Mya	100,000 years ago		
		3000 — Iron tools	
		6000 — Domestication of horse	
		8000 — Domestication of cattle	

Wolf/dog split
120,000 — *Homo* language possible

12,000 — Early agriculture
14,000–10,000 — Domestication of dog · Megafauna extinction in North America

1.6 — *Homo erectus* in Asia
250,000–160,000 — *Homo sapiens*

1.8 — *Homo habilis* out of Africa
200,000 — Anatomically modern human

20,000 years ago

2.5 — Tool use

30,000 — Human migration from Asia to North America

Panama Isthmus rises
3.6 — Lucy fossil
Australopithecus afarensis footprints
355,000 — *Homo heidelbergensis* footprints

4 Mya
400,000
40,000

45,000 — Megafauna extinction in Australia

Formation of current Galápagos Islands
5–4 — Last common ancestor of chimps and humans
500,000 — *Homo erectus* use of fire

50,000 — Humans migrate from Asia to Australia

6 Mya — Common ancestor of human/chimp/gorilla
600,000 — Human/Neanderthal split
60,000

6–7 — *Sahelanthropus*

79,000–15,000 — Start of Wisconsin glaciation

8 Mya
800,000
80,000

10 Mya
1 Mya
100,000

EVOLUTION

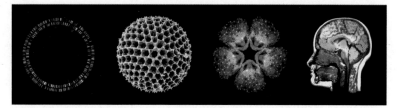

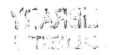

EVOLUTION

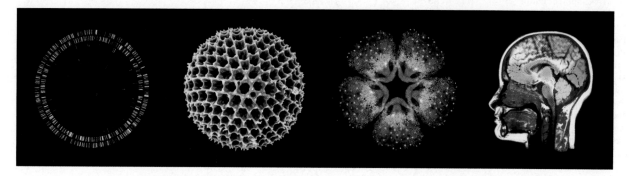

Nicholas H. Barton
University of Edinburgh

Derek E.G. Briggs
Yale University

Jonathan A. Eisen
University of California, Davis

David B. Goldstein
Duke University

Nipam H. Patel
University of California, Berkeley

Cold Spring Harbor, New York • http://www.cshlpress.com

EVOLUTION

Publisher	John Inglis
Acquisition Editor	Alexander Gann
Director of Book Development, Marketing, and Sales	Jan Argentine
Text Development Editor	Michael Zierler
Art Development Editor	Judy Cuddihy
Project Coordinator	Maryliz Dickerson
Permissions Coordinators	Carol Brown, Nora McInerny, and Maria Fairchild
Production Manager	Denise Weiss
Production Editor	Kathleen Bubbeo
Desktop Editor	Susan Schaefer
Marketing Manager	Ingrid Benirschke Perkins
Sales Account Manager	Jane Carter
Interior Designer	Denise Weiss
Art Studio	Electronic Illustrators Group
Cover Designer	Michael Albano

Front Cover and Title Pages Artwork: 1st Image: A map of the genome of the bacterium *Deinococcus radiodurans* R1. A portion of Fig. 1 from White O. et al., *Science* **286:** 1571–1577, © 1999 American Association for the Advancement of Science. **2nd Image:** Mineral skeleton of a radiolarian (marine). The hard skeleton is composed of silica. SEM ×145. © Dr. Dennis Kunkel/Visuals Unlimited. **3rd Image:** Grasshopper embryo (image duplicated and rotated fivefold). Courtesy of Sabbi Lall and Nipam H. Patel, University of California, Berkeley. **4th Image:** Photo of enhanced brain MRI image, © Scott Camazine.

Preface Artwork: Earthrise—Apollo 8. Courtesy of NASA.

Back Endpaper Artwork: Courtesy of Sandie Baldauf, University of York.

Library of Congress Cataloging-in-Publication Data

Evolution / Nicholas H. Barton ... [et al.].
    p. cm.
Includes bibliographical references and index.
ISBN 978-0-87969-684-9 (hardcover : alk. paper)
  1. Evolution (Biology)--Textbooks. I. Barton, Nicholas H. II. Title.

QH366.2.E847 2007
576.8--dc22

2007010767

1006453959

10 9 8 7 6 5 4 3 2 1

Contents

About the Authors x
Preface xi
Aim and Scope of the Book 1

 PART I

AN OVERVIEW OF EVOLUTIONARY BIOLOGY 7

1 The History of Evolutionary Biology: Evolution and Genetics 9
2 The Origin of Molecular Biology 37
3 Evidence for Evolution 65

 PART II

THE ORIGIN AND DIVERSIFICATION OF LIFE 85

4 The Origin of Life 87
5 The Last Universal Common Ancestor and the Tree of Life 109
6 Diversification of Bacteria and Archaea. I: Phylogeny and Biology 137
7 Diversification of Bacteria and Archaea. II: Genetics and Genomics 169
8 The Origin and Diversification of Eukaryotes 195
9 Multicellularity and Development 225
10 Diversification of Plants and Animals 253
11 Evolution of Developmental Programs 287

 PART III

EVOLUTIONARY PROCESSES 323

12 Generation of Variation by Mutation and Recombination 325
13 Variation in DNA and Proteins 355
14 Variation in Genetically Complex Traits 381
15 Random Genetic Drift 413
16 Population Structure 439

17 Selection on Variation 457
18 The Interaction between Selection and Other Forces 489
19 Measuring Selection 521
20 Phenotypic Evolution 555
21 Conflict and Cooperation 585
22 Species and Speciation 619
23 Evolution of Genetic Systems 657
24 Evolution of Novelty 695

PART IV

HUMAN EVOLUTION 725

25 Human Evolutionary History 727
26 Current Issues in Human Evolution 755

Glossary 783
Figure Credits 802
Index 811

ONLINE CHAPTERS (http://www.evolution-textbook.org)

27 Phylogenetic Reconstruction
28 Models of Evolution

Detailed Contents

About the Authors x
Preface xi

Aim and Scope of the Book 1

Evolutionary Biology Describes the History of Life and Explains Why
  Organisms Are the Way They Are 1

Evolutionary Biology Is a Valuable Tool 3

Molecular Biology and Evolutionary Biology Are Overlapping Fields
  of Study 5

 PART I

AN OVERVIEW OF EVOLUTIONARY BIOLOGY 7

1 **The History of Evolutionary Biology: Evolution and Genetics 9**

A Brief Summary of Modern Genetics and Evolution 9

Evolution before Darwin 10

Charles Darwin 16

The Eclipse of Natural Selection 18

The Evolutionary Synthesis 30

2 **The Origin of Molecular Biology 37**

The Beginnings of Molecular Biology 38

Evolutionary and Molecular Biology: A New Synthesis? 58

3 **Evidence for Evolution 65**

Evidence for Evolution 66

Objections to Evolution 76

Science and Society 81

 PART II

THE ORIGIN AND DIVERSIFICATION OF LIFE 85

4 **The Origin of Life 87**

When Did Life Begin on Earth? 87

How Did Life Begin on Earth? 91

5 The Last Universal Common Ancestor and the Tree of Life 109

Tracing Early Evolutionary History 109

Universal Homologies, LUCA, and the Tree of Life 115

6 Diversification of Bacteria and Archaea. I: Phylogeny and Biology 137

Introduction to the Bacteria and Archaea 137

Phylogenetic Diversification of Bacteria and Archaea 142

Biological Diversification of Bacteria and Archaea 151

7 Diversification of Bacteria and Archaea. II: Genetics and Genomics 169

The Nature of Archaeal and Bacterial Genomes 169

Lateral Transfer of DNA 182

8 The Origin and Diversification of Eukaryotes 195

Introduction to the Eukaryotes 195

Endosymbioses Have Played a Key Role in the Evolution of Eukaryotes 202

Structure and Evolution of the Nuclear Genome 214

Eukaryotic Diversification 221

9 Multicellularity and Development 225

How Multicellularity Happens 225

Division of Labor through Differentiation 230

Diversity of Body Plans 238

Genetics of Building a Body Plan 244

10 Diversification of Plants and Animals 253

Fossilization and Geological Time 253

The Flow of the Evolution of Life-forms 254

The Next 500 Million Years—Life Since the Cambrian Period 272

Patterns of Evolution 280

11 Evolution of Developmental Programs 287

Anterior–Posterior Patterning: *Hox* Gene Regulation of Development 288

Hox Genes Are Shown to Be Involved in Evolutionary Change 293

Skeletal Evolution in Sticklebacks 307

Evolution of Maize from Teosinte 313

Universality of Developmental Systems 318

 **PART III**

EVOLUTIONARY PROCESSES 323

12 Generation of Variation by Mutation and Recombination 325

Mutations and Mechanisms for Their Generation 325

Protection, Prevention, and Correction Mechanisms Limit the Number of Mutations Caused by DNA Damage and Replication Errors 334

Rates and Patterns of Mutations 343

Generation of Variation by Mixing: Sex and Lateral Gene Transfer 349

13 Variation in DNA and Proteins 355

Genetic Variation 355

Types of Genetic Variation 367

14 Variation in Genetically Complex Traits 381

Introduction to Quantitative Traits 381

Analyzing Quantitative Variation 385

The Genetic Basis of Quantitative Variation 399

Generation of Quantitative Variation 409

15 Random Genetic Drift 413

Evolution Is a Largely Random Process 413

Random Drift of Allele Frequencies 415

Coalescence 420

The Neutral Theory 425

Recombination and Random Drift 427

16 Population Structure 439

Gene Flow 439

Gene Flow Interacts with Other Evolutionary Forces 444

Genealogies in Structured Populations 449

17 Selection on Variation 457

The Nature of Selection 457

Selection on Quantitative Traits 476

Selection on Multiple Genes 485

18 The Interaction between Selection and Other Forces 489

Selection and Random Drift 489

Selection and Gene Flow 496

Balancing Selection 505
Mutation and Selection 510

19 Measuring Selection 521
Direct Measurement of Selection 521
Indirect Measurement 530
Selection on Linked Loci 536
Selection on Noncoding DNA 542
The Extent of Selection 547

20 Phenotypic Evolution 555
Evolutionary Optimization 556
Aging 561
Evolutionary Games 567
Sexual Selection 573

21 Conflict and Cooperation 585
Social Evolution 586
Conflict between Genes 587
Interactions between Relatives 599
Evolution of Cooperation 611

22 Species and Speciation 619
Defining Species 620
The Genetics of Speciation 630
Mechanisms of Speciation 640
The Geography of Speciation 644

23 Evolution of Genetic Systems 657
Studying the Evolution of Genetic Systems 657
Evolution of Mutation Rates 659
Evolution of Sex and Recombination 663
Consequences of Sex 683
Evolution of Evolvability 689

24 Evolution of Novelty 695
The Basic Features of Novelty 696
Changes in the Activity of Gene Products 700

Changes in Gene Regulation and Interactions
within a Network: Targeting, Differentiation,
and Development 704
Redundancy 708
Robustness, Modularity, and
Compartmentalization 711
Acquiring New Functions from Other Species:
Gene Transfer and Symbiosis 717
Natural Selection over Long Periods of Time
Leads to the Origin of Novelty 719

PART IV

HUMAN EVOLUTION 725

25 Human Evolutionary History 727
Placing Humans on the Tree of Life 727
The Evolution of Hominins 731
Genetics and Human Evolution 741
Genomics and Humanness 749

26 Current Issues in Human Evolution 755
The Genetic Basis of Disease 755
Understanding Human Nature 769

Glossary 783
Figure Credits 802
Index 811

ONLINE CHAPTERS (http://www.evolution-textbook.org)

27 **Phylogenetic Reconstruction**
28 **Models of Evolution**

About the Authors

NICHOLAS H. BARTON'S early research was on narrow zones of hybridization that subdivide many populations, with work on a variety of species, including grasshoppers, butterflies, and toads. More recently, his research, which has been mainly theoretical, is attempting to understand the influence of selection on complex traits, models of speciation, the evolution of sex and recombination, and the coalescent process. He is Professor of Evolutionary Genetics at the Institute of Evolutionary Biology, University of Edinburgh.

DEREK E.G. BRIGGS works on preservation and the evolutionary significance of exceptionally preserved fossils, including those of the Cambrian Burgess Shale of British Columbia. His current research focuses on the chemical changes that occur during the transformation from living organism to fossil. He is Frederick William Beinecke Professor of Geology and Geophysics at Yale University and Director of the Yale Institute for Biospheric Studies.

JONATHAN A. EISEN uses a combination of genomic sequencing and evolutionary reconstruction methods to study the origin of novelty in microorganisms. Previously, he applied this phylogenomic approach to cultured organisms, such as those from extreme environments. Currently he is using phylogenomic methods to study microbes in their natural habitats, including symbionts living inside host cells and planktonic species in the open ocean. He is Professor in the Department of Medical Microbiology and Immunology and the Department of Evolution and Ecology at the University of California, Davis.

DAVID B. GOLDSTEIN'S principal interests include human genetic diversity, the genetics of neurological disease, population genomics, and pharmacogenetics. His laboratory currently investigates how human genetic variation influences the response to drug treatments for common neurological and cardiovascular disorders. He is Director of the Center for Population Genomics and Pharmacogenetics at the Duke University Medical Center.

NIPAM H. PATEL initially studied the development of several model and nonmodel species, including cows, chickens, grasshoppers, and *Drosophila melanogaster*. His research group studies the evolution of development, with a focus on the evolution of segmentation, neurogenesis, appendage patterning, and gene regulation. He is Professor in the Department of Molecular and Cell Biology and the Department of Integrative Biology at the University of California, Berkeley and an Investigator of the Howard Hughes Medical Institute.

Preface

EVOLUTIONARY BIOLOGY IS ONE OF THE CORNERSTONES of modern science. The publication of *On the Origin of Species*, by Charles Darwin in 1859, changed forever the way that people view life on our small planet. Over the 150 years since, the study of evolution has itself evolved, taking a quantum leap in the past 20 years through the exciting discoveries that have occurred at the interface of molecular and evolutionary biology. The powerful tools available through the use of molecular markers, genome sequences, and genetic manipulation have provided extraordinary insights into how evolution works. In addition, a fascinating new realm of natural history has opened up, revealing a remarkable variety of molecular adaptations and an array of fundamental molecular mechanisms that are unexpectedly conserved throughout the living world.

Yet, these remarkable developments are scarcely reflected in the way that evolutionary biology is taught. Recent advances in topics such as molecular evolution and developmental evolution are often relegated to minor chapters in other textbooks. Our aim here is to integrate these and other modern breakthroughs into a complete evolutionary perspective that forms the central theme of the book.

Fundamental principles are illustrated repeatedly across all levels of organization (from molecular to organismal) and the entire diversity of life (from microbial to human). We illustrate the basics of natural selection using in vitro selection on RNA molecules and using the Galápagos finches (Chapter 17). We show how selection can be measured in laboratory populations of yeast and in natural populations of deer (Chapter 19). We illustrate evolutionary optimization using mating behavior in dung flies and metabolic flux in *Escherichia coli*, and we illustrate evolutionary games using production of toxins by bacteria and mating behavior in lizards (Chapter 20). We show how evolutionary conflicts occur between transposable elements and their host genomes and also between bees in a hive (Chapter 21).

The interdisciplinary approach that we adopt will be valuable to students from a whole-organism background (the majority of students now taking evolution courses), because it will widen their perspective on the scope of evolutionary biology and show them the new methods that are so important in current research. The integration of molecular and evolutionary biology also will be attractive to students whose background and interests are primarily molecular and to faculty in molecular biology departments who develop modern evolution courses.

Structure of the Book

The book is divided into four parts. In Part I, we summarize the history of evolutionary and molecular biology and the evidence for evolution by natural selection. As well as giving a historical overview, this provides a compact summary of our current understanding, which guides the reader through the rest of the book. Part II describes the origin and diversification of life, including biochemical as well as morphological complexity, and gives a balanced treatment of bacterial, archaeal, and eukaryotic diversity. This section includes some of the most exciting recent discoveries in biology, such as the identification and characterization of particular genes and genetic pathways that play key roles in the development of plants and animals. In Part III, we explain the fundamental processes of evolution: mutation, random drift, recombination, gene flow, and, most importantly, natural selection. We then bring this understanding of the evolutionary process to bear on major questions: Why do organisms age? How can conflicts between organisms be resolved? How do new species originate? Why is sexual reproduction widespread? How do novelties evolve? In the final two chapters (Part IV), we turn to our own species, describing our evolutionary history and discussing the implications of evolution for humanity today.

This book provides the basis for a full course on evolutionary biology: building from a historical overview of evolutionary and molecular biology, through an account of the origin and diversification of life, to an explanation of the processes of evolution and how they interact, and ending with a discussion of human evolution. Parts of the book can also be used as a text for smaller courses.

- Chapters 1–3 give a concise overview of the subject.

- Part II provides material for a course on the origin and diversity of life. More specifically, Chapters 5–8 form the foundation for a course on microbial evolution.

- Part III treats the topics found in standard texts on the processes of evolution but with a strong molecular emphasis. Ideally, this would follow from a course based on Part II.

- Chapters 14, 17, 19, and 26 provide a course in quantitative genetics that goes further, and is more up-to-date, than any other textbook at this level.

Web Resources

Our book is accompanied by an extensive Web site (http://www.evolution-textbook.org), which will be especially valuable for instructors and advanced students.

- Many of the illustrations and tables are available in a form that can be used in teaching.

- The glossary, which is included in the printed book, is also available online. All glossary terms are highlighted in **bold** at their first appearance within a chapter. The glossary possesses a search function.

- Each chapter is accompanied by discussion topics, which will allow the student to review ideas and should stimulate further thought.

- Chapters in Part III are accompanied by problems. It is important that the student work through these, so as to reinforce the quantitative material in these chapters. Some of the problems are more advanced and will require knowledge of the material in the online chapters (see next bullet).

- The main text describes the basics of quantitative and population genetics. The Web site includes two supplementary chapters, which allow students to build on the basics and explore more quantitative aspects of the subject. Chapter 27 explains methods for inferring phylogenetic relationships, and Chapter 28 shows how the evolutionary process can be modeled. (These chapters are not indexed, but they can be searched, and are accompanied by Web Notes, as explained below.)

Exploring the Literature

It is important that the reader can use this book as a platform to explore the literature of evolutionary biology. The success of science depends on being able to trace the evidence and arguments that support each statement and on this web of knowledge being open to all. We have not cluttered the main text with references to our sources, but have provided several routes into the literature. At the most basic level, each chapter ends with a Further Reading section, which lists reviews that cover the material. On the Web site, we include a section that describes the most useful journals in the field and explains how online databases such as the ISI Web of Knowledge and Google Scholar can best be used to search the literature. Each section of the book is accompanied by Web Notes, which give the primary sources for the material in the main text and expand on some points. Together, the Web resources make it straightforward for the reader to go beyond the introduction that we provide in the main text and to explore the subject in depth.

Evolutionary biology is an active and thriving field. Much is well established, but there are many open questions, and new discoveries are continually opening up new questions, which are highlighted throughout the book. We aim to set out clearly what *is* known, but also to lay out the evidence and arguments upon which this knowledge is based and to show how we can set about answering new questions. Our book is unique in integrating molecular biology with evolutionary biology—an approach that reflects the convergence of these fields in recent years.

Acknowledgments

First, we would like to thank Jim Watson, who conceived the idea of a textbook that would integrate molecular and evolutionary biology. Cold Spring Harbor Laboratory Press has given outstanding support to the project throughout its long gestation: John Inglis and Alex Gann, in particular, showed exceptional skill in guiding their sometimes troublesome authors through to completion of the book. During the development and production of the book, Michael Zierler, Judy Cuddihy, and Hans Neuhart have shaped the writing and artwork into a coherent and attractive text. We also thank Jan Argentine, Elizabeth Powers, Maryliz Dickerson, Carol Brown, Mary Cozza, Maria Fairchild, Nora McInerny, Denise Weiss, Susan Schaefer, and Kathleen Bubbeo for their help throughout production of the book. Thanks are also due to Mila Pollock, Director of Cold Spring Harbor Laboratory Library and Archives and her staff, especially Clare Clark, Gail Sherman, Claudia Zago, and Rhonda Veros.

We have received invaluable comments on draft chapters from our colleagues, whose efforts in correcting our errors and improving our writing were much appreciated. We thank Peter Andolfatto, Maria-Inés Benito, Brian and Deborah Charlesworth, Satoshi Chiba, Nick Colegrave, Jerry Coyne, Angus Davison, Laura Eisen, Michael Eisen, Andy Gardner, Paul Glenn, Ilkka Hanski, Amber Hartman, Bill Hill, Holly Huse, Saul Jacobson, Chris Jiggins, Mark Kirkpatrick, John Logsdon, Hanna Miedema, Erling Norrby, Josephine Pemberton, Mihai Pop, Rosie Redfield, Jay Rehm, Jeffrey Robinson, Denis Roze, Michael Turelli, Craig Venter, Peter Visscher, Naomi Ward, Stu West, and Merry Youle. We also thank the librarians at the Darwin and New College Libraries at the University of Edinburgh for their assistance.

We would also like to thank the reviewers who gave detailed and helpful comments on early drafts of the book: Tiffany M. Doan, David H.A. Fitch, Joerg Graf, Rick Grosberg, Thomas Hansen, Kevin Higgins, Trenton W. Holliday, Robert A. Krebs, David C. Lahti, Richard E. Lenski, Michael P. Lombardo, James Mallet, Jennifer B.H. Martiny, Rachel J. Waugh O'Neill, Kevin J. Peterson, Michael Petraglia, Ray Pierotti, Richard Preziosi, David Raubenheimer, Mark D. Rausher, Gary D. Schnell, Eric P. Scully, David Smith, Steve Tilley, Martin Tracey, John R. Wakeley, and Susan Wessler.

Finally, we are all immensely grateful to our friends and families who provided us with such strong support, despite our frequent absences and distractions.

Nicholas H. Barton
Derek E.G. Briggs
Jonathan A. Eisen
David B. Goldstein
Nipam H. Patel

Aim and Scope of the Book

"Nothing in biology makes sense except in the light of evolution"

—Theodosius Dobzhansky

EVOLUTIONARY BIOLOGY DESCRIBES THE HISTORY OF LIFE AND EXPLAINS WHY ORGANISMS ARE THE WAY THEY ARE

OUR WORLD IS FILLED WITH AN extraordinary diversity of living organisms (Fig. A&S.1). The Sun's light is harvested by bacteria, algae, and plants, and every feasible source of chemical energy, from hydrogen gas to carbon monoxide, is exploited by some microbe. Life thrives in the most extreme environments, from Antarctic rocks to scalding undersea vents to crevices miles beneath the Earth's crust. Organisms are sensitive to the slightest variations in their environment: Bacteria find their way by sensing the Earth's magnetic field, moths find their mates from a few molecules' scent, and owls see their prey from afar on a moonless night. Birds can use specialized tools to help them extract food, bees coordinate the activity of their hive, and human societies use language and technology to overcome the limitations of their biology.

All of these diverse functions are achieved by the same fundamental biochemical system. Genetic information stored in the sequence of bases in DNA is **transcribed** into RNA and then **translated** into sequences of amino acids that construct and maintain the organism. The engineering of organisms is as remarkable as its consequences. DNA sequences can be replicated with less than one error in a billion bases. Proteins at ambient temperatures can catalyze reactions that human chemists can achieve only under extreme conditions and with much coarser specificity. Interactions between DNA and proteins regulate precise patterns of gene expression that allow the reliable construction of complex organs, including, most impressively, the human brain, which generates and controls elaborate behavior. All of this is determined by a remarkably small amount of information; the human DNA sequence, for example, encodes less information than is stored in a personal computer.

We know, at least in outline, how the DNA sequence is translated into proteins, how these proteins affect metabolism and regulate gene expression, and how multicellular organisms develop reliably from a single cell. We also know, at least in outline, how the diverse organisms now living came to be. All of them (including us) descended from one or a few "universal ancestor(s)" that lived more than 3 billion years ago. That ancestor or ancestors descended from much simpler organisms in which RNA carried out the present roles of both DNA and protein. Thus, all of the biochemical, morphological, and behavioral diversity around us descended from a few original genes.

The exquisite biological devices that we now see appear as though carefully designed for their present purposes, and this appearance of design was long taken as evidence of

FIGURE A&S.1. Diversity of adaptation. (*Top row*) Life in extreme environments. Organisms can live in the very cold environment of Antarctic dry valleys (*left*) or in hydrothermal vents deep within the sea (tube worm community; *right*). (*Middle row*) Diversity of morphological adaptation. The fine structure of moth antennae (*left*) allows the detection of just a few molecules of pheromone. The large eyes of the vermiculated screech owl (*Otus guatemalae; right*) allow it to find prey at night. (*Bottom row*) Behavioral adaptation. The New Caledonia crow (*Corvus moneduloides; left*) fashions tools for specific purposes, here to extract insects from holes in a tree branch. Honey bees (*Apis mellifera; right*) live and work in an organized "society" with specific tasks assigned to maintain the hive.

an intelligent creator. We now know that biological function is constructed and maintained by natural selection: the gradual accumulation of variations that arise by chance and that are preserved because they aid the survival and reproduction of their carriers. The theory of evolution is a synthesis of Darwinian natural selection and Mendelian genetics. It allows us to ask not just how life evolved, but why it is as it is: Why do organisms develop from a single cell? Why is the genetic code as it is? Why is there sexual reproduction?

EVOLUTIONARY BIOLOGY IS A VALUABLE TOOL

Our understanding of the history of life and the mechanism by which life has evolved has influenced virtually every aspect of human society from literature to medicine. However, evolutionary biology is not simply a historical science. Information on evolution and the application of principles learned from the study of evolution also have many practical uses in fields as diverse as geology, computer science, and epidemiology. We give here three examples.

Because closely related organisms tend to have similar features, evolutionary classification can help predict the biology of an organism through comparison with its relatives. This is best seen in studies of microbes. Traditionally, microorganisms are studied by isolating them in culture. However, the great majority of microbes cannot at present be grown in the laboratory. By combining tools from molecular and evolutionary biology, genes from unculturable species can be cloned and analyzed to provide the information that places them into the tree of life. This allows scientists to infer much about these unculturable microbes, such as details of their metabolic processes or their interactions with other organisms.

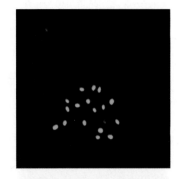

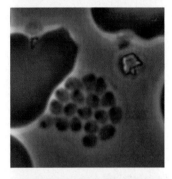

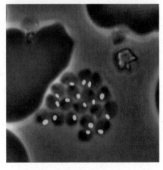

FIGURE A&S.2. Apicoplasts are visible through the use of green fluorescent protein (*top*); round balls in center are malaria parasites (*middle*); and green apicoplasts inside parasites (*bottom*).

This approach has led to the development of new drugs for treating malaria and related diseases. Malaria is caused by the infectious parasite *Plasmodium falciparum.* Traditional antimalarial drugs that kill *P. falciparum* are frequently quite toxic to humans, because they attack pathways common to eukaryotes. In the 1970s, scientists found an unusual organelle in *P. falciparum* that contained its own DNA. Much to their surprise, evolutionary trees based on genes in this DNA indicated that this organelle (the **apicoplast**) was closely related to plant **chloroplasts** (Fig. A&S.2). Plant chloroplasts originated from an **endosymbiosis** between ancestors of plants and a bacterium related to modern day **cyanobacteria** (see p. 203). Evolutionary analysis has shown that the *Plasmodium* lineage obtained this plastid by entering into a symbiosis with another eukaryote (most likely an alga) that contained a chloroplast. Although many of the features of this alga-like symbiont have since been lost, the plastid was maintained and became the apicoplast. Although *Plasmodium* does not carry out photosynthesis, the metabolic functions of the apicoplast are essential for its survival. The absence of the apicoplast metabolic pathways in mammals has suggested new targets for antiparasite medications. A variety of antibiotics, enzyme inhibitors, and herbicides that target the apicoplast show promise at killing *P. falciparum* and relatives that cause toxoplasmosis and cryptosporidiosis.

Evolutionary biology has also been very useful in enhancing our understanding of macromolecular structure. This is perhaps best shown through studies of **ribosomal RNA** (rRNA). In all organisms, rRNA is involved in the translation of messenger RNA into proteins. rRNA molecules form intricate structures, driven in part through internal base pairing, much like the base pairing that forms the DNA double helix (Fig. A&S.3). Biochemical analysis of rRNA molecules has been unsuccessful in solving the rRNA folding problem, partly because there are simply too many base-pair combinations to test. Until recently, structural studies have also been unsuccessful, because rRNA molecules do not readily form the crystals needed to solve structures by X-ray diffraction. However, the base pairings within rRNA molecules have been determined by tracing the evolutionary history of nucleotide changes and identifying cases in which changes occurred in two separate parts of the molecule at the same time. By identifying thousands of these correlated changes, researchers have been able to determine which nucleotides base pair with each other (see p. 544).

Numerous human endeavors have benefited from our understanding of how evolution works. Animal and plant breeding programs are continually improved through the application of principles of population genetics. Similar approaches are being used in biotechnology, through the use of so-called molecular breeding (see Fig. 17.2). One ex-

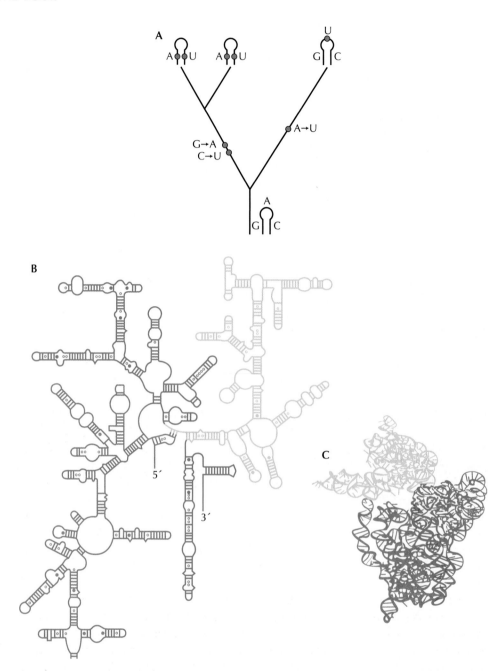

FIGURE A&S.3. The pattern of evolutionary change in an RNA molecule can be used to determine its structure. (A) Where two parts of the RNA molecule pair, a change in one base must be accompanied by a change in its partner if pairing is to be maintained (e.g., G→A [*red*] and C→U [*blue*]). This maintains the pairing between these bases. In contrast, changes in unpaired regions can occur independently (e.g., the single change A→U [*green*]). (B) The secondary structure of the 16S ribosomal RNA molecule was inferred using this method, based on comparisons between more than 7000 sequences from different bacterial species. Ladders indicate paired helices. (C) The full three-dimensional structure of crystals of the molecule was determined directly in 2000 and confirms the inferences made by the comparative method: More than 97% of the predicted pairs were found in the crystal structure. Different colored regions in *B* and *C* correspond to different domains of the molecule.

FIGURE A&S.4. Molecular evolution can be used to solve a complex practical problem. Here in vitro selection of a synthetic green fluorescent protein (GFP) gene (*B*) was used to increase its signal over wild type (*A*).

ample is the creation of designer proteins that have new or enhanced functions. Fluorescent proteins are very useful tools in biological research (e.g., Fig. A&S.2). They are detected using fluorescence microscopy and other light-sensitive instruments through the light emitted by the proteins when placed under the proper conditions. To make the best use of fluorescent proteins, it is helpful to have different forms that emit different colored lights. The original green fluorescent protein was isolated from a jellyfish species, *Aequorea victoria*. Using in vitro molecular breeding methods, scientists have altered the green fluorescent protein, creating new versions that fluoresce at other wavelengths. Molecular breeding methods rely on our understanding of the mechanisms of evolution to enable designer macromolecules to be created (Fig. A&S.4). In nature, the evolution of new functions for, say, a protein may be limited by the likelihood that the appropriate genes carry the necessary combination of mutations. **Recombination** (which is used deliberately in molecular breeding) brings together different mutations and greatly increases the efficiency of in vitro selection (see Fig. 23.21).

MOLECULAR BIOLOGY AND EVOLUTIONARY BIOLOGY ARE OVERLAPPING FIELDS OF STUDY

Molecular biology and evolutionary biology are both thriving fields. Over the last half-century, however, they have largely developed independently of each other. Of course, evolutionary biology uses many molecular techniques and investigates molecular as well as morphological and behavioral variation. Conversely, molecular biology uses (at least implicitly) evolutionary methods to establish relationships between organisms and between genes and to distinguish functional sequences. Nevertheless, the two fields remain as largely separate communities and ask rather different questions. Our aim in this book is to explain evolutionary biology in a way that will be accessible—and interesting—to molecular biologists and to show how evolutionary methods bear on some of the most recent molecular discoveries.

Part I provides a historical discussion of the two fields, outlining key advances and concepts. The overwhelming evidence for evolution is presented, including arguments supporting natural selection as its central mechanism. Part II presents an overview of the diversification of life. Beginning with the origins of life more than 3.5 billion years ago, this part shows how the three domains of life—bacteria, archaea, and eukaryotes—have diversified. We describe the evolution of multicellular organisms, as seen both in the fossil record and through our recent understanding of developmental programs. Part II lays the groundwork for Part III, which explains the nature of genetic variation, the mechanisms of evolution, and their consequences. Almost all populations contain abundant variation, which is generated by mutation and recombination. This variation is seen both at the level of DNA and protein sequence and in the morphology and behavior of the whole organism—traits that depend on the interaction between large numbers of genes. Natural selection acts on this variation and is responsible for the remarkable adaptations that we see in the living world. We show how the various evolutionary mechanisms interact with each other to generate these adaptations, to produce diverse species, and to shape the genetic system itself. Part IV closes the book with a discussion of the evolution of our own species and the role that evolution and evolutionary biology now play in human endeavors.

This textbook is supported by a comprehensive Web site, which which will be valuable for instructors and students. Many of the textbook illustrations and tables, as well as the glossary, are available. Discussion questions, designed to review concepts and aid the student in thinking more deeply about the topics, are presented for every chapter. Problems (with answers) are provided for Chapters 12–24. The Web site also includes chapter notes, which give full references, expand on some of the arguments in the main text, and provide links to other useful Web sites. Combined with the recommended reading list found at the end of every book chapter, this will guide the student into the extensive literature of evolutionary biology and will provide useful background material for the instructor. The Web site also contains two chapters not found in the book that introduce the quantitative methods that are used to study evolution. The first (Chapter 27) describes in detail how phylogenetic trees are inferred and how they are used to reveal evolutionary relationships. The second (Chapter 28) is a detailed introduction to evolutionary modeling.

We hope that, as well as providing a clear explanation of biological evolution, our book will help to bring molecular and evolutionary biology closer together. The past century has seen extraordinary advances in both branches of biology, and the coming decades promise to be still more exciting.

FURTHER READING

Meagher T.R. and Futuyma D. 2001. Evolution, science, and society. *Am. Nat.* **158:** 1–46.
    Examples of how evolutionary biology can be applied in diverse ways. Also available from http://evonet.sdsc.edu.evoscisociety/.
Working Group on Teaching Evolution, National Academy of Sciences. 1998. *Teaching about evolution and the nature of science.* National Academies Press, Washington, D.C.

Web Sites

http://evonet.org is a Web site that provides a good overview of evolutionary biology.
AAAS Press Room—Evolution on the Front Line: http://www.aaas.org/new/press_room/evolution/
National Center for Science Education: http://www.natcenscied.org/
Understanding Evolution: http://evolution.berkeley.edu/

PART I

AN OVERVIEW OF EVOLUTIONARY BIOLOGY

Evolutionary biology seeks to explain the origin and diversification of life. Our subject formally began when Charles Darwin demonstrated that all organisms share a common ancestry and that the key mechanism of evolution is natural selection. It emerged in its modern form after the rediscovery of Mendel's work established classical genetics, providing the tools for quantitative analysis of evolutionary processes. Chapter 1 traces this history and outlines how these key questions are now being answered using a combination of experimentation, theoretical approaches, and the analysis of genomes and gene sequences.

Chapter 2 sketches the emergence of molecular biology. The molecular basis of genetics and evolution was largely established in the decade following the discovery of the structure of DNA in 1953. Molecular techniques have transformed the study of evolution. Molecular and evolutionary biology have become closely intertwined, in a way that is reminiscent of the synthesis between Darwinian evolution and Mendelian genetics almost a century ago.

Chapter 3 summarizes the evidence for evolution by natural selection. The arguments here go back to Darwin and were overwhelmingly convincing in his time. Throughout the book, we will see many more examples that reinforce the points summarized in this chapter.

As you read more of this book, it will be useful to return to Part I to see how the detailed arguments presented elsewhere fit into the basic framework provided here and to see how far we have come toward answering the key questions in evolutionary biology.

This is an exciting time for evolutionary biology. We are awash in a flood of new information about phylogenetic relationships, newly discovered organisms, the nature of genetic variation, and the fundamental mechanisms of development. However, the present-day debates all trace back to the origins of evolutionary biology. Thus, a historical perspective is invaluable for understanding the subject today.

1

The History of Evolutionary Biology: Evolution and Genetics

ALL LIVING ORGANISMS ARE DESCENDED FROM a much smaller number of organisms that lived about 3.5 billion years ago. All the genes in every living organism descend from just a few ancestral genes. This modern understanding of the living world was established only quite recently. In 1859, Charles Darwin's *On the Origin of Species by Means of Natural Selection* set out the evidence for "descent with modification" and for the key evolutionary mechanism of **natural selection**. Gregor Mendel explained inheritance in 1865, although his work was overlooked until 1900, when its rediscovery triggered the growth of classical genetics. In 1953, James Watson and Francis Crick's determination of the structure of DNA established the physical basis of genetics.

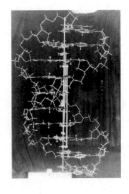

Until these recent advances, human ideas of heredity and of the origin of species were quite different—indeed, so different that it is hard for us to comprehend past thinking. Even now, many misconceptions about genetics and evolution, and many controversies in present-day evolutionary biology, can be traced back to these older ideas. Thus, to understand present debates and to appreciate just how much our view of the world has changed, we must consider at least a brief history. Since ideas on heredity and evolution are so closely intertwined, we consider both together.

A BRIEF SUMMARY OF MODERN GENETICS AND EVOLUTION

Before we begin our history of evolutionary biology, it will be helpful to highlight the key features of our current understanding of genetics and evolution.

- *DNA interacts with the cell and the environment to determine the phenotype.* The DNA sequence alone is meaningless. It must be **transcribed** into **messenger RNA** (mRNA), **ribosomal RNA**, and other functional RNA sequences. The mRNA is then translated into proteins by the **ribosomes**. These proteins must then interact with each other, various RNA molecules, and the DNA to determine the timing of

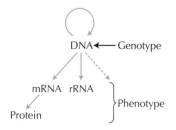

FIGURE 1.1. The Central Dogma: DNA makes RNA makes protein.

gene expression within the cell, the response to the outside world, and the development of morphology and behavior (Fig. 1.1; see p. 53). This sequence of interactions determines all the characteristics that we observe—that is, the **phenotype**.

- *It is populations that evolve.* A species is not a single homogeneous unit: It is a population of diverse individuals. A species evolves as the proportions of different kinds of individuals within it change, rather than in abrupt, discontinuous transitions.

- *Evolution is like a branching tree.* Populations change, split into separate species, and often become extinct. Tracing forward, most individuals leave no descendants and most species go extinct. Tracing backward, closely related species merge in a recent common ancestor, and more distantly related species share ancestry further back in time (Fig. 1.2).

- *Evolution does not progress toward a goal.* Populations evolve in response to chance variations and to their changing environment. This leads to erratic change and to diversification, rather than progress toward any particular end point.

- *All adaptation is caused by natural selection.* Although there are many causes of evolutionary change, only natural selection can lead to **adaptation**—that is, to the complex and finely tuned structures that allow organisms to survive and reproduce in diverse environments. Organisms are not designed for their present way of life; rather, their ancestors accumulated variations that were favorable for reproduction.

We elaborate these themes later in the book. For the moment, however, you should keep them in mind and contrast them with the quite different ideas that were held in earlier times.

EVOLUTION BEFORE DARWIN

Past Ideas of Inheritance and the Origin of Species Were Fundamentally Different from Ours

Mankind has long had a practical and religious interest in inheritance and natural variation. Those that live by hunting and gathering have a detailed classification of the plants and animals on which they depend—a classification that is usually close to that of modern biologists (see Chapter 23). Agriculture began with the domestication of crops and livestock; it relies on the selection (conscious or unconscious) of improved varieties and some understanding of the inheritance of their traits. Quite apart from the practicalities, all societies have more or less elaborate beliefs about their own origins.

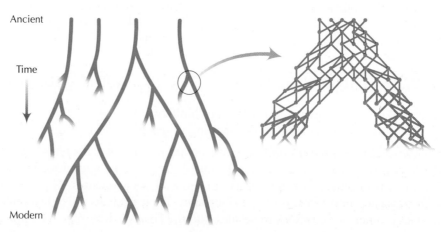

FIGURE 1.2. The evolution of species through time. (*Left*) Each line indicates a population of individuals. (*Inset*) Sexual individuals that make up the population.

The first to record their systematic search for natural explanations were the Greek philosophers. Plato taught that the world we perceive is but a shadow of underlying essences that remain fixed. This view had immense influence on Western thought, but it was incompatible with any gradual transformation of one species into another. Aristotle studied with Plato, but developed rather different views: He emphasized the importance of direct observation, instead of abstract reasoning, and gave much attention to biological issues. Aristotle contrasted two possible mechanisms for reproduction and development. A complex embryo might form by differentiation of homogeneous material (**epigenesis**) or, alternatively, the structure might be present from the start, so that development simply entails an unfolding (**preformation**). These alternatives run through the whole history of developmental biology: Our present view of development as a process of differentiation guided by a genetic program contains elements of both. Aristotle ordered organisms into a linear sequence from least to most complex (the *Scala Naturae*; Fig. 1.3). Like Plato, however, he held that species have fixed essences and that the world is eternal and unchanging—views that precluded any kind of evolution.

Aristotle had a profound influence on medieval Christian thought: Indeed, the authority given to his writings inhibited further enquiry. Aristotle's emphasis on the fine adaptations of organisms led St. Thomas Aquinas to the **argument from design**. According to this idea, the order seen in the living world proved that it was created by an intelligent and benevolent designer. Right up until the time of Darwin, such **natural theology** motivated the study of biology through a belief in the revealed wonders of creation. The one fundamental difference introduced by Christianity was the proposal that the creation was recent. However, the world was not thought to develop after its creation, and so this view was just as incompatible with any idea of evolution as was Aristotle's belief in an unchanging and eternal world.

Mechanical Explanations of the World Did Not Lead to an Understanding of Evolution

In the late 15th century, the Renaissance saw the beginnings of modern science: study of the world by observation and experiment, rather than by introspection or by interpretation of sacred texts. The new astronomy of Nicolaus Copernicus and Galileo Galilei removed man (i.e., the Earth) from the center of the universe and revealed a world far larger and stranger than had previously been imagined. The discovery of simple physical laws culminated in Isaac Newton's *Principia* (1687), in which one universal law accounted for the movements of the planets, for the tides, and for gravity itself. These findings stimulated the search for entirely material explanations, an outlook propounded most influentially by René Descartes' mechanical philosophy. This reduction of complex phenomena to simple mechanisms finds its modern expression in molecular biology, which sets out to explain all aspects of life in terms of physics and chemistry.

Although this new view of the physical world opened up the possibility of scientific inquiry into the living world, it was not as helpful to biology as one might suppose. God was thought to have established fixed laws of nature, which then generated the universe in a mechanical way, with no further divine intervention (a philosophy termed **deism**). Thus, the emphasis was on understanding the divine plan by study of the adaptations of organisms to their environment and of their relationships with each other. Natural theologians continued to study the details of individual adaptations in the belief that they gave an insight into the majesty of creation (as is apparent from the title of John Ray's [Fig. 1.4] *The Wisdom of God Manifested in the Works of the Creation* [1691]). Species were still seen as fixed and part of Aristotle's *Scala Naturae*, now known as the "Great Chain of Being" (Fig. 1.3). New observations soon complicated this idea of a simple linear

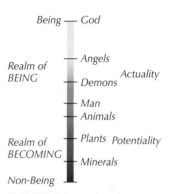

FIGURE 1.3. Medieval version of the Great Chain of Being based on Aristotle's ideas.

FIGURE 1.4. John Ray provided one of the first systematic descriptions of plants.

FIGURE 1.5. The polymath Robert Hooke clarified the relationship between living organisms and fossils. This engraving shows fossils from the Isle of Wight, compared with living species.

progression, but they were still seen as revealing a fixed plan. Thus, fossils of organisms now extinct forced the "Great Chain" to be seen as unfolding through time, but it was still seen as having a fixed underlying structure, with humans just below the angels. Carolus Linnaeus' *Systema Naturae* (1735) established the modern hierarchical system of biological nomenclature, in which species are grouped within genera and orders. However, whereas we now see this as revealing ancestry, Linnaeus saw it as a manifestation of the divine order, which related the separately created groups.

The New Science of Geology Revealed an Old and Changing Earth

FIGURE 1.6. The geological ideas of Vulcanism and Neptunism continued to develop in the 18th century. (*Top*) Cone-shaped mountains identified as volcanoes in central France. (*Bottom*) James Hutton collecting fossils. He proposed a theory that combined sedimentation and Vulcanism.

Christian theology supposed that the Earth was recently created: In 1650, Archbishop James Ussher dated the creation at 4004 B.C. by counting back through the biblical chronology. The scientific discoveries that followed the Renaissance encouraged material explanations for the origins of our planet. In 1644, Descartes proposed that the Earth had formed from a star that cooled to ash.

By the 18th century, it was clear that most sedimentary rocks had been laid down under ancient oceans. Fossils were known to be genuine relics of earlier organisms, and so the presence of marine fossils in most sedimentary rocks confirmed their origin (Fig. 1.5). The great thicknesses of sediment demanded either a drastic catastrophe or a very much longer expanse of time than was apparently allowed by the Bible. These alternatives defined opposing camps of "catastrophists" and **"uniformitarians"** who continued to dispute into the mid-19th century: Indeed, we hear echoes of this debate in current arguments over whether the fossil record shows **"punctuated equilibrium"** and whether mass extinctions are due to asteroid impacts (see Chapter 10).

The presence of sedimentary rocks on dry land was a key problem for the new science of geology. Either the ocean had retreated ("Neptunism") or volcanic activity had raised the sediments up from the sea ("Vulcanism"). The identification in midcentury of cone-shaped mountains in central France as extinct volcanoes encouraged Vulcanism (Fig. 1.6). In 1788, James Hutton argued that eroded rock was continually being washed into the sea where it was laid down as sediment, a process counterbalanced by the uplifting of new land by earthquakes that were driven by the inner heat of the Earth. This view was supported

by Hutton's finding that granite and basalt are igneous rocks and by experiments that showed how the slow cooling of molten rock could produce crystalline minerals. Hutton thought that there was a steady balance between erosion and uplifting and that there was "no vestige of a beginning—no prospect of an end."

Fossils Played a Crucial Part in the Development of Geology

Georges Cuvier (Fig. 1.7), who was responsible for the collections of vertebrates at the newly founded Muséum d'Histoire Naturelle in Paris, made careful comparisons of the anatomy of different living species and was the first to use such knowledge to reconstruct fossil forms. This allowed him to show that many fossil forms, such as the woolly mammoth and mastodon, were now extinct and to use fossils to identify a sequence of strata of different ages. Cuvier interpreted the abrupt transitions between strata as reflecting catastrophes, although he supposed that these were usually only local. His emphasis on the intricate relations between different body parts, which his comparative anatomy had demonstrated, led him to a firm belief in the fixity of species and to oppose the idea of evolutionary transformation.

FIGURE 1.7. Georges Cuvier expanded Linnaeus' taxonomy by grouping both living species and fossils in classes. He believed that function was the basis for classification of species and that new species were created after catastrophic extinction events.

The methods for dating rock strata that were developed by Cuvier and others revealed how life had changed through time. In 1824, William Buckland discovered a giant carnivore that he named *Megalosaurus*—the first known dinosaur (Fig. 1.8). In the following years, Adam Sedgwick and Roderick Murchison established the chronology of the ancient rocks of Wales and so outlined the sequence of geological ages back to the Cambrian (see Fig. 10.1). The new geology gave the first glimpse of the history of life and showed the successive appearance of different groups through time—first fish, then reptiles, and then mammals. The dominant interpretation was of a series of catastrophes separating distinct eras.

In 1830, Charles Lyell published his *Principles of Geology*, which set out a quite different view—uniformitarianism. Lyell revived Hutton's steady-state theory and emphasized that observable processes such as erosion and volcanic activity, acting at more or less their present rates, could cause great changes if continued for long enough. Lyell supported his argument by showing that Mount Etna, in Sicily, had been built up by a long sequence of eruptions, most occurring before recorded history; yet, the volcanic rocks were geologically young and lay over sediments that contained fossils of forms still found in the Mediterranean. Lyell's uniformitarianism and his *Principles of Geology* (1830; Fig. 1.9) greatly influenced Darwin, who also sought to explain biological evolution through the accumulated effects of present-day processes. However, Lyell's steady-state view of the world was incompatible with the long-term changes that were beginning to be revealed from the fossil record and, indeed, with evolution.

Diverse Ideas on Evolution Emerged in the Years before Darwin

The 18th century, known as the "Age of Enlightenment," saw a fascination with biology. There was a new confidence in scientific explanation, founded on the suc-

FIGURE 1.9. The frontispiece from Charles Lyell's *Principles of Geology* (1830) shows the Roman Temple of Serapis, near Naples, which Lyell used as evidence that the level of the land had changed in historical times. After its construction 2000 years ago, the temple was partly submerged and later uplifted. Evidence for its submersion is shown by the roughened areas on each column, caused by the action of the marine rock–boring clams; below, the columns are smooth, having been protected by sediment.

FIGURE 1.8. The jaw of Buckland's *Megalosaurus* (1824)—the first published description of a dinosaur.

FIGURE 1.10. Map of Cook's voyage around the world in the *Endeavour* (1768–1771).

FIGURE 1.11. Joseph Banks was the naturalist on the *Endeavour*. He and his colleagues identified more than 2400 species of plants and animals not known in Europe.

cess of Newton's physics. At the same time, systematic improvement of agriculture began in Northern Europe and exploration brought Europeans into contact with the full diversity of life. The new biology was of great economic importance as new crops and new markets were discovered. Governments sponsored expeditions such as those of James Cook, which involved naturalists whose task was to make systematic collections of the newly discovered flora and fauna (Figs. 1.10 and 1.11).

A new faith in social and economic progress suggested the possibility of biological change. Pierre de Maupertuis (Fig. 1.12A) discussed ways by which humans in the tropics might have become dark-skinned, either by inheritance of chance variations in skin color or by a direct influence of the sun on the hereditary particles. Buffon suggested that closely related species might have shared a common ancestry, although he held that each family of species shared the same "internal mould." Linnaeus also allowed for the evolutionary transformation of species within genera, emphasizing the importance of hybridization in generating new forms. Erasmus Darwin (Charles Darwin's grandfather) wrote several evolutionary speculations in poetry and prose.

Jean-Baptiste de Lamarck set out what was the first clearly evolutionary theory (Fig. 1.12B). He came to question the fixity of species only in 1800, soon after he was appointed curator of invertebrates at the Muséum d'Histoire Naturelle. One factor that

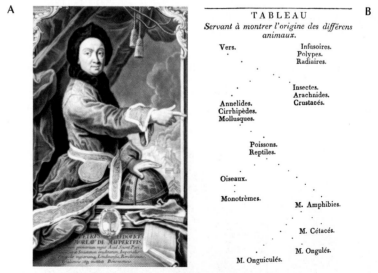

FIGURE 1.12. (*A*) Pierre de Maupertuis devised a theory of inheritance that involved hereditary particles and the spontaneous generation of life. (*B*) Jean-Baptiste de Lamarck saw evolution as a progression of separately generated lineages; his chain of animal life was published in his *Philosophie Zoologique* (1809).

influenced him was the continuous gradation he saw in fossil mollusks—then and now among the most complete records of evolution. But Lamarck's theory was quite different from Darwin's. He supposed that primitive organisms were generated spontaneously and then changed progressively along a linear Chain of Being. Because new forms were constantly being generated and ascending toward greater complexity, we now see a linear series of species, each evolving separately. Lamarck's theory, set out in his *Philosophie Zoologique* (1809), had little influence at the time: It was dismissed by Cuvier and was out of keeping with the conservative reaction against the French Revolution. In the late 19th century, an influential "neo-Lamarckian" school emphasized the inheritance of acquired characteristics in evolution. However, the idea was not original to Lamarck; such inheritance was widely accepted in his time.

After 1800, the Western Belief in Progress Strengthened as the Pace of Economic and Technological Change Accelerated

The belief in progress took different forms in different countries and led to a very different reception for evolutionary ideas. In France, social progress was seen as the inevitable consequence of increasing knowledge. However, human nature was seen as fixed, and so these ideas were not compatible with biological evolution. In Germany, biology was dominated by *Naturphilosophie*, which arose as a reaction against the mechanical worldview of Newton and Descartes. The idea of evolution was popular, but it was seen in mystical terms, as an unfolding analogous to individual development. In England, a pragmatic, utilitarian philosophy emerged, which was strongly influenced by Adam Smith's (Fig. 1.13) *An Inquiry into the Nature and Causes of the Wealth of Nations* (1776). Individual competition was seen as leading to an efficient economy that functioned for the good of all. This certainly influenced Darwin: However, the "invisible hand" of the market envisaged by Smith gave a more optimistic and benevolent outlook than did the "struggle for existence" emphasized by Darwin.

In 1844, the *Vestiges of the Natural History of Creation* was published anonymously (Fig. 1.14). This book set out an evolutionary theory akin to Lamarck's, in which organisms arose by spontaneous generation and were then progressively transformed into ever more complex forms. The book was a popular success, but

FIGURE 1.13. Scots economist Adam Smith's theories of economics influenced Darwin.

VESTIGES

OF

THE NATURAL HISTORY

OF

CREATION.

WITH A SEQUEL.

NEW YORK:
HARPER & BROTHERS, PUBLISHERS,
329 & 331 PEARL STREET,
FRANKLIN SQUARE.

FIGURE 1.14. (A) Robert Chambers believed that nature's laws could explain the material universe. (B) His *Vestiges of the Natural History of Creation* (published anonymously) described a theory of everything, from the origin of the universe to the origin of life and of humanity.

was fiercely criticized—primarily because Chambers included man within the evolutionary series, in direct conflict with Christian teaching. The significance of the *Vestiges* was not its scientific content, but rather its demonstration of the intense popular interest in evolution and the strength of the reaction against it.

CHARLES DARWIN

Darwin's Voyage in the *Beagle* Provided the Observations that Led Him to the Theory of Evolution

FIGURE 1.15. Charles Darwin as a young man in 1840.

Charles Darwin was born in 1809, into a prosperous and intellectually active family. He was grandson to Erasmus Darwin and to Josiah Wedgwood, who had brought the Industrial Revolution to the pottery trade. Although Darwin was undistinguished at school and loathed his classical education, he was from a young age an ardent naturalist (Fig. 1.15). Sent to medical school in Edinburgh, he recoiled from the horrors of the dissecting table and instead devoted himself to studying marine invertebrates with Robert Grant, a prominent exponent of Lamarck's evolutionary ideas. After 2 years, he gave up his medical studies and began to train for religious orders in Cambridge—more in the hope of pursuing natural history as a country parson than from religious zeal. The Cambridge degree required Darwin to master William Paley's *Natural Theology* (1802), which presented the exquisite adaptations of living organisms as evidence of their supernatural design. As at Edinburgh, however, Darwin's unofficial studies gave him a far more important training. He was gripped by the then-current craze for collecting beetles (Fig. 1.16), attended John Stevens Henslow's course in botany and his scientific soirees, and accompanied Adam Sedgwick on a geological expedition to North Wales in 1831. Later that year, his connections with scientific circles in Cambridge and his reputation as a naturalist brought him an invitation to join what was to be a 5-year survey of South America on the *H.M.S. Beagle*.

FIGURE 1.16. "Go it Charlie!" cartoon drawn by one of Darwin's Cambridge schoolmates. Beetle collecting was a passion for Darwin during his years at Cambridge. As he described in his *Autobiography*, "But no pursuit at Cambridge was followed with nearly so much eagerness or gave me so much pleasure as collecting beetles. It was the mere passion for collecting, for I did not dissect them, and rarely compared their external characters with published descriptions, but got them named anyhow."

At the start of his voyage, and despite acute sea sickness, Darwin read Lyell's recently published *Principles of Geology*. This, together with his recent training from Sedgwick, focused his attention on geological questions. In the Cape Verde Islands, he saw beds of sedimentary rock that had been lifted up from the ocean floor by volcanic action and then gradually curved downward as the land around subsided. Arriving in Chile just after the earthquake of 1835, Darwin saw that the land had been raised by several feet. In his first major contribution to geology, he realized that coral reefs could be explained by the gradual growth of coral as the shallow bed on which they had first formed slowly sank deeper beneath the sea (Fig. 1.17). All this convinced Darwin of Lyell's uniformitarian view—that the continued action of present-day processes could account for the geological form of the Earth.

It was natural for Darwin to ask whether present processes could also explain the "mystery of mysteries"—the origin of diverse and well-adapted species. He found fossil mammals in South America that were similar to the armadillos, sloths, and llamas presently living there, which suggested continuity of descent (Fig. 1.18). More crucial, however, were his observations of geographic distribution. On the open plains of Argentina, Darwin found (and ate) two subtly different species of rhea (flightless birds) that seemed to be competing to occupy the same territory. On the Galápagos, he found that each island held distinct species of mockingbirds, which all appeared to be related to species on the mainland.

On the Origin of Species Set Out the Argument for Evolution by Natural Selection

On his return to England in 1836, Darwin was convinced of the transmutation of species. In "the busiest two years of his life," he began to search for natural

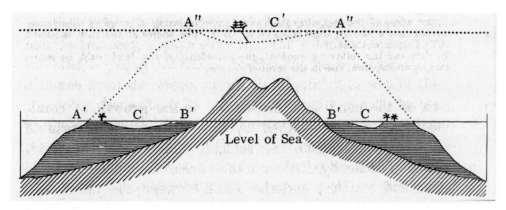

FIGURE 1.17. Darwin's explanation of coral reefs as shown in *The Voyage of the Beagle*. He explained that as the "new encircling barrier-reef" slowly sinks down, the corals continue to grow upward. The sea encroaches on the island until only the atoll remains (*dotted curve*).

mechanisms that would explain this evolutionary process. He was sure that evolution must be based on the gradual accumulation of slight individual variations. The idea of natural selection came to him in September 1838, as he read Thomas Robert Malthus' *Essay on Population*, which argued that because resources are finite, and yet reproduction is potentially exponential, there is an inevitable struggle for existence between individuals. Darwin saw that if individuals vary in their ability to survive this struggle and if such variations are inherited, then they will be "naturally selected":

> If variation be admitted to occur occasionally in some wild animals, and how can we doubt it, when we see thousands of organisms, for whatever use taken by man, do vary. If we admit such variations tend to be hereditary, and how can we doubt it when we remember resemblances of feature and character … . If we admit selection is steadily at work, and who will doubt it, when he considers amount of food on average fixed and reproductive powers act in geometric ratio. If we admit that external conditions vary, as all geology proclaims, they have done and are now doing—then, if no law of nature be opposed, there must occasionally be formed races, slightly differing from the parent races.

These words, written in 1842, come from a sketch of Darwin's theory, to be published if he were to die. However, Darwin was well aware that his theory would be controversial, an awareness reinforced by the hostility that had greeted Chambers' *Vestiges*. Darwin therefore spent 20 years amassing supporting evidence and told only a few close friends of his ideas. He began investigations into artificial selection and experiments on inheritance, and he spent much effort on a study of barnacles, which would show how evolutionary principles could be applied to classification. He also developed the theory further. In particular, he explained the divergence of species by analogy with the division of labor in economics, as expounded by Adam Smith. Just as an individual worker gains an economic advantage by specializing in a new trade, so individual organisms gain over their competitors by exploiting their environment in a different way.

In 1858, as Darwin was laboring over his massive work, he received a paper from Alfred Russel Wallace (Fig. 1.19) that set out essentially the same theory. Wallace was a professional collector then in the East Indies on his second major expedition. Like Darwin, he had been convinced of evolution by the geographic distribution of plants and animals and had come to natural selection after reading Malthus. His paper was read alongside extracts from Darwin's work at a meeting of the Linnaean Society of London. These papers had little immediate effect—indeed, the President of the Linnaean Society described 1858 as a year "which has not, indeed, been marked by any of those striking discoveries which at once revolutionise … the department of science on which they bear." In the following year, Darwin published an "abstract" of

FIGURE 1.18. Darwin's fossil finds during the voyage of the *Beagle* included the bones of this *Megatherium*. (The Wellcome Library, London.)

FIGURE 1.19. Alfred Russel Wallace sent Darwin a manuscript describing a theory almost identical to Darwin's.

his projected work, *On the Origin of Species by Means of Natural Selection.* All 1250 copies of the book were sold on the first day.

Within 15 Years of the Publication of the *Origin,* Most Educated People Accepted Evolution as a Fact

This rapid change in worldview was in large part due to Darwin's skill in amassing evidence and anticipating objections: As he put it, the *Origin* is "one long argument" and a most convincing one. His success was also due to his identification of a definite mechanism for evolutionary change. As we shall see, in the following half century, few agreed with Darwin that natural selection was the primary cause of adaptation and diversification. However, by proposing a testable and coherent mechanism, Darwin brought evolution into the realm of science rather than speculation.

The strength of Darwin's argument was in showing how evolution could explain many disparate facts: geographic distribution, the fossil record, morphological relationships, and comparative embryology. Acceptance of evolution opened up "a grand and almost untrodden field of enquiry" and allowed biologists to address questions that had hitherto been inaccessible. Paleontology and comparative anatomy received a particularly strong stimulus, and the search for evolutionary relationships dominated late 19th century biology. However, there was little progress in furthering Darwin's insights into the mechanism of evolution.

THE ECLIPSE OF NATURAL SELECTION

Natural Selection Was Not Widely Accepted as the Mechanism of Evolution

The *Origin* argues that evolution is by means of natural selection: specifically, by the accumulation of small variations, each established because they increase the reproduction of one individual relative to another. However, although Darwin's arguments for evolution were widely accepted, few took natural selection to be its main cause. Even Darwin's strongest supporters played down the role of selection based on slight differences between individuals. Wallace was perhaps the strongest advocate of natural selection, but tended to see it as being between varieties and species rather than between individuals. Thomas Henry Huxley, famous for his public defense of Darwin (Fig. 1.20), did not use the idea of natural selection in his own research on comparative anatomy and thought that evolution was based on variations of large effect (saltations or "sports"). In Germany, Ernst Haeckel was a strong supporter of "Darwinismus," but, again, did not see natural selection as its main mechanism. Haeckel saw selection as sifting among variants that had been produced by directed processes and saw evolution as progressing toward higher forms such as man.

Darwin's theory of evolution was most strongly supported by naturalists, who best appreciated the wide range of natural variation and the struggle for existence. Henry Walter Bates' (Fig. 1.21) and Fritz Müller's studies of **mimicry** provided what is still one of the best demonstrations of natural selection: Butterflies of different species evolve a closely similar appearance because individuals that resemble distasteful forms are less likely to be eaten by birds (Fig. 1.22 and pp. 475 and 506). However, this work was not followed by any systematic study of selection in nature. In the 50 years after publication of the *Origin*, there were only two studies of natural selection. In 1899, Herman Bumpus demonstrated that sparrows that survived a New England storm differed in shape from those that died and, around the same time, W.F.R. Weldon found that the shape of crabs in Plymouth Sound (England) changed as a result of selection (Fig. 1.23).

FIGURE 1.20. *Vanity Fair* magazine caricatures of Bishop Wilberforce (*left*), Thomas Henry Huxley (*middle*), and Charles Darwin (*right*). Wilberforce attacked Darwin's ideas at an 1860 British Association meeting, and Huxley—"Darwin's bulldog"—defended Darwin at this debate.

The Theory of Natural Selection Faced Real Obstacles, Which Were Only Resolved in the 20th Century

Because transitional forms were rarely found, evidence for both evolution and the mechanisms of evolution had to be indirect. Darwin emphasized the analogy with artificial selection—for example, the diverse plumage and behaviors that had been bred by pigeon fanciers (Fig. 1.24). However, it was often objected that no new species had been produced in this way and that selection could only produce changes within certain limits. A stronger objection was that the precursor of a complex organ, such as the vertebrate eye, could not have given any initial advantage. Darwin's response was that

FIGURE 1.21. Henry Walter Bates was a friend of Alfred Russel Wallace and encouraged his ideas on evolution. He is best known for explaining **Batesian mimicry** of color patterns in butterflies.

- even a limited function might be advantageous (e.g., the ability to sense light without forming a clearly focused image);
- functional intermediates could in fact be identified (e.g., eyes of varying degrees of complexity);
- organs might switch from one function to another.

These issues are discussed in more detail in Chapters 3 and 24.

FIGURE 1.22. Batesian mimicry. Nonpoisonous butterflies from the same species (*bottom row*) mimic the warning coloration of different species of poisonous butterflies (*top row*) to improve their chances of survival.

FIGURE 1.23. W.F.R. Weldon studied natural selection in the crab, *Carcinus maenas*.

A difficulty that was harder to counter came from calculations of the age of the Earth, made by the physicist William Thomson (later Lord Kelvin). He argued that the Earth had cooled from a molten state as heat was radiated into space and that this would have taken only about 25 million years—much less than Darwin thought necessary for gradual geological change and for gradual evolution by natural selection. The error in Thomson's calculations became apparent only in the next century when it was discovered that radioactive decay provides enough heat to offset the Earth's cooling. Radioisotopes now date the Earth at about 4.5 billion years, in rough agreement with Darwin's estimates based on rates of geological processes.

The problems that gave Darwin the greatest difficulty were those related to inheritance. In a review of the *Origin* published in 1867, the engineer Fleeming Jenkin distinguished small variations from sports. He believed that the former would respond to selection, but only within fixed limits. In contrast, if a single distinct "sport" were picked out by selection, it would necessarily be mated with the original population, and so its offspring would blend back toward the original state: Blending inheritance would eliminate inherited variation, making selection ineffective. Francis Galton, Darwin's cousin, also emphasized the importance of "sports," although for different reasons. He devised the statistical method of **regression** (Chapter 28 [online]) as a tool for quantifying the inheritance of continuous variation. Galton mistakenly believed that the offspring regressed back toward the mean of the original population, which therefore could not change (Fig. 1.25).

Jenkin's and Galton's arguments were based on an ignorance of heredity. As we shall see, although Mendelian genetics did eventually settle these difficulties, it took 20 years from the rediscovery of Mendel's laws for the confusion to be resolved.

Alternative Mechanisms of Evolution Were Popular

Cultural preconceptions were more important than these scientific objections in preventing acceptance of natural selection. Darwinian evolution is clearly incompatible with a literal interpretation of the Bible, but such views had in any case lost force by the mid-19th century. The religious and moral objections to natural selection were more subtle. On Darwin's view of evolution, there is no overall tendency for life to progress toward "higher" forms and man has no special place in nature. Darwinian evolution is consistent with deism, in which God works through regular laws rather

FIGURE 1.24. Varieties of pigeons illustrate the results of artificial selection for specific traits. Darwin kept a collection of such pigeons.

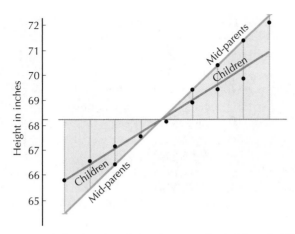

FIGURE 1.25. Galton's original data on the relation between the heights of children (*y* axis) and the average height of their parents (*x* axis). Galton realized that the children deviate less from the average of the population than do their parents and termed this a "regression to mediocrity." He believed that the offspring of a selected group of parents would regress toward a fixed mean value, which would limit the change caused by selection. In fact, the regression is toward the *current* mean of the population, which may change with time (see Chapters 4 and 28 [online]).

than by miracle (see p. 82). However, natural selection, which is based on random death and extinction, was widely felt to be an unacceptable mechanism.

Alternative mechanisms of evolution seemed more plausible. The direct effects of use and disuse of organs and the inheritance of acquired characters (known as **Lamarckism**) were thought to allow faster evolution and to introduce a tendency to progress. In the face of the objections summarized above, Darwin gave such mechanisms greater weight in later editions of the *Origin*, although he never thought them as important as natural selection. August Weismann argued strongly (and correctly) that, in animals at least, the **germ line** is separate from the **soma** and that inherited characteristics therefore cannot be influenced directly by the environment; he was the only major figure to rely exclusively on natural selection as an evolutionary mechanism. "Neo-Lamarckian" inheritance was seen as a key evolutionary force well into the 20th century.

A rather different class of theories involved some kind of **orthogenesis**, in which lineages have an inherent tendency to change in some direction. Prominent supporters of "Darwinism," such as Ernst Haeckel and Herbert Spencer, relied on such mechanisms and invoked natural selection only to sift among variations generated by other forces. Those such as Huxley and Galton, who emphasized the role of "sports," can be seen in the same light. Such theories are fundamentally in conflict with Darwin's view of natural selection, as they require complex adaptations to be produced directly, rather than gradually assembled and refined by selection.

In Britain, the debate over whether evolution was based on continuous variation or on "sports" came to an acrimonious head. On one side, Weldon (Fig. 1.26) began an energetic research program into selection on continuous variation (e.g., Fig. 1.23) and stimulated the mathematician Karl Pearson to develop the necessary statistical methods. Weldon and Pearson gained Galton's support in setting up a committee of the Royal Society to sponsor their **biometric** researches. On the other side, in 1894 William Bateson published *Materials for the Study of Variation*, a massive catalog that argued for the importance of discontinuous evolution (**saltation**). Bateson vigorously criticized Weldon and Pearson's biometry and, in 1896, was added to the Evolution Committee. Four years later, and after much argument, the biometricians resigned, and the committee became exclusively a vehicle for Bateson and his supporters. The disputes remained unresolved when Weldon died in 1906.

FIGURE 1.26. W.F.R. Weldon (*left*), Karl Pearson (*middle*), and Francis Galton (*right*).

Rediscovery of Mendelian Heredity Strengthened the Divide between Gradualists and Saltationists

FIGURE 1.27. Gregor Mendel and his peas.

The dispute at the end of the 19th century over the mechanism of evolution stemmed largely from ignorance of the mechanism of heredity. Ironically, the essentials of our modern understanding of inheritance had been established in 1866 by a monk from Moravia, Gregor Mendel (Fig. 1.27). If his work had been properly understood at the time, half a century of confusion might have been avoided. Yet, its rediscovery in 1900 had the paradoxical effect of deepening the divide between saltationists, such as Bateson, who emphasized the importance of discontinuous variations like those studied by Mendel, and gradualists, who held to the Darwinian reliance on continuous variation. It is important to understand the history of this time, both because it shaped the development of biology over the next century and because many present-day issues concerning evolution trace back to the old arguments between gradualists and saltationists.

Mendel was ordained as a priest at the age of 25 and then trained as a science teacher at the University of Vienna. When Mendel returned to his monastery in Brunn (now Brno), he began 15 years of systematic experiments on heredity in peas and other species. His key innovations were to concentrate on particular characteristics that remained distinct (e.g., round or wrinkled seeds, or the position of flowers on the stem) and to make careful counts of these characteristics in the offspring. Working with 22 **true-breeding** varieties of pea, Mendel observed more than 30,000 plants. He found that the offspring of a cross between two true-breeding varieties (termed the F_1) were uniform: For each character, they showed only one of the two alternative types. If these plants were self-fertilized to produce an F_2 population, then three-quarters of the plants showed the **dominant** character that had been seen in the F_1, but the **recessive** character reappeared in one-quarter of the plants. Mendel then self-fertilized these plants and found that of the three-quarters that showed the dominant character, one-third bred true, but two-thirds produced 3:1 ratios in their offspring, in just the same way as the F_1 (Fig. 1.28). If two varieties differing in several characters were crossed, then each character segregated independently in subsequent generations.

Mendel's explanation was that each characteristic in his experiments was determined by a pair of elements, one derived from the mother and one from the father. A plant that received two different elements took on the appearance characteristic of one of them, but nevertheless produced two different kinds of egg cell and two kinds of pollen grain, each carrying a single element, and in equal proportions. The simple ratios seen in later generations arose because these elements combine at random. The crucial feature of Mendelian inheritance is that it is

	Round	Wrinkled
1.	45	12
2.	27	8
3.	24	7
4.	19	16
5.	32	11
6.	26	6
7.	88	24
8.	22	10
9.	28	6
10.	25	7
Total	**336**	**107**

FIGURE 1.28. The wrinkled versus round seed produced by ten plants of the F_1 generation of Mendel's peas showing the overall 3:1 ratio.

particulate: The elements (which we now call **genes**) remain intact. Thus, variation is not dissipated by "blending"; there is no direct influence of the environment (heredity is "hard," as Weismann argued); and enormous variety can be generated by the random combination of different elements. Previous theories, put forward by Darwin, Weismann, and others, had also postulated particles. However, their models had many particles for each characteristic, and so no simple ratios were expected.

Why was Mendel's work ignored for so long? He was hardly a prolific author: His 8 years of experiments with peas were summarized in a single paper "Experiments in plant hybridisation," read at the Brunn Natural History Society in the spring of 1865 and published in the journal of the Society in the following year. However, the journal was reputable and was distributed to more than a hundred libraries across Europe. A more fundamental reason for Mendel's neglect, however, was that the times were not ready for such a theory. In the mid-19th century, there was little interest in studying heredity for its own sake: Inheritance was seen in relation to the origin of species and to the development of the individual, and Mendel's work had no obvious bearing on these larger questions.

By the Early 1880s, Advances in Microscopy Had Made Clear the Basic Biology of the Cell

Much of the groundwork for the rediscovery and elaboration of Mendel's work was laid by the development of **cytology**. This showed that organisms are made up of cells whose internal structure is remarkably similar across a wide range of organisms. The nucleus was shown to determine the form of the cell—most elegantly by Theodor Boveri, who found that when the nucleus of one species of sea urchin was transplanted into the egg of another, the larva had all the characteristics of the parent nucleus. The nuclear material was organized into long, threadlike chromosomes, which divided in a regular way. Every adult cell of a species had the same number of chromosomes, whereas the gametes had half that number (Fig. 1.29). A new individual was formed by sexual union,

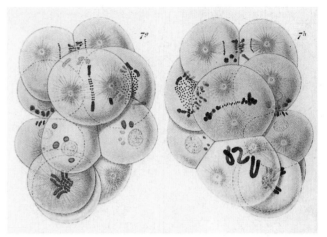

FIGURE 1.29. Theodor Boveri's studies of chromosomes in the roundworm *Ascaris'* eggs answered many questions about the number and behavior of chromosomes in germ and somatic cells.

with an equal contribution from father and mother. All of these findings, which we now take for granted, were established over little more than a decade. Summarized in books such as Hugo de Vries' *Intracellular Pangenesis* (1889) and Weismann's *Germ Plasm* (1892), they spurred new interest in inheritance. de Vries began breeding experiments and found that striking new variants sometimes arose in the evening primrose (*Oenothera lamarckiana*). In his influential *Theory of Mutation* (1900), he proposed that new species originated from just such discontinuous variations.

By the turn of the century, breeding experiments had shown clear 3:1 segregation ratios in many plant species. This led Carl Correns, de Vries, and Erich Tschermak von Seysenegg independently to the correct theory of particulate inheritance. When they came to write up their results, they found that Mendel had anticipated them by 35 years. These independent rediscoveries were immediately seen as a major advance and, within the next decade, the essentials of classical genetics were established. Mendelian inheritance was found to apply to diverse traits in both animals and plants, and **linkage**, **incomplete dominance**, and multiple **alleles** were discovered. Bateson, in particular, coined the term "genetics" and took up the new Mendelian theory with enthusiasm. He presented it as evidence for the importance of discontinuous variation in evolution. The biometricians responded by arguing that Mendel's laws did not apply to the continuous variation that they saw as driving evolution.

Morgan Introduced the Fruit Fly *Drosophila melanogaster* as a Model Organism for Genetic Studies

Mendel had carefully chosen varieties and characters that would give the simplest patterns. In the decade after his work was rediscovered, it was extended to essentially its present form. In modern terms, it was found that each gene could take many alternative forms or alleles; that these did not always show simple dominance relations, so that **heterozygotes** could be distinguished; that several genes may affect each trait and that several traits might be affected by each gene; and that genes did not always segregate independently, but instead could be **linked**. In 1909, Thomas Hunt Morgan's group, working in the "fly room" at Columbia University (Fig. 1.30), began to make large-scale searches for mutations and used linkage relations to show that genes could be arranged in a linear order.

The main result in the following years was the demonstration that these linkage groups correspond to chromosomes. Walter Sutton had proposed as early as 1902 that the chromosomes consist of **homologous** pairs, one coming from each

FIGURE 1.30. T.H. Morgan in his Columbia University fly room. Morgan and his coworkers, who included Calvin B. Bridges, Alfred Sturtevant, Theodosius Dobzhansky, and Hermann Muller made seminal discoveries regarding mutations, linkage groups, and genes and established paired chromosomes as the physical basis of Mendel's laws of heredity.

parent, and that each pair carries distinct information: These "constitute the physical basis of the Mendelian law of heredity." Morgan's group confirmed this hypothesis, by documenting the close correlation between chromosomal variants and the patterns of inheritance they determined. Most important was the discovery of sex-linked genes: In male *Drosophila*, which carry a single X chromosome, only one copy of such genes is found, whereas in females, which carry two X chromosomes, there are two copies of each sex-linked gene. This strongly suggests that sex-linked genes are carried on the sex chromosomes. There was surprisingly strong resistance to the chromosomal theory; indeed, Bateson never accepted it. Even after it was established, there was still no understanding of the physical basis of heredity. As we shall see in the next chapter, that would come only half a century later.

The early Mendelians doubted the efficacy of selection on discontinuous variation and instead emphasized mutation as the primary cause of evolutionary change. The Danish plant breeder Wilhelm Johannsen was particularly influential. He made the first clear distinction between **genotype** and **phenotype** and argued, quite correctly, that selection could not be effective on a "pure line," which consists of only a single genotype. His selection experiments on brown beans appeared to confirm this view: Although selection could change a population that consisted of a mixture of supposedly pure lines, each propagated by self-fertilization, it seemed to have little effect on any one line. Johannsen then argued that selection could only sort among the genotypes present in a mixed population and could not create anything new. We can now see that this argument is quite misleading: Because parents usually differ in many genes, there are so many potential gene combinations in their F_2 offspring that only a tiny fraction can actually be realized (Fig. 1.31). Selection can thus be effective for many generations, even if mutation introduces no new variation. At the time, however, Johannsen's "pure line theory" was widely accepted as a proof that Mendelian heredity and Darwinian natural selection were incompatible.

Discrete Mendelian Genes of Small Effect Can Account for Continuous Variation

By 1910, Mendelian genetics was established as a thriving field of research. However, it was almost universally regarded as being incompatible with Darwinian selection, which depended on the selection of slight variations. Over the following

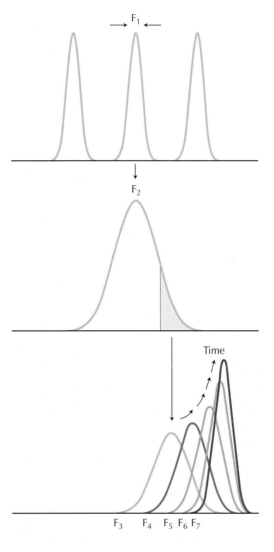

FIGURE 1.31. An example showing how the cumulative effect of selection can produce phenotypes that are never seen in the original population. Two true-breeding lines differ substantially in some trait, in this example by 20 standard deviations. This difference is due to ten unlinked genes, with equal and additive effects. The F_1 of a cross between two lines is intermediate between the two parents (*top*). If the F_1 individuals are crossed with each other to make an F_2 generation, recombination produces genetic variation (*middle*). In principle, every possible genotype is present in this population. However, the chance of recovering one of the parental genotypes is extremely small—2^{-20}, or less than one in a million. Thus, the distribution in the F_1 generation does not appreciably overlap either parental phenotype. Nevertheless, a few generations of selection can reconstruct either parental genotype. The bottom panel shows that if the largest 10% of the population is selected in every generation, the original parental distribution is recovered after only five rounds of selection.

decade, this divide was bridged by both experiment and theory. Continuous variation was explained as a consequence of multiple Mendelian factors, and selection on such variation was shown to be remarkably effective. This reconciliation laid the foundations for the wider synthesis that was to follow in the middle of the century.

We now know that multiple Mendelian factors, each making a small contribution to the character of interest, combine with random environmental effects to give apparently continuous variation. This explanation was suggested by Mendel himself to explain a gradation in flower color in the F_2 of a cross between beans with red or white flowers. The idea was known to Bateson, Pearson, and others and was laid out in some detail in 1902 in a paper by George Udny Yule. However, it was at first neglected in the shadow of the bitter dispute between Mendelians and biometricians. In 1908, the Swedish plant breeder Herman Nilsson-Ehle reported his work on oats and wheat. He showed that a careful analysis of slight variations in grain color revealed underlying Mendelian ratios. For example, in the F_2 of a cross between brown and white chaffed winter wheat, he found 15 brown grains to each white grain and continuous variation in color within the brown class. This he explained by two Mendelian factors: The double recessive homozygote was white, and at a frequency of one-sixteenth, and the other gene combinations gave various shades of brown. Figure 1.32 shows how variation in three factors can lead to virtually continuous variation of the kind studied by Nilsson-Ehle.

As you will see in Chapter 14, it is difficult even today to demonstrate the exact genetic basis of a continuous trait—that is, to show which genes contribute and how they do so. Nevertheless, Nilsson-Ehle's work, together with Morgan's discovery of mutations in *Drosophila* with minute effects, quickly convinced geneticists that essentially all kinds of variations could be explained in Mendelian terms. Such work also emphasized the enormous variety that could be generated by sexual reproduction among individuals that differed at many genes. The full significance of the variation generated by sex is only now becoming understood (Chapter 23).

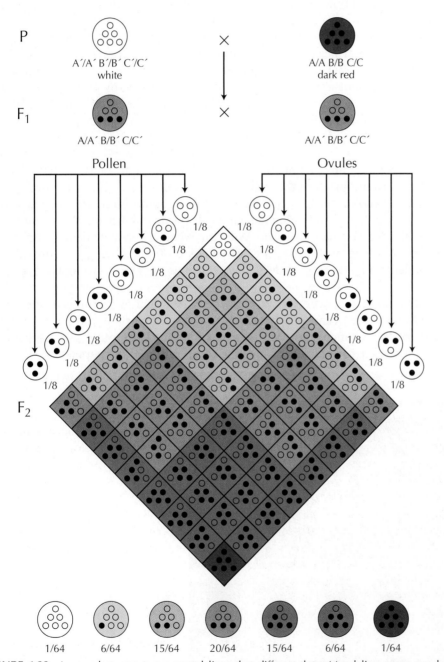

FIGURE 1.32. A cross between two parental lines that differ at three Mendelian genes can lead to almost continuous variation. One parent is homozygous for alleles A, B, C and has dark red seeds; the other is homozygous for alleles A', B', C' and has white seeds. The F_1 has intermediate-colored seeds, and in the F_2, there is wide variation. If seed color depends on the number of alleles inherited from one or the other parent, as shown here, then the 64 different F_2 genotypes show seven different average phenotypes; in practice, this would appear virtually continuous. We discuss this kind of variation in detail in Chapter 14.

Following Johannsen's experiments on "pure lines," it was widely held that selection could not take a population beyond the extremes seen in an F_2 population—after all, every possible gene combination could in principle be generated there. However, selection experiments soon showed this argument to be flawed. For example, lines of maize had been selected since 1889 for high and low oil content and by 1910 had passed beyond anything seen in the original stock. (This experiment is still continuing, and the lines are still diverging [see Fig. 17.31B].) William E. Castle's experiments on rats had great influence. He crossed "hooded" rats, which were homozygous for a recessive allele that gives a black stripe on the back, to the "Irish" strain, which is entirely black on the back and sides (Fig. 1.33). When these were crossed back to the parent strain to recover the recessive homozygote, the stripes were larger. Castle then selected for larger or smaller stripes and, after 4 years, saw extreme divergence. Castle at first thought that he was sorting among different alleles of the hooded gene. However, by 1919, further crosses had shown that the dramatic divergence was due to differences at many genes. It is this accumulation of multiple minor variants that allows the population to move far beyond anything actually seen in the original population (Fig. 1.31).

There was considerable confusion at first over how Mendelian variation would be distributed within a population. It was clear that with random mating, an F_2 population would produce the same 1:2:1 ratios in subsequent generations. However, the generalization to a population of any composition was not made until 1908. The accomplished mathematician G.H. Hardy published a short note in *Science*, while the German physician Wilhelm Weinberg independently published a detailed (and long neglected) series of papers on the mathematical consequences of Mendelian inheritance. What is now known as the **Hardy–Weinberg law** establishes that after a single generation of random mating, the genotypes $QQ:PQ:PP$ will be found in proportions $q^2:2pq:p^2$, where the initial proportions of the two alleles Q and P are q and p, respectively. Its most important consequence is that Mendelian inheritance does not dissipate variation: Thus, other processes such as selection are free to gradually change the composition of the population (Box 1.1).

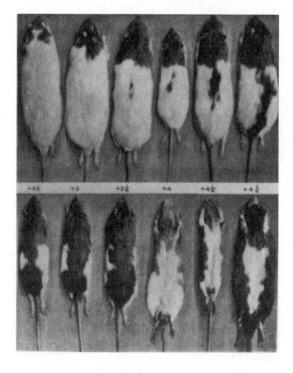

FIGURE 1.33. William E. Castle showed the effect of selection on coat color through experiments on hooded rats. Instead of a 3:1 ratio, they demonstrated a graded series, caused by variation of expression of the *hooded* gene. (*Top*) Selection for white coat. (*Bottom*) Selection for dark coat.

Box 1.1   Hardy–Weinberg Law

How do the proportions of different genotypes change over the generations under Mendelian inheritance? This is the most basic question in population genetics, and it is answered by the Hardy–Weinberg law. We explain the law here for the simplest case of a single gene that comes in two forms, or alleles. We refer to these different alleles as P and Q. (Genetic terminology is summarized in Box 13.1.)

Imagine a population of diploid organisms that release male and female gametes, which then come together at random to form the next generation of diploid zygotes (many marine invertebrates have this kind of life cycle). We will assume that the proportion of P and Q alleles in the pool of gametes is p and q, respectively. Because there are only two alternative alleles, it must be true that $p + q = 1$. For simplicity, assume that allele frequencies are the same in male and female gametes.

Now, the chance that a new diploid individual will have genotype QQ is just the chance that it has a Q allele from the male parent and also a Q allele from the female parent, so $q \times q = q^2$. The chance of being heterozygous, PQ, is the chance of getting Q from the male gamete and P from the female or getting P from the male and Q from the female: $(q \times p) + (p \times q) = 2pq$. Finally, the chance that the genotype will be PP is just p^2. Thus, random union of gametes gives genotype frequencies $q^2:2pq:p^2$. Reassuringly, these add up to 1, because $q^2 + 2pq + p^2 = (p + q)^2 = 1$.

This is illustrated in Figure 1.34A. The lengths along the vertical direction represent the allele frequencies in female gametes, and those along the horizontal axis, the allele frequencies in male gametes. The areas of the four rectangles represent the frequencies of the different diploid genotypes formed by random union of gametes: q^2 for QQ (top left), $2pq$ for PQ (top right and bottom left), and p^2 for PP (bottom right).

What happens if diploid adults mate at random? Then, there are nine kinds of mating, as illustrated by the nine squares in Figure 1.34B. Matings between QQ males and QQ females produced entirely QQ offspring (top left); mating between PQ males and QQ females produce half QQ and half PQ offspring (top center); matings between PQ males and PQ females produce $QQ:PQ:PP$ in proportions 1:2:1 (center); and so on. We can see from the diagram that QQ offspring occupy a square (shaded red, top left) that has edges with length equal to the frequency of QQ plus half the frequency of PQ. This is just the frequency of the Q allele in the adult diploid population, q. So, the proportion of QQ offspring is just q^2, and by the same argument, the proportions of PQ and PP are $2pq$ and p^2, respectively.

We also see that even if the diploid population is *not* in Hardy–Weinberg proportions, a single round of random mating restores Hardy–Weinberg proportions in the new generation of zygotes. This is easy to see for organisms that release gametes into a pool (as we assumed in Fig. 1.34A), because all that matters is the allele frequency in these gametes. However, as we can see from Figure 1.34B, the same is true when diploid individuals mate and produce offspring by internal fertilization.

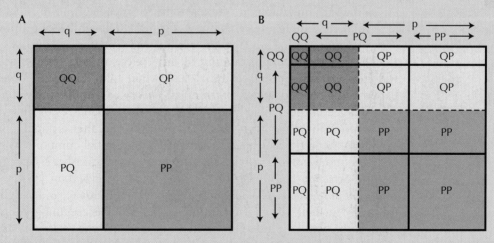

FIGURE 1.34. The Hardy–Weinberg formula gives proportions of the diploid genotypes formed by random union of haploid gametes (*A*) or by random mating between diploids (*B*). (*A*) Areas of the boxes are proportional to the frequencies of the diploid genotypes that are formed by the random union of haploid gametes; distances along the sides represent frequencies of alleles q and p. (*B*) Frequencies of diploid genotypes are indicated along the sides, and areas of the nine boxes (bounded by *solid lines*) indicate the frequencies of the nine mating combinations. Adding up the areas occupied by the genotypes shows that the genotype proportions are still given by the Hardy–Weinberg formula (compare *colored squares* with those in *A*).

THE EVOLUTIONARY SYNTHESIS

The Ideas of Mendel and Darwin Were Reconciled in Classical Population Genetics

FIGURE 1.35. R.A. Fisher (*top*), Sewall Wright (*middle*), and J.B.S. Haldane (*bottom*). Their work reconciled Mendelian genetics, biometry, and natural selection.

The full theoretical reconciliation between Mendelian genetics, biometry, and natural selection came primarily through the work of R.A. Fisher, Sewall Wright, and J.B.S. Haldane (Fig. 1.35). The first and most important contribution came in 1918, with Fisher's paper on "The correlation between relatives on the supposition of Mendelian inheritance." This set out to show that the correlations between relatives that had been measured by Galton, Pearson, and the other biometricians could be explained by multiple Mendelian factors, together with random nongenetic influences. However, Fisher did much more than this. He defined the **variance** and showed that it could be expressed as a sum of components due to genetic and nongenetic causes. The genetic component could itself be separated into components due to additive effects, dominance, and interactions between different genes. Fisher's 1918 paper established much of the present theory of quantitative genetics, as well as the key statistical technique of **analysis of variance** (see Chapter 14).

Sewall Wright began his research as a graduate student with Castle and devised the crosses that showed that the response to selection on hooded rats was due to numerous modifying factors. Using guinea pigs, he went on to study how genes interacted to determine traits such as coat color. His main theoretical contribution was to the understanding of **inbreeding** and the closely related process of random **genetic drift**. Mendel had shown that in a population of separate, self-fertilizing lines, the proportion of heterozygotes falls by one-half in every generation. However, despite the practical importance of inbreeding, it was not known how other systems of mating altered the genetic composition of a population. In a series of papers published in 1921, Wright set out a general analysis of inbreeding and went on to show how the matings between relatives that are inevitable in a small population cause random fluctuations in allele frequency. He termed this process genetic drift and showed how it interacts with selection, mutation, and migration (see Chapters 15 and 18).

The theoretical understanding of how selection acts on Mendelian variation was not seriously addressed until 1924, when J.B.S. Haldane published the first paper in his series on the *Mathematical Theory of Natural and Artificial Selection*. This was a comprehensive treatment of the way differences in survival and reproduction due to one or two Mendelian genes would alter a population.

These theoretical advances established the essential components of population genetics and were summarized in three key works: Fisher's *Genetical Theory of Natural Selection* (1930); Haldane's *The Causes of Evolution* (1932); and Wright's *Evolution in Mendelian Populations* (1931). Fisher, Haldane, and Wright differed significantly in their views of evolution. Fisher emphasized the gradual change of a single large population due to selection on many minor variations. Haldane placed greater emphasis on strong selection on single genes. Trained as a biochemist, he was also more concerned with the physical basis of genetics: For example, in 1923, he published one of the first suggestions as to how life might have originated from prebiotic chemistry (Chapter 4). Finally, Wright argued that adaptation would be most effective if species were subdivided into many small subpopulations, so that there could be a **shifting balance** between different evolutionary processes (p. 607). Despite these differences, however, all three founders of population genetics saw their main achievement as the reconciliation between the ideas of Mendel and Darwin, which allowed a wider synthesis of diverse fields of biology.

The "Evolutionary Synthesis" Brought Together Diverse Fields

The division in the early 20th century was not simply between Mendelians and biometricians. Rather, a divide had opened up between experimental biologists, who worked with model organisms in the laboratory, and field naturalists, who observed natural populations in the wild. (As we will see at the end of Chapter 2, a similar divide is seen today in the different perspectives of evolutionary biologists who work with molecular variation and those who investigate whole organisms.) Theories of evolution that were popular among experimentalists, such as de Vries' mutation theory, concentrated on how variation was generated. They provided no explanation of how organisms came to be adapted to their environment and supposed that new species appear instantaneously. In contrast, naturalists thought of variation as ever present and considered species as breeding populations; they asked how these populations came to be separated into new species and how they adapted to changing circumstances. Thus, the emergence of modern evolutionary biology required more than the reconciliation between Mendelism and Darwinism that was completed by the early 1920s. The new genetic Darwinism had to be shown to apply in nature, and different fields of biology had to be brought together into a consistent framework. This integration, which took place in the years around 1940, is known as the **Evolutionary Synthesis.**

Russian researchers played an important part in taking genetics out of the laboratory. Breeding experiments were cheap and so could be carried out in the turbulence that followed the Russian Civil War (1918–1920). Moreover, there was already a strong Darwinian tradition in Russia, with an emphasis on ecological questions. Sergei Chetverikov showed that natural populations of *Drosophila* harbor considerable genetic variation in the form of recessive alleles, whose effects are revealed when the flies are inbred and could be taken up by natural selection when conditions change. In 1927, Theodosius Dobzhansky (Fig. 1.36) joined Morgan's laboratory as it moved to the California Institute of Technology and took the ideas of the Russian school to the West. He began field collections in the Sierra Nevada and again found substantial hidden variability. He also began to investigate the genetic basis of hybrid sterility between *Drosophila* species. Dobzhansky was impressed by Wright's summary of his shifting balance theory of evolution at the Sixth International Congress of Genetics, in 1931. Six years later, Dobzhansky published *Genetics and the Origin of Species*, which combined Wright's theoretical ideas with Chetverikov's perspective on natural populations. This book had wide influence, because it applied the new genetics to the questions that concerned naturalists—especially those concerning the nature and origin of species.

Parallel developments were taking place in Great Britain. E.B. Ford (Fig. 1.37) applied Fisher's ideas to study selection on visible polymorphisms in moths and butterflies. In 1924, Haldane had shown that the spread of the melanic form of the moth *Biston betularia* could be explained if the melanic moths had a 50% greater chance of survival than nonmelanic moths in polluted areas, possibly because these darker forms were better camouflaged against darker backgrounds (Fig. 1.38). Ford confirmed that such strong selection pressures are common and, with Fisher, argued against any significant role for Wright's random genetic drift. His students went on to demonstrate that natural selection acts strongly on even apparently minor differences, such as the banding patterns on snails or the beak shape of the Galápagos finches. Similarly, Dobzhansky moved toward greater emphasis on natural selection. He found that chromosomal **inversions** in *Drosophila pseudoobscura* fluctuated regularly over the seasons, which implied the action of selection rather than drift (Fig. 1.39). Overall, there was a move

FIGURE 1.36. Theodosius Dobzhansky provided a fusion of Darwinism and genetics in his book *Genetics and the Origin of Species* (1937).

FIGURE 1.37. Edmund Brisco ("Henry") Ford studied selection of visible polymorphisms in moths and butterflies using Fisher's ideas.

FIGURE 1.38. *Biston betularia,* the peppered moth, in its typical (*top*) and melanic (*bottom*) forms. Haldane theorized that the melanic form had a greater chance of survival because it was better camouflaged against the darkened tree trunks in polluted areas.

through the 1950s toward the view that most differences within and between species were due to adaptive natural selection rather than chance.

Following the theoretical framework laid out by Fisher, Haldane, and Wright in the 1920s and its application to natural organisms in Dobzhansky's 1937 book, other fields were brought into the evolutionary synthesis. Ernst Mayr's *Systematics and the Origin of Species* (1942) discussed the nature of species, their origin, and their geographic distribution; George Gaylord Simpson's *Tempo and Mode in Evolution* (1944) showed that the patterns of change seen in the fossil record are consistent with population genetics (Fig. 1.40); and, rather later, G. Ledyard Stebbins's *Variation and Evolution in Plants* (1950) brought botany into the synthesis. Evolutionary biology began to emerge as a professional discipline, with lecture courses, grant support, and, in 1946, its own journal, *Evolution.*

Adaptation Is the Result of Natural Selection on Mendelian Variation

After a century of confusion and dispute, Darwin's theory of evolution by natural selection had been combined with Mendelian genetics to give a synthetic theory that could make sense of very different aspects of biology. There was agreement on two key issues.

- *Mendelian genetics is conducive to natural selection.*
  - Heritable variation is preserved from one generation to the next, rather than being dissipated by "blending inheritance."

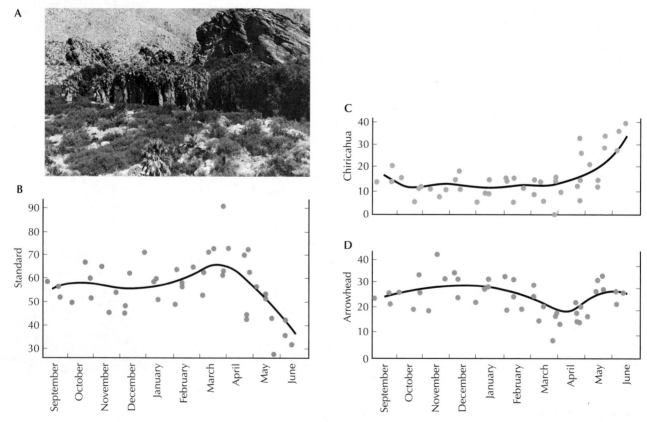

FIGURE 1.39. Inversion frequencies over time as studied by Dobzhansky. (*A*) Study site in the Sierra Nevada, California. (*B–D*) Frequencies of three chromosome types changed systematically through the year. These results implied the action of selection rather than of genetic drift.

- Variation is not influenced directly by the environment; in other words, there is no "Lamarckian" inheritance.

- Mutation is random with respect to adaptation; there is no bias toward well-adapted variations.

• *Adaptation is solely due to natural selection.*

- Although many processes affect evolution (e.g., random drift, migration, and mutation), none of these tends to produce organisms with better survival and reproduction.

- Even minor differences are influenced by selection, and even weak selection can be effective.

We set out the evidence for these basic tenets in Chapters 12–24. Although accepted by evolutionary biologists, they are still sometimes disputed, often with the same arguments that were raised in the early 20th century; we summarize these objections in Chapter 3.

Among the founders of the evolutionary synthesis, there was much enthusiasm for the wide prospects that had been opened up and a sense of triumph over past confusions. However, much remained unknown, and the central importance of the new "evolutionary synthesis" was not yet appreciated by most biologists. In 1955, for example, Mayr, Simpson, and Wright were dismayed to attend a National Research Council meeting at which none of the other biologists named natural selection as one of the basic principles of biology. In the following section, we outline the major issues that remained open.

The "Evolutionary Synthesis" Left Major Questions Unresolved

At the time of the synthesis, genetic variability could not be observed directly and the physical basis of the gene was essentially unknown. There was a consensus that genetic variability was extensive and that long-term evolution could be explained entirely by the processes seen in present-day populations. However, a more detailed understanding of genetic variability across diverse taxa, past and present, had to await advances in molecular biology.

With hindsight, we can see several major questions that remained almost entirely open at the time of the synthesis. The likely solutions to these questions were understood, at least by some, but there was little way of directly investigating them. As you will see from the rest of this book, even now we often do not have definitive answers.

What Is the Nature of Genetic Variation?

As Darwin argued, and as was confirmed in the early years of genetics, there is abundant variation in almost all aspects of the organism. However, we still do not know just how genes combine to influence the organism as a whole or what maintains variation in these genes. How many genes affect traits such as the shape of a *Drosophila* wing, the yield of a crop of wheat, or human intelligence? How many alleles are found at each gene? Do their effects simply add up or do they combine in more complicated ways? What kind of genes are involved? The present activity in human genetics concentrates on these kinds of questions: Our health is influenced by a complicated interaction between our genes, the genes of our pathogens, and environmental conditions. We come back to these issues in Chapters 14 and 26.

What Maintains Genetic Variation?

In a sense, these are purely genetic questions, concerning the relation between genotype and phenotype. However, the answers depend on how genetic variation

FIGURE 1.40. (*Top*) Ernst Mayr (on the *right* in photo) worked on the nature of species, their origin, and their geographical distribution. (*Bottom*) George Gaylord Simpson showed that patterns of change seen in the fossil record are consistent with population genetics.

has evolved and is maintained. At the time of the synthesis, there were two distinct views on this issue. The **classical view**, which traces back to Morgan and to his student Hermann Muller, is simple: There is typically a **wild-type** allele that has greatest fitness in any particular environment, and genetic variation around this optimum is due to deleterious mutations. In contrast, the **balance view**, propounded by Dobzhansky and his students, takes a more positive view of variation. Different alleles are maintained by **balancing selection**, and this variability allows rapid adaptation to ever-changing conditions. As we will see in Chapter 19, these two views have evolved over the years, but the debate about their relative importance remains unresolved.

What Does Natural Selection Act on?

Darwin insisted that natural selection acted on individuals: A population changed because some individuals left more offspring than others. Fisher, Haldane, and Wright all followed Darwin in holding that the differential survival of groups or of whole species is of negligible importance. The rate of selection on groups must be much weaker than that on individuals, because groups reproduce less often. Moreover, adaptations favorable to only the group or species are always vulnerable to cheats, who gain an individual advantage at the expense of the group. However, during the evolutionary synthesis, as greater emphasis was placed on adaptation rather than random drift and mutation, this insistence was often forgotten. Those working with natural populations often failed to distinguish whether adaptations were "for the good of the individual" or "for the good of the species." For example, V.C. Wynne-Edwards argued that animals may have evolved to reduce their individual reproduction, in order for the species not to overexploit its resources. This stimulated strong arguments in favor of the primacy of selection among individuals, such as George C. Williams's *Adaptation and Natural Selection* (1966), and W.D. Hamilton's extension of individual-based selection to allow for social interactions between relatives. The important point here is that, until the 1960s, there was no general appreciation that adaptations must usually be explained by natural selection among individuals. The level at which selection acts is still a controversial issue, with implications for such questions ranging from the origin of life to social behavior (Chapter 21).

How Did the Major Evolutionary Transitions Occur?

How did the very different organisms we see today evolve from a common ancestor, and how did life itself evolve on the early Earth? Darwin privately speculated about the origin of life in some "warm little pond," and, in 1923, Haldane postulated that a simple virus-like organism had emerged from prebiotic chemistry. Alexander Oparin's *Origin of Life* (1936) went further and suggested that there was a stage of chemical evolution involving natural selection among droplets that concentrated small organic molecules. Haldane and Oparin's idea that simple organic molecules could be generated spontaneously under the oxygen-free conditions of the early Earth were confirmed by experiments such as Stanley Miller's, in 1953, in which a simulated lightning bolt generated amino acids. However, little further progress could be made without knowing how genetic information was encoded in nucleic acids (see Chapter 4). Similarly, major changes such as the evolution of the eukaryotic cell with all its organelles, or of vertebrate body form, could not be understood without knowing the common genetic machinery shared by all organisms, or the way this machinery leads to the development of complicated morphologies (Chapters 9 and 24).

Molecular Biology Has Transformed Research in Evolutionary Biology

Some progress could be made with these questions solely with the aid of classical genetics and in ignorance of molecular details; indeed, much of the progress that has been made over the last 50 years has been based on classical methods that were familiar to the founders of evolutionary biology. For example, the biometric methods developed by Pearson and Fisher are the basis of modern efforts to identify **quantitative trait loci** (**QTLs**), genes that are involved in plant and animal improvement and complex human diseases. However, the development of molecular biology has fundamentally changed research in evolutionary biology.

- *Molecular biology provides powerful new techniques.* The availability of large numbers of **genetic markers** allows evolutionary relationships among all kinds of organisms to be more accurately established.

- *Molecular biology allows old questions to be answered, often in unexpected ways.* To give one example, we now have objective measures of levels of genetic variability, which can be applied in the same way to all organisms (Chapter 13). To take another (Chapter 9), we now have a clear understanding of what it means for organs to be homologous. For example, in what sense are the eyes of insects and vertebrates homologous through common ancestry, rather than having independently evolved to fulfill the same function (p. 110)?

- *Molecular biology raises wholly new questions.* The discovery of the molecular world opens up an entirely new realm of natural history. For example, we can now ask how entities can coexist, despite replicating independently and thus having different evolutionary interests (Chapter 21). How can the eukaryotic nuclear genome survive, despite conflicts with mitochondria and among "selfish" **transposable elements**? These questions parallel questions about how groups and species can be well adapted, even though selection acts on individual self-interest. We can also ask entirely new questions about the fundamental genetic machinery (Chapters 4 and 23): Why is the genetic code as it is? Why are genes regulated in the way that they are? Why do multicellular organisms develop in the way that they do? Each molecular fact that is established raises a corresponding evolutionary question.

In the next chapter, we trace how molecular biology emerged in the second half of the 20th century, its relation with evolutionary biology over this time, and the way it is now coming together in a new synthesis.

SUMMARY

Several kinds of evidence led to our present understanding of evolution and its mechanism. The development of geology in the 18th century showed that the Earth is old and that processes such as erosion and volcanic action could eventually account for long-term geological change. Fossils were understood to be remains of species now extinct, and present-day groups appeared successively through time. Darwin's arguments for evolution rested especially on the geographic patterns that he saw during his voyage on the *Beagle*—for example, the diverse yet related species found on remote oceanic islands. *On the Origin of Species* convinced most educated people of evolution, but its primary mechanism—natural selection—remained disputed well into the 20th century. One of the main obstacles was ignorance of heredity. Even though Mendel's work was rediscovered in 1900, leading to the rapid establishment of classical genetics, there was continuing dispute between those who emphasized the importance of mutations of large effect as opposed to natural selection on slight variations. This dispute was resolved by the 1920s, when population genetics showed both that apparently continuous variation can be caused by Mendelian variants and how selection can act effectively on such variation. This Evolutionary Synthesis was extended in the following decades to include all aspects of biology. Many open questions remained; in particular, the nature and causes of genetic variation were obscure, and the implausibility of selection for species or groups was only slowly appreciated. Now, evolutionary biology is being transformed by techniques and data from molecular biology, which helps to answer old questions and opens up extraordinary new phenomena to investigation.

FURTHER READING

Bowler P.J. 1989. *Evolution: The history of an idea*. University of California Press, Berkeley; Ruse M. 1996. *Monad to man: The concept of progress in evolutionary biology*. Harvard University Press, Cambridge, Massachusetts.

Two excellent histories of evolutionary thinking, up to the present.

Browne E.J. 1996, 2002, 2 volumes. *Charles Darwin: A biography*. Princeton University Press, Princeton, New Jersey; Desmond A. and Moore J.R. 1991, 2 volumes. *Darwin*. Michael Joseph, London; Viking Penguin, New York.

Two good recent biographies of Darwin.

Provine W.B. 2001. *The origins of theoretical population genetics*, 2nd ed. University of Chicago Press, Chicago.

A fascinating account of the bitter disputes over genetics and evolution in the early 20th century and their resolution in the "evolutionary synthesis."

Classic Works

Darwin C. 1839. *Journal of researches into the geology and natural history of the various countries visited by H.M.S. Beagle, under the command of Captain FitzRoy, R.N., from 1832 to 1836*. Henry Colburn, London.

Both an excellent travel book and an account of the observations and thinking that led Darwin to the theory of evolution by natural selection.

Darwin C. 1859. *On the origin of species by means of natural selection*. John Murray, London.

There are many reprints of this key work. We recommend reading reprints of the first edition.

Dobzhansky T. 1937. *Genetics and the origin of species*. Columbia University Press, New York.

The textbook that applied population genetics to the study of natural populations; crucial in establishing the "modern synthesis."

Fisher R.A. 1930. *The genetical theory of natural selection*. Clarendon Press, Oxford.

A concise summary of population genetics followed by Fisher's views on **eugenics**.

Haldane J.B.S. 1932. *The causes of evolution*. Longmans, Green, New York.

A clear and readable summary of basic population genetics, and its relevance to evolution.

CHAPTER

2

The Origin of Molecular Biology

URING THE TIME WHEN GENETICS AND evolution were coming together in the **Evolutionary Synthesis**, there was virtually no knowledge of the material basis of the genes or of how genes directed the function and development of the organism. The basic biochemistry of small molecules was established during the first half of the 20th century, but there was much confusion over the nature and role of proteins and nucleic acids. Between 1940 and 1965, there was a dramatic revolution in our understanding of biology, in which the physical bases of heredity, of gene regulation, and of enzyme function were established and revealed a surprising simplicity to life.

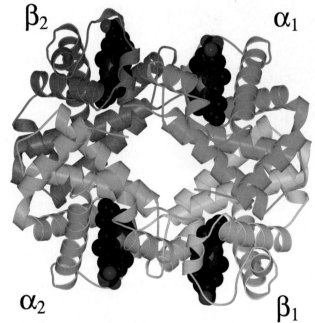

With the birth of molecular biology, an entirely mechanistic account of living organisms became feasible. This new discipline concentrated on the flow of information rather than of energy and on the three-dimensional structure of macromolecules rather than on organic chemistry. Substantial technical advances were later made, particularly between 1972 and 1980, which made DNA sequencing and genetic manipulation quick and simple. However, the basics of classical molecular biology were laid down during a brief period in midcentury.

In this chapter, we give an overview of the development of molecular biology, which complements the history of evolutionary biology in Chapter 1 and helps us understand the relationship between the two fields. It also serves as a summary of the key principles established by molecular biology, which will help you understand the detailed examples that we will discuss later. In the rest of the book, we seek to explain how and why organisms have evolved to share the fundamental structures that were revealed by classical molecular biology.

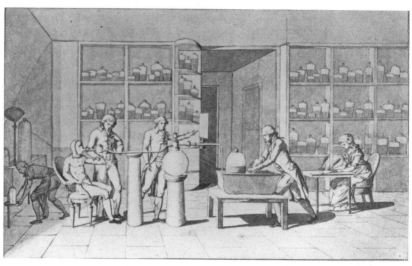

FIGURE 2.1. Antoine-Laurent Lavoisier performing one of his experiments on human respiration. Lavoisier quantified the oxygen consumed and the carbon dioxide produced by metabolism and identified respiration as a very slow combustion process. This drawing was done by Lavoisier's wife Marie-Anne Paulze, who appears at the right.

Free enzyme + Free substrate

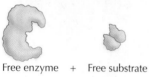

Free + Free
enzyme products

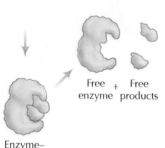

Enzyme–
substrate
complex

FIGURE 2.2. Jöns Jakob Berzelius (*top*), in addition to discovering several chemical elements and suggesting the present chemical symbols, identified and named the process of catalysis, essential for understanding the mechanism of enzyme action (*bottom*).

THE BEGINNINGS OF MOLECULAR BIOLOGY

The Physical Basis of Life Was Not Understood until the 1950s

Attempts to explain life in terms of physics and chemistry began in the late 18th century, long before fully materialistic explanations were possible or even plausible. They were stimulated by the success of Newtonian mechanics in explaining the movements of earthly and heavenly bodies and by the rational philosophies of Gottfried Wilhelm Leibniz and René Descartes. Around 1770, Antoine-Laurent Lavoisier (Fig. 2.1) showed that respiration and combustion caused similar chemical changes and that both living and nonliving material could be analyzed into its component elements in the same way. In the early 1830s, "ferments" (i.e., enzymes) were found in malt, almonds, and gastric juices; these accelerated chemical reactions but were not themselves consumed. Jöns Jakob Berzelius (Fig. 2.2) called this **catalysis** and applied the term to both organic and inorganic substances.

A better understanding of enzyme action did not come until the end of the 19th century. There was considerable dispute as to whether fermentation required living cells or was instead a matter of simple chemistry. Louis Pasteur (Fig. 2.3) showed by a variety of experiments around 1861 that fermentation took place only in the presence of living cells. (For example, blood or urine drawn sterile from the body did not take up oxygen over a 3-year period, yet quickly did so when exposed to microbial infection.) However, in 1897, Eduard (Fig. 2.4) and Hans Buchner showed that a cell-free extract of yeast would catalyze fermentation. By the turn of the century, it was understood that enzymes could catalyze reactions either inside or outside cells and did not depend on any "vital force."

One property that seemed unique to organic substances is their asymmetric effect on light. For example, polarized light passed through a solution of tartaric acid derived from plants is twisted to the right, whereas paratartaric acid (chemically identical, but synthesized artificially) has no such effect. In 1848, Pasteur found that paratartaric acid forms crystals of two types, which are mirror images of each other. Pasteur separated these by hand and showed that they rotate polarized light in opposite directions and that they react quite differently with other substances of organic origin. In 1874, Jacobus Hendricus van't Hoff and Joseph-Achille Le Bel independently gave an explanation that did not require any vital force: The optical asymmetry reflects the asymmetry of one of the carbon atoms in tartaric acid, which can be in a right-handed or a left-handed configuration (Fig. 2.5A). Molecules that differ in this way are now known as **stereoisomers**. In the 1890s, Emil Fischer confirmed their hypothesis, by synthesizing the 16 possible stereoisomers of glucose, which differ in the configuration of their four asymmetric carbon atoms (Fig. 2.5B). Fischer found that enzymes had strong specificities for particular stereoisomers, which he explained by supposing that catalysis was only effective when enzyme and substrate fit "like lock and key." The central importance of three-dimensional structure was only fully recognized half a century later, with the coming of molecular biology.

Although it was clear, from Berzelius' time, that enzymes consisted largely of protein, the structure of proteins and the site of catalytic activity remained obscure. In 1902, Fischer and Franz Hofmeister independently proposed that proteins are made up of amino acids, joined end to end by peptide bonds. However, it was widely held that proteins consist of short peptides, bound together in some kind of **colloid**, and that this colloidal state was responsible for their biological activity. Colloids had been defined as substances that formed amorphous solids rather than well-defined crystals. The term was later defined more precisely as a mixture of two distinct phases (e.g., an emulsion of oil and water). The idea that the "dynamical" features of life and, in particular, the catalytic function of enzymes were due to their colloidal organization was attractive, in part because many colloids do have some catalytic activity as a result of the extensive interface between their phases.

The colloid theory was demolished in the 1920s by three lines of evidence. First, clear X-ray diffraction patterns were produced from fibers of cellulose and silk, implying a regular molecular structure (see Box 2.1). Second, Theodor Svedberg used the **ultracentrifuge** to show that most proteins have a high molecular weight, which is characteristic of each kind of protein (Fig. 2.6). For example, Svedberg showed that each of the four subunits of the hemoglobin molecule has molecular weight 16,700 (see opening figure [p. 37]). Finally, Hermann Staudinger's studies of natural and synthetic polymers gave convincing chemical evidence that these were long chains of subunits, bound by ordinary covalent chemical bonds, rather than by the weaker bonds postulated in the colloid theory. Staudinger termed such long chains macromolecules.

The clearest evidence against the colloid theory came in 1926, when James Sumner crystallized the enzyme urease (Fig. 2.7) from jack beans; 3 years later, John Northrop grew extremely pure crystals of pepsin and demonstrated that they had enzyme activity.

By the 1940s, biochemists had a good understanding of the basic metabolic and synthetic pathways and of the flow of energy through the cell. Both nucleic acids (DNA and RNA) and protein were known to be macromolecules made up of long chains of nucleotides and amino acids, respectively. However, little was known about the structure and function of these macromolecules. The sequences of macromolecules were thought to be in some way regular: In particular, the "tetranucleotide hypothesis" supposed that nucleic acids consist of regularly repeated sets of the four different bases, adenine, thymine, guanine, and cytosine for DNA (e.g., ATGCATGCATGC. . .). It was not known that either nucleic acids or proteins were stable: Indeed, influential experiments suggested that amino acids are rapidly exchanged between proteins. It was widely thought that in vivo individual macromolecules have no stable conformation, but rather are in some kind of dynamic equilibrium. For example, Linus Pauling proposed in 1940 that antibodies recognize particular antigens by folding around them and thus taking up a complementary shape that could then recognize other antigen molecules of the same kind (Figs. 2.8 and 2.9). What was

FIGURE 2.3. Louis Pasteur's experiments showed that fermentation was a physiological process that took place only in living cells. Pasteur's wide-ranging accomplishments established stereochemistry (see Fig. 2.5), recognized the role of microorganisms in infectious disease, and discovered a vaccine against rabies.

FIGURE 2.4. Eduard Buchner's work on noncellular fermentation, which linked the processes of catalysis and fermentation, won him the Nobel Prize in Chemistry in 1907. His cell-free fermentation studies using yeast allowed him to separate the activities of the cells and to study their contents, particularly enzymes, which he described as acting as "the overseers" in the cells of plants and animals.

FIGURE 2.9. Linus Pauling and Sir Lawrence Bragg examining a model of nucleic acid structure in 1953. Bragg developed X-ray crystallography of atoms in a crystal, a key tool for determining the structures of molecules.

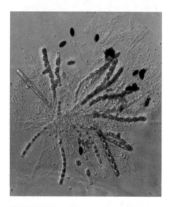

FIGURE 2.10. George Beadle and Edward Tatum established the relationship between genes and proteins, the one gene–one enzyme (or polypeptide) principle, with their experiments using the mold *Neurospora*.

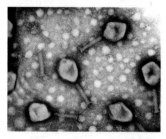

FIGURE 2.11. Great advances in molecular biology, from detecting rare mutations and recombination to producing maps showing the locations of mutations within single genes, were made by studying bacteriophage, viruses that infect bacteria.

themselves they gave no way of establishing the material structure of the genes and were limited by the slow generation time and complex physiology of organisms such as *Drosophila*, maize, and mice. Advances came on two fronts. First, the close correspondence between genes and enzymes was gradually established. Archibald Garrod had proposed in 1908 that human inherited diseases are due to "inborn errors of metabolism," involving the failure of specific metabolic steps. During the 1920s, J.B.S. Haldane and Rose Scott-Moncrieff identified genes that controlled particular steps in pigment synthesis in *Primula*. Such work led up to George Beadle and Edward Tatum's development of biochemical genetics in the bread mold, *Neurospora crassa*, which gave a general method for the genetic analysis of biochemical pathways. In 1945, Beadle proposed that one gene corresponded to one enzyme (Fig. 2.10).

The second key to progress was the development of new model organisms. As well as introduction of the genetics of bacteria and fungi, in which mutants were identified by biochemical defects, a distinctive program of research into the genetics of **bacteriophage** began (Fig. 2.11). Groups centered on Delbrück in Pasadena and Lwoff in Paris developed efficient methods for studying these bacterial viruses, which allowed the detection of extremely rare mutations and recombinations. By the early 1950s, these techniques allowed Seymour Benzer to produce the first maps showing the location of mutations within single genes and, eventually, down to single nucleotides.

Biochemistry

As we shall see, the structure of DNA was discovered through a combination of X-ray crystallography and physical chemistry. However, the subsequent establishment of the genetic code, by which DNA sequence determines amino acid sequence and hence protein structure, eventually came primarily through the refinement of classical biochemistry, which allowed the construction of well-defined cell-free systems that could synthesize proteins. This can be seen as a development of work begun by the Buchner brothers' demonstration of fermentation by yeast extracts, at the turn of the century.

Two other contributions of biochemistry stand out. In 1951, Sanger (Fig. 2.12) published the complete sequence of the 30 amino acids of the insulin B chain. Erwin Chargaff (Fig. 2.13) accurately measured the proportions of the four bases in DNA. By 1950, he had shown that there are wide variations in composition between different species, but that there are always equal proportions of adenine and thymine and of guanine and cytosine. Sanger's work with proteins and Chargaff's with DNA refuted the idea that each consisted of a regularly repeated sequence of amino acids or nucleotides.

Conceptual and technical advances came together to create the distinct discipline of molecular biology, which then fed back into the rest of biology. The rapid progress made in midcentury was not the result of a single breakthrough—nor was it entirely a matter of chance. The Rockefeller Foundation, founded in 1913 by the oil magnate John D. Rockefeller, had a deliberate policy of promoting molecular biology—a term coined by its director, Warren Weaver, in 1938. The Foundation funded work by Lwoff, Thomas Hunt Morgan, Pauling, Svedberg, and many others.

The Structure of DNA Established the Physical Basis of Heredity

Before the 1940s, genes were thought to consist of protein or perhaps of some kind of complex of nucleic acids with protein. DNA itself was seen as a "stupid molecule" monotonously repeating the four bases, whereas proteins were known to be extraordinarily versatile in function. DNA came to be seen as the genetic material primarily through work on the transformation of *Streptococcus pneumoniae*. This bacterium is found in various virulent S strains and also in an R strain that is not infectious. (The strains are named after the *smooth* or *rough* appearance of colonies.) Frederick Griffith had found in 1928 that the R strain can be permanently transformed into the virulent form by dead bacteria of an S strain. Moreover, the new virulent form is of the same strain as the dead bacteria. Avery's (Fig. 2.14) laboratory showed by a variety of painstaking tests that the transforming factor consists of DNA (Fig. 2.15). For example, it is sensitive to enzymes that degrade DNA, but not to those that degrade protein.

This evidence, published most definitively by Avery, Colin McLeod, and Maclyn McCarty in 1944, is not by itself evidence that DNA is a general hereditary material: The DNA might simply switch bacteria between alternative forms. However, other evidence accumulated. Penicillin resistance was also shown to be acquired by transformation in the same way; the DNA content of sperm and eggs was found to be half that of other cells; and DNA was found to turn over much more slowly than proteins, suggesting a role as the stable carrier of genetic information. By the end of the 1940s, DNA was widely regarded as the substance of the genes. Further support came from experiments using radioactively labeled bacteriophage. Upon infection of bacteria by the phage, the results showed that only the DNA was transferred into the bacteria, whereas the phage protein coat remained on the outside of the bacteria.

James Watson arrived in Bragg's X-ray crystallography unit at Cambridge University in October 1951, having just completed a thesis on phage genetics with Luria. There, he met Francis Crick, 12 years older, but still researching for his thesis on protein structure. This meeting between the English school of structural biology and the American phage genetics stimulated both Watson and Crick to find a structure for DNA (Fig. 2.16) by the model-building approach that had been introduced by Pauling.

The year before, Maurice Wilkins at King's College in London had begun X-ray diffraction studies on DNA and had recruited Rosalind Franklin to work on the problem (Fig. 2.17). By the fall of 1951, Franklin had recognized that DNA fibers can take on two different forms (A and B) depending on their water content and that their density implied two or perhaps three chains per molecule. She presented her results at a seminar in November 1951, which Watson attended. Based on Watson's somewhat inaccurate recollections of the seminar and on theoretical results on the diffraction patterns expected from helical molecules, which Crick had recently derived, Watson and Crick proposed a model with three ribose-phosphate backbones on the inside and the bases sticking out. When the King's group saw the model a few days later, they immediately saw that it could not be correct: The backbones would be packed too tightly to hold the known water content and, in any case, the strong helical diffraction pattern required that the heavy phosphate atoms of the backbone be on the outside. After this debacle, Bragg forbade further work on DNA structure by Watson and Crick.

The following spring, Franklin took her best picture yet of the B form; this was to prove crucial for the final solution (Fig. 2.18). The pattern showed that the spacing between nucleotide units is 3.4 Å, that the molecule repeats after ten of these units, and that its diameter is about 20 Å. But Franklin chose not to work further on the B form, instead turning to a laborious calculation aimed at direct interpretation of the diffraction pattern from the A form.

Around the same time, Crick asked a mathematician, John Griffith, whether like bases might attract like. Griffith thought it more likely, based on tentative quantum mechanical calculations, that adenine would attract thymine and guanine would attract cytosine. Chargaff visited Cambridge soon after and pointed out to Crick that the proportions of adenine and thymine and of guanine and cytosine are always the same (i.e., A = T, G = C). Crick saw that this fitted with Griffith's calculation and, moreover, suggested a mechanism of complementary replication. However, at this stage it was not clear how this idea could be fitted into a structure.

Watson and Crick resumed work on DNA in earnest in January 1953 when they saw the draft of a paper by Pauling and Robert Corey that proposed a model similar to their own first attempt—and that failed for similar reasons. Watson visited

FIGURE 2.12. Frederick Sanger's contributions to sequencing proteins (insulin in particular) and nucleic acids won him two Nobel Prizes (1958, 1980).

FIGURE 2.13. Erwin Chargaff's determination of the proportions of the four bases—adenine = thymine, cytosine = guanine—in DNA was critical for determining the structure of DNA.

FIGURE 2.14. Oswald Avery's chemical analyses, with Maclyn McCarty and Colin MacLeod, in the 1940s showed that DNA and not protein was the "transforming principle"—the material of inheritance—in pneumococcal cells. Avery wrote that "nucleic acids are not merely structurally important but functionally active substances in determining the biochemical activities and specific characteristics of cells—and that by means of a known chemical substance it is possible to induce predictable and hereditary changes in cells."

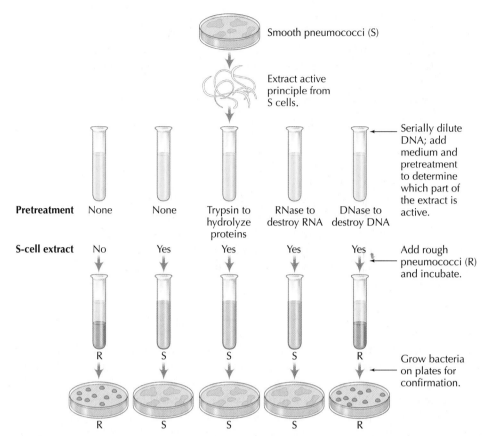

FIGURE 2.15. Oswald Avery and his colleagues used transformation between the smooth and rough properties of *Streptococcus pneumoniae* colonies, as well as enzymes that destroy polysaccharides, lipids, DNA, protein, and RNA, to determine the nature of the transforming principle. Transforming activity was abolished by pretreatment with an extract of dog intestinal mucosa that had been shown to destroy DNA. Thus, DNA was the genetic material, a result that was viewed with skepticism by many in the field at the time. Previous to these experiments, most scientists believed that genes were made up of protein.

FIGURE 2.16. All of the preceding work lead up to the discovery of the structure of DNA by James D. Watson (*left*) and Francis H.C. Crick (*right*) in 1953. Their publication of two *Nature* papers that year established the double-helical structure of the molecule and described how base pairing in the molecule was the basis for replication of the molecule—the molecular basis for heredity. Watson and Crick are shown with their original model of the double helix.

FIGURE 2.17. Rosalind Franklin (*left*) and Maurice Wilkins (*right*) provided key physical evidence regarding the structure of the DNA molecule.

King's College the following month and saw Franklin's photograph of diffraction from the B form (Fig. 2.18). Talking to Wilkins later, he first learned of the distinction between the A and B forms and of their correct dimensions. These were confirmed by a brief report in which the King's group had summarized their results from the previous autumn. Crick saw that the symmetry of the diffraction pattern implied that there must be two chains running in opposite directions so that the structure would remain identical when rotated through 180°.

The final piece of the puzzle came when Watson realized that he and Crick had assumed that the bases were in a chemical form that was incorrect. Substituting the correct forms, he found immediately that adenine pairs naturally with thymine, and guanine with cytosine, each being held together by hydrogen bonds (Fig. 2.19). This immediately explained Chargaff's ratios, and moreover, because the larger purines (A, G) fit with the smaller pyrimidines, each base pair fits neatly within the double helix. Watson completed the first rough model in late February 1953; the structure was published by Watson and Crick in *Nature* at the end of April, accompanied by reports of the experimental evidence from the Kings' group.

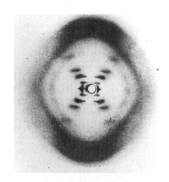

FIGURE 2.18. The crucial X-ray diffraction image of the B form of DNA (Photograph 51) taken by Rosalind Franklin and Ray Gosling. The X-shaped pattern in this image meant that the DNA molecule was a helix, critical information for Watson and Crick's model building.

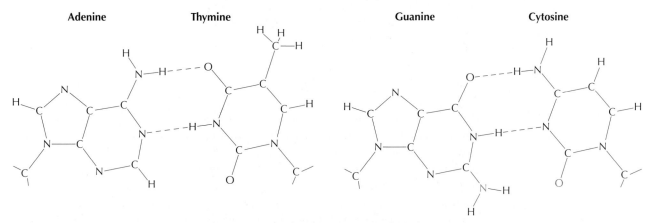

FIGURE 2.19. Pairing of bases in the DNA double helix based on the structures originally shown in Watson and Crick's 1953 paper. Hydrogen bonds are indicated by dashed lines. Pauling later showed that guanine and cytosine share a third hydrogen bond between the nitrogen (*blue*) and the oxygen (*red*).

The DNA Sequence Is Expressed through the Genetic Code

Watson and Crick closed their paper by saying that "[i]t has not escaped our notice that the specific pairing we have postulated immediately suggests a possible copying mechanism for the genetic material." That is, the two strands of the double helix unwind, and new bases then assemble in the correct sequence by complementary base pairing.

By 1958, Matthew Meselson and Franklin Stahl had shown, by labeling DNA with heavy nitrogen, that DNA replication is **semiconservative**: The first-generation offspring molecule carries one heavy strand inherited from the labeled parent and one light strand, newly synthesized (Fig. 2.20). By the same time, Arthur Kornberg had purified a DNA polymerase from *Escherichia coli*. He showed in vitro that the proportions of adenine and thymine and of guanine and cytosine matched in the new strand and that the composition of the new DNA mirrored that in the original molecules.

Establishment of the structure of DNA as a linear sequence of four alternative bases immediately raised the question of how this sequence encodes and expresses the genetic information carried by the DNA. Although the mechanism of replication was rapidly elucidated, how the genetic code is translated took many years to be fully understood.

By this time, evidence existed linking genetic variation with variation in proteins (i.e., linking genotype with phenotype; see Chapter 14). For example, in 1949, sickle cell anemia had been shown by Pauling and Harvey Itano to be due to a defective hemoglobin and, in the same year, was shown to be inherited as a simple Mendelian recessive trait. In 1955, Vernon Ingram used paper chromatography of peptide fragments to show that sickle cell hemoglobin differed from the normal variety by a single amino acid (Fig. 2.21). This was the first evidence that a genetic variant was expressed through a specific change in a protein.

It was clear that DNA must somehow direct protein synthesis. As Crick wrote in 1957, "once the central and unique role of proteins is admitted, there seems little point in genes doing anything else." However, there were widespread misconceptions about the nature of proteins and their synthesis. It was thought that proteins are unstable, continually exchanging amino acids with each other; enzymes were believed to be made from precursor molecules; and protein synthesis was thought to involve a reversal of the well-understood pathways of protein degradation. The elucidation of the actual mechanisms of protein synthesis—that is, of the primary mechanism by which genes are expressed—came by three routes: the abstract analysis of the genetic code, formal genetics, and classical biochemistry. Although analysis of the coding problem and of the genetics of model organisms such as phage produced some elegant arguments and evidence, it was the detailed biochemistry that proved decisive.

The Coding Problem

Shortly after publication of the structure of DNA, George Gamow, the physicist who in 1946 had put forward the Big Bang theory for the origin of the Universe, wrote to Crick with a simple proposal. He had noticed that the arrangement of bases within the double helix created a series of pockets, each surrounded by four bases. He suggested that particular amino acids might fit directly into particular holes (Fig. 2.22). Moreover, Gamow pointed out that there are 20 kinds of holes, which corresponds to the number of commonly occurring amino acids.

Gamow's model was biochemically flawed, but it did force a clear definition of the problem; it initiated attempts to deduce the code from logical principles. The most immediate effect was to spur Crick and Watson to draw up the correct list of 20 amino acids, discarding unusual types that are formed by modification of proteins after translation. (For example, brominated tyrosine is found in coral

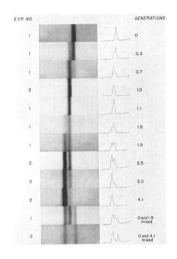

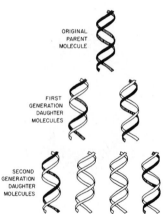

FIGURE 2.20. Matthew Meselson and Franklin Stahl's demonstration that DNA replication is semiconservative—the first-generation DNA molecule carries one parental strand of DNA and one newly synthesized strand. (*Top*) In this elegant experiment, a population of bacteria was made to reproduce synchronously in a medium containing the heavy isotope of nitrogen (^{15}N). The bacteria were then transferred to unlabeled medium and the density of the DNA molecules was measured in an ultracentrifuge. This photo shows the DNA bands that resulted. Regions of equal density occupy the same horizontal position on each photograph. Adjacent are microdensitometer tracings of the DNA bands. (*Bottom*) In generation 0, all of the DNA is labeled. In generation 1, all of the DNA consists of one labeled strand and one unlabeled strand. In generation 2, half the DNA is hybrid and the other half unlabeled.

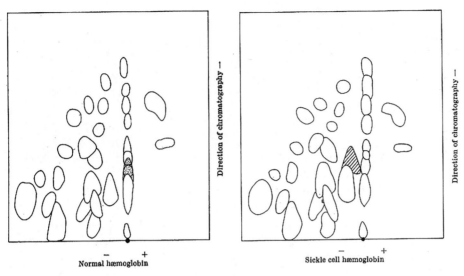

FIGURE 2.21. Vernon Ingram showed that the difference between normal hemoglobin (HbA) and sickle cell hemoglobin (HbS) was at the #6 position in the β-globin chain. (*Left*) Normal and (*right*) sickle cell hemoglobin fingerprints using electrophoresis.

and iodinated tyrosine in the thymus gland.) Remarkably, their initial list of 20 standard amino acids turned out to be exactly correct.

Gamow's was an overlapping code, because the four bases that determine each amino acid also contribute to the code for neighboring amino acids (Fig. 2.22). This puts constraints on the possible amino acid sequences that could be encoded and thus would be found in proteins. For example, the insulin B chain of sheep and cows differs at only one position, occupied by glycine in the first species and serine in the second. With an overlapping code such as Gamow's, a single base change would alter more than one amino acid, and multiple changes would only cause further complications. By 1956, enough proteins had been sequenced to allow Sydney Brenner to show that most of the 20 × 20 = 400 possible pairs of neighboring amino acids had been found. So, any overlapping triplet code is impossible.

Crick soon realized that a direct interaction between a nucleic acid template and the protein chain, of the kind proposed by Gamow, is unlikely. Some amino acids are chemically very similar (e.g., phenylalanine and tyrosine differ by a single hydroxyl group), whereas others have side chains with very different size and charge (Fig. 2.23). Thus, it is hard to see how *any* direct scheme could work. The chemical homogeneity of DNA contrasts with the heterogeneity of proteins and reflects their respective roles as the stable carrier of information and as the versatile chemical agent that executes the genetic program. The implausibility of a direct mechanism for coding led Crick to propose an alternative: the adapter hypothesis. According to this idea, the sequence of bases would be recognized by binding to one end of an adaptor, whose other end carried the amino acids. At least 20 such molecules would be required, each coupling a specific amino acid to a specific sequence of bases. Such molecules were discovered around the same time that Crick proposed them on theoretical grounds; now, we know them as **transfer RNAs** (tRNAs) (Fig. 2.24).

The Genetic Code Was Deciphered Primarily by Painstaking Biochemistry

Once it was realized that the code is nonoverlapping and is read by adapter molecules, it became obvious that it could not be deduced from first principles. Although genetic arguments and experiments contributed to our understanding of the genetic code, it

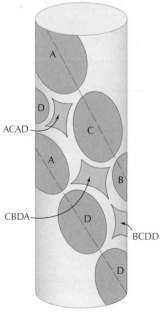

FIGURE 2.22. George Gamow suggested a physical basis for the genetic code. The double helix contains pockets (*brown*) that are surrounded by four bases (*blue*; labeled A, B, C, D). Gamow showed that these pockets can be surrounded by 20 combinations of bases and so could potentially code for 20 amino acids.

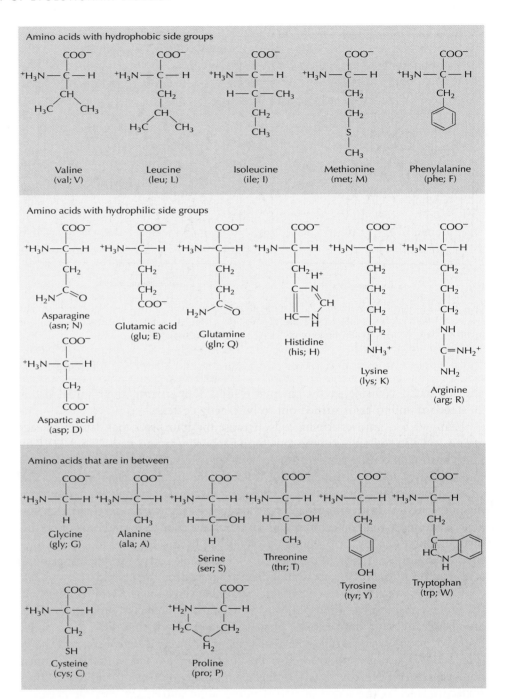

FIGURE 2.23. The 20 commonly found amino acids carry a wide variety of side groups. The three-letter and one-letter abbreviations are given below each name.

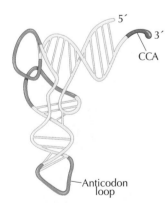

FIGURE 2.24. The transfer RNA (tRNA) acts as an adaptor molecule in the process of translation. The anticodon binds to the triplet codon at one end and the CCA terminus to the corresponding amino acid at the other. The genetic code is determined by the tRNAs together with the specific enzymes that bind the amino acid to them.

is more or less arbitrary, and thus the code could be solved only by laborious biochemical experiments that established the code for each amino acid one by one.

Evidence that RNA is involved in protein synthesis came in the early 1940s. Torbjörn Caspersson in Stockholm and Jean Brachet in Brussels found that high RNA levels are correlated with rapid growth; that although DNA is contained in the chromosomes, RNA is predominantly found in the cytoplasm; and that this cytoplasmic RNA is mostly associated with small protein particles, now known as **ribosomes**. In 1946, Brachet was the first to suggest that these particles are the site of protein synthesis.

By 1955, Paul Zamecnik and Mahlon Hoagland in Boston had shown that the first steps in protein synthesis take place in the soluble fraction, away from the ribosomes. Each amino acid is coupled to adenosine monophosphate (AMP) by a covalent bond that will later supply the energy required for formation of the peptide bond. The amino acid and AMP are then bound to a small RNA molecule, and the whole complex moves to the ribosome. In the summer of 1956, Zamecnik and Hoagland isolated an RNA molecule that bound specifically to radioactively labeled leucine, and when this was added to an extract containing ribosomes, the leucine was incorporated into protein. Thus, Crick's adaptor molecules (i.e., tRNAs) had been identified shortly after he had proposed them on theoretical grounds.

Thus far, the in vitro work had not demonstrated the synthesis of a specific protein, directed by a specific nucleic acid. The genetic code was finally broken by Marshall Nirenberg (Fig. 2.25, top) and Johann Matthaei, working at the National Institutes of Health in Bethesda, Maryland. They found that if a bacterial cell-free system was treated with RNase, protein synthesis immediately halted. But, protein synthesis resumed if RNA was added back. In early 1961, Nirenberg and Matthaei used RNA from tobacco mosaic virus to stimulate copious production of protein. However, this protein proved difficult to characterize. The breakthrough came in May 1961: When Matthaei added synthetic polyuracil to the system, polyphenylalanine was produced. Assuming a triplet code, UUU must encode phenylalanine (Fig. 2.25, bottom).

This first decoding was announced at the International Congress of Genetics in Moscow, in August 1961. Despite intense competition, it took another 6 years for the code to be fully uncovered. The full code was set out in its now-familiar form by Crick (see Fig. 2.26). A cartoon illustrating the essential steps of protein synthesis is shown in Figure 2.27.

The First Gene Regulation Mechanism Was Discovered by Jacob and Monod

Knowing how genes code for proteins is not enough; we must also know how their *expression* is regulated. For all organisms, it is the timing and location of protein production that matters: If all genes were expressed all the time, organisms would be a formless mass of protein. Genes must be expressed at the appropriate time in the cell cycle and in response to changing conditions—for example, to DNA damage or changes in food source.

The first understanding of the mechanism of gene regulation came from two

TABLE 6
TENTATIVE SUMMARY OF CODE WORDS *

C¹⁴-amino acid	M-codewords			
Alanine	CCG			
Arginine	CGC			
Aspartic acid³⁶	ACA			
Asparagine³⁶	UAC or UAA†			
Cysteine	UUG or UGG‡			
Glutamic acid³⁶	ACA	AGA	AGU§	
Glycine	UGG			
Histidine	ACC			
Isoleucine	UUA			
Leucine	GUU	CUU	AUU†	(UUU)
Lysine	AAA	AAC	AAG	AAU
Methionine	UGA§			
Phenylalanine	UUU			
Proline	CCC	CCU	CCA	CCG
Serine	UCG	UCU		
Threonine	CAC	CAA		
Tryptophan	UGG			
Tyrosine	UAU			
Valine	UGU			

FIGURE 2.25. Marshall Nirenberg (*top*) presented this summary of the genetic code (*bottom*) in a 1962 paper.

2nd→ 1st↓	U	C	A	G	3rd↓
U	Phe	Ser	Tyr	Cys	U
	Phe	Ser	Tyr	Cys	C
	Leu	Ser	Stop	Stop	A
	Leu	Ser	Stop	Trp	G
C	Leu	Pro	His	Arg	U
	Leu	Pro	His	Arg	C
	Leu	Pro	Gln	Arg	A
	Leu	Pro	Gln	Arg	G
A	Ile	Thr	Asn	Ser	U
	Ile	Thr	Asn	Ser	C
	Ile	Thr	Lys	Arg	A
	Met	Thr	Lys	Arg	G
G	Val	Ala	Asp	Gly	U
	Val	Ala	Asp	Gly	C
	Val	Ala	Glu	Gly	A
	Val	Ala	Glu	Gly	G

FIGURE 2.26. The actual code, written in the standard form proposed by Crick.

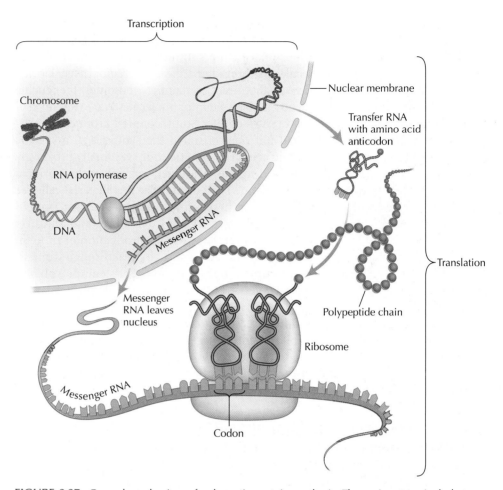

FIGURE 2.27. General mechanism of eukaryotic protein synthesis. The major steps include transcription of the DNA gene sequence into the messenger RNA template in the nucleus of the cell, translation of the DNA codons of that gene into amino acids, and their assembly into polypeptides in the cytoplasm. Important mediators of this process include transfer RNAs, splicing elements, and ribosomes.

model systems, both developed in the attic of the Institut Pasteur in Paris. The first was Jacques Monod's study of how *E. coli* exploit the sugar lactose as a source of energy and carbon; the second was the phenomenon of **lysogeny** in bacteriophage, unraveled by André Lwoff, Elie Wollman, and François Jacob (Fig. 2.28).

The lac System

As he completed his doctoral thesis in 1940, Monod discovered that when *E. coli* is given a mixture of glucose and lactose, it grows in two steps: rapidly at first, plateauing out as glucose is used up, and then growing rapidly again as it turns to use lactose. This is an example of what was termed "enzymatic adaptation," which was thought to involve a direct effect of the substrate on enzyme activity. Moreover, some kind of dynamic equilibrium was envisaged in which unstable proteins are converted between alternative forms. Monod's work was to overturn both misconceptions.

The enzyme that breaks lactose into its component sugars, glucose and galactose, cleaves the molecule at a particular site—the β carbon of the galactose ring. Hence, the enzyme is known as β-galactosidase. Working with Melvin Cohn, Monod found that other substances that contain this site could also be cleaved by β-galactosidase and sometimes more efficiently (Fig. 2.29).

FIGURE 2.28. (*Left to right*) François Jacob, Jacques Monod, and André Lwoff won a Nobel Prize for their determination of the nature of gene regulation using *Escherichia coli* and bacteriophage.

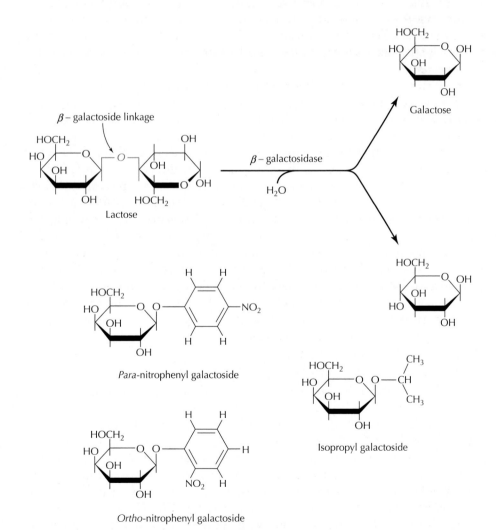

FIGURE 2.29. The regulation of β-galactosidase could be studied separately from its catalytic function by the use of chemical analogs. This enzyme cleaves at a particular position (marked in *red*). *para-* and *ortho-*nitrophenyl galactoside are cleaved much more efficiently than lactose, but do not induce production of the enzyme. Conversely, isopropyl galactoside is a much more powerful inducer of enzyme production, but it cannot be broken up by the enzyme.

This work with lactose analogs was crucial both practically, by providing a powerful set of experimental tools, and conceptually, by demonstrating that the stimulation of enzyme production was an entirely separate phenomenon from the catalytic activity of the enzyme itself. This independence was confirmed by the discovery in 1955 of the lactose permease, which allows the uptake of lactose and is regulated in the same way as β-galactosidase. To define the phenomenon more clearly and to distinguish it from other kinds of evolutionary **adaptation**, the Pasteur group renamed it **induction**.

The λ System

Bacteriophage typically grow lytically. That is, upon infection they make new phage and lyse the cell. However, some phage can become dormant within the host and replicate with it. Bacteria that contain phage in this kind of latent state are known as **lysogenic**. They occasionally produce new phage, and so there is always a low level of phage in a lysogenic culture. This phenomenon had been known since the 1920s and remained quite mysterious. By culturing single individuals of the giant *Bacterium megaterium*, Lwoff showed that the free phages are produced by the occasional lysis of a small fraction of cells in the usual way. Surprisingly, the remaining lysogenic bacteria contained no trace of bacteriophage. In 1949, he proposed that the phage genes are incorporated into the host genome, where they suppress both their own expression and the expression of any phage genes from new infections.

In 1951, Esther Lederberg identified lysogeny in the K12 strain of *E. coli*, infected by what she named phage λ. Two years later, the *E. coli Hfr* strain was discovered, which donates genes to other bacterial cells (i.e., bacterial mating) at high frequency. In 1954, Wollman used a blender to abruptly interrupt mating at set times and showed that the *Hfr* chromosome was transferred from a fixed starting point and at a constant rate, so that the time at which a gene is transferred indicates its position on the chromosome. This technique made possible the formal genetics of phage infection. For example, Jacob and Wollman found that if a lysogenic bacterium transferred its genes into a nonlysogenic cell, the recipient cell soon lysed to release a hundred or so bacteriophage. This implied that something in the cytoplasm of the lysogenic bacterium had repressed the expression of phage genes. The studies of lysogeny and of β-galactosidase were now being done in the same bacterial strain, K12, and using the same techniques, and both phenomena were known by the same term: induction. Yet, the close parallels between the two systems were not yet appreciated.

Mutations That Revealed Regulation

The key experiment that revealed the mechanism by which β-galactosidase was regulated and that led to a general theory of gene control was performed in 1958, by Arthur Pardee, Jacob, and Monod. (It is known from their initials as the PaJaMo experiment; Fig. 2.30.) Mutations had been found that were **constitutive**— that is, bacteria expressed both β-galactosidase and the lactose permease even in the absence of an inducer. By analogy with Jacob and Wollman's work on phage λ, genes from a wild-type bacterium were transferred into a constitutive strain lacking a functional β-galactosidase. Thus, in the absence of an inducer, neither the donor nor the recipient would express β-galactosidase. Three minutes after the mating, β-galactosidase started to be synthesized by the recipient cells; but, after 90 minutes or so, enzyme production declined. Just as for the parallel experiment with phage λ, this implied that, in the absence of an inducer, a repressor molecule present in the cytoplasm prevented gene expression. Immediately after the transfer of the functional β-galactosidase gene into a cytoplasm lacking the repressor, gene

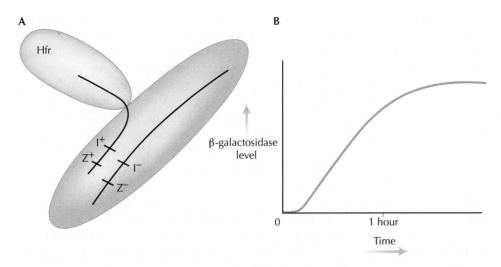

FIGURE 2.30. The PaJaMo (Pardee, Jacob, Monod) experiment. (*A*) Wild-type *Escherichia coli* (I^+, Z^+) were mixed with bacteria that lacked the normal β-galactosidase enzyme (Z^-) and that also carried a constitutive mutation that caused expression of the *lac* operon even in the absence of lactose (termed I^- because they were not inducible by lactose). The wild-type bacteria came from the *Hfr* strain and so acted as males, donating genes to the double-mutant strain. (*B*) A few minutes after mating, β-galactosidase began to be produced at a high rate, even in the absence of lactose. However, enzyme production later declined.

expression started; however, a functional repressor gene had also been transferred, so that the β-galactosidase gene was eventually turned off again.

Jacob realized the close similarities between the two model systems in the summer of 1958: In both cases, a repressor controlled the expression of multiple genes. There followed a fertile period in which the model was exhaustively tested in both systems. For example, one prediction is that as well as constitutive mutations that lacked the repressor molecule (as in the PaJaMo experiment), there should be constitutives that lacked the site to which the repressor bound. Such mutations were indeed found.

The Central Dogma: DNA Makes RNA Makes Protein

In 1958, Crick presented the **Central Dogma** (Fig. 2.31) of molecular biology: "Once 'information' has passed into protein, *it cannot get out again...* . Information here means the *precise* determination of sequence, either of bases in the nucleic acid or of amino acid residues in the protein." As Crick later said, this was "an idea for which there was *no reasonable evidence*"; rather, it was a statement of what is logically possible. It can be justified in two ways. First, we now know (and it was suspected then) that the genetic code is degenerate: Several triplets code for each amino acid, and so it is impossible to precisely reconstruct DNA sequence from protein sequence. However, the stronger argument is that back-translation from protein to nucleic acid sequence would require a machinery at least as complex as the system of transfer RNA, amino-acyl transferases, ribosomes, and so on. "There is no trace of such machinery, and no reason to believe that it may be needed."

Although the slogan "DNA makes RNA makes protein" had been put forward in 1947, the actual role of RNA in carrying information from DNA to protein remained obscure for another 13 years. It was widely assumed that the RNA molecules found in each ribosome (now known as **ribosomal RNA** or **rRNA**) were a transcript of the DNA that coded for a particular protein. Thus, there would be a different set of ribosomes for each different protein. This view raised two difficulties

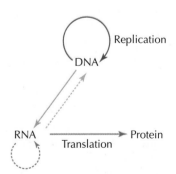

FIGURE 2.31. The Central Dogma of molecular biology is that information flows from DNA to RNA to protein. *Solid arrows* indicate information flows that occur in all cells, through DNA replication, transcription of DNA into mRNA, and translation of mRNA into protein. *Dotted arrows* indicate flows that are seen occasionally, through reverse transcription and replication of RNA. Crucially, information cannot flow from protein back into nucleic acid sequence.

FIGURE 2.32. Max Perutz (*right*) spent much of his professional lifetime studying the structure and function of the hemoglobin molecule. John Kendrew (*left*) determined the structure of myoglobin.

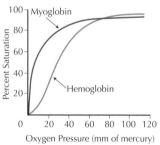

FIGURE 2.33. (*Top*) The hemoglobin molecule structure as presented by Perutz. An alternative presentation of the hemoglobin molecule is shown in the opening illustration (p. 54). (*Bottom*) Interactions between the four subunits of hemoglobin make it more efficient at transporting oxygen. Myoglobin, a closely related protein with only one subunit, shows the usual hyperbolic relation between the amount of oxygen bound and the partial pressure of oxygen (*upper curve*). In contrast, hemoglobin shows a sigmoidal relation, which allows it to unload oxygen in the tissues at a much higher partial pressure.

that were acutely felt by 1960. First, how could genes be turned on and off as quickly as was seen in the PaJaMo experiment? Second, how could the ribosomal RNA code for proteins of widely differing sizes when it was known to be found in only two lengths? Worse, Andrei Belozerskii and Alexander Spirin had shown in 1958 that although the composition of bacterial DNA varied greatly from species to species, there was no corresponding variation in the composition of their RNA.

These puzzles were resolved at a meeting between the Paris and Cambridge groups, on Good Friday, 1960, at which the PaJaMo results were discussed. It was realized that an unstable **messenger RNA** (mRNA) carried the information from the DNA and that the ribosomes are merely a "reading head" that translates any message. This immediately explained the metabolic stability and uniform composition of the bacterial RNA—this was mainly ribosomal RNA, which is highly conserved. It also explained the speed with which genes could be switched on and off: The repressor molecule could bind directly to the gene, preventing its **transcription** into messenger RNA. Within a few months, the messenger RNA had been isolated and the hypothesis substantiated.

Jacob and Monod set out both the idea of the messenger and of regulation of clusters of genes (**operons**) in 1961. By this stage, all the basic principles of molecular biology had been established. However, the actual mechanism by which proteins catalyzed chemical reactions, and by which they were regulated, remained obscure.

Protein Function Depends on Changes in Shape: Allostery

The discovery of the structure of DNA quickly led to the understanding that DNA sequence determines protein sequence and hence, presumably, the three-dimensional structure of the protein. However, the structure of proteins (or even, of any one protein) was not known; nor was it known how proteins carry out their catalytic and regulatory functions. The answers were to come primarily through 40 years of work on hemoglobin by Max Perutz (Fig. 2.32).

In 1936, Perutz came to Cambridge to work with Bernal, who had just taken the first clear X-ray diffraction images of crystallized proteins. Perutz chose to study hemoglobin because it formed good crystals, with a symmetry that somewhat simplified X-ray analysis. Although Perutz did not appreciate this at the time, the biological function of hemoglobin was the best understood of any protein. It acts as a "molecular lung," carrying oxygen to the tissues. As oxygen is released, hemoglobin's affinity for the remaining oxygen falls, so that oxygen can be unloaded almost completely. This behavior, which is reflected in the sigmoid relation between oxygen pressure and saturation (Fig. 2.33, bottom), depends on interactions between the four subunits of the molecule, each with one heme group (Fig. 2.33, top). Once one O_2 molecule is released from one heme group, the remaining three O_2 molecules become less tightly bound. Such biochemical subtleties were all known by the 1920s. However, it took half a century for them to be explained in terms of the structure of the hemoglobin molecule.

Perutz's early hopes for a solution rested on the assumption that proteins share some regular arrangement, so that simple and general models might explain the diffraction patterns. This view was encouraged by Pauling's discovery in 1949 of the α helix, a regular structure for the peptide chain that happens to be particularly abundant in hemoglobin. However, these hopes were dashed by Crick's demonstration in a seminar (entitled "What Mad Pursuit") that hemoglobin must be fundamentally irregular: If the structure had contained repeated elements, these would generate stronger diffraction patterns than were in fact observed.

Nevertheless, the weak scattering of hemoglobin was to prove an advantage. In the method of "isomorphous replacement," the effect of a heavy atom inserted into the molecule is used to deduce the phases of the X rays and hence the full structure. Perutz had believed that changing one atom within such a large molecule would have a negligible effect. However, because the reflections from the irregularly spaced atoms of the protein mostly cancel, one heavy atom does in fact have a large influence on the diffraction pattern.

This method was first applied in 1953 and led to slow but steady progress. Methods for automatic measurement of spot intensity and computer programs for their interpretation were also required. By 1957, Perutz's colleague John Kendrew obtained a crude structure of myoglobin, a homolog of hemoglobin with a single subunit found in muscle. Two years later, Perutz had the first rough structure of hemoglobin; by 1968, he had resolved the molecule to a high level of detail. By that time, the structures of several enzymes had also been obtained.

Knowing the structure did not, however, give much clue as to the function. Indeed, the first results were puzzling: The heme groups were found to be far apart, which made it hard to see how they interacted to give the crucial cooperation in the binding of oxygen. A comparison between the structures of the oxygenated and deoxygenated forms showed that the interaction was caused by the moving apart of the β chains when oxygen is released. However, it was not until 1970 that Perutz understood the mechanism. When oxygen is released, the iron atom at the center of the heme group pushes out by 0.6 Å; this in turn displaces the histidine to which the iron is attached, and the motion of this molecular lever breaks the salt bridges that link the two halves of the molecule (Fig. 2.34).

In 1962, Jacob, Monod, and Jean-Pierre Changeux brought together the ideas that were emerging from studies of hemoglobin with their own work on gene regulation in the *lac* genes and in phage λ. In all these examples, molecules with quite different shapes bound to the same protein and interacted via a mechanical change in conformation. (In hemoglobin, there is an interaction between the binding of oxygen at the heme group and the binding of hydrogen ions elsewhere in the molecule, as well as between heme groups; and in the *lac* repressor, the binding of lactose loosens binding to DNA at a different site.) They termed this phenomenon **allostery** and emphasized its fundamental importance: It allows the coupling of

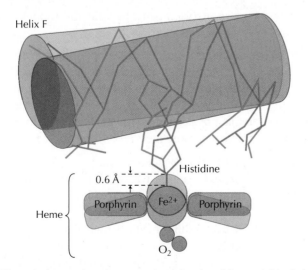

FIGURE 2.34. Allosteric interactions between the subunits of hemoglobin help it load and unload oxygen efficiently. When oxygen is bound to the iron atom at the center, the heme group is flat (*red*). When oxygen is released, the heme group pushes upward, displacing the attached histidine (*arrow*). This moves the protein chain (*upper cylinder*), and breaks the links between the salt bridges that link two halves of the molecule. This in turn distorts other heme groups, making it easier for them to unload oxygen.

arbitrary molecular structures into molecular circuits that execute the logic of metabolic and genetic regulation. Monod wrote that this principle "is of such value to living beings that natural selection must have used it to the limit."

Direct Observation of DNA Revealed Some (but Surprisingly Few) New Phenomena

During the decade that followed the birth of molecular biology, the basic findings were consolidated, but no fundamentally new discoveries were made. Many of the founders moved on to other questions: developmental mechanisms, study of the nervous system, eukaryotic cell biology, and so on. The next period of rapid change was driven by technical refinements, rather than by new discoveries about the nature of life.

Several key techniques were introduced, or became widely available, during the 1970s.

FIGURE 2.35. The 30*S* subunit of the *Thermus thermophilus* ribosome to 3 Å resolution. The ribosome, which translates the genetic information encoded in mRNA into protein, is a large nucleoprotein complex made up of a 30*S* and 50*S* subunit in bacteria.

- *Structure.* Better methods for deducing molecular structure by X-ray crystallography were introduced gradually: better monochromatic X-ray sources, freezing crystals to slow their degradation by irradiation, and faster computers. Application of nuclear magnetic resonance to structural studies introduced a new method that complemented X-ray studies. By the end of the century, the structure of the entire ribosome had been determined (Fig. 2.35).

- *Sequencing.* In the 1970s, Sanger, who had been the first to sequence a protein, developed a rapid method for determining the sequence of DNA. Allan Maxam and Walter Gilbert invented an alternative DNA sequencing method during the same period. This initially relied on labor-intensive separation of DNA molecules of different lengths by electrophoresis. Later, automation of the Sanger method made large-scale sequencing feasible.

- *Manipulation.* In 1972, Paul Berg at Stanford showed how the enzymes that had evolved to manipulate the cell's own DNA could be used to artificially recombine DNA taken from different species. Bacterial restriction enzymes were especially useful, because these cut DNA at specific sequences, leaving single-stranded ends that can be rejoined in new combinations. After a brief moratorium in which the possible risks of the new technology were assessed, the new recombinant DNA techniques were rapidly applied to allow genetic manipulation of a wide range of organisms, beyond the model systems to which genetics had previously been restricted.

- *PCR.* The most recent major innovation was Kary Mullis's invention of the **polymerase chain reaction** (PCR) in 1983. Repeated cycles of denaturation and reannealing allow the amplification of a specific DNA sequence from as little as a single source molecule. In many cases, this removes the need to **clone** genes in bacterial culture and makes it possible to determine the genetic makeup of single cells—for example, from unculturable microorganisms or from human embryos.

These new techniques revealed new phenomena and elaborated on old findings. Especially in eukaryotes, gene regulation proved more complex and more diverse than had been indicated by the simple models of β-galactosidase and phage λ. Genes could be activated as well as repressed, and regulation could occur at many stages: Besides binding of a regulatory molecule to the gene itself, there could be processing or degradation of the messenger, regulation of translation, localization or modification of the protein product, and so on. Signals might be passed along chains of interacting proteins and multiple transcription factors might interact at a gene's promoter. With hindsight, such complexity is not so surprising; natural selection can exploit any available mechanism to alter the way genes are expressed to produce the phenotype.

The most striking finding to come from the first direct studies of DNA sequence was that in eukaryotes the coding sequence is usually interrupted by **introns**, which are spliced out from the messenger RNA before translation. Most eukaryotic DNA does not code for protein and, indeed, may have no function that benefits the whole organism. For example, much of the genome consists of transposable elements that replicate at the expense of host fitness (see pp. 217 and 596). Genetic manipulation revealed that in organisms such as yeast, *Drosophila*, or mice, the majority of individual genes can be eliminated without drastic ill effects. Such apparently redundant genes could not easily have been detected by the methods of classical genetics.

Challenges were made to the core of molecular biology—that DNA makes RNA makes protein. **Prions**—proteins that can switch between alternative, stably inherited, conformations—were found in yeast and mammals. Enzymes were found that methylate DNA, in such a way that this modification is stably inherited. These are examples, albeit rare and limited, in which heredity is not based on base pairing in nucleic acids. In 1970, **reverse transcriptase** was discovered. This enzyme allows transposable elements and RNA-based viruses to synthesize DNA that can be incorporated into the host genome. Examples were found in which the cell manipulates its own genetic material—for example, in switches between the two mating types of the yeast *Saccharomyces cerevisiae* or in cells of the vertebrate immune system.

None of these examples challenge the Central Dogma: In no case is protein sequence translated back into nucleic acid sequence. They can best be seen as exceptional cases in which the basic genetic machinery has been subverted to give some local advantage. Indeed, it seems likely that one could, in principle, construct an organism based only on the simple principles of replication, transcription, translation, and allosteric regulation identified by 1962.

The discovery of catalytic RNA by Thomas Cech and Sidney Altman in 1982 is arguably the one fundamentally new finding made since the establishment of molecular biology (Fig. 2.36). Although it had long been known that many pro-

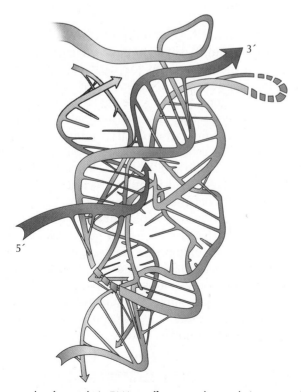

FIGURE 2.36. An example of a catalytic RNA or **ribozyme**: the catalytic center of the *Tetrahymena* group I intron (*blue*). The substrate RNA (*red*) is cleaved by the active site at the junction of the two helical domains (*red arrow* at *center*). Remarkably, this RNA molecule catalzyes its own cleavage, cutting itself out of the pre-mRNA to produce the processed mRNA, which will be translated.

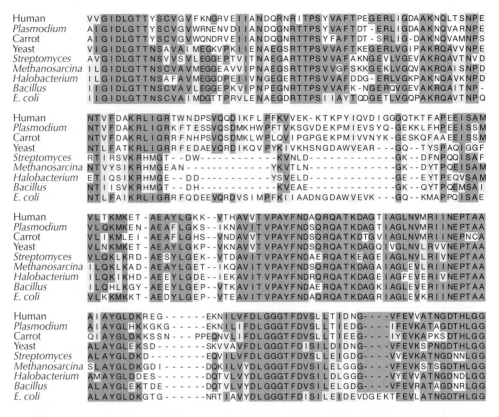

FIGURE 2.37. An alignment of highly conserved heat shock proteins in the HSP70 family, which is found in all cellular organisms. The *shading* highlights conserved amino acid residues (*red* means identical and *blue* means similar), and amino acids are identified using the standard single-letter code (see Fig. 2.23).

teins are associated with small RNA molecules (most obviously, the ribosomal RNA), these RNAs were thought to play a passive structural role. However, it is now clear that many of them carry out the basic catalytic steps—for example, synthesis of the peptide bond by rRNA—and are relics of an early world in which RNA acted as both gene and enzyme (pp. 101–104).

From an evolutionary perspective, the most striking outcome of the past 40 years of molecular biology has been the extraordinary conservation of basic biology. It is perhaps not surprising that the genetic code and the basic machinery for replication, transcription, and translation are so similar across all living organisms. What is more remarkable is that the molecules used in other processes such as development, the cell cycle, cell movement, and defense against pathogens have retained both similar sequence and similar function across a wide taxonomic range (Fig. 2.37). Examples include transporter proteins; chaperonins, which guide the folding of proteins; cytoskeletal components such as tubulins; proteases, which aid in regulating protein activity (among other roles); and kinases and phosphatases that regulate cellular proccesses. This molecular conservation makes biology much simpler, because the same explanations and techniques apply to many organisms. However, it makes it much harder to explain the most fundamental features of life, because evolution has provided us with only a single example.

EVOLUTIONARY AND MOLECULAR BIOLOGY: A NEW SYNTHESIS?

Modern evolutionary and molecular biology were born around the same time, and both have since grown to become large and thriving fields. However, they have remained surprisingly independent of each other, asking different kinds of questions about different aspects of biology. In this section, we outline the relation between the two fields, the differences between them, and the ways in which they interact most fruitfully.

Molecular Biology Opened Up Study of How All Life Evolves

The remarkable conservation of molecular structures emphasizes the common origin of all life. Its practical effect has been to make it possible for evolutionary biologists to study many genes in many organisms and to make such studies in nature rather than in the laboratory. Early work on evolutionary genetics (e.g., by Theodosius Dobzhansky and E.B. Ford; see p. 31) was restricted to genes that happened to vary in a detectable way: blood groups in humans, the patterns of butterfly wings or snail shells, or chromosomal inversions in *Drosophila*. Genetics required crosses, and so was limited to species that could be bred artificially. Now, it is possible to choose any kind of gene and to use techniques such as PCR to take samples from natural populations: from the droppings of wild primates, from Neanderthal fossils, or from microbes that cannot be cultured in the laboratory.

Beginning with protein electrophoresis in the mid-1960s and, more recently, using direct surveys of DNA, such work revealed unexpectedly high genetic variation within species. Similarly, Sanger's first studies revealed differences in amino acid sequence between species and made it possible to compare the same protein across a wide taxonomic range. Moreover, it was soon found that the rate of sequence divergence of a protein, even across very different organisms, is similar. In other words, there is a **molecular clock** (Fig. 2.38; see pp. 371 and 531).

This wealth of genetic variability made it possible to study the evolutionary processes that shape variation within species and to reconstruct the phylogenetic relations that connect different species with their common ancestor. The existing questions posed by evolutionary biologists could be asked about a wide range of new genes and organisms. This ability to apply common techniques to obtain comparable data has brought a great simplification and unification to the subject. Evolutionists had always attempted a universal explanation of life, but without the unification of mechanism provided by molecular biology, this was hard to realize in practice.

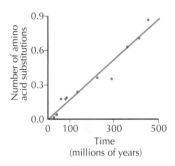

FIGURE 2.38. The molecular clock ticks steadily for any one protein. The graph shows the number of amino acid differences per site relative to the human sequence. This increases with divergence time, and the rate of change is roughly constant across very different lineages. Sequences are from α-globin of shark, carp, newt, chicken, echidna, kangaroo, dog, and several primate species (*right* to *left*).

The Pattern of Molecular Variation Suggested that Most of It Does Not Affect Fitness: The "Neutral Theory"

The first observations of molecular variation and divergence prompted Motoo Kimura (Fig. 2.39) to propose, in 1968, the **neutral theory**. Kimura argued, first, that the molecular clock was best explained by the steady accumulation of mutations that had no significant effect on fitness. Second, he argued that there is too much genetic variation within species for more than a small fraction to be subject to natural selection. This theory was (and is) controversial, especially because, at the time Kimura proposed it, evolutionary biologists were placing greater emphasis on the role of selection in accounting for even slight phenotypic differences (see Chapter 1). As we will see in Chapter 19, this issue is still unresolved. It is clear that in eukaryotes at least, most noncoding variation must be neutral simply because there is so much DNA in their genomes. However, it is not known what fraction of amino acid differences between individuals or species are maintained by selection, and, conversely, there is good evidence of selection on some **synonymous** changes (i.e., changes that do not alter protein sequence but may affect the way genes are expressed).

The neutral theory transformed evolutionary biology by making it possible to apply existing theory in a quite different way. Originally, population genetics had been used to show that observations are consistent with the action of natural selection and other evolutionary processes—for example, that rates of change in the

FIGURE 2.39. Motoo Kimura championed the neutral theory of molecular evolution.

fossil record can easily be accounted for by slight fitness differences or that differences from place to place in the frequency of deleterious mutations in humans or fruit flies could be explained by **random genetic drift**. The neutral theory set out a simple and definite null hypothesis, which made predictions that depended only on population size and mutation rate and not on complicated and unknown details of selection (Chapter 15). Thus, it was possible to devise statistical tests that could reject neutrality and to make estimates of population size and mutation rates from genetic data (at least, where variation can be assumed neutral). The neutral theory also justifies a simple model of molecular evolution and leads to rigorous methods for testing and estimating phylogenetic relationships. Such methods have become especially powerful with the availability of DNA sequence data.

Discovery of the Physical Basis of Life Did Not by Itself Answer the Questions Asked by Evolutionary Biology

The rediscovery of Mendelian genetics transformed evolutionary biology; combined with Darwin's theory of natural selection, it established the Evolutionary Synthesis, which still stands as the core of the subject. In contrast, the discovery of the substance and mechanism of the genes has had remarkably little effect on the basic structure of the field. Molecular methods have greatly changed the practice of evolutionary biology, but they have not immediately answered any of the central questions or even altered the nature of those questions. The new genetic markers have in the main been used to solve existing problems. For example, we can now determine how individual birds are related and so better study their social behavior, and we can use sequence differences rather than external morphology to determine phylogenetic relationships. We may ask questions about molecules in the same way that we would ask about more readily observed features of the organism. For example, we can ask how the biochemistry of the hemoglobin molecule aids survival and reproduction in the same way as we would try to understand the function of the gills or lungs. The arguments over the neutral theory can be seen as a recasting of the old arguments between the **classical** and the **balance views** of genetic variation (see p. 34): To what extent is genetic variation of any kind maintained by selection, rather than being deleterious or neutral? Thus, the formal structure of the subject has not been greatly affected by the molecular revolution, and up till now only a minority of modern developments in evolutionary biology have been driven by the rise of molecular biology.

Molecular Biology Asks "How?"; Evolutionary Biology Asks "Why?"

To some extent, the fields have remained separate because they ask different kinds of questions. One can ask *how* any feature of an organism works: For example, how does hemoglobin bind oxygen? Given that organisms have evolved, we can ask: What was the historical sequence that led to the present state of affairs? For example, from what kind of genes, in what kind of creatures, did hemoglobin descend? Finally, we can ask *why* the feature is as it is. For example, why does hemoglobin consist of four subunits rather than one? Sometimes, answers to such questions can be given without regard to the evolutionary process: In principle, we can trace the replacement of one kind of globin gene by another without knowing *why* that replacement occurred, and we can explain the present mechanism by showing that it is an efficient piece of engineering. (Recall that hemoglobin unloads oxygen efficiently as a result of interactions between its different subunits.)

However, such nonevolutionary explanations are often misleading, incomplete, or inadequate. Later in the book, we explain the possible pitfalls and how they can be avoided (Chapter 20).

We can illustrate the different kinds of questions that can be asked by taking the influenza virus as an example (Fig. 2.40). A molecular biological approach might ask how its RNA is introduced into the host cell, how it is replicated, and how the virus-encoded RNA polymerase replicates the viral genome rather than the host's RNA. An evolutionary biological approach can ask about the evolutionary history of the virus. When and where did different strains share a common ancestor? Was that ancestor in pigs, ducks, or humans? Finally, we can ask why the virus has the properties it has. Why is its genome divided into three separate RNA molecules? Why does it usually cause mild disease rather than rapid death? Such questions cannot safely be answered by any kind of optimality argument: In this example, there is a conflict between what is best for the virus and what is best for its host. These kinds of questions about the evolution of life histories and genetic systems, which we examine in Chapters 20, 21, and 23, require that we trace the fate of individual genes through the evolutionary process.

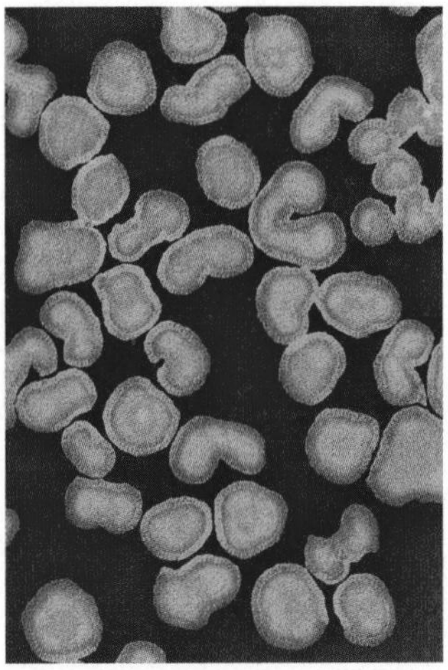

FIGURE 2.40. Human influenza virus type A (original magnification, 191,700x).

Evolutionary Questions Are Difficult to Answer

Questions about past history, and about why that history was as it was, are necessarily harder to answer than questions about present-day mechanisms. Indeed, many such questions may be impossible to answer. Understanding evolutionary relationships relies on making inferences about past ancestors. The fossil record is of little help here because only a tiny fraction of species and traits have been preserved, and those that have are unlikely to lie on a direct line of descent to living organisms. We therefore rely on comparing present-day organisms. If too few species survive, or if there has been too much divergence between those that do survive, then information about the past may be irretrievably lost.

Still harder is explaining why life is as it is. We must imagine possible alternatives and understand how the evolutionary process sorts among these alternatives. We cannot assume, as did Voltaire's Dr. Pangloss, that we live in "the best of all possible worlds." Rather, we must find the constraints on what is possible, and we must replace a simple idea of what is "best" by a knowledge of how the evolutionary process determines the chance that one or another possibility will succeed in competition with others. For example, in the case of the influenza virus, it is not at all obvious what degree of pathogenicity or what genomic arrangement is "best" or, in any case, whether any kind of "optimal" solution would in fact evolve.

We can use a variety of methods to address evolutionary questions. To give a specific example, suppose that we wish to explain the elaborate antlers of red deer stags (Fig. 2.41). We can make theoretical arguments to show that a preference of hinds for stags with large antlers or an advantage in competition between males would favor increased antler size despite their obvious disadvantages. We can observe populations in nature and look for a statistical correlation between antler size and mating success. We might be able to make experimental manipulations of the antlers to show whether any such correlation is due to antler size itself or instead reflects a correlation with some underlying factor such as nutrition. We can compare different species of deer to find whether antler size is correlated with the degree of competition for mates. In principle, we could use artificial selection to find whether various kinds of selection would in fact lead to different antler size. Thus, a combination of such methods is often needed—population genetic theory, field observation, experimental manipulation, comparisons across species, and artificial selection.

FIGURE 2.41. The evolution of antlers in red deer can be studied in a variety of ways.

FIGURE 2.42. One of Kipling's "Just So Stories" explains how the elephant got his trunk.

Because inferences are indirect and predictions are often counterintuitive, considerable care is needed. Indeed, after the early enthusiasm that followed the Evolutionary Synthesis, there has been a strong reaction within evolutionary biology against loose reasoning and Panglossian explanation. Plausible but untestable explanations have often been (and still are) put forward to explain why this or that trait has evolved. These are known derisively as "**just-so stories**," after Rudyard Kipling's (nonevolutionary) accounts of, for example, how the leopard got its spots or how the elephant got its trunk (Fig. 2.42).

Hypotheses should be framed clearly, with a coherent explanation set out in terms of populations of individual genes, and with some means of testing against observations. In particular, it is not enough to show that some feature evolved "for the good of the species." To take two examples from molecular biology, it has been claimed that introns evolved in order to allow new proteins to evolve by a shuffling of different functional domains, thus giving a long-term evolutionary advantage to species possessing split genes. More recently, it has been proposed that groups of genes that are expressed at the same time are clustered in the genome in order that new regulatory patterns can evolve more readily. However, even if new proteins have in fact evolved by exon shuffling or new regulatory patterns have evolved by switching the expression of whole blocks of genes, that does not show that these features have evolved for that reason. What is required is a consistent model that shows how the reproduction of individual genes could lead to these outcomes. Furthermore, some method is needed to distinguish such adaptive explanations from the alternatives—for example, that split genes arise from the random insertion of introns or that coordinated gene expression evolves simply because individual fitness is greatest when certain sets of genes are expressed at certain times.

Molecular and Evolutionary Biology Interact Especially Fruitfully in Several Areas

Although we have emphasized the separate development of molecular and evolutionary biology and the different kinds of questions that tend to be asked by the two fields, they do interact fruitfully in many areas. These interactions go beyond shared techniques to ask questions that link what have been very different lines of research. Here, we outline three examples; each will be discussed in much more detail later.

- *Evolution of development.* (See Chapters 9 and 24.) Following publication of *On the Origin of Species*, the studies of development and evolution were closely intertwined: Comparative embryology was thought to be the key to evolutionary relationships, and the process of evolution was thought to be guided by "laws of form." However, the fields separated during the 20th century. Mendelian genetics seemed to give more direct insight into evolution but, at the same time, offered no obvious solution to developmental problems. Now that molecular biology has shown how the same kinds of genes and mechanisms are responsible for the development of very different forms, it has become possible to apply the new developmental genetics to evolutionary questions. In particular, we can ask how major changes in morphology might have occurred via stepwise genetic substitutions and to understand how variation in individual genes can account for variation in morphology within populations.

- *Comparisons between genomes.* Direct observation of DNA has revealed an astonishing diversity of species and of the habitats they occupy (Fig. 2.43). This brings together questions of molecular mechanism (how can an organism

FIGURE 2.43. Tropical rain forests illustrate the extraordinary biodiversity that has evolved.

function at temperatures above 100°C?), ecology (how can seawater support so many species of bacteria and their viruses?), and evolution (how do new metabolic functions evolve to allow species to occupy extreme environments and what are the limits to such adaptation?). Comparison of large stretches of sequence—in the limit, of whole genomes—has revealed extensive gene transfer, ranging from single genes to whole organelle genomes (pp. 131–133 and 182–191). This challenges our view of what a species is (Chapter 22). In the extreme, given the possibility of extensive genetic exchange, we can ask why we see species at all, rather than a pool of genes that come together in temporary coalitions—in an alternative, termed by John Maynard Smith, the "football team" model of evolution.

- ***Why is life as it is?*** Our ability to at least attempt to reconstruct the deep tree of life emphasizes the complexity of the last universal ancestor. This must have had a genetic code, with the associated machinery for transcription and translation, as well as an elaborate metabolism. Although it is hard to know how that ancestor evolved, since the time roughly 3500 Mya when life itself began, it has proved possible to make surprising progress. In particular, the discovery of catalytic RNA makes it likely that the universal ancestor was preceded by a world in which RNA acted as both enzyme and hereditary material. The study of this lost world depends on an intimate combination of evolutionary arguments about what could have evolved and of molecular studies of the properties of RNA. Evolutionary biology has developed methods for understanding why RNA molecules work as they do (e.g., what limits the size of the genome, whether sexual reproduction is required for efficient evolution), but these must be complemented by a thorough understanding of the properties of actual molecules. Experimentally, the selection of RNA molecules in vitro both tells us about the possible nature of the RNA world and allows us to evolve molecules with new and useful properties.

SUMMARY

Our understanding of the physical basis of life developed gradually, from the birth of organic chemistry in the late 18th century to the emergence of modern biochemistry by the 1930s. However, although the chemical composition of macromolecules (especially, proteins and nucleic acids) was known, and basic metabolism was well understood, essential concepts were missing: What is the importance of macromolecular three-dimensional structure and sequence specificity? What is the role of information flow, in addition to the flow of energy and matter, in biological systems? These questions became the domain of molecular biology. The key step came in 1953, when Crick and Watson showed that the structure of DNA is a double helix, in which a sequence of four alternative bases (A, T, G, or C) pair with the complementary sequence, according to specific base-pairing rules. Over the following decade, the foundations of molecular biology were laid. Intense efforts revealed how DNA sequence specifies protein sequence via the genetic code; Jacob and Monod showed how genes can be regulated; and the intermediary role of messenger RNA was established. In the culmination of three decades' effort, Perutz used X-ray crystallography to determine the three-dimensional structure of hemoglobin. This revealed how allosteric changes in the protein's shape allow flexible regulation by connecting different active sites. Later developments in molecular biology led to powerful techniques that enabled DNA to be amplified by PCR, manipulated using restriction enzymes, and sequenced. Despite these rapid advances, the basic findings established in the early years of molecular biology remain intact. Arguably, the one fundamental discovery since then was that RNA acts as an enzyme in some key biological pathways.

Evolutionary and molecular biology are now closely intertwined. The extraordinary conservation of molecular mechanisms allows all living organisms to be studied by the same techniques. The observation that sequences evolve at a roughly constant rate and the large sizes of many genomes suggest that most DNA sequence variation, both within and between species, has negligible effect on fitness. Later in the book, we will see how this neutral theory provides a powerful null model that allows sophisticated analysis of sequence variation. We also will see how evolutionary and molecular approaches come together in understanding the evolution of development, in establishing the phylogenetic relation between species, and in understanding why organisms are the way they are.

FURTHER READING

Crick F. 1970. Centra dogma of molecular biology. *Nature* **227**: 561–563.
  An update and response to critics of the Central Dogma, by its original author.
Hunter G.K. 2000. *Vital forces: The discovery of the molecular basis of life.* Academic Press, New York.
  A history of biochemistry from its earliest years.
Judson H.F. 1995. *The eighth day of creation.* Penguin, London.
  An impressively detailed account of the origins of classical molecular biology based on interviews with the key figures. It includes in-depth descriptions of the seminal papers.
Monod J. 1971. *Chance and necessity: An essay on the natural philosophy of modern biology.* Alfred A. Knopf, New York.
  An elegant discussion of the nature of life by the codiscoverer of the mechanism of gene regulation.
Morange M. 1998. *A history of molecular biology.* Harvard University Press, Cambridge, Massachusetts.
  Summarizes the history of more recent advances in molecular biology.

CHAPTER

3

Evidence for Evolution

EVOLUTIONARY BIOLOGY HAS PROFOUNDLY altered our view of nature and of ourselves. At the beginning of this book, we showed the practical application of evolutionary biology to agriculture, biotechnology, and medicine. More broadly, evolutionary theory underpins all our knowledge of biology, explains how organisms came to be (both describing their history and identifying the processes that acted), and explains why they are as they are (why organisms reproduce sexually, why they age, and so on). However, arguably its most important influence has been on how we view ourselves and our place in the world. The radical scope of evolutionary biology has for many been hard to accept, and this has led to much misunderstanding and many objections. In this chapter, we summarize the evidence for evolution, clarify some common misunderstandings, and discuss the wider implications of evolution by natural selection.

Biological evolution was widely accepted soon after the publication of *On the Origin of Species* in 1859 (p. 17). Charles Darwin set out "one long argument" for the "descent with modification" of all living organisms, from one or a few common ancestors. He marshaled evidence from classification of organisms, from the fossil record, from geographic distribution of organisms, and by analogy with artificial selection. As we saw in Chapter 1, the detailed processes that cause evolution remained obscure until after the laws of heredity were established in the early 20th century. By the time of the **Evolutionary Synthesis**, in the mid-20th century, these processes were well understood and, crucially, it was established that adaptation is due to **natural selection** (p. 33). Now, evolution is accepted as a fact, and active research is extending our understanding of the processes responsible for it.

Despite this strong scientific consensus, many people do not accept that living organisms have evolved by purely natural processes. This skepticism has several roots. For some, it may arise from a conflict with prior religious beliefs. In others, it may come from doubts that the astonishing diversity of the living world could descend from one simple ancestral organism or that complex adaptations—especially, the human mind—could be built up by natural selection acting on random variation. The conflict with religious belief is most sharply focused for those who believe in the literal truth of their sacred texts. Such believers must reject much of science—physics, astronomy, and geology, as well as biology—and, indeed, must reject the very methodology of science.

There is a continuous range of beliefs about the origin of life, from a literal belief in one of the creation myths in Genesis through to the purely material account given in this book. Some hold that each of the 6 days of the biblical account corresponds to many millions of years. Thus, they accept an old Earth but still hold that species (or at least, higher taxa) were separately created. Others accept evolution but invoke an "intelligent designer" to explain complex adaptation. The predominant position of the main Christian churches is **theistic evolution**, in which God works through natural laws, with little or no direct intervention. The Catholic Church accepts physical evolution via natural selection and other evolutionary processes, but it invokes a supernatural introduction of the human soul. Most of the clashes between evolutionary biology and religion have come from people who believe in a single divine creator. Religions that do not assume such a deity, for example, Buddhism and Hinduism, have generally seen evolutionary ideas as compatible with their belief in a world that is in continual transformation.

Later in this book, we set out the detailed history of evolution (Chapters 4–11); explain the mechanisms of natural selection and other evolutionary processes (Chapters 12–17); and explain how this accounts for adaptation, speciation, and the emergence of novel features (Chapters 18–24). We also devote considerable discussion specifically to human evolution (Chapters 25 and 26). In this chapter, we summarize the evidence that natural selection is responsible for the appearance of design. We discuss how the scientific process works and its relationship with the religious and moral beliefs that fuel much of the opposition to evolution by natural selection. Many of the points we make about the nature of evolutionary thinking are considered in more detail later in the book. This chapter serves as a concise summary of the key arguments.

EVIDENCE FOR EVOLUTION

Patterns of Relationship Provide the Most Powerful Evidence for Evolution

Although direct observation and the fossil record each provides powerful support, the most compelling evidence for evolution comes from the patterns of similarity between present-day organisms, which reveal features that are shared across all organisms: a nested pattern of groups within groups, consistent across many different traits, and a correspondence between biological relationship, geological history, and geographical distribution.

Universally Shared Features

Even in Darwin's time, the similarity of all living organisms was clear enough for him to suppose that all life descended from one or at most a few ancestors (Fig. 3.1). However, the full extent of this similarity was revealed when the universal principles of molecular biology were discovered in the middle of the last century. Almost all organisms use DNA to encode their genetic information, which is transcribed into RNA and then translated by a single universal genetic code into protein sequence. (Some viruses are based on RNA, not DNA, and there are slight variations in the genetic code [Table 5.3], but these are minor exceptions.) Many molecular functions have been conserved across widely different taxa. For example, yeast that are defective in genes that control the cell cycle can be rescued by human genes that carry out the same function. Indeed, the basic machinery of replication, transcription, and translation is conserved across all living organisms. The success of molecular biology lies in the essential universality of its mechanisms (Chapter 2).

This shared biochemistry is largely arbitrary: These universal features are not constrained to be the way they are by physics or chemistry. For example, proteins are always made from L-amino acids and never from their mirror-image D-stereoisomers

FIGURE 3.1. Charles Darwin. "There is grandeur in this view of life, with its several powers, having originally been breathed by the Creator into a few forms or into one" (from conclusion to *On the Origin of Species*).

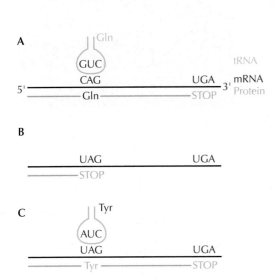

FIGURE 3.2. **Suppressor mutations** show that the universal genetic code is not constrained to be exactly as it is. **Nonsense mutations** generate stop codons that prematurely terminate translation of the protein. They can be suppressed by mutations in a transfer RNA that enables the tRNA to recognize the stop codon as an amino acid and so allow translation of the protein to be completed. In effect, these mutations have changed the genetic code. (A) Translation of the wild-type sequence is terminated by a UGA stop codon. (B) A mutation changes CAG (which coded for glutamine, Gln) to a stop codon, UAG, causing premature termination of translation. (C) A mutation in one of the transfer RNAs that codes for tyrosine (Tyr) changes the anticodon to AUC. This recognizes the UAG nonsense mutation, and so a full-length protein is produced, with tyrosine substituted for glutamine. (Recall that U in RNA corresponds to T in DNA.)

(Fig. 2.5)—even though a left- or a right-handed biochemistry would function equally well. This universal handedness is readily explained by descent from a single common ancestor. The genetic code is to a large extent a "frozen accident": Any code that maps the 64 possible triplet codons onto the 20 amino acids would work and could be implemented just as easily by an appropriate set of transfer RNAs. This point is illustrated by occasional natural variants to the code (Table 5.3) and by laboratory mutations that alter the code (Fig. 3.2). (Later, we discuss some regularities in the genetic code that indicate that it is not *entirely* random.) As a final example, we saw in Chapter 2 that RNA molecules carry out key catalytic functions—most notably, the joining of amino acids by peptide bonds to form a protein. These are explained as relics of an **RNA world**, in which RNA molecules instead of proteins were responsible for the chemical work of the cell, as well as for carrying hereditary information (pp. 101–104).

Hierarchical Classification

The naturally hierarchical classification of organisms into groups within groups, reflected in the Linnaean system of species, genera, and families, is immediately explained by "descent with modification"; the classification directly reflects shared ancestry. Moreover, the characteristics that are most useful in classification are not those that adapt species to their individual way of life, but rather those that retain their ancestral state throughout a group. This distinction between **analogy** and **homology** was appreciated before Darwin (Fig. 3.3). For example, the streamlined shape of fish and whales is an **analogous** feature: In other words, it is a consequence of their convergent ways of life. In contrast, the structure of the mammalian limb has remained the same, even though it is used for very different purposes in bats, humans, and porpoises. Such **homologous** structures are readily explained by common descent. Certain embryonic stages tend to be similar, even between species that have very different adult forms (Fig. 3.4). Darwin explained this pattern by pointing out that selection would diversify the adult form but would act against potentially disruptive changes in the embryo.

This pattern of groups within groups leads to a nested classification, with each group defined by sharing a unique set of characters. For example, vertebrates (the

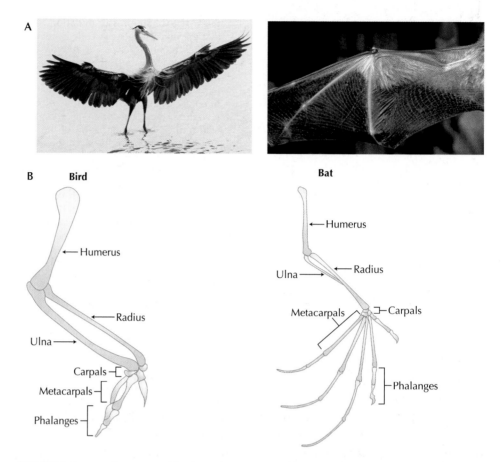

FIGURE 3.3. (*A*) The wings of birds and bats are *analogous*, because they carry out the same function, but are not descended from a common ancestral structure having that function (*left*, great blue heron; *right*, male red bat) (*B*) The skeletal forelimbs of birds and bats are *homologous*, because they descend from the same structure in their common ancestor.

subphylum Vertebrata) have a backbone and limbs built to the same five-fingered plan. Within the Vertebrata are mammals (class Mammalia), who produce milk and are covered in hair. Ruminants (e.g., antelope, sheep, and cattle; a suborder of Mammalia) share cloven hooves and a specialized digestive system. This method of classifying nested groups of organisms continues down to the level of individual species, such as *Bos taurus* (the domesticated cow) (Fig. 3.5).

Notably, molecular characteristics support the same classification. For example, vertebrates share a particular arrangement of *Hox* genes, and classifications based on multiple DNA or protein sequences yield the same nested pattern. This nested classification of groups within groups, and its consistency across traits, is most easily explained as a reflection of the tree-like pattern of descent from a common ancestor. It is quite different from the pattern seen in designed artifacts. For example, although cars of the same brand share some superficial resemblance, different features are scattered across different makes wherever they are found useful.

As we explain in Chapter 27 (online), the trees inferred from any one set of characters cannot be identified with certainty; rather, they are statistical estimates of the actual relationship and so will usually not be perfectly accurate. Nevertheless, the consistency across entirely unconnected traits—both molecular and morphological—is striking and is strong evidence for common descent.

The frequent difficulty in deciding whether two forms rank as species or mere

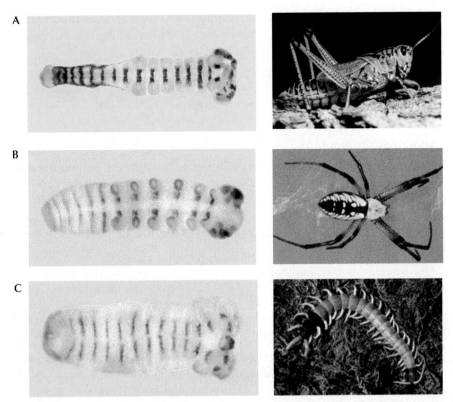

FIGURE 3.4. Similarity of embryos during early stages of development. Embryos of a grasshopper (insect; *A*), spider (chelicerate; *B*), and centipede (myriapod; *C*) are juxtaposed with their adult forms. Despite the differences between the adults of these three arthropods, the embryos are remarkably similar at this stage of development. (The embryos are stained for a gene product that highlights their conserved segmental nature.)

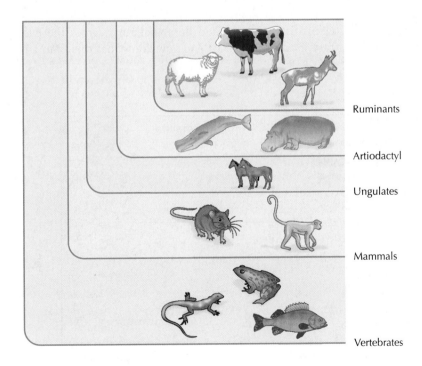

Ruminants

Artiodactyl

Ungulates

Mammals

Vertebrates

FIGURE 3.5. The vertebrates can be classified in a series of nested groups-within-groups, each group sharing a set of unique features. This pattern is explained by the underlying phylogeny that connects each species (see Fig. 9.18). The artiodactyls (even-toed ungulates) are a diverse group that includes the ruminants (shown here) but also highly modified groups such as the whales.

varieties was seen by Darwin as further support for "descent with modification." Varieties are incipient species, and a clear dividing line between them is not expected (see pp. 620–622). In *On the Origin of Species*, Darwin wrote:

> Systematists will be able to pursue their labours as at present; but they will not be incessantly haunted by the shadowy doubt whether this or that form be in essence a species. This I feel sure, and I speak after experience, will be no slight relief.

Geographic Distribution

Darwin's most compelling evidence for evolution came from the geographic distribution of myriad plants and animals that he saw during his travels on the *Beagle* (see p. 16). Wherever he looked, Darwin found that organisms were related by their proximity to one another, even across diverse habitats. The marsupials in Australia and the toothless mammals in South America are but two of the many examples that led to Darwin to write:

> . . . the naturalist in travelling, for instance, from north to south never fails to be struck by the manner in which successive groups of beings, specifically distinct, yet clearly related, replace each other. He hears from closely allied, yet distinct kinds of birds, notes nearly similar, and sees their nests similarly constructed, but not quite alike, with eggs coloured in nearly the same manner.

This pattern is especially striking on oceanic islands. For example, the mockingbirds that Darwin found in the Galápagos Islands differ between islands, but they share an underlying resemblance with each other and, to a lesser extent, with the mainland birds from which they are derived. We now have many such examples of dramatic **adaptive radiations** on oceanic islands, including many species found on the Hawaiian Islands, and the cichlid fishes within the African Great Lakes (pp. 650–652). Darwin explained the presence of the same species, or closely allied species, of alpine plants on different mountaintops across Europe and North America as a consequence of the retreat of glaciers, which stranded these organisms on isolated peaks. Without invoking evolution, such distributions cannot be explained except through an arbitrary number of separate creations.

Some of the most striking support for evolution comes from the correspondence between geographical distribution and geological history. Alfred Russel Wallace identified a sharp boundary between distinct fauna and flora that runs across the East Indies in an apparently arbitrary location (Fig. 3.6). We now know that this is an an-

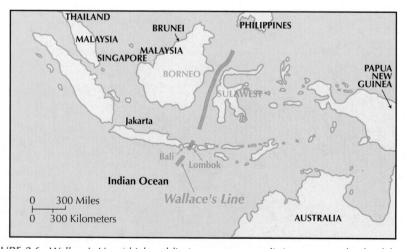

FIGURE 3.6. Wallace's Line (*thick red line*) separates two distinct present-day land faunas.

cient deep-water passage between two land masses, which remained separate from each other even when sea levels were about 100 m lower, during the Pleistocene (see p. 285). A still more striking example is the distribution of species such as marsupials, lung-fishes, and the southern beech (*Nothofagus*) across the southern continents. Although now widely separated, these were all part of the supercontinent of Gondwana, 120 Mya (see Fig. 10.5). Such patterns are immediately explained in terms of evolution, if species gradually disperse away from their point of origin.

Evolutionary Processes Can Be Observed Directly

The astonishing success of artificial selection played a large part in shaping Darwin's ideas on evolution (p. 19). Agriculture relies on an extraordinary variety of domesticated animals and plants, which have been shaped simply through the continued selection of those individuals with desirable characteristics. Often, one species has yielded radically different varieties. The different breeds of dogs differ much more in morphology and behavior than do typical mammalian species. Similarly, one species of *Brassica* has been selected to produce apparently quite different crops (Fig. 3.7).

Rapid evolutionary change is also seen in nature. Good examples are the change in morphology of sparrows as they spread across North America since their introduction in 1852 or the response of soapberry bugs to the introduction of a new host plant in the 1920s (Fig. 3.8). On a quite different scale, when people infected by the human immunodeficiency virus (HIV) are treated with antiviral drugs, the population of HIV that infects them evolves multiple amino acid substitutions that confer resistance. These evolutionary changes are consistently seen across different infections (Fig. 3.9).

The origin of new species is usually too slow to be seen directly, but we do have several striking examples. There are cases where insects have been seen to shift to use a new host plant, thus producing populations that are on their way to becoming fully

 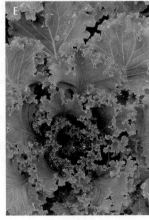

FIGURE 3.7. Diverse varieties of *Brassica oleracea* include (*A*) cabbage; (*B*) broccoli; (*C*) cauliflower; (*D*) brussels sprouts; and (*E*) flowering kale.

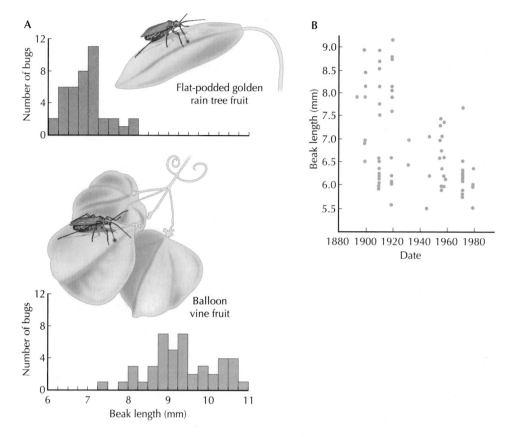

FIGURE 3.8. Soapberry bugs in Florida originally fed on the native balloon vine (*A, lower*), using their sharp beaks to penetrate the fruit. In the 1920s, the flat-podded golden rain tree (*A, upper*) was introduced from Asia. This has thinner-skinned fruit, and, correspondingly, soapberry bugs evolved shorter beaks after they switched to feed on this new host plant. (*B*) Each dot in the scatter plot shows the beak length of an individual bug taken from museum collections.

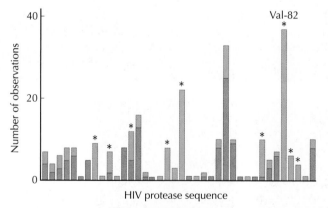

FIGURE 3.9. HIV evolves resistance to the antiviral drug ritonavir through multiple substitutions in the drug's target, HIV protease. The evolution of this enzyme was followed in 42 patients. The *red bars* show the variants observed in the base population. The *blue bars* show variants that emerged after drug treatment, most of them on multiple occasions. The nine variants marked by an asterisk contributed to resistance, but significant resistance required several substitutions. For example, a change to valine at position 82 appeared first in most patients but does not give resistance by itself.

FIGURE 3.10. *Primula kewensis* (*left*) was created artificially by crossing *Primula verticillata* (*middle*) and *Primula floribunda* (*right*). It has twice as many chromosomes as its parent species and so can interbreed with neither.

separate species; we discuss such examples on page 653. Most striking is the rapid origin of new species following hybridization. An F_1 individual occasionally doubles up its genome to produce a new **polyploid** species that cannot interbreed with either parental species. A substantial proportion of plant species have formed in this way; in many cases, their hybrid origin has been confirmed by artificially crossing the presumed parents and generating the hybrid species anew (pp. 631–633; Fig. 3.10). Indeed, this process is used routinely to generate new horticultural varieties, especially in orchids (Fig. 3.11).

As we will see later in this book, we now have abundant and detailed observations of just how the various evolutionary processes work both in the laboratory and in nature.

The Fossil Record Provides Several Lines of Evidence for Evolution

Although our knowledge of the fossil record is much fuller than it was in Darwin's time, there are still gaps—as is to be expected. The chance that any individual will be preserved and discovered is extremely small, and species that are soft-bodied or that lived in restricted areas may be lost entirely. In addition, the chance that a particular fossil will be found is remote—the world is a very large place. Even those fossils that are found are unlikely to be on the direct line of descent to living species, which makes reconstruction of phylogenetic relationships difficult (Fig. 3.12). Few species, and few individuals within species, are actually ancestors of today's organisms. We now have many examples where more or less continuous evolutionary change can be traced (e.g., Fig. 3.13), but these are exceptional. Fossils do provide strong evidence for evolution, but for the most part, this is not through direct observation of evolutionary change.

As we saw in Chapter 1, arguments in the 19th century that the Earth is only a few million years old seemed among the strongest objections to evolution (p. 20). The discovery of radioactivity resolved these arguments by permitting the age of fossils and geological samples to be measured accurately. We now know, through multiple lines of evidence, that our planet formed 4.65 billion years ago and that conditions suitable for life were present relatively soon afterward (p. 90). Both morphological and molecular changes (seen directly and in the fossil record, and inferred from comparisons between living species) can be rapid, and so there has been ample time for evolution to occur.

We would predict that groups that have diverged more recently should appear later in the fossil record—just as is observed (recall Fig. 3.5). Thus, the first chordates and fishes appear approximately 525 Mya, the first amphibians approximately 247

FIGURE 3.11. Orchids made by hybridization and polyploidy. (*Top*) Hybrid lady slipper orchids. (*Bottom*) Polyploid orchid *Ionocidium* Popcorn.

Time

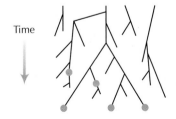

FIGURE 3.12. Fossils (*blue*) will rarely be on the direct line of descent to present-day species (*red*).

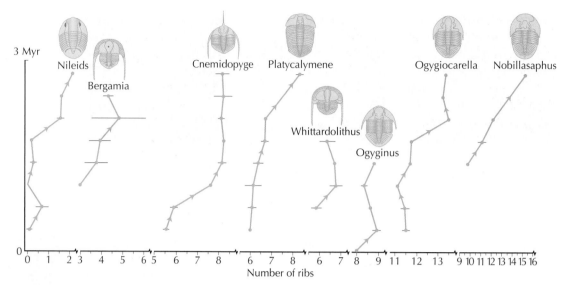

3 Myr

Nileids   Bergamia   Cnemidopyge   Platycalymene   Whittardolithus   Ogyginus   Ogygiocarella   Nobillasaphus

0

0 1 2 3   4 5   6 5   6 7 8   6 7 8   6 7 8 9 11 12 13   9 10 11 12 13 14 15 16

Number of ribs

FIGURE 3.13. These eight lineages of Ordovician trilobites show gradual, rather than punctuated, change. The plot shows the mean number of ribs, with standard error.

FIGURE 3.14. Darwin collected fossils of the extinct *Glyptodon* (*top*), a giant edentate (toothless mammal), and realized that it is related to armadillos (*bottom*), which live in the same region of South America.

Mya, and the first mammals approximately 225 Mya. It is important to realize that this prediction is made from comparisons between present-day species, which tell us the phylogenetic relationships and hence the order of appearance that we expect to see in the fossil record. J.B.S. Haldane made this point in a characteristically pithy way. When asked what observation might refute evolution, he replied "a pre-Cambrian rabbit."

Another pattern that supports evolution is what Darwin called the "Law of Succession," that is, fossils in any one region are related to that region's present-day inhabitants (e.g., Fig. 3.14). The positions of the continents and their climates change over time, and so this pattern holds only for relatively recent fossils. Over longer periods of time, we must take account of geological changes. For example, fossils of marsupials have been found in Antarctica, just as predicted from their distribution across the southern continents (Fig. 3.15; also Fig. 10.5).

Natural Selection Causes the Appearance of Design

The cumulative selection of slight variations has an astonishing power to create complex adaptations. HIV evolves resistance through multiple changes in the proteins that are targets for drugs (Fig. 3.8), and bacteria acquire antibiotic resistance from plas-

FIGURE 3.15. Marsupial fossils in Antarctica. This upper Eocene fossil example of a lower jaw of a polydolopid was found on Seymour Island of the Antarctic Peninsula. An artist's reconstruction of this 20-cm-long marsupial is shown at *right*.

mids that carry multiple resistance genes. Artificial selection among molecules in vitro is used to produce efficient functions that we could not design from first principles (recall the example of fluorescent proteins in Fig. A&S.4). Most obviously, the extraordinary range of domesticated plants and animals has been shaped by selection over the past few thousand years (e.g., Fig. 3.7). In computing, evolving programs often produce novel solutions to difficult problems. We will examine these and other examples of selection later in the book (especially in Chapters 17 and 24).

Of course, a sufficiently intelligent designer would find all these solutions directly. Such an ideal mechanism for adaptation would avoid the **genetic load** that accompanies natural selection. If one type of gene is to replace another, very many individuals must die or fail to reproduce over the many generations that such a replacement requires. We discuss the limits to natural selection further in Chapter 19. Here, we simply note that natural selection is an imperfect mechanism. Thus, evidence that natural selection is responsible for the appearance of design in the living world comes from characteristic imperfections in adaptation.

Vestigial structures, such as the rudimentary pelvis of snakes and whales (Fig. 3.16; also see Fig. 3.18, below), teeth that remain hidden under the gum in the upper jaws of calves, or the remnant eyes of blind cave fish, are extremely puzzling if organisms are rationally designed or are constructed according to some universal law. However, they are to be expected if the structures are relics that functioned in the ancestors. At the molecular level, extra copies of genes are generated by random mutation; these almost always accumulate deleterious mutations and lose their function. These **pseudogenes** are functionless by-products of a process that occasionally leads to the evolution of new functional genes (see p. 184).

Natural selection must act on existing variation, and so adaptations are based on co-option of structures that evolved for other purposes. For example, vertebrates have employed the basic pentadactyl (five-fingered) limb for a variety of purposes (e.g., Fig. 3.3). In *On the Origin of Species*, Darwin discussed this example along with several others:

> How inexplicable are these facts on the ordinary view of creation! Why should the brain be enclosed in a box composed of such numerous and such extraordinarily shaped pieces of bone? . . . Why should similar bones have been created in the formation of the wing and leg of a bat, used as they are for such totally different purposes? Why should one crustacean, which has an extremely complex mouth formed of many parts, consequently always have fewer legs; or conversely, those with many legs have simpler mouths? Why should the sepals, petals, stamens, and pistils in any individual flower, though fitted for such widely different purposes, be all constructed on the same pattern?

Molecular examples include the use of a metabolic enzyme, lactate dehydrogenase, as the transparent material in the lens of the vertebrate eye (p. 710) and the use of the same basic set of *Hox* genes to direct early development in all animals (see pp. 288–307). In Jacques Monod's memorable phrase, these are examples of "evolution by tinkering."

Natural selection acts through competition between individuals. As we shall see in Chapter 21, this conflict drives much of evolution. Some of the most striking examples involve **sexual selection**, in which males compete with each other to fertilize females. For example, many male insects produce a "mating plug" to prevent their mate being fertilized by other males (see p. 577). Most known examples of rapid molecular evolution involve sexual selection, the spread of selfish genetic elements, or the struggle between host and parasite. Evolutionary conflicts such as these are to be expected from natural selection but not if adaptations are designed in some optimal way.

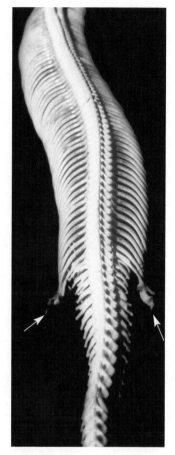

FIGURE 3.16. Python skeleton showing vestigial pelvic limbs (*arrows*).

▐ OBJECTIONS TO EVOLUTION

Objections to the Fact of Evolution

In this section and the next, we list some of the objections that have been made to evolution by natural selection and give a brief summary of the arguments that refute them. We do not include those that are directly and immediately refuted by science—for example, claims that the Earth is only a few thousand years old—because these imply rejection of science as a whole. Also, we do not consider here objections based on the supposedly pernicious social effects of belief in human evolution; we discuss such objections at the end of this chapter. Here, we refute arguments against evolution; in the next section, we refute arguments against natural selection as its chief mechanism.

Argument: Evolution cannot be observed and so cannot be proved.

Rebuttal: Just as in other areas of science, evolutionary biology does not for the most part rely on direct observation. In physics, we do not observe the gravitational attraction between Earth and Moon directly (how could we?), but instead we bring together diverse lines of evidence—both in the laboratory and from astronomy—to build a simple and satisfying explanation. Similarly, in geology, we can observe the slow action of processes such as erosion and continental drift directly, but must then extrapolate to produce a consistent account of large-scale change. In just the same way, evolutionary biologists use direct observation to thoroughly understand the underlying processes, but rely on many lines of indirect evidence to explain evolution on a larger scale.

No hypothesis can be "proved" with absolute certainty. Indeed, a productive theory suggests new investigations that lead to its revision and modification. For example, Newton's laws of gravitation make extraordinarily accurate predictions about the motions of celestial bodies, but slight deviations from them support Einstein's general theory of relativity—a still more accurate theory, and one that brings a more profound understanding of space and time. Mendel's laws of heredity were refined to include phenomena such as linkage after their rediscovery in 1900; half a century later, the discovery of their physical basis in DNA led to further refinements in our understanding of the gene. Evolutionary biology has developed in just the same way as the rest of science—changing and becoming richer as an ever-wider range of phenomena become understood.

Argument: Evolutionary theory is not testable.

Rebuttal: A scientific theory is successful if many predictions are made based on the theory, those predictions are tested, and the theory survives those tests. These tests need not involve laboratory experiments. In evolutionary biology they more often involve predictions about patterns across living species, in the fossil record, or in the structure of the genome. We have already noted several key tests of evolution: the consistency of phylogenies inferred from different characters, the order of appearance of taxa in the fossil record, and patterns of geographic distribution. We will see much more evidence of this sort in the rest of the book.

Argument: There are no transitional forms.

Rebuttal: Continuous transitions can be seen only in the most favorable cases (e.g., Fig. 3.13). However, there are many striking examples of intermediate forms, which carry ancestral combinations of characters that had been predicted to exist from reconstructed phylogenies. Among the most striking examples are feathered dinosaurs, discovered in China in 1996 (Fig. 3.17), and the series of intermediates between whales and hippos that shows how mammals adapted to life in the sea (Fig. 3.18). Fossils show

FIGURE 3.17. Feathered dinosaurs, like those from which birds evolved. These fossilized vertebrae of a tyrannosauroid from the Early Cretaceous show filamentous structures (*upper portion of rock*) thought to represent protofeathers.

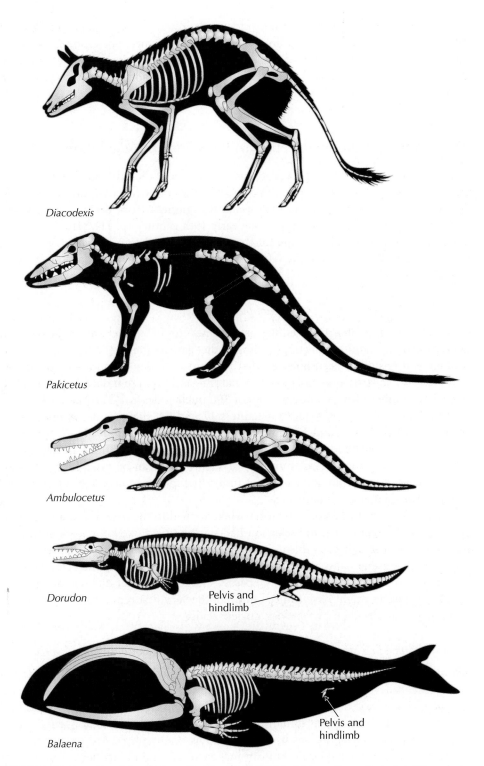

Diacodexis

Pakicetus

Ambulocetus

Dorudon   Pelvis and
                   hindlimb

Balaena   Pelvis and
                   hindlimb

FIGURE 3.18. A series of fossils from the Eocene (~50 Mya) hippo-like artiodactyl (*Diacodexis*, *top*) to a skeleton of the modern whale (e.g., *Balaena, bottom*) shows how mammals adapted to life in the sea. Among the most important changes, the pelvis and hindlimbs were reduced, the tail was lengthened for swimming, and the jaws were modified for feeding on plankton.

clearly that different characters evolve independently, rather than appearing together all at once or accumulating as a linear, progressive sequence. For example, we see in Chapter 25 that hominid fossils show this pattern of mosaic evolution—increased brain size, sexual dimorphism, changes in dentition, and adaptations for walking upright all change more or less independently.

Objections to Natural Selection as the Cause of Adaptation

Until Darwin discovered the process of natural selection, the existence of complex and functional structures was taken to imply creation by an intelligent designer. As we saw in Chapter 1 this long-standing philosophical **argument from design** was especially influential in the **natural theology** of the early 19th century (p. 11). Natural selection provides an alternative explanation for the appearance of design, which relies solely on natural causes. In this section, we refute some of the objections to the adequacy of this process.

FIGURE 3.19. R.A. Fisher. "Natural selection is a mechanism for generating an exceedingly high degree of improbability."

Argument: Chance cannot generate complexity.

Rebuttal: The influential 19th century astronomer John Herschel objected to the random element in natural selection, calling it the "law of the higgledy-piggledy." Subsequently, it has often been objected that random mutations cannot lead to ordered complexity. The structures assembled by natural selection are indeed highly improbable: The number of possible sequences of 100 nucleotides is 4^{100}, or more than 10^{60}. However, as explained in detail in Chapters 17 and 24, the cumulative effect of selection is precisely to build highly improbable structures (Fig. 3.19). Although the reproduction of each individual and the generation of new mutations are each highly random, the outcome of large numbers of such events can be essentially deterministic—just as the random movements of individual molecules average out to give the precise laws of thermodynamics.

This objection does have some force when applied to the origin of the first reproducing system—before natural selection had started to act. We know very little about how the very first living organisms originated, but as we will see in Chapter 4, there are several plausible hypotheses. In particular, the first replicating molecule need not have required a *precise* sequence of 100 bases: A large number of different and perhaps shorter sequences might have sufficed. Indeed, the very first **replicators** may well have been much simpler than present-day nucleic acids.

Argument: The first step toward complex adaptation could not have been favored.

Rebuttal: Darwin felt this was one of the strongest objections to evolution by natural selection and devoted a section of *On the Origin of Species* to refuting it; as an example, he used the vertebrate eye. More recently, the term **irreducibly complex** has been used to describe systems that cannot function if any one of their components is missing: The bacterial flagellum has been proposed as one example. We discuss such arguments in detail in Chapter 24 but can make two general points here. First, the initial steps need only give some slight selective advantage. As the eminent and extremely nearsighted evolutionist John Maynard Smith put it, his imperfect vision was far better than complete blindness. Second, the initial stages may have evolved for a quite different purpose than that of the final complex structure (see Chapter 24). Finally, although the present function may be entirely destroyed by changes to most of its components, there only needs to be one path of increasing fitness that connects the ancestor with the present structure (Fig. 3.20). This point is discussed in more detail on pages 642–644.

Greenfield Medical Library - Issue Receipt

Customer name: Omar, Nazri

Title: Principles of genetics.
ID: 1001699967
Due: 05 Jan 2013 16:30

Title: Evolution / Nicholas H. Barton ... [et al.].
ID: 1006453959
Due: 05 Jan 2013 16:30

Total items: 2
12/07/2012 17:41

All items must be returned before the due date
and time.
The Loan period may be shortened if the item is
requested.

WWW.nottingham.ac.uk/is

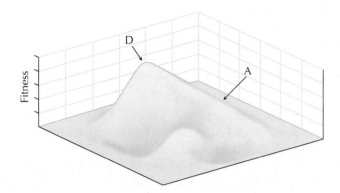

FIGURE 3.20. A complex structure can evolve even if every change to it reduces fitness: All that is required is that there was at least one path from ancestor (*A*) to descendant (*D*), along which fitness was high and increasing. The diagram shows an **adaptive landscape** (see p. 472), which plots average fitness against the state of the population (morphology, allele frequencies, etc.). In reality, the landscape fluctuates, giving further pathways for evolutionary change. Moreover, evolution can proceed along many axes, not just the two shown here.

Argument: Natural selection creates nothing new.

Rebuttal: A single round of selection does indeed just pick from existing variation. However, successive rounds of selection increase the frequency of each favorable variant, so that very soon new combinations are seen that were vanishingly rare in the original population (Fig. 1.31). As we explain in Chapter 17, this criticism is no more valid than saying that an author merely rearranges existing letters but creates nothing new (p. 465).

Argument: Natural selection violates the second low of thermodynamics.

Rebuttal: In a *closed* system, disorder must necessarily increase. More precisely, the **entropy** of any closed system—a quantitative measure of its disorder—will almost certainly increase. However, living systems are *open:* They take in nutrients and free energy (ultimately from sunlight or from some source of chemical energy) and export waste products and heat. Overall, entropy increases in open systems. Organisms grow and reproduce in an orderly way, but this is more than offset by the heat and chemical waste that they produce.

Argument: Human intellect could not have evolved by natural selection.

Rebuttal: We have already considered arguments that selection can create complex adaptations such as human language, which have obvious selective value. (In Chapter 25, we consider the detailed steps that could have led to complex language.) A distinct objection is that many human abilities—musical and mathematical talent, religious feelings, and so on—could have no survival value and so could not have been selected. (Wallace, who discovered natural selection independently of Darwin [Fig. 1.19], held to this view and so never accepted a fully natural explanation of human nature.) There are two responses to this argument. First, we know that apparently quite useless traits can increase fitness—for example, by influencing social status or mate choice (Chapter 20). Second, traits that have no direct effect on fitness readily evolve as side effects of direct selection on other traits (see p. 479). Thus, mathematical abilities might be a side effect of a general reasoning ability and intellectual curiosity.

Abstract reasoning, facilitated by symbolic language, is clearly of enormous value to individual humans, both directly (allowing better hunting, toolmaking, and much more) and indirectly (by enhancing social skills and attractiveness to mates). As was

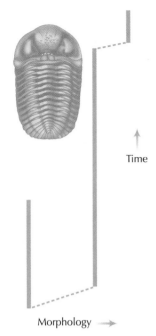

Time

Morphology →

FIGURE 3.21. An example of punctuated equilibrium. Shown schematically is the evolution of the Devonian trilobite *Phacops rana* (one of the organisms used to develop this theory). Time is on the vertical axis (*solid lines*) and morphology is on the horizontal axis (*dotted lines*). Time ranges of three species are illustrated (the three *vertical lines*). Punctuated equilibrium is evidenced by the abrupt changes in morphology that occur during the relatively short periods of time indicated by the *dotted lines*.

just noted and is covered in more detail in Chapter 17, a curiosity about the world and an interest in explaining subtle patterns could lead to intellectual abilities that have no direct effect on fitness.

Argument: Punctuated equilibrium implies that natural selection within species is ineffective.

Rebuttal: In 1972, the paleontologists Niles Eldredge and Stephen Jay Gould proposed a theory known as **punctuated equilibrium.** They emphasized that species often persist unchanged for millions of years but then shift abruptly to a new form (Fig. 3.21). If these sudden shifts correspond to formation of a new species, then extinction and speciation—in effect, selection among species—would be more effective than selection within species in shaping **macroevolution.**

Eldredge and Gould accepted that natural selection within species shapes adaptation. They were instead concerned with large-scale macroevolutionary trends—increases in body size, for example. However, their theory has been misinterpreted by others as a criticism of evolution by natural selection. In fact, the pattern of punctuated equilibrium is entirely consistent with change through natural selection. It simply implies that selection maintains the same phenotype for long periods, but that when change does occur, it can be rapid. Although "punctuations" are fast on an evolutionary timescale, they span thousands of years and can easily be accounted for by natural selection (see Fig. 17.30). Conversely, as we show in Chapter 21, selection among species is very slow, simply because species originate and go extinct so much less often than individuals are born or die. Thus, although species selection may lead to macroevolutionary trends, it cannot build complex adaptations.

Argument: The human genome is too simple to account for such a complex organism.

Rebuttal: When the draft sequence of the human genome was published in 2001, many scientists were surprised that only about 30,000 genes were identified; this estimate has since fallen to about 25,000 protein-coding genes. Previously, it had been estimated that our genome contains 80,000 or more genes, although there was no reliable data on which to base such an estimate. The unexpectedly low number of genes caused surprise that such a complex organism could be encoded by so little genetic information. It is indeed remarkable that less than 20 Mb of information can instruct human development and, especially, can determine the structure of our brain with so many intricate connections between its 100 billion neurons. However, this puzzle was apparent even before sequence data gave such a surprisingly low estimate of gene number. As we discuss in Chapters 21 and 24, gene number is a poor guide to complexity; for one thing, alternative splicing of RNA transcripts gives many more proteins than genes (Box 13.1). In any case, many biological differences involve changes in gene regulation, determined by sequences that do not code for protein.

As we will see in Chapter 14, there is no simple relationship between genotypes and phenotypes; each gene affects many traits, and each trait is influenced by many genes. The DNA sequence does not *code* for the phenotype. Rather, a large number of genes interact with each other and with the cellular machinery to allow the organism to develop. The arrangement of neurons in our brains is not specified exactly, but rather develops through interactions among neurons and with our environment (see pp. 715–717). The DNA sequence itself is meaningless—it must be expressed in the context of the cell. Most of the information required to build an organism is contained in the cellular machinery, rather than in the genome itself.

A clearer comparison is to ask whether the *differences* between (say) human and chimpanzee can be accounted for by the differences in their genomes. There are approximately 40,000 amino acid differences, perhaps one-third established by selection, plus a similar number of differences in noncoding regions that are maintained by selection and so are likely to be functional (see pp. 535 and 542–547). As we will see later, these many differences can readily be established by selection over 6 million years of divergence (see pp. 549–553). It does not seem implausible that our differences from our closest relatives are due to these many thousands of differences.

SCIENCE AND SOCIETY

The Fact of Evolution Is Explained by Evolutionary Theory

Popular debate over evolution and its mechanism is marked by confusion over the scientific use of the terms "fact" and "theory." Sometimes, it is said that evolution is "only a theory," suggesting that it is a mere speculation, with little support. In science, however, a **theory** means a web of interconnected hypotheses, which makes predictions that are consistent with what we see and makes new predictions that stimulate further research. A theory that has survived many different tests may be so well supported that we take it as a *fact*. This applies to the theories of gravitation, of plate tectonics, of quantum mechanics, and of evolution—all are treated as thoroughly established facts.

The fact of evolution is explained by a sophisticated body of theory that shows how it has come about. Much is firmly established. We understand how all the evolutionary processes work, and we have very many examples where we understand how they have generated adaptation and divergence. We also know a great deal about the history of life, through paleontology and phylogenetics. As we emphasize throughout this book, many questions remain open. Just what fraction of variation is selected, and how? What is the genetic basis of complex phenotypes? What role does the exchange of genes among different evolutionary lineages play in the diversification of bacteria and archaea? Why do most eukaryotes reproduce sexually? How do originally free-living organisms come together to cooperate in the eukaryotic cell, in multicellular organisms, and in social colonies? Evolutionary biologists are making rapid progress in answering such questions, but no doubt new puzzles will arise. The strength of evolutionary theory and the reason why it was so rapidly accepted is that it explains a wide range of phenomena in terms of a few simple principles. It continues to be a fertile source of ideas that can be tested in nature or in the laboratory.

Of course, everything we see could be consistent with special creation—species might be created with features *as if* they had evolved, and fossils might have been placed into geological strata in the same order *as if* they had evolved. However, such a perverse alternative explains nothing. Each species would be just the way it is, and we would have no explanation for any of the patterns described above. Special creation only explains the facts of biology by making arbitrary assumptions that can fit any observations. The hypothesis of special creation cannot be tested and so is not considered to be scientific.

Many accept evolution, and many admit that natural selection accounts for evolution within species or perhaps larger taxonomic groupings. However, they suppose that occasional interventions by a supernatural being (often referred to as an "intelligent designer") have created some particularly complex adaptations. Again, if such interventions are arbitrary, then this hypothesis cannot be tested—anything might happen. On the other hand, if the designer is supposed to have certain properties (always giving optimal designs or designs favorable to humans, say), then the hypothesis *can* be refuted. As we saw in the previous section, adaptations in the natural world show just the kinds of imperfections that we would expect from natural selection but not from an omnipotent designer. In any case, invoking a "God of the Gaps" is unsatisfactory from a theological point of view, because as science advances, it explains more, which causes the gaps to become more and more restricted in scope.

Understanding Nature and Humanity

Objections to evolution by natural selection have come from those who see a clash with their religious beliefs or, more broadly, from those who object to a materialist worldview. Objections also come from those who worry that evolution could be used to justify moral positions that they object to or that belief in evolution might undermine morality. (As one woman said at hearings on the teaching of evolution in the Louisiana Senate in 1981, "I think that if you teach children that they are evolved from apes, then they will start acting like apes.")

Plainly, evolution does contradict *literal* readings of the Bible and other sacred books. However, a literal interpretation contradicts science as a whole—not just evolution—and in any case, Genesis includes two different accounts of creation. Most major religions see no inconsistency between their beliefs and the scientific account of evolution. For example, the Catholic view now accepts evolution by natural selection, including the physical evolution of our own species. A common position is that God works through natural law, including natural selection and other evolutionary processes. At an individual level, many evolutionary biologists have no difficulty in reconciling their various religions with their scientific beliefs. Many hold the view eloquently expounded by Gould that science and religion are separate domains—one concerned with explaining the natural world, the other with interpreting the meaning of human life.

The ideas of evolution and natural selection have sometimes been used to justify bad policies. Racial segregation has been justified by supposed innate differences; **positive eugenics** (e.g., compulsory sterilization of the mentally ill) has been justified as aiding natural selection; and **Social Darwinism** extolled the "survival of the fittest" in economic and social policy. However, the scientific justification for such policies is now seen as baseless. Indeed, our knowledge of evolution has quite properly supported many policies that most would agree with. Hermann Muller campaigned to ban atmospheric tests of nuclear weapons on the grounds that radiation-induced mutations place a severe load on future generations; the close genetic similarity between human populations and the absence of distinct "races" supports equal treatment of all humans (see p. 769); and the importance of cooperation in evolution (Chapter 21) has been emphasized as a contrast to the naive view that emphasizes conflict (in Tennyson's words, "nature red in tooth and claw").

More fundamentally, neither evolution nor its mechanism in themselves justify any particular moral position. Philosophers term this the **naturalistic fallacy**, an argument that claims that *what is* justifies *what should be*. The same point applies, of course, to religious beliefs about how the world is—beliefs about how the world began or the existence of God do not in themselves tell us how we should behave. Nevertheless, both scientific and religious beliefs do change our perspective on our place in the world and so may indirectly influence the values that we choose. For example, an awareness of our continuity with the living world may make us value the existence of other species, rather than seeing them purely as useful to ourselves. Similarly, an awareness of the importance of variation for evolution may make us set more value on diversity, both of different species and within our own species. We share with Darwin the sentiment with which he closed his private *Sketch* of his theory, written in 1842:

> There is a simple grandeur in the view of life with its powers of growth, assimilation and reproduction, being originally breathed into matter under one or a few forms, and that whilst our planet has gone circling on according to fixed laws, and land and water, in a cycle of change, have gone on replacing each other, that from so simple an origin, through the process of gradual selection of infinitesimal changes, endless forms most beautiful and most wonderful have been evolved.

SUMMARY

Evolution is accepted among biologists as explaining the diversification of life, and natural selection is accepted as the sole cause of adaptation. Yet, many outside biology still do not accept this scientific consensus.

Evidence for evolution comes from the nested classification of group within group, consistent across varied molecular and morphological traits, from direct observation in the laboratory or on the farm, and from the fossil record. Evidence for natural selection comes from analogy with artificial selection, from the characteristic imperfections of natural adaptations (e.g., vestigial organs), and from the recruitment of ancestral structures for new purposes. We briefly summarize arguments against various misunderstandings and objections here and discuss these further later in the book (especially in Chapters 17 and 24).

The fact of evolution is explained by evolutionary theory—which is as well established as other scientific theories, such as quantum mechanics, plate tectonics, or molecular genetics. This theory is consistent with the major theologies and with the varied religious beliefs held by evolutionary biologists. Although it does not justify any particular morality (and cannot be held responsible for any lack of morality), the theory of evolution does give us a radically new perspective on the place of humanity in nature. As Darwin wrote at the end of *On the Origin of Species,* "There is grandeur in this view of life... ."

FURTHER READING

Eldredge N. 2005. *Darwin: Discovering the tree of life.* W.W. Norton, New York.
    Written as an accompaniment to the 2005–2006 Darwin exhibition at the American Museum of Natural History (www.amnh.org), it combines an account of the development of Darwin's ideas with their present place in biology and in society.
Futuyma D.J. 1995. *Science on trial: The case for evolution.* Sinauer Associates, Sunderland, Massachussetts.
    A counter to creationism, but written before arguments for "intelligent design" emerged. For a more recent concise summary of the arguments, see Futuyma D.J. 2005. *Evolution.* Sinauer Associates, Sunderland, Massachussetts.
Pennock R. 1999. *The tower of Babel: Evidence against the new creationism.* MIT Press, Cambridge, Massachussetts; Young M. and Edis T. (eds.). 2004. *Why intelligent design fails.* Rutgers University Press, Piscataway, New Jersey.
    Two out of several recent books refuting arguments for "intelligent design."
Pope John Paul II. 1996. *Message to the Pontifical Academy of Sciences.* Reprinted in *Q. Rev. Biol.* **72:** 381–406, together with four commentaries, which discuss the relation between science and religion more generally.

Web Resources

www.NationalAcademies.org/evolution/
    Resources and statements from the National Academy of Sciences.
www.amnh.org/exhibitions/darwin/
    A guide to the exhibition celebrating the bicentenary of Darwin's birth (see Eldredge 2005).
www.pbs.org/wgbh/evolution/
    A wide range of material accompanying the PBS television series, *Evolution.*
www.talkorigins.org
    A newsgroup devoted to the discussion and debate of biological and physical origins.
www.pamd.uscourts.gov/kitzmiller/kitzmiller_342.pdf
    The decision in a 2005 case against the Dover district school board in Pennsylania; summarizes the nature of science and the scientific arguments against "intelligent design."

THE ORIGIN AND DIVERSIFICATION OF LIFE

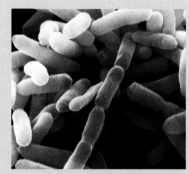

Part II explores the history of life, from its origin billions of years ago to the fantastic diversity of life today. Major advances in recent years, most notably the discovery that primordial life likely used RNA to carry out the tasks now done by DNA and protein, have led to a much better understanding of how life began (Chapter 4). Chapter 5 introduces the grand phylogenetic tree that connects all living organisms and discusses the inferences that can be made about their (and our) common ancestor.

Chapters 6–8 describe the three domains of life: bacteria, archaea, and eukaryota. Throughout time, single-celled organisms have dominated and shaped the biosphere. Molecular techniques have revealed the relationships between microorganisms and highlighted their extraordinary biochemical versatility.

The remainder of Part II examines how multicellular organisms evolved and diversified. Chapter 9 explains how cells interact to construct complex organisms and describes the highly conserved genetic mechanisms that underlie the extraordinary diversity of life-forms. Chapter 10 introduces the history of plants and animals as seen through the fossil record. Fossils provide us with precise dates for evolutionary history and make clear that the great majority of species have left no descendants. Without geological evidence, we would have no idea that dinosaurs or trilobites once lived, or that the Earth has gone through a series of mass extinctions. Chapter 11 explains how changes in development lead to both major and minor changes in plant and animal morphology. All this sets the stage for the more detailed examination of evolutionary processes in Part III.

Part II raises questions that are addressed in later chapters. How does one species split into two? Why do most eukaryotes reproduce sexually? How can cells in a multicellular organism, or members of an insect or human society, cooperate? How do novel biochemical, morphological, and behavioral adaptations evolve? As you survey the history of life presented in Part II and then read on in the book, keep these key questions in mind.

C H A P T E R

4

The Origin of Life

I T IS EXTRAORDINARY TO CONSIDER that all living creatures on Earth share a common ancestry. What was the Earth like billions of years ago, when life began? How did life begin? And where?

To investigate the origin of life requires an understanding of the key features that set living systems apart from nonliving ones. For our purposes, we use the following definition: *Life is composed of organized matter that is capable of undergoing reproduction and natural selection.* This definition allows us to distinguish living from nonliving systems and identifies some of the functions that we should expect to find in any living system. To reproduce, a living system must be able to accumulate biomass and to organize it into coherent, inheritable biological structures. These processes require energy, a network of chemical reactions (i.e., metabolism), and some memory of pattern. Furthermore, for natural selection to operate on a system, that system must generate inheritable variation (see p. 17).

In this chapter, we discuss the origin and very early evolution of life on Earth. We focus on two main questions: *When* and *how* did life begin?

WHEN DID LIFE BEGIN ON EARTH?

Determining when life began on Earth is important because it allows us to locate specific events in the origin of life within the context of the Earth's history. This information is critical for understanding how life began because it tells us how much time was available for these processes to take place and what the Earth might have been like during those periods.

The general approach for determining when life began on Earth is to place an upper and a lower boundary on this origination. The lower boundary (the first point at which life could have existed) is determined by geological studies of the origin and formation of the Earth itself. The upper boundary (when life was definitely present) is determined through studies of the fossil and geochemical record. Both approaches are discussed in the following sections.

Estimates of the Age of the Earth Place a Lower Boundary on When Life Began

There are two approaches for determining the age of the Earth. Through astronomical and planetological studies, the age of the solar system has been calculated. Geological studies have been used to estimate the age of the oldest known rocks and minerals on Earth.

Geological Approach

A variety of methods are used to date ancient rocks. The most commonly used method is **radioisotope dating**, in which the amounts of different isotopes in a rock sample are analyzed (Box 4.1). Any radioactive isotopes that were present in the rock when it was formed will decay over time. The decay results in a decrease in the radioisotope with a concomitant increase in a stable product. Because the radioactive decay of isotopes is uniform over time, the age of the sample can be estimated (see Fig. 4.1). Thus, a search for the oldest rocks on Earth involves looking for materials that are likely to be old (e.g., those at or near the bottom of layers of rocks already known to be old) and then estimating their age through radioisotope analysis.

Box 4.1   Radioisotope Dating

Chemical elements are defined by their atomic number, which equals the number of protons present in the nucleus. Most elements come in a variety of isotopes, which have different numbers of neutrons present in the nucleus. Some **isotopes** are unstable and can change their nature over time by emitting particles and thus are known as **radioactive isotopes**. Every radioactive isotope has a unique and extraordinarily uniform decay pattern, which can be described by its "half-life," the time required for half of the population of nuclei to decay (Fig. 4.1). This regular pattern of decay is the basis for **radiometric dating**, a method of determining the absolute age of objects.

To understand radiometric dating, it is useful to start with a hypothetical example. Imagine being given a sealed box, into which 1 mole of a radioisotope (X) was placed an unknown time (*t*) ago. Knowing that X converts to the daughter isotope Y with a half-life of 1 billion years, how can *t* be calculated? One way would be to measure the amount of X present in the box now and then use the half-life to calculate *t*. Suppose, for example, that there was 1/8 mole of X present. Because we started with 1 mole, this means that three half-lives have elapsed (1/2 x 1/2 x 1/2) or 3 billion years. To check

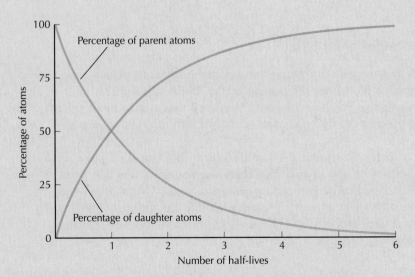

FIGURE 4.1. Radiometric dating. Graph showing the decline in the number of atoms of a radioactive isotope (*blue line*) and the increase in the number of atoms of daughter isotopes (*red line*).

Box 4.1. Continued.

this calculation, the amount of Y can be measured, which should equal the amount of X that has decayed—7/8 mole. This type of calculation can be made to estimate the age of a sample, as long as the half-life of the radioisotope in question and the amount of starting material are known.

In practice, however, the amount of X (continuing with the nomenclature in the previous paragraph) at the start is usually unknown. In such cases, one way to calculate the age of the sample is by examining the ratio of X to Y. After one half-life the ratio will be 1:1, after two half-lives 1:3 (1/4 the original amount of X remains), and after three half-lives 1:7. Knowing the half-life, t can be calculated from these ratios.

Age estimates of real-world samples are usually determined by analyzing the absolute amounts of X and Y and the ratios of X and Y at different times. The analysis, however, is frequently more complicated than in the example of the sealed box. For example, the two methods described above assume that Y is not present in the original sample, which is often incorrect. It is also assumed that no material has entered or left the sample. In addition, the sample must contain a radioactive isotope with a half-life that is similar in timescale to the age of the item in question. To understand in more detail how these issues are dealt with, we consider two examples of radiometric dating.

Example 1: Using Carbon-14 to Date Recent Biological Samples: Although this chapter focuses on the origin of life, an event that happened billions of years ago, it is instructive to consider a common method for dating recent biological samples based on carbon-14 (^{14}C) decay. Carbon exists as three principal isotopes: ^{14}C, ^{13}C, and ^{12}C. ^{14}C is radioactive with a half-life of 5730 years, making it useful for dating samples up to about 60,000 years old. ^{14}C is produced constantly in the atmosphere by cosmic rays, and when carbon is fixed, fresh ^{14}C enters the food chain. As ^{14}C decays, it produces nitrogen-14 and a β particle. Unfortunately, the X:Y ratio method does not work well in this case, because no reliable method exists to estimate the amount of Y (i.e., ^{14}N) in the original sample. The success of ^{14}C dating depends on accurately estimating the amount of X (i.e., ^{14}C) in the original sample. Such estimates are reasonably accurate because the

ratio of the three carbon isotopes is relatively uniform in the carbon of living organisms today (~98.89% ^{12}C, 1.11% ^{13}C, and 0.0000000001% ^{14}C). It is assumed that these ratios were about the same in living organisms in the recent past. Importantly, ^{12}C and ^{13}C are stable isotopes. Thus, for biological samples, the starting amount of ^{14}C can be estimated from the current amounts of ^{12}C and ^{13}C present. The age of the sample can then be calculated using the amount of ^{14}C currently in the sample and the half-life of ^{14}C.

Example 2: Using Radiometric Dating of Zircon Crystals to Estimate the Age of the Earth: With a half-life of 5730 years, ^{14}C dating is not useful in dating the age of the Earth; instead isotopes with very long half-lives are used (see Table 4.1). Of course, as the ^{14}C example showed, just having the right half-life is not sufficient; the right combination of materials and circumstances is also necessary to accurately estimate the age of the Earth. For example, when certain rocks are made molten by volcanic activity, many of the isotopes present in the starting material will reequilibrate with their surroundings and thus cannot be used to estimate the age of the starting material. Here we describe how analysis of zircon crystals can be used to estimate the age of the Earth, despite the challenges.

Zircon is a mineral with chemical formula $ZrSiO_4$. Crystals of zircon are formed when molten igneous rock cools. Even if the crystal is heated during its lifetime, it retains a signature from its first crystallization. Frequently trapped within zircon crystals are radioactive isotopes of uranium, including uranium-238 (^{238}U), which decays to lead-206 (^{206}Pb) with a half-life of 4510 million years. Fortunately for purposes of radiometric dating, lead is frequently absent in the initial crystals. If any lead was present, the total amount of ^{206}Pb that was present originally in the sample can be estimated from the current amount of ^{204}Pb (a stable isotope of lead) in the sample, much as ^{12}C is used to estimate the amount of ^{14}C. Thus the age of a zircon crystal can be estimated from the ratio of ^{238}U to ^{206}Pb. Other radioisotopes, such as ^{235}U, which also are found trapped in zircon crystals, can be used to date the crystals. ^{235}U decays to ^{207}Pb with a half-life of 704 million years. Using multiple radioisotopes for dating provides a set of internal controls for the age estimates.

TABLE 4.1. Elements with long isotopic half-lives

Parent Isotope	Stable Daughter Product	Half-Life
Uranium-238	Lead-206	4.5 billion years
Uranium-235	Lead-207	704 million years
Thorium-232	Lead-208	14.0 billion years
Rubidium-87	Strontium-87	48.8 billion years
Potassium-40	Argon-40	1.25 billion years
Samarium-147	Neodymium-143	106 billion years

Source: From http://interactive2.usgs.gov/learningweb/teachers/geoage.htm.

The oldest rocks known, which are estimated to be approximately 4 billion years old, are gneisses from Great Slave Lake in Canada. Many other rocks have been found that are approximately 3.8 billion years old. In addition to these rocks, some minerals have been found that are even older. For example, zircon crystals have been found in Australia that are approximately 4.3 billion years old. These crystals are found embedded in rocks that are much younger than the crystals, and the ultimate source of the crystals is still not known. If they are from Earth, the age of these crystals indicates that the Earth is at least 4.3 billion years old.

Planetological Approach

It is thought that the Earth formed through a process involving the accretion of meteorites and other material present in the early solar system. Thus, a lower boundary for the age of the Earth can also be estimated by finding and dating material that was formed at about the same time as the Earth but did not participate in the accretion. There are multiple sources of such material, including moon rocks, meteors, and meteorites. Radiometric dating of meteorites that were formed just before the formation of the Earth indicates that they are approximately 4.5 billion years old. Thus, the accretion and formation of the Earth likely happened between 4.5 and 4.3 billion years ago.

Many studies suggest that the very early Earth was inhospitable to any form of life. As the Earth was being formed, it was bombarded by meteors and other large objects, some as big as planets (e.g., when the moon was formed). These collisions would have generated massive amounts of heat, evaporating any water on the Earth's surface and possibly sterilizing the planet. Even if life had evolved, it might have been destroyed. Thus a key question is: When did these major collisions end? Although this is difficult to estimate, most studies suggest that major sterilization events ended approximately 4 billion years ago.

Fossil Evidence and Comparisons of Modern Organisms Suggest That Life Evolved Soon after Earth Was Hospitable

To find evidence for the earliest existence of living organisms, we turn to fossils (the preserved remains of organisms or their activities). Studies of fossils are discussed in detail in Chapter 10. Here we focus on one particular issue: What are the oldest remnants of life that can be found?

Studies searching for the remnants of early life have looked mainly for two types of traces: fossilized cells and the residue of ancient cells (either chemical remains or geochemical effects of such cells). One potential example of ancient fossils is found in ancient rocks that resemble stromatolites. Stromatolites are formed by colonies of **cyanobacteria**, a type of photosynthetic bacteria (Fig. 4.2). Ancient stromatolite-like

FIGURE 4.2. Stromatolites are rock-like mounds formed by mats of cyanobacteria or other photosynthetic microbes.

structures have been found that date to approximately 3.5 billion years ago. Because these resemble modern stromatolites, it has been proposed that they are remnants of ancient life. Furthermore, electron microscopy has revealed structures in some ancient stromatolites that resemble modern cyanobacterial cells (Fig. 4.3). Although there are certainly very old fossilized stromatolites, it is unclear if these 3.5-billion-year-old structures are remnants of living organisms, because the mounded structures and cell-like microscopic forms can be created by purely abiotic processes. If these are genuine fossilized cells, and given their apparent complexity and diversity, then life would have been evolving for some time prior to 3.5 billion years ago, thus placing the origin of life into the even more distant past.

There is solid evidence that life was present on Earth more than 3 billion years ago. This comes from studies of chemical fossils, which are the chemical residue left behind by processes unique to living organisms. One chemical fossil is **kerogen**, an organic substance produced by the decay and transformation of living organisms. Kerogen has been found in sedimentary rocks in Greenland that are 3.85 billion years old. To confirm that the kerogen in the Greenland rocks came from living organisms, scientists have compared the ^{12}C to ^{13}C ratio in the kerogen with the same ratio in abiotic carbon compounds found in the same rocks. As discussed in Box 4.1, both ^{12}C and ^{13}C are stable isotopes of carbon (unlike ^{14}C), and thus the ratio of ^{12}C to ^{13}C does not change once a material is formed. However, the ratio is not uniform for all carbon-containing material. In particular, in the fixation of carbon by living organisms (e.g., the conversion of CO_2 to sugars), lighter carbon (i.e., ^{12}C) is incorporated preferentially. In the Greenland rocks, the ^{12}C to ^{13}C ratio in kerogen is consistent with the ratio expected for carbon of biological origin. However, isotope ratios may have been reset during deformation of these rocks, and more reliable signatures date from approximately 3.5 billion years ago.

Summary

Putting all of this information together, life probably originated during the 200 million years between the last major accretion impacts (~4 billon years ago) and the formation of the earliest known chemical fossils (~3.8 billion years ago). This is a relatively short period of time compared to both the age of the Earth and the history of life (Fig. 4.4).

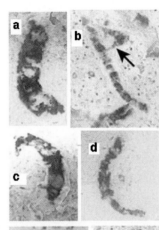
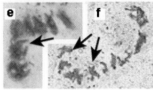

FIGURE 4.3. Ancient cyanobacteria? Structures from the ~3.465-billion-year-ago Apex **cherts** of the Warrawoona Group in Western Australia were originally thought to be organic microfossils, an interpretation that is now disputed. Names assigned to these putative organisms: (a) *Primaevifilum laticellulosum*; (b) *Primaevifilum delicatulum* showing a side branch; (c) *Primaevifilum conicoterminatum*; (d) narrow filament; and (e,f) trichomes (small outgrowths) showing apparent bifurcations (*arrows*), which are a result of interspersed quartz grains.

HOW DID LIFE BEGIN ON EARTH?

More fundamental than determining when life began is determining how it began. How did living systems originate from nonliving ones? What key steps and innovations were required? In modern organisms DNA encodes proteins and proteins copy

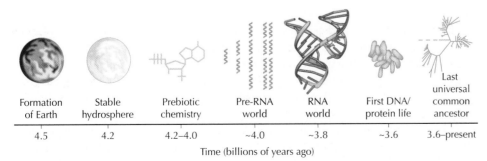

FIGURE 4.4. Steps in the origin of life.

DNA; so how did one evolve without the other? In general terms, scientists have adopted two strategies to address how life began and how it evolved the common features that we recognize in all modern living organisms. In one approach, research takes off from the point when Earth was hospitable for life to evolve. Investigations focus on finding out how features of living organisms could have come into existence and utilize both experimental and theoretical simulations of conditions on the early Earth or in early protoliving systems.

In the alternative strategy, modern organisms are compared in order to ascertain what systems are ancient and which are new. Reaching as far back into the past as possible, the goal is to determine what the earliest life-forms were like. What were their genetic systems? What metabolic processes did they have?

The two approaches are not independent of each other. Each one generates information that leads to refinements of the other approach. By using the methods in concert, scientists are developing a better understanding of how life began. From such studies, a number of critical steps have been identified that appear to be necessary for the origin of life. Possible ways that these steps may have occurred have also been identified. In the remainder of the chapter, we discuss seven of these steps:

1. the generation of simple organic molecules from inorganic molecules

2. chemical "evolution" to produce more complex organic molecules and primitive metabolic networks

3. the origin of self-replication and the creation of "genotypes"

4. compartmentalization and the creation of cells

5. the linking of **genotype** and **phenotype**

6. the origin of the genetic code

7. the takeover of early replication systems by one involving DNA

Although each step was essential for life to originate, the order of the steps is uncertain. Some processes may have occurred simultaneously. In the following sections, we view these seven steps as separate components in the origin of life.

Many Molecules Required for Life Can Be Created by Chemical or Physical Means

It is often said that all the conditions for the first production of a living organism are now present, which could ever have been present. But if (and Oh! what a big if!) we could conceive in some warm little pond, with all sorts of ammonia and phosphoric salts, light, heat, electricity, etc., present, that a protein compound was chemically formed ready to undergo still more complex changes, at the present day such matter would be instantly devoured or absorbed, which would not have been the case before living creatures were formed.

Darwin in a letter to the botanist Joseph Hooker

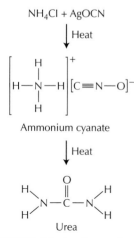

FIGURE 4.5. Synthesis of urea from inorganic salts by the Wöhler reaction.

A question that has been the subject of much research is, where did the molecules come from that were required for life to begin? Experiments in organic chemistry (dating back to the 19th century) have proven that complex organic molecules can be synthesized from simple inorganic components. In the first of these experiments (performed in 1828), the organic compound urea ($CO(NH_2)_2$) was synthesized by heating two inorganic salts, ammonium chloride and silver cyanate (Fig. 4.5), which form the unstable intermediate, ammonium cyanate, which then forms urea. This experiment helped disprove the theory of vitalism, which claimed that the chemistry of living systems is fundamentally different from that of nonliving systems (see p. 38). In the 1850s, the amino acid alanine was produced by exposing formamide and water to UV irradiation and sparks. Shortly thereafter it was shown that mixing formaldehyde and sodium hydroxide produces sugars. Thus around the time *On the Origin of Species* was published, it was already known that organic molecules could be made from inorganic compounds by relatively simple chemical reactions.

In the 1920s and 1930s, A.I. Oparin and J.B.S. Haldane independently suggested that early conditions on Earth might have led to the accumulation of organic molecules that could have been precursors for life. However, these and other theoretical suggestions simply took the results of chemical studies and tried to connect them to the origin of life. They did not try to address specifically what could have been made on the early Earth.

It was not until the middle of the 20th century that scientists attempted to simulate, in the laboratory, conditions that might have existed on the early Earth. A groundbreaking experiment, performed by Stanley L. Miller, changed how most scientists regarded these **prebiotic synthesis** experiments. Miller, a student working in Harold C. Urey's lab, created a simulated ocean–atmosphere interface in which methane, ammonia, and hydrogen gas were mixed with boiling water and circulated past an electrical discharge (simulated lightning) (Fig. 4.6). After only a few days under these conditions, reaction products included high yields of amino acids, including many found in modern proteins (Table 4.2). In subsequent experiments using similar conditions, a diverse array of organic molecules were created, including adenine, guanine, cyanoacetylene (a possible precursor of uracil and cytosine), and a variety of sugars. One example is ribose, the key sugar for nucleic acids, which can be produced through a version of the formose reaction (Fig. 4.7) under prebiotic conditions. When additional conditions were added, such as cycles of desiccation and evaporation (which certainly occurred on the prebiotic Earth), the variety of organic products that were formed increased greatly (Fig. 4.8).

The Miller–Urey experiment and many of the follow-up studies were performed under reducing conditions, which, in essence, means that no O_2 was present, and compounds that were present were oxygen poor and hydrogen rich (e.g., CH_4 not CO_2 and NH_3 not NO_2). Miller used strongly reducing conditions because, at that time, a number of scientists believed that the early atmosphere was a highly reducing environment. More recent studies suggest this may not have been true. Instead, conditions near the Earth's surface appear to have more closely resembled a weakly reducing mix of CO_2, N_2, water, and other components. Under these conditions, many fewer kinds of organic

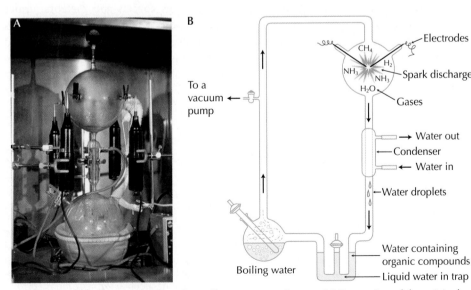

FIGURE 4.6. The apparatus used in the Miller–Urey experiments. (*A*) Recreation of the original apparatus. (*B*) Diagram of the apparatus.

TABLE 4.2. Amino acids from the Miller–Urey experiment

Amino Acid	Amount Produced			Amino Acid	Amount Produced		
	Exp.1	Exp. 2	Exp. 3		Exp.1	Exp. 2	Exp. 3
Protein amino acids				β-alanine	4.3	—	—
Glycine	100	100	100	α-amino-n-butyric	61	—	—
Alanine	180	2.4	0.87	α-aminoisobutyric	7	—	—
Valine	4.4	0.005	<0.001	β-amino-n-butyric	0.1	—	—
Leucine	2.6	—	—	β-aminoisobutyric	0.1	—	—
Isoleucine	1.1	—	—	γ-aminobutyric	0.5	—	—
Proline	0.3	—	—	N-methyl-β-alanine	1.0	—	—
Aspartic acid	7.7	0.09	0.14	N-ethyl-β-alanine	0.5	—	—
Glutamic acid	1.7	0.01	<0.001	Pipecolic	0.01	—	—
Serine	1.1	0.15	0.23	α-hydroxy-γ-amino			
Threonine	0.2	—	—	butyric	17	—	—
Nonprotein amino acids				α,β-diaminobutyric	7.6	—	—
Sarcosine	12.5		—	α,β-diaminopropionic	1.5	—	—
N-ethylglycine	6.8	—	—	Isoserine	1.2	—	—
N-propylglycine	0.5	—	—	Norvaline	14	—	—
N-isopropylglycine	0.5	—	—	Isovaline	1	—	—
N-methylalanine	3.4	—	—	Norleucine	1.4	—	—
N-ethylalanine	trace	—	—	Allothreonine	0.2	—	—

From Tables 3.1 and 3.2 of Maynard-Smith J. and Szathmáry E. 1995. *The major transitions in evolution.* Oxford University Press, Oxford; and based in part on Miller S.L. 1987. *Cold Spring Harbor Symp. Quant. Biol.* **52:** 16–27.
Mole ratios normalized to glycine as 100.
Exp., experiment. Exp. 1 contained $CH_4/N_2/NH_4Cl$ at 1:1:0.05 mole. Exp. 2 contained $CO/N_2/H_2$ at 1:1:3 mole. Exp. 3 contained $CO_2/N_2/H_2$ at 1:1:3 mole.

molecules are created. Thus, scientists have looked to other possible locations where simple reducing conditions may have been present.

One such location is outer space, which, in general, is a highly reducing environment. Objects in space that lack an atmosphere, such as comets and meteors, under the proper conditions, could serve as a substrate where Miller–Urey-type reactions occur. Billions of years ago, the organic reaction products could have been delivered when these objects collided with Earth. This scenario is supported by analyses of me-

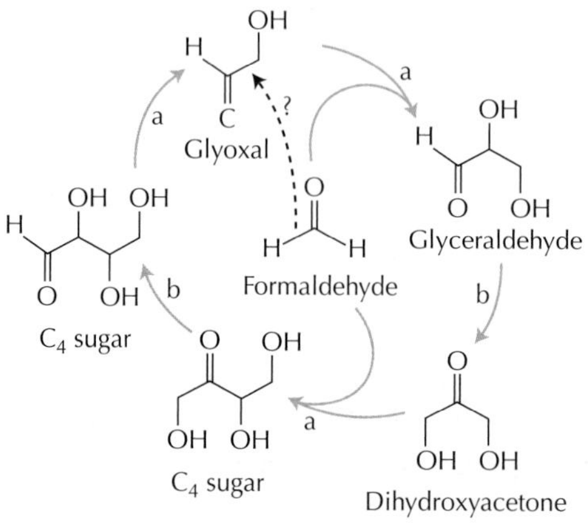

FIGURE 4.7. The chemical pathways that make up the formose reaction. In this reaction, formaldehyde can be polymerized to produce longer-chain sugar molecules. The reaction products (e.g., the C_4 sugar) can be readily converted to ribose. *Arrows* labeled with b represent ketone-alcohol isomerizations. *Arrows* labeled with a represent aldol/retroaldol reactions.

$$CH_2O \xrightarrow{Ca(OH)_2} \text{Sugars, including a small amount of ribose}$$

$$CH_4 + NH_3 + H_2O \xrightarrow[\text{discharge}]{\text{Electric}} \text{Amino acids, including glycine}$$

$$HCN \xrightarrow[\text{ammonia}]{\text{Aqueous}} \text{Adenine}$$

$$HC \equiv C - C \equiv N \xrightarrow[\text{or urea}]{\text{Cyanate}} \text{Cytosine}$$

FIGURE 4.8. Schematic diagram of pathways for the synthesis of some key molecules required for the origin of life. Formaldehyde (CH_2O) can polymerize to produce various sugars (i.e., the formose reaction; Fig. 4.7). This polymerization is aided by the availability of reactive groups like $Ca(OH)_2$. Methane (CH_4), ammonia (NH_3), and water (H_2O) if mixed in the presence of electrical discharges (in Miller–Urey-like settings) can lead to the production of amino acids (and other compounds). Hydrogen cyanide (HCN) when in the presence of aqueous ammonia can produce adenine. Finally cytosine can be produced from cyanoacetylene when mixed with urea.

teorites and comets, such as the meteorite that landed in Murchison, Australia in 1969. The Murchison meteorite is 4.6 billion years old and contains many of the products formed in the Miller–Urey-type experiments, including dozens of amino acids.

Space is not the only place we might look for the origin of life. Even if the atmosphere of the early Earth was not a reducing environment, there are other places on Earth that almost certainly were. Perhaps the best examples are the hydrothermal vents found deep beneath the sea (Fig. 4.9). These vents are sites where the seafloor is spreading apart, because of the drift of the tectonic plates under the ocean, and from which large upwellings of very hot, highly reduced compounds are released. Four billion years ago, the molecules necessary for the origin of life could have been synthesized at these deep-sea sites. Importantly, when the vent outflows hit the cold surrounding deep-sea water, the unstable organic molecules generated at high temperatures become "fixed" by the lower-temperature water. In addition, if life did originate in hydrothermal vents, those molecules and primitive organisms would be protected from much of the damaging radiation found on the surface of the Earth.

Although the idea that life originated deep within the ocean may have once seemed far-fetched, this changed with the discovery in 1977 of teeming oases of life at hydrothermal vents located thousands of meters below the surface. These oases are the sites of some of the most diverse and unusual ecosystems on the planet (Fig. 4.9). The basis of the food chain at the vent oases is chemosynthesis—the fixation of carbon using chemical energy (pp. 159–161). Thus this system does not rely on energy from the Sun.

FIGURE 4.9. Life at a hydrothermal vent oasis. The photo shows a community of animals at a vent site.

Chemical Reactions Can "Evolve" to Produce Complicated Molecules and Primitive Metabolism

Even if a suitable reducing environment did exist billions of years ago, the results of the prebiotic synthesis experiments are not sufficient to explain life's origins. First, there are some important compounds that have never been produced in any prebiotic synthesis experiment, regardless of the conditions used. The most notable of these are **nucleotides**, the building blocks of nucleic acids. Second, many reaction products are created only in minuscule amounts that probably were insufficient for the origin of life. Third, many of the compounds important for life are very unstable, especially in aqueous solutions. They would not accumulate over time but would instead degrade. For example, sugars such as ribose have short half-lives, especially at high temperatures. Protoliving organisms would have to use these unstable compounds very rapidly or convert them to more stable compounds. Other limitations of the prebiotic synthesis experiments, such as limited production of long polymers and production of right- and left-handed forms of compounds, are discussed in subsequent sections.

The limitations of prebiotic synthesis experiments can be explained in several ways.

Perhaps we have yet to try the right experimental conditions. Maybe the temperatures or pressures need to be higher, or the mixtures of simple starting chemicals need to be different, or some particular source of energy needs to be added. Alternatively, maybe all that is needed is enough time for molecules to accumulate. Or perhaps the molecules made were sufficient for primitive living systems to evolve. For example, the absence of nucleotides could be explained if the early evolving systems did not use nucleotides for their genetic material. Although these explanations may have some truth to them, many scientists consider another possibility more plausible: Additional chemical reactions utilized the molecules produced by prebiotic synthesis experiments to produce the full suite of molecules needed for the origin of life. The variety of compounds that could be produced becomes very large if we accept the highly plausible scenario that environments on the early Earth varied over time and space. In addition, and perhaps most importantly, long periods of time were available for reactions to occur. Scientists have thus sought ways to predict what types of chemicals could be made through longer-term reactions of the molecules produced by prebiotic synthesis experiments. These longer-term reactions have come to be called **chemical evolution**.

One example of chemical evolution is the formation of biologically important compounds in reactions driven by commonly occurring metal sulfides. Compounds like iron sulfide and hydrogen sulfide can provide ample amounts of free energy that drive reactions without the need for external energy sources like light or heat. This chemical energy can be used to convert single-carbon compounds (e.g., CO_2 or CO) into interesting organic compounds and to generate complex polymers. Rocks and clays that are rich in metal sulfides may have served as both reaction centers and sites where compounds accumulated at high concentrations (Fig. 4.10), providing suitable conditions for the formation of a primitive metabolism, without the presence of organisms. Metal sulfides are found in abundance in hydrothermal vents, which, as we discussed above, are one of the places where organic molecules necessary for the origin of life could have been synthesized.

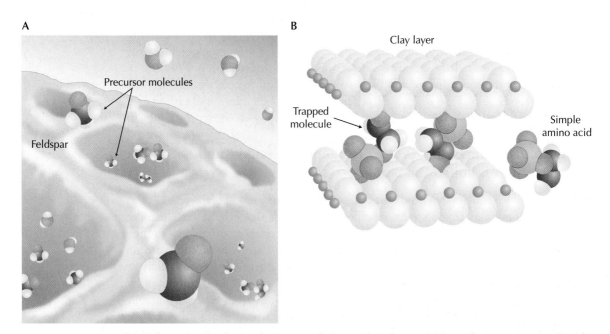

FIGURE 4.10. Minerals may have assisted chemical evolution. (*A*) Irregularities are present on the surfaces of many minerals, such as the feldspar shown here. The small pits could have protected molecules from the environment (e.g., UV irradiation) and served as "concentrating" pockets. (*B*) The layers within clays can serve as traps for compounds, increasing the probability of intermolecular reactions by forcing molecules into small spaces. Clays themselves are frequently reactive, which can further aid chemical reactions.

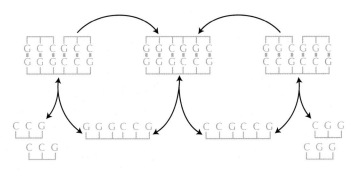

FIGURE 4.11. Example of an autocatalytic network involving DNA replication. Prior to the origin of DNA polymerase enzymes, DNA molecules could have served as scaffolds for the assembly of their complementary molecules from a mix of oligonucleotides present in a sample. Here the molecule made up of six Gs serves as a template to bring together two molecules of CCC. If these are ligated together they can then serve as the template to assemble the original GGGGGG molecule.

Chemical evolution theories frequently hypothesize that some of the chemical reactions form closed cycles. As with modern metabolic cycles (such as the Krebs cycle), the components of closed cycles do not disappear from the system. Maintaining these closed cycles requires a continued input of energy, along with input and output of some chemical compounds. Such a system is known as an **autocatalytic network** (Fig. 4.11). If autocatalytic networks were able to form on the early Earth, they may have served as the core chemical engines leading to the origin of life.

Perhaps most importantly, chemical evolution could have contributed to the creation of complex polymers. Of particular interest is the creation of polymers of amino acids and nucleotides, because they are essential to modern living systems. As discussed above, amino acid polymers can be formed in prebiotic synthesis experiments without the need for chemical evolution. Even more diverse polymers can be created via chemical evolution. However, it is nucleotide polymers that are most important, because they may have the highest capacity for evolution (see later discussion). But how nucleotide polymers might have formed is still unclear. On the one hand, in experiments that include nucleotides among the starting materials, diverse chemical evolution reactions can extend them into polymers. One example is the autocatalytic network shown in Figure 4.11. Another example is the ligation of nucleotides and small nucleotide oligomers into long chains by UV irradiation (Fig. 4.12). On the other hand, it is problematic to form useful long chain nucleotide polymers in experiments in which nucleotides are not included in the initial reactants. Prebiotic synthesis experiments generally have not yielded any nucleotides that could then be used to make nucleic acid polymers. Perhaps the most important issue is the formation of polymers in which every subunit is the same stereoisomer (see Fig. 4.18; also pp. 39 and 104).

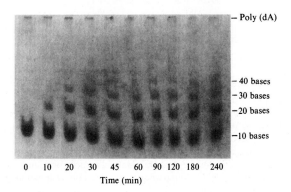

FIGURE 4.12. Polymerization of oligonucleotides by UV cross-linking. A solution of a DNA oligomer (10 bases long) was exposed to UV irradiation for different lengths of time. Longer exposures generated longer polymers.

Self-Replication Is Necessary for Evolution

The production of complex chemicals and chemical cycles, by whatever means, is not sufficient for life. Also absolutely critical is the ability to replicate, and when variants of a replicating entity arise and are then replicated, the copies are of the variant, rather than the original. These are the properties of **self-replication** of a **genotype**. In this section, we consider how a self-replicating system may have evolved.

To understand some of the issues associated with the origin of replication, it is useful to consider simple replicating systems. Perhaps the simplest involves nucleation, which occurs when, in the assemblage of a complex or in the formation of a polymer from monomers, the first steps in the assembly are energetically less favorable than continued growth. For the first steps to occur, the reaction needs to be seeded by a small portion of the complex or polymer. Nucleation can be a form of self-replication when the nature of the polymer or complex is determined by the nature of the seed. An example of a seed-driven nucleation is given in Figure 4.13A. Such a system can produce different types of polymers based on different pairings of the seeding monomers (Fig. 4.13B). **Prions**—the infectious protein molecules responsible for bovine spongiform encephalopathy (BSE)—may replicate by nucleation.

Although nucleation-based systems satisfy some of the requirement for self-replication, they are limited in many ways. For example, because they are seed driven and lead to the production of repetitive polymers that are multiples of the seed, nucleation systems do not lead to the production of diverse polymers. In addition, if a variant was introduced into the polymer, it would not be heritable, because it is the seed that drives the formation of the polymer. Thus nucleation systems are insufficient to allow for the significant amounts of evolution needed to lead to early life-forms.

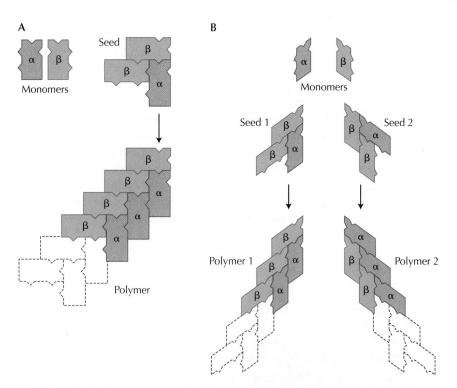

FIGURE 4.13. Nucleation of monomers. (*A*) Monomers can form polymers by a nucleation process. Monomers α and β represent two different but related chemicals. The monomers bind to each other through the triangular notches (internal and external). If the binding is only possible when a monomer binds to two other monomers, then a seed molecule may "nucleate" the polymerization process. (*B*) Nucleation can lead to multiple polymers if the initial seeds are different. Using one α and two βs (from *A*), two different seeds can form depending on how the triangular notches are bound. These seeds can catalyze polymerization, but the polymers produced are different.

In an alternative self-replicating system, initial "replicators" could have operated via guided polymerization in which the chemical or physical environment influenced the order of monomers assembled into a chain. One intriguing way that monomers can "self-organize" into polymers is through interactions with mineral substrates, such as clays. This has been demonstrated experimentally using monomers of nucleotides, which are "activated" or ready to react with other monomers or oligomers to form small chains.

A self-organizing system that produces oligomers of any monomer can "evolve." Although the population of oligomers will be predominantly the most stable molecules or those that form the most rapidly, this selection by "stability and speed" is not evolution by natural selection, because it lacks generation of variation and the differential survival of variants (see Chapters 5 and 17).

The key to life is the ability for evolution by natural selection to occur. This would require polymers that can make copies of themselves and a mechanism of heritable variation, in which the replicator makes mistakes (creating variants) and the new variants make copies of themselves. Even though a variety of chemical compounds may be capable of nucleation or self-organizaton, nucleic acids have attracted the most attention as potential early replicators because of their ability to base-pair, which allows for template-based polymerization. For an oligomer of nucleotides to utilize base pairing as a replication mechanism, all that is needed is a system to ligate the monomers that base pair with the polymer. Whatever the mechanism, the entity performing the self-replication can be called a **replicase**. Although it is easy to imagine how a replicase might evolve, difficulties arise when considering issues such as the accuracy of replication, and compartmentalization, which are discussed next.

Compartmentalization Facilitates Self-Organization and Accelerated Evolution

Researchers studying the origins of replication systems have identified bottlenecks that restrict the ability of simple systems to become more complex. Many of these obstacles could have been overcome if early replication systems were compartmentalized. In this section, we introduce three of these problems, indicate how compartmentalization could have provided a way around them, and discuss how early compartmentalization systems may have evolved.

The first problem with early replication systems concerns mutation rates and the size of genomes (this problem is called Eigen's paradox, and was proposed by Manfred Eigen in 1971). Mutation rates of these systems were certainly very high, because the enzymes involved were inaccurate. To lower mutation rates, complex machinery for error correction (see pp. 334–338) was required. But this machinery would have to be encoded within large genomes. Large genomes, however, could not evolve until the mutation rate was lower, because only small genomes could be replicated accurately. Eigen proposed that without error correction enzymes, a replicating molecule could be no larger than 100 base pairs, too small to encode an enzyme that corrects replication errors.

A second difficulty is that early replication systems were easily parasitized. If multiple systems interacted with one another in, for example, the primordial ocean, the products of one system could have been co-opted by a second system. If a self-replicating system also coded for other (nonreplicating) molecules, then a third problem arose. In an environment that was undergoing frequent mixing (e.g., the primordial ocean), the connection between the encoded information and the products made would be difficult to sustain, because mixing could have interrupted ongoing reactions and molecular interactions.

One way around all of these problems is to partially isolate the replicating system from its surroundings. Compartmentalizing a replicating system reduces parasitism because it restricts the flow of molecules and permits energy to be sequestered. In addition, local concentrations of metabolites can become quite high, which in turn can help drive reactions. Perhaps most importantly, compartmentalization creates a framework in which different entities (i.e., reactions occurring within different compartments) can cooperate. It even suggests a solution to Eigen's paradox: Each compartmentalized replicating system is sufficiently small to minimize catastrophic mutation rates, but the separate systems cooperate at another level to create more complex structures and processes. We discuss selection of multiple levels in Chapter 21.

What was the first compartmentalization system? One possibility is a system based on the same types of molecules that make up modern cell compartments—lipids. Lipids spontaneously form compartments because of their **amphipathic** structure. One end of the molecule, the polar head group, readily interacts with water (**hydrophilic**) and the other end, the long hydrocarbon tails, avoids interactions with water (**hydrophobic**) (Fig. 4.14A). Lipids can spontaneously form a bilayer, a structure in which two layers associate with each other, with the hydrophobic portions facing inward and the hydrophilic portions facing the water (Fig. 4.14B). Scientists have found that with a bit of vigorous mixing, lipid bilayers in an aqueous solution can be induced to adopt a cell-like structure, in which the bilayer folds over onto itself to form a sphere, trapping water within the sphere, which is itself bathed in the aqueous environment. These structures are known as **liposomes** (Fig. 4.14C).

So, were liposomes the first compartments? Maybe not. Neither lipids nor the long-chain fatty acids that form the hydrocarbon tails are common products of prebiotic synthesis experiments. Although it is possible that processes akin to "chemical evolution" could have formed such compounds, alternatively, there may have been a different kind of compartmentalization system used by prebiotic and early life, which was later replaced by lipids. One possibility is protein microspheres, which form easily from amphipathic peptides that are common products of prebiotic synthesis experiments.

When we introduced compartmentalization, we noted that the replicating system should be only *partially* isolated from its surroundings. There must be some degree of movement of molecules in and out of the compartment. Today, cellular membrane proteins carry out this transport. A few theories have been proposed to explain how primitive compartments may have allowed such movement in the absence of transport proteins. For example, the compartments may have been discontinuous, with gaps created by objects such as mineral surfaces or large macromolecules (see Fig. 4.10). These gaps could function much like a cave entrance—things could move in and out through the

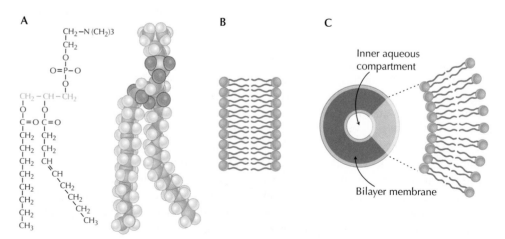

FIGURE 4.14. Lipids. (*A*) General structure of phospholipids. Phospholipids are made up of fatty acids, glycerol, and a phosphate group. They are amphipathic, with one hydrophobic end and one hydrophilic end. (*B*) Bilayers form when phospholipids spontaneously aggregate in water. The hydrophobic ends attract each other in the center of the layer and the hydrophilic ends are surrounded by water. (*C*) Liposomes are formed when a lipid bilayer folds over itself.

hole, but overall the insides were kept relatively separate from the outsides.

Whatever the molecular basis of the compartment, its critical advantages were that a primitive self-replicating entity could keep much of the products of its metabolic processes to itself and defend its replication machinery against parasitism, thus allowing evolution by natural selection to begin and eventually to proceed more rapidly.

The Chicken-and-Egg Problem Solved: RNA Can Serve a Dual Role as Information Carrier and Catalyst

Today, all cellular organisms use DNA as the molecule of heredity and proteins (along with RNA) to carry out the instructions. Working backward from the cellular organization of today, we encounter a chicken-and-egg problem when conceiving how life originated. DNA encodes the information for making proteins, which are incapable of self-replication. But proteins are required to catalyze the replication of DNA. It seems unlikely that the entire system could have evolved all at once from scratch. Could proteins have once been self-replicating? This seems unlikely. It also appears improbable that DNA—a relatively inert chemical—ever possessed catalytic activity.

In the 1960s, Francis Crick suggested a possible solution to this conundrum—maybe RNA could have served as both the information carrier and the catalyst as life was originating. His arguments were mostly theoretical, but the idea that RNA could have been both "genotype and phenotype" received an enormous boost when experimental studies revealed that even now, RNA functions as a catalyst (see p. 57). Here we outline the theoretical and experimental evidence that supports the possibility that an "RNA world" existed at one time in the past.

In modern organisms, RNA has a diversity of cellular roles (Table 4.3), whereas the role of DNA is largely limited to being the genetic material. The most well-known role of RNA is as a messenger (mRNA), relaying the message in the DNA to the protein translation apparatus (see p. 54). In addition, RNA is a key player in translation, in which transfer RNA (tRNA) and ribosomal RNA (rRNA) have critical roles. Other RNA functions include DNA replication and gene regulation. The multifunctional nature of RNA is evidence that it may have played an even more central role in the past.

TABLE 4.3. Examples of modern RNA roles

Function	Type of RNA	Role of RNA
Translation	mRNA	Product of DNA transcription
	tRNA	Involved in translation of the genetic code
	rRNA	Serves as part of a ribosomal subunit
DNA replication	RNA primers	Replication of the lagging DNA strand initiates with an RNA primer
	Telomerase RNA	Needed at the ends of linear chromosomes
Splicing and RNA processing	Small nuclear RNA (snRNA)	Involved in splicing
	Small nucleolar RNA (snoRNA)	Required for posttranscriptional processing of rRNA
	RNase P	Essential for tRNA processing
Translation quality control	tmRNA	Targeting aberrant protein products for degradation in bacteria
Protein translocation	Signal recognition particle (srpRNA)	A component of the signal recognition particle (SRP)
RNA interference (RNAi)	Many types	Involved in regulating RNA stability and translation in euykaryotes
Transcription regulation	6S	Regulates the function of bacterial RNA polymerase

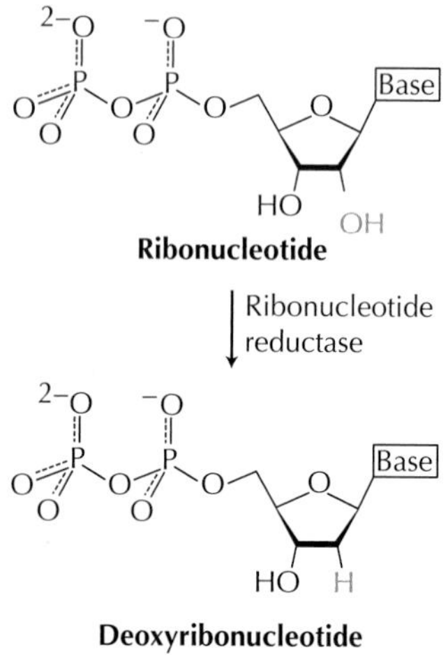

Ribonucleotide

↓ Ribonucleotide reductase

Deoxyribonucleotide

FIGURE 4.15. Ribonucleotide reductase. In modern organisms, deoxyribonucleotides are synthesized from ribonucleotides using the enzyme ribonucleotide reductase. This is consistent with deoxyribonucleotide synthesis being a more recent invention than ribonucleotide synthesis.

Furthermore, consider the modern molecules and processes in which RNA and ribonucleotides are involved. Most of them are ubiquitous and many are ancient. For this reason these molecules and processes have been called molecular fossils. For example, many essential coenzymes (e.g., NAD^+ and FAD) used in all organisms are derivatives of a ribonucleotide. Deoxyribonucleotides are synthesized from ribonucleotides (Fig. 4.15). DNA replication uses an RNA primer. ATP, the global energy carrier, is a ribonucleotide. These ancient roles of RNA suggest it was serving a crucial role early in the evolution of life.

RNA has been identified as a catalytic component of many modern cellular processes (Table 4.4). These functions include **splicing** of some RNA transcripts, replication of DNA ends, protein translation, and cleavage of tRNA precursors. It seems likely that many other examples of RNA molecules with catalytic function remain to be discovered. In addition, in vitro evolution experiments have shown that RNA molecules can be created with enzymatic activities capable of catalyzing diverse reactions, and exhibiting **allostery** (Chapters 2 and 17).

The catalytic activity of RNA is due in large part to two features. First, the molecule is able to fold into diverse three-dimensional structures. This folding takes advantage of base pairing to form what are called secondary structures (stems and loops), which can then fold to produce complex three-dimensional structures (Fig. 4.16). Although single-stranded DNA can fold into diverse structures, it lacks the level of reactivity that would allow it to catalyze reactions.

The second critical component of RNA catalysis is its reactivity. RNA is more chemically reactive than DNA because of the presence of the OH group on the 2′ carbon. This extra reactivity is in part what allows RNA to have more catalytic potential. (Conversely, the limited chemical reactivity of DNA makes it a better information-storage molecule.)

TABLE 4.4. Ribozymes

Ribozyme	Description
Self-splicing introns	Some introns splice themselves by an autocatalytic process. There is also growing evidence that the splicing pathway of GU-AG introns includes at least some steps that are catalyzed by snRNAs.
Ribonuclease P	This enzyme creates the 5′ ends of bacterial tRNAs. It consists of an RNA subunit and a protein subunit, with the catalytic activity residing in the RNA.
Ribosomal RNA	The peptidyl transferase activity required for peptide bond formation during protein synthesis is associated with the 23S rRNA of the large subunit of the ribosome.
Virus genomes	Replication of the RNA genomes of some viruses involves self-catalyzed cleavage of chains of newly synthesized genomes linked head to tail. Examples are the plant viroids and virusoids and the animal hepatitis delta virus. These viruses form a diverse group with the self-cleaving activity specified by a variety of different base-paired structures, including a well-studied one that resembles a hammerhead.
Telomeres	In some species, replication of DNA ends is catalyzed by an RNA subunit of its telomerases.

From Brown T.A. 2002. *Genomes*, 2nd ed., Table 10.4, BIOS Scientific Publishers Ltd., Oxford. snRNA, small nuclear RNA; tRNA, transfer RNA.

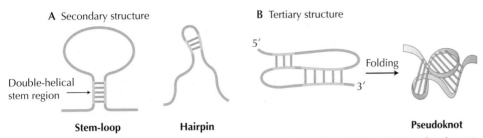

A Secondary structure

B Tertiary structure

Double-helical stem region →

5′

3′

Folding →

Stem-loop   **Hairpin**   **Pseudoknot**

FIGURE 4.16. RNA structure. Structural motifs that can be formed by folding RNA molecules. (*A*) Stem-loops, hairpins, and other secondary structures form by base pairing between distant complementary segments of an RNA molecule. In stem-loops, the single-stranded loop (*blue*) between the base-paired helical stem (*red*) may be hundreds or even thousands of nucleotides long, whereas in hairpins, the short turn may contain as few as 6–8 nucleotides. (*B*) Interactions between the flexible loops may result in further folding to form tertiary structures such as the pseudoknot. This tertiary structure resembles a figure-eight knot, but the free ends do not pass through the loops, so no knot is actually formed.

Considering all of this evidence has led to the generally accepted conclusion that early life was an RNA world in which RNA functioned as both genotype and phenotype.

The Chicken and Egg, Part 2: How Could an RNA Replication System Have Evolved?

How did an RNA replication system originate? Perhaps activated RNA monomers simply reacted to form chains. This would have been sufficient for a variety of RNA chains to form. If one of these RNA chains had RNA ligase activity, it could create many long chains. The formation of such a ligase is supported by in vitro evolution experiments in which RNA ligases form from random sequences after a few rounds of mutation and selection. In vitro evolution experiments are described on pages 458–460.

With base pairing, this ligase activity could serve as a primitive form of replication. It is not a large biochemical leap to go from an RNA ligase to an RNA polymerase. Once such a polymerase came into being, it could copy itself (Fig. 4.17). However, the replicase could also copy other RNA molecules, perhaps overwhelming the RNA polymerase with parasitic RNAs (Fig. 4.17). As we have seen, compartmentalization would solve this problem, by allowing for the evolution of cooperation and limiting the evolution of parasitism (also see pp. 614–618).

One limitation in the evolution of an RNA replicator is that nucleic acid monomers can come in two distinct forms based on the orientation of particular bonds within

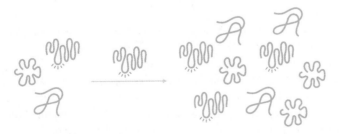

FIGURE 4.17. RNA replication. (*Left*) Replicase. A hypothetical RNA molecule that can catalyze its own replication. This hypothetical process would require catalysis of the production of both a second RNA strand of complementary nucleotide sequence and the use of this second RNA molecule as a template to form many molecules of RNA with the original sequence. The *small red lines* represent the active site of this hypothetical RNA enzyme. (*Right*) Cheating. A hypothetical mix of RNAs in which one is a replicase. It will copy the other RNAs unless they are isolated from each other.

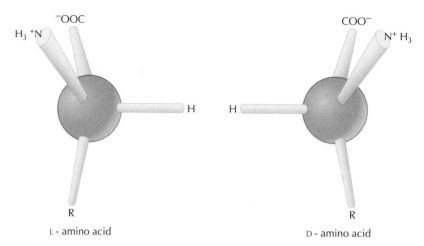

L - amino acid

D - amino acid

FIGURE 4.18. L and D amino acids. The L and D forms are stereoisomers (i.e., having the same chemical composition but with mirror images in bond positions). Building polymers (of amino acids or nucleotides) usually requires using only one form for the whole polymer.

the molecules: a D form and an L form (e.g., see Fig. 4.18). Mixing the L and D forms of the nucleotides would disrupt the replication of nucleic acid polymers if such a system was based on base pairing. Thus scientists have sought experimental ways to produce only one of the two forms (L or D) or to identify conditions under which the mechanism of self-replication would preferentially use one of the two.

Leaving the RNA World Required the Evolution of the Translation System and the Genetic Code

When we began this chapter, we identified two interdependent approaches scientists are using to determine how life began—top-down and bottom-up. If, as seems likely, an RNA world previously existed, we can ask, how did things move from that RNA world to a world of RNA, DNA, and proteins? More specifically, how did a common genetic code and its associated translation apparatus come into being?

To understand the theories concerning the origin of the translation system, it is useful to review some features of modern protein translation. Transfer RNA molecules (tRNAs) and amino acids are joined in a process known as aminoacylation or charging. Each tRNA is charged with its own specific amino acid. A charged tRNA aligns with the first codon of a protein-coding gene through base pairing of its anticodon sequence. The second charged tRNA aligns with the next codon and the amino acid from the first is added to the amino acid from the second through the formation of a peptide bond. Additional charged tRNAs bind to the next codons and the peptide is extended until a signal is received to stop the translation process. Recall that RNA is the catalytic component of the modern translation system (Table 4.4). Thus proteins did not have to exist before some type of translation system could have evolved.

In considering the origin of the genetic code, it is important to recognize that although the genetic code is redundant, the redundancy is not random (see Figs. 2.23 and 2.26). For example, among the amino acids encoded by multiple triplets, the triplets (**codons**) are usually variable only in the third position. (There are three exceptions to this: serine, arginine, and leucine, which are each encoded by six codons.) Another important aspect of the code is that chemically related amino acids are, in some cases, encoded by similar codons. For example, GAU and GAC code for aspartic acid, whereas the other acidic amino acid, glutamic acid, is encoded by GAG and

GAA. A third critical feature is that, despite many references to the "**universal genetic code**," the code is not in fact universal. Instead, there is a code used by many species, which can be called the **standard code** or **canonical code**, and exceptions to the standard code are found in various organisms and organelles (see Table 5.3). Patterns in the modern genetic code provide insight into what the RNA world may have been like and how translation evolved.

Based on the information outlined above, four stages in the origin of the genetic code have been identified:

1. the origin of the coding system, including both the use of codons and the use of tRNA adaptors

2. expansion of the coding system to add additional amino acids

3. adaptation of the coding system

4. modification of the system in some lineages

Here we focus on the origin of a coding system. (Expansion of the code is not discussed in further detail; adaptation of the code is discussed in Chapter 20, and modification of the code is presented in Chapters 7 and 10.) To address the origin of a coding system, we need to investigate (1) RNA-guided protein synthesis, (2) the mapping of particular RNA sequences to particular amino acids, and (3) the origin and role of tRNA adaptors.

Before there was translation, RNA molecules and amino acids were likely interacting in other ways. Amino acids may have stabilized RNA structures, much as proteins do for RNAs today. Amino acids may also have been used as enzyme cofactors, enhancing the catalytic activities of RNA molecules or facilitating the development of new functions. Eventually, RNAs may have evolved functions that formed the underpinnings of a translation system. They may have developed amino acid synthesis and stabilization pathways; mechanisms to acquire and sequester particular amino acids; the ability to form small chains of amino acids, much as is done by nonribosomal peptide synthesis pathways today; and as aminoacyl-tRNA synthetases do today, the ability to attach amino acids to other RNAs.

The one-to-one association of an amino acid with a tRNA molecule determines much of the specificity of our current protein translation system. How these associations formed is a key origin-of-life question. There are two prevailing theories (Fig. 4.19). The first suggests that the linkage was due to the development of tRNA adapter molecules used in the making of small peptides. It is plausible that some RNA molecules evolved the ability to "charge" other RNAs with amino acids. This possibility is supported by evolution of such

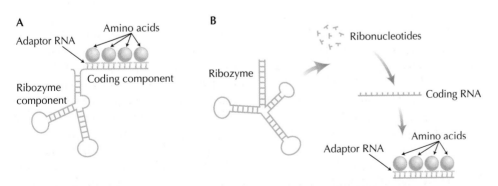

FIGURE 4.19. RNA to amino acids. Two possible scenarios for the evolution of the first coding RNA. A ribozyme could have evolved to have a dual catalytic and coding function (*A*), or a ribozyme could have synthesized a coding molecule (*B*). In both examples, the amino acids are shown attaching to the coding molecule via small adaptor RNAs, the presumed progenitors of today's tRNAs.

systems in in vitro selection experiments. Over time, such a system would evolve greater precision if selection favored one RNA for each (or, at most, a few) amino acid(s). In the modern protein translation system, tRNAs bind to mRNA at the ribosome. A primitive translation system could have achieved a similar result if it had an analog of mRNA to guide placement of the charged RNA molecules and a mechanism for linking amino acids and peptides. Through extensive fine-tuning, a triplet codon–based guiding system could form. If this model is correct, then the mapping of codons to particular amino acids could in essence be random. This theory is known as the "frozen accident" hypothesis.

An alternative model hypothesizes that the original interaction of amino acids was tied to the guide RNA sequence through stereochemical constraints (and is thus known as the stereochemistry theory). Carl Woese originally proposed this idea based simply on looking for structural fits between codons and amino acids. It has gained much support from experiments showing that RNA molecules can be selected for based on their preferential binding to certain amino acids. In many cases, the selected RNA contains a codon that matches the canonical code for that amino acid. Furthermore, these codons appear to be the likely binding site for the amino acid. So far, results suggest that perhaps half of the 20 amino acids used in modern proteins preferentially bind to one of their corresponding triplet codons over other possible RNA triplets. Thus the genetic code may have originated as an extension of sequence-specific RNA–amino acid interactions. The apparent ability of specific RNA sequences to interact with specific amino acids may have contributed to the initial evolution of RNA–amino acid interactions as well. This may explain, for example, why some amino acids that were present on the early Earth are not used in the genetic code. They may have lacked the ability to form highly specific interactions with RNA sequences. The stereochemistry theory still requires a way of using adaptor molecules instead of direct interaction with guide RNAs. Perhaps the original interaction of amino acids with specific RNAs was via tRNA-like molecules, and the guide RNAs were added later, as in the frozen accident theory.

DNA Replaces RNA

In the RNA world, RNA was genotype and phenotype. This was a critical step in the origin of life. However, once a translation system evolved, proteins rapidly took over most catalytic functions—a role they have retained to this day. There are many advantages to having proteins instead of RNA as enzymes. Proteins have a more complex alphabet and thus can form a more varied array of polymers. This allows them to have far more diverse catalytic activities than RNA. For example, no RNA molecule has been found that can carry out oxidation–reduction reactions or break carbon–carbon bonds. As RNA ceased to be the exclusive domain of phenotype, evolutionary forces may have begun to select for new genetic molecules. Eventually DNA took over (at least in cellular organisms).

There are many possible reasons why DNA replaced RNA as the information center. The lower reactivity of DNA makes it a more stable carrier of the genetic material. The higher reactivity of RNA is in part what allows it to be such a good catalyst. However, this also makes RNA more prone to chemical reactions that could damage the genetic material (see Chapter 12). The DNA takeover may have provided advantages by dividing the labor up between RNA and DNA. With RNA carrying out a variety of chemical reactions and metabolic processes, it may have been less available for replication processes or may have been more exposed to potential damaging chemicals. Thus the sequestering of the genetic material in DNA would allow RNA to carry out its catalytic and structural functions with fewer restrictions. The use of different nucleic acids for the information center (DNA) and the information carrier (RNA) would make it easier to distinguish what molecules in the cell are supposed to be doing what processes (DNA and RNA have multiple chemical differences).

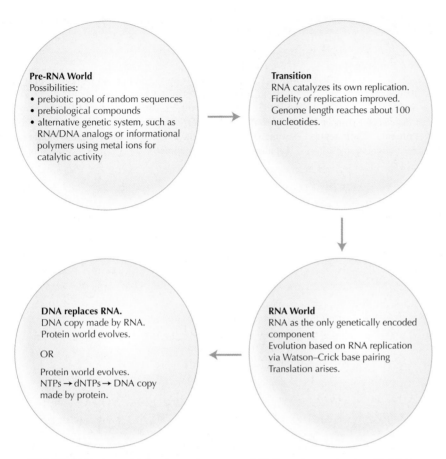

FIGURE 4.20. A model for the replacement of RNA as the genotype with DNA.

Given that there are many reasons why DNA would have replaced RNA as the hereditary material, most research in this area has focused on how the change occurred. Recall that in modern organisms, DNA precursors are made from RNA precursors, suggesting that DNA could have been made with relative ease once organisms were making RNA. Still one needs to have a way to convert the information in RNA into DNA. Today, reverse transcription enzymes can do this. They are found in many viruses and transposons and possibly are relics from the time when the RNA world was converted to a DNA world. Perhaps the RNA-to-DNA conversion happened only after proteins (like reverse transcriptase) were being synthesized (Fig. 4.20). Alternatively, perhaps the simple RNA replicases evolved to become primitive reverse transcriptases, able to convert RNA molecules into DNA.

SUMMARY

In this chapter, we have discussed some of the ways in which primitive life-forms could have been generated from abiotic systems to eventually become evolving organisms. It is important to realize that as soon as evolution by natural selection began occurring, the pace of change was massively accelerated. In addition, the same principles of evolution discussed in later chapters of this book should apply to these early protoliving systems. Recombination should accelerate scanning of the adaptive landscape. Cooperation and conflict will occur. The genetic systems themselves will evolve. What is of fundamental importance is that there is a relatively clear path from simple chemicals to the complex chemicals required for life, and a moderately simple path from there to evolving systems.

SUGGESTED READING

Dalrymple G.B. 1991. *The age of the earth*. Stanford University Press, California.

Fenchel T. 2002. *The origin and early evolution of life*. Oxford University Press, Oxford.

Knoll A.H. 2003. *Life on a young planet: The first three billion years of evolution on earth*. Princeton University Press, Princeton, New Jersey.

Maynard-Smith J. and Szathmáry E. 1998. *The major transitions in evolution*. Oxford University Press, Oxford.

McKay C.P. 2004. What is life—And how do we search for it in other worlds? *PLoS Biol.* 2(9): e302. Epub 2004 Sep.

Miller S.L. 1987. Which organic compounds could have occurred on the prebiotic earth? *Cold Spring Harbor Symp. Quant. Biol.* **87:** 17–27.

Newman W.L. 1997–2001. *Geologic time*, Online Edition. http://pubs.usgs.gov/ gip/geotime/.

CHAPTER

5

The Last Universal Common Ancestor and the Tree of Life

I N CHAPTER 4, WE DISCUSSED HOW LIVING systems on the early Earth could have evolved from nonliving ones. In this chapter, we focus on the early evolution of life and consider, in particular, two related issues: the possible existence of a common ancestor of all modern life-forms and the Tree of Life, which shows the evolutionary history of all life. As we examine what this **last universal common ancestor (LUCA)** may have been like, some key concepts in evolutionary biology are introduced, including **vertical descent**, **homology**, **divergence**, and **phylogenetic trees**. These concepts are used to illustrate how the properties of extinct organisms, such as LUCA, can be inferred. The chapter also discusses why a tree that shows the relationship among *all* life-forms is critical for such reconstructions. It was the analysis of such a tree that revealed that early in evolution life diversified into three main lineages of cellular organisms: the **bacteria**, **archaea**, and **eukaryotes**. The chapter closes with a discussion on why one of the key assumptions about the evolution of life—that of vertical descent—does not always apply and how the exchange of genes among organisms affects our understanding of the origin of life.

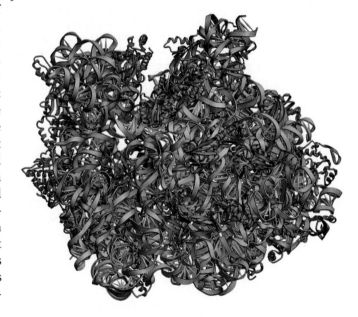

TRACING EARLY EVOLUTIONARY HISTORY

Evolution Can Be Represented by a Branching Process

One of Charles Darwin's fundamental insights into biology was that the evolution of species over time can be represented as a branching process in which individual species "split" to produce two or more distinct "daughter" species (pp. 17 and 67). These daughter species evolve as distinct units until one or both split again, and this process continues with more and more branches being created.

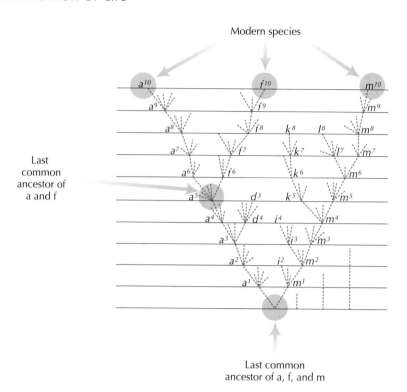

FIGURE 5.1. Vertical inheritance as represented in the only figure in *On the Origin of Species*. The evolution of species over time is represented by the branching of the tree with a single species giving rise to one or more descendent species. The descendent species inherit traits from the parental species in the form of "vertical" inheritance. Modern species are represented by a^{10}, f^{10}, and m^{10} with all other lineages having become extinct. a^{10} and f^{10} can trace their ancestry to a common ancestral species a^5. In turn, a^{10} and f^{10} can jointly trace their common ancestry with m^{10} to the species at the bottom of the figure.

Although Darwin did not understand the exact mechanisms of heredity, he was able to extend to the evolution of species what was known about the transmission of traits between parent and offspring. In fact, Darwin viewed this concept of **vertical inheritance** as so important that a diagram of this branching process was the only illustration he included in *On the Origin of Species* (Fig. 5.1). Such diagrams, which are known as **phylogenetic trees** or evolutionary trees, are described in more detail in Box 5.1.

The Concept of Descent with Modification Can Be Used to Infer Evolutionary Relationships

In the evolution of species, for some time after a lineage splits, the two "daughter" lineages will resemble each other, just as offspring resemble each other and their parents. The fact that offspring resemble their parents, or that new species resemble the ancestral species from which they formed, is of course due to the inheritance of genetic material. This sharing of traits via a common ancestry is known as **homology** and the shared traits are **homologous**. Homology can manifest itself in a number of ways, including among genes, morphology, and behavior (see Table 5.1). For example, the arm bones of humans and front leg bones of quadruped mammals are considered homologous, as are the different types of globin proteins in human blood. Over time, though, similarities between daughter lineages will decrease, as a result of a process known as **divergence** or **descent with modification** (p. 67).

Box 5.1   Evolutionary Trees

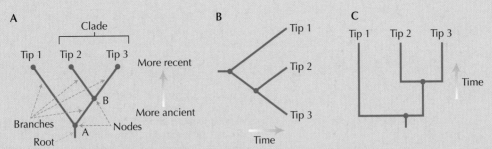

FIGURE 5.2. Model phylogenetic trees showing tips, nodes, and branches. (*A*) All the key elements of a phylogenetic tree. This tree shows the evolutionary history of three operational taxonomic units (OTUs). OTUs can represent species, individuals, genomes, genes, or other entities having an evolutionary history. (*Thick blue lines*) The branches represent the evolution of the OTUs over time. In this tree, evolutionary time is shown progressing from bottom to top, and thus this is known as a vertical tree. When considering evolutionary time from the past to the present, nodes (*blue circles*) represent the points at which one lineage separated into two. When considering evolutionary time from the present to the past, nodes represent the common ancestor of the organisms above the node. In this case, Tip 2 and Tip 3 share a common ancestor at node B. All descendants from node B, including Tip 2 and Tip 3, can be considered a clade or monophyletic group (see Fig. 5.3 for more detail on clades). The separation of taxa on the horizontal axis and the angles of the branches have no real meaning; it is done to be visually pleasing. (*B*) The tree in *A* has been rotated 90°. Such horizontal trees contain the same information as vertical trees. In this case, evolutionary time progresses from left to right, and the separation on the *y*-axis has no meaning. (*C*) A T-branching tree. It too has the same information as the trees in *A* and *B*, but the branches are drawn with a T-shaped junction instead of a V-shaped junction. In each of these trees, the "root" of the tree is the branch leading up to the common ancestor of all taxa shown in the tree.

Phylogenetic trees contain three main components: **branches**, nodes, and tips (Fig. 5.2). The branches (each distinct line in the tree) represent evolution over time. **Nodes**, which are the points at which multiple branches are joined, represent the separation of one lineage into multiple distinct lineages. The **tips** (also called leaves, **operational taxonomic units [OTUs]**, or terminal nodes) are the end points for particular branches. They can represent taxonomic units present today (e.g., modern species) or those that have become extinct. Nodes also represent hypothetical common ancestors of the species from different branches that descended from that branch point. Tree-like diagrams were originally applied to species evolution, but they are also used to represent the evolution of other features, such as genes and individuals within populations. The exact structure of the diagram varies depending on what is being represented. Figure 5.2 and its legend provide additional details about the key features of phylogenetic trees.

It is frequently useful to refer to groups by how they relate to each other in a phylogenetic tree (Figs. 5.3 and 5.4). The simplest grouping is that of a **monophyletic** group, or

clade, which consists of an ancestor and all of its descendants. Sometimes, species that share a common ancestor are grouped together, but some descendants of that common ancestor are excluded from the grouping. This is known as a **paraphyletic** group. One example is the reptiles (see Fig. 5.4), which all share a common ancestor, but because birds evolved from within the reptile clade, not all the descendants of the reptile common ancestor are reptiles. In other situations, species are treated as a group because of some shared biological features, even though they do not share a common ancestor to the exclusion of other species. Such a collection of species is a **polyphyletic** group (derived from many [poly] ancestors; Fig. 5.3). Examples include gliding mammals (made up of species related to both foxes and squirrels), gram-negative bacteria (see Fig. 6.2), and algae (see p. 198).

The trees shown so far contain information only about the branching patterns among the tips and nodes. They are referred to as **cladograms**. In many cases, it is useful to represent the amount of evolutionary change that has occurred along different branches. This is done by making the **branch**

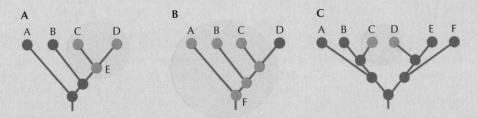

FIGURE 5.3. Different types of phylogenetic groups. In each panel, the phylogenetic group is indicated by a *green shaded circle*. (*A*) Monophyletic group. All species (C and D) in the group share a common ancestor (E) not shared by any of the other species. (*B*) Paraphyletic group. All species in the group share a common ancestor (F), but some species (D) have been excluded from the group. (*C*) Polyphyletic group. A grouping of lineages each more closely related to other species not in the group than they are to each other.

Box 5.1. Continued.

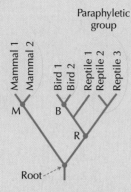

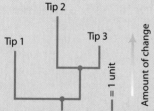

FIGURE 5.4. An example of monophyletic and paraphyletic groups in vertebrate evolution. A phylogenetic tree of the relationships among two mammals, two birds, and three reptiles is shown. Mammals share a common ancestor (at node M) to the exclusion of birds and reptiles and therefore are a monophyletic group. Birds are also a monophyletic group (sharing an ancestor at node B). Although reptiles share a common ancestor (at node R), not all descendants of this common ancestor are reptiles; some are birds. Reptiles are not a monophyletic group; they are a paraphyletic group.

length proportional to the amount of change. When all branches are allowed to be variable length, a tree is referred to as an **additive** tree, or a **phylogram** (Fig. 5.5). In such trees, the distance along the time axis corresponds to the amount of change and a scale bar can be shown. In some cases, it is useful to restrict the branch length variation and force all the tips of the tree to be equidistant from the root (e.g., such a tree can be used to represent evolutionary time since divergence from common ancestors). These are referred to as **dendrograms** or **ultrameric** trees.

The relative positions of branches in a tree diagram can

FIGURE 5.5. A phylogram (shown) is a phylogenetic tree that indicates the amount of evolution in addition to the branching order. The amount of evolution is represented by the branch lengths along the time axis (in this example, the vertical axis). In this tree, Tip 2 and Tip 3 share a common ancestor to the exclusion of Tip 1. However, during the time since they diverged from their common ancestor, Tip 2 has undergone more change. If Tip 2 and Tip 3 are modern organisms, this means that the rate of evolution in the lineage leading up to Tip 2 was greater than that leading up to Tip 3. Differences in rates of evolution are common and can be due to many factors such as different mutation rates, different population sizes, and different selective forces. Regardless of the cause, it is frequently very useful to incorporate such differences into evolutionary trees. In phylograms, a scale bar defines how much change is represented per unit length.

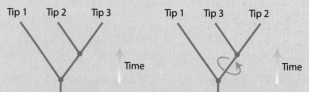

FIGURE 5.6. Branch rotation does not change the information in a phylogenetic tree. The two trees are the same except the branch leading up to the ancestor of Tip 2 and Tip 3 has been rotated such that Tip 3 is on the left and Tip 2 is on the right. The tree drawings on *left* and *right* are simply different forms of the same tree. It is useful to consider a phylogenetic tree like a mobile: The branches are free to rotate, but the branching patterns never change.

usually be changed without changing the meaning of the tree, because the informative components of a tree are the branching patterns and branch lengths. Thus, for example, the positions of Tip 2 and Tip 3 in Figure 5.6 can be rotated without changing the meaning of the tree. The tree still indicates that Tip 2 and Tip 3 share a common ancestor at the node that joins them and thus form a clade to the exclusion of Tip 1.

Although most trees are drawn with two branches emerging from a single node (i.e., two lineages diverging from a common ancestor), nodes can be connected with more branches. This is referred to as a **polytomy** (Fig. 5.7). Polytomies are sometimes used to represent evolutionary

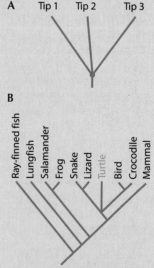

FIGURE 5.7. A phylogenetic tree drawn with three branches emerging from a single node represents a polytomy. (*A*) A generic polytomy. There are many uses for polytomies in phylogenetic trees. In some cases they are used to represent starburst-like evolutionary patterns when many lineages evolve from a single common ancestor over a very short period of time (i.e., essentially simultaneously). (*B*) Polytomies are also used to represent ambiguity in a model, such as this tree, which was used in a study that concluded it was not possible to determine the order of branching among turtles, snakes/lizards, and birds/crocodiles.

Box 5.1. Continued.

events—as in the possible occurrence of simultaneous divergence of lineages from a single common ancestor. Alternatively, they can be used to represent uncertainty in phylogenetic inference—meaning that the order of divergence of lineages cannot be inferred with reasonable certainty.

The phylogenetic trees examined thus far are examples of rooted trees, which have a root branch at the base. All entities in the tree share a common ancestor represented by the node connecting the root to the rest of the tree. It is not always possible, however, to identify the root of a phylogenetic tree. A tree without a root is unrooted. The assumption is that the organisms represented by an unrooted tree do share a common ancestor, and thus the tree has a root. The problem is that the position of the root is unknown and could be located at any point in the diagram. Figure 5.8 shows an example of an unrooted tree with three tips. Interpreting unrooted trees is challenging, and the placement of the root has profound implications on the conclusions that can be drawn from a phylogenetic tree. The issues associated with rooting trees are discussed in more detail on pages 124–129 and in Chapter 27 (online).

The uncertainty represented by polytomies raises an important issue in representing evolution by phylogenetic trees, which in almost all cases is based on inference rather than observation. A phylogenetic tree is a model of evolution, and some parts of the tree may be based on stronger evidence than other parts. There are numerous measures of the support for particular parts of a tree. One of the commonly used measurements is the **bootstrap** value, which is a statistical method for estimating how much a particular grouping in the tree is supported by *all* the data used to build the tree. For example, in a tree built with data on many types of bones, some groupings within the tree might be

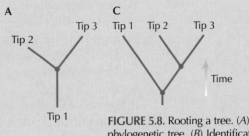

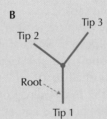

FIGURE 5.8. Rooting a tree. (*A*) Unrooted phylogenetic tree. (*B*) Identification of the position of the root of a tree. (*C*) Redrawing the tree in *B* to now include the root. In this case, with the root on the branch leading to Tip 1, Tip 2 and Tip 3 are more closely related to each other than either is to Tip 1. Other placements of the root would lead to different conclusions about the relationships among the tips.

supported by all of the bones, whereas other groupings might be supported by only a few bones. Bootstrap values are calculated by creating new datasets (based on the original dataset) that are the same size as the original, but differ from the original because some traits are included multiple times and others are not included at all. Next, a new tree is inferred for each new dataset. Finally, all of the datasets (i.e., trees) are analyzed by asking, in what fraction is a grouping the same as in the original dataset? The fraction is the bookstrap value, and the greater it is, the more a **clade** is supported by the data. Bootstrap value methodology is discussed further in Chapter 27 (online). Bootstrap values are shown in a phylogenetic tree by placing the value next to the appropriate node in that tree.

TABLE 5.1. Homology and analogy

Type of Trait	Homologous Example	Analogous Example
Morphology	Human hand and bat wing	Bird and insect wings
Physiology	Ion transport via ATP-binding cassette transporter proteins	Growth in high salt of algae and halophilic archaea
Biochemistry	Capture of light by chlorophyll in cyanobacteria and chloroplasts	Capture of light by chlorophyll vs. rhodopsins
DNA composition	High genome A + T content of mitochondrial genomes from humans and chimps	High genome A + T content of mitochondrial and mycoplasmal genomes

Homologous features are defined as those that share a common descent, whereas analogous features are those that have a common form or function but are derived separately. Homology and analogy can be seen in every type of biological feature, and a few examples are given in this table.

Using the concepts of descent with modification and the branching pattern of species evolution, it is possible to compare modern organisms and **infer** their evolutionary relationships. Such inferences are important because in most cases evolutionary history is not observed directly. It would seem that such inferences could be made by comparing organisms (or genes or other features) and converting levels of similarity to a tree where similar species are grouped together. All of the branches in a tree could be filled in based on these measures of similarity. In essence, one could work backward in time using similarity to determine where to join together species branches in the tree. However, evolutionary inference is not this straightforward; starting with measures of similarity and working backward to infer relatedness, unfortunately, does not always work.

One obstacle to this method is that rates of evolution may vary between different lineages. These differences are caused by variation in features like mutation rate, population size, generation time, and strength of selection. (These processes are examined in more detail in Chapters 12, 15, and 19.) Whenever rate differences occur, the overall level of similarity will not necessarily be directly related to evolutionary history. For example, the tree in Figure 5.9 shows the evolutionary history of three species, one of which (species 3) has had a higher rate of evolution than the other two. In such a tree, similarity between any two species can be approximated by measuring the total branch length that must be traversed to connect them. In Figure 5.9, the length from species 2 to species 1 is less than that from species 3 to species 2, meaning that species 2 is more similar to species 1 than it is to species 3. Yet species 3 and 2 share a common ancestor to the exclusion of species 1. Thus levels of similarity in this case are not reflective of the evolutionary relationships.

Although some traits resemble each other, they may have independent origins, and thus their similarity does not reflect homology. This type of relationship, which is known as **analogy**, poses a second complication when attempting to infer relatedness from measures of similarity. There are two main processes that can lead to the generation of analogous features: **convergence** and **parallel evolution**. In convergence, similarities arise from distinct (nonhomologous) features in different lineages as a result of adapting to similar environments or strategies of life. For example, selection for improved flight has led to the convergent evolution of wings and flight mechanics in birds, bats, and some insects. The structures involved in flight in these distantly related organisms have superficial similarities but are not homologous. In parallel evolution, similarities arise in different lineages through separate modifications of the same features. For example, there have been many separate origins of gliding across the mammals including in lemurs, foxes, squirrels, and marsupials. The gliding mechanism usually involves very similar modifications of the forelegs, but it arose separately in each distinct group. Another example of parallel evolution is shown in Figure 5.10. In this example, different lineages of arthropods have separately evolved a crab-like appearance, most likely through simple changes of the same developmental pathways. Additional examples of homologous and analogous traits are given in Table 5.1 and Figure 3.3.

The imperfect connection between levels of similarity and patterns of evolutionary relatedness cannot be overemphasized. This is one of the critical concepts to understand in studies of the evolutionary history of organisms. For example, to infer the evolutionary history of organisms, traits must be chosen that are not prone to convergent evolution and that have relatively uniform rates of change in different evolutionary branches. Evolutionary reconstruction methods and the issues that result from unequal rates of evolution and convergence and other phenomena are discussed in more detail in Chapter 27 (online).

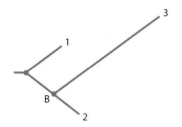

FIGURE 5.9. Unequal rates of evolution. A hypothetical phylogenetic tree showing relationships among three species. The rate of evolution of the lineage leading up to species 3 has accelerated relative to that of species 2, as indicated by the longer branch leading from their common ancestor (node B).

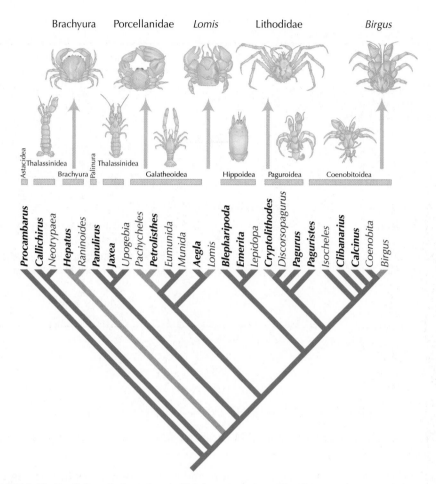

FIGURE 5.10. Parallel evolution of crab-like form in arthropods. The tree shows an inferred phylogeny of a diversity of arthropod groups that include species with crab-like form. The general appearance of each group is shown above the taxon names. Crab-like forms are shown at the top of the figure. The *arrows* indicate which genera have that form. *Red branches* represent multiple origins of crab-like structure. Multiple origins are likely the result of the same or very similar simple changes in shared developmental pathways (see Chapter 11), and this is considered an example of parallel evolution.

UNIVERSAL HOMOLOGIES, LUCA, AND THE TREE OF LIFE

The Existence of Universal Homologies Implies a Single Origin of Life and the Existence of LUCA

In considering the origin and early evolution of life, two fundamental questions arise: How many times did life evolve from nonliving systems? Do all modern life-forms share a common ancestor? Although it is widely believed that a common ancestor of all modern life did exist, it could certainly be otherwise. What then is the evidence for this common ancestry? The evidence is the existence of **universal homologies**, which were introduced in Chapters 2 and 3. Examples of universal homologies (Table 5.2) include:

1. DNA as the genetic material

2. copying DNA using a template and base-pairing mechanism

3. transcribing DNA into RNA using an RNA polymerase with a homologous catalytic mechanism

TABLE 5.2. Universal homologies of life

Feature	Trait
Core cellular features	
Genetic material	DNA
Bases used in DNA	A, C, T, G
Bases used in RNA	A, C, U, G
Genetic code	Three letter
Cell envelope	Lipoprotein membrane
Protein composition	20 core amino acids
Cellular complexes found in all organisms (examples)	
Translation	Small-subunit rRNA
	Large-subunit rRNA
	Multiple ribosomal proteins
	Aminoacyl-tRNA synthetases
	tRNA
Transcription	RNA polymerase
Membrane transport systems	ABC transporters

rRNA, ribosomal RNA; tRNA, transfer RNA; ABC, ATP-binding cassette.

4. translating RNA into proteins using a three-letter genetic code

5. the use of a mix of ribosomal RNAs, transfer RNAs (tRNAs), and ribosomal proteins in translation

6. the use of ATP as a cellular energy store and as a building block for making DNA and RNA

7. enclosing cells in a plasma membrane across which nutrients and waste must flow

Assuming that traits are transmitted only by vertical descent, then the sole explanation for the existence of universal homologies is that all modern organisms can trace their ancestry to a single common ancestor. The logic behind this argument is as follows. If *two* species share homologous features (and the transmission of those features is only by vertical descent), then those two species must have a common ancestor from which the homologous feature came. Those species could not be part of distinct lineages that represent separate origins of life. By extending this logic from two species to all species, it follows that if all species share some homologous features (i.e., the universal homologies), then all species must have a common ancestor from which these universal homologies came. It is then possible to draw a tree representing the evolutionary history of all species descending from a single ancestral node. This tree is known as the Tree of Life; and the organism represented by the ancestral node, this putative forebear of all life, is LUCA (Fig. 5.11). LUCA, the last universal common ancestor or the last universal cellular ancestor, is the organism that Darwin was referring to in *On the Origin of Species*, when he wrote, "... probably all the organic beings which have ever lived on this earth have descended from some one primordial form, into which life was first breathed." Later in this chapter, we will see that **lateral gene transfer**, which violates the rules of vertical descent, creates problems for this concept of LUCA.

It is important to understand that LUCA is not the only common ancestor of all of life, but rather the most recent one. That is, the branch leading up to LUCA includes other common ancestors of life that lived further back in the past. Thus LUCA

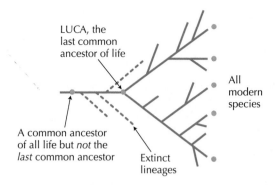

LUCA, the
last common
ancestor of life

All
modern
species

A common ancestor
of all life but *not the
last* common ancestor

Extinct
lineages

FIGURE 5.11. Evolutionary tree showing the most recent (also known as last) universal common ancestor (LUCA) of life. The tree also shows lineages (with *dashed branches*) that branched off before the existence of LUCA but subsequently went extinct and a common ancestor of life but not the most recent one.

is by no means the first living organism. It is the common ancestor of modern species and the predecessor to the three domains: eukaryotes, bacteria, and archaea. In addition, it is important to recognize that LUCA was not the only life-form around at the time. LUCA was one species or population among a diverse array of life-forms. These other forms, however, did not leave modern descendants because their lineages went extinct, whereas the lineage of LUCA gave rise to all modern forms (Fig. 5.11).

The Features of LUCA Can Be Inferred from Universal Character States or by Evolutionary Reconstructions

What was LUCA like? What type of cell structure did it have? What type of metabolism did it possess? What was the structure of its genome? To answer these questions, we need to use **evolutionary character state reconstruction** methods, which are described in more detail in Chapter 27 (online). These methods work by first taking any group of organisms of interest and overlaying their characteristics onto an evolutionary tree of those organisms (Fig. 5.12). Then the tree can be used to infer the likely characteristics of organisms in the past at nodes in the tree. As one works further and further back in time, the likely characteristics of deeper and deeper nodes can be inferred, until the common ancestor of all the organisms being studied is reached. Character traits are particular *parts* of an organism. They can be structures (e.g., webbed feet), organs (e.g., the heart), or genes (e.g., ribosomal RNA genes). The particular *form* that a trait takes is known as its character state (e.g., a three-chambered vs. a four-chambered heart). When comparing organisms, we can study the origin of character traits (e.g., the origin of the heart) or the change in character states of a particular trait (e.g., the evolution from a three-chambered to a four-chambered heart). In the most straightforward character state reconstruction methods, the assignment of the character state or trait to the ancestral nodes makes use of the principle of **Occam's razor**, or **parsimony**, which says that if all else is equal, choose the simplest explanation. Using this logic, the history that should be chosen is the one requiring the fewest changes in characteristics to produce the observed patterns.

In terms of character state reconstruction for LUCA, the simplest scenario is one in which a trait is universal and it takes the same state in all organisms. If the trait is overlaid onto the tree and traced back in time, the most straightforward possibility is that all individuals past and present had the trait. Thus all the universal homologies described previously (see Table 5.2) would be inferred to be present in LUCA. For example, because homologous small-subunit rRNAs are present in all species, it can be inferred that LUCA had small-subunit rRNAs.

Things become more complex when traits are universal but come in different states. We can infer that LUCA had the trait, but we would also like to know what state it took.

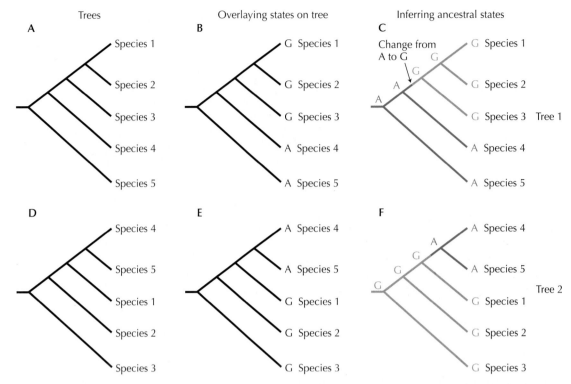

FIGURE 5.12. Evolutionary character state reconstruction. The trees show the evolutionary relationships between five species for which data about a single homologous trait are available. This trait comes in two states, A or G (e.g., nucleotides in a conserved position in a gene). Species 1, 2, and 3 have the G state and species 4 and 5 have the A state. Character state reconstruction methods allow one to infer which character state is ancestral and which is derived. (*A,D*) Two different trees relating the five species. (*B,E*) The character states are overlaid onto the trees. (*C,F*) Character state reconstruction methods are used to infer the likely states present in ancestral nodes. In *C* the ancestral state is A and in *F* it is G. Thus, the two different species trees lead to different inferences regarding which state is ancestral.

A good example involves the genetic code. Most species use what is called the **standard genetic code** (see Fig. 2.23). However, a significant number of species use variations of this three-letter code (Table 5.3). For example, in mycoplasma (see Table 6.2), the UGA codon codes for tryptophan, but in the standard code, UGA is a stop codon, signaling the termination of translation. UGA is also used in many **mitochondria** to code for tryptophan. (Note that mitochondria, which are the energy-producing organelles of eukaryotes, evolved from free-living bacteria, still have their own genomes and still carry out functions such as replication, transcription, and translation [see pp. 163 and 202–210

TABLE 5.3. Alternative genetic codes

Codon	Standard Code	Other Codes in Cellular Genomes				Mitochondrial Codes		
		Mycoplasma	Ciliates	Euplotes	Many Species	Yeast	Protozoa	Mammals
UGA	Stop	Tryptophan	s	Cysteine	Selenocysteine	Tryptophan	Tryptophan	Tryptophan
UAA/UAG	Stop	s	Glutamine	s	s	s	s	s
AUA	Isoleucine	s	s	s	s	Methionine	Methionine	Methionine
CUA	Leucine	s	s	s	s	Threonine	s	s
AGA/AGG	Arginine	s	s	s	s	s	s	Stop

Data in part from Madigan M.T., Martinko J., and Parker J. 1997. *Brock biology of microorganisms*, 8th ed. Prentice Hall, New York, Table 6.7, p. 222. s, same as standard code.

A

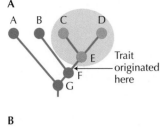

B

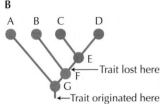

FIGURE 5.13. Ancestral and derived traits. (*A*) Derived traits. Consider a clade of species ABCD, with a trait present in species C and D but not A and B. If the trait evolved in the common ancestor of C and D (labeled E), then this is a derived trait in the group of ABCD because it was not present in the common ancestor of all four species. (*B*) Ancestral traits. Consider a clade of species ABCD, with a trait present in species A and B but not C and D. If the trait evolved in the common ancestor of A and B (labeled G), then this is an ancestral trait in the group of ABCD because it was present in the common ancestor of all four species. (Incidentally, the trait was subsequently lost in the lineage ECD.)

and Fig. 7.5].) In the standard genetic code, UAA and UAG are the other stop codons, but in many ciliates (such as *Tetrahymena thermophila*) these code for glutamine. Some species have even expanded the genetic code to more than 20 amino acids. For example, species from across the tree (including humans and *Escherichia coli*) use a modified tRNA to insert selenocysteine, the "21st" amino acid, into proteins.

Given that there are different states of the genetic code being used, how do we figure out which state was used in LUCA (and thus can be called an **ancestral character state**) and which represent more recent modifications (and are referred to as **derived character states**) (Fig. 5.13)? Without any other information it seems plausible that the standard code was ancestral and that the exceptions are derived, but it is also plausible that one of the exceptions was ancestral and that the standard code and the other exceptions are derived. To make such an inference, we need one critical piece of data—a rooted tree showing the relationships among all life-forms. This is the Tree of Life, which was introduced previously in this chapter (also see Box 5.2).

Box 5.2   The Tree of Life: A Brief History

Classification and the Tree of Life are intimately related: Prior to acceptance of the idea that organisms are related to each other by descent with modification, many classification schemes were developed. Most of these were quite confusing and focused on either small groups of organisms (e.g., birds) or organisms from only a few locations. Completely different schemes were used for the same types of organisms from different places and for different types from the same place. In particular, the naming of species was poorly organized, and some species had dozens of different names. These classification schemes were almost entirely replaced by a system introduced by Carolus Linnaeus, a Swedish botanist. The Linnaean classification scheme had two main components. First, all species were given two-part names, the first part of which was the name for the genus and the second part was the species-specific name. This **binomial nomenclature** system was rapidly adopted and is still in use today. Linnaeus also formalized a system in which all species were organized into a series of groups of different levels. This **hierarchical classifi-** cation system is also still in use today. For many years the classification levels used were kingdom, phylum, class, order, family, genus, and species. (Many mnemonics have been created to make the series easier to remember. One example is "kings play chess on fine grained sand.") Recently, domain has been added (see below) as a level above kingdom.

Although Linnaeus' scheme for classification was revolutionary, the methods for determining the classification hierarchy itself were not generally agreed on. This was particularly apparent in regard to classifying microorganisms when they were first observed with microscopes. Antoni van Leeuwenhoek referred to the microorganisms he discovered as "wee animalcules," clearly implying they belonged to the animal group. Others believed that they should be grouped with plants. This debate is, of course, based on an artificial premise, because most microbes are neither plants nor animals. It was only after the evolution of species became more generally accepted that classification hierarchies began to be based on branching patterns in an evolutionary tree

Box 5.2. Continued.

rather than on phenotypic characteristics that the observer believed to be important. In other words, the question for classification of microbes should be "How does this microbe relate, within an evolutionary tree, to animals or plants?" To figure this out, one needs a tree of life.

Concepts of the Tree of Life have gone through many changes: The first well-known attempt to draw a tree of life was by Ernst Haeckel (Fig. 5.14). This tree branched into three main lineages—the animals, the plants, and the protists. Bacteria were grouped into the Monera lineage within the protist branch and the root of the tree was also labeled as Monera, implying that bacteria were ancestral to other organisms. A major refinement of this tree came from Edouard Chatton, who, in 1937, divided life into two main groups—the **prokaryotes** and the **eukaryotes**—on the basis of whether or not an organism possessed a nucleus, the organelle that contains the bulk of the DNA within eukaryotic

cells. Another refinement came in 1959 in the five-kingdom tree of Robert Whittaker (Fig. 5.15). In this tree the distinction between prokaryotes and eukaryotes is still there, and the prokaryotes are still at the base of the tree. In addition, a new eukaryotic group was added—the fungi.

In general, as new depictions of the Tree of Life were developed, the relationships within the eukaryotic branch became more and more detailed, in particular for multicellular organisms, and the relationships among microbes in general and prokaryotes in particular were not highly detailed. Thus these trees of life were in essence incomplete—either they did not include many microorganisms or the placement in the tree of microorganisms relative to other species was highly suspect. There are two major reasons why it was difficult to include microorganisms in trees of life. First, there was a dearth of traits that could be observed for most microbial species. In contrast, many eukaryotes have traits like bones, organs, or leaves that could be readily measured and compared. In addition, there were few traits that could be used to link microorganisms to other species. This is important because to build an evolutionary tree, it is necessary to compare homologous traits (those with a common evolutionary history) among the taxa for which one wants to infer the history (for more details, see Chapter 27 [online]). For example, to infer the history of vertebrates, one could compare bone structures and patterns. However, to infer the relationship of vertebrates to invertebrates, other characteristics are needed because invertebrates lack bones. In general, the deeper the branches, the more universal the characters need to be. And to study the earliest branches in the tree of life, one thus needs universal homologies.

Molecular analysis revolutionized our view of the Tree of Life, leading to the modern "three-domain tree": Before the era of molecular biology, few, if any, universal homologies had been identified, and thus determining the early branching points on the tree of life was more based on intuition than on analysis of actual data. The advent of molecular biology changed this and led to the discovery of many new universal homologies, such as those described in Table 5.2. Another critical step in studying the tree of life was the development of techniques that use sequence data to study evolutionary history, an approach that is now known as molecular systematics. In its standard form, molecular systematics allows one to use the rich information of DNA and protein sequences to infer phylogenetic history. This sequence-based molecular systematics works by lining up the genes from different species, or different forms of a gene within a species, in the form of a multiple sequence alignment. Each site in the aligned genes can be considered a separate character trait and the particular residue used as a different character state (Fig. 5.16).

The culmination of molecular systematics in studies of the Tree of Life came with the work of Woese and colleagues in the 1970s. They focused on analyzing the sequences of rRNA molecules. rRNA is a component of the ribosomes—the machines that synthesize proteins in all species.

Woese's evolutionary analysis of the rRNA data revealed

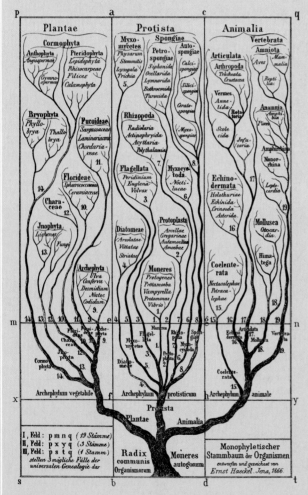

FIGURE 5.14. Tree of life constructed by Ernst Haeckel in 1866. This represents one of the first attempts to draw an evolutionary tree that included all known life-forms.

Box 5.2. Continued.

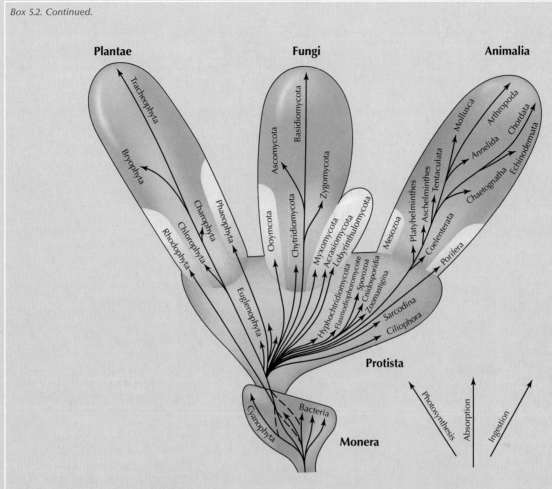

FIGURE 5.15. Whittaker's five-kingdom tree. This system contains five kingdoms based on three levels of organization: prokaryotic (kingdom Monera), eukaryotic unicellular (kingdom Protista), and eukaryotic multicellular and multinucleate (kingdoms Fungi, Animalia, and Plantae). The three kingdoms at the top of the figure are distinguished mainly by differences in nutrition (see the *inset*).

a striking finding—the Tree of Life was divided into three main lineages, and one of these lineages had not previously been recognized (Fig. 5.17). One group corresponded to the organisms whose cells have nuclei (eukaryotes). A second group contained most of the well-known bacterial species including all the known pathogens (e.g., *E. coli*, *Vibrio cholerae*, *Mycobacterium tuberculosis*, and *Haemophilus influenzae*) as well as many of the known free-living organisms (such as cyanobacteria). The third group contained mostly unusual, poorly characterized, extremophilic species, such as those that grow at high salinity, temperature, or pressure. Woese named this group *archaebacteria* with the "archae" referring to the similarity of the niche of these species to the conditions that some people believed existed on the primitive Earth. One of the most striking things about this finding was that the prokaryotes (organisms without nuclei) were split into two distinct lineages (the **bacteria** and the archaebacteria). Thus the grouping of all species without

nuclei turned out to be incorrect. Another striking thing about this finding was the existence of a previously unrecognized major branch on the Tree of Life. Subsequent analysis of other molecular sequence data supported this trichotomy of life, and in 1990 the three groups were assigned a new taxonomic status (the domain) and the archaebacteria were renamed **archaea** to emphasize their uniqueness.

Although the exact relationship between the domains is still being debated, Woese's work showed that molecular systematic techniques provide a picture of the microbial world very different from that obtained with traditional morphological methods. The discovery of a third domain of life is one of the great triumphs of modern molecular evolutionary biology, and the use of rRNA for molecular systematics continues to this day. Although other genes, and even entire genomes, are now used to infer the phylogeny of microbes, molecular systematics is viewed as the only reliable way to include all organisms in a tree of life.

Box 5.2 Continued.

```
- - - - - - - - - -    - - - MAIDENK    Q ALAAALGQ    I EKQFGKGS I    MRL GEDRSM-    Escherichia coli
- - - - - - - - - -    - - - - -MDENK    KRALSAALSQ    I EKQFGKGSV    MRMGDRYI E -    Xanthomonas campestris
- - - -MSQNSL    RLYEDKSVDK    SKALEAALSQ    I ERSFGKGS I    MKLGSNENV I    Rhizobium phaseoli
- - - - - - - - - -    - - - MSKLAEK    LKAVAAAVAS    I EKQFGKGSV    MTLGGEAREQ    Myxococcus xanthus
- - - - - - - - - -    - - - MAIDEDK    QKAISLAIKQ    IDKVFGKGAL    VRLGDKVQE -    Helicobacter pylori
- - - - - - - - - -    MAINTDTSGK    QKALTMVLNQ    I ERSFGKGA I    MRLGDATRM-    Anabaena variabilis
- - - - - - - - - -    - - - -MAGTDR    EKALDAALAQ    I ERQFGKGAV    MRMGDRTNE -    Steptomyces lividans
- - - - - - - - - -    - - - - - -MSDR    QAALDMALKQ    I EKQFGKGS I    MKLGEKTDT -    Bacillus subtilis
- - - - - - - - - -    - -MANIDKDK    LKAI EMAMGQ    I EKQFGKGSV    MKLGEQGAP -    Clostridium perfringens
MSKLKEKREK    AVVGIERASK    EEAI ELARVQ    I EKAFGKGSL    I KMGESPVGQ    Borrelia burgdorferi
- - - - - - - - - -    - - - -MSVPDR    KRALEAAI AV    I EKQFGAGS I    MSLGKHSSAH    Chlamydia trachomatis
- - - - - - - - - -    - - - -MASSEK    LKALQAAMDK    I EKSFGKGS I    MKMGE- EVVE    Bacteroides fragilis
- - - - - -MAEE    KI PTVQDEKK    LQALRMATEK    I EKTFGKGA I    MNMGANTYE -    Porphyromonas gingivalis
- - - - - - - - - -    - - -MPEEKQK    KSYLEKALKR    I EENFGKGS I    MI LGDETQVQ    Thermotoga maritima
- -MSKDATKE    ISAPTDAKER    SKAI ETAMSQ    I EKAFGKGS I    MKLGAESKL -    Deinococcus radiodurans
- - - - - - - - - M    ARVSENLSEK    MKALEYALSS    I EKRFGKGAV    MPLKAYETV -    Aquifex pyrophilus
```

```
DVETISTGSL    SLDIALGAGG    LPMGRIVEIV    GPESSGKTTL    TLQVIAAQR     Escherichia coli
AVEVIPTGSL    MLDIALGIGG    LPKGRVVEIV    GPESSGKTTL    TLQAIAECQK    Xanthomonas campestris
EIETISTGSL    GLDIALGVGG    LPKGRIIEIV    GPESSGKTTL    ALQTIAESQK    Rhizobium phaseoli
KVAVIPSGSV    GVDRALGVGG    VPRGRVVEVF    GNESSGKTTL    TLHAIAQVQA    Myxococcus xanthus
KIDAISTGSL    GLDLALGIGG    VPKGRIIEIV    GPESSGKTTL    SLHIIAECQK    Helicobacter pylori
RVETISTGAL    TLDLALG-GG    LPRGRVIEIV    GPESSGKTTV    ALHAIAEVQK    Anabaena variabilis
PIEVIPTGST    ALDVALGVGG    IPRGRVVEVV    GPESSGKTTL    TLHAVANAQK    Steptomyces lividans
RISTVPSGSL    ALDTALGIGG    VPRGRIIEVV    GPESSGKTTV    ALHVIAAAQQ    Bacillus subtilis
QMDAVSTGCL    DLDIALGIGG    VPKGRIIEIV    GPESSGKTTV    ALHVVAAAQK    Clostridium perfringens
GIKSMSSGSI    VLDEALGIGG    VPRGRIIEIF    GPESSGKTTL    TLQAIAEVQK    Borrelia burgdorferi
EISTIKTGAL    SLDLALGIGG    VPKGRIVEIF    GPESSGKTTL    ATHIVANAQK    Chlamydia trachomatis
QVEVIPTGSI    ALNAALGVGG    VPRGRIIEIV    GPESSGKTTL    AIHAIAEAQK    Bacteroides fragilis
DVSVIPSGSI    GLDLALGVGG    VPRGRIIEIV    GPESSGKTTL    AIHAIAEAQK    Porphyromonas gingivalis
PVEVIPTGSL    AIDIATGVGG    VPRGRIVEIF    GQESSGKTTL    ALHAIAEAQK    Thermotoga maritima
DVQVVSTGSL    SLDIALGVGG    IPGGRITEIV    GPESGGKTTL    ALAIVAQAQK    Deinococcus radiodurans
EVETIPTGSI    SLDIATGVGG    IPKGRITEIF    GVESSGKTTL    ALHVIAEAQK    Aquifex pyrophilus
```

```
EGKTCAFIDA    EHALDPIYAR    KLGVDIDNLL    CSQPDTGEQA    LEICDALARS    Escherichia coli
LGGTAAFIDA    EHALDPIYAA    KLGVNVDDLL    LSQPDTGEQA    LEIADMLVRS    Xanthomonas campestris
KGGICAFVDA    EHALDPVYAR    KLGVDLQNLL    ISQPDTGEQA    LEITDTLVRS    Rhizobium phaseoli
AGGVAAFIDA    EHALDVSYAR    KLGVRVEELL    VSQPDTGEQA    LEITEHLVRS    Myxococcus xanthus
NGGVCAFIDA    EHALDVHYAK    RLGVDTQNLL    VSQPDTGEQA    LEILETITRS    Helicobacter pylori
EGGIAAFVDA    EQALDPTYAS    ALGVDIQNLL    VSQPDTGESA    LEIVDQLV S    Anabaena variabilis
AGGQVAFVDA    EHALDPEYAK    KLGVDIDNLI    LSQPDNGEQA    LEIVDMLVRS    Steptomyces lividans
Q- RTSAFIDA    EHALDPVYAQ    KLGVNIEELL    LSQPDTGEQA    LEIAEALVRS    Bacillus subtilis
LGGAAAVIDA    EHALDPVYAK    RLGVNIDDLV    VSQPDTGEQA    LEITEALVRS    Clostridium perfringens
EGGIAAFIDA    EHALDPVYAK    ALGVNVAELW    LSQPDTGEQA    LEIAEALIRS    Borrelia burgdorferi
MGGVAAVIDA    EHALDPNYAA    LIGANINDLM    ISQPDCGEDA    LSIAEALARS    Chlamydia trachomatis
AGGIAAFIDA    EHAFDRFYAA    KLGVDVDNLF    ISQPDNGEQA    LEIAEQLIRS    Bacteroides fragilis
AGGLAAIIDA    EHAFDRTYAE    KLGVNVDNLW    ISQPDNGEQA    LEIAEQLIRS    Porphyromonas gingivalis
MGGVAAFIDA    EHALDPVYAK    NLGVDLKSLL    IAQPDHGEQA    LEIVDELVRS    Thermotoga maritima
AGGTCAFIDA    EHALDPVYAR    ALGVNADELL    VSQPDNGEQA    LEIMELLVRS    Deinococcus radiodurans
RGGVAVFIDA    EHALDPKYAK    KLGVDVDNLV    ISQPDVGEQA    LEIAESLINS    Aquifex pyrophilus
```

FIGURE 5.16. Example of a multiple sequence alignment, showing alignment of a portion of the RecA proteins from different bacterial species. Amino acids conserved across most or all of the species are highlighted in *red*. *Letters* are the abbreviations of different amino acids (see Fig. 2.23).

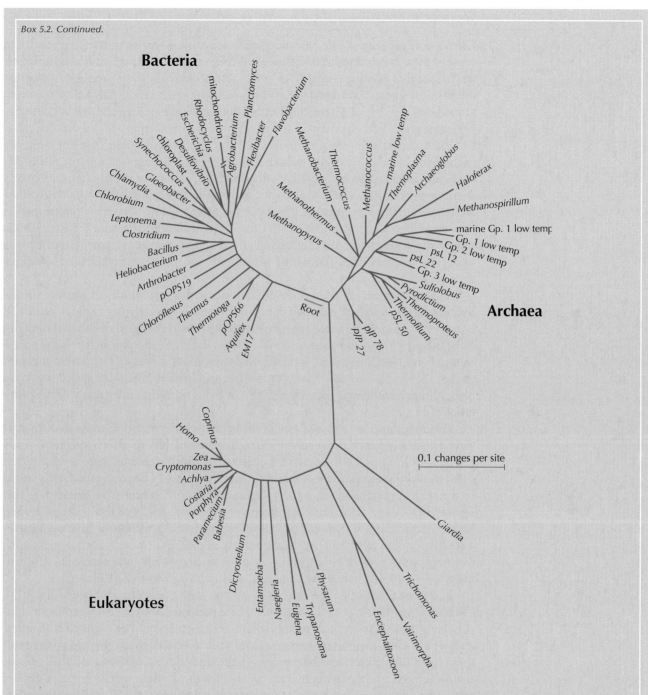

Box 5.2. Continued.

FIGURE 5.17. A phylogenetic tree with representatives of each of the three domains of life. The tree is based primarily on analysis of sequences of small-subunit rRNA molecules. It is believed that the positions of some organisms in this tree do not accurately reflect their true phylogenetic position (e.g., see Box 8.1 for a more thorough discussion of the evolution of eukaryotes and the back endpaper for a more upt-to-date tree). Nevertheless, the division of life into three domains is supported by analyses of many other characteristics.

To Infer Properties of LUCA, We Need a Rooted Tree of Life

Our view of the Tree of Life has undergone many drastic revisions over the years, and some of these are summarized in Box 5.2. Perhaps the most important realization came when scientists starting using DNA and protein sequence data from genes that are found in all species (**universal genes**) for molecular systematics. This led to the modern view of the Tree of Life with its three domains (see Fig. 5.17 and the back endpaper). One group corresponded to the eukaryotes, the organisms with nuclei. The other two groups contained only prokaryotes (i.e., species without nuclei). One of these, the bacteria, included most of the well-known prokaryotic lineages. The third, and previously unrecognized group, now known as archaea, contained mostly unusual, poorly characterized, extremophilic prokaryotic species, such as those that grow at high salinity, temperature, or pressure (see Box 5.2 for more detail). The term "prokaryote" is now viewed by many scientists as antiquated and misleading because the two distinct prokaryotic lineages (and the eukaryotes) are understood to be separate but equally important groups. Thus the words prokaryote and prokaryotic are used sparingly throughout the book.

Although the rRNA-based three-domain tree of life revolutionized our view of the total extent of biological diversity, it is difficult to use for inferring features of LUCA, because it is an unrooted tree (compare with the one in Fig. 5.8), and thus it does not show the position of the ancestral node. In this case (i.e., the Tree of Life), the ancestral node is LUCA itself (Fig. 5.11). More details about unrooted trees, what it means to root them, and how this relates to the Tree of Life are shown in Figure 5.18. What is most important here is that a *rooted* tree of life is needed to infer many of the properties of LUCA.

The importance of a rooted tree of life is best seen in regard to traits that are conserved within but not between the three domains of life, of which there are many (Table 5.4). In these cases, we need to determine which character state is ancestral in order to determine what LUCA was like. To illustrate this analysis, consider the origin of the nuclear membrane or, more specifically, the trait identified as "membrane that separates DNA and DNA functions from the rest of the cell." For this trait, eukaryotes have a character state that can be called "present" and prokaryotes have a character state that can be called "absent." The question is, did LUCA have a nuclear membrane (meaning state "present" is ancestral) or did a nuclear membrane evolve later (meaning it is derived)? To test these scenarios, we need a rooted tree of life.

In theory, the root of the three-domain tree could be in many places (Fig. 5.18). Three different rootings of the Tree of Life are presented, and for each rooting a comparison is made between how many evolutionary events would be required if the presence of a nuclear membrane were ancestral or if it were derived. For two of the rootings (Fig. 5.18D, middle and bottom rows), the simplest explanation (and thus the one that is favored using parsimony criteria) is that the nuclear membrane is a derived trait. This is the simplest explanation because it requires only a single evolutionary event—the origin of the nuclear membrane in the branch leading to eukaryotes. The alternative, that the nuclear membrane is ancestral, would require three evolutionary events: the origin of the nuclear membrane prior to the existence of LUCA and then the loss of the nucleus in the archaeal and the bacterial branches separately. Although such parallel evolutionary events are possible, unless there is other evidence relating to this feature, we would generally rely on the principle of parsimony and conclude that the simpler scenario is more likely to be correct.

If the root of the Tree of Life is on the eukaryotic branch (meaning bacteria and archaea share a common ancestor to the exclusion of eukaryotes; Fig. 5.18, top row), then loss of the nuclear membrane need occur only once for the nucleus to be ancestral (Fig. 5.18E, top row). Thus, the derived scenario is still favored but less so than with the other rootings.

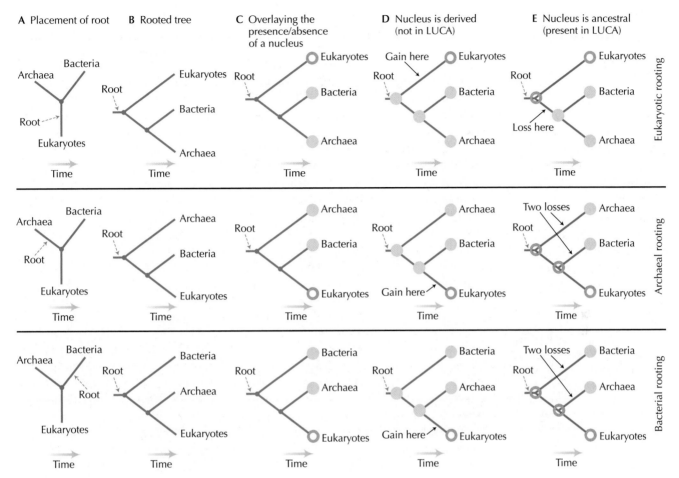

FIGURE 5.18. The effect of differentially rooting the three-domain Tree of Life on inferred properties of LUCA. (*A*) Three different rootings of the Tree of Life. (*B*) The rooted forms of these trees. (*C*) The presence (*red circle*) or absence (*blue filled circle*) of the nuclear membrane overlaid onto the rooted trees. (*D,E*) Evolutionary character state reconstruction can be used to test different hypotheses about the origin of the nuclear membrane. (*D*) A reconstruction of the evolutionary gain and/or loss of the nuclear membrane if it is assumed that the membrane is derived. (*E*) A reconstruction of the evolutionary gain and/or loss of the nuclear membrane if it is assumed that the membrane is ancestral. For each rooting, *D* and *E* can be compared to see if one is more parsimonious. When the root is in either the bacterial or archaeal lineage, the derived-membrane scenario requires fewer evolutionary events and thus is more parsimonious. When the root is in the eukaryotic lineage, the derived scenario is also more parsimonious but only slightly so because the ancestral scenario simply requires the loss of the nuclear membrane in the ancestor of the bacteria and archaea.

Initial Attempts to Root the Tree of Life Suggested That Archaea and Eukaryotes Are Sister Groups

How then do we infer the placement of the root on the Tree of Life? The root of any phylogenetic tree is usually inferred using an **outgroup**, a lineage that branched off of the tree prior to the existence of a common ancestor of all the taxa of the group being studied (Fig. 5.19). For example, in a study of primate evolution, one could use any rodent species as an outgroup (see Chapter 27 [online] for details on how an outgroup is used to root a tree). Unfortunately, there is no outgroup to the entire Tree of Life, because all known species are within the tree and the only lineages that branched off prior to the common ancestor of all living organisms have since gone extinct (Fig. 5.11).

In the 1980s, an ingenious method was worked out to root the Tree of Life. It was based on analysis of ancient gene duplications. To understand this analysis, we first

TABLE 5.4. Examples of major differentiating features among bacteria, archaea, and eukaryotes

Feature	Bacteria	Archaea	Eukarya Nucleus/Cytoplasm
Chromosome structure	Usually circular	Usually circular	Usually linear
Operons	Present	Present	Absent
mRNA introns	Absent	Absent	Present
Membrane-bound nucleus	Absent[a]	Absent	Present
Membrane lipids	Ester-linked, unbranched hydrocarbons	Ether-linked, sometimes branched hydrocarbons	Ester-linked, unbranched hydrocarbons
Initiator tRNA	Formyl-methionine	Methionine	Methionine
Plasmids	Common	Occasional	Rare
tRNA introns	Rare	Present	Present
Ribosome size based on sedimentation	70S	70S	80S
Capping and poly(A) tailing of mRNA	Absent	Absent	Present
Methanogenesis	Absent	Present	Absent
N_2 fixation	Present	Present	Absent
Reduction of S_0 to H_2S	Present	Present	Absent
Sensitivity to chloramphenicol, streptomycin, kanamycin	Yes	No	No
Chlorophyll-based photosynthesis	Present	Absent	Present (organellar)
RNA polymerase type	I	II	I, II, III

Based in part on Zillig W. 1991. *Curr. Opin. Genet. Dev.* **1:** 544–551.
[a]A nucleus-like structure has been found in one group of bacteria.
mRNA, messenger RNA; tRNA, transfer RNA.

provide some background on gene duplication. Gene duplication occurs when a copy of a particular gene is added to the genome of an individual. Duplicated genes are homologous to each other (because they share a common ancestry), and they have been given a special name, **paralogs**, because they evolve in *parallel* within that species. To distinguish them from paralogs, homologous genes that diverge as a result of speciation events are called orthologs (ortho- means "same"). Orthologs are the same form of a gene but in different species. Figure 5.20 shows evolutionary trees of orthologs and paralogs (one tree shows the evolution of the genes within the species tree and the other tree simply shows the evolution of the genes themselves).

The proteins that make up hemoglobin are a good example of orthologs and paralogs. Hemoglobins are composed of four subunits of distinct globin proteins, and different forms of hemoglobin (e.g., maternal vs. fetal) are made up of different combi-

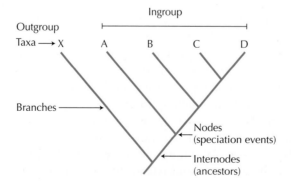

FIGURE 5.19. Rooting a tree with an outgroup. One of the ways to determine the root of a tree is to use an outgroup. An outgroup is a taxonomic unit that branched off in evolution prior to the existence of the common ancestor of all the taxa being studied in an ingroup. For example, salmon is an outgroup relative to humans, chimps, gorillas, and lemurs. This phylogenetic tree shows an outgroup and four taxa in an ingroup.

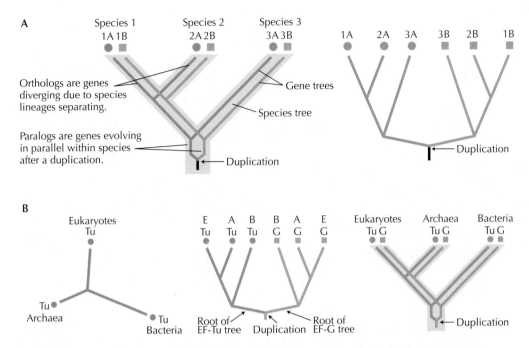

FIGURE 5.20. Orthologs, paralogs, and rooting the tree of life. (A) Evolutionary trees of species and genes representing gene duplication events. (Left) The tree includes a species tree (thick gray lines) and gene trees (blue and red lines). A gene duplication event, leading to the coexistence of the blue and red paralogs in the root of the species tree, is labeled. The species tree subsequently splits twice, producing three species, each of which has inherited the blue and red paralogs. (Right) The gene tree is extracted from the species tree and untangled. The red forms of the gene, which are orthologs of each other, are more closely related to each other than to any of the blue forms. The same is true for the blue forms of the gene. Note that the species relationships among the two groups of orthologs (red and blue) are the same. (B) The same types of trees as in A, but these correspond to the evolution of elongation factors Tu and G across the three domains of life. The red and blue branches in the rightmost tree each correspond to a Tree of Life, and each is rooted by the paralogous elongation factor.

nations of subunits. In humans, there are at least six distinct globin proteins that can be used to make different hemoglobins including, for example, α- and β-globin. The globin proteins are homologous to one another and are members of one large globin family of proteins. Evolutionary studies have shown that the genes encoding the distinct globin proteins arose in a series of gene-duplication events, most likely from a gene encoding an ancestral myoglobin-like protein. Thus the distinct globin proteins are paralogs of each other. It is useful to consider just two of the globin proteins and trace their evolutionary history. Sometime during vertebrate evolution, α- and β-globin arose as the result of a duplication event. It was at this point that the genes started to diverge from each other. Any species that descended from the organisms with the distinct α- and β-globin then inherited both α- and β-globin. In any species that inherited both genes, its α-globin is an ortholog of other α-globins and its β-globin is an ortholog of other β-globins. Any and all α-globins are paralogs of any and all β-globins.

The finding that some genes were duplicated prior to the existence of LUCA allows for a special method of rooting the Tree of Life. One example of such anciently duplicated genes involves proteins used for the elongation step of translation. Elongation factor Tu (EF-Tu) and elongation factor G (EF-G) are both involved in protein translation and are related to each other by an ancient duplication event; thus they are paralogs (Fig. 5.20). All species contain EF-Tu and EF-G genes. Thus an EF-Tu tree can be constructed that is rooted using EF-G as an outgroup. In addition, an EF-G tree can be made and rooted using EF-Tu as an outgroup. These two constructions

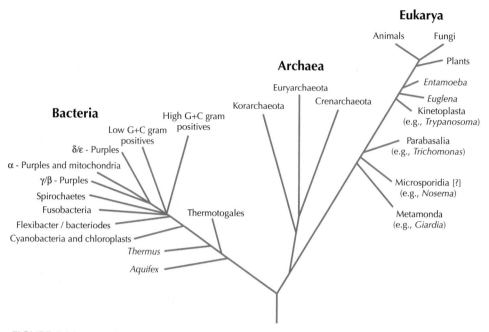

FIGURE 5.21. Rooted tree of life. This tree was constructed by rooting the rRNA-based tree (e.g., see Fig. 5.17) based on the results of studies of anciently duplicated genes. Note that many of the positions of particular organisms within each domain in this tree are now thought to be inaccurate (see Chapters 6–8).

yielded the same tree rooted in the same location. In the first studies using analyses of such anciently duplicated genes, the root of the Tree of Life was found to lie between the bacteria on the one hand and the archaea and eukaryotes on the other (Fig. 5.20).

One of the implications of the rooted tree in Figure 5.20 is that archaea and eukaryotes share a common ancestor to the exclusion of bacteria. Thus not only are the archaea not the same as bacteria, but they are in fact more closely related to eukaryotes. In other words, although archaea and bacteria share phenotypic similarities (e.g., absence of a nucleus), these two groups do not have the closest evolutionary relationship among the three domains (recall that similarity is not a direct measure of relatedness [Fig. 5.9]). Thus, prokaryotes are a polyphyletic group. This rooted tree was well received by many people studying the molecular biology of archaea, who were discovering many similarities of core molecular features between archaea and eukaryotes. An important use of the rooted tree in Figure 5.20 (which is of elongation factors) is to root other trees, such as the rRNA tree of life (Fig. 5.21). This is important because the rRNA tree is much more filled in as a result of the availability of hundreds of thousands of rRNA sequences.

Perhaps most importantly for studies of LUCA, rooting the Tree of Life allows us to better infer which characteristics that differ across taxa are ancestral and which are derived. For example, considering again the evolutionary history of the nuclear membrane (recall Fig. 5.18), rooting the Tree of Life as shown in Figure 5.21 supports the conclusion that the nuclear membrane is most likely derived, evolving in eukaryotes some time after LUCA because alternative scenarios require more "gain" and "loss" events. In addition, having a rooted tree makes it possible to infer what *states* are ancestral for universal traits. For example, it is possible to infer that the use of the standard genetic code is an ancestral state and alternatives found in species like mycoplasma and in mitochondria are derived. This conclusion is reached because other scenarios (e.g., the mycoplasma code being ancestral) would require the occurrence of many more evolutionary changes. The evolution of the genetic code is discussed in more detail in Chapter 20 (p. 561).

LUCA and Modern Cells Share Features of Protein Translation

Using character state reconstruction methods and the analysis of universal homologies, we are beginning to get a good idea of what LUCA was like. In many ways LUCA resembled a modern single-celled organism surrounded by a membrane and possibly a cell wall. The basic molecular machinery of life was present: a DNA-based genome, transcription of RNA, and translation of proteins from the RNA by a ribosome made up of proteins and noncoding RNAs. DNA was replicated with a DNA polymerase, and a multitude of DNA repair enzymes were present to remove DNA damage and limit the occurrence of mutations. Other possible features of LUCA are summarized in Table 5.5.

Translation in LUCA was probably much as it is in modern organisms. Many of the structures and genes used today were present including rRNAs, tRNAs, elongation factors, ribosomal proteins, and aminoacyl-tRNA synthetases. Although some components of modern protein translation were absent in LUCA (e.g., some additional ribosomal proteins), overall the process has remained largely unchanged.

Although protein translation in modern organisms is very similar to translation in LUCA, mechanisms of DNA replication and transcription appear to have undergone many changes since the time of LUCA. It is not clear whether LUCA possessed well-developed systems for replication and transcription that were modified later in evolution or whether LUCA utilized very simple forms of these processes that were separately elaborated by the different domains. For example, it is possible to infer that transcription in LUCA was carried out by an enzyme made up of homologs of RpoA, RpoB, and RpoC, which are used in all modern RNA polymerases. However, most of the molecular details of initiation, regulation, elongation, and termination appear to have evolved separately in the three domains.

TABLE 5.5. Inferred molecular processes and gene families present in LUCA

Feature	Details	Associated Gene Families
Protein translation	Ribosomes made up of RNA and protein Charged tRNAs used	16S/18S rRNA 23S/28S rRNA 5S rRNA Many ribosomal proteins Elongation factor Tu Elongation factor G Multiple aminoacyl-tRNA synthetases
Transcription	At least a three-subunit DNA-dependent RNA polymerase	RpoA RpoB RpoC
DNA replication	DNA was the genetic material.	Pol1-type exonuclease DnaN (sliding clamp)
DNA recombination		RecA
Protein insertion in membranes		SecY FtsY

Data based in part on Harris J.K. et al. 2003. *Genome Res.* **13:** 407–412.
rRNA, ribosomal RNA; tRNA, transfer RNA.

Why should translation in LUCA have been relatively complex compared to DNA replication and transcription? There appears to be a relatively simple explanation. Because the invention of translation was required for transition from the RNA world to the protein world (see p. 104), it appears that this occurred before DNA replaced RNA as the genetic material. Thus at the point when the lineages of the three domains began to diverge from one another, translation had had much more time to evolve than replication or transcription. Thus it is possible that LUCA was a rudimentary cell in many ways. Such a possibility was proposed originally by Woese in a theory he titled the **progenote hypothesis**.

The Bacterial Rooting of the Tree of Life Has Been Challenged by Some Analyses

Evolutionary studies generally become more difficult as one tries to infer events that occurred further and further into the past. This is certainly true for attempts to root the Tree of Life, because the branches involved occurred a very very long time ago. Although most of the published studies support the bacterial rooting of the Tree of Life, there are a reasonable number that have challenged this rooting. We believe it is important to discuss the arguments here. A major criticism is that a methodological artifact of many phylogenetic studies, known as **long-branch attraction**, may be biasing the results. In this phenomenon, species at the ends of long evolutionary branches will tend to group together in a tree, regardless of whether they are actually most-closely related (Fig. 5.22). The reasons that long branches tend to group together are discussed in more detail in Chapter 27 (online). What is important here is that long-branch attraction may have caused the bacteria to be artificially pulled toward the outgroups in

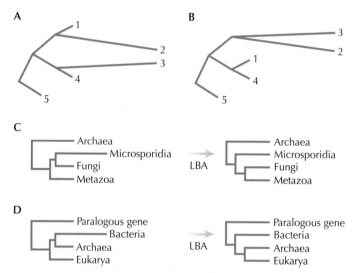

FIGURE 5.22. Long-branch attraction is a methodological artifact that can cause phylogenetic trees to inaccurately portray evolutionary history. The phenomenon causes errors in phylogenetic reconstruction when two (or more) of the entities being studied lie on the end of long branches in their "real" tree but are not sister taxa. (A) In this hypothetical "real" tree of five species, species 2 and 3 (which are not sister taxa, as indicated) have undergone higher rates of evolution than the other three, and thus sit at the end of longer branches. Many phylogenetic reconstruction methods used to infer the evolution of species will cause the long branches to appear to be closely related and thus produce an incorrect tree (as shown in B). (C) In studies of the evolution of microsporidia (a relative of fungi, *left tree*), long-branch attraction (LBA) is believed to have erroneously identified them as deeply branching eukaryotes (*right tree*). (The evolution of microsporidia is discussed in more detail on p. 198.) (D) In trees of anciently duplicated genes, long-branch attraction might have pulled bacteria down to the paralogs used to root the tree, because the paralogs are at the end of a long branch (*right tree*). This would occur if bacteria evolved at a higher rate than archaea and eukaryotes (as suggested in the *left tree*).

the tree, thus making them appear to be the deepest branch (Fig. 5.22D). When attempts are made to eliminate or reduce the influence of this artifact, the bacterial rooting is no longer as strongly supported.

There is another complication to rooting the Tree of Life that is far more important than long-branch attraction: The assumption of vertical descent may not apply.

The Occurrence of Gene Transfer between Species Means That Different Genes May Have Their Own LUCAs

A key assumption when attempting to infer properties of LUCA from the properties of modern organisms is that species evolve by vertical descent. However, vertical descent is not the only mode of inheritance. Closely related species can hybridize with each other, sometimes producing fertile offspring. Occasionally, distantly related species will hybridize too. Perhaps most dramatically, DNA can be passed from one evolutionary lineage to another, by a process known as lateral gene transfer (Fig. 5.23A), which is described in more detail on pages 182–191. Lateral transfer is a form of sex, and the significance of sex and gene transfer is discussed more in **Chapter 23**. For our purposes, what is important is that lateral gene transfer creates chimeric organisms—organisms in which different parts of the genome have different histories. For example, in the his-

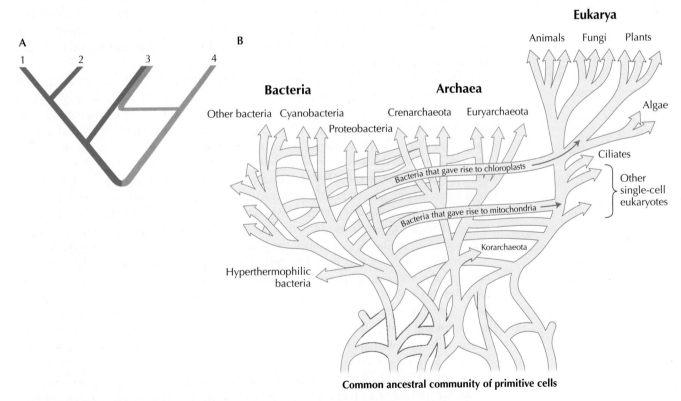

FIGURE 5.23. Lateral gene transfer and the tree of life. (*A*) Gene transfer diagrammed on a hypothetical tree of four species. The *horizontal red line* shows the transfer of genes from the lineage of species 4 to that of species 3. The transferred genes then merge with the endogenous genes in the lineage of species 3, creating a chimera, where some genes have the "blue" ancestry and others have the "red" ancestry. (*B*) Gene transfer in the Tree of Life. The figure shows a reticulated network following the same general scheme as the rooted Tree of Life in Fig 5.21 with the bacteria as the deepest branching domain. However, many branch-crossing events (e.g., gene transfer from organelles to the nuclear genomes of eukaryotes) create a network, not a tree. Note that there is no single LUCA shown, i.e., no single common ancestor of all modern organisms, because gene transfer prevents the species tree from tracing back to a single organism. However, individual gene trees may still have a single LUCA.

tory of eukaryotes, it is clear that mitochondria and chloroplasts were once endosymbiotic bacteria (see p. 163). These organelles retain relics of their evolutionary past in the form of cellular structure, function, and DNA. Many of the genes that once were encoded in the organelle genomes have since been transferred to the nuclear genomes of their hosts (pp. 202–207). Many other examples of lateral gene transfer are known, such as the movement (especially in hospitals) of antibiotic resistance genes between bacterial species. It therefore follows that there cannot be a single "Tree of Life." That is, a single tree cannot accurately represent the evolution of life. It may be better to represent species evolution as a reticulated network (e.g., Fig. 5.23B) with interconnecting branches.

This possibility of lateral transfer is a component of the progenote hypothesis mentioned above. If the rudimentary cells in the distant past freely exchanged genetic information with each other, then the three domains of life may not have originated by a tree-like evolutionary pattern. Instead, they may have separately sampled from the pool of genetic information available at the time. Rather than tracing their origin back to a single LUCA, the base of the Tree of Life may have multiple nodes (see Fig. 5.23).

Lateral gene transfer also confounds our ability to automatically use universal homologies or character state reconstruction methods to infer the properties of LUCA. For example, suppose that the use of rRNA evolved in the bacterial branch of the Tree of Life and then was transferred laterally to the archaeal and eukaryotic branches, spreading eventually to all species. If such an event were to occur, then the presence of ribosomal RNA in all species could give the mistaken impression that it was present in LUCA. Theoretically, any feature could have a chimeric history. Therefore, to infer the properties of LUCA, it is necessary to determine if lateral transfer has occurred for those features.

The degree of difficulty in reconstructing the features of LUCA and studying early evolutionary history depends in part on how much lateral transfer has occurred and on the details of the genes that have been transferred. Genome sequence analyses are beginning to answer these questions. The general picture that is emerging is that organisms, especially microbes, appear to readily take up DNA from all sorts of foreign sources. However, this DNA rarely spreads into the populations of the recipient organism (probably because the DNA produces few advantageous mutations). Furthermore, even when foreign genes are able to spread into a recipient population, they rarely stay in an organism's lineage for extended periods of evolution, probably because newly acquired DNA is frequently used to adapt to specialized environmental conditions (e.g., presence of an antibiotic), and once the environment changes the DNA disappears.

This is not to say that lateral transfer does not occur—only that it is somewhat rare and requires strong selective advantages for foreign DNA to be maintained. One result of this is that there tend to be "core" portions of genomes that do not undergo much lateral gene transfer (e.g., housekeeping genes, like ribosomal proteins) and there are peripheral portions of genomes that undergo a greater amount of transfer. Thus, to study the evolutionary history of a species, phylogenetic studies of "core" genes can be used to obtain a picture of the "average" history of a species. An example of this is provided in Figure 5.24, right, which shows a phylogenetic tree of genomes based on combining the alignments of a set of core genes from across the bacterial species shown into one mega-alignment. This "core" tree can then be used to ask questions regarding ancestral traits and the occurrence of gene transfer, loss, and duplication events. With this approach, researchers are thus able to study the Tree of Life, even though it is, in part, a network, not a tree. In addition, using these genome-level approaches, researchers have found that the "three-domain" view of the Tree of Life appears valid. As more genomes become available, it should be possible to more accurately test questions concerning the relationships among these domains and the early events in the history of life.

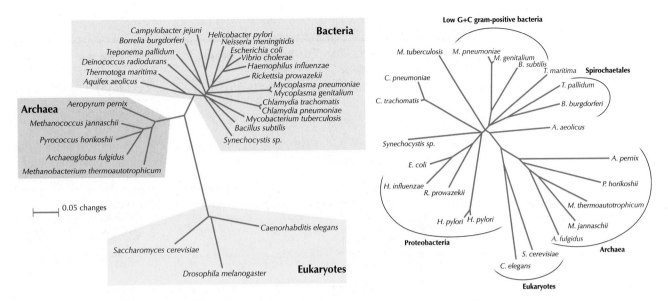

FIGURE 5.24. Once many genomes became available, it was possible to compare phylogenetic trees of rRNA (*left*) with those based on comparing whole genomes (*right*). The general topology of the two trees is very similar, indicating that whatever lateral transfers have occurred did not swamp out the core phylogenetic signal for these species.

Where Do Viruses Sit on the Tree of Life?

Viruses are acellular obligate parasites, and, therefore, are often not included with cellular organisms when discussing the Tree of Life. However, given that viruses have their own genomes, this is probably a mistake. Mitochondria and chloroplasts, which once were free-living bacteria but are now obligate host-dependent organelles, are active topics in studies of the Tree of Life. Cellular obligate parasites, such as *Plasmodium*, the causative agent of malaria, are also granted full status in Tree of Life studies. In addition, although some viruses depend on their host for nearly everything, others are quite complex and encode many functions previously found only in cellular organisms. For example, the recently discovered mimivirus has a 1.2-Mb genome that encodes more than 900 proteins.

Viruses evolve, more or less, as independent entities, allowing them to be treated in phylogenetic studies much like cellular organisms. For example, studies of the evolution of HIV (human immunodeficiency virus) or influenza-causing viruses are based on phylogenetic analysis of genes encoded by these viral genomes. However, this is only true for the recent evolution of viruses; further back in time, viral evolution gets murkier. In this section, two questions concerning early viral evolution are addressed. First, how are viruses related to each other? Second, how did viruses originate? In other words, can we build a tree of viruses and can we link that tree to the Tree of Life?

It is useful to first introduce the diversity of viruses. Viruses vary in shape and size, in the types of interactions they have with hosts, in the content of their genomes, in their life cycles, and in their patterns of evolution. Viruses are commonly classified on the basis of their general mechanisms of replication and the structure of their genomes. Using these criteria, six classes have been identified (Table 5.6). Further classification of viruses into families has been done on the basis of a variety of viral features. More than 60 major families are known today.

The study of *recent* viral evolution is relatively straightforward because, within any one family, different viruses are moderately closely related and can thus be compared using standard techniques, such as alignment-based phylogenies (see Chapter 27 [online]). There are, however, three major challenges when studying *older* events in viral

TABLE 5.6. Types of viruses

Virus Class	Described Families	Examples of Viruses
Double-stranded DNA	20	SV40, phage T4, herpesviruses, smallpox
Single-stranded DNA	5	Phage φX174, phage M13
DNA–RNA reverse transcribing viruses	5	HIV, hepatitis B
Double-stranded RNA viruses	6	Rice dwarf virus
(–) sense single-stranded RNA viruses	7	Ebola, mumps, rabies, influenza
(+) sense single-stranded RNA viruses	19	Polio, SARS, dengue, hepatitis C

Based on Table 8.1 in Mindell D.P., Rest J.S., and Vallareal L.P. 2004. Viruses and the tree of life, pp. 107–118. In *Assembling the Tree of Life* (ed. J. Cracraft and M.J. Donoghue), Oxford University Press, New York. For more information, see the Web site of the International Committee on the Taxonomy of Viruses (ICTV) at http://phene.cpmc.columbia.edu/index.htm.
HIV, human immunodeficiency virus; SARS, severe acute respiratory syndrome.

evolution. First, viruses generally evolve more rapidly than cellular organisms. This makes it difficult to identify homologous genes between viral types even when they are present. Second, viruses appear to be particularly prone to lateral gene transfer, which means that the different genes within one viral genome frequently have different histories. Third and most importantly, there are no genes shared by all viruses that would allow them to be easily placed into a single tree. Most of the "universal" genes used to study the early events in the evolution of cellular organisms either are absent from viruses entirely or are present in only a few viral families.

Despite these difficulties, the recent availability of complete genome sequences of many viruses is greatly improving our understanding of viral evolution by enabling new types of phylogenetic analysis not possible previously. One example of this is the "phage proteomic tree," shown in Figure 5.25. This tree is based on analysis of the overall similarity of the complete genomes of many phage (which are viruses that infect bacteria and archaea). This tree is quite revealing and informative. For example, it shows that in a single tree, different classes of viruses (single-stranded DNA [ssDNA], single-stranded RNA [ssRNA], and double-stranded DNA [dsDNA] phage) can be linked together, even though there is no single gene shared by all of these viruses. The phage proteomic tree has also revealed that some previous classifications of viruses are inconsistent with their evolutionary history; thus, these classifications probably need to be changed (much as rRNA-based trees revolutionized the classification of microorganisms). Genome-level analysis has also helped reveal that many viral lineages undergo recombination and lateral gene transfer at a very high rate, suggesting that, perhaps more than in most cellular organisms, viral evolution should be considered more as a network rather than as a single tree.

Whole-genome analyses and phylogenies like the phage proteomic tree are greatly improving our ability to classify viruses but do little to help us answer the question: "Where did viruses come from in the first place?" In the remainder of this section, three theories concerning the ultimate origins of viruses are presented. Given the diversity of viruses, keep in mind that different viruses may have separate origins.

The first possibility is that viruses are relics from the precellular world. Viruses may have existed early in the origin of life and have hung around ever since. This possibility is supported by the presence of some processes and genes in many viruses that are thought to be ancient. For example, many viruses encode their own forms of ribonucleotide reductase, the enzyme that converts ribonucleotides into deoxyribonucleotides. Other common and ancient enzymes found in many viruses include DNA repair enzymes, reverse transcriptases, and DNA polymerases. In addition, many of these viral genes appear to be *distantly* related to their cellular counterparts, suggesting that they were not recently ac-

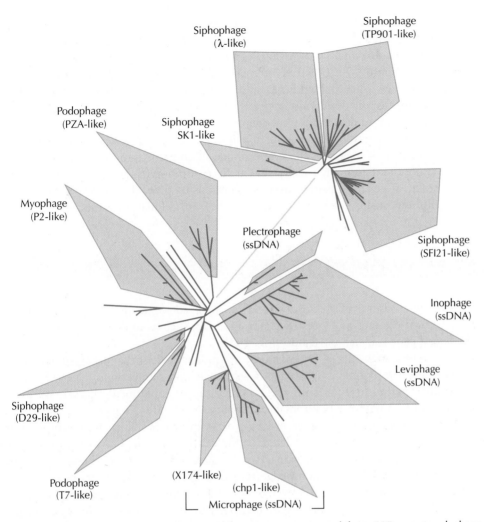

Siphophage
(λ-like)

Siphophage
(TP901-like)

Podophage
(PZA-like)

Siphophage
SK1-like

Myophage
(P2-like)

Plectrophage
(ssDNA)

Siphophage
(SFI21-like)

Inophage
(ssDNA)

Leviphage
(ssDNA)

Siphophage
(D29-like)

(X174-like)

(chp1-like)

Podophage
(T7-like)

Microphage (ssDNA)

FIGURE 5.25. The phage proteomic tree. This tree was constructed from 105 sequenced phage genomes. To make the figure easier to read, the siphophage group has been manually moved away from the other groups. The *beige* line indicates where it connects to the other groups. The *green shapes* highlight different groups.

quired from host organisms. A difficulty with this theory is the current dependence of viruses on cellular organisms; how could viruses have existed before cellular organisms originated? In the precellular world, one of the keys to evolution was the ability to replicate. Many viruses encode their own replication machinery; thus, it is plausible that they existed as RNA replication machines, where RNA performed catalytic functions (see p. 101). As cellular organisms evolved, some of the primitive RNAs could have become parasites of these cellular organisms and are today what we call viruses.

A second theory proposes that viruses are "escaped" portions of cellular organisms. Consistent with this, many viruses resemble **transposable elements**. Known as the "escaped transcript" model, this theory postulates that viruses were once normal components of a cell, similar in many ways to models concerning the origin of transposable elements, which also may be derived from normal cellular components. According to this theory, to become a virus, certain cellular components would have had to acquire a protein coat and some means of regulating their own replication. This may be easiest to imagine for a retrovirus, which could have started as a cellular RNA and later linked up with a reverse transcriptase. Alternatively, a virus could have started as a mobile DNA element, like a **transposon** or a **plasmid**. To convert a mobile plasmid to a virus requires little more than the addition of a protein coat. The evolution of transposable elements is discussed in Chapter 21 (pp. 595–599).

The final possibility is that viruses may be remnants of once free-living organisms. Until recently this idea seemed far-fetched. However, two lines of evidence suggest otherwise. First, genome analysis of organelles and intracellular organisms has shown that genome streamlining can be taken to great extremes. Some mitochondrial genomes encode only a few of their own genes, and, in terms of genome content, some intracellular pathogens and symbionts appear to be on the verge of becoming organelles. In addition, as more and more viral genomes are sequenced, viruses with very large genomes and an incredible diversity of activities are being found. The previously mentioned mimivirus is currently the best example. It contains more genes than any organelle and more than many bacteria and archaea. Although large viral genomes could be the result of gene acquisition over long periods of time, their existence is consistent with the possibility that viruses are derived from cellular ancestors.

SUMMARY

In this chapter, we have discussed the early evolution of life. We have laid out the concept of vertical descent, as well as the complications created by lateral gene transfer. The existence of universal homologies, coupled with the idea of vertical evolution, has been used to show that all life should have a single common ancestor, referred to as LUCA. We have discussed the importance of rooting evolutionary trees and of rooting the Tree of Life in particular. Ideas about the Tree of Life have changed over time, and in the current view, the Tree of Life is divided into three domains—bacteria, archaea, and eukaryotes. There is some agreement among scientists that archaea and eukaryotes are more closely related to each other than either is to bacteria. In addition, we have seen how evolutionary trees can be used to aid in the inference of ancestral character states, such as to determine the possible features of LUCA.

Inferring evolutionary events that occurred billions of years ago is very challenging and cannot always be done with perfect precision. In subsequent chapters, as more recent evolutionary events are discussed, our ability to infer events greatly improves. Thus, for example, our knowledge of the relationships within each of the three domains (see Chapters 6–8) is much better than our knowledge regarding the very early events of evolution. However, as more and more genome sequence data are gathered, our ability to infer early events in the evolution of life will also improve.

FURTHER READING

Rooting the Tree of Life

Brown J.R. and Doolittle W.F. 1997. Archaea and the prokaryote-to-eukaryote transition. *Mol. Microbiol. Biol. Rev.* **61:** 456–502.

Doolittle W.F. 2000. Uprooting the tree of life. *Sci. Am.* **282 (2):** 90–95.

Forterre P. and Philippe H. 1999. Where is the root of the universal tree of life? *BioEssays* **21:** 871–879.

Philippe H. and Forterre P. 1999. The rooting of the universal tree of life is not reliable. *J. Mol. Evol.* **49:** 509–523.

Universal Homologies and LUCA Features

Doolittle W.F. 2000. The nature of the universal ancestor and the evolution of the proteome. *Curr. Opin. Struct. Biol.* **10:** 355–358.

Kyrpides N., Overbeek R., and Ouzounis C. 1999. Universal protein families and the functional content of the last universal common ancestor. *J. Mol. Evol.* **49:** 413–423.

Penny D. and Poole A. 1999. The nature of the last universal common ancestor. *Curr. Opin. Genet. Dev.* **9:** 672–677.

Woese C.R. 1998. The universal ancestor. *Proc. Natl. Acad. Sci.* **95:** 6854–6859.

Viruses and the Tree of Life

Mindell D.P., Rest J.S., and Villareal L.P. 2004. Viruses and the Tree of Life. In *Assembling the Tree of Life* (ed. J. Cracraft and M.J. Donoghue), pp. 107–118. Oxford University Press, New York.

Rohwer F. and Edwards R. 2002. The Phage Proteomic Tree: A genome-based taxonomy for phage. *J. Bacteriol.* **184:** 4529–4535.

History of Trees of Life and the Three-Domain Tree

Pace N.R. 1997. A molecular view of microbial diversity and the biosphere. *Science* **776:** 734–740.

Woese C.R. and Fox G.E. 1977. Phylogenetic structure of the prokaryotic domain: The primary kingdoms. *Proc. Natl. Acad. Sci.* **74:** 5088–5090.

Woese C.R., Kandler O., and Wheelis M.L. 1990. Towards a natural system of organisms: Proposal for the domains Archaea, Bacteria, and Eukarya. *Proc. Natl. Acad. Sci.* **87:** 4576–4579.

6

Diversification of Bacteria and Archaea
I: Phylogeny and Biology

A S WE DISCUSSED IN CHAPTER 5, organisms without nuclei comprise two distinct evolutionary lineages: the bacteria and the archaea, each with their own unique features. Despite their evolutionary distinctness, the archaea and bacteria share many biological features (Table 6.1). Indeed, it is these similarities that led us to group them together as "prokaryotes" for so long. Most importantly, there are some aspects of their biology, such as haploidy, binary fission, significant **lateral gene transfer**, and coupling of transcription to translation, that cause them to have similar patterns and mechanisms of diversification and evolution relative to eukaryotes. In this chapter, we focus on these two domains of life together, highlighting both their similarities and differences. First, we introduce the bacteria and archaea and outline their phylogenetic groups. Next, we present a

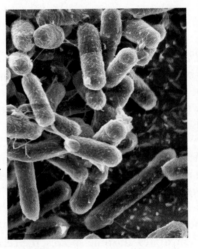
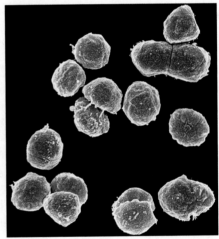

few examples of their biological evolution: their ability to thrive in extreme conditions, the array of mechanisms they possess for acquiring energy and carbon from the environment, and their interactions (both beneficial and detrimental) with other organisms. In Chapter 7, we discuss the genomic basis for this diversification.

INTRODUCTION TO THE BACTERIA AND ARCHAEA

Bacteria Share Many Key Features

The phylogenetic and biological diversity of the bacterial domain is immense. Within this group are model organisms used for genetic studies (e.g., *Escherichia coli* and *Bacillus subtilis*), the causative agents of many diseases (e.g., *Vibrio cholerae* [cholera], *Mycobacterium tuberculosis* [tuberculosis], *Bacillus anthracis* [anthrax], and *Yersinia pestis* [the plague]), organisms of agricultural or industrial importance (e.g., *Lacto-*

TABLE 6.1. Bacteria and archaea share many features to the exclusion of eukaryotes

Feature	Archaea and Bacteria	Eukaryotes
Nuclear membrane	Absent	Present
Genomes	Usually one circular chromosome; frequently one or more small DNA circles (plasmids)	Multiple linear chromosomes
Mitosis	Absent	Present
Introns	Rare	Common
Cytoskeleton composed of microtubules	Absent	Present
Internal membranes	Simple and rare	Complex and common
Primary mode of genetic recombination	Lateral gene transfer	Sexual reproduction
Size	Small, 1–5 μm	Larger, 20–100 μm
Organelles	No	Yes
Ploidy	Haploid	Diploid or more

Based on Madigan M.T., Martinko J., and Parker J. 1997. *Brock biology of microorganisms*, 8th ed. Prentice Hall, New York, Table 3.5.
What we list here are features of most of the members of these groups.

coccus lactis used in cheese production and *Thermus aquaticus*, from which the first heat-stable DNA polymerase for the polymerase chain reaction [PCR] was purified), some of the major players in global nutrient cycles (such as the nitrogen-fixing bacterial symbionts of legumes), and even the mitochondria and chloroplasts of eukaryotes (Fig. 6.1).

Despite the enormous diversity among bacteria, there are a variety of features characteristic of most species. A typical bacterium is a single-celled organism with cells that are much smaller than those of most eukaryotes (usually 0.8–2 μm along the narrowest axis). The cells are frequently rod-like in shape (known as bacilli) or spherical (known as cocci), but many other shapes are found. Whatever the shape, the cell is surrounded by a phospholipid membrane encased in a cell wall composed of protein and carbohydrates. In some species, the cell wall is thick and tough; in others it is thin and covered by a porous outer membrane. This difference in cell wall composition is the basis for the differential staining mechanism of the **Gram stain** (Fig. 6.2).

Attached to the outer wall or membrane of many bacteria is a whip-like appendage known as a flagellum (Fig. 6.3). Flagella are used to move cells through liquid environments. This movement is generated through rotation of the flagella, making use of a molecular motor embedded in its base. In most species of bacteria that have flagella, the movement is linked to systems that allow the bacteria to respond to environmental cues such as chemical gradients (this is known as chemotaxis) or light (known as phototaxis).

In a typical bacterium the majority of the genome is maintained as a large circular chromosome. Sometimes the chromosome is condensed into a coiled bundle called a nucleoid, although there is no separate compartment for the DNA like the nucleus of eukaryotes. Additional DNA is frequently found in the form of smaller circles called **plasmids** (Fig. 6.4) (see p. 169). Most species of bacteria grow and divide by binary fission and live life as haploid organisms (Fig. 6.5). Unlike eukaryotes, transcription of

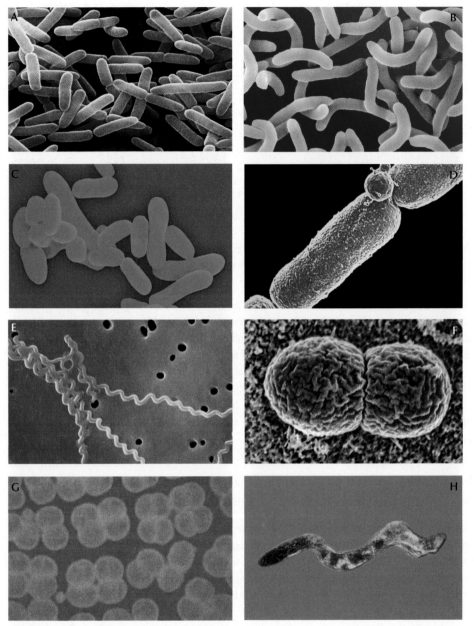

FIGURE 6.1. Examples of bacteria. (*A*) *Bacillus subtilis*; (*B*) *Vibrio cholerae*; (*C*) *Mycobacterium tuberculosis*; (*D*) *Bacillus anthracis*; (*E*) *Leptospira interrogans*; (*F*) *Lactococcus lactis*; (*G*) *Deinococcus radiodurans*; and (*H*) *Borrelia burgdorferi*.

Gram–positive cell wall

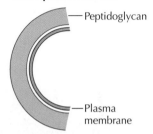

— Peptidoglycan

— Plasma membrane

Cell wall 20–80-nm thick
One–layer cell wall
>50% peptidoglycan content
Teichoic acids are present
0–3% lipid and lipoprotein content
0% protein content
0% lipopolysaccharide content

Gram–negative cell wall

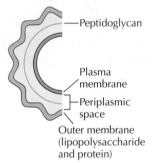

— Peptidoglycan

— Plasma membrane

— Periplasmic space

— Outer membrane (lipopolysaccharide and protein)

Cell wall 10-nm thick
Two–layer cell wall
10–20% peptidoglycan content
Teichoic acids are absent
58% lipid and lipoprotein content
9% protein content
13% lipopolysaccharide content

FIGURE 6.2. (*Top*) Gram stain of mixed bacteria. The *dark purple* indicates gram-positive bacteria and the *cerise* indicates gram-negative bacteria. (*Bottom*) Schematic diagram and characteristics of the cell wall and membrane of gram-positive and gram-negative cells.

protein-coding genes is coupled to translation. As a gene is being copied into RNA, the protein synthesis machinery attaches to the RNA and begins making protein. In many cases, multiple genes arrayed in tandem are transcribed into a single RNA. Such an arrangement is called an **operon** (see pp. 54 and 172–174).

As with almost everything in biology, there are interesting exceptions to these generalities. Although most motile bacteria use a flagellum, other mechanisms of bacterial movement include gliding, twitching, and inflation/deflation of gas vesicles. Rather than being single celled, the cells of some species remain together in clusters after division. Others, such as *Chondromyces crocatus* (Fig. 6.6A), which produce a fruiting body, are truly multicellular with differentiated cells and divi-

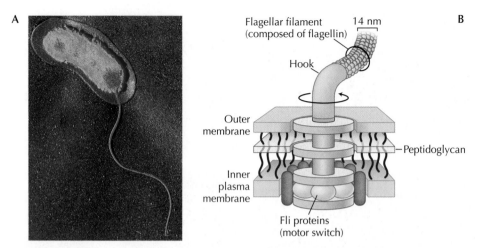

FIGURE 6.3. (*A*) The γ-proteobacterium *Proteus mirabilis* with attached flagellum. (*B*) Schematic diagram of a bacterial flagellum (from a gram-negative bacterium).

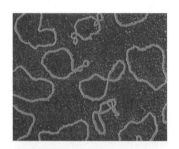

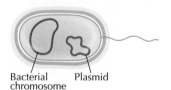

Bacterial   Plasmid
chromosome

FIGURE 6.4. (*Top*) Bacterial plasmids. (*Middle*) Lysed bacterial DNA. (*Bottom*) Cartoon of a bacterial cell with a circular chromosome and a circular plasmid. (The DNA is not drawn to scale.)

sion of labor. There are some bacteria, such as *Gemmata obscuroglobus,* that have a nucleus-like structure (Fig. 6.6B). There are also bacteria with very large cells. *Epulopiscium fischelsoni,* a symbiotic species that lives inside the gut of surgeonfish, is greater than 500 μm in length. When *E. fischelsoni* was first identified, it was misclassified as a protist because of its large size. *Thiomargarita namibiensis,* a sulfur-metabolizing bacteria found on the ocean floor off the coast of Namibia, is even larger. *T. namibiensis* can grow up to a millimeter in length, with much of the space being filled with large numbers of storage vesicles holding sulfate and nitrates for consumption during lean times (Fig. 6.6C).

The Archaea, although Unique in Many Ways, Share Some Features with Bacteria and Others with Eukaryotes

The first species recognized as archaea were found only in so-called "extreme" environments—for example, places with very high salt concentrations or very high temperatures. The "archae" part of their name, which means "ancient," was chosen to express the idea that these species might be relics from the ancient Earth, when such harsh conditions may have been the rule. Although many additional archaeal species have been found living in extreme environments, the choice of the term "archae" is somewhat misleading, because archaea have also been found in more mundane habitats, including ordinary soil, the open ocean, and animal digestive tracts. Examples of

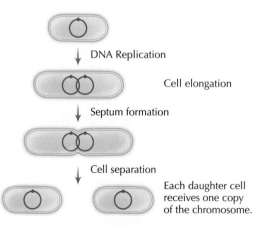

DNA Replication

Cell elongation

Septum formation

Cell separation

Each daughter cell receives one copy of the chromosome.

FIGURE 6.5. Binary fission in bacteria.

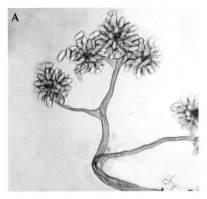
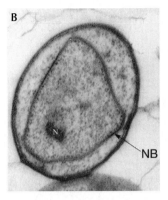
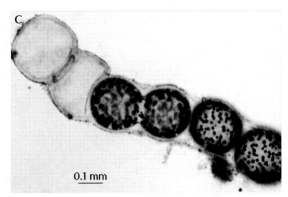

FIGURE 6.6. Diverse bacterial forms. (*A*) *Chondromyces crocatus*. (*B*) *Gemmata obscuroglobus* contains a nucleus-like structure. In this cross section of a *Gemmata*-like cell, the double-membrane-bounded nuclear body is labeled as NB and the nucleoid as N. (*C*) *Thiomargarita namibiensis* giant cells.

archaea include *Pyrobaculum aerophilum*, an aerobic species that grows optimally at 100°C, *Haloferax volcanii*, which grows in water nearly saturated with NaCl, and *Thermoplasma acidophilum*, which lives at a pH of approximately 0.

Phenotypically, the average archaeon resembles the typical bacterium (Fig. 6.7). Archaea are usually single celled and small, with sizes ranging from 0.3–1 X 0.3–6 μm. Most archaeal cells are surrounded by a membrane and a cell wall. As with bacteria, most archaea divide by binary fission and are haploid organisms. Archaea frequently use flagella for motility. The structure and organization of archaeal genomes outwardly resemble those of bacteria. Most species have a single circular chromosome that is copied bidirectionally from a single replication origin. Many species also contain extrachromosomal genetic elements, such as plasmids. Genes are frequently organized in operons and are transcribed by a single RNA polymerase. Translation is coupled to transcription as in bacteria.

Despite these similarities to bacteria, the basic components of archaeal replication, transcription, and translation more closely resemble eukaryotic processes at the molecular level. For example, in some archaeal species, the DNA is packaged into a coiled, condensed structure that resembles the chromatin of eukaryotes more than the nucleoids of bacteria. This packaging is achieved using homologs of the eukaryotic histone proteins (Fig. 6.8). In addition, although archaeal genes can be organized into operons as in bacteria, the mechanism of transcription is generally more similar to that of eukaryotes than to that of bacteria. Archaea have TATA-box-based promoters that are controlled by homologs of the eukaryotic TATA-binding protein. The single archaeal RNA polymerase is most closely related to the eukaryotic RNA polymerase II.

Although archaea share some features with bacteria and others with eukaryotes, they also possess many unique features. For example, in the archaeal cell membrane, the bonds between the glycerol and hydrocarbon chains are ether linkages, whereas those of eukaryotes and bacteria are ester linkages (Fig. 6.9). The hydrocarbon chains are made up of highly methyl-branched isoprenyl chains, instead of the predominantly straight fatty-acid chains of bacteria and eukaryotes. In some species, the side chains from two glycerols are bound together tail to tail, leading to the creation of a lipid monolayer, instead of a bilayer (Fig. 6.9C). Such monolayers may be important in allowing archaea to thrive in extremely high temperatures, because the two faces of the membrane cannot separate as in bilayers. These distinctive membranes allow archaeal species to be rapidly identified within samples of microorganisms. In addition, the distinct properties of these membranes can be used as "chemical fossils," biomarkers that provide evidence for the presence of archaea in ancient sedimentary rocks.

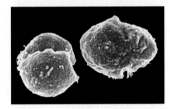
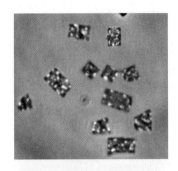
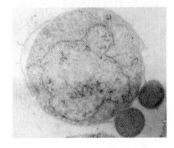

FIGURE 6.7. Diverse archaeal forms. (*Top*) A square haloarchaea. (*Middle*) *Methanococcus jannaschii* (see also opening figure, p. 137, *right*). (*Bottom*) Symbiosis between *Nanoarchaeum equitans* (smaller cells at lower right) and host *Ignicoccus*; both are archaeons.

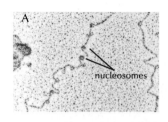

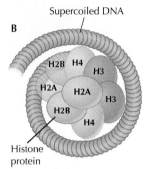

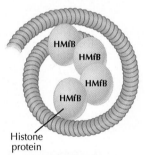

FIGURE 6.8. Some archaeal species fold their chromosomal DNA into structures resembling the nucleosomes of eukaryotes. (*Top*) Electron micrograph of nucleosomes from *Methanobacterium thermoautotrophicum*. (*Bottom*) Proteins homologous to eukaryotic histones are used in the core of these nucleosomes, indicating that these structures are homologous with nucleosomes of eukaryotes.

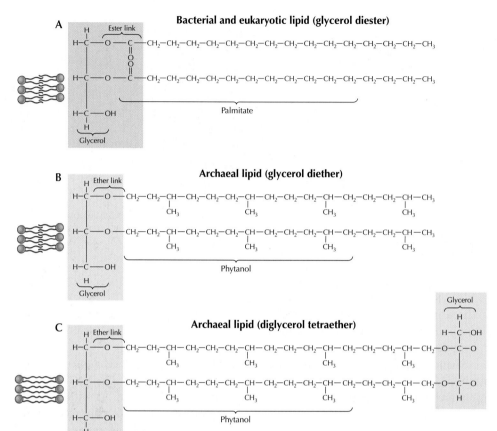

FIGURE 6.9. Comparison of cell membrane lipids. (*A*) In bacteria and eukaryotes, membrane lipids contain ester lineages between glycerol at one end and linear fatty acids at the other. (*B,C*) In archaea, lipids contain ether linkages between glycerol and branched hydrocarbon chains. In some (*B*), glycerol is only at one end; in others (*C*), glycerol is on both ends, resulting in a monolayer membrane.

PHYLOGENETIC DIVERSIFICATION OF BACTERIA AND ARCHAEA

The Only Reliable Way to Infer the Phylogenetic History of Bacteria and Archaea Is by Molecular Systematics

Phylogenetic studies of archaea and bacteria today rely almost entirely on molecular analysis. A variety of molecular data types have been used for such studies, including DNA–DNA hybridization, **restriction fragment length polymorphisms**, gene and genome sequence data, and gene presence/absence. In contrast, phylogenetic studies of many eukaryotes, especially multicellular ones, can be done reliably with morphological or molecular approaches or a combination of the two. The reason that bacterial and archaeal phylogenetic studies rely so heavily on molecular approaches is simple. Phenotypic features such as morphology, physiology, chemical composition, and disease-causing properties, all of which have been used in the past, are unreliable for inferring relationships among bacteria or archaea. These features generally lack sufficient information to be used for phylogenetic reconstructions or they evolve too rapidly and are too prone to **convergence** to be reliable phylogenetic markers (see Chapter 27 [online]).

The importance of molecular approaches to studies of the phylogeny of bacteria and archaea cannot be overstated. The premier discovery from these studies has been the recognition that the archaea are a separate group. In addition, the use of molecular systematics has led to the reclassification of a very large number of well-established bacterial and archaeal groupings. In some cases organisms that were grouped into the same genera are now known to be in different phyla. In other cases, close relationships among species were completely missed because of phenotypic differences. These reclassifications were not just esoteric exercises—they have had profound impacts on biology, medicine, and agriculture. Molecular systematics has changed how we perceive and study microbes by giving us more precise information on microbial relationships. We consider two examples.

The first one involves a bacterium that used to be known as *Micrococcus radiodurans*. This species was originally isolated from canned meat that had been exposed to what was believed to be a sterilizing dose of ionizing radiation. The microbe was assigned to the genus *Micrococcus* based on appearance and the knowledge that some other *Micrococcus* species were radiation resistant. *M. radiodurans* became the subject of extensive experimental studies after it was shown to be the most radiation-resistant organism known. Its similarity to other *Micrococcus* species turned out to be limited, however, preventing researchers working on *M. radiodurans* from using what was known about other *Micrococcus* species to help guide their work. Molecular systematic studies revealed that *M. radiodurans* is not closely related to *Micrococcus* species. *M. radiodurans* was removed from the genus *Micrococcus* and renamed *Deinococcus radiodurans*. *D. radiodurans* is not even a member of the same phylum as *Micrococcus*. It is related to *Thermus aquaticus*, the source of much of the DNA polymerase used in PCRs (Fig. 6.10). Many thermophiles are also highly resistant to radiation, which aids our understanding of the mechanism of radiation resistance in *D. radiodurans*.

The second example of an important historical misclassification comes from studies of intracellular pathogenic bacteria. In many cases, such organisms have been difficult or impossible to grow in isolation in the laboratory. Therefore, their classification was based primarily on morphological characteristics such as membrane structure, cell shape, and staining patterns. It was often difficult to be certain that such an organism was even a species of bacterium. When it was certain (e.g., by detecting ribosomes), more detailed phylogenetic classifications were frequently inaccurate. For example, *Rickettsia prowazekii*, the causative agent of Q-fever, and *Chlamydia trachomatis*, the causative agent of sexually transmitted chlamydia, were for some time grouped together because of phenotypic similarities. Molecular systematic studies have shown, however, that *R. prowazekii* is in the α subgroup of the phylum Proteobacteria and is related to species in the genera *Ehrlichia*, *Wolbachia*, *Rhizobium*, and *Caulobacter*. *C. trachomatis* is in its own phy-

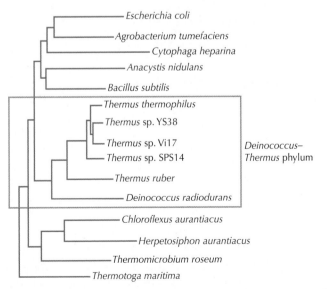

FIGURE 6.10. Phylogenetic tree showing the relationship of the *Deinococcus* and *Thermus* genera. The tree is based on analysis of rRNA gene sequences.

lum, very distant from any of the proteobacteria. Although many Rickettsias cannot be grown in the laboratory, experimental studies of other α-proteobacteria can be used to make reasonable predictions about the biology of Rickettsias.

There are many other examples of revisions in bacterial and archaeal classification based on molecular phylogeny. Molecular systematic approaches have also been very important in phylogenetic studies of eukaryotes, especially microbial eukaryotes. This is discussed in more detail on pages 198–202.

Bacteria and Archaea Have Been Split into Many Distinct Subdivisions

Using molecular systematic approaches, scientists have divided the bacterial domain into more than 40 major phylogenetic subgroups. A phylogenetic tree based on ribosomal RNA (rRNA) sequences is shown in Figure 6.11A. The bacterial phyla include many

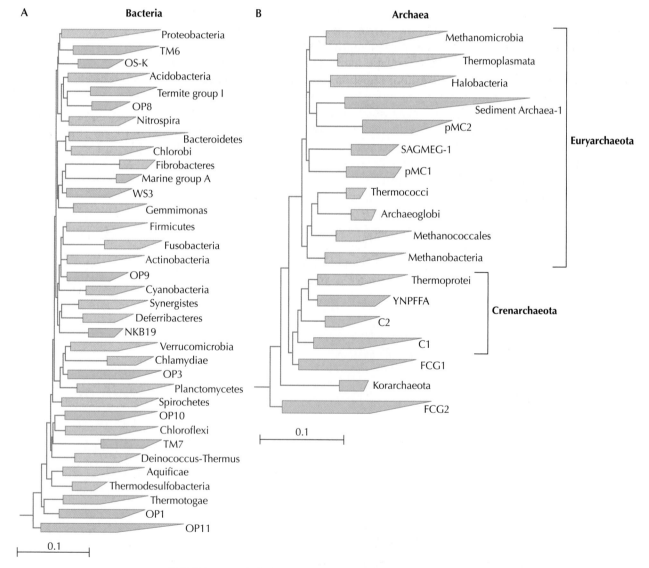

FIGURE 6.11. Phylogenetic trees of bacteria (*A*) and archaea (*B*) showing major lineages (phyla) as *wedges*, with horizontal dimensions reflecting the known degree of divergence within that lineage. Phyla with cultivated representatives are in *orange*. Phyla known only from environmental sequences are in *blue*. They are usually named after the first clones found from within the group. The scale bar represents 0.1 changes per nucleotide.

groups familiar to the nonmicrobiologist, such as spirochetes and cyanobacteria (also known as blue-green algae). Many of the other groups include well-known species, although the groupings themselves are not widely known. For example, *E. coli* is in the γ subgroup of the phylum Proteobacteria, and *B. anthracis*, the causative agent of anthrax, is in the phylum Firmicutes (also known as the low-GC gram-positive bacteria). Some of the key features of these phyla are described in Table 6.2 (additional details on metabolism and other properties of these microorganisms are discussed later in this chapter).

Molecular approaches have allowed scientists to subdivide the archaeal domain too. The major lineages have been recognized as three kingdoms: the Euryarchaeota, the Crenarchaeota, and the Korarchaeota—each containing distinct phyla. A phylogenetic tree based on rRNA sequences of the major archaeal groups is shown in Figure 6.11B. Some of the features of each subgroup are described in more detail in Table 6.3. The phylogeny of archaea appears to be simpler than that of bacteria, but that may be an artifact of the far greater amount of research done on bacteria and the relatively recent discovery that archaea are distinct from bacteria.

The phylogenetic trees of bacteria and archaea can be used in many ways. As discussed in the previous section, they can be used to improve the classification of species in these two domains of life. The trees are also essential for understanding the patterns of diversification of bacteria and archaea. For example, the finding that there are pathogenic species intermixed with nonpathogens throughout the bacterial tree indicates that pathogenicity is a relatively labile, or plastic, trait. In contrast, it seems that the trait of being gram-positive only originated once in bacterial evolution, appearing

TABLE 6.2. Descriptions of some of the major bacterial phyla that have cultured isolates

Phyla	Biology, Physiology, and Evolution	Species and Genera
Aquificae	Most species are hyperthermophiles, growing with an optimal growth temperature of ~85°C. Most are aerobic chemolithoautotrophs fixing carbon by the reductive TCA cycle. The membranes are made up of diethers, instead of phospholipids, although the diethers are distinct from those of archaea. Some studies suggest that this group is the deepest branching of the bacteria.	The phylum is composed of two main genera, *Hydrogenobacter* and *Aquifex*.
Thermotogae	Most species are hyperthermophiles with optimal growth temperatures exceeding 80°C. Most have an unusual "toga-like" structure that wraps around one end of the rod-shaped cell. The membranes are made up of diethers, like those in the Aquificae.	The best-characterized member of this phylum is *Thermotoga maritima*, a species that was originally isolated from a volcano on the seafloor near Italy.
Deinococcus-Thermus	Most are aerobic heterotrophs. All species in the *Deinococcus* branch of this group are radiation resistant. The *Thermus* group includes some of the first- and best-characterized thermophilic species.	*Deinococcus radiodurans* is the most radiation-resistant species known. *Thermus aquaticus* is the source of the thermostable "Taq" DNA polymerase protein used for PCR.
Chloroflexi/green nonsulfur bacteria	Many species are photosynthetic, filamentous, and motile, using a specialized gliding mechanism to move. The photosynthetic species fix CO_2 by the hydroxypropionate pathway. They grow best using electrons from organic compounds and thus are called nonsulfur bacteria to contrast with the green sulfur bacteria.	*Chloroflexus aurantiacus*, a photosynthetic species, is one of the best-studied species in this group.
Chlorobi/green sulfur bacteria	All species in this group are strictly anaerobic phototrophs. In contrast to the green nonsulfur bacteria, these species prefer to use sulfur compounds as the source of electrons. Carbon is fixed by the reverse TCA cycle.	*Chlorobium tepidum* and relatives are model systems for this group.

(Continued on following page.)

TABLE 6.2. *(Continued)*

Phyla	Biology, Physiology, and Evolution	Species and Genera
CFB/Bacterioidetes	This phylum is composed of three main lineages, the *Cytophaga* and the *Flavobacteria* (which are somewhat closely related to each other) and the more divergent *Bacteroides* group. Each subgroup includes pathogenic species interspersed with nonpathogenic lineages. Thus, it appears that the pathogenicity of these species has evolved separately. This phylum is related to the green sulfur bacteria.	*Porphyromonas gingivalis* is the cause of a large fraction of the gingivitis in the world.
Cyanobacteria	Phenotypically and physiologically heterogenous group best known for their photosynthetic species. All are oxygenic phototrophs. Many species fix nitrogen. Also known as blue-green algae. Chloroplasts are derived from bacteria in the cyanobacterial lineage.	*Spirulina* species are a common food source. *Prochlorococcus* species are responsible for a large fraction of the primary production in the world's oceans.
Chlamydiae	All are very small, even for bacteria, ranging in size from 0.2 to 0.7 μm in diameter. Most species are obligate intracellular pathogens and are considered energy parasites—they do not synthesize their own ATP but instead steal it from their host. Many species were originally misclassified as viruses because of their small size and other phenotypic features	*Chlamydia trachomatis* is the leading cause of blindness in the world and the causative agent of a major sexually transmitted disease.
Proteobacteria	The largest (in terms of numbers of known species) and most physiologically diverse bacterial phylum. Made up of five major divisions—the α, β, γ, δ , and ε. Phototrophs, heterotrophs, and chemilithotrophs are scattered throughout these divisions. Many species are symbiotic, including both pathogens and mutualists. Mitochondria are derived from bacteria in the α-proteobacterial lineage.	The pathogens include the causative agents of cholera (*Vibrio cholerae*, a γ), meningitis (*Neisseria meningitidis*, a β, and *Haemophilus influenzae*, a γ), and stomach ulcers (*Helicobacter pylori*, an ε). The model organism *Escherichia coli* is in this group.
Spirochetes	Most are relatively long compared to many other bacteria and are slightly helical in shape. The group is metabolically very diverse and includes a variety of free-living species as well as many major animal and human pathogens. *B. burgdorferi* has dozens of genetic elements—almost all of which are linear.	The pathogenic species include the causative agents of syphilis (*Treponema pallidum*) and Lyme disease (*Borrelia burgdorferi*).
Actinobacteria	This is one of two phyla of gram-positive bacteria. Most species have high G + C contents of their genomes and thus they are also known as high-GC gram-positive bacteria. The pathogens are interspersed with nonpathogenic lineages suggesting that pathogenicity has evolved multiple times in this group of bacteria.	Many well-known pathogens are in this group including *Mycobacterium leprae* (leprosy) and *Mycobacterium tuberculosis* (tuberculosis). Also includes many soil microbes, like *Streptomyces*.
Firmicutes	This is the other major phylum of gram-positive bacteria. Because most species have low G + C contents, the group is also known as the low-GC gram-positive bacteria. This group contains *Heliobacterium*, a photosynthetic lineage. The pathogens are interspersed with nonpathogenic lineages suggesting that pathogenicity has evolved multiple times in this group of bacteria. The mycoplasmas lack cell walls and many use modified genetic codes.	Many well-known pathogens are in this group including *Bacillus anthracis* (anthrax), *Staphylococcus aureus*, *Streptococcus pneumoniae*, and *Clostridium perfringens*.
Planctomycetes	Members of this group divide by budding and frequently are found in stalk-like structures of multiple cells. They are frequently found floating in aquatic habitats. All species lack peptidoglycan. Some species have their DNA packaged in a nucleus-like structure.	*Gemmata obscuroglobus* has a nucleus-like structure.

TCA, tricarboxylic acid; rRNA, ribosomal RNA; mbp, megabase pair; PCR, polymerase chain reaction; ATP, adenosine triphosphate.

TABLE 6.3. Descriptions of the three major divisions of archaea and some of the phyla with cultured isolates

Division/Phylum	Biology, Physiology, and Evolution	Species and Genera
Euryarchaeota/Halobacteria	These species differ in many ways from halophilic bacteria and eukaryotes: They are the only extreme halophiles (able to grow and thrive in salt solutions >4 M), they cannot grow at low-salt concentrations, and their mechanisms of salt tolerance are very different from those of bacteria and eukaryotes. They also use the energy of light to operate proton pumps that are used for a specialized form of phototrophy. The extreme halophiles were the first archaea for which genetic tools became available.	*Haloferax volcanii* and *Halobacterium cutirubrum* are among the models for this group.
Euryarchaeota/Methanogens Methanococcales Archaeoglobi Thermococci Methanomicrobia Thermoplasmata	Most species of Euryarchaea are methanogens (species that produce methane as a by-product of metabolism) or thermophiles or both. The methanogens produce methane either by reduction of CO_2 or from methylated substrates such as methanol (CH_3OH). All of these lineages are dominated by thermophilic species and all except the *Thermoplasmales* contain methanogens. Many species within these lineages are not extreme thermophiles. These lineages do not correspond to a single monophyletic group. Instead, there are multiple methanogen and thermophile lineages that branch off separately from the base of the euryarchaeal tree.	*Methanococcus jannaschii*, the first archaeon to have its genome sequenced, was isolated from a hydrothermal vent in the deep sea.
Crenarchaeota/Thermoproteales	This group generally consists of hyperthermophiles (organisms that grow optimally at temperatures >85°C) and thermoacidophiles (thermophiles that thrive in low-pH environments).	*Sulfolobus sulfotaricus* was isolated from a hot acidic spring in Yellowstone National Park.
Korarchaeota	The Korarchaea are a very poorly understood group, represented only by rRNA sequences from uncultured species. They are a group defined solely on the basis of DNA samples that have been isolated from environmental samples.	

in a common ancestor of the Actinobacteria and Firmicutes phyla. This is based on three findings: Gram-positive species are limited to only these two phyla, these phyla are closely related to each other (and may in fact be "sister" phyla), and nearly all members of these phyla stain positively (with the only exceptions being organisms, like mycoplasmas, that have severely reduced genomes). Throughout this and the next chapter, examples are given of how reliable phylogenetic trees of bacteria and archaea improve our understanding of the biology and diversification of these groups. The trees shown in Figure 6.11 are models of the evolutionary history of these organisms. They almost certainly are imperfect, and thus alternative trees have been proposed and others will continue to be proposed. Therefore, where it is appropriate, we discuss how some of the inferences based on these trees would change if the trees were inaccurate.

Molecular Approaches Allow Microbes to Be Studied in the Environment

Studies of bacteria and archaea in nature can be quite challenging. Most of the organisms are microscopic and can be difficult to identify accurately based solely on physi-

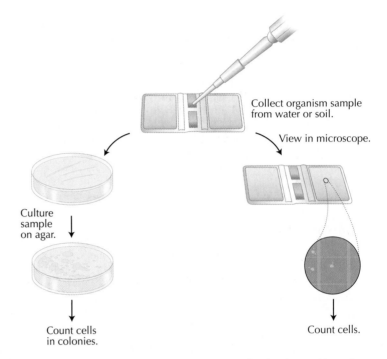

Collect organism sample from water or soil.

View in microscope.

Culture sample on agar.

Count cells in colonies.

Count cells.

FIGURE 6.12. The great plate count anomaly. Plate counts of cells obtained by cultivation are usually much lower, sometimes by orders of magnitude, than those from direct cell counts under a microscope. Possible reasons are (1) the differing nutritional requirements of the organism, (2) the organism may enter a noncultivatable resting state, or (3) the organism may rely on other organisms and thus cannot be cultivated in isolation.

cal features and ecology. One alternative is to **culture** the organisms in the laboratory, with the goal of growing a "pure culture" of each organism of interest, uncontaminated by any other organisms. If an organism can be grown in pure culture, then its physiology, biochemistry, molecular biology, and genetics can be studied in great detail.

Many microorganisms, however, cannot be grown as pure cultures in the laboratory. This is exemplified by something known as the **great plate count anomaly**. If a sample of pond water is diluted and spread onto a dish of solid growth media (such dishes are commonly known as plates), and the organisms are allowed to grow until visible colonies appear, then the number of colonies that grow on each plate can be counted. Each colony is an aggregate of cells representing a single starting cell from the sample. This provides an estimate of the number of colony-forming units (CFUs) present in the pond sample. Comparing the number of CFUs in a sample to the number of cells from the same sample when seen through a microscope reveals a striking difference: Many more cells (typically, orders of magnitude more) are visible under the microscope than could form colonies. This plate count anomaly (Fig. 6.12) is evidence that most microbes have never been grown in pure culture. Such organisms are known as **uncultured microbes**.

To appreciate how little we currently know about most uncultured microorganisms, consider the following thought experiment. Imagine you are studying carbon cycling in a tropical rainforest, where you know a large amount of carbon fixation is occurring, but rather than knowing to look at the trees for much of the photosynthesis, you do not know which organisms to consider. Imagine also that you are able to examine the forest only from 1000 miles away in low-resolution images from which you cannot tell a squirrel from a tree. Supplementing your studies are a few pieces of forest material that you were able to bring back to the lab and keep alive. Over time, some details of the carbon cycle would be worked out. But the pace would be slow, the task difficult and expensive, and the picture would remain incomplete. Environmental microbiology was in this sort of predica-

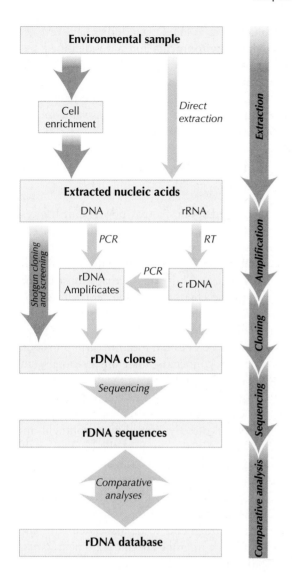

FIGURE 6.13. A flowchart illustrating the methods available for characterizing an environmental sample using rRNA sequence analysis. RT, reverse transcriptase.

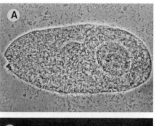

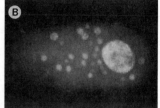

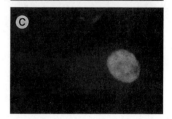

FIGURE 6.14. In situ hybridization can be used to label specific organisms in a mixed sample. This is shown here with a *Paramecium* (a eukaryote in the Ciliate phylum; see Box 8.1) that harbors a bacterial symbiont. (*A*) Phase-contrast micrograph. (*B* Fluorescent micrograph using an rRNA probe that detects all bacteria. The probe lights up multiple locations in the cell because *Paramecium* eat bacteria. (*C*) Fluorescent micrograph with a probe specific for the symbiont *Halospora obtusa*, which is detected only in the **macronucleus**.

ment for a long time, because (as noted above) few microorganisms can be studied in detail in the laboratory and field studies alone provide limited information.

Environmental microbiology was revolutionized, beginning in the 1980s, when rRNA-based molecular systematic studies (as described in Box 5.2) began to be applied to environmental questions. Using advances in DNA and RNA sequencing methods (especially the PCR), it was shown that 16S rRNA gene sequences from uncultured microbes could be determined (Fig. 6.13). From rRNA sequence data, the microbes present in an environmental sample can be identified. For example, phylogenetic analysis of sequences could be used to determine what phylogenetic types (or **phylotypes**) of organisms are present. Analysis of the numbers of clones found can be used to estimate the abundance of particular phylotypes and the total number of types present. Furthermore, these sequences could serve as species-specific "tags," enabling organisms to be identified in an environmental sample by staining them using probes that hybridize only to the rRNAs of interest. The morphology of those microbes that "light up" with particular probes can be examined using a microscope, allowing scientists to connect the morphology of particular microbes to an rRNA type and to estimate the abundance of different forms (Fig. 6.14).

Such studies have provided striking results. For example, of the more than 40 recognized phyla of bacteria, more than half have been identified only via molecular systematics analyses (the groups shaded in blue in Fig. 6.11). Other studies have revealed that archaea are found in many "nonextreme" environments and have identified specific rRNA phylotypes (i.e., species) that are abundant throughout the oceans, yet have never been studied directly. A very important discovery is that many of the uncultured microbes are more abundant and more ecologically significant than those that can be grown in pure culture. Thus, studies of cultured microbes may be of less value for understanding particular environments than originally believed.

Genome Sequencing Allows the Biology of Uncultured Microbes to Be Predicted

Although analysis of rRNA genes from uncultured species is very powerful, it has certain limitations. First, the number of copies of rRNA genes differs between species. This hampers attempts to use rRNA gene analysis to quantify the abundance of different species. Second, the slow rate of change of rRNA genes means that analysis of rRNA is not very useful for comparing closely related species. Third, and perhaps most important, rRNA-based classification of microbes in the environment does not tell us what these organisms are doing. This is because microbial physiology, biochemistry, and other processes change very rapidly, so closely related microbes may differ significantly in their biology. Although in situ experimental studies of microbes can measure carbon fixation, sulfur metabolism, and other metabolic processes, connecting these observations to individual microbial types identified by rRNA surveys can be very difficult.

A solution to this problem has come from the application of genome sequencing methods to uncultured microbes, an approach known as **environmental genomics** or **metagenomics**. In this approach, DNA is isolated directly from the environment and then fragments are sequenced. In principle, one can then predict the biology of the organisms in the sample by analyzing the sequence data. In practice, there is one major difficulty—determining which DNA fragments came from which organisms in the sample. The simplest way to do this is to isolate and sequence fragments that contain rRNA genes. Each rRNA gene serves as a **phylogenetic anchor** for some of the genome pieces (Fig. 6.16A), enabling particular genome fragments to be associated with specific types of microbes. The other genes found on a microbe-specific genome fragment can then be identified, providing an indication of what roles that organism plays in its environment. In principle, any gene can be used as an anchor, but rRNA genes are the most useful because a database of approximately 200,000 rRNAs from different organisms is available.

The phylogenetic anchor–based approach has led to fundamental insights into environmental microbes. For example, PCR surveys had shown that a few types of proteobacteria (defined by rRNA sequence) were abundant in all surface waters of most of the world's marine ecosystems. Yet, because none of them had been cultured, nobody could determine what roles the organisms played in these waters, although photosynthesis was assumed. However, when metagenomic methods were used to sequence a fragment of the genome containing the 16S rRNA gene from one of these organisms, it was discovered that they encoded a homolog of rhodopsin previously known only in archaea and single-celled eukaryotes. These "proteorhodopsins" function as light-mediated proton pumps—an important energy-generating mechanism that was missed by searching for photosynthesis processes already known in bacteria.

Although the phylogenetic anchor method has yielded many important results, it only provides a snapshot of the cell from which the large fragment is derived (Fig. 6.15A). An alternative approach (Fig. 6.15B) applies the random **shotgun sequencing** methods, which are used to sequence genomes, to sequence DNA from environmental samples. This approach has enormous potential because all of the organisms present in an environment can be randomly sampled. Analysis of the data generated by environmental shotgun sequencing has been very fruitful. For example, organisms not

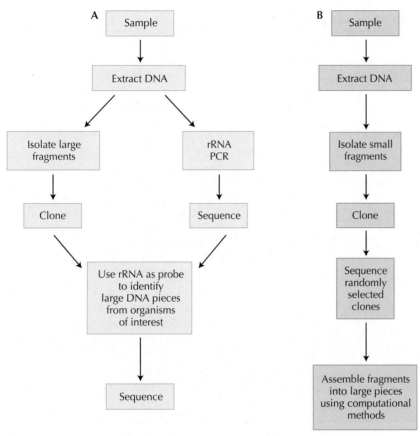

FIGURE 6.15. Methods used in metagenomics. (*A*) Phylogenetic anchor method. (*B*) Environmental shotgun sequencing.

detected by PCR have been identified by this method. In addition, this method provides a wealth of data for population genetic studies of uncultured species.

BIOLOGICAL DIVERSIFICATION OF BACTERIA AND ARCHAEA

The extraordinary diversification of bacteria and archaea is exemplified by the wide range of biological processes that different species carry out. In this section, we discuss three aspects of this biological diversification: the ability to live and thrive in extreme environments, the acquisition of energy and carbon from the environment, and interactions with other species.

Many Bacteria and Archaea Thrive in Extreme Environments

Some organisms have evolved to occupy niches that are intolerably hostile to most other species. For example, many archaea and bacteria live at temperatures greater than 50°C, and a few thrive at temperatures above 100°C. Although acidic or basic conditions will kill many species, there are some microorganisms that thrive at pH 0 and others that thrive at pH 14. Organisms that live at the fringes of particular environmental conditions (such as temperature, pH, pressure, radiation, water potential, and salinity) are known as **extremophiles** (Table 6.4).

Studying extremophiles has produced numerous practical and scientific benefits. Products from extremophiles are used in many industries; for example, the DNA polymerases found in thermophiles are used in the biotechnology industry, and proteases are used in detergents. Studies of organisms that grow at different temperatures have

TABLE 6.4. Tolerances of life-forms

Type of Environment	Examples of Environments	Mechanism(s) of Survival	Practical Uses
High temperature (thermophiles)	Hydrothermal vents, hot springs, volcanoes	Amino acid changes, increase H-bonds, metal cofactors	Thermostable enzymes
Low temperature (psychrophiles)	Antarctic Ocean, glacier surfaces	Antifreeze proteins, solutes	Enhancing cold tolerance of crops
High hydrostatic pressure (barophiles)	Deep sea	Solute changes	
High salinity (halophiles)	Evaporating ponds and seas, salt evaporators	Solute changes, ion transport, protein amino acid adaptation	Industrial enzymes; soy sauce production
High pH (alkaliphiles)	Soda lakes	Transporters	Detergents
Low pH (acidophiles)	Mine tailings	Transporters	Bioremediation
Desiccation (xerophiles)	Evaporating ponds, deserts	Spore formation, solute changes starvation tolerance, DNA repair, scavenge free radicals	Freeze-drying additives
High radiation (radiophiles)	Nuclear reactors or waste sites, high-altitude surfaces	Absorb radiation, enhance DNA repair, scavenge free radicals	Bioremediation

yielded insights into how proteins fold. Research on organisms that grow at high salinities (halophiles) has led to the discovery of novel mechanisms for dealing with osmotic stress. Probing the biology of radiation-resistant organisms has revealed novel mechanisms for repairing damaged DNA. Perhaps most important for our purposes in this book, because of the strong selective pressures present under extreme conditions, extremophiles are excellent models for studying mechanisms of evolution.

Life at Extreme Temperatures

That organisms are able live at all in extreme temperatures might seem surprising. Low temperatures are harmful to "normal" organisms because of an increased risk of freezing, decreased flexibility of proteins, reduced chemical reaction rates, a drop in solubility of most compounds, and an increase in the viscosity of membranes. High temperatures accelerate the denaturation of proteins, cause excessive membrane fluidity, reduce the stability of DNA and RNA base pairing, and increase rates of DNA damage. For these reasons, it was long held that life could be found only within a relatively narrow range of temperature, between 0°C and 60°C. Today we know of organisms with optimal growth temperatures as low as −12°C and as high as 113°C. These extremes would rapidly kill most organisms.

It is important to distinguish between survival temperature and optimal growth temperature. Many organisms can survive extreme conditions for a limited period of time. However, we are interested in those species that grow best at extreme temperatures. The optimal growth temperature of any species is determined by plotting its rate of growth against temperature (Fig. 6.16). Organisms with optimal growth temperatures below 15°C are called **psychrophiles** and those that grow best above 50°C are called **thermophiles**. In between are the **mesophiles**. Organisms that grow and thrive at temperatures above 80°C are rare and have been assigned to a special category called **hyperthermophiles**.

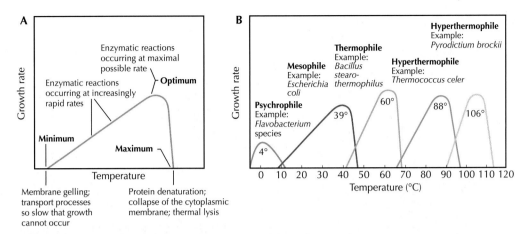

FIGURE 6.16. Optimal growth temperatures. (*A*) Hypothetical curve of temperature vs. growth rate. The optimal temperature for growth is at the peak. (*B*) Optimal growth temperatures of different types of organisms.

Among the bacteria and archaea, there are many thermophilic species. However, few thermophilic eukaryotes are known, and all have optimal growth temperatures just above 50°C (Table 6.5). Thermophilic environments are dominated by bacteria and archaea, and at the highest temperatures, environments are almost exclusively populated by archaea.

Among the few bacterial hyperthermophiles known, all species are in either the phylum *Thermotogae* or *Aquificae*. The fact that these two phyla are the deepest-branching bacterial lineages in the rRNA tree of life (Fig. 6.17) has led to the proposal that the common ancestor of bacteria and archaea (which in most theories would be the last

TABLE 6.5. Presently known upper temperature limits for growth of living organisms

Group	Upper Temperature Limits (°C)
Animals	
Fish and other aquatic vertebrates	38
Insects	45–50
Ostracods (crustaceans)	49–50
Plants	
Vascular plants	45
Mosses	50
Eukaryotic microorganisms	
Protozoa	56
Green algae	55–60
Fungi	60–62
Prokaryotes	
Bacteria	
Cyanobacteria	70–74
Anoxygenic phototrophic bacteria	70–73
Chemoorganotrophic bacteria	90
Archaea	
Hyperthermophilic methanogens	110
Sulfur-dependent hyperthermophiles	113

Reprinted from Madigan M.T., Martinko J., and Parker J. 1997. *Brock biology of microorganisms*, 8th ed. Prentice Hall, New York.

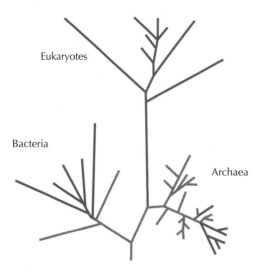

Eukaryotes

Bacteria

Archaea

FIGURE 6.17. Rooted rRNA tree of life showing the distribution of hyperthermophilic lineages (in *red*).

universal common ancestor [LUCA]) was hyperthermophilic. However, as discussed on pages 130–131, the deep branches in the rRNA tree of life are not considered completely resolved. In other trees of life, the *Thermotogales* and *Aquificales* are not always the deepest-branching bacterial lineages. If that is the case instead, hyperthermophily in these phyla likely evolved separately from that in archaea, possibly aided by the lateral transfer of some portions of the archaeal genomes to these bacteria (see pp. 182–187).

Adaptations to Growth at Extreme Temperatures Can Be Identified through Comparative Analysis

What are the adaptations that allow some organisms to survive temperature extremes that would kill most living things? Perhaps the best studied are adaptations to the direct kinetic effects of temperature on cellular reactions and structure. As temperature increases, if all else is held constant, membranes become more fluid, proteins become more flexible, and chemical reactions occur more rapidly. To understand how organisms respond to these kinetic effects, studies have been performed in which the optimal growth temperatures from a wide variety of organisms and biological features are compared. The data are analyzed to determine if any features correlate with optimal growth temperature. Initial results revealed that organisms compensate for the kinetic effects of temperature. For example, when compared with mesophiles grown at the same temperatures, thermophiles had lower membrane fluidity, stiffer proteins and noncoding RNAs, and slower reaction rates. The opposite was true among psychrophiles, whose membranes were more fluid, proteins and noncoding RNAs were more flexible, and chemical reactions occurred at higher rates.

The above observations, although important, do not reveal the mechanisms by which these evolutionary changes occurred. To uncover those, we need to examine more closely the membranes, proteins, and RNAs of the extremophiles to try to understand how they changed. Again, comparative studies have been very useful in this regard. For example, comparisons of amino acid composition of proteins reveals that, at higher temperatures, charged residues (Asp, Glu, Lys, and Arg) are preferred over uncharged polar residues (Ser, Thr, Asn, and Gln). In addition, hydrophobic residues are preferred over hydrophilic ones. These observations led to studies that showed that these amino acid differences lead to an increase in the number of salt bridges and hydrophobic interactions, which in turn make these proteins more stable at higher temperatures (Fig. 6.18). These discoveries have allowed researchers to engineer proteins from mesophilic organisms to withstand high temperatures.

A

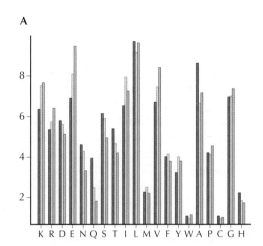

B

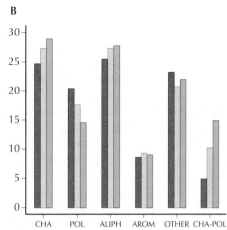

FIGURE 6.18. Amino acid content vs. optimal growth temperature. (*A*) Plot of the percentages of the various amino acids (see Fig. 2.23) in mesophiles (*blue*), thermophiles (*yellow*), and hyperthermophiles (*red*). (*B*) Plot of the various amino acid classes; colors as in panel *A*. CHA, charged; POL, polar; ALIPH, aliphatic; AROM, aromatic.

Comparative studies have also led to a very important finding about DNA and RNA thermostability. Based on what is known about base pairing in DNA and RNA, it is reasonable to assume that organisms that grow at high temperatures should have more Gs and Cs in their genome than organisms that grow at low temperatures because G:C base pairs have an extra hydrogen bond compared to A:T base pairs. The extra hydrogen bonds would assist in stabilizing genomes of thermophiles. Surprisingly, this turns out not to be the case. The genome G + C content is not correlated with optimal growth temperature, which suggests that other factors keep the DNA from unwinding in thermophiles. DNA stabilization factors, such as the use of high intracellular salt concentrations, have been identified. Interestingly, there is a strong correlation between growth temperature and the G + C content of genes for rRNA. The organisms that have higher optimal growth temperatures have more G + C. This is probably because the folding of rRNAs (and thus their function) is more dependent on base pairing than is DNA.

The correlation between growth temperature and G + C content of rRNA has many implications. First, it means that there is convergent evolution in rRNA in distantly related thermophiles. This can result in errors in phylogenetic reconstruction based on rRNAs. For example, the first molecular systematic studies of *Thermus* species incorrectly grouped these species with other thermophilic bacteria like *Thermotoga maritima* and *Aquifex pyrophilus*. As mentioned above, we now know that *Thermus* species are related to the radiation-resistant organism *D. radiodurans*. This misgrouping of *Thermus* occurred because the 16s rRNA gene was used and there was convergence of G + C content among the thermophiles. Only when phylogenetic methods that corrected for G + C content variation were used did the relationship of *Thermus* and *Deinococcus* become apparent (Fig. 6.10; also see Chapter 27 [online], for a discussion of this correction).

Comparative Analysis Has Limitations

Although the comparative analyses described above have great utility, they do not take into account the relatedness among the organisms being studied. That is, all pairwise comparisons are counted equally, whether the organisms are distantly related or closely related. A variety of methods have been developed to account for the relatedness of organisms when doing correlation studies. These are discussed in more detail in Chapter 27 (online).

Comparative studies, such as those described above, have another limitation: They look at the end points of evolution (in this case, the current state of temperature adaptation) across a wide range of taxa. When comparing widely divergent organisms, many differences (e.g., pathogenicity or differing metabolic processes) exist that are unrelated to temperature. An alternative approach compares closely related species that differ mainly in growth temperature. Because the organisms being compared are closely related, the difference in temperature adaptation must have evolved relatively recently. In addition, the close relationship means that, presumably, there will be less "noise" from other evolutionary adaptations confusing the picture. Furthermore, by using **evolutionary character state reconstruction** methods (see p. 117) when comparing closely related species, it is possible to infer the actual evolutionary changes that occurred as the growth temperature changed. For example, one study, which analyzed changes in the amino acid composition of proteins that accompanied changes in growth temperature (Table 6.6), found that changes from mesophily to thermophily were specifically accompanied by increases in the isoleucine, arginine, lysine, and glutamic acid content and by decreases in serine, threonine, asparagine, glutamine, and methionine. This is similar to the result of the correlation analysis described above. However, this evolutionary analysis was also able to determine that not all hydrophobic amino acids were increased in the shift to thermophily (methionine is not and valine and leucine are kept at roughly equal levels). Thus the evolutionary analysis can reveal additional or complementary results to the correlation studies.

TABLE 6.6. Differences in amino acid composition between mesophilic and thermophilic proteins

Amino Acid	Gains	Losses	P^a	Change (%)
Ile	842	658	2.2×10^{-6}	9.5
Glu	739	562	1.0×10^{-6}	9.1
Arg	383	214	4.5×10^{-12}	16.5
Lys	789	620	7.4×10^{-6}	8.3
Pro	167	96	0.000014	7.0
Tyr	224	177	0.021	5.8
Ala	504	458	0.15	2.8
Trp	23	11	0.058	8.3
Leu	560	548	0.74	0.6
Cys	72	69	0.87	0.9
Phe	200	202	0.96	−0.3
Asp	429	432	0.95	−0.2
Val	666	670	0.93	−0.2
His	80	92	0.40	−2.8
Gly	201	264	0.0040	−3.4
Met	174	248	0.00037	−11.3
Gln	158	234	0.00015	−13.1
Thr	336	431	0.00068	−8.4
Asn	313	481	2.7×10^{-9}	−15.9
Ser	271	664	9.5×10^{-39}	−31.7

Reprinted from Haney P.J. et al. 1999. *Proc. Natl. Acad. Sci.* **96:** 3578–3583.
[a]The random probability of a directional bias greater than or equal to that observed (calculated using the two-tailed **binomial distribution**, indicating statistical significance).

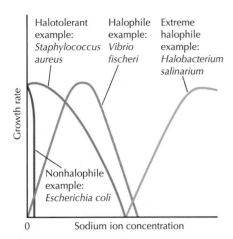

FIGURE 6.19. Sodium ion concentration vs. growth rate, comparing nonhalophiles, halotolerant species, halophiles, and extreme halophiles.

The fact that some forms of extremophily have evolved multiple times also constrains the conclusions that can be drawn from correlation analysis. When organisms that have a particular form of extremophily are placed on the tree of life, in many cases they are interspersed with nonextremophilic species, suggesting that there have been multiple origins of the extremophily. This raises the possibility that there are multiple solutions to the problem of growth in that particular extreme. If so, including organisms with the different solutions in the comparative analysis may result in a spurious explanation for the adaptation. Evolutionary reconstructions allow each origin of extremophily to be treated separately.

It Is Difficult to Determine the Underlying Adaptive Mechanisms Used by Organisms with Multiple Extremophilies

Some organisms thrive in multiple types of extreme conditions. For example, halophilic archaea are also resistant to radiation and desiccation. *D. radiodurans* and its relatives are resistant to radiation and desiccation, but they cannot grow at high salinity. In this subsection, we discuss the complex interplay in the evolution of three types of extremophiles—halophiles, xerophiles, and radiophiles.

First, let us look at the adaptive mechanisms that **halophiles** use (Fig. 6.19). Growth at high salinity is challenging because there is significant osmotic pressure exerted on a cell. Osmotic pressure causes water inside the cell to diffuse across the cell membrane in order to reach equilibrium with the high-salt solution outside.

All halophiles have a mechanism to increase their internal osmolarity and thus reduce the force of the water trying to leave the cell. Bacterial and eukaryotic halophiles usually do this by increasing the concentrations of organic solutes (such as proteins, carbohydrates, and amino acids) inside the cell. For example, halophilic species in the genus *Staphylococcus* use proline, whereas many yeasts use glycerol. Halophilic archaea use inorganic salts to counterbalance the high salt in their environment. Thus rather than excluding salts from the cell, their cellular machinery has adapted to function at high concentrations of salt. Halophilic archaea are the only extreme halophiles—i.e., they are the only species able to grow at salinities greater than 4 M (about ten times the salt concentration of seawater). In fact, they are obligate halophiles, unable to grow at normal salinity.

Halophiles must often combat desiccation, because the high-salinity environments they inhabit are frequently created by evaporation of water. During desiccation, the loss of water from inside the cell increases internal solute concentrations, which actually aids the halophile in compensating for the high salinity outside of the cell. The mechanisms halophiles have evolved that enable their cellular processes to function in high internal solute concentrations are also well suited for dealing with the initial stages

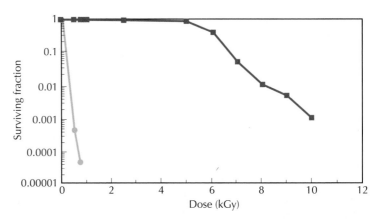

FIGURE 6.20. *Deinococcus radiodurans* radiation resistance. Representative survival curves of *D. radiodurans* (*red*) and *Escherichia coli* (*blue*) are shown following exposure to different doses of γ irradiation (kGy = kiloGrays).

of desiccation. Not all desiccation-resistant organisms (**xerophiles**), however, are halophilic. Xerophily has also arisen in environments that are of normal salinity, but prone to drying out. Many microorganisms form spores in response to drought but sporulation does not necessarily protect against high salinity.

Many halophiles and xerophiles are also resistant to the damaging effects of UV and γ-radiation. For many years, the belief was that resistance to radiation damage (radiophily) must have evolved independently of halophily or xerophily. However, genetic studies of *Deinococcus radiodurans* refuted this belief. *D. radiodurans* is approximately 1000-fold more resistant to γ-radiation than *E. coli*, placing it among the most radiation-resistant species known (Fig. 6.20). *D. radiodurans* is also resistant to desiccation. While screening *D. radiodurans* mutants to identify genes responsible for radiation resistance, scientists discovered that the genes that made *D. radiodurans* sensitive to radiation also increased its sensitivity to desiccation. For example, genes involved in DNA repair processes, which are discussed on pages 334–336, protect cells from both radiation- and desiccation-induced damage. Thus, a mechanistic connection exists between xerophily and radiophily, at least in this microbe. Based in part on these studies, scientists are now aware that for some cells, the cellular damage caused by desiccation and rehydration is similar to the damage caused by radiation.

Bacteria and Archaea Have Evolved an Enormous Variety in Biochemical Reactions

The biochemical processes seen in bacteria and archaea are staggering in their variety. Whether it is the capture of energy to run cellular processes or the conversion of raw materials into cellular structures, bacteria and archaea have found ways to use virtually every chemical and energy source available on Earth. From inorganic single-carbon compounds to complex organic molecules, there are bacteria and archaea that can use them, manipulate them, and convert them into biomass. For just about every form of nitrogen, from hard-to-crack nitrogen gas to pure nitric acid, there is a bacterium or archaeon that can convert it to a usable form and feed it into its metabolic networks. The same is true for sulfur, phosphorus, manganese, magnesium, and many other elements.

To appreciate this diversity, it is useful to understand how we categorize energy and carbon metabolism. There are three broad categories. The first relates to the ultimate source of energy. Those processes that get energy directly from chemical reactions are *chemo*trophic, whereas those that use the energy of light are *photo*trophic (-trophy is derived from the Greek word that means "to feed"). The second category

TABLE 6.7. Examples of bacterial and archaeal energy and carbon metabolism

Type of Metabolism	Energy Source	Electron Donor	Carbon Source
Photolithoautotrophs	Light	Misc.	CO_2
Photolithoheterotrophs	Light	Sulfide	Acetate, fumarate
Photoorganoautotrophs	Light	Lactate	CO_2
Photoorganoheterotrophs	Light	Misc.	Misc.
Chemolithoautotrophs	H_2S, CH_4, H_2, NH_4^+, NO_2^-	H_2S, H_2	CO_2
Chemolithoheterotrophs	H_2	H_2, acetate	Acetate
Chemoorganoautotrophs	Sugars	Misc.	Misc.
Chemoorganoheterotrophs	Sugars	Sugars	Sugars

considers the source of electrons used to convert the primary energy to forms of energy that cells can use. Processes that use an inorganic compound as an electron donor (e.g., H_2S) are *litho*trophic, whereas those that use organic compounds (e.g., acetate) are *organo*trophic. The third category considers the source of carbon for cellular growth. Processes that use inorganic, single-carbon compounds (e.g., CO_2) and assemble these into larger molecules are *auto*trophic. Those that use organic compounds (e.g., sugars) are *hetero*trophic (Table 6.7). Among bacteria and archaea there are species that use every possible combination of these different processes. In addition, within each category, the details of the processes (in terms of both the compounds and the mechanisms involved) are also very diverse. To illustrate the biochemical diversity among bacteria and archaea, in the next two subsections we discuss some of the carbon fixation pathways and phototrophic mechanisms used by these organisms.

Carbon Fixation by Bacteria and Archaea Is Carried Out in Nearly Every Imaginable Way

Bacteria and archaea are able to fix just about any form of single-carbon compound including methane (CH_4), methanol (CH_3OH), methylamine (CH_3NH_2), formate ($HCOO^-$), formamide ($HCONH_2$), carbon monoxide (CO), and carbon dioxide (CO_2). Carbon "fixation" is a process in which inorganic single-carbon compounds are combined into carbon chains. The fixation of carbon dioxide is, of course, what green plants and their relatives do, through the use of chloroplasts, which are derived from free-living cyanobacteria (see p. 203).

The diversity of carbon fixation pathways is simply too great to consider in this textbook, and we focus here on the fixation of carbon dioxide (Fig. 6.21). Perhaps the simplest pathway is the hydroxypropionate pathway, in which two molecules of CO_2 are converted into glyoxylate (CHO-COOH). The discovery of this pathway in species in the Chloroflexi phyla, which is deeply branching in the rRNA tree of life, has led to the suggestion that it may have been the first carbon fixation pathway to evolve on Earth. An alternative to the hydroxypropionate pathway is the reductive tricarboxylic acid (rTCA) cycle. This pathway is essentially the reverse of the well-known oxidative TCA cycle, which is used in humans and many other species to generate energy from the end products of glycolysis. The rTCA cycle was originally discovered in species in the Chlorobi phyla but has subsequently been found in the *Aquificae*, many archaeal lineages, and a variety of single-celled eukaryotes. It is important to note that the use of the term "reverse" to describe this pathway does not mean that the oxidative pathway evolved first. In fact, because of its presence in many deeply branching species, it is possible that the rTCA cycle evolved before the "forward" form of the cycle.

Perhaps the best-studied CO_2 fixation process is the Calvin cycle, which is also known as the reductive pentose pathway. In the first step in this process, CO_2 and a

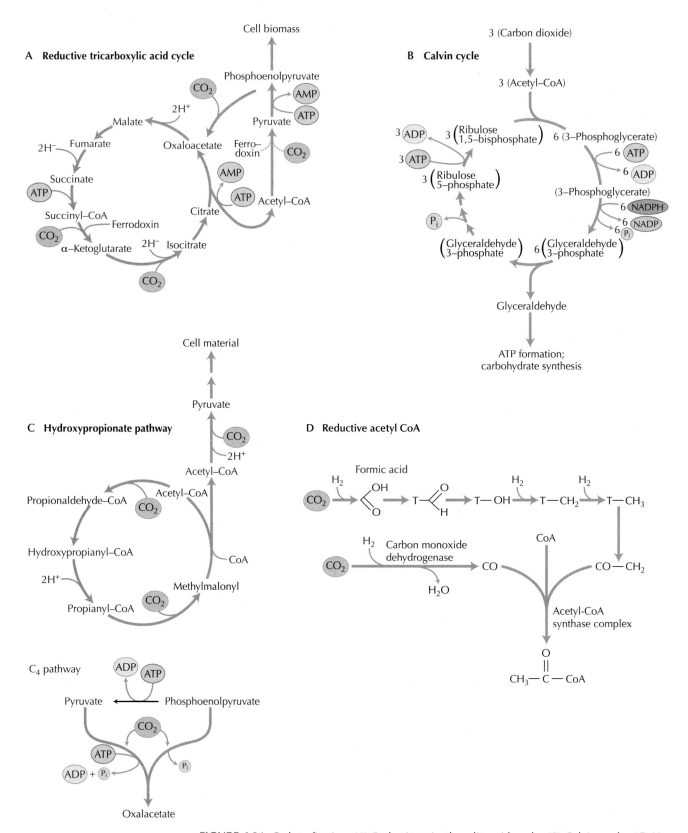

FIGURE 6.21. Carbon fixation. (*A*) Reductive tricarboxylic acid cycle. (*B*) Calvin cycle. (*C*) Hydroxypropionate and C4 pathways. In the 3-hydroxypropionate cycle, CO₂ is fixed by acetyl-coenzyme A (CoA) and propionyl-CoA carboxylases eventually forming Malyl-CoA. This is split into acetyl-CoA, to replenish the cycle, and glyoxylate, for use in cell carbon. (*D*) Reductive acetyl-CoA pathway. The reductive acetyl-CoA pathway is a noncyclic pathway. One CO₂ is captured on a special cofactor (tetrahydrofolate: T in the figure) and reduced to a methyl group. The other CO₂ is reduced to a carbonyl group (C=O) by carbon monoxide dehydrogenase and this enzyme-bound carbonyl group is combined with the methyl group to form acetyl-CoA by a collection of enzymes termed the acetyl-CoA synthase complex.

five-carbon compound (ribulose bisphosphate) are combined and the resulting six-carbon compound is split to form two molecules of 3-phosphoglyceric acid. This first step is catalyzed by the enzyme ribulose bisphosphate carboxylase, known as rubisco. Using a series of steps and by combining the products of multiple rubisco reactions, carbon compounds are built and ribulose bisphospate is regenerated. These steps utilize the **adenosine triphosphate (ATP)** and NADPH generated by phototrophic or chemotrophic reactions. The Calvin cycle is found in many photoautotrophs (the photosynthetic cyanobacteria, plants, algae, and the α-proteobacteria) as well as many chemoautotrophs. Some plants use a slight variant of this process in which the CO_2 is "stored" in the form of a four-carbon compound (the C4 pathway). Eventually the carbon is fed into the reactions of the Calvin cycle.

A specialized carbon assimilation pathway is found in one group of archaea—the methanogens. These species produce methane either by reduction of CO_2 (with electrons from various sources, including H_2) or from methylated substrates such as methanol. There are also many cases where a species has only some of the components of a pathway. For example, some organisms use parts of the methanogenesis pathway to fix carbon monoxide instead of carbon dioxide.

Bacteria and Archaea Possess Many Ways to Convert Light Energy to Cellular Energy (Phototrophy)

Phototrophs convert the energy found in visible light to cellular energy. Most phototrophs can make use of the energy they acquire from light to drive carbon fixation reactions. Thus these species are photoautotrophic and are also known as **photosynthetic**. Some of them can also obtain carbon from organic compounds and thus are heterotrophic as well.

If the mechanisms of photosynthesis in these different species are compared, we find that they all follow the same general schema (Fig. 6.22). Light energy is captured by light-antennae complexes, which channel the energy to generate excited electrons. The energy in the excited electrons is passed from the antennae to one of the proteins in a "reaction center." Each electron is passed through the reaction center, where the electron energy is slowly dissipated and the extracted energy is used to drive proton pumps across the membrane. Thus the reaction centers are frequently known as *electron transfer chains*. The proton gradient that is generated by these pumps can be used to generate cellular forms of energy (e.g., ATP), which are used to run carbon fixation reactions (among other things).

Comparing the mechanistic details of energy conversion in different photoautotrophs, we discover a great variety of photosynthetic processes. For the light absorption step, each species uses its own particular mix of "antenna" molecules including different types of chlorophyll, phycobilin, phycocyanin, phycoerythrins, and carotenoids (Fig. 6.23). All use some form of chlorophyll as the final step in the electron excitation process, but four major forms are known. The reaction centers are also highly diverse and come in two major forms, called Photosystem I and Photosystem II. Reaction centers differ in the types of molecules used for electron transport and in the molecules that are oxidized during the course of the transfer. Last, as discussed above, the mechanism of carbon fixation varies greatly.

Photosynthesis is sporadically distributed within the Tree of Life. It occurs in five groups of bacteria: the cyanobacteria, the Chlorobi, the Chloroflexi, the heliobacteria (members of the gram-positive lineage), and the α-proteobacteria. The only other known photosynthetic organisms are those eukaryotes (e.g., plants, algae, and euglenas) that have acquired photosynthesis through symbioses with chloroplasts (see pp. 202–208). Because plastids are derived from cyanobacteria, this means that all known forms of photoautotrophy evolved in bacteria. It is unclear why no separate form of photoautotrophy evolved in archaea or eukaryotes. Phylogenetic studies indicate that the photosynthetic processes in different bacterial groups did not evolve separately from scratch but rather that some amount of lateral gene transfer was involved.

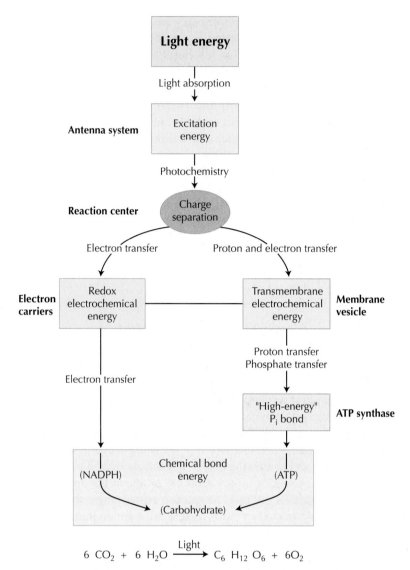

$$6 \ CO_2 \ + \ 6 \ H_2O \ \xrightarrow{\text{Light}} \ C_6 \ H_{12} \ O_6 \ + \ 6O_2$$

FIGURE 6.22. Photosynthesis is shown here as a series of reactions that transform energy from one form to another. The different forms of energy are shown in *boxes*, the direction of energy transformation is shown by the *arrows*, the energy-transforming reaction is shown by the type intersecting the arrows (e.g., electron transfer), and the site at which the energy is stored is shown in *boldface type* outside the boxes. The primary photochemical reaction, charge separation, is shown in the *oval*.

FIGURE 6.23. Light-harvesting pigments. A simplified scheme shows light absorption in antenna pigments followed by excitation energy transfer to a reaction center. The antenna and reaction center chlorophyll molecules are physically located in different proteins. Primary photochemistry (electron transfer from the primary electron donor to the primary electron acceptor) takes place in the reaction center.

Although photo*auto*trophy is a bacterial invention, other forms of phototrophy evolved outside the bacteria. One example is a system originally described in the halophilic archaea, which uses the energy of light to pump protons out of the cell and then uses the proton gradient to synthesize ATP. Unlike the complex systems of photosynthesis, only a single protein, known as bacteriorhodopsin, is required for this reaction (Fig. 6.24). (Bacteriorhodopsin was named before the archaea had been recognized as a separate domain.) It is similar in many ways to the rhodopsin that functions in human eyes; i.e., light is absorbed using a retinal chromophore that is bound in the membrane-spanning region of the bacteriorhodopsin protein. It is unclear, however, if this similarity is due to homology or convergence. Interestingly, extreme halophiles use homologs of this bacteriorhodopsin to pump chloride ions into the cell to maintain their high internal potassium ion concentration. Homologs of this bacteriorhodopsin have been discovered in uncultured oceanic bacteria (see pp. 150–152).

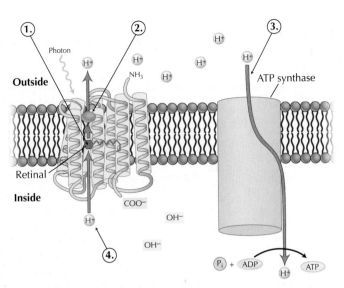

FIGURE 6.24. Bacteriorhodopsin function. (*1*) The protein absorbs a photon and retinal becomes protonated. (*2*) The proton is transferred through a series of intermediates and released outside the cell membrane. (*3*) Protons outside the cell reenter via ATP synthase, which generates ATP. Thus, sunlight is converted directly to energy. (*4*) The cycle begins again when deprotonated retinal takes up another H^+ from the cytoplasm.

These bacteria apparently use their rhodopsin to generate energy when food sources are limited. This form of phototrophy may account for up to 20% of the energy budget of all oceanic microoorganisms.

Bacteria and Archaea Have Diverse Types of Interactions with Other Species: Mutualism, Parasitism, and Sociality

The social and symbiotic interactions of bacteria and archaea play critical roles in almost every ecosystem. In the broadest sense, symbiosis includes any interaction between two or more organisms that persists for a significant portion of the partners' lifetimes. Such interactions range from parasitic (e.g., pathogen and host) to mutualistic (where both partners benefit). There are notable differences between archaea and bacteria in their symbioses. For example, there are no known archaeal pathogens. In this section we discuss some of the inter- and intraspecies interactions that exemplify the diversity of bacteria and archaea.

The Metabolic Diversity of Bacteria and Archaea Has Been Harnessed by Eukaryotes through Mutualistic Symbioses

The diversity of metabolic processes in bacteria and archaea is much greater than in eukaryotes. However, eukaryotes have been able to capitalize on this diversity through mutualistic symbioses with bacteria and archaea. The most profound examples of mutualistic symbiosis are the mitochondria and chloroplasts of eukaryotes, which once were free-living bacteria. The ancestors of chloroplasts were free-living cyanobacteria that were internalized by eukaryotic host cells. This symbiosis led to the development and diversification of all photosynthetic eukaryotes. The ancestors of mitochondria were free-living α-proteobacteria, whose metabolic capabilities were harnessed early in eukaryotic evolution (the evolution and functions of these organelles are discussed further on pp. 202–210). Mitochondria and chloroplasts are highly derived systems that have become so integrated with their host that they are now classified as organelles. There are hundreds of other types of mutualistic symbioses between bacteria or archaea and eukaryotes that are in different stages of evolution (Table 6.8). We describe some of them here.

TABLE 6.8. Symbioses

Type of Symbiosis	Host Species	Symbionts	Description
Autotrophic	Giant tube worm, *Riftia pachyptila*	γ-proteobacteria	Chemoautotrophic bacteria feed the host sugars just as plastids do for plants.
	Tunicates	*Prochloron*	Photosynthetic bacteria feed the host animal sugars.
	Fungus	Algae or cyanobacteria	Lichens are able to photosynthesize.
Nutritional	Aphids	*Buchnera*	Symbionts synthesize amino acids for insect species that feed on amino-acid-poor sap.
	Humans	*Escherichia coli, Bacteroides* spp.	Microbes in the gut aid in digestion and the production of vitamins and cofactors.
Nitrogen fixation	Legumes	Rhizobia	Rhizobia infect nodules in legume roots and fix nitrogen for host.
	Azolla (water fern)	Cyanobacteria	Cyanobacteria fix nitrogen for host.
Digestive	Termites	Bacteria, archaea	Symbionts help digest cellulose.
	Cattle	Ciliates, bacteria, archaea	Symbionts help digest cellulose.
Luminescent	Flashlight fish	Photobacterium	Bacteria emit light in specialized fish organs.

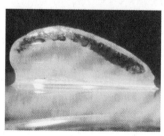

FIGURE 6.25. *Prochloron didemni* symbiosis with the ascidian *Lissoclinum patella*. (*Top*) An *L. patella* colony (~3 cm long). (*Bottom*) Cross section of *L. patella*, showing location of *Prochloron* (green).

Photoautotrophic symbioses. The chloroplast is an example of a photoautotrophic symbiosis. Others include the symbiosis between cyanobacteria and fungi to form lichens, and one of the most unusual, involving *Prochloron*, a cyanobacterium that lives inside the tunic region of sea squirts, which are primitive marine chordates (Fig. 6.25). In addition to carrying out photosynthesis, *Prochloron* may function as a rudimentary immune system for its host by producing secondary metabolites that protect the tunicate from pathogens.

Chemoautotrophic symbioses. A completely novel form of autotrophic symbiosis was discovered in the 1980s from studies of deep-sea organisms inhabiting hydrothermal vent communities. Chemoautotrophic bacteria live inside various invertebrates, such as giant tube worms, providing their hosts with fixed carbon. Unlike photoautotrophic symbionts, the bacteria inside the tube worms use chemical energy (e.g., the oxidation of H_2S) to drive the carbon fixation. In many ways, the host animals resemble plants. For example, the tube worms are immobile in the adult stage, have no digestive system, and are highly adapted to providing chemicals and CO_2 to the symbionts (see Fig. A&S.1). Other chemoautotrophic symbioses have been found widely in nature, especially in the oceans—generally wherever there are large amounts of highly reduced chemicals. These types of symbioses have been described in more than 200 species from at least five invertebrate phyla and one ciliate and involve multiple kinds of bacteria.

Nitrogen fixation in legumes. Leguminous plants (such as peas, beans, and clover) are able to utilize soil bacteria to increase their access to usable forms of ni-

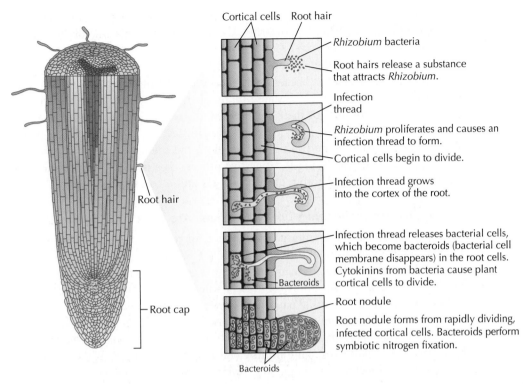

Cortical cells   Root hair

- *Rhizobium* bacteria
- Root hairs release a substance that attracts *Rhizobium*.

Infection thread

- *Rhizobium* proliferates and causes an infection thread to form.
- Cortical cells begin to divide.
- Infection thread grows into the cortex of the root.
- Infection thread releases bacterial cells, which become bacteroids (bacterial cell membrane disappears) in the root cells. Cytokinins from bacteria cause plant cortical cells to divide.

Bacteroids

- Root nodule
- Root nodule forms from rapidly dividing, infected cortical cells. Bacteroids perform symbiotic nitrogen fixation.

Root hair

Root cap

Bacteroids

FIGURE 6.26. Symbiosis of *Rhizobium* bacteria with legumes.

trogen. The plant roots are colonized by *Rhizobium* or *Bradyrhizobium* species, which form colonies within special root nodules. The host provides the proper environmental conditions for the bacteria to carry out nitrogen fixation—the reduction of gaseous nitrogen to ammonia (Fig. 6.26). This symbiotic relationship is of enormous agricultural importance, as it is used to replenish nitrogen-depleted soils. Leguminous "cover crops," such as clover or alfalfa, are planted and then plowed into the earth, thus greatly enriching the nitrogen content of the soil. Clearly, this symbiosis also can provide a selective advantage to the host legumes as it allows them to grow in nitrogen-poor soil much more effectively than other plants.

Nutritional symbioses. Another form of nutritional symbiosis is found in many sap-feeding insects, which obtain most or all of their nutrients from the fluid inside the circulatory system of plants. In many plant species this fluid is very low in nutrients including amino acids essential to insects. Many sap-feeding insects, such as aphids, harbor intracellular symbiotic bacteria in their guts, which synthesize these essential amino acids for the host. Because many of these insect lineages have been feeding on sap for millions of years, it is not surprising to find that the symbionts appear to have been vertically transmitted for millions of years. The existence of such co-evolving symbioses is very important for studies of bacterial evolution, because the fossil records of the hosts can be used to date the evolutionary history of the symbionts. In addition, these symbioses may explain why these bacteria have highly reduced genomes, similar to those in organelles (p. 175).

Digestive and other gut symbioses. Many animals, including humans, live on foods that they cannot digest. They make use of symbiotic microbes that live in their gastrointestinal system to carry out many of the steps required for digestion. The best-characterized example involves microbes that aid in the digestion of cellulose, a complex carbohydrate that makes up the bulk of many plants. Ruminants live almost entirely on grasses rich in cellulose. Lacking the enzymes necessary to digest cellulose, they instead rely upon a diverse collection of microbes in their gut to slowly digest cellulose and convert it into sugars that can be absorbed into the bloodstream. Other cellulose-digesting animals, including termites and shipworms, also rely on symbiotic microbes to digest the cellulose.

A key component of the ecology of ruminant digestion is the formation of methane by methanogenic archaea. These species are autotrophic, fixing CO_2 and using H_2 as an electron acceptor. Because many of the cellulolytic microbes produce CO_2 and H_2 as a by-product of their metabolism, the utilization of these compounds by the methanogenic archaea helps drive cellulolytic reactions. Thus, archaea play a critical, if indirect, role in the process. In addition, the methane produced, which is released by the ruminants during the digestive process, is a major source of methane in the atmosphere, which is both a greenhouse gas and a contributor to smog.

Many Bacteria Are Pathogenic, but No Archaea Are Known to Be

Of course, not all symbioses are mutualistic. In fact, microorganisms are best known for their detrimental associations with eukaryotic hosts. Such microbes are generally known as pathogens. Although there are many bacteria that are pathogenic, interestingly, there are no known archaeal pathogens. It is not clear whether this is because archaea cannot be pathogenic or because archaeal pathogens have not yet been identified.

The best-studied bacterial pathogens are those that cause human diseases, such as cholera (*Vibrio cholerae*), plague (*Yersinia pestis*), Lyme disease (*Borrelia burgdorferi*), syphilis (*Treponema pallidum*), and anthrax (*Bacillus anthracis*) (see Fig. 6.1). Bacterial pathogens also infect plants, fungi, other animals, and protists. Much can be learned about bacterial pathogens from an examination of their evolution. Most striking is the intermixing in the bacterial tree of pathogens and nonpathogens, which suggests that either the ability to cause disease (also known as pathogenicity) has evolved multiple times or the genes required for pathogenicity have been transferred between species. This is further supported by the discovery that the mechanisms of pathogenicity and **virulence**, which is the degree of pathogenicity, differ significantly between pathogens. For example, many bacterial pathogens produce virulence factors that enhance their ability to cause disease. The diversity of known virulence factors is incredible. Some function by breaking down cellular structures such as collagen, clots, or hyaluronic acid. Others lyse host cells by punching holes in the cell membrane (e.g., hemolysins). Still others alter cell–cell signaling by binding to and/or degrading neurotransmitters (in animals) or hormones (in plants and animals).

In hosts with immune systems, infecting pathogens often have evolved ways of avoiding or delaying the immune response. Some do this by "hiding" inside host cells; others rapidly change their surface proteins by recombination and mutation (pp. 662–663), thus delaying the host's ability to mount an effective immune response.

The genes required for virulence and pathogenicity frequently cluster within the genomes of pathogenic species. These clusters are often significantly different from other regions of the genome in terms of GC content, codon usage, and other signals, and thus have become known as **pathogenicity islands** (see p. 187). Many studies indicate that these islands can be transferred relatively easily between strains or species, which may explain the "sudden" appearance of pathogenicity in a group of otherwise harmless organisms.

In considering the evolutionary origin of pathogenicity, it is helpful to examine parallels between how pathogens and mutualistic symbionts interact with their respective hosts. For example, many mutualistic symbionts have "symbiosis" islands in their genomes that encode genes necessary for the mutalistic partnership. In addition, for those mutualistic symbionts that infect host cells (e.g., *Rhizobium* nodulation of legumes), the infection mechanisms (which are genetically encoded) are similar to those used by pathogens. The secretion systems (a set of proteins working in a coordinated fashion) used to export toxins from pathogens are the same systems used to export symbiosis factors in mutualistic symbionts. Perhaps the transition from pathogen to mutualist or vice versa is not very complicated. Some pathogens may be "runaway" mutualists that added toxins to their repertoire to the detriment of their hosts. Some mutualists may be "tamed" pathogens that have lost virulence factors but provide some benefit to their hosts. In fact, there are many organisms that appear to exist as both mutualistic symbionts and pathogens depending on the stage in their life cycle. For example, pathogenic bacteria in the group Rickettsiaceae are dependent on invertebrate vectors for their transmission between animal hosts. Although they cause major diseases in various animals (e.g., Rocky Mountain spotted fever, typhus, and ehrlichiosis), many appear to be beneficial symbionts for the vector (e.g., ticks) that transmits them between hosts.

Bacteria and Archaea Have Evolved Many Ways of Interacting with Other Members of Their Species

Contrary to what is commonly thought, bacteria and archaea may have complex social interactions with other members of their species. These social interactions allow them to function in many ways like multicellular organisms. Here we describe one of the key aspects of this sociality—the ability to regulate their biological processes in response to changes in population density. This is known as **quorum sensing**.

Quorum sensing occurs in many species of bacteria, and although it has been suggested to be present in archaea, no conclusive proof has been found. Where it occurs, individual cells continuously secrete a chemical signal into their environment. As the population density increases, the concentration of the chemical signal increases. Individual cells monitor the signal concentration in the environment, and when it reaches a threshold level, a regulatory cascade is induced. This is known as **autoinduction**, and the chemical signal is the **autoinducer**. Quorum sensing enables bacteria to coordinate the behavior of a group of cells and function in many ways like multicellular organisms (Fig. 6.27).

In many mutualistic symbionts, quorum sensing is used to determine when the

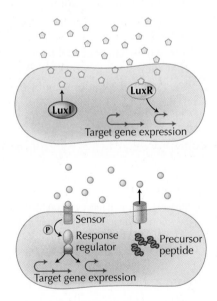

FIGURE 6.27. There are two major variants of quorum sensing. In many gram-negative species, the autoinducer is an acyl homoserine lactone (*pentagons*). In gram-positive species, it is often a peptide known as AIP (autoinducing peptide). A specific strain of bacteria produces its own form of autoinducer molecule, which then diffuses freely into the environment as well as into other cells. Each strain also has a receptor protein that recognizes its own autoinducer. When the autoinducer concentration crosses a particular threshold, the receptor becomes a transcription activator, leading to the induction of a suite of genes that have the activator-binding site as part of their promoter. An important aspect of the evolution of the autoinducer-based quorum sensing is that all known autoinducer-based processes are homologous and use members of one gene family (LuxI) to produce the autoinducer and members of another gene family (LuxR) to encode the receptor.

bacteria should activate particular processes used by the host. For example, the bacterium *Vibrio harveyi* produces light in the light organs of many animal hosts. The bacteria use quorum sensing to detect when they are in a confined space (the autoinducer concentration is able to cross a high threshold). This indicates that the light organ of the host has been colonized and a regulatory cascade is turned on that eventually leads to the production of light. In many pathogenic bacteria, quorum sensing is used to activate virulence factors, probably because they are likely to be ineffective if the pathogen is not present in sufficient numbers. In *Streptococcus pneumoniae* and many of its relatives, quorum sensing is used to enable cells to take up DNA from their environment, a process known as competence. In other species quorum sensing is used to regulate the formation of biofilms, the development of fruiting bodies, and the production of spores. The processes that cause the evolution of social traits, like quorum sensing, are discussed in Chapter 21.

SUMMARY

Bacteria and archaea play fundamental roles in every major ecosystem and biogeochemical process on Earth. Although their importance as pathogens and in industrial processes, like fermentation, has been appreciated for the past 100 years, it was not until the recent development of molecular survey methods that we have begun to fully appreciate their importance and understand their diversity. The best example of this is the recognition in the 1970s that bacteria and archaea are separate domains of life. In addition, molecular methods have allowed a thorough understanding of the diversity within these domains. For example, PCR surveys of rRNA genes have led to the identification of dozens of phyla of each of these groups and have shown that many phyla are composed entirely of species that have never been grown in the laboratory.

These rRNA surveys, coupled with genome sequencing methods and experimental studies, have started to reveal many details in the lives of these important microbes. For example, features have been identified that are universal in bacteria but absent in archaea, and vice versa. This adds support to the classification of these organisms as distinct domains of life. In addition, we have begun to understand the incredible phenotypic and physiological diversification of these species. In this chapter, three aspects of this diversification were highlighted: the ability to thrive in extreme environments, the ability to carry out nearly any conceivable biochemical reaction, and the diverse ways that these species interact with others (e.g., mutualistic symbioses and pathogenic interactions). Although bacteria and archaea are still somewhat neglected in evolutionary studies, we believe that they serve as excellent model systems for addressing fundamental questions about the evolution of life.

FURTHER READING

Phylogenetic Diversification

Hugenholtz P. and Pace N.R. 1996. Identifying microbial diversity in the natural environment: A molecular phylogenetic approach. *Trends Biotechnol.* **14:** 190–197.

Pace N.R., Olsen G.J., and Woese C.R. 1986. Ribosomal RNA phylogeny and the primary lines of evolutionary descent. *Cell* **45:** 325–326.

Woese C.R. 1987. Bacterial evolution. *Microbiol. Rev.* **51:** 221–271.

Biological Diversification

Extremophiles

Horikoshi K. and Grand W.D., eds. 1998. *Extremophiles: Microbial life in extreme environments.* Wiley-Liss, New York.

Rothschild L.J. and Mancinelli R. 2001. Life in extreme environments. *Nature* **409:** 1092–1101.

Metabolism

Shively J.M., van Keulen G., and Meijer W.G. 1998. Something from almost nothing: Carbon dioxide fixation in chemoautotrophs. *Annu. Rev. Microbiol.* **52:** 191–230

Xiong J. and Bauer C. 2002. Complex evolution of photosynthesis. *Annu. Rev. Plant. Biol.* **53:** 503–521.

Inter- and Intraspecies Interactions

Bassler B.L. 2002. Tiny conspiracies. *Nat. Hist.* **110:** 16–21.

Losick R. and Kaiser D. 1997. Why and how bacteria communicate. *Sci. Am.* **276:** 68–73.

Ochman H. and Moran N.A. 2001. Genes lost and found: Evolution of bacterial pathogenesis and symbiosis. *Science* **292:** 1096–1099.

CHAPTER

7

Diversification of Bacteria and Archaea II: Genetics and Genomics

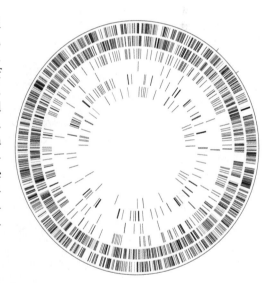

HE RECOGNITION THAT ORGANISMS WITHOUT nuclei are composed of two distinct phylogenetic groups, the bacteria and the archaea, is one of the most important successes of molecular systematics. Strikingly, despite their phylogenetic separation, there are many aspects of genome structure and genetic processes that are similar in the two groups. These similarities include circular chromosomes, genomes densely packed with genes, limited use of introns, and the presence of many operons. Both archaea and bacteria couple transcription and translation, possess a diversity of phage and plasmids, and exhibit an extensive amount of lateral DNA transfer across wide evolutionary distances. There are also some significant differences between bacterial and archaeal genetic processes, including mechanisms of DNA replication and transcription. In this chapter, we discuss how these similarities and differences have shaped the diversification of bacteria and archaea.

THE NATURE OF ARCHAEAL AND BACTERIAL GENOMES

Bacteria and Archaea Vary Greatly in the Size and Types of Different Genetic Elements

Bacteria and archaea exhibit great variation in the size and types of genetic elements they contain. In many cases, the entire genome is contained in a single circular piece of DNA. For example, *Haemophilus influenzae*, the first organism to have its genome completely sequenced, has a single circle of 1830 kilobase pairs (kb; 1 kb = 1000 base pairs, 1000 kb = 1 Mb). More than half of the bacteria and archaea for which complete genomes are available at this time have all of their DNA in a single circular chromosome.

In addition to their chromosomes, many bacteria and archaea contain small circles of DNA referred to as **plasmids**. Examples include the pathogenic bacterium *Escherichia coli* O157:H7 (one chromosome of 5.5 Mb, and plasmids of 92 kb and 3 kb) and the endosymbiotic bacterium *Buchnera aphidicola* APS (one chromosome of 640 kb, and plasmids of 14 kb and 7 kb). We include the strain designations here (O157:H7 and APS) and throughout the chapter when it is known that there are significant differences in genome content or structure among strains within that species.

In many organisms, there is a clear distinction in size between the chromosomes and the plasmids. However, as more complete genome sequences are determined, size is no longer an infallible criterion for distinguishing different types of genetic elements. For example, the halophilic archaeon *Haloferax volcanii* has five circular DNA elements with sizes of 2.92 Mb, 690 kb, 442 kb, 86 kb, and 6.4 kb (other examples are in Table 7.1). If size were the only criterion, we might consider the 690-kb element in *H. volcanii* to be a second chromosome because it is larger than the chromosome of *B. aphidicola* APS, for example. However, size is just a property that helps distinguish plasmids from chromosomes. More importantly, there are significant biological differences between plasmids and chromosomes. Some of these differences are discussed in the following sections.

Plasmids, unlike chromosomes, are generally "accessory" elements, carrying genes that are required only under certain conditions (Table 7.2). For example, the *B. aphidicola* APS plasmids encode genes needed to synthesize tryptophan and leucine, two of the amino acids that the bacteria provide for their host. The *B. aphidicola* APS chromosome encodes all the information for DNA replication, transcription, translation, cell-membrane and cell-wall formation, and the other genes required to assemble the core machinery of the cell. In *E. coli* O157:H7, the 92-kb plasmid encodes many virulence factors that contribute to the disease caused by this bacterium, whereas the chromosome encodes all the housekeeping functions. Because plasmids typically have only

TABLE 7.1. Examples of bacteria with multiple genetic elements

Species	Form	Size (kb)	Shape
Streptomyces coelicolor	Chromosome	8667	Linear
	Plasmid	356	Linear
	Plasmid	31	Circular
Agrobacterium tumefaciens	Chromosome	2842	Circular
	Chromosome	2057	Linear
	Plasmid	543	Circular
	Plasmid	214	Circular
Borrelia burgdorferi	Chromosome	911	Linear
	Plasmid ($n = 11$)	9–54	Circular/Linear
Brucella melitensis	Chromosome	2117	Circular
	Chromosome	1178	Circular
Clostridium acetobutylicum	Chromosome	3941	Circular
	Plasmid	192	Circular
Deinococcus radiodurans	Chromosome	2649	Circular
	Plasmid	412	Circular
	Plasmid	177	Circular
	Plasmid	46	Circular
Ralstonia solanacearum	Chromosome	3716	Circular
	Chromosome?	2095	Circular
Salmonella typhi	Chromosome	4809	Circular
	Plasmid	218	Circular
	Plasmid	107	Circular
Sinorhizobium meliloti	Chromosome	3654	Circular
	Plasmid	1683	Circular
	Plasmid	1354	Circular
Vibrio cholerae	Chromosome	2941	Circular
	Chromosome	1072	Circular
Yersinia pestis	Chromosome	4654	Circular
	Plasmid ($n = 3$)	10–96	Circular

Based on Bentley S.D. and Parkhill J. *Annu. Rev. Genet.* **38**: 771–792, as adapted from Ohmachi M. 2002. *Curr. Biol.* **12**: R427–428.

TABLE 7.2. Plasmid functions

Genetic Function of Plasmid	Gene Functions	Examples
Resistance	Antibiotic resistance	Rbk plasmid of *Escherichia coli* and other bacteria
Fertility	Conjugation and DNA transfer	F plasmid of *E. coli*
Killer	Synthesis of toxins that kill other bacteria	Col plasmids of *E. coli*, for colicin production
Degradative	Enzymes for metabolism of unusual molecules	TOL plasmid of *Pseudomonas putida*, for toluene metabolism
Virulence	Pathogenicity	Ti plasmid of *Agrobacterium tumefaciens*, conferring the ability to cause crown gall disease on dicotyledonous plants

accessory functions, an organism can usually survive without them provided it is not exposed to the specialized conditions for which the plasmids are needed. In turn, this means that plasmids are commonly lost from particular bacterial and archaeal strains.

In most species, only one copy of the chromosome (or, at most, a few) is present per cell. Plasmids, however, are frequently present in much greater copy number; sometimes there are hundreds of copies per cell. Allowing the copy number of plasmids to increase (while controlling chromosome copy number) in essence means that all of the genes on the plasmids undergo substantial gene duplication. For example, in *B. aphidicola* APS, the ratio of tryptophan and leucine plasmid number to chromosome number is greater than 10:1. This difference in copy number arises because plasmid and chromosomal replication are not coupled. Furthermore, the two frequently use entirely separate replication mechanisms. In addition, because plasmids and chromosomes use different replication systems, they frequently have different rates and patterns of mutation.

From an evolutionary point of view, the most important distinction between plasmids and chromosomes is the ease with which plasmids move between strains and even species. The mobility of plasmids plays a critical role in **lateral gene transfer** (see below). This transfer of plasmids results in very sporadic plasmid distribution patterns when different strains of one species or different species are compared.

In almost all species, there is only one chromosome and all other genetic elements are plasmids. There are, however, a few notable exceptions of bacteria with more than one chromosome. The causative agent of cholera, *Vibrio cholerae*, has two large genetic elements (2.9 and 1.1 Mb in size; see Table 7.1). Both encode multiple housekeeping genes and are found in all close relatives of this species (and thus do not have sporadic distribution patterns). Therefore, both elements qualify as chromosomes.

Agrobacterium tumefaciens, which causes crown gall tumors in plants, has an unusual pair of chromosomes: One is circular (as is typical for bacteria), but the other is linear. Once thought to be the exclusive province of eukaryotes, linear genetic elements have now been found in several species of bacteria.

Linear chromosomes are faced with a unique problem: DNA polymerases cannot replicate the ends of the chromosome, because the enzymes cannot replace the terminal RNA primer of the lagging strand (see Box 12.1). Without another mechanism for replicating the ends (i.e., the **telomeres**), linear chromosomes would become shorter with each round of replication. Eukaryotes use a specialized enzyme, **telomerase**, which adds a repeating DNA motif to the telomeres (see Fig. 8.17). Bacteria, like *A. tumefaciens*, appear to use a similar mechanism for preserving the ends of their linear chromosomes. Furthermore, it appears that these replication systems arose independently in bacteria and eukaryotes and, thus, are an interesting example of convergent evolution.

Bacterial and Archaeal Genomes Are Much Smaller and More Compact Than Those of Eukaryotes

Bacterial and archaeal genomes are smaller than the vast majority of eukaryotic genomes (Fig. 7.1). Among bacteria, genomes range in size from 160 kb (the obligate symbiont *Carsonella ruddii*) to more than 13 Mb (the δ-proteobacterium *Sorangium cellulosum*). Archaeal genomes range from 490 kb (*Nanoarchaeum equitans*, a symbiotic species [Fig. 6.7]) to 5.7 Mb (the methanogen *Methanosarcina acetivorans*). The median genome size for both archaea and bacteria is approximately 2 Mb.

When comparing bacteria and archaea with eukaryotes, the difference in genome size is much greater than the difference in the number of genes. This is because the density of genes is very great within bacterial and archaeal genomes (Fig. 7.2). For example, the human genome is approximately 1000 times bigger than the *E. coli* K12 genome, yet humans have only about ten times as many protein-coding genes (Fig. 7.3). In fact, a number of bacteria and archaea have more protein-coding genes than some eukaryotes. Almost all species of Myxobacteria (a subgroup of fruiting-body-forming δ-proteobacteria, including *S. cellulosum* with the 13-Mb genome) have greater than 8000 protein-coding genes, which is more than in the model yeast species *Saccharomyces cerevisiae* and *Schizosaccharomyces pombe*.

The great density of genes within the genomes of bacteria and archaea is due to the paucity of noncoding DNA compared with that in eukaryotic genomes. **Introns** and intergenic regions (i.e., the DNA located between genes) are rare and generally small in bacteria and archaea. Instead, as mentioned in Chapter 6, many bacterial and archaeal genes are organized into **operons**, clusters of cotranscribed genes that use only a single promoter for the entire gene cluster. This organization helps to create a compact genome. The genes found in a single operon are usually involved in similar functions (e.g., the same metabolic pathway or a single-protein complex) (Fig. 7.4). Operons are a critical feature of the genomes of bacteria and archaea. For example, it is estimated that *E. coli* K12 has about 700 operons in its genome.

Eukaryotic genomes are bulky in part because they contain large numbers of repetitive DNA elements (Fig. 7.2). Common eukaryotic repetitive DNA elements include sim-

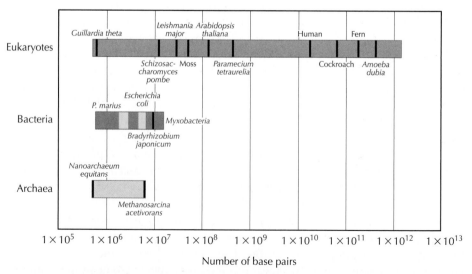

FIGURE 7.1. Genome sizes in the three domains of life. A selection of genome sizes and size ranges from specific groups of organisms is indicated.

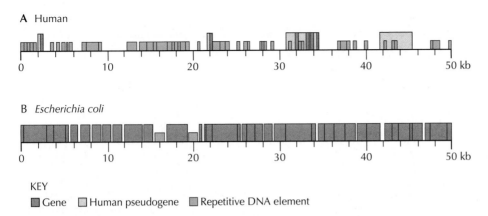

KEY
■ Gene □ Human pseudogene ■ Repetitive DNA element

FIGURE 7.2. Genome density. Comparison of the genome density and content of humans and *Escherichia coli*. Each segment is 50 kb in length and represents (*A*) a portion of the human β T-cell receptor locus and (*B*) a region of the *E. coli* K12 genome. Note the much greater proportion of genes (*red boxes*) in *E. coli* compared to humans.

ple sequence repeats (e.g., **microsatellites** and **minisatellites**), gene duplications (both tandem arrays and **pseudogenes**), and **transposable elements**. Although bacterial and archaeal genomes contain repetitive DNA, the total amount is relatively small. For example, hundreds of thousands of copies of transposable elements are present in many eukaryotic genomes, yet in bacteria and archaea it is rare to have even 100 copies.

Pressure to Streamline Genomes Causes Bacteria and Archaea to Lose Genes Not Actively Maintained by Selection

To understand the evolution of bacterial and archaeal genomes, it is useful to ask why there is so much more noncoding DNA in most eukaryotic genomes. Clearly, some of the extra DNA in eukaryotes has important functions such as gene regulation. However, much of the noncoding DNA in eukaryotic genomes has been classified as either **junk DNA** or **selfish DNA**. Junk DNA appears to provide little benefit or no function to the organism. (In some cases this designation is a misnomer resulting from a lack of infor-

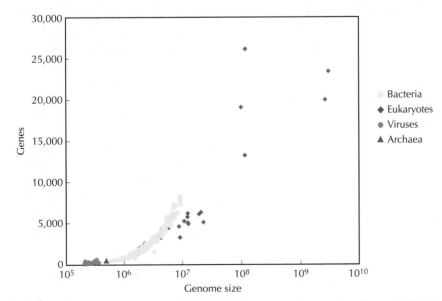

FIGURE 7.3. Genome size vs. number of protein-coding genes. The number of genes is highly correlated to genome size for bacteria, archaea, and viruses, but less so for eukaryotes. Many archaeal points (*blue triangles*) are hidden under bacterial ones (*yellow squares*).

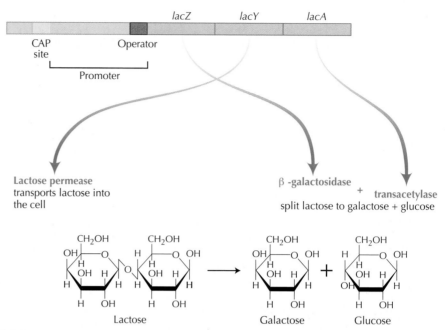

FIGURE 7.4. Lac operon from *Escherichia coli*. This operon consists of three genes whose transcription is regulated by a single promoter. The genes encode proteins involved in utilizing lactose, including a permease (encoded by *lacY*), which brings lactose into the cell from the outside, and two enzymes (encoded by *lacZ* and *lacA*), which split lactose into glucose + galactose (see pp. 52–53).

mation. Some stretches of "junk DNA" have been determined to be involved in gene regulation, chromatin organization, centromere activity, and other functions.) Selfish DNA is composed of mobile DNA elements that facilitate their own duplication, even if it is to the detriment of the host.

All of the many theories that have been proposed to explain why junk DNA and selfish DNA are less abundant in bacteria and archaea agree that there is some global pressure to keep total genome size small. This global pressure is most likely selection, although there may also be a bias toward deletion of DNA. Indeed, such a mechanism in bacteria and archaea could be responsible for keeping introns both small and rare, holding transposable elements in check, maintaining operons, and culling junk DNA. This global pressure and other theories on the evolution of genome size are discussed in more detail in Chapter 21. Here we discuss its effects on the general patterns of genomic evolution in bacteria and archaea.

The limited occurrence of introns in bacteria and archaea has many important consequences. For example, although eukaryotes can make thousands of protein products from a single gene by **alternative splicing**, this is not seen in bacteria and archaea. In addition, mixing and matching of protein domains is less common in bacteria and archaea than in eukaryotes, possibly because such events are caused mainly by **recombination** in introns.

The extensive use of operons also has significant consequences. In some respects, operons are a major constraint; mutations that break up the operon (e.g., by causing a rearrangement in the middle of the operon) may be quite detrimental. In other ways, operons facilitate rapid acquisition of new features by bacteria and archaea because they allow complete pathways to be transferred readily between strains or species (see the discussion of lateral transfer later in this chapter). In contrast, in many eukaryotes, with genes involved in the same pathway scattered around the genome, it is unlikely that all of the genes would be transferred at one time to another strain or species.

In bacteria and archaea, the pressure to streamline genomes (whether caused by mutation bias or selection for small genomes or both) means that genes that provide no advantage are rapidly lost (see Box 18.2). Thus, although vestigial genes may linger for long periods in eukaryotes, they do not linger in bacteria and archaea. For exam-

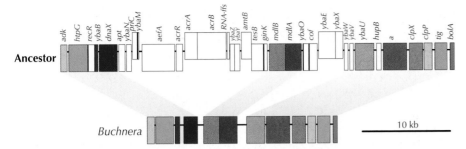

FIGURE 7.5. Genome reduction in *Buchnera* endosymbionts of aphids. A fragment of two genomes is shown. (*Top row*) The putative ancestor of all aphid endosymbionts in the *Buchnera* genus. (*Bottom row*) The genome of the symbionts today. The massive amounts of gene loss are indicated by the genes colored *white* in the ancestral genome that are missing from the modern genome below. Orthologous genes between the two genomes are shown in the same color. Note the conservation of gene order between the two genomes despite the gene loss. The direction of gene transcription is indicated by the gene box being shifted above or below the *black line*.

ple, *B. aphidicola* APS has undergone a massive reduction in its genome since it shared a common ancestor with *E. coli* (Fig. 7.5). This symbiont lives inside aphid cells where many genes required for the free-living lifestyle of *E. coli* are not needed.

Gene Content Is in Constant Flux in Bacteria and Archaea

The availability of hundreds of complete genome sequences enables scientists to examine how gene content evolves. The first analysis of this sort was performed using the first two sequenced genomes: *M. genitalium* and *H. influenzae*. Despite the fact that both species have very small genomes, hundreds of homologous genes were identified. These shared genes were proposed to be the "minimal gene set" of a bacterium; that is, they might represent the genes that are essential for making a bacterium (Fig. 7.6A).

However, as more genomes from different phylogenetic groups have been sequenced, the number of "core" homologous genes has diminished (Fig. 7.6B). The reason for this became clear when genomes from different strains of the same species were compared. This was first done with the pathogenic strain *E. coli* (O157:H7) and the *E. coli* K12 laboratory strain. Although these strains share approximately 4000 highly conserved genes, O157:H7 has more than 1000 genes not found in K12, and K12 has approximately 500 genes absent from O157:H7 (Fig. 7.7). Similar patterns of

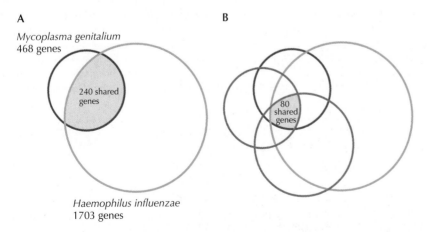

FIGURE 7.6. (*A*) Comparison of predicted protein-coding genes in the first two completed genomes *Haemophilus influenzae* and *Mycoplasma genitalium*. Approximately 240 genes are shared between the two species. (*B*) Comparison of the predicted protein-coding genes of the first 25 bacterial genomes (not all 25 circles are shown). Note that only about 80 genes can be identified as being shared among all of these species.

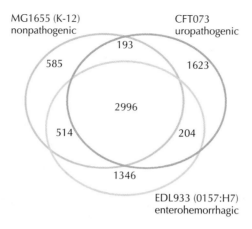

FIGURE 7.7. Number of shared proteins between strains of *Escherichia coli*. Note the large number of genes found in one strain but not the others (seen in the *outer* portions of each *circle*).

substantial variation in gene content among members of *the same species* have been reported in other lineages of bacteria and archaea. Thus, the diminishing number of core **orthologous genes** is simply an extension of something happening among close relatives.

How do such extensive differences in gene content among close relatives originate? One of the most important clues comes from comparing the genome structures of related species. (A graphical method for aligning circular genomes is introduced in Box 7.1; see Figs. 7.8 and 7.9.) In comparing *E. coli* K12 and O157:H7, the genes that are shared between the two strains not only are highly conserved at the sequence level; but

---

Box 7.1 Graphical Alignment for Comparing Circular Genomes Using Dotplots

Comparing the arrangement of genomes is a critical tool for understanding how they evolve. This enables scientists to identify and characterize genome rearrangements (e.g., inversions and translocations) and to search for patterns and associations that may explain how and why certain events occur. For example, differences in gene order between species are frequently at sites where repetitive DNA is found, which suggests that recombination at the repetitive DNA may have led to rearrangements. One of the more useful methods compares two genomes on an *x–y* plot, a procedure commonly referred to as a dotplot.

Dotplots let people use their visual pattern-recognition skills to identify similarities. Their power and simplicity have made them a valuable analytical tool in fields beyond biology, including electrical engineering and computer sci-

ence. Let us illustrate the method using some text-based examples. Figure 7.8A plots a familiar quotation against itself. The central diagonal line is the axis of identity. The outlying points represent text that repeats. A quick examination can distinguish a pattern that is repeated in its entirety (Fig. 7.8B) from one with some unique elements (Fig. 7.8C).

Because most bacterial and archaeal chromosomes are circular, a chromosome must first be "opened" before laying it out on the *x*- or *y*-axis. Although the circle can be linearized at any point, it is preferable to open each chromosome at its origin of replication (Fig. 7.9A). One linearized chromosome is then aligned along the *x*-axis with the origin of replication placed at the graphical origin. The other chromosome is similarly arranged along the *y*-axis. The two chromosomes are com-

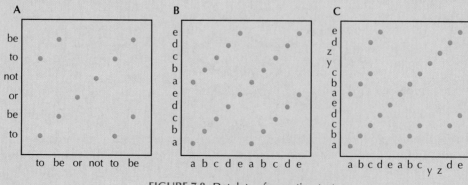

FIGURE 7.8. Dotplots of repeating text.

Box 7.1. Continued.

pared by searching for some designated pattern of similarity or conservation; for example, homologous or orthologous genes can be identified, and a dot is used to represent each pair of such genes. The dotplot (in this example) is a visualization of conserved gene order between the two chromosomes.

Alternatively, a sliding-window approach can be used to compare two DNA sequences. Each gene is divided into segments of a specified number of base pairs. Pairs of segments with sequence similarity are then represented by a dot. Analyzing these dotplots reveals patterns beyond conservation of gene order and features such as conserved noncoding DNA or conserved locations of repeats.

Figure 7.9 provides some examples of how genome dot-

plots are used. Figure 7.9B–D shows how dotplots can be used to investigate the internal architecture of a single genome by comparing the chromosome to itself. Figure 7.9B shows a nonrepeating genome plotted against itself. All of the dots are along a single diagonal, which indicates that each plotted region is similar only to itself. Repeated genes or DNA sequences appear as rows of dots off the primary diagonal (Fig. 7.9C). A greater number of repeats create a more complex pattern (Fig. 7.9D). When two different genomes are compared, dotplots can provide insight into how genomes have diverged. For example, Figure 7.9E shows a dotplot for two genomes that differ by an inversion and Figure 7.9F shows a dotplot for genomes that differ by an insertion or deletion.

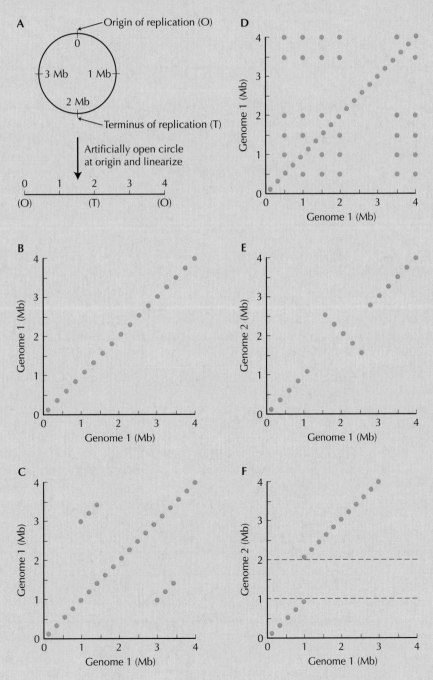

FIGURE 7.9. Using dotplots to compare circular genomes. See box text for explanations.

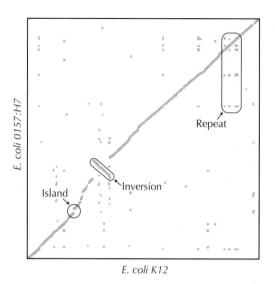

FIGURE 7.10. Conserved gene order in the backbone of *Escherichia coli* K12 and 0157:H7. The two genomes were aligned with each other and the matching regions were plotted. The conserved order of genes in the backbone of the two *E. coli* strains is indicated by the *diagonal line*. Three important genomic regions are *circled*. An island present in one of the two strains causes a slight shift in the position of the main diagonal.

they also occur in virtually the same order in both strains (Fig. 7.10). The genes unique to each strain are clustered into "islands" interspersed among the stretches of common genes. Similar patterns of DNA "islands" within a conserved genome backbone have been found among other related bacteria or archaea.

How do these islands originate? These are two possibilities: insertion of DNA into the strain with the island or deletion of DNA in the strain without the island. Gene loss is very common and frequently very rapid in bacteria and archaea (e.g., Fig. 7.5). However, relying on gene loss alone to explain genomic islands becomes untenable as more and more species are compared. For example, when the genome of a third strain of *E. coli* was determined, it was found to have many additional islands that are absent from both K12 and O157:H7 (Fig. 7.7). For gene loss to explain all the islands in the various *E. coli* strains, their common ancestor would have required an enormous genome from which different regions were lost in different lineages. Indeed, such a mechanism would require ancestral species to have had bigger and bigger genomes further back in time. Thus genes must be acquired to offset gene loss. Acquisition of genes is one of the hallmarks of bacterial and archaeal evolution and is discussed on pages 182–191.

Gene Order Changes Rapidly but with Strong Constraints

In addition to studying the location of genes found in one organism but not another, it is useful to compare the order of genes and other genomic features that are conserved between species. These comparisons reveal how genomes evolve and what the constraints are on the relative positioning of genes.

As with gene content, there is little conservation in the gene order between distantly related species (Fig. 7.11). Some sets of genes, however, are strongly conserved. The best example is the genes that encode many of the ribosomal proteins (Fig. 7.12). When such conservation occurs across such large evolutionary distances, it suggests that tightly coordinated regulation of transcription and translation is necessary for functionality. This is probably due in part to the coupling of transcription and translation in bacteria and archaea. In turn, the lack of coupling in eukaryotes may explain why there are few examples of gene-order conservation across such large distances.

When gene-order comparisons are made among closely related strains or species, rearrangements are frequently observed at sites of repetitive sequences such as transposons or duplicated genes (Fig. 7.13). Although repetitive DNA is less abundant in bacteria and archaea, it still plays a major role in genome evolution.

Comparing gene order among multiple sets of close relatives has revealed what types of rearrangements are most common. In bacteria and archaea, one of the most com-

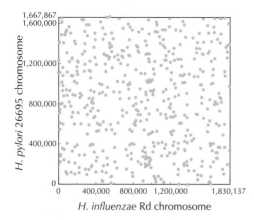

FIGURE 7.11. The lack of conservation of gene order between *Haemophilus influenzae* and *Helicobacter pylori* is illustrated. Linearized chromosomes of *H. influenzae* and *H. pylori* are plotted on the horizontal and vertical axes, respectively. Each *dot* represents a single pair of orthologous proteins. Genes in similar operons, which do exist, are too close together to give separated points on the scale used.

mon is symmetric inversion around the origin of replication (Fig. 7.14). Such inversions are seen in almost every comparison of moderately closely related strains or species. Although other rearrangements occur, the symmetric inversions serve as a useful tool for understanding some features of general evolution and we focus on them here.

Symmetric inversions around the origin are due to a combination of mutation bias and selection bias. To understand how mutation bias could cause this, it is helpful to understand some of the features of circular chromosome replication in bacteria and archaea. Replication of circular chromosomes almost always begins at a single region—referred to as the origin of replication. DNA replication proceeds bidirectionally from this origin, continuing until the replication forks collide on the other side of the DNA circle at the terminus of replication (Fig. 7.15). It is thought that the replication complex stands relatively still and the DNA is threaded through this complex, which would place the two replication forks close to each other. This threading can thus lead to symmetric inversions. If the DNA replication complexes were to slip and drop the DNA strands, they might restart replication by using the template from the opposite side of the origin to extend the recently replicated DNA, thereby causing an inversion. As the two replication forks should

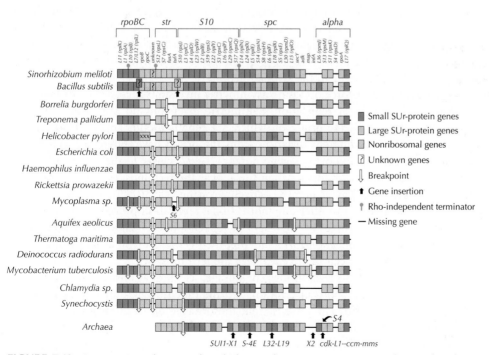

FIGURE 7.12. Conservation of gene order of ribosomal protein operons across bacterial and archaeal species.

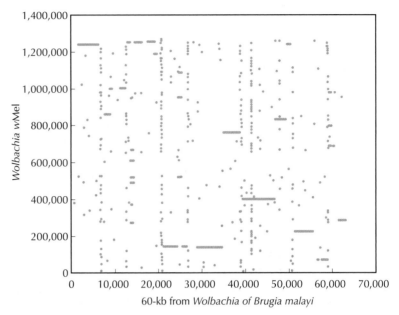

FIGURE 7.13. Breaks in gene order conservation between close relatives occur frequently at sites of repetitive DNA. The figure shows a dotplot of a fragment of the genome of one *Wolbachia* (an intracellular α-proteobacteria) with the complete genome of another. Note the different scales, which mean that regions of conserved gene order appear as *nearly horizontal lines*. The regions of conserved order are noncontiguous, indicating rearrangements between the two genomes. The breaks in gene order are associated with *vertical columns of dots*, which indicate the presence of repetitive DNA elements.

proceed at roughly the same speed, this would lead to a symmetric inversion.

In addition to this mutation bias, there is a clear selection bias against other types of inversions. In particular, for highly expressed genes, there may be selection against gene transcription in the direction opposite to the direction of DNA replication. If the two processes were to occur in opposite directions, the result might well be a collision between the RNA and DNA polymerases. Symmetric inversions do not change the relative orientation of transcription versus replication for any genes inside the inversion, although asymmetric inversions do. Selection against colliding DNA and RNA polymerases is thought to be partially responsible for an unusual genome pattern in firmicutes. In these species, more than 90% of the genes are oriented such that transcription and replication occur in the same direction.

A second selective force opposing asymmetric inversions involves the distance of a gene from the origin of replication. Symmetric inversions do not change this distance—just the side of the origin on which the gene is found. This is important because genes that are different distances from the origin have different effective copy numbers per cell (also known as **gene dosage**). Genes closer to the origin of replication have higher effective dosages than those near the terminus, especially when cells are actively growing. This is because actively growing cells usually make serial copies of the genome; that is, just after one bidirectional replication process starts, two more start on the duplicate copies of the origin of replication left behind. If there are many active bidirectional replication forks moving, then genes near the origin will be present in more copies per cell than genes near the terminus.

The combination of selection for gene dosage, selection against changes in the orientation of replication and transcription, and the mutation bias for symmetric inversions is a powerful trio. The result is that the distance of a gene from the origin of replication is highly conserved even when no gene-order conservation is present (Fig. 7.14C). As a consequence, because the mutation rate is influenced in part by the distance a gene is from the origin, mutation rates for different genes may be conserved across large evolutionary distances in bacteria and archaea.

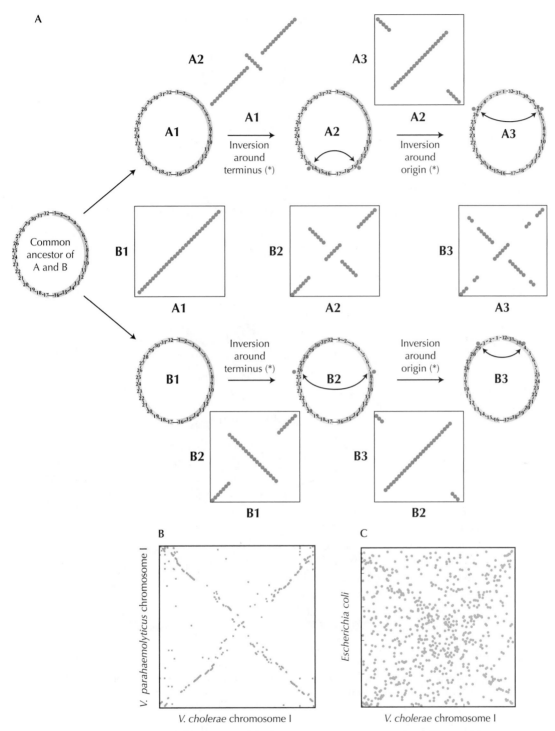

FIGURE 7.14. X-alignments. (*A*) Schematic model of symmetric genome inversions. The model shows an initial speciation event, followed by a series of inversions in the different lineages (A and B). Inversions occur between the asterisks (*). Numbers on the chromosome refer to hypothetical genes 1–32. At time point 1, the genomes of the two species are still colinear (as indicated in the scatterplot of A1 vs. B1). Between time point 1 and time point 2, each species (A and B) undergoes a large inversion about the terminus (as indicated in the scatterplots of A1 vs. A2 and B1 vs. B2). This results in the between-species scatterplot looking as if there have been two nested inversions (A2 vs. B2). Between time point 2 and time point 3, each species undergoes an additional inversion (as indicated in the scatterplots of B2 vs. B3 and A2 vs. A3). This results in the between-species scatterplots beginning to resemble an X-alignment. (*B*) X-like alignment in dotplot of the main chromosomes of *Vibrio cholerae* (*x*-axis) and *Vibrio parahaemolyticus* (*y*-axis). (*C*) A weak X-like pattern exists even when comparing more distantly related species, in this case *V. cholerae* and *E. coli*. An X-like pattern indicates that the distance of a gene from the origin is conserved, but the side of the origin on which it is located is not conserved.

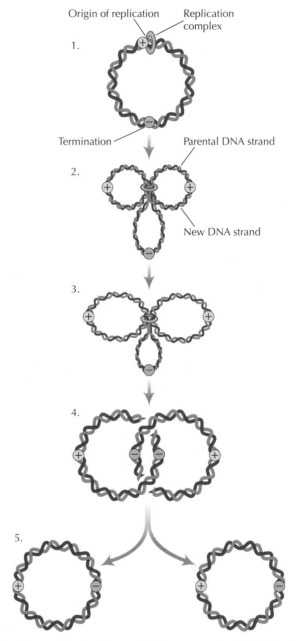

FIGURE 7.15. Bidirectional replication. The replication process for circular bacterial genomes is diagrammed. (*1*) Origin (+) and terminus (−) of replication are shown. The replication machinery is shown as a ring around the DNA. (*2*) Replication proceeds in both directions from the origin. (*3*) Replication continues. (*4*) Replication is completed and the two circles disentangle by an enzyme that cuts, moves, and then ligates one of the circles. (*5*) Two daughter molecules.

LATERAL TRANSFER OF DNA

Lateral DNA Transfer Is Pervasive in Bacteria and Archaea

Recombination greatly increases the patterns of variation in a population by mixing and matching different variants. In eukaryotes, recombination primarily occurs within species through sexual reproduction when DNA from two parental lineages mixes to produce new combinations. Because sexual recombination of the form seen in eukaryotes does not occur in bacteria and archaea, for many years it was believed that **genetic recombination** was largely absent from these organisms. We now know that this is not the case. Bacteria and archaea undergo recombination, both within and between species, even very distantly related ones. This recombination comes in many forms and is generally referred to as lateral gene transfer; it is also known as lateral DNA transfer, horizontal DNA transfer, and horizontal gene transfer.

As discussed on page 131, the terms lateral and horizontal are meant to contrast this form of DNA transmission with the normal "vertical" transmission of DNA from parent to offspring. Lateral gene transfer is different from sexual recombination in many ways, although some of the effects are similar. In lateral gene transfer, the exchange is usually unidirectional—with DNA going from one organism to another, rather than mixing the DNA from two individuals. In addition, the evolutionary distance between the partners participating in lateral transfer can be great—it can even occur between species from different domains (e.g., bacteria and archaea).

The importance of lateral gene transfer in bacteria and archaea is reflected in the diverse array of both active and passive processes that lead to lateral DNA transfer. This is not to say that there is no lateral gene transfer in eukaryotes; it certainly does occur. However, its extent appears to be much more limited, with most of the well-documented cases involving transfer of genes from organelle genomes to the nucleus (see p. 163). Lateral DNA transfer may be the main reason that bacterial and archaeal diversification differs from that in eukaryotes. In the remainder of this section we introduce different mechanisms as well as specific examples of lateral gene transfer.

Transformation

In transformation, naked DNA is taken up directly from the environment, a process also known as **competence** (Fig. 7.16). Only certain species are naturally competent, although scientists have developed laboratory methods to induce competence in species in which it does not normally occur. Naturally competent species include the radiation-resistant bacterium *Deinococcus radiodurans* and the pathogens *H. influenzae* and *Streptococcus pneumoniae*. It is believed that naturally competent species may be more likely to experience lateral gene transfer. It was a form of competence that was responsible for the uptake of DNA in Avery's experiments showing that DNA is the genetic material (see Fig. 2.15).

Conjugation

Gene transfer between microbes also occurs by a form of mating called **conjugation** (Fig. 7.17). During conjugation, a physical linkage is made between cells, and DNA is passed from one cell to another. Conjugation is frequently orchestrated by self-replicating genetic elements known as conjugative plasmids, which have been found throughout the bacteria and archaea. These plasmids encode all the proteins necessary to carry out the conjugation process. As with competence, the DNA can then be maintained as a self-replicating element or integrated into the genome of the recipient cell.

Conjugation can occur across large evolutionary distances, even into eukaryotes. For example, the 543-kb element in *A. tumefaciens* (see Tables 7.1 and 7.2) encodes proteins that mediate the injection of the plasmid DNA into plant cells (Fig. 7.18). Once inside the plant cells, this DNA is integrated into the plant nuclear genome, causing other plasmid genes to be expressed, and leading to the production of gall tumors on the plant. The tumors serve as feeding reservoirs for the bacteria, because the transferred genes cause the plant to excrete nutrients and other metabolites. This gene transfer process has been very valuable in plant molecular biology studies. The *A. tumefaciens* plasmid is used in the laboratory as a vector to transfer a wide variety of genes into plants, allowing for a better understanding of their genetic functions.

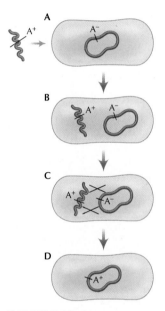

FIGURE 7.16. Competence. (*A*) A naturally competent cell is exposed to DNA in its environment. (*B*) The cell takes up the DNA. (*C*) The DNA recombines with the chromosome. (*D*) A section of the new DNA replaces the homologous segment (in this case, resulting in the A⁻ allele replacing the A⁺ allele).

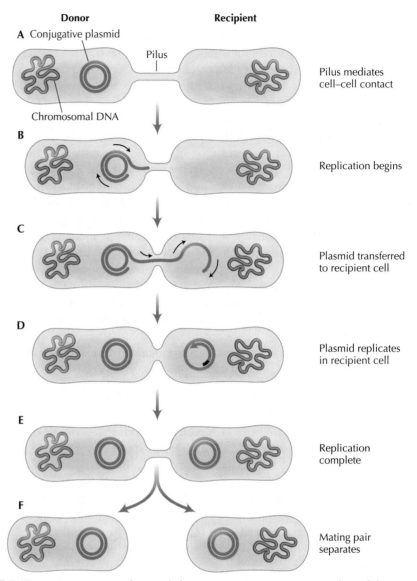

FIGURE 7.17. Conjugation. (*A*) A bacterial donor containing a conjugative plasmid forms a connection (pilus) with a neighboring cell. (*B*) A single-stranded copy of the plasmid genome is made. (*C*) The single-stranded copy is sent through the pilus to the recipient cell. (*D*) The complementary strand of plasmid DNA is made. (*E*) DNA replication is completed. (*F*) The pilus is broken.

Conjugation plays a key role in imparting antibiotic resistance to pathogenic bacteria. Many kinds of organisms (microbes, animals, plants, and protists) produce **antimicrobial** or antibiotic compounds that kill, harm, or inhibit the growth of microorganisms. This has led to the evolution of diverse strategies in microbes to limit the harmful and toxic effects of these antibiotics. In many cases, the genes responsible for resistance are found on conjugative plasmids, which can be readily transferred to other strains or species. Thus, via conjugation, antibiotic resistance can be spread widely (Fig. 7.19).

Transduction

Viruses (which are generally known as **phage** when they infect bacteria or archaea) are a very important vector for lateral DNA transfer. Phage-mediated gene transfer occurs by **transduction**, of which there are two main types (Fig. 7.20). In generalized transduction, phage first infect a cell and then concurrently produce copies of their genome and phage **capsids**. The phage DNA is packaged into the capsids. Occasionally, frag-

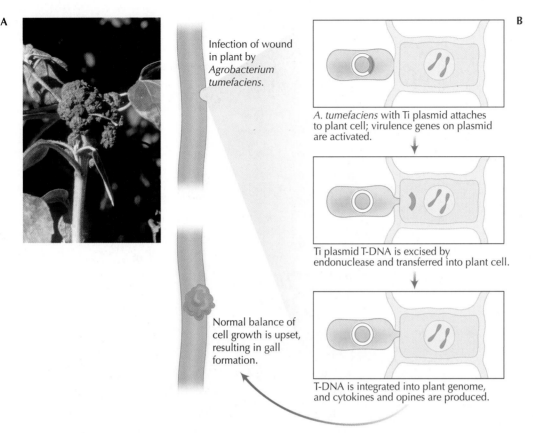

Infection of wound in plant by *Agrobacterium tumefaciens*.

A. tumefaciens with Ti plasmid attaches to plant cell; virulence genes on plasmid are activated.

Ti plasmid T-DNA is excised by endonuclease and transferred into plant cell.

T-DNA is integrated into plant genome, and cytokines and opines are produced.

Normal balance of cell growth is upset, resulting in gall formation.

FIGURE 7.18. Conjugation in *Agrobacterium tumefaciens*. (*A*) Crown gall tumor formed by *A. tumefaciens*. (*B*) Mechanism of conjugation.

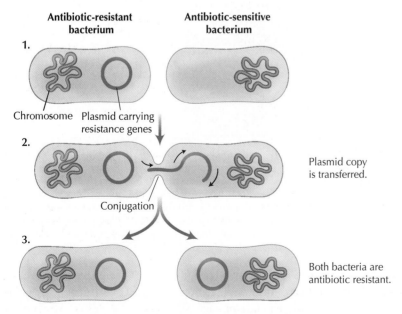

Antibiotic-resistant bacterium   **Antibiotic-sensitive bacterium**

1.

Chromosome   Plasmid carrying resistance genes

2.

Conjugation

Plasmid copy is transferred.

3.

Both bacteria are antibiotic resistant.

FIGURE 7.19. Lateral transfer of antibiotic resistance via plasmids. (*1*) Antibiotic-resistant and -sensitive bacteria are shown. (*2*) Bacteria "mate" via conjugation, during which a copy of the plasmid carrying antibiotic resistance genes is transferred. (*3*) Both bacteria are now antibiotic resistant.

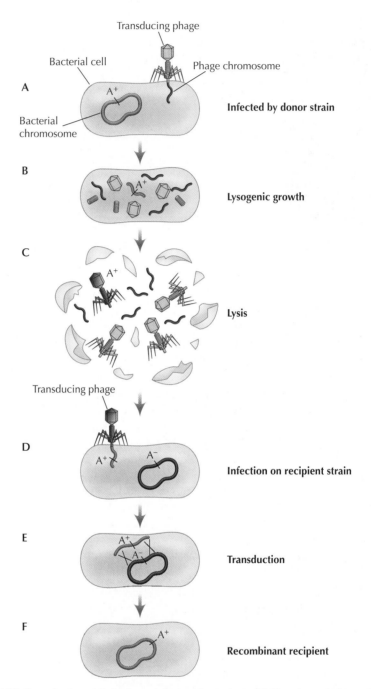

FIGURE 7.20. Transduction. (*A*) A phage infects a bacterium. (*B*) The phage induces the host to synthesize multiple copies of the phage genome and structure components. (*C*) Multiple phage are created with DNA inside the phage head. One phage (*red*) has picked up a portion of the host genome, including an antibiotic resistance gene, A$^+$. (*D*) The A$^+$ phage infects an A$^-$ bacterium. (*E*) The phage DNA and part of the chromosome are homologous, which enables recombination to occur (i.e., transduction). (*F*) This recipient has converted to A$^+$. Phage transduction occurs most often within a species but can also facilitate DNA exchange between distantly related species.

ments of the host genome are incorporated into some of the phage particles as they are assembled. If these particles infect another host cell, the DNA from the original host is released into the new host. The fate of such DNA is similar to that which occurs with competence. It can be maintained in the new cell or degraded and have no long-term consequences. In specialized transduction, phage first integrate their genome into the host genome. Following capsid production, the integrated DNA is excised and packaged into capsids. Occasionally, DNA excised from the host genome may include neighbor-

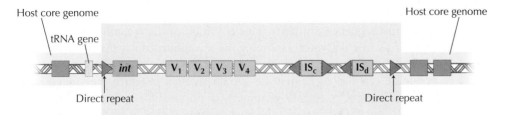

FIGURE 7.21. Diagram of a pathogenicity island (within the *large green shaded region*). Flanking the island are core host genes (*small blue shaded regions*). Immediately flanking the island on the *left* is a tRNA gene—many islands are found near or within tRNA genes. At the edges of the island are direct repeats (*red triangles*). Inside the island are virulence genes (*orange boxes*) and some insertion sequences (*purple boxes*).

ing portions of the host DNA. When these phage infect a new host and this DNA integrates into the new genome, the result is the transfer of bacterial or archaeal DNA from one microbe to another using the phage as a vector.

Because many phage can infect multiple species, phage transduction can move DNA across wide evolutionary distances. It may be responsible for interspecies movement of genes involved in pathogenicity and virulence. Pathogenicity and virulence genes are often clustered in contiguous stretches of the genomes of pathogenic bacteria, called **pathogenicity islands** (Fig. 7.21). In many cases, the islands found in a particular strain are anomalous in terms of GC content, codon usage, and phylogenetic history, which suggests that they may have been introduced from another species. The occurrence of these genes in a single stretch of the chromosome allows them to be readily transferred between strains or species. Although the exact transfer mechanism is not well understood, it is probably mediated by transduction.

Barriers Limit the Occurrence of Gene Transfer in Natural Populations

Despite the mechanisms of lateral DNA transfer, DNA does not flow freely among and between bacteria and archaea. There are barriers to its movement, some active and some passive, which greatly limit the amount of lateral gene transfer experienced by species.

One barrier to the movement of DNA between species are enzymes that recognize and specifically degrade foreign DNA. These **restriction enzymes** have been found in many species of bacteria and archaea. (Restriction enzymes are essential tools in modern biological research and helped make possible recombinant DNA technology and genetic engineering [see p. 56].) They work by cleaving DNA molecules that contain specific sequences of bases (often referred to as restriction sites). Organisms are protected from their own enzymes either by the absence of all such restriction sites in their own genome or, more commonly, by methylation of their DNA. Some species do not have any restriction enzymes and may be more prone to the successful transfer of foreign DNA. Other species have many different restriction enzymes, each recognizing a unique sequence of DNA, thus enabling the organism to restrict the type of DNA that can be acquired.

Even if foreign DNA does escape a host's restriction system, it must contain the necessary regulatory elements recognized by the DNA polymerases and DNA segregation machinery to ensure that it is transmitted reliably from parent to offspring. One way to ensure reliable transmission is for a piece of DNA to carry its own replication and segregation instructions. This is why plasmids are such an important vector for lateral DNA transfer—they are self-replicating genetic elements.

Alternatively, to ensure propagation, the foreign DNA needs to integrate into the host genome, using one of two possible paths. One route is via **homologous recombination**. If the foreign DNA includes a portion that is very similar at the sequence level to some of the host DNA, it can align with a host genetic element and recombine at the point of sequence similarity. This is similar to the recombination that occurs during meiosis in eukaryotes. Homologous recombination is discussed in more detail in Chapter 12.

Recombination can also occur in a site-specific manner in which the foreign DNA is simply inserted into the recipient's genome with no need for homology. For example, many phage insert their DNA into the host genome at sites of transfer RNA genes, making use of the highly conserved sequences and structures of the tRNA genes to guide their insertion. The lack of a requirement for sequence similarity allows phage to move DNA between even distantly related species.

Even after the foreign DNA acquires the means to propagate within the new host, it may still be lost if it does not provide some selective advantage to the host organism, and even this may be insufficient. There is great variation between species in the subtle details of how their genes function. For example, different species have codon usage biases; that is, they differ in their preference for which codons are used for particular amino acids (see p. 543). This **codon usage bias** can lead to inefficient or inaccurate translation if the nonpreferred codons are used. If the codon usage bias differs in the foreign DNA and the host, a protein encoded by foreign DNA may be translated inefficiently. Similarly, differences in GC content, promoter sequences, targeting signals, and other features may also limit the effectiveness of foreign DNA.

All of the barriers described above are typically circumvented if lateral DNA transfer occurs between closely related strains or species. Such DNA will usually have the same restriction sites, enough similarity for homologous recombination to occur, and all the right genome features for any genes to function properly. However, mismatch repair to correct DNA replication errors, which is one of the primary mechanisms organisms use to limit mutations, can also limit lateral transfer from close relatives. Mismatch repair works by recognizing incorrect base pairs (or small loops) in double-stranded DNA (see pp. 336–339). Such mismatches and small loops are also created when DNA from a close relative recombines with native DNA. In many cases the mismatch repair proteins will "correct" these mismatches—thus preventing lateral transfer.

There are two ways to overcome barriers to lateral DNA transfer from a distantly related organism. A selfish DNA element can replicate sufficiently to outrun the streamlining pressure for deleting vestigial/nonfunctioning DNA. This is why transposable elements can move rapidly between species. Another way is to provide a selective benefit to the host even in the face of poor genome content. This can occur if there are strong selective pressures favoring the acquisition of the new DNA. For example, if some members of a bacterial or archaeal population take up a plasmid that encodes genes for penicillin resistance, then even if the resistance genes are poorly transcribed or translated, the plasmid may provide a selective advantage on exposure of the organisms to penicillin.

Once some of the barriers to lateral gene transfer are overcome and foreign DNA "sticks" in a new population, an interesting process occurs. The gene(s) and other portions of the DNA acquire more of the subtle features of the host that enable them to function well in this new environment. This process, known as **amelioration**, slowly erases the unusual features of a gene that may have limited its functionality (Fig. 7.22). This is "good" for the foreign DNA but conceals evidence of lateral gene transfer between species.

Lateral Transfer Is Common but Most Species Appear to Have a "Core" of Genes That Evolve Primarily by Vertical Evolution

Given the prevalence of lateral gene transfer among bacteria and archaea, it is natural to ask how much of any given genome is acquired and how much follows traditional vertical evolution. The increasing number of genome sequences that are available for

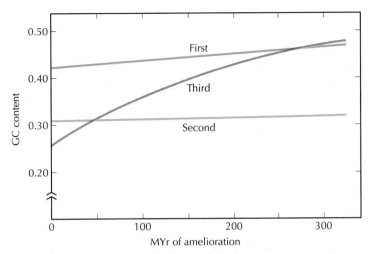

FIGURE 7.22. Amelioration in coding regions. The plot shows the results of a simulation of amelioration of genes from *Shigella flexneri* as they shift in composition to become more like genes from *Salmonella enterica*. Both of these are species of γ-proteobacteria. The different lines correspond to first, second, and third positions in codons. MYr, millions of years. The change is more rapid in the third position because the genetic code is redundant.

bacteria and archaea has helped to answer these questions by allowing genome-wide quantification of the relative amount of foreign DNA.

One strategy for such quantification examines variation in the nucleotide composition within the genome. Regions of the genome that have recently been inserted from some other source will likely differ in composition (e.g., codon usage or GC content). Although processes other than lateral transfer—such as selection—can lead to compositional differences in regions of a genome, it is clear that for many species, a significant percentage of their genome was acquired recently (Fig. 7.23).

Examination of DNA composition simply takes a snapshot of one genome. However, just because strains are around to possess some foreign DNA does not mean that such DNA stays found for a long time. The best way to determine the long-term fate of acquired DNA is to trace the gain and loss of DNA onto a phylogenetic tree of species.

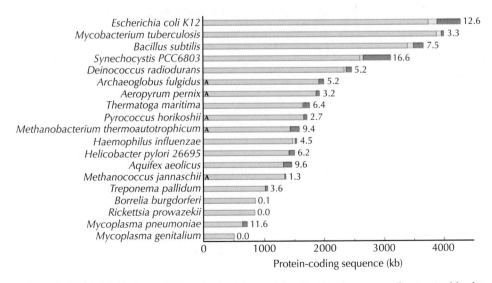

FIGURE 7.23. Estimation of the percentage of the genome that has been recently acquired by lateral transfer for different species of bacteria and archaea. *Blue,* "native" DNA (not acquired by transfer); *yellow,* known mobile DNA elements; *red,* other foreign DNA; and A, archaeal species.

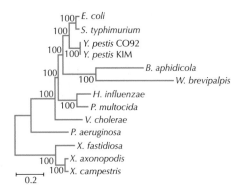

FIGURE 7.24. Tree of 205 core proteins from γ-proteobacterial genomes. The topology shown agrees with almost all individual gene alignments. The same tree is obtained after removing the two genes showing evidence for lateral gene transfer. The position of the root corresponds to the one obtained repeatedly using SSU (small subunit) rRNA.

There is, however, one major problem with trying to trace the history of genes onto a **species tree.** How do we determine a species tree if different genes in a genome have different histories? Or even more daunting—what if there really is no species tree? Fortunately, there appears to be a set of genes ("core genes") within most genomes that are resistant to gene transfer. These genes can be used to create something akin to a species tree, by treating all of the core genes as a single entity and building evolutionary trees with this "concatenated" unit (Fig. 7.24). Concatenation-based trees can then be used as a surrogate for a species tree in studies of lateral transfer. An example of this is shown in Figure 7.25 for the organisms in Figure 7.24. When this kind of analy-

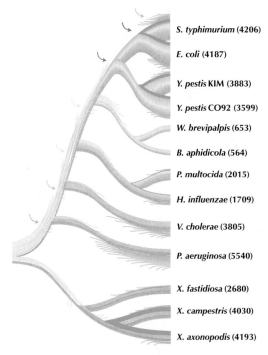

FIGURE 7.25. Lateral gene transfer and genome evolution in γ-proteobacteria. Only a small proportion of genes have been retained since the common ancestor of γ-proteobacteria (in *orange*). Under the assumption that ancestral and contemporary genome sizes are similar, most of the genes present in this ancestral genome (in *gray*) have been replaced by nonhomologous genes (*yellow* to *green*), usually via lateral gene transfer from organisms outside of this clade. Once a new gene is acquired, its transmission follows vertical inheritance. The abundance of genes unique to a species (in *blue*) indicates that these bacteria (with the exception of the endosymbionts *W. brevipalpis* and *B. aphidicola*) constantly acquire new genes, most of which do not persist in the long term within lineages. (Numbers of protein-coding genes, excluding those corresponding to known insertion sequence elements and phage, are in parentheses for each genome.)

sis has been done, researchers have found that genes in a variety of evolutionary groups that encode components of the information-processing machinery (replication, transcription, and translation) frequently have low rates of lateral transfer. Known as **informational genes**, they may resist gene transfer because they interact with too many other genes in the genome. The genes that are more prone to lateral transfer tend to be those involved in metabolic processes and peripheral functions and are known as **operational genes**.

The Balance between Vertical Evolution, Gene Loss, and Lateral Transfer Varies in Different Bacteria and Archaea

The general pattern of evolution in bacteria and archaea is determined largely by the balance between vertical evolution, gene loss (e.g., due to streamlining), and acquisition of DNA by lateral gene transfer. Each of these forces varies between taxa (see Figs. 7.23 and 7.25), which, in turn, can cause different groups to have major differences in patterns of evolution. For example, the exceptionally small genomes of some intracellular species may be due in part to their inability to acquire genes to compensate for those lost as a result of streamlining forces.

The importance of lateral transfer should not be underestimated. The vast array of genetic possibilities that can be rapidly created as a result of lateral gene transfer is daunting. Consider the genetic diversity present in phage and plasmids, which are ready-made vectors for genetic exchange. The number of different genes present in plasmids and phage within *E. coli* species is greater than the number of genes in any single *E. coli* genome. The number of genes contained in all of the phage and plasmids on Earth is staggering. This enormous shared genomic resource may contain more genes than all the eukaryotes alive today. We discuss the consequences for the evolution of novelty in Chapter 24.

Because of the extensive gene loss and gene transfer among bacteria and archaea, entire pathways can have sporadic distribution patterns across species. This may explain the presence of similar mechanisms of photosynthesis in only a few distantly related bacterial lineages (see pp. 161–163). The gain and loss of entire pathways provide a useful means of studying bacterial and archaeal genomes. Genes can be grouped by their distribution patterns across species, which aids both in identifying genes that have the same distribution patterns and in predicting gene function. For example, if an unidentified gene is found consistently with a set of genes known to be involved in sporulation, then it is reasonable to hypothesize that the unidentified gene is also involved in sporulation (Fig. 7.26).

The combination of gene loss and gene transfer in bacteria and archaea has also resulted in some very closely related strains or species that show significant differences in gene contents even when they have a shared genome backbone, such as in the *E. coli* example presented earlier in this chapter (Fig. 7.6). This gene-content variation, as well as the chimeric nature of most bacterial and archaeal genomes, makes it very difficult to define what is a bacterial or archaeal species. We discuss the nature of species further in Chapter 22.

C4-dicarboxylate response regulator
degV family protein
putative DNA-binding protein
sigF RNA polymerase sigma-F factor
conserved hypothetical protein
spoVK stage V sporulation protein K
cotJC cotJC protein
SCP-like extracellular protein
putative phosphoesterase
spo0A stage 0 sporulation protein A
conserved hypothetical protein
conserved hypothetical protein
putative sporulation protein
sigG RNA polymerase sigma-G factor
pheB ACT domain protein pheB
putative sporulation specific protein SpoIIGA
YabG peptidase,U57 family
putative membrane protein
putative sporulation protein
spore maturation protein B
stage II sporulation protein D
putative stage II sporulation protein E
putative membrane protein
putative stage II sporulation protein M
putative sporulation protein
putative sporulation protein
putative sporulation protein
gpr spore protease
conserved hypothetical protein
conserved hypothetical protein
putative membrane protein
stage IV sporulation protein A
stage V sporulation protein AE
stage V sporulation protein AD
putative stage IV sporulation protein B
stage V sporulation protein AC
stage II sporulation protein R
sigE RNA polymerase sigma-E factor
conserved hypothetical protein
gerKA spore germination protein GerKA
spore germination protein
putative sporulation protein
spoVT stage V sporulation protein T
transcriptional regulator, AbrB family
putative sporulation protein
small acid-soluble spore protein
small acid-soluble spore protein
PRC-barrel domain protein
putative sporulation protein
stage III sporulation protein D
sigK RNA polymerase sigma-K factor
rpoZ DNA-directed RNA polymerase, omega subunit
conserved hypothetical protein
putative ATP-dependent protease La
spoVS stage V sporulation protein S
putative lipoprotein
RNA polymerase sigma factor
conserved hypothetical protein
conserved hypothetical protein
putative germination protein GerM
putative membrane protein
conserved hypothetical protein
putative membrane protein
putative glycosyl transferase
ATP:guanido phosphotransferase domain protein
uvrB/uvrC motif domain protein
glpP glycerol uptake operon antiterminator regulatory protein
anti-sigma F factor
CBS domain protein
conserved hypothetical protein
putative membrane protein
transcription regulator, Fur family
conserved hypothetical protein
conserved hypothetical protein
conserved hypothetical protein
aspartate kinase, monofunctional class
N-acetylmuramoyl-L-alanine amidase
vanW domain protein
spore maturation protein A

FIGURE 7.26. Phylogenetic profile analysis (grouping genes by their distribution patterns across species) helps predict their functions. This method analyzes complete genome sequences to identify genes that are found in similar sets of species. First, complete genomes are searched for the presence of a gene that is found in a particular (reference) genome. Second, a binary profile is created of the gene's presence and absence across the species. These two steps are repeated for all the genes in the reference genome. Third, a clustering method is used to identify genes with similar profiles (and thus similar distribution patterns across species). This figure shows one major cluster based on an analysis using the genome of the bacterium *Carboxydothermus hydrogenoformans* as the reference genome. In the *center* of the figure, profiles are shown for a subset of genes from this species: Genes are in *rows*, species are in *columns*, *red boxes* mean a gene is present in a species, and *black* means it is absent. Gene names are shown on the *right*; species names are not shown. The tree at the *left* represents the cluster diagram showing the similarity of the profiles for these genes. The columns of *red* in the central portion of the figure indicate that those species possess all or nearly all of the genes shown. Most of the genes listed, whose functions are known, have some role in sporulation (see the gene descriptions on the *right*), because species that sporulate use basically the same set of genes, and those that do not sporulate lack these genes (as indicated by the *mostly black spots* in the *central* figure). In turn, it can be inferred that genes found in the cluster, but with no known function, probably have some as yet unknown role in the sporulation process.

SUMMARY

In this chapter, we have presented an overview of the general patterns of genome evolution in bacteria and archaea. These general patterns include

- pressure to keep genomes streamlined
- the rapid loss of genes that are not actively under positive selection
- constraints on gene order within operons but few constraints on global gene order
- significant gene-content variation among closely related organisms
- significant amounts of gene transfer between species

Taken together, these and other genome-level processes may explain why bacteria and archaea have similarities within their patterns of diversification, even though archaea appear more closely related to eukaryotes than to bacteria. The key feature of bacterial and archaeal genome evolution is the extensive amount of lateral gene transfer that has taken and does take place. Despite the absence of recombination by sexual reproduction (as is found in eukaryotes), bacterial and archaeal recombination methods allow genes to move across much larger evolutionary distances. This has resulted in highly mosaic genomes; it allows them to sample genes from a collective gene pool and permits evolution of new processes.

FURTHER READING

Genome Reduction

Mira A., Klasson L., and Andersson S.G. 2002. Microbial genome evolution: Sources of variability. *Curr. Opin. Microbiol.* **5:** 506–512.

Moran N.A. 2002. Microbial minimalism: Genome reduction in bacterial pathogens. *Cell* **108:** 583–586.

Gene Order

Eisen J.A., Heidelberg J.F., White O., and Salzberg S.L. 2000. Evidence for symmetric chromosomal inversions around the replication origin in bacteria. *Genome Biol.* **1:** RESEARCH0011.

Lathe W.C., 3rd, Snel B., and Bork P. 2000. Gene context conservation of a higher order than operons. *Trends Biochem. Sci.* **25:** 474–479.

Rocha E.P. 2004. The replication-related organization of bacterial genomes. *Microbiology* **150:** 1609–1627.

Genome Structure

Bentley S.D. and Parkhill J. 2004. Comparative genomic structure of prokaryotes. *Annu. Rev. Genet.* **38:** 771–792.

Casjens S. 1998. The diverse and dynamic structure of bacterial genomes. *Annu. Rev. Genet.* **32:** 339–377.

Lateral Transfer

Bushman F. 2002. *Lateral DNA transfer: Mechanisms and consequences.* Cold Spring Harbor Laboratory Press, Cold Spring Harbor, New York.

Doolittle W.F. 1999. Phylogenetic classification and the universal tree. *Science* **284:** 2124–2129.

Ochman H., Lawrence J.G., and Groisman E.A. 2000. Lateral gene transfer and the nature of bacterial innovation. *Nature* **405:** 299–304.

Syvanen M. and Kado C. 2002. *Horizontal gene transfer,* 2nd ed. Academic Press, New York.

CHAPTER

8

The Origin and Diversification of Eukaryotes

I N CHAPTERS 6 AND 7, WE DISCUSSED TWO OF THE THREE main branches on the tree of life: the **bacteria** and **archaea**. These organisms are united by a variety of features, the most striking of which is the absence of **nuclei**. It is the presence of a cell nucleus that defines the **eukaryotes**—the third branch on the tree of life. Eukaryotes differ from bacteria and archaea in many other ways as well—from molecular processes like DNA replication to sexual reproduction to cellular organization to the presence of many multicellular forms.

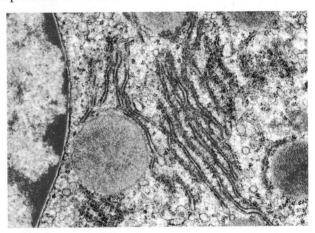

In this chapter, we discuss the early events that created this dramatic change in cellular organization—the origin of eukaryotes and their diversification. As in the chapters on bacterial and archaeal diversification, it is not possible to cover all of the interesting aspects of early eukaryotic diversification in a single chapter. We focus on three topics: the classification of eukaryotes, the origin of eukaryotic organelles and the nucleus, and the structure and evolution of the nuclear genome.

INTRODUCTION TO THE EUKARYOTES

Eukaryotes Are Very Diverse, but They Share Many Unique Molecular and Cytological Features

The eukaryotic domain is an assemblage of organisms with incredible diversity of form and function. There are carnivorous plants, animals that are **photosynthetic** because of symbioses with algae, fungi that grow over thousands of acres, obligate intracellular pathogens, diatoms with their intricate cell walls made from silicon, and **amoebas** that can rapidly mold their cells into any shape. Despite this diversity, eukaryotes share many cellular and molecular features that distinguish them from bacteria and archaea (Fig. 8.1). Probably the most important of these features is the location of the bulk of the cell's genome within a membrane-bound nucleus. Nuclear DNA, when not being transcribed or replicated, is kept tightly wound as **chromatin**, a coiled structure of DNA and proteins. These proteins guide the **supercoiling** of each DNA molecule to form a visible chromosome during some stages of the cell cycle (Fig. 8.2).

In eukaryotes, RNA is transcribed from the DNA in the nucleus and is subsequently transported to the cytoplasm where translation and protein synthesis take place (Fig. 8.3). Proteins that function in the nucleus, such as transcription factors, must then be

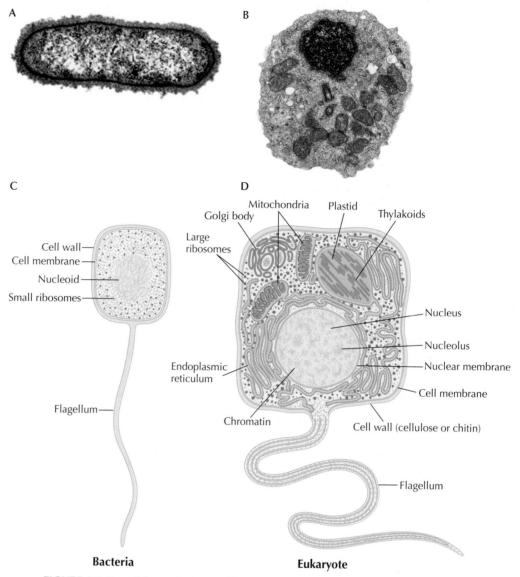

FIGURE 8.1. Bacterial vs. eukaryotic cell (not drawn to scale). (*A*) Photo of a bacterial cell. (*B*) Photo of a eukaryotic cell. (*C*) Diagram of a bacterial cell. (*D*) Diagram of the main features of eukaryotic cells. Note that both diagrams are composites; not all bacteria or eukaryotes have all such features.

transported back through the nuclear membrane into the nucleus. In contrast, in bacteria and archaea, which lack a nuclear compartment, transcription and translation are coupled (see pp. 138–141). As a messenger RNA (mRNA) is being transcribed from the DNA, the ribosomes attach to the nascent RNA to translate it immediately into proteins.

In addition to the nucleus, all eukaryotes have a dynamic and complex system of internal membranes and vesicles, called the **endomembrane system**, which consists of the Golgi apparatus, the **endoplasmic reticulum**, lysosomes, and **peroxisomes** (see Fig. 8.1). The endomembrane system functions in protein processing, intracellular transport, secretion, acquisition and degradation of nutrients from outside the cell, and cell-to-cell signaling. The endomembrane compartments are in constant flux as membranes and vesicles bud and fuse with one another. Many eukaryotes have other specialized membrane-bound compartments, such as **mitochondria**, **hydrogenosomes**, and **chloroplasts**, that are used for energy conversion processes.

Eukaryotic cells also contain a **cytoskeleton** that guides the transport of materials within the cell, gives structure to the cell, organizes and moves **organelles**, and directs

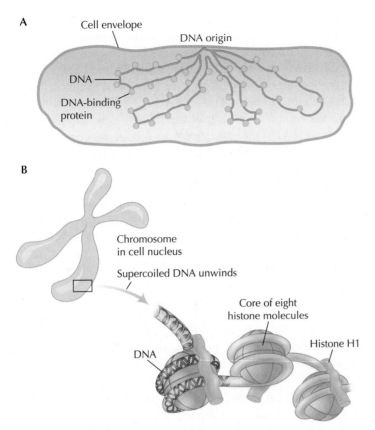

FIGURE 8.2. Models of the DNA structure in bacterial vs. eukaryotic cells. (*A*) In the bacterial cell, DNA is packaged in loops from a protein core that is attached to the cell wall. The chromosome is usually circular and there is no nuclear membrane. (*B*) Eukaryotic cells have their DNA packaged around cores of histone proteins to form nucleosomes. This DNA is then supercoiled into long bundles. Eukaryotic cells have a nuclear membrane.

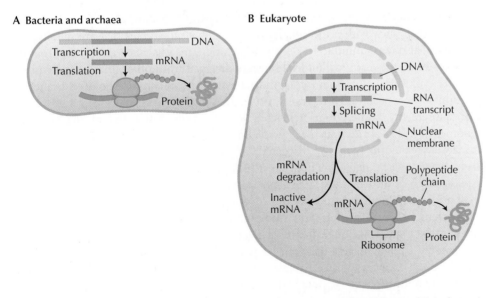

FIGURE 8.3. Eukaryotic vs. bacterial and archaeal transcription and translation. (*A*) In bacteria and archaea, transcription and translation happen in the same location in the cell and are frequently coupled to each other with translation occurring while an RNA is still being made. (*B*) In eukaryotes, transcription and splicing occur in the nucleus and then RNA is exported out of the nucleus where translation occurs. Coding regions are in *red*.

cell division and chromosome segregation. The cytoskeleton is made from a complex set of protein components, including **microtubules**, **microfilaments**, and molecular motors. Eukaryotic flagella, which can be considered a component of the cytoskeleton, are structurally distinct from flagella of bacteria and archaea (see Fig. 6.3). Although some cytoskeleton-like structures and proteins have been found in bacteria and archaea, such as the tubulin homolog FtsZ, no archaea or bacteria are known that possess anything like the complex cytoskeletal machinery found in eukaryotes.

As discussed on pp. 125–131, many core molecular functions of eukaryotes are more similar to processes in archaea than to those in bacteria. For example, the archaeal RNA polymerase is most closely related to the eukaryotic RNA polymerase that synthesizes mRNA. In addition, many of the components of large protein complexes that carry out transcription, replication, and repair in eukaryotes function in similar complexes in archaea but not in bacteria (see pp. 140–142). In each case, however, the eukaryotic process is much more complex than the analogous process in archaea.

Eukaryotes Can Be Divided into Eight Major Lineages, or Kingdoms

In early classification systems, all unicellular photosynthetic organisms were classified as "algae" and considered to be primitive plants, and all motile unicellular nonphotosynthetic eukaryotes were called "protozoa" and considered to be primitive animals. Many other classifications of eukaryotes have been proposed since that time, focusing on morphological and/or physiological differences and similarities. However, as with bacteria and archaea (see pp. 142–146), studies of morphological and physiological traits have been unable to resolve the deepest branches in the eukaryotic tree.

Modern molecular phylogenetic studies have revolutionized our interpretation of the deep relationships among eukaryotes, which have led to major changes in classification of eukaryotes. Initial studies relied on single genes, often those encoding ribosomal RNA (rRNA); more recently, multiple protein-coding genes and entire genome sequences have been compared.

The molecular phylogenetic studies of **microsporidia** are an excellent example of such a reclassification. Microsporidia are obligate intracellular parasites that lack mitochondria and peroxisomes and infect a variety of animal species, including humans (Fig. 8.4A). They were previously classified as a separate phylum, which was considered to be a deep branch within the eukaryotic tree. Other eukaryotes that are missing mitochondria also were found to be deeply branching. This was taken as evidence that the ancestors of all these amitochondrial lineages diverged from other lineages early in eukaryotic evolution before the mitochondrial symbioses occurred. The deep branching position of these taxa was based mostly on analysis of rRNA sequences (Fig. 8.4B). However, subsequent studies of protein-coding genes (Fig. 8.4C) and of the complete genomes of microsporidia, such as *Encephalitozoon cuniculi*, showed that microsporidia are closely related to fungi. As discussed on pages 130–131, the misclassification was likely due to long-branch attraction. Their reclassification into the fungal kingdom helps explain much of their biology, such as their infectivity via a spore stage (Fig. 8.4A). Microsporidial infections are now controlled using antifungal drugs.

Current molecular phylogenetic studies indicate that there are at least 40 major lineages of eukaryotes. Although the exact relationships among these groups are still debated, some studies suggest they can be placed into eight supergroups, or kingdoms. These are the **excavates**, **discicristates**, **heterokonts**, **alveolates**, **rhizaria**, plants, **Amoebozoa**, and **opisthokonts**. We note that whether these kingdoms are monophyletic is still debated. The characteristics of these eight eukaryotic kingdoms are summarized in Box 8.1. Because these studies are incomplete, the current divisions and their relationships may change as more information becomes available. A phylogenetic tree showing the relationships among the groups is shown in Figure 8.5. Notably this tree appears to lack

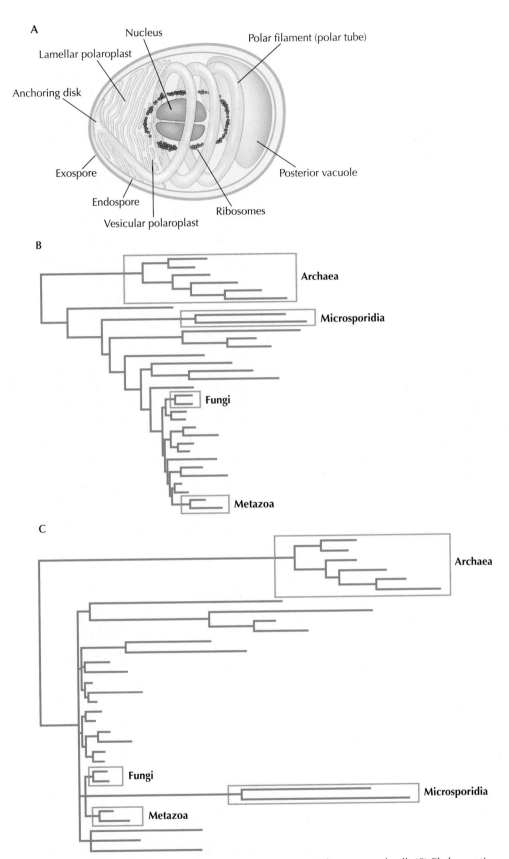

FIGURE 8.4. Microsporidia are related to fungi. (*A*) Microsporidia spore and cell. (*B*) Phylogenetic tree of many eukaryotic groups (four groups are identified). The tree was based on analysis of rRNA sequences. Note the deep position of microsporidia. (*C*) Phylogenetic tree, believed to be more accurate than the one in *B*, based on multiple protein-coding genes. Note the grouping of microsporidia with metazoa and fungi. Names of species and other eukaryotic groups are left out for clarity.

Box 8.1   Major Evolutionary Lineages of Eukaryotes

Most **excavates** are heterotrophic flagellated single-celled organisms. There are four major subdivisions within this kingdom (Fig. 8.5). The best-studied members include the human parasites *Giardia lamblia* (diplomonad) and *Trichomonas vaginalis* (parabasalid). Interestingly, none of the members of the excavates are known to have mitochondria, although some have other organelles, such as hydrogenosomes, that resemble mitochondria (Fig. 8.6A).

The discicristates include four major lineages of single-celled organisms, the best studied of which are the **kinetoplastids** and **euglenoids**. Most species in this kingdom have mitochondria with a unique morphology made up of disk-shaped internal structures. The kinetoplastids include the leishmanias and trypanosomes (Fig. 8.6B), mostly parasitic organisms that include the causative agents of human diseases such as Chagas disease, African sleeping sickness, and leishmaniasis. They have a single large mitochondrion that is usually associated with a unique organelle, the **kinetoplast**. The euglenoids include many photosynthetic organisms. Some, such as members of the genus *Euglena,* serve as model systems for studying eukaryotic photosynthesis. Their photosynthesis is carried out by a plastid-like organelle. Some studies suggest that the discicristates and excavates are sister lineages (i.e., they are more closely related to each other than either is to other eukaryotic groups).

Heterokonts are a diverse assemblage of single-celled organisms including at least six subgroups (see Fig. 8.5). Most of the members of this kingdom have hollow straw-like flagella and mitochondria with tubular **cristae**. Before the development of molecular systematics, most of these subgroups were

considered unrelated. Many had been grouped with fungi because of similarities in life cycle and traits that now are considered to be convergent similarities. For example, oomycetes are also known as downy mildews and water molds, including *Phytophthora infestans*, which caused the potato blight that was responsible for the mid-19th century famine in Ireland. The **labyrinthulids** (Fig. 8.6C), also called marine **slime molds**, are mostly parasitic and include *Labyrinthula zosterae*, the causative agent of eelgrass wasting disease.

Alveolates, which are sometimes grouped into a superkingdom with heterokonts, include three major lineages (Fig. 8.5). All known **apicomplexans** are animal parasites, including the causative agents of malaria (*Plasmodium* spp.) and toxoplasmosis (*Toxoplasma* spp.). They are a sister group to the **dinoflagellates**, a group of flagellated and (frequently) photosynthetic microbes commonly found in aquatic environments (Fig. 8.6E). Dinoflagellates are the cause of "red tides" in which blooms of cells become so abundant that the color of the water changes. Many dinoflagellates are bioluminescent. The **ciliates**, which include both parasitic and free-living species, use their cilia to move and feed. The best-studied examples include *Paramecium* and *Tetrahymena*. All ciliates have two distinct types of nuclei within a single cell: a **macronucleus**, which is the location of gene expression, and a **micronucleus**, which serves as the equivalent of germ cells in animals (see Fig. A&S.2).

Rhizaria (sometimes called Cercozoa) include five major lineages (Fig. 8.5). The **radiolarians** (Fig. 8.6F), the deepest branching members of this kingdom, are radially symmetric, mostly planktonic organisms, with a substantial fossil record.

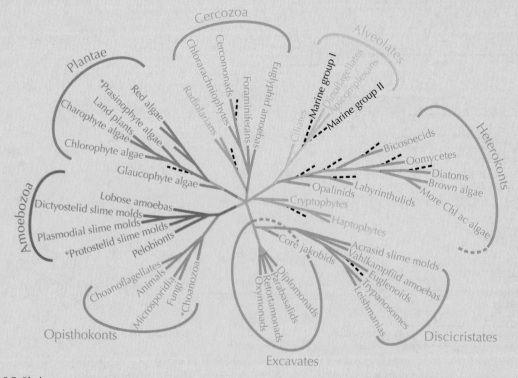

FIGURE 8.5. Phylogenetic tree of major evolutionary groups of eukaryotes. Branches are colored based on proposed supergroups, with *gray* indicating significant ambiguity. *Dashed lines* indicate major uncultured lineages known only from environmental surveys. *Starred* groups are possibly paraphyletic. Note the relationships among groups and the nature of the supergroups are still greatly debated.

Box 8.1. Continued.

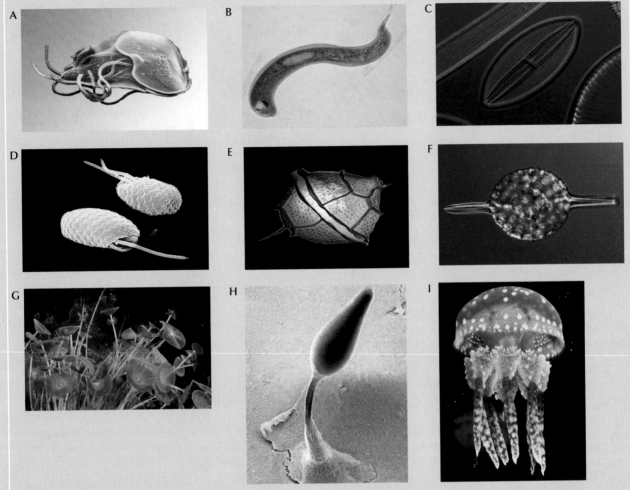

FIGURE 8.6. Diversity of eukaryotes. (*A*) *Excavates: Giardia.* (*B*) *Discicristates:* Euglena. (*C*) *Heterokonts:* Diatom. (*D*) *Cryptophyte: Cryptomonas.* (*E*) *Alveolates:* Dinoflagellate. (*F*) *Rhizaria:* Radiolarian protozoa. (*G*) Plants: *Acetabularia.* (*H*) *Amoebozoa:* Slime mold. (*I*) *Opisthokonts:* Jellyfish.

The marine foraminiferans also possess a significant fossil record because of their mineralized shells. The photosynthetic chlorarachniophytes contain chloroplasts. Interestingly, many other members of the Rhizaria contain algal symbionts, and thus the chlorarachniophytes' plastid is thought to be a relic of an ancestral algal symbiont. The chlorarachniophytes also contain a **nucleomorph** with its own genome, thought to be a relic nucleus left over from that algal symbiont.

The **Plantae** lineage includes at least six major subgroups: red algae (Rhodophytes), two groups of green algae (**Chlorophytes** and Prasinophytes), glaucophyte algae, charophyte algae, and land plants (Fig. 8.6G). Most of these organisms are free-living and contain chloroplasts; many species in the group are multicellular. The chloroplasts of red algae contain phycobilin pigments, which are responsible for the organisms' red color. The green color of charophyte algae, green algae, and plants comes from chlorophyll, their dominant chloroplast pigment.

The **Amoebozoa** are mostly heterotrophic organisms that move using pseudopods. They are composed of three major lineages (lobose amoebas, mycetozoan slime molds, and pelobionts) as well as many smaller groups (e.g., *Entamoeba*). Some members of this group, such as pelobionts and enta-moebas, have no mitochondria. The slime molds exhibit a type of multicellularity by aggregating and differentiating to form a stalked fruiting body structure containing spores (Fig. 8.6H).

Opisthokonts include three major groups: **choanoflagellates**, fungi (including microsporidia), and **metazoa** (animals). The choanoflagellates are single-celled organisms, which even prior to molecular phylogenetic studies were considered by many to be distantly related to animals. The fungi include well-known fungal organisms like mushrooms and yeast, and fungal pathogens like the Aspergilli, and micro-sporidia. The metazoa include animals (Fig. 8.6I) as well as the myxozoa, unicellular spore-forming parasites.

As with bacteria and archaea, many of the eukaryotic microorganisms found in nature cannot be grown in pure culture in the laboratory (see pp. 147–151). As a result, we know little about many of them. However, researchers have been able to characterize the phylogenetic diversity of microbial eukaryotes using molecular methods, similar to those used to study bacteria and archaea (see pp. 147–151). For example, rRNA surveys have found completely novel lineages of alveolates in marine samples (see Marine groups I and II in Fig. 8.5). It is likely that as more studies are done, new phyla or even kingdoms of eukaryotes will be found.

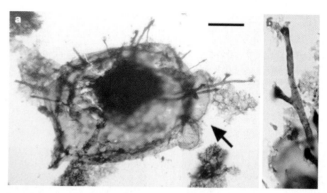

FIGURE 8.7. Early eukaryotic fossils (acritarchs). (a) A fossil of *Tappania plana*. (b) A zoom-in of a region with dichotomous branching. For more detail see Chapter 9. The scale bar is 35 μm for a, and it is 10 μm for b.

the "crown" that had been seen in previous studies, in which the fungi, animals, and plants clustered together into a single group. There appears, instead, to be an early radiation into many lineages. This tree allows one to interpret many eukaryote features in a phylogenetic context. For example, it suggests that multicellularity has evolved many times in different lineages (see p. 226).

Early Eukaryotic Fossils, although Limited, Suggest That Eukaryotes May Have Arisen Long after Bacteria and Archaea

In the fossil record, eukaryotes do not appear until long after bacteria- and archaea-like cells. The earliest convincing cellular fossils of eukaryotes are those of **acritarchs**, found in rocks of the late **Paleoproterozoic** (~1.9–1.7 billion years ago) (Fig. 8.7). The term acritarch is used for a diverse collection of microfossils. Many of them likely represent cysts of primitive single-celled photosynthetic eukaryotes (e.g., the ancestors of dinoflagellates). Although it is not always possible to determine their taxonomic status, all of them are likely derived from eukaryotic cells with thick decay-resistant organic walls, which give them a high preservation potential.

Additional evidence that eukaryotes existed almost 2 billion years ago comes from "molecular fossils." These "fossils" are molecular signatures derived from compounds known to exist in eukaryotic membranes but not in those of modern bacteria or archaea. Examples of these molecular signatures include **steranes**, which are derived from **sterols**, molecules that are common in eukaryotes but rare in bacteria and archaea. In particular, finding molecular fossils such as these in association with acritarch fossils strengthens the evidence of their eukaryotic origin. It is difficult to estimate the amount of eukaryotic evolution that had occurred by this point because acritarchs cannot be placed into any of the known modern eukaryotic groups.

Extensive diversification of eukaryotic forms is seen in fossils from the **Neoproterozoic** era (see pp. 259–260). By 1300–1000 Mya, many lineages were present, based on reliable fossils of plantae (green algae and red algae) alveolates (dinoflagellates), and heterokonts (brown algae).

ENDOSYMBIOSES HAVE PLAYED A KEY ROLE IN THE EVOLUTION OF EUKARYOTES

In Chapter 6 we discussed how eukaryotes have made extensive use of the biochemical and metabolic diversity of bacteria and archaea through mutualistic symbioses. Mutualistic symbioses have been taken even farther in the evolution of mitochondria and **plastids.** In this section we discuss theories concerning the origin and evolution of these organelles as well as evidence suggesting that the origins of the nucleus and the eukaryotic cell were the result of symbiotic events.

Many Lines of Evidence Indicate That Mitochondria and Chloroplasts Were Derived from Bacteria

Mitochondria are present in most eukaryotes; chloroplasts are found in plants and many single-celled eukaryotic lineages (including red and green algae, dinoflagellates, euglenoids, cryptomonads, and **glaucocystophytes**). Although they are now so integrated with the cells in which they are found that they are called organelles, the evidence is overwhelming that both mitochondria and chloroplasts are the descendants of endosymbiotic bacteria.

The first indication that these organelles were once free-living microbes came from microscopic investigations. Each organelle resides inside the cytoplasm of the cell and is surrounded by multiple (from two to five) membranes, giving the appearance of a bacterial cell living inside the cytoplasm (compare Fig. 8.11 to Fig. 6.2).

Many other similarities link mitochondria and plastids to bacteria. Both organelles contain a single circular genome. Both carry out transcription and translation within their compartments, using bacteria-like enzymes and components. Both replicate by dividing in a manner akin to bacterial binary fission and do so independently of the nuclear-controlled division of the host cell. The electron transport system of mitochondria closely resembles that of free-living bacteria. Chloroplasts are similar to photosynthetic **cyanobacteria.** Both transform light energy into ATP and **NADPH** (and O_2), using these to fix carbon dioxide. The proteins, pigments, and processes used by plastids to carry out these reactions are very similar to those of some cyanobacteria. As we learned in Chapters 5 and 6, however, appearances can be deceptive. Conclusive proof that mitochondria and chloroplasts are derived from bacteria comes from phylogenetic analysis of the genes encoded in their genomes.

The Existence of Hundreds of Other Types of Endosymbiosis Supports the Theory That Organelles Arose through Endocytosis

Eukaryotes have formed a diverse array of mutualistic symbioses with intracellular bacteria (see pp. 163–166). These relationships are found in nearly every major evolutionary group of eukaryotes. They range from those in which the symbionts spend only part of their life cycle in association with the host to the highly codependent organelle systems.

The pervasiveness of intracellular symbioses throughout the eukaryotes is a reflection of the relative ease with which they can evolve. This in turn is likely due to two features of eukaryotic biology: (1) The diversity of structures inside the cell enables intracellular organisms to occupy highly specialized niches, and (2) eukaryotes are able to carry out **endocytosis,** the process of enveloping objects and internalizing them in a vesicle surrounded by a membrane. Endocytosis is used by eukaryotes for many functions, such as nutrient acquisition, cell-to-cell signaling, and killing pathogens. Many mutualistic symbionts that are not transmitted from parent to offspring use endocytosis to "invade" their host cells, with some even living inside the vesicles after internalization. These pathways are also used by many pathogenic species. Thus, endocytosis, which has encouraged the formation of diverse mutualistic associations and contributed to the development of the organelles, also makes eukaryotes vulnerable to intracellular pathogens.

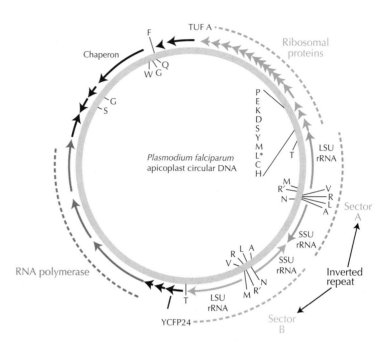

FIGURE 8.8. Apicoplast genome from *Plasmodium falciparum*. Recall from Aim and Scope that the apicoplast is related to chloroplasts. Genes are indicated by *arrows*, which show the direction of transcription. *Broken lines* highlight functional regions. Ribosomal RNA genes are indicated by LSU (large subunit) and SSU (small subunit). Each tRNA is labeled with the single-letter abbreviation of the corresponding amino acid.

Organelle Genomes Are Highly Reduced Because Most of Their Genes Were Transferred to the Nuclear Genomes of Their Hosts

In Chapter 7, we saw that the gene content of **endosymbionts** is much smaller than that of their free-living relatives. This reduction has been taken to the extreme in mitochondria and chloroplasts, whose genomes lack many of the genes common to all bacteria. The greatest number of protein-coding genes found in organelle genomes is approximately 100, which is fewer than that seen in any known bacterium, even endosymbiotic ones. Many organelles encode many fewer than 100 (e.g., see Fig. 8.8).

The reduced content of organelle genomes reflects the stable environment in which they reside, rendering unnecessary many of the genes that are essential to other bacteria. This genome reduction has been accompanied by a general simplification in the biology of organelles. Many lack cell walls, have simpler membrane structures, have reduced protein synthesis machinery, and have a highly reduced set of **transfer RNAs** (**tRNAs**) that make use of **wobble pairing** in the third position, which enables many of the tRNAs to recognize multiple codons. The organelles even use alternative genetic codes (Table 8.1).

Despite this reduction in their genome content, the biological complexity of organelles remains high. This is true because organelles import many materials from their hosts, including metabolites, noncoding RNAs, and proteins. For example, many of the ribosomal proteins required for protein synthesis in the organelles are encoded in the nuclear genome, synthesized in the cytoplasm, and imported by the organelle as peptides. Most striking, these and many of the other imported proteins, although now encoded in the nuclear genome, were originally encoded in the organelle genome, but the genes that encode them were transferred over time to the nuclear genome.

TABLE 8.1. Mitochondrial genetic codes

Codon	Standard Code: Nuclear-encoded Proteins	Mitochondria		
		Mammals	*Drosophila*	*Neurospora*
UGA	Stop	Trp	Trp	Trp
AGA, AGG	Arg	Stop	Ser	Arg
AUA	Ile	Met	Met	Ile
AUU	Ile	Met	Met	Met
CUU, CUC, CUA, CUG	Leu	Leu	Leu	Leu

Reproduced from Lodish H. et al. 2000. *Molecular cell biology*, 4th ed., Table 9-4, p. 335. W.H. Freeman, New York.

Analysis of gene and genome sequences has revealed that a staggering number of nuclear genes are derived from organelle genomes. It is useful to consider how such genes are identified within a nuclear genome. For example, suppose we wanted to know how many genes in a plant nuclear genome were derived from chloroplasts. We could examine the phylogenetic history of all the nuclear-derived protein-coding genes by building their evolutionary trees. Chloroplast-derived genes would be expected to be most closely related to genes from chloroplasts or cyanobacteria. When such an analysis was performed on the genome of the model plant *Arabidopsis thaliana,* it was discovered that approximately 18% of all nuclear protein-coding genes appeared to be derived from chloroplasts. Applying the same type of analysis to look for mitochondrially derived genes in the nuclear genomes of other eukaryotes has revealed similarly high numbers. It is extraordinary that a substantial fraction of our genome is bacterial rather than eukaryotic in origin.

Evolutionary Analyses Help Reveal the Mechanisms of Gene Migration from Organelle Genomes to the Nucleus

How do organellar genes move to the nuclear genome? The process is thought to require a series of steps (Fig. 8.9A). The first step is for the organelle DNA to come into contact with the nuclear DNA. This may be facilitated by the dissolution of the nuclear membrane during cell division. The DNA next needs to integrate into the nuclear genome. This can happen by homologous recombination (see the discussion on lateral transfer on pp. 182–187). However, experimental studies suggest that integration occurs most frequently when there are double-strand breaks in the nuclear genome, enabling the organellar DNA to be inadvertently incorporated by DNA repair processes. Although typically only small fragments of organelle DNA are incorporated, sometimes entire organelle genomes are integrated into the nuclear genome.

From an evolutionary perspective, the frequency of DNA migration from organelles to the nucleus is high, and in some extreme cases it occurs at approximately the same rate as point mutations. Most of the time, however, the integrated DNA is nonfunctional. Only rarely does the DNA integrate in such a way as to have the correct transcription and regulatory signals to work with the eukaryotic machinery of the nucleus to produce functional transcripts. This still does not mean that a gene will encode a product that will function in the organelle. Products must also acquire the necessary signals to be transported from the cytoplasm into the appropriate organelle.

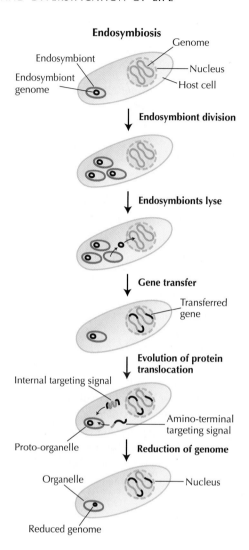

FIGURE 8.9. Organelle to nuclear gene transfer. Multiple steps in the transfer of DNA from an organelle genome to that of the nucleus are shown. First, an endosymbiosis forms. Next the endosymbiont replicates itself. Then some symbionts lyse and their DNA migrates to the nucleus where it is integrated into the nuclear genome. Genes that were integrated acquire targeting and translocation signals allowing the gene product to be sent to the symbiont. This allows the symbiont version of this gene to be deleted without functional consequences. Many consider this the point at which the symbiont becomes an organelle.

When a gene migrates from the organelle genome to the nuclear genome and acquires all of the signals necessary for its product to end up in the organelle, selective pressures are reduced on the copy of the gene within the organelle. This makes it more likely that the gene in the organelle may be lost or degraded. As there is virtually no gene flow from the nucleus to organelles, once a gene is lost from the organelle genome, it has little chance of being replaced by the nuclear copy. Thus gene movement may be a selectively neutral process, providing no advantage or disadvantage to the genes involved or to the organism.

A variety of lines of evidence, however, suggest that gene migration is not neutral. It may be advantageous for organisms to have these genes encoded in their nuclear genome rather than the organelle genome. For example, nuclear genomes typically undergo recombination during sexual reproduction, which provides greater variation for natural selection to act upon and helps prevent the accumulation of deleterious mutations. We discuss this issue in detail in Chapter 23.

Why, then, are any genes left in the genomes of organelles? Some proteins may not be transported easily from the cytoplasm into the organelles, and thus there is pressure to retain their genes within the organelle. Some resident DNA may be required for organelle replication. An organellar gene that uses alternative codons would be difficult to integrate into the nuclear genome because the protein translation machinery in the cytoplasm does not recognize those codons. Importantly, the rapid and coordinated regulation of some genes is critical for proper organelle function. The delay inherent in im-

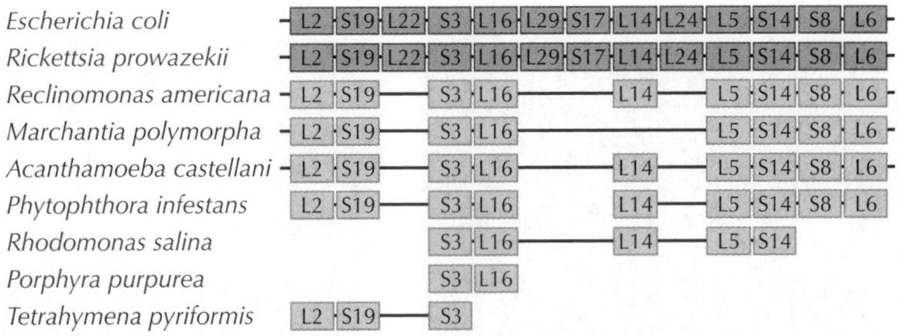

FIGURE 8.10. Mitochondria and bacteria have similar ribosomal protein gene order. The order of genes encoding some of the ribosomal proteins is shown for two bacterial species (*red*) and for multiple mitochondrial genomes (*orange*). Although some genes were lost in the mitochondrial genomes (and likely moved to the nuclear genomes in those species), the order of the other genes is retained.

porting nuclear-encoded proteins may be a selective pressure that causes those genes to remain in the organelle. Perhaps this is why organelles retain genes for many of their ribosomal proteins and rRNA molecules, and why the arrangement of ribosomal protein genes is highly conserved between organelles and free-living bacteria (Fig. 8.10).

Phylogenetic Studies Reveal the Complex History of Plastids and Their Hosts

Plastids can be classified as primary or secondary according to their evolutionary history. We use the general term plastid to refer to all organelles that trace their ancestry to the primary plastid and chloroplast to specifically refer to photosynthetic forms of these organelles. Phylogenetic analysis of extant plastid genomes and nuclear-encoded genes derived from plastids suggests that all primary plastids trace their origin to a single endosymbiotic event between a cyanobacterium and a eukaryotic cell (Fig. 8.11A). Some of the characteristic structural features of cyanobacteria have been retained by these plastids. For example, the cyanobacterial cell wall surrounds the inner and outer membranes in the plastids of glaucocystophytes. Primary plastids are found in green plants, red and green algae, and glaucophytes.

Secondary plastids are thought to be the result of subsequent symbioses in which plastid-containing eukaryotes became intracellular symbionts in other eukaryotic lineages (Fig. 8.11B). Many lines of evidence support this theory. First, some plastids are surrounded not by two, but by three or four membranes. Second, many eukaryotic species that contain plastids also possess a nucleomorph, i.e., a second nucleus with a degenerate genome that is unrelated to the functional "major" nuclear genome. Third, molecular phylogenetic analyses have shown that the nucleomorph genomes are related to those of eukaryotic organisms that possess primary plastids, whereas the major nuclear genomes are related to other eukaryotic lineages. For example, in cryptophytes, the nucleomorph and plastid genomes are related to those of red algae, whereas the nuclear genome is from a distinct phyla, possibly related to haptophytes (Fig. 8.5). A wide variety of secondary endosymbioses have now been discovered. In some cases, the organism has retained all of the genomes present when the secondary symbiosis occurred, that is, the nuclear and mitochondrial genomes of the original host, as well as the nuclear, mitochondrial, and plastid genomes of the original symbiont. In other cases, such as the apicomplexan parasites (e.g., *Plasmodium* and *Toxoplasma*), the nuclear and mitochondrial genomes of the symbiont were discarded and only its plastid genome was retained (see Aim and Scope).

All Mitochondria Appear to Be Derived from One Symbiosis That Occurred at or near the Origin of Eukaryotes

As with plastids, the ancestry of extant mitochondrial genomes has been traced by phylogenetic studies. These phylogenetic reconstructions show that all mitochondria have

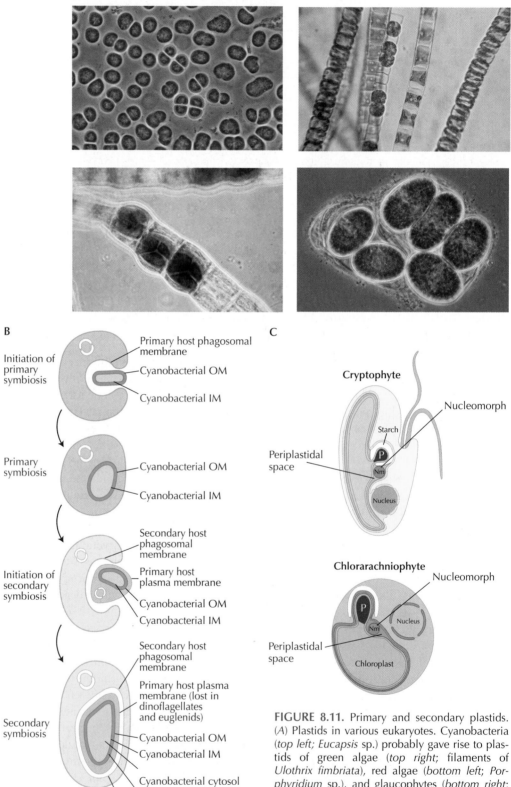

FIGURE 8.11. Primary and secondary plastids. (*A*) Plastids in various eukaryotes. Cyanobacteria (*top left; Eucapsis* sp.) probably gave rise to plastids of green algae (*top right*; filaments of *Ulothrix fimbriata*), red algae (*bottom left; Porphyridium* sp.), and glaucophytes (*bottom right; Glaucocystis* sp.) through a single symbiotic event that took place more than 1.2 billion years ago. (*B*) Model of primary and secondary plastids. OM, outer membrane; IM, inner membrane. (*C*) Two examples of eukaryotes with plastids formed by secondary endosymbiosis.

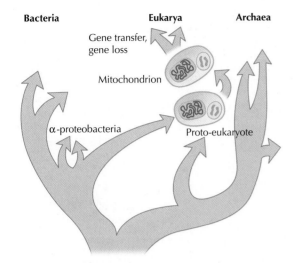

FIGURE 8.12. Mitochondria are α-proteobacteria. The drawing shows a model for the origin of the mitchondria. The overall tree is that discussed in Chapter 5 with the archaea being closer to eukaryotes than to bacteria.

a common ancestry and that they evolved from within the α subclass of the bacterial phylum Proteobacteria (Fig. 8.12). The exact placement of mitochondria within the α group is still unknown. Early studies suggested that mitochondria are related to *Rickettsia*, α-subclass bacteria with highly reduced genomes and an intracellular parasitic lifestyle. However, this putative phylogenetic position has not been supported by analysis of the complete genomes of species in the α subclass.

Some eukaryotes, such as *Trichomonas*, *Giardia*, and *Entamoeba*, do not have mitochondria. Can this tell us anything about the origin of mitochondria? Early molecular systematic studies, based primarily on rRNA analysis, suggested that these amitochondrial eukaryotes branched off near the root of the eukaryotic tree (see Chapter 5). Thus, it was theorized that the symbiotic event that introduced the mitochondrial precursor into an ancient eukaryote occurred after the branching of these amitochondrial lineages. The realization that these amitochondrial lineages live in anaerobic environments led to the suggestion that the mitochondrial symbiosis may have allowed other organisms to exploit aerobic environments better. However, the new picture of the tree of eukaryotes discussed in Box 8.1 is inconsistent with this model. In this tree, these three lineages do not group together, and, most likely, mitochondria are ancestral to *all* eukaryotes.

What was the fate of mitochondria in these amitochondrial lineages? In many cases, the mitochondria have not been lost but instead have been converted into other organelles. For example, the hydrogenosome in *Trichomonas* and the mitosomes in *Giardia* and *Entamoeba* lack their own genomes but appear to be derived from mitochondria. The mitochondrial ancestry of these organelles is supported by their appearance

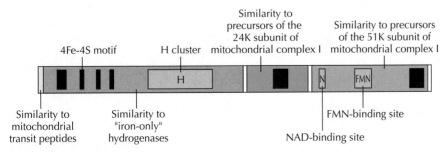

FIGURE 8.13. Are hydrogenosomes relic mitochondria? An open reading frame from a nuclear-encoded hydrogenosome-targeted gene is shown. The open reading frame encodes a putative [Fe]-hydrogenase, which has many similarities to known mitochondrial-derived proteins.

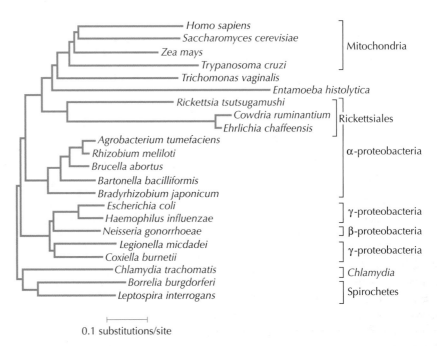

FIGURE 8.14. Mitochrondrial-derived genes may be found in amitochondrial species. A phylogenetic tree of bacterial- and nuclear-encoded cpn60 showing that genes from *Trichomonas* and *Entamoeba*, which are amitochondrial species, group with mitochondrial genes from other eukaryotes.

and biochemistry. In addition, within the nuclear genomes of these species, there are genes that are evolutionarily related to mitochondria-derived genes (Figs. 8.13 and 8.14). Although it is possible that these genes were acquired by lateral transfer from free-living α-**proteobacteria**, the finding is consistent with the possibility that these species once had mitochondria with genomes and that some of the mitochondrial genes migrated to the host's nuclear genome before the organelle was lost.

The Origin of the Nucleus Remains Unclear

The key feature that distinguishes eukaryotes from bacteria and archaea is the cell nucleus. Although the nucleus has been extensively studied for more than 200 years, its origin remains a mystery. A sampling of the theories regarding the origin of the nucleus and the nuclear genome is given in Table 8.2. These theories can be divided into two categories. The karyogenic (or autogenous) models postulate that the nucleus evolved within a single evolutionary lineage, in which the cellular genome became enclosed within a membrane. This structure evolved into the nucleus we see today. The endokaryotic models invoke some type of endosymbiosis (or fusion) between cells from different lineages, in which the endosymbiont becomes the nucleus. All of the models attempt to be consistent with two key features of the nucleus: the current structure of the nuclear membrane, and the fact that some genes in the nuclear genome are more closely related to archaeal genes, whereas others are closer to bacterial genes.

Karyogenic Models

Most karyogenic models include some type of invagination of the cell membrane to create a compartment that eventually became the nucleus. The invagination is akin to phagocytosis as seen in modern eukaryotes. The ancestor of eukaryotes must have sep-

TABLE 8.2. Schematic summary of various models for the origin of the nucleus

Schematic Model	Membrane That Nuclear Membrane Is Derived from and Is Homologous to	Compartment That the Nuclear Compartment Is Derived from and Is Homologous to
A Gram-positive bacterium (actinobacterium) Archaea → Amitochondriate eukaryote	Plasma membrane of a bacterium	Bacterial cytoplasm
B Gram-positive bacterium Endospore formation → Amitochondriate eukaryote	Plasma membrane of a bacterium	Bacterial endospore
C Gram-negative bacterium, Endokaryosis Crenarchaeote → Amitochondriate eukaryote	Plasma membranes of a bacterium and an archaea	Archaeal cytoplasm
D Archaeal (methanogens), δ-proteobacterial fusion H₂-producing δ-proteobacteria → Amitochondriate eukaryote	Plasma membranes of several bacteria	Archaeal cytoplasm
E Archaeal (methanogens), δ-proteobacterial fusion H₂-producing δ-proteobacteria, α-proteobacterial anaerobic methane oxidizer, Mitochondria → Mitochondriate eukaryote	Plasma membranes of several bacteria	Archaeal cytoplasm
F Archaeal (methanogens) H₂-producing α-proteobacteria, Archaeal host with mitochondrial symbiont, Vesicle accumulation, Facultative anaerobic mitochondriate eukaryote	Vesicles of bacterial lipids synthesized in an archaea cytoplasm	Archaeal cytoplasm around the chromosome
G *Thermoplasma* Spirochaete → Amitochondriate eukaryote	Plasma membranes of a bacterium and an archaea	Spirochaete cytoplasm
H Complex-enveloped DNA virus *Methanoplasma*-like methanogen, Bacterial syntrophs, Consortium → Eukaryote (mitochondriate?)	Viral coat	Viral lumen

Reproduced from Martin W. 2005. *Curr. Opin. Microbiol.* **8:** 630–637 (Table 1, p. 632) (© Elsevier).

arated from the ancestors of bacteria and archaea at some point to form a third major branch in the tree of life. Most likely, such an ancestor would have had a cell wall, as modern bacteria and archaea do. Thus, to have invaginations of the membrane as in phagocytosis, this cell wall would have to have been lost. Importantly, this theory helps explain the structure of the nuclear envelope (see Fig. 8.1). In karyogenic models, it is postulated that the lineage in which karyogensis happened is more closely related to archaea than to bacteria. The bacterial genes would have come from the symbiosis with the mitochondrial ancestor (see Fig. 8.12). This symbiosis could have occurred before or after the nucleus was created, but it must have happened after the cell wall was lost and some type of endomembrane system had been developed.

One critical difficulty for the karyogenic models is postulating what the selective pressures would have been for the nucleus to evolve. One possibility is that a membrane around the DNA protected the genome from the waste products of metabolism and from toxins, such as oxygen. Inside cells, oxygen radicals, which are formed from oxygen, react with DNA causing strand breaks and lesions in the DNA bases, which can be mutagenic and toxic (e.g., see Fig. 12.12). The nucleus may also protect the genetic material from genetic parasites like viruses and transposable elements. Encapsulating the DNA and the replication machinery in a membrane would make it more difficult for genetic parasites to integrate into the genome or to co-opt the replication machinery for their own purposes. These theories concerning the selective pressure for the origin of the nucleus depend on when in evolution the nucleus originated. The models above consider the origin to have been after cellular life was evolving. An alternative theory proposes that the nucleus evolved as part of the transition from the RNA world to the dominance of DNA as the genetic material. In this model, the nucleus is considered to be the single-cell equivalent of germ cells in animals: a way to protect a copy of the genome. In this case, the DNA would be a copy of the functioning RNA genome and eventually DNA would have taken over as the sole genomic copy.

In considering the karyogenic models, it is important to realize that same bacteria and archaea possess features that, if present in the ancestor of eukaryotes, would have made the creation of a nucleus relatively straightforward. For example, in most bacteria and archaea, the DNA is not simply floating around loose in the cytoplasm; it is packaged into one region of the cell, and it is usually attached to the cell membrane. This allows DNA and RNA polymerases to concentrate around the genome. It is a small evolutionary leap to surround this collection of DNA and proteins with a membrane. In fact, this is what some species of bacteria in the Planctomycete phyla have done. For example, the DNA in *Gemmata obscuriglobus* is enclosed in a membrane compartment isolated from the rest of the cell (Fig. 8.15).

Endokaryotic Models

The endokaryotic hypothesis postulates that the nucleus is simply the by-product of endosymbiotic events, much like the membrane around mitochondria and plastids. Given the origins of mitochondria and plastids from endosymbiosis, the symbiotic origin of the nucleus is not hard to imagine. In many of these models, the symbiont's genome became the nuclear genome and the original host's genome eventually migrated to the nuclear genome. This explains the chimeric ancestry of nuclear genes.

In endokaryotic models, who were the symbiotic partners and what drove the symbiosis? As with the karyogenic models, the endokaryotic models require that the cell wall of the original host be lost prior to the symbioses. This is why in some models the original host was from a lineage whose extant species lack a cell wall (e.g., *Thermoplasma*). Most models, however, propose that the endosymbiont that became the nucleus was an archaeon and the original host was some type of bacterium. This ex-

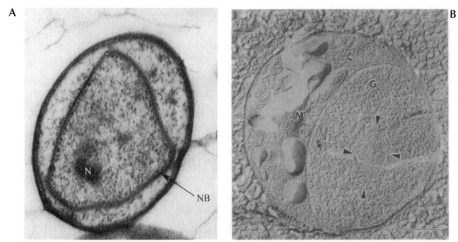

FIGURE 8.15. DNA-enclosing membrane in a bacterium. (*A*) Cross section of a *Gemmata*-like cell. The double-membrane-bounded nuclear body is labeled as NB and the DNA nucleoid as N. (*B*) An electron micrograph of a cross section of *Gemmata obscuriglobus*. The membrane-bound nuclear body is labeled M: The cytoplasm outside the nuclear body (C) is a different texture than the cytoplasm located inside the nuclear body (G). Bar = 0.5 μm.

plains why many genes encoding nuclear-associated functions in eukaryotes (e.g., replication and transcription) are more closely related to archaeal genes than bacterial genes, and why many genes for cytoplasmic functions (e.g., metabolism) are closer to those of bacteria. Because, as was discussed above, the prevailing theory regarding mitochondrial evolution is that they evolved at or near the beginning of eukaryotic evolution, many endokaryotic models propose that the same α-proteobacteria use the original host of the symbiosis (Fig. 8.16). If instead, a separate host was involved, then the mitochondrial symbiosis was an independent event, and some of the genes in the original host's genome migrated to the nuclear genome, whereas the rest of the host's genome disappeared. The question of what drove the symbiosis is one of the more interesting aspects of the endokaryotic theories. Most of the models propose that the driving forces involved some type of metabolic partnership, as is seen in many modern symbioses (see p. 163).

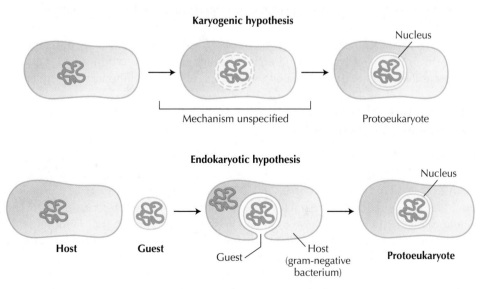

FIGURE 8.16. Origin of nucleus.

Separate Genomes Can Evolve Differently

There are evolutionary consequences to locating some genes in separate organellar genomes. The nuclear, mitochondrial, and plastid genomes are under different selective pressures and thus frequently evolve in different ways. For example, in sexually reproducing eukaryotes, organelle genomes do not undergo sexual recombination, but instead they are frequently transmitted primarily through the maternal lineage. This makes them useful as genetic markers for tracing maternal lineages in populations, much as the Y chromosome can be used to trace paternal lineages in humans (see pp. 745–748). However, lack of recombination makes selection less effective and may cause organelle genomes to eventually degenerate (see pp. 680–683).

Rates of evolution differ between nuclear genomes and nonnuclear genomes. For example, in animals, genes in the mitochondrial genome evolve more rapidly than nuclear genes. In many plants, the opposite is true; mitochondrial genes evolve more slowly than nuclear genes. The reasons for the differences in rates are unclear, but they need to be taken into account when using genes from different genomes for population and evolutionary studies.

Summary: Eukaryotes Are Genomic Chimeras

Having organelles each with their own genome has resulted in the "colonization" of eukaryotic genomes with genes derived from bacteria. However the nucleus originated, extant nuclear genomes are chimeras of genes with significantly different evolutionary histories. In eukaryotes that are the product of secondary symbioses, one cell can possibly contain genes from the original host's nuclear and mitochondrial genomes as well as the symbiont's nuclear, mitochondrial, and plastid genomes.

In addition, although lateral gene transfer in eukaryotes is probably not as frequent or extensive as in bacteria and archaea, there are indications that it does occur. Among eukaryotes, it is probably most extensive in single-celled organisms. There is also some evidence suggesting that viruses can move genes between eukaryotic species in much the same way that phage do in bacteria and archaea. Given all of these potential mechanisms for moving genes, it is more accurate to envision the evolution of a eukaryotic nuclear genome not as a single tree but rather as multiple trees with branches that occasionally interlace.

STRUCTURE AND EVOLUTION OF THE NUCLEAR GENOME

The structure and content of eukaryotic nuclear genomes is significantly different (on average) from those of bacterial and archaeal genomes. In this section, some examples of these differences are presented along with some of their evolutionary consequences.

The Structure of Eukaryotic Chromosomes Has Important Evolutionary Consequences

Eukaryotic nuclear DNA is typically in the form of linear chromosomes. As discussed on page 171, DNA replication enzymes are unable to replicate the lagging strand of linear DNA to the end of the chromosome. This is because there is no site to make a primer to bind that is required for the terminal **Okazaki fragment**. Failure to address this problem would cause the ends of the chromosome to be shortened a little in each round of replication (Fig. 8.17A). Diverse solutions to this problem have evolved. In some organisms, the ends of the chromosomes, termed **telomeres**, are composed of highly repetitive DNA. These organisms use a specialized enzyme called **telomerase** to cap the chromosome ends with the specific telomere sequence, thus reversing this end degradation (Fig. 8.17B). Many telomerases are related to the reverse transcriptases that are used by RNA viruses and retro-

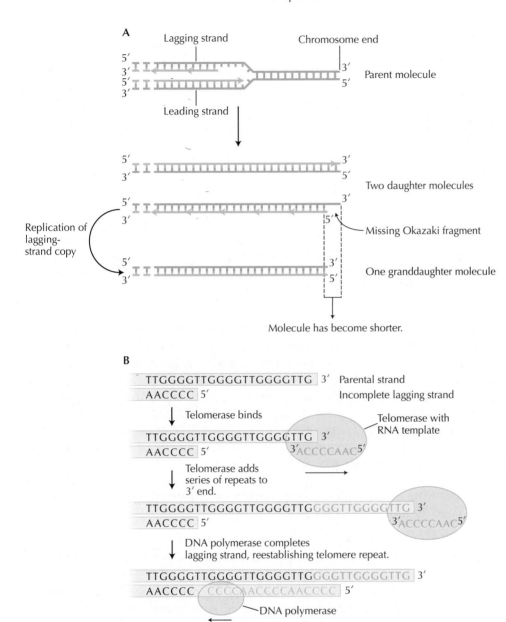

FIGURE 8.17. (*A*) The inherent problem in replicating telomeres. (*Top panel*) DNA is being replicated near the chromosome end (*right*). The leading strand can be completed (*top* molecule in *middle panel*), but the lagging strand is incomplete (because a primer cannot be made for the last piece at the end). When this is then replicated, it leads to a shortening of the chromosome. (*B*) Telomerase enzymes allow for replication of the ends of the lagging strands. In this example, an RNA molecule inside the telomerase serves as a template for adding DNA to the chromosome end.

transposons to copy their RNA genomes into DNA. This has led to the theory that telomerases in eukaryotes were captured from a transposon or a virus. Other eukaryotes, including some that also have a telomerase, add DNA to their chromosomal ends by other methods, including using unequal recombination and transposable elements.

A prominent feature of most eukaryotic chromosomes is the **centromere** (Fig. 8.18A). The centromere serves as the point of attachment connecting each chromosome to microtubules during cell division. During meiosis and mitosis, this attachment directs the chromosomal movements that result in chromosome segregation and the distribution of one complete set of chromosomes to each daughter cell.

The presence of centromeres has important evolutionary consequences. For exam-

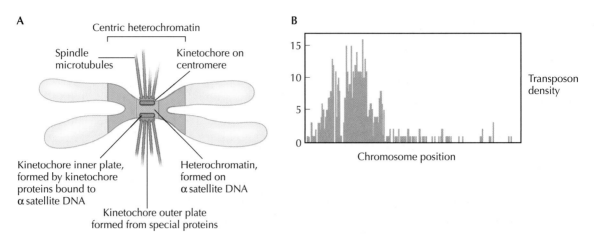

FIGURE 8.18. Transposons accumulate in centromeres. (A) Diagram of a eukaryotic centromere. (B) Transposon density in chromosome II of *Arabidopsis thaliana*. The peak of transposon density is in the area of the centromere.

ple, there is a recombination gradient within eukaryotic chromosomes, with higher rates at telomeres and lower rate at centromeres (see Fig. 19.21). The lower recombination rate near centromeres leads to less efficient natural selection in those areas of the genome (see pp. 540 and 680–683). As a result, DNA of little functional importance or DNA that may be slightly deleterious is often relegated to the centromeric region of the chromosome. For example, in the model plant species *Arabidopsis thaliana*, the density of transposable elements is approximately 20 times higher at the centromere than at the telomeres (Fig. 8.18B), and multiple copies of the entire mitochondrial genome are located in the centromeric region of chromosome II.

Genome Size and Relative Frequencies of Different Types of DNA Vary Greatly among Eukaryotes

Compared to archaea and bacteria, eukaryotic genomes can be quite large (see Fig. 7.1). The variation in genome size within the eukaryotes is even greater than the differences between eukaryotes and both archaea and bacteria. When haploid genome sizes (i.e., the number of base pairs per version of the genome) are compared, we find that some eukaryotic genomes are smaller than those found in some bacteria and archaea, whereas others are enormous. For example, the nuclear genome of the intracellular pathogen *Encephalitozoon cuniculi* is less than 2×10^6 bp in length, whereas the genomes of some plants, such as *Lilium longiflorum*, are more than 9×10^{10} bp or 30 times as large as the human genome. The genomes of some amoebae are estimated to be 6.7×10^{11} bp. Because the number of chromosomes does not always increase in proportion to genome size, some organisms have extremely large chromosomes, which require multiple origins of replication to reduce the time required for replication of the entire genome. Similarly, eukaryote genome size does not correlate closely with the number of genes.

Much of the "extra" DNA has been labeled junk or selfish DNA (see p. 173), implying that it plays no role in the normal biology of the organism in which it resides. Although some of this DNA may have important functions in eukaryotic cells (see pp. 542–547), much of it is probably of little benefit to the host cell. The evidence for this view and the reasons why eukaryotes seem to accumulate so much extra DNA are discussed in more detail on pages 595–598. In the following sections we examine the nature and consequences of this extra DNA, because it plays a critical role in the variation in genome structure among eukaryotes. Specifically, we discuss three classes of DNA that vary either in size or copy number among eukaryotes: **tandem DNA repeats**, transposable elements, and **introns**.

Tandem DNA Arrays Can Be Quite Extensive in Eukaryotes

Many eukaryotic genomes contain long stretches of DNA in which the same or very similar sequences are repeated. Known as tandem DNA repeats or tandem DNA arrays, such features can be quite extensive and come in several characteristic forms. One is known as **satellite DNA**, so named because it was originally identified as "satellite" bands during DNA purification. Because of its repetitive nature and different nucleotide composition, this DNA has a different buoyant density than the rest of the DNA in that genome. Thus, when the DNA from an entire genome is purified using density gradient centrifugation, this DNA forms distinct "satellite" bands. The term satellite DNA is now used to refer to any noncoding tandem repeat that is about 100–1000 bp in length, regardless of whether its buoyant density differs from that of the rest of the genome. Experimental and genomic studies have shown that satellite DNA is highly abundant in many eukaryotic genomes, especially in centromeric regions and in heterochromatin.

Tandem repeats smaller than 100 bp are known as **minisatellites** (where the repeat is ~10–100 bp) and **microsatellites** (where the repeat is 2–10 bp). These are valuable as genetic markers because of their high mutation rate (Fig. 12.9 and Box 13.3). Many eukaryotes also have extensive tandem arrays of complete genes (e.g., Fig. 12.3).

Transposable Elements and Their Relics Account for Much of the DNA in Many Eukaryotes

In contrast to the streamlined genomes of bacteria and archaea, eukaryotic genomes are often greatly enlarged by the presence of large numbers of various types of mobile DNA elements (termed transposons). More than half of the human genome is composed of transposons of some type. Because such mobile DNA elements can have significant negative effects on the host genome, they are considered to be selfish DNA. Explanations for why selfish DNA varies among eukaryotic genomes are given in Chapter 21. Here we discuss the diversity of transposable elements found in eukaryotic genomes and their effects on the genome.

Transposable elements can be divided into three categories based on their mechanism of replication: retroelements, replicative DNA transposons, and conservative (or nonreplicative) DNA elements (see Table 12.2). Retroelements make copies of themselves through an RNA intermediate using a reverse transcriptase. To replicate, they first transcribe their DNA into RNA, then reverse transcribe this RNA into a copy of the original DNA which is inserted at another location in the genome (Fig. 8.19). Retroelements come in two main forms defined by whether or not they possess long repeats in their flanks. These long terminal repeats (LTRs) are present in **retrotransposons** and absent from **retroposons** (see Fig. 8.19). Retrotransposons are related to **retroviruses**, which contain LTRs and also replicate via an RNA intermediate followed by genomic insertion. The most familiar example of a retrovirus is the human immunodeficiency virus (HIV), which causes acquired immunodeficiency syndrome (AIDS).

The human genome, like many eukaryotic genomes, contains millions of different retroelements (see p. 597). There are two important groups of retroposons: **LINEs** (long interspersed nuclear elements) and **SINEs** (short interspersed nuclear elements) (Fig. 8.19). LINEs are present in multiple forms, some with hundreds of thousands of copies. SINEs are present in more than a million copies per cell and also include ALU elements. SINEs do not encode a reverse transcriptase but instead make copies of themselves by co-opting reverse transcriptases from other retroelements.

Eukaryotic genomes also contain **replicative DNA transposons** (which make a copy of themselves that is then inserted at a new genomic location) and **conservative DNA transposons** (which move within the genome, excising from one location and reinte-

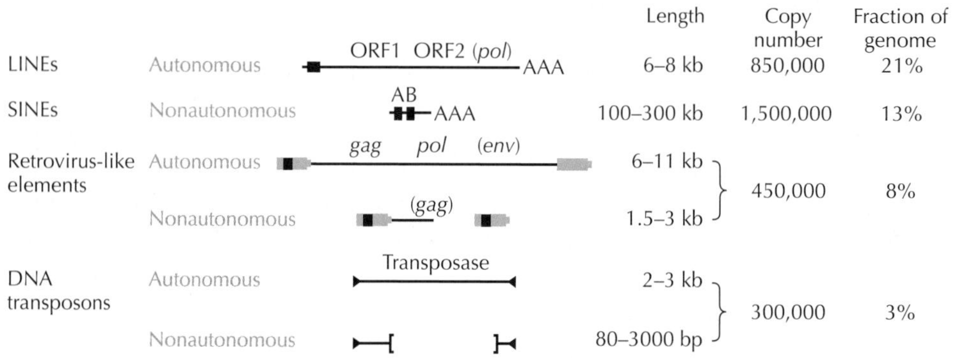

			Length	Copy number	Fraction of genome
LINEs	Autonomous	ORF1 ORF2 (*pol*) ▪━━━━━AAA	6–8 kb	850,000	21%
SINEs	Nonautonomous	AB ━▪▪━AAA	100–300 kb	1,500,000	13%
Retrovirus-like elements	Autonomous	*gag pol (env)* ▪▪━━━━━▪	6–11 kb	450,000	8%
	Nonautonomous	*(gag)* ▪▪━▪	1.5–3 kb		
DNA transposons	Autonomous	Transposase ◂━━━━━▸	2–3 kb	300,000	3%
	Nonautonomous	▸━┤ ├━◂	80–3000 bp		

FIGURE 8.19. Examples of the more abundant repetitive sequences (all of which are some form of transposon) found in the human genome. These can be classified by the type of element and whether they are autonomous (encode the means of transposition) or nonautonomous (depend on exogenous factors for transposition). Shown are four types of elements, their general structure, length, copy number, and fraction of the genome they represent. The first three listed, LINEs, SINEs, and retrovirus-like elements, are forms of retroelements that transpose through an RNA intermediate. LINEs and SINEs are related forms of retroposons; LINEs are autonomous (encoding a pol protein) and SINEs are nonautonomous. Retrovirus-like elements, which include retrotransposons, have long terminal repeats at each end (shown in *blue*) with autonomous elements encoding a pol protein. DNA transposons are frequently flanked by inverted repeats (*small triangles*) with autonomous elements encoding a tranposase protein.

grating at another) (Fig. 8.20). DNA transposons are typically less abundant in eukaryotic genomes than are retroelements.

This plethora of transposons in eukaryotes is responsible for generating a large number of mutational events. They frequently disrupt the host genome by inserting into coding genes and regulatory sequences or by participating in genome rearrangements via homologous recombination (see pp. 342–343). In addition, the excision step during replication of conservative DNA transposons is usually inexact and thus can produce a mutation at the excision site.

The reverse transcriptases encoded by active retroelements can also operate on cellular RNAs. When endogenous RNA undergoes reverse transcription, the resulting **complementary DNA (cDNA)** is sometimes inserted into the genome. The cDNA copy often differs from the original gene that encoded the copied RNA because the cellular RNAs that are reverse transcribed have already undergone RNA processing. RNA processing removes any introns and splices together the remaining exons, so that the cDNAs lack the intron sequences. Free from strong selective pressure, these gene copies accumulate mutations rapidly and become nonfunctional **pseudogenes**. Pseudogenes, which are regions of the genome that originated from genes but have become nonfunctional because

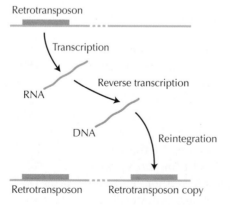

FIGURE 8.20. Retrotransposition. A retrotransposon is transcribed, and then the RNA is reverse transcribed into DNA. This DNA is then integrated into another location in the host genome.

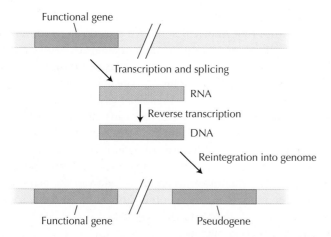

FIGURE 8.21. Reverse transcriptase–mediated creation of pseudogenes. A gene (in the *top* row) is transcribed and spliced into a final RNA, which is then reverse transcribed back into DNA. This intronless DNA can be integrated into the genome. It will most likely become a pseudogene. Note the similarities to the process of retrotransposition in Fig. 8.20.

of the accumulation of mutations, can be created by other processes (Fig. 8.21). Despite their lack of function, pseudogenes can have major effects on the genome by serving as sites for intragenomic recombination and causing chromosome misalignment during DNA repair and meiosis (see pp. 339–342). Researchers have found that they serve as useful "neutral" evolutionary markers because pseudogene sequences are similar to the corresponding functional genes but are nonfunctional. Analysis of pseudogenes has been used to characterize mutation rates and patterns in genomes. They have also served as molecular fossils for phylogenetic studies. Pseudogenes that are shared between species can be a useful derived characteristic in phylogenetic reconstructions.

Eukaryotic Genomes Contain Various Types of Introns

The abundance of introns among eukaryotes varies greatly. Some species, such as the model yeast *Saccharomyces cerevisiae*, have very few and those that are present are short. In other species, genes are filled with dozens of introns, some larger than entire bacterial genomes. Introns are classified into four major types based on their location and mechanism of action: tRNA, group I, group II, and pre-mRNA or spliceosomal introns (Table 8.3). Some tRNA genes contain introns, which must be removed from the transcribed pre-tRNA. Processing the pre-tRNA requires an endonuclease to excise the intron, as well as a **kinase** and ligase to rejoin the remaining RNA segments to form the functional tRNA.

TABLE 8.3. Major types of introns

Intron Type	Where Found
Spliceosomal	Eukaryotic nuclear pre-mRNA
Group I	Eukaryotic nuclear pre-rRNA, organelle RNAs, few bacterial RNAs
Group II	Organelle RNAs, some prokaryotic RNAs
Pre-tRNA	Eukaryotic nuclear pre-tRNA

Reproduced, with modifications, from Brown T.A. 2002. *Genomes*, 2nd ed., Table 10-2. Wiley-Liss, New York.

mRNA, mitochondrial RNA; rRNA, ribosomal RNA; tRNA, transfer RNA.

Group I introns were the first RNAs shown to be catalytic (Chapter 4). In vitro, they can excise themselves from RNA molecules; in vivo they do so in conjunction with various proteins. Group I introns are found in some organellar tRNAs and rRNAs, as well as in some mRNAs of various organisms. Their structure and function are highly conserved across all species in which they are found. The group I introns present in nuclear genomes may have been transferred from organellar genomes (see pp. 204–207).

Group II introns are also catalytic RNAs that can self-excise in vitro. Like group I introns, their structure and function are highly conserved in all species in which they have been found. However, their mechanism of action differs from that of group I introns and is instead similar to that of spliceosomal introns. Many group II introns encode proteins that transpose the intron to another region in the genome. Group II introns are found most commonly in the genomes of bacteria and eukaryotic organelles, and thus were likely present in the bacterial ancestors of organelles.

Spliceosomal introns (also known as pre-mRNA introns or nuclear mRNA introns) are so named because the splicing process requires a complex of proteins known as the spliceosome. Some eukaryote species, such as the yeast *S. cerevisiae*, have few spliceosomal introns and most of those are quite small. Few yeast genes have more than one spliceosomal intron. In contrast, human spliceosomal introns are common and very large.

Introns are frequently considered to be junk DNA. However, comparative sequence analysis has revealed that the sequence of some introns is highly conserved, suggesting that functional constraints have played a role in the evolution of intron sequences (see Chapter 19, where the conservation of noncoding DNA is discussed). Some of the observed conservation can be attributed to those sequences required for RNA splicing. Other conserved sequence elements regulate the timing of splicing, a property that enables multiple proteins to be produced from a single protein-coding gene through the process of **alternative splicing** (Fig. 8.22). If all possible alternative splicing possibilities are taken into account, species such as humans can make millions of different proteins, even though they each have only 25,000 protein-coding

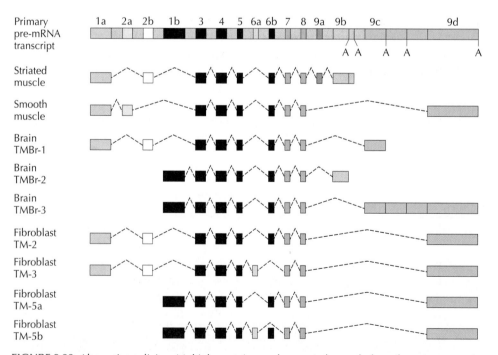

FIGURE 8.22. Alternative splicing. Multiple proteins are known to be made from the α-tropomyosin gene by alternatively splicing in different cell types. The *orange boxes* in the primary transcript represent introns; all other boxes are exons. For each tissue type the splicing form is shown by *lines* connecting the exons that are used.

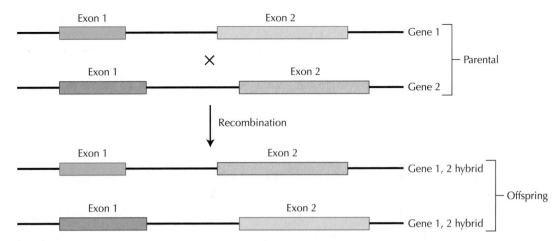

FIGURE 8.23. Exon shuffling occurs when exons from different genes are mixed and matched by recombination in the region between the exons.

genes. Alternative splicing provides a significant source of novelty for diversification (see pp. 712–713).

The presence of introns has the potential to affect the evolution of genes and genomes in various ways. For example, the exon theory of genes proposes that introns were present very early in the history of life and were lost from bacterial and archaeal lineages early in their evolution (possibly because of selection favoring compact genomes). The theory also proposes that the presence of introns early in evolution allowed for the diversification of proteins by facilitating the reshuffling of exons (Fig. 8.23). Most studies suggest that the exon theory of genes is not correct regarding the *very* early origin of introns and that many, if not most, introns arose later in evolution. Nevertheless, once introns are present, they do contribute to increased protein family diversification (see Chapter 24).

It is important to recognize that introns are not static entities. They are added to or deleted from genes relatively frequently. The loss of introns is often the result of reverse transcription of mRNAs that have already been spliced. The addition of introns may be due in part to the ability of some introns to act as transposable elements.

EUKARYOTIC DIVERSIFICATION

Single-celled Eukaryotes Show an Incredible Diversity of Cellular Forms

By far the greatest share of the phylogenetic diversity of eukaryotes is represented by single-celled organisms. They also exhibit extensive biological diversity, especially in their cellular structures and form (see, e.g., Fig. 8.6). This diversity of form is accompanied by a comparable diversity of function. For example, single-celled amoebae can change their shape to accommodate their environment, can envelop and "swallow" prey and food particles, and can glide along surfaces to explore their environment. Some algae can form plant-like structures that consist of a single cell. The diversity of cellular form exhibited by single-celled eukaryotes is also present in the various cell types within multicellular species (see Chapter 9).

Underlying this variety of cellular shapes and functions are two dynamic features of eukaryotic cells: their elaborate cytoskeleton and their internal transport network. The eukaryotic cytoskeleton allows cells to maintain diverse shapes and structures but also to change shape rapidly. The cytoskeleton also subdivides the cell into diverse microenvironments. The nucleus, organelles, other membrane-bound compartments, and other components are kept separated, yet accessible. The extensive intracellular transport system directs the movement of molecules into and out of the cell and can target them to specific intracellular locations. These active processes enable eukaryotic cell size to be less constrained by the rate of diffusion than is the case for bacterial and archaeal cells.

Complex Regulatory Networks Are a Key Aspect of Eukaryotic Diversification

The increased diversity of form in eukaryotes is accompanied by the need for more elaborate gene and protein regulatory networks than is required in bacteria and archaea (Fig. 8.24). It is important to realize that this is as true for single-celled eukaryotes as for multicellular ones. The complexity of these systems is seen at almost every level in most eukaryotes. Transcription is regulated by chromatin structure, DNA methylation, DNA supercoiling, complex promoters, interfering RNA molecules, and many other factors. Transcriptional products are modified in many ways, including RNA splicing, polyadenylation, sequence editing, and degradation. Translation products are also regulated in a variety of ways, most importantly by posttranslational modification of the proteins. These modifications include acetylation, phosphorylation/dephosphorylation, protease cleavage, regulated degradation in proteosome complexes, glycosylation, and **prenylation**. Although protein modifications are also seen in bacteria and archaea, they are less elaborate than those used by eukaryotes.

Another important eukaryotic regulatory mechanism is the ability to orchestrate the targeting of proteins, metabolites, organelles, and other components within the diverse architecture of the cell. Sorting and trafficking of proteins are assisted by the protein modifications described above. Likewise, the cytoskeletal transport systems described earlier are essential components of this mechanism. These complex regulatory processes are one feature that has enabled eukaryotes to develop diverse cellular forms and to carry out diverse processes, despite their limited range of biochemical capabilities.

Sexual Reproduction Is a Key Invention in Eukaryotic Evolution

From an evolutionary point of view, perhaps the most important benefit of the eukaryotic nuclear genome is that it made possible a new specialized reproductive process—sexual reproduction. Sexual reproduction combines the genomes of two individuals to generate the genome of an offspring (Fig. 8.25).

Essentially, sexual reproduction is a cycle involving an alternation between **haploid** and **diploid** (one or two copies of each chromosome per cell) states. Diploid individuals (**zygotes**) produce haploid cells by **meiosis** (Fig. 8.25). These cells give rise to the haploid **gametes** that fuse to form a new diploid zygote. The cycle then repeats.

Some aspects of sexual reproduction are highly conserved across most or all eukaryotes, whereas other features vary widely. For example, the general scheme of meiosis is highly conserved. First, each chromosome in the diploid cell is replicated producing two sister **chromatids**. Thus there are four chromatids for each pair of homologous chromosomes. The four chromatids then align based on sequence similarity and undergo crossing over, a process in which segments are exchanged between chromatids. Two subsequent cell divisions yield four gamete cells, each getting one chromatid from each chromosome. This process is described in more detail in Figure 12.24.

Within this conserved framework there are also some important variations. For example, in *Drosophila* and many other Diptera, crossing over during meiosis occurs only in females, not in males. However, for all species that have sexual reproduction, crossing over always occurs in at least one of the sexes.

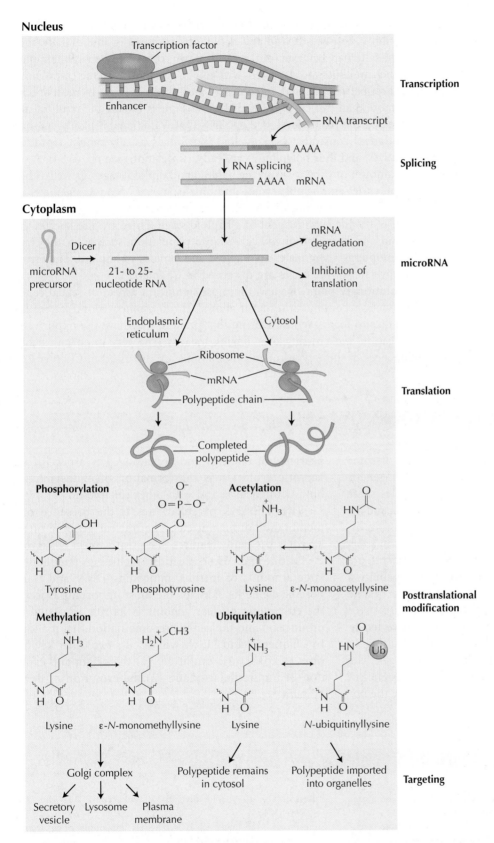

FIGURE 8.24. Eukaryotic gene regulation is generally much more elaborate than gene regulation in bacteria and archaea. Regulatory steps include transcription, RNA splicing, RNA stability (microRNAs), translation, posttranslational modifications, and protein targeting. Although each step is present in some bacteria and archaea, in eukaryotes the processes are more complex and diverse.

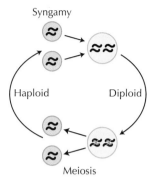

Syngamy

Haploid   Diploid

Meiosis

FIGURE 8.25. Sex in eukaryotes. The basic eukaryotic life cycle involves the coming together of haploid cells to form diploids (**syngamy**) and the production of haploid cells from diploid by meiosis. The diagram shows a genome arranged on two chromosomes. Recombination occurs at meiosis by both segregation and crossing over (as shown by *red crosses*).

Another variable aspect is the degree of ploidy in the zygotes and gametes. The prototypical alternation is between diploid and haploid, as in humans. Interestingly, there are also species that alternate between four copies and two copies (i.e., between tetraploid and diploid). Other multiples are also used, with one of the most extreme cases known being an alternation between 32 and 16 copies of each chromosome. In most species, the presence of more than two copies of each chromosome is usually the result of mutations that increased the ploidy level (see pp. 338–339). Variations in ploidy can have significant evolutionary consequences. Polyploid individuals usually cannot reproduce with diploid relatives, and thus polyploidy can lead to speciation (see pp. 631–633).

The relative amount of time spent in the haploid and diploid stages, as well as how that time is spent, varies among eukaryotes. In humans, meiosis produces gametes directly: The haploid ovule does not divide. However, in other groups (e.g., fungi), the primary portion of the life cycle is spent in a haploid stage. There are also species that have both haploid and diploid individuals within a population, such as in the "haplodiploid" Hymenoptera, where males are haploid and females are diploid. The factors that shape this diversity of life cycles are discussed in Chapter 23.

From an evolutionary point of view, the most significant aspect of sexual reproduction is the genetic recombination that it allows. Chromosomes from different lineages come together to form zygotes. During that process, they have the opportunity to mix by segregation and crossing over during meiosis. The very important evolutionary consequences of these types of recombinations are discussed in Chapter 23.

SUMMARY

In this chapter, we have presented an overview of some of the general features of eukaryotes, their phylogenetic diversity, and their genomic structures. Eukaryotes share many features that distinguish them from bacteria and archaea. They possess organized intracellular structures including diverse organelles, a cytoskeleton, and a transport system. Transcription and translation mechanisms are more complex, and with increased regulatory capabilities, thus enabling eukaryotes to develop diverse forms despite their limited range of biochemical capabilities.

Modern molecular phylogenetic studies have led to major changes in eukaryote classification, resulting in their current division into eight primary lineages. The fossil record provides strong evidence for the existence of eukaryotes almost 2 billion years ago, long after the appearance of bacteria and archaea. A key aspect of eukaryotic evolution is the formation of endosymbioses both with bacteria and with other eukaryotes.

The hallmark of eukaryotes is the presence of a membrane-bound nucleus. The nuclear genome of eukaryotes has many features that greatly affect eukaryotic evolution and diversification including (frequently) large amounts of introns, noncoding DNA, and transposable elements. Perhaps the most important aspect of the eukaryotic nuclear genome is its role in sexual reproduction and the genetic recombination that it allows. In Chapters 9 and 11, we will discuss two other key features of eukaryotic evolution: (1) the origin and evolution of multicellularity and (2) the evolution of development.

FURTHER READING

Archibald J.M. and Keeling P.J. 2002. Recycled plastids: A green movement in eukaryotic evolution. *Trends Genet.* **18:** 577–584.

Brown J.R. and Doolittle W.F. 1997. Archaea and the prokaryote-to-eukaryote transition. *Microbiol. Mol. Biol. Rev.* **61:** 456–502.

Patterson D.J. 1999. The diversity of eukaryotes. *Am. Nat.* **154:** S96–S124.

Simpson A.G. and Roger A.J. 2004. The real "kingdoms" of eukaryotes. *Curr. Biol.* **7:** R693–R696.

CHAPTER

9

Multicellularity and Development

I N PREVIOUS CHAPTERS, WE DESCRIBED THE EVOLUTION of various unicellular organisms and the remarkable levels of biochemical and genomic diversity they have achieved. In this chapter, we explore how multicellularity may have come about and emphasize the ability of organisms to produce multiple cell types. More complex levels of multicellular organization require a process of development in which an initial, single cell generates a multitude of different cell types in an organized and reproducible fashion.

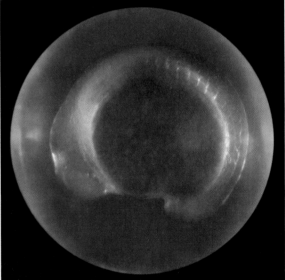

By positioning these different cell types in specific locations during development, animals and plants are able to generate characteristic morphologies. These morphologies constitute what we call the **body plan** and are key criteria for the classification scheme we use to describe all animals and plants. Understanding the relationships of these body plans has guided our understanding of the evolutionary history of these organisms.

More recently, a variety of approaches have allowed us to understand the development of multicellular animals and plants at the genetic and molecular level. These studies help us to understand how processes such as differential gene expression and cell–cell signaling act to guide development, and in this chapter, we introduce the basic mechanisms that guide the remarkable process that transforms a single egg cell into a complex organism such as ourselves. In the following two chapters, we describe the paleontological record that reveals the evolutionary history of animals and plants and then present some examples of how development has evolved to lead to these changes.

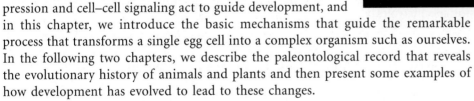

HOW MULTICELLULARITY HAPPENS

The Advent of Multicellularity Represents a Significant Evolutionary Step in the History of Life

As we saw in Chapters 6–8, the diversity of biochemical processes in unicellular organisms far exceeds that of multicellular organisms, and a number of unicellular organisms

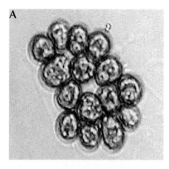

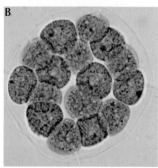

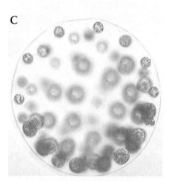

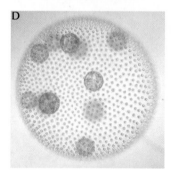

FIGURE 9.1. Various genera of Volvocales possess different types of multicellular organization. (A) *Gonium*; (B) *Pandorina*; (C) *Pleodorina californica*; and (D) *Volvox carteri*. In *Gonium* and *Pandorina*, cells are aggregated in specific geometries, but all the cells are identical. *Pleodorina* and *Volvox*, however, contain differentiated cell types (an external layer of somatic cells and internal reproductive cells) with distinctly different morphologies and functions.

show remarkable levels of morphological complexity. The emergence of multicellular organisms, however, allowed for entirely new levels of organization. Multicellularity arose many times during evolution: It is thought to have evolved independently, for example, in animals and plants. With multicellularity came the evolution of **differentiation**—the allocation of cells to different functions during the development of an organism.

In reality there is a continuum from unicellular to multicellular organization. Many organisms we consider to be unicellular can form colonies, and within these colonies, cells are exposed to different environments and respond appropriately. For example, the cells in the center of a bacterial colony growing on an agar plate see different nutrients and wastes relative to those on the edge of the colony, and thus different genes are **transcribed**, and different metabolic pathways activated, in cells of different parts of the colony. Other unicellular organisms can be considered colonial because they always occur as organized colonies of apparently identical individuals; however, again environmental influences can create differences between individuals of the colonial organism. In "true" multicellular organisms, the distinction between different cells is controlled in a more predictable way, so that there is a reproducible generation of different cell types within the multicellular organism, and the survival of the whole organism depends on having several distinctly different types of cells.

Multicellularity Is Achieved by Cells Either Coming Together or Not Separating after Division

Multicellularity, in the simplest colonial sense, can arise in two principal ways. First, as cells divide, they can remain attached instead of separating. Depending on the regularity and orientation of the divisions, this mode of becoming multicellular can easily form any of a number of macroscopic shapes, from long filaments to flat plates to perfect spheres. Second, multicellularity can be achieved through the aggregation of unicellular individuals. Such an aggregation can take on specific shapes and behaviors and in many cases can be triggered by environmental conditions.

Even in its simplest manifestation, where individual cells remain identical, multicellularity can have several benefits, including protection against predation, improved attachment to substrates, and greater buffering against fluctuating environments. Once multicellularity has been attained, the process can be taken one step further: Cells within the aggregation can differentiate and become specialized for specific roles.

In *Volvox*, Two Different Cell Types Are Maintained

Within the **chlorophyte** algae (see Fig. 8.5) is the order the Volvocales, which can be used to illustrate the type of multicellularity that arises when dividing cells stay together after cell division. Members of this group have achieved various levels of differentiation of cell types within the multicellular form.

Within the Volvocales are examples of unicellular ciliated eukaryotes such as *Chlamydomonas*. There are also simple multicellular forms, such as members of the genus *Oltmannsiella*, which contain four attached unicellular ciliated cells in a row. In *Gonium*, the different species contain a flat plate of 4–16 ciliated cells, whereas in *Pandorina*, they consist of a sphere of 16 cells (Fig. 9.1A,B). In these examples, the organisms are multicellular with varying geometries, but the cells remain identical.

Pleodorina californica and *Volvox carteri* have evolved an added level of complexity, having differentiated cell types within the multicellular body. In *Pleodorina californica*, the colony size is usually 64 or 128 cells, and 24 or 48 of these are somatic cells that do not participate in reproduction (Fig. 9.1C). The remaining cells, which are reproductive, are organized on the posterior side of the colony. In *Volvox carteri*, this division between somatic and reproductive cells remains, but the reproductive cells do not develop flagella and thus are not involved in locomotion. Adult *V. carteri* are composed of a sphere of 2000 or so small nonreproductive cells, with two flagella per cell. Within the sphere, and toward one side, is an internal population of 16 reproductive cells called the gonidia (Fig. 9.1D). These reproductive cells are set aside early in the development of the colony and during asexual reproduction form juvenile animals within it, each with its own set of somatic and gonidial cells. These juvenile colonies are eventually released, and the resulting sphere of purely somatic cells (the "parent") undergoes programmed cell death (Fig. 9.2). As with all the other volvocaceans, *Volvox* can also reproduce sexually. Under certain environmental conditions, *Volvox* will produce a sexual inducer protein that causes the colony, and all those colonies surrounding it, to undergo sexual development. Depending on whether they are male or female, the gonidia now produce sperm or eggs, and fertilization results in zygotes that can develop into mature adults (Fig. 9.3).

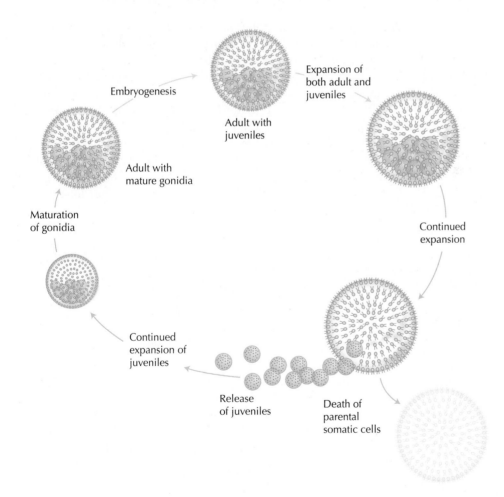

FIGURE 9.2. *Volvox carteri* asexual reproduction. The adult has external somatic cells that contain flagella and move the organism about, whereas the internal gonidial cells form juvenile animals. Juveniles develop their own gonidia and somatic cells, and when the juveniles are released, the somatic parent cells die.

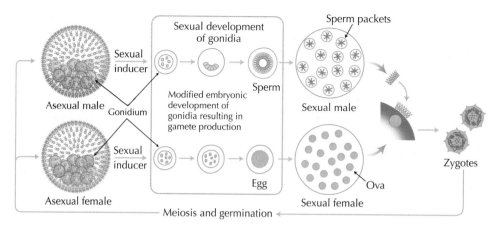

FIGURE 9.3. *Volvox carteri* sexual reproduction. Sexual inducer protein results in the development of gonidia into sperm in males and ova in females. The sperm then fertilize the ova of adult females, and the resulting zygotes (embryos) complete meiosis and development to produce adult asexual males and females.

In *Dictyostelium*, Environmental Cues Trigger Aggregation

The unicellular, amoebal slime mold, *Dictyostelium discoideum* (see Fig. 8.5), represents the type of multicellularity that arises from the aggregation of individual cells (Fig. 9.4). When food (bacteria) is abundant, the haploid amoebas reproduce by fission and exist as dispersed cells. When food becomes scarce, the amoebas stream together into a large aggregation of cells. This aggregation is a response to the production of cyclic adenosine monophosphate (cAMP), which prompts cells to move toward the source and make their own cAMP. This results in waves of cells that move toward a stochastically formed center.

Once aggregated into a mass of many thousands, the cells form a slug that begins to migrate over the substrate. The anterior part of the slug will eventually form a stalk and the remaining cells will form spores. Once lifted into the air by the stalk, the spore cells disperse to form new amoebas, and the stalk cells die. Thus *Dictyostelium* is only

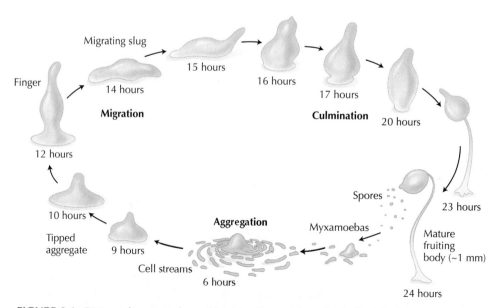

FIGURE 9.4. *Dictyostelium* sexual reproduction. Cues such as the lack of food cause the amoebas to aggregate together to form a crawling slug. Cell differentiation then occurs within the slug and results in part of the slug forming a stalk while the remaining cells form spores that disperse.

multicellular for a portion of its life and only in response to conditions in the environment. More importantly, once aggregated, *Dictyostelium* does not just remain a homogeneous mass of cells. It differentiates distinct cell types (stalk vs. spore) and is capable of generating a morphologically complex multicellular structure designed to improve its chance of surviving adverse conditions.

Cells Must Recognize and Communicate with Each Other

Although it seems likely that the type of multicellularity that led to modern plants and animals arose from the *Volvox*-like mode, both *Dictyostelium* and *Volvox* illustrate the basic principles involved in the evolution of multicellularity and differentiation. In both cases, communication between cells is required for the colony to achieve its organization. Cells must be able to attach, or remain attached, to one another, and the colony must establish an overall polarity. Finally, individual cells must take on different fates, and the continued survival of the organism depends on the reproducible acquisition of these multiple cell fates. These basic principles of growth, communication, and differentiation are recapitulated in the development of all animal and plant embryos today.

Some recent insights into the evolution of animal multicellularity have come from the study of **choanoflagellates**, which are unicellular eukaryotes that appear to be the closest surviving unicellular organisms to all the multicellular animals (Fig. 9.5A; see Box 8.1). Choanoflagellates have an ovoid cell body with a single flagellum surrounded by a collar of fine tentacles (Fig. 9.5B,C). They use the flagellum to create a water current that propels food (bacteria) into the collar for capture. Some species appear to always be unicellular, but other species of choanoflagellates can become colonial in response to certain environmental cues such as the density and type of bacterial food present in the surroundings. Remarkably, more than 135 years ago, biologists realized that a very similar cell morphology was seen in a specialized cell type of **sponges**. These sponge cells, called choanocytes (Fig. 9.5D), are the food-gathering cells of the sponge, and their flagellum and collar are used for similar food-gathering purposes. These observations led to the first suggestion that choanoflagellates were closely related to multicellular animals, and this has been borne out by molecular phylogenetic studies (see Box 8.1).

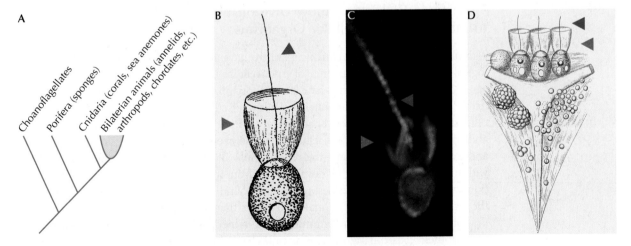

FIGURE 9.5. (*A*) Choanoflagellates are the closest living unicellular relatives to all extant multicellular animals. The single choanoflagellate cell has a collar (*red arrowhead* in *B* and *C*) and a single flagellum (*green arrowhead* in *B* and *C*). *B* is a drawing of a choanoflagellate, whereas *C* is an actual photo in which the cell body and flagella are stained green, the nucleus blue, and the collar red. Cells, called choanocytes, with a very similar morphology are seen in sponges (*D*, *green arrowhead* points to a flagellum and the *red arrowhead* points to a collar of a single choanocyte cell).

Interestingly, choanoflagellates contain many genes that encode for proteins that we might otherwise have thought unique to multicellular animals. For example, cadherins are a family of transmembrane proteins that plays an important role in sticking cells together in a wide variety of animals, and these cadherins are characterized by extracellular repeat structures known as cadherin domains. Proteins containing very similar cadherin domains are present in choanoflagellates. Similarly, proteins called receptor tyrosine kinases (RTKs) in animals have an extracellular domain that binds an extracellular ligand and an intracellular kinase domain that becomes activated by ligand binding. These RTK molecules play key roles in cellular signaling and the control of cell shape in multicellular animals, but again have also been discovered in choanoflagellates. What are these proteins doing in unicellular choanoflagellates? This remains an unanswered question, but they could certainly be involved in sensing and adhering to bacteria or other choanoflagellates. Thus, the types of proteins key to the advent of multicellularity probably arose for somewhat different functions in the common unicellular ancestor of animals and choanoflagellates.

Evolution of the First Multicellular Organisms Involved Environmental Inputs

Dictyostelium, *Volvox*, and choanoflagellates all illustrate the importance of environmental inputs into the evolution of multicellularity. The ability of cells to respond to environmental cues, a fundamental property of all living organisms, is a prerequisite for achieving multiple, differentiated cell fates within a single organism. One theory holds that the initial asymmetry of multicellular organisms was a simple response to external stimuli acting differentially on an otherwise homogeneous aggregation of cells. For example, a cellular aggregation might have become large enough to rest on the substrate in an aquatic environment. Those cells on the substrate would experience a different environment than those on top. It might become advantageous to differentiate the bottom cells for adhesion (if there were a benefit to staying in one place on the substrate) or to develop flagella only in those cells at the bottom (if moving along the substrate were beneficial).

DIVISION OF LABOR THROUGH DIFFERENTIATION

Specialized Cell Types Allow Division of Labor: The Key to Success of Multicellular Organisms

Although a number of unicellular organisms have attained remarkable levels of morphological complexity, the emergence of multicellular organisms well over 600 million years ago (Mya) allowed for entirely new levels of organization. The simplest examples of multicellularity include organisms such as *Volvox* (discussed above) in which the two types of cells, somatic and reproductive, are segregated and distinguished by different functions and morphologies. With multicellularity came evolutionary advances in the complexity and extent of differentiation—the reproducible allocation of cells to different functions within an organism during its development from an embryo to an adult. In turn, evolutionary changes in the processes, mechanisms, and patterns of differentiation have led to the remarkable diversity of animals and plants that we see on Earth today.

Embryos of plants and animals have gained control of the microenvironment around their cells during development. What was probably once a stochastic process acting to make colonial cells slightly different from one another has evolved into a system of generating a controlled environment to achieve reproducible patterns of cell differentiation. Animals and plants ensure that the proper cues are received at the appropriate times and locations by embryonic cells, resulting ultimately in remarkable levels of cell-type differentiation and large-scale structural complexity. This process of

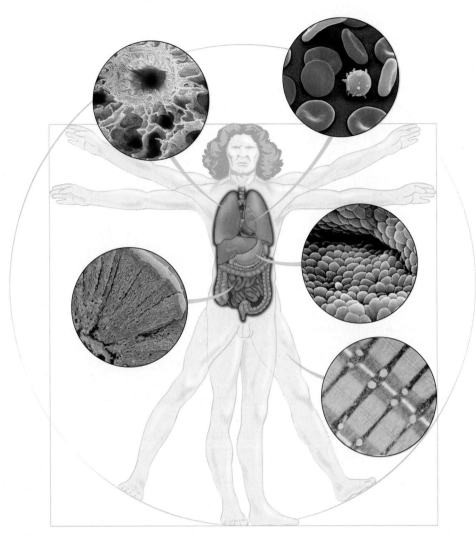

FIGURE 9.6. Multicellular animals, like us, can have thousands of uniquely differentiated cells. Spatially organizing these cell types allows animals and plants to have organs with specific functions and to take on characteristic morphologies. For example, humans have (*clockwise from top left*) alveoli of the lung with cells specialized for gas exchange, blood cells that transport oxygen and carbon dioxide as well as carry out defensive functions, a stomach composed of cells that secrete digestive enzymes, muscles composed of cells with highly arrayed proteins that generate force efficiently, and a small intestine composed of cells that can absorb nutrients.

differentiation has reached greater and greater levels of elaboration, generating organisms such as humans in which there are many thousands of differentiated cell types, each with its own set of functions that contribute to the success of the organism as a whole (Fig. 9.6). Each differentiated cell type contains unique combinations of proteins and biochemical activities. None of these cells could survive on its own, but as part of an integrated organism together they achieve things that no aggregation of homogeneous cells ever could. This specialization of cells, with its consequent division of labor, is the key to the success of multicellular organisms.

Differentiated Cells within an Organism Contain the Same Genetic Information

How do cells become differentiated from one another? As we will see, a multitude of mechanisms are used to enable a fertilized egg from a plant or animal to develop

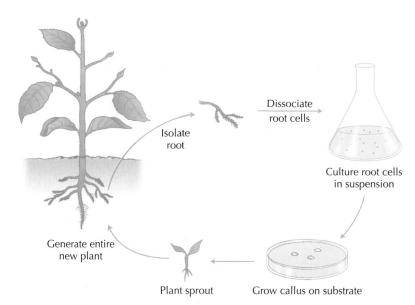

FIGURE 9.7. An entire plant can be produced from differentiated cells when grown under the appropriate conditions. In this example, isolated cells from a root can be grown to produce a callus, which then can form a normal plant capable of its own reproduction.

into an organism composed of many different cell types. What has become apparent, however, is that in virtually all cases the differentiated cells all contain the same genetic information (one notable exception, the antibody-producing B cells, is discussed below). The cells of our skin contain exactly the same genes as the cells of our liver.

This constancy of genetic material between differentiated cells within an organism has been illustrated in a number of ways. The clearest cases come from plants, where an entire plant can be regenerated from a single differentiated cell. For example, isolated root cells can be cultured in a medium where they will eventually form a callus of cells in suspension, which then settle on a substrate and generate an entire, mature plant capable of its own reproduction (Fig. 9.7).

Differentiation in animal cells can also be reversed, although not as easily as in plants. The first indication of the genomic equivalence of cells within an animal came from regeneration studies. For example, when the lens of an amphibian eye is removed, it eventually grows back. This does not require a pool of undifferentiated cells that were maintained in the adults; rather the lens is regenerated from adjacent cells of the iris. These iris cells, which are clearly differentiated (e.g., containing specialized pigments), begin to dedifferentiate, lose many of their specialized aspects, proliferate, and then redifferentiate as lens cells with all the appropriate specializations associated with this cell type (Fig. 9.8). It should be noted, however, that in most cases adult organisms will replace missing cells not through a process of local dedifferentiation and redifferentiation, but rather through the use of stem cells that remain relatively undifferentiated, but can differentiate into a limited number of cell types when required. It is in this way that, for example, we replace missing skin cells.

Another clear illustration of the **pluripotency** of differentiated cells comes from animal-cloning experiments. For a number of animal species, it is now possible to isolate a differentiated cell from an adult tissue, inject that cell's nucleus into an enucleated egg that lacks its own genetic material, and grow this egg under the appropriate conditions. Once implanted into a donor mother, the egg can develop into a newborn, grow into a juvenile animal, reach adulthood, and have its own offspring. This proves that differentiated cells still contain all of the genetic information needed to make all

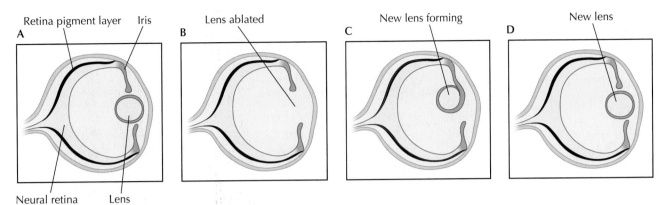

FIGURE 9.8. Regeneration in the amphibian eye. (A) Normal eye showing the lens and pigmented iris. (B) Animal in which the lens has been ablated (removed), but the iris is still present. A week after the ablation, cells at the edge of the iris dedifferentiate and proliferate, then begin forming a small, new lens (C), which eventually grows into a normal-looking lens (D) about a month after the ablation.

other cell types (Fig. 9.9). It should be noted that this is far from the normal mode of reproduction for these organisms, of course, and requires significant intervention to trigger dedifferentiation.

Furthermore, we know of exceptions to the general rule that differentiated cells contain exactly the same genetic material. A well-understood example of such an exception comes from the vertebrate immune system. One way in which infections are fought off is through the production of antibody molecules, which are produced by mature B cells of the immune system. The effectivenesss of the response relies on the production of an enormous diversity of antibodies, such that almost all foreign proteins can be recognized. This diversity is generated by several mechanisms, one of which involves combining together different portions of genomic DNA (Fig. 9.10). Only by joining together various DNA segments that encode certain peptide domains can a complete immunoglobulin gene be made. There are, however, many DNA segments that can be used in the process—an enormous number of combinations of joining are possible, which is in part why the immune system can generate such a huge variety of different antibody molecules (also see Chapter 23, p. 663). However, after the rearrangement is complete, an individual B cell can generate only one type of antibody molecule. Be-

FIGURE 9.9. (Left) Dolly, the first cloned sheep; (right) Tetra, a cloned monkey.

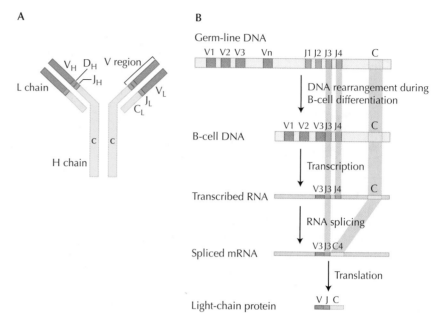

FIGURE 9.10. **Genomic rearrangements** are required to produce a functional immunoglobulin-encoding gene. Immature B cells contain a number of different DNA segments that encode many potential alternative portions of an antibody molecule. These portions, however, are not joined together through exon/intron splicing but instead are combined through an enzymatically mediated recombination process, called V(D)J recombination, that results in permanent genomic rearrangement in each cell. (*A*) Antibodies consist of four protein chains, two identical short (light) and two identical long (heavy) chains. Within these chains, the variable regions produced through recombination are V (*red*), D (*purple*, heavy chain only), and J (*blue*). (*B*) During the differentiation of B cells, a random combination of VL and JL sequences is brought together with the constant (CL) region to produce the light chain. Similarly, a random combination of VH, DH, and JH sequences codes for the heavy chain (not shown).

cause the DNA rearrangment (and deletion of intervening regions) is not reversible, the immune system of any animal derived (cloned) from a single mature B-cell nucleus would be limited to producing a single antibody variant. Although an adult animal might result, it would be doomed in the real world, because its immune system could not generate the diverse antibodies needed to successfully fight off infection.

Diverse Mechanisms Make Cells Different from One Another

Given that, with rare exceptions, all differentiated cells within an individual contain the same genetic information, how do the different cell types arise? Each differentiated cell type takes on a characteristic morphology and carries out a certain set of biochemical processes requiring a specific combination of active proteins. For example, why do blood cells contain hemoglobin, but neurons do not? And why do embryonic red blood cells produce a different form of hemoglobin from adult blood cells? It turns out that multiple regulatory mechanisms are involved in the process of cell differentiation, and these mechanisms act at various biochemical levels (see Fig. 8.24).

The majority of the decisions about which proteins to produce in a given cell are made at the level of gene regulation through the control of **transcription**—as, for example, in the classic case of the Lac operon (see pp. 52–53). An individual gene is composed of a transcribed region (including the coding region for the gene product) and regulatory domains that specify when and in which cells the gene will be transcribed. These regulatory DNA regions can lie 5′ (upstream) or 3′ (downstream)

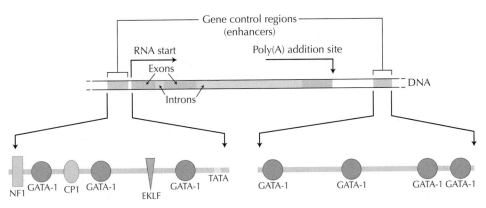

FIGURE 9.11. Hemoglobin gene regulation. The (so-called) gene control regions that regulate when and where transcription occurs for the hemoglobin gene lie both 3' and 5' of the region of DNA that is transcribed to produce the hemoglobin mRNA. These gene control regions, also commonly referred to as enhancers, contain the sites where DNA-binding proteins, known as transcription factors, bind to the genomic DNA. Once bound, these transcription factors can act as either activators or repressors of mRNA transcription. In the case of the hemoglobin gene, binding sites have been identified for several transcription factors including GATA-1, EKLF, NF1, and CP1 that help to activate transcription of the hemoglobin gene. It is the presence of these transcription factors in maturing red blood cells that causes the hemoglobin gene to be transcribed in these cells, whereas the absence of one or more of these activating transcription factors (or the presence of inhibitory transcription factors) in other cells of the body prevents the transcription of the hemoglobin gene.

of the transcribed region, or even within **introns.** A variety of nuclear proteins, known as transcription factors, bind to these regulatory domains and act to either promote or repress transcription of the associated gene. The genes encoding the hemoglobin proteins are not transcribed in neurons, but they are transcribed in developing blood cells (in many animals, mature blood cells lose their nuclei, thus becoming transcriptionally silent) (Fig. 9.11). The distinction between embryonic versus adult forms of hemoglobin is also regulated at the transcriptional level. A particular hemoglobin gene is transcribed in a particular cell type and at a particular time because the appropriate combination of transcription factors required to activate its transcription is present (or biochemically active) only at specific times and places during development. Of course these transcription factors are regulated at many levels, and the entire process of development involves a highly orchestrated cascade of transcriptional activation.

Cell differentiation is also controlled by posttranscriptional regulation of **messenger RNA (mRNA) processing.** For example, the gene, *CyIIIa*, which encodes a specific type of actin, is transcribed in the **ectoderm, mesoderm,** and **endoderm** of sea urchin larvae, but the protein is made only in the ectoderm, because the mRNA splicing required to make a functional mRNA occurs only in that tissue. Nuclear transcripts of *CyIIIa* in the mesoderm and endoderm do not have their introns spliced out and never enter the cytoplasm. Other examples of regulation by splicing include alternative tissue-specific splice variants. For example, the thyroid hormone calcitonin and the neuropeptide calcitonin gene-related peptide (CGRP) are encoded by the same gene, but alternative tissue-specific splicing generates different processed mRNAs, which are then translated into the functionally different proteins (Fig. 9.12; recall Fig. 8.22). Alternative splicing is very frequent in multicellular organisms, and some molecules, such as cell adhesion proteins, can have many different forms generated in this way. These alternative forms are regulated temporally and spatially during development, and the regulated production of the various forms is thought to play a role in complex morphogenetic movements of many tissues.

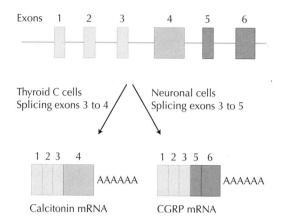

FIGURE 9.12. Alternative splicing can generate functionally distinct proteins. The thyroid hormone calcitonin and the neuropeptide CGRP are produced by alternative splicing from the same gene. In thyroid cells, exons 1, 2, 3, and 4 are used, whereas in neurons, exons 1, 2, 3, 5, and 6 are used. Also see Fig. 8.22.

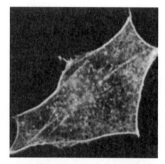

FIGURE 9.13. Localization of actin, a key cytoskeletal protein, in cells. The top cell is a quiescent (resting) cell and shows relatively little organization of its actin cytoskeleton, whereas the bottom cell is one in which the GTPase, Rho, has been activated, which, in turn, has led to the dramatic reorganization of the actin cytoskeleton into cable-like structures. Many external stimuli can cause a rapid change in cell morphology by activating enzymatic pathways like this one and do so directly without the need to control gene transcription.

Sequences within the mRNA can also affect the stability of the message, and these sequences often have different effects in different cell types, leading to differential protein expression. There are also genes encoding **microRNAs** that function by either triggering the degradation of target mRNAs or by blocking their translation. There are also examples where cell differentiation is regulated by posttranslational events. For example, external stimuli, such as binding of ligands on the cell surface, can cause the clustering of transmembrane proteins called integrins. This in turn leads to the activation of focal adhesion kinase (kinases add phosphates onto other molecules, and focal adhesion kinase is so named because this particular protein accumulates inside cells near sites where cells adhere to one another or to the substrate). The resulting cascade of protein phosphorylation causes rapid changes in cell morphology. Regulated activity of monomeric GTPases can regulate the polymerization of actin within the cell, which in turn regulates the formation of filopodia and lamellipodia—cellular structures important in controlling cell shape and movement (Fig. 9.13). See Figure 8.24 for examples of additional levels of regulation.

Mechanisms of differentiation are often interrelated and feed back on one another, and the process of cellular specialization is often a progressive one. For example, within a vertebrate spinal cord, motoneurons and interneurons represent two distinct cell types. Normally, both these types of neurons differentiate only from cells already committed to a neural fate within the developing neural tube (Fig. 9.14). In turn, the neural tube cells differentiate only from the dorsal embryonic ectoderm. Likewise, the development of cells within the immune system is characterized by the progressive restriction of lineages to more and more restricted cell types.

Maintenance of Differentiation Is Achieved by Several Mechanisms

During development, cells may move along their appropriate differentiation pathway relatively quickly, but usually at some point the state of cellular specialization becomes relatively stable. Various biochemical feedback loops contribute to this stability. For example, transcription factors often autoregulate their own transcription, and signaling pathways often form stable circuits between small groups of adjacent cells. In addition, some control of transcription is also established by higher-order structuring of chromatin regions, and this appears to be stably passed along to the daughter cells formed through cell division. Finally, highly specialized cells will also often cease division. A stably maintained differentiated state is essentially a new kind of inheritance but one based on states of gene expression rather than DNA sequence.

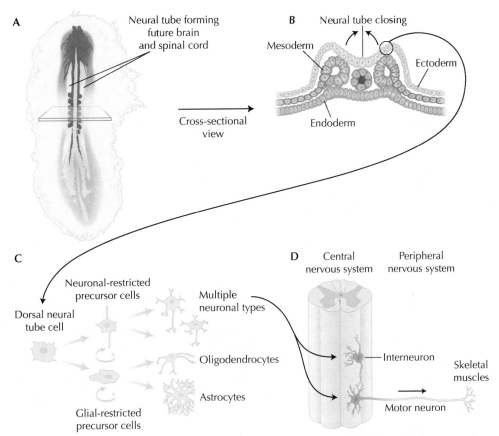

FIGURE 9.14. The progressive nature of cell differentiation can be illustrated with the events that lead to the formation of different types of neurons in the chick embryo. The dorsalmost ectodermal cells pull together to start the process of neural tube formation (*A*; seen in cross section in *B*). The cells of this tube will form the brain and spinal cord, whereas the rest of the ectoderm will form the skin. Depending on cues received within the neural tube, an individual neural tube cell (*C*) may become either a neural precursor cell or a glial precursor cell (glia are support cells within the nervous system and include oligodendrocytes and astrocytes). Spatial cues within the neural tube act on the neural precursor cell to cause it to differentiate into any of a number of distinct neurons, such as interneurons that send signals with the spinal cord or motor neurons that innervate the muscles of the body.

Cancer Is One Manifestation of the Loss of a Differentiated State

The regeneration and animal cloning experiments described above show that cells can dedifferentiate. Being able to dedifferentiate cells, and then redifferentiate them in a controlled manner, holds great promise for repairing damaged tissues and organs in the future.

Unfortunately, however, inappropriate reversals of differentiation can lead to cancer. Most cells of the adult are highly controlled in their growth—a key aspect of their differentiation. In cancerous tumors, this growth control is lost. If the cells grow and divide quickly, but otherwise retain their specialized properties, the tumor can often be treated relatively easily. If, however, the cells also lose other aspects of their differentiated state, the cancerous cells can pose a more serious problem, especially if they begin to spread. Embryonic cells are often fairly motile, whereas differentiated cells generally stay put. Cancerous cells that lose most of their previous specialized properties, and become highly mobile, pose a great danger. Left unchecked, they may ultimately kill the organism as they disrupt normal tissues in which they come to reside.

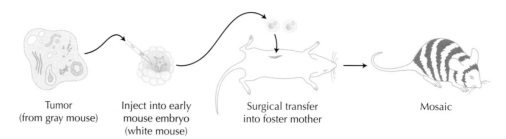

Tumor
(from gray mouse)

Inject into early
mouse embryo
(white mouse)

Surgical transfer
into foster mother

Mosaic

FIGURE 9.15. Cancer cells have the ability to redifferentiate. If tumor cells derived from a gray mouse are injected into an early embryo (derived from white mice), and then these injected embryos are implanted into a host female, mice that are mosaic for gray and white coat color will be born. The areas of gray coat color are derived from cancer cells that have been able to differentiate into normal skin cells within the host embryo.

Remarkably, however, cancer cells can be forced to redifferentiate. This can be beautifully illustrated by injecting murine carcinoma cells into an early mouse embryo (Fig. 9.15). The cancerous cells will end up contributing to the formation of normal differentiated cells in tissues throughout the host. The environment of the embryo can clearly promote the redifferentiation of these cells, just as it controls the normal differentiation of pluripotent embryonic cells. It should be noted that in some cancers, however, tumor progression is marked by the deletion of certain genes and chromosomal regions. Just as with B cells (Fig. 9.10), such genetically altered cells will be limited in their capacity, and such cancer cells are unlikely to have the capacity to form normal differentiated cell types.

DIVERSITY OF BODY PLANS

We have talked so far about the generation of different cell types. Most multicellular organisms not only contain a variety of differentiated cell types, but also organize these differentiated cells in a highly reproducible manner. This pattern of organization leads to the characteristic morphology that we associate with each individual species. Indeed the process of differentiation and morphological organization are intimately tied together during development—precise spatial control over the pattern of differentiation is essential. For example, the formation of a heart in a location where it could not be connected to the developing system of veins and arteries or the development of flowers on the ends of plant roots would be of little use. Indeed, we classify multicellular organisms according to the ways that they have arranged differentiated cell types and organs in space to construct their characteristic body plans. Over time, the process of development itself has evolved to craft the overall shape of individual animal and plant species. Thus, extant body plans can be seen as a clear example of **descent with modification**.

A Body Plan Is the Morphology Characteristic of a Group of Organisms

Morphological characteristics have long been used to try to understand the evolutionary relationships of various plant and animal groups. For example, **arthropods** are united by the possession of a segmented exoskeleton and jointed appendages. Vertebrates, like ourselves, are also segmented but possess an internal skeleton, and it is the

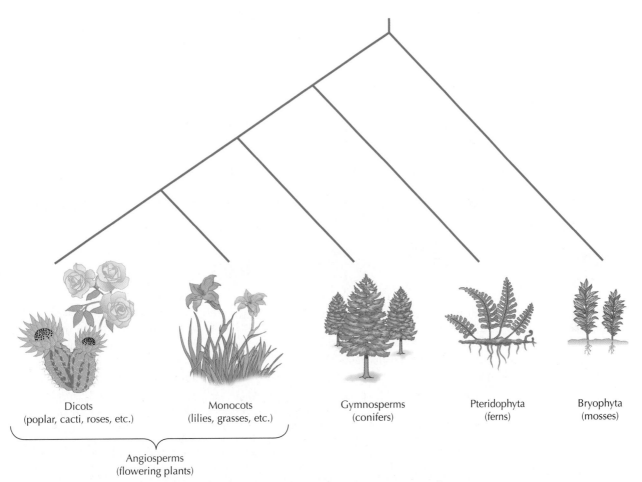

FIGURE 9.16. Phylogenetic relationship of some major plant groups. The angiosperms (flowering plants) are subdivided further into two groups, the monocots and dicots, based on their pattern of embryonic leaves. This morphological phylogeny is also supported by molecular analyses.

vertebral column (backbone) that is the main defining characteristic of the group. Flowering and nonflowering plants make up two large subdivisions within the plants; monocotyledons and dicotyledons are a further subdivision of flowering plants based on the initial pattern of embryonic leaves (Fig. 9.16). Many of these morphological criteria do accurately classify monophyletic groups; but of course some shared body features turn out not to be homologous but instead the result of convergent evolution—for example, the possession of wings by bats, birds, and insects. In other cases, a feature is lost by a subset of an evolutionary lineage. For example, fleas are holometabolous insects (**holometabolous** insects are the **monophyletic** group that undergoes complete metamorphosis and includes flies, bees, and butterflies), but unlike most holometabolous insects, fleas lack wings, because of the evolutionary loss of this feature (Fig. 9.17). Likewise, snakes and whales both derive from four-legged tetrapod ancestors, but the extant forms of these groups contain only vestiges of limbs (Fig. 3.16).

Multicellular Organisms Can Build a Diversity of Body Plans

Each of the major groups of multicellular organisms possesses a characteristic body plan. In this chapter, we focus on the development of animals and, in Chapter 10, on their paleontological record, so it is useful to introduce these animal groups, at least up to the level of major subdivisions of the better-studied phyla (Fig. 9.18).

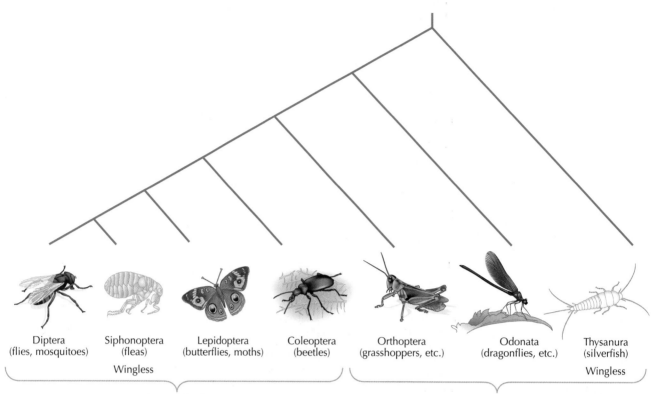

Diptera
(flies, mosquitoes)

Siphonoptera
(fleas)

Lepidoptera
(butterflies, moths)

Coleoptera
(beetles)

Orthoptera
(grasshoppers, etc.)

Odonata
(dragonflies, etc.)

Thysanura
(silverfish)

Wingless

Holometabolous

Hemimetabolous

Wingless

FIGURE 9.17. The phylogenetic relationship of several insect orders. Of the groups shown here, only two, the Thysanura and Siphonoptera, are wingless. Based on the fossil record and relationship to other arthropods, Thysanura are thought to have evolved from the last common wingless ancestor of all the insects shown here. Thus, Thysanura retain the wingless condition of the presumed ancestor. On the other hand, the Siphonoptera lineage is thought to have become wingless as it is nested within multiple orders of winged insects. Thus the common ancestor of fleas and flies was winged, but wings were lost in the lineage leading to fleas.

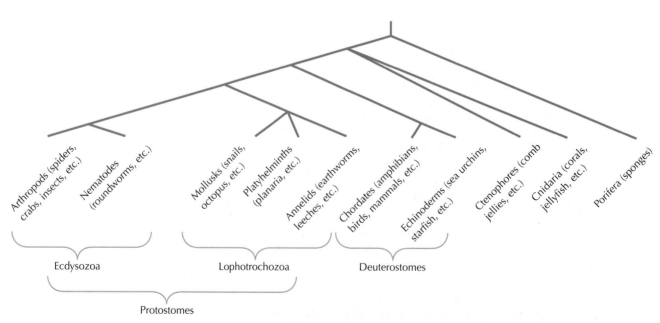

Arthropods (spiders, crabs, insects, etc.)

Nematodes (roundworms, etc.)

Mollusks (snails, octopus, etc.)

Platyhelminths (planaria, etc.)

Annelids (earthworms, leeches, etc.)

Chordates (amphibians, birds, mammals, etc.)

Echinoderms (sea urchins, starfish, etc.)

Ctenophores (comb jellies, etc.)

Cnidaria (corals, jellyfish, etc.)

Porifera (sponges)

Ecdysozoa

Lophotrochozoa

Deuterostomes

Protostomes

FIGURE 9.18. Phylogenetic relationship of some of the major animal (**metazoan**) groups. The deuterostomes plus protostomes combined make up the so-called Bilaterian animals, which have the property of bilateral symmetry (although it can be argued that Cnidaria have clear indications of bilateral symmetry as well). Based primarily on molecular phylogenetics, the Protostomes are further divided into two large clades known as the Ecdysozoa and the Lophotrochozoa.

Within the animals, the Porifera (sponges; Fig. 9.19A) have one of the simplest body plans. Although these animals contain a variety of cell types, they are not organized into distinct organs. They have relatively simple tube-shaped bodies and feed on suspended food particles by using flagellated cells, called choanocytes, to establish a flow of water through their bodies. The Cnidaria (e.g., corals, jellyfish, sea anemones; Fig. 9.19B) and Ctenophora (e.g., comb jellies; Fig. 9.19C) are regarded as very ancient animal lineages and possess recognizable organ systems such as a mouth and a digestive cavity. They are generally referred to as having radial symmetry (although this is not quite true), are soft bodied, and feed using a system of tentacles.

The remaining animal phyla can be separated into two large groups known as **deuterostomes** and **protostomes** (based on the embryonic origins of their mouth vs. anus). Within the deuterostomes are the **echinoderms** and the **chordates**. The Echinodermata, which include starfish and sea urchins (Fig. 9.20A,B), have a mesodermal skeleton composed of calcite plates located just underneath their outer layer of epidermis. They have bilateral symmetry as larvae, but the adults generally have pentameral (fivefold) radial symmetry. They move using small tube feet that are powered by the animals' internal water vascular system. The other major deuterostome phylum is Chordates, which includes fish, amphibians, birds, reptiles, and mammals (Fig. 9.20C–G). The members of this phylum have an internal skeletal structure, which includes an embryonic notochord that is sometimes replaced by vertebrae in the adults. They also have a hollow nerve cord on the dorsal side and an embryonic tail that may or may not persist into adulthood.

The protostomes are further subdivided, mostly by molecular studies, into two large clades, the Ecdysozoa and the Lophotrochozoa (Fig. 9.18). The Lophotrochozoa (Fig. 9.20) include the **mollusks**, annelids, and platyhelminths, along with a number of less well-studied, but very interesting, phyla. The mollusks (Fig. 9.21A–C) include gastropods (e.g., snails and limpets), bivalves (e.g., clams and mussels), and cephalopods (octopus and squid). The annelids (Fig. 9.21D–F) are subdivided into three main groups: oligochaetes (e.g., earthworms), polychaetes (e.g., bristle worms), and hirudinids (e.g., leeches). Annelids show an obvious external segmentation along their anterior–posterior axis appendages, and this subdivision is also seen for their internal cavities,

 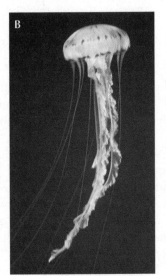 

FIGURE 9.19. Animal phyla outside the traditional Bilaterians include the Porifera (sponge; *A*), the Cnidaria (jellyfish; *B*), and the Ctenophores (comb jelly; *C*).

FIGURE 9.20. Representative Deuterostomes include echinoderms such as (*A*) crinoid, (*B*) starfish, and chordates such as (*C*) clownfish, (*D*) frog, (*E*) bird, (*F*) snake, and (*G*) elephant.

allowing them to localize hydrostatic pressure to individual segments and giving them excellent burrowing abilities. In the polychaetes, this burrowing ability has been compromised somewhat as they have developed pauropodia (simple appendages) for crawling across surfaces.

The other protostome subdivision, the Ecdysozoa, includes a number of phyla as well, but the two best studied, and most numerous, are the Arthropoda and the Nematoda. The extant arthropods (Fig. 9.22) have four major subdivisions: hexapods (springtails plus the insects, e.g., flies and ants), myriapods (centipedes and millipedes), chelicerates (e.g., spiders and horseshoe crabs), and crustaceans (e.g., lobsters and copepods). All these are characterized by the possession of jointed appendages and a tough, chitinous exoskeleton that is clearly segmented. Unlike annelids, however, they do not have the extensive internal segmentation of their body cavity and, given the strength of their exoskeleton, generally no longer need a hydrostatic system for locomotion. Nematodes are relatively simple from a morphological perspective, generally looking like very small, clear worms. Some are free-living and others are parasitic. They are remarkably abundant and cover just about every surface on Earth.

FIGURE 9.21. Representative Lophotrochozoa include (A) nudibranch (mollusk), (B) cowrie (mollusk), (C) Pacific octopus (mollusk), (D) bristleworm (annelid), (E) Christmas tree worm (annelid), (F) leech (annelid), (G) planaria (platyheminth), (H) Brachiopod, and (I) **Bryozoan**.

FIGURE 9.22. Representative Ecdysozoa include (A) butterfly (arthropod insect), (B) crayfish (arthropod crustacean), (C) millipede (arthropod myriapod), (D) spider (arthropod chelicerate), (E) chaetognath, (F) priapulid, and (G) onychophoran (velvet worm).

The History of Body Plans Is Revealed in Part by the Fossil Record

Although limited by the poor preservation of many organisms, an examination of the fossil record provides some very important information regarding the emergence of the major body plans seen in multicellular animals and plants. For example, from the fossil record we know that most of the major animal body plans appeared very rapidly during a period known as the Cambrian explosion some 530–510 Mya (see pp. 263–270). Before this time, there were multicellular organisms, but they are believed to have been relatively small and simple and many may not have possessed morphologies that we can easily relate to extant groups. Further details of the fossil record for the major animal and plant groups are presented in the next chapter.

GENETICS OF BUILDING A BODY PLAN

Evolutionary biologists have long sought to understand certain aspects of development, because it is clear that evolutionary changes in body morphology are due to changes in the events of development. Indeed, Darwin devotes a chapter in *On the Origin of Species* to a discussion of animal embryology and development. Likewise, developmental biologists have used comparative studies in an attempt to understand the relationships between different groups and thus how development is related between different organisms. During the past decade there has been a renaissance in attempts to understand the evolution of development. The renewed interest stems primarily from understanding significant aspects of development at the genetic and molecular levels and the remarkable discovery that developmental pathways are often conserved even between widely divergent organisms. These common underpinnings of development allow us to understand how evolutionary changes at the genetic level lead to morphological diversity.

Genetic Approaches Have Revealed the Genes Involved in Development

A significant breakthrough in understanding development has come through genetic approaches to the problem. Developmental biology has a long and impressive history of experimental manipulations, many of them involving delicate surgical procedures on early embryos. Unfortunately, such approaches revealed relatively little about the genes involved in development. These were, however, identified in genetic screens used to isolate mutants that affect development. Many of these first genetic screens to identify developmental mutants were undertaken in the fruit fly, *Drosophila melanogaster*, and the nematode, *Caenorhabditis elegans*.

The most common approach is to randomly mutate genes in the cells that make sperm by exposing males to mutagens (chemicals or X rays are generally used) and then mating these males to normal females (Fig. 9.23 shows the scheme for a zebrafish mutagenesis screen). This results in a first generation (F_1) in which individuals carry single copies of mutated genes. In rare cases, these mutations may create a dominant effect and can be recognized at this step. More commonly, a series of controlled crosses must be carried out to eventually create offspring (F_3) that are homozygous for a mutation. These F_3 are screened for the phenotype of interest. For example, in zebrafish, such screens led to the isolation of a mutation that prevents the formation of somites, which are the initial condensations of tissue that form the segmentally repeated vertebrae and muscles of the fish (Fig. 9.23B). In addition, mutations that cause dramatic alterations in the color patterns of the fish were isolated (Fig. 9.23C).

One of the early intensive genetic screens in *Drosophila* led to the identification

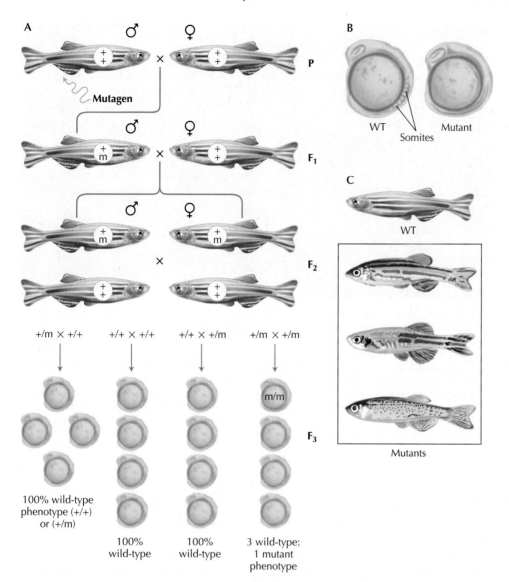

FIGURE 9.23. Mutagenesis screen in zebrafish. (A) Males are treated with a chemical mutagen and then mated to a wild-type female. Some F₁ progeny will carry induced mutations but will not be homozygous, so for recessive mutations, no phenotypes will be visible. By controlled mating, some homozygous mutant animals will be generated in the F₃ generation (see Box 13.2 for a similar technique in *Drosophila*), and these can be screened for phenotypes of interest. Examples of phenotypes include (B) embryonic lethal mutations that result in embryos that fail to form somites (*arrows* point to somites in the wild-type [WT] embryo) and (C) viable adult phenotypes that alter the color patterns of the fish.

of a large number of genes responsible for generating the correct pattern of segments in the larvae (Fig. 9.24). The subsequent characterization of these mutant phenotypes, and the identification of the corresponding genes, has given us a remarkably detailed picture of the early developmental program that regulates the process of segmentation in this insect. To date, dozens of other developmental programs have been studied through genetic screens and have identified the genes involved in such events as diverse as flower patterning in *Arabidopsis* and neural development in *C. elegans*.

Although at first these various developmental processes might appear obscure, having little to do with human development, we now know that this is not the case. As we will demonstrate in Chapter 11, the genes and molecular pathways they have helped

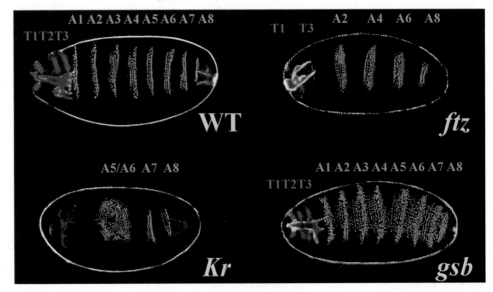

FIGURE 9.24. Mutagenesis screens in *Drosophila* have led to the identification of genes involved in the process of segmentation. At the *upper left* is a wild-type (WT) larva that is just about to hatch from the egg. Anterior is to the *left* and the ventral surface is visible. Each segment is clearly delineated by bands of denticles, stiff hairs that help the animals crawl through their food. The three thoracic segments (T1–T3) and eight of the abdominal segments (A1–A8) are clearly visible. In *Krüppel* (*Kr*) mutants (meaning embryos homozygous for **null alleles** of the *Kr* gene), the thoracic and anterior abdominal segments are deleted. In *fushi tarazu* (*ftz*) mutant embryos, every other segment is deleted. In *gooseberry* (*gsb*) mutant embryos, there is the correct number of segments, but patterning within each segment is altered so that the denticle bands have expanded into the regions that are normally free of denticles. Starting with these and other similar mutants, developmental biologists now understand in great detail how the *Drosophila* embryo becomes segmented.

uncover are widely used in the development of all organisms. Indeed, Nobel Prizes in Physiology or Medicine have been awarded to scientists who have focused on understanding the development of *Drosophila* and *C. elegans* because their discoveries have been so instrumental in understanding human development and disease. Likewise, the analysis of *Arabidopsis* (a common weed) has been invaluable in improving the crops we depend on.

Although remarkably powerful, such genetic screens are generally limited to a relatively small number of model organisms that are tractable to such approaches. For these "forward" genetic approaches to work efficiently, the animal or plant being studied must be easy and economical to raise, it must have a reasonably short generation time, and the phenotype of interest must be relatively simple to identify.

Reverse Genetic Approaches Expand Our Understanding of Development

Development in animals and plants not amenable to classical forward genetic screens can nevertheless be studied genetically using a variety of approaches that can be characterized as **reverse genetics**. For example, it is possible to selectively mutate genes in mice through a targeted mutation method that takes advantage of **homologous recombination**. In a number of organisms, including both plants and animals, gene function can be "knocked out" (i.e., eliminated) by taking advantage of a phenomenon known as **RNA interference (RNAi)** (Fig. 9.25). In one version of this approach, many double-stranded RNA copies of a gene of interest are synthesized in vitro and then injected into the developing embryo. This double-stranded RNA leads to the degradation of the corresponding endogenous mRNA. Similarly, in animals such as

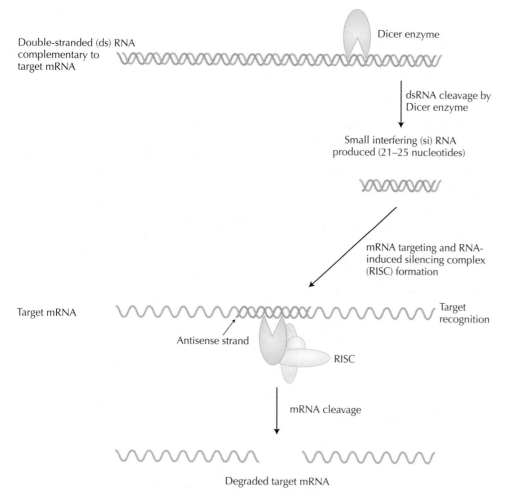

Double-stranded (ds) RNA
complementary to
target mRNA

Dicer enzyme

dsRNA cleavage by
Dicer enzyme

Small interfering (si) RNA
produced (21–25 nucleotides)

mRNA targeting and RNA-
induced silencing complex
(RISC) formation

Target mRNA

Target
recognition

Antisense strand

RISC

mRNA cleavage

Degraded target mRNA

FIGURE 9.25. In many organisms, the introduction of double-stranded RNA complementary to a target mRNA is an efficient way to test the function of a gene. The double-stranded RNA can be introduced into an organism by any of several means, including injection, soaking, or even feeding. Once inside the cell, the double-stranded RNA is cleaved by the Dicer enzyme into 21–25-nucleotide fragments called small interfering RNAs (siRNAs). The antisense strand of the siRNA complexes to the target mRNA sequence along with the RNA-induced silencing complex (RISC) protein complex. The target mRNA is then cleaved and degraded. In this way, gene function can be "knocked down" and the effect on the organism observed. Using this technique, it is possible to screen through thousands of genes for developmental phenotypes and to begin to carry out genetic analyses in animals and plants not amenable to standard forward genetic approaches.

Xenopus, **antisense oligonucleotides** can also be used to block gene function by interfering with mRNA translation. Misexpression of a gene can also be used to elucidate gene function. For example, a muscle-specific transcription factor may be isolated via biochemical approaches, and the potential of this gene to specify cell muscle fate during development could then be tested by injecting the mRNA encoded by this gene into other cell types of the embryo to see if this then alters their differentiation.

Approaches such as these make it possible to carry out detailed genetic analyses without using forward genetic screens. In principle these "reverse genetic" approaches could be carried out with genes picked at random, but in practice the genes are often chosen for analysis for specific reasons. For example, a gene named *tinman* was first isolated through forward genetic screens because of its role in *Drosophila* heart formation. It was subsequently found that an ortholog of *tinman* was expressed in the mouse heart. Using gene targeting, a mouse mutant for this gene was generated, and it too showed defects in heart development. In this way, a genetic analysis of mouse heart development could be initiated using information from forward genetic screens in another animal.

Both Lineage and Cell–Cell Interactions Play a Role in Patterning the Embryo

Embryological studies have shown that both cell lineage and cell–cell interactions are important in the process of embryogenesis. For example, each sensory bristle (hair) on the surface of an adult fly, and its underlying structure, is made up of four cells derived from a common precursor cell. The lineage (cell division and differentiation) pattern generating these four cells from the single precursor is shown in Figure 9.26A,B. An obvious question is whether or not there is an intrinsic asymmetry during cell division— that is, does one daughter cell inherit something different from another daughter during cell division that is responsible for the two sibling cells having different fates? Genetic analyses of neural development in *Drosophila* isolated several mutations that affected this lineage pattern, including a mutation called *numb*. In *numb* mutants, the cell lineage pattern is altered as shown in Figure 9.26C, resulting in the production of just socket cells and the elimination of the other cell types (hence the name *numb*, because the bristles and neuron were missing). It turns out that the numb protein is asymmetrically inherited during cell division as shown in Figure 9.26D–F. The precursor cell

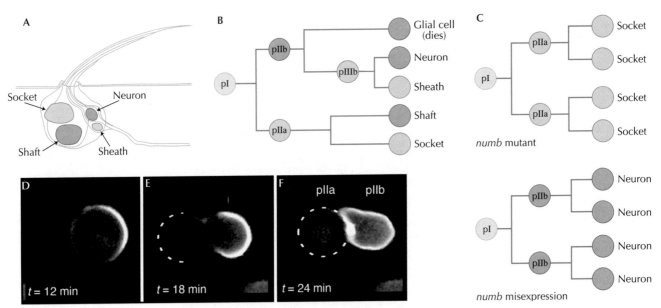

FIGURE 9.26. (A) The sensory hair structures (known as bristles) in *Drosophila* are made up of four cells: the socket, shaft, neuron, and sheath cells. (B) These cells are all derived from a single precursor cell called pI through a particular division pattern. In *numb* mutants (upper half of C), the cell differentiation program is altered so that all the progeny become socket cells. When *numb* is overexpressed (lower half of C), all the progeny become neurons. (D) Analysis of numb protein distribution reveals that numb is localized to one side of the pI cell. (E,F) As the cell begins to divide, the asymmetry in numb protein distribution results in all the numb protein going to the pIIb daughter (the *dotted line* in E and F helps to outline the edge of the pIIa cell) and numb protein is also asymmetrically distributed in subsequent divisions as well. The normal pattern of numb distribution is reflected by the coloring of the cells in B (numb-containing cells are *red*; those without numb protein are *blue*). This asymmetric distribution can be used to explain the phenotypes seen in *numb* mutants and when *numb* is misexpressed. When *numb* is absent, all the progeny behave like the cells that would normally lack numb protein (pIIa and then the socket cell). When *numb* is misexpressed, all the progeny behave like the cells that have the numb protein (pIIb and then the neuron).

initially expresses numb protein uniformly, but just before it divides, the numb protein becomes localized to one side of the cell (Fig. 9.26D,E). In the end, the numb protein is inherited by only one daughter, which will be pIIb (Fig. 9.26F). Later in development, *numb* will be expressed in pIIIb and pIIa but is again asymmetrically inherited so that it ends up in the cells that will be the socket and sheath.

Further evidence that this pattern of asymmetric inheritance is crucial to proper patterning was obtained by experimentally misexpressing the numb protein (Fig. 9.26G, bottom). If all the cells are forced to uniformly express the numb protein, then the resulting phenotype is essentially the opposite of that seen in a *numb* loss-of-function mutant—that is, there are an excess of neurons and a deficit of the other cell types. Subsequent studies have shown that the numb protein regulates a signaling pathway in the cells that leads to transcriptional differences between the various daughter cells.

The importance of cell–cell interactions can be illustrated by studies on the formation of the vulva in the nematode, *C. elegans*. The vulva forms by the specialization of a specific set of ectodermal cells that produce the opening and lining of the vulva (Fig. 9.27). Laser ablation studies showed that the initiation of this developmental program required the presence of an internal cell, called the anchor cell. In the absence of the anchor cell, the ectodermal cells do not form a vulva but instead behave like adjacent ectodermal cells that form the skin of the worm. Subsequent genetic screens in the worm identified many genes involved in this process, and we now understand this signaling pathway in great detail.

The anchor cell expresses a gene called *lin-3*, which encodes a secreted protein very similar to the epidermal growth factors (EGFs) of vertebrates. This protein diffuses away from the anchor cell and acts as a ligand that binds to an EGF receptor–like transmembrane tyrosine kinase on the ectodermal cells (the receptor is the product of the *C. elegans let-23* gene). The ectodermal cell closest to the anchor cell adopts what is known as a primary fate, whereas the two ectodermal cells that are a bit further away adopt what is termed a secondary fate (Fig. 9.27). Various experiments have shown that this difference in fate is due to the slightly different levels of lin-3 protein product present at these different distances away from the source (the anchor cell). This differential response is also amplified by another signaling cascade initiated by the central cell inhibiting the more lateral cells from adopting the primary fate.

Of course, transcriptional regulation still has a key role in virtually all developmental processes including this one. We know that previous patterning events in the embryo led to the transcription of the *lin-3* gene in the anchor cell and in no other nearby cell. This allows the anchor cell to be the sole source of the signal controlling the fate of the adjacent ectodermal cells. When the lin-3 protein binds the extracellular receptor portion of the let-23 protein, this activates the enzymatic activity of the intracellular tyrosine kinase domain of the let-23 protein. Acting through what is known as the RAS pathway, this ultimately controls the transcription of a number of target genes in the ectoderm that affect the pattern of growth, division, and morphology of the ectodermal cells, so that they form the correct vulval structure.

Patterning Is Controlled by a Variety of Molecular Mechanisms

We have quite a detailed understanding of many steps in the development of various plants and animals. A full account would be beyond the scope of this book, but there are a number of generalizations, most of which we have already presented in this chapter, that will be useful to our discussion of the evolution of morphological and developmental diversity. Let us review them.

1. One of the main hallmarks of multicellularity is the differentiation of cell types within an organism. This is precisely what development accomplishes.

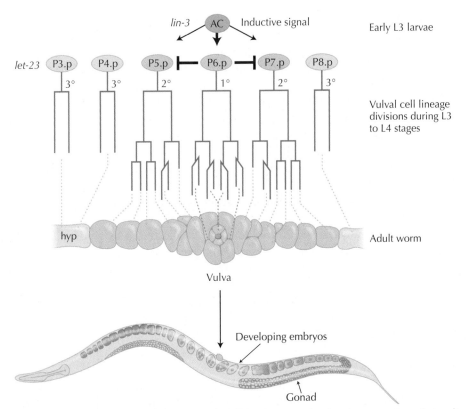

FIGURE 9.27. The formation of the *Caenorhabditis elegans* vulva involves a series of signaling events. In the L3 stage larvae, the anchor cell (AC) secretes the lin-3 protein, which controls the development of the three closest ectodermal cells P5.p, P6.p, and P7.p. The cell that receives the highest level of signal, P6.p, adopts the primary fate and the other two cells adopt a secondary fate. This signaling pathway ultimately affects the transcription of a number of genes within these ectodermal cells and thus controls their fate so that they undergo a specific pattern of division and differentiation to ultimately form the vulva in the proper position and with the proper morphology.

2. A single-cell embryo forms a complex multicellular animal or plant and does so by having each cell express a particular set of gene products that then defines that cell's function within the organism. In addition, developmental processes also assure that different cell types are located in the appropriate places within the organism.

3. The expression of a particular set of gene products is accomplished through regulation at the transcriptional, posttranscriptional, and posttranslational levels. The three are intertwined and often act to feed back on one another (see Fig. 8.24).

4. The process of patterning is a progressive one, and each patterning event is intimately tied to previous ones. In general, as the embryo goes from a single, pluripotent cell to a fully formed organism composed of many different cell types, the fate of cells becomes progressively restricted. At the same time, the position of the cells must be precisely coordinated and is intimately tied to the entire process of cell differentiation.

SUMMARY

Multicellularity arose on several occasions, and, through the controlled process of cell differentiation, organisms evolved in which individual cells carried out subsets of functions required by the organism as a whole. In modern-day animals and plants, the allocation of different cell types is achieved through a precise and reproducible pattern of development. This process of development employs many mechanisms that have been uncovered through genetic approaches.

Through repeated steps of differential gene regulation and cell–cell signaling, embryos are able to both generate different cell types and position these cells in precise lo-

cations. The remarkable process of development is not so much the reading of a blueprint as it is the interactive unfolding of a series of events that can reliably and robustly produce complex patterns. In doing so, individual organisms generate a characteristic morphology, or body plan. In Chapter 10, we describe the evolutionary history of the main groups of animals and plants from the fossil record, and then in Chapter 11, we illustrate several examples of how evolutionary changes in development lead to changes in animal and plant morphology. In Chapter 21, we discuss how cells in a multicellular organism have come to sacrifice their reproduction for the sake of other cells.

FURTHER READING

Gilbert S. 2003. *Developmental biology*, 7th ed. Sinauer Associates, Inc., Sunderland, Massachusetts.

This book covers the details of animal and plant development.

King N. 2004. The unicellular ancestry of animal development. *Dev. Cell* **7:** 313–325.

This review provides further details on choanoflagellates and their relationship to multicellular animals.

Nusslein-Volhard C. 2006. *Coming to life. How genes drive development*, 1st ed. Kales Press, Carlsbad, California.

Watson J., Baker T., Bell S., Gann A., Levine M., and Losick R. 2004. *Molecular biology of the gene*, 5th ed. Benjamin Cummings, San Francisco.

This comprehensive book gives details of the various mechanisms of transcriptional regulation.

Wolpert L., Jessell T., Smith J., Lawrence P., Robertson E., and Meyerowitz E. 2006. *Principles of development*, 3rd ed. Oxford University Press, New York. In press.

An excellent book on the general principles of developmental biology.

CHAPTER

10

Diversification of Plants and Animals

THE HISTORY OF LIFE IS WRITTEN in the macro-molecules of organisms. It is also revealed in the bones, the patterns of development, and the behavior of ancient and modern creatures. As we have learned so far in this section, the stage for life was set more than 3 billion years ago. The tree of life has grown and shifted since then as organisms have evolved, diversified, and been extinguished. In this chapter, we explore the remarkable evolution of plants and animals. Although some of the evidence we present is based on phylogenetic analysis and evolutionary developmental biology, most of our story is told through the fossil record. Thus, we begin with an introduction to fossils and the concept of geological time.

FOSSILIZATION AND GEOLOGICAL TIME

Most fossils are the remains of shells, bones, and teeth—the hard parts of animals that are mineralized in life. Plants also have an extensive fossil record because of their decay-resistant tissues such as wood and leaf cuticles. Some fossils provide evidence of behavior in the form of tracks, trails, and burrows, and even chemical signatures of the composition or diet of organisms.

The chances of an organism becoming fossilized are vanishingly small. The familiar hierarchical food chain and the biological cycles that sustain life are based on the breakdown and recycling of nutrients that dead organisms contain. To become a fossil, at least some part of the animal or plant must avoid decay and physical destruction, and with few exceptions, such as trapping of insects in resin exuded from trees (e.g., amber), it must be buried in sediment. Because most sedimentary deposition occurs in the sea, marine organisms have a much higher chance of becoming entombed and fossilized than the animals and plants that live on the land. Likewise organisms that live in lakes and rivers, or close to them, are better represented than animals and plants that live away from water. Burial, however, is only the beginning of the story—the sedimentary rocks that contain a potential fossil must survive geological processes such as the addition of overlying layers of sediment (with its attendant rise in pressure and temperature), uplift and erosion, and the destruction of ocean floor as tectonic plates descend into trenches at the margin of continents. Because of this last process there is no oceanic crust more than about 180 million years old flooring today's oceans. Older marine rocks that have survived were deposited in basins on continental crust or thrust onto the continents during episodes of mountain building. Finally the fossil must be found and described.

About 250,000 species of fossil marine animals have been named, slightly more than the number of described living marine animal species. The great majority of organisms that lived in the past, however, do not occur as fossils either because they were soft-bodied and prone to decay or because they lived in environments or during intervals of time that are poorly represented in the fossil record. Thus 250,000 may represent only 2–4% of the total number of marine animal species that ever existed. If we extrapolate this fraction to estimates of the total number of species living on Earth (between 3 and 50 million), it indicates that between 75 and 2500 million species lived, and became extinct, in the past. These figures are very imprecise because estimates of the number of species on Earth today are very approximate. It is clear, however, that the fossils known today represent only a tiny fraction of all the organisms that once lived. Nonetheless, a number of important evolutionary phenomena are revealed by the fossil record.

1. The fossil record provides evidence of events in the history of life. Although relationships among living organisms reveal the order of branching of the tree of life, without the fossil record we would have no inkling that extinct organisms such as trilobites and dinosaurs once existed.

2. The fossil record reveals the pattern of biodiversity through time, based on the timing of the origination and extinction of taxa. Geological time is subdivided largely on the basis of events in the history of life (Box 10.1). The end of the Paleozoic is marked by the Permian mass extinction, and the end of the Mesozoic, widely known as the K–T (i.e., Cretaceous–Tertiary) boundary, by the extinction that followed the asteroid impact at the end of the Cretaceous (K is used as an abbreviation for Cretaceous to distinguish it from Cambrian and Carboniferous).

3. The fossil record provides the dimension of time, essential for calibrating the **molecular clock**, which uses genetic differences between organisms to estimate the timing of evolutionary splits, and for analyzing rates of evolution (see p. 59).

4. Our understanding of large-scale processes in evolution (**macroevolution**) relies largely on evidence from the fossil record. This includes determining rates of speciation and analyzing the effects of mass extinctions.

THE FLOW OF THE EVOLUTION OF LIFE-FORMS

Biologists commonly use a hierarchical system to classify organisms: phylum, class, order, family, genus, species (see p. 67 and Box 5.2). Such concepts are based on similarity and are constructed by taxonomists in an attempt to analyze and understand the evolutionary history of life on earth. A fundamental assumption is that all members of a group, at whatever level in the hierarchy, share a common ancestor—that is, the group is **monophyletic**. Phyla, at the highest level in the hierarchy, represent distinct **body plans**, morphologically distant one from another (see pp. 238–244). Most estimates indicate that there are 30–35 animal phyla among the living biota. Molecular sequencing provides new ways to determine the interrelationships of phyla; morphological similarities are a product of evolution and convergence over the past 500 million years.

Box 10.1   Geological Time

Geological time (Fig. 10.1) is divided up mainly on the basis of events in the history of life. Our major focus here is the Phanerozoic, which began 542 Mya (millions of years ago) with the Cambrian period, defined essentially by the appearance of the first animals with shells. The Phanerozoic is subdivided, primarily on the basis of fossil appearances and extinctions, into the Paleozoic, Mesozoic, and Cenozoic eras, representing only about 15% of the Earth's history. Although the Earth is approximately 4.5 billion years old, no rocks survive from the first 800 million years. Time prior to the Cambrian is referred to as the Precambrian, subdivided at 2.5 billion years ago into the earlier Archean and the later Proterozoic. Eons of geological time are calibrated in years based on an analysis of the radioactive elements in certain minerals, which decay at a constant rate (Box 4.1).

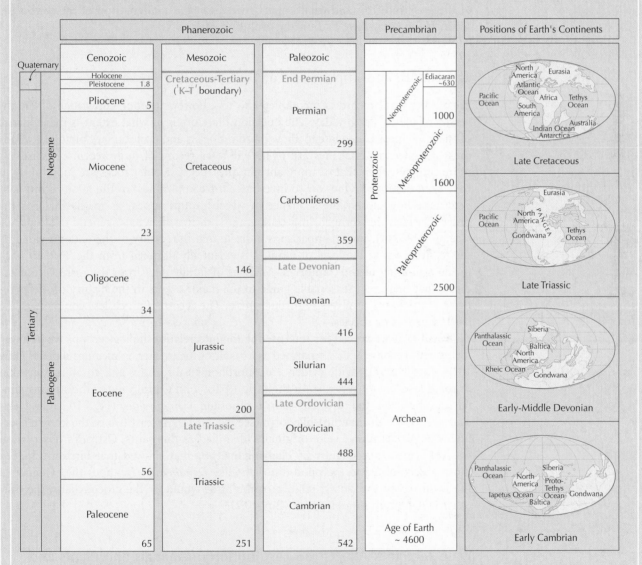

FIGURE 10.1. The geological time scale. Numbers are millions of years ago. *Red* represents mass extinctions.

The earliest fossil evidence for life on Earth was considered in Chapter 4. The major innovation that is represented by the formation of a cell nucleus and the evolution of eukaryotes was described in Chapter 8. In Chapter 9, we considered the evolution of multicellularity and the way in which body plans develop. In this chapter, we are concerned with the origin and diversification of many-celled animals and plants. The first large organisms appear about 565 Mya but microfossil evidence of many-celled algae and animals is known from the late Neoproterozoic Doushantuo Formation of China more than 600 Mya. Most of the phyla originated during the Cambrian explosion. No new animal phyla appeared even during the major ecological shift represented by the colonization of land; this was the time of diversification of the major divisions of the plant kingdom.

Before we embark on this story we must first consider how reliable the evidence of the fossil record is and what impact changes in the configuration of the continents through time may have had on the evolution of life.

The Nature and Importance of Data from the Fossil Record

Clearly the fossil record is incomplete. So too, however, is our knowledge of the diversity of life on Earth today. The question therefore is to what extent is the evidence of the fossil record adequate to allow paleontologists and evolutionary biologists to address the questions that they ask of it? The fossil record of more recent organisms is more complete than that of more ancient ones, because older rocks are more likely to have been destroyed. One way to investigate the completeness of the fossil record is to determine how closely the sequence in which groups appear as fossils matches the order in which they branch in a cladogram showing their phylogenetic relationships (see pp. 238–244). Such comparisons indicate that the quality of the data provided by the fossil record at the level of families is essentially uniform from the Tertiary back to the beginning of the Cambrian. Analyses of family data from the fossil record indicate that they are a reasonable guide to the major events in the history of life as reflected in patterns of diversity through time. The record is much less complete at the level of genera or species.

Fossil taxa are crucial for interpreting the interrelationships of groups when living forms have become widely separated as a result of the extinction of intermediates (Box 10.2). Fossils may provide evidence of morphological features and combinations that in some cases no longer exist. In the early 1970s, John Ostrom noted the striking similarities between the dromaeosaurid dinosaur *Deinonychus* (Fig. 10.3) and *Archaeopteryx*, the earliest bird-like form (see Fig. 10.32). At least two of the known specimens of *Archaeopteryx* were originally identified as dinosaurs. Ostrom's observation was confirmed subsequently by **cladistic** analyses that showed that birds are indeed most closely related to theropod dinosaurs like *Deinonychus*. Without the evidence of the fossil record the closest relative of the birds would be the crocodylians, the only other living group of archosaurs (Box 10.2).

Plate Tectonics Affects the Distribution of Animals and Plants and Generates Barriers That Lead to Diversification

In the early 20th century Alfred Wegener noted the reciprocal similarity between the coastlines on the eastern side of South America and the western side of Africa and documented similarities in fossil faunas and floras in Brazil and West Africa. He reconstructed a large continent Pangea, the result of the collision of the Earth's major plates during the Permian, which subsequently broke up and drifted apart (Fig. 10.4).

| Box 10.2 | Fossil Taxa Are Important in Determining the Evolutionary History and Relationships of Living Groups |

Fossils are critical to determining not just the evolutionary history, but also the relationships of living taxa. A classic example is provided by the tetrapods (animals with four limbs); analyses of the living taxa alone yield a spurious solution, which is remedied when fossils are included. Analysis of living taxa results in a sister-group relationship between mammals and archosaurs (crocodiles plus birds); turtles and lizards are successively lower on the cladogram. When fossils are added, turtles lie below lizards, which are the sister group to crocodiles and dinosaurs (including birds). Mammals are the sister group to all other tetrapods. The supposed relationship between mammals and archosaurs was an artifact of similar specializations for locomotion that are the result of convergence rather than a common origin (Fig. 10.2).

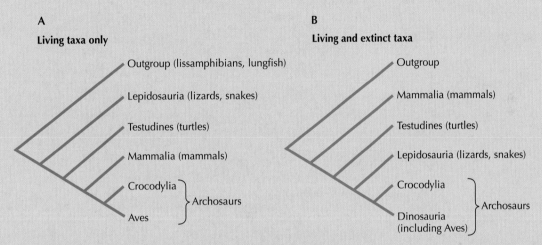

A
Living taxa only

Outgroup (lissamphibians, lungfish)
Lepidosauria (lizards, snakes)
Testudines (turtles)
Mammalia (mammals)
Crocodylia }
Aves } Archosaurs

B
Living and extinct taxa

Outgroup
Mammalia (mammals)
Testudines (turtles)
Lepidosauria (lizards, snakes)
Crocodylia }
Dinosauria (including Aves) } Archosaurs

FIGURE 10.2. Cladograms of the evolutionary relationships of tetrapods based on living forms alone (*A*) and on living and extinct taxa (*B*) yield two different results. The correct relationships are revealed in *B*. This comparison illustrates the importance of fossils in determining relationships. Note particularly the different position of mammals in the two examples.

FIGURE 10.3. The Cretaceous dromaeosaurid dinosaur *Deinonychus*—"terrible claw"—is a 3-m-long agile theropod like *Velociraptor*.

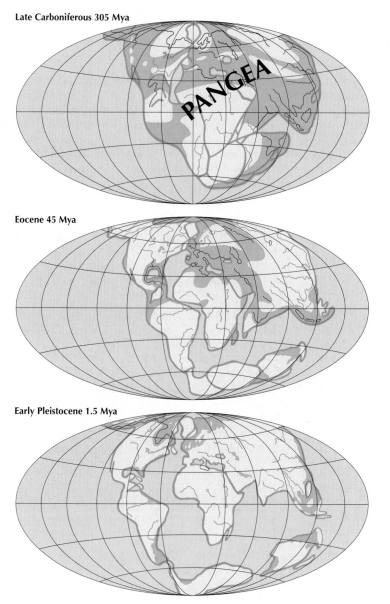

FIGURE 10.4. Alfred Wegener's reconstruction of the breakup and movement of the Earth's land masses (the *darker blue* areas represent shallow seas). Originally known as the theory of continental drift, plate tectonics has had major consequences for many scientific fields.

Although Wegener's ideas were rejected at the time, the movement of land masses through time is now incontrovertible and can be explained by seafloor spreading. The breakup of Pangea and the formation of the Atlantic Ocean eventually gave rise to the configuration of continents today.

Indeed, the Earth is a dynamic planet and the geography of the continents and oceans has changed through geological time as a result of the processes of **plate tectonics**. Oceanic crust is generated along the mid-ocean ridges and sinks along subduction zones changing the configuration of the continents on the Earth's surface. The implications for the history of life extend to dramatic changes in biogeography and climate, which in turn affect the number of separate faunal provinces on the Earth's surface and the extent to which some of them support unique biotas. The configuration of the continents was a factor influencing biodiversity in the past, just as it is at present.

A

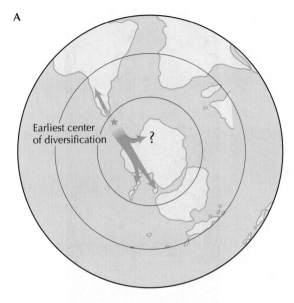

FIGURE 10.5. Long-distance land dispersal route of *Nothofagus*. (*A*) Initial routes of *Nothofagus* from its earliest center of diversification. Position of the land masses during the Late Cretaceous (~70 Mya) is shown. (*B*) Dispersal of *Nothofagus* to New Zealand and route of *Brassospora*, a subgenus of *Nothofagus*, northward through Australia to New Guinea, and from New Zealand to New Caledonia, during the Early Miocene (~20 Mya). (*C*) *Nothofagus gunnii*.

B

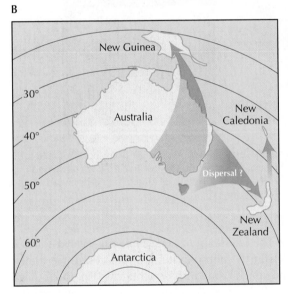

C

Discontinuous distributions of organisms generally can be explained in one of two ways: either they migrated across some barrier or their original geographic range was fragmented by geological processes. Australia and Antarctica, for example, separated about 40 Mya and are now some 20° latitude apart. *Nothofagus*, the southern beech, occurs mainly in New Guinea, New Caledonia, New Zealand, and Australia, but fossil evidence shows that it was formerly much more widespread, extending into South America and Antarctica (Fig. 10.5). Its distribution is in part the result of the breakup of the southern continent Gondwana during the early Tertiary (~65–40 Mya), but long-distance dispersal also played a role—for example, in its spread to New Zealand.

The First Multicellular Animals Appear in the Neoproterozoic

Some of the earliest evidence for multicellular animals is found in the Neoproterozoic Doushantuo Formation of the South China platform, which ranges from 635 to 551 Mya (Fig. 10.6). The Doushantuo Formation yields exceptionally preserved fossils from three rock types: **chert** (siliceous nodules), carbonaceous shales, and **phosphorites** (a sedimen-

FIGURE 10.6. Location of the Doushantuo Formation: Weng'an, in the Guizhou Province of southern China where phosphatized embryos were found.

FIGURE 10.7. (*A*) Spiny acritarch *Meghystrichosphaeridium reticulatum* from the Neoproterozoic Doushantuo Formation, China. Scanning electron micrograph. (*B*) Fossilized tetrads of the red alga *Paratetraphycus* found in the Doushantuo Formation. (*C*) Fossilized animal embryos in early stages of cleavage preserved in the Doushantuo Formation.

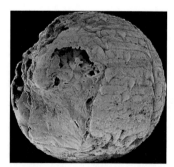

FIGURE 10.8. Cambrian phosphorites have yielded embryos such as *Markuelia*. The example illustrated here, from the early Cambrian of Siberia, is 555 μm in diameter.

tary rock rich in calcium phosphate). The Doushantuo Formation has yielded some 25 different multicellular algae. The fossils in the phosphorites are preserved three dimensionally in apatite (a calcium phosphate mineral) and include acritarchs (Fig. 10.7A), algae showing cellular preservation (Fig. 10.7B), and spectacularly preserved animal embryos displaying the earliest stages of cleavage (Fig. 10.7C). Because these embryos are in the earliest stages of division, it is difficult to assign them to specific groups. Phosphorites in younger rocks have yielded similarly preserved embryos, including *Markuelia* (Fig. 10.8) from the Cambrian of South China (~500 Mya), which are sufficiently developed to allow them to be placed in the stem group of the clade that includes the priapulids (a group of worm-like animals, sometimes known as Introverta, which are rare today) (see Fig. 9.22F). This indicates that Precambrian phosphorites have the potential to provide evidence of the early stages of metazoan evolution. This type of material may also allow evidence from developmental biology to be integrated with fossil data in unravelling the early history of metazoans. The fossil record of embryos, however, is likely to be biased toward forms that acquire cuticle early in their embryonic development, because they resist decay long enough to become mineralized in calcium phosphate as a result of microbial activity.

The Earliest Large Organisms Are the Ediacara Forms

The first large multicellular organisms that occur abundantly in the fossil record are often referred to collectively as Ediacara organisms after the Ediacara Hills in the Flinders Ranges of South Australia (Fig. 10.9), where they were first discovered in 1946.

They are found in rocks that range in age from 575 to approximately 542 million years, during the last part of the interval of time known as the Ediacaran Period. Many of the Ediacara fossils are found in sandstones laid down in shallow water, where the relatively coarse grain size limits the morphological detail that is preserved compared to that in finer-grained mudstones (a phenomenon akin to the loss of detail in images displayed on a computer screen in a larger pixel size). Some of the Ediacara organisms look very different from anything living today, so it is perhaps not surprising that the nature and affinities of these fossils remain controversial.

The Ediacara fossils show significant variability in size—from just a few millimeters to several tens of centimeters in maximum dimension. They are known from more than 30 localities (most notably in Newfoundland, South Australia, the White Sea region of northern Russia, and Namibia) from every continent apart from Antarctica, and they show a trend in increasing diversity and apparent complexity from the oldest to the youngest occurrences. Nonetheless they lack shells or other hard parts. They are usually preserved on the underside of a sandstone layer that cast an impression of the external surface. Their widespread fossilization despite a lack of hard parts may reflect the paucity of grazing, scavenging, and burrowing organisms at this time, which ensured that the microbial films that colonized the sediment surface were not consumed or destroyed. These microbial films created conditions conducive to very early mineralization of the layer of sediment immediately in contact with the decaying organism, preserving an impression in the sediment referred to as a death mask.

About 100 species of Ediacara organisms have been described, the most spectacularly preserved examples coming from the Avalon Peninsula in Newfoundland. Their affinities have been difficult to interpret because of the unusual nature of these forms. Traditionally they have been interpreted as early metazoans, either as forms ancestral to modern groups or early offshoots from the metazoan line. Many appear to lack dis-

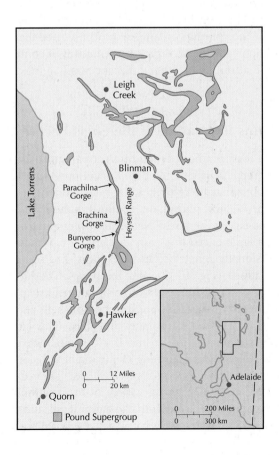

FIGURE 10.9. Location of the Ediacara site near Adelaide in southern Australia. The fossils occur in the Pound Supergroup, a major thickness of sedimentary rocks.

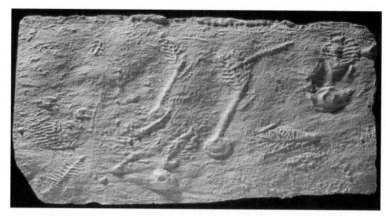

FIGURE 10.11. Ediacara fossils (*Charniodiscus* and spindle-shaped forms) about 565 million years old from Mistaken Point, Avalon Peninsula, Newfoundland. Field of view is 70 cm across.

FIGURE 10.10. The Ediacara organisms *Kimberella* (*top*) and *Dickinsonia* (*bottom*) are considered by some investigators to be complex metazoans. Magnifications: (*top*) x0.5; (*bottom*) x0.44.

crete internal organs and have been compared to cnidarians, particularly medusae and sea pens. Others, such as *Yorgia*, *Kimberella*, and *Dickinsonia*, are more complex and are compared to more complex metazoans (Fig. 10.10). An alternative interpretation of the Ediacaran organisms, developed mainly by Adolf Seilacher, is that many of them display a quilt-like structure lacking internal organs, and absorbing oxygen and nutrients through the body surface, a type of organization that largely disappeared before the Cambrian. Seilacher called these unfamiliar creatures vendobionts (Fig. 10.11), and argued that they existed alongside metazoans, the latter represented mainly by small trace fossils. More recently Seilacher has developed this interpretation to argue that many of the problematic Ediacara organisms represent giant protozoans and that more familiar metazoans (there are convincing examples of **sponges**, cnidarians, and **mollusks**) were smaller and rare. The controversy about the affinities of the Ediacara fossils illustrates the difficulty of interpreting unfamiliar morphologies that are preserved in unusual ways. What is important is that early representatives of some metazoan clades *are* present, whereas the affinities of other Ediacara fossils remain problematic.

The Appearance of Shells Transforms the Nature of the Fossil Record

The morphology of a number of Ediacara fossils indicates that different parts of these organisms were more or less rigid than others. It is clear that some developed tubes, carapaces, and even tooth-like structures but that these organs were not biomineralized (i.e., strengthened by minerals like the calcium phosphate [apatite]) in vertebrate bone). Rare biomineralized skeletons appear in the latest Neoproterozoic approximately 550 Mya, including *Namapoikia*, a large encrusting early cnidarian or sponge, *Cloudina* (a tube-like fossil, which is occasionally penetrated by holes, providing some of the earliest evidence for drilling predation) (Fig. 10.12), and the poppy-like *Namacalathus*. Biomineralization appeared in different lineages almost simultaneously at the beginning of the Cambrian (Fig. 10.13) and involved the construction of shells and other hard parts in a variety of minerals: phosphates, carbonates and silica. This led to a dramatic increase in the preservation potential of organisms: It is no coincidence that the base of the Cambrian (542 Mya) defines the beginning of the major division of geological time known as the Phanerozoic, the interval of "revealed life." The reasons for the onset of biomineralization are unknown and a variety of explanations have been offered including changes in ocean chemistry. Initially biomineralization may have been a response to a requirement for sequestering harmful ions, but it was

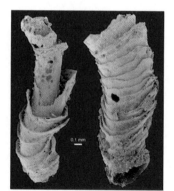

FIGURE 10.12. Phosphatized *Cloudina* specimens from the Neoproterozoic Dengying Formation in Shaanxi, China. Hard shelly parts appeared in metazoans during the Ediacaran, and the fossil on the *right* shows evidence of drilling predation.

FIGURE 10.13. Early Cambrian shelly fossils, illustrating the range of morphology. Sizes of these fossils are just a few millimeters in maximum dimension.

rapidly adopted for other functions. It was particularly important as a defense against predators. The appearance of biomineralized skeletons marks an ecological rather than a phylogenetic event; it reveals nothing about the timing of the origin of major groups.

The Cambrian Explosion

Occasionally circumstances conspire to preserve extraordinary details of an ancient community, including soft-bodied animals. The Cambrian Burgess Shale of British Columbia is perhaps the most famous example (Fig. 10.14). The Burgess Shale (505 Mya) and the older Chengjiang faunas of Yunnan Province in China (525 Mya) provide critical evidence of the results of the Cambrian radiation during which all the major body plans appeared. They preserve not only animals with hard skeletons (trilobites, echinoderms, mollusks, and brachiopods), but also a range of soft-bodied animals that are normally lost through decay (lobopods, most **arthropods**, annelids, priapulids, and chordates) (Fig. 10.15; for some examples of living representatives of these groups, see Figs. 9.20–9.22). Deposits like the Burgess Shale provide a wealth of information on the temporal range and morphology of extinct forms, but they also emphasize how biased the rest of the fossil record is. A census of the Burgess Shale fossils in the Smithsonian Institution (>65,000 specimens) showed that 86% of the genera and 98% of the individuals present would not be preserved in a normal deposit yielding only fossils with hard parts.

Some of the details of why and how the Burgess Shale fossils were preserved are not fully understood, but in outline the process was as follows: The organisms were swept up in a thick cloud of suspended mud. Turbulent transport led to the burial of carcasses in a variety of orientations. The sediment was low in oxygen, and the overlying mud protected the carcasses from burrowing scavengers. The carcasses collapsed as a result of decay. Specimens preserved in different orientations to the planes in the shale are equiv-

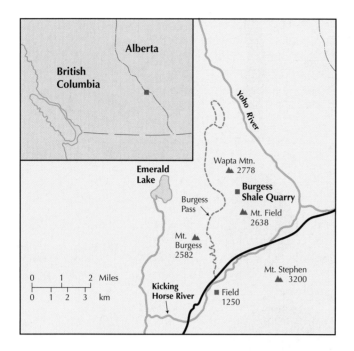

FIGURE 10.14. (*Left*) Location of the Burgess Shale Quarry in British Columbia in Yoho National Park between Wapta Mountain and Mount Field (heights are in meters). Discovered by Charles Walcott in 1909, the fossils comprise a complete Cambrian marine ecosystem that was caught in a series of mud slides. (*Right*) Charles Walcott and others at the Burgess Shale Quarry site.

alent, in a sense, to photographs of the animal taken from different angles. Data from these different specimens assist in restoring the original three-dimensional appearance of the animal. As the carcass decayed, minerals formed on the surface; analysis of the fossils shows that different minerals formed on different tissues. The more decay-resistant parts of the animal (e.g., the exoskeleton of arthropods) were slowly transformed from their original composition to a stable organic material that survived over geological time. A number of Cambrian animals, like *Anomalocaris* (Fig. 10.15), represent forms different from those around today. The morphology of these early invertebrates may provide clues to the evolutionary history and relationships of living organisms.

Sponges have the simplest type of organization among multicellular organisms.

FIGURE 10.15. Restoration of the Burgess Shale fauna. Key to the animals is as follows. Sponges: *Pirania* (*1*), *Vauxia* (*2*), *Wapkia* (*3*). Lobopods: *Aysheaia* (*4*), *Hallucigenia* (*5*). Anomalocaridids: *Anomalocaris* (*6*), *Laggania* (*7*). Arthropods: *Marrella* (*8*), *Odaraia* (*9*), trilobite *Olenoides* (*10*), *Sanctacaris* (*11*), *Sarotrocercus* (*12*). Priapulid: *Ottoia* (*13*). Polychaete annelid: *Canadia* (*14*). Chordate: *Pikaia* (*15*). Animals of disputed or uncertain affinity: *Amiskwia* (*16*), *Dinomischus* (*17*), *Eldonia* (*18*), *Odontogriphus* (*19*), *Opabinia* (*20*), *Wiwaxia* (*21*).

They form an aggregation of cells that are differentiated to perform different functions, but they lack the internal organs of more advanced invertebrates. The earliest examples are known from the Ediacara assemblages. **Ctenophores** and **cnidarians** (see Figs. 9.18 and 9.19) have just two layers of cells, the **endoderm** and **ectoderm**, and a single opening. The ctenophores (known as comb jellies or sea gooseberries) are gelatinous, mainly floating, creatures propelled by rows of cilia. They have a very poor fossil record because they are soft bodied, but Cambrian examples are known. The cnidarians include polyps (e.g., sea anemones and corals) and medusae. The colonial corals build the largest organic structures on Earth today, reefs like the Great Barrier off the east coast of Australia, and they were also important reef builders in the past.

The more complex forms of animal, the **bilaterians**, fall into two major clades: **protostomes** and **deuterostomes**, which diversified during the Cambrian (Fig. 10.16). The distinction between protostomes and deuterostomes is based mainly on differences in their embryology. The protostomes have been subdivided into **ecdysozoans** and **lophotrochozoans** on the evidence of molecular data. Ecdysozoans are characterized by molting (i.e., they periodically shed their cuticle in order to grow). This clade is dominated by the arthropods, but includes the **lobopods**, and worm-like **nematodes, kinorhynchs, priapulids,** and possibly the **chaetognaths** (see Fig. 9.22). Arthropods have been the most diverse group on Earth since the Cambrian (they represent ~40% of the genera in the Middle Cambrian Burgess Shale) when they were confined to the sea. They are even more dominant today, because of the huge variety of insects, representing about 80% of all known living species of animal. The four major living groups of arthropods are the hexapods (insects), crustaceans, myriapods, and chelicerates (horseshoe crabs and arachnids) (see Fig. 9.22). The interrelationships of these groups have long been controversial, with conflicts between hypotheses based on fossil and living forms and on morphological and molecular data (Fig. 10.17). It appears that the insects and crustaceans together form a clade; the interrelationships of myriapods and chelicerates are less certain. The fossil trilobites, which appeared in the Cambrian and died out by the end of the Permian, have a particularly abundant fossil record because of their heavily calcified exoskeleton. The other groups of ecdysozoans are poorly represented in the fossil record because they are essentially soft bodied, but lobopods (belonging to the same clade as living **onychophorans**), priapulids, and chaetognaths first appear in the Lower Cambrian Chengjiang deposits of China.

The lophotrochozoans are the nonmolting protostomes, and they include the lophophore-bearing organisms (e.g., **brachiopods** and **bryozoans**) and those with a **trochophore** larva such as the annelids and mollusks (but not all lophotrochozoans have a lophophore or a trochophore larva; see Fig. 9.21). Clades of lophotrochozoans with a biomineralized skeleton (brachiopods, bryozoans, mollusks) have a substantial fossil record that allows changes in their biodiversity through time to be documented. Brachiopods, for example, are a major component of Paleozoic marine communities, but they were severely depleted by the end-Permian extinction, after which bivalves became more dominant. The deuterostomes include the **echinoderms, hemichordates,** and **chordates**. The echinoderms are characterized by a calcitic skeleton, and this ensures a significant fossil record from the Cambrian onward. Echinoderms have a unique water vascular system and fivefold symmetry. The hemichordates (acorn worms and pterobranchs) have a poor fossil record, but the graptolites, extinct colonial organisms that floated in the oceans, are very common in Paleozoic shales where their robust organic skeletons accumulated in very large numbers. The chordates include the vertebrates and their relatives. The earliest vertebrates have now been recorded from rocks of Early Cambrian age from China—the jawless fish *Haikouichthys* (Fig. 10.18) from the Chengjiang assemblage. The fishes give rise to the tetrapods, which include the most familiar animals: the synapsids, which branched off early and gave rise to the mammals, and the diapsids, which gave rise to the lizards, crocodilians, and birds.

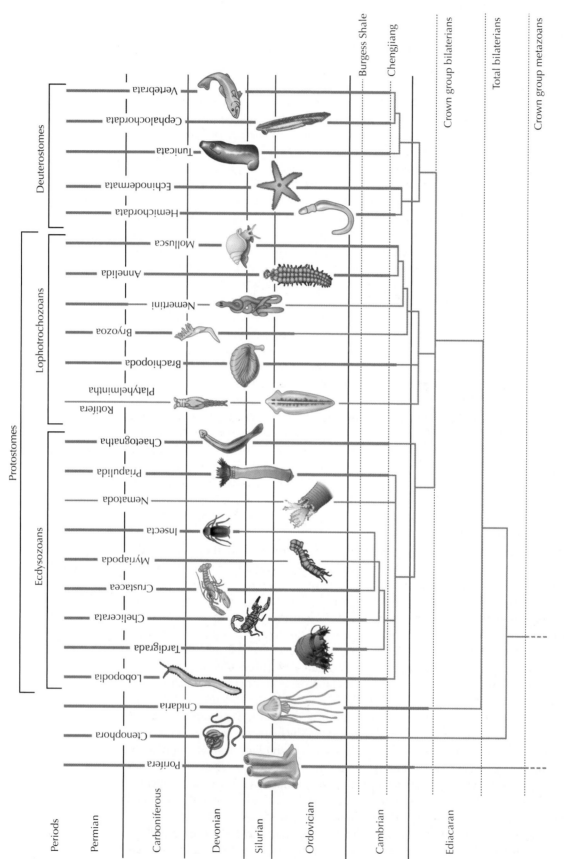

FIGURE 10.16. The fossil record and metazoan phylogeny. *Dark lines* represent the temporal range of phyla from their first appearance in the fossil record to the present. Extrapolation into the Ediacaran is based on molecular clock data. Some relationships, particularly among the arthropods, remain controversial (see Fig. 10.17).

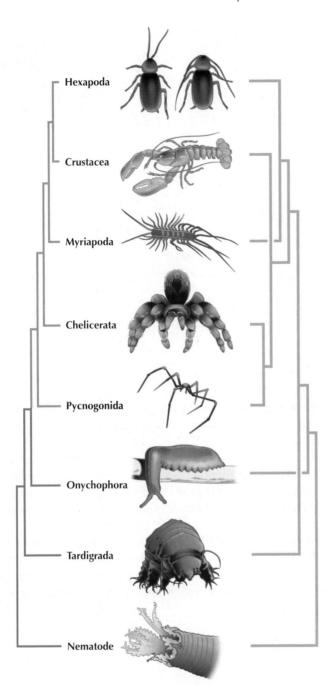

Hexapoda

Crustacea

Myriapoda

Chelicerata

Pycnogonida

Onychophora

Tardigrada

Nematode

FIGURE 10.17. Two views of the interrelationships of arthropod groups based on morphology and molecular data (*left; red* indicates Arthropoda) and morphology including fossil data (*right*).

How Large Was the Cambrian Explosion?

The increase in diversity that characterizes the Cambrian was detected first in the sudden appearance of shelly fossils at the beginning of this period and the steep increase in numbers of taxa plotted against geological time (Fig. 10.19). The suggestion that this might just be a reflection of the evolution of mineralized shells was countered by the demonstration that the diversity of soft-bodied organisms (as evidenced by exceptional fossil occurrences like those of Chengjiang and the Burgess Shale) and of trace fossils (e.g., trackways, feeding traces, burrows), increases in parallel.

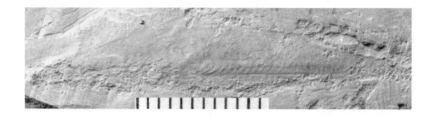

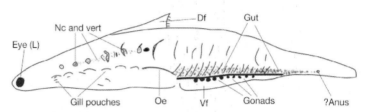

FIGURE 10.18. Early Cambrian jawless fish *Haikouichthys ercaicunensis* from Haikou, Kunming, Yunnan in southern China. The anterior end is to the *left*. This organism has a small lobate extension to the head including eyes, possible nasal sacs, and otic capsules, and a notochord with separate vertebral elements. These features indicate well-advanced vertebrate evolution by the Early Cambrian. Nc and vert, notochord with vertebral elements; Df, dorsal fin; Oe, esophagus; Vf, ventral fin fold. The divisions in the scale are mm.

The first detailed calibration and analysis of life's diversity through time was made by Jack Sepkoski. His database was taxonomic; he used the literature to document the occurrence of families of marine invertebrates through geological time (because seashells provide the best continuous record). He subsequently generated an expanded database of genera. Plotting the numbers of families that are represented in successive intervals of geological time revealed an increase in diversity overall, interrupted by major extinctions. The nature of the database raised a number of questions: How representative a sample does the fossil record provide of life in the past? How do biases in the record (marine organisms better represented than terrestrial; organisms with hard parts better represented than those without) vary through time? How valid are the fossil taxa used in the plot (many of the groups had not been subjected to cladistic analyses)? How do other factors, such as differences in the amount of exposed rock of different ages, or even the degree to which paleontologists have investigated particular groups of fossils or intervals of geological time ("paleontological interest units"), affect the pattern? To what extent

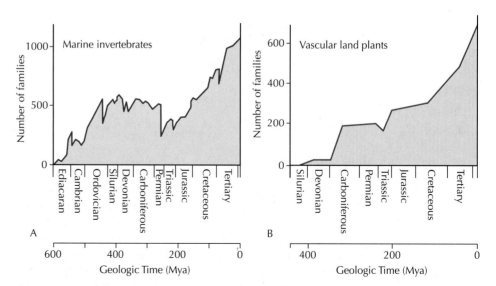

FIGURE 10.19. Patterns of diversity through time of marine invertebrates (*A*) and vascular land plants (*B*). Mya, million years ago.

do plots of diversity based on families or even genera represent true diversity, which normally is enumerated in species? Notwithstanding these concerns, the pattern of diversity revealed by Sepkoski's research has largely withstood subsequent refinement and augmentation of the database. The patterns illustrated in Figure 10.19 appear to be real.

Cladistic methodology ensures that the classification of organisms is much more rigorous than it used to be, but higher taxonomic categories (e.g., families and orders) may not be equivalent from one group to the next (e.g., fossil ammonites have received more attention and are more finely split than fossil gastropods; birds are better understood than nematodes). Plots of taxa through time provide one way of considering patterns of evolution through time, but changes in the range of morphology provide an alternative measure of evolution. Stephen Jay Gould emphasized this distinction between range of morphology (termed disparity) and numbers of taxa (termed diversity) and prompted attempts to quantify morphology as an approach to understanding the Cambrian radiation. Gould argued that the organisms of the Cambrian displayed greater disparity than those at any time since, including the present day (Fig. 10.20). He pictured the distribution of disparity from the Cambrian to the present as an inverted cone, flared at the base and tapering toward the top as a result of the loss of body plans through extinction (Fig. 10.20). In formulating this model Gould was influenced by some of the extraordinary animals discovered in the Burgess Shale (Fig. 10.15), which he called "weird wonders," regarding them as body plans (taxa of phylum-level status) that have become extinct and are unknown today.

Gould's approach emphasized the *differences* between taxa rather than considering their similarities. Subsequent analyses of morphology in arthropods and priapulids have shown that the range of morphology in the Burgess Shale fauna is similar to or slightly less (in the case of priapulids) than that of the present day (Fig. 10.20). It appears that groups of organisms realized much of their potential for evolving different

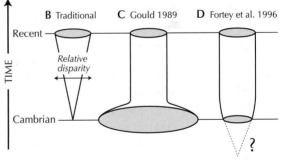

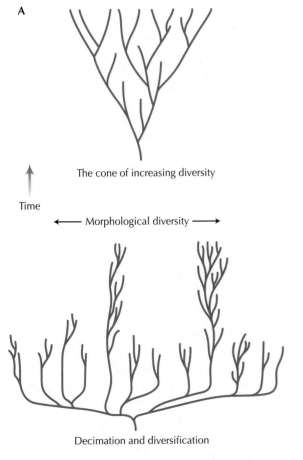

FIGURE 10.20. (*A*) The cone of increasing diversity (*top*) and Gould's model of diversification and decimation based on the evidence of the Burgess Shale (the *vertical axis* is time and the *horizontal* represents range of morphology). In this latter model, there is an initial rush of diversification of anatomical forms that are then restricted to a few surviving models, which then produce variants. (*B–D*) Models of Cambrian radiation. (*B*) Traditional model in which metazoans originate as simple forms and become more complex through the Phanerozoic. (*C*) Gould's model, in which Cambrian organisms are perceived as weird wonders that appeared rapidly, with disparity diminishing in the early Phanerozoic as a result of extinction. (*D*) Multivariate methods of quantifying morphology indicate that the Cambrian and recent disparities were similar.

morphologies early in their history during the Cambrian and that subsequent evolution involved variation on the themes established at that time. The distribution of morphology from the Cambrian onward can be modeled as a tube of uniform diameter.

The Timing of the Cambrian Explosion

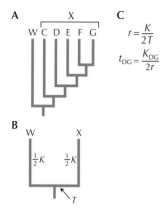

FIGURE 10.21. Use of the molecular clock, in which molecular differences are used to determine when two lineages diverged. In a simple example, the molecular clock is calibrated using the split between two taxa (W and X) (see A) which occurred at time T. (B) The number of nucleotide differences between W and X can be used to calculate the rate of nucleotide substitution for other taxa inside or outside the given clade. This result is used to calculate the time elapsed between the split of taxa D and G. In the equations in C, K is the number of nucleotide substitutions between two taxa, r is the rate of substitution, T is the calibration time, and t is the calculated time of split between two taxa.

We cannot rely on a literal reading of the fossil record to determine when the different groups of metazoans originated. First appearances as a fossil give only a minimum age for the origin of a group. The oldest recognizable fossil representative of a lineage is unlikely to represent the time at which it branched from an ancestral lineage. The advanced morphology of some of the animals that first appear in the lower Cambrian indicates that significant evolution had taken place by this time, but the Precambrian fossil record offers little evidence of the timing and nature of this process. An independent estimate of when groups originated is provided by the molecular clock (Fig. 10.21; also see Chapter 27 [online]). The clock is based on comparisons between gene sequences, such as that for small subunit RNA, for a range of organisms. Molecular differences can be used to estimate the time at which two lineages diverge. The rate at which sequences change can be estimated by dating the point at which two living taxa diverged, based on the fossil record. Those rates can then be extrapolated back in time to provide an indication of the timing of earlier events. Perhaps not surprisingly, different genes, and alternative methods of measuring differences between sequences and determining time, have resulted in widely varying estimates of the timing of the origin of groups—some of them long before the first appearance of examples in the fossil record. Many early estimates of branching times were based on differences between sequences in model organisms, mainly vertebrates. Now more sequences are available from invertebrates that are relevant to particular branching points in the phylogenetic tree and have a good fossil record. But a further complication is the likelihood that rates of genetic change varied through the history of life; we discuss this issue in Chapter 19 (p. 531).

Improvements in molecular data and methods have resulted in younger estimates of the age of branching points early in metazoan evolution, but a number of fundamental splits in the history of life still lie well within the Precambrian. Molecular evidence suggests that metazoans may date from more than 650 Mya, leaving more than 100 million years before the appearance of diverse fossil metazoans at the base of the Cambrian (Fig. 10.16). This begs the question as to why there is no fossil evidence of this early history. Various explanations have been offered centering on the argument that most Precambrian metazoans were soft bodied and tiny. The timing of splits between major groups remain controversial; some estimates indicate that the cnidarians may have split from other taxa just more than 600 Mya, but the protostome–deuterostome lineage may have diverged about 580 Mya, which would certainly imply very rapid evolution. As techniques continue to improve and more discoveries are made, the convergence of dates obtained from molecular and fossil data indicates that estimates of the timing of the origins of phyla are becoming more precise. Most major groups originated in a relatively short interval in the late Neoproterozoic.

One of the challenges posed by the Cambrian radiation is to explain how such a wide range of morphology could have evolved so quickly. Developmental biology provides one possible solution to this difficulty. Genetic studies of development have identified mutations in a number of key regulatory genes that can cause quite remarkable morphological alterations. The majority of these laboratory mutations are highly deleterious, and it is implausible that single mutations in key developmental genes could generate new body plans in a single step. However, developmental mechanisms could create relatively rapid morphological change in the time frame of the Cambrian explosion. The nature of these developmental systems and specific examples of their potential role in evolution are presented in Chapters 9 and 11, and their evolutionary consequences are presented in Chapter 24.

Cambrian Life Was Confined Mainly to the Sea, but There Is Some Evidence That the First Steps onto Land Were Taken by the End of this Period

Although the diversity of life on land today is much higher than it is in the sea, multicellular life originated in the sea and only subsequently invaded land. The early land surface was a barren environment that provided no protection for potential colonizers. All organisms that moved onto land faced similar challenges: They needed to adapt for respiration, for new sensory functions, for reproduction, and to provide support in air as opposed to water. Above all, they needed to develop strategies to avoid desiccation.

The earliest evidence of activity on land is trackways of a large arthropod in dune sands of late Cambrian to early Ordovician age in Ontario, Canada (Fig. 10.22). Although the trackways were made on land, the nature of the gait suggests that the arthropod was more at home in water, and it may have been amphibious. The arthropods are one of only four phyla that evolved a large diversity of terrestrial forms; the others are mollusks, annelids, and vertebrates. Some other phyla invaded at very small size, or as internal parasites of other organisms, but there is little or no fossil evidence of this transition. No phylum is known to have originated on land.

Although the fossil record of life on land is less complete than that of marine organisms, it shows that different terrestrial groups invaded at different times. They also used a variety of routes—via the seashore, rivers, or salt marshes. The arthropods made inroads at least by the early Ordovician: An oribatid mite has been described from early Ordovician marine deposits on the island of Öland, Sweden, that may have come from an intertidal setting. The earliest definitive terrestrial animal (i.e., with clear evidence of air breathing) is a millipede *Pneumodesmus newmani* from the Silurian of Scotland, but its ancestors probably lived in water (Fig. 10.23). Trigonotarbids (Fig. 10.24), an extinct group of arachnids that possessed sensory and respiratory structures for life in air, occur in slightly younger Silurian rocks from Shropshire in England.

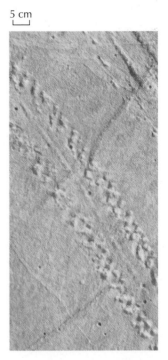

5 cm

FIGURE 10.22. Trackways of a possibly amphibious arthropod found in rocks representing dune sands of late Cambrian to early Ordovician age near Kingston in Ontario, Canada. This find dates early footprints on land to about 500 Mya.

A

B

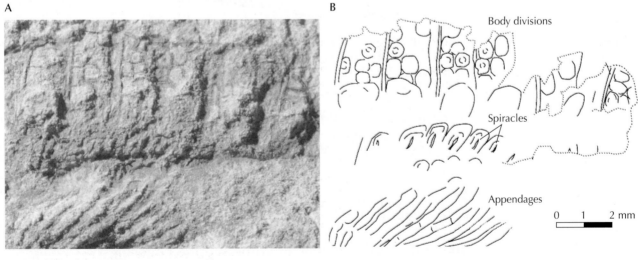

Body divisions

Spiracles

Appendages

0   1   2 mm

C

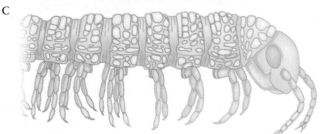

FIGURE 10.23. The millipede *Pneumodesmus newmani* is the earliest definitive terrestrial animal capable of air breathing, as evidenced by the spiracle-like structures found in these fossils (*A,B*). This fossil was discovered in the Silurian Cowie Formation, Stonehaven, Scotland. A portion of the body with appendages is shown facing to the right. The reconstruction (*C*) shows what a younger form, *Palaeodesmus* from the Devonian of Ayrshire, Scotland, might have looked like.

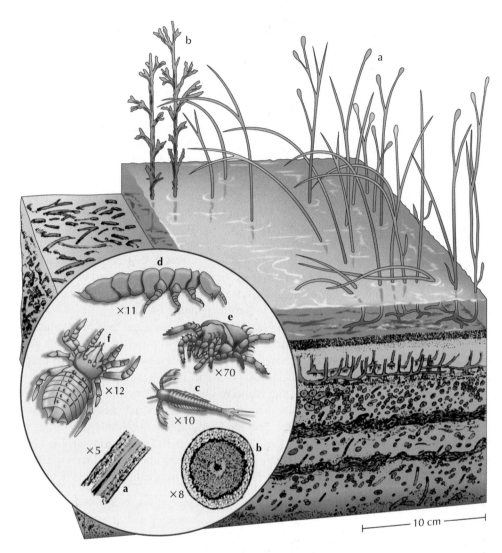

FIGURE 10.24. Lower Devonian terrestrial community of the Rhynie Chert. Life-forms shown include *Rhynia major* (*a*) and *Asteroxylon* (*b*) (both pteridophytes), *Lepidocaris* (*c*) (crustacean), *Rhyniella* (*d*) (hexapod), and *Protacarus* (*e*) and *Palaeocharinus* (*f*) (a mite and a trigonotarbid, both chelicerates).

Scorpions moved from the sea onto land in the Devonian (400 Mya), as evidenced by the discovery of examples with lungs rather than gills. A diversity of terrestrial forms had evolved by the Devonian (Fig. 10.25).

THE NEXT 500 MILLION YEARS—LIFE SINCE THE CAMBRIAN PERIOD

The Ordovician and Silurian Witnessed the Greening of the Landscape

Bacteria, followed by algae, probably invaded land during the Precambrian. These early colonizers relied on damp conditions and they were essential to breaking down the substrate to form the first soils. The major groups of algae, including diatoms and brown, green, and red algae, appeared in the Neoproterozoic. The green algae are a sister group to the plants: liverworts, mosses, ferns, **lycophytes**, and **sphenopsids** (other nonseed plants), and seed plants. The mosses (bryophytes) are a monophyletic group

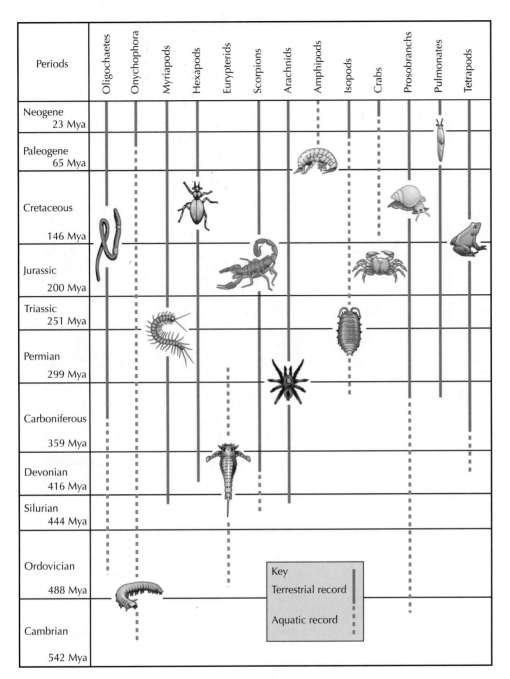

FIGURE 10.25. Development of diversity in terrestrial animals. The terrestrial fossil record is indicated by *solid lines* and the aquatic fossil record by *dashed lines*. Mya, millions of years ago.

of low creeping plants that lack a fully developed transport system and rely on damp conditions for survival. The earliest record is Late Devonian, but this is likely to reflect the low preservation of these plants and of the environment in which they lived; mosses very likely were present at an earlier time.

A key innovation in the evolution of higher land plants was the development of **tracheids,** the defining characteristic of vascular plants (Fig. 10.26). Tracheids are elongate cells that transport fluids within plant tissues so that water can be carried from the roots to other parts of the plant. The earliest generally accepted vascular land plant is *Cooksonia* from the Lower Silurian rocks of Wales. It is a naked, dichotomously branch-

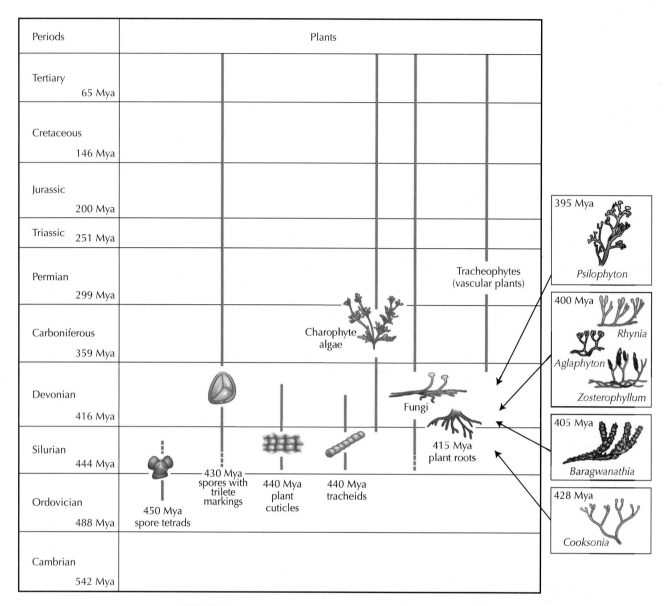

FIGURE 10.26. Innovation in plants leading to terrestrial plant forms. An important innovation was the development of structures such as tracheids to carry water from the roots to other plant parts. Mya, million years ago.

ing stem no more than a few centimeters high with spore-bearing structures at the tips (terminal sporangia). Land plants radiated in the Early Devonian with the appearance of the **rhyniophytes, zosterophylls, lycopods,** and **trimerophytes.** The Rhynie Chert in Scotland is one of the most important sources of data on the earliest land plants and animals (Figs. 10.27 and 10.24). Cellular details survive because the fossils are preserved in the highly siliceous deposits of a hot spring, which led to the precipitation of mineral in the plant tissue. Five of the Rhynie Chert plants preserve evidence of tracheids, demonstrating that they were vascular plants. A number of the plants, like *Rhynia*, had **sporangia** that split to disperse the spores. The Rhynie Chert also preserves algae, fungi, and lichens, and some of the plants preserve evidence of the decay activities of fungi and bacteria. The evidence of gut contents and damage to plant tissues indicate that a number of the associated arthropods made up a plant-litter community that included mites, insects, and a myriapod. Others, including **trigonotarbids** and a centipede, were carnivores.

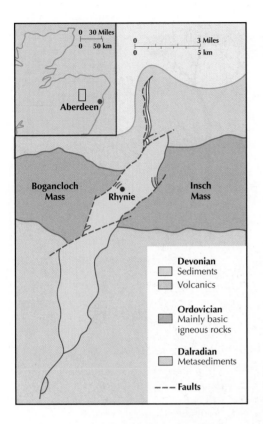

FIGURE 10.27. Location of the Rhynie Chert, near Aberdeen, in Scotland. Evidence of vascular plants was discovered in this area.

The ferns are **vascular plants** with independent **gametophytes** and motile sperm that first appeared in the Devonian. The first seed plants were seed ferns (pteridosperms), which appeared first in the late Devonian and produced large trees by the Permian. During the Carboniferous the dominant plants were the **lycopsids** (giant club mosses), which produced trees up to 30 m tall. Other swamp plants included the cordaites, tree ferns, pteridosperms, and calamites (sphenopsids) (Fig. 10.28). They formed huge thicknesses of coal in Europe and North America as a result of organic matter accumulation in wet environments. Different elements of the flora dominated in different environments and at different times. The coal swamp flora disappeared in large measure in the Late Carboniferous, except in China where it persisted through the Permian. The Carboniferous coal swamp forests accommodated the 2-m-long *Arthropleura*, a giant herbivorous myriapod. Since the beginning of the Mesozoic, plant assemblages have been dominated by seed plants. They include the cycads, which are important in the Carboniferous; conifers and gingkos, which appeared in the Triassic; and angiosperms, the flowering plants, which appeared in the Cretaceous.

Vertebrates Moved onto Land

The transition from water to land in vertebrates is linked to the evolution of tetrapods from fish. The classic picture is of air-breathing fishes (lobe-finned crossopterygians) crawling from a shrinking pool to a larger water body to avoid desiccation in an arid Devonian landscape. However, it appears that most tetrapod characters evolved in shallow water before they moved onto land. The late Devonian *Ichthyostega* combined an otherwise tetrapod morphology with a fish-like brain case and a tail fin, but its hindlimb was like the flipper of a seal. Many of these critical morphological changes took place rapidly in a relatively short interval in the late Devonian when forms leading to the modern tetrapods first appeared (Fig. 10.29). Subsequent events are obscured by a 25-million-year hiatus in the lowermost Carboniferous fossil record of

FIGURE 10.28. Late Carboniferous coal swamp forest. Plants from *left* to *right*, a calamite tree, scrambling cordaite, tree fern, lycopsid, seed fern, lycopsid, and mangrove cordaite.

tetrapods, during which time amphibians and amniotes (mammals, birds, and reptiles) diverged. One of the most unexpected discoveries was the presence of a larger number of digits (eight in *Acanthostega*, seven in *Ichthyostega*, and six in *Tulerpeton*) in the limbs of some of the stem taxa (Fig. 10.29). The evolution of five or fewer digits appears to have been one of the later changes in the evolution of tetrapods.

Arthropods and Vertebrates Evolved the Ability to Fly

The Rhynie Chert yielded the earliest known hexapod, a springtail called *Rhyniella*. A second taxon, *Rhyniognatha*, appears to be a more advanced insect (Fig. 10.30). Although the single known specimen preserves details of only the jaws, these jaws are similar to those known in winged insects, suggesting that flying insects may have been present by this time. There are no insect fossils preserved for about 55 million years from the Middle Devonian to the Early Carboniferous, after which there appears a spectacular diversity of winged forms. Evidence from Permian and Carboniferous fossils suggests that wings evolved from the gill, the outer of the two branches that made up the trunk limbs in an aquatic precursor, and that early insects bore wing-like structures along the length of the body. The hypothesis that wings evolved from gills is also supported by developmental data that reveal similarities in gene expression patterns between crustacean gills

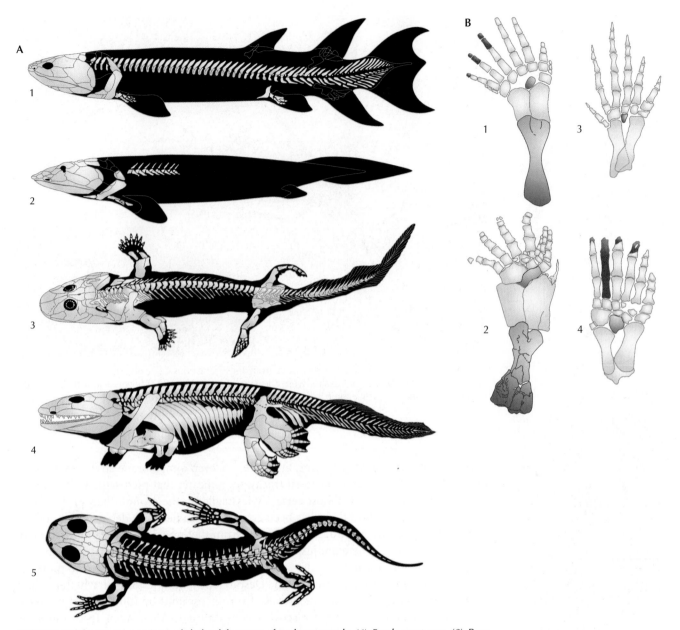

FIGURE 10.29. (*A*) Comparison of skeletal features of early tetrapods. (*1*) *Eusthenopteron*, (*2*) *Pan-derichthys*, (*3*) *Acanthostega*, (*4*) *Ichthyostega* (*1–4* all Devonian), and (*5*) *Balanerpeton* (Carboniferous). Not all skeletal features are shown; those that are shown emphasize the retained skeletal features among these genera and clarify their comparison. (*B*) Pectoral fin and limb patterns of early tetrapods. (*1*) *Acanthostega* hindlimb, (*2*) *Ichthyostega* hindlimb, and (*3,4*) *Tulerpeton* fore-limb and hindlimb. *Blue* indicates homologous structures; *dark brown* indicates restored structures.

and fly wings. The subsequent radiation of insects produced forms with wings on three or fewer segments of the thorax. Suppression of wings on certain segments was presumably necessary to produce an insect that was capable of flight.

Three groups of vertebrates evolved the ability to fly: pterosaurs, bats, and birds. Pterosaurs were the first group to evolve flapping flight, in the Triassic, and they include *Quetzalcoatlus* from the Cretaceous of Texas, with a wingspan of up to 12 m, the largest flying animal known (Fig. 10.31). It is likely that pterosaurs also soared, particularly the larger varieties, in order to conserve energy. The wings were formed of membranes stretched between the limbs. The arrangement of these membranes dif-

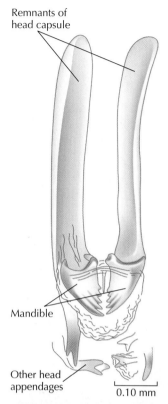

Remnants of head capsule

Mandible

Other head appendages

0.10 mm

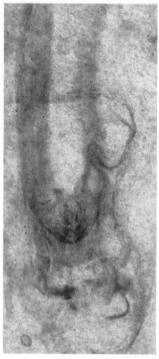

FIGURE 10.30. *Rhyniognatha hirsti*, the oldest fossil insect, found in the Rhynie Chert of Scotland. This fossil has derived characters that are shared with winged insects and thus suggests a Devonian origin of wings.

FIGURE 10.31. *Quetzalcoatlus*, a pterosaur from the Cretaceous of Texas, had a wingspan of up to 10–12 m. The smallest pterosaurs were the size of sparrows. An elongated fourth digit in the wing skeleton provided the main support for the wing. The evolution of flight in the pterosaurs was separate from that in birds and bats—an example of convergent evolution.

fered between taxa; in some pterosaurs there was even a membrane stretched between the hindlimbs. The evidence of fossil trackways indicates that pterosaurs walked on all fours; most of them were almost certainly less agile on the ground than in the air.

The evolution of bird flight has been extensively studied. The debate has long centered around the earliest bird, *Archaeopteryx*, from the Jurassic Solnhofen limestone of Bavaria (Fig. 10.32). There are just ten known specimens of this animal (plus an isolated feather), and it is unlikely that any suite of fossils has been subjected to more intense investigation and interpretation. Despite this the question of whether flight evolved via gliding from trees, or directly from the ground by running and flapping, remains unresolved. The recent discovery of a dinosaur *Microraptor* from the Early Cretaceous of Liaoning, China, with two pairs of wings, has rekindled the debate (Fig. 10.33). The feathers of the forelimbs, and of the hindlimbs and tail, are asymmetric and appear to be adapted for flight; both pairs of wings plus the tail make up an airfoil that was suitable for gliding. The long feathers on the hindlimbs would have inhibited running in this dinosaur, indicating that it could not get airborne directly from the ground. *Microraptor* belongs to a group of theropod dinosaurs that lies basal to *Archaeopteryx*, even though it occurs in younger rocks. As a gliding animal *Microraptor* provides support for an arboreal-gliding origin of flapping flight, which was followed by the loss of flight feathers on the hindlimbs (in effect a reduction to a single pair of wings). As ever, of course, questions remain to be addressed. *Microraptor* was clearly adapted for gliding, but was it also capable of powered flight? Was its four-winged morphology sufficiently widespread to suggest that it was the ancestral condition, or did it just evolve in a group of small gliding dinosaurs?

Bats were the last group to evolve flapping flight. The earliest known examples are from the Eocene Green River Formation of Wyoming (48 Mya). X-radiographs of slightly younger Eocene bats from the Messel Oil Shale of Germany revealed details of the ear bones that indicate that these bats were capable of echolocation.

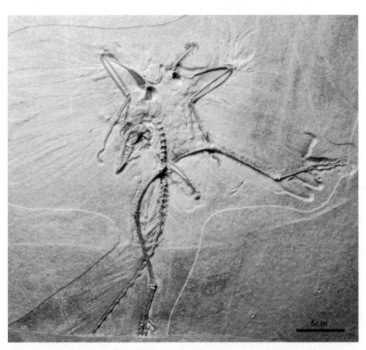

FIGURE 10.32. The Thermopolis specimen of *Archaeopteryx*—"ancient wing"—from the Jurassic Solnhofen Limestone in Bavaria, southern Germany. Dating from 150 Mya, this fossil has features that are both absent (teeth, flat sternum, long bony tail, claws on wings) and present (feathers, wings, wishbone, reduced fingers) in modern birds. This specimen had a wingspan of about 0.4 m.

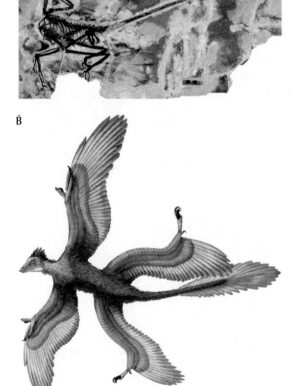

FIGURE 10.33. *Microraptor*, a dinosaur from the Early Cretaceous of Liaoning, China, with two pairs of wings and feathers, provides clues about the origins of flapping flight. This species is about 77 cm long. The fossilized form is shown in *A* and a reconstruction is shown in *B*.

PATTERNS OF EVOLUTION

The Fossil Record Provides Evidence of the Pattern of the Diversity of Life through Time

Estimates of the diversity of life today range between 3 and 50 million species. Such figures are based, in part, on the number of living species that have been cataloged, approximately 1.7–1.8 million. If it is so difficult to estimate the number of species living today, how can patterns of biodiversity through geological time be determined from an imperfect fossil record? Accurate values for biodiversity at any time in the past are impossible to obtain. However, our estimates are consistent enough to provide figures that are reliable from one relative to another, and this allows us to identify significant fluctuations in species diversity throughout geological time. Once such fluctuations are identified, their possible causes can be explored.

Animal diversity is much greater on land today than it is in the marine realm, even though life in the sea has been diversifying for much longer. About 85% of described living species live on land, most of them insects, and this diversity has evolved over the past 450 million years. Nonetheless a much higher proportion of fossils (95%) are described from marine settings than from land.

Many biologists define species in terms of **reproductive isolation** (see Chapter 22). In practice, however, species of living organisms are usually identified on morphological grounds, the same criteria that are used to identify fossil species. The problem with fossils is that they are part of an evolving lineage, raising the question of where does one fossil species end and the next begin? In the 1970s Niles Eldredge and Gould reconsidered the traditional paleontological view that speciation is a gradual process by positing their model of punctuated equilibrium (Fig. 10.34).

The origin of new species is often explained by the isolation of small populations on the fringe of their main distribution range (pp. 641–642). The chances of organisms from such peripheral isolates being preserved as fossils are very small. The pattern that is evident in the fossil record, where morphology remains the same for long periods only to be punctuated by an abrupt shift, could be explained by a similar process. The ideas of Eldredge and Gould prompted a search for complete sections in the sedimentary record where their model of speciation could be tested. Although punctuated equilibrium appears to be widespread, examples of gradual change also occur, particularly among planktonic Foraminifera that inhabit the more stable environment of the oceans.

Models of speciation are largely irrelevant to considerations of ancient biodiver-

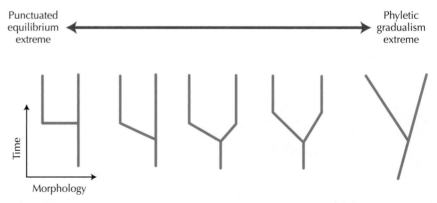

FIGURE 10.34. Continuum of models of speciation including the Eldredge and Gould model of punctuated equilibrium at one end and the model of phyletic gradualism at the other. Shown are the varying patterns of change along this continuum.

sity, because the fossil record of species is far too incomplete. Biodiversity through time is normally estimated based on the temporal ranges of genera, or more often families, as a proxy for species. Sepkoski's plots of families of marine animals with readily preserved hard parts (the "shelly" fossil record) (Figs. 10.19A and 10.35) revealed a plateau in diversity in the Cambrian that lasted about 40 million years, and a much longer one from the Ordovician to the Permian that lasted about 250 million years. Following the end-Permian extinction there was a steep increase in diversity through the Mesozoic and Cenozoic that continues today. The diversity curve based on marine families has acquired a near iconic status as the pattern of the history of life.

Rates of taxonomic turnover are higher on land than in the marine realm. The potential for evolution into new habitats in the terrestrial realm may have yet to be fully realized. The history of diversity of shelly marine invertebrates is paralleled by that of vascular plants on land (Fig. 10.19B), although they did not appear until the Silurian and the first significant increase was in the Early Devonian. Lycopsids, ferns, sphenopsids, and cordaites diversified in the Carboniferous (where they dominated the coal swamp floras) and Permian. Gymnosperms radiated in the Triassic to Jurassic, and angiosperms mainly from the end-Cretaceous extinction onward. Land-dwelling tetrapods originated in the Devonian and remained at a relatively low level of diversity through the Mesozoic. Their major expansion has been since the end-Cretaceous extinction with the diversification of modern groups including the birds and mammals.

Extinctions

The rate of change of diversity is the difference between rates of origination and of extinction. Certain intervals of time in the past were characterized by much higher rates of extinction than the normal background. These mass extinctions (Fig. 10.36), particularly the five largest ones (end-Ordovician, end-Devonian, end-Permian, end-Triassic, and end-Cretaceous), had a profound impact on the history of life. They were a response to a variety of different global perturbations, sometimes acting together. Initial analyses suggested that major extinctions might have occurred with a regular periodicity, and a variety of extraterrestrial agents, such as a cloud of comets bombarding the Earth, were hypothesized to explain this pattern, but this inference is no longer accepted.

The greatest extinction of all was that at the end of the Permian (which defined the end of the Paleozoic era). Recent field work has allowed more precise data to be accumulated on the temporal ranges of the organisms involved. A middle Permian event mainly affected marine organisms in tropical latitudes and was followed by a recovery. The more short-lived end-Permian event was much more severe and global in impact. Some major taxa became extinct (tabulate and rugose corals and trilobites) and others were severely decimated (crinoids, of which perhaps only one genus survived, and brachiopods). Large, sessile, surface-dwelling forms were replaced as the dominant taxa in seabed communities by bivalves and gastropods that lived in the sediment (Fig. 10.37). Terrestrial groups, including insects, tetrapods, and plants, also were affected severely. More than 60% of families and perhaps 90% of species of marine and terrestrial organisms died out. A number of agents have been implicated in this largest of all biodiversity crises.

The Permo-Triassic boundary is dated at 251 Mya. A major rise in sea level and associated reduction in circulation led to a widespread development of anoxic water in the oceans. This is evidenced by the occurrence of pyrite, which is generated by sulfate reduction, in fine-grained sediments. Anomalous levels of helium and argon in sections in China suggest the possibility of an asteroid impact, but the interpretation is equivocal. However, substantial volcanic eruptions in Siberia produced about 2 million km^3 of lava and extensive ashfalls within about 600,000 years at this time. Such sustained activity would have resulted in elevated levels of CO_2 and global warming.

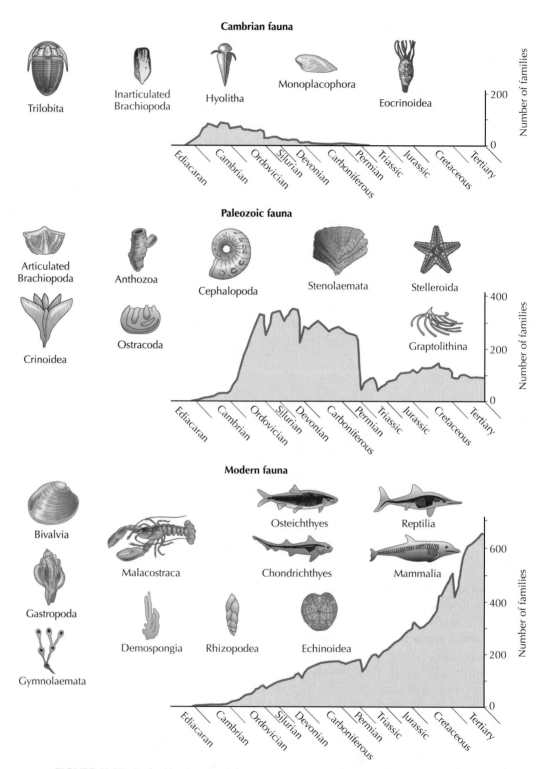

FIGURE 10.35. Sepkoski recognized three great marine evolutionary faunas: Cambrian, Paleozoic, and Modern. Here the main groups of animals that made up these faunas are illustrated and their diversity is plotted as numbers of families through geological time.

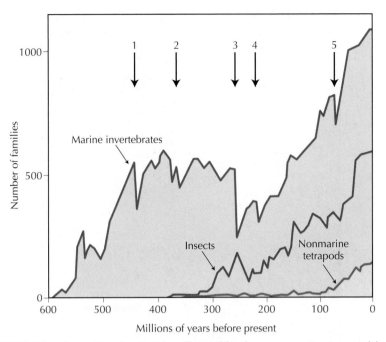

FIGURE 10.36. Diversity profiles showing the effects of five large mass extinctions on life on Earth: (1) end of the Ordovician, (2) Late Devonian, (3) end of the Permian, (4) end of the Triassic, and (5) end of the Cretaceous.

It has been suggested that melting of frozen **gas hydrates** may have released large volumes of methane, evidenced by a pronounced carbon isotope shift, producing a greenhouse effect. The dramatic climatic warming that followed could account for the very high levels of extinction at the boundary. Rates of recovery after the end-Permian extinction were variable. Diversity on land recovered relatively rapidly, but preextinction levels were not achieved in the marine realm for more than 100 million years.

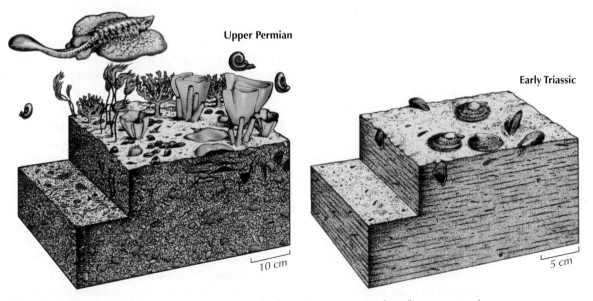

FIGURE 10.37. Effects of the Permian extinction. The latest Permian tropical seafloor compared with that of the Early Triassic, based on the section at Meishan, China, showing the loss of reef-dwelling organisms.

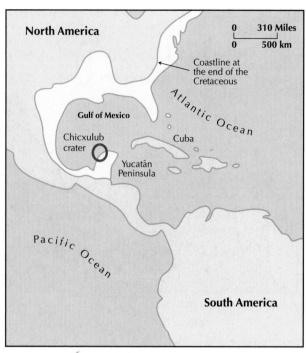

FIGURE 10.38. Location of the Chicxulub crater on the Yucatán Peninsula of Mexico.

Although evidence for an extraterrestrial impact associated with the Permo-Triassic boundary is questionable, there is clear evidence that such an impact was the primary agent that caused the mass extinction at the end of the Mesozoic era. Evidence that this end-Cretaceous (K–T) extinction was the product of a major asteroid impact first came to light in the early 1970s at Gubbio in Italy where Walter Alvarez found a layer of clay that contained a concentration of the element iridium more than 300 times the normal level. Iridium is very rare on Earth but occurs in significant concentrations in extraterrestrial material. Subsequently, evidence of a giant crater of the correct age, the Chicxulub Crater, was discovered on the Yucatán Peninsula in Mexico (Fig. 10.38). A major cause of the K–T extinction was the cooling effect of the enormous cloud of dust generated by the impact, which also blocked out much of the light from the Sun. This was followed by global warming and a substantial increase in precipitation. Longer-term climate change may also have been a contributory factor.

The K–T extinction was much more selective than the end-Permian. It is estimated to have led to the demise of 40–76% of species. Some major groups, such as dinosaurs and ammonites, died out, whereas others, such as deep-sea benthic Foraminifera, were virtually unscathed. Plant extinctions are evident in the record of pollen and spores and in leaf taxa, and disruption of the food chain led to the extinction of some vertebrates. There is evidence of reduced nutrient levels in the seas following the extinction, and the representatives of many marine groups became smaller in size. Both plants and mammals underwent major radiations during the Paleocene, the interval immediately following the extinction, but for some groups it was more than 10 million years before they returned to their Cretaceous diversity levels.

The factors that contribute to survival during normal background levels of extinction may not apply during mass extinctions. Organisms died out in the K–T extinction as a result of the aftereffects of the impact rather than the collision itself. Although organisms cannot evolve strategies to avoid being extinguished by an asteroid, the effects of extinctions on taxa are not random. Broad geographic distribution, for example, increased the likelihood of survival during the K–T event.

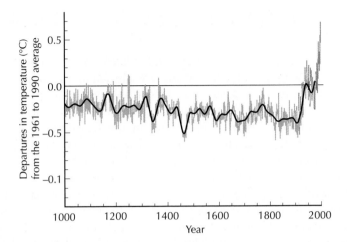

FIGURE 10.39. Variations of the Earth's surface temperature over the last millennium. The year-by-year (*blue* and *red curves*) and 50-year average (*black curve*) variations of the average surface temperature of the Northern Hemisphere for the past 1000 years are shown. The rate and duration of warming of the 20th century has been much greater than in any of the previous nine centuries. It is likely that the 1990s have been the warmest decade and 1998 the warmest year of the millennium. Data from thermometers (*red*) and from tree rings, corals, ice cores, and historical records (*blue*).

Modern Extinctions

The K–T event was followed by a recovery to diversity levels well in excess of those prior to the extinction. Life today is more diverse than ever before. Human activities, however, are having a major impact on biodiversity. Habitat destruction, particularly of forests and coral reefs in the tropics, is rapidly eliminating species. The loss of larger birds and mammals is more striking but may be no more significant than the elimination of invertebrates and plants. Overall, it has been estimated that about half of present-day species of plants and animals are at risk of extinction this century—a magnitude comparable with the K–T event.

The effects of global warming are potentially disastrous. Levels of carbon dioxide are now higher than at any time in the past 20 Myr, and, as a consequence, temperature over the past half-century has risen by more than any other time in the past 1300 years (Fig. 10.39). If emissions continue unchecked, global temperature is projected to rise around 2–5°C this century. The last time that temperatures were this high for a prolonged period, about 125,000 years ago, sea level was 4–6 m (13–20 feet) higher.

Although populations can adapt to changing conditions, and new species eventually evolve to replace those that are lost, when measured on the timescale of the human species, these processes are exceedingly slow. As a result, specialist species, like the koala, the panda, or desert cacti, tend to be replaced by invaders like rats and cockroaches. After previous mass extinctions, it took millions of years for global biological diversity to recover. We need to take steps to halt the progress of extinction before unique habitats are eliminated together with their inhabitants.

SUMMARY

The fossil record provides a time scale for the diversification of plants and animals. Some of the oldest evidence for multicellular life comes from South China where remarkably preserved microfossils of algae and animal embryos occur in rocks more than 600 million years old. The earliest large organisms, the Ediacara forms, appear before the base of the Cambrian 542 Mya. Their unusual morphology has made their affinities difficult to determine. The appearance of diverse shells in the early Cambrian heralds a major increase in the diversity of life on Earth, the Cambrian explosion. The extraordinary preservation in deposits like the Burgess Shale provides evidence that all the major body plans evolved at this time. But estimates based on molecular differences, together with the evidence of fossils, indicate that some of the major splits in the evolution of animals occurred much earlier (sponges branched off more than 600 Mya, cnidarians around 600 Mya). Arthropods made the first forays onto land during the late Cambrian, but the earliest vascular land plants are Silurian in age (about 425 Mya), and a diversity of land plants and arthropods is not known until the late Silurian and Devonian. Vertebrates also moved onto land in the Devonian, as early tetrapods evolved from fish. The coastal swamps of the Carboniferous were dominated by giant plants that gave rise to enormous accumulations of coal. Changes in the configuration of the continents have had a major impact on diversity. The history of life is punctuated by major extinctions, the greatest of which was at the end of the Permian, around 250 Mya, when perhaps 90% of species died out. A less severe, but no less dramatic, extinction occurred at the end of the Cretaceous (65 Mya) when an asteroid impact saw the demise of the ammonites and dinosaurs. The diversity of life today is at a maximum, but habitat destruction and global warming threaten another mass extinction.

FURTHER READING

Briggs D.E.G. and Crowther P.R., eds. 2001. *Palaeobiology II.* Blackwell Science, Oxford.

Briggs D.E.G., ed. 2005. *Evolving form and function: Fossils and development.* Yale Peabody Museum of Natural History, New Haven.

Clack J.A. 2002. *Gaining ground: The origin and evolution of tetrapods.* Indiana University Press, Bloomington, Indiana.

Erwin D.H. 2006. *Extinction: How life on earth nearly ended 250 million years ago.* Princeton University Press, Princeton.

Hou X.-G., Aldridge R.J., Bergström J., Siveter David J., Siveter Derek J., and Feng X.-H. 2004. *The Cambrian fossils of Chengjiang, China. The flowering of early animal life.* Blackwell Science, Oxford.

Narbonne G.M. 2005. The Ediacara biota: Neoproterozoic origin of animals and their ecosystems. *Ann. Rev. Earth Planet. Sci.* **33:** 421–442.

Peterson K.J., McPeek M.A., and Evans D.A.D. 2005. Tempo and mode of early animal evolution: Inferences from rocks, Hox, and molecular clocks. *Paleobiology* (suppl. Part 2) **31:** 36–55.

11

Evolution of Developmental Programs

I N THE PREVIOUS TWO CHAPTERS, WE PROVIDED AN introduction to the diversity of multicellular creatures and what we know about their history from the fossil record. We also discussed the basic principles that guide the development of multicellular organisms and ultimately shape their **body plans**. Now we will explore how we can combine these approaches to understand how development itself has evolved to yield the remarkable diversity we see.

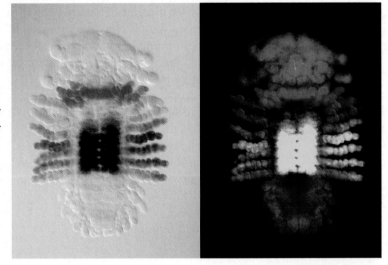

In *On the Origin of Species*, Darwin devoted a chapter to the subject of embryology, realizing that morphological changes must be the result of changes in development. Although a number of early embryologists were interested in evolutionary questions, the fields of evolution and developmental biology nevertheless remained fairly separate for quite a long time. In the past two decades, however, there has been renewed interest in uniting these two fields, leading to the renaissance of evolutionary developmental biology, often called "EvoDevo." Much of this renewed interest has come about because we now understand development at the molecular and genetic levels and can combine this information with our understanding of the process of evolution at these same molecular and genetic levels. We can consider evolution to be the change in populations and species over time, whereas development represents the change of an individual over time. By tying evolution and development together, we gain better insight into both processes.

In this chapter, we present several examples of how developmental data can be used to understand morphological evolution. We begin with the *Hox* genes, which are known to play an ancient and well-conserved role in anterior–posterior patterning during animal development. Recent studies indicate that changes in the function and expression of these genes can also lead to evolutionary changes in morphology at both macroevolutionary and microevolutionary scales. In the next set of examples we show how biologists, taking advantage of the ability to interbreed closely related species or morphologically distinct populations within a species, have been able to identify genes responsible for microevolutionary changes in stickleback fish and maize. In these cases, the identified genes have turned out to be those that were already known for the roles they play during the normal course of development. We end the chapter with an example of how comparative developmental data from different phyla can even help us to reconstruct the development and morphology of ancestors for which we have little fossil data.

ANTERIOR–POSTERIOR PATTERNING: *Hox* GENE REGULATION OF DEVELOPMENT

Homeotic Mutations Transform One Body Region into Another

William Bateson used the term homeosis in 1894 to describe the transformation of one body part into another (see p. 21). Bateson was looking at naturally occurring aberrations, such as humans in which one cervical (neck) vertebra was transformed into a thoracic vertebra or insects in which one antenna was replaced with a leg. In many of these cases, Bateson was not observing genetic mutants, but anomalies that resulted from environmental pertubations during development. Nevertheless, he concluded that many animals were composed of repeating units that were somehow made distinct from one another during development.

Almost 60 years ago, Edward Lewis started characterizing a series of mutations in *Drosophila melanogaster* that caused one body region to be transformed into another adjacent region. These particular mutations did not change the total number of segments, but rather caused segments to take on different identities. For example, a series of mutations could transform posterior abdominal segments into segments with the characteristics of more anterior abdominal segments. He also studied mutations, in a gene named *Ultrabithorax* (*Ubx*), that could transform the third thoracic segment into the second thoracic segment, resulting in a fly with two second thoracic segments and no third thoracic segment. What made this mutant phenotype very striking was that the second thoracic segment normally possesses the fly's single pair of wings, whereas the third thoracic segment contains just a small pair of balloon-like projections called **halteres**. Thus, the fly with two second thoracic segments possessed two pairs of wings instead of the normal single pair (Fig. 11.1). Lewis also found that these various mutations defined a series of loci positioned next to each other on one of the fly's chromosomes. As a class they were called **homeotic genes** in keeping with Bateson's term homeosis.

Homeotic Genes Are Specifically Expressed in the Body Regions That They Control

Eventually, it was shown that an additional region of the same chromosome harbored a second cluster of homeotic genes responsible for correctly patterning the more anterior segments of the fly. Interestingly, for both clusters, the order of genes along the chromosome correlated with the anterior–posterior (A-P) position of the segments affected by the mutations in the genes. Although these two gene clusters, called the *Bithorax* and *Antennapedia* complexes (*BX-C* and *Antp-C*), were first characterized by **alleles** that produced adults with homeotic transformations, it soon became clear that **null** (i.e., complete loss-of-function) **alleles** for all but one of these genes were recessive embryonic lethal (killing the animal as a larva just before hatching from the egg) and resulted in the homeotic transformation of segments.

In all, there are eight homeotic (*Hox*) genes in *Drosophila*, five in *Antp-C* and three in *BX-C* (Fig. 11.2). To illustrate the general function of homeotic genes, we focus on the phenotypes associated with embryonic lethal alleles of the three *BX-C* genes: *Ultrabithorax* (*Ubx*), *abdominal-A* (*abd-A*), and *abdominal-B* (*Abd-B*). As can be seen in

FIGURE 11.1. Wild-type fly (*top*) and fly homozygous for reduced-function alleles of *Ubx* (*bottom*). In wild-type flies, the second throacic segment bears a pair of wings, whereas the third thoracic segment has a pair of halteres (*inset* shows a magnified view; *arrow* points to a haltere). In the *Ubx* mutant, loss of *Ubx* expression in the third thoracic segment results in a fly with two pairs of wings as the halteres have been transformed into an additional wing pair.

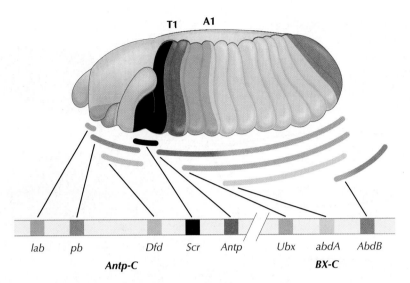

lab   pb   Dfd   Scr   Antp   Ubx   abdA   AbdB

Antp-C   **BX-C**

FIGURE 11.2. Homeotic gene organization in *Drosophila*. *Lower* portion shows a schematic of the fly *Hox* genes and their organization into two clusters, known as *Antp-C* and *BX-C*. The *upper* portion shows a side view of a fly embryo (anterior is to the *left*; ventral is *down*; T1, first thoracic segment; A1, first abdominal segment). The *Hox* genes are expressed in specific domains along the anteroposterior (A-P) axis, and the relative position of each gene on the chromosome is also reflected in the order of expression along the A-P axis, a correlation known as colinearity. The *colored bars* below the embryo give a better representation of the entire expression domain for each gene and the regions of overlapping expression.

Figure 11.3, the loss of any one of the *BX-C* genes causes a particular group of segments to take on the identity of a more anteriorly located segment (e.g., in embryos homozygous for null alleles of *Ubx*, the T3 and A1 segments are transformed into T2 segments). The anterior limit of the transformed segments in the mutant animals corresponds to the anterior boundary of the gene's expression domain in a wild-type animal (Figs. 11.2 and 11.3). Thus, there is a clear correspondence between the spatial order of the *BX-C* genes on the chromosome, their relative anterior boundaries of expression along the anterior–posterior axis of the animal (*Ubx* is expressed more anteriorly than *abd-A*, which in turn is expressed more anteriorly than *Abd-B*), and the relative order of the segments that are transformed by loss-of-function mutations in the corresponding genes (see Figs. 11.2 and 11.3). The same relationship between chromosome position, domain of expression, and affected segments also holds true for the *Antp-C* genes. This three-way correspondence between the relative position of the gene within the complex, its expression domain, and the region affected in the corresponding mutants is known as colinearity and is a striking feature of the homeotic genes.

Inappropriate expression (misexpression) of a homeotic gene outside of its normal domain can create the reverse effect of loss-of-function alleles, which results in body transformations just as striking as those seen in loss-of-function mutations. Misexpression of a homeotic gene anterior to its normal expression domain causes a more anterior segment to adopt the identity of a more posterior segment. A spectacular illustration of this misexpression effect is seen when the *Antennapedia* (*Antp*) gene is misexpressed in the head of a developing *Drosophila* larva (Fig. 11.4). Normally, *Antennapedia* is expressed in the thoracic region of the embryo and larvae and is responsible for the correct specification of the anterior thoracic segments. The original alleles of *Antennapedia*, however, turned out to be regulatory mutations (due to a chromosomal rearrangement) that caused the gene to be expressed in the developing head of the fly larvae. As a result, imaginal disk cells (larval cells that are precursors to adult

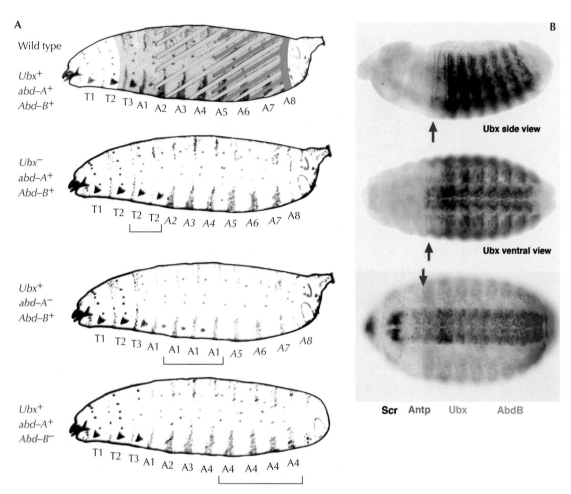

FIGURE 11.3. (*A*) (*Top*) A wild-type larvae with schematic of gene expression patterns (*Ubx, blue; abdA, yellow; AbdB, brown*). Lower larvae illustrate the homeotic transformation in segment identity that occurs in null alleles of *Ubx, abdA*, and *AbdB*. In all cases, the total number of segments is unchanged, but segment identity is altered. (*B*) Antibody staining reveals Hox protein distribution in wild-type embryos (*blue arrows* point to the boundary between T2 and T3). By about mid-embryogenesis (*top* two embryos), Ubx protein is detected in T3 and throughout most of the abdomen. *Bottom* embryo (ventral view) shows the simultaneous detection of four different *Hox* gene protein products (*Scr, black; Antp, red; Ubx, blue; AbdB, brown*). Note the sequential expression of the proteins along the anteroposterior axis of the embryo.

appendages) in the head, which would normally make the antennae, now behave as if they were in the thorax and produce a leg instead.

Hox Genes Contain a Conserved DNA-binding Domain Called the Homeobox

In the early 1980s, the first homeotic genes from *Drosophila* were cloned and analyzed. It was soon discovered that these cloned *Hox* genes could be used as molecular probes to isolate the remaining genetically defined homeotic genes, as well as a number of other genes that played important patterning roles during development. This approach succeeded because these genes all share a highly conserved sequence, called the **homeobox**. This homeobox sequence of 180 base pairs encodes a 60-amino-acid protein domain known as the **homeodomain**, which is the DNA-binding motif of this family of transcription factors (Fig. 11.5).

Given the common developmental role of the eight homeotic genes in patterning the *Drosophila* body axis, it is not surprising that they encode similar proteins. What was unexpected, however, was that similar proteins would be found in species spanning a number of animal phyla. The homeobox-containing fragments of *Drosophila* homeotic genes could also be used to identify genes with similar homeobox-containing sequences from other animals. Within a short period of time, a number of homeobox-containing genes were isolated from organisms quite distantly related to *Drosophila*, including *Xeno-*

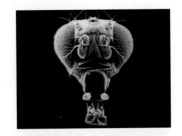

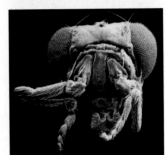

FIGURE 11.4. (*Top*) Wild-type fly head. (*Bottom*) *Antp* mutant fly head showing legs growing from where the antennae should be due to inappropriate expression of *Antp* more anterior than normal.

A

Scr group

Fruit fly	TKRQRTSYTRYQTLELEKEFHFNRYLTRRRRIEIAHALCLTERQIKIWFQNRRMKLKKEH
Grasshopper	TKRQRTSYTRYQTLELEKEFHFNRYLTRRRRIEIAHALCLTERQIKIWFQNRRMKWKKEH
Beach hopper	TKRQRTSYTRYQTLELEKEFHFNRYLTRRRRIEIAHALCLTERQIKIWFQNRRMKWKKEH
Centipede	TKRQRTSYTRYQTLELEKEFHFNRYLTRRRRIEIAHSLCLSERQIKIWFQNRRMKWKKEH
Mite	TKRQRTSYTRYQTLELEKEFHFNRYLTRRRRIEIAHSLCLSERQIKIWFQNRRMKWKKEH
Leech	NKRTRTSYTRHQTLELEKEFHFNRYLSRRRRIEIAHVLNLSERQIKIWFQNRRMKWKKDH
Sea urchin	SKRSRTAYTRYQTLELEKEFHFNRYLTRRRRIEIAHALGLTERQIKIWFQNRRMKWKKEH
Zebra fish	GKRARTAYTRYQTLELEKEFHFNRYLTRRRRIEIAHALCLSERQIKIWFQNRRMKWKKDN
Mouse	GKRARTAYTRYQTLELEKEFHFNRYLTRRRRIEIAHALCLSERQIKIWFQNRRMKWKKDN
Human	GKRARTAYTRYQTLELEKEFHFNRYLTRRRRIEIAHALCLSERQIKIWFQNRRMKWKKDN

Antp group

Fruit fly	RKRGRQTYTRYQTLELEKEFHFNRYLTRRRRIEIAHALCLTERQIKIWFQNRRMKWKKEN
Grasshopper	RKRGRQTYTRYQTLELEKEFHFNRYLTRRRRIEIAHALCLTERQIKIWFQNRRMKWKKEN
Beach hopper	RKRGRQTYTRYQTLELEKEFHFNRYLTRRRRIEIAHALCLTERQIKIWFQNRRMKWKKEN
Centipede	RKRGRQTYTRYQTLELEKEFHFNRYLTRRRRIEIAHALCLTERQIKIWFQNRRMKWKKEN
Spider	RKRGRQTYTRYQTLELEKEFHFNRYLTRRRRIEIAHALCLTERQIKIWFQNRRMKWKKEN
Leech	QKRTRQTYTRYQTLELEKEFYSNRYLTRRRRIEIAHSLALSERQIKIWFQNRRMKWKKEN
Sea urchin	GKRGRQTYTRQQTLELEKEFHFSRYVTRRRRFEIAQSLGLSERQIKIWFQNRRMKWREH
Zebra fish	GRRGRQTYTRYQTLELEKEFHFNRYLTRRRRIEIAHALCLTERQIKIWFQNRRMKWKKEN
Mouse	GRRGRQTYTRYQTLELEKEFHYNRYLTRRRRIEIAHALCLTERQIKIWFQNRRMKWKKES
Human	GRRGRQTYTRYQTLELEKEFHYNRYLTRRRRIEIAHALCLTERQIKIWFQNRRMKWKKES

B

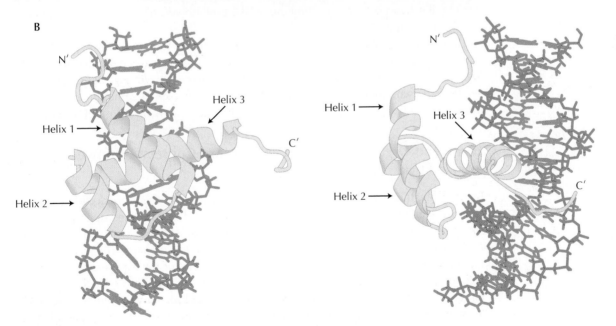

FIGURE 11.5. (*A*) Amino acid sequence alignment of the the *Scr* orthologs and *Antp* orthologs reveals a high level of sequence conservation across a wide range of phyla. Conserved amino acids are *aqua*, unique amino acids for *Scr* family members are *yellow*, and unique amino acids for *Antp* family members are *red*. (*B*) Two views of the structure of the homeodomain (*yellow*) and associated DNA (*red* and *blue*). The unique amino acid sequence of each homeodomain-containing protein allows it to bind to specific DNA sequences, which in turn allows homeodomain-containing proteins to regulate the transcription of target genes that contain these homeodomain-binding sequences.

pus (the South African clawed frog) and humans. It was quickly realized that many orthologous homeobox-containing genes could be characterized and used to define evolutionarily conserved gene families in distantly related animals.

Hox Genes Are Present in All Animals

What is the significance of this conservation during animal evolution? Initially, one interpretation was that there was an underlying commonality to pattern formation in disparate animal phyla. An alternative interpretation was that the presence of these conserved homeobox sequences implied nothing more than that all animals used transcription factors of this class for any of a number of unrelated gene regulation events. We now know that homeobox-containing genes are widespread and serve many different roles in development and gene regulation in all organisms, including both animals and plants. Indeed *Drosophila* has approximately 150 homeobox-containing genes. However, we now have ample evidence that particular homeobox-containing genes do have conserved functions in developmental patterning, even between different phyla. Indeed, the homeotic genes turn out to be a striking example of how the function of some developmental genes has been conserved during animal evolution.

As mentioned earlier, a large number of genes with very close sequence similarity to fly homeotic genes were isolated from a variety of animals. A careful analysis of their sequences, aided by the fact that their chromosomal organization has been remarkably well conserved (at least in those few animals where this can be examined in detail), reveals that an **orthologous** complex, and sometimes multiple paralogous complexes, of genes can be found in almost all animals. We refer to these genes as *Hox* genes (as opposed to the much broader category of all genes containing a homeobox). The common ancestor to deuterostomes (**chordates**, echinoderms, etc.) and protostomes (arthropods, annelids, etc.) probably contained approximately eight *Hox* genes organized in a single cluster. Figure 11.6 illustrates our current thinking on the evolutionary history of the *Hox* complex within these phyla.

In the lineage leading to vertebrates, the *Hox* complex underwent two rounds of duplication, and some of the resulting duplicate copies of individual genes were subsequently lost. In other lineages, degeneration of the complex has occurred (e.g., although the *Hox* genes are present, they are no longer organized in a complex in some tunicates), and in other lineages (e.g., nematodes) several of the *Hox* genes have been lost altogether.

Hox Gene Organization and Function Are Conserved between Animal Phyla

Remarkably, these *Hox* genes are not just well conserved at the sequence level, but also serve similar developmental functions in a variety of distantly related animals. Researchers discovered that the expression patterns of *Hox* genes in vertebrates and *Drosophila* were surprisingly similar. In both cases these genes are expressed in specific regions of the early embryo, and the anterior boundaries of the expression domains are colinear with their position on the chromosome (Fig. 11.7). Mutant mice that contained targeted disruptions for individual *Hox* genes displayed phenotypes remarkably similar to the homeotic transformation phenotypes seen in *Drosophila*. For example, a loss-of-function mutation in the mouse *Hoxb4* gene results in homeotic transformation of the second and third cervical vertebrae into ones that look like the first cervical vertebra (Fig. 11.8A). The duplicated *Hox* genes in vertebrates have partially redundant functions, and, thus, relatively mild phenotypes are often seen for null mutations in any one *Hox* gene, but deletions of the multiple paralogous copies often result in far more striking homeotic transformation (Fig. 11.8B). It is now apparent that in both protostomes and deuterostomes, the *Hox* genes serve a role in regional specification of the body plan.

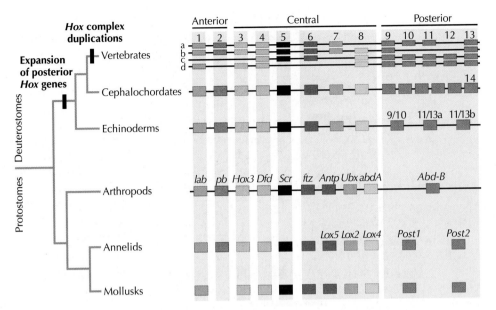

FIGURE 11.6. Evolutionary history of the *Hox* gene complex in metazoan phyla. The common ancestor of the phyla shown here probably possessed at least eight *Hox* genes organized into a single complex. Various expansions of the complex have occurred in multiple lineages. For example, an expansion of the Posterior group appears to have occurred early in deuterostome evolution, and several duplications of the entire cluster occurred during vertebrate evolution, which led to four clusters (a, b, c, and d; note that some *Hox* genes are absent from some of the clusters because of gene loss after duplication) on four different chromosomes. The precise evolutionary relationships for the Posterior class genes and the *Hox6, -7,* and *-8* group genes between phyla are still unresolved (e.g., independent duplications may have led to *Hox7* and *-8* in deuterostomes and *Antp* and *Ubx* in protostomes). The coloring pattern shown here for these genes is an oversimplification.

This suggests that in the common ancestor to deuterostomes and protostomes, *Hox* genes were already involved in specifying the regional organization of the body plan.

This striking level of conservation of *Hox* gene structure and function led to the suggestion of the "zootype" in animal development. The zootype is defined as the stage where all animal embryos display regional expression of their various *Hox* genes. Thus, although embryos of animals from different phyla look quite different morphologically, they look remarkably similar (Fig. 11.7) when viewed as a pattern of *Hox* gene expression. These observations all point toward a very ancient and highly conserved function of *Hox* genes in shaping the body plans of all animals.

Hox GENES ARE SHOWN TO BE INVOLVED IN EVOLUTIONARY CHANGE

Evolution of Two-Winged Flies from Four-Winged Ancestors

If the *Hox* genes show such a remarkable degree of evolutionary conservation in structure and function, why do distantly related animals still look so different from one another? Furthermore, why should we suspect that changes in the *Hox* genes themselves, or changes in the way that they function, contribute to evolutionary changes in body morphology seen during animal evolution? We saw earlier that null alleles of homeotic genes result in homeotic transformations of embryonic structures, which also lead to embryonic lethality. Some alleles of homeotic genes, however, result in transformations that, at least superficially, are similar to the types of changes that have occurred during

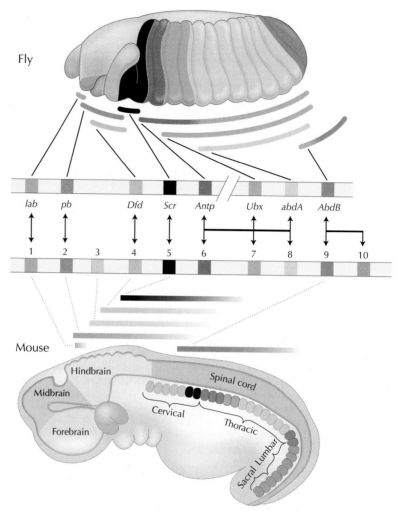

FIGURE 11.7. Comparison of *Drosophila* and mouse *Hox* gene organization and expression. In both species, the *Hox* genes are clustered. As described previously, there are two clusters (complexes) in *Drosophila*. In mice, as in other mammals, the *Hox* complex is in a single, compact cluster with four copies of the cluster in the genome (see Fig. 11.6; each complex is on a different chromosome; only one copy is shown here). In both species, *Hox* genes are expressed along, and control the patterning of, the anteroposterior axis. In mice, expression patterns are seen both in the nervous system (*colored bars* indicate the domains in the brain and spinal cord) and in the somites (*coloring* indicates *Hox* gene expression domains in the somites that will form the cervical, thoracic, and lumbar vertebrae). The *Hox11–13* genes are not shown here, but they are expressed in various patterns of the lumbar and sacral vertebrae.

animal evolution. Recall that *Ubx* plays a role in giving unique identities to the different thoracic segments during embryogenesis. Within these segments are also groups of cells, called imaginal disks, that grow during the larval stages and go on to form the adult appendages of the corresponding adult segments. In *Drosophila* larvae, *Ubx* is normally expressed in the imaginal disks of the third thoracic segment but not in the imaginal disks of the second thoracic segment. The four-winged fly shown in Figure 11.1 is the phenotype that results from a *Ubx* mutation that does not alter early *Ubx* expression but does result in the loss of *Ubx* expression from the larval T3 imaginal disks. The result is a homeotic transformation of the halteres of the third thoracic segment into the wings normally associated with the second thoracic segment, giving rise to a four-winged fly.

But, of course, there are many insects, such as dragonflies, butterflies, and bees, which normally have four wings (Fig. 11.9). Both our knowledge of the phylogenetic relation-

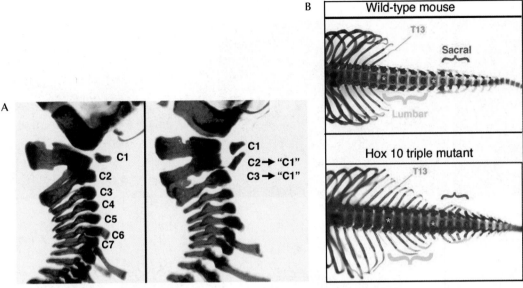

FIGURE 11.8. Mutation of *Hox* genes causes regional transformations in mice. (*A*) (*Left*) Wild-type mouse skeleton with normal first, second, and third cervical vertebrae (C1, C2, C3); (*right*) the skeleton of the mouse lacking *Hoxb4* gene (the *Hox4* gene of the b cluster; see Fig. 11.6) showing transformed cervical vertebrae so that what should be C2 and C3 look more like a normal C1 vertebrae. (*B*) Ventral views of wild-type (*above*) mouse lumbar and sacral vertebrae (the last rib bearing vertebra is thoracic 13, T13; *yellow bracket*, lumbar vertebrae; *green bracket*, sacral vertebrae) and a *Hox10* triple mutant version (*below*; deleted for the *Hox10* gene from clusters a, c, and d; there is no group 10 gene on cluster b). This complete elimination of *Hox10* genes leads to the transformation of the vertebrae of the sacral and lumbar region into rib-bearing thoracic-type vertebrae.

ship of various insect groups and evidence from the fossil record indicate that the ancestor of all these winged insects was one with four wings. During the lineage leading to flies, the wings of the third thoracic segment were modified into halteres. This knowledge, combined with the initial observations on the function of *Ubx* in *Drosophila*, led to three hypotheses to explain the evolutionary transition from four-winged to two-winged insects (Fig. 11.9D–F). The first was put forward by Lewis, who carried out much of the early genetic characterization of the *Bithorax* complex of *Hox* genes. He initially hypothesized that the four-winged common ancestor of flies and dragonflies lacked the *Ubx* gene. Somewhere in the evolutionary lineage leading to flies, *Ubx* appeared and resulted in the transformation of the third thoracic wings into halteres (Fig. 11.9D). At the time, Lewis had little information on the molecular nature of *Ubx*. After the cloning and characterization of the *Drosophila Ubx* gene, along with the discovery that a *Ubx* gene was present throughout the insects, this first hypothesis could be rejected.

A second hypothesis was based largely on the discovery that the alleles of *Ubx* that resulted in adult homeotic transformation were caused by mutations that affected the regulation of *Ubx*. According to this hypothesis, the *Ubx* gene is present in all insects and carries out its many different functions, but in those insects with four wings, *Ubx* expression is regulated so that it is not expressed in the appendage primordia of both the second and third thoracic segments. During the lineage leading to flies, the regulation of *Ubx* was altered such that it was now expressed in the appendage primordia of the third thoracic segment, resulting in the evolutionary appearance of two-winged insects such as flies. Thus it was hypothesized that changes in the regulation of *Ubx* expression might underlie the evolutionary changes in insect wing patterning (Fig. 11.9E).

But when the Ubx expression pattern was examined in a number of different insects, it was found to be highly conserved (Fig. 11.10). For example, in butterflies and moths,

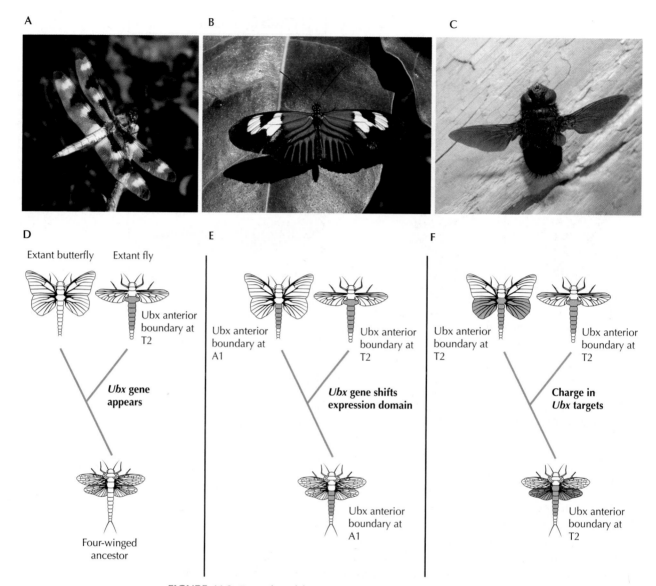

FIGURE 11.9. Examples of four-winged insects include dragonflies (*A*) and butterflies (*B*), whereas flies (*C*) are two-winged insects. (*D–F*) Three hypotheses for the evolution of two-winged insects from a four-winged ancestor. In *D*, there was no *Ubx* gene in the ancestor, and it arose only in the lineage leading to flies. In *E*, the anterior expression boundary of *Ubx* shifted anteriorly from A1 to T3. In *F*, the pattern of *Ubx* did not change, but the targets of *Ubx* regulation did. As explained in the text, the data available suggest that the third hypothesis, shown in *F*, is the most likely explanation.

which are collectively called Lepidoptera and are clearly four-winged insects, Ubx protein is expressed in the wing primordia (imaginal disks) of the third thoracic segment, but not the wing primordia of the second thoracic segment (Fig. 11.11). Indeed, the boundary of Ubx protein expression between the third and second thoracic segments seems well conserved throughout the insects (Fig. 11.10). This point is made especially clear by examining *Ubx* expression in firebrats, which belong to an insect lineage that is very basal in the insect phylogenetic tree (see Fig. 9.17) and that retains the wingless trait associated with the very first insects. In the firebrat, just as in *Drosophila*, *Ubx* is expressed throughout the third thoracic segment, but not throughout the second thoracic segment. Thus, it would appear that *Ubx* has always played a role in distinguishing the third thoracic segment from the second thoracic segment irrespective of the presence or absence of wings, and so the second hypothesis could also be rejected.

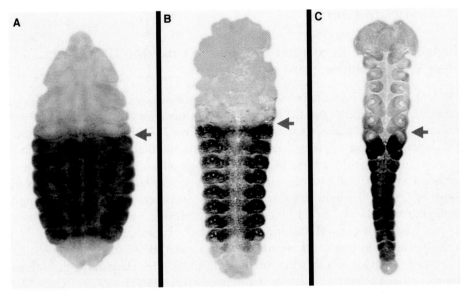

FIGURE 11.10. Ubx protein expression in *Drosophila* (*A*), beetle (*B*), and grasshopper (*C*) embryos. each approximately one-third of the way through embryogenesis. In all three, the anterior boundary of Ubx expression extends through T3 at this time. The *red arrow* indicates the boundary between T2 and T3; all embryos are oriented with anterior up.

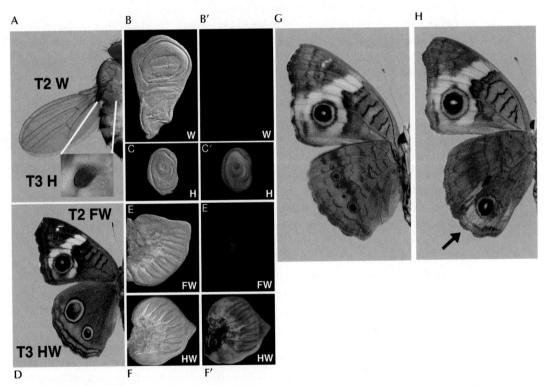

FIGURE 11.11. (*A–F*) Differential Ubx protein expression makes the appendages of the second and third thoracic segments morphologically distinct. In T2 of flies, the wing (T2 W in *A*) derives from the larval wing imaginal disk (*B*), which does not express Ubx (lack of green stain in *B'*). In fly T3, the haltere (T3 H in *A*) derives from the larval haltere disk (*C*), which does express Ubx (green staining in *C'*). In butterflies, the T2 forewing (T2 FW in *D*) derives from the larval forewing disk (*E*), which does not express Ubx (lack of green stain in *E'*). The T3 butterfly hindwing (T3 HW in *D*) derives from the larval hindwing disk (*F*), which does express Ubx (green stain in *F'*). (*G*) Wild-type butterfly (underside of the species *Precis coenia*) shows different coloration patterns between the forewing and hindwing. If *Ubx* expression is removed from small patches of the hindwing during development, the resulting adult wing (*H*) displays regions of the hindwing that now contain patterns normally associated with the forewing. Note the lighter color and presence of the large eyespot in the hindwing (indicated by *arrow*), which are patterns normally seen in the forewing.

Changes in Gene Regulation by *Ubx* Are Linked to Changes in Insect Wing Evolution

If we look closely at butterfly or moth wings, it is obvious that the forewings (from the second thoracic segment) and the hindwings (from the third thoracic segment) are not identical. The overall shape, coloration, and sometimes even the morphology of the individual scales of these two wings are clearly different, and thus it was suggested that *Ubx* has a role in distinguishing these two types of wings. Subsequent manipulation of Ubx expression in studies of butterfly wings showed that removal of *Ubx* expression from groups of cells in the hindwing during pupal development resulted in forewing-like patches in the hindwing (Fig. 11.11G,H), and misexpression of *Ubx* in groups of cells in the forewing during pupal development resulted in hindwing-like patches in the forewing. Thus, *Ubx* makes the appendage of the second thoracic segment—whether it is a butterfly forewing or a fly wing—different from the appendage of the third thoracic segment, either a butterfly hindwing or a fly haltere (Fig. 11.11).

Given that *Ubx* expression appeared to be the same in butterflies and flies, those trying to understand the developmental difference between halteres and hindwings turned their attention to the genes "downstream" of *Ubx*. Because *Ubx* is a transcription factor, these "downstream" genes are those whose transcription is controlled by the DNA-binding activity of Ubx protein. In the case of flies, the wing and haltere differ in a number of morphological criteria. For example, the wings have veins and sensory hairs, whereas the halteres do not. Furthermore, the wing is far larger and contains a greater number of cells than does a haltere. The genetic pathways that make veins and sensory hairs and regulate cell proliferation causing growth of the wing disk have been identified in the course of genetic studies of wing formation. It was then shown that *Ubx* regulated one or more steps in each of these pathways. Thus, the presence or absence of *Ubx* expression in fly haltere and wing disks leads to the inactivation or activation of specific downstream genetic pathways that lead to the morphological differences between adult wings and halteres.

This has led to a third, and currently favored, hypothesis to explain the appearance of two-winged insects (Figs. 11.9F and 11.12). According to this hypothesis, *Ubx* has maintained its differential expression between forewing and hindwing disks, but what has changed during evolution is the way in which various genes respond to *Ubx* expression. The pathways resulting in vein and sensory hair formation appear to be activated in both the forewing and hindwing disks of butterflies, despite the presence of *Ubx* in the hindwing disk. This was presumably also the case in the common four-winged ancestor of both flies and butterflies. During the evolution of two-winged insects, such as flies, various genes involved in processes such as vein formation, sensory hair growth, and cell proliferation came under the control of *Ubx*. We would expect that these same genes in butterflies are not regulated by *Ubx* (at least not in the same way). On the other hand, genes controlling the coloration and scale morphology differences between the hindwing and forewing are regulated by *Ubx* in Lepidoptera. Presumably then, the acquisition of Ubx-binding sites in genes of particular genetic pathways is the evolutionary basis for the transition from four-winged insects to two-winged insects (Fig. 11.12). In this case, then, evolution has not acted on the *Ubx* gene itself to bring about morphological change, but knowledge of how *Ubx* functions has allowed us to suggest that evolutionary gains and losses of Ubx-binding sites in target genes have played a major role in the evolutionary changes in insect wing morphology.

Changes in Gene Regulation by *Ubx* Are Linked to Changes in Larval Appendage Evolution

Another clear example of the changing role of *Ubx* function during evolution comes from the analysis of appendage formation during insect embryogenesis. Insects that re-

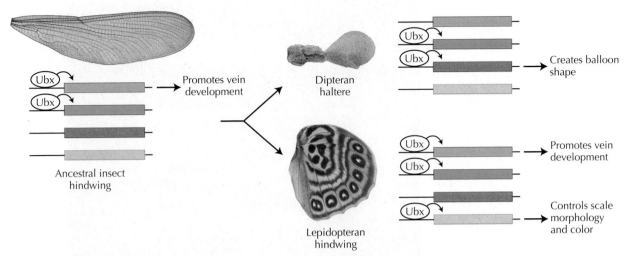

FIGURE 11.12. Evolutionary changes in the targets of *Ubx* regulation have led to the diversification of insect hindwings and halteres (T3 appendages). Given that *Ubx* is expressed in similar spatial domains in the primordial of various hindwings and halteres, the diversification of these structures would appear to involve changes in the genes that are regulated by *Ubx* (through the binding of Ubx protein [*ovals*] to the enhancers of these genes). For example, *Ubx* may have promoted vein development in the ancestral hindwing and continues to do so in extant Lepidoptera, but no longer does so in the veinless halteres of diptera. On the other hand, a novel function of *Ubx* in regulating scale and color patterning appears to have evolved in the lepidopteran lineage; similarly, function in creating the balloon-like shape of the haltere may have emerged during the evolution of diptera.

tain many of the ancestral characteristics of development form legs during embryogenesis and hatch with functional legs for locomotion (e.g., grasshoppers). Although *Drosophila* embryos do not form obvious legs during embryogenesis, various morphological and molecular markers reveal that remnants of larval legs are present during embryogenesis. One particularly useful molecular marker in this regard is a gene called *Distal-less*, which is expressed in all manner of arthropod appendages, including legs, antennae, and most jaw parts. *Distal-less* was first genetically defined from hypomorphic alleles (mutations that reduce, but do not eliminate, gene function) that cause stunted legs. It was subsequently shown to encode a transcription factor involved in the establishment and patterning of appendages as well as a number of other developmental processes. During *Drosophila* embryogenesis, distinct spots of *Distal-less* expression appear in the head and thorax, marking the position of the embryonic appendage primordia in these segments. No *Distal-less* expression is visible in the abdomen. It has been shown that this lack of *Distal-less* expression in the abdomen is due to the direct repression of *Distal-less* transcription by both the *Ubx* and *abd-A* homeotic gene products (Fig. 11.13). At first this appears to be contradictory to the previously discussed role of *Ubx* in distinguishing T3 from T2. The solution to this contradiction lies in the temporal complexity of *Hox* gene expression. The initial anterior expression boundary for *Ubx* is at A1, and then later moves anteriorly to T3. Initially, *Ubx* acts to distinguish T3 from A1, and it is at this time that *Ubx* represses *Dll*. Subsequently, *Ubx* expression will move forward to T3, and it is at this later time in development that *Ubx* acts to distinguish T3 from T2. By the time *Ubx* expression begins in T3, *Dll* expression is no longer regulated by *Ubx*.

Like most insects, adult flies lack abdominal legs. However, an examination of the embryos of many other insects, including those of butterflies, moths, beetles, grasshoppers, and crickets, reveals that there is an appendage, called a pleuropod, that develops in the first abdominal segment of these insect species. The pleuropod starts off looking like a smaller version of a thoracic leg, but then it develops a distinctly dif-

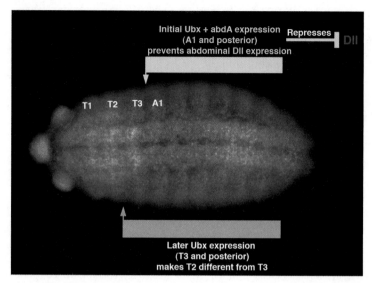

FIGURE 11.13. *Ubx* and *abd-A* regulate *Distal-less* (*Dll*) expression during *Drosophila* embryogenesis. Dll protein (*red*) is expressed in specific cells in the head and thorax that mark the appendages of these segments, but Dll is not expressed in the abdomen. Although *Ubx* is eventually expressed in T3 and the abdomen (*blue bar*), initial expression of *Ubx* is restricted to the abdomen (*yellow bar*), and this initial domain of *Ubx* expression plays a role, along with *abdA*, in keeping transcription of the *Dll* gene off in the abdomen. In *Ubx* mutants, Dll expression expands to include A1, and in *Ubx/abdB* double mutants, expression of Dll expands to include A1–A8. Thus, the early domains of expression of *Ubx* and *abdA* are essential to prevent Dll expression and appendage development in the abdomen. Later in development, *Dll* is no longer regulated by *Ubx* or *abdA*, and the expansion of the *Ubx* domain to include T3 allows *Ubx* to help establish the distinctions between T2 and T3.

ferent morphology. This appendage is not used for locomotion; instead it helps the animal to hatch from the egg and is then left behind. Nevertheless, clearly the pleuropod is an appendage and it does express *Distal-less* just like the appendages of the thorax. What is surprising, however, is that *Distal-less* expression occurs in this appendage even though Ubx protein is present throughout the cells of this segment. One possible explanation for this is that, in these insects, *abd-A* is able to repress *Distal-less*, but *Ubx* is not (Fig. 11.14).

This hypothesis is supported by looking at the effects of removing *Ubx*, *abd-A*, or both *abd-A* and *Ubx* from the developing embryos of the beetle *Tribolium castaneum*. Without *Ubx*, the pleuropod of the first abdominal segment is transformed into a thoracic leg. Without *abd-A*, all the abdominal segments develop pleuropods, and without both *Ubx* and *abd-A*, thoracic legs develop all along the abdomen (Fig. 11.15). This indicates that *Ubx* does not repress *Distal-less* in *Tribolium*, but does modify the type of appendage that appears in the first abdominal segment. *abd-A*, however, just as in *Drosophila*, does repress *Distal-less* and is responsible for repressing limb formation posteriorly from the second abdominal segment. It may be that the *Distal-less* enhancer has gained a binding site for *Ubx* in the lineage leading to the Diptera. This new binding site allows *Ubx* to repress *Distal-less* in Diptera, something that does not appear to occur in insects such as beetles, which have retained the ancestral character of possessing pleuropods.

There is, however, some indication that evolutionary changes in the Ubx protein itself may be at work here. When the Ubx protein of *Onychophora*, a lobe-limbed arthropod very distantly related to *Drosophila*, is expressed in *Drosophila*, it can carry out some functions of *Drosophila* Ubx, such as repressing wing development in the third thoracic segment, but is unable to repress *Distal-less* expression during embryogenesis. Likewise, the Ubx protein of the crustacean, *Artemia salina*, is also unable to

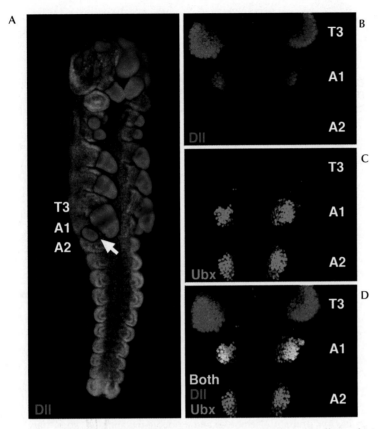

FIGURE 11.14. In grasshopper embryos Distal-less protein (*red* staining in all panels) is expressed in the appendages of the head, thorax, and the first abdominal segment. The *white arrow* in *A* points to the Dll-expressing pleuropod appendage, which is part of the A1 segment. As in *Drosophila*, *Ubx* in grasshopper embryos is expressed throughout the entire abdomen during the time that appendages are first formed. Fluorescent double labeling for Dll protein (*red* in *B* and *D*) and Ubx protein (*green* in *C* and *D*) shows that *Dll* expression in the A1 appendage occurs despite the expression of Ubx in these same cellls (overlap of *green* and *red* creates the *yellow* color seen in *D*), and thus *Ubx* does not appear to repress *Dll* in grasshoppers.

repress *Drosophila Distal-less* expression. Therefore, the ability of *Ubx* to repress *Distal-less* in Diptera may reflect an evolutionary change in the Ubx protein instead of in the *Distal-less* enhancer.

At least within insects, the expression pattern of *Ubx* along the anteroposterior axis is relatively well conserved. Evolutionary changes in morphology instead seem to occur because downstream genes respond to *Ubx* differently in different species. At least some of this evolutionary change may be due to alterations in the enhancer sequences of direct targets of the Ubx protein, whereas other changes in downstream responses may be due to changes within the Ubx protein itself.

Alterations in *Ubx* Regulation May Have a Role in the Macroevolution of Crustacean Morphology

Crustaceans, which include animals such as lobsters, crabs, brine shrimp, barnacles, and copepods, are closely related to the insects. But in these arthropods, unlike in insects, the A-P boundary of *Ubx* expression varies significantly between species and, more importantly, this variation is well correlated with evolutionary changes in body plans. Crustacean species all have the same basic pattern of head segments and head appendages. Just anterior to the thorax, crustaceans have three pairs of appendages (one pair per segment) that form the jaw apparatus. The segments are known as the mandibular, maxillary 1, and

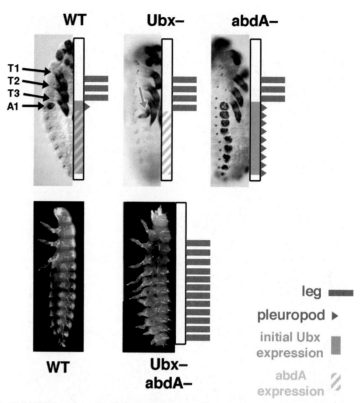

FIGURE 11.15. Wild-type beetle (*Tribolium castaneum*) embryos (*above*) and larvae (*below*) and the phenotypes in *Ubx*⁻ and *abdA*⁻ embryos and a double-mutant *Ubx*⁻ *abdA*⁻ larva. The embryos are labeled for Distal-less protein to highlight the appendages. The adjacent schematics show the number of legs and pleuropods along with the expression domains of *Ubx* and *abdA*. In wild-type embryos, Distal-less-expressing legs are present on T1–T3, whereas a Distal-less-expressing pleuropod is present on A1. Elimination of *Ubx* transforms the pleuropod into a leg, whereas the elimination of *abdA* leads to the formation of pleuropods on the abdominal segments. Elimination of both *Ubx* and *abdA* creates legs along the length of the abdomen. These results indicate that *abdA* represses *Dll* and appendage development, whereas *Ubx* does not repress *Dll* but instead initially distinguishes the leg-bearing T3 segment from the pleuropod-bearing A1 segment. Later in development, *Ubx* expression will expand to include T3 and then function to distinguish T2 from T3.

maxillary 2 segments. The appendages of these segments are generally compact, held near the mouth opening, and specialized for feeding. Crustaceans, however, do vary in the number of thoracic segments, from as few as 6 in copepods, to 8 in lobsters, and 11 in brine shrimp. In any given species, most of the thoracic appendages are morphologically distinct from the jaw appendages and usually function in locomotion. For example, in the brine shrimp (*Artemia salina*), all 11 thoracic appendages are morphologically similar swimming appendages. In *Artemia*, Ubx protein expression is seen in the first thoracic segment (T1) and extends posteriorly throughout the entire thorax. This same anterior boundary, beginning at T1, is seen in all crustacean species that possess more or less identical appendages throughout the thorax (Fig. 11.16).

In some crustaceans, however, the anteriormost thoracic limbs have a morphology that is more reminiscent of the jaw appendages of the head (mouth segments). These modified thoracic appendages are termed maxillipeds ("jaw feet"). Not only are these appendages morphologically more similar to jaw appendages than the rest of the thoracic appendages, but they are also generally used for feeding instead of locomotion. The number of maxillipeds also varies between species. For example, mysid shrimp possess one pair of maxillipeds (on T1), whereas cleaner shrimp have three pairs of maxillipeds (on T1–T3). Remarkably, the A-P expression boundary of *Ubx*, beginning at stages well before thoracic limbs are morphologically visible and continuing through embryogenesis, predicts where along the A-P axis the transition from maxillipeds to lo-

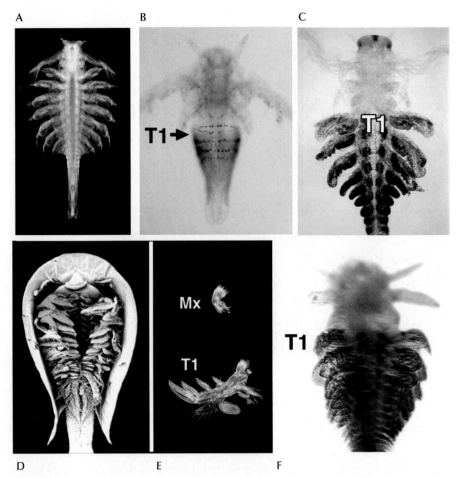

FIGURE 11.16. In crustaceans, proteins such as *Artemia* (A) and *Triops* (B), that have swimming appendages on all thoracic segments, Ubx expression begins in the T1 segment. *Artemia* (A) have 11 pairs of feathery thoracic swimming appendages, and during development Ubx expression (*brown* staining in B; *black* staining reveals the boundary between adjacent segments) begins in T1. Later, when the limbs are visible, Ubx expression (*black* staining in C) continues in all thoracic segments and appendages. *Triops* (D) have around 60 pairs of thoracic appendages (the T1 appendage on one side is colored *red* and T2 is colored *green*). These thoracic appendages are all very similar (isolated T1 appendage shown in E) and are clearly different from the jaw appendages (the jaw appendage of the maxillary [Mx] segment is also shown in E). Ubx expression (*black* staining in F) is present in all thoracic segments and appendages, but absent in all the head segments (just anterior to T1).

comotory thoracic appendages occurs. In mysids, *Ubx* has its anterior expression boundary at T2, whereas in cleaner shrimp embryos, the anterior boundary is at T4 (Fig. 11.17). Various lines of evidence suggest that the ancestral crustacean had a body plan like *Artemia*—one in which there were no maxillipeds. Thus, it would appear that the boundary of *Ubx* expression has shifted posteriorly during the evolution of several crustacean lineages. Not only does this shift correlate with the apparently independent appearance of maxillipeds in copepods and malacostracan crustaceans (Fig. 11.18) and to the subsequent appearance of additional maxillipeds within the malacostracans, but the genetic data from model systems, such as *Drosophila* and mice, suggest that shifting the pattern of *Ubx* expression could have a causal role in the evolutionary change in morphology. Recall that in *Drosophila* expression loss of a *Bithorax* complex gene transforms segments at the anterior of the normal wild-type expression domain into more anterior segments. Thus, what occurs as a mutant phenotype in *Drosophila* may be seen as an evolutionary change in the crustacean lineage, although the changes that have occurred during crustacean evolution involve changes in the expression domain of *Ubx* and not the total loss of *Ubx* function.

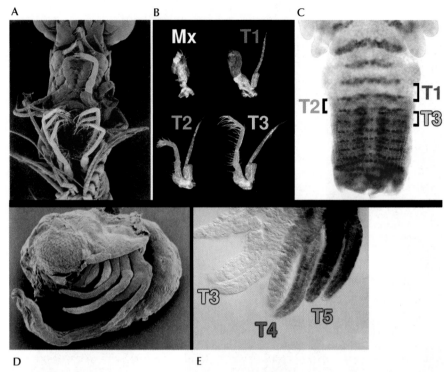

FIGURE 11.17. In adult mysid shrimp (scanning electron micrograph in *A*, and dissected appendages in *B*), the T1 appendage (*red*) is a maxilliped and morphologically similar to the more anteriorly located jaw appendages, such as the maxillary appendage (Mx). T3 and more posterior thoracic appendages (*yellow, pink,* and *orange* in *A*) are all morphologically similar to each other and used for swimming, whereas the T2 appendage (*green*) is morphologically intermediate between T1 and T3. During mysid embryogenesis (*C*), Ubx protein is not expressed in T1 and more anterior segments, is expressed at low levels in T2, and is expressed at high levels in T3 and more posteriorly. In cleaner shrimp (scanning electron micrograph of hatchling shown in *D*), the first three thoracic appendages are maxillipeds (T1, T2, and T3 are colored *red, green,* and *yellow,* respectively, in *D*). During embryogenesis (*E*), Ubx expression is absent from T3 and all segments more anteriorly, is weakly expressed in T4, and is strongly expressed in T5 and more posteriorly.

Clearly the developmental role of *Ubx* is not identical in *Drosophila* and crustaceans. For example, we have already discussed that one of the functions of *Drosophila Ubx* is to repress expression of the appendage patterning *Distal-less* gene, but this is not true for crustacean *Ubx*, possibly because of differences in the structure of the Ubx protein itself. Nevertheless, the general nature of regional transformations resulting from altered *Hox* gene expression in *Drosophila*, mice, and various other organisms can help us predict the functional consequences of altering *Ubx* expression in crustaceans. The loss of *Ubx* expression from the first thoracic segment of an *Artemia*-like ancestor would presumably result in the production of an animal with one pair of maxillipeds through the transformation of the locomotory T1 appendage into one characteristic of the jaws located just anteriorly in the head. It should be noted that our understanding of *Ubx* regulation in *Drosophila* suggests that such shifts are indeed possible through mutations in the regulatory elements of *Ubx*. The regulatory region of *Ubx* spans more than 100 kb in *Drosophila* and is composed of modular elements that regulate *Ubx* expression in specific segments, and apparently even specific cells within a segment, at precise times in development. Indeed the famous four-winged *Drosophila* is the result of combining two regulatory mutant alleles of *Ubx*, resulting in loss of *Ubx* expression only in the larvae (not during embryogenesis) and only from specific portions of the haltere disk. Furthermore, these proposed shifts in *Ubx* expression need not have occurred instantaneously. Because of the highly refined nature of *Ubx* regulation, *Ubx* expression in crustaceans could have been altered gradually over evolutionary time through the gradual

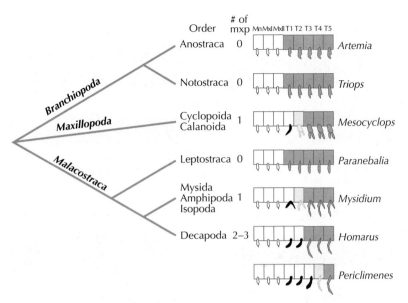

FIGURE 11.18. Phylogenetic distribution of different patterns of segmental specialization and *Ubx* expression among various crustaceans. *Dark blue,* high Ubx protein expression; *light blue,* weaker Ubx expression; *white,* no Ubx expression during early embryonic development; *black,* maxillipeds. For each organism, three mouth segments are followed by the first five thoracic segments. The number of maxillipeds (mxp) is given just to the right of the family names. The early embryonic pattern of *Ubx* expression correlates with the morphology of the appendages. *Ubx* is expressed in segments that will produce locomotory legs of the thorax but not in maxillipeds of the thorax. The phylogenetic distribution suggests that maxillipeds may have evolved more than once, but each change in maxilliped numbers correlated with shifts in the anterior boundary of *Ubx* expression.

accumulation of regulatory mutations, yielding a series of subtle limb variants for selection to act upon, rather than the sudden appearance of a "hopeful monster." Indeed, the T2 limb of the mysid shows exactly what such a limb might look like (Fig. 11.17). The mysid T2 limb is not classified as a maxilliped, but it clearly possesses a morphology intermediate between the T1 maxilliped and the T3 swimming leg (Fig. 11.17). During development, the T2 segment expresses Ubx protein, but at a lower level than the T3–T8 segments. Furthermore, as the T2 limb grows, it shows a mosaic appearance of Ubx expression. The proximal part of the limb does express Ubx, whereas the distal part does not (Fig. 11.17). This correlates with the fact that the proximal part of the T2 limb is more like the T3 limb, whereas the distal part is more like the T1 limb.

Although the data from crustaceans are quite compelling, it will be important to experimentally show that perturbations of *Ubx* expression in these arthropods are actually capable of causing the types of homeotic transformations postulated in these studies. Also, these types of macroevolutionary studies do not directly address whether the change in *Ubx* expression is really caused by changes in the *Ubx* regulatory elements (*cis* changes) or by changes in any number of upstream factors (*trans* changes) that regulate *Ubx* expression. Microevolutionary studies (i.e., genetic experiments using closely related species that can still be interbred in the lab), on the other hand, can overcome these limitations, and one such study is presented in the next section.

Changes in *Ubx* Have a Role in the Microevolution of Fly Bristle Patterns

Although loss of *Ubx* function is lethal, and many well-known *Ubx* alleles cause very obvious morphological changes, other alleles can create remarkably subtle phenotypes that would be missed by someone not familiar with the details of *Drosophila* morphology. *Ubx* plays a major role in body patterning early in development and in imag-

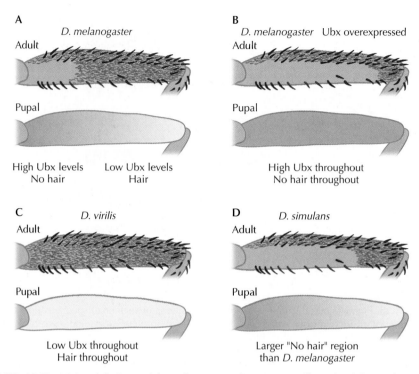

FIGURE 11.19. (A) in adult *Drosophila melanogaster*, there is a small patch of the T2 femur that is devoid of hairs (bristles), and this hairless (naked) region corresponds to the region of the femur that expresses high levels of Ubx during pupal development. (B) If high levels of Ubx expression are experimentally generated in the *D. melanogaster* T2 leg during the pupal stage, the hairless region in the adult T2 leg is greatly expanded. (C) In *Drosophila virilis* adults, there is no naked region on the T2 femur, and *Ubx* is expressed at low levels in the pupal T2 femur. (D) In *Drosophila simulans*, the naked region of the T2 femur is larger than in *D. melanogaster*. Currently available techniques for detecting Ubx protein are not sufficiently accurate to measure differences in expression between the pupal legs of *D. simulans* and *D. melanogaster* (hence the similarity in the *Ubx* profiles shown in *A* and *D*), but genetic experiments suggest that there are functionally significant differences in the level of *Ubx* expression between the two species.

inal disk development during larval life; however, it also functions to regulate additional fine patterning processes. For example, the expression of *Ubx* in particular cells late in imaginal disk development controls the positioning and fate of specific neural structures in the halteres. Thus *Ubx* is both a "macromanager" and a "micromanager."

A relatively subtle role of *Ubx* is to control the pattern and density of bristles (fine hairs) on the femur of the T2 leg (Fig. 11.19). Although Ubx is not expressed in the T2 wing imaginal disk during larval development, Ubx expression is found later in the femur of the T2 leg during pupal development and acts to repress bristle formation. A graded expression pattern of Ubx protein, with the highest levels seen at the proximal base of the femur, correlates roughly with the pattern of bristles seen later on the adult leg of *Drosophila melanogaster*—a "naked" area free of bristles is located at the proximal end of the femur where Ubx expression is highest and bristles are found in the areas where Ubx expression is low (Fig. 11.19A). Experimental manipulations that create high levels of Ubx expression at this stage lead to an expansion of the naked area, whereas lowering the levels of Ubx expression decreases the size of the naked region (Fig. 11.19B).

As it turns out, the size of this naked region also varies between different closely related *Drosophila* species. In *Drosophila virilis*, the naked region is absent, and this is correlated with a lowered level of Ubx expression in the T2 pupal leg of this species relative to *D. melanogaster*. In *Drosophila simulans*, the naked region is expanded (Fig. 11.19D). The currently available techniques lack the quantitative accuracy to detect significant differences in Ubx protein expression between these two *Drosophila* species, but because *D.*

melanogaster and *D. simulans* can be mated to produce hybrid progeny (albeit sterile progeny), it is possible to carry out a genetic analysis of the contribution of *Ubx* to this phenotypic difference. Using such an approach, it has been shown that it is the *Ubx* locus that does indeed contribute significantly to this evolutionary difference between the two fly species. Given that the Ubx-protein-coding regions are identical between these two flies, it is assumed that the difference must lie in regulatory elements controlling the pattern and levels of *Ubx* expression in each species. Thus, variation in *Ubx* clearly has a role in generating the morphological differences seen between closely related species.

Ubx Variation within Populations Contributes to Morphological Evolution

One additional observation underscores the potential role of *Ubx*, and other such regulators of development, in evolutionary change. Genetic variation within the population must be present, and this variation must be able to manifest itself in some way. If all flies always carried identical copies of the *Ubx* locus, then there would be nothing for selection to act on. Researchers relied on a classic embryological perturbation to show that there are various functionally distinguishable *Ubx* alleles with fly populations, and they can be acted on by selection. Conrad Waddington found that treating *Drosophila* embryos with ether created a few adult flies with slightly enlarged halteres. If these flies were selected and bred together and then their embryos were subjected to ether and again selected for progeny with enlarged halteres, the frequency and severity of the phenotype could be increased. What was remarkable, however, was that eventually the flies would display this phenotype even when not treated with ether. This suggested that the ether treatment was revealing some underlying genetic variation, which increased the sensitivity of flies to the ether treatment. Through selection and breeding, these various alleles could be combined until a point was reached when the phenotype would manifest itself in a significant number of flies without the perturbations caused by the ether (and to this day there is not a definitive explanation for why ether results in this phenotype). The trait of having enlarged halteres mimics the phenotype seen in some alleles of *Ubx*. Is it possible that one of the alleles being selected for was one or more allelic variants of *Ubx* present in *Drosophila*? Through various mapping techniques it was shown that this was the case. A significant fraction of the phenotypic difference could be mapped to the *Ubx* locus. Thus variation does exist at the *Ubx* locus, and at least some of these variants can manifest themselves phenotypically under the appropriate conditions.

Although the *Hox* genes provide an excellent example of how individual genes, or at least very specific pathways, can play an important role in morphological evolution, we do not want to leave the impression that they explain all, or even most, of the morphological evolution of body plans. Although certain apsects of *Hox* gene function may make them well poised to play a role in morphological evolution, there are, no doubt, thousands of genes that contribute to morphological diversity. The next several examples from sticklebacks and maize reveal additional genes that contribute to morphological evolution, and although they primarily focus on transcription factors, alterations at all levels of developmental pathways probably can and do play roles in evolution.

SKELETAL EVOLUTION IN STICKLEBACKS

Sticklebacks Vary in Their Pelvic Skeletal Anatomy

Sticklebacks are a common and abundant group of fish with a wide distribution in both freshwater and marine habitats; they have been well studied by fishery biologists. They also possess remarkable behaviors, especially during courtship and reproduction, and these courtship behaviors were the subject of study by the Nobel Prize–winning behaviorist, Nikolaas Tinbergen.

Stickleback fish have undergone a particularly impressive and rapid evolutionary radiation. The present-day populations of sticklebacks arose from a common ancestral population that lived exclusively in the ocean approximately 15,000 years ago. With the end of the last ice age, this ancestral population found itself fragmented into many isolated populations, including many in inland seas that became freshwater lakes. Sticklebacks today live in both deep and shallow freshwater lakes as well as various ocean environments.

Having been separated for some 15,000 years and adapted to varied environments, different populations of sticklebacks have evolved many morphological and behavioral variations. For example, the three-spine stickleback, *Gasterosteus aculeatus*, is one of the recognized species of sticklebacks, and it gets its name from the three prominent spines on its back. The marine populations of this species also have an especially prominent pair of pelvic spines projecting from a well-developed pelvic skeleton. In contrast, some of the freshwater populations of three-spine sticklebacks, especially those that live at the bottom of shallow lakes (called benthic fish), have reduced or completely absent pelvic spines and pelvic skeletons (Fig. 11.20).

What are the functions of these pelvic spines and why do they occur in some populations but not in others? It is believed that the pelvic structure of the marine sticklebacks protects the fish from certain types of predators encountered in the relatively open ocean environment. The pelvic girdle and spines effectively enlarge the fish, making them too big to fit into the mouths of some would-be predators. In addition, the spines pierce the soft mouths of predators that do manage to grab them, further discouraging preda-

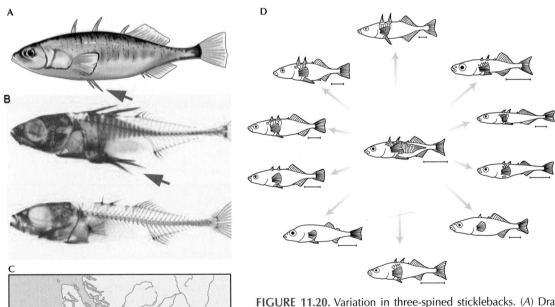

FIGURE 11.20. Variation in three-spined sticklebacks. (*A*) Drawing of a typical three-spined stickleback, which gets its name from the three spines on its back. This specimen also possesses pelvic spines (*red arrow*). (*B*) Skeleton preparation showing a marine stickleback (*above*) with a pelvis and pelvic spines (*red arrow*), but a lack of these structures in the benthic freshwater sticklebacks of Paxton Lake (*below*). (*C*) Map showing the location of Paxton Lake (*red dot*). (*D*) Over a relatively short period of time, an ancestral population of marine sticklebacks with pelvic spines (*center* fish) diversified into many morphologically distinct populations of freshwater fish with or without pelvic spines (*surrounding* fish).

tion. There are at least two possible explanations for why many freshwater stickleback populations lack these pelvic spines. First, less calcium is available in these lakes for the production of skeletal material. Second, although the lakes inhabited by many of these stickleback populations contain few predatory fish, dragonfly larvae prey on the stickle-backs swimming at the bottom of the lake, and the presence of pelvic spines makes it easier for the dragonfly larvae to grasp their prey.

Although the adaptive advantages of these variations have been extensively studied, scientists have long been intrigued as to the developmental basis of these morphological variations. Only recently have techniques been developed that allow us to explore the molecular and genetic bases for some of the variations in morphology. We discuss these in the next section.

Genetic Polymorphisms Underlie Stickleback Skeletal Evolution

Fifteen thousand years is actually a remarkably short period of evolutionary time. Although the morphologically distinct populations of fish are isolated and so have no opportunity to interbreed, they will do so if brought together in the laboratory. This ability to interbreed sticklebacks and analyze the traits in the progeny allows us to associate specific chromosomal regions (and ultimately individual genes) with the traits being studied. This technique, called **quantitative trait locus** (QTL) analysis, depends on genetic recombination to re-sort alleles during the formation of sperm and eggs, but to do so in a way that is dependent on the distance between genes. The technique itself is described in much greater detail in Chapter 14. One essential step in QTL analysis is the creation of a genetic map. For sticklebacks, this was created initially using numerous **microsatellite** markers (see Box 13.3 for more information on microsatellites).

In one experiment, a female marine three-spine stickleback (with the typical well-developed pelvis and pelvic spines) was crossed with a male fish from a benthic population from Paxton Lake, Canada (these fish have reduced spines and pelvises; Fig. 11.20B). The resulting progeny (i.e., the F_1 generation) had pelvises and spines that looked like those of the marine mother (Fig. 11.21). When two of the F_1 progeny were mated to each other, it was found that 75% of the F_2 progeny showed complete pelvises, but 25% of the F_2 progeny had at least some measurable change in the pelvis (size reduction, loss, or some asymmetry in the shape of the pelvis) (Fig. 11.21). The occurrence of the Mendelian 3:1 ratio suggests that a single locus plays a substantial role in the variation of this trait. Indeed, QTL mapping revealed that 13.5–43.7% of all the measured pelvic parameters (such as the length of the spines or the length of the underlying pelvic girdle) showed a strong linkage to a single chromosomal region (Fig. 11.22). To further refine this interval down to the level of a single gene would require very extensive crosses and molecular analyses. A more rapid way to identify the candidate gene came from something already known about the genetics of pelvis development in mice.

The *Pitx1* Gene Has a Role in the Evolution of Stickleback Pelvic Morphology

Developmental biologists have found two genes, *Pitx1* and *Tbx4*, that are specifically expressed in the developing hindlimbs, but not forelimbs, of mouse embryos. Mice missing the *Pitx1* gene develop asymmetric hindlimbs with the right femur shorter than the left (Fig. 11.23). Remarkably, pelvic reduction in sticklebacks is also asymmetric, with the right side of the pelvis often smaller than the left one. The phenotype in mice is further enhanced by also mutating the gene *Pitx2*, which is closely related to *Pitx1*. *Tbx4* is also involved in mouse hindlimb development and appears to be downstream of *Pitx1* in the genetic hierarchy controlling this structure (Fig. 11.23). Thus these three genes were regarded as candidates for the control of pelvic traits in the three-spine sticklebacks.

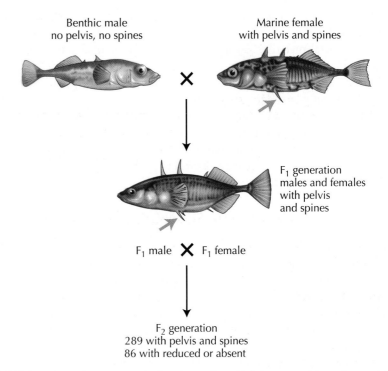

FIGURE 11.21. Mating of a Paxton Lake benthic male, which has no pelvis and no pelvic spines, with a marine female, which does have a pelvis and pelvic spines, results in male and female F_1 progeny that do have a pelvis and pelvic spines. Mating these F_1s to each other produces an F_2 generation in which about 3/4 of the fish have a complete pelvis and pelvic spines, but the remaining 1/4 have at least some measurable reduction of the pelvis and pelvic spines.

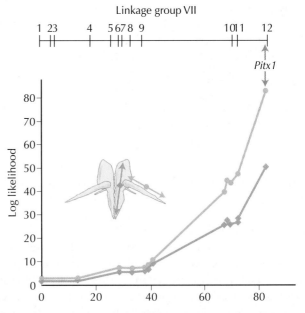

FIGURE 11.22. QTL mapping showing strong linkage of measured stickleback pelvic parameters to a single chromosomal region. Inset shows diagram of the pelvis and the measurements that were made of the pelvis (*blue line*) and the pelvic spines (*green line*). Mapping these traits onto the genetic map shows a strong linkage for both traits to the chromosome position of the *Pitx1* gene.

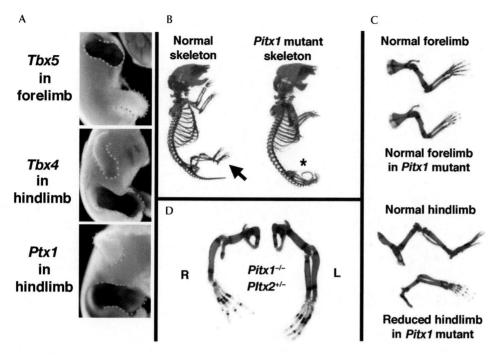

FIGURE 11.23. Molecular distinctions between forelimbs and hindlimbs. (*A*) *Tbx5* is expressed (*blue* staining) in the mouse forelimb (outlined in *pink*), but not the hindlimb (outlined in *green*). The reverse is true for *Tbx4* and *Pitx1*. (*B*) On the left is a wild-type mouse skeleton and on the right is a *Pitx1* mutant mouse skeleton (position of the hindlimb indicated by the *arrow* in the wild-type and asterisk in the mutant). (*C*) Isolated forelimb and hindlimb bones from wild-type and *Pitx1* mutant mice. Deletion of the *Pitx1* has no effect on the forelimbs but results in the reduction of the size of the hindlimb. (*D*) The effect of *Pitx1* deletion is left–right asymmetric, and the effect is further enhanced by also removing one copy of the closely related *Pitx2* gene. In the *Pitx1*$^{-/-}$ *Pitx2*$^{+/-}$ embryo, both left and right hindlimbs are reduced, but the effect is always much more severe for the right hindlimb.

To see whether any of them truly associated with the trait, the stickleback orthologs of *Pitx1*, *Pitx2*, and *Tbx4* were isolated from the ocean and Paxton Lake sticklebacks populations described above. By sequencing these genes and the flanking genomic DNA from a number of individual sticklebacks, DNA polymorphisms were found that distinguished the starting marine and Paxton Lake parents. These polymorphisms were then used to place the *Pitx1*, *Pitx2*, and *Tbx4* genes onto the genetic map, and one of them, *Pitx1*, mapped to the same region as the major QTL identified by the analysis of the F$_2$ progeny (Fig. 11.22). These data suggest that it is indeed a difference in the *Pitx1* gene that is responsible for most of the phenotypic difference in pelvic spine and pelvic skeleton morphology. It should be noted, however, that given the resolution of the QTL technique in these particular experiments, the phenotypic difference might really be due to polymorphism in a different gene that lies very close to the *Pitx1* gene.

Further evidence to bolster the argument that it is *Pitx1* itself, however, comes from a comparative analysis of the expression of *Pitx1* in the sticklebacks. Using **in situ hybridization** techniques to look at messenger RNA (mRNA) distribution in the tissues of developing fish, *Pitx1* expression is seen in the prospective region of the pelvis in the marine sticklebacks but not in the Paxton Lake sticklebacks (Fig. 11.24B,D). This difference in expression is specific to the pelvic region—that is, levels and patterns of expression of *Pitx1* are similar in both fish in other tissues. For example, *Pitx1* also functions in the development of head structures, and similar patterns and levels of expression of *Pitx1* are seen in this region in both fish (Fig. 11.24A,C). Thus the difference in the two *Pitx1* alleles most likely lies at the level of their **cis**-regulation. In the

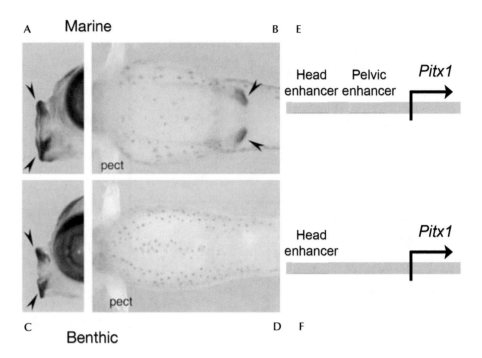

FIGURE 11.24. *Pitx1* mRNA expression in marine vs. Paxton Lake benthic sticklebacks. *Pitx1* expression (*blue* staining in all panels) is seen around the mouth of both fish (*arrowheads* in *A* and *C*). In the pelvic area, however, *Pitx1* expression is seen in the marine fish (*arrowhead* in *B*), but is absent from the benthic fish (*D*; the label "pect" shows the position of the anterior pectoral fins). This suggests that changes in the regulation of the *Pitx1* gene underlie the morphological differences seen in the pelvis and pelvic spines between these fish. It has been predicted that the marine fish possess both a head (mouth) enhancer and pelvic enhancer that cause *Pitx1* to be expressed in these two locations (*E*). In the benthic fish, the pelvic enhancer is presumably absent, and thus xpression of *Pitx1* is seen in only the mouth (*F*).

benthic fish there is presumably a *cis*-regulatory alteration specifically leading to reduced expression only in the pelvic area compared to the marine three-spine sticklebacks (Fig. 11.24E). The finding that the amino acid sequence encoded by the two *Pitx1* alleles is identical further supports this interpretation. Thus the two alleles do not differ in the proteins they encode; rather differences in the expression of the two alleles underlie the different phenotypes conferred by the two alleles.

These experiments all suggest that evolutionary variation in the regulation of the *Pitx1* gene explains the changes in pelvic morphology between the presumably ancestral marine population and the Paxton Lake population of sticklebacks. But what about other freshwater stickleback populations? It appears that other freshwater populations have convergently (i.e., independently) evolved reduced pelvic structures similar to those seen on Paxton Lake fish. Are these also the result of changes at the *Pitx1* locus? Genetic crosses using sticklebacks living in other lakes suggest that pelvic reductions in these fish populations likewise involve variation at the *Pitx1* locus. Either these *Pitx1* variants have arisen by independent mutation on several occasions or the variant allele was present in the ancestral marine population and selected for during the evolution of each lake population. Distinguishing between these possibilities will require a more detailed analysis of the molecular nature of *Pitx1* alleles and their distributions within various populations. These data from sticklebacks provide an excellent illustration of how a single gene (in this case, *Pitx1*), identified initially from its role in development, can play a critical role in the evolutionary changes in morphology within and between populations during evolution.

Another trait that varies between many marine and freshwater benthic populations is the extent of body armor (hard plates) on the sides of the fish. Marine populations have extensive armor, whereas the freshwater fish have relatively little armor. Using QTL analysis, a single gene was shown to have a major effect on this trait. Eventually the gene was identified as the fish ortholog of a human gene called *Ectodysplasin* (*Eda*). Mutations in the human gene had been identified in patients with a disease known as ectodermal dysplasia, which is characterized by the loss or reduction of teeth, hair, sweat glands, and eyebrows. *Eda* encodes a secreted molecule that is involved in cell–cell signaling in many tissues. The difference between the *Eda* alleles of the two fish populations again appears to be regulatory, resulting in lowered expression in the skin of freshwater versus marine fish. This example illustrates that it is not just transcription factors that are involved in the evolution of morphological change, but all levels of developmental pathways can and do play roles in evolution.

EVOLUTION OF MAIZE FROM TEOSINTE

More than 600 million metric tons of corn are produced worldwide each year, making it a crop of significant global economic importance. The plant that produces corn is called maize, and although it is easy to find acres upon acres of cultivated maize, it is unusual to find it growing wild. Unlike tomatoes, for example, for which wild tomato plants are essentially smaller-fruited versions of the domesticated plant, no equivalent wild plants exist for maize. Various lines of evidence suggest that maize has been cultivated for the past 6,000 to 10,000 years, but where did maize come from?

Maize Is Derived from the Teosinte of Mexico

More than 100 years ago, botanists first suggested that maize might have been derived from a plant known as teosinte, which is found throughout parts of Central and North America. This claim was contentious, however, as the two plants are morphologically distinct and others have suggested that teosinte is more closely related to rice than maize (see Fig. 11.28). In the 1930s, George Beadle and Ralph Emerson provided new evidence to support the hypothesis that an ancestral teosinte was a forerunner to maize. Teosinte, which is in the genus *Zea*, includes several described species. Beadle reported that maize and a particular form of teosinte called the Mexican annual teosinte, now known as *Zea mays* ssp. *mexicana*, have chromosomes that appear identical under the microscope. Furthermore, maize and *Zea mays* ssp. *mexicana* can be interbred and the resulting hybrid offspring are fully fertile. This led Beadle to suggest that maize and Mexican annual teosinte are the same species and that maize is simply a domesticated form of teosinte.

Maize and Teosinte Differ in Very Significant Ways

Although Beadle's work provided compelling evidence that teosinte is the ancestor of maize, there remained the question of how domestication had been able to change this plant so radically. The overall architecture of the two plants is quite different. Teosinte has many lateral branches containing tassels (the male reproductive part of the plant), whereas maize tends to have just a single long branch with ears of corn radiating from the sides and a tassel on top (see Fig. 11.28).

Even more striking are the differences in the ears (the female reproductive part of the plant). Teosinte ears have just 5–12 kernels, and each is encased in a hard fruit case that is not easy to break open. Once mature, the ear shatters, dispersing the kernels. The kernels may be eaten, but they are not destroyed as they pass through the digestive tracts of birds and mammals and are further dispersed by these animals. Maize ears (corn), on the other hand, have as many as 500 kernels per ear. The kernels have little external protection and are easily digested by animals (including humans). The kernels also remain firmly attached to the ear and do not disperse when the ear matures. Indeed, if the ears are not harvested, the ear will fall to the ground and all the

FIGURE 11.25. (*Left to right*) Teosinte, maize–teosinte hybrid, and maize ears. Note that the hybrid ear (in the *center*) is intermediate in many respects to the ears of its two parents.

kernels will sprout so close together that almost none survive. Thus maize cannot maintain itself without human intervention.

A Simple Cross Suggests That a Small Number of Genes Are Responsible for the Differences between Teosinte and Maize

Beadle focused on the morphology of the ears and carried out a very simple cross to estimate how many genes contributed to the differences that are seen between the ears of the two plants. Beadle found that the ears of the hybrid offspring looked intermediate between teosinte and maize (Fig. 11.25). Beadle then crossed the hybrids (the F_1 generation) together and produced 50,000 plants (the F_2 generation) from this cross. An analysis of the F_2 plants revealed that approximately 1 in 500 looked like maize and another 1 in 500 looked like teosinte. Beadle surmised that this was the expected outcome if just four or five genes were responsible for the differences between maize and teosinte ear morphology (Fig. 11.26).

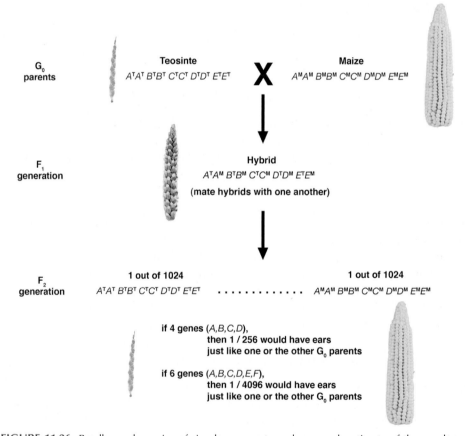

FIGURE 11.26. Beadle used a series of simple crosses to make a rough estimate of the number of genes that distinguish teosinte from maize ears. To begin, imagine that five genes (genes *A, B, C, D,* and *E*) control the difference between the two. Teosinte contains all teosinte alleles of the genes (A^TA^T, B^TB^T, etc.) and maize contains all maize alleles (A^MA^M, B^MB^M, etc.). The F_1 hybrids are heterozygous for all these alleles (A^TA^M, B^TB^M, etc.) and morphologically intermediate between the parents. When the F_1 hybrids are crossed to one another, many different genotypes and phenotypes are generated, but 1/1024 ($4 \times 4 \times 4 \times 4 \times 4 = 1024$) F_2 progeny will have the same genotype at the five loci as teosinte, and another 1/1024 will have the same genotype as maize. These ears are expected to look just like teosinte and maize, respectively, if these five genes are responsible for all the differences. If there had been four genes, the number would have been 1/256, and if six genes, 1/4096. Because Beadle got a number roughly 1/500, he estimated that about five genes were responsible for the difference between teosinte and maize ears. This is, of course, a very crude estimate as it does not take into account potential dominant/recessive relationships between alleles and genetic interactions between loci, but it did prove to be a remarkably accurate first estimation.

To Beadle these genetic results further supported the notion that maize had been produced from teosinte through human domestication. Early farmers, who selected for traits that made the plant more useful to humans, might have originally cultivated once-wild teosinte. Beadle's genetic experiments suggested that these early farmers appear to have selected for various alleles, in just a small handful of genes, that led to the rapid transformation of teosinte into modern day maize. But what are these genes?

QTL Mapping Identifies *teosinte branched 1* as the Gene Responsible for the Difference in Lateral Branching Patterns

As with sticklebacks, QTL analysis was used to map the chromosomal locations of specific genes controlling particular trait differences between teosinte and maize. (QTL analysis is described in more detail in Chapter 14.) Six chromosomal intervals were identified that appeared to account for many of the differences that could be scored between maize and teosinte (Fig. 11.27). One mapped trait was the number of lateral branches coming off the main branch of the plant. Recall that in teosinte there are many of these branches, whereas in maize there are only a few. A significant QTL for this trait was mapped to a particular portion of the left arm of chromosome 1, and the teosinte trait of "many lateral branches" was found to be recessive to the maize trait of "few lateral branches."

As chance would have it, scientists who were using genetic approaches to study the development of maize isolated a recessive maize mutant that had lateral branching patterns more like teosinte (Fig. 11.28). Interestingly, this mutation, which had been named *teosinte branched 1* (*tb1*), also mapped to the same interval of chromosome 1 as the QTL associated with the lateral branch pattern.

To prove that the QTL analysis and *tb1* mutation identified the same gene, complementation tests were carried out. Through many generations of crosses and selection, plants were produced with a genome entirely of maize origin, except for the small interval on chromosome 1 containing the lateral branch QTL, which was of teosinte origin. These plants were mostly maize-like but had the lateral branching property of teosinte. When these plants were crossed to maize *tb1* mutants, the resulting progeny still had many lateral branches. Thus, the *tb1* mutation failed to complement the lateral branch QTL, confirming that *tb1* and the mapped QTL were part of the same gene. (This is a variation of the complementation test described in more detail in Box 13.2.)

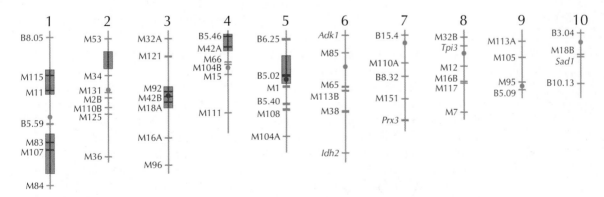

FIGURE 11.27. Estimated locations (*red*) of quantitative trait loci (QTLs) on the maize genetic map that distinguish teosinte from maize. This distribution of QTLs suggests that six chromosomal regions spread over five of the ten chromosomes account for most of the differences between maize and teosinte. The resolution of the mapping data narrows down the search to regions that still contain many genes, and each individual QTL could contain several linked genes that contribute to trait differences. Additional experiments showed that one of the QTLs (near genetic marker M107 on chromosome 1) was due to differences at a single gene—the *tb1* gene.

FIGURE 11.28. (A) Teosinte plant, (B) maize plant, and (C) maize *tb1* mutant plant. The teosinte (A) and maize (B) plants are quite different looking (e.g., the teosinte plant has many lateral branches, whereas maize has few or none), but the *tb1* mutant maize (C) plant has several morphological features that make it look more like teosinte, including the presence of lateral branches.

The Normal Function of *tb1* Is to Repress Organ Growth

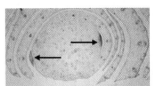

FIGURE 11.29. *tbl* expression patterns in maize. Cross section through a growing plant shows that *tb1* mRNA is expressed in axillary meristems (*arrows* in *top*) and in the primordia of husk leaves (h, *purple* staining in *bottom*), In both cases *tb1* is expressed in maize tissues whose growth is suppressed relative to the same tissues in teosinte.

When the maize *tb1* gene was cloned and characterized, it was found to encode a transcription factor of the TCP family (the family is named after the founding members, *T*eosinte branched, *C*ycloidea, and *P*CF2). *tb1* was subsequently shown to be expressed in a number of different maize organ primordia (precursor cells of the final organs), including the axillary (lateral) buds, branches, and husk leaf primordia (Fig. 11.29). What all these organ primordia have in common is that their growth is either stopped or reduced during normal maize development relative to the growth of the equivalent organs in *tb1* mutants. Other TCP family members are known to positively regulate a variety of genes that promote cell proliferation and growth by binding to their *cis*-regulatory regions and activating or increasing their level of transcription. The current hypothesis is that the Tb1 protein may bind competitively to these same *cis*-regulatory regions recognized by other TCP proteins, but the Tb1 protein is unable to activate transcription, and thus *tb1* represses growth and proliferation by competing with these other TCP-class transcription factors.

When the coding regions of *tb1* from multiple maize and teosinte populations are compared, no amino acid changes are found that are exclusive to either maize or teosinte. This finding suggests that the difference between the maize and teosinte alleles is not due to amino acid differences between the proteins encoded by the respective alleles. Thus, it appears that the difference between the teosinte and maize phenotypes is controlled instead at the level of gene regulation. Consistent with this, the maize allele is expressed at about twice the level of the teosinte allele in lateral branches.

Finally, sequencing the *tb1* gene from many populations of maize and teosinte revealed equivalent levels of variation within the coding region for both maize and teosinte alleles. In contrast, sequencing the DNA immediately 5′ of the start site for *tb1* transcription indicated that the maize DNA exhibits very low levels of nucleotide variation, whereas the teosinte DNA 5′ to the transcription start site of *tb1* had greater amounts of variation. The DNA immediately 5′ of the *tb1* transcription start site is a likely location for regulatory domains controlling the transcription of *tb1*. These results can be interpreted as evidence of selection for regulatory changes in *tb1* during the domestication of this plant. (See Chapter 19 for further details on detecting selection at the sequence level.) Thus, the earliest farmers of maize appear to have selected for particular regulatory mutations in the *tb1* gene during the course of maize domestication.

Several other traits differing between maize and teosinte have been mapped to individual genes as well. These include a gene associated with the presence of a hard fruit case around teosinte kernels, which is absent in maize. A single QTL for this trait has been linked to a gene named *teosinte glume architecture 1* (*tga1*), which appears to control a number of aspects of fruit-case development including the deposition of silica (a mineral that contributes to the hardness of the fruit case) and lignification (the formation of a complex plant polymer).

It should be noted, however, that other traits do not behave in such a simple genetic manner. For example, the disarticulation trait (the ease with which the ear falls apart) is distributed over many QTLs, which show complex patterns of genetic interactions. Thus, although the example provided by *tb1* and *tga1* clearly shows that single genes can have a major effect on some traits, other traits have clearly come about through the selection of many interacting alleles.

Teosinte Was Quickly Domesticated to Produce Maize

More detailed molecular analyses have identified a form of teosinte called *Zea mays* ssp. *parviglumis*—commonly called the Balsas teosinte—as the closest living relative to modern-day maize (*Z. m.* ssp. *parviglumis* forms interfertile hybrids with maize just as *Z. m.* ssp. *mexicana* does, but other genetic analyses indicates that maize is more closely related to *Z. m.* ssp. *parviglumis* than to *Z. m.* ssp. *mexicana*). Phylogenetic reconstruction based on microsatellite data suggests that there was a domestication process, which occurred only once and in a defined location within Mexico (Fig. 11.30). In addition, molecular dating places the divergence of *Z. m.* ssp. *parviglumis* and maize at about 9000 years ago, which is in close agreement with archaeological data.

Intriguing results have also been obtained from the analysis of "ancient" DNA extracted from maize as old as 4400 years preserved at archaeological sites throughout Mexico and the southwestern United States (Fig. 11.31). As might be expected, molecular studies suggest that the *tb1* allele associated with maize was already present in even the oldest samples. On the other hand, the analysis of another gene, called *su1*, indicates that selection for the modern day *su1* allele occurred approximately 2000 years ago. The *su1* gene encodes a starch debranching enzyme expressed in kernels, and the

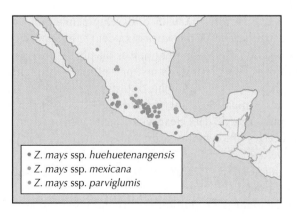

• *Z. mays* ssp. *huehuetenangensis*
• *Z. mays* ssp. *mexicana*
• *Z. mays* ssp. *parviglumis*

FIGURE 11.30. Domestication of *Zea mays* in Mexico and Central America. The various subspecies of *Z. mays* (teosinte) that are most closely related to maize grow in the southwest part of Mexico and parts of Guatemala. Of the three subspecies shown here, it is *Z. mays* ssp. *parviglumis* that seems to be most closely related to maize.

FIGURE 11.31. Maize cob from the Ocampo Cave in Venezuela, dated to ~3890 years before present. Its length is 47 mm.

presence of the modern allele is associated with the textural properties required for tortilla production.

All of these lines of evidence suggest that in a relatively small amount of time, a small community of farmers living in a restricted geographical area were able to turn teosinte into modern maize. Although the occurrence of new spontaneous mutations within the breeding population may have played some role, it seems likely that these farmers were mostly selecting for variation that already existed within their teosinte populations. Even today, maize-like alleles for genes such as *tb1* can be found within wild teosinte populations. Through selection, these farmers increased the frequency of specific alleles in genes such as *tb1* and *tga1* that play key roles in developmental processes and could by themselves produce fairly dramatic changes. At the same time, they also must have unveiled completely new variants by generating novel allele combinations, which show up today as complex multilocus QTLs.

The examples from stickleback and maize illustrate how morphological and developmental changes during evolution can be accomplished through relatively subtle modifications in one or a small number of genes. Remarkably, this applies both to a case of rapid human domestication (maize) and to a more gradual example of natural selection (stickleback). Furthermore, we have focused on examples that highlight the role of regulatory evolution, but evolutionary changes in protein-coding regions also play important roles. For example, the evolutionary diversification of the *Hox* gene cluster involved changes in both regulation and the encoded proteins. There is little doubt that if we study enough specific examples of evolutionary change, we will find that alterations in all aspects of developmental pathways have played a role and that the changes will involve both regulatory and protein-coding modifications. Nevertheless, it is interesting to see how many examples to date have implicated regulatory changes in transcription factors as a major factor in morphological and developmental evolution. This has lead many to speculate that these transcription factors are particularly well poised to play roles in evolution, because they can coordinate the activity of many downstream targets and their own regulation is modular, allowing changes to affect individual tissues at precise times in development (see Chapter 24).

UNIVERSALITY OF DEVELOPMENTAL SYSTEMS

One of the most significant findings to come from the fields of developmental biology, genetics, and genomics in the past several decades is the surprising degree of similarity in the molecular and genetic bases of animal development across all phyla. The *Hox* gene example discussed at the beginning of this chapter is just one of many examples of the remarkable similarities that exist even between animals of distantly related phyla. The evolutionary interpretations of these similarities, however, have fueled considerable debate. Although the molecular and phylogenetic data clearly indicate that many of the genes involved in developmental processes are evolutionarily ancient, it is not always clear what this tells us about the development and morphology of the ancestors that possessed these genes. We close this chapter with an example of a remarkable molecular similarity that underlies the development of a complex organ together with possible interpretations of what this tells us about the evolution of this and other organs.

Pax6 and the Evolution of Eyes

Eyes, the organs animals use to see the world, have long been thought to have evolved multiple times. Indeed, eye structures vary in such fundamental ways that some morphologists have argued that eyes have evolved independently as many as 40–65 times! There are three basic types of eyes: camera eyes, like those found in vertebrates and cephalopods (the class of mollusks that includes octopus, squid, cuttlefish, and nautilus); compound eyes, like those found in arthropods; and mirror eyes, like those found in mollusks such as scallops (Fig. 11.32). Given these remarkable differences in design, it was assumed that these eyes had independent evolutionary origins. But these views were challenged by the discovery of the developmental role of a gene called *Pax6*.

Pax6 belongs to a family of transcription factors that contain a PRD-type DNA-binding domain. The PRD domain takes its name from its founding member, the *Drosophila* paired (prd) protein, which is involved in the process of segmentation.

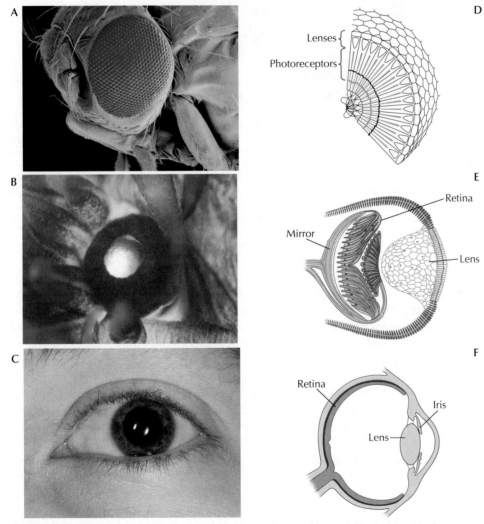

FIGURE 11.32. Examples of three types of eyes. (*A*) Flies, and many other arthropods, have compound eyes, which are composed of a repeating array of hexagonal units called ommatidia (*D*). Each ommatidium contains a lens and cluster of photoreceptors. (*B*) Scallops have mirror-type eyes that contain a reflective mirror at the very back (*E*). The photoreceptors of the retina detect light as it enters the eye and as it is reflected by this mirror. (*C*) Humans and many other organisms have camera-type eyes in which the lens focuses an image onto a layer of photoreceptors in the retina at the very back of the eye (*F*).

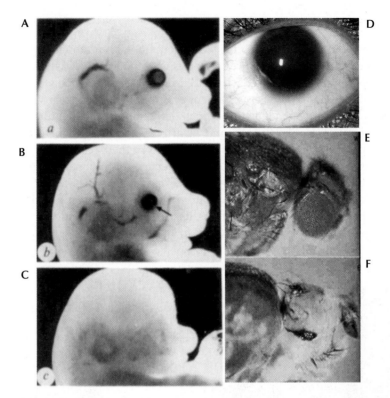

FIGURE 11.33. The *Pax6* gene controls eye development in many animal species. (*A*) Wild-type mouse embryo with a normal eye. (*B*) In an embryo heterozygous for a mutant allele of *Pax6* (the gene is called *Small-eye* [*Sey*] in mice), the eye is reduced in size. (*C*) In an embryo homozygous for the *Sey* mutation, the eye, along with the nose and other parts of the face, is completely absent. (*D*) Humans heterozygous for a mutation in *Pax6* (the gene is called *Aniridia* in humans) have eye defects that include the loss of the iris. A few rare human embryos homozygous for *Pax6* mutations show a complete loss of the eye, nose, and facial structures just as seen in mice. (*E*) Wild-type *Drosophila* with a normal eye. (*F*) Mutation of *Pax6* (the gene is called *eyeless* in flies) results in severe disruption of the eye. Complete elimination of *eyeless* results in headless flies.

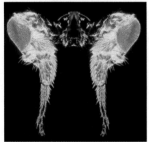

FIGURE 11.34. Ectopic expression of *Pax6* creates ectopic eyes. (*Top*) Inappropriate expression of *Drosophila eyeless* in the wing imaginal disk results in the formation of an eye (*red* structure) on the wing. Flies also contain a second *Pax6* gene, called *twin-of-eyeless*, which can also create ectopic eyes. (*Bottom*) Ectopic expression of *twin-of-eyeless* in leg imaginal disks results in eyes (*red* structure) on the adult legs. Even misexpression of the mouse *Pax6* (*Sey*) gene in *Drosophila* leads to the formation of ectopic eyes—these are, of course, ectopic fly eyes and not mouse eyes on the flies! Likewise, ectopic expression of *Drosophila eyeless* gene in frogs leads to ectopic frog eye structures in frogs.

Drosophila paired actually contains two DNA-binding motifs, a homeodomain and a PRD domain. Using the *Drosophilia* PRD domain, it was possible to clone related family members from a multitude of animals. Some of the cloned proteins contain both a homeodomain and a PRD domain, whereas others contain only the PRD domain.

Mutations in *Pax6* were found to be associated with defects in eye development in both mice and humans (Fig. 11.33). In humans, a genetically dominant syndrome, called aniridia, is characterized by defects in the iris, whereas the mouse mutant phenotype, called *Small-eye*, showed reduction in the size of the eyes. In both cases, the cause of the defect was traced back to mutations in the *Pax6* gene, and the phenotypes were due to haploinsufficiency (loss of one allele) at the *Pax6* locus, hence the dominant nature of the mutations. Homozygous loss of the *Pax6* gene results in embryonic lethality with a complete lack of eyes, nose, and parts of the brain. At the time that *Pax6* was characterized in vertebrates, no ortholog was known from *Drosophila*.

When the *Drosophila Pax6* gene was finally isolated and characterized, it came as somewhat of a shock to find that it corresponded to a mutation known as *eyeless* (Fig. 11.33). As the name implies, these flies lack eyes. Even more remarkably, the *Pax6* gene, when expressed in other parts of the animal, is capable of creating ectopic (i.e., extra) eyes (Fig. 11.34), suggesting that the gene plays a role as a "master regulator" of eye development. Subsequent studies have suggested that *Pax6* is involved in the development of all animal eyes. From an evolutionary point of view, what is so remarkable is that the same gene is involved in the development of very different types of eyes—such as those of vertebrates and arthropods—that were thought to have evolved independently.

Perhaps eyes have not evolved independently, but were instead present, and being patterned by *Pax6* function, in the common ancestor of all bilaterian animals. During the course of evolution, modifications in eye structure occurred that led to the diversity of eye types we see in extant animals. Thus, this common ancestor, which lived before the Cambrian explosion, would be expected to have eyes. An alternative interpretation is that *Pax6* was independently recruited multiple times to make eyes each time they evolved. The evolutionarily ancient function of *Pax6* might be one of patterning anterior neural structures in general. Given that eyes tend to evolve at the anterior end of animals for functional reasons, *Pax6* may have been recruited independently on several occasions to serve as the transcription factor guiding eye development.

Yet a third, somewhat intermediate, explanation is possible. The common ancestor may have had a very simple light-sensing structure composed of a photoreceptor (the cell that actually senses photons of light using a molecule called rhodopsin) and a pigment cell that shades light from at least one side so that the photoreceptor receives light only from certain directions. It may be that this very simple structure, which certainly could be called an eye, was patterned by *Pax6*. In support of this idea, it has been found that among the many transcriptional targets of *Pax6* are the rhodopsin genes. As this primitive eye independently evolved in different animal lineages into the many morphologically complex eye types, *Pax6* continued to serve as a key regulator of eye development and acquired additional transcriptional targets, all of which play a role in eye development in extant eyes. The ultimate arbitrator, however, should be the fossil record. What sort of eye, if any, is found in the common ancestor of all bilaterians? Unfortunately, the fossil record has yet to provide an answer.

SUMMARY

Developmental genetics has led to the identification of the genes and pathways involved in the development of many different animal species. Somewhat surprisingly, the molecular underpinnings of development are well conserved, and many of the genes involved in development are evolutionarily ancient. The way in which these genes are used, however, appears to underlie many of the evolutionary differences that do occur during development and thus leads to the morphological diversity we see in extant organisms.

The *Hox* genes, which play a role in anterior–posterior axis patterning, provide an excellent example of an evolutionarily ancient set of genes that play a similar role in a wide variety of animals. Despite their remarkable level of conservation, evolutionary changes in the expression and specific functions of this group of genes do help explain a number of morphological changes. These extend from macroevolutionary changes in overall body plan to microevolutionary changes in such structures as leg bristles in closely related species of flies. Other genes identified for their role in development, such as *Pitx1*, *Eda*, and *tb1*, have also been shown to play a role in the evolution of such diverse morphologies as the pelvic spines of sticklebacks and the morphological changes that occurred during the domestication of maize from teosinte.

Although these comparative developmental studies have led to important insights into the mechanisms of evolutionary change, caution should still be employed in using them to reconstruct the development and morphology of ancestors. On some occasions, these developmental genes may have been independently co-opted for similar roles on multiple occasions. In other instances, the morphological structures we see in extant animals may have been derived from morphologically simple precursors.

FURTHER READING

Hox *Gene Evolution*

Carroll S.B, Grenier J.K., and Wetherbee S.D. 2005. *From DNA to diversity. Molecular genetics and the evolution of animal design.* Blackwell Publishing, Oxford.

Stickleback Evolution

Tickle C. and Cole N.J. 2004. Morphological diversity: Taking the spine out of three-spine stickleback. *Curr. Biol.* **14:** R422–R424.

Evolution of Maize from Teosinte

Doebley J. 2004. The genetics of maize evolution. *Annu. Rev. Genet.* **38:** 37–59.

Eye Evolution

Gehring W. 2002. The genetic control of eye development and its implications for the evolution of the various eye-types. *Int. J. Dev. Biol.* **46:** 65–73.

Nilsson D.-E. 2004. Eye evolution: A question of genetic promiscuity. *Curr. Opin. Neurobiol.* **14:** 407–414.

PART III

EVOLUTIONARY PROCESSES

  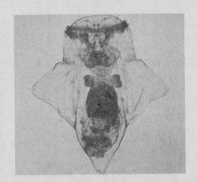

As we saw in Chapter 1, one of the most striking observations in evolutionary biology is that almost all populations contain abundant genetic variation, which is the basis for adaptation by natural selection. Part III explains the nature of genetic variation, the processes that act on it, and the way these processes interact in evolution. Ultimately, this variation is generated by mutation and recombination, which are detailed in Chapter 12. It is straightforward to describe variation in DNA and proteins (Chapter 13), but the effect of this variation on the whole organism is harder to determine. Chapter 14 explains how the methods of quantitative genetics can be used to analyze genetically complex traits.

Chapters 15–17 explain the key processes of evolution: random genetic drift, the flow of genes from place to place, and, most important, selection. Interactions between these evolutionary forces are explored in Chapter 18. In Chapter 19, diverse lines of evidence are brought together to assess the extent of selection. We know that there is abundant genetic variation, but how is it maintained, and what are its consequences?

In the remaining five chapters, we apply our understanding of the processes of evolution to specific problems: the evolution of phenotype, including aging and mating behavior; conflicts and cooperation between genes and organisms; the origin of species; and the evolution of the genetic system itself (in particular, the evolution of sexual reproduction). Finally, we ask in Chapter 24 how fundamentally novel features—be they biochemical, morphological, or behavioral—evolve.

Throughout Part III, as is the emphasis throughout this book, molecular and organismal approaches are combined and integrated. Almost all of the phenomena that are discussed can be studied at different levels. For example, the evolution of social interactions can be studied using microbes as well as insects, birds, and mammals; and sexual reproduction is important for all organisms, from viruses to humans. Evolutionary biology is experiencing a resurgence as long-standing questions are studied in a variety of new ways.

12

Generation of Variation by Mutation and Recombination

A WALK THROUGH A FOREST HIGHLIGHTS the extraordinary variation among living creatures. Dozens of species of trees, birds, vines, flowers, fungi, amphibians, and small mammals are readily seen and heard. Within the leaf litter live a wide array of worms, ants, and mites. Examination of just a spoonful of soil reveals a rich microscopic world of viruses, bacteria, archaea, and single-celled eukaryotes. Perhaps more stunningly, a walk through a city park highlights the variation that can be seen even within species. People, dogs, apple trees, and roses all come in a myriad of shapes, colors, and forms. In this chapter, we explore the sources of variation both between and within species—**mutation** and **genetic recombination**.

Mutation, formally defined as a heritable change in the genetic material (DNA or RNA) of an organism, is the ultimate source of all variation. Without mutation, there would be no evolution. Variation is also generated when mixing occurs between genetic material from different lineages, a process known as genetic recombination.

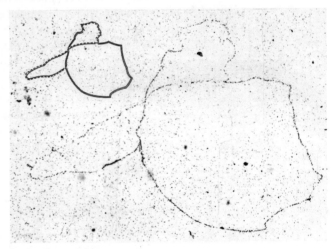

Much of this chapter focuses on mutations: the different types, how they are generated, and the variation seen in patterns of mutation within and between species. The chapter closes with a discussion of how genetic recombination amplifies the variation existing in populations. We look first at the most familiar form, genetic recombination that occurs in sexually reproducing species, such as ourselves; we then consider other types of recombination, such as mitotic recombination and lateral gene transfer.

MUTATIONS AND MECHANISMS FOR THEIR GENERATION

As we begin our discussion on mutations, it is useful to consider the life cycle of a simple organism, such as a haploid single cell that reproduces asexually by binary fission. Reproduction requires that the cell first copy its genome and then divide, giving each daughter cell one of the genome copies. Within this life cycle, there are three major sources of mutations: errors in replication, errors in **segregation** of the replicated genome to the daughter cells, and modification of the genome by other processes including transposition, DNA damage, and aberrant homologous recombination.

Of course, few biological systems are as simple as the one described above, but in all cases, mutations fall into the same three categories. These categories will help to organize our discussion. First, the different classes of mutations are identified. Next the more straightforward causes of mutation are described: errors in DNA replication and physical damage to the DNA. Then mechanisms that limit the mutagenic effects of replication errors and DNA damage, including DNA **proofreading** and DNA repair processes, are discussed. Next, more complex mutation-generating mechanisms are presented, including errors in chromosome segregation and processes that occur during normal cellular growth, such as aberrant homologous recombination and transposition. The section ends with a discussion of how and why patterns and rates of mutation vary.

Mutations Come in Many Forms

Mutations can be classified according to the type of change to the DNA sequence. For simplicity, the focus will be on organisms that use DNA as their genetic material, but the same principles apply to RNA-based genomes, such as some viruses and hypothetical early life-forms. Base substitutions, or **point mutations**, occur when there is a change from one base pair to another at a single position in the DNA sequence. Using one DNA strand as a reference, there are 12 possible point mutations (each base can change to three other bases). Such substitutions can be classified by whether they are within or between the two forms of DNA bases, **purines** (G and A) and **pyrimidines** (C and T). Changes from one purine to another or one pyrimidine to another are referred to as **transitions**. Changes from a purine to a pyrimidine or vice versa are called **transversions** (Table 12.1).

Although point mutations change only a single base in the sequence of the genome, they can have profound effects. For example, a single transversion (from A to T) in the coding strand for the β-globin protein results in the substitution of valine for glutamic acid in the amino acid sequence in the β-globin protein. Heterozygotes benefit by having resistance to malaria, but homozygotes suffer from sickle-cell anemia (Fig. 12.1), a devastating disease. When a substitution occurs that leads to a change in a single amino acid in a protein (e.g., Figs. 12.2 and 12.3), this is known as a **missense mutation**. On the other hand, because of the redundancy of the **genetic code**, point mutations may not change the amino acid sequence. These are termed **synonymous mutations**.

In other mutations, the daughter DNA has a different number of base pairs in a particular region than does the parental DNA. These insertions and deletions range in length from a single base, to a gene fragment, to thousands of genes. The smaller insertions and deletions are often referred to as **indels** (Fig. 12.2). An indel within a region of DNA that codes for a protein results in a messenger RNA (mRNA) with the corresponding insertion or deletion. If the number of base pairs inserted or deleted is not divisible by 3, the indel will alter the reading frame used during translation of that mRNA for all regions downstream of the indel. Thus, these particular indels are known as **frameshift mutations** (Fig. 12.2).

Most insertions, whether large or small, are not made up of random DNA sequence. Rather the inserted sequence is usually derived from other DNA that has been copied or moved from elsewhere within the cell. If the source DNA was copied, that sequence is now present in both its original location and the new location, resulting in a duplication. If the new copy is adjacent to the original sequence, it is termed a **tandem duplication** (Fig. 12.3).

Entire chromosomes and even whole genomes can be duplicated. In diploid species, chromosomes are in pairs (humans, for instance, have 23 pairs). The loss of one member of a chromosome pair leads to **monosomy** and a duplication leads to **trisomy**. In humans, most changes in somy level have dire consequences; all result in pre-

TABLE 12.1. Transitions and transversions

Transitions	Transversions	
A→G	A→T	T→A
G→A	A→C	T→G
T→C	G→T	C→A
C→T	G→C	C→G

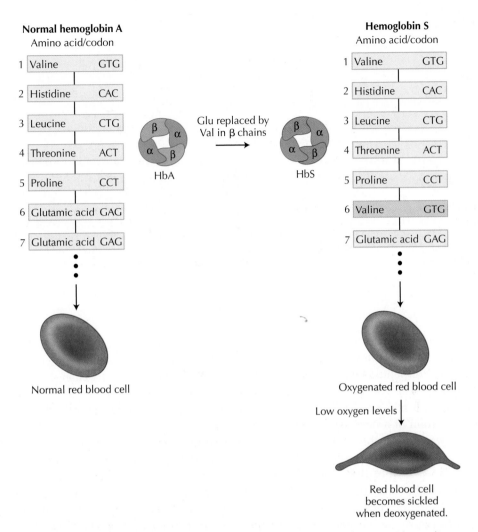

Normal hemoglobin A
Amino acid/codon

1	Valine	GTG
2	Histidine	CAC
3	Leucine	CTG
4	Threonine	ACT
5	Proline	CCT
6	Glutamic acid	GAG
7	Glutamic acid	GAG

HbA

Glu replaced by Val in β chains

Hemoglobin S
Amino acid/codon

1	Valine	GTG
2	Histidine	CAC
3	Leucine	CTG
4	Threonine	ACT
5	Proline	CCT
6	Valine	GTG
7	Glutamic acid	GAG

HbS

Normal red blood cell

Oxygenated red blood cell

Low oxygen levels

Red blood cell
becomes sickled
when deoxygenated.

FIGURE 12.1. Sickle-cell anemia can be caused by a single-base change in the coding sequence of the gene that encodes one of the proteins (hemoglobin A) that makes up hemoglobin, resulting in the glutamic acid to valine substitution at the 6th amino acid position in the mutant-form hemoglobin S in sickle-cell disease. A schematic diagram of this is shown in which the normal β subunit (in *blue*) is present in the normal hemoglobin (HbA) and an abnormal β subunit (in *red*) is in the sickle-cell form of hemoglobin (HbS). When present, the altered hemoglobin molecule causes red blood cells to occasionally form a sickle shape, especially when under low oxygen levels. This in turn can lead to major phenotypic consequences such as the blocking of small blood vessels.

Category	Wild type	Missense	Frameshift by insertion
DNA	5′ AAA-GCT-ACC-TAT-CGG-TTA 3′ 3′ TTT-CGA-TGG-ATA-GCC-AAT 5′	5′ AAT-GCT-ACC-TAT-CGG-TTA 3′ 3′ TTA-CGA-TGG-ATA-GCC-AAT 5′	5′ AAA-GCT-ACC-ATA-TCG-GTT 3′ 3′ TTT-CGA-TGG-TAT-AGC-CAA 5′
mRNA	5′ UUU-CGA-UGG-AUA-GCC-AAU 3′	5′ UUA-CGA-UGG-AUA-GCC-AAU 3′	5′ UUU-CGA-UGG-TAU-AGC-CAA 3′
Protein	N PHE-ARG-TRP-ILE-ALA-ASN C	N LEU-ARG-TRP-ILE-ALA-ASN C	N PHE-ARG-TRP-TYR-SER-GLY C
	Amino Carboxyl	Amino Carboxyl	Amino Carboxyl

FIGURE 12.2. Indels and frameshifts. In column 2, a region of DNA corresponding to a protein is shown, as is the mRNA and the protein encoded in that region (see Figs. 2.23 and 2.26 for the genetic code and amino acid abbreviations). In column 3, a missense mutation and the resulting change in the protein sequence are shown (with differences to wild type shown in *red*). In column 4, a frameshift mutation (an addition of an A-T base pair) and the resulting changes in the protein sequence are shown.

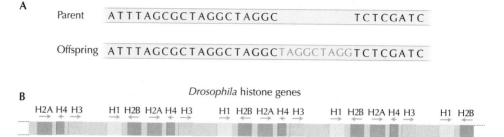

FIGURE 12.3. Tandem duplications. These are duplication mutations in which the new copies ("offspring") of the DNA are adjacent to the source DNA ("parental"). Tandem duplications come in a range of sizes, from single bases to many genes. (*A*) A schematic diagram of 9-bp tandem duplication (*red*). (*B*) A tandem array of histone genes in *Drosophila melanogaster*. *Arrows* indicate the direction of transcription. Such tandem arrays, which are common in eukaryotic genomes, are usually generated via multiple tandem duplication events.

natal or very early death, except trisomy at chromosome 21, which causes Down syndrome. Because of these severe consequences, somy mutations are usually not maintained within populations. Duplication of the entire genome leads to a change in the **ploidy** level. For example, in many eukaryotes, the chromosome copy number can be doubled with relative ease, although this often has profound effects—even leading to the formation of a new species (see pp. 631–633).

Rearrangements of DNA result in mutations without any change in the total DNA content or local DNA sequence. For example, **inversions** are mutations in which the order of a section of the genome (ranging from a few base pairs to large segments of a chromosome) has been reversed (Fig. 12.4). In chromosomes with **centromeres** (most eukaryotic chromosomes), inversions are classified as either **pericentric** (if they span the centromere) or **paracentric** (if they do not span the centromere) (Fig. 12.5). **Translocations** occur when sections of two chromosomes (or two ends of one chromosome) are exchanged (Fig. 12.6). Unless the break points disrupt genes, inversions and translocations frequently have no immediate phenotypic consequences. However, these rearrangements can create problems during genetic recombination (i.e., sexual recombination in eukaryotes and mating in bacteria and archaea). When one chromosome contains an inversion or translocation that is absent from the homologous chromosome, **synapsis** may not occur correctly and **crossing over** during meiosis may lead to deletions. Thus it is common for individuals that are heterozygous for inversions or translocations to be partially or completely sterile (see Box 12.2).

Many Mutations Are Caused by Spontaneous Errors in DNA Replication

When DNA is "copied," DNA polymerases do not always reproduce the original sequence with perfect accuracy (DNA replication is described in Box 12.1). Most commonly, errors involve violation of the base-pairing rules (A pairs with T and G pairs with C) leading to a base substitution. The most common errors are base transitions; one major cause of transitions is base **tautomerization**. Each of the bases in DNA exists in two alternative tautomeric states (Fig. 12.8). More than 99% of the time, the bases are in the forms commonly seen in DNA, the *keto* forms of G and T and the *amino* forms of C and A. However, each base has a low (<1%) probability of being in the alternative form (*enol* and *imino* forms, respectively). These alternative states can create base pairings if they occur at a critical moment during DNA replication. For example, the *enol* tautomer of G pairs with T rather than C. The *imino* form of A pairs with C rather than T (Fig. 12.8A).

A

Parent   A T T T A G C G C T A G G C T A G G C T C T C G A T G

Offspring A T T T A T C G G A T C G C G A G G C T C T C G A T G

B

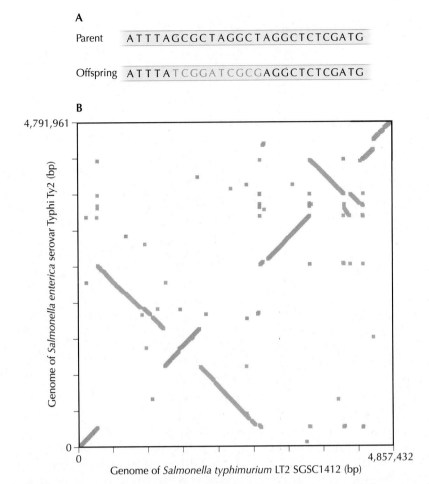

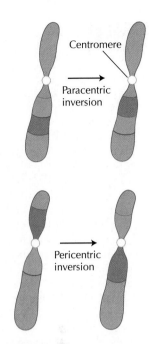

FIGURE 12.5. Paracentric vs. pericentric inversion. Paracentric inversions are outside of the centromere; pericentric inversions span the centromere. The *colored bands* highlight the inversions.

FIGURE 12.4. Inversions. (*A*) Hypothetical example of an inversion in a small section of DNA. Parental sequence is shown *above* and the offspring sequence is shown *below* with the inverted region highlighted in *red*. (*B*) Comparison of the genomes of two strains of the bacterial genus *Salmonella* showing the occurrence of multiple large inversions. The diagram shows a genome dot-plot (as in Box 7.1). The genome of one strain is on the *x*-axis and the other is on the *y*-axis (with the replication origins at (*x, y*) = (1, 1)). Conserved regions between the two genomes are indicated by a *dot*. If the two genomes showed the same total orientation, all the dots would be on the *y* = *x* diagonal. The *blue* segments are inversions between the two strains.

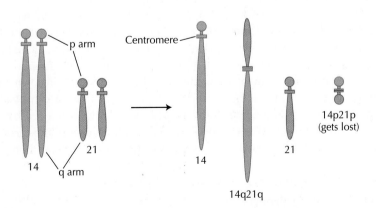

FIGURE 12.6. Translocation. Diagram of translocation between human chromosomes 14 and 21. The short (p) arms of chromosomes 14 and 21 can recombine, causing the two translocations (14q21q and 14p21p).

Box 12.1 DNA Replication

It is essential to understand the mechanisms that organisms use to replicate their DNA in order to understand how replication errors can lead to mutations.

A DNA molecule is a double helix made from two strands of nucleotides. The backbone of each strand is built of phosphate units, which form phosphodiester bridges between pairs of sugar molecules; each phosphate linking the 3'-OH of one deoxyribose to the 5'-OH of the adjacent one (Fig. 12.7; also see pp. 42–46). The orientation or polarity of a DNA strand is the result of a free 5'-OH group on one end and a free 3'-OH group on the opposite end. The two strands of a double helix have opposite polarity (referred to as antiparallel).

The opposing polarity of the two strands has implications for how DNA replication can proceed. The DNA polymerase complex, which catalyzes DNA replication, initiates replication at one point and then proceeds in one direction along the double-stranded DNA, copying both strands at the same time. DNA polymerases, however, add new nucleotides only to the 3' end of growing DNA chains, not to the 5' end. This works smoothly when the direction of strand elongation is the same as the direction of movement of the DNA polymerase complex. This occurs on the strand referred to as the **leading strand** of DNA replication. For the other strand, called the **lagging strand**, the DNA polymerase complex moves in a direction opposite to that of chain elongation. To compensate for this, the DNA polymerase jumps a little bit ahead and then copies the intervening segment in the reverse direction by adding nucleotides as usual to the 3' end of the growing fragment. Then it jumps ahead again, copies the next segment, and so on. The newly synthesized fragments, termed Okazaki fragments, are ligated together to form a continuous strand. Synthesis of the leading strand is regulated to match the slower pace of lagging-strand synthesis.

A further complication arises because DNA polymerases are unable to initiate the replication process them-

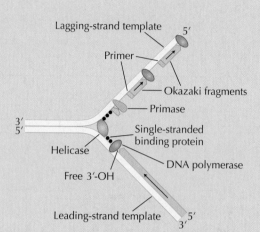

FIGURE 12.7. A DNA replication fork. The parental DNA (*blue*) serves as the template for synthesis of the two new strands of DNA (*green*). The DNA to the right of the helicase (*purple oval*), which has been unwound, is being replicated, whereas to the left of the helicase, which unwinds the DNA, the DNA is not yet replicated. Additional details are explained in the box text.

selves. Initiation requires the assistance of a **primase** that synthesizes a short complementary sequence (of RNA or DNA, depending on the species), which serves as the replication primer. A primer is required each time replication is initiated. Replication of the leading strand requires only one primer, but replication of the lagging strand requires a new primer for each segment, which further slows the synthesis of the lagging strand. These differences in replication mechanism for the two antiparallel strands can create asymmetries in the pattern of mutation.

DNA replication systems also make mistakes that generate indels, typically via a mechanism termed **slip-strand mispairing** (SSM). In SSM, the DNA replication machinery "slips" as it proceeds along the DNA template strand, causing the template and the newly synthesized DNA strands to temporarily separate. If the strands are one or several bases out of register, then when they reassociate, a loop will form on one of those strands. As replication continues, the result is the addition of extra bases to the new strand (if the new strand contains the loop) or the deletion of a few bases (when template contains the loop) (Fig. 12.9). This phenomenon occurs most often in regions with short repetitive sequences, such as **microsatellite DNA** (see Box 13.3).

Physical Damage to DNA Can Generate Mutations, either Directly or Indirectly

The DNA of organisms is not inert in between cycles of replication and segregation. It is constantly being wound and unwound, bound by various proteins, and even chemically altered, all as part of normal cellular processes. These processes are integral to life, but

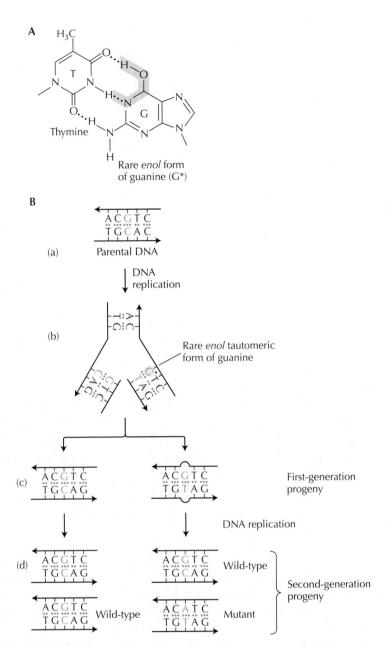

FIGURE 12.8. Tautomers and mutations. (*A*) Base pairing of the *enol* tautomer of guanine with thymine. (*B*) Mutation generated by tautomeric shifts in the bases of DNA. In (*a*), parental DNA is shown. In (*b*), DNA replication is proceeding on both strands. A guanine in one parental strand (*arrow*) undergoes a tautomeric shift to its rare *enol* form (G*). This leads to a T being placed opposite it, rather than a C. In (*c*), the first-generation progeny are shown. The G* has shifted back to a normal G. However, there is now a mismatched G-T base pair in the DNA. This can be repaired by mismatch repair (see main text). If it is not repaired, this will become fixed as a mutation when the DNA is replicated again because an A will be placed opposite the T. In (*d*), the DNA sequences in this region are shown after another round of replication has occurred.

also endanger genome stability. DNA is also continually subjected to chemical and physical forces, which can chemically alter the DNA or cause other types of damage. These alterations can cause mutations directly by changing the base sequence or indirectly by increasing the frequency of replication or segregation errors.

A common form of base alteration in DNA is the deamination of cytosine (Fig. 12.10). This occurs spontaneously at a high rate and can also be induced by mutagenic chemicals or heat. The loss of the amino group converts the cytosine to uracil, a base that is nor-

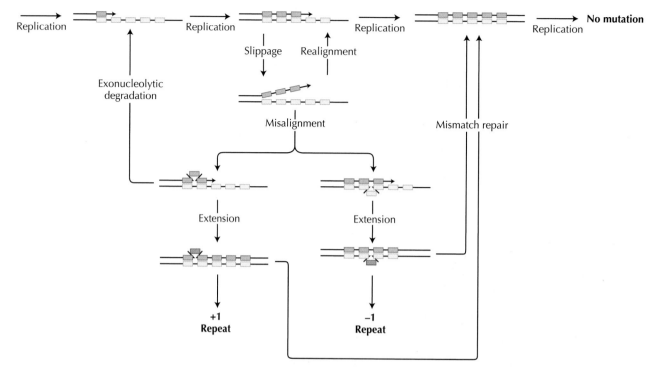

FIGURE 12.9. Slip-strand mispairing (SSM). Model of the SSM process that can occur during DNA replication. SSM is most common in tandemly repeated DNA. In the figure, cartoons of double-stranded DNA containing a tandem repeat (e.g., GAGAGAGAGAGA) are shown at different stages of the replication and mutation process. In the cartoons, DNA strands are represented by *thin lines*, individual copies of repeats by *small boxes*, and ongoing replication by *small arrows*. Starting from the *left*, a section of parental DNA (*blue*) containing five repeats is being replicated (new DNA is *green*). In the *middle*, the possibility of slippage is indicated in the panel *below*. After slippage, the DNA can realign in register and return to the correct form or realign out of register, forming a loop (shown *further below*). The loop can form on the template or the strand being synthesized. These can be repaired by mismatch repair (*arrows* to the *right*) or by proofreading of the DNA polymerase (*arrow* to the *left*). If they are not repaired, an indel mutation will occur.

mally found in RNA but not in DNA. For most purposes, uracil in DNA functions as though it were thymine and will base-pair with adenine. Thus, during DNA replication, a uracil in a template strand will cause adenine to be inserted in the opposite strand, thus converting the original C-G base pair to a U-A base pair, which is converted to a T-A base pair by the next replication cycle. Cells recognize that uracil is not a normal constituent of DNA and have special repair processes to remove any uracil found in the DNA, thus limiting the damage caused by the deamination of cytosine (see p. 336). However, in many species, some of the cytosine bases are modified by the addition of a methyl group (see Fig. 12.10). The deamination of methylcytosine, which also occurs spontaneously at a high rate, produces thymine. If uncorrected, this results in the change of a C-G base pair to a T-A base pair. Because thymine is a normal DNA base, cells cannot distinguish between a normal thymine and one resulting from the deamination of methylcytosine and thus cannot correct this error in a straightforward manner (although some species possess systems to specifically recognize and repair the G:T mismatches produced by this deamination). Thus, the deamination of methylcytosine contributes to the high mutation rate of cytosines observed in the methylated regions of genomes (see below).

UV irradiation can damage DNA by forming covalent bonds between two adjacent pyrimidine bases in the same DNA strand (Fig. 12.11). Replication of sites containing UV-induced pyrimidine damage is more error prone than replication of normal DNA; thus UV irradiation can cause mutations. Damage to DNA, which leads to replication errors, can also be caused by DNA **intercalating agents**, which are able to wedge them-

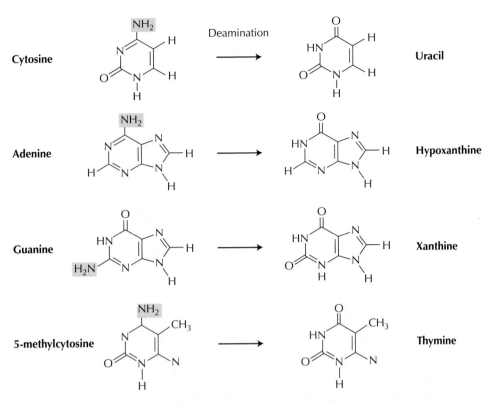

FIGURE 12.10. Deamination of DNA bases. The removal of the NH_2 group from DNA bases occurs spontaneously and can be induced by heat and various chemical agents. It happens most commonly to cytosine and 5-methylcytosine. In many species, the deamination of 5-methylcytosine to thymine is particularly troublesome because, as a naturally occurring nucleotide, thymine cannot be recognized as an abnormal DNA base (although the abnormal G-T mismatches can be recognized in some species).

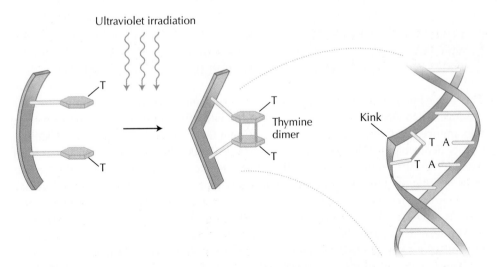

FIGURE 12.11. Cyclobutane pyrimidine dimer (CPD) induced by UV irradiation. The figure shows a CPD involving two thymines, which is also known as a thymine dimer. CPDs such as this are mutagenic and toxic because they cause distortions in the DNA helix, as shown, which in turn disrupts replication, transcription, and other processes.

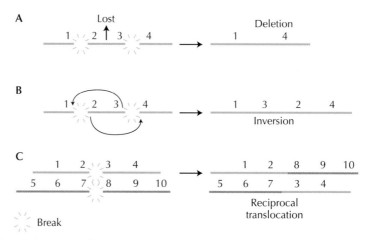

FIGURE 12.12. Chromosomal rearrangements generated by incorrect rejoining of DNA double-strand breaks. (*A*) Two strand breaks (indicated by *blue starbursts*) occur flanking regions 2 and 3. Religation between the flanking regions leads to the deletion of the region containing 2 and 3. (*B*) A different ligation of the same double break as in the first panel leads to an inversion. (*C*) Single breaks in two chromosomes are incorrectly religated leading to a translocation.

selves into the space between two adjacent base pairs. Perhaps the most significant cause of base damage in most organisms is the presence of oxygen radicals, which can damage purines and pyrimidines.

A number of different factors can induce breaks in the DNA strands by breaking the phosphodiester bond of the DNA backbone. These factors include oxygen radicals, γ-irradiation, chemicals such as the antibiotic mitomycin C, and incomplete transposition. Strand breaks cause mutations in a variety of ways. For example, double-strand breaks can cause deletions or create inversions and translocations when the broken ends are joined incorrectly (Fig. 12.12). In addition, in eukaryotes containing chromosomes with centromeres, double-strand breaks that exist during chromosome segregation will prevent the part of the chromosome without the centromere from segregating correctly. Thus, one of the daughter cells may not receive that section of the genome.

PROTECTION, PREVENTION, AND CORRECTION MECHANISMS LIMIT THE NUMBER OF MUTATIONS CAUSED BY DNA DAMAGE AND REPLICATION ERRORS

The mutagenic potential of DNA replication errors and DNA damage is immense. For example, when DNA polymerases are studied in vitro, their base misincorporation error rate approaches 1 mistake per 1000 bases. Some form of base damage (e.g., deamination or depurination) occurs at about 1 of every 10,000 bases per day, even under ideal conditions. Yet in most organisms, the mutation rate is remarkably low. For example, in *Escherichia coli* the rate of point mutations is approximately 1×10^{-9} and in humans it is approximately 1×10^{-10} per site per replication (see Fig. 12.23). That is, in humans, there is only 1 error for every 10,000,000,000 sites in the genome per replication. As will be seen in Chapter 19, the mutation rate has to be extremely low because even a few mutations per genome per replication can drastically reduce fitness.

This low mutation rate, despite the high rates of polymerase error and DNA damage, is due to processes that protect DNA from damage, prevent the transmission of errors to offspring by repairing DNA damage, and correct DNA replication errors.

Together these protection, prevention, and correction processes reduce the mutational load by many orders of magnitude. These processes are discussed in the following sections.

Protection Mechanisms Limit the Effects of Mutagens

If environmental mutagens are unable to reach cells and their DNA, then the organism is protected against the damage they can cause. For example, many organisms shield themselves with pigments that absorb UV irradiation. Organisms also use diverse mechanisms to scavenge oxygen radicals within cells, which limits both the mutagenic and toxic effects of these radicals.

DNA Repair Processes Prevent Mutations by Repairing Damaged DNA

DNA repair processes work by recognizing damaged, altered, or abnormal DNA and restoring the DNA to an undamaged, unaltered, or normal state. Based on their general mechanism of action there are three classes of repair processes: direct repair, excision repair, and recombinational repair.

Direct repair mechanisms simply reverse alterations in the structure of DNA. In one form of direct repair, **DNA ligation** (Fig. 12.13A), the phosphodiester bonds of a broken DNA backbone are religated together. This can be used to repair both single- and double-strand DNA breaks. A specialized form of DNA ligation, called nonhomologous end joining, is used in many organisms to repair double-strand breaks. Other examples of direct repair include photoreactivation, in which inappropriately formed covalent bonds between adjacent pyrimidines (caused by UV irradiation) are broken (Fig. 12.13B), and alkyl transfer, in which abnormal alkyl groups are removed from DNA by transferring them to proteins.

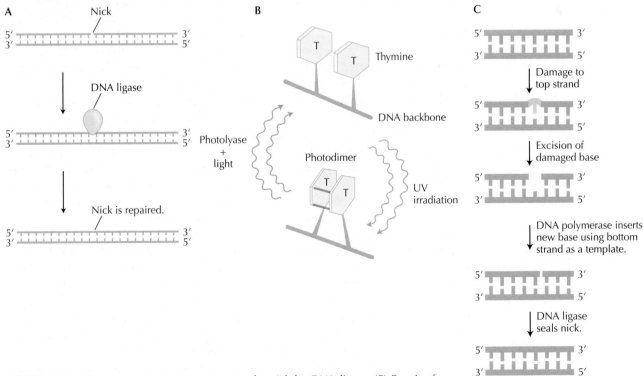

FIGURE 12.13. DNA-repair processes. (*A*) Repair of a nick by DNA ligase. (*B*) Repair of a UV-induced pyrimidine dimer by photolyase. (*C*) Base excision repair.

Excision repair mechanisms share a common strategy. First, the section of the DNA strand containing the abnormality is removed. Second, the other strand is used as a template to resynthesize the removed section. Third, the newly synthesized patch is ligated into place (Fig. 12.13C). There are three major forms of excision repair, distinguished by the type of abnormality removed and by the mechanism of recognition and removal: Base excision repair replaces abnormal single bases (e.g., uracil, as discussed on p. 332); nucleotide excision repair restores a damaged region of DNA by replacing a large section of DNA that includes the damage; and mismatch excision repair repairs base-pair mismatches and indel errors generated by replication mistakes. Mismatch excision repair (also known as mismatch repair) is discussed in the next section.

Recombinational repair is a specialized form of **homologous recombination**. In homologous recombination, two pieces of DNA that are identical or nearly identical in sequence (e.g., two copies of a chromosome) are aligned, and a section of DNA is exchanged between the two. The mechanics of recombination are discussed in more detail on pages 339–342. For the purposes of DNA repair, the important point is that homologous recombination can be used to exchange a section of one chromosome that has been damaged or altered with an undamaged section present in a homologous copy of the same chromosome. This is most commonly used to repair single- and double-strand breaks or to allow DNA replication to proceed past the point of DNA damage. Homologous recombination is also used in sexual reproduction, and the connections between sex and DNA repair are discussed in Web Notes for Chapter 23.

Error-correcting Mechanisms Reduce the Number of Mutations Resulting from Replication Errors

Even under the best situations (e.g., when all DNA damage has been removed and cells are healthy), DNA replication errors occur on the order of once every few thousand base pairs. Two mechanisms for detecting and correcting errors, proofreading and **mismatch repair**, greatly reduce this error level and achieve a remarkable overall accuracy of replication. Both of these processes are based on a simple premise: Whenever the wrong base is inserted during DNA replication or a few extra bases are added or deleted, the normal structure of the double helix is distorted.

During proofreading, the DNA replication machinery itself checks the recently replicated DNA for these structural distortions. If abnormal base pairing is detected, the newly replicated base is removed and DNA replication continues from the position of the base that was removed (Fig 12.14). If the error was due to a transient tautomerization, it is unlikely that it will occur again at the same site, and thus the polymerase is unlikely to make a mistake the second time. DNA proofreading can sometimes correct slip-strand mispairing (SSM) mistakes by recognizing a distortion in the double helix due to the incorporation of extra DNA in one strand.

Proofreading is imperfect, leaving many mispaired bases and most SSM errors. However, until the DNA has been replicated again, the possibility remains of recognizing mismatched base pairs or unpaired regions that result from SSM errors. Many of these errors are corrected by the mismatch repair process (Fig. 12.15). Base-pairing mismatches or unpaired loops are recognized, and the base that is on the newly replicated DNA strand is removed. The region that contained the excised base is then resynthesized by DNA polymerase (frequently one specialized for this repair) using the opposite (and presumably correct) strand as a template.

Two critical features enable mismatch repair to reduce mutations caused by replication errors. First, it recognizes alterations in the general structure of the genome. It is not necessary that each type of mismatch and small loop of DNA be recognized. Sec-

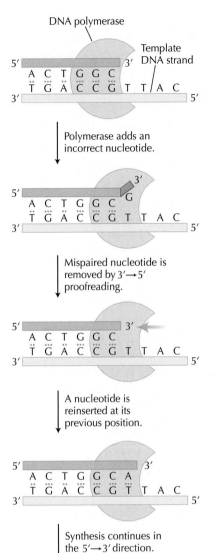

DNA polymerase

Template
DNA strand

Polymerase adds an
incorrect nucleotide.

Mispaired nucleotide is
removed by 3′→5′
proofreading.

A nucleotide is
reinserted at its
previous position.

Synthesis continues in
the 5′→3′ direction.

FIGURE 12.14. Proofreading during DNA replication. The DNA polymerase is shown as a *crescent*. The template strand (*blue*) is on the bottom and the newly replicated strand (*green*) is on the top. In the *second* panel, the polymerase makes an error and places the wrong base opposite the T, leading to an abnormal DNA structure (indicated by the *red kink*). In the *third* panel, the proofreading function of polymerases recognizes this abnormality and backs up by removing the recently added base. Replication then restarts from this position (*fourth* panel).

ond, the mismatch repair system correctly identifies which strand is newly synthesized. This is critical because the replication error will be on the new strand. The mechanisms of this strand recognition vary between species. In *E. coli*, this is done through a methylation-based process in which methyl groups are attached to the adenines in regions of the genome where the sequence GATC is present (Fig. 12.15). This methylation occurs a short time (i.e., within minutes) after DNA replication occurs. A newly synthesized strand will thus be transiently unmethylated, whereas the template will be methylated fully. It is the unmethylated state of the newly synthesized strand that is recognized by the mismatch repair system. The mismatch repair process can then accurately remove any newly synthesized segments that contain replication errors. The repair process must be complete before the next round of replication, otherwise the mismatch will disappear after the strand containing the error is used as a template for DNA replication.

Mismatch repair is a bit of a misnomer, because the systems repair base-pair mismatches and small loops (e.g., those generated by SSM error). SSM errors are not readily corrected by proofreading, so mismatch repair is the primary way these mistakes are corrected before they become mutations. Thus, when mismatch repair systems are absent or defective, there is a large increase in indel mutations.

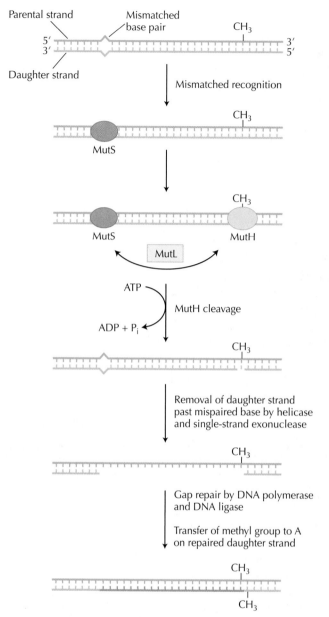

FIGURE 12.15. Mismatch repair in *Escherichia coli*. A recently replicated section of DNA containing a mismatch (indicated by a distortion in the DNA backbone) is shown. The parental (template) strand is on the *top* and the daughter (newly replicated) strand is on the *bottom*. The parental strand contains a methylated adenine (at a GATC sequence, indicated by the CH_3) as a result of the action of a methylase enzyme. The daughter strand has not yet had a chance to be methylated because replication just occurred. The MutS enzyme binds to the mismatch, and then in conjunction with the MutL protein causes the MutH endonuclease to cleave the DNA backbone of the newly replicated strand. MutH cleaves the unmethylated DNA strand at sites where one strand has a methyl group on the A at GATC sites. The DNA surrounding the MutH-generated break is removed and then resynthesized. Mismatch repair in other species works in similar ways, except the strand recognition system is sometimes different.

Mutations Can Arise from the Missegregation of DNA

The process of transmitting traits to offspring requires more than simply copying the genomes. Once replication is completed, the two copies of the genome need to be properly partitioned to daughter cells, with each daughter cell getting exactly one complete genome. This process is known as chromosome segregation.

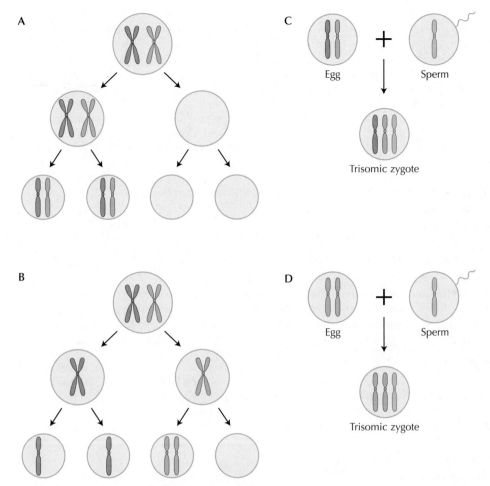

FIGURE 12.16. Segregation errors during meiosis. Chromosomes are supposed to be partitioned equally to the daughter cells. When this does not occur, it is known as nondisjunction. (*A*) Nondisjunction in the first division of meiosis (meiosis stage I). (*B*) Nondisjunction in the second division of meiosis (meiosis stage II). (*C*) What happens if an egg with the pattern from the *left* part of *A* participates in sexual reproduction. (*D*) What happens if an egg with the pattern from the *right* part of *B* participates in sexual reproduction.

In diploid or polyploid species, errors in chromosome segregation can produce daughter cells with different numbers of copies of individual chromosomes (somy) or all the chromosomes (ploidy). These and other types of segregation errors are somewhat common during meiosis (Fig. 12.16). Thus sexual reproduction not only generates variation through recombination, but it also contributes to the origin of mutations (see Box 12.2).

Duplicate Regions of the Genome Can Catalyze Major Rearrangement Events and Other Mutations

The form of homologous recombination most familiar to us occurs between two perfectly aligned homologous chromosomes during meiosis in sexually reproducing eukaryotes (discussed more below). However, recombination can occur between any two regions of a genome that have significant levels of sequence similarity. Known as **ectopic recombination**, such events can generate mutations. To understand how, it is necessary to understand the basic mechanism of homologous recombination (Fig. 12.18A). (In this broader context, "homologous" means that the participating DNA molecules are very similar to each other in sequence.)

In sexually reproducing eukaryotes, when there are polymorphisms in the structure of the genome (e.g., indels and inversions), there can be significant effects on meiotic recombination and chromosome segregation (Fig. 12.17). Here we describe several kinds of polymorphisms and their effects.

Insertion/deletion: In a heterozygote, an insertion cannot pair and thus loops out. Clearly, there can be no recombination in this region.

Fusion/fission: Chromosomes with terminal centromeres can fuse together to produce a single chromosome with the centromere in the middle. In a heterozygote, the fused chromosome will pair with the two unfused chromosomes. Usually, these segregate properly at meiosis, so that each gamete receives a complete haploid genome. However, occasional errors produce **aneuploid** gametes that lack one copy of each region of genome.

Multiple fusions: When a heterozygote contains multiple fusions, the chromosomes will pair in a single complex structure. This rarely segregates properly at meiosis, and so most such heterozygotes are sterile.

Reciprocal translocation: A segment from one chromosome may be exchanged with a segment from another. A complex structure involving four chromosomes forms at meiosis in the heterozygote. Proper segregation is rare, usually leading to sterility.

Inversion: If there is an inversion in a large section of chromosome, a loop will form at meiosis in the heterozygote. The entire chromosome pairs, and provided there is no recombination within the inverted region, the chromosomes will segregate properly at meiosis. However, if a cross-over occurs within the inversion loop, aberrant gametes are formed. One copy carries two centromeres and a duplication of the region near the centromere, whereas the other carries a chromosome fragment with no centromeres and a duplication of the other end of the chromosome. Because cross-overs are relatively common in inversion loops, inversion heterozygotes are usually sterile or have reduced fecundity, and large inversions are usually absent as polymorphisms within populations.

Inversion polymorphism in Diptera: In Diptera (i.e., flies, including *Drosophila*), special mechanisms limit sterility in inversion heterozygotes. Because there is no recombination in males, inversion heterozygotes suffer no problems in male meiosis. In females, only one of the four products of meiosis goes on to produce the egg. Thus, although recombination in female heterozygotes does produce aberrant chromosomes, these are shunted into meiotic products that have no future. Only intact nonrecombinant chromosomes segregate into the future egg cell. The consequence is that inversion heterozygotes are fully fertile as males and females; offspring show no recombination within the inverted region. In effect, these two chromosome arrangements form separate pools, which cannot exchange genes with each other. *Drosophila* inversions are valuable tools in the laboratory and inversion polymorphisms are common in nature.

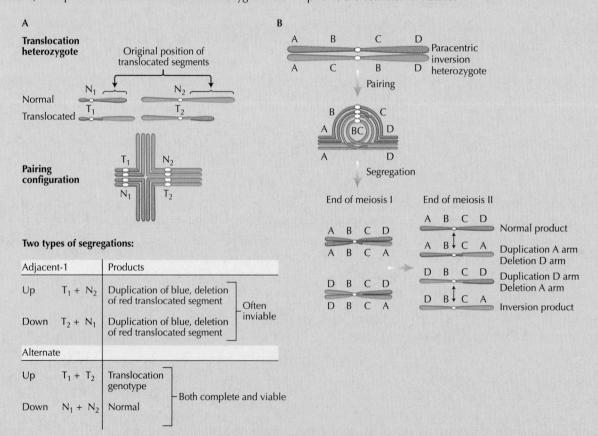

FIGURE 12.17. (*A*) The meiotic products resulting from the most commonly encountered chromosome segregation patterns in a reciprocal translocation heterozygote. (*B*) Meiotic products resulting from a meiosis with a single cross-over within a paracentric inversion loop (*top*, different chromosomes; *middle*, pairing of these; *bottom*, segregation results).

1. Two separate regions of DNA (either in different molecules or at different locations in the same molecule) that are identical or nearly identical are physically aligned. The frequency of alignment increases with the length of similarity; regions less than 50 base pairs long rarely align.

2. Nicks are introduced into one or both DNA strands (if they are not already present). The DNA near the nick is frequently degraded.

3. One strand from one of the DNA molecules base-pairs with the complementary strand in the other molecule. This "strand invasion" results in the formation of a cross-over of DNA strands known as a **Holliday junction**.

4. The position of the Holliday junction can "migrate" by a zipper-like mechanism, unpairing the DNA strands ahead of it and zipping up the strands behind it.

5. The Holliday junction is cleaved (a process known as resolution).

The outcome of a recombination event depends on a variety of factors, including whether the participating DNA molecules are circular or linear, where the duplicated regions are located relative to elements that control chromosome segregation (e.g., centromeres), whether the recombination is intra- or intermolecular, and how the Holliday junction is resolved. For example, when duplicated regions are arranged in tandem within a chromosome, recombination between two copies of that chromosome can lead to mutations if the tandem repeats are not aligned accurately (Fig. 12.18B). The result of this **unequal crossing over** is that one offspring receives too many copies of the duplicated sequence and one receives too few. This outcome is similar to the indel mutations generated by SSM replication errors that were discussed earlier. Unequal crossing over (and SSM) can result in the homogenization of repeats within a genome, because one form of the repeat can spread and eliminate other forms (Fig. 12.18C). The result is that all of the duplicates within a genome remain similar to each other in sequence, even as they diverge from homologs in other species, a phenomenon known as **concerted evolution**.

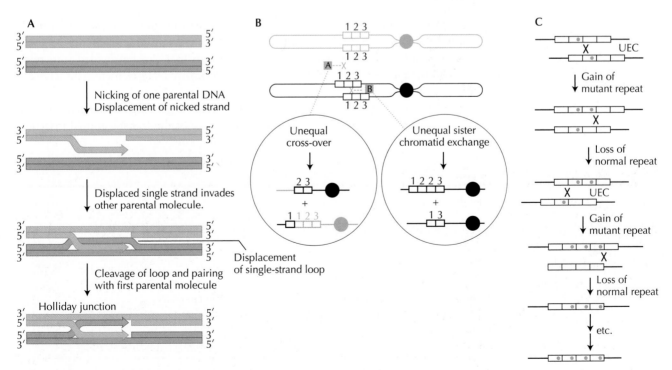

FIGURE 12.18. Steps in homologous recombination. (*A*) Homologous recombination. (*B*) Unequal cross-over. (*C*) Homogenization of repeats by unequal cross-over (UEC).

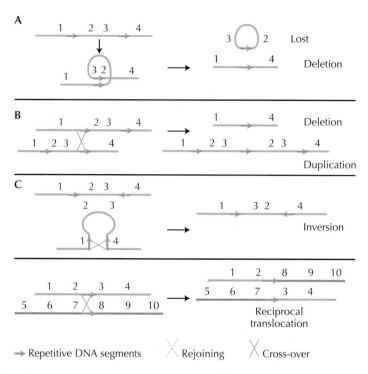

FIGURE 12.19. Chromosomal rearrangements due to crossing over between repetitive DNA. (*A*) An intrachromosomal recombination leads to a deletion. (*B*) Homologous chromosomes recombine out of register at duplication regions, leading to a deletion in one molecule and a duplication in the other. (*C*) Intrachromosomal recombination at inverted repeats leads to an inversion.

Other types of mutations that can be generated by recombination at repeats include translocations, inversions, duplications, and deletions (Fig. 12.19).

Many Mutations Are Caused by Transposition

Another important source of mutation is **transposition**, the movement within the genome of mobile DNA elements (also known as **transposable elements** or **transposons**). Transposable elements are a diverse group that can be divided into retrotransposons, replicative DNA elements, and nonreplicative DNA elements (pp. 217–218; Table 12.2; Fig. 12.20).

The mutagenic effects of transposition are varied, depending on the genomic sites that are affected, the type of transposon, the accuracy of the transpositional insertions

TABLE 12.2. Major classes of transposable elements

Type	Characteristics
Retroelements (RNA elements)	Original element is transcribed into RNA, reverse transcribed into DNA, and then inserted into the genome.
Retrotransposon (LTR)	Retroelement that possess LTRs
Retroposon (non-LTR)	Retroelement that does not possess LTRs
DNA elements	Found in DNA genomes. Do not function through an RNA intermediate
Replicative	Original element is copied at the DNA level and inserted elsewhere in the genome, resulting in a duplication.
Nonreplicative (conservative)	Element moves via excision from the original location and insertion into a new location.

LTR, long terminal repeat

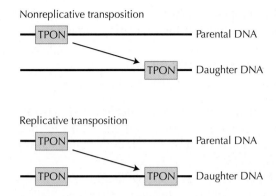

FIGURE 12.20. DNA transposition. Schematic diagram of two classes of transposition. In nonreplicative transposition *(top)*, a transposon moves from one location to another. In replicative transposition *(bottom)*, a transposon adds a copy of itself to a new location while maintaining its original copy.

and excisions, and other factors. Many transposable elements contain transcription terminator signals and stop codons in all reading frames. Thus, the insertion of a transposon into a coding region of a genome often prevents full transcription and translation. Transposons can also have indirect effects, such as changing transcription levels of genes near the insertion site (e.g., p. 405). Because they often contain promoters and splicing signals, transposable elements are important contributors to the evolution of new genes and of new patterns of expression and splicing for existing genes.

Transposition is not the only way that transposable elements can cause mutations. The frequent duplication of retrotransposons or replicative DNA transposons creates opportunities for mutation through ectopic recombination (described in the previous section). Insertion of all types of transposons is frequently accompanied by a small duplication of the site flanking the insertion. In essence, transposons function as portable regions of homology, and their proliferation can have profound effects on genome evolution. The more copies there are of a particular element within a genome, the more likely that they will promote genomic rearrangements, duplications, and deletions. Evidence for the effects of transposition can be found by aligning the genome sequences of closely related species. Transposable elements are often found at inversions and other breaks in the conservation of genome structure, suggesting that the elements caused the rearrangements (Fig. 7.13).

Transposons are almost universal among living organisms. However, the impact of transposition varies greatly within and between species, caused in part by the total number of transposable elements within the genome as well as the amount of transposition. Approximately 50% of the maize genome is transposons. There are dozens of classes of transposons represented, and some are present in thousands of copies. One familiar effect of transposition in maize is the color variegation seen in the kernels of some strains. In *Drosophila*, transposition accounts for a large fraction of the spontaneous mutations that occur in its genome (e.g., Fig. 13.11). Although the human genome contains a large number of transposable elements, their transposition rarely causes mutation because most of those elements are inactive (see p. 598). However, they are the sites for many ectopic recombination events. Some species, such as intracellular bacteria, lack transposons, and so transposition is not a cause of mutation for them.

RATES AND PATTERNS OF MUTATIONS

Mutation rates and patterns vary in many ways. Different types of mutations occur at different rates. The same types of mutations can occur at different rates in different regions of the genome or in different individuals. Both the rate and the pattern of mutation differ greatly in different species. Environmental conditions such as exposure to irradiation and stress influence the rates and types of mutations seen. In other words, the generation of variation by mutation is variable. In the following sections we discuss some of the key factors that are known to influence patterns and rates of mutation. In Chapter 23, we will ask whether mutation is just a side effect of imperfect replication or is it sometimes selected as an adaptation for generating variation.

The Occurrence of Mutations Is Independent of Their Consequences for Fitness

Modern evolutionary theory presumes that mutation and selection are separate processes. Mutations (and recombination) generate variation without regard to their consequences for fitness (i.e., survival and reproduction); selection and other forces then act on the variation, leading to adaptation.

This was not always the commonly held view. For example, before 1900 it was widely believed that the changes in phenotype that occur during an organism's lifetime could be passed on to its offspring. This **inheritance of acquired characteristics** implies that the characteristics acquired by the parents produce genetic changes in the offspring. This theory is inconsistent with the general tenets of molecular biology, including the **Central Dogma**. Nevertheless, it is quite difficult to either disprove the possibility of "Lamarckian" evolution or prove that mutation and selection are not connected. A conclusive answer requires knowing whether mutations already exist in populations or whether they arise only in response to particular selective pressures. In particular, if a population were exposed to stressful conditions and only a subset survived the stress, what had enabled that subset to survive? Did they adapt to the stress without the need for mutational change? Did the stress generate new mutations? Did the stress select individuals with preexisting adaptive mutations, who were then able to reproduce?

The first compelling experimental test of when mutations occurred relative to selection came from replica-plating experiments with bacteria (Fig. 12.21). In this procedure a master plate was prepared by inoculating an agar plate with a small number of bacte-

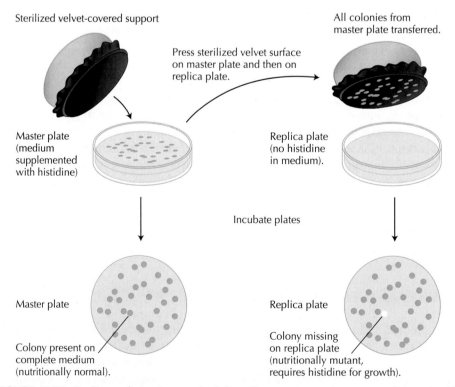

FIGURE 12.21. Replicate plating is a method that allows one to make a copy of the microbial colonies on one plate by carefully placing a piece of material that lifts a small sample from the original plate and can then place this sample onto another plate. In the example shown here, the method is used to identify colonies that require histidine for growth.

ria and allowing them to grow to produce a number of individual colonies. Using a small number of bacteria ensured that each colony would be composed of a clone of bacterial cells all derived from one individual bacterium. A portion of every colony on the master plate was transferred to a series of fresh plates such that each new plate was a replica of the master plate (i.e., a small number of bacteria from each colony were present in the same spatial arrangement as on the master plate). Next, selective pressure was applied uniformly to each replica plate, and the bacteria were given an opportunity to grow under these selective conditions. Each visible colony represented a clone of bacteria that could grow under selective pressure due to the presence of a mutation.

This experimental protocol was able to distinguish between two possible processes. If the mutant bacteria already existed in the population prior to selection, then visible colonies would be found in identical locations on each replica plate. Alternatively, if the mutations arose during selection in response to the selection conditions, similar numbers of bacterial colonies would be visible on each replica plate, but their arrangement would vary from plate to plate. The results showed visible colonies in identical locations on each replica plate. This indicates that the mutations existed in the bacterial population prior to selection and thus were not induced by the selection.

However, replica-plating experiments left unanswered some questions about the origin of mutations. A different experimental approach was applied by Salvador Luria and Max Delbrück in the 1940s, who devised a simple, yet elegant, test known as a **fluctuation test** (Fig. 12.22). A culture of a **bacteriophage**-sensitive strain of bacteria was grown in liquid medium. Equal volumes from that culture were then removed and grown in separate subcultures. Next, equal volumes of each subculture were transferred onto agar plates containing a phage capable of infecting and killing the bacteria. The plates were incubated to allow any phage-resistant bacteria to grow and produce a visible colony. The number of colonies on each plate was counted to determine the number of phage-resistant bacteria in each subculture. If mutations conferring phage resistance arose in response to selection, then every plate should contain approximately the same number of resistant colonies. On the other hand, if phage resistance arose prior to selection, then the samples from different subcultures should have vastly different numbers of colonies— a result Luria and Delbrück called the "jackpot" pattern. This variation is observed because mutations that arose early in the growth of the subculture would have produced many resistant offspring, whereas mutations arising much later would have had little time to produce resistant offspring. Luria and Delbrück's experimental results displayed the "jackpot" pattern, indicating that the mutations arose prior to selection.

From these results, the conclusion was that mutations arose randomly without respect to their adaptive value, because they occurred before selection was applied. This is what would be expected based on the mechanisms of mutation and repair discussed earlier in this chapter. For example, when proofreading a newly synthesized strand, DNA polymerase cannot tell the difference between coding and noncoding regions, nor can it distinguish between a gene promoting cell division and one preventing it. Similarly, the enzymes repairing mismatched bases are unaware if the error being corrected would benefit or harm a particular cellular function (see pp. 336–338).

To many scientists, Luria and Delbrück's work answered all the major questions about the origin of mutations. However, a few people noticed a flaw in this experiment that left one question unanswered: Do mutations that result in increased fitness arise both before and after selection is applied? The selective regime imposed by Luria and Delbrück did not allow mutations to arise after selection, because the phage killed any bacteria that were not resistant. To address this question, subsequent experiments were performed using a nonlethal selective agent. The results demonstrated that mutations that result in increased fitness arise both before and after selection is applied.

In one of these experiments, bacteria containing a mutation in the *lac* gene, which prevented them from utilizing lactose as a carbon source, were transferred to plates containing lactose as the only carbon source. Under these conditions, the mutant bac-

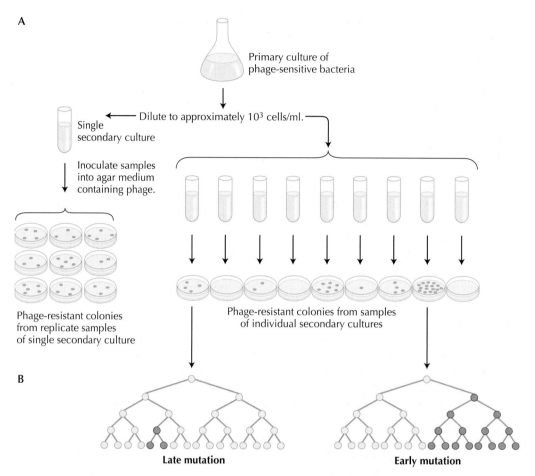

FIGURE 12.22. Fluctuation test. (*A*) The protocol as used by Luria and Delbrück to study the origin of mutations. Their test focused on mutations in bacteria that conferred resistance to killing by phage. The test was designed to determine if such mutations arose prior to exposure to the phage or specifically in response to exposure. A primary culture of bacteria was grown that had been inoculated with a phage-sensitive strain. Multiple secondary cultures were made by transferring small amounts of the primary culture to new growth media. The secondary cultures underwent many rounds of replication. Subsamples of each culture were removed, mixed with phage, and placed onto growth plates. After several days, the number of colonies on each plate was counted. This number is a measure of the number of cells in the subsample that were able to resist killing by the phage. There were two key results in this experiment. First, replicate subsamples from a single secondary culture gave similar numbers of phage-resistant colonies (shown on the *left*). Second, and more importantly, different secondary cultures of the same primary culture yielded wildly different numbers of colonies (shown on the *right*). They concluded that this "jackpot" pattern could only occur if mutations in the bacteria arose prior to their exposure to the phage. (*B*) Mutations that occurred late in the growth of the secondary culture would yield few mutant cells (shown in *red*), and thus few colonies, at the end of the growth of the culture (tree on the *left*). Mutations that occurred early would yield many mutants, and thus many colonies (tree on the *right*).

teria were not killed, but nor could they grow. A mutation (e.g., a reversion in the defective *lac* gene) was required for growth on lactose. The pattern of mutation observed was a mix between the jackpot model expected for preexisting mutations and the equal distribution expected for mutations that arose after applying the selection. Although there was some initial controversy over these results, it has now been shown that the selection pressure did not selectively increase the rate of mutation in the *lac* gene, but rather increased mutation rates across the entire genome. These results indicate that although mutations continue to occur after selection pressure is applied, the mutations occur randomly with respect to adaptation.

It is important to realize that just because mutation is random with respect to adaptation does not mean that organisms have no control over mutation processes. In fact, mutation is actively regulated and manipulated in organisms in many ways. For example, mutation rates can be induced under stress or increased in specific regions of the genome. These processes are discussed in greater depth in Chapter 23. For now, note that even when there is some control over the rate of mutation, the particular mutations that are seen are not directly linked to their adaptive value.

Different Types of Mutations Occur at Different Rates

Mutation rates vary depending on the type of mutation. Perhaps the best example is the difference between transitions and transversions. In almost all species that have been examined, transition mutations occur at a much higher rate than transversions, in part because of the tautomerization of DNA bases described above (see pp. 326 and 328). Another reason is that base misincorporations that lead to transitions involve substituting one purine for another or one pyrimidine for another. This distorts the DNA double helix less than when substituting a purine for a pyrimidine or vice versa. For example, substituting a T for a C in a G-C base pair can lead to a transition mutation. The G-T base pair that is formed by the misincorporation will still be a purine–pyrimidine pair. In contrast, any misincorporation that would lead to a transversion would create either a purine–purine or a pyrimidine–pyrimidine pair. Pyrimidine–pyrimidine and purine–purine pairs are less likely to occur in the first place and are more likely to be recognized by the DNA proofreading and repair machinery. In turn, this reduces the rate of transversions compared to transitions.

Mutation Rates Can Vary across the Genome

Mutation rates at different genomic locations can vary because of the local sequence context. For example, regions containing long runs of the same base have a higher rate of indel mutations than do other regions, because of the increased likelihood of SSM errors during DNA replication. If the DNA template contains ten adenines in a row (AAAAAAAAAA), the chance of an SSM error producing a new complementary strand with a short indel would be much greater than if the template was ATCGCTGATT. Similarly, SSM-induced mutations are more frequent for **microsatellite DNA,** which contains repeat elements of 2–10 base pairs long (such as ATATATATAT) or **minisatellite DNA** with repeats of 11 to approximately 100 base pairs (see Box 13.3). The SSM-induced mutation in these regions involves changes in the number of copies of the repeat motif (e.g., AT above). Thus these regions are generally known as **variable number tandem repeats** (VNTRs). The high mutation rate of these short repeats makes microsatellite repeats excellent markers for population genetic studies (see Box 13.3).

Another example of sequence context altering mutation rates can be seen in mammalian genomes. In mammals, normal DNA methylation mechanisms preferentially methylate the cytosines in the DNA sequence 5′-CG-3′ (also known as CpG). Thus, a C next to an A will be less likely to be methylated than a C next to a G. As discussed above, when a methylated cytosine is deaminated, it becomes a thymine, which, if replicated, leads to a mutation. If the cytosine is not methylated, the deamination creates uracil, which can be easily recognized and repaired (see pp. 331–332). Thus, because they are more likely to be methylated, cytosines next to guanines have higher rates of mutation than do cytosines in other locations.

Other factors that affect mutation rates include proximity to the origin of replication, repetitive sequences, the presence of transposable element insertion signals, and proximity to telomeres and centromeres. The realization that mutation rates vary with location cautions us not to apply universally a mutation rate that is measured in only one part of the genome. It also reminds us that a gene's sequence context and genomic context can profoundly affect its rate and pattern of evolution.

Mutation Rates and Patterns Vary Enormously between Species

The variation of mutation rates between species has been the subject of intense study over many years, and some interesting patterns have emerged. The mutation rates characteristic of organisms with different genetic systems are shown in Figure 12.23. The mutation rates are expressed as the number of mutations per base pair per replication, and as such they reflect the molecular fidelity of the enzymes that replicate and repair the genome. Notice that smaller genomes have higher mutation rates than larger ones. Thus, on average, there is a negative correlation between genome size and mutation rate per base pair. If mutation rates are measured in terms of number of mutations per genome per generation, less variation between species is observed.

The highest mutation rates are seen in viruses such as HIV (human immunodeficiency virus), which have small, single-stranded RNA genomes. Their high error rate is thought to be partly caused by the lack of proofreading by the RNA polymerases and reverse transcriptases that replicate them and by the inherent difficulty of repairing single-stranded polynucleotides. This property of RNA viruses has been extensively studied in HIV. If samples of HIV taken from one individual at different times or from different locations during the course of an infection are compared, large genetic differences are frequently found, even when it is known that the individual has been infected with only a single strain of the virus (see p. 447). This enormous amount of genetic diversity is generated during the course of a single infection because the mutation rate in HIV is extraordinarily high—approximately 1,000,000 times higher than the mutation rate for DNA replication in human cells. The mutation rate in HIV reflects the high rate of base misincorporation by its reverse transcriptase (~1 error per 2000 nucleotides) and the absence of proofreading and mismatch repair processes.

Predicting evolutionary consequences from mutation rates per cell division can be misleading because different species, and even different sexes within a species, experience different numbers of cell divisions per generation. For example, in humans all eggs are produced in the developing female fetus where they undergo approximately 30 cell divisions before entering stasis. In contrast, human sperm are produced throughout the lifetime of adult males and their lineages have usually undergone more

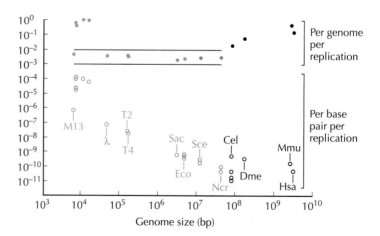

FIGURE 12.23. Mutation rates versus genome size. Mutation rates are shown for a variety of organisms relative to genome size. On the *bottom* portion of the graph, mutation is plotted per base pair per replication event. Note the downward trend, with mutation rate decreasing as genome size increases. On the *top* portion, mutation rate is plotted per genome per replication. Note the relative uniformity in the values across diverse organisms and genome sizes. RNA viruses (*red*): rhinovirus, poliovirus, vesicular stomatitis virus, and measles virus. DNA viruses (*green*): MI3, λ, T1, and T2. Archaea (*blue*): Sac, *Sulfolobus acidocaldarius*. Bacteria (*blue*): Eco, *Escherichia coli*. Eukaryotic microbes (*purple*): Sce, *Saccharomyces cerevisiae*; Ncr, *Neurospora crassa*. Metazoa (*black*): Cel, *Caenorhabditis elegans*; Dme, *Drosophila melanogaster*; Mmu, mouse; Hsa, human.

than 100 cell divisions. This difference in the number of cell divisions may explain why the germ line of human sperm accumulates new point mutations more rapidly than does the germ line of human eggs.

GENERATION OF VARIATION BY MIXING: SEX AND LATERAL GENE TRANSFER

Sexual Recombination Mixes Gene Combinations within Species through Assortment and Crossing Over

In species that reproduce sexually, most of the genetic differences between individuals are the result of sexual recombination rather than recent mutations. To understand how sexual recombination contributes to variation, it is necessary to understand the process of recombination as it occurs during sexual reproduction (as described next). The selection that shapes sexual reproduction is discussed in Chapter 23.

Sexual reproduction mixes the DNA from two haploid gametes, typically from different lineages, to produce a diploid offspring. Whereas a haploid cell contains one of each chromosome and has a DNA content designated as $1n$, a diploid cell contains two of each chromosome and has a DNA content of $2n$. Thus if the genome of an organism consists of four chromosomes, a diploid cell would contain eight chromosomes, four from one parental lineage and four from the other. The matching chromosomes from the two parents are termed homologous chromosomes.

The process of sexual reproduction starts with the formation of haploid gametes by **meiosis** (Fig. 12.24A). At the beginning of meiosis, a diploid cell replicates its DNA so that its DNA content now equals $4n$. The products of replication of each chromosome, termed **sister chromatids**, remain together. Next, the sister chromatids from one chromosome pair up and align with the sister chromatids of the homologous chromosome. The matched chromosomes can then exchange genetic information through the process of crossing over.

On the molecular level, crossing over involves a series of reciprocal homologous recombination events that produce recombinant chromosomes containing DNA segments from the two parental lineages (Fig. 12.24B). During the first meiotic division, the pairs of homologous chromosomes segregate randomly, one homolog of each chromosome normally going to each daughter cell. The sister chromatids then separate during the second nuclear division, thus generating four gametes each containing a randomly assorted haploid set of chromosomes (Fig. 12.24C). Thus, meiosis utilizes two mechanisms—homologous recombination and **random segregation**—to generate genetic variation.

The process of sexual recombination can generate vast amounts of diversity. Imagine an organism with only one pair of homologous chromosomes that has 1000 single-nucleotide differences (also known as single-nucleotide polymorphisms [SNPs]) between the two. In this example, 2^{1000} ($\sim 10^{301}$) different forms of each chromosome can be generated by sexual recombination. Although crossing over will not generate all these combinations (for closely linked polymorphisms, a cross-over between them will be rare), the *potential* for variation is enormous.

Recombination and Segregation Occur Independently of Sexual Reproduction

The production of genetic mixing by recombination is not limited to the meiotic divisions that occur during sexual reproduction. In diploid organisms, a low level of recombination occurs between homologous chromosomes during mitosis. When mitotic recombination occurs in single-celled organisms or in the germ line of multicellular organisms, it can lead to genetic mixing in much the same way as meiotic recombination does. Likewise, homologous recombination in conjunction with mating or lat-

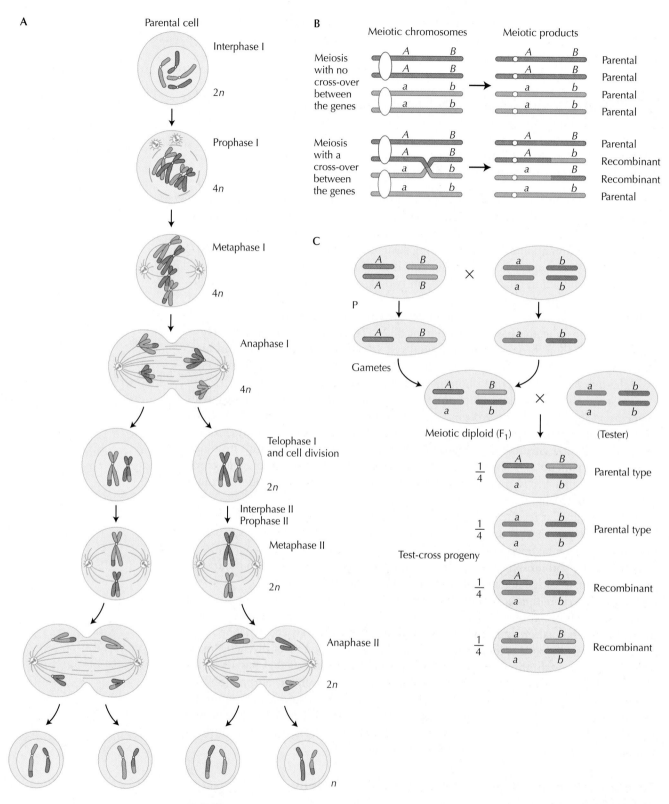

FIGURE 12.24. (*A*) Meiosis in a diploid organism. First, the genome is copied to produce a *4n* cell. Then crossing over occurs and cell division produces two *2n* cells. These go through another round of division to produce haploid (*n*) gametes. (*B*) Recombination due to cross-over. One region of the genome is shown for a *4n* cell. (*Top*) No cross-over; (*bottom*) a single cross-over. (*C*) Recombination due to segregation. A cross between gametes from two parents is shown. The assortment can be readily detected using a tester cross (not recombinants).

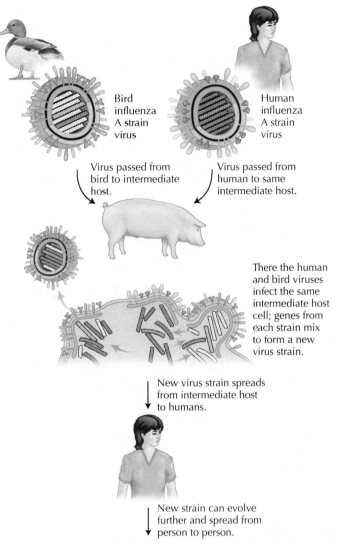

Bird influenza A strain virus

Human influenza A strain virus

Virus passed from bird to intermediate host.

Virus passed from human to same intermediate host.

There the human and bird viruses infect the same intermediate host cell; genes from each strain mix to form a new virus strain.

New virus strain spreads from intermediate host to humans.

New strain can evolve further and spread from person to person.

FIGURE 12.25. Antigenic shift in influenza.

eral gene transfer in bacteria and archaea also leads to the mixing of the genetic sequences from different lineages.

Viral genomes that are composed of multiple chromosomes provide another opportunity for recombination. When multiple viruses infect a host cell, genetic exchange can occur during viral replication. The participating viruses can be different lineages of the same virus or different types of viruses. The flu virus, whose genome is composed of eight chromosomes, provides an example of great importance to humans. During infection of a host cell, all eight chromosomes are replicated in preparation for packaging into new viral particles. When a single host cell is infected with two strains of the virus, the new viral particles produced contain a mixture of chromosomes from the two original flu strains (Fig. 12.25). In addition, recombination events similar to those occurring in sexually reproducing organisms can occur between the two versions of each chromosome.

Lateral Gene Transfer Moves Genes between Species

As discussed in Chapter 5, **lateral gene transfer** (LGT) refers to the movement of genes from one evolutionary lineage to another (see Fig. 5.23). It introduces variation into a population in a manner that is reminiscent of both sexual recombination and mutation. It is useful to compare LGT with meiotic sex in order to understand the effects of LGT on evolution and how LGT itself might evolve.

Both meiosis and LGT create variation by combining genetic material from two sources. Some types of LGT require homologous recombination, the same mechanism as used in sexual recombination. When a segment of foreign DNA is very similar to a region of the recipient's genome, it can integrate into the genome by homologous recombination, replacing the corresponding segment of the recipient's DNA. Because a close similarity between donor and recipient DNA is required, successful integration is possible only for closely related cells.

In some bacteria and archaea (see pp. 183–184), conjugation transfers large sections of chromosome from a donor cell to a recipient of the same species. The subsequent homologous recombination creates new variants in much the same way as meiotic sex does. Thus conjugation is frequently referred to as bacterial sex.

LGT also differs from meiotic sex in several significant ways.

- LGT is a unidirectional transfer between a donor and a recipient.
- Whereas meiotic sex reshuffles two complete genomes in a single step, each LGT event typically moves only one small segment of a genome.
- LGT is not a regular part of any life cycle and usually occurs much less frequently than meiotic sex.
- Perhaps most importantly, LGT can occur across vast evolutionary distances. Thus, LGT can deliver entirely novel genes to an organism, not merely new alleles of existing genes.

It is important to realize that LGT has had a great impact on evolution in bacteria and archaea. The effects of LGT in bacteria and archaea in generating variation that selection can act upon may be similar to the effects of recombination by meiosis that occur in sexually reproducing organisms. We return to this issue in Chapter 23.

SUMMARY

Genetic variation, which is the grist of the mill of evolution, is generated through a combination of mutation and genetic recombination. Mutation, which is a change in the genome between parent and offspring, is the ultimate source of all variation. Genetic recombination generates novel patterns by mixing and matching variation from different lineages.

Mutations range from single base changes to small insertions, duplications, and deletions to chromosomal changes such as translocations and the formation of polyploids. Although there are many kinds of mutations, all of them can be traced to three types of events: errors that occur during replication of the genome, errors during segregation of the genome to offspring, and modifications to the genome that occur between rounds of replication (e.g., transposition). Overall, the rate of such events is quite high; yet in most species, the rate of mutation is extremely low. This is because repair and error checking mechanisms work together to limit the occurrence of mutations.

The processes that generate mutations occur without any direct regard for the consequences of mutations. This does not mean, however, that all mutations occur at equal rates. Mutation rates vary within genomes, under different conditions, and between individuals and species. This variation has important consequences for evolutionary processes and patterns (see Chapter 23).

The majority of variation is the result of genetic recombination—primarily, sex in eukaryotes and lateral gene transfer in bacteria and archaea. These processes mix and match mutations that occur in different lineages, greatly increasing the amount of variation that can be acted upon by evolutionary forces, like natural selection.

FURTHER READING

General Mutation

Drake J.W. 2006. Chaos and order in spontaneous mutation. *Genetics* **173:** 1–8.

Miller J.H. 2005. Perspective on mutagenesis and repair: The standard model and alternate modes of mutagenesis. *Crit. Rev. Biochem. Mol. Biol.* **40:** 155–179.

Sniegowski P.D., Gerrish P.J., Johnson T., and Shaver A. 2000. The evolution of mutation rates: Separating causes from consequences. *BioEssays* **22:** 1057–1066.

Mismatch Repair and Proofreading and Replication Errors

Kunkel T.A. and Erie D.A. 2005. DNA mismatch repair. *Annu. Rev. Biochem.* **74:** 681–710.

Schofield M.J. and Hsieh P. 2003. DNA mismatch repair: Molecular mechanisms and biological function. *Annu. Rev. Microbiol.* **57:** 579–608.

DNA Repair

Cline S.D. and Hanawalt P.C. 2003. Who's on first in the cellular response to DNA damage? *Nat. Rev. Mol. Cell Biol.* **4:** 361–372.

Fuss J.O. and Cooper P.K. 2006. DNA repair: Dynamic defenders against cancer and aging. *PLoS Biol.* **4:** e203.

Recombinational Repair and Recombination-induced Mutations

Kanaar R., Hoeijmakers J.H., and van Gent D.C. 1998. Molecular mechanisms of DNA double strand break repair. *Trends Cell Biol.* **8:** 483–489.

Adaptive Mutation

Luria S. and Delbrück M. 1943. Mutations of bacteria from virus sensitivity to virus resistance. *Genetics* **28:** 491.

Roth J.R., Kugelberg E., Reams A.B., Kofoid E., and Andersson D.I. 2006. Origin of mutations under selection: The adaptive mutation controversy. *Annu. Rev. Microbiol.* **60:** 477–501.

Mutation Rates

Denamur E. and Matic I. 2006. Evolution of mutation rates in bacteria. *Mol. Microbiol.* **60:** 820–827.

CHAPTER

13

Variation in DNA and Proteins

ONE OF THE MOST IMPORTANT OBSERVATIONS in evolutionary biology is that populations are highly variable. The development of techniques for sequencing proteins and for separating them by electrophoresis made clear the extent of variation at the molecular level (see p. 59). Almost all populations conceal an enormous wealth of genetic variability. For example, two randomly chosen human genomes differ at about 2.3 million positions within their DNA sequence. At the level of the expressed protein sequence, there will be approximately 11,000 amino acid differences. When we compare species, we see a more-or-less steady accumulation of molecular differences over time, known as the **molecular clock** (see p. 59). Thus, our understanding of evolutionary processes must begin with an appreciation of the nature and extent of inherited variation.

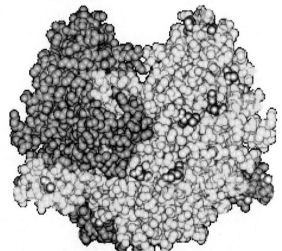

In this chapter, we describe the extent of variation in different kinds of genes and in different kinds of organisms. After summarizing the evidence from classical genetics, which first suggested that there is abundant cryptic variation hidden within populations, we concentrate on our current knowledge of the variation in DNA and protein sequence found within and between species. In the next chapter, we will see how variation in the traits that we observe directly depends on this underlying genetic variability. Later in the book, after explaining the different evolutionary processes that act on variation, we ask why such extensive variation exists, and what are its consequences. However, we begin this chapter by explaining just why so much of evolutionary biology is devoted to the study of variation between individuals within a population.

GENETIC VARIATION

Evolution Requires Genetic Variation

Populations evolve primarily through changes in the proportions of different kinds of individuals. All evolutionary processes depend on genetic variability. To see this, think of a completely homogeneous population, in which all individuals have exactly the same DNA sequence. Unless mutation introduces variability, the offspring must be identical to their parents, and there can be no evolutionary change (Fig. 13.1).

Darwin saw clearly that variation is essential to evolution. Over time, variation

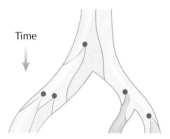

Time

FIGURE 13.1. Variability is necessary for evolution. The diagram shows a single ancestral species that splits into three present-day species (as indicated by the three branches). Evolution consists of the occasional origin of new alleles by mutation (*red dots*), followed by changes in their frequency in the population, represented here by different colors.

within populations is transformed into variation between populations and ultimately between species. Thus, much of *On the Origin of Species* is concerned with documenting variation, and many of the difficulties faced by Darwin's theory stemmed from ignorance of the nature of inherited variation (see p. 20).

We are especially concerned with natural selection, because it is this that creates the complex, well-adapted structures characteristic of life. Natural selection requires inherited variation in **fitness**—that is, there must be inherited variation that affects an organism's survival and reproduction. In this chapter, we examine the most fundamental kind of variation, which is seen in the sequences of DNA and of protein. In the next chapter, we show how this is related to variation in complex traits, such as body shape and behavior, and ultimately in fitness. We will then be in a position, in the rest of Part III, to examine natural selection and its consequences. In particular, in Chapter 19 we consider the evidence on variation in fitness itself.

To describe genetic variation, we need some basic terminology; this is explained in Box 13.1, as well as in the Glossary. Two key distinctions need to be kept in mind throughout the rest of the book. First, we distinguish between **genes** and **alleles**. A gene is a stretch of DNA sequence that codes for a protein or RNA molecule, together with the associated regulatory elements. Genes vary in sequence, and these different variants are called alleles (Box 13.1). Thus, it is more accurate to refer to the frequencies of different kinds of genes in a population as **allele frequencies**, rather than gene frequencies.

Second, we will refer to the combination of alleles carried by an individual as its **genotype**. Ultimately, this is the complete DNA sequence, but usually we need to focus only

Box 13.1 Genetic Terminology

Gene, locus, site: The fundamental concept of a gene appeared well-defined within classical genetics, as an indivisible factor that segregates according to Mendelian rules and that could recombine with other such factors. However, we now understand a gene as a loosely defined stretch of DNA that includes sequences that code for a protein or a functional RNA molecule (such as ribosomal RNA) together with associated regulatory sequences (Fig. 13.2A). Complications such as overlapping genes, genes-within-genes, and alternative splicing (Fig. 13.2B) make a watertight definition impossible. Nevertheless, the term gene has a straightforward meaning in most cases.

We use the term **locus** (plural loci) to refer in a general way to a location in the genome. This may refer to a long stretch of DNA that includes several genes or to a few hundred bases within a gene. The term is useful, because when our information comes from laboratory crosses, studies of human families, or population genetic analyses, we often do not know whether a locus corresponds to several genes, to one gene, or to a fragment of a gene. In the next chapter, we see how the loci responsible for variation in quantitative traits (e.g., crop yields, morphology, and disease susceptibility) can be identified. These will be referred to as **quantitative trait loci** (**QTLs**) because we usually do not know which gene is responsible and hardly ever know which bit of sequence within that gene is responsible. Similarly, in population genetics we often refer to loci rather than genes, because what matters is the effect of some small region of

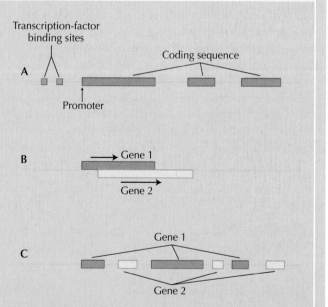

FIGURE 13.2. (*A*) A gene includes regulatory sequences (e.g., transcription-factor binding sites and the coding sequence, which may be broken up into exons (here, three). Complications to this definition include those illustrated in *B* and *C*. (*B*) Genes that may overlap. The sequence is translated in one reading frame to give one protein and in another reading frame to give a different protein. (*C*) Exons coding for two proteins may be interspersed. The messenger RNA must be spliced in two different ways to give two different proteins.

Box 13.1. Continued.

genome on fitness—not whether that region corresponds to the sequence coding for one functional protein or RNA molecule.

Ultimately, the fundamental and indivisible unit of genetics and evolution—the atom of biology—is the single base pair or **nucleotide**. Base pair refers to a pair of bases such as A:T or G:C within a DNA or RNA double helix, whereas nucleotide refers to the (deoxy)ribose-base subunit of the DNA or RNA polymer. In evolutionary genetics, the two terms are often used interchangeably to refer to a single unit of genetic information. Often, the smallest unit is referred to as a **site** (sometimes, a **nucleotide site**) rather than a locus.

To summarize: A gene refers to a functional unit, coding for a single protein or RNA molecule, whereas a locus refers to a region of genome that we treat as an approximate unit of inheritance. Ultimately, genes and loci are made up of strings of nucleotide sites.

Alleles and haplotypes: Different varieties of a gene are called alleles. In classical genetics, these were identified through their visible effects on the organism: An allele might cause wrinkled peas or white-eyed flies when present in two copies in a homozygote. Mutations at several different genes can cause the same phenotypic effect—any defect in the pathway that makes eye pigment can cause the white-eyed phenotype. The most important way of showing that different mutations are alternative alleles at the same locus is the complementation test (Fig. 13.3; recall pp. 244–246). If the recessive variants are brought together in the same individual, then if they are alleles at the same locus, the recessive phenotype will appear, just as if either of the two alleles were made homozygous. However, if the alleles are variants of different **genes**, then their different defects will be complemented by the functional **wild-type** alleles at each locus; thus, the wild-type phenotype will appear. This test is not foolproof—most obviously, it applies only to **recessive** alleles. However, it is the most widely used and straightforward test.

A gene can have many alleles—for example, the *white* locus in *Drosophila melanogaster* has at least 1517 different alleles (Fig. 13.4). The conventional notation for denoting alleles varies somewhat from organism to organism, and homologous genes often have different names in different organisms. In population genetics, and in this book, we often use a compact notation, with P, Q denoting alternative alleles at a single gene. If we have several loci, labeled A, B, C, ..., then the alternative alleles at locus A would be denoted by A^P, A^Q, the alternatives at locus B by B^P, B^Q, and so on.

The genotype of an individual is the complete set of alleles that it carries. For example, $A^P A^P B^P B^Q C^P C^Q$ represents the genotype of a diploid, carrying alleles P or Q at three genes (A, B, C), indicating a homozygote at gene A and heterozygotes at genes B and C. However, when genes are linked, this does not show how the alleles are combined on chromosomes. Is allele B^P combined on the same chromosome as C^P, and B^Q with C^Q, or is B^P with C^Q and B^Q with C^P? The full genotype can be written as $A^P B^P C^P / A^P B^Q C^Q$,

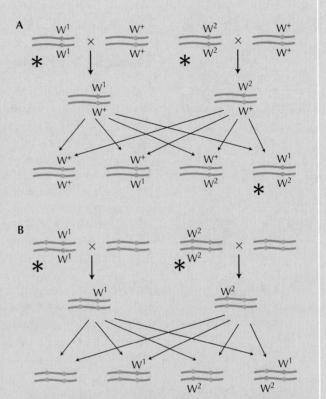

FIGURE 13.3. The complementation test is used to show whether two recessive alleles that produce the same phenotype (*asterisk*) when homozygous are alleles of the same gene. Homozygous individuals from each mutant strain are crossed to wild type to produce heterozygotes (middle row in *A* and *B*). (*A*) When heterozygotes are mated with each other, 25% of the offspring will show the recessive phenotype if the alleles are variants at the same gene. (*B*) However, if the alleles are at different genes, all the offspring will be wild type. This is because individuals that carry the two recessive alleles (one at each gene) will also carry the wild-type alleles at both genes (bottom right in *B*).

showing that A^P, B^P, and C^P are carried together on the same chromosome and were inherited from the same parental gamete. The haploid genotype that is inherited from one gamete is called the **haplotype**; in this example, the diploid genotype is made up of the two haplotypes $A^P B^P C^P$ and $A^P B^Q C^Q$. A difficult problem in the genetics of diploid organisms is that the haplotype cannot be determined solely by listing the diploid genotypes, locus by locus.

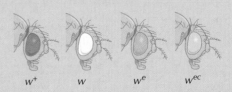

FIGURE 13.4. Eye phenotypes of alleles of the *white* gene in *Drosophila melanogaster*. The wild-type allele is denoted by +.

A

Pink unbanded   Yellow unbanded

Pink banded   Yellow banded

B

Nonmelanic   Melanic

FIGURE 13.5. Until the development of molecular markers, there were few examples of natural polymorphism with a simple genetic basis. Examples included (A) shell pattern in the snail *Cepaea nemoralis* and (B) melanism in the ladybird beetle, *Adalia bipunctata*.

on a set of genes of interest for the problem in hand. This contrasts with the individual's **phenotype**, which is all of the characteristics that we actually observe. In the next chapter, we consider how variation in genotype is related to variation in phenotype.

Classical Genetics Revealed Cryptic Variation

Darwin documented extensive inherited variation in phenotype—in particular, in traits that affect survival and reproduction—but the genetic basis of this variation was obscure. In contrast, the first geneticists concentrated on variants that followed simple Mendelian rules. However, these arose as spontaneous mutations in the laboratory or greenhouse or were found as rare aberrations. Examples of genetic variation in nature were hard to find, and the overall degree of variation could not be quantified. A few exceptionally variable systems were known (e.g., Fig. 13.5) and were termed **polymorphisms**. However, most examples of genetic variation came from rare deleterious alleles. In humans, although a substantial proportion of individuals are affected by some inherited condition, each particular genetic defect is very rare. For example, the most common inherited disease, cystic fibrosis, has an incidence of only 1 in 2000 Europeans (see p. 755). In large surveys of natural *Drosophila* populations, variants similar to laboratory mutations were found, but only at very low frequency. All of this was consistent with the simple view that a gene usually carries a common wild-type allele, with deleterious variant alleles being rare.

The first evidence that populations in fact contain extensive genetic variation came from studies in which wild-caught *Drosophila* were crossed to specially constructed laboratory strains, such that whole chromosomes could be made homozygous (Box 13.2). Remarkably, 10–30% of *Drosophila* chromosomes were found to carry **recessive lethals**, so that flies homozygous for these chromosomes die; the remaining chromosomes carry recessive alleles with more or less severe effects (Table 13.1). Thus, if a randomly chosen fly were made completely homozygous for all its chromosomes, it would be unlikely to survive. This is an instance of the well-known phenomenon of **inbreeding depression**, which we discuss in more detail in Chapter 18 (pp. 515–518).

These early studies also showed that recessive lethals are scattered over many genes, so that, at any one gene, they are rare. One can find out whether two lethals that were isolated from different flies are in fact alleles of the same gene (i.e., are **allelic**) by using the complementation test (Box 13.1; recall pp. 244–246). If the alleles are brought together in the same fly, and the fly survives, then the defects are at different genes. If the fly dies, however, then the two are alleles of the same gene and so produce a lethal homozygote when they come together (Box 13.2). In *Drosophila pseudoobscura*, the chance of allelism between two recessive lethal alleles on the second chromosome isolated from different places is approximately 1/300. This implies that there are approximately 300 genes that can mutate to recessive lethals on the second chromosome. Dividing the total frequency of second-chromosome recessive lethals of approximately 15% by this number gives an estimated allele frequency at each locus of approximately 0.0005. Thus, although this technique reveals extensive hidden variation due to recessive alleles, the results are still consistent with low levels of variation within each gene.

Cryptic genetic variation can be revealed when normal development is perturbed either by an unusual environment or by the presence of a mutation. For example, if early *Drosophila* embryos are briefly exposed to ether, a fraction will develop abnormally to show adult phenotypes similar to those produced by the **homeotic** mutation *Ultrabithorax* (see p. 288), in which body segments change their identity. In extreme cases, the thoracic segment that usually carries **halteres** instead carries an extra pair of wings (Fig. 13.7). As we saw in Chapter 11, classic experiments by Waddington showed that the expression of this phenotype was largely genetically determined, because selection could rapidly increase or decrease it. Recently, it has been shown that much of

Box 13.2	Balancer Chromosomes and Tests for Allelism

Balancers are specially constructed *Drosophila* chromosomes, which can be used to keep single wild-type chromosomes intact in laboratory culture. Thus, one particular wild-type chromosome can be "extracted" from a natural population and its effects studied under controlled conditions (Fig. 13.6A,B). A **balancer chromosome** carries multiple **chromosomal inversions**, recessive lethals, and dominant marker alleles. *Drosophila* that are heterozygous for an inverted chromosome do not produce recombinant offspring (Box 12.2). Moreover, because balancers carry recessive lethals, they cannot become homozygous, and because they carry dominant marker alleles (e.g., mutations that alter eye color when in a single copy), their presence can easily be seen.

A population heterozygous for a balancer and one particular wild-type chromosome can be maintained indefinitely (Fig. 13.6A). The +/+ homozygote is much less fit than the B/+ heterozygote, and so a polymorphism is maintained by an extreme form of **heterozygote advantage** (Box 17.2). A single chromosome can be extracted from a natural population by crossing a single wild-caught male to a female from the balancer stock (Fig. 13.6B). (Because there is no recombination in male *Drosophila*, using a wild-caught male ensures that his chromosome will be passed on intact, without being broken up in meiosis.) A single offspring carrying a balancer is then chosen and used to establish a new balancer stock that carries the "extracted" chromosome. If no +/+ individuals appear in this stock, then the wild-caught chromosome must have carried one or more recessive lethals. As is illustrated in Figure 13.6C, if two different balancer stocks are crossed, then any wild-type individuals in the offspring must carry two different wild-type chromosomes (one from each parent stock). If no wild-type flies survive, then the two stocks must have carried recessive lethals

at the same gene. If wild-type flies do survive, then the two strains must have carried recessive lethals at different genes. (This is the **complementation test** explained in Box 13.1.)

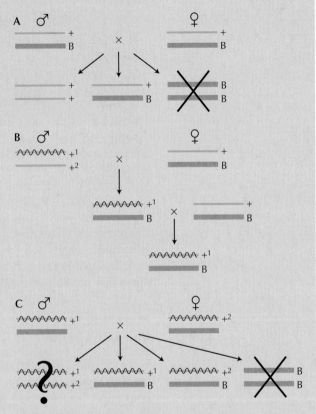

FIGURE 13.6. (A) Maintaining a balancer stock. (B) "Extracting" a wild-type chromosome. (C) Testing for allelism.

TABLE 13.1 Frequencies of recessive variants in Brazilian populations of *Drosophila willistoni*

Effect	Second Chromosome	Third Chromosome
Lethal	28.6	19.7
Semilethal	12.6	12.4
Sterile	31.0	27.7
Visible mutants	15.9	16.1

Reprinted with permission from Dobzhansky T. 1937. *Genetics and the Origin of Species*, Table 4, p. 66. Columbia University Press, New York (©1982 Columbia University Press, Reissue Edition October 15, 1982; ISBN 0231054750—ppbk), and references therein.
"Semilethals" are defined as chromosomes that kill more than half of individuals.

FIGURE 13.7. Flies exposed to ether as early embryos show phenotypes similar to those of homeotic mutations of *Ultrabithorax*. (*Left*) Wild-type fly with tiny haltere indicated by arrow in inset. (*Right*) Fly in which haltere has been partially transformed into wing (*inset*) as a result of weak mutation of Ubx. This phenotype is also generated by ether treatments.

this variation is due to cryptic polymorphism in the *Ultrabithorax* gene itself, which is revealed only by this environmental perturbation (see p. 307). As with the recessive variation that is revealed by inbreeding, these observations hint at the extensive variation that normally remains hidden within the population.

Members of a Population Differ Extensively in Protein and DNA Sequence

Until the mid-1960s, the amount of genetic variation was not known for even the best-studied organisms (see p. 59). The methods of classical genetics and biochemistry could find variation only fortuitously, in particular traits (e.g., blood groups or visible polymorphisms; Fig. 13.5) and in particular organisms (e.g., humans and *Drosophila*). Thus, it was only possible to guess at the overall pattern of variation across genes and organisms in general.

The key innovation was **protein electrophoresis**, a method that could be used with any organism and that measured variation in proteins (and thus indirectly in genes) that were chosen solely because they could be scored on gels, rather than because they were known to be polymorphic. The proteins that can most easily be scored are abundant enzymes—for example, those involved in basic metabolism. Different allelic variants of such enzymes are known as **allozymes**. The first applications of this technique surprised geneticists by revealing far more genetic variation than had been imagined. In both flies and humans, about one-third of enzymes were polymorphic (i.e., had more than one allele at appreciable frequency), and any randomly chosen individual would have a 5–10% chance of being heterozygous at a gene chosen at random. Protein electrophoresis was taken up with enthusiasm by population geneticists, who set out to describe levels of variation across the whole living world—in other words, to "find them and grind them." Since then, many new techniques for measuring variation have been devised, but the basic observation of abundant genetic variability has remained unchanged.

Electrophoresis detects changes in charge and conformation, which affect the speed at which a protein moves through a gel. These reflect changes in amino acid sequence and hence in DNA sequence. However, not all amino acid changes are detected, and sometimes proteins are modified after they have been translated, which can alter their mobility within a gel. Other kinds of genetic markers give us simple, but more or less indirect, ways of observing the DNA sequence itself. **Microsatellite** markers count the number of repeated elements in a tandemly repeated sequence. **Restriction fragment length polymorphisms (RFLPs)** detect whether a piece of DNA is cut by a **restriction enzyme**, which cleaves a particular sequence of typically four to six bases. Variation in the length of restriction fragments is due to changes in the cleavage site and also to insertions and deletions between cleavage sites. In a recently developed technique, it is possible to search for variation across the genome in a single experiment by hybridizing DNA from an individual to an array (or "chip") made of thousands of **oligonucleotides** corresponding to different regions of the genome.

In principle, we could make an exact survey of genetic variation by directly comparing complete DNA sequences. Something approaching this has been done for the human genome. When the first draft of the human genome sequence was published in 2001, it was accompanied by a survey of 1.4 million **single-nucleotide polymorphisms** (SNPs, pronounced "snips"). These were detected by finding differences in sequence between genomes from 24 people of diverse origins. The aim was to identify genetic markers that could be associated with diseases rather than to describe genetic variation in the human species; for the latter, a much larger number of individuals would need to be sampled. For most purposes, direct surveys of DNA sequence would be extremely inefficient. We now have a wide variety of genetic markers that reflect variation in DNA sequence and that can be chosen to suit different kinds of problems. The details of these and other techniques are set out in Box 13.3.

DNA sequences show many kinds of variations. These are illustrated in Figure 13.11, which shows variation in a 13-kb region around the alcohol dehydrogenase gene (*Adh*) of *D. melanogaster*. There are eight sites at which single bases vary; nine small insertions and deletions, each of a few bases; and seven larger insertions, of hundreds of base pairs, which are caused by **transposable elements** (see pp. 217–219). These variations occur mainly in **introns** and flanking sequences, where they do not affect amino acid sequence. There is only a single variant that affects the protein sequence, a shift from A to C, which causes a change from lysine to threonine. This protein polymor-

---

Box 13.3   Genetic Markers

Ideally, a **genetic marker** is an inherited variant that can be easily scored, varies within or between populations, and has negligible effect on phenotype and, in particular, on fitness. In principle, we could sequence segments of DNA from multiple individuals. However, this would be prohibitively expensive for large samples and, in any case, is usually unnecessary.

We are not interested in the marker itself but rather wish to use it to indicate an individual's genotype. As we will see, genetic markers have a wide variety of uses—for example, finding the genes responsible for variation in quantitative traits or for disease (Chapters 14 and 26), inferring paternity or geographic origin (Chapter 16), or detecting the effects of selection (Chapter 19).

In general, genetic markers require a way of detecting specific genetic loci and a way of distinguishing variation at these loci. Here, we summarize the most important techniques.

Allozyme electrophoresis: Allozymes are allelic variants of an enzyme coded by a single locus. Crude protein extract is pulled through a gel (starch, polyacrylamide, or cellulose acetate) by an electric field. One particular protein is then detected with a stain that detects a specific enzyme activity. The position of bands on the gel reflects differences in the rate of movement of enzyme molecules through the gel, which depends on both their charge and their size. Figure 13.15A (p. 367) provides an example in which homozygotes show a single band, whereas heterozygotes show three bands. This is because this enzyme has two subunits, so that there can be three kinds of **dimer** in a heterozygote. Allozyme electrophoresis is cheap and simple. However, it does not reveal all amino acid variation, and it can detect only abundant proteins for which there is a specific stain. Typically, up to 50 genes can be scored easily.

Restriction fragment length polymorphisms (RFLPs): DNA is cut with a restriction enzyme and pulled through a gel by an electric field (Fig. 13.8). Single-stranded probe DNA, labeled radioactively or with a fluorescent dye, is then applied to the gel. The probe anneals to complementary sequences, allowing detection of just those fragments that share homology with the probe sequence. The position of the bands within the gel reflects the lengths of the fragments. Variation in length can be caused either by substitutions of

Box 13.3. Continued.

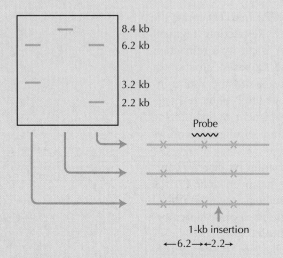

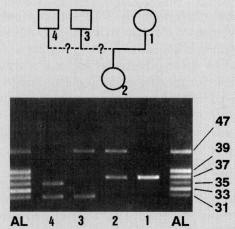

FIGURE 13.8. Restriction fragment polymorphism.

FIGURE 13.9. Minisatellite loci can be used to accurately identify relationships. This diagram shows genotypes at a single minisatellite locus for a mother (1), a daughter (2), and two putative fathers (3, 4). The daughter is heterozygous for two alleles, with 37 and 47 repeats. The shorter allele must come from the mother, who is homozygous for the 37-repeat allele. Individual 4 does not carry the 47-repeat allele and so cannot be the father. However, individual 3 is heterozygous for that allele and so could be the father. The ladders on either side show a reference mixture of DNA that includes alleles that are common in the population.

single bases (causing loss or gain of restriction sites) or by insertions or deletions between restriction sites. Thus, only a small fraction of sequence variation is seen, but almost all length variation is detected.

Variable number tandem repeats: Many markers involve repeated sequences that vary in copy number. These **variable number tandem repeat [VNTR]** loci are classified according to the length of the repeat involved and the number of copies.

Minisatellites are arrays of many copies of short sequences, which can be from nine base pairs long up to a few hundred base pairs. These arrays are scattered over multiple sites in the genome, and typically each consists of 10–100 repeats. (In contrast, **satellite DNA** consists of extremely large numbers of repeats [see p. 217]. Although this also varies, it is difficult to score and so is not a suitable genetic marker.) By probing electrophoretic gels with one particular minisatellite repeat sequence, large numbers of bands are revealed. Alternatively, individual loci can be detected by choosing two primer sequences that recognize unique sequences on either side of the array and using these in the **polymerase chain reaction** (PCR) (e.g., Fig. 13.9). Each band corresponds to one of the two alleles at a **minisatellite** locus, and its position on the gel corresponds to the number of repeats at that locus. The number of repeats is highly variable, because there is an exceptionally high mutation rate that generates new length variants (see pp. 339–342). The extremely high variability of minisatellite loci makes them useful for identifying individuals and determining relationships (Fig. 13.9).

Microsatellites are arrays of short tandem repeats, each unit being just a few base pairs long. For example, there are about 300,000 tri- and tetranucleotide repeats in the human genome—one every 10 kb—and about 50,000 two-base repeats, ...CACACA... . Individual microsatellite loci are detected using primers that recognize unique sequences, as described above for minisatellites. Microsatellites have a high mutation rate and show correspondingly high variation within populations (Fig. 13.10).

Single-nucleotide polymorphisms (SNPs): Sequencing

segments of genome from multiple individuals reveals variation at single-nucleotide sites (e.g., see Fig. 13.15D). At any one site, variation is uncommon—in humans, for example, the chance that a particular site will be heterozygous is $\pi \sim 0.0008$. Therefore, if variation is found, there are usually two alternative bases (and only very rarely three or four). It is estimated that the human population contains approximately 10 million variants with a minor allele frequency of at least 1%; approximately 90,000 of these are within coding regions and, of those, approximately half alter amino acid sequence. A substantial fraction of these human SNPs have already been discovered in large-scale sequencing projects. **Microarrays** can be designed that will efficiently score very large numbers of SNPs. Short sequences, 10–20 bp long, surrounding each SNP are attached to a chip, such that only DNA carrying one of the SNP alleles will hybridize to the chip.

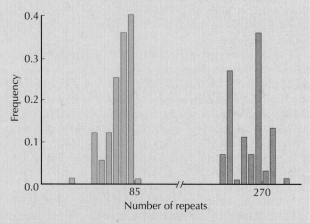

FIGURE 13.10. Variation in repeat number at two human microsatellite loci. These are both dinucleotide repeats.

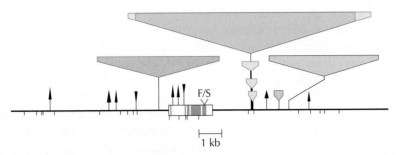

1 kb

FIGURE 13.11. Different kinds of variations are seen around the *Adh* gene of *Drosophila melanogaster*. Forty-eight chromosomes were sampled from four populations across the United States, and variation was detected through differences in the lengths of fragments cut by restriction enzymes (Box 13.3). *Colored triangles* show large insertions, caused by insertion of transposable elements. *Small triangles* show insertions or deletions, defined relative to the most common haplotype in the sample. *Dashes* show single-base changes, which cause the presence or absence of the site at which the restriction enzyme cuts. Only a small fraction of single-base changes is detected using this method. Allowing for this fraction, it is estimated that 2.7% of sites are polymorphic in this sample and that the nucleotide diversity is $\pi = 0.0064$. The box at the center shows the *Adh* gene, with exons *pink* and introns *white*; the position of the site that causes the F/S polymorphism in amino acid sequence is also shown.

phism was detected by electrophoresis: The allele carrying threonine migrates faster through the gel and so is known as the *fast* (F) allele, in contrast to the *slow* (S) allele. It was this difference in electrophoretic mobility that first encouraged detailed sequence studies of the *Adh* gene.

Variation Can Be Described by Allele Frequencies

The patterns revealed by different kinds of genetic markers must be summarized to give measures of variability that can be compared across methods, across genes, and across species. From the perspective of classical genetics, all methods detect different kinds of genes (i.e., different alleles). Thus, observations can be reduced to lists of allele frequencies. For example, the data of Figure 13.11 show that 16 of the *Adh* sequences carry the F allele, and 32 carry the S allele, giving a frequency of allele F in the sample of 0.333. We could calculate the frequencies of all the sequence variation in Figure 13.11 in a similar way. However, long lists of allele frequencies are not by themselves particularly informative. To make sense of large datasets, we need ways of summarizing lists of allele frequencies in simple statistics.

The simplest measure is whether a locus is polymorphic, that is, whether it carries more than one allele. This criterion clearly depends on sample size. A common definition is therefore that a locus is polymorphic if the commonest allele is at a frequency of (say) no more than 95%. A related measure is just the number of alleles that are found in the sample; again, this will tend to be larger for larger samples.

Another measure would be to take the proportion of heterozygotes in the sample. However, this is influenced by the system of mating, as well as by allele frequency. For example, a field of wild oats, which reproduces mainly by self-fertilization, typically contains many different alleles at each locus, and yet almost all genes in any individual are homozygous. More seriously, such a measure cannot be applied to haploid organisms, such as bacteria. Usually, therefore, one calculates the **expected heterozygosity**, which is the proportion of heterozygotes that would be seen in an idealized population of random mating diploids. Suppose the frequency of the ith allele is p_i ($\Sigma_i p_i = 1$). From the **Hardy–Weinberg formula** (Box 1.1), the chance that both copies carry the ith allele (i.e., the frequency of homozygotes for the ith allele) is p_i^2. Therefore, the expected heterozygosity is $H = 1 - \Sigma_i p_i^2$. Because this measure of variability is often applied to haploid or self-fertilizing organisms, it is best referred to as the genetic diversity, and it is best thought of as the chance that a pair of genes chosen at random from the population carries different alleles.

These measures of allele frequency variation can be applied to sequence data, provided that we take individual nucleotide sites as the unit. There is little chance that any particular site will be polymorphic (see Fig. 13.11). The number of alleles at a nucleotide site is therefore usually 1, and occasionally 2, rather than the maximum of 4 set by the structure of DNA. If the genetic diversity is calculated base by base, we obtain a measure, which is denoted by π and is usually termed the **nucleotide diversity**. Similarly, the chance that any one site is polymorphic is just the number of sites that show any variation in the sample (the number of **segregating sites**), divided by the length of sequence.

Describing variation in this way, by treating each nucleotide site separately, captures very little of the information in a sample of sequences. There are an enormous number of possible combinations of even a small number of segregating sites, yet typically, only a small number of combinations are found. (Each particular combination of variants—in full, each particular sequence—is known as a haplotype.) For example, the sample of 48 *Adh* sequences shown in Figure 13.11 contains 24 variable sites within 13 kb, but only 29 different haplotypes are present—far fewer than the $2^{24} = 17$ million that are possible.

However, it would be foolish to treat haplotypes simply as alleles and count their frequency. In a sufficiently long sequence, every sampled sequence will typically be different, yielding as many alleles as we have sequences. The problem with summarizing sequence data in terms of allele frequencies is that this ignores information about differences between sequences. If sequences differ at many positions, then there is more genetic variability in the sample. The simplest way to take this information into account is to count the average number of sites that distinguish a pair of sequences. This is just the length of sequence multiplied by π, the nucleotide diversity defined above. As we shall see in the next section, it is the way variants are combined with each other in the sample that contains the fullest information about the evolutionary history of the sampled sequences. This requires a more sophisticated description of variation.

Genetic Variation Is Most Fully Described in Terms of Genealogies

If the sequence harbors enough variation, and if there has been no recombination in the history of the sample, then we can do much more: The full genealogical relationship between the sequences can be estimated using the same methods as for reconstructing the phylogeny of species (Box 5.1; Chapter 27 [online]). (The term **genealogy** refers to the relationship between genes within a species, whereas the term **phylogeny** usually refers to the relationship between different species.) Often, it is reasonable to assume that each sequence variant arose just once and so is found within a single set of descendant sequences (known as a **clade**). Then, the genealogy can be at least partially reconstructed from the sequence data (Fig. 13.12).

In practice, however, it is not easy to reconstruct relationships from genetic data (Chapter 27 [online]). There are two main problems: convergence and recombination. If a site evolves rapidly or if the sequences are very distantly related, then a particular variant may arise several times, and the different clades that descend from each cannot then be distinguished. (For example, if the mutations shown in Figure 13.12 were not distinguished by different colors, then the genealogy could not be reconstructed.) A more serious difficulty is that in the nuclear DNA of sexually reproducing species, there is an appreciable rate of recombination within stretches of a few kilobases. Even within bacterial species, where recombination occurs only sporadically, it may be frequent enough over the history of the sample to make it impossible to assume a single genealogy (Fig. 5.23 and Chapter 27 [online]). With recombination, different stretches of the sequence have different genealogies (Fig. 13.13). For example, there is clear evidence of recombination in the ancestry of the *Adh* gene of *D. melanogaster* (Fig. 13.11) and in this case, likely recombination events can be identified (Fig. 13.14).

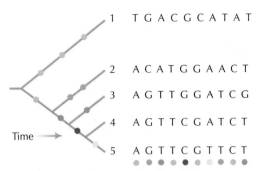

FIGURE 13.12. Unique mutations in a nonrecombining DNA sequence can be used to reconstruct their genealogy. This example shows four sequences (nos. 2–5) sampled from a population, together with a more distantly related sequence (1) termed an **outgroup**. Each unique mutation defines a group of related sequences, or clade; for example, sequences 3, 4, and 5 all share an A→T substitution (*green*), which shows that they form a single clade, and sequences 4 and 5 share a G→C substitution (*dark blue*), which shows that they also form a clade. In this example, enough mutations have occurred that the genealogy can be reconstructed unambiguously. Mutations that occurred between the outgroup and sequences 2–5 are represented by *brown dots*. However, these are of no use in reconstructing the genealogy that relates 2–5. Only variable positions in the sequence are shown: In the full sequence, the great majority of sites would not vary. See Chapter 27 (online) for a full discussion.

In general, however, it is extremely difficult to allow for recombination in a genealogical analysis. We return to this issue on pages 427–432.

Variation at individual genetic loci—in the limit, at individual bases—can be summarized in terms of simple measures such as the genetic diversity. It is much harder to summarize the information contained in complete haplotypes, and so the estimated genealogy is often simply presented as a tree (e.g., Fig. 13.12 and 13.14). Where recombination is suspected, a set of genealogies might be depicted (e.g., Fig. 13.13). Although this conveys more information than a single statistic, it is important to realize that such diagrams give only an *estimate* of the actual relationships among the sampled genes. Moreover, the true genealogy is itself based on a random sample of genes from the population and of genetic loci from the whole genome. For example, even if we knew the true relationships among the 48 *Adh* sequences depicted in Fig-

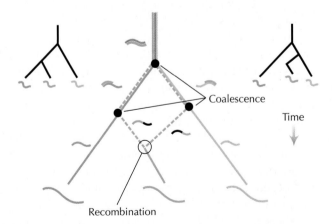

FIGURE 13.13. Different parts of a recombining genome have different genealogies. The diagram shows a simple example, with three genomes (*red, blue, green*). Tracing back through time, the *red* and *green* lineages **coalesce** (*top filled circle*) without experiencing recombination. However, the *blue* genome traces back to a recombination event (*open circle*) so that the left part is inherited from one parental genome and the right part from another. The left lineage coalesces with the *red* lineage, whereas the right lineage coalesces with the *green* lineage (*dashed blue lines*). As a result, the two sections of genome have different genealogies (*top left* and *top right* diagrams). This process of coalescence with recombination will be discussed in more detail on pp. 427–432.

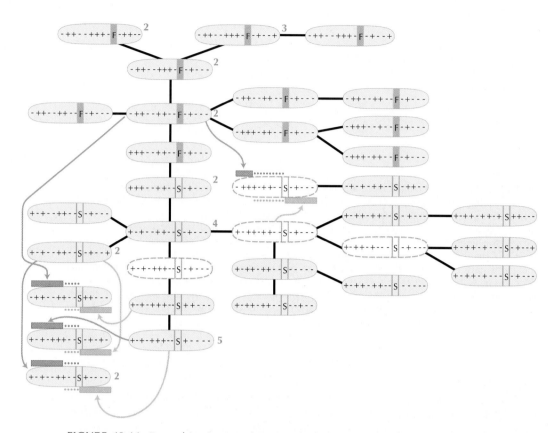

FIGURE 13.14. Recombination can be seen in the ancestry of the *Adh* gene of *Drosophila melanogaster*. Each haplotype is written as a sequence of + and –, which indicate the presence or absence, respectively, of the polymorphic variants shown in Fig. 13.11. The diagram shows the likely relationship between the different haplotypes, with adjacent haplotypes differing by a single change. Hypothetical intermediates are *dashed*, and the number of times each was found is indicated on the *right* (when >1). The fast and slow alleles (F, S) fall into two clusters. However, four haplotypes have clearly been formed by recombination. The likely sources of the left and right parts of these recombinant haplotypes are shown by *arrows*. Note that this diagram cannot be drawn as a conventional genealogy, because the **root** is not known (see Chapter 27 [online]).

ure 13.14, at some particular site, this would be only a random subset of the relationships among all the *Adh* genes in the population. In turn, the relationships among genes at some other locus would be quite different. The genealogy that we estimate is subject to random errors in estimating the actual relationships, in sampling a small fraction of individuals from the population, and in sampling one **genetic locus** from the whole genome. Similar considerations apply to statistics based on allele frequencies. These different levels of randomness demand that we sample many individuals and many genetic loci to find out about the evolutionary process.

In this chapter, we have only outlined a few ways of summarizing the basic pattern of genetic variation. The complete information contained in allele frequencies and genealogical relationships can be used more fully if we have in mind a definite model of the evolutionary process. Alternative models can be tested and their parameters estimated (e.g., mutation rates, recombination rates, and population size). This approach is discussed later in the book (e.g., Box 25.2). In particular, the **neutral theory** of molecular evolution, in which the primary processes are mutation and random genetic drift, allows sophisticated models to be fitted (see p. 536). If our evolutionary model is correct, then this is a powerful method for interpreting genetic data. However, it is wise to be aware that if the assumptions do not hold, then the results can be quite misleading. It is best to begin with a simple examination of the patterns in the data of the kind outlined in this section.

TYPES OF GENETIC VARIATION

Most Populations Contain Abundant Genetic Variation

The most striking observation in evolutionary genetics is that almost all populations of all kinds of species contain abundant genetic variation (Figs. 13.15 and 13.16). Some features of organisms (e.g., their population size and habitat) do correlate with levels

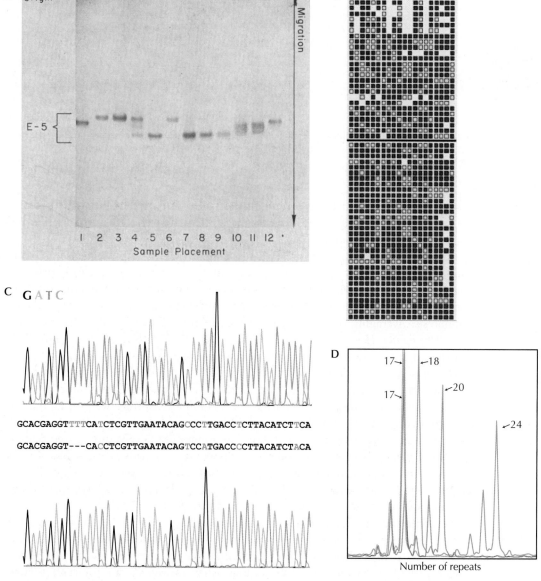

FIGURE 13.15. Examples of several kinds of molecular variability (see Box 13.3). (*A*) Variation at the *Esterase 5* locus of a population of *Drosophila pseudoobscura*, detected using allozyme electrophoresis. (*B*) Single-nucleotide polymorphisms (SNPs) in a sample of 20 human chromosomes. (*Yellow*) Variant bases; (*light blue*) missing data. Only 59 of the 147 sites that vary in this 31-kb region of chromosome 21 are shown. (*C*) A 51-bp section of the cytochrome oxidase 3 gene from two earthworms, *Lumbricus rubellus*. The two traces show the output from an automatic sequencing machine. There are a single insertion/deletion of a TTT triplet and five single-base changes. (*D*) Microsatellite variation in red deer (*Cervus elaphus*). Electrophoretic traces are shown for three individuals; peaks indicate the length of the microsatellite allele, which varies with the number of AC repeats. These three individuals are heterozygous for 17 and 20 repeats (*red*), homozygous for 17 repeats (*blue*), and heterozygous for 18 and 24 repeats (*green*).

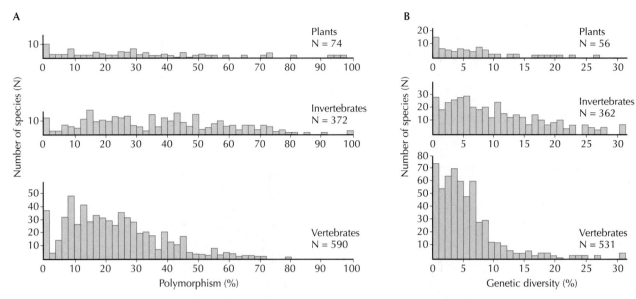

FIGURE 13.16. The distribution of (A) polymorphism and (B) genetic diversity at allozyme loci, based on a survey of nearly 1000 species. (A locus is defined as polymorphic if the commonest allele is at <0.99.)

of genetic variation, but these correlations are weak and give us little clue as to the causes of variation. What is most important is the high level of genetic variation that is found in almost all populations.

As assayed by gel electrophoresis, a substantial fraction of proteins are polymorphic: 19% in mammals, 48% in *Drosophila*, and 90% in the bacterium *Escherichia coli* (Fig. 13.16A). Similarly, if we take two proteins at random from any population, there is a substantial chance that they will carry different alleles as detected by electrophoresis. Averaging over all allozyme loci, mammals have genetic diversity $H \sim$ 4.1%; *Drosophila, H* $\sim$ 12%; and *E. coli, H* $\sim$ 47% (Fig. 13.16B). At the level of DNA sequence, data are available for many fewer species. In humans, the chance that a randomly chosen nucleotide site will be heterozygous is $\pi \sim 0.0008$, whereas within *D. melanogaster*, this measure of nucleotide diversity is an order of magnitude higher, at $\pi \sim 0.01$. (Note that the gene diversity per nucleotide site is much lower than the corresponding diversity measure for whole proteins, simply because proteins are typically coded by many hundreds of nucleotides. The protein diversity is due to that fraction of DNA sequence diversity that codes for different amino acids.) Over a broad range of species, nucleotide diversity is highest for prokaryotes, lower for single-celled eukaryotes, and lowest for multicellular eukaryotes (Fig. 13.17A). Organisms with larger genomes and more genes tend to have lower diversity (Fig. 13.17B).

For almost all species, the bulk of variation is held within local populations. Two proteins or two nucleotides are somewhat more likely to differ if they come from different places than if they come from the same local population, so that diversity is greater at the level of the whole species than when measured within localities. However, as we will see in Chapter 16 (p. 448), the difference is small—typically 10–20%

Genetic Variation Is Greater in More Numerous Species, but the Relationship Is Weak

The most obvious pattern that we see is that genetic diversity is higher in more abundant organisms. Not surprisingly, extremely small populations may have little or no variation. For example, the elephant seal, which was hunted almost to extinction at the end of the 19th century, shows no variation at 24 protein-coding loci (Fig. 13.18).

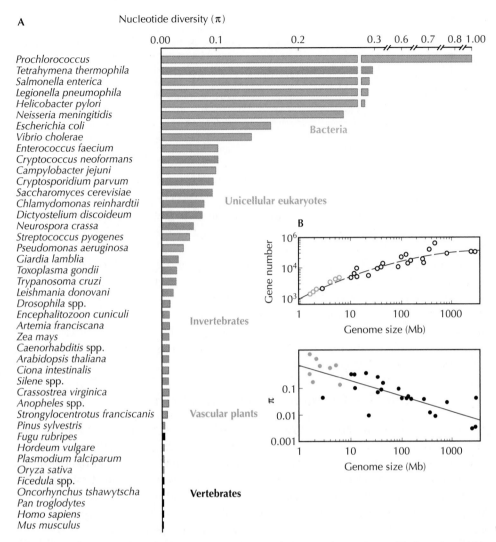

FIGURE 13.17. (*A*) Nucleotide diversity π across a wide range of organisms. (*B*) Organisms with larger genomes tend to have more genes (*top*). They also have lower within-species diversity (*lower curve*). Data for bacteria and archaea are shown in *blue*.

The increasing diversity from mammals, through *Drosophila*, to *E. coli* is consistent with an effect of increasing population size. Nevertheless, small populations can be highly variable. For example, a population of a few hundred Soay sheep on the isolated Hebridean island of Hirta contains allozyme and microsatellite variation that is not much lower than in wild sheep populations on the mainland (Fig. 13.19).

The relation between genetic diversity and population size is difficult to discern, in part, because it is extremely hard to estimate the numbers of most species and because the number that matters is an average back into the distant past. Figure 13.20 shows the relation between heterozygosity at enzyme loci and crude estimates of population size. Heterozygosity does increase with population size, but the relationship is remarkably weak, even though population sizes differ by many orders of magnitude. Moreover, the apparent increase is mainly due to a group of carnivores, which are rare and genetically depauperate, and at the other extreme a set of *Drosophila* species that are abundant and genetically diverse. For the species in between, there is actually no significant relation between estimated population size and genetic variability. Similarly, there are puzzles within the well-studied *Drosophila:* Several Hawaiian species, which are restricted to specialized habitats within remote islands, have almost as much al-

FIGURE 13.18. Northern elephant seals (*Mirounga angustirostris*) were hunted to near extinction in the late 19th century, with as few as 20 seals surviving. As a consequence, they contain exceptionally little genetic variation.

FIGURE 13.19. Soay sheep on Hirta, in the St. Kilda archipelago.

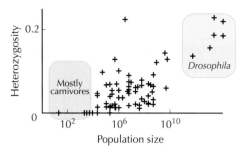

FIGURE 13.20. Heterozygosity increases only weakly with population size. The figure shows the relation between heterozygosity, measured by electropheretic assay of enzyme loci, and estimated population size. There are 76 species of multicellular eukaryotes in the survey. The group of abundant species with high heterozygosity are all *Drosophila*, and the group of rare species with low heterozygosity are mostly carnivores.

lozyme variation as cosmopolitan species such as *D. melanogaster*, which are far more abundant. Finally, Figure 13.17B shows that nucleotide diversity decreases by a factor of approximately 100 from single-celled prokaryotes to large mammals; yet population sizes must vary over a far wider range.

We examine the effects of such factors as population size, breeding system, and environmental heterogeneity in detail later in the book. In particular, we will see why genetic variation is expected to increase in proportion to population size (see pp. 425–426), and then in Chapter 15 (p. 426) we discuss the various reasons why the relation is in fact much weaker than this expectation. For the moment, the key point is that both abundant and rare species contain substantial genetic variability, whether measured at the level of protein or DNA sequence.

Sequence with Less Function Usually Varies More

Comparisons between variation in different parts of the genome can be understood more easily than comparisons between species. Moreover, for completely sequenced organisms, we can use data for the entire genome. Although measures of variation for any one site may be crude, because so many sites are available, statistical errors become small. The primary observation is that changes that have the least effect on the organism are most likely to be seen as polymorphisms. This one rule accounts for almost all the patterns that we see in cross-genome comparisons.

The clearest and most important contrast is between changes that alter amino acid sequence and those that do not; that is, between **nonsynonymous** and **synonymous** changes. For human coding sequences, the average nucleotide diversity for those changes that alter the protein sequence is less than a third that of synonymous sites within the coding sequence (Fig. 13.21). Of course, some synonymous changes may affect the functioning of the organism—for example, by altering the binding of RNA polymerase, the processing of messenger RNA, or the binding of transfer RNAs during translation. (Note that diversity in noncoding regions adjacent to genes is substantially lower than at synonymous sites, which suggests that these have some function in gene regulation; see Fig. 13.21, bottom.) Conversely, many changes in amino acid sequence may have negligible effects on the protein's function. However, it is clear that, overall, changes that alter the protein's sequence are much more likely to alter its function and hence affect fitness. As we will see in Chapter 19, many statistical analyses of the causes of DNA sequence variation are based on comparisons between synonymous and nonsynonymous variation.

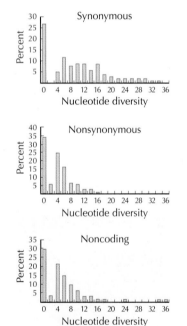

FIGURE 13.21. The distribution of nucleotide diversity for humans. These data come from a survey of 106 genes and compare the distribution of diversity for synonymous changes (which do not alter amino acid sequence), nonsynonymous diversity (i.e., changes that do alter amino acid sequence), and diversity in noncoding sequence. The mean pairwise diversity for these three classes is π = 0.00107, 0.00028, and 0.00052, respectively.

The close relation between functional significance and amount of variation can be seen in many other comparisons. Some amino acid changes have rather little chemical effect, whereas others cause large differences in charge and/or size. Such major changes are much less likely to be seen as polymorphisms. Genes that are present as single copies tend to be less variable than members of gene families, which code for many closely related proteins that may have overlapping functions. Similarly, "housekeeping" genes, which are present in all tissues, tend to be less variable than genes that have a narrower function. Coding sequences tend to be less variable than introns and other noncoding sequences (Fig. 13.22). The most striking example is with **pseudogenes**. These are genes that have lost their function and have begun to degenerate. Often, they are recognized because they have acquired one or more stop codons that prematurely halt translation. They show levels of polymorphism even higher than those for synonymous changes.

Rather than comparing the levels of different kinds of variations (e.g., synonymous vs. nonsynonymous), we can compare diversity along the genome. By far the most extensive data are for humans, which show substantial variations in gene diversity over scales of around 100 kb (e.g., Fig. 13.23). Over a larger scale, there are significant differences in the average diversity carried by different chromosomes (Fig. 13.24). The sex chromosomes and mitochondrial DNA harbor less variation than the autosomes, which may be in part because they are present in fewer copies in the population. (The human X chromosome is carried in two copies in females and one in males; the Y is present as a single copy and only in males; and the mitochondrial genome is passed down only in the female line.) However, there are also significant differences between the autosomes, which cannot be explained in this way. We discuss the causes of genetic variation in more depth in Chapters 15 and 19.

Differences between Species Accumulate Steadily: The Molecular Clock

Evolutionary biologists spend much effort documenting genetic variation, because it is this variation that makes it possible for populations to evolve through time. We cannot observe change through time directly, for DNA and proteins are far too fragile to be preserved in the fossil record. (The most ancient fragments of DNA that have been recovered are from the mitochondrial DNA of Neanderthals, who lived 20,000–30,000 years ago.) Instead, we measure rates of change by counting the genetic differences between species that diverged from each other at a known date (see Figs. 10.21 and 13.25 and Chapter 27 [online]).

One of the most remarkable findings that came from the first comparisons of protein sequences is the existence of a molecular clock; that is, any given protein evolves at a steady rate, even though it is evolving within very different organisms (see p. 59). For example, α-globin evolves at approximately 1.2×10^{-9} amino acid changes per year, a rate that holds steady across the whole range of vertebrate lineages (Fig. 13.26). The

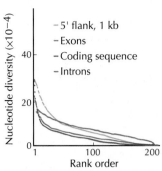

FIGURE 13.22. Regions under higher functional constraint show lower diversity in a survey of 213 human genes. Genes are ranked in order of nucleotide diversity, with most diverse on the left and least diverse on the right. The 5′ flanking region and introns are much more diverse than coding sequences and exons.

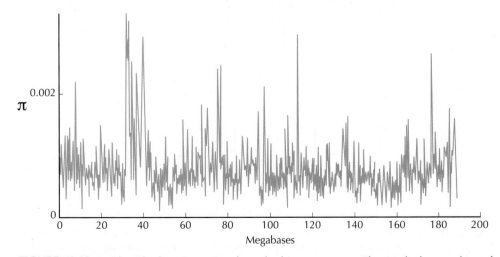

FIGURE 13.23. Nucleotide diversity varies along the human genome. The graph shows values of π for each block of 200 kb along chromosome 6. The region of high diversity near 34 Mb corresponds to the exceptionally diverse region around the major histocompatibility complex, which presents antigens to the immune system (see pp. 233 and 541).

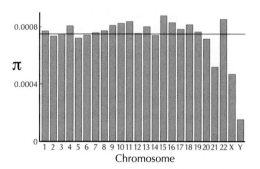

FIGURE 13.24. Nucleotide diversity, π, varies between human chromosomes. The *horizontal line* shows the genome-wide average. These estimates are from a sample of 24 ethnically diverse individuals.

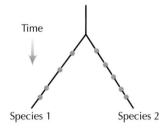

FIGURE 13.25. Rates of sequence divergence can be measured by counting the number of differences between two species that diverged at known time *T* in the past. In this example, species 1 and 2 differ by nine changes (*red dots*), and so the rate of divergence is estimated as 9/2*T*; the factor of 2 arises because divergence occurs down both lineages. See Chapter 27 (online).

α- and β-globin genes evolved from an ancestral globin gene early in the divergence of the vertebrates. Therefore, the α- and β-globins within any vertebrate species have been diverging for exactly the same time. Correspondingly, they show the same level of divergence, consistent with a molecular clock (Fig. 13.27). Another test of the constancy of the molecular clock comes from comparisons between pairs of plant species, which were separated from each other by climatic change at about the same time, around 5 Mya. Chloroplast DNA sequence from most pairs of sister species differs by about the same amount, reflecting similar rates of divergence.

Sequence divergence across the whole genome can be measured from the temperature at which different single-stranded DNA molecules anneal to form a double helix.

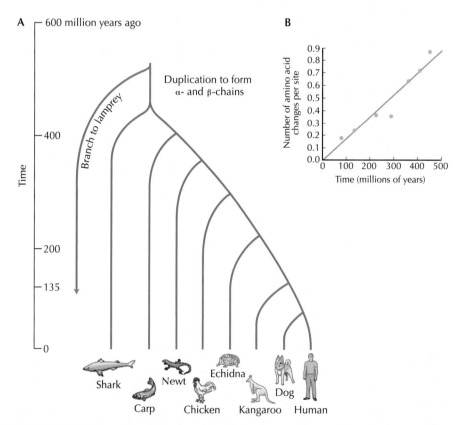

FIGURE 13.26. Amino acid substitutions in α-globin accumulate steadily with time. (*B*) The number of amino acid differences per site against time since divergence, based on sequences from eight vertebrate species, whose relationship is shown by the phylogenetic tree (*A*). Comparisons are between shark and the average of the other seven species (*top right closed circle*), carp and the other six, and so on down to human versus dog (*bottom left closed circle*). These values are corrected for multiple substitutions (Chapter 27 [online]). Estimates of time are from the fossil record.

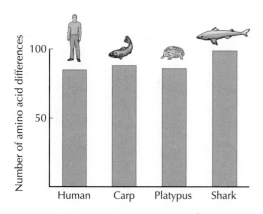

FIGURE 13.27. Comparisons between the α- and β-globin genes show that the molecular clock ticks steadily; that is, divergence occurs at almost the same rate in different lineages.

Strands with complementary sequences will base-pair at all sites and so will melt at a higher temperature than hybrid sequences from different species, which fail to pair at some fraction of sites. Thus, sequence divergence of 1.5–2% corresponds to a decrease in temperature for melting and reannealing of 1°C. This overall measure of genetic divergence increases steadily with divergence time (Fig. 13.28).

Rates of divergence vary greatly between different genes and between different kinds of changes within genes (Table 13.2). For example, histone H4 is highly con-

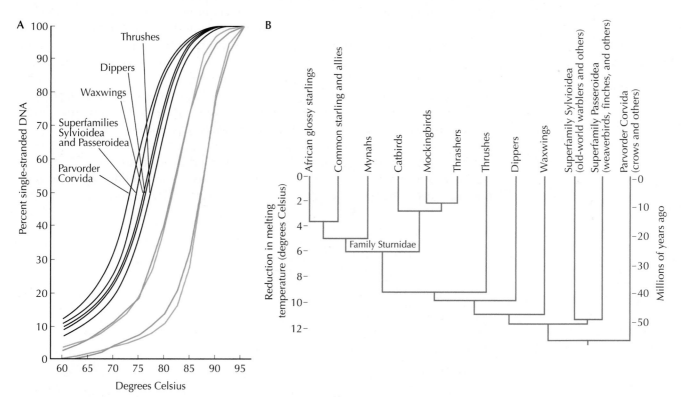

FIGURE 13.28. Overall sequence divergence can be measured from the melting temperature of double-stranded DNA. (*A*) These curves show the proportion of DNA that has melted into single strands as a function of temperature. As the divergence between the two strands increases, they melt at a lower temperature, and the curves shift to the left. The rightmost pair of curves shows DNA where both strands come from starlings or from mockingbirds. The next pair shows hybrid duplexes between species of starlings and mockingbirds, respectively. The *curves* to the left are for molecules with one strand of mockingbird or starling DNA and more distantly related groups. (*B*) The phylogeny derived from these data. The vertical scale shows the temperature at which half the DNA melts and the divergence time estimated from this temperature, assuming a molecular clock.

TABLE 13.2 Rates of divergence of different proteins

Protein	Amino Acid Substitutions per Site per Billion Years
Fibrinopeptides	8.3
Pancreatic ribonuclease	2.1
Lysozyme	2.0
α-globin	1.2
Myoglobin	0.89
Insulin	0.44
Cytochrome c	0.3
Histone H4	0.01

Data from Kimura M. 1983. *The neutral theory of molecular evolution*, Cambridge University Press, Cambridge, Table 4.1, and Li. W.-H. 1997. *Molecular evolution*. Sinauer, Sunderland, Massachusetts, Table 7.1. Estimates are based on protein sequences from a wide range of multicellular eukaryotes.

served across all eukaryotes and differs by only 2 out of 100 amino acids between peas and humans. In contrast, fibrinopeptides diverge hundreds of times more rapidly. Patterns such as these are explained primarily by differences in the degree to which they are constrained by their function. Histones are responsible for packaging eukaryotic DNA and have retained the same essential function; on the other hand, fibrinopeptides are cleaved away from the fibrinogen molecule during blood clotting, and their detailed sequence seems irrelevant. The same explanation holds for different kinds of changes within genes. The rate of divergence of synonymous changes in coding sequences is higher on average, and much less variable, than the rates of nonsynonymous change in different genes (Fig. 13.29). To take another example, **pseudogenes**—which presumably have lost all function—diverge at a faster rate than even synonymous changes within functional coding sequences. Just as for variation within species, which we discussed in the previous section, patterns of variation between species are shaped primarily by the degree of functional constraint.

Later in the book, we will see that the rate of molecular evolution of a given gene does vary from lineage to lineage (see p. 531). In particular, the adaptation of mole-

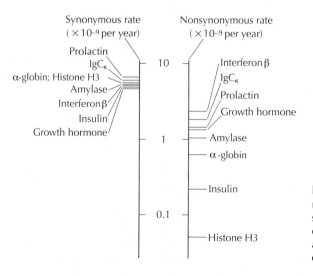

FIGURE 13.29. Rates of synonymous nucleotide substitution are similar across genes, whereas rates of nonsynonymous change, which alter amino acid sequence, are generally lower and more variable.

cules to a new environment or a new function may lead to a burst of substitutions. However, even when there is statistically significant variation in rate, the clock can still keep quite good time. If the rate were absolutely steady, then the number of substitutions that separate two species would follow a **Poisson distribution**. This distribution has a variance equal to its mean. Thus, if we expected to see, on average, differences at 200 sites in a long DNA sequence, the variance in number would be 200, and the standard deviation would be $\sqrt{200}$ (i.e., ~14). So, 95% of the time we would see between 172 and 228 differences, a range of fluctuation of only ±14% (blue curve in Fig. 13.30). Even if the variance in rate were four times this baseline level, the rate of the clock would still lie within a range of ±28% (green curve in Fig. 13.30), a reasonably small range, given the usual uncertainties in evolutionary inferences.

Gene Arrangement Evolves More Erratically

We have seen that mutations include many complex changes, such as chromosomal rearrangements, in addition to changes at single base pairs (see pp. 326–328). The net rate of chromosomal rearrangements is high: The net rate of detectable chromosomal mutations is typically one in a few thousand per generation in animals. In this section, we first consider variants that are large enough to be detected by direct **cytological** examination of the chromosomes. We then summarize what is known about smaller rearrangements, which are detected using genetic or molecular techniques.

Even closely related species usually differ by one or more chromosomal rearrangements, reflecting the gradual accumulation of chromosomal changes over time. For this reason, cytology is important for taxonomy, because it provides a clear way of distinguishing species. Moreover, because many chromosomal rearrangements are unique, they provide a powerful way to reconstruct the phylogenetic relationships between species. Rates of chromosomal evolution vary considerably between groups. For example, rodents have a net rate of chromosomal change of 17.8 changes per million years, whereas whales change an order of magnitude more slowly, at 1.7 changes per million years (Table 13.3). Within species, different chromosome arrangements are found in different parts of a species' range, but they are rarely found together within the same local population. This is simply because chromosomal heterozygotes suffer a variety of problems in meiosis and so are partially or completely sterile (e.g., Fig. 12.17; see Box 12.2). We discuss the consequences of this kind of selection against heterozygotes in more detail in Chapters 18 and 22 (pp. 496, 641–642, and 645).

There are exceptions to the general lack of chromosomal polymorphism within species. In the Diptera, heterozygotes for large **inversions** are fully fertile, and inversion polymorphisms are common. These can be conveniently seen in the giant **polytene** chromosomes (Fig. 13.31). There is in effect no recombination within the inverted region (Box 12.2), which facilitates genetic analysis (e.g., using balancer chromosomes; see Box 13.2) and has important evolutionary consequences for speciation and for the evolution of the genetic system, which we explore in Chapters 22 (p. 647) and 23 (p. 682), respectively.

Over smaller scales, that of a few base pairs or more, insertions, deletions, duplications, and inversions are commonly seen when we compare closely related species. Such rearrangements are rare within the coding sequence, because they disrupt gene function, most seriously by shifting the reading frame. However, they accumulate in the noncoding sequence at a rate similar to that of single-base changes, as, for example, in Figure 13.11. This makes it very difficult to properly align the noncoding sequences (Chapter 27 [online] and p. 546). Mutations are especially frequent in tandem repeats, because **unequal crossing over** and slip-strand mispairing (Fig. 12.9) increase or decrease copy number (see p. 341). This high mutation rate leads to high

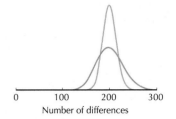

FIGURE 13.30. The *blue curve* shows the distribution of the number of differences between two sequences, given that the expected number is 200 and the actual number follows a Poisson distribution. The *green curve* shows a distribution with four times the Poisson variance. Even with this increased variance in rate, the molecular clock is roughly steady.

TABLE 13.3 Rates of chromosomal evolution vary substantially between groups

Group	No. of Genera Examined	Average Age of Genera	Net Chromosomal Changes per Myr
Placental mammals			
Rodents	42	4.6	17.8
Primates	12	4.4	14.2
Rabbits	3	9.0	12.8
Ungulates	14	4.3	11.5
Insectivores and			
edentates	8	11.0	6.5
Carnivores	11	11.6	4.5
Bats	17	10.7	3.3
Whales	3	6.3	1.7
Average	—	7.7	9.1
Other vertebrates			
Marsupials	8	1.9	1.3
Snakes	12	12.4	2.6
Lizards	15	23.0	2.4
Turtles and			
crocodiles	13	51.0	0.21
Frogs	12	16.7	1.8
Salamanders	9	21.5	0.6
Teleost fishes	23	18.8	2.6
Average	—	20.6	1.7
Mollusks			
Prosobranch snails	16	64.7	0.3
Other snails	15	49.0	0.4
Bivalves	3	77.0	0.1
Average	—	64.0	0.3

Adapted from Wilson A.C. et al. 1975. *Proc. Natl. Acad. Sci.* **72:** 5061–5065 (©1975 Wilson et al.). Myr, millions of years.

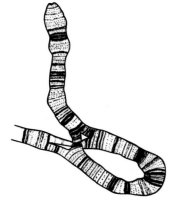

FIGURE 13.31. Chromosomes from a *Drosophila pseudoobscura* that is heterozygous for an inversion. This shows the giant polytene chromosomes that are found in the salivary glands. The chromosomes are present in very many copies, so that they are clearly visible. Homologous chromosomes are paired, and so heterozygotes for inversions form a characteristic loop, as explained in Box 12.2.

levels of polymorphism, making microsatellites and minisatellites excellent genetic markers for variation within a population (Box 13.3).

Recent surveys have revealed surprisingly high polymorphism in gene arrangement. For example, an average of 11 differences were found between two randomly chosen human genomes (Fig. 13.32). Most rearrangements were rare, being found only once in a panel of 20 individuals. However, many were found multiple times, several had been previously reported, and some were close to regions that are known to mutate at high frequency to give rearrangements responsible for genetic diseases (regions A–D in Fig. 13.32). This shows that some regions of the human genome are prone to rearrangements (some of which are seriously deleterious) that persist at appreciable frequency for a considerable time.

Although chromosome arrangements differ substantially between sister species, and small rearrangements are often polymorphic within populations, the net rate of chromosomal evolution is still very slow. The chance that any particular region will suffer a rearrangement is therefore small, and so linkage relations are conserved for surprisingly

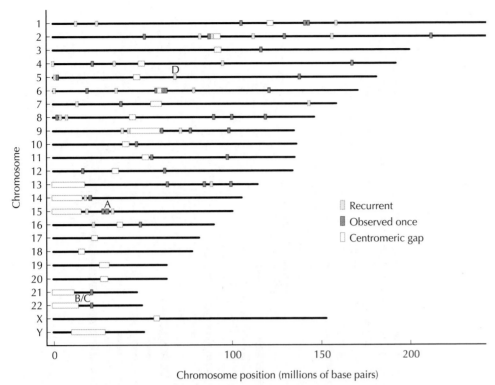

Chromosome position (millions of base pairs)

FIGURE 13.32. There is extensive variation in gene content within the human population. DNA from 20 individuals revealed genome-wide variation in copy number at 76 loci, which reflects the presence of deletions and duplications. These polymorphisms each average 465 kb in length and contain 70 known genes in total. Some polymorphisms coincide with loci that have a high rate of chromosomal rearrangement, causing inherited diseases (A, Prader–Willi and Angelman syndromes; B, cat eye syndrome; C, DiGeorge/velocardiofacial syndrome; D, spinal muscular atrophy). This survey should detect most large-scale deletions and insertions but will miss smaller rearrangements.

long times. There is substantial conservation of linkage between mice and humans, so that large blocks of chromosome can be seen to be homologous (Fig. 13.33). We will see in Chapter 23 that selection can act on recombination rates and therefore also on gene order; and we have already seen on pp. 178–181 that there are constraints on gene order in bacteria and archaea (p. 178). However, it may well be that in eukaryotes, gene order is a purely arbitrary consequence of random chromosome rearrangements.

We have looked at variation in the content of genomes over increasing scales, from the sequence of individual base pairs, through amino acid sequence, and on to chromosomal rearrangements of various sizes. We have already seen that at the largest scale the size of the whole genome varies enormously: 0.16 Mb in the bacterial symbiont *Carsonella* to more than 13 Mb in δ-proteobacteria *Sorangium cellulosum*, and 2.3 Mb in the parasitic microsporidian *Encephalitozoon cuniculi* to 133,000 Mb in lungfish; humans are intermediate, with 2,900 Mb (Fig. 7.1). This variation in DNA content does not correspond to variation in the number of distinct genes or in the amount of sequence that codes for protein (Fig. 7.3). Some variation in genome size is due to **polyploidy** (i.e., multiplication of the whole genome, both coding and noncoding; pp. 328 and 631) or duplication of large segments of genome. However, most variation in gene size is not due to polyploidy, but instead reflects extreme variation in the amount of noncoding DNA. Substantial variation is seen within genera and even within species. For example, DNA content varies 1.6-fold within the genus of kangaroo rats, *Dipodomys*, and within maize (*Zea mays*), genomes vary by approximately 40% in size. In Chapter 21, we will see that variation in genome size largely reflects the nonadaptive accumulation of "selfish" and "junk" DNA.

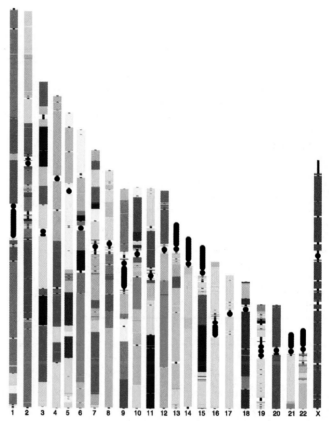

FIGURE 13.33. Gene order is conserved across wide evolutionary distances. The colored segments show blocks of genome that have maintained the same order between mouse and humans. Each color corresponds to a mouse chromosome, overlaid onto the human chromosomes. Note that gene content on the X chromosome is completely conserved (far right).

SUMMARY

Evolution requires genetic variation. Populations must contain different DNA sequences, coding for different proteins and causing differential expression of these proteins. In this chapter, we have seen that populations typically contain abundant variation in DNA and proteins, which can be measured in several ways. In the next chapter, we will see how the overall phenotype is influenced by this sequence variation.

Until the development of molecular biology, the extent of genetic variation was almost entirely unclear. Cryptic variation was revealed by inbreeding and by environmental and genetic perturbations, which suggested that substantial variation was hidden behind a more-or-less uniform phenotype. Electrophoretic detection of allozyme variation and, more recently, direct observation of the DNA sequence made it clear that there is an enormous range of molecular variation of all kinds: differences in the sequence of bases, insertions and deletions of various sizes, chromosomal rearrangements, and variation in the size of the whole genome.

This variation can be described in various ways. The simplest is in terms of allele frequencies, which can in turn be summarized by measures such as the genetic or nucleotide diversity (H and π, respectively). However, such measures do not capture the frequencies of combinations of alleles at different sites. The most fundamental description of genetic variation is based on the genealogies of each site.

Multicellular eukaryotes have allozyme diversity of up to $H \sim 15\%$ and nucleotide diversity up to $\pi \sim 0.01$. However, diversity is typically higher for single-celled eukaryotes and higher still for bacteria. More abundant species tend to be more diverse, but the relationship is weaker than expected. Sites that are constrained by their function tend to be less variable. Most important, changes that alter amino acid sequence are less common than synonymous changes that do not.

We see similar patterns when we measure differences

between species. The rate of divergence is surprisingly steady across different lineages—there is a molecular clock, albeit a crude one. However, different parts of the genome diverge at very different rates, reflecting the degree of functional constraint.

Gene order shows a different pattern from sequence variation. Large chromosomal rearrangements are rarely found as polymorphisms within populations, and the rate of chromosomal evolution varies greatly between taxonomic groups. However, smaller-scale rearrangements (e.g., small inversions) are common wherever they do not disrupt function.

FURTHER READING

Bentley D.R. 2003. DNA sequence variation of *Homo sapiens*. *Cold Spring Harbor Symp. Quant. Biol.* **68:** 55–63; Hinds D.A., Stuve L.L., Nilsen G.B., Halperin E., Eskin E., Ballinger D.G., Frazer K.A., and Cox D.R. 2005. Whole-genome patterns of common DNA variation in three human populations. *Science* **307:** 1072–1079.

Recent summaries of the pattern of variation across the human genome, concentrating on SNPs.

Eichler E.E. and Sankoff D. 2003. Structural dynamics of eukaryotic chromosome evolution. *Science* **301:** 793–797.

Reviews recent findings on chromosomal evolution, based on comparisons between sequenced genomes.

Gibson G. and Dworkin I. 2004. Uncovering cryptic genetic variation. *Nat. Rev. Genet.* **5:** 681–690.

Reviews of both the older literature on cryptic variation and our more recent understanding of its molecular basis.

Kimura M. 1983. *The neutral theory of molecular evolution.* Cambridge University Press, Cambridge.

A very clear summary of the extent of variation within and between species, in both protein and DNA sequence. Although written at the very beginning of studies of DNA sequence, the basic patterns still hold.

Lewontin R.C. 1974. *The genetic basis of evolutionary change.* Columbia University Press, New York.

A classic text, written soon after the discovery of extensive genetic variation, that describes the extent of variation detected using electrophoresis and the difficulties in explaining this variation in terms of the neutral theory.

Nevo E. 1988. Genetic diversity in nature—Patterns and theory. *Evol. Biol.* **23:** 217–246.

A comprehensive review of the within-species diversity revealed by electrophoresis.

C H A P T E R

14

Variation in Genetically Complex Traits

MENDEL DISCOVERED THE BASIC LAWS OF GENETICS by studying clear-cut differences that were inherited in a simple way—peas that were round or wrinkled, yellow or green. Similarly, molecular biology focuses on simple phenotypes—bands on gels that directly reflect the DNA sequence. Yet, simple traits such as these are not necessarily interesting in themselves. For both practical and scientific reasons, we also seek to understand the properties of whole organisms—the yield of a crop, the human life span, the variety of organismal forms, or the diversity of behaviors. In the first section of this chapter, we introduce ways of thinking about the genetics of the whole organism, which allow for the immensely complicated relation between genotype and phenotype—that is, between the simple sequence of nucleotides and the complicated traits that we usually observe.

Quantitative genetics describes the inheritance of traits that depend on many genes. It does not require any knowledge of the actual genetic mechanisms that are responsible for the **variation** and, indeed, it was developed before we had any knowledge of these mechanisms (pp. 20–21 and 25–28). However, now that genetic markers are available in abundance, quantitative genetics can be used to identify the genes responsible for variation. Quantitative genetics now thrives both in its traditional application to plant and animal breeding and in studies of natural populations and of human disease.

INTRODUCTION TO QUANTITATIVE TRAITS

Genes Influence Phenotype through a Complex Web of Interactions

A DNA sequence is, by itself, meaningless. The information in the double helix is interpreted through its interactions with the rest of the cell. The first steps, and the simplest, are the transcription of DNA into RNA, and its translation into protein. The simplicity of the **Central Dogma** that DNA makes RNA makes protein, and the fact that the **genetic code** is identical or nearly identical in all species, allowed for the spectacular success of molecular biology. However, this central machinery does not take the organism very far: A formless mass of proteins can hardly function. Protein and RNA must fold into the correct three-dimensional structures; often they may be modified so that their final sequence does not exactly reflect the sequence specified by the DNA, and so that nonstandard side groups are attached. Genes must be expressed at the right time and their products transported to the right place, such as to different compartments within the cell. In multicellular organisms, cells must reliably differentiate, so that the correct structure, with its specialized organs, develops. In some organisms, yet more layers of complexity are added—for example, the differentiation of ants into castes with distinct roles in the colony (Box 21.2). All this complexity must be robust against disturbance, yet flexible enough to respond to changing circumstances.

In principle, we understand how DNA interacts with the cell to produce an organism: how gene expression can be regulated, how proteins can be assembled into organelles, and how delicately shaped organs develop. In a few cases, we have a more or less complete qualitative understanding of which molecules bind to which and how the consequent interactions are responsible for overall function: for example, the reproduction of phage λ (p. 52) or the early development of the *Drosophila melanogaster* embryo (Fig. 9.24). However, we do not know how to actually compute even the simplest organism from the DNA sequence, even though cells themselves routinely carry out this computation. Proteins usually fold into unique three-dimensional structures, determined by their amino acid sequence, yet protein structure can only be predicted a priori for small molecules. Similarly, we cannot predict which transcription factors will bind to which DNA sequences—even though rapid, accurate, and specific binding is essential to gene regulation.

Some scientists are attempting to build a quantitative model of the whole cell. This seems a distant prospect, but even if it were successful, such a detailed mechanical description—following the interaction between each of several thousand kinds of molecule through time and space—would not in itself give us much understanding: The model would be as complex as what is being modeled (Fig. 14.1). What is needed is some way of condensing the complex relationship between an organism and its genome into a few comprehensible parameters.

Quantitative Genetics Is Concerned with Variation between Individuals

Fortunately, we do not need a full model of the organism. For most purposes, we are concerned with variation between individuals, populations, or species that share a common ancestry. We want to know why some people get sick and some do not; why some breeds of dog instinctively herd, whereas others hunt or track; or why each species of fig wasp breeds inside a particular species of fig. Similarly, evolution itself is concerned with variation; change is incremental and consists of a long series of steps that eventually lead to very different organisms. At each step, it is the *change* in the organism that matters: Thus, we need to know both how the change affects reproductive success and how it is produced by genetic variation in DNA.

The distinction here, between the biochemical mechanism that builds an organism and the effect on that organism of a *change* in its genome, is an important one. Popular accounts of genetics often refer to genes that "determine" a trait. What that usually means is that a particular change in a gene (i.e., replacing one **allele** by another) has an effect on the trait. That does not mean that it is the only allele or gene that affects the trait or that it would have the same effect in a different population or in a different environment. Also, it does not necessarily mean that the gene involved is in the direct pathway that constructs the trait. Often, alleles have effects due to some general disruption of the organism, rather than being attributable to the function it-

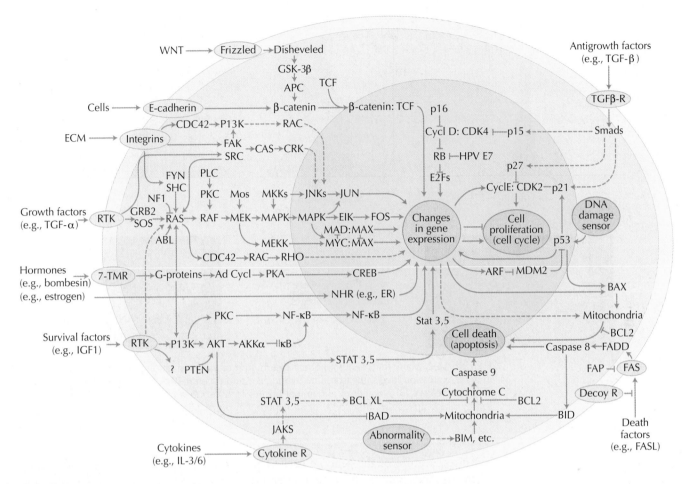

FIGURE 14.1. A qualitative model that illustrates some of the complex interactions that sustain a functioning cell. Shown is a model of a human cell with a cytoplasmic membrane, cytoplasm, and nucleus. In cells, a network of proteins causes changes in cell behavior by changing gene expression, protein translation, and protein function. External signals (e.g., growth factors, hormones, and cytokines) influence cells via membrane-bound receptors (e.g., Frizzled and E-cadherin), which in turn trigger interconnected signaling pathways that lead to changes in cell behavior. *Arrows* and *lines* represent interactions (e.g., positive and negative regulation of one protein by another or binding of proteins to nucleic acids).

self. (For example, screens for mutations that affect learning in *Drosophila* often pick up defects in control of movement, because the tasks that the flies are required to learn require movement.) It is similarly misleading to talk of genes that "code" for a trait, because there is no simple code that relates DNA sequence to its consequences. This issue is especially important in human genetics, and so we discuss it in more depth in Chapter 26.

Most Traits Are Influenced by Many Genes

As we will see, variation in most traits is **polygenic**. That is, differences between individuals or species are usually due to several genetic differences, so that there is no simple pattern of Mendelian inheritance. The figure that opens this chapter and Figure 14.2 provide some examples of the variety of characters we will be concerned with in this chapter. Some are strictly determined by the genome. For example, the total DNA content can be seen as a continuously varying trait, which depends on a large number of sites in the genome. Thus, it is both polygenic and

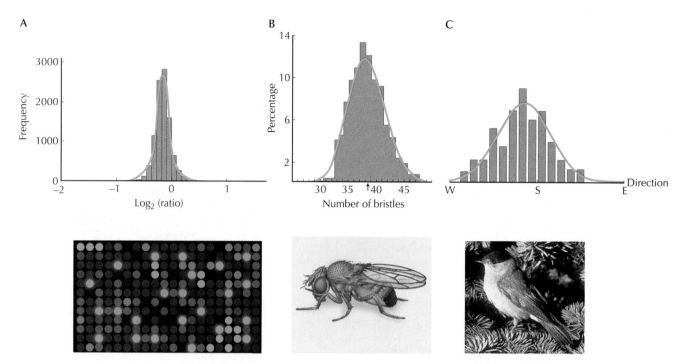

FIGURE 14.2. Many traits show a normal distribution of variation among individuals. (*A*) Variation in gene expression levels, measured by hybridizing fluorescent cDNA to oligonucleotides on a microarray (*below*). (*B*) Number of abdominal bristles in *Drosophila melanogaster*. (*C*) Direction chosen by blackcaps (*Sylvia atricapilla*) in an experiment on migratory behavior.

is a direct feature of the genome itself. The amount of messenger RNA produced by a gene depends both on the promoter and other regulatory sequences at that gene and on variation at other genes, for example, those that code for the relevant transcription factors. Moving to the opposite extreme, behavioral variation—for example, the direction taken by a migrating bird—depends both on the structure of the brain and on the early experiences that shape the behavior. However complex the mechanisms involved, all these examples can be treated as **quantitative traits** and are studied using the methods of **quantitative genetics**.

One of the most important applications of quantitative genetics, and one that has grown rapidly over the last decade, is in human genetics, an application that is discussed further in Chapter 26. Many inherited diseases are caused by defects in single genes and have been well characterized using the classical methods of Mendelian genetics. However, such diseases cause only a small proportion of illness relative to diseases with more complex inheritance. In Western societies, only approximately 1.5% of individuals will be affected by a disorder caused by a single gene (e.g., cystic fibrosis), whereas most will suffer at some time in their lives from one or more of the many diseases that are strongly heritable, but genetically complex (e.g., coronary heart disease, diabetes, or schizophrenia). Thus, understanding the genetic basis of genetically complex traits might make a considerable contribution to human health.

In the remainder of this chapter, we set out statistical methods for analyzing variation in traits that depend on many genes. We show how the individual **phenotype** can be separated into components that are inherited in different ways and how the variation in the whole population can be separated in a similar way. We show how these components of variance can be measured and survey the nature and extent of quantitative genetic variation. Last, we show how the statistical description of quantitative traits is related to the underlying genes and how these genes can be identified.

ANALYZING QUANTITATIVE VARIATION

Traits Often Follow a Normal Distribution

Many features of organisms follow a characteristic **normal distribution**—a bell-shaped curve illustrated for various examples in Figure 14.2 and discussed in more detail in Chapter 28 (online). The prevalence of the normal distribution is important for two reasons. First, it can be described by just two parameters: the **mean** and the **variance**. Thus, these two parameters can represent *any* normal distribution: Once those are given, the rest of the distribution follows. Second, the normal distribution is expected whenever observations are the sum of many independent and random effects. Thus, the observation that biological traits are often normally distributed suggests that these are due to the sum of effects of the many genetic and environmental perturbations that influence each individual. This simple **additive model** is the foundation of quantitative genetics. However, we will see that even when genes interact in complicated ways, normal distributions may still be observed and, whatever the distribution, variability can still be described in terms of variances.

A normal distribution may appear only when traits are measured on the appropriate scale. For example, Figure 14.3 shows the distribution of weights of tomatoes from a **backcross** between two varieties with very different fruit weights. There is a very wide spread, approaching the small cherry of the wild variety at the lower extreme. The distribution is far from normal: Indeed, a normal curve with the observed variance would imply that approximately 2% of hybrid tomatoes should weigh less than nothing (Fig. 14.3, top). However, if the weights are plotted on a logarithmic scale (Fig. 14.3, bottom), tomatoes necessarily have a positive weight, and the distribution of log(weight) is close to normal. This is as expected if fruit weights are the *product* of many independent factors, each multiplying the weight by some random factor. One factor increases weight by 10%, a second decreases it by 5%, and so on, so that the final weight is 1.1 × 0.95 × ⋯ . Now, the log(weight) is the *sum* of many independent effects (i.e., log(weight) = log(1.1 × 0.95 × ⋯) = log(1.1) + log(0.95) + ⋯) and so is expected to be approximately normally distributed.

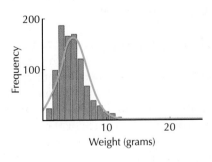

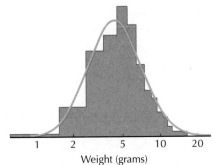

FIGURE 14.3. Distributions are often normal when plotted on a log scale. The *top panel* shows the distribution of fruit weight, in a backcross between two varieties of tomato (Red Currant, with mean weight 0.45 g, and Danmark, mean 10.4 g). This raw distribution is strongly skewed to the right and deviates significantly from a normal distribution (*blue bell-shaped curve*). On a log scale, however, the distribution fits well to a normal curve (*bottom panel*).

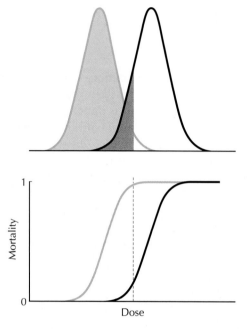

FIGURE 14.4. Under the threshold model, survival depends on an underlying normally distributed trait. (*Top*) Suppose that individual insects vary in the threshold dose of an insecticide. Thus, individuals with a threshold below the actual dose die (*shaded areas*). In this example, the population on the left has a distribution of thresholds that is lower, and so 97.7% die (*blue*); the right-hand population has generally higher thresholds, and so only 16% die (*red*). (*Bottom*) When survival is plotted against dose for these two populations, a characteristic sigmoid curve is seen. The *dashed line* shows the dose given in the *top* panel.

Traits restricted to discrete values, such as counts, cannot strictly follow a normal distribution, because that distribution is defined for truly continuous variation. However, when observations span a moderately large range of values, they can be approximated to a normal curve. For example, Figure 14.2B shows the distribution of the number of bristles on the abdomen of *D. melanogaster* (a popular trait, because it is easy to count on live flies). Even traits that can take on only two values (e.g., alive or dead) can be described by a normal distribution, using a **threshold model**. Here, the assumption is that there is some underlying quantity, which is normally distributed, and that the trait changes when this exceeds a threshold. Thus, if 2.5% of a population survives, the threshold score for survival must be two standard deviations above the population mean. Figure 14.4 shows an example in which the proportions of insects surviving different doses of insecticides fit the threshold model.

The Normal Distribution Describes Variation in Multiple Traits

Multiple traits can also be described by the normal distribution (Chapter 28 [online]). For example, Figure 14.5A shows the distribution of leg length and body length of a population of fire-bellied toads. The scatter of points fits well to a multivariate normal distribution, shown by the contours. This distribution is now defined by the means of the two traits, by the variances of each trait (16.0 mm^2 for body length and 3.4 mm^2 for leg length), and by the **covariance** between them. In this example, toads with large bodies tend to have longer legs, reflected in the clustering of points along the diagonal in Figure 14.5A. Correspondingly, the covariance between the traits is positive. (The covariance has units equal to the product of the units of the two traits—here 6.1 mm^2.) Often, it is convenient to express the covariance as a dimensionless **correlation coefficient**, which lies between –1 and +1. In this example, the correlation between body length and femur length is 0.83.

This extension of the normal distribution to describe more than one variable is important for several reasons. First, it gives a compact description of the joint variation of any set of traits. Figure 14.5B shows the growth rates of various **genotypes** of aphid, grown on either peas or alfalfa. In this figure, points scattered away from the diagonal indicate a tendency to specialize on one or the other host. Thus, we can meas-

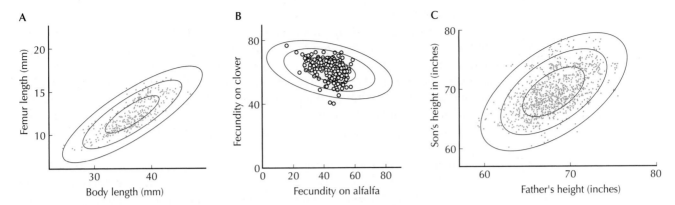

FIGURE 14.5. (*A*) The distribution of body length and leg length in a population of fire-bellied toads (*Bombina*). (*B*) Relative fecundities of aphids reared on different host plants. Aphids derived from alfalfa are more fecund on alfalfa, whereas aphids derived from clover are more fecund on clover. The scatter shows the distribution of fecundities of an F₂ cross between populations derived from alfalfa and from clover. (*C*) Father's height plotted against son's height. Contours show the region expected to contain 50%, 90%, and 99% of a multivariate normal distribution.

ure different traits under the same circumstances (e.g., Fig. 14.5A) or the same trait under different circumstances (e.g., Fig. 14.5B). Second, and most important from a genetical point of view, a trait can be measured on a set of related individuals. For example, the points in Figure 14.5C show the heights of fathers and sons.

In general, the pattern of inheritance can be described by the distribution of the trait across relatives and, in particular, by the **covariance** of the trait between relatives. (This method was devised by Darwin's cousin, Francis Galton; Fig. 1.26.) Use of a normal distribution condenses all the complexities of inheritance into just a few parameters. Moreover, we shall see that these covariances depend in a simple way on the fraction of genes that are shared by the different relatives (see Box 14.2) and do not depend on details of the underlying genetics.

The Phenotype Can Be Analyzed as a Sum of Independent Effects

Even if genetically identical individuals were reared under standard conditions, traits would inevitably vary between them. Any particular genotype must have a characteristic distribution of each trait. Individuals with the same genotype have the same set of genes but nevertheless do not have identical phenotypes. Such variation between genetically identical organisms is termed **environmental variation**. It includes the effects of factors that we can identify (e.g., temperature and nutrition) and those we cannot (e.g., random fluctuations in early development).

Quantitative genetics is based on the separation of the variance in a population into a component due to genetic differences among individuals and a component due to nongenetic environmental variation. To understand how the variance can be separated in this way, think of the distribution of the trait across different individuals with the same genotype (Fig. 14.6). The mean of this distribution is written G and is termed the **genotypic value**. An individual's phenotype can then be written as $P = G + E$, where E is the random **environmental deviation**. Because G is defined as the average of P, E must average 0. With species such as maize, we can actually measure both G and the distribution of E simply by rearing large numbers of identical genotypes and measuring them under standard conditions.

It is often reasonable to assume that the environmental deviation E is normally distributed and that it has the same distribution for all genotypes. With these assumptions, a genotype can be described simply by its **environmental variance** V_E and

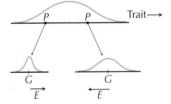

FIGURE 14.6. An individual's phenotype can be separated into a genotypic and an environmental component: $P = G + E$. The *curve at the top* shows the distribution of phenotype in the whole population. If we take any one individual and rear very many genetically identical copies, these will have a distribution (*lower curves*). The mean is the genotypic value G and the difference between this and the actual phenotypic value P is the environmental deviation E.

A

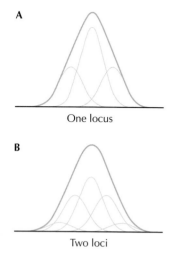

One locus

B

Two loci

FIGURE 14.7. The additive model generates an approximately normal distribution, even when only a few genes are involved. (A) A single genetic locus carries two alleles at equal frequency. The *thin curves* show the trait distributions of the three genotypes and the *thick curve* shows the overall distribution. (B) Two genetic loci, each with two alleles at equal frequency. Both genes have the same size of effect on the trait. In both examples, 62% of the total variance is due to genetic differences.

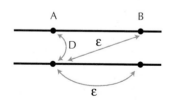

FIGURE 14.8. Dominance is an interaction between homologous genes at the same locus (D), whereas epistasis is an interaction between genes at different loci (ε). Lines show the two genomes in a diploid individual, with loci A, B.

genotypic value G. However, the separation of phenotype into genotypic value and environmental deviation does not depend on assuming that environmental deviations follow a fixed normal distribution and, indeed, genotypes might well vary in their sensitivity to environmental perturbations. For example, inbred individuals often show an increased environmental variance, and some mutations disrupt developmental stability. We examine the possible implications of such inherited variation in sensitivity in Chapter 23 (pp. 689–692). For the moment, though, we assume the simplest case, in which all genotypes follow a normal distribution with a fixed environmental variance V_E.

If organisms were asexual, we would now have a complete statistical model of inheritance: Barring mutation, G would be passed unchanged from mother to daughter. With sexual reproduction, however, genotypes are not passed on intact. Instead, offspring carry a mixture of genes from the two parents. So, we must find how the **genotypic value** of the offspring depends on the genes carried by its two parents. It is simplest to begin by assuming that the effects of each gene add up. For example, a diploid genotype $A_1A_2B_1B_1$, heterozygous at locus A and homozygous at B, would have phenotype $P = \alpha_{A_1} + \alpha_{A_2} + 2\alpha_{B_1} + E$, where the αs are known as the additive contributions of each allele. (This simple scheme is known as the **additive model**.) Now, if we could observe the genotypic value directly, we would see distinct values for each genotype. However, addition of a little environmental variance yields a smooth and approximately normal distribution, even when only a few genes are involved (Fig. 14.7).

Genes Interact through Dominance and Epistasis

Of course, genes do not have effects that simply add up. It would be extraordinary if such a simple model could describe the complexity of gene action. Two kinds of interactions must be distinguished: **dominance**, which is the interaction between the two homologous genes at a single genetic locus, and **epistasis**, which is an interaction between genes at different loci (Fig. 14.8). Dominance can occur when individuals carry two or more copies of each gene (i.e., in diploids and tetraploids), whereas epistasis can occur in any organism, including haploids. Sometimes, the terms dominance and epistasis are used to refer to an interaction in which one gene completely masks variation at another. (For example, a defect at one step in a biochemical pathway will mask the effects of any variation earlier in the pathway.) In quantitative genetics, however, the terms refer to any deviation from the additive model. Thus when heterozygote B_1B_2 has an average value midway between the homozygotes B_1B_1 and B_2B_2, we say that there is no dominance. Similarly, if $A_1A_2B_1B_2$ were the sum of contributions from the two loci (i.e., from A_1A_2 and from B_1B_2), we would say that there is no epistasis.

These ideas can be visualized by diagrams in which the phenotype is plotted against the genotype. Figure 14.9 shows an example from a study of the effects of **transposable elements** (Chapter 8, p. 217) on metabolic traits in *D. melanogaster*. Flies that differed by the presence or absence of transposable elements at two loci were crossed, and the activities of all nine possible genotypes were measured. Eight pairs of transposable element insertions, and 16 traits, were examined in this way. For each combination, the phenotypes of the nine different genotypes are plotted in a diagram that would give three parallel lines if there were no epistasis and that would give equally spaced straight lines if there were no dominance (Fig. 14.9A). A wide variety of patterns were seen, suggesting extensive deviations from simple additive gene action. Some of this variation is due to measurement error, but statistical tests confirmed widespread epistasis; 27% of tests showed statistically significant interactions between the effects of different transposable element insertions.

Even when there are gene interactions, the overall distribution can still appear approximately normal, as would be the case for the examples shown in Figure 14.9. If

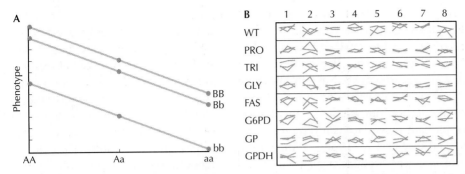

FIGURE 14.9. Epistasis and dominance for metabolic characters in *Drosophila*. For each of eight pairs of transposable element insertions, the nine different genotypes were measured. The measured traits included weight (WT), soluble protein (PRO), triglyceride and glycogen contents (TRI, GLY), and the activities of 12 metabolic enzymes, of which four (fatty acid synthase [FAS], glucose-6-phosphate dehydrogenase [G6PD], glycogen phosphorylase [GP], and α-glycerol-3-phosphate dehydrogenase [GPDH]) are shown. For each trait and each pair of transposable elements, the nine phenotypes were plotted as in *A*. If the genes had independent effects, the three lines would be parallel. Dominance between alleles A and a would show as deviations from a straight line, whereas dominance between B and b would show as an uneven spacing of the three lines. (In the example shown in *A*, there is no epistasis or dominance for A and a, but allele B is more or less dominant over allele b.) The results show a wide variety of patterns (*B*). For example, cross 2 shows strong **overdominance** for several traits, because the heterozygote Aa is often outside the range of the homozygotes. In cross 1, GPDH shows epistasis, because differences between bb, Bb, and BB are seen only when combined with genotype aa (*bottom left*).

there is dominance, but no epistasis (i.e., there are only interactions between genes at the same locus), then the genotype is still the sum of independent contributions from each locus, and so we still expect a normal distribution. Even with epistasis, there may still be an approximately normal curve, because interactions in different directions cancel out to leave the overall distribution unaffected. Regardless of the shape of the distribution, we shall see that quantitative genetics provides a description of genetic variation that applies to any kind of gene interaction.

With Dominance or Epistasis, the Effect of a Gene Depends on Its Genetic Background

How can we analyze genetic variation in complex traits when the effects of the genes do not add up? The crucial idea is to focus on the effect of one particular gene, averaging over the variation at all the other genes with which it finds itself and over random variations caused by the environment. We can define this marginal effect of a gene precisely for any particular population and for any particular set of environmental conditions. However, if there is dominance or epistasis, this effect will change as the composition of the population changes.

In Chapter 17 (p. 462), we will see that the immediate outcome of natural selection depends primarily on the marginal effect of a gene on reproductive success, averaged over all the circumstances in which it finds itself. Thus, although the ideas developed in this chapter apply to any kind of trait (e.g., body shape or behavior), they are particularly important when applied to the trait that determines the propagation of its genes: its reproductive success or **fitness**.

The effect of a particular gene on a trait can be defined in several ways. Although there are subtle differences between these definitions, for our purposes they can be taken as equivalent. We give two definitions: **average excess**, based on comparing the averages of individuals that carry different alleles, and **average effect**, based on a statistical **regression** of phenotype on genotype (see p. 20).

As a simple example, involving just one gene, we consider the effect of the sickle-cell hemoglobin polymorphism on human mortality (Box 14.1). In populations from parts of Africa where malaria is common, there are two alternative alleles at the β-globin locus, Hb^S and Hb^A. This polymorphism is maintained by a balance between the severe anemia experienced by Hb^SHb^S homozygotes and the susceptibility to malaria of Hb^AHb^A homozygotes. In Box 17.1, we consider in more detail how this kind of natural selection can maintain genetic variation, but for the moment, we use it as an example of a polymorphism responsible for substantial quantitative variation in survival and that involves strong dominance interactions—in other words, where the effect of one gene depends strongly on its homologous partner.

Box 14.1   Calculating Components of the Genetic Variance: Sickle-Cell Hemoglobin in Humans

Table 14.1 shows data from a Tanzanian population in which the Hb^S and Hb^A alleles are at frequencies $p = 0.207$ and $q = 0.793$ at birth. The frequencies of the three genotypes at birth are in **Hardy–Weinberg** proportions (Box 1.1), on the assumption that mating is random (i.e., $q^2:2pq:p^2$). Relative survival was estimated from the increase in frequency of Hb^AHb^S heterozygotes between childhood and adulthood. (Only the effects of malaria and sickle-cell anemia are included; of course, there are many other causes of mortality that reduce the actual survival below these relative values.)

Genotypic values and genotypic variance: Figure 14.10 shows the distribution of genotypic values in the population: On the left, the small proportion of Hb^SHb^S homozygotes are certain to die ($G = 0$), whereas on the right, the heterozygotes Hb^AHb^S have slightly higher survival than the Hb^AHb^A homozygotes (1 vs. 0.837). The total **genotypic variance** is just the variance of this distribution. It is calculated as the average of the squared deviations from the mean (Chapter 28 [online]): $V_G = 0.629(0.837 - 0.855)^2 + 0.328(1 - 0.855)^2 + 0.043(0 - 0.855)^2 = 0.0384$.

Average excess: An Hb^A allele has a chance of 0.793 of

finding itself with another Hb^A allele in a homozygote, and 0.207 of finding itself with an Hb^S allele in a heterozygote. So, its average survival is $(0.793 \times 0.837) + (0.207 \times 1) = 0.871$. Similarly, the average survival of the Hb^S allele is $(0.793 \times 1) + (0.207 \times 0) = 0.793$. The average excess is just the difference between the average survival of each allele and the overall average, 0.855.

In this example, the average excesses are quite small (0.016 for Hb^A and –0.062 for Hb^S), because the mortality from sickle-cell anemia almost exactly counterbalances the reduced mortality from malaria. For the polymorphism to be stable, the average excess fitness of each allele must be 0 (Chapter 17). The slight deviations from 0 here are due to slight errors in the estimates.

The overall genotypic value, G, can be written as a sum of two components. One contribution comes from the additive effects of each allele and is known as the **breeding value**, A. For example, each Hb^A allele is predicted to increase survival by +0.016, and so the predicted survival of an Hb^AHb^A homozygote is the overall mean plus twice the contribution from each Hb^A allele: $0.855 + (2 \times 0.016) =$

TABLE 14.1. Statistics on sickle-cell genotypes in a Tanzanian population

Genotype	Hb^AHb^A	Hb^AHb^S	Hb^SHb^S	Overall Mean	Variance
Frequency at birth	0.629	0.328	0.043		
Relative survival, G	0.837	1	0	0.855	$V_G = 0.0384$
	Hb^A		Hb^S		
Allele frequency	0.793		0.207		
Mean survival	0.871		0.793	0.855	
Average excess	0.016		–0.062	0	
Breeding value, A	0.887	0.809	0.731	0.854	$V_A = 0.0020$
Dominance deviation, D	–0.050	0.191	–0.731	0	$V_D = 0.0364$

Data from Allison A.C. 1956. *Ann. Hum. Genet.* **21:** 67; and Allison A.C. 1965. Population genetics of abnormal haemoglobins and glucose-6-phosphate dehydrogenase deficiency. In *Council for International Organizations for Medical Science Symposium on abnormal haemoglobins in Africa* (ed. J.H.P. Jonxis), pp. 365–391. Blackwell Scientific Publications, Oxford.

Box 14.1 Continued.

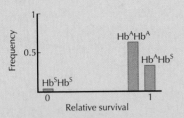

FIGURE 14.10. Genotypic values.

0.887. The deviation between this additive prediction and the actual survival is 0.837 – 0.887 = –0.050; this is known as the **dominance deviation**. In this example, dominance deviations are large, because the main differences in survival are between heterozygote and homozygotes.

Average effect: Figure 14.11 shows how relative survival depends on the number of copies of the Hb^S allele. The red bars show the actual genotypic values, with the heterozygote having the highest survival (middle bar). The gray bars show the breeding values, which necessarily change linearly with genotype. The breeding values can be calculated by a regression of trait against genotype, which finds the straight-line relationship that minimizes the squared deviation from the actual values (dashed line). As explained in the text, the slope of this regression is known as the average effect; when alleles are combined at random, as here, this is identical with the average excess.

Because the average effects of the alleles are quite small in this example, breeding values are similar to each other. In other words, any parent can expect his or her offspring to have about the same chances of survival, regardless of his or her genotype. As long as mating is random, the family will contain a mixture of homozygotes and heterozygotes. Figure 14.12 shows the distribution of breeding values. Its variance is necessarily equal to the **additive genetic variance**, $V_A = 0.629(0.887 – 0.855)^2 + 0.328(0.809 – 0.855)^2 + 0.043(0.731 – 0.855)^2 = 0.0020$. The difference between the total genotypic variance and the additive variance is equal to the variance of the dominance deviations: $V_D = V_G – V_A = 0.0364$. We see that in this example, almost all the inherited variance in survival is due to dominance deviations, rather than to the additive effects of the separate alleles.

Heritability: In quantitative genetics, we usually deal with traits that are more or less normally distributed. However, the methodology applies to a trait with any kind of distribution, including survival. If we treat survival as a quantitative trait, we can set a value of 1 for a child that survives and 0 for one that dies from malaria or sickle-cell anemia. For simplicity, we assume that all the heterozygotes survive. Looking at the population as a whole, the average survival is then 0.855. In other words, a proportion 0.855 have value 1 and 0.145 have value 0. So, the variance of the distribution is $V_P = 0.855(1 – 0.855)^2 + 0.145(0 – 0.855)^2 = 0.1242$.

There are two ways to define heritability; each is useful for a different purpose. The **broad-sense heritability** is defined as the proportion of the total variance due to inherited differences: $H^2 = V_G/V_P = 0.0384/0.1242 = 0.31$. The **narrow-sense heritability** is defined as the proportion of the total variance due to additive differences: $h^2 = V_A/V_P = 0.0020/0.1242 = 0.016$. In this example, we have considered the effects of only one gene on survival and have found that it accounts for a substantial fraction of total variation. Presumably, many other genetic differences also contribute to variation in survival, so that the actual broad-sense heritability may be considerably higher.

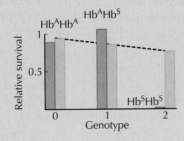

FIGURE 14.11. Relative survival.

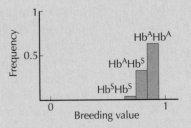

FIGURE 14.12. Breeding values.

The average excess of an allele is defined as the difference between the average trait value of individuals who carry that allele at a particular gene relative to the overall average of the population (Fig. 14.13A). Imagine that we pick out a particular gene, such as the β-globin gene inherited from the father. Now, we can compare mortality in individuals who inherit the Hb^S allele from the father with the population mean. This will be an average over those who also inherit Hb^S from the mother (and hence suf-

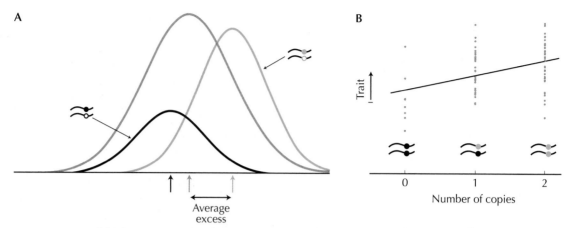

FIGURE 14.13. (A) The average excess of an allele is the difference between the trait value of individuals carrying that allele and the overall mean. Here, we imagine that the population (red) is divided into those that carry a particular allele (blue) and those that do not (black). The average excess is an average over all the environments and all the other alleles with which the allele in question finds itself. (*Open circles* indicate alleles chosen randomly from the population.) (B) The average effect of an allele is the slope of a regression of trait value on the number of copies of the allele. This example shows the distribution of trait values for the three diploid genotypes, carrying zero, one, or two copies of the allele in question; the *line* shows the best fit to these values, and its slope is the average effect.

fer sickle-cell anemia) and those who inherit Hb^A from the mother (and hence gain partial resistance to malaria). If mating is random, then these two possibilities will occur in proportions equal to the frequencies of the alleles. For a population in an area with no malaria, the average excess mortality of the Hb^S allele is high. However, in areas where malaria is prevalent, the average excess is close to 0, because the advantage of malaria resistance in the heterozygote counterbalances the disadvantage of sickle-cell anemia in the $Hb^S Hb^S$ homozygote. It is this balance that maintains the polymorphism (Box 14.1). Note that the average excess of an allele depends on genotype frequencies (because they determine the genotypes over which we average) and the environment (which determines the trait values of each genotype).

The average effect of a gene is defined in a slightly different way, which is equivalent to average excess when the genes are combined at random in the population. Suppose that we plot the phenotype as a function of the number of copies of the various kinds of allele—here, as a function of the number of Hb^S alleles (Fig. 14.10B). We can then find the additive model that best fits the actual, more complicated relationship, in the sense that it minimizes the mean squared deviation between actual and predicted genotype (Box 14.1). The slope of this regression is the average effect of one allele relative to another. For a random-mating population, this is the same as the average excess. In general, however, average effect and average excess may differ slightly. The idea of average excess is perhaps intuitively grasped more easily: It is simply the average trait value of individuals carrying one particular allele relative to the overall population mean. However, average effect and its extensions, discussed below, have simpler statistical properties and are the basis of more elaborate quantitative genetic analyses.

Quantitative Genetics Describes Interactions among Any Number of Genes

The ideas developed in the previous sections can be extended to describe interactions between two, three, or any number of genes. The average excess of an allele is defined in the same way, by the mean trait value of all those who carry that allele. Each gene is embedded in a whole genome, and so we must average over all the different combinations of genes with which it might find itself—in other words, over all possible genetic backgrounds. The average excess of an allele depends on the genetic background in two ways. Our particular allele may tend to be found with other genes that also affect the trait. Moreover, if there is dominance or epistasis, then the trait depends on the combination of alleles and not just on the sum of their individual effects.

The idea of an average effect also extends to any number of genes. We simply write the phenotype as a sum of contributions from each gene independently, of each pair of genes, of each triplet of genes, and so on. Then, we choose contributions that minimize the squared deviation from the actual trait values. (As for a single gene, the average effect of each individual gene is the same as the average excess, provided that the genes are combined at random.) However complicated the system, we can write the value of an individual phenotype as a sum:

$$P = G + E = A + D + I + E. \tag{14.1}$$

Here, the genotypic value G is the sum of the average effects of each individual gene, A; the effects of dominance interactions between pairs of homologous genes, D; and, finally, more complicated epistatic interactions between genes at different loci, I. Of course, each of these terms is itself a sum of the particular effects of individual genes and combinations of genes. However, we will see that for many purposes we need only this straightforward separation into different kinds of contribution.

The first contribution, A, is called the breeding value. Although we have defined it in terms of the sum of the additive contributions from each gene (i.e., the sum of the average effects), it has a straightforward biological interpretation, which does not depend on knowing about any of these individual genes (Fig. 14.14). Imagine that an individual is mated at random to the rest of the population. His or her offspring will have some average trait value, which will differ from the overall average because of the contribution of the parent's genes. The breeding value is defined as twice this difference: The factor of 2 arises because only half the genes come from the one parent in question. This breeding value can be measured directly if one parent can produce very many offspring. For example, the breeding value of a bull with respect to milk yield can be measured from the average milk yield of all his daughters. For a dairy farmer, the only concern when buying a bull is its breeding value.

Causes of Variation Can Be Separated from Each Other

We have shown how an individuals' phenotype can be written as a sum of components with different causes: additive, dominance, epistatic, and environmental. In practice, it is rarely possible to measure these components. To do so requires that we know the genotype of each individual at all the loci that affect the trait and that we also know the average trait value for each possible genotype. If we knew all this, a quantitative genetic analysis might seem superfluous. The value of a separation of phenotype into its components comes when we use it to analyze the variance in the population as a whole. This variance can be separated into distinct components corresponding to components of the mean phenotype. Moreover, the variance can be measured without knowing anything about individual genes.

The variance of a sum of independent values is equal to the sum of the variances of each considered separately (Chapter 28 [online]). So, because the phenotype is the sum of the genotypic value and the environmental deviation, the variance of phenotype is the sum of the variance of genotypic values and the variance of environmental deviations. This can be written as $\mathrm{var}(P) = \mathrm{var}(G) + \mathrm{var}(E)$ or more compactly,

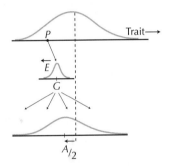

FIGURE 14.14. The breeding value of an individual can be measured from the average trait value in offspring produced when it mates randomly with the rest of the population. The *top curve* shows the distribution of the trait in the population. An individual with phenotype P is chosen. It has an underlying genotypic value G, which is the average trait value of a set of identical genotypes, reared under the same conditions (*middle curve*). The chosen individual is mated with others at random, producing the distribution of offspring shown in the *bottom curve*. The difference between the mean of these offspring and the population mean (*dashed line*) is half the breeding value ($A/2$); the factor of 1/2 arises because offspring get only one-half their genes from the chosen parent.

$V_P = V_G + V_E$. These variances are termed the **phenotypic variance**, genotypic variance, and environmental variance, respectively. Similarly, because the genotypic value can be split into additive, dominance, and interaction components ($G = A + D + I$), the genotypic variance can be split into additive, **dominance**, and **interaction variances** ($V_G = V_A + V_D + V_I$). The last term can itself be split into different kinds of interactions. For example, the variance due to interactions between the additive effects of pairs of genes is written V_{AA} and that due to interactions between the dominance effects of sets of three genes V_{DDD}. Overall, $V_I = V_{AA} + V_{AD} + V_{DD} + V_{AAA} + \cdots$. (Analyses into separate components of variance are possible because the components are defined in a special way, so that they are statistically independent of each other.) A simple calculation, involving just one gene (β-globin), is illustrated in Box 14.1.

The proportion of the total phenotypic variance that is due to genetic differences is termed the **heritability**. Because the heritability is a ratio of variances, it is dimensionless, and so comparison can be made between traits that are measured in different units. There are in fact two distinct versions of the heritability. Broad-sense heritability is the proportion of variance due to genetic differences of any kind; it is written as $H^2 = V_G/V_P$. In contrast, the narrow-sense heritability is the proportion due to strictly additive effects and is written as $h^2 = V_A/V_P$. This latter measure is more important, especially in an evolutionary context, because it is this term that allows a response to selection, as we discuss further in Chapter 17 (pp. 476–479).

Variance Components Are Estimated by Comparing Relatives

We have already seen (Fig. 14.6) the simplest method for estimating components of the variance, in which a collection of **inbred lines** is measured. If we regard the whole collection as the "population," then the variance between line means is V_G and the variance within lines is V_E. In human genetics, it is considered unethical to establish inbred lines or to clone large numbers of genetically identical individuals. However, comparisons between **identical twins** provide a similar method of estimation. A problem with studies on identical twins, however, is that the similarities shared by the siblings may be due to the family environment they share, as well as to the same set of genes they inherited. A partial solution is to compare them with nonidentical twins, whose genetic relationship is the same as brothers or sisters (Fig. 14.15). Alternatively, comparing twins who were separated and reared in separate families from an early age can reduce the effects of a shared environment.

The resemblance between a pair of relatives can be measured by the covariance between them. The simplest case is the relation between parents and offspring. If we assume the additive model, the offspring's phenotype can be written as $P = A + E = A_1 + A_2 + E$. Here, the additive effect, A, has been divided into the components inherited from the mother, A_1, and from the father, A_2. The father's phenotype can be written as $P^* = A_1 + A_3 + E^*$, where A_1 is the contribution from genes shared with the offspring and A_3 is the remainder (Fig. 14.16). Assuming that these components are independent, the covariance between parent and offspring is $\text{cov}(P, P^*) = \text{cov}(A_1 + A_2 + E, A_1 + A_3 + E^*)$. This expands to a sum of covariances: $\text{cov}(A_1, A_1) + \text{cov}(A_1, A_3) + \cdots + \text{cov}(E, E^*)$. Because all these components are independent, only the covariance involving *shared* genes contributes, and so the formula reduces to $\text{cov}(A_1, A_1) = \text{var}(A_1)$. The total additive variance is the sum of equal contributions from the two genomes in a diploid ($V_A = \text{var}(A) = \text{var}(A_1 + A_2) = \text{var}(A_1) + \text{var}(A_2)$). Thus, the covariance is just $\text{var}(A_1) = \frac{1}{2} V_A$ if the parents contribute equally. The factor of $\frac{1}{2}$ is the fraction of genes that a parent shares with its offspring (Box 14.2). Therefore, we can measure the additive genetic variance simply as twice the covariance between parent and offspring, using data such as those illustrated in Figure 14.5C.

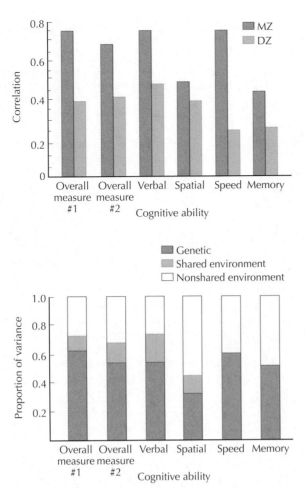

FIGURE 14.15. Components of variance can be estimated by comparing similarities between identical and nonidentical twins. A sample of 110 pairs of identical (i.e., **monozygotic**, MZ) twins, and 130 pairs of nonidentical (i.e., **dizygotic**, DZ) twins, from Sweden, all more than 80 years old, were tested for several measures of cognitive ability. (The *two leftmost columns* are alternative measures of overall cognitive ability.) (*A*) Correlations between genetically identical twins are significantly higher than between nonidentical twins. (*B*) About half of the variance was estimated to be genetic; with these data, it is not possible to distinguish additive from nonadditive variation. A small fraction of variance was attributed to shared environment (*gray*); this could be distinguished by comparing twins reared apart with those who grew up in the same family.

Estimating Variance Components Is Difficult in Practice

The covariance between relatives is simply a mixture of the different components of variance in the population as a whole (Box 14.2). Thus, in principle **variance components** can be estimated by comparing the covariances among different kinds of relatives. This method is used, for example, in animal breeding, where measurements are taken from very large numbers of individuals with a known relationship.

However, there are difficulties—for example, the effects of shared environments must be taken into account. Related animals may be on the same farm, and allowance must be made for known factors such as variation in nutrition. More problematic is that very large samples are needed if complex interactions are to be detected. Distinguishing different variance components depends on measuring *differences* in the degree of similarity between different kinds of relatives. For example, if there is dominance variance (V_D), full siblings are more similar to each other than parents are to their offspring. This is because siblings can inherit the same gene from the mother and

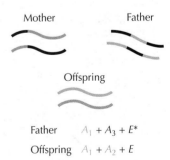

FIGURE 14.16. The additive genetic variance can be estimated from the covariance between parent and offspring. The offspring carries one genome from the father (*blue*) and one from the mother (*red*); these contribute A_1 and A_2 to the breeding value, respectively. The covariance between father and offspring depends on only the genes that they share (*blue*), and so is equal to var(A_1).

Box 14.2 | Contributions of Variance Components to Covariances between Relatives

The resemblance between relatives depends on the components of genetic variance in a simple way. In Table 14.2, each row shows how the covariance between each kind of relative depends on the additive variance (V_A), the dominance variance (V_D), and the various higher-order components (V_{AA}, V_{AD}, V_{DD}, ...). For example, the covariance between half-siblings is $\frac{1}{4} V_A + \frac{1}{16} V_{AA}$. Only pairwise interactions are shown, because higher-order variance components (e.g., V_{AAA}) make little contribution. It is assumed that mating is random and that relatives do not share environmental influences.

The simplest case is a pair of identical twins, who share exactly the same genotype. We can write the phenotypes of each of the twins as $P_1 = G + E_1$, $P_2 = G + E_2$. Assuming that the random environmental deviations E_1, E_2 are independent (i.e., $cov(E_1, E_2) = 0$), the covariance equals the genotypic variance: $cov(P_1, P_2) = cov(G, G) + cov(E_1, E_2) = V_G = V_A + V_D + V_{AA} + \cdots$. (Recall that $cov(x, x) = var(x)$, so that $cov(G, G) = var(G) = V_G$.) The covariance between genetically identical twins is therefore equal to the total genotypic variance, and all components of that variance contribute equally (first row).

In general, the contributions of the various components are just the fraction of sets of genes that are shared by the relatives (see Box 15.3). The second through sixth rows show relationships in which the relatives share only one ancestor, so that genes are shared down a single line of descent. In such cases, dominance components make no contribution, because individuals never share *pairs* of genes at the same locus. As an example, Figure 14.17, top panel, shows a pair of half-siblings who share one parent. If we choose one gene at random from each half-sibling, the chance that they are the same, and can therefore contribute to a resemblance between the two relatives, is 1/4. However, the *pairs* of genes cannot be identical, because the half-siblings do not share both parents. Thus, dominance deviations, which involve interactions between the two genes at a single locus, cannot contribute. In contrast, when individuals share more than one ancestor, they can share the same *pair* of genes, and so dominance components do contribute. For example, the chance that the pairs of genes carried by two full-siblings are identical is 1/4, and there is a corresponding contribution from the dominance variance, V_D (Fig. 14.17, bottom panel).

TABLE 14.2. The components of genetic variance determine resemblance between relatives

Relationship	V_A	V_D	V_{AA}	V_{AD}	V_{DD}
Identical twins	1	1	1	1	1
Parent/offspring	$\frac{1}{2}$		$\frac{1}{4}$		
Grandparent/grandchild	$\frac{1}{4}$		$\frac{1}{16}$		
Great-grandparent/ great-grandchild	$\frac{1}{8}$		$\frac{1}{64}$		
Half siblings	$\frac{1}{4}$		$\frac{1}{16}$		
First cousins	$\frac{1}{8}$		$\frac{1}{64}$		
Full siblings	$\frac{1}{2}$	$\frac{1}{4}$	$\frac{1}{4}$	$\frac{1}{8}$	$\frac{1}{16}$
Double first cousins	$\frac{1}{4}$	$\frac{1}{16}$	$\frac{1}{16}$	$\frac{1}{64}$	$\frac{1}{256}$

Adapted from Lynch M. and Walsh J.B. 1998. *Genetics and analysis of quantitative traits.* Sinauer Press, Sunderland, Massachusetts.

Numbers show the contribution of each variance component to the covariance between relatives.

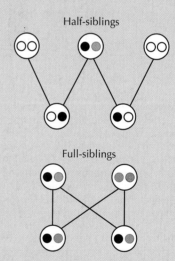

FIGURE 14.17. Sharing of genes between half- and full-siblings.

from the father and thus share identical genotypes. In contrast, an individual inherits only one copy of each gene from a given parent and so cannot share the combination of genes responsible for the interaction (Box 14.2). The dominance variance can therefore be estimated from the difference between the covariance between siblings, and the covariance between parent and offspring. However, such comparisons are statistically inaccurate, because estimates of both covariances are themselves inaccurate. The key point to take from Table 14.2 is that complex interactions, especially those involving dominance, hardly ever make much contribution to the similarity between relatives: This depends most strongly on the additive genetic variance, V_A, in the population.

Separating the trait variance into its components gives a powerful method for analyzing genetic variation when (as is almost always the case) we do not know the detailed genetic basis of the trait. However, it is important to realize that there is no simple relation between the magnitudes of the various components and the underlying interactions between the genes. Although genes interact in complicated ways to build the organisms that we observe, we will see that most variation in most traits can nevertheless be ascribed to additive effects of the genes. Why this is so is illustrated by the simple example of a single gene with two alleles (Fig. 14.18). When the effects of these alleles are additive (i.e., when the heterozygote is midway between the homozygotes), the genetic variance is entirely additive and is greatest at intermediate allele frequency (Fig. 14.18A). However, even if one allele is completely recessive (P, say, so that PQ and QQ are indistinguishable), most of the variance is still additive, provided that this recessive allele is common (right-hand side of Fig. 14.18B). This is because when allele P is common, the population consists mainly of PQ and PP, which have different phenotypes: The difference in phenotypes is thus associated with the additive effect of P versus Q. When the recessive allele is rare, a different phenotype is expressed only when the rare genotype PP is formed, and so dominance variance is larger than the additive variance ($V_D > V_A$; see left of Fig. 14.18B). However, in this situation both variance components are small, and so the total variance of a trait determined by many genes is still likely to be primarily additive. The dominance variance is only likely to be large in exceptional cases in which the heterozygote is outside the range of the two homozygotes (e.g., Box 14.1). The same kind of arguments carry over to interactions between multiple genes (epistasis)—even if the trait depends on strong interactions between genes, its variance can still be ascribed mainly to the additive effects of the separate genes.

Variance components and heritabilities are not constants of nature. They depend on the genetic diversity in the population and on how nongenetic "environmental" factors influence the trait. Nevertheless, we will see in the next section that there are some surprisingly consistent patterns in the quantitative genetic variation found within populations—whether natural, domesticated, or human.

There Is Extensive Inherited Variation in Most Traits

Just as there is extensive variation in DNA and protein sequences within most populations (Chapter 13), there is also abundant heritable variation in quantitative traits. This is shown in two ways: directly, by the resemblance between relatives for most characters, and indirectly, by the ready response to artificial selection. We examine selection on quantitative traits in detail in Chapter 17. For the moment, note that if continued selection for breeding of individuals with high values of some trait (e.g., milk yield) leads to an increase in that trait, then there must be a correlation between parent and offspring—in other words, there must be additive genetic variance.

The abundance of heritable variation has long been understood at a qualitative level. Darwin's strongest evidence for its existence was the clear similarity between parents and offspring and the striking success of artificial selection (see pp. 17–19). However, the methods outlined in this section allow a quantitative survey to be made. Figure 14.19 shows a compilation of estimates of narrow-sense heritability, h^2, which includes a wide range of traits—the response to light of polychaete worms, heat resistance of isopods, spine morphology of sticklebacks, growth rates of rainbow trout, and so on. The values span the whole range of possible values, but in most cases, heritabilities are high—that is, much of the overall phenotypic variance is due to additive genetic variance.

Some traits show consistently high heritability. In humans, the number of ridges on the fingers (one measure of the fingerprint) has broad-sense heritability of about 0.96. Presumably, almost all the variance is genetic because finger ridge count is determined early in development and so is not much perturbed by the external envi-

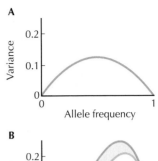

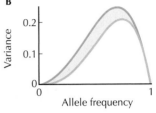

FIGURE 14.18. Components of variance for a single locus with two alleles. (*A*) The plot shows the genetic variance ($V_G = V_A$) when gene action is entirely additive. (Phenotypes QQ, PQ, and PP are 0, 0.5, and 1, so that each P allele adds 0.5 to the trait.) (*B*) The plot shows the components of variance when there is complete dominance (i.e., phenotypes 0, 0, and 1). The *upper curve* shows the total genotypic variance V_G and the *lower curve*, the additive genetic variance V_A; the difference between them (*yellow*) is the dominance variance V_D. The horizontal axis shows the frequency of the recessive allele.

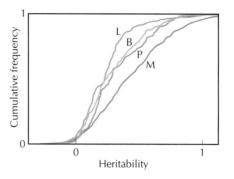

FIGURE 14.19. The distribution of narrow-sense heritabilities (V_A/V_P). Traits are classified as life history (L), behavioral (B), physiological (P), or morphological (M). The curves show the cumulative frequency distribution (Chapter 28 [online]). (The few estimates that are less than 0, or greater than 1, arise from statistical error.) Data are from 1120 estimates from 75 animal species.

ronment, and also selection does not act to eliminate genetic variation. At the other extreme, several experiments have failed to select for any left–right asymmetry in *Drosophila*; this is one of very few traits for which there is no detectable genetic variance. Even traits that appear invariant at the phenotypic level may be influenced by cryptic genetic variation (see p. 358). Under normal conditions, all wild-type flies have similar venation patterns. However, after a brief heat shock at a particular stage of pupal development, about 40% of adults emerge with a gap in a cross-vein (Fig. 14.20A). When flies carrying such a gap were selected, the proportion of flies that expressed a gap after a heat shock increased (Fig. 14.20B). Eventually, even flies reared under normal conditions show the cross-veinless phenotype. (Recall a similar example on pp. 307 and 358–360.) This phenomenon was termed **genetic assimilation**, because a trait that is initially induced by an extreme environment comes under genetic

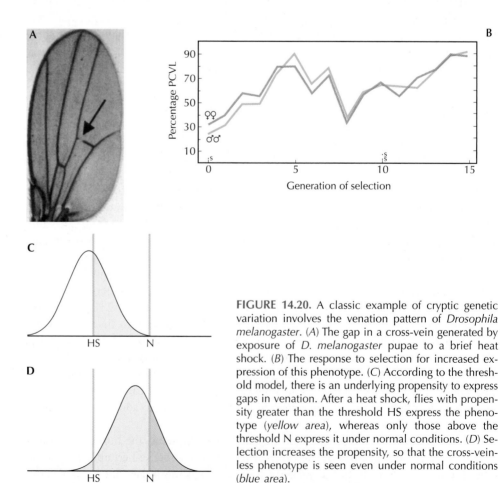

FIGURE 14.20. A classic example of cryptic genetic variation involves the venation pattern of *Drosophila melanogaster*. (*A*) The gap in a cross-vein generated by exposure of *D. melanogaster* pupae to a brief heat shock. (*B*) The response to selection for increased expression of this phenotype. (*C*) According to the threshold model, there is an underlying propensity to express gaps in venation. After a heat shock, flies with propensity greater than the threshold HS express the phenotype (*yellow area*), whereas only those above the threshold N express it under normal conditions. (*D*) Selection increases the propensity, so that the cross-veinless phenotype is seen even under normal conditions (*blue area*).

control even in a normal environment. It can be understood in terms of the threshold model discussed above—even though the cross-veinless phenotype is invariant in the normal environment, there is cryptic variation in an underlying propensity to lose the cross-vein (Fig. 14.20C,D). We discuss the relevance of this kind of experiment to the evolution of novel features in Chapter 24.

It Is Hard to Measure the Components of Natural Genetic Variation and the Numbers of Genes Involved

There are systematic differences between different kinds of traits. Adult morphology tends to show higher heritability than **life-history traits**—that is, traits that are associated with survival and reproduction. Physiological and behavioral traits show intermediate heritabilities (compare curves in Fig. 14.19). Interpretation of this pattern is not straightforward. Narrow-sense heritability is defined as the *ratio* of additive genetic variance to total phenotypic variance ($h^2 = V_A/V_P$). Although it was initially thought that life-history traits have low heritability because they have low additive genetic variance, closer examination showed that, in fact, the pattern arises because life-history traits have higher environmental variance and hence higher overall phenotypic variance. An alternative way of comparing genetic variances between traits with different scales of measurement is to express them relative to the trait mean $\bar{z}$ (i.e., to compare $V_A/\bar{z}^2$). On this measure, life-history traits in fact tend to have *higher* additive genetic variance than morphological traits.

The number of genes responsible for variation in quantitative traits is extremely hard to determine using classical genetic methods. We have seen that just a few genes can give more or less normal trait distributions (Fig. 14.7); indeed, the power of quantitative genetics is that it can make predictions that do not depend on genetic details, such as the numbers of genes and their individual effects. Nevertheless, some information about the number of genes involved comes from the variance seen in the F_2 of a cross between two divergent populations. Qualitatively, if parental phenotypes are rarely seen in the F_2 generation, then large numbers of genes must be involved (recall the maize example of Fig. 11.26). A quantitative estimate of the number of genes responsible for trait differences was introduced by Wright and Castle (see p. 28). This has been widely applied and tends to give estimates in the range of 1 to 15 genes (Fig. 14.21). Unfortunately, many factors conspire to make the Wright–Castle estimator a serious underestimate of the actual number of genes involved. We now turn to more powerful methods for identifying the actual genes responsible for polygenic variation.

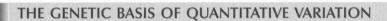

FIGURE 14.21. Estimating the number of genes from the variance of an F_2 cross. In the F_1 generation, all individuals have half their genes from one parent and half from the other. In the F_2, the distribution of the trait depends on the number of genes. If there is one gene, with intermediate heterozygote, then the distribution is broad (*lower left*). If there are five genes, the distribution is narrower and the parental genotypes only arise rarely (frequency $2^{-10} = 1/1024$; *lower right*). The number of genes involved is inversely proportional to the genetic variance in the F_2 relative to the difference between the parental lines.

THE GENETIC BASIS OF QUANTITATIVE VARIATION

Classical Genetics Can Identify Quantitative Trait Loci

In classical genetics, genes are mapped to positions on the chromosome by seeing how their alleles segregate in a cross. If alleles of different genes tend to stay together as they are passed from parent to offspring, then they must be physically linked, and their distance apart can be measured by the rate of recombination between them. It was this method that was used by T.H. Morgan's group at the beginning of the last century to first map genes onto chromosomes (see p. 24). The same method can be used to map the genes responsible for variation in a quantitative trait. If Mendelian markers are correlated with different trait values within a family, then one or more genes that affect the trait must be linked to the marker.

This approach identifies so-called **quantitative trait loci** or QTLs. The term refers not to an actual gene, but rather to a chromosome region that affects the trait. As we shall see, it is not easy to go from a statistical demonstration of a QTL to find the actual gene or genes involved—and, ultimately, to find how individual sequence variants affect the trait. We begin by discussing how the approximate position of a QTL can be estimated and then move on to methods that can locate them more precisely.

Mapping QTLs simply requires that a suitable population be scored for a large number of Mendelian markers and for the trait of interest (Box 14.3). In principle, any group of related individuals can be used—most often, an F_2 or backcross between two species or between a domesticated plant or animal and its wild relative. This is the most favorable case, because the cross may involve a difference of many standard deviations, so that the population used for mapping will have high heritability (i.e., $V_G \gg V_E$). Also, the exact genotype of the markers in parents is known and is the

---

Box 14.3   Mapping Quantitative Trait Loci

Suppose that we cross two inbred lines that differ for a quantitative trait and also for different alleles (P, Q) at a marker locus (Fig. 14.22, top panel). In the F_1, all individuals are heterozygous for the marker (PQ) and also for all the QTLs that affect the trait. Thus, the distribution of the trait is intermediate, and its variance is solely environmental (V_E; Fig. 14.22, middle panel).

In the F_2, there is increased genetic variance for the trait, and the marker segregates in Mendelian ratios (Fig. 14.22, bottom panel). In this example, there is an association between the trait and the marker. PP individuals have larger values than PQ, which in turn are larger than QQ. Each additional P allele is associated with an increase of 0.4 units in the trait. This could be due to a direct effect of the marker on the trait, but it is more likely due to linkage between the marker and a QTL.

If the QTL and the marker were completely linked, then we would estimate that the additive effect of the QTL allele associated with P is +0.4. Because the PQ heterozygotes are halfway between the two homozygotes in this example, we estimate that there is no dominance at the QTL. Note that the original line carrying PP is *smaller* than that carrying QQ (Fig. 14.22, top panel) and yet, in the F_2, individuals carrying P alleles are *larger*. This indicates that the lines differ by several QTLs, with mostly *negative* effects in the low line, but that the P marker allele happens to be associated with a *positive* QTL allele.

The difference between the lines in this example is two trait units; the difference between QQ and PP marker homozygotes is 0.8 or 40% of the total (albeit in the opposite direction). In the F_2 generation, the total genetic variance is 0.25. The marker is associated with some of this variance: One quarter of the population is QQ with mean –0.4, one half is PQ with mean 0, and one quarter is PP with mean +0.4. The variance of this distribution is $\frac{1}{4}(0.4)^2 + \frac{1}{4}(0.4)^2 = 0.08$, which is 32% of the total genetic variance in the F_2.

Of course, it is unlikely that the QTL and marker locus are completely linked. Suppose that the QTL recombines at a rate c with the marker. What difference in trait do we expect to see associated with allele P? In the gametes that go to make up the F_2, the four genotypes are at frequencies as follows.

Genotype	Frequency	Effect on trait
Q –	$(1 - c)/2$	0
Q +	$c/2$	α
P –	$c/2$	0
P +	$(1 - c)/2$	α

The mean effect of gametes carrying Q is αc, and the mean effect of gametes carrying P is $\alpha(1 - c)$. So, the difference in trait associated with the marker is $\alpha(1 - 2c) = 0.4$. If there is no linkage, then $c = 1/2$, and there is no association. With complete linkage, $c = 0$ and the difference is just α. Thus, information from a single marker can demonstrate linkage to a QTL, but it is impossible to disentangle the effect of the QTL, α, from the recombination rate. There could be a tightly linked QTL of small effect or a loosely linked QTL of large effect. With two markers, it is possible both to locate the QTL and to estimate the size of its effect. In practice, QTLs are detected using large numbers of markers spread over the genome. However, the basic method is essentially the same as in the single marker case considered here. For further details, see Web Notes.

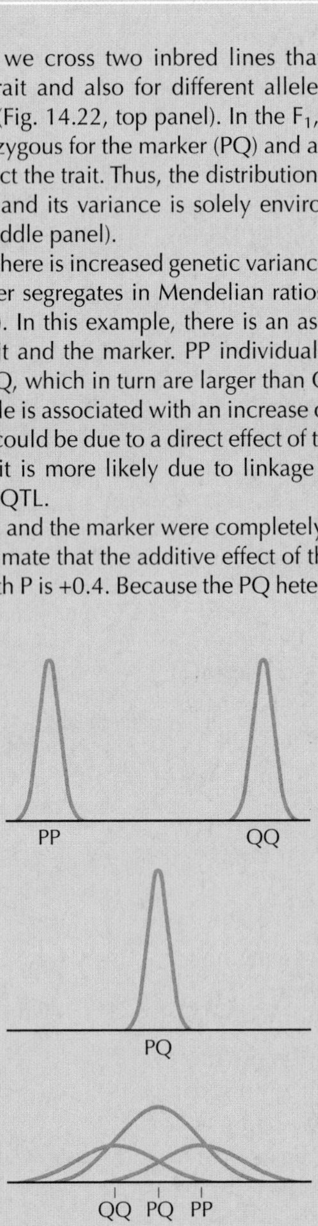

FIGURE 14.22. The distribution of a quantitative trait in individuals with different marker genotypes (PP, PQ, QQ) in parental, F_1, and F_2 generations.

same for all the individuals. For example, an F_1 would be heterozygous for all the marker alleles that distinguish the species or lines being crossed. If we label alleles from the two parental populations as 1 or 2, and the loci as A, B, C, ..., then the F_1 carries them in the coupling arrangement $A^1B^1C^1 \cdots /A^2B^2C^2 \cdots$.

The same method can be used within a single outbred population—for example, in studies of human families or dairy cattle. For several reasons, the statistical power is then much lower. First, there is less genetic variation in the study population. Second, most families are not informative, because the parents are not heterozygous at both QTLs and marker loci. Finally, the arrangement of the marker and QTL alleles in the parents is not known and varies from family to family. For example, a family that shows 1:1 segregation at two marker loci might have parents with any of the six genotypes

$$\frac{A^1B^1}{A^2B^1} * \frac{A^1B^1}{A^1B^2} \text{ or } \frac{A^1B^1}{A^2B^2} * \frac{A^1B^1}{A^1B^1} \text{ or } \frac{A^1B^2}{A^2B^1} * \frac{A^1B^1}{A^1B^1} \text{ or } \frac{A^1B^1}{A^1B^1} * \frac{A^1B^1}{A^2B^2} \text{ or } \frac{A^1B^1}{A^1B^1} * \frac{A^1B^2}{A^2B^1} \text{ or } \frac{A^1B^1}{A^1B^2} * \frac{A^1B^1}{A^2B^1} .$$

There must be a single A^2 allele and a single B^2 allele, but they could be in different parents, in the same parent but inherited from different grandparents, or in the same parent and from the same grandparent.

The recent proliferation of QTL mapping studies has been stimulated by the availability of easily scored molecular markers (see Box 13.3). Such markers can be found in any organism, can be scored on a large scale, and almost never have a direct effect on the trait being studied. A final requirement is that the trait be measured accurately. Here, there is a strong advantage to using organisms where lines of genetically similar individuals can be produced, so that environmental effects average out and the genotypic value can be measured precisely. For example, in dairy cattle a bull might sire hundreds of daughters, allowing extremely accurate measurement of his breeding value for milk yield.

The Principles of QTL Mapping Are Illustrated by a Study of Differences in Wing Shape in *Drosophila*

We can illustrate these methods using an example from *D. melanogaster*. Two lines of flies were selected in opposite directions for differences in the positions of landmarks defined by the wing veins (Fig. 14.23). Within and between natural populations of *Drosophila*, different parts of the wing vary together in a characteristic way. This phenomenon is known as **allometry** and has been thought by some to reflect underlying constraints on how organisms can develop (see Chapter 24, p. 696, for further discussion). Selection was applied so as to distort this allometric relation, and was remarkably successful. After 20 generations, the "high" and "low" lines (call them H and L) differed by 20 standard deviations, to give wing shapes quite different from anything found in nature (Fig. 14.23A,B).

The genetic basis of this difference was investigated by constructing more than 500 stocks of flies, each carrying a particular recombinant chromosome. We can concentrate on differences due to the third chromosome, which makes up 40% of the genome (the effects of the second chromosome were similar). Wing shape increased smoothly with the proportion of chromosome inherited from the H line. Moreover, the same pattern was seen whether the genetic material from the H line came from the left or right end of the chromosome (Fig.14.23C). This is consistent with there being an indefinitely large number of genes, each with a small additive effect. Statistically, at least eight genes are required to explain this pattern, because otherwise a step-like pattern would be seen, rather than a smooth change in wing shape with the proportion of genome from "high" versus "low" lines.

A more sophisticated analysis was made, which included data from double and triple recombinants. For each interval, a model in which there is a single QTL was fitted (see Box 14.3). A plot of the **likelihood** of this model against the position of the QTL identified 11 statistically significant loci (Fig. 14.23D). Moreover, a further sig-

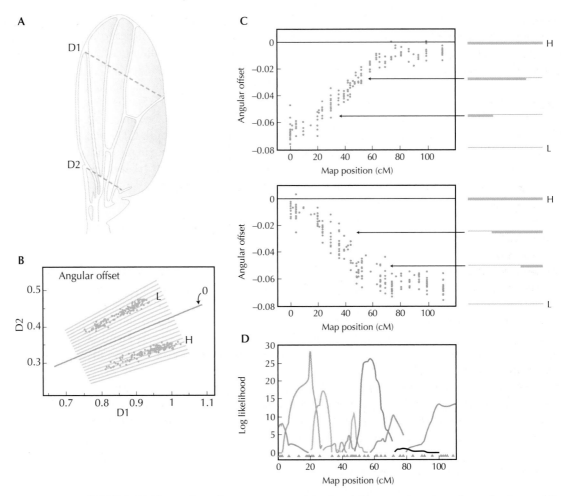

FIGURE 14.23. Finding the QTL responsible for differences in wing shape in *Drosophila melanogaster*. (*A*) The distances D1 and D2 were measured in two strains of flies that had been selected in opposite directions. (*B*) In the base population, these measurements fell on the straight line at the center. The two clusters of dots show measurements of the two selected lines, denoted L and H. The difference between them is measured by the "angular offset," whose value is indicated by the series of thin lines. (*C*) Flies were produced that were homozygous for a recombinant third chromosome, which was derived partly from the H line and partly from the L line. The diagrams show how wing shape changes with the proportion of third chromosome derived from the H lines. (*D*) A statistical model was fitted in which 11 QTLs influenced wing shape. The 11 curves show the likelihood for the position of each QTL: Peaks indicate the best estimate for its location. *Triangles* at the bottom show the positions of the markers.

nificant improvement was made by including epistatic interactions between some pairs of genes. Thus, there is some statistical evidence for interactions between individual genes, but these tend to cancel out, so that overall the pattern fits an additive model remarkably well. It is important to realize that in situations such as this, where many genes contribute to genetic variation, a statistical analysis cannot reliably identify individual gene effects and interactions, because there are simply too many parameters to estimate from a limited amount of data.

It Is Difficult to Find the Precise Position of QTLs

Studies such as these have yielded many estimates of the numbers of QTLs and their effects. However, there are several reasons to be cautious about taking such results at face value.

1. QTLs of small effect are likely to be missed. In a study of a few hundred individuals, only effects larger than approximately $0.2 \sqrt{V_E}$ are likely to be detected.

2. Conversely, the effects of those QTLs that are correctly identified tend to be overestimated, simply because effects that happen to be larger than expected are more likely to be detected. This bias is especially serious if they are close to the threshold of significance (e.g., the effects of those detected at a significance level of $P =$ 1% are overestimated by a factor of around 2).

3. Usually, the statistical test compares the hypothesis that there is a single QTL in the region of interest with the alternative that there is no QTL. If there are in fact a very large number of loci, spread evenly along the genome, then the usual statistical analysis still gives localized peaks like those in Figure 14.23D, just by chance. This may suggest that there are just a few QTLs, when in fact there are many.

4. Alleles whose effect varies with environment or with genetic background (i.e., that show epistasis) may be missed.

5. A standard QTL experiment can locate effects only to within 20 cM or so. Such regions may contain many loci that affect the trait, and, as we will see, apparent QTLs tend to "break up" on closer investigation.

This last limitation is the most fundamental. It implies that a typical study of a few hundred individuals can detect only at most about 12 QTLs and cannot locate them to anything like the precision needed to find the actual gene involved. The limitation is primarily due to the limited number of recombination events that occur in each experiment: If some block of genome is never split apart by recombination, then the effects of different genes within that block can never be distinguished, regardless of how many genetic markers are scored. To go beyond the limits of the standard experimental design requires either that we have some clue about likely **candidate genes** or that the effects of larger numbers of recombination events can be measured.

How can we locate QTLs more precisely? A powerful method is to backcross one line (say, one that produces a "low" trait value) repeatedly with the "high" line to generate an array of **nearly isogenic lines** or NILs (Fig. 14.24). Each NIL carries a unique set of introgressed genetic material, which can be identified by scoring large numbers of genetic markers. If the trait value of one of these NILs differs significantly from the "high" line, this difference must be due to a gene (or genes) in one of the introgressed segments, and because repeated generations of backcrossing have divided the genome into very short segments, any such genes can be located accurately.

In a few cases, this method has allowed identification of individual genes responsible for trait differences. For example, the much larger size of cultivated tomatoes compared to their wild relatives is due to multiple QTLs. One of these, *fw2.2*, causes a weight difference of up to 30% (Fig. 14.25A). A cross between two nearly isogenic lines located the QTL to within a region of about 4 cM, which was small enough to be characterized by constructing **transgenic** tomatoes. Ultimately, the effect on fruit weight was shown to be caused by differences in the regulatory region upstream of a gene, which is expressed early in fruit development. Because of the effort required, such methods for the statistical location of QTLs have only just begun to identify individual genes.

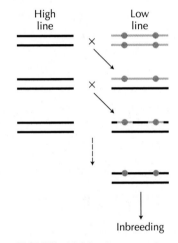

FIGURE 14.24. Construction of nearly isogenic lines (NILs). After several generations of backcrossing a "low" line to a "high" line, each backcross will carry a small fraction of genome from the "low" line, interspersed as short segments within a "high" background. If we select for low trait values during the experiment, these will be enriched for QTLs that reduce the trait (*red dots*). Genetically identical NILs are then derived by inbreeding—for example, by repeated mating between siblings.

Mutations with Major Effects on a Trait Show Us Which Candidate Genes Are Likely to Influence Milder Variation

Classical Mendelian genetics is based on mutations that cause obvious and major effects on the organism; for example, albino mice, wingless *Drosophila*, or severe human diseases with Mendelian inheritance. The first geneticists took such major mutations

A

B

FIGURE 14.25. The *fw2.2* QTL is responsible for fruit weight in tomatoes. (*A*) Wild and cultivated tomatoes can differ by 1000-fold in fruit weight. (*Left*) *Lycopersicon pimpinellifolium*; (*right*) *Lycopersicon esculentum*. (*B*) All cultivated varieties of tomato carry the large-fruit allele of *fw2.2*, and all wild varieties examined so far carry the small-fruit allele. The two tomatoes shown here differ by the insertion of a small-fruit allele (*left*) into a cultivated variety.

to be the stuff of evolution, a view that, as we saw in Chapter 1.15, caused an enduring controversy with those studying quantitative variation in nature. Major mutations are in fact unlikely to contribute to evolution. Individuals carrying them are usually at a serious competitive disadvantage even under laboratory conditions. However, genes that can mutate to cause major changes in a trait may also have alleles that have milder effects on the trait and less drastic effects on fitness. Thus, major mutations indicate which genes are likely to contribute to quantitative variation in a trait: that is, they indicate candidate genes. In the last section, we saw that the simple statistical association between markers and trait within a mapping experiment can only locate QTLs quite crudely. To identify the specific genes responsible requires that we narrow the search from many hundreds down to a few likely candidates.

Mapping studies do often indicate QTLs that lie in the same region as some obvious candidate gene. We have seen two examples in Chapter 11: the *Pitx1* gene, which influences pelvic morphology in sticklebacks, and *teosinte branched 1*, which is one of several genes responsible for differences between maize and its wild relative teosinte. In Chapter 26 (pp. 758–763), we will see examples of how candidate genes have been implicated as causes of diseases with complex inheritance.

Associations between Genetic Markers and Quantitative Traits in the Population as a Whole Can Help Map QTLs

The most obvious way to show that some candidate gene influences a quantitative trait is to demonstrate an association between particular sequence variants and differences in trait value. Studies of this kind have successfully identified a substantial fraction of the variance in *Drosophila* bristle number in natural populations. In flies, the bristles are motion-sensing organs. Thus, major changes in bristle pattern are produced by mutations to genes that specify development of these organs. QTL mapping experiments, of the kind described above, suggested that loci influencing bristle number often lay near to plausible candidate genes. In a series of studies, individual chromosome regions around these loci were isolated. Measurements of large numbers of flies from inbred lines thus gives an accurate estimate of the homozygous effect of each region, which could be compared with sequence variation in the 50–100 kb surrounding the candidate gene. There were significant differences in bristle number between chromosomes carrying different variants at the candidate loci. In the example shown in Figure 14.26, the presence of insertions in the *achaete-scute* region accounted for about 5% of the genetic variance in bristle number, a result that has been replicated

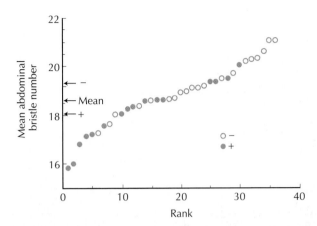

FIGURE 14.26. Insertions in the *achaete-scute* region of the *Drosophila melanogaster* genome reduce bristle number. Thirty-six X chromosomes were extracted from natural populations. For each chromosome, the mean number of abdominal bristles was measured, along with the location of insertions in a 106-kb-wide region. In the figure, the lines are arranged from smallest to largest bristle number, and lines carrying one or more insertions are marked by *closed circles*. These lines have significantly fewer bristles (compare +, – at *left* of figure).

in a separate study. Similar results have been found at two other candidate genes, *Delta* and *scabrous*, which are also involved in development of the peripheral nervous system. Crucially, at both loci much of the variation was associated with the presence of transposable elements (see pp. 217 and 342), which suggests that these elements actually cause the differences in bristle number. If the differences were caused by variation at sites that happened to be linked to the observed loci, then one would not expect a particular *kind* of variation to be involved.

These studies are encouraging, because they show that we know enough about genetics to identify the genes that influence a trait in nature. However, there are several difficulties that make it hard to apply this approach to other traits and organisms and, in particular, to human disease. First, in *Drosophila*, accurate measurements of genotypic value of a locus can be made by backcrossing onto a standard genetic background and by measuring large numbers of genetically identical individuals in a standard environment. This is only practical in a few model organisms. Very large sample sizes are needed to establish a correlation between marker and trait using measurements on *individuals*. It may be impossible to do this if allelic effects depend on environment and on other genes and, as we will see in the next section, such interactions are common. More fundamentally, we saw in Chapter 13 that in almost all populations there is abundant genetic variation, which is often clustered into distinct haplotypes (pp. 364–366)—that is, combinations of sequence variants that arise by chance and that are only slowly broken up by recombination. It is impossible to show, from a purely statistical analysis, which site in a haplotype is actually responsible for trait differences.

Genetic Manipulation Can Confirm the Genetic Basis of Trait Variation

The only way to conclusively demonstrate the *causes* of trait variation is by genetic manipulation in which individuals that differ only in specific ways are constructed. In one of the few experiments in multicellular eukaryotes in which this has been done, the results are sobering. Variation in alcohol dehydrogenase activity in *D. melanogaster* is strongly influenced by polymorphism in the *Adh* gene, which codes for the alcohol dehydrogenase enzyme. Nearly all natural populations of *D. melanogaster* are polymorphic for two alleles that are distinguishable by their properties during elec-

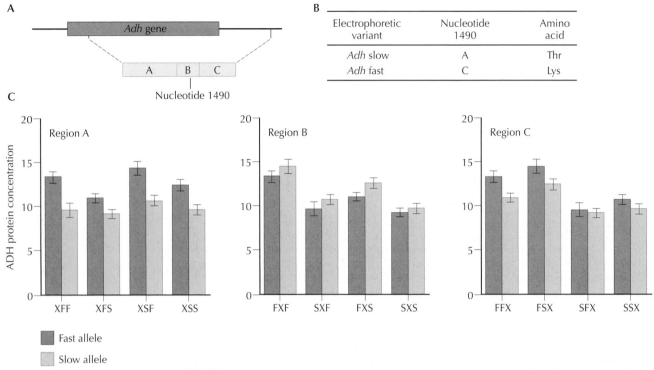

FIGURE 14.27. Multiple sites interact to determine the concentration of protein produced by the *Adh* gene. (A) Eight different constructs were made by combining regions (A, B, and C) from either a *fast* (F) or *slow* (S) allele. These alleles are distinguished by a nucleotide change at position 1490, which causes the amino acid change shown in the table (*B*). (C) For each of these eight constructs, accurate measurements were made of the *Adh* protein concentration in adults. Thus, when region A is derived from the F allele, concentration is consistently increased by 32%. In region B, sequence derived from F *decreases* concentration by 10% (*middle* histogram), whereas in region C, F-derived sequence increases concentration only when region A is also derived from F (*right* histogram). *Bars* show standard errors.

trophoresis. The fast (F) and slow (S) alleles are defined by a single amino acid difference (Fig. 14.27A; see also Fig. 13.11), which is entirely responsible for the different properties of the alcohol dehydrogenase enzyme. Specifically, the enzyme coded by the F allele has a 2.5-fold greater activity. However, the F allele also causes a 50% greater level of gene expression, and there is substantial genetic variation in expression within the F and S allelic classes. This variation is presumably due to noncoding sequence differences that affect gene regulation. However, it is extremely difficult to determine which of the many variable nucleotide sites is responsible.

This problem was partially solved by dividing the *Adh* locus into three regions and constructing flies carrying eight possible combinations of sequences from this region, derived from F or S reference alleles. It was shown that each region had a significant effect on the quantity of enzyme expressed and that there were epistatic interactions between these regions. Thus, although *Adh* activity at first sight appears to have a very simple genetic basis with segregation of two alleles accounting for much of the genetic variance, variation at this locus in fact results from interactions among at least three very tightly linked sites—or more likely, some much larger number of sequence variants. This kind of complexity may be typical. In humans, for example, multiple sequence variants at the β-globin locus affect oxygen binding and malaria resistance in a complex way (Fig. 18.4) and, similarly, many different alleles of genes such as *apoE* interact to determine cholesterol levels (see Chapter 26 [pp. 758–763]). Thus, it may not be possible to identify which sites cause quantitative variation from statistical arguments alone.

Although approaches based on candidate genes have shown remarkable successes, they are of limited applicability. As we discuss in the next section, variation in most quantitative traits may usually be due to alleles of such small effect that even if we knew which genes were involved, it would be impossible to demonstrate an association with trait values, even in organisms where accurate measurements are possible. A second difficulty is that even in the best-studied organisms, most genes have no known function: Indeed, in eukaryotes around two-thirds of genes show no obvious phenotypes when deleted and so are inaccessible to classical genetics. If one of these genes were responsible for substantial variation in a quantitative trait, we would not know of its existence as a candidate. The converse problem is that for many traits, a long list of potential candidates can be generated. For example, more than a hundred genes have been suggested as candidates that may influence cardiovascular disease.

The loci that influence a quantitative trait can, in principle, be found without knowing candidate genes, simply by searching for associations with markers distributed across the entire genome. This approach is now being applied on a large scale to search for the variants responsible for human disease. We discuss genome-wide association studies in Chapter 26.

Fisher's Geometrical Argument Emphasizes Alleles of Small Effect

Darwin held that the evolution of complex and well-adapted organisms required an abundant supply of variants with small effects. He drew an analogy with a telescope, which requires many independent and precise adjustments to reach its optimum performance. R.A. Fisher quantified this argument in his 1930 book, *The Genetical Theory of Natural Selection*. He imagined an organism as being described by a large number n of variables—for example, these might describe the shape of the skeleton or measures of behavior. Fitness is pictured as declining away from the optimum combination of these variables (Fig. 14.28). Individuals are unlikely to be perfectly adapted and so lie some distance from the optimum d. A mutation might move the population a distance r, in a completely random direction. Fisher made a simple geometrical argument: Mutations of small effect ($r << d$) are as likely to improve fitness as to reduce it, whereas mutations of large effect ($r > 2d$) must reduce it. (If we imagine a sphere enclosing the optimum with radius d, then *any* movement of more than a distance $2d$ must take the population outside the sphere and reduce its fitness.) Fisher

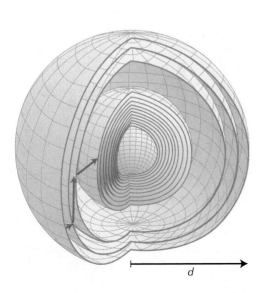

FIGURE 14.28. Fisher's argument that changes of small effect are most likely to contribute to adaptation. The optimal organism lies at the center of the diagram, and a population starts with all the individuals a distance d away, on the outer sphere. Thus, any change of more than a distance $2d$ must take the population further from the optimum, whereas small changes ($<<d$) are as likely to increase fitness as to decrease it. The *red arrows* show a sequence of evolutionary steps, in which random mutations that take the population toward the optimum are fixed by natural selection. The first successful mutation has magnitude $r = 0.137d$ and takes the population 8.7% of the way to the optimum (*first red line*, leading to the *second sphere*). The third successful mutation has the largest magnitude, $0.271d$; it is followed by smaller steps that, on average, follow a geometric series (*inner spheres*). This simulation uses ten dimensions, although only three can be shown.

d

went on to show that the chance that a mutation of size *r* is favorable falls away with $r\sqrt{n}/d$, so that favorable mutations are typically of size approximately $d/\sqrt{n}$ or smaller. This quantifies Darwin's intuition that in a complex organism (i.e., *n* large), only small changes are likely to be adaptive.

Fisher's argument was extended by Motoo Kimura, who allowed for the chance that a favorable mutation would actually be picked up by selection (Fig. 18.2), and by H. Allen Orr, who considered the process of evolution toward the optimum, as a sequence of favorable mutations arises and is fixed by selection (Fig. 14.28). Orr found that in such a sequence, the distribution of effects of the mutations that are fixed is approximately exponential (Chapter 28 [online]). Thus, successful mutations will have a range of effects, with some contributing much more to adaptation than others (Fig. 14.28). Nevertheless, Fisher's basic point remains valid: In a complex organism, in which many different features must work together, adaptation is expected to involve alleles of small effect.

It is not clear how far this kind of geometrical argument applies to actual traits. Some kinds of traits often evolve through changes at one or a few key genes. For example, resistance to pesticides often involves specific changes to the target molecule, which prevent the toxin binding to it. Similarly, resistance to malaria in humans is in large part due to changes in the globin genes, which code for the predominant protein within the red blood cells in which the malaria parasite lives. In such examples, there is a very strong selective challenge, which outweighs the disruption of existing function by a new mutation.

Quantitative Variation Is Based on Alleles with a Range of Effects and Interactions

Fisher's argument has been extremely influential in the **Evolutionary Synthesis** and led to the widely held belief that evolution proceeds by small steps. Orr's extension, to show that the accumulation of adaptive mutations leads to an exponential distribution of effects, is intuitively plausible and leads to testable predictions. These arguments apply to adaptive differences, such as those between artificially selected populations, or closely related species in nature. It seems plausible that such variation also involves a distribution of effects, mostly small but some large. What do we actually see?

As yet, QTL mapping has not given a clear answer to this question. Figure 14.29 shows two examples, one from studies of pigs and dairy cattle and one from a classic study of *Drosophila* bristle number that used visible markers. These distributions are at least consistent with an exponential distribution, although, of course, QTLs of small ef-

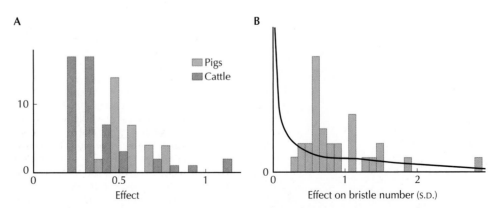

FIGURE 14.29. Distributions of effects of QTLs, scaled relative to the phenotypic standard deviation. (*A*) A compilation of studies of milk yield in cattle and growth and meat quality in pigs. (*B*) The distribution of effects of 31 QTLs on the third chromosome on bristle number in *Drosophila*. Effects smaller than about 0.5 s.d. are undetectable. The *curve* shows a hypothetical exponential distribution.

fect go undetected. Also, as explained above, the effects of those QTLs that are detected tend to be overestimated. There are many examples where individual QTLs cause substantial effects (one or more standard deviations, say) and where associations with genetic markers can account for an appreciable fraction of the genetic variance (e.g., Fig. 14.26). We discuss the genetic basis of species differences in Chapter 22.

We can also ask about the pattern of interaction between genes. Again, the evidence is as yet somewhat contradictory. For many traits, we have seen that estimates of variance components suggest that most genetic variance is additive. More detailed genetic analyses of quantitative traits can also show a surprisingly good fit to the additive model (e.g., Fig. 14.23). On the other hand, when individual QTLs of large effect are isolated, they have been shown to interact strongly with each other (e.g., Figs. 14.9 and 14.27). Such interactions need not maintain much nonadditive variance in the whole population and may average out so that overall the trait appears additive (recall the example of wing shape in *Drosophila*, discussed above, and see Fig. 14.18).

GENERATION OF QUANTITATIVE VARIATION

Mutation Generates Abundant Heritable Variation

Ultimately, quantitative genetic variation is generated by mutation. We can measure the rate at which genetic variance increases as a result of mutation by taking a set of genetically identical populations and propagating them for many generations. The rate of increase in the variance between lines, per generation, is called the **mutational variance** V_M. Often, this is scaled relative to the environmental variance to give a quantity that does not depend on the units of measurement. This is the **mutational heritability** V_M/V_E; it is the heritability produced by mutation in one generation, within a genetically homogeneous line.

Although this procedure is simple in principle, it is laborious in practice and has only recently been carried out on a wide scale. Crucially, the diverging lines must be allowed to accumulate mutations without opposition from natural selection. Most new mutations are deleterious and so tend to be eliminated (see previous section, and pp. 510–513). For organisms that can reproduce without sex, selection can be avoided simply by propagating lines asexually over many generations. For sexually reproducing organisms, self-fertilization or mating between siblings allows propagation of a line in which genetic variation is so low that natural selection is ineffective (e.g., Fig. 14.30). In *Drosophila*, **balancer chromosomes** can be used to preserve single chromosomes intact (Box 13.2). Obviously, mutations that kill or sterilize heterozygotes cannot be propagated, but mutations of smaller effect will accumulate with negligible opposition from natural selection.

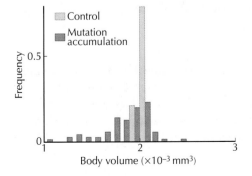

FIGURE 14.30. Estimation of mutational variance V_M for body size in the nematode *Caenorhabditis elegans*. Replicate lines were propagated by self-fertilization for 152 generations. *Gray bars* show the mean body size in control lines that were frozen; *red bars* show the mean body size in lines that had been allowed to accumulate mutations. The variance in body size between lines increased, giving an estimate of $V_M = 0.004$ per generation.

TABLE 14.3. Estimates of the mutational heritability, V_M/V_E

Organism	Trait	$h_M^2 = V_M/V_E$
Drosophila melanogaster	Abdominal bristle number	0.0035
	Enzyme activities	0.0022
	Wing dimensions	0.0020
	Body weight	0.0047
Tribolium castaneum	Pupal weight	0.0091
Daphnia pulex	Life-history traits	0.0017
Mouse	Limb length	0.0234
	Skull measures	0.0052
	6-week weight	0.0034
Arabidopsis thaliana	Life-history traits	0.0039
Maize	Plant size	0.0112
	Reproductive traits	0.0073
Rice	Plant size	0.0030
Barley	Life-history traits	0.0002

From Lynch M. and Walsh J.B. 1998. *Genetics and analysis of quantitative traits*. Sinauer Press, Sunderland, Massachussetts, Table 12.1, p. 338.

Mutation generates variation surprisingly rapidly. Moreover, the rate of increase in variation is remarkably consistent across a wide range of traits and organisms. The mutational heritability is in the range $V_M/V_E = 0.001 - 0.01$ (Table 14.3). This figure has a simple interpretation: It would take 100–1000 generations for mutation alone to replenish high heritability ($h^2 \sim \frac{1}{2}$ or $V_G \sim V_E$). If the quantitative trait is under selection, then favorable mutations will be picked up much sooner than this; conversely, if most mutations are deleterious, the buildup of genetic variance will be slower. The short timescale associated with mutation of quantitative traits contrasts with the much slower rate of mutation at single genes (roughly 10^{-5} per generation) or single base pairs (roughly 10^{-9} per generation; Chapter 12).

The mutational variance introduced in each generation can be written as $V_M = 2\Sigma_i \mu_i E[\alpha_i^2]$. This is a sum over all the loci i that can mutate to affect the trait; μ_i is the mutation rate at locus μ_i and $E[\alpha_i^2]$ is the average squared effect of mutations at that locus. We can see from the formula that the surprisingly high mutational variance could be due to large effects of each mutation ($E[\alpha_i^2]$) or to a high overall mutation rate, $2\Sigma_i \mu_i$, and a high mutation rate could be due to a large number of genes influencing the trait or to high mutation rates at each gene. We will see that all three factors play a role in explaining the high rate of generation of polygenic variation by mutation.

High Mutational Variance Could Be Due to Mutations of Large Effects or a High Mutation Rate

Earlier in the chapter, we saw that QTLs can have large effects. This does not by itself tell us the size of effects of new mutations, because selection will tend to pick up those mutations with the largest effects and because a series of substitutions at the same locus can accumulate to cause a large effect. (Recall that at least three mutational changes at the *D. melanogaster Adh* locus affect enzyme activity; Fig. 14.27.) More direct evidence comes from experiments in which new mutations are generated by random insertion of transposable elements. These mutations are tagged by the presence of the transposable element, and so they can be isolated and their individual effects can be measured. For bristle number in *Drosophila*, this method gives estimates of $E[\alpha^2]$ of around $0.1–0.2V_E$; that is, the standard deviation of effects ($\sqrt{E[\alpha^2]}$) is around 0.3–0.45 of the standard deviation of environmental effects, $\sqrt{V_E}$. This figure may not be representative of the mutations that affect bristle number in nature. However, transposable elements have been shown to be the cause of a substantial part of the natural variation in bristle number in *Drosophila* (e.g., Fig. 14.26), and so these experiments show that mutations of fairly large effect do contribute, at least to this particular trait.

Even if most of the mutations responsible for new genetic variance have effects around $E[\alpha^2] = 0.1V_E$, the total mutation rate must still be high: $2\Sigma_i\mu_i \sim 0.01–0.1$ to account for $V_M \sim 0.001–0.01V_E$. A few experiments have attempted to directly estimate the total mutation rate by counting the number of new variants that arise. In mice, two experiments gave estimates of $2\Sigma_i\mu_i \sim 0.02$ for skeletal traits and, in maize, two experiments gave figures of 0.05–0.1 for reproductive characters. These estimates are certainly too low, because mutations of small effect will be missed. Nevertheless, they suggest that total mutation rates are high enough to account for high values of mutational variance.

The most obvious explanation for the high total rate of mutation to polygenic variation is that very many genes are involved. For example, consider the overall viability of the organism as a quantitative trait. In eukaryotes, around one-third of the genes can mutate to give recessive lethals, and many more genes can have milder effects on survival (see pp. 510–515). Thus, many thousands could contribute to this trait. However, traits such as bristle number seem likely to depend on a small number of genes involved in specific developmental pathways. The discovery that a few candidate genes contribute much of the genetic variation in such traits suggests that the total number is not large, although it is still possible that many more genes contribute smaller effects through their general effects on physiology.

Finally, we can ask whether mutation rates at genes contributing to quantitative genetic variation are unusually high. Classical estimates come from rates of mutation to recessive lethals or, at least, mutations that have major phenotypic effects (e.g., eye color in *Drosophila*). However, quantitative traits may be influenced by changes with smaller effects. Changes in the amino acid sequence, or even complete deletion, may have no obvious effect under laboratory conditions. Moreover, changes outside the coding region may alter gene expression. (Most of the QTLs identified so far lie in noncoding regions.) In Chapter 19 (pp. 542–547), we examine evidence that suggests that at least as much noncoding as coding sequence is of functional importance. Thus, classical estimates of mutation rate may be too low because they miss changes with individually small effects, both inside and outside coding regions.

Some loci may also be exceptionally mutable. For example, tandemly repeated sequences are often replicated inaccurately, generating mutations with different numbers of copies (see p. 341 and Box 13.3). Variation in the number of copies of the genes that code for ribosomal RNA is responsible for some variation in bristle number in *Drosophila* and is associated with yield in several cereal species. Mutations of this sort may be an important source of quantitative variation.

SUMMARY

Most quantitative traits are influenced by many interacting genes. They often follow a normal distribution, which can be described simply by its mean and variance.

The phenotype is the sum of effects of genetic and environmental effects ($P = G + E$). The total genotypic value G is made up of the breeding value A (which represents the net additive effect of individual genes) and two kinds of interactions—dominance, due to interactions between genes at the same locus, and epistasis, due to interactions between genes at different loci. Thus, $G = A + D + I$, where I is the contribution to phenotype from epistatic interaction. The phenotypic variance V_P can be analyzed in the same way: $V_P = V_G + V_E = V_A + V_D + V_I + V_E$. The fraction of phenotypic variance due to all genetic causes is called the broad-sense heritability ($H^2 = V_G/V_P$), whereas the fraction due to additive effects is the narrow-sense heritability ($h^2 = V_A/V_P$). The covariance between relatives depends on these variance components and on the fraction of genes that they share. Thus, components of variance can be estimated by comparing different kinds of relatives (e.g., identical and nonidentical twins or parents and offspring).

For most traits, in most populations, heritabilites are high, and most genetic variation is additive. Traits associated with fitness tend to show lower heritability, but this is because they have higher environmental variance V_E rather than because they have lower genetic variance.

The approximate location of quantitative trait loci (QTLs) on the genetic map can be found using associations between traits and genetic markers. However, it is much harder to identify the precise genetic difference that causes trait variation; knowledge of candidate genes helps here.

Fisher set out an influential geometrical argument that adaptive variation in complex traits should be due to alleles of small effect. However, recent QTL studies have revealed a range of effects, with some large and many small.

Mutation makes a surprisingly high, and surprisingly consistent, contribution to trait variation: $V_M \sim 0.001–0.01 V_E$ per generation. It is not yet known to what extent this is due to large numbers of genes, large allelic effects, or high mutation rates per gene.

FURTHER READING

Barton N.H. and Keightley P.D. 2002. Understanding quantitative genetic variation. *Nat. Rev. Genet.* **3:** 11–21.

   Reviews results from QTL mapping in an evolutionary context.

Falconer D.S. and Mackay T.F.C. 1995. *Introduction to quantitative genetics.* Longman, London.

   The classic textbook on quantitative genetics, which covers most of the material in this chapter.

Lynch M. and Walsh J.B. 1998. *Genetics and analysis of quantitative traits.* Sinauer Press, Sunderland, Massachussetts.

   An advanced reference on modern methods in quantitative genetics.

Mackay T.F.C. 2001. Quantitative trait loci in *Drosophila. Nat. Rev. Genet.* **2:** 11–20.

   A clear explanation of methods for finding the genes responsible for quantitative variation.

Roff D.A. 1997. *Evolutionary quantitative genetics.* Chapman and Hall, New York.

   A straightforward textbook, somewhat more advanced than Falconer and Mackay and emphasizing application to natural populations.

15

Random Genetic Drift

EVOLUTION IS NEITHER DIRECTED NOR DETERMINISTIC. There is no necessary trend toward ever more elaborate or "perfect" structures, and the evolutionary process is not in general smooth or predictable. Seen over the long span of the Earth's history, we are amazed by the variety of extraordinary organisms that have evolved and by the structures that sustain them. However, if we follow any one lineage we see an erratic and fluctuating pattern of change with the usual outcome being extinction. Overall, the diversification of different species and competition between them have led to more complicated and more elaborate organisms (Chapters 6, 9, and 10). Seen in detail, however, the evolutionary process is fundamentally random. This is so even for our recent evolution from our primate relatives. We are impressed by our use of tools, upright gait, large brains, and so on. However, these traits have evolved erratically across several **hominin** lineages, of which only one happens to remain (Chapter 25).

Random events enter in many ways. Mutations occur at random, because of inevitable errors in copying the genetic material (Chapter 12). The environment changes at random over different scales: from the sporadic flux of nutrients experienced by marine bacteria as organic particles fall through the sea, to the patchy and transient spread of diseases, and up to large-scale changes in climate that affect whole species. Most dramatically, occasional collisions with asteroids may have caused the catastrophic extinction of entire groups, such as the dinosaurs (see p. 284).

In this chapter, we are concerned with the most basic source of randomness: the random outcome of individual reproduction. Imagine a diploid, sexually reproducing population, which is growing at 10% in each generation (Fig. 15.1). On average, each individual must leave 2.2 offspring. However, any particular individual will leave zero, one, two, or more offspring in a distribution that only *averages* 2.2. Even if each individual left exactly the same number of offspring, any particular gene would have only a 50% chance of being passed on at meiosis, giving a further source of randomness. This fundamental randomness of the reproduction of individuals, and of the genes they carry, inevitably leads to a process called **random genetic drift**.

EVOLUTION IS A LARGELY RANDOM PROCESS

Because it is the only natural process that can lead to adaptation, we pay most attention in this book to natural selection. However, we begin our consideration of the processes responsible for evolution by emphasizing the randomness of evolution. The importance

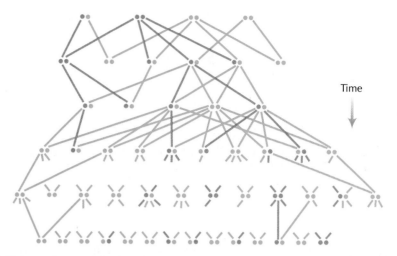

Time

FIGURE 15.1. Random growth of a sexual population. The average number of offspring is 2.2, so that the population grows on average by 10% in each generation. However, the actual number of offspring is 0, 1, 2, ..., and so the actual growth of the population is erratic (here, 4, 6, 5, 10, 12, 14, ...). Allele frequencies (red vs. blue) also fluctuate erratically ($\frac{3}{8}, \frac{5}{12}, \frac{4}{10}, \frac{5}{20}, \frac{6}{24}, \frac{7}{28}$, ···) both because of the random number of offspring from each individual and because of the randomness of meiosis.

of random processes is seen most clearly at the molecular level. Most eukaryotic DNA does not code for protein and is unlikely to possess any precise function. The sheer quantity of even those sequences that do code for protein suggests that they are only weakly constrained and mostly evolve at random. Indeed, the growth of molecular studies over the past half century has brought a greater emphasis on **neutral evolution**, in which change is largely due to mutation and random reproduction (see p. 59).

We should make clear here what we mean by "random." The dictionary definition is "of unknown cause." There may be definite physical causes of a changing climate or of whether some individual leaves a certain number of offspring, and with enough effort we might (in principle) be able to predict the outcome. In practice, though, we are ignorant of such complex processes and treat the outcome as "random." In biology, however, there is another, more crucial, meaning of the term: We often mean that change is random *with respect to adaptation* (recall pp. 344–347). Mutations occur as a result of copying errors, regardless of any effect they might have on function. (There are some examples where the *rate* of mutation is adjusted in an adaptive way—for example, in the vertebrate immune system. However, the mutations themselves still occur in a random direction [see pp. 233–234 and 663].) Similarly, whether or not an individual reproduces is largely independent of whether that individual carries particular genes that, on average, would improve its function. Variation in reproduction that is independent of genotype drives random genetic drift, whereas any consistent dependence of reproduction on the presence or absence of particular alleles drives the systematic, cumulative process of **natural selection.**

In physics, the basic processes are random, both because particles are subject to the uncertainties of quantum mechanics over small scales and because of chaotic dynamics over larger scales. However, for many purposes physics appears deterministic. For example, there is a precise relation between the temperature, volume, and pressure of a gas, even though on a small scale the gas consists of randomly colliding molecules. Indeed, the importance of chance processes only became clear with the emergence of the atomic theory at the end of the 19th century. Chance plays a much more obvious role in evolution for two reasons. First, while populations may consist of large numbers of individuals, their number is nowhere near as large as the number of molecules in even a small mass: There are approximately 6×10^9 humans on the planet compared to 6×10^{23} hydrogen atoms in a gram of gas. Thus, individual fluctuations are relatively more important in evolution than in physics—at least, over familiar scales. Second, and more importantly, reproduction amplifies individual events. A single mutation, involving a change in a single DNA molecule, can spread through an entire population. As a consequence of both random mutation and random reproduction, populations can diverge into distinct species even if they experience exactly the same physical conditions.

In this chapter, we begin by showing how random fluctuations in individual reproduction lead to ever increasing fluctuations in the frequencies of different alleles. Tracing forward in time, populations become more and more different from each other and eventually will fix different alleles. We then show how the same process of random genetic drift describes how genes come to share common ancestors, as we trace backward in time. We then introduce mutation and recombination and show how the process of neutral evolution leads to a simple framework for understanding molecular data. In the following chapters, we will extend this framework to include gene flow and natural selection and, in Chapter 19, we will use it to find ways to analyze molecular variation.

RANDOM DRIFT OF ALLELE FREQUENCIES

Allele Frequencies Change Whenever Individuals Have Different Numbers of Offspring

Random genetic drift is almost inevitable. A population could only stay the same if every gene produced exactly the same number of copies in the next generation. In large populations, fluctuations average out and so accumulate slowly. Nevertheless, random genetic drift occurs in any finite population, however large. We first examine the simplest case of asexual reproduction, starting from a single gene, and then go on to see how the same process occurs in a sexually reproducing population and how it influences quantitative traits as well as Mendelian loci.

Imagine a population of bacteria that reproduces asexually and maintains a steady size. Thus, the average rate of cell division must equal the average rate of cell death. Now, focus on a single cell, and its descendants—that is, a **clone**. The cell might carry a unique mutation, which would allow us to trace its offspring. However, we assume that any such mutation is **neutral**—that is, it has no effect on survival or reproduction.

Now, any lineage that starts with a single individual cell is likely to be lost quite quickly. The first event is as likely to be death as fission, and so there is only a 50% chance of surviving past this first event. If the cell does divide before dying, then there is still a 25% chance that both offspring will die before reproducing. In the long run, any neutral lineage is almost certain to die out (Fig. 15.2). This conclusion may seem paradoxical, because the expected number of offspring left after some time is just 1. This is so because the population as a whole is assumed to be stable, fission and death are equally likely, and so the expected number of descendants stays the same indefinitely far into the future. The explanation is that although the chance that the clone survives decreases toward 0, the average size of those lucky clones that *do* survive becomes larger and larger (Fig. 15.2). Thus, the distribution of the number of surviving offspring keeps the same mean of 1, but develops a larger and larger variance (see Chapter 28 [online]).

Every population experiences random genetic drift, whether it reproduces sexually or asexually, whether it is **haploid** or **diploid**, and whether it reproduces in discrete generations or continuously in time. In a sexual population, even if each

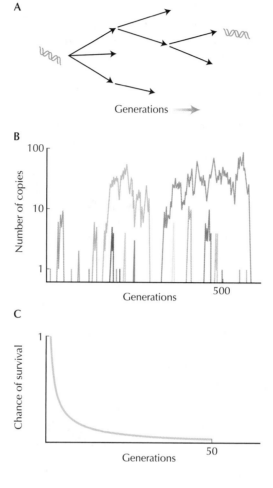

FIGURE 15.2. On average, in a population of constant size, each gene leaves one copy in the next generation. However, the actual number varies randomly, and so the lineage will eventually go extinct. (*A*) In this example, one gene leaves three daughters; one daughter leaves two offspring, one none, and the third daughter leaves one; only one of the granddaughters reproduces, and none of her offspring reproduce, so that this genetic lineage has gone extinct after three generations. (*B*) This shows 20 random lineages, each starting with one copy. Most die out in a few generations and are never present in more than a few copies. However, one lineage (*blue*) rises to 53 individuals before dying out after 138 generations; another lineage (*orange*) reaches 83 individuals and persists for 255 generations. (Numbers are plotted on a log scale because they range so widely: usually only a few copies are present, but numbers can occasionally grow large.) (*C*) The chance that the descendants of a single gene copy will survive decreases over time: For example, the chance of surviving for 50 generations is only 3.8%. (Numbers of offspring genes are assumed to follow a Poisson distribution with average 1.)

mating produced exactly the same number of offspring, the random segregation of genes at meiosis would lead to variation in the reproductive success of the individual genes, because only one of the two copies is passed on at meiosis.

The Wright–Fisher Model Is a Standard Representation of Random Drift

The simplest model of random genetic drift was developed independently by Sewall Wright and R.A. Fisher and so is known as the **Wright–Fisher** model. There are N diploid individuals, which each produce a very large number of **gametes**. These gametes come together at random to form a large pool of **zygotes**; N of these are then sampled at random to form the new generation. (Equivalently, one could sample $2N$ gametes and pair these at random to form the next generation.) Individuals are hermaphrodites (there are no separate sexes) and have a small chance ($1/N$) of fertilizing themselves. Now, imagine that there are two different alleles in the population (P and Q, say), and that they are at frequencies p and q, respectively. The number of copies of allele P in the next generation (j_P, say) will follow a **binomial distribution**, which averages $2Np$, but has variance $2Np(1 - p) = 2Npq$ (see Chapter 28 [online]). So, the new allele frequency, $p^* = j_P/2N$, will equal p on average—there is no directional trend—but will have variance $\mathrm{var}(p^*) = \mathrm{var}(j_P)/4N^2 = pq/2N$. We can think of the fluctuations in allele frequency as being due to the fluctuating numbers of two competing clones, P and Q, each behaving in a similar way to the bacterial example that we began with. In Box 15.1, we show how the variance in allele frequency increases over time.

Box 15.1 Random Drift Increases Variance between Populations and Decreases Variation within Them

Suppose that a population has allele frequency p_0. According to the Wright–Fisher model, the next generation is formed by sampling $2N$ genes at random from this population, to form N diploid individuals. As explained in the text, the new allele frequency p follows a binomial distribution. The mean stays the same ($E[p] = p_0$) but random drift has

generated a variance in allele frequency in the first generation of $V_1 = p_0 q_0/2N$. What happens as this process continues over many generations?

It is easiest to think of a large number of replicate populations as in the real experiment shown in Figure 15.3. The variance in allele frequency between these replicates increases at first at a rate $p_0 q_0/2N$. Eventually, it approaches a maximum variance $p_0 q_0$ when a fraction p_0 of the populations have fixed allele P and the remainder have fixed allele Q (see Chapter 28 [online]).

The rate of increase of variance, which is equal to $pq/2N$, must slow down over time (Fig. 15.3C). It does so because although the average allele frequency stays constant at $E[p] = p_0$, the average of pq decreases as the allele frequencies spread out. Ultimately, when all populations are fixed, $E[pq] = 0$, because either $p = 0$ or $q = 0$. To see how the average product of allele frequencies decreases over time, write $p = p_0 + \delta$. Then, $E[pq] = E[(p_0 + \delta)(q_0 - \delta)] = E[p_0 q_0 + \delta(q_0 - p_0) - \delta^2]$. Because $E[\delta] = 0$, and by definition $E[\delta^2] = \text{var}(p) = V$, $E[pq] = p_0 q_0 - V$. So we see that there is an exact trade-off between the variance within populations and the variance between populations, V.

The increase in variance of allele frequency from generation t to generation $t + 1$ is

$$V_{t+1} = V_t + \frac{E[pq]}{2N} = V_t + \frac{(p_0 q_0 - V_t)}{2N} = V_t\left(1 - \frac{1}{2N}\right) + \frac{p_0 q_0}{2N}.$$

It is easy to show from this that the variance after t generations is

$$V_t = p_0 q_0\left(1 - \left(1 - \frac{1}{2N}\right)^t\right) \sim p_0 q_0\left(1 - \exp\left(-\frac{t}{2N}\right)\right).$$

This formula fits poorly to the experiment if N is taken to be the actual number of flies ($N = 16$). However, it fits well if we assume that the **effective population size** is $N_e = 11.5$ (Fig. 15.3C).

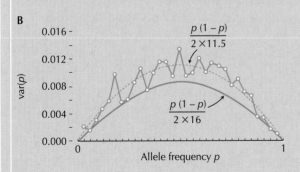

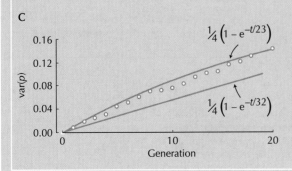

FIGURE 15.3. Random drift in experimental populations of *Drosophila melanogaster*, propagated with eight males and eight females. (*A*) The distribution of allele frequencies across replicate populations, all started at $p = 0.5$. Populations that had fixed one or the other allele are shown at *left* and *right*. (*B*) The variance in allele frequency generated by drift in a single generation. The experimental data (*circles*) show the variance in allele frequency among populations that had frequency $p = \frac{1}{32}, \frac{2}{32}, ..., \frac{31}{32}$ in the previous generation. (*C*) The accumulated increase in variance of allele frequency over 19 generations. (This is the increase in variance of the distributions shown in *A*.) In *B* and *C*, the *lower curve* shows the variance that would be expected from the actual number of flies ($\text{var}(p) = pq/2N$, $N = 16$), whereas the *upper dashed curve* shows the variance assuming an effective population size $N_e = 11.5$.

Fluctuations in allele frequency are well illustrated by an experiment on *Drosophila melanogaster*. Approximately 100 replicate vials of *Drosophila* were set up, each containing eight males and eight females. Initially, all of the vials were set up with a 50% frequency of an allele, bw^{75}, that altered their eye color slightly. Over 19 generations, frequencies in the different vials drifted apart, and 43 vials lost one or the other allele completely. Thus, genetic variability between populations increased at the expense of variation within populations (Fig. 15.3A). Overall, however, the allele frequency stayed roughly constant, implying that the bw^{75} allele had no strong effect on survival or reproduction. The simple prediction that the variance of changes in allele frequency is $pq/2N$ can be tested by following the distribution of allele frequencies through time. There is a good fit, but only if the number N is taken to be 11.5, instead of the actual number, 16 flies (Fig. 15.3B,C). That is, the variance in allele frequency increases faster than expected from the number of flies in the vials. This is because the variation in reproductive success of each fly is greater than is assumed under the idealized Wright–Fisher model.

Random genetic drift influences quantitative traits that are determined by multiple genetic loci; as the frequencies of the underlying alleles fluctuate, so too does the distribution of the trait itself (Fig. 15.4). What is remarkable is that the variance in the trait mean increases in proportion to the **additive genetic variance** ($\mathrm{var}(\bar{z}) = V_A/N$) and does not depend on the detailed genetic basis of the trait. As variation between populations increases, so variability within populations decreases. On average, the additive genetic variance decreases by a fraction $1/2N$ in every generation (i.e., $\mathrm{E}[\delta V_A] = -V_A/2N$). The variance of the trait also fluctuates randomly around this expectation, but here there is no simple prediction. The sensitivity of the genetic variance to random drift does depend on the genetic basis of the trait, which in most cases is unknown (Chapter 14).

The Rate of Random Drift Is Determined by the Variance in Fitness

At this point, we introduce a crucial concept: **fitness**. The fitness of an individual is simply the number of offspring it leaves after one generation. (If reproduction can occur at any time, then fitness would be defined as the net rate of increase in offspring numbers.) Similarly, the fitness of a particular gene is the number of copies that it leaves

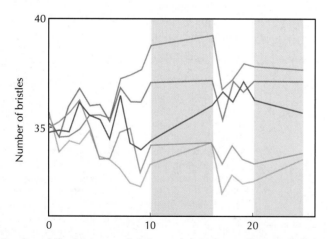

FIGURE 15.4. Random drift of the mean bristle number in five replicate populations of *Drosophila melanogaster*. There were $N = 20$ breeding individuals per generation. The additive genetic variance in the base population was $V_A = 6.0$, and so random drift is expected to generate variance between lines of $\sim V_A/N = 0.3$ per generation. Over the first ten generations, the rate of increase of variance between line means is greater than this (~ 0.5 per generation). This may be because the effective population size is smaller than the actual (cf. Fig. 15.3C), but the number of populations is so small that the difference is not significant. The *shaded area* indicates unscored generations.

after one generation. When we talk of the fitness of an *allele*, we mean the average fitness of genes carrying that allele. For example, in a human population polymorphic for two alleles Hb^A, Hb^S of β-globin, the fitness of Hb^A would simply be the total number of copies that genes carrying Hb^A pass on to the next generation divided by the number of genes carrying Hb^A in the present generation. Similarly, the fitness of a genotype (that is, a particular combination of alleles carried by an individual) is the average fitness of the individuals carrying that combination. For example, the fitness of $Hb^A Hb^S$ heterozygotes is simply the average number of offspring produced by such genotypes. Obviously, fitness depends on the environmental conditions and perhaps also on the genetic composition of the rest of the population. (For example, the fitness of Hb^S alleles depends on the incidence of malaria, because $Hb^A Hb^S$ heterozygotes are partially resistant to this disease, and it also depends on the frequency of Hb^A in the population, which determines the chance that an Hb^S allele will find itself in a heterozygote.)

Sometimes, we will distinguish between **absolute fitness**—the actual numbers of offspring produced—and **relative fitness**—the number of offspring produced *relative* to some other gene or genotype. The absolute fitness, averaged over the whole population, determines how fast the population grows, whereas the composition of a population—that is, the relative proportions of different genotypes—depends only on the relative fitnesses. Even in the literature of evolutionary biology, the term fitness is sometimes used in a vague way to refer to general vigor or "evolutionary potential." In this book, we keep to a precise and straightforward definition of fitness, which applies to individual genes and organisms as well as to genotypes and alleles. Together with mutation, gene flow, and recombination, it is variation in fitness that determines how the composition of a population changes.

In this chapter, we are concentrating on neutral alleles, which all have the same average fitness—that is, the reproduction of a gene is not affected by which allele it carries. However, *individual* genes vary in fitness, and it is this random variation in individual fitness that drives random genetic drift. In the idealized Wright–Fisher model, each gene leaves (to a good approximation for large *N*) a **Poisson distribution** of offspring. That is, if the average number of copies produced is 1, the chance of leaving 0, 1, 2, 3, ... is 0.37, 0.37, 0.18, 0.06, ... (Chapter 28 [online]). The variance of this distribution is 1 and, as we saw above, the rate of increase of the variance of allele frequency is $pq/2N$. In general, the fitness of genes might vary with some different distribution, with a variance *v*. Then, the variance of allele frequency increases as $vpq/2N$. The greater the variance in individual fitness, the faster the rate of random drift.

A variety of factors, including unequal sex ratio, fluctuating population size, and sex linkage, influence the rate of random genetic drift (Box 15.2). The effects of such factors are often summarized by defining an effective population size, written as N_e. This is simply the size of the ideal Wright–Fisher population that would lead to the same rate of drift as in the actual, more complicated population. Thus, an increase in the rate of genetic drift over the expectation from the Wright–Fisher model is described by saying that the **effective size** of the population has been reduced. For example, the excess rate of drift observed in the experiment described in Box 15.1 relative to what would be expected for a Wright–Fisher population is described by saying that the effective size of these populations is $N_e = 11.5$ (Fig. 15.3).

The rate of random drift relative to that expected from the actual number of individuals varies greatly from case to case. However, one expects that substantial random variation in individual fitness will typically accelerate the rate of drift substantially. In large populations, the rate of drift may be much faster than would be expected from the number of individuals; this is reflected in the observation that levels of genetic variability are not extremely high in abundant species such as humans, fruit flies, or bacteria (see pp. 368–370). One explanation may be that there are occasional drastic **population bottlenecks**, which dominate the process of random drift in the long term. We discuss this issue in more detail in Chapter 19 (pp. 536–540).

Box 15.2 Effective Population Size

The effective population size N_e is defined as the size of the ideal Wright–Fisher population that would give the same rate of random drift as the actual population. Many factors can influence the rate of random drift and hence N_e.

Varying population size: The total variance in allele frequency that accumulates over a series of k generations with population sizes $N_1, N_2, ..., N_k$ is (approximately)

$$pq \left(\frac{1}{2N_1} + \frac{1}{2N_2} + \cdots + \frac{1}{2N_k} \right).$$

Thus, a single generation with a very small population size can produce a burst of random drift that contributes more than many succeeding generations with large N. The effective population size is given by

$$\frac{1}{N_e} = \frac{1}{k} \left(\frac{1}{N_1} + \frac{1}{N_2} + \cdots + \frac{1}{N_k} \right).$$

Unequal sex ratio: When the numbers of males and females (N_m, N_f) differ,

$$\frac{1}{N_e} = \frac{1}{4} \left(\frac{1}{N_m} + \frac{1}{N_f} \right).$$

This assumes random mating, a Poisson distribution of progeny number, and that N_m and N_f are large. N_e is closest to the numbers of the rarer sex.

Variance in fitness of individual diploids: Suppose that the number of offspring from each diploid individual averages 2 (so that the population stays constant on average) but has variance V. Then, the effective population size is given by

$$\frac{1}{N_e} = \frac{2 + V}{4N}.$$

In the Wright–Fisher model, progeny number is approximately Poisson and so the variance equals the mean ($V = 2$), giving $N_e = N$. This formula assumes an equal sex ratio and large N. Usually, progeny number varies more than in the Poisson and so effective size is smaller. If all individuals have the same number of offspring ($V = 0$), random drift proceeds at half the rate ($N_e = 2N$), because there is still random segregation of genes at meiosis. (Note that in the text, we give the much simpler formula for the effect of variance in the fitness of individual genes, v.)

Sex-linked genes: Maternally inherited genes, such as mitochondrial DNA in mammals, have an effective size that depends only on the female population. If there are N_f females, with Poisson progeny number, then $N_e = N_f/2$, because each female carries only one copy of each gene, not two. Similarly, paternally inherited genes, such as those on the mammalian Y chromosome, have $N_e = N_m/2$. X-linked genes spend two-thirds of their time in females and one-third in males. Again assuming Poisson progeny number and large N_m and N_f, for X-linked genes,

$$\frac{1}{N_e} = \frac{4}{9N_f} + \frac{2}{9N_m}.$$

With an equal sex ratio ($N_m = N_f = N/2$), this simplifies to $N_e = (3/4)N$.

COALESCENCE

Tracing Backward in Time, an Inbred Population Has Fewer and Fewer Ancestors

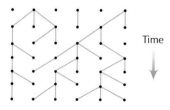

FIGURE 15.5. As we trace back in time, the number of ancestors inevitably decreases. The diagram shows the reproduction of individual genes.

So far, we have thought of the process of random genetic drift as running forward in time, by following the number of descendants of a particular gene or by following fluctuations in the frequency of an allele. We can also think of the process as running backward in time and consisting of a narrowing of the number of ancestors. The number of ancestral genes cannot be larger than the total number of genes in the ancestral population and will usually be much smaller, simply because not all past individuals reproduced (Fig. 15.5).

This way of thinking about random drift is especially powerful when we are trying to understand the ancestry of a particular collection of genes, rather than of the whole population—for example, when we are analyzing samples taken from a population. In Chapter 13 (pp. 364–366), we showed how samples of genes can be described by their **genealogy**. Here, we explain how this genealogy is shaped by random drift.

The narrowing ancestry of a population can be quantified by the idea that genes become **identical by descent** (**IBD**; Box 15.3); if we trace back far enough, all genes become identical by descent from a single ancestral gene. The idea of identity by de-

Box 15.3   Inbreeding and Identity by Descent

The extent of inbreeding is quantified by the probability of identity by descent. Two genes are IBD if they descend from the same gene in some ancestral population. For example, in Figure 15.6 the genes a, b are identical by descent relative to the ancestral population, three generations back, because they both descend from gene x. However, genes a and c are not IBD, because they trace back to different genes (x and y, respectively) (Fig. 15.6).

To understand the idea of identity by descent, three points need to be kept in mind.

1. Identity is defined with respect to a particular ancestral population. If we trace any homologous genes back far enough, they will be identical by descent from a common ancestor.

2. It is simplest to think of identity by descent as purely a description of ancestry of the genes, which is independent of their allelic state—that is, of the sequences they carry. In the literature, these two ideas are often confused. However, we will use the term IBD to refer solely to the relationships between genes, so that it is unaffected by (for example) mutations that change the allelic state of the genes as they descend through the population.

3. Often, we are interested in the chance that two genes within the same individual are identical by descent. This is called the **inbreeding coefficient**. The idea of identity by descent can also be used to describe the relationship between two individuals. The chance that two genes, one from each individual, are IBD is called the **coancestry** or the **coefficient of kinship**. It is this that determines similarities in quantitative traits (Box 14.2).

The relationship between the genes carried by sexually reproducing individuals depends on how those genes are passed on through meiosis. Figure 15.7 (left-hand diagram) shows a mating between brother and sister and shows how each gene passes through meiosis. The two genes in the offspring (bottom) are *not* identical by descent relative to the first generation (top), because they come from *different* grandparents: one gene comes from the grandfather, whereas the other comes from the grandmother.

Usually, however, we do not know which genes are passed on at meiosis (Fig. 15.7, right-hand diagram). Now, there is a chance of $f = \frac{1}{4}$ that the two genes in the offspring are identical by descent. There is a chance of $\frac{1}{2}$ that the first gene comes from the grandmother, and then there is a chance of $\frac{1}{4}$ that the second gene also descends from that same gene in the grandmother. Thus, there is a chance of $\frac{1}{2} \times \frac{1}{4} = \frac{1}{8}$ of identity by descent via the grandmother. Adding the same chance of identity via the grandfather, we find that $f = \frac{1}{4}$.

There is a simple rule that makes it possible to calculate the probability of identity by descent (written as f) for complicated pedigrees. The two genes in a diploid individual can become identical only if there is a *loop* in the pedigree,

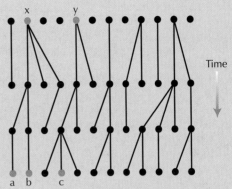

FIGURE 15.6. Identity by descent.

such that it can receive the same gene down two different paths. The probability of identity is a sum over all the loops:

$$f = \sum_{loops} \left(\frac{1}{2}\right)^{n-1} (1 + f_A),$$

where n is the number of individuals in each loop and f_A is the probability of identity of the ancestral individual. In Figure 15.7 (right), there are two loops, each with $n = 4$ individuals. The chance that the same gene passes down both sides of the loop, through the $n - 1 = 3$ meioses, is $(\frac{1}{2})^3 = \frac{1}{8}$, and so summing over the two loops, we have $f = \frac{1}{4}$, as before. Here, we have calculated identity relative to the first generation in the pedigree, so that the two genes within each grandparent are assumed to be unrelated ($f_A = 0$). If the diagram is part of some larger pedigree, then the grandparents might themselves be inbred, relative to some earlier generation. This is taken into account in the formula through the term $(1 + f_A)$, which allows for the possibility that two different genes from the ancestral individual are passed on (one down one side of the loop, one down the other) and that these two genes were identical by descent (i.e., $f_A > 0$).

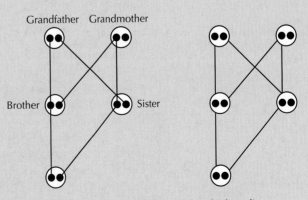

FIGURE 15.7. A simple example of inbreeding.

Box 15.3. Continued.

So far, we have supposed that the relationships are known, so that the only randomness comes through segregation at meiosis. If we only know the general system of mating (e.g., random mating under the Wright–Fisher model), then probabilities of identity depend on the vagaries of reproduction as well as segregation. It is easy to find how inbreeding accumulates, because probabilities of identity increase from one generation to the next. Under the Wright–Fisher model, each gene "chooses" an ancestor at random from the $2N$ genes in the previous generation. Thus, there is a chance $1/2N$ that two genes in generation t trace back to the *same* ancestral gene in generation $t-1$. If they do, they must be identical by descent; if they do not, they have a chance f_{t-1} of being identical. Thus,

$$f_t = \frac{1}{2N} + \left(1 - \frac{1}{2N}\right) f_{t-1}.$$

Applying this relationship over t generations, starting at $f_0 = 0$,

$$f_t = \left(1 - \left(1 - \frac{1}{2N}\right)^t\right).$$

This is the same relationship as for the variance in allele frequency (Box 15.1). Looking at fluctuations in allele frequency or at the increase in identity by descent are two ways of viewing the same underlying process of random drift.

scent is important because it measures the degree of **inbreeding** in a population. An individual is said to be inbred if its parents are related to each other, and the degree of inbreeding is measured by the chance that two homologous genes within that individual are IBD.

Understanding inbreeding is important because inbred individuals often suffer **inbreeding depression** (see p. 515). Plant and animal breeders and conservation biologists therefore aim to minimize inbreeding. We will also see that the concept of identity by descent is important for understanding how selection shapes the social behavior of groups of relatives (see pp. 601–603).

Genetic Ancestry Is Described by a Process Known as Coalescence

In an ideal Wright–Fisher population of N diploid individuals, ancestral genes are picked at random from the $2N$ genes in the previous generation. So, there is a chance $1/2N$ that any two genes will share a common ancestor in the previous generation. This shared ancestry reflects a **coalescence** of the two lineages in one common ancestor. As we trace back the ancestry of any sample of genes, we follow a genealogy that consists of a series of coalescence events. In other kinds of populations, the rate of coalescence will differ from $1/2N$, but the pattern of ancestry can be described in qualitatively the same way.

In general, the rate of coalescence is given by the inverse of twice the effective population size, $1/2N_e$. This is the same effective size as was defined above, which determines the accumulation of variance in allele frequency. The rate of coalescence increases with the variance in fitness between genes in just the same way as does the variance in allele frequency (Box 15.2). Thus, the same factors that increase the rate of drift of allele frequencies also narrow the ancestry of a population by accelerating the coalescence of lineages as we trace backward in time.

Despite its simplicity, the coalescent process has some surprising properties. Most important, the ancestry of a sample of genes rapidly narrows down to a small number of lineages, which may then coexist for a long time before merging in the single ancestral gene. This is because at first, there are a large number of pairs that might coalesce—with 20 genes, for example, any of 190 pairs of lineages might coalesce. Thus, coalescence is initially rapid, but slows down as the number of lineages decreases. Overall, it takes on average $2N_e$ generations to go from a large number of genes down

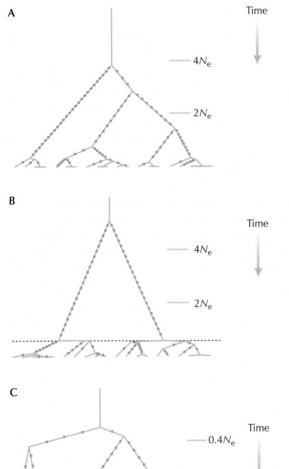

A

B

C

Time

FIGURE 15.8. (*A*) A typical genealogy relating 20 genes sampled from a population of constant size N_e. Note that most coalescence events occur in the recent past, so that for most of the history of the sample, there are just a few ancestral lineages. In this example, unique mutations (*red dots*) occur at a rate $\theta = 4N_e\mu = 20$. (*B*) A genealogy sampled from a population that had experienced a sharp bottleneck at $0.6N_e$ generations in the past (*dashed line*). This causes a rapid burst of coalescence: Eight lineages present after the bottleneck trace back to just two lineages immediately before the bottleneck. (*C*) A genealogy sampled from a population growing exponentially. Coalescences now tend to occur relatively further back, when the population was much smaller. The whole sample shares common ancestry $0.45N_e$ generations back, an order of magnitude more recently than expected if population size had stayed at N_e throughout. At that time, the population was only 10% of the size when the sample was taken, N_e.

to two ancestral lineages, and then takes a further $2N_e$ generations for these two to coalesce at the common ancestor (Fig. 15.8). However large the initial number of genes, a typical genealogy will have just two lineages for half of its history.

This structure has two important consequences. First, sampling more and more genes from a population gives surprisingly little extra information—extra genes are likely to trace back to a quite-recently shared ancestor and so are unlikely to have accumulated novel genetic variants that have not already been found in the rest of the sample. Partly for this reason, current surveys of human variation are based on the genomes of just a few individuals. (For example, the first large-scale survey of **single-nucleotide polymorphisms** [SNPs] in humans identified 1.4 million SNPs, based on a panel of only 24 people.) Second, the deeper parts of genealogies are highly variable, depending on when the two oldest lineages happen to coalesce. Although these two will coexist for an average of $2N_e$ generations, there is a 2.5% chance that they will coexist for more than $7.4N_e$ generations, and a 2.5% chance that they will coalesce in less than $0.05N_e$ generations. (That is, the 95% **confidence interval** spans from $0.05N_e$ to $7.4N_e$, more than two orders of magnitude.) So, however many genes are sampled, their genealogical relationships will remain highly variable.

Genealogies Can Be Inferred from Sequences

We cannot observe a genealogy directly, but we can infer it from the pattern of mutations that have accumulated on the genealogy (Chapter 27 [online]). If we are comparing DNA sequences from within a population, then **nucleotide diversity** is typically low (see p. 368), and so we can assume that each variant base is due to a unique mutation. Under this **infinite-sites model**, inferring a genealogy is straightforward (recall Fig. 13.12). In the example in Figure 15.8A, the mutation rate is high enough that most branches of the genealogy carry at least one unique mutation. Thus, the shape of the genealogy can be reconstructed almost completely from the pattern of mutations in the sample. Moreover, the number of mutations on each branch gives an estimate of the time spanned by that branch.

The shape of a genealogy directly reflects past population size. In principle, N_e could be read from the rate of coalescence at each time in the past. For example, a drastic population bottleneck would cause a sudden burst of coalescence, giving a genealogy, with one or more "stars," in which several lineages coalesce at the bottleneck (Fig. 15.8B). In a rapidly growing population, coalescence tends to occur relatively further back in the past when the population was much smaller (Fig. 15.8C). These differences in shape can be seen in the genealogies of viral populations with different histories (Fig. 15.9). In Chapter 25, we will see what this approach can tell us about the history of our own species.

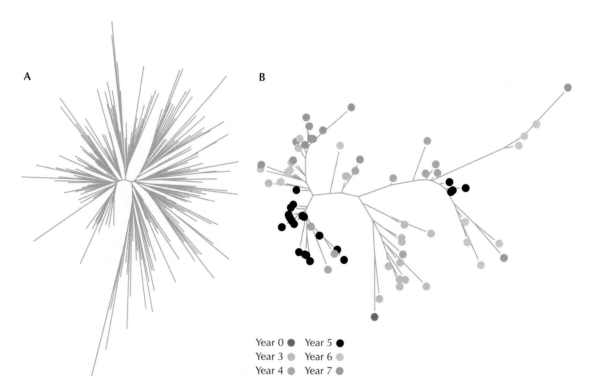

Year 0 ● Year 5 ●
Year 3 ● Year 6 ●
Year 4 ● Year 7 ●

FIGURE 15.9. Genealogies relating genes from the HIV virus reflect different population histories. (*A*) Relationship between sequences from the *gag* gene, sampled from 200 individuals from northern Britain in 1993. Coalescence tends to occur back in the past (toward the center of the diagram), a pattern characteristic of a rapidly growing population (cf. Fig. 15.8C). (*B*) Relationship between *env* genes sampled from within a single patient over 7 years. This example is unusual in that genes are sampled through time. The divergence of genes through time can be seen by following out from the original sequence (*red*) through to the most diverse population at year 7 (*magenta*). In this example, most coalescence occurs relatively recently, a pattern characteristic of a steady population (cf. Fig. 15.8A). (Both these genealogies are drawn **unrooted**, whereas Fig. 15.8 shows **rooted** genealogies.)

We now have a simple model of the process of random genetic drift, which we can use to explain both the future evolution of a population and its past ancestry. In the rest of this chapter, we combine this model with the processes of mutation and recombination to give a complete picture of neutral molecular evolution.

THE NEUTRAL THEORY

The Rate of Neutral Divergence Equals the Mutation Rate

The time when two genes last shared a common ancestor cannot be observed directly, but can be inferred from the number of mutations that distinguish them. If the two lineages coalesce at a shared ancestor who lived t generations ago, then there have been $2t$ generations during which mutations might have occurred. We therefore expect that if mutations occur at a rate μ per gene per generation, then the two genes will differ by $2\mu t$ mutations. This is a key result for understanding molecular evolution: Provided that those mutations are neutral, the rate at which pairs of genes diverge is equal to the total rate of mutation.

We can see the same result by tracing alleles forward in time. Almost all lineages will go extinct and, in the long run, the whole population will consist of descendants of just one gene (Fig. 15.10). Therefore, the chance that any particular gene in a diploid population of $2N$ genes will eventually survive is just $1/2N$. All $2N$ genes have the same chance of being the lucky survivor, provided that they are assumed to be neutral and therefore equivalent to each other. Now, in every generation, a total of $2N\mu$ mutations enter the population, that is, $2N$ genes, each of which mutates at a rate μ. Because only a fraction $1/2N$ of these mutations survive in the long run, the net rate of divergence of the population is μ per gene per generation. The argument is really the same as before—because the whole population traces back to a single lineage, the rate of change of the population is in the long run equal to the rate of change of that single lineage (Fig. 15.10).

It was the observation in the 1960s of a steady **molecular clock** in protein sequences that led to the proposal of the **neutral theory** of molecular evolution by Motoo Kimura (see p. 59). It was found that any particular protein accumulates changes in amino acid sequence at a steady rate, even when the protein is evolving in very different organisms. For example, α-globin accumulates about one amino acid change every 6 million years, whether it is evolving in fish, birds, or mammals (Figs. 2.28, 13.26, and 13.27). We saw on pages 373–375 that different proteins evolve at different rates, reflecting different degrees of functional constraint. For example, histone H4 differs by only two amino acids between plants and mammals, whereas fibrinopeptides evolve three orders of magnitude more rapidly (Table 13.2). Crucially, the rate of change of weakly constrained proteins or of nonfunctional sequences, such as **pseudogenes**, approaches the mutation rate, which is measured directly. The more-or-less steady ticking of the molecular clock, at a rate similar to the mutation rate, is the strongest support for the neutral theory. In Chapter 19 (p. 531), we examine evidence for its accuracy more closely.

In the Absence of Selection, Neutral Variation Is Determined by a Balance between Mutation and Drift

We have seen that lineages accumulate neutral mutations at a rate equal to the mutation rate. Thus, two genes taken from separate species that diverged a time t ago will, on average, differ by $2\mu t$ mutations, where μ is the total mutation rate for the gene. We now ask: How different are genes from within the same population? That is, how much genetic variation will we see within populations as a result of the joint action of mutation and random genetic drift?

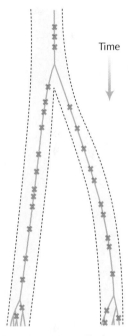

Time

FIGURE 15.10. All the genes within a species descend from a single ancestral lineage, which traces back into the distant past. Thus, the rate at which two species diverge over time is equal to the rate at which their two ancestral lineages accumulate mutations. The *shaded area* shows two species, which descend from a single ancestral species (*left*), and the *solid lines* show the ancestry of all the genes presently in the two species. Mutations are indicated by *red crosses*.

The answer is simple: Two genes share a common ancestor on average $2N_e$ generations in the past and therefore on average differ by $4N_e\mu$ mutations per base pair. When we deal with DNA sequences, we can count these differences directly. The proportion of nucleotides that differ between two randomly chosen sequences is called the nucleotide diversity π (see p. 364). Under the neutral theory, it is expected to equal $4N_e\mu$. This crucial parameter is denoted as θ.

We may not be able to count directly the number of mutations that have occurred. For example, if we score proteins by electrophoresis, mutations will typically produce a new allele, but we cannot tell whether two alleles differ by one or by many mutations. Now the appropriate measure of diversity is the **gene diversity** H, which is the chance that two randomly chosen genes carry different alleles (see p. 363). Under the neutral theory, we expect $H = 4N_e\mu/(1 + 4N_e\mu)$. Many other models of mutation have been studied. Whatever the details, however, the amount of genetic variation depends on the key parameter, four times the product of effective population size and mutation rate, or $\theta = 4N_e\mu$. (Note that the mutation rate is measured per base pair if we are dealing with sequence data, or per gene if we are counting differences between whole genes.) θ measures the relative rates of mutation and random drift, and hence it determines the balance between them.

The most obvious prediction of the neutral theory is that genetic variation should increase in proportion to population size. The larger the population, the more slowly variation is eliminated by drift. One test of this prediction is to compare levels of diversity on chromosomes with different modes of inheritance, but within the same population. Because the X and Y chromosomes have smaller effective population sizes than the autosomes (see Box 15.2), we expect them to show correspondingly lower nucleotide diversity. The analysis of 1.4 million single-nucleotide polymorphisms in humans, mentioned above, gave estimates of $\pi = 0.000765$ for the autosomes, 0.000469 for the X, and 0.000151 for the nonrecombining region of the Y. Thus, the X has 61% of the diversity of the autosomes and the Y has 20% of the diversity. These figures are quite close to the values predicted solely on the basis of effective population size (75% and 25%, respectively). Other factors are also involved and may account for the lower diversity of the sex chromosomes relative to the simple neutral prediction. The distribution of progeny number differs between the sexes; lack of recombination on the Y chromosome can lead to reduced diversity (see pp. 536–540 and 676); and the mutation rate is higher in males than in females (see pp. 348–349). Nevertheless, differences in effective population size seem to account for most of the differences in diversity among chromosomes with different patterns of inheritance (Fig. 13.24).

Abundant Species Have Less Genetic Diversity Than Expected from the Neutral Theory

The mutation rate can be estimated both directly and from the degree of sequence divergence between species whose time of divergence is known. Thus, it is straightforward to estimate the effective population size that is needed to account for the genetic variation that we see. For example, in most autosomal regions of the *Drosophila melanogaster* genome, diversity at silent and noncoding sites averages $\pi \sim 0.01$. Taking $\mu = 3 \times 10^{-9}$ per base per generation, and assuming $4N_e\mu = \pi$, we estimate an effective population size $N_e \sim 10^6$. This is far lower than the actual number of *Drosophila* living now, or the numbers likely to have lived in the past.

Similar discrepancies are seen in wider surveys. In Chapter 13 (pp. 368–370), we saw that genetic diversity does tend to increase with the abundance of the organism, but only weakly: Diversities differ by less than an order of magnitude over a range of actual population sizes that spans many orders of magnitude (Fig. 13.17). Fewer data are available for DNA sequence variation, but a similar pattern is seen. For example, nucleotide diversity in humans is about tenfold lower than in *D. melanogaster*, yet population numbers surely differ by a much larger factor than this.

Two explanations have been put forward for why extremely abundant species such as *Escherichia coli* or *D. melanogaster* do not show correspondingly high levels of genetic diversity. The first is that occasional population bottlenecks greatly reduce variation, far below that expected from census numbers. Suppose that a population is extremely large most of the time, but drops to N^* individuals in 1 out of 100 generations. Then, its effective size would be only $100N^*$ (Box 15.2). Drastic bottlenecks play some part in explaining the wide discrepancy between the effective population sizes required to explain genetic diversity and the actual population size. For example, the *E. coli* inhabiting our guts consist of a few predominant clones that reproduce asexually and that are occasionally displaced by new clones. Thus, the effective size of an *E. coli* population may be closer to the number of mammalian hosts than to the vast numbers of individual bacteria. Nevertheless, it is hard to believe that population bottlenecks alone can bring effective population sizes down low enough to fit the facts or that they can account for the moderate ratios between the diversities of organisms with such different population sizes.

The second explanation is that selection sweeps variation out of populations. When a new favorable mutation arises and spreads to high frequency by natural selection, it carries with it part of the genome in which it first arose; this region of genome therefore becomes homogeneous, and variation is lost from the population. This phenomenon, known as a **selective sweep** or **hitchhiking**, could greatly reduce genetic variability, so that even extremely large populations could not accumulate high levels of variation. We explain this process in more detail later in this chapter and examine evidence for its importance in Chapter 19 (pp. 536–538). For the moment, though, note that both population bottlenecks and selective sweeps cause lineages to coalesce much sooner than expected from the actual numbers of individuals. In fact, when the problem is seen in terms of coalescence times, a naive application of the neutral theory becomes absurd. In a population with an effective size of a billion individuals, genes are expected to coalesce 2 billion generations ago, which in many cases will be long before the species, or anything like it, existed. For many genes that have arisen by duplication, the gene itself may not have existed $2N_e$ generations ago. The simple process of random drift cannot be extrapolated back over the very long timescales of molecular evolution.

RECOMBINATION AND RANDOM DRIFT

Recombination Breaks Up the Genome into Regions with Different Ancestry

Any set of **homologous** genes traces back to a single ancestral gene. In a strictly asexual population, every gene shares the same ancestry simply because all genes are passed on together from mother to daughter. However, if there is any kind of sexual reproduction (e.g., the occasional transfer of genes between bacteria or regular meiotic sex in eukaryotes), then ancestry will vary from gene to gene. We can see this most clearly by thinking of the ancestry of the human mitochondrial DNA or the human Y chromosome. In mammals, because mitochondria are always inherited from the mother, they all must trace back to a single ancestral female—the so-called "mitochondrial Eve." Similarly, all Y chromosomes descend from a single male, the "Y chromosome Adam." Now, "mitochondrial Eve" and "Y chromosome Adam" lived at different times and places. Moreover, there is nothing special about these two individuals, except that by chance they contributed a small part of their genome to the future human population. Indeed, it is possible that they contributed nothing else to future generations. Conversely, if we look back in time, we see that the present-day human genome is divided into many blocks that trace back to a large number of different ancestors (Fig. 15.11).

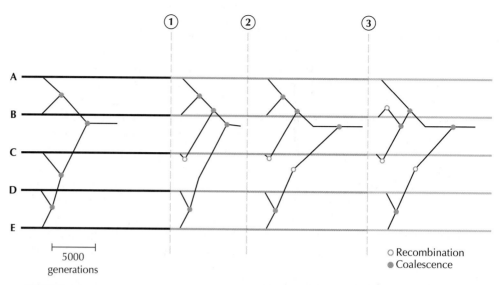

FIGURE 15.11. In a sexual population, different parts of the genome have different ancestry. This example shows a small region from five genomes (labeled A, B, C, D, and E). In the *leftmost section*, A and B share a common ancestor 2000 generations back; D and E share an ancestor 1000 generations back, and lineage (D, E) traces back to a shared ancestor with C 2000 generations into the past. The whole sample shares a common ancestor at 5000 generations. The genome C descended from an ancestral genome that underwent a recombination event 500 generations back; thus, the section to the *right* of position ① in the genetic map has a slightly different ancestry: C is more closely related to A and B than to D or E. Moving to the *right*, the next event is a recombination event at ② in the genetic map, which occurred 3000 generations back in the ancestor of D, E. This event did not change the qualitative relationships, but now the five genomes share a common ancestor 8000 generations into the past. Finally, moving to the *rightmost section*, a recombination event at 1000 generations makes B more closely related to C than to A.

We can better understand inheritance in a sexual population by thinking not of the ancestry of single genes, but of the ancestry of individuals in the everyday sense—that is, of the **pedigree** of the population. Each individual has two parents, four grandparents, and so on. As we go further back, the number of ancestors increases exponentially (e.g., $2^{10} = 1024$ ancestors 10 generations back, $2^{20} \cong 10^6$ ancestors 20 generations back), so that even in a large population, ancestors must start to be repeated; that is, some ancestors leave offspring via several different lines of descent (Fig. 15.12). Of course, matings between close relatives will produce this kind of inbreeding, but even with random mating in a large population some degree of inbreeding is inevitable just because the number of ancestors grows so rapidly as we go back in time. Remarkably, if one traces a pedigree back for a few tens of generations, then there will be an individual who is an ancestor of the *entire* current population; he or she might have passed on one or more blocks of his genome to every living individual and so might potentially be an "Adam" or "Eve" for some segment of DNA. If we go back about twice as far, then we find that about 20% of the population has left *no* descendants, whereas the remaining 80% of individuals appear as ancestors of *every* living individual. Some may appear only once in the pedigree, but others may appear many times and so potentially contribute much more (Fig. 15.12).

It is important to realize that we are considering here ancestry in the everyday sense of a pedigree relationship: An individual might appear as a relative and yet might pass on no genetic material at all. Each individual receives one set of genes from the mother and one set from the father (sex chromosomes and mitochondria aside). Each of these sets is a mixture of genetic material inherited from one or the other grandparent through the random process of meiosis. In principle, it is possible that one of the grandparents might by chance pass on no genetic material at all through meiosis. That

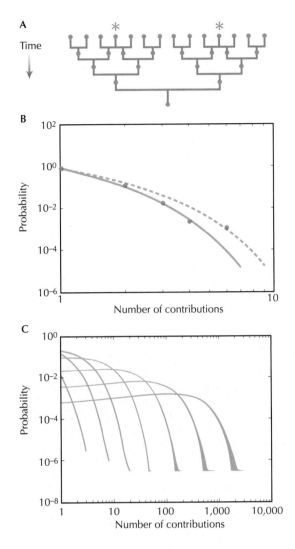

A

Time

B

C

FIGURE 15.12. (*A*) Inbreeding is inevitable even in a large population. Going back in time, there are potentially 2, 4, 8, 16, ... ancestors in the pedigree. Eventually, some individuals must contribute several times. In this example, the 2 individuals marked with a *red asterisk* each contribute twice to the pedigree, whereas the other 12 ancestors each contribute once. Thus, there are 14 ancestral individuals instead of 16. (*B*) The dots show the distribution of contributions of the ancestors of the English King Edward III (1312–1377), traced back for about ten generations. Some ancestors contribute once, but others up to six times. The lines show simulations assuming that mating is random, within a closed population of 2048 (*upper curve*) or 4096 (*lower curve*), representing the English nobility. (*C*) Tracing back further in time, ancestors may make contributions down many lines of descent. The curves show the distribution of numbers of contributions in a simulated population of 2^{15} = 32,768 individuals, traced back 9, 11, 13, ..., 23 generations (*left to right*). Twenty-three generations back, ancestors typically make up to several hundred contributions to the pedigree, but rarely more than 1000 (see peak in *rightmost curve*).

is extremely unlikely, at least for organisms such as ourselves with many chromosomes and a high rate of recombination. Indeed, if a human being has any pedigree descendants at all, he or she will almost certainly pass on some genetic material to at least some descendants thousands of generations into the future. However, this contribution is likely to consist of only a few small blocks of genome (Fig. 15.13).

We see, then, that the genome of a sexually reproducing organism consists of a mosaic of blocks, each with a different ancestry and a different fate (Figs. 15.11 and 15.13). So far, we have argued forward in time. When we analyze samples from present-day populations, however, we need to trace their ancestry back through time. This can be done by extending the **coalescent process** to include recombination (Box 15.4). We will now see how this process determines the patterns of genetic variation that are shaped by mutation, recombination, and drift.

The Pattern of Ancestry Depends on the Number of Recombination Events per Generation, $N_e c$

The genealogical structure is reflected in the pattern of variation seen in samples of DNA sequences. In one especially thorough study, 20 copies of human chromosome 21 were sequenced. These were sampled from a panel of people of different races in an attempt to represent overall variation in our species. In the 21 Mb of unique sequence,

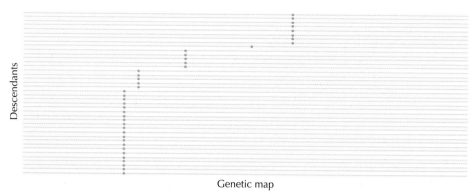

Descendants

Genetic map

FIGURE 15.13. A single individual is likely to pass on only a small part of his genome to future generations. The figure shows a simulation of a single genome, with map length 35.7 Morgans (as for humans) for 50 generations. By then, the single ancestor has 4.3×10^{14} pedigree descendants. (In a real population, there would be far fewer than this, because of matings between relatives; see Fig. 15.12.) Only 91 of this enormous number of descendents inherit any genetic material, 40 of which are represented here by the 40 lines, with the ancestral material in *red*. On average, each block of ancestral material is 2.2 cM long. Only five blocks survive anywhere in the population, making up 0.3% of the ancestor's genome.

Box 15.4   The Coalescent Process with Recombination

Neighboring sections of genome come to have a different ancestry because of recombination. At some time, a gamete carrying two genes was derived via a recombination event at meiosis, so that one gene came from the paternal genome and one from the maternal genome. Tracing back further in time, their ancestry follows independent paths, until by chance the two genes are again found next to each other in the same genome. Over a long period of time, the two genes spend some time in different genomes and some time within the same genome.

The ancestry of two genes can be quantified very simply, as an extension to the standard coalescent process (see Fig. 15.8). Any two lineages have a chance $1/2N_e$ per generation of coalescing with each other, as before. In addition, any lineage that carries both genes has a chance c per generation of recombining, such that further back in time the two genes were present in different ancestors. The proportion of time that two linked genes spend in the same individual or in different individuals depends on the relative rates of these processes, $N_e c$. If this value is large, then the two genes spend most of their ancestry in different individuals.

The process of coalescence with recombination is illustrated in Figure 15.14. Six genomes are sampled, each carrying three genes (labeled black, blue, and red and in that order along the genome). The genealogy of the first gene is shown in black and follows the standard coalescent process. The genealogy of the blue gene is almost the same, except that a single recombination event (indicated by a circle on the right) has occurred such that the blue and black genes descend from different individuals. The blue lineage then coalesces with a different black lineage (circle on the left), so that the genealogies again coincide. Going back still further before the common ancestor of the whole sample (filled circle), the lineages recombine and coalesce so that the blue and black genes spend some time together and

some apart. These ancient events are undetectable. More recent recombination events can still be hard to detect. For example, the blue and black genealogies have the same shape and only differ in the time to common ancestry (earlier for the blue gene than for the black). The red genealogy is for a gene somewhat further away, which differs by three recombination events and has a different shape as well as a different depth. As one moves further along the genome, the genealogy becomes more and more different until for large $N_e c$ it is almost uncorrelated.

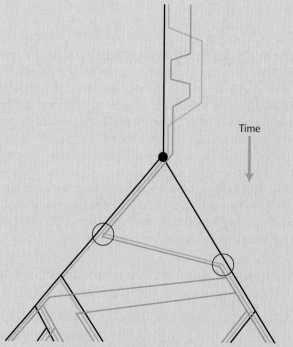

Time

FIGURE 15.14. Coalescence with recombination.

35,989 SNPS were identified; these are sites at which two (or occasionally, more) different bases were found. Potentially, these alternative bases could appear in an enormous number of different combinations. However, in any one region of genome the different SNPs are closely associated and fall into a small number of combinations or **haplotypes**. For example, in one 19-kb region, there were 26 SNPs, giving $2^{26} = 6.7 \times 10^7$ possible combinations. Thus, if each base pair had an independent ancestry, we would expect every chromosome to have a different haplotype. In fact, only seven haplotypes were found (inset in Fig. 15.15). Overall, the sequence variation could be described by dividing the chromosome into a set of blocks, each containing just a few haplotypes. These strong associations between different SNPs, in which each block is organized into a limited number of alternative combinations, reflect the shared ancestry of closely linked regions of genome (Fig. 15.11). From a practical point of view, this block structure allows most genetic variation to be captured by scoring relatively few SNPs; for example, scoring just 2793 SNPs would describe 81% of the total variation. The blocks identified in this analysis do not correspond directly to regions of shared ancestry (see Fig. 15.11). They will usually be much larger, because most recombination events will have undetectably small effects on the genealogy. However, this and other studies show that neighboring sequence variants tend to be strongly associated, primarily as a result of the random genetic drift involved in the coalescent process.

In sequences from sexual eukaryotes, such as humans or *Drosophila*, it is not usually possible to see exactly where recombination events have occurred because the sequences are too similar for their distinct genealogies to be discerned. However, in prokaryotes recombination is much less frequent and may involve the transfer of genes between very distantly related species (see p. 184). Thus, the boundaries between regions with different

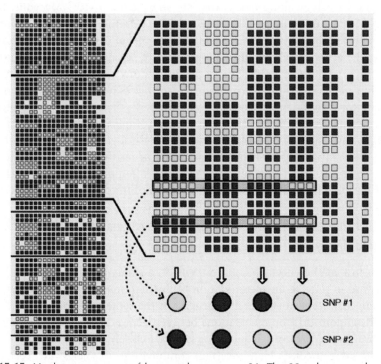

FIGURE 15.15. Haplotype structure of human chromosome 21. The 20 columns to the *left* represent variation in a sample of 20 human chromosomes. The rows correspond to 69 single-nucleotide polymorphisms (SNPs) spanning 50 kb; *yellow* indicates a variant allele and *light blue* indicates missing data. The inset on the *right* shows a block of 26 SNPs, which define seven distinct haplotypes. The first five columns represent the commonest haplotype, the next four the next most common haplotype, and so on. Most of these haplotypes could be distinguished by scoring only two of the SNPs, as indicated by the two rows at the *bottom right*: these two SNPs define four alternative arrangements, which capture variation among the four most common haplotypes in this block.

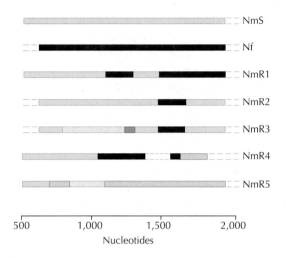

FIGURE 15.16. Some strains of the pathogenic bacterium *Neisseria meningitidis* (Nm) have evolved resistance to penicillin by acquiring a short segment of DNA (*black*) from the naturally resistant (but harmless) species *Neisseria flavescens* (Nf). The high level of divergence between these species allows the different ancestry of different parts of the genome to be seen clearly.

histories can sometimes be seen clearly. For example, resistance to penicillin in *Neisseria meningitidis* is due to a short segment of DNA that has been transferred from a naturally resistant *Neisseria* species—a transfer between species that differ in sequence by more than 20% (Fig. 15.16). Here, the transfer of a block of sequence that confers antibiotic resistance has been driven by the strong advantage that it confers upon the recipient.

In Box 15.5, we show that correlation in ancestry between two linked genes depends on the product of effective population size and recombination rate, $N_e c$. Putting this another way, we expect to see correlations between genes that are $c \sim 1/N_e$ map units apart. For example, the effective population size for *Drosophila melanogaster*, inferred from observed nucleotide diversity, is $N_e \sim 10^6$ (see p. 426). Thus, we expect to see the same genealogy along blocks that are shorter than $\sim 10^{-6}$ Morgans. Recombination rates vary considerably along the *Drosophila* genome, but average about 2×10^{-8} per base pair per generation. Thus, a map length of $1/N_e$ corresponds to only 50 bp. Of course, the actual block length will vary greatly, because the actual lengths of genealogies vary, recombination events occur at random on those genealogies, and recombination rates vary along the genome. Nevertheless, we expect that even within a gene coding for a single protein (say, a few kilobases), there will be many blocks with different ancestry (e.g., Fig. 13.14), which greatly complicates the analysis of DNA sequence. It is impossible to reconstruct a single ancestral genealogy for the whole gene, and there is usually too little sequence variability to reconstruct the separate genealogies for each region. This is one reason why so much population genetic work has concentrated on mitochondrial DNA and on Y chromosomes, which do not recombine.

Associations between Pairs of Alleles Are Measured by Linkage Disequilibrium

We have seen that the effect of random drift on a single genetic locus can be viewed either forward in time, as an accumulation of fluctuations in allele frequency, or backward, as the coalescence of different lineages. In the same way, the effect of recombination and random drift on associations between genes can be viewed either forward or backward. In the previous section, we saw that closely linked sites ($c \sim 1/N_e$) tend to share the same ancestry. Hence, certain combinations of alleles (i.e., certain haplotypes) tend to be found in association with each other. In this section, we see how these associations evolve as we go forward in time.

Imagine that a new mutation occurs at locus A. We can label the new allele that is produced as A^P; initially, its frequency will be $p_A = 1/2N$. It will be associated with the particular combination of alleles with which it happens to arise—that is, with a particular genetic background. For simplicity, focus on one linked site (locus B) at which

Box 15.5   Measuring Associations between Genes: Linkage Disequilibrium

Much of population genetics assumes that alleles at different loci are combined at random. That is, the chance of finding an allele at one locus is independent of which alleles are present at other loci. Then the frequency of a genotype is just the product of the allele frequencies. (For example, a gamete carrying $A^P B^P C^Q D^P$ would be at frequency $p_A p_B q_C p_D$.) This state is called **linkage equilibrium**. It brings an enormous simplification. For example, with two alleles at ten loci, there are $2^{10} = 1024$ haploid genotypes; at linkage equilibrium, these can be described by just ten allele frequencies.

How can we describe associations between alleles at different loci? Suppose we have two loci (A, B), the first segregating alleles A^P, A^Q and the second, B^P, B^Q. The corresponding allele frequencies are $p_A + q_A = 1$, $p_B + q_B = 1$. The coefficient of linkage disequilibrium is defined as the difference between the actual genotype frequency and the frequency expected if alleles are combined at random. Thus, we can write the frequencies of the four genotypes as follows:

$A^Q B^Q$	$q_A q_B + D$
$A^Q B^P$	$q_A p_B - D$
$A^P B^Q$	$p_A q_B - D$
$A^P B^P$	$p_A p_B + D$

The value of linkage disequilibrium varies over a range that depends on allele frequencies. Clearly, D cannot become so large that the frequency of $A^Q B^P$ or $A^P B^Q$ becomes negative, and it cannot become so negative that the frequency of $A^P B^P$ or $A^Q B^Q$ becomes negative. These constraints on D make it difficult to compare between populations with different allele frequencies. Sometimes, D is expressed relative to the maximum possible value, given the allele frequencies. This measure is defined as $D' = D/D_{max}$ and must lie between −1 and +1. Alternatively, a kind of correlation coefficient is sometimes used:

$$r = \frac{D}{\sqrt{p_A p_B q_A q_B}}.$$

This also lies within the range −1 to +1. It has the useful property that in a balance between recombination and random drift, the mean of r is 0, and its variance is $1/(1 + 4N_e c)$.

Linkage disequilibrium describes the strength of associations between different alleles in a population. If the presence of a particular allele at one locus increases the chance that some allele will be found at another locus, then we say that the alleles are in linkage disequilibrium (Fig. 15.17). The term is misleading in two ways. Alleles at loci on different chromosomes can be associated with each other in "linkage disequilibrium" even though they are genetically unlinked. Moreover, populations can reach a stable equilibrium in which various evolutionary forces maintain "linkage disequilibrium" at a constant level. The term refers to a statistical association, which describes how the frequencies of gene combinations deviate from the expectation based on allele frequencies alone.

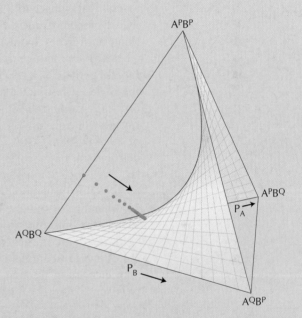

FIGURE 15.17. The four vertices of the tetrahedron correspond to populations fixed for each of the four haploid genotypes, and points within the tetrahedron show populations polymorphic for all four genotypes. Populations in linkage equilibrium lie on the surface; those with positive disequilibrium (i.e., A^P associated with B^P, A^Q with B^Q) lie to the left of the surface, and those with negative disequilibrium, to the right. If the linkage disequilibrium is complete, the population lies on the line connecting $A^P B^P$ with $A^Q B^Q$ (i.e., only these two genotypes are present), and similarly for complete negative disequilibrium. The series of *red dots* shows successive generations of a population that initially is in complete linkage disequilibrium and evolves toward linkage equilibrium; $c = 20$ cM.

---

an allele B^P is found at frequency p_B. If the mutation A^P happens to occur alongside allele B^P, then at first all copies of A^P will be associated with B^P and the frequency of the combination $A^P B^P$ will be the same as the frequency of the new allele A^P. Now, if the alleles were combined at random, we would expect to see $A^P B^P$ at frequency $p_A p_B$. The difference between the frequency we observe and what we expect is termed the coefficient of **linkage disequilibrium**, denoted by D (Box 15.5). In this case, the linkage disequilibrium is just $D = (p_A - p_A p_B) = p_A q_B$, where $p_B + q_B = 1$.

Recombination has a simple effect on the coefficient D: After one generation of random mating and recombination, the linkage disequilibrium among the gamete population is reduced by a factor $1 - c$. Thus, disequilibrium between unlinked loci ($c = \frac{1}{2}$) is halved in every generation and rapidly approaches 0. For linked loci, recombination dissipates linkage disequilibrium over a time of approximately $1/c$ generations. Thus, strong linkage disequilibrium will be seen only when it is generated rapidly enough to be maintained despite recombination. (Note that the symbol r is sometimes used for recombination rates rather than c; we use c, because r also denotes a measure of linkage disequilibrium; Box 15.5.)

The rate of decay of linkage disequilibrium can be used to find the age of a new allele. A nice example of this method is given by a 32-base pair deletion to the human CCR5 gene (CCR5-Δ32). The deletion causes loss of a chemokine receptor, which is used by the human immunodeficiency virus (HIV) to invade cells. As a consequence, homozygotes for the deletion are resistant to infection by HIV, and heterozygotes take several years longer to develop acquired immunodeficiency syndrome (AIDS) than do wild-type homozygotes. The CCR5-Δ32 allele is not found in African or Asian populations, but reaches about 10% frequency in Caucasian populations. The key question is whether it rose to high frequency slowly by random drift or increased in frequency more rapidly, perhaps as a result of selection for resistance to some past epidemic. The age of the allele was estimated by examining two closely linked microsatellite markers—one 0.72 cM away and the other 0.93 cM away in the same direction. One particular pair of microsatellite alleles was found with 86% of the CCR5-Δ32 alleles, but with only 36% of the wild-type CCR5 alleles. This strong association is most easily explained if the deletion arose as a single mutation in a genome carrying those two microsatellite alleles. The association has now partly dissipated as a result of recombination at a total rate of 0.0093 per generation (Fig. 15.18). These data suggest that the mutation arose about 30 generations (roughly 700 years) ago and increased rapidly as a result of strong selection; the linked microsatellite markers have increased by hitchhiking.

Just as random genetic drift causes random fluctuations in allele frequencies, so too does it cause random fluctuations in linkage disequilibrium. Certain combinations may increase in frequency simply because the individuals carrying them happen to leave more offspring. We have just looked at the extreme case, where a single new mu-

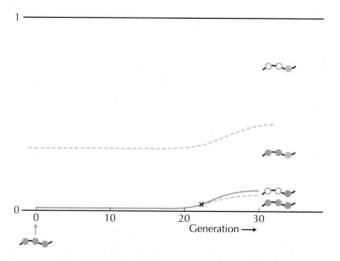

FIGURE 15.18. The age of the CCR5-Δ32 allele can be estimated from its association with two microsatellite markers. This allele (*red*) arose by mutation on a genetic background that carried two microsatellite markers that were then at a frequency of about 36% (*green filled circles; arrow, lower left*). The mutation increased to about 10% in present European populations and is still largely associated with the original two markers. However, one or more recombination events (*black cross*) have occurred, so that some CCR5-Δ32 alleles are now associated with different microsatellites.

tation arises on a random genetic background and so at first is completely associated with a random combination of alleles. These chance associations are just the same as those described in terms of the coalescent in the previous section: Alleles tend to be found together because they happen to be in a region of genome that shares the same ancestry. In a balance between recombination and random drift, the average linkage disequilibrium D between two alleles is 0: There is no reason why one combination should be systematically favored over another. However, the value of D fluctuates up and down and has a variance proportional to $1/(1 + 4N_ec)$ (see Box 15.5). As we saw when we examined the ancestry of a pair of linked genes, the key parameter is the product of recombination rate and population size, N_ec, which is proportional to the *number* of recombination events that occur in every generation in the whole population. This is the ratio between the rate at which random drift builds up random associations ($\sim 1/N_e$) and the rate at which recombination breaks them down ($\sim c$). We will see the same kind of dependence on the relative rates of different processes when we consider other forces such as selection and gene flow.

The Amount of Linkage Disequilibrium Varies Greatly along the Genome

The simple proportionality between the variance of linkage disequilibrium and $1/(1 + 4N_ec)$ can be misleading. Actual associations fluctuate greatly around this expectation, both because of the limited samples that we observe and because of the chance evolutionary history of each region of the genome (Figs. 15.11 and 15.14). The picture of blocks of genome with different ancestry is more satisfactory than measures of association between discrete genetic loci. Viewed in this way, we can see that the level of nucleotide diversity π can vary abruptly as we move along the genome. Although the average diversity equals $\theta = 4N_e\mu$, the actual level depends mainly on the depth of the genealogy at each point—that is, on the time when the sample of genes at each locus last shared a common ancestor. Because a single recombination event can alter the depth of the genealogy, the observed diversity may change abruptly. This need not reflect any real biological differences such as a change in mutation rate. For the same reason, levels of linkage disequilibrium fluctuate greatly in a way that does not accurately reflect variation in recombination rate (Fig. 15.19). The extreme randomness of the evolutionary process frustrates attempts to make simple inferences from DNA sequence data.

The dependence of linkage disequilibrium on the evolutionary history of each section of genome can be seen in surveys of the human genome. Nucleotide diversity averages $\pi = 0.0008$, but varies over a tenfold range across different regions of the genome. The amount of diversity seen in neighboring regions, a few kilobases apart,

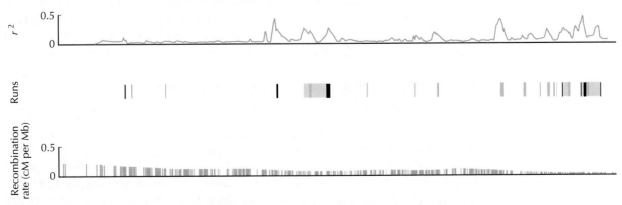

FIGURE 15.19. Variation in linkage disequilibrium (LD) along the short arm of human chromosome 19. The *upper panel* shows r^2 (see Box 15.5), measured in a "sliding window" 500 bp wide. The density of shading in the *middle panel* indicates the statistical significance of runs of excess LD (gray: $p < 10^4$). The *bottom panel* shows the recombination rate estimated from these data.

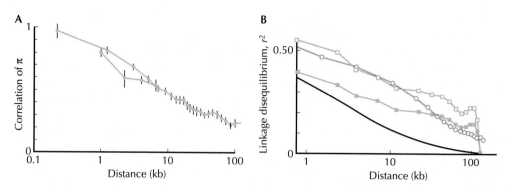

FIGURE 15.20. Variation in evolutionary history predicts patterns of linkage disequilibrium along the genome. (*A*) The correlation between nucleotide diversity in regions separated by different distances. The two lines show two subsets of the single-nucleotide polymorphism data, which agree well. (*B*) The *red line* shows the observed linkage disequilibrium (r^2) as a function of the distance between sites. This is compared with predictions based on the sharing of evolutionary history derived from *A* (upper and lower bounds, *blue*). The theoretical prediction in *B* (*black curve*) is based on a simple model assuming constant population size and uniform recombination rates.

is strongly correlated (Fig. 15.20A). Now, we have seen that neutral diversity is expected to equal $2\mu\tau$, where τ is the average time to common ancestry of two genes. Under the simple model of coalescence in a single population, τ averages $2N_e$, but all kinds of factors can distort this relationship (selection, population structure, and so on). Local variation in nucleotide diversity largely reflects variation in the coalescence time τ rather than the mutation rate μ.

The correlation in coalescence times between nearby regions of genome (Fig. 15.20A) reflects the similarity in genealogy between closely linked sites (Box 15.4). As expected, the correlation is stronger and extends over more base pairs in regions of the genome with lower recombination rates. The pattern of linkage disequilibrium that would be expected given the observed correlation in coalescence times fits well with the actual levels of linkage disequilibrium measured in a separate study of 2745 SNPs (Fig. 15.20B [red, blue]). However, both the correlation in coalescence time τ and the extent of pairwise linkage disequilibrium r^2 extend over much larger regions of genome than predicted by a simple model of a single well-mixed population with uniform recombination rates. Both extend over scales of approximately 20 kb, rather than just a few kilobases, as predicted by the simple relation $r^2 \sim 1/(1 + 4N_ec)$ (black line in Fig. 15.20B). A similar discrepancy is seen in studies of *D. melanogaster*. It may be explained if recombination is confined to "hot spots," so that long regions of genome can maintain strong linkage disequilibrium without being broken up at meiosis. Alternatively, the human population may have been strongly subdivided until recently; mixing of genetically divergent populations can generate strong linkage disequilibria (see p. 452). At present, it is not entirely clear which of these two explanations is more important.

Linkage Disequilibrium Both Helps and Hinders the Search for QTLs

Understanding the pattern of linkage disequilibrium is crucial to attempts to find the genetic basis of quantitative traits—and, in particular, to find the genetic basis of human disease by searching for associations between the disease and genetic markers (see pp. 404 and 758–773). Associations between different alleles both help and hinder the search for the genetic changes that cause differences in quantitative traits. On the one hand, if genetic variation is organized into blocks of genome that are found in a small number of combinations (i.e., in just a few haplotypes), then only a few sites within these blocks need to be scored (recall Fig. 15.15).

On the other hand, if an association *is* found between some haplotype and (say) an inherited disease, it is impossible to find which particular site is responsible for the differences. There are (at least) two difficulties here. First, spurious associations may arise with sites that are far from the actual causative site, because of sampling error, because of the randomness of the evolutionary process, and because mixing of genetically different populations can generate strong associations over wide map distances (see p. 452). Second, several sites might interact in a complicated way to determine the overall phenotype (recall the example of *Adh* in *Drosophila*; Fig. 14.27). These difficulties are discussed further in the human context in Chapter 26 (pp. 758–763).

Just as linkage disequilibria hinder our search for the causes of phenotypic variation, so they also hinder natural selection. If a particular variant increases fitness, then all the alleles that are associated with it will increase—even if those alleles are slightly deleterious. In Chapter 23 we examine how the chance associations generated by random genetic drift interfere with selection and how this gives an important explanation for the prevalence of sex and recombination.

SUMMARY

Random genetic drift is caused by random variation in fitness. We define fitness as the number of offspring left after one generation. Because the fitness of individual genes is necessarily random, the frequencies of alleles also fluctuate at random. In the Wright–Fisher model, the variance of allele frequency increases by $pq/2N$ in each generation, where $p + q = 1$ are the allele frequencies and there are N diploid individuals in the population. In other models, the variance of allele frequency increases at a rate $pq/2N_e$, where N_e is the effective population size.

In any finite population, mating between relatives (i.e., inbreeding) is inevitable, and so the number of ancestors decreases as we look further back in time. Thus, genes become identical by descent from fewer and fewer ancestral genes as we go back. Ancestral lineages coalesce, and ultimately trace back to a single common ancestral gene. This is described by the coalescent process, in which each pair of lineages has a chance $1/2N_e$ of coalescing in each generation. Typically, a sample of genes will quickly coalesce down to two ancestral lineages in $2N_e$ generations and, on average, these two trace to their common ancestor a further $2N_e$ generations back. However, the process is highly variable.

Under the neutral theory, molecular evolution is shaped by mutation and random genetic drift. The rate at which lineages diverge by mutation is just equal to the mutation rate μ—providing a simple explanation for the "molecular clock." The amount of variation within populations depends on $\theta = 4N_e\mu$, which measures the relative rates of drift and mutation. In fact, variation within abundant species is much lower than predicted by the simple neutral theory. This may be because of drastic population bottlenecks or because the substitution of favorable mutations reduces variation at linked loci (hitchhiking).

In a sexual population, recombination breaks the genome up into blocks with different ancestry. Correspondingly, sequence variation shows a blocklike pattern, such that only a few haplotypes segregate within each block. These strong associations between alleles are known as linkage disequilibria; they are important because they are responsible for hitchhiking, and because they are the basis for attempts to find the alleles responsible for quantitative variation and for human disease.

FURTHER READING

Crow J.F. and Kimura M. 1970. *An introduction to population genetics theory*. Harper & Row, New York.

The classic textbook that sets out the population genetics of random drift. The most accessible reference for drift of allele frequencies and inbreeding, but does not cover the recent development of the coalescent process.

Felsenstein J. 1978–2007. *Lecture Notes in Population Genetics*. http://evolution.gs. washington.edu/pgbook/pgbook.html

An excellent introduction to random genetic drift.

Gould S.J. 2000. *Wonderful life: Burgess Shale and the nature of history*. Vintage, New York.

The essential randomness of evolution was a strong theme in Stephen Jay Gould's writings; this is one of several relevant references. (However, note that Gould's interpretations of the Burgess Shale fauna are disputed.)

Hudson R. 1990. Gene genealogies and the coalescent process. *Oxf. Surv. Evol. Biol.* **7:** 1–44.

A clear explanation of the coalescent process, including its extension to include recombination.

Kimura M. 1983. *The neutral theory of molecular evolution*. Cambridge University Press, Cambridge.

Summarizes Kimura's arguments for the neutral theory, and also gives a clear introduction to the interaction between random drift and mutation.

Reich D.E., Schaffner S.F., Daly M.J., McVean G.A.T., Mullikin J.C. et al. 2002. Human genome sequence variation and the influence of gene history, mutation and recombination. *Nat. Genet.* **32:** 135–142.

An elegant analysis of the very large dataset of human SNPs that was published with the draft human genome.

Rosenberg N.A. and Nordborg M. 2002. Genealogical trees, coalescent theory and the analysis of genetic polymorphisms. *Nat. Rev. Genet.* **3:** 380–390.

A short introduction to the way the coalescent process describes genealogies, and its application to understanding genetic variation.

Wakeley J. 2006. *Coalescent theory: An introduction*. Roberts and Company, Englewood, Colorado.

An excellent summary of the coalescent process.

Wall J.D. and Pritchard J. 2003. Haplotype blocks and linkage disequilibrium in the human genome. *Nat. Rev. Genet.* **4:** 587–597.

Reviews recent work on linkage disequilibrium in human populations, such as that shown in Figs. 15.15 and 15.19.

16

Population Structure

S O FAR, WE HAVE THOUGHT OF populations as single homogeneous pools of genes. Thus, we have needed to follow only the proportions of different kinds of genes or gene combinations (i.e., allele frequencies or genotype frequencies). We have assumed that every individual is equally likely to mate or interact with every other and that all individuals experience the same conditions. A population of this sort is called **panmictic**.

Real populations are not like this. They are spread over large areas, with barriers to movement and with varying density and environmental conditions. As well as this spatial distribution, there can be subtler deviations from **panmixis**. Individuals may tend to meet and mate with others living in the same local habitat, social grouping, or those of the same age. We say that any population that deviates from ideal panmixis is **structured**. In this chapter, we will concentrate on **spatial structure**, but most of what we say will apply to other kinds of structures as well.

Many key questions about evolution depend on population structure. Do species adapt to local environments? Can a single population split into separate species? How do favorable alleles spread over a wide area? These questions involve selection, which we consider in the next chapter. In this chapter, we introduce the concept of gene flow, explain how it acts, and explain how it interacts with mutation, recombination, and, especially, random drift to shape neutral variation. We will see that the way genes are distributed through space can tell us much about populations, including their history, rates of movement, and population density. In Chapter 25, we will see how the principles explained here can help us understand the structure of our own species.

GENE FLOW

Natural Populations Are Not Single Well-Mixed Gene Pools

All kinds of organisms live in **structured populations**. Spatial structure is most obvious in organisms with limited dispersal abilities, which can only move a small distance compared to their overall range (e.g., snails or flightless insects). However, even species that regularly migrate across enormous distances may nevertheless return to breed in the same place year after year (e.g., monarch butterflies, salmon, or swallows). On a smaller scale, microbes may be confined to live in thin films on petri dishes, or

Pink unbanded Yellow unbanded

Pink banded Yellow banded

FIGURE 16.1. (*A*) The snail, *Cepaea nemoralis*, is highly polymorphic for shell color and banding. In the Pyrenees (on the border of France and Spain), these traits vary erratically from place to place (*B*). Darker shells tend to be found in colder places, because they warm up more in the sun. (*C*) In contrast, allele frequencies at allozyme loci vary over large scales: Eastern, central, and western populations have distinct combinations of allele frequencies, separated by relatively sharp clines. These reflect the historical origins of these populations. *B* shows the frequency of the allele for yellow shell color and *C* shows the frequencies of four alleles at the indophenol oxidase locus, *Ipo-1*.

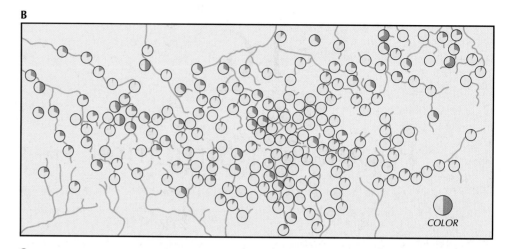

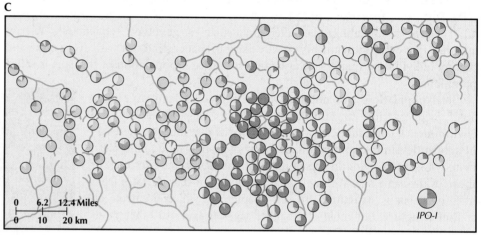

on the surface of rocks. Later in this chapter, we will look at spatial structure in HIV virus populations over a scale of a few millimeters within the human spleen.

Genetic variation shows all kinds of spatial patterns. Often, allele frequencies or quantitative traits vary erratically from place to place (e.g., Fig. 16.1B). Alternatively, allele frequencies or traits may vary gradually, in patterns known as **clines**. These may extend over thousands of kilometers or over a few hundred meters or less (Fig. 16.1C; also see Figs. 18.13–18.17). Spatial patterns, whether apparently random or smooth and clinal, may show no correlation with any obvious environmental feature (e.g., Fig. 16.1C). Alternatively, there may be a close relation with some environmental variable. We have already seen the example of the sickle-cell hemoglobin allele in humans, which is found in areas where malaria is common. In *Cepaea nemoralis*, there are fine-scale correlations between banding pattern and type of vegetation, such that the snails tend to be well camouflaged. In addition, snails living in cooler places tend to have darker shells, because these warm up more in the sun (Fig. 16.1B). Coat color in many mammal species changes in relation to habitat in a similar way and can give sharp clinal patterns (e.g., see Fig. 18.12).

To understand patterns such as these, we must take account of an evolutionary force that we have not yet discussed: **gene flow**. Animals move from place to place in search of food or mates; pollen and seeds may be blown by the wind or carried by animals; bacteria can be dispersed by air or by water or carried about by their animal hosts. Whatever the mechanism, a gene can change its location from one generation to the next. It is this movement that is termed gene flow.

Gene Flow Homogenizes Populations

Movement of genes from place to place has a simple effect: It makes different parts of a population more similar to each other. Acting alone, it will eventually make the population homogeneous. Here, we show how the rate of gene flow can be measured and its effects quantified.

The simplest case is the **island model**: A single local population receives migrants from some outside "mainland" population. The local population is often referred to as a **deme**. In each generation, a fraction m of the genes are derived from the mainland, and the remaining $1 - m$ come from the local deme. This mixing has the simple effect of reducing differences in allele frequency between mainland and island by a factor $1 - m$ in every generation. For example, when, in each generation, 10% of the genes are derived from the mainland (i.e., migration rate $m = 0.1$), differences in allele frequency will decrease by a factor 0.9, 0.81, 0.729, ... over successive generations, and will decrease by a factor ~0.35 after ten generations. In general, differences are expected to become substantially smaller over a timescale of approximately $1/m$: that is, over 100 generations if $m = 0.01$, 1000 generations if $m = 0.001$, etc. (Box 16.1).

This simple island model can easily be extended to describe movement between any number of demes. We just need to know the proportion of genes in every generation that came from all the potential source demes. The effect of gene flow between multiple demes is qualitatively similar to the one-island case; that is, differences decay at a rate inversely proportional to the fraction of genes that move. These kinds of models can be used when we have detailed information on migration rates—for example, from historical records of the villages where people were born and where they died. In most cases, however, it is not feasible to gather this kind of detailed information.

Fortunately, there is a simple approximation that describes the rate of gene flow through a population that extends over a wide region of space. Imagine a set of genes that all start in the same place. Each will be carried to a new location, so that after one generation, the genes will have spread out in some distribution. This might have a complicated shape, with most individuals typically staying near home and some moving further away (Fig. 16.3). A full description of movement over one generation requires knowledge of the whole distribution. However, as we follow the progress of the

---

Box 16.1   Allele Frequencies in the Island Model

In each generation, a fraction m of genes in the island comes from a "mainland" population. The remaining $1 - m$ come from the island itself. The island allele frequency at time t is simply a mixture of the allele frequencies in the two sources:

$$p_t = mp_m + (1 - m)p_{t-1},$$

where p_m is the allele frequency on the mainland and p_{t-1} is the allele frequency in the island in the previous generation (Fig. 16.2). The difference in allele frequency between island and mainland just decreases by a factor $1 - m$ in every generation:

$$(p_t - p_m) = (1 - m)(p_{t-1} - p_m).$$

Thus, the difference in allele frequency decreases approximately exponentially at a rate m per generation:

$$(p_t - p_m) = (p_0 - p_m)(1 - m)^t$$
$$\sim (p_0 - p_m)\exp(-mt).$$

(Note that $(1 - m)^t$ is approximately $\exp(-mt)$ for small m [Chapter 28 (online)].)

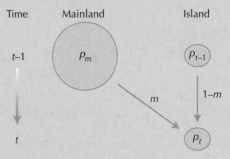

FIGURE 16.2. The island model.

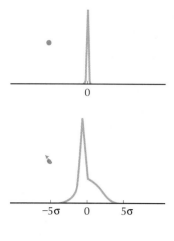

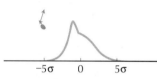

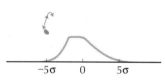

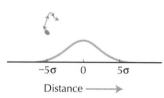

Distance ⟶

FIGURE 16.3. The red arrows show a gene moving randomly through a two-dimensional habitat, over 0, 1, 2, 3, 4 generations. In the first generation, the distribution of positions might be complicated (*blue, top*), but soon the sum of several random movements approaches a normal distribution (*blue, bottom*). Only movements along the *x*-axis are shown in the distributions.

genes over many generations, the distribution approaches a **normal distribution** (see Chapter 28 [online]). If there is no tendency to move in any particular direction, this distribution has mean 0 and a variance that increases in proportion to time (Fig. 16.3).

Instead of thinking about the distribution of a collection of genes, we can think instead of the movement of a single gene (Fig. 16.3, inset). Its position after t generations is the sum of random movements over each of those generations: $x_1 + x_2 + \cdots + x_t$. If these movements are independent of each other, then their sum approaches a normal distribution, with variance $\sigma^2 t$ (see Chapter 28 [online]). This approximation of the long-term movements of genes is called the **diffusion approximation**. It is the same as the approximation that is used to describe the diffusion of a chemical. Over large scales, the concentration appears to spread smoothly, but on a fine scale, we see that it is due to many random motions of individual molecules. In the same way, the long-term changes in allele frequency due to gene flow in a large population appear smooth, but are actually the cumulative effect of many individual movements.

The Rate of Diffusion of Genes Is Measured by σ^2

The rate of diffusion of genes through a population is measured by a single parameter: the variance of the distance between parent and offspring over one generation, denoted σ^2. (In two dimensions, this distance is measured along the direction of interest.) In principle, this is easy to measure: We must simply follow individuals and find out where their offspring are located one generation later. One of the first measurements of the rate of diffusion of genes was made by Theodosius Dobzhansky and Sewall Wright using *Drosophila pseudoobscura*. In a series of experiments, they released a total of 14,026 flies marked by an orange eye-color mutation and followed them over the next few days (Fig. 16.4). In one experiment, the variance of distance moved along a line of traps increased at a rate of 4600 m² per day. Assuming that only adults disperse, and that they have a life span of 4.5 days, the rate of gene flow is estimated as $\sigma^2 = 21{,}000$ m² per generation, corresponding to a standard deviation of $\sigma = \sqrt{21{,}000} = 145$ m in one generation (blue line in Fig. 16.4B).

In practice, direct measurements of this kind suffer from many difficulties. For example:

- Rates of movement are sensitive to local conditions (e.g., temperature, as in Fig. 16.4B).

- Marking or observing individuals may disturb them; the eye-color mutation used by Dobzhansky and Wright presumably altered the flies' vision and hence their behavior.

- Even if very large samples are followed, and large areas surveyed, occasional long-range migrants will be missed; yet these contribute most to the variance of distance moved.

- It is difficult to follow an entire generation to find whether the offspring of migrants survive and themselves reproduce.

For these reasons, direct measurements of gene flow are at best inaccurate. In the rest of this chapter, we will see how indirect estimates can be made, by inferring the rate of gene flow from the pattern of genetic variation.

Diffusion Is a Slow Process

We have seen that after t generations, genes will have spread out into a normal distribution with variance $\sigma^2 t$. Thus, a gene will typically have moved a distance with standard deviation $\sigma \sqrt{t}$ in each direction. For example, the flightless grasshopper *Podisma pedestris* (see Fig 16.5) moves about 20 m in one generation (more precisely, $\sigma^2 = 400$ m² per generation). Thus, over 100 generations a gene will have moved about

A

B

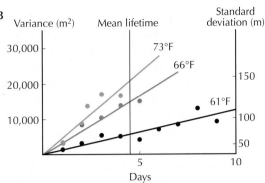

FIGURE 16.4. Dobzhansky and Wright (1943) measured the rate of dispersal of *Drosophila pseudoob-scura* by releasing marked flies at sites in the Sierra Nevada, California (*A*). Over the following days, flies were caught in a series of traps. The graphs (*B*) show how the variance of the distribution of marked flies increased over time. The three sets of points show results from experiments at different times during the summer: Rates of movement increase strongly with temperature. The rate of diffusion of genes is estimated by assuming a mean lifetime of 4.5 days (*vertical line*).

FIGURE 16.5. *Podisma pedestris* (adult female).

$20\sqrt{100}$ = 200 m, and over 10,000 generations, about $20\sqrt{10,000}$= 2 km. The process is slow because genes move in a random walk and are as likely to move forward as backward (see Fig. 16.3, inset).

The slowness of diffusion can be seen by thinking of the mixing of two genetically different populations, rather than the spread of genes out from a point. A smooth cline will form, which will gradually become broader. After a time *t*, the width of the cline will be $\sqrt{2\pi\sigma^2 t}$. As this increases only with the square root of time, mixing becomes slower and slower over longer times (Fig. 16.6). For example, the two chromosomal races of *P. pedestris* met in the Maritime Alps after the last glaciation, less than 10,000 generations ago (this species has an annual life cycle, with one generation per year). If they had simply been mixing together, the cline separating them would be only $\sqrt{2\pi 400)(10,000}$ = 5 km wide. The actual cline, at approximately 800 m, is much narrower than this, indicating that some

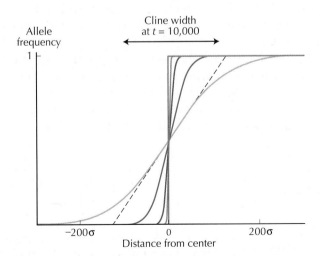

FIGURE 16.6. After two populations fixed for different alleles meet at a sharp boundary, gene flow will gradually broaden the transition between them. The figure shows the clines after 1, 10, 100, 1000, and 10,000 generations, when the clines are $\sqrt{2\pi\sigma^2 t}$ = 2.5σ, 7.9σ, 25σ, 79σ, and 250σ wide, respectively (*red, green, purple, blue,* and *orange lines*). The cline width is defined by drawing a tangent to the steepest part of the cline (*dotted line*), as shown by the arrows for the cline at *t* = 10,000 generations.

faster process must be keeping the two types separate. In Chapter 18, we will see how narrow clines can be maintained and, in Chapter 19, we will use these ideas to distinguish neutral mixing from other processes.

This kind of calculation shows that diffusion is an extremely slow process. We know that the Alps were covered in ice 10,000 years ago, and so the organisms living there now must have colonized them relatively quickly, because a passive diffusion northward would have been far too slow. The diffusion approximation works well over short timescales (a few hundred generations, say), but over the longer term, the spread of genes depends on unpredictable large-scale expansions and contractions of the population as a whole, not on random local movements. Later in this chapter, we will see how genetic patterns can help us to infer this ancient history.

GENE FLOW INTERACTS WITH OTHER EVOLUTIONARY FORCES

Geographic Variation Is Generated by Random Genetic Drift

Gene flow alone will eventually lead to homogeneity. Thus, interesting patterns will be seen only if some other process generates divergence. In the remainder of this chapter, we will look at how random genetic drift can cause divergence. Random drift can be seen in terms of fluctuating allele frequencies or as a process of coalescence of ancestral lineages (Chapter 15). In this section, we look at how drift and gene flow balance each other to generate variance in allele frequency and how this variance can be used to estimate the relative rates of these two processes. In the next section, we will look at random drift in a different way, in terms of ancestral lineages that wander across the species' range.

Imagine a set of small populations or demes, which all start out with the same allele frequencies. If these are isolated from each other, then they will drift apart until eventually different demes will be fixed for different alleles (recall Box 15.1 and Fig. 15.3). Now, suppose that each deme receives a fraction m of its genes from some other deme in every generation. (This is an extension of the single-island model of Box 16.1; migrants come not from a large "mainland" population, but from a pool of migrants contributed equally by all the demes. A set of demes of this sort is known as a **metapopulation**.) Exchange of genes will tend to make the demes more similar and will balance the diversifying effects of random drift. The population as a whole will settle to an equilibrium distribution of allele frequencies (Fig. 16.7).

The variance in allele frequency between demes depends on the product of the **effective population size** of each deme and the migration rate, $N_e m$ (Box 16.2). This measures the rate of gene flow relative to the rate of random drift, in much the same way that $N_e \mu$ measures the rate of mutation relative to drift (see p. 426) or $N_e c$ measures the rate of recombination relative to drift (see p. 429). The parameter combina-

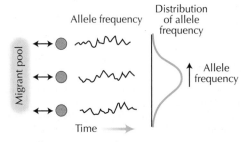

FIGURE 16.7. In the island model, demes exchange genes with a pool of migrants (*left*). Allele frequencies in individual demes fluctuate at random over time (*middle*), but the population as a whole reaches a steady statistical distribution (*right*).

tion $N_e m$ has a simple interpretation: It is the effective *number* of individuals that migrate into the population in every generation. Thus, very low *rates* of migration can still keep populations genetically similar despite random drift, simply because drift is a slow process in reasonably large populations. If more than a few individuals are exchanged per generation, the variance in allele frequencies among demes will be low. (For example, if $N_e = 10,000$ and m is only 0.001, then $N_e m = 10$, which is enough to keep allele frequencies quite similar across demes.)

Box 16.2   Variance of Allele Frequency in the Island Model

Imagine a very large number of demes that all start at the same allele frequency p_0. The average allele frequency stays constant (or at least, fluctuates very slowly) because the population as a whole is large. However, there will be a distribution of allele frequency across demes (Fig. 16.8A), with variance V_t at time t. We saw in Box 15.1 that, in each generation, random drift increases this variance to

$$V_{t+1} = V_t + \frac{E[pq]}{2N_e} = V_t + \frac{(p_0 q_0 - V_t)}{2N_e},$$

where $q_0 = 1 - p_0$. A fraction m of genes come in from elsewhere; on average, these immigrants have allele frequency p_0. The deviation in allele frequency in any one deme is reduced by a factor $1 - m$ (Box 16.1). Because a variance is defined as a mean *squared* deviation, gene flow reduces the variance by $(1 - m)^2$. Overall, we have

$$V_{t+1} = (1 - m)^2 \left(V_t + \frac{(p_0 q_0 - V_t)}{2N_e} \right).$$

The variance between demes increases toward an equilibrium that we can find by setting $V_{t+1} = V_t = V$. To a good approximation for small m and large N_e,

$$V = \frac{p_0 q_0}{1 + 4N_e m}.$$

The variation between demes, V, relative to the maximum possible variance for these allele frequencies, $p_0 q_0$, is measured by Wright's F_{ST}:

$$F_{ST} = \frac{V}{p_0 q_0} = \frac{1}{1 + 4N_e m}.$$

We have ignored mutation here, even though some mutation must be present if variation is to be maintained in the long term. Usually, however, mutation is much rarer than migration ($\mu << m$) and has a negligible effect on variation from deme to deme. Over approximately $1/m$ generations the collection of demes soon settles to a steady-state distribution, with variance determined by $N_e m$. The allele frequency in the whole population then evolves very slowly over a timescale set by the mutation rate and by the overall effective population size (Fig. 16.8C).

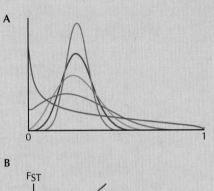

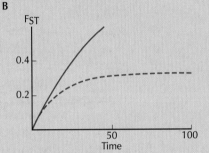

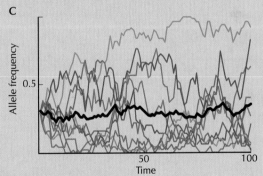

FIGURE 16.8. (*A*) The distribution of allele frequencies across a set of demes, each with $N_e = 25$ and exchanging migrants at a rate $m = 0.02$. The demes all start at $p_0 = 0.3$; the distribution is plotted at 1, 2, 4, 10, and 100 generations (*red, dark green, green, blue,* and *purple lines,* respectively). (*B*) The increase in variance of allele frequencies over time, measured by $F_{ST} = \text{var}(p)/p_0 q_0$. The *solid line* shows how F_{ST} would increase toward 1 if there were no migration; the *dashed line* shows the increase toward an equilibrium F_{ST} of $1/(1 + 4N_e m)$ = 0.333, in a balance between drift and gene flow. (*C*) An example showing allele frequencies in each of ten demes. The average over the whole population of ten demes drifts much more slowly (*black line*).

> ## Box 16.3   Example Calculation of F_{ST}
>
> Suppose that an allele has average frequency 0.5 and is at frequencies 0.4, 0.5, 0.5, and 0.6 across four demes. Then, var(p) = 0.005 and
>
> $$F_{ST} = \frac{var(p)}{\overline{p}(1-\overline{p})} = \frac{0.005}{0.5 \times 0.5} = 0.02.$$
>
> If another allele has average frequency 0.1 and has frequencies 0.04, 0.1, 0.1, and 0.16 across the four demes, then its variance in allele frequency is much lower, at var(p) = 0.0018. However, $F_{ST} = 0.0018/(0.1 \times 0.9) = 0.02$ is the same.
>
> A quantitative trait with **narrow-sense heritability** of $V_A/V_P = 50\%$ and phenotypic variance $V_P = 400$ (i.e., standard deviation $\sqrt{400} = 20$) would have additive genetic variance $V_A = 200$. In a population with $F_{ST} = 0.02$ for Mendelian loci, the analogous measure $Q_{ST} = var(\overline{z})/(var(\overline{z}) + 2V_A)$ should also equal 0.02. We therefore expect to see a between-deme variance in the trait of $var(\overline{z}) \cong 2Q_{ST}V_A = 2 \times 0.02 \times 200 = 8$; that is, a standard deviation between demes of $\sqrt{8} \cong 2.8$.

F_{ST} Is a Standardized Measure of the Genetic Variation between Demes

The degree of genetic divergence between populations is measured by the coefficient F_{ST}. Under any model of random drift and gene flow, the variance in frequency of an allele that overall is at mean frequency $\overline{p}$ is proportional to $\overline{p}(1 - \overline{p})$, so that we can write $var(p) = \overline{p}(1 - \overline{p}) F_{ST}$. One can think of F_{ST} as the variance in allele frequency, relative to the maximum possible variance of an allele that has an overall mean $\overline{p}$. The maximum variance is $\overline{p}(1 - \overline{p})$, which occurs when a fraction $\overline{p}$ of demes are fixed at $p = 1$ and the remainder are fixed at $p = 0$ (Chapter 28 [online]). F_{ST} can also be thought of as a measure of the proportion of genetic variance that is between subpopulations, relative to the total; hence Wright's use of the ST subscript.

The standardized coefficient F_{ST} allows us to combine information from many different alleles, with different frequencies. Alleles at intermediate frequencies will show higher variance from place to place than those that are nearer fixation, but the ratio F_{ST} is predicted to be the same for all of them (e.g., Box 16.3). Indeed, a similar measure can be defined for quantitative traits, as the variance in trait mean between demes relative to the average additive genetic variance within them, V_A: $Q_{ST} = var(\overline{z})/(var(\overline{z}) + 2V_A)$. In a balance between gene flow and random drift, this is predicted to be the same for all traits and to be equal to F_{ST}. The differences in F_{ST} between loci, and differences in Q_{ST} between traits, give us a powerful way to detect the action of selection (see Web Notes).

The Population as a Whole Drifts Slowly

Thus far, we have discussed how local subpopulations diverge from each other as a result of random drift. We can also ask about the evolution of the species as a whole. This will be far slower than changes in individual demes. Allele frequencies in a single deme of effective size N_e will fluctuate over a timescale of approximately N_e generations, whereas a population made up of n such demes, connected by migration, will drift over the much longer timescale of approximately nN_e generations (Fig. 16.8C).

How quickly does the overall mean allele frequency, $\overline{p}$, drift? For the island model, the answer to this question is surprisingly simple: The effective size of the whole population is just $nN_e/(1 - F_{ST})$. Thus, subdivision of the population into separate demes has increased the overall effective size (i.e., it has reduced the overall rate of genetic drift). This can be understood by thinking of the extreme case where the number of migrants (N_em) is very low. Then, demes will mostly be fixed for one allele or another

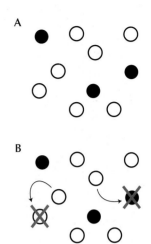

($F_{ST} \sim 1$), and so there can be no further change: Genetic variation is preserved in many isolated subpopulations (Fig. 16.9A). This idea extends to similar models: Provided that migration is symmetrical, so that exchange of genes does not alter the overall allele frequency, the rate of loss of genetic variation from the whole population is reduced by a factor $1 - F_{ST}$, because variation is locked up as differences between local demes.

In reality, local populations fluctuate in size. Alleles that happen to find themselves in demes that are expanding will increase in frequency, whereas alleles in demes that face extinction will become rarer. Migration is now asymmetric, so that gene flow is predominantly out of large demes into small demes and favors the spread of genes from the larger demes. These kinds of fluctuations can greatly accelerate the overall rate of genetic drift and so greatly reduce genetic diversity in the population as a whole. In the extreme case, again suppose that demes are near fixation for different alleles ($F_{ST} \sim 1$). If demes occasionally go extinct and are recolonized from another deme, then the effective size of the whole population is close to the number of demes rather than to the number of individuals (Fig. 16.9B).

Extinction and Recolonization in HIV Populations

The key obstacle to treatment of HIV infection is that over time the virus evolves resistance to antiviral drugs. The rate of this response to selection depends on levels of genetic variation in the viral population. This is a good example of how loss of variation due to fluctuating population size can easily outweigh its preservation through population subdivision. Individuals infected by HIV carry 10^7 or more virus particles, and the mutation rate is high ($\mu \sim 3 \times 10^{-5}$ per site per generation). Although nucleotide diversity is high (0.01–0.05; Table 16.1), it is much lower than expected from the product of census numbers N and mutation rate μ ($N\mu > 300$, so that $4N\mu/(1 + 4N\mu) \sim 1$; p. 426). HIV lives within small clusters of lymphoid cells in the spleen. There can be substantial genetic divergence between these clusters over distances of just a few millimeters (Table 16.1). This is due to a rapid turnover of infected clusters such that local clusters go extinct, whereas uninfected cells are colonized by viruses from elsewhere. This process of extinction and recolonization greatly accelerates the rate of genetic drift and goes some way toward explaining why sequence diversity in abundant organisms such as HIV is usually far lower than suggested by their actual numbers, a pattern that we emphasized before (see pp. 368–370 and 426–427).

FIGURE 16.9. (*A*) If a population is so strongly subdivided that demes are fixed for different alleles (i.e., $F_{ST} = 1$), then as long as these demes survive, there can be no further change: Genetic variation is preserved indefinitely. (*B*) If demes sometimes go extinct (*red* ×s) and are recolonized from a surviving deme, then allele frequencies change. Demes are analogous to individuals that die and reproduce, and so the effective size of the whole metapopulation is close to the number of demes, rather than the number of individuals.

TABLE 16.1. Spatial variation in HIV		
Patient	Nucleotide Diversity, π	F_{ST}
B	0.041	0.594
L	0.040	0.369
M	0.045	n.s.
N	0.013	0.078
P	0.039	0.215
S	0.029	0.090

Adapted from Frost S.D.W. et al. 2001. *Proc. Natl. Acad. Sci.* **98:** 6975–6980 (© National Academy of Sciences, U.S.A.).

Tissue samples were taken from the spleens of six patients. Several islands of lymphoid tissue were microdissected from each, and samples of the HIV *env* gene were sequenced. The second column shows the average pairwise nucleotide diversity, π, within each island, and the third column shows the fraction of genetic variation between islands, F_{ST}. In patient M, there was no significant (n.s.) difference between islands, but in others, there was extensive divergence.

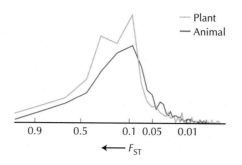

FIGURE 16.10. The overall distribution of F_{ST} values across more than 1000 studies of plants and animals.

Rates of Gene Flow Can Be Estimated from F_{ST}

It is unusual to find genetic differentiation as high as is seen in the HIV example (Table 16.1). A survey of F_{ST} estimates shows a wide range of values, with values of F_{ST} typically around 0.1–0.2 (Fig. 16.10). Overall, the pattern is consistent with the idea that geographic variation is generated by random drift and dissipated by gene flow. In plants, values of F_{ST} are higher for organelles, which are transmitted by pollen or by seed only, than for nuclear genes, which can move in both ways. Here, an additional factor is that the effective population size is smaller for organelles than for nuclear genes (Table 16.2A). In animals, a similar pattern is seen: Mitochondrial genes show about twice as much geographic divergence as nuclear genes, presumably because their effective population size is about half as great.

Studies such as these are usually made with the aim of estimating the rate of gene flow, using the relationship $F_{ST} = 1/(1 + 4N_e m)$ (Box 16.2). However, it is extremely difficult to validate such estimates directly. There are several reasons why a good deal

TABLE 16.2. Estimates of F_{ST} by (A) mode of transmission and (B) spatial scale

(A) Mode of Transmission	Number of Studies	Mean F_{ST}
Plant		
Pollen	8	0.39
Seed	29	0.46
Both	294	0.32
Animal		
Mitochondrial	150	0.45
Nuclear	781	0.20
(B) Scale of Sampling	**Number of Studies**	**Mean F_{ST}**
Plant		
Local	43	0.25
Regional	52	0.39
Species-wide	36	0.33
Animal		
Local	65	0.10
Regional	116	0.28
Species-wide	48	0.23

Adapted from Morjan C.L. and Rieseberg L.H. 2004. *Mol. Ecol.* **15:** 1341–1356, Table 1, p. 1346 and Table 3, p. 1347, and references therein (© Blackwell Publishing).

of caution is needed. First, this formula applies to the island model, in which immigrants come at random from the rest of the population. A similar relationship holds for populations spread smoothly over two dimensions, but does not hold for other kinds of population structure (e.g., fish in a branching river system or shells along the seashore). Second, the pattern of spatial differentiation must be due to a balance between gene flow and genetic drift. If it is generated by selection or by the vagaries of past history, then obviously one cannot infer $N_e m$. F_{ST} is high over broad geographic scales, and this divergence is most likely due to past history rather than to a short-term balance between drift and gene flow (Table 16.2B). Third, what is being estimated is the rate of gene flow relative to the rate of random drift, which depends on the effective *numbers* of migrants. Later in this chapter, we will see how the *rate* of gene flow, m or σ^2, can be measured.

GENEALOGIES IN STRUCTURED POPULATIONS

Genealogies Are Distorted by Population Structure

As we trace lineages back through time, they move from place to place. Each of our genes comes from a parent, a grandparent, and so on—a chain of ancestors that each lived in more or less distant places. The coalescent process extends to the **structured coalescent**, in which lineages move randomly as we trace back in time and may coalesce if they find themselves in the same place.

This picture is especially simple in the island model (Fig. 16.11). Suppose we sample two genes from within the same deme, having an effective size N_e. There is a chance $1/2N_e$ that the genes share a common ancestor in the previous generation, but also a chance of $\sim 2m$ that one of the two will have an ancestor outside the deme. Tracing back several generations, the lineages will either coalesce soon (in $\sim 2N_e$ generations) or escape and spend a very long time wandering through the entire population of n demes. Surprisingly, the average **coalescence time** for two genes from within the same deme, T_w, is equal to the total effective number of genes in the whole population and does not depend on the migration rate at all:

$$T_w = 2nN_e.$$

Although the average time to coalescence is not affected by subdivision of the population into n demes, the distribution of coalescence times is much more variable (Fig. 16.12). This variability in **genealogy** is typical of structured populations. As we shall see in the next section, it makes it hard to interpret data from one or a few genetic loci.

If we sample two genes from different demes, they cannot coalesce until they meet in the same deme. The chance that one or the other will change position is $\sim 2m$ per generation, and the chance that if one of the lineages does move, it lands in the same deme as the other is $1/n$. So, the expected time to coalescence for two genes from different demes, T_d, is $n/2m$ generations for them to land in the same place, plus a further $T_w = 2nN_e$ generations for them to coalesce once they are in the same deme:

$$T_d = \frac{n}{2m} + 2nN_e.$$

This description in terms of coalescing lineages can be related to the description in terms of allele frequencies through Wright's coefficient, F_{ST}:

$$F_{ST} = \frac{\overline{T} - T_w}{\overline{T}},$$

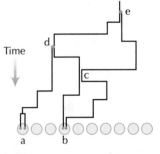

FIGURE 16.11. Coalescence in the island model. Genes that are in the same deme (a) may trace back to a common ancestor within the same deme in the recent past. However, one or another lineage may escape (b) and wander through the whole population for a long time. Eventually, lineages come together in the same deme. They may then move apart again (c) or coalesce (d,e).

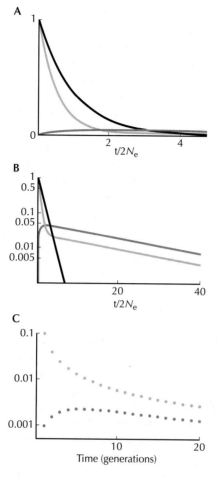

FIGURE 16.12. Distribution of coalescence times in structured populations. (*A*) The *black curve* shows the exponential distribution expected in a single panmictic population. The *blue curve* shows the distribution of times for two genes that start in the same deme, under the island model. (There are ten demes in all, and $4N_em = 1$; time is scaled relative to $2N_e$ generations.) The chance that the two genes share a recent common ancestor is reduced, whereas the chance that they are distantly related is increased. With $4N_em = 1$, there is about the same chance that the genes are related recently, via an ancestor in the same deme, as that they are related much more distantly. (*B*) The very long tail to the distribution can be seen more clearly on a log scale. The *red curves* in *A* and *B* show the distribution of coalescence times for two genes that are in different demes. (*C*) The distribution of coalescence times in a two-dimensional population for genes that are near each other (*upper blue points*) and that are 5σ away from each other (*lower red points*). Neighborhood size is Nb = 5. As for the island model, this distribution is highly skewed. For nearby genes, there is a chance (1/2Nb = 10%) of common ancestry in the previous generation and a 27% chance of common ancestry within 20 generations. However, the mean coalescence time is twice the total effective number of genes in the whole population, which could be very large (e.g., 800,000 generations for a population ranging over 1000σ by 1000σ).

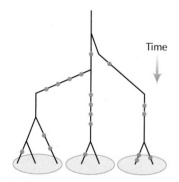

FIGURE 16.13. Wright's F_{ST} is related to the mean coalescence time between pairs of genes within demes, compared with the mean coalescence time between randomly chosen pairs: $F_{ST} = (\overline{T} - T_w)/\overline{T}$. These coalescence times, and hence F_{ST}, can be estimated from the number of mutations that separate each pair of genes (assuming the infinite sites model; see p. 424). In this example, seven genes are sampled from three demes; mutations are indicated by *red circles*. On average, there are 8.1 differences between pairs of genes sampled at random compared with 2.0 differences between genes within the same deme. Hence, F_{ST} is estimated to be (8.1 − 2.0)/8.1 = 0.753.

where $\overline{T}$ is the average coalescence time between two genes chosen at random from the whole population. If there are very many demes ($n \gg 1$), $\overline{T} \sim T_d$, and we see that $F_{ST} = 1/(1 + 4N_em)$, as before (Box 16.2). Thinking in terms of coalescence times rather than fluctuating allele frequencies is appropriate when we deal with DNA sequence variation, because the average number of differences between a pair of sequences is then proportional to their coalescence time. Thus, F_{ST} can be estimated from the number of differences between sequences from within the same deme, compared with that between sequences sampled from different demes (Fig. 16.13). (F_{ST} for HIV was estimated in this way [Table 16.1].)

Genealogical patterns are similar for genes that live in a large two-dimensional population—the most usual kind of population structure. Two genes that are close to each other may share a common ancestor very recently (i.e., within the previous few generations). If they do not, their ancestral lineages are likely to wander randomly over the whole species' range, before eventually meeting in a common ancestor in the distant past (Fig. 16.14). As in the island model, the average time for two nearby genes to share a common ancestor is equal to the total effective number of genes ($2nN_e$ for n demes each of N_e), but the distribution is extremely skewed (Fig. 16.12C). In fact, with symmetrical migration the average coalescence time between nearby genes is independent of subdivision in any population. This result applies under the same conditions as for the total effective size to be increased by $1 - F_{ST}$ (see p. 446). Again, however, if the population structure fluctuates (e.g., with local extinction and recolonization), then coalescence times can become much shorter. (Recall the example of sequence variation in HIV [Table 16.1].)

Neighborhood Size

In a two-dimensional population, the importance of gene flow relative to genetic drift depends on the **neighborhood size**. Sewall Wright defined this as $Nb = 4\pi\rho\sigma^2$, where ρ is the effective population density and σ^2 is the rate of diffusion of genes, which we defined on page 442. This can be thought of as the number of individuals living in a circular neighborhood of radius 2σ. In an ideal population, the chance that two nearby genes share a common ancestor in the previous generation is $1/2Nb$ and, more generally, the chance of common ancestry within the previous few generations is of this order (Fig. 16.12C). Wright's F_{ST}, which measures the proportion of genetic variation between populations relative to the total, also depends on the neighborhood size: $F_{ST} \sim 1/(1 + CNb)$, where C is a number that depends on the details of the model, but is usually in the range 2–10. Thus, the neighborhood size is closely related to $N_e m$ in the island model: Both are products of population density and migration rate, and both determine F_{ST} through similar formulas (Box 15.2). In two dimensions, neighborhood size can be inferred from F_{ST} in the same way that $N_e m$ can be inferred in the island model.

The Relation between Genes from Different Places Reflects Their History: Phylogeography

Over large distances, the simple theory we have outlined breaks down. In a species that lives in a habitat that spans a thousand dispersal distances, σ, it will take approximately a million generations for ancestral lineages to wander throughout the species' range, as assumed by the theory. In reality, populations expand and contract over much shorter timescales. This is obviously so for organisms that live in high latitudes in areas that were covered by ice during recent glaciations. However, even in the tropics, climate changes caused drastic changes in distribution. Now, of course, human activity is causing great changes both directly and through our disruption of climate.

The advent of DNA-based genetic markers and, especially, the ease of sequencing **mitochondrial DNA** have stimulated many studies that infer species' historical movements from genealogical patterns. (Recall that mitochondrial DNA is inherited maternally, without recombination, and so has a single genealogy.) This approach is known as **phylogeography**. We have already seen how the expansion and contraction of single populations are reflected in genealogies (see Fig. 15.8). When genes are sampled from different places, rather than from just one place, we have considerably more information. In a population with large neighborhood size and a stable history, we expect that ancestral lineages would wander across the species' range, so that genealogies would show no strong geographic patterns (e.g., Fig. 16.14). In contrast, if genes in different places descend from different ancestral populations, their relationships may reflect the spread of those populations (Fig. 16.15). For example, think of the northward expansion of populations from southern European refuges after the last glaciation (Fig. 16.16).

Many phylogeographic studies of mitochondrial DNA variation have been carried out in the southeastern United States. For example, the majority of coastal species show a deep separation between two families of sequences, one found in the Gulf of Mexico and the other along the Atlantic Coast (Fig. 16.17). There is a sharp transition between these types along the East Coast of Florida. This suggests that two ancestral populations had been separated, possibly by lower sea levels during past glaciations. Similarly concordant east-to-west separations have been found in freshwater fishes of the same region.

Examples such as these are compelling, because of the consistent patterns seen across several species and because there is a plausible explanation for the genetic patterns. Information from just a single gene in a single species, however, must be treated cautiously. We have seen that genealogies in a structured population have extremely variable histories (Fig. 16.12), so that patterns are quite likely to arise by

FIGURE 16.14. In a two-dimensional population, nearby genes may be closely related through a coalescence event in the recent past. Otherwise, lineages wander over a wide area as they trace back to a common ancestor in the distant past.

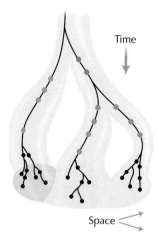

Time

Space

FIGURE 16.15. If a population has a long history of geographic isolation, then genealogies will tend to reflect this history. In this example, the present population formed through the merging of three parts, which had been isolated from each other for a long time. If genes are sampled from across the species' range (*black dots*), the genealogy that relates them falls into three divergent lineages, reflecting this history. *Colored dots* on the lineages indicate mutations. All individuals derived from the "red" population share four mutations (*red dots*), and similarly for the "green" and "blue" populations.

chance, even without an extrinsic cause such as a barrier to gene flow or a historical separation. For example, mitochondrial DNA sampled from the greenish warbler showed two deep genealogical separations. One coincided with a transition between two subspecies, defined by numerous morphological traits. However, the other occurred at an apparently arbitrary location and with no evidence for genetic divergence in other traits or loci (Fig. 16.18). Computer simulations showed that such deep divisions may often arise by chance, even in a population with a stable and uniform distribution. Of course, we know that the range of these warblers must in fact have shifted as a result of climatic changes. However, one cannot reliably infer such changes from genealogies sampled from a single locus. As we discuss in Chapter 25, the wealth of information on multiple genetic loci can help us reconstruct human history in more detail than is possible in other organisms.

Mixing of Populations Generates Linkage Disequilibria

We have seen that the effect of gene flow on single genetic loci is simply to reduce genetic differences from place to place. When populations differ at multiple loci, gene flow has another effect. In a mixed population, there will be an excess of gene *combinations* characteristic of the various source populations—in other words, there will be **linkage disequilibrium** (Box 15.5). These associations will be broken down by recombination, and so the population will tend to an equilibrium in which the strength of linkage disequilibrium depends on the relative rates of migration and recombination ($D \sim m/c$). Because migration rates can be high, gene flow is a potent source of linkage disequilibrium.

The effects of gene flow can be seen most clearly when genes come in from a source that is genetically very different. We have already seen one example in *Neisseria*, where segments of genome responsible for penicillin resistance can be identified because multiple sequence variants derived from another species are found in association, i.e., in linkage disequilibrium (see Fig. 15.16). An influx of multiple genes by hybridization with a distinct population or species is known as **introgression**. It can be thought of as leading to gene flow, as described by the simple island model of Box 16.2.

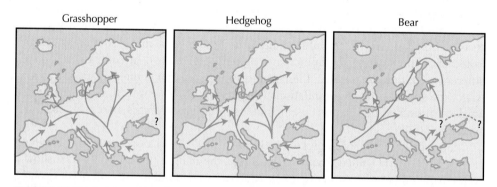

Grasshopper   Hedgehog   Bear

FIGURE 16.16. Patterns of recolonization of Europe after the most recent glaciation, reconstructed on the basis of DNA sequence variation. The grasshopper, *Chorthippus parallelus*, shows little variation across northern Europe, suggesting expansion primarily from a Balkan refuge. The hedgehog is classified into two species, *Erinaceus europeus* and *Erinaceus concolor*, with subdivisions within these. The genealogy suggests expansion from three main refuges—in Iberia, Italy, and Greece. The bear, *Ursus arctos*, is thought to have expanded from two main refuges, in Iberia and in the Caucasus/Carpathian area; distinct sequences characteristic of these two expansions abut in Sweden. These histories are based on single loci (a noncoding nuclear sequence for *Chorthippus* and mitochondrial DNA for the others). However, they are supported in *Chorthippus* by the existence of sharp **hybrid zones** in the Pyrenees and Alps and in the hedgehog by the division into two species.

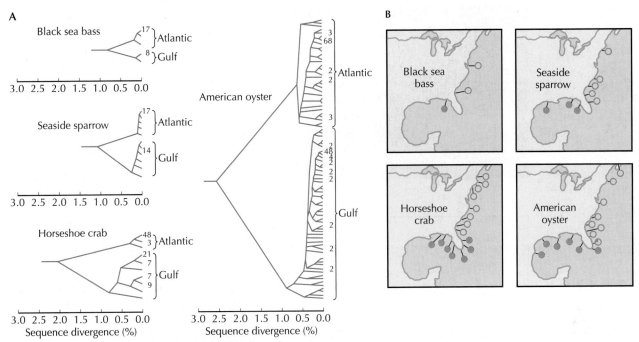

FIGURE 16.17. (*A*) Relationships among mitochondrial DNA sequences from four species that live along the coast of the southeastern United States. Each divides into two distinct clades. Numbers of individuals of each haplotype are indicated by numbers; unnumbered lineages are found in single individuals. (*B*) These clades are largely confined to the Atlantic vs. Gulf coasts, with a sharp transition between them. The pie diagrams show the proportion of haplotypes typical of the Gulf coast.

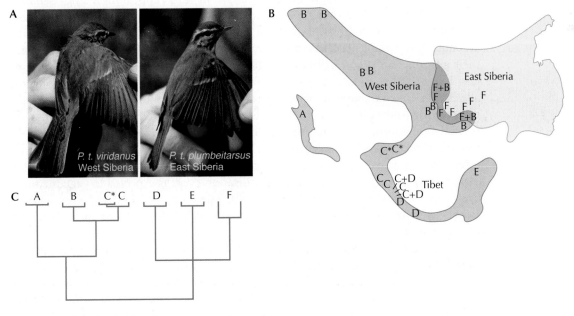

FIGURE 16.18. The greenish warbler, *Phylloscopus trochiloides*, has a broad range across Siberia and the Himalayas (*A,B*). The mitochondrial genealogy is divided into two clades (*C*), which are found in different areas. These clades meet in two places. In Siberia, type B meets type F in a zone that coincides with differences in plumage, song, and migratory behavior; these are named *P.t. viridanus* and *P.t. plumbeitarsus* (*A*). In Kashmir, types C and D abut; however, in this case there are no indications of divergence in other traits.

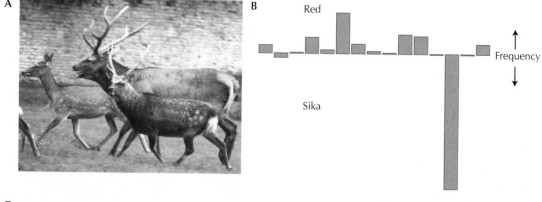

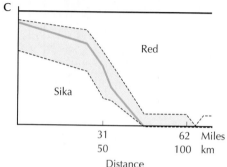

FIGURE 16.19. (A) Red and sika deer look quite different and yet can hybridize. The large stag in the *middle* is a red deer (*Cervus elaphus*), whereas the smaller stag at *front* is a sika deer (*Cervus nippon*). At rear, a juvenile F₁ hybrid. (B) Distributions of allele frequencies at 1 of 11 microsatellite loci in deer from Kintyre, Scotland. Alleles are classified along the *x*-axis according to the number of repeats they contain (see pp. 361–362). (C) The proportion of phenotypically sika and red deer along the Kintyre peninsula. The *shaded area* shows deer containing one or more apparently introgressed alleles (phenotypically sika deer below the *thick line*, phenotypically red above).

Another example of introgression is the hybridization between sika and red deer in Scotland. Sika deer escaped from deer parks about a century ago and have increased in numbers since. Although quite different in size, appearance, and behavior (Fig. 16.19A), sika deer sometimes mate with the native red deer. The two species carry different distributions of alleles at 11 microsatellite loci; sika deer in Scotland are much less variable, because of the severe population bottleneck that accompanied their introduction (Fig. 16.19B). In Kintyre, where the two species overlap, almost all deer appear to be either red or sika. However, up to 40% of deer carry one or more alleles that are characteristic of the opposite species (Fig. 16.19C). These alleles derive from hybridization over the past century or may have been present in the ancestral populations. However, deer that carry several "foreign" alleles are likely to be recent hybrids. A backcross between an F₁ and a red deer will be heterozygous for foreign alleles at half its loci, a second generation backcross will be heterozygous at a quarter of its loci, and so on; each hybridization event is expected to generate many backcross hybrids in subsequent generations. Three out of 246 deer were identified as backcross hybrids in this way, implying a rate of hybridization of about 1 in 1000 per generation.

We can view the population in another way, by saying that there is linkage disequilibrium between alleles derived from sika. Because the marker loci are unlinked ($c = 1/2$), linkage disequilibria would halve in every generation and so must be caused by recent hybridization. Linkage disequilibrium can be used to measure migration rates even when differences are not as marked as in these examples. Although human populations differ rather little in allele frequency ($F_{ST} \sim 0.15$ worldwide), it is nevertheless possible to assign individuals to particular populations and hence to estimate rates of mixing if enough loci are assayed (see pp. 741–745).

We know very little about the history of most populations. Even in humans, with a written history and extensive archaeology, we have little idea of patterns of population movement more than a few thousand years back. Yet, what we know of the present implies that populations of our own and other species have had complex patterns of divergence, movement, and mixing. Thus, the effects of population structure give a plausible explanation for the excess linkage disequilibrium observed in humans (see p. 436). Our ignorance of population structure is a serious obstacle to using linkage disequilibria to locate the genes responsible for complex traits (Chapters 14.13 and 26.3).

SUMMARY

Populations are usually structured. They may be divided into discrete demes or spread out over a more or less continuous region. Gene flow tends to homogenize populations. For discrete demes, the rate of gene flow is measured by m, the fraction of immigrant genes; demes become more similar over a timescale of $\sim 1/m$ generations. In a continuous population, gene flow can be approximated by diffusion at a rate σ^2, the variance of distance between parents and offspring. Diffusion is slow: In t generations, genes typically move a distance $\sqrt{\sigma^2 t}$.

Populations tend toward a statistical equilibrium in which random drift increases divergence and gene flow reduces it. The extent of divergence is measured by F_{ST}; at equilibrium, it depends on the effective number of migrants, $N_e m$, with discrete demes, or the neighborhood size, Nb $=$ $4\pi\rho\sigma^2$ in two dimensions. These parameters can be estimated from F_{ST}, provided that genetic divergence is due to a balance between drift and gene flow. The whole population drifts slowly, at a rate proportional to the total number of individuals. This overall rate is reduced by population subdivision (specifically, by a factor $1 - F_{ST}$), but may be greatly increased by fluctuations in local population size.

Genealogies in a structured population are described by the structured coalescent, in which ancestral lineages wander randomly from place to place and can coalesce only when they meet in the same place. If populations have a history of splitting and reunion, this history will be reflected in sampled genealogies. However, genealogical structure is highly random, which makes phylogeographic inference difficult.

FURTHER READING

Avise J.C. 2004. *Molecular markers, natural history and evolution.* Sinauer Press, Sunderland, Massachussetts.
A comprehensive overview of phylogeography.
Charlesworth B., Charlesworth D., and Barton N.H. 2003. The effects of genetic and geographic structure on neutral variation. *Annu. Rev. Ecol. Syst.* **34:** 99–125.
Reviews the behavior of genealogies in structured populations.
Hey J. and Machado C.A. 2003. The study of structured populations—new hope for a difficult and divided science. *Nat. Rev. Genet.* **4:** 535–543.

A critical review of how population history can (and cannot) be inferred from genealogies.

WWW Resources

http://evolution.gs.washington.edu/index.html Felsenstein J. *Lecture notes in population genetics.*
An excellent introduction to the way random drift interacts with gene flow.

CHAPTER

17

Selection on Variation

THE MOST STRIKING FEATURE OF THE LIVING WORLD is its variety of intricate adaptations. Enzymes catalyze specific chemical reactions under mild conditions; in contrast, chemists use extreme temperatures and pressures to achieve much less precise results. Moreover, the many chemical reactions in a cell are delicately arranged in a well-organized traffic of molecules between cellular compartments. Organisms have evolved to exploit an extraordinary range of habitats—from swifts that spend most of their lives aloft, to tube worms that live on microbes around deep-sea vents. Behaviors have evolved that mediate elaborate interactions between organisms. Single cells can aggregate to form the fruiting bodies of slime molds, and colonies of social insects (ants, termites, and bees) use a complex division of labor to forage efficiently, to build their shelter, and to fight other colonies. Human children have evolved the ability to associate words and phrases with subtle meanings simply by listening to those around them. A central task of evolutionary biology is to understand how these kinds of complex adaptations have come about.

In this chapter, we explain how natural selection leads to adaptation. We begin by showing that natural selection is the inevitable consequence of inherited variation in fitness—an idea that is made precise by Fisher's **Fundamental Theorem**. We discuss the qualitative features of natural selection and, in particular, the way that complex adaptations can be built up as a series of small steps. We show how the changes in allele frequency and in quantitative traits caused by selection can be predicted and, finally, we show how interactions and associations among multiple genes can be understood. This chapter focuses on the most important evolutionary process—selection; later chapters will explore its wider consequences.

THE NATURE OF SELECTION

Natural Selection Depends on Inherited Variation in Fitness

Natural selection is the only process that leads to adaptation—a crucial point that we explain in the next section. It is therefore the most important evolutionary process, and the one to which we pay most attention. It is simple and straightforward, and yet, despite its simplicity, it is often misunderstood. Surprisingly for such a fundamental principle, natural selection was discovered only relatively recently, by Darwin in 1838.

Moreover, it was not widely recognized as the key evolutionary process until a century or so later, as we saw in Chapter 1.

Darwin first explained natural selection in the *Origin of Species*, using these words:

> How will the struggle for existence ... act in regard to variation? ... Let it be borne in mind in what an endless number of strange peculiarities our domestic productions, and, in a lesser degree, those under nature, vary; and how strong the hereditary tendency is. ... Let it be borne in mind how infinitely complex and close-fitting are the mutual relations of all organic beings to each other and to their physical conditions of life. Can it, then, be thought improbable, seeing that other variations useful ... in some way to each being in the great and complex battle of life, should sometimes occur, in the course of thousands of generations? If such do occur, can we doubt (remembering that many more individuals are born than can possibly survive) that individuals having any advantage, however slight, over others, would have the best chance of surviving and of procreating their kind? ... This preservation of favourable variations and the rejection of unfavourable variations, I call Natural Selection.

That is, if some individuals tend to leave more offspring than others because they carry certain inherited traits, then those traits will tend to increase. Natural selection is the inevitable consequence of inherited variation in reproductive success. The question is not whether it operates, but, rather, how effectively it can accumulate adaptations when combined with other evolutionary processes such as mutation, random drift, recombination, and migration (Fig. 17.1).

Selection Acts on Replicating Molecules

We begin with a simple example of how natural selection operates—not among organisms in nature, but among RNA molecules selected in the laboratory. The experiment is based on the self-splicing **intron** from the ciliate *Tetrahymena* (Fig. 17.2A), which catalyzes its own cleavage during maturation of the final functional **ribosomal RNA** molecule. (Such **ribozymes** are relics of an early **RNA world**, in which RNA acted as both enzyme and hereditary material [see p. 101].) Normally, the ribozyme requires magnesium ions (Mg^{2+}). This experiment set out to select a molecule that would instead function in the presence of calcium ions (Ca^{2+}). It exploited the fact that the ribozyme emerges from the reaction joined to part of its substrate, which makes it possible to select for those sequences that have successfully carried out the reaction. Variation was deliberately introduced into the initial population, and further variation arose later by mutation. Although activity in the presence of calcium ions was extremely low at first, 12 generations of selection led to a new ribozyme that was almost as active as the original, but that had adapted to a new chemical environment (Fig. 17.2B). This change was due to seven substitutions of one base for another (Fig. 17.2C).

This kind of in vitro selection experiment differs from more familiar examples of artificial selection in that it involves an extremely large population of approximately 10^{13} RNA molecules. Also, the trait that is selected is a direct consequence of the nu-

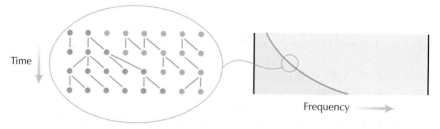

FIGURE 17.1. If some individuals tend to leave more offspring than others because they carry certain inherited traits, then those traits will tend to increase. Individual reproductive success varies randomly (*left*), but when averaged over a large population, slight differences in fitness cause steady evolutionary change (*right*).

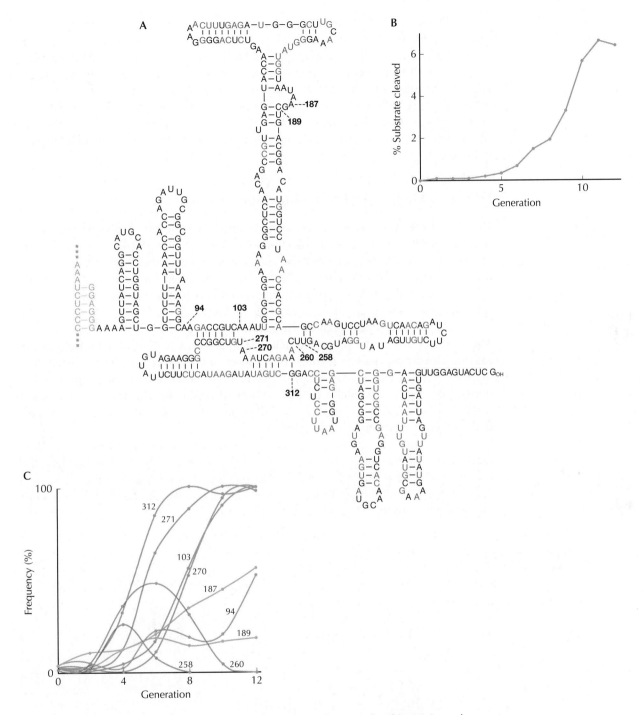

FIGURE 17.2. Selection on catalytic activity of a population of RNA molecules. (*A*) Secondary structure of the *Tetrahymena* ribozyme. Sites in *red* were never seen to vary during the course of the experiment. The RNA substrate is shown in *purple* (*left*). (*B*) The increase in activity over 12 generations. (*C*) The change in frequency of the nine commonly seen variants; numbers give the position in the RNA sequence, as shown in *A*. Two of these variants (258, 260) increased at first, but were later displaced by other variants that resulted in greater activity.

cleic acid sequence, rather than being the complicated outcome of metabolism and development. Nevertheless, the principles are the same: Those variants that reproduce most successfully (in this case, because they catalyze the required reaction) increase in the population. Within a few generations, a particular combination of variants with high catalytic activity comes to dominate.

In Chapter 15, we introduced the idea of **fitness:** the number of offspring left after one generation or after some chosen time interval. We have seen that random noninherited variation in fitness between individuals leads to random fluctuations in allele frequency, that is, to **random genetic drift.** Natural selection arises from the systematic accumulation of inherited variations in fitness. If some characteristic (e.g., replacement of A by G at position 270 in the in vitro selection experiment) leads to increased fitness, then that characteristic will increase. Because random variations in fitness tend to cancel out, the systematic effects of selection will dominate in large populations (Fig. 17.1).

Fitness Is Made Up of Separate Components

Fitness is made up of various **components,** which depend on the replicating organism that we consider. In the in vitro selection experiment of Figure 17.2, the **replicator** is the individual RNA molecule. Its fitness is the chance that a molecule will catalyze the desired reaction, hence picking up the tag that is required for its replication, multiplied by the number of copies of each tagged molecule that is then generated. In modeling the spread of infectious diseases, epidemiologists define fitness as R, the number of new infected cases generated by each initial infection: If $R > 1$, the disease spreads. The unit here is taken to be the infected host rather than the individual infectious organism. The fitness measure R equals the number of infectious particles produced during the course of the illness times the chance that one of these is successfully transmitted (e.g., Fig. 17.3). Finally, to take the most familiar case, the fitness of humans is made up of the chance of surviving to adulthood, the chance of finding a mate, and the number of offspring of each couple.

When we work in discrete generations, different fitness components are multiplied together to give the overall fitness, which is usually denoted by W. For example, the average number of offspring produced after one generation is equal to the probability of surviving to adulthood multiplied by the average number of offspring that an adult can expect to produce. When we work in continuous time, fitness components are

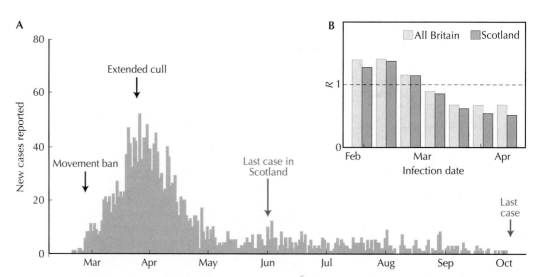

FIGURE 17.3. (*A*) Change in fitness of the foot-and-mouth disease virus during the 2001 epidemic in the United Kingdom. At least 6 million animals were culled in order to contain the spread of the virus. (*B*) The estimated number of new infected farms that are generated by each infected farm, *R*. Despite a ban on animal movements, *R* remained above 1 until an extended cull was introduced in late March. *R* then fell below 1, and the epidemic declined. It died out much sooner in Scotland, where faster culling reduced *R* more effectively (see *B*). The long tail of the epidemic is due to geographically localized pockets of infection.

added together to give an overall measure, often denoted by r. Thus, if we are considering a continuously growing bacterial culture, then fitness would be defined as the rate of cell division per unit time, minus the rate of cell death. The arithmetic of selection is explained in more detail in Chapter 28 (online).

In these various examples, the term **fitness component** can be used in a precise sense to refer to quantities that can be multiplied or added to give the overall fitness. Sometimes, however, the term is used more loosely to refer to quantities that are closely related to fitness, such as life span or age at first reproduction.

For discrete generations, we have defined fitness as the absolute number of progeny. This **absolute fitness** determines the rate of growth or decline of the population and is a crucial quantity for ecology (the study of the distribution and abundance of organisms). In the long run, absolute fitness must be close to 1 for an asexual population. With sexual reproduction, each offspring must have two parents, and so fitness must be 2 on average if the population size is to remain the same. Even a slight deviation from this equilibrium fitness would eventually lead to extinction or explosion of the population. Similarly, in continuous time the rate of change in population size is equal to the average fitness and must be close to 0. There must always be some **density dependence**, so that as the population becomes more crowded, its absolute fitness declines, thus maintaining more or less constant numbers over time.

In population genetics, however, we are concerned not with the growth of the whole population, but with competition between different types within the population. Thus, we often deal with **relative fitness**—the *ratio* between the absolute fitnesses in discrete time or the *difference* in growth rates in continuous time. For convenience, we might assign one genotype an arbitrary fitness of (say) 1 and measure the fitness of other genotypes relative to it ($1+s$, say). Alternatively, we might measure fitness relative to the **mean fitness** in the whole population.

Differences in relative fitness are referred to as **selection coefficients** and are usually denoted by s. This is a useful convention, because the proportions of different types in the population only depend on relative, and not on absolute, fitnesses. Also, the selection coefficients may often be very small; as we shall see, very slight differences in fitness eventually have large effects. It is more convenient to work with selection coefficients of 10^{-4} than with fitnesses of 1.0001, even though these are equivalent to each other. It is important to realize that assigning a *relative* fitness of, for example, 1 to a genotype does not make any assumption about *absolute* fitness: The population as a whole might be steady, shrinking, or growing. The separation of absolute from relative fitnesses corresponds to a separation of ecology from evolution.

We have so far supposed that the population is homogeneous, so that all its members are equivalent to each other. In reality, individuals might be of different sexes or ages or live in different places. In these cases, we say that the population is **structured** (Chapter 16). A full description of a structured population can be quite complicated. For example, we might need to keep track of the rates of survival and reproduction of individuals of different ages. However, we can still summarize the fitness of each genotype in a single number provided that the population is in a steady state. Then, the age distribution of the variant will also settle to a steady state. Its numbers will then grow or decline exponentially, and this steady growth rate is our measure of its fitness. This approach is especially powerful when the genotype we are interested in is so rare that its presence does not disturb the rest of the population, so that it settles into a steady state before it becomes common (Chapter 28 [online]).

We will see later that the way reproduction at different ages contributes to fitness is crucial for understanding the evolution of aging (see pp. 561–566). More generally, a common way of understanding the evolution of complicated systems is to look at the invasion of a new type, after it is introduced at low frequency (Box 20.4). Then, the fitness measure just described is what determines whether the invasion will succeed or fail.

FIGURE 17.4. R.A. Fisher believed that his Fundamental Theorem "held the supreme position among the biological sciences."

The Increase in Mean Fitness Equals the Additive Genetic Variance in Fitness: Fisher's Fundamental Theorem

We have emphasized that natural selection is based on inherited variation in fitness. R.A. Fisher made this idea precise, in what he called the Fundamental Theorem of Natural Selection (Fig. 17.4). We explain Fisher's theorem here for two reasons. First, it gives a deeper understanding of exactly how selection increases the fitness of a population and, second, it provides us with a key measure of the overall extent of selection.

Fisher's theorem states that $\Delta\overline{W} = \text{var}_A(W)/\overline{W}$. In words, the increase in mean fitness due to natural selection ($\Delta\overline{W}$) is equal to the **additive genetic variance in fitness**, $\text{var}_A(W)$, divided by the current mean ($\overline{W}$). Now, $\text{var}_A(W)$ is not the total variance in fitness, which includes random noninherited differences in reproductive success between individuals. Rather, it is the variance in fitness that can be attributed to the **average effect** of each allele (recall Chapter 14 [pp. 387–394]). Every allele has a fitness, which is simply the average fitness of all the individuals that carry that allele. The additive genetic variance in fitness is just the variance of these average effects summed over all the genetic loci in the population (Fig. 17.5). Because fitness is defined as the number of offspring copies produced in the next generation, the frequency of each allele must increase in direct proportion to its average fitness. These changes in allele frequency in turn cause an increase in mean fitness, which Fisher showed is directly proportional to the additive genetic variance in fitness.

The Fundamental Theorem is not an equation that makes predictions about how absolute fitness changes over time. In the long run, mean absolute fitness, $\overline{W}$, must stay around 1 (for an asexual population with discrete generations) and cannot keep increasing, as suggested by the Fundamental Theorem. Rather, this theorem identifies that *part* of the increase in mean fitness caused by changes in allele frequency that result from natural selection. This increase is offset by decreases in mean fitness caused by other processes. Usually, mean fitness decreases as the environment changes. A population adapted to current conditions will not reproduce as successfully in new conditions. Moreover, other species will themselves be evolving in a way that tends to reduce the fitness of their competitors. Similarly, mean fitness will tend to decrease if the additive effects of genes change. As we saw in Chapter 14 (p. 389), the effect of any one gene on fitness depends both on environmental conditions and on the frequencies of other genes in the population. As these change, additive effects change, and mean fitness will

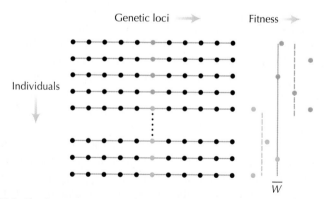

FIGURE 17.5. The increase in mean fitness caused by selection on allele frequencies equals the additive genetic variance in fitness. The diagram shows a population consisting of many genomes, each with its own individual fitness (shown at right). Each allele has its own average fitness; for example, the allele shown in *red* has fitness slightly higher than the population mean $\overline{W}$ (*red dashed line*). It is each allele's average fitness that determines changes in its frequency and that contributes to the additive genetic variance in fitness. (In Chapter 14, we defined these averages in two ways—**average excess** and average effect—which for our purposes are equivalent.)

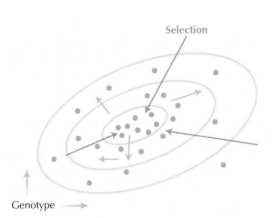

Selection

Genotype ——→

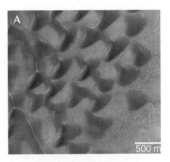

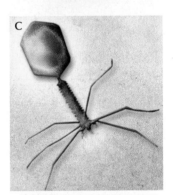

FIGURE 17.6. Natural selection increases mean fitness, whereas all other evolutionary processes tend to decrease it. The diagram represents a population of genotypes clustered around the most fit genotype at the center. Selection increases the frequency of fit genotypes and so pushes the population toward this central optimum (*red arrows*). Other processes (mutation, random genetic drift, recombination, etc.; *green arrows*) act in arbitrary directions and so overall tend to move the population away from the fittest genotype. (The two dimensions here stand for the very many dimensions along which a population can evolve; *contours* show fitness increasing toward the center.)

tend to decrease. Mutation, migration, and recombination alter the frequencies of genotypes in arbitrary ways and so will also tend to decrease the fitness of a well-adapted population. Fisher's Fundamental Theorem represents the continual struggle between natural selection, which increases mean fitness, and other processes that tend to reduce it (Fig. 17.6). In Chapter 19 (pp. 547–549), we review evidence on the overall strength of selection, as measured by the additive genetic variance of fitness: Does natural selection typically increase mean fitness by 0.1%, 1%, or 10% in each generation?

Natural Selection Is the Only Process That Causes Adaptation

Although many processes shape evolution, natural selection is special because it alone creates complex, functioning organisms. All other processes tend to degrade what has been built up by natural selection, simply because these processes act at random with respect to function. Mutation makes random changes in DNA sequence that, if they have any effect at all, tend to disrupt function. Migration introduces genes from elsewhere, which tend not to be adapted to their new environment. Similarly, recombination and random drift will, on average, disrupt genotype frequencies that have been built up by selection so as to increase fitness. They may, by chance, occasionally increase fitness, but overall tend to reduce it. Only natural selection has a systematic tendency to build up the elaborate functions that are needed to maintain reproductive success.

Many physical processes can build complex structures, often through simple underlying rules. A well known example is the way regular crystals form by the spontaneous aggregation of simple molecules. Here, the geometry of the crystal directly reflects the geometry of the molecules. More generally, simple processes can lead to complicated and well-ordered patterns, even when the processes themselves are everywhere the same. Figure 17.7 shows some of the patterns that can be produced by relatively simple processes.

This phenomenon, known as self-organization, is essential for life. The linear sequence of bases in the DNA, expressed in a more or less homogeneous egg, must somehow generate an elaborate organism built from many precise patterns in space and time. Compact genomes could only code for elaborate organisms with the aid of self-organization. For example, the multitude of behaviors mediated by our brains, including the complexities of language, are based on the expression of the genes that specify the nervous system. Although thousands of genes are involved, they encode only a small amount of information (perhaps a few megabytes). Complex behaviors are generated from limited information through the repetition of a limited number of

FIGURE 17.7. Complex structures can be produced by self-organization. (*A*) Regular patterns produced by the wind blowing across a sandy desert on Mars. (*B*) Spirals generated spontaneously from a mixture of inorganic chemicals. (*C*) The T4 bacteriophage is formed by the spontaneous aggregation of protein molecules. (*D*) Aggregation of the slime mold *Dictyostelium discoideum*.

cell types and structures and through the use of general rules for learning from the environment (see Chapters 11 and 24).

Mechanisms of self-organization have received much attention in recent years. It is sometimes claimed that the form of living organisms is largely shaped by **developmental constraints** rather than by natural selection. (These ideas trace back to the 18th and 19th centuries when it was widely held that organisms are shaped by intrinsic laws, which can be understood from the similarities in morphology between different species [see p. 21].) Clearly, physical laws, along with the genetics and biochemistry shared by all organisms, do limit what kinds of organisms can evolve; we discuss the way adaptation interacts with constraint in Chapter 20. For the moment, the key point is that natural selection exploits self-organization to produce well-adapted organisms. Thus, although the same basic physical processes are involved in formation of inorganic crystals as in the formation of the head capsule of a virus, only the latter is directed toward a goal—the reproduction of the virus. It is crucial to realize that complex structures will function to aid reproduction only to the extent that they are shaped by natural selection.

We have seen that complicated structures can be produced without the aid of natural selection. Conversely, natural selection does not necessarily tend to produce more and more complicated organisms. This can be seen most clearly where functions that are no longer necessary are lost, as in eyeless cave fishes (Fig. 17.8), or in our own inability to synthesize vitamins and other nutrients that are abundant in our diet. The most extreme examples are seen in parasites. For example, viruses depend on their hosts for almost all functions and may retain only a few genes. Some viruses even have parasites of their own: **defective interfering viruses** that have deleted regions coding for RNA polymerase (for example) and thus reproduce faster, provided that they can exploit the polymerase coded by intact viruses infecting the same cell (see p. 590).

Of course, many processes may facilitate natural selection. Mutation generates random genetic variation that may be material for adaptation by natural selection. Recombination shuffles that variation in ways that can accelerate adaptation by increasing the inherited variance in fitness. The movement of genes from place to place can allow the spread of favorable alleles. In Chapter 23, we will see that genetic systems can evolve under natural selection in such a way that adaptation occurs more effectively. That is, organisms have evolved so as to be **evolvable**. Nevertheless, it is crucial to realize that any such adaptations for better **evolvability** are driven by natural selection in essentially the same way as are simpler adaptations, which function to increase fitness in more obvious ways.

FIGURE 17.8. The fish, *Astyanax mexicanus*, has evolved eyeless forms in caves. There have been multiple origins of eyelessness—an example of convergent evolution that has eliminated features no longer favored by selection.

Natural Selection Works Step by Step by Assembling Multiple Favorable Variations

A recurrent criticism of natural selection is that it creates nothing new, but rather just sifts among a predetermined range of possibilities. It is true that, in principle, all genetic combinations are present in a sufficiently large population. For example, in the F_2 of a cross between two species, every possible combination of the genes that distinguish them will be found if there are enough individuals. Selection is sometimes seen as simply picking out the fittest genotype from these, without itself generating anything novel. This argument was raised in the early years of genetics and has often reappeared since (Fig. 1.31). However, it is quite misleading once one considers more than a handful of genes, simply because there are so many possible gene combinations. For example, suppose that two diploid populations differ at 20 genetic loci and that, at each locus, one allele slightly increases fitness. The fittest genotype will carry a particular combination of 40 alleles (two copies at 20 loci) and so will be present at a frequency of 2^{-40}, or less than 1 in 10^{12} in the F_2. In general, selection acts not by picking out one vanishingly rare combination of genes, but rather by gradually raising the frequency of several favorable alleles. Eventually, the fittest genotype becomes common and ultimately fixes in the population (Fig. 17.9).

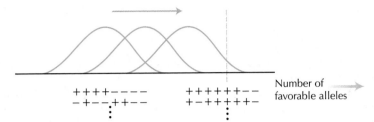

FIGURE 17.9. The chance of finding a genotype that brings together very many fit alleles (++++ ⋯) is vanishingly small (*left curve*). Natural selection works by increasing the frequency of alleles that are individually favorable. As each becomes more common, the chance of finding fit combinations increases (*rightmost curves*).

We have considered here a sexual population in which variation arises through recombination between genomes from different populations. However, the same argument applies even with no sex or recombination, when variation arises by mutation, as in the in vitro selection experiment discussed above. In this example, the initial pool of 10^{13} RNA molecules was so large and variable that an extremely rare combination of four variants was in fact picked out from the initial pool, so that these variants increased more or less at the same time (green curves in Fig. 17.2C). However, the other three variants that completed the adaptation probably arose later as separate mutations (blue curves in Fig. 17.2C). With smaller populations, or when more variants must be combined, the chance that all these mutations will arise in the right combination becomes vanishingly small even in the largest of populations. Instead, the fittest genotype arises because as one mutant lineage increases in frequency, other mutations can arise in combination with it, so that a progressively fitter combination is built up.

Jorge Luis Borges' story *The Library of Babel* well illustrates the absurdity of viewing selection as simply picking the fittest genotype out of all possible genotypes (Fig. 17.10). He tells of a vast library whose volumes contain all possible sequences of letters, and hence all possible books. This library contains every truth but also every falsehood, both immersed in an ocean of nonsense. Picking out some particular book from the incomprehensibly large space of possibilities is essentially impossible. Figure 17.11 illustrates this point using a simple word game.

Of course, we do not imagine that a book is written by an author who merely chooses a sequence of words from all possible sequences. Rather, he or she is seen as creating something new. This creation is achieved by building on previous cultural traditions, using ideas and styles generated by others before. Natural selection works in the same way by assembling separate variations into a sequence of bases in the DNA that allows the organism to reproduce efficiently. The power of natural selection is its ability to accumulate a series of changes eventually leading to a highly improbable outcome. This argument of course requires that there is a series of simple changes, each of which increases fitness, so that selection is able to act cumulatively. In Figure 14.28, we saw Fisher's argument that adaptation is most likely to proceed through many small steps. We return to this issue in Chapter 24.

FIGURE 17.10. The Tower of Babel.

Selection Is Often More Effective Than Rational Design

It is striking that even for humans selection is often the most effective way to build complex structures. For example, although we know the general principles that govern the folding of proteins and RNA and determine their catalytic activity, we cannot design efficient catalysts from first principles. (Naturally occurring enzymes can be altered to fulfill slightly different functions, but that is very different from the design of an enzyme de novo with no help from natural analogs.) Thus, there is considerable interest in using in vitro selection such as that described in Figure 17.2 to generate molecules for specific purposes. (Recall the example of novel fluorescent proteins [p. 5].)

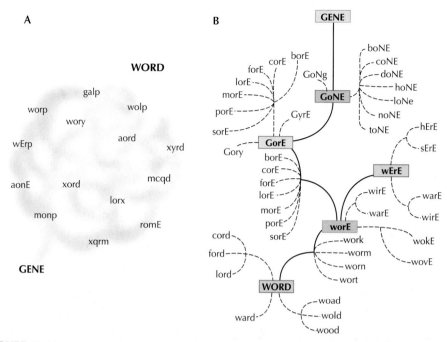

FIGURE 17.11. (*A*) It would take an extremely long time to change WORD to GENE by making random changes to each letter: The desired set of four letters must be picked out from nearly half a million possibilities ($26^4 = 456,976$). (*B*) In contrast, the new word can quickly be reached by accumulating a series of simpler changes. If only valid English words can survive, and if at each stage the words that share most letters with GENE are selected, then the target is reached after only four steps.

Selection is also used to solve difficult optimization problems—for example, deciding the arrangement of components on a silicon chip. Random changes are made to a trial solution, and the chance that the variant survives is determined by an appropriate measure of fitness. If the only changes that are accepted are those that improve fitness, then the solution will reach a local optimum, but may never find the globally optimal solution. (Imagine climbing uphill on a rough fitness landscape, as in Fig. 17.12.) However,

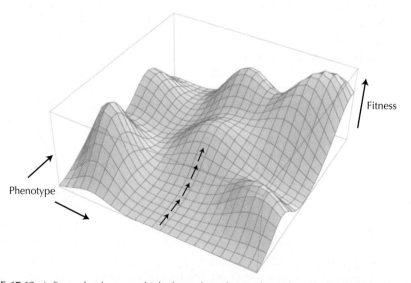

FIGURE 17.12. A fitness landscape, which shows how fitness depends on individual phenotype (horizontal axes). Unlike Fig. 17.6, this landscape has multiple peaks. Thus, a series of stepwise improvements (*arrows*) may lead to a local optimum (*center*) rather than to the highest peak (*far right*). Here, the phenotype is represented by just two dimensions. In reality, there might be very many dimensions, and fitness might be plotted against a discrete genotype rather than a continuous phenotype.

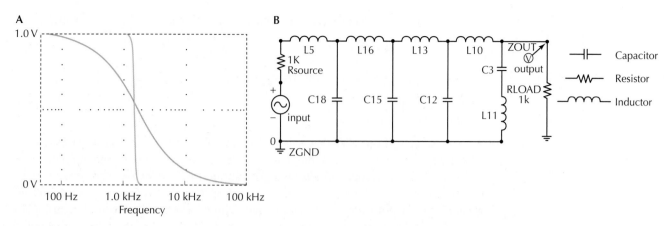

FIGURE 17.13. Low-pass filters designed by selection among alternative designs. Selection was for full transmission of frequencies below 1 kHz and zero transmission above 2 kHz. (A) The frequency response of the best of a random initial population (*red*) compared with the sharp frequency response achieved after 49 generations of selection (*blue*). (B) The filter that gave the sharp frequency response had been patented in 1917.

by sometimes choosing inferior changes, the algorithm can explore a wider range of possibilities. This method, known as simulated annealing, can be a simple way to find the best out of a large number of local optima. We discuss this kind of interaction between selection, mutation, and random drift in a biological context in Chapter 18 (pp. 494–496).

Engineers and computer scientists use a similar approach to solve difficult design problems. Alternative designs are competed against each other, and variation is generated by mutation and recombination. Ingenious and unexpected solutions often emerge that would be difficult or impossible to produce by rational thought. Figure 17.13 shows a simple example. Low-pass filters are constructed from capacitors and inductors; their function is to block transmission of frequencies above a sharp threshold. It is not at all easy to design such circuits. Yet, fewer than 50 generations of selection on random variation produced excellent solutions, including several patented designs. Unlike simulated annealing, this example used recombination as well as mutation; such approaches are known as **genetic algorithms** or **genetic programming**. In Chapter 23, we examine how recombination can improve the response to selection.

Even Slight Differences in Fitness Eventually Become Significant

Let us now look at the very simplest form of selection. Two alternative alleles (call them P and Q) compete in the population. We count the frequencies of the two alleles in the pool of gametes and call the allele frequencies in generation 0 p_0 and q_0, respectively. In a typical sexual eukaryote, the haploid gametes meet at random to form diploid zygotes. These grow to adulthood and reproduce, giving the new generation of haploid gametes. The fitness of an individual gene is the number of offspring that it produces after one generation, counting from some definite point in the life cycle (say, from haploid gamete to haploid gamete). The fitness of each allele (W_P, W_Q) is the average fitness of genes carrying that allele. As we have just seen, this might be made up of many components (e.g., survival of the haploid stage, chance of union with another gamete, survival of the diploid stage, and so on). We might also imagine other life cycles: Diploid individuals might mate at random instead of casting out gametes that combine randomly or there might be no diploid stage at all, with only asexual reproduction of haploids. Regardless of these complications, however, changes in the composition of the population depend only on W_P and W_Q (Fig. 17.14).

In every generation, the numbers of each allele change in proportion to their fitness. If we call the numbers of type P and Q at time t $N_{P,t}$ and $N_{Q,t}$ then $N_{P,1} = W_P N_{P,0}$, $N_{Q,1} = W_Q N_{Q,0}$, and so on for successive generations. So, the ratios of numbers, and

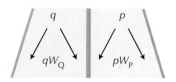

FIGURE 17.14. The ratios of allele frequencies $q{:}p$ change in proportion to the fitnesses of the alleles, W_Q, W_P.

hence the ratios of allele frequencies, change in proportion to the ratio of fitnesses:

$$\frac{N_{P,1}}{N_{Q,1}} = \frac{p_1}{q_1} = \frac{W_P}{W_Q}\frac{p_0}{q_0}. \tag{17.1}$$

Over many generations, the ratio changes by

$$\frac{p_t}{q_t} = \left(\frac{W_P}{W_Q}\right)^t \frac{p_0}{q_0}. \tag{17.2}$$

For example, if one allele has twice the fitness of the other (e.g., $W_P = 2W_Q$), and the allele frequencies are initially in the ratio $p{:}q = 1{:}1000$, then after one generation, the ratio will have risen to 2:1000, and after ten generations, to $2^{10}{:}1000$, or 1024:1000. In terms of frequencies, p will have risen as $\frac{1}{1001}$, $\frac{2}{1002}$, ..., $\frac{1024}{2024}$, or 0.001, 0.002, ..., 0.506 over ten generations. The time course of this simplest form of selection is shown in Figure 17.15.

This calculation shows several important features that also apply to more complicated models.

1. *Relative fitness is what matters.* If the *absolute* fitness of each allele (i.e., the actual number of offspring) were constant, then the rate of population growth would increase over time, as the fitter allele replaced the less fit (Fig. 17.15B). This suggests that, in general, the population size would increase at a faster and faster rate as selection acted to increase mean fitness. However, fitness must decrease with population size if the population is to stay within bounds. The population will increase to a higher equilibrium size when a fitter allele replaces a less fit allele, but will not increase indefinitely (Fig. 17.15D,E). Provided that crowding affects both alleles in the same way, the same formula for allele frequency will apply. It depends only on the relative, not the absolute, fitnesses (Fig. 17.15C).

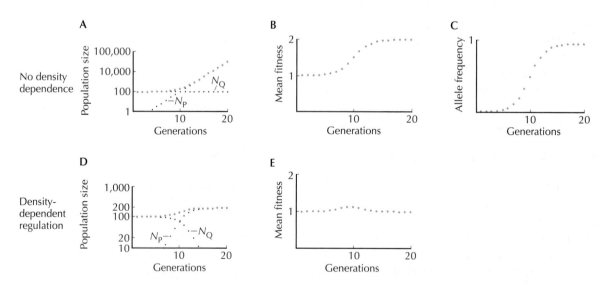

FIGURE 17.15. The time course of selection in the simplest case of two alleles (P, Q), with and without density-dependent regulation of population size. The *top row* is for the case where fitnesses are constant, regardless of population size. (A) If $W_Q = 1$, numbers of type Q stay constant at $N_Q = 100$. If type P is twice as fit ($W_P = 2$), then its numbers increase geometrically, as 2^t. On a logarithmic scale, this appears as a straight line. The total population size ($N_P + N_Q$; *red dots*) remains close to 100 until around ten generations, when P becomes common and the population as a whole increases geometrically in size. (B) The rate of population growth (i.e., the slope of the *red dots* in A) is equal to the mean fitness. This increases from 1 to 2 as P replaces Q. (C) The allele frequency, $p = N_P/(N_Q + N_P)$. (D) If the fitness of both types decreases geometrically with population size, then the population stays within bounds ($W_Q = 2^{1 - N/100}$, $W_P = 2W_Q$, say), and equilibrates when $\overline{W} = 1$. Type Q reaches an equilibrium at $N_Q = 100$ (*left*); type P is twice as fit and in this example reaches an equilibrium at $N_P = 200$ (*right*). (E) Now, the mean fitness increases above $\overline{W} = 1$ as P replaces Q, but then returns toward equilibrium as population size increases. However, because the *ratio* in fitness between the two types is constant ($W_P/W_Q = 2$), the time course of allele frequency change is exactly the same (C).

2. *The timescale of change is inversely proportional to the selection coefficient.* Because only relative fitness matters, we can set W_Q to 1 and W_P to $1+s$. Now, we see that, given enough time, even slight differences in fitness will cause one allele to replace another. The ratio of allele frequencies changes with $(W_P/W_Q)^t = (1+s)^t$, which is close to $\exp(st)$ for small s (Fig. 17.16). For example, a difference in relative fitness of 1 in 1000 (i.e., $s = 0.001$) will change the ratio of allele frequencies by $e \sim 2.718$ in 1000 generations, and by $e^{10} \sim 22,000$ in 10,000 generations. In general, the timescale over which selection acts is roughly $1/s$. In Chapter 19, we review evidence on just how strong are the selection coefficients actually found in nature.

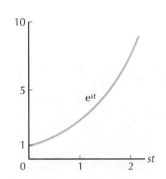

FIGURE 17.16. An allele with selective advantage s increases as $\exp(st)$ or e^{st} while it is rare.

3. *The frequency of favorable alleles increases in a characteristic sigmoid curve.* An allele that is initially very rare (e.g., a single copy of a new mutation) will increase exponentially ($p \sim \exp(st)$). That is, it will take $2.3/s$ generations to go from $p = 10^{-6}$ to 10^{-5}, from 10^{-5} to 10^{-4}, from 10^{-4} to 10^{-3}, and so on. There will be a long lag before the allele actually appears at a detectable frequency in the population. (This lag is around $(1/s) \log(1/p_0)$ generations, where p_0 is the initial frequency.) The new allele then sweeps to high frequency relatively quickly, in $\sim 1/s$ generations. There is then another long lag before the original allele is finally eliminated (Fig. 17.15C).

In this simple example, we supposed that relative fitnesses stayed the same over time. In general, however, relative fitnesses change for a variety of reasons. Nevertheless, the change in any one generation, t, still depends on the fitness of one allele relative to the other in that generation ($W_{P,t}/W_{Q,t}$), and, in the long run, whether an allele increases or decreases depends on whether the product of the relative fitnesses is greater than 1 or less than 1. It is therefore helpful to think of the **geometric mean** fitness, which is defined as

$$\left(\frac{W_{P,1}}{W_{Q,1}} \frac{W_{P,2}}{W_{Q,2}} \ldots \frac{W_{P,t}}{W_{Q,t}} \right)^{1/t}.$$

In the long run, the allele with the highest geometric mean fitness will win (Fig. 17.17).

A

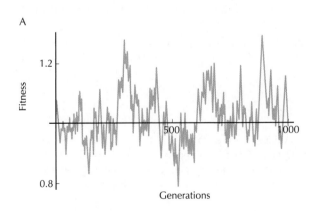

B

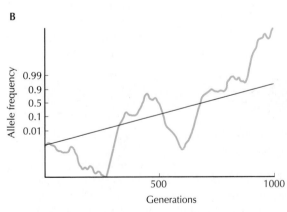

C

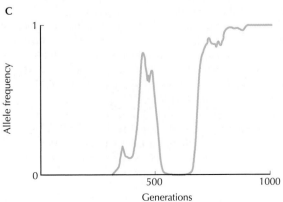

FIGURE 17.17. In the long run, the rate of increase of an allele depends on its geometric mean fitness. (*A*) In this example, the fitness of allele P relative to Q fluctuates randomly, with a geometric mean 1.01. (*B*) The ratio of allele frequencies, on a logarithmic scale. An allele with a constant advantage of $s = 0.01$ would increase steadily, with a frequency increasing as a straight line on this scale. The actual allele frequency fluctuates greatly, but in the long term increases at a rate given by $s = 0.01$. (*C*) Allele frequency on the original scale.

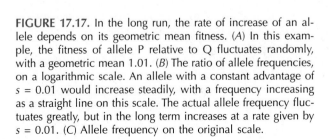

The Fitness of a Gene Depends on Interactions with the Physical and Biotic Environment and with Other Genes

FIGURE 17.18. The peppered moth, *Biston betularia*, is polymorphic for pale (*top*) and melanic (*bottom*) forms. The melanic form is more common in polluted areas, where tree trunks are darker and the paler form is more conspicuous (see p. 31).

Fitnesses may vary for many reasons—most obviously as external environmental conditions vary. We have already seen several examples and will see many more in later chapters. In humans, homozygotes for the common Hb^A allele of β-globin suffer considerable childhood mortality from malaria, but heterozygotes for the sickle-cell allele (Hb^{S}Hb^A) survive the disease much better. Thus, both absolute and relative fitnesses of these genotypes depend strongly on the incidence of malaria (Box 14.1). Similarly, rates of predation of the melanic and nonmelanic forms of the peppered moth, *Biston betularia*, vary with the degree of pollution, which alters the effectiveness of their camouflage (see p. 31; Fig. 17.18).

Most important, conditions may change as a direct or indirect result of changes in the population; there is feedback between evolution and environment. We have already seen that absolute fitnesses must decline with population size if the population is to stay within bounds. If, as we have assumed so far, this decline is the same for all genotypes, then it has no effect on relative fitnesses and hence no effect on evolution. However, different genotypes might have different sensitivities to changes in crowding, in which case we say that selection is **density dependent**. More generally, fitnesses may depend on the frequencies of other genotypes in the population, for example, if different genotypes exploit different resources. We will see in Chapter 18 (pp. 505–508) how this kind of **frequency-dependent selection** can maintain genetic variation in the population.

In a diploid organism, there is a more direct influence of the composition of a population on the fitnesses of the genes within it: The effect of a gene depends on its partner. For example, an Hb^S allele gives resistance to malaria when it is paired with the Hb^A allele, but causes severe anemia when it is partnered with another Hb^S allele in an Hb^{S}Hb^S homozygote. With random mating, the fitness of an allele is an average over all the fitnesses when combined with other possible alleles, weighted by the frequency of those alleles in the population. Thus, a recessive allele that only increases fitness in the homozygote will increase slowly when rare, because it will only rarely be found in the fitter homozygous form. Conversely, a dominant favorable allele will increase quickly at first, but will then be slow to completely displace the original recessive allele (Box 17.1).

Box 17.1   Selection on Diploids

In a diploid population that contains two alleles (Q, P), there are three possible genotypes (QQ, PQ, PP). However, we can still understand the evolution of the population by focusing on the frequencies of the two alleles (q, p), rather than keeping track of the frequencies of all three genotypes. This is because, provided that there is random mating, the proportions of the three diploid genotypes among newly formed zygotes are given by the **Hardy–Weinberg** proportions (q^2:$2pq$:p^2), which depend only on the allele frequencies (Box 1.1). Similarly, we can understand the effects of selection by focusing on the average fitness of the two kinds of genes, W_P and W_Q. A P allele has a chance p of finding itself paired up with another P in a PP homozygote, and a chance q of being paired with a Q allele in a heterozygote. Thus, its average fitness is $W_P = pW_{PP} + qW_{PQ}$. Similarly, $W_Q = pW_{PQ} + qW_{QQ}$. These fitnesses can be inserted directly into Equations 17.1 and 17.2 to give the change of allele frequency over time.

The detailed formulas for the effects of selection are summarized in Table 17.1 (see Chapter 28 [online]). As explained in the text, the change in allele frequency depends on the fitness of each gene (W_P, W_Q), which is an average over the different diploid genotypes in which the allele might find itself. The net selection coefficient is proportional to the difference between these genic fitnesses, $W_P − W_Q$.

When selection is weak (e.g., $s < 20\%$), the change of allele frequency is approximately continuous in time and only depends on the product st. Thus, the change produced by 100 generations of selection at $s = 1\%$ is almost exactly the same as the change produced by 1000 generations of selection at $s = 0.1\%$, and so on. This approximation for the rate of change in allele frequency is shown in the last column of the table.

Figure 17.19A compares three forms of directional selection in which P is favored over Q. With additive selection, each extra copy of P increases fitness by s. Allele frequency increases rapidly and changes from the initial $p =$

Box 17.1. Continued.

TABLE 17.1 Change in allele frequency for some simple models of selection

	Fitnesses of Diploid Genotypes			Genic Fitnesses		Selection Coefficient	Rate of Change
	W_{QQ}	W_{PQ}	W_{PP}	W_Q	W_P	$W_P - W_Q$	dp/dt
Haploid				1	$1 + s$	s	spq
Additive	$1 - s$	1	$1 + s$	$1 - sq$	$1 + sp$	s	spq
Dominant P	1	$1 + s$	$1 + s$	$1 + sp$	$1 + s$	sq	spq^2
Recessive P	1	1	$1 + s$	1	$1 + sp$	sp	sp^2q
Overdominant	$1 - s_1$	1	$1 - s_2$	$1 - \bar{s}\,p_e q$	$1 - \bar{s}\,q_e p$	$\bar{s}\,(p_e - p)$	$\bar{s}\,pq(p_e - p)$
Underdominant	$1 + s_1$	1	$1 + s_2$	$1 + \bar{s}\,p_e q$	$1 + \bar{s}\,q_e p$	$\bar{s}\,(p - p_e)$	$\bar{s}\,pq(p - p_e)$

Here $p_e = s_1/(s_1 + s_2)$ and $\bar{s} = s_1 + s_2$.

1% to $p = 99\%$ in a time $st = 9.2$ (e.g., in 9200 generations if $s = 0.001$). If the P allele is dominant, it initially increases in the same way, because the difference in fitness between a PQ heterozygote and the original QQ homozygote is still equal to s. However, as it becomes common, the population consists mostly of PQ and PP genotypes, which have the same fitness. The Q allele is eliminated slowly, because it reduces fitness only when expressed in the rare QQ homozygote. If the P allele is recessive, the pattern is reversed. It increases very slowly at first, because its advantage is expressed only in the rare PP homozygote. In this example, it takes a time $st = 108.2$ to increase to 99%, an order of magnitude slower than with additive selection.

Figure 17.19B shows the typical pattern with heterozygote advantage, or **overdominance**. (In this example, the equilibrium is at $p_e = 0.7$, and the fitnesses are $1 - 0.7\bar{s}$:1:1 $- 0.3\bar{s}$.) If the P allele is at first very common, it will be found mostly in homozygotes and so will decrease (top curve). On the other hand, if it is rare, it will be found mostly in heterozygotes and will increase (bottom curve). An equilibrium will be reached after a time $t \sim 1/\bar{s}$ generations.

Figure 17.19C shows the opposite case of **underdominance**, in which heterozygotes are less fit. Now, whichever allele is more common tends to increase, and so there is an *unstable* equilibrium. In this example, fitnesses are $1 + 0.7\bar{s}$:1:1 $+ 0.3\bar{s}$, so that the unstable state is at $p_e = 0.7$. Populations with p very slightly above this threshold will eventually fix for P, whereas those below the threshold will lose this allele.

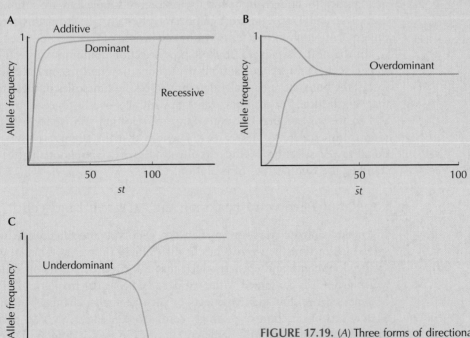

FIGURE 17.19. (*A*) Three forms of directional selection in which P is favored over Q. (*B*) Selection on diploids in which the heterozygotes have the advantage. (*C*) Selection on diploids in which the heterozygotes are less fit.

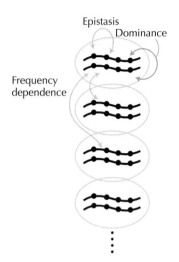

Epistasis
Dominance

Frequency
dependence

FIG. 17.20. Fitness depends on several kinds of interaction. Dominance is an interaction between the two homologous alleles within a diploid individual. Epistasis is the interaction between alleles at different genetic loci within an individual. Frequency dependence is an interaction between genes in different individuals at the same or at different loci. The diagram shows a population of diploids, each carrying two genomes with four loci (*dots*).

If heterozygotes are fitter than either homozygote, as in the sickle-cell hemoglobin example, then whichever allele is rarer will most often meet the alternative allele in a heterozygote and so will on average be fitter. Because each allele increases when rare, a stable equilibrium will be reached in which both kinds of gene are maintained in the population. The classic example of this kind of polymorphism is the maintenance of the Hb^S allele in areas where malaria is endemic (see Box 14.1). In the following chapters, we will look at other examples and examine the general importance of heterozygote advantage in maintaining variation (see pp. 505 and 541).

So far, we have looked at the consequences of interactions between the two homologous copies of a gene at the same locus in a diploid organism. (These are referred to as **dominance** interactions.) Similar effects arise even in haploids, through interactions between genes at different loci, that is, through **epistasis** (Figs. 14.8 and 17.20). For example, in the in vitro selection experiment on RNA molecules, two variants increased at first, but then decreased (red curves in Fig. 17.2C). This pattern occurred because although these two variants do increase catalytic activity, relative to the original type, they do not increase it as much as the four-variant combination, which increased slightly later (green curves in Fig. 17.2C). Crucially, these variants cannot all be brought together successfully. For example, an artificial construct carrying variants 260 (red) and 270 (green) functions poorly. It is these epistatic interactions that prevent the two transient variants from establishing in the final population.

Frequency-dependent selection is yet another kind of feedback, mediated by interactions between genes in *different* individuals. Again, the important point is that although the fate of an allele is determined in a very simple way by its fitness, that fitness depends on an averaging over all the circumstances in which the allele finds itself—which allele is present at the homologous gene in a diploid, which genes are present in the rest of the genome, and the environmental conditions. Because the environment itself depends on the size and genetic make-up of the whole population, as well as on interactions with other species, the course of natural selection can be very complicated.

Despite the complex interactions that determine fitness, the basic population genetic principles underlying natural selection are simple and are summarized by Equations 17.1 and 17.2. The most important point to take from this section is that the timescale of selection is determined by the inverse of the selection coefficients ($t \sim 1/s$).

In the next chapter, we look at how selection combines with other evolutionary forces, such as mutation, random drift, migration, and recombination. For all these forces, the outcome is basically determined by the timescales of these forces relative to that of selection. For example, random drift acts over a timescale of N_e generations, and so we will see that the importance of selection *relative* to drift depends on the ratio of timescales, $N_e/(1/s) = N_e s$. The idea that the strength of selection is proportional to the selection coefficients involved gives a simple qualitative way of understanding the complexities of evolution.

Populations Evolve to Peaks on the Adaptive Landscape

Given the complex relationships between genotype and phenotype, we expect widespread interactions between genes in their effects on fitness. That is, we expect widespread dominance and epistasis, such that the effect of a gene on fitness depends on which genes it is combined with (see p. 388). So, as the frequencies of different kinds of genes change, the additive effects of each gene also change. This makes the long-term effects of selection surprisingly complex (e.g., Fig. 17.2C). In particular, populations exposed to exactly the same conditions can evolve different gene combinations, depending on the initial allele frequencies.

Sewall Wright, one of the three founders of population genetics, devised an attractive way to picture the effects of gene interaction on allele frequency evolution. He showed that the change in allele frequency is proportional to the rate at which mean fitness is increased by raising the frequency of that allele. One can picture mean fitness as a function

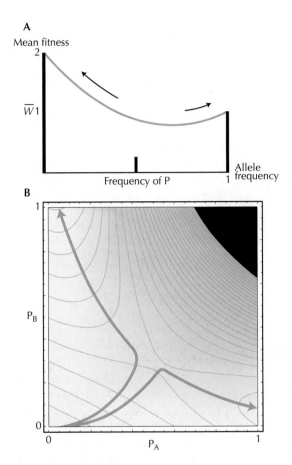

A

FIGURE 17.21. The effects of selection can be understood by plotting an adaptive landscape, which is a plot of mean fitness against allele frequency. (*A*) If heterozygotes are less fit than the homozygotes, the population will evolve uphill (*arrows*) to fix one or the other allele. The genotypes QQ, PQ, and PP have fitnesses 2, 0.25, and 1 (*vertical bars*). (*B*) An example with two loci, each with two alleles (A^P, A^Q at locus A; B^P, B^Q at locus B). Each copy of allele A^P adds 5% to fitness, and each copy of B^P adds 10%. However, the double homozygote $A^P B^P B^P B^P$ is lethal. Thus, there are two alternative adaptive peaks, one near fixation for $A^Q B^P$ with mean fitness 1.2 (*top left*) and the other near fixation for $A^P B^Q$ with mean fitness 1.1 (*bottom right*). The *arrows* show the trajectory of populations that start with predominantly $A^Q B^Q$. These may evolve to either adaptive peak depending on exactly where they start. Contours are spaced at intervals of mean fitness of 0.02. (These landscapes plot mean fitness against the state of the population, whereas the landscape in Fig. 17.12 plots individual fitness against individual phenotype.)

of all the allele frequencies, giving an **adaptive landscape**, which has peaks corresponding to the combination of allele frequencies that give the population the highest mean fitness. The steeper the gradient on the adaptive landscape, the faster the rate of movement.

When genes interact to determine fitness, there can be multiple **adaptive peaks**. Populations can evolve to several alternative combinations of allele frequencies, and so the outcome depends on where they start out. The simplest example is where the heterozygote is less fit (known as underdominance, in contrast to overdominance; Box 17.1). This is typical of many chromosome rearrangements in which the heterozygote suffers problems in meiosis. The different arrangements may not segregate properly, so that gametes do not receive one copy of each gene (Box 12.2). The population evolves so as to increase its mean fitness, and therefore will move toward one of the two adaptive peaks, which in this example correspond to fixation of one or the other allele (Fig. 17.21A). Different parts of the species' geographic range may reach different adaptive peaks, leading to a patchwork of regions separated by sharp boundaries (Fig. 17.22). This is known as a **parapatric** distribution.

In this first example, the interaction is between the two copies of homologous genes at the *same* locus. Similar behavior can be seen when genes at *different* loci interact. For example, suppose that at each of two loci (A and B), the P allele gives a slight selective advantage, but the double homozygote, $A^P A^P B^P B^P$, is lethal (Fig. 17.21B). Thus, the population tends to establish one or the other of them, but not both. In Chapter 22 (p. 643), we will see how this kind of epistasis can lead to the evolution of separate species.

The Metaphor of an Adaptive Landscape Can Be Misleading

Although Wright's metaphor of an adaptive landscape is a helpful way to picture the joint effects of different evolutionary processes, it can be misleading. First, it applies

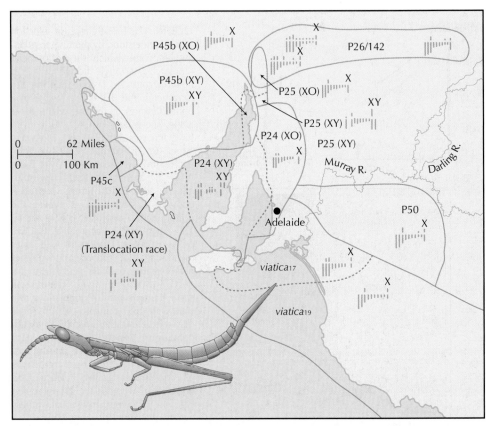

FIGURE 17.22. *Vandiemenella* grasshoppers in southern Australia are divided into a parapatric mosaic of different geographic races and species, which are distinguished by different chromosome arrangements (indicated by small diagrams). Boundaries between taxa regarded as species are delimited by *solid lines*, and between races, by *dotted lines*. These are separated by narrow **hybrid zones** a few hundred meters wide. The ranges are shown extending into the areas of the Southern Ocean that were dry during the Pleistocene; this accounts for the distribution of the three taxa found on Kangaroo Island (*center*). *Inset:* a male *Vandiemenella pichirichi*.

only if the fitnesses of each genotype are constant. If fitnesses change with a changing environment or with changing allele frequencies, then the landscape itself changes over time and cannot be used to predict the evolution of the population. Wright's formula also requires that alleles be randomly associated (i.e., in **linkage equilibrium**). If certain alleles tend to be associated with each other, then we cannot predict genotype frequencies just by multiplying together all the allele frequencies (Box 15.5). The population can be described only by tracking all the genotype frequencies, and its evolution will depend on how recombination breaks up **linkage disequilibria**. The outcome cannot then be predicted by using Wright's simple picture, which only describes the evolution of allele frequencies.

A second difficulty is that the average fitness of a population depends on many genes, and so Wright's diagram should really be drawn in as many dimensions, not just two or three. The usual three-dimensional picture can be misleading, because it suggests that the population can easily be trapped on a local adaptive peak. In fact, the adaptive landscape has thousands of dimensions, and so there are very many directions in which the population can evolve—many kinds of changes can be made to many different genes. Typically, there may be several changes that will increase mean fitness and that will be exploited by selection. It is not clear whether populations are typically trapped at local fitness peaks.

Finally, two quite different kinds of fitness landscape are commonly used. Here, we have thought of the mean fitness of the population as depending on the allele frequencies that describe the state of the population. Often, however, individual fitness is plotted against individual genotype (e.g., Fig. 17.12). Indeed, Wright himself used both versions without making an explicit distinction between them. Both population- and individual-based fitness landscapes can be useful, but in this book we mainly use the first.

Despite these caveats, Wright's adaptive landscape is a useful way to understand how selection acts on allele frequencies in the presence of epistasis. In the next chapter, we will see how Wright extended it to include the effects of mutation, migration, and random drift. Wright's idea has had a major influence on evolutionary thinking, not so much through its mathematical development, but rather because many biologists have found it to be a useful metaphor for thinking about the way different factors combine in the evolutionary process.

Müllerian Mimicry Favors Common Color Patterns

Wright's adaptive landscape can only be used to predict the effects of selection on allele frequencies if genotypes have constant relative fitnesses. Thus, it is a useful way to understand dominance and epistasis, but it cannot be used when relative fitnesses depend on the composition of the population. However, multiple stable equilibria can also occur if fitnesses are frequency dependent, even though it is mathematically incorrect to visualize these as adaptive peaks. If common alleles are fitter than rare alleles, then the population will tend to fix whatever allele happens to be most common at the outset.

Müllerian mimicry provides the best example of this kind of frequency-dependent selection. Many species that are distasteful or toxic to predators have evolved conspicuous warning patterns that advertise their distastefulness. There is strong selection for all individuals to share the same common pattern, because this is what is recognized by predators. A rare variant may not be recognized as being distasteful and will be eaten. Often, many different distasteful species will mimic each other closely and thus gain greater protection by converging on a single pattern. For example, species of butterflies in the genus *Heliconius* have evolved bright wing patterns and, in any one region, all these species share the same pattern (Fig. 17.23). Indeed, more distantly related butterflies and moths also gain protection by mimicking this one pattern. How-

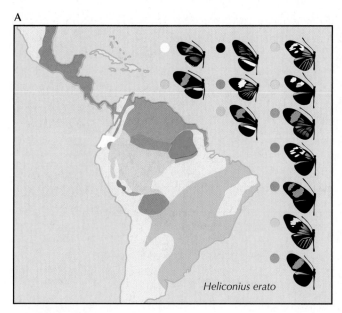
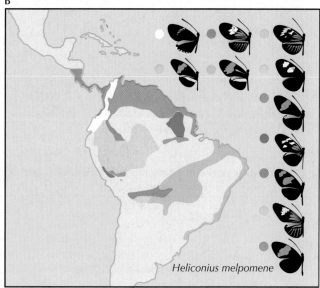

FIGURE 17.23. Müllerian mimicry in *Heliconius* butterflies. Within any one area, *Heliconius erato* (*A*) and *H. melpomene* (*B*) share the same warning pattern. However, patterns differ considerably across South and Central America.

ever, which pattern is used as a warning is arbitrary and, in different places, different patterns are found (Fig. 17.23). This parapatric distribution of different patterns is maintained by selection against rare phenotypes, which takes several forms. The fitness of each individual genotype increases as that genotype becomes more common (frequency-dependent selection). Heterozygotes may differ from either of the common parental forms and so will be predated more often (underdominance). Finally, where several genes are involved, recombinant genotypes will also show unusual patterns and will also survive less well (epistasis).

SELECTION ON QUANTITATIVE TRAITS

Selection Acts in a Simple Way on Quantitative Traits

How will a crop plant respond to artificial selection? How does selection shape the morphology or behavior of an organism? It would be extremely difficult to answer such questions if we needed to follow the allele frequencies at all the genes that influence a trait and the effects of each allele on fitness. The problem would be utterly hopeless if we had to go further and account for the frequencies and effects on fitness of every combination of alleles (i.e., for linkage disequilibrium and epistasis). Fortunately, the evolution of at least the mean value of a quantitative trait can be understood without knowing its genetic basis, by using the methods introduced in Chapter 14. We will see that the effect of selection on the average value of a quantitative trait

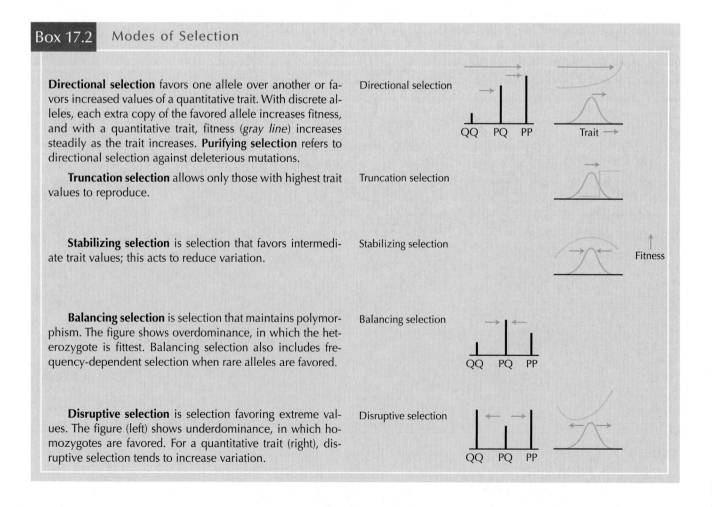

Box 17.2   Modes of Selection

Directional selection favors one allele over another or favors increased values of a quantitative trait. With discrete alleles, each extra copy of the favored allele increases fitness, and with a quantitative trait, fitness (*gray line*) increases steadily as the trait increases. **Purifying selection** refers to directional selection against deleterious mutations.

Truncation selection allows only those with highest trait values to reproduce.

Stabilizing selection is selection that favors intermediate trait values; this acts to reduce variation.

Balancing selection is selection that maintains polymorphism. The figure shows overdominance, in which the heterozygote is fittest. Balancing selection also includes frequency-dependent selection when rare alleles are favored.

Disruptive selection is selection favoring extreme values. The figure (left) shows underdominance, in which homozygotes are favored. For a quantitative trait (right), disruptive selection tends to increase variation.

is similar to its effect on a single gene and can be described in just as simple a way.

Suppose that there is directional selection on the trait; that is, individuals with a larger trait value are more likely to reproduce (Box 17.2). A classic example is provided by the Galápagos finches, which have been studied for more than 30 years by Peter and Rosemary Grant. These birds are remarkable, because a single founding population has diversified to produce 14 species, which fill a wide range of ecological niches on the Galápagos Islands. Species vary greatly in their diets and show corresponding variations in beak shape (Fig. 17.24). There is also considerable variation in beak shape within species (Fig. 17.24C,D), and this variation is under strong selection in relation to diet. The Galápagos finches are one of the best examples in which the causes of natural selection within species are understood and where selection of this kind has led to adaptive divergence between species.

The medium ground finch, *Geospiza fortis*, feeds on seeds, which it cracks with its bill. After a severe drought in 1977, the average size of seeds increased, as the more easily handled seeds were eaten (Fig. 17.25B). Those birds with deeper and stronger beaks could crack these large seeds more effectively and so survived much better. Almost all of the smaller birds died of starvation (Fig. 17.25C,D). The net effect was an increase in the average beak depth of the survivors relative to the original population. This change, which is directly caused by selection, is termed the **selection differential** *S*. In this example, beak depth increased by 0.30 standard deviations (S.D.) in the first half of 1977 and by 0.23 S.D. in the second half, giving a total selection differential of 0.53 S.D. over the year.

This process of selection is straightforward when we look at changes within a generation—survivors had deeper beaks than those that died. But what effect will this selection have on the next generation? We know from Chapter 14 that the covariance between a parent and its offspring is half the additive genetic variance. This leads to a simple relationship between the average of the trait in the offspring and the average of the two parents (Fig. 17.26) The slope of the best-fitting **regression** line is equal to the **narrow-sense heritability** ($h^2 = V_A/V_P$) and equals the proportion of variance that is due to additive genetic effects (p. 394). It immediately follows that the change in mean in the next generation, termed the **selection response**, is just the change in the parents multiplied by the heritability: $R = h^2S$. Thus, a fraction of the selection differential is passed on to the next generation, and that fraction is equal to the proportion of variance that is inherited through the additive effects of genes.

In the *G. fortis* example, the heritability of bill depth was estimated as 0.80, so that a change of approximately $R = 0.80 \times 0.53 = 0.42$ S.D. is predicted. The observed difference in bill depth between offspring born in 1978 and in 1976 was approximately 0.6 S.D., which is in reasonable agreement. This simple calculation omits several important factors. First, selection acts at several stages of the life cycle. Young birds with shallower beaks survive slightly better when conditions are good, perhaps because they can handle small seeds more efficiently. On the other hand, males with deeper bills are more likely to find a mate. Thus, the net selection differential, including all stages in the life cycle, will differ from that due just to survival of the drought. Second, bill depth depends on environmental conditions (e.g., nutrition) and so changes may be nongenetic. (For example, offspring born in 1976 were smaller than their parents, whereas those born 2 years later were the same size on average; dashed versus solid lines in Fig. 17.26.) Finally, a trait may change as a result of selection on other traits to which it is correlated (see p. 479).

FIGURE 17.24. Diversity in diet and beak shape in species of Galápagos finches. (*A*) Large ground finch, *Geospiza magnirostris*, which can crack large seeds. (*B*) Woodpecker finch, *Cactospiza pallida*, using a tool to extract insect larvae. (*C,D*) Different beak shapes of two *Geospiza fortis* from the island of Daphne Major.

Selection Changes Quantitative Traits at a Rate Proportional to Their Additive Genetic Variance

When we are dealing with artificial selection, it is reasonable to think in terms of the selection differential *S*. For example, a crop breeder chooses to sow seeds from those

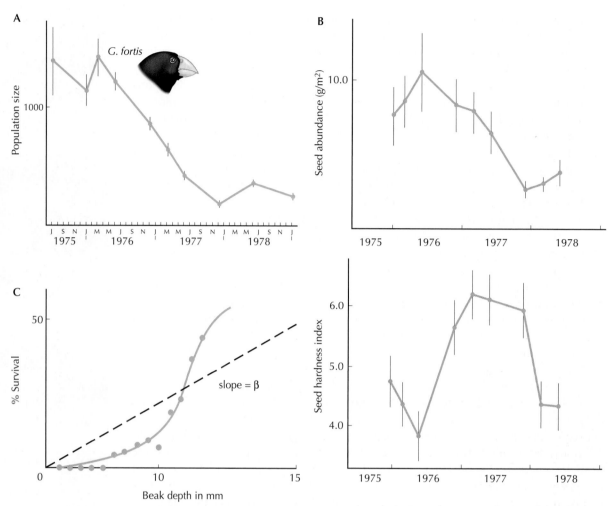

FIGURE 17.25. Selection on medium ground finches, *Geospiza fortis*, on Daphne Major during the drought of 1977. (*A*) Population size. (*B*) Seed abundance (*top*) and hardness (*bottom*). (*C*) Birds with deeper beaks survived the drought much better. The *dashed line* indicates the regression, which has slope equal to the selection gradient β. In *A* and *B*, 95% confidence intervals are shown.

plants that gave the greatest yield and calculates *S* from the difference between the mean yield of those selected and the yield of the original population. However, for natural populations, it is often more appropriate to think in terms of the relation between the trait and fitness. The slope of this relationship is called the **selection gradient**, written β. (Actually, the slope must be divided by mean fitness; what matters is relative fitness.) To understand how β is calculated, look at Figure 17.25C, which shows how survival depends on beak depth in *G. fortis*. The straight line that best approximates this curve is called a regression line, and its slope divided by the mean survival of the population gives the selection gradient.

Provided that the trait is normally distributed, the change in mean due to selection is equal to the selection gradient multiplied by the phenotypic variance ($S = V_P\beta$). Thus, the change in mean due to selection is also equal to the selection gradient multiplied by the additive genetic variance:

$$R = h^2 S = \frac{V_A}{V_P} V_P \beta = V_A \beta. \tag{17.3}$$

This set of formulas tells us that selection will change a quantitative trait at a rate

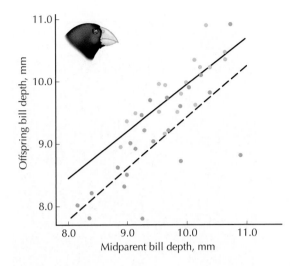

FIGURE 17.26. The relation between bill depth of offspring and parents, for the *Geospiza fortis* on Daphne Major. The *solid line* gives the best fit for 1978 (*blue circles*), whereas the *dashed line* gives the same for 1976 (*red circles*). For both years, heritability was about 0.8 as shown by the slopes of the regression lines.

equal to the product of the heritability and the selection differential, which is equal to the product of the additive genetic variance and the selection gradient.

How can such a simple result emerge when the underlying genetics may be very complicated? The formula $R = h^2S$ can be seen as simply an empirical observation. For example, if larger parents give larger offspring with a regression gradient h^2, then selection to breed from parents larger by S must produce offspring that are larger by h^2S (Fig. 17.26). A more detailed argument shows the relation between selection on the trait and changes in frequency at the many genes that affect the trait. Remarkably, genetic details, such as the number of genes and the sizes of their various effects, cancel out. The accumulated effect of all the changes in allele frequency is $R = V_A\beta$.

We have seen that populations evolve uphill on the adaptive landscape. This we defined as the surface of mean fitness considered as a function of allele frequencies (Fig. 17.21). When we deal with quantitative traits, we can think of the mean fitness as a function of the means of all the traits. The slope of this surface is equal to the selection gradient, and so Equation 17.3 shows us that populations will evolve uphill on this surface at a rate proportional to the additive genetic variance. Just as for allele frequencies, selection may push populations toward alternative adaptive peaks that represent different favorable combinations of traits (Fig. 17.27).

Selection on One Trait Causes Changes in Other Traits That Are Genetically Correlated with It

We have seen that the response to selection on the trait mean is proportional to the additive genetic variance of the trait. The difficulty, of course, is that the genetic variance as well as the mean may change, so that it becomes impossible to predict how fast the mean will respond to selection in future generations. If genetic variance is based on a few alleles of large effect, we expect directional selection to fix them, and so genetic variation will soon be lost. If these alleles are rare at first, they will contribute a transient burst of variation as they increase to intermediate frequencies. However, variation will be eventually be lost as the alleles rise to fixation (see Fig. 17.33A). As we shall see in the next section, simply assuming that the genetic variance stays constant seems to be a surprisingly good approximation over a few tens of generations. However, as we saw in Chapter 14 (p. 408), we do not know enough about the genetic basis of trait variation to predict evolution in the longer term.

Different traits may be correlated with each other, so that selection on one trait increases other traits. For example, the finches that survived the 1977 drought on the island

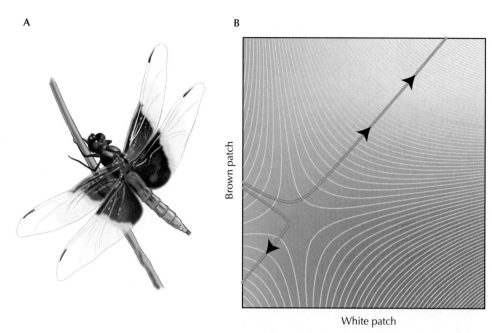

FIGURE 17.27. (*A*) A male dragonfly, showing the brown and white patches that act as sexual signals (Chapter 20). (*B*) Male mating success as a function of the sizes of patches of each color. Sexual selection favors either large patches of both colors (*top right*) or neither (*bottom left*). Populations could evolve to either adaptive peak (*arrows*).

of Daphne Major had deeper beaks (Fig. 17.25C), but were also larger overall. This is mainly because beak depth is correlated with body size.

Correlations between different traits have no long-term effect if they are due entirely to the **environmental** component and so are not passed on to the next generation. (For example, well-fed individuals may be larger by several measures. These measures will therefore be correlated as a result of random variation in nutritional environment.) However, if there are **genetic correlations**, either because the same genes are responsible for several traits or because of linkage disequilibrium between genes responsible for different traits, then selection on one trait will cause changes in other traits in the next generation. Such genetic correlations can be seen as constraints on evolution: It is difficult to change one trait without changing another. However, the optimal combination of traits will eventually be reached provided that correlations are not absolute; that is, as long as there is some genetic variation in the direction favored by selection and only a single optimum (Fig. 17.28). Moreover, genetic correlations may themselves change under selection.

Artificial Selection Usually Produces a Rapid and Continued Response

The improvement of plants and animals by selection was necessary for agriculture to be established—and with it, the explosive growth in human population and the surplus production that has allowed elaborate civilizations to form. These developments began when the major crop plants and animals were first domesticated, approximately 7000 years ago, and, more recently, with agricultural reforms in late 18th-century Europe. Since then, yields have continued to improve with the aid of better statistical and genetic methods (Fig. 17.29).

The dramatic success of artificial selection in Darwin's time gave major support to his theory of evolution by natural selection (see pp. 17–19). Artificial selection can

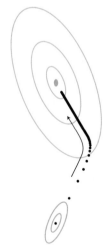

FIGURE 17.28. Correlations between traits will change the direction of the selection response, but will not prevent the population from reaching the optimum. The *blue* contours show mean fitness increasing toward an optimum (*red*). The *green* contours show the genetic variation for two traits, which are positively correlated. The *dots* show the evolution of this population toward the optimum, over more than 100 generations.

FIGURE 17.29. Examples of major crop plants and animals, with their wild relatives. Pictures of (*A*) maize and teosinte and (*B*) domestic and wild sheep.

change populations far more rapidly than the rates seen in the fossil record (Fig. 17.30), and an almost limitless variety of traits will respond to selection. Thus, although Darwin had no direct evidence for the action of selection in nature, the analogy with artificial selection gave a strong argument that the same process, extended over much greater time spans and acting on much larger populations, would be as effective in nature as on the farm.

A classic example of the power of artificial selection comes from an experiment on maize that was begun in Illinois in 1895 and is still being continued. One line was started using the 24 ears with highest oil content and another using the 12 ears with lowest oil content. These two lines were selected in a similar way each year, with the most extreme 20% of the population being chosen to breed. Remarkably, oil content has continued to change over more than 100 generations (Fig. 17.31A).

Although selection experiments have been carried out in the laboratory for more

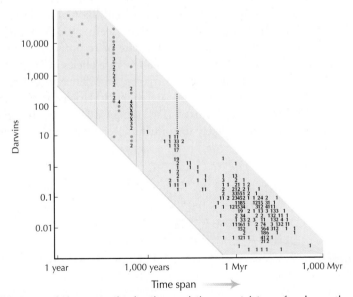

FIGURE 17.30. Rates of change in the fossil record (*bottom right*) are far slower than are seen in the selection experiments (*top left*). The scatter diagram plots the rate of morphological change against the time interval over which it was measured. Rates are in **darwins;** 1 darwin is defined as a change of 1 part per million per year. Myr, millions of years.

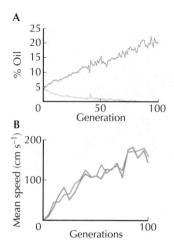

FIGURE 17.31. (A) Long-term selection for oil content in maize. (B) Selection on *Drosophila* for ability to fly upwind. The response to selection in two replicates is shown.

than a century (e.g., Castle's experiment on hooded rats; Fig. 13.3), these have usually been restricted to small numbers of individuals and a limited number of generations. Recently, however, Ken Weber has devised a variety of ingenious devices for handling large numbers of *Drosophila*, which have provided some remarkable illustrations of the power of sustained artificial selection. Figure 17.31B shows the response to selection for the ability to fly upwind toward a light. Initially, the flies could only fly against a wind of 2 cm/sec. But, after 100 generations in which the best 4.5% of flies were selected, the average flying speed increased 85-fold, to 170 cm/sec. The response to selection was remarkably similar in two replicate lines (Fig. 17.31B).

One of the longest running selection experiments to date has been carried through by Lenski and colleagues, who have propagated replicate populations of the bacterium *Escherischia coli* for more than 20,000 generations. The initial stock was adapted to live on a rich medium, but during the experiment it was propagated on minimal medium with only glucose as a carbon source. Conditions were such that the population grew exponentially, so that growth rate in a novel environment was selected. The growth rate relative to the ancestral strain increased substantially during the experiment, although mostly during the first few thousand generations (Fig. 17.32). Interestingly, different replicates diverged in fitness, presumably because different combinations of favorable mutations were acquired by chance. This example differs from Weber's *Drosophila* experiment in that the bacterial population is extremely large and the bacteria are reproducing asexually. As we shall see in Chapter 23, selection is considerably less efficient in asexual than in sexual populations. On the other hand, the very large population size makes it much more likely that favorable mutations will be picked up.

Additive Genetic Variance Remains High Despite Continued Selection

These examples are typical of many selection experiments that have been carried out on a wide variety of organisms and traits. Although selection responses soon slow down in small populations, selection on moderately large sexual populations (a few hundred or more individuals) sustains a steady response for as long as the experiment is continued (e.g., Fig. 17.31). Moreover, this response is often closely similar between replicate lines. The initial response to selection is to be expected, given that high additive genetic variance V_A is seen for most traits in most populations (see p. 397). What is surprising is that the additive genetic variance, and hence the rate of response, *remain* high despite continued selection. If variation were based on a few rare alleles of large effect, then we would expect to see a transient increase in variance, and hence in the response to selection, as rare favorable variants rose from low frequency. However, the response would soon peter out as all favorable alleles fixed (Fig. 17.33A). There are two possible solutions to this puzzle, and both are likely to contribute to the observed results.

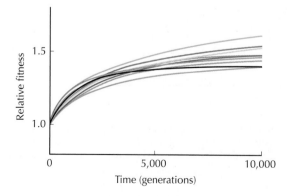

FIGURE 17.32. Average fitness of nine replicate *Escherichia coli* populations relative to the ancestral strain during 10,000 generations of adaptation to a glucose-limited medium.

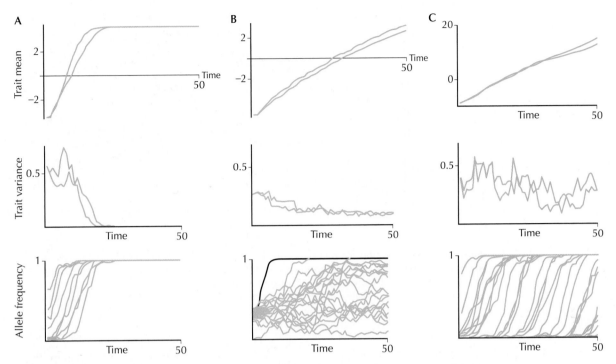

FIGURE 17.33. Simulations of artificially selected populations, comparing three scenarios. The environmental variance is set to $V_E = 1$. In each generation, the 50 highest-scoring individuals were chosen from 250, heritability started at 30%, and major alleles had effects equal to 0.5 units. *Upper row:* trait mean over time (two replicates). *Middle row:* genetic variance (trait variance in the figure) V_G. *Bottom row:* allele frequencies (one replicate). (A) Ten unlinked loci with major alleles. (B) 400 unlinked loci with minor effects (0.03 units) apart from two major loci (allele frequencies shown by the *black line* in *bottom panel*). (C) 100 unlinked loci; mutation rate such that mutational variance $V_M = 0.0023$.

The first and simpler is that additive genetic variance is largely due to many alleles of very small effect. Then, selection on each individual allele would be so weak that allele frequency would hardly change, and the genetic variance would stay almost constant. The mean does shift, because of the accumulated effects of many small changes in allele frequency, all causing slight increases in trait mean. However, the changes in allele frequency at different loci have fluctuating effects on the genetic variance, which therefore stays approximately constant when enough loci contribute (Fig. 17.33B).

This so-called **infinitesimal model** has an attractive simplicity, but conflicts with evidence that alleles of large effect do contribute to selection response; for example, between wild and cultivated tomatoes or maize and teosinte (see pp. 307–318 and 403–407). In some laboratory experiments, the selected trait may increase sharply over only a few generations. In some cases, such jumps have been shown to be due to an allele of large effect (Fig. 17.34). This is often found to be an allele that has large effects on the trait in the heterozygote, but is inviable or sterile in the homozygote. (Such **recessive lethals** reach a balance between artificial selection for the heterozygote and natural selection against the homozygote. Variation is maintained by heterozygote advantage, as discussed in Box 17.1.)

If alleles of major effect do contribute, why are they not quickly fixed and variation lost? The answer is that such alleles arise as new mutations within the selected lines. Although individual genes mutate very rarely (about once in 10^5 generations), mutation does generate considerable genetic variance in quantitative traits, which depend on many genes. Typically, the additive genetic variance increases by at least one-thousandth of the environmental variance in each generation ($V_M \sim 10^{-3} V_E$; Table

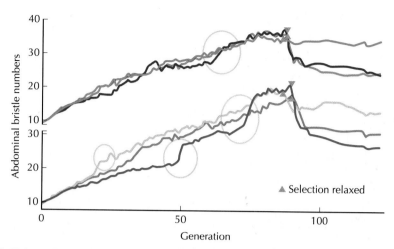

FIGURE 17.34. Selection for increased bristle number in *Drosophila* produced a continued response in six replicate lines. Sharp jumps (*pale circles*) were shown to be due to selection of recessive lethals, which added up to 11 bristles when heterozygous. When selection was relaxed at generation 90, the bristle number fell as the recessive lethals were eliminated by natural selection.

14.3). If this input of **mutational variance** were due to alleles of small effect, they would take so long to be picked up by selection that they would not contribute much to the response seen in laboratory experiments. However, mutations of large effect can be picked up immediately and can explain selection responses of the kinds seen in Figures 17.31 and 17.34.

It is still puzzling that in large populations, selection response can be so similar between different replicates (e.g., Figs. 17.31A and 17.34). It may be that enough mutations accumulate in large populations to smooth out the random fluctuations that are often seen in smaller experiments, as in the simulation of Figure 17.33C. Another difficulty with attributing long-term response to mutational input is simply that populations that initially contain no genetic variation respond much more slowly than those that contain standing variation. A review of six selection experiments that started with inbred lines found that the average response over the first 50 generations was 0.07 phenotypic standard deviations per generation (i.e., $\sqrt{V_P}$). In contrast, the response to selection on outbred populations is typically much higher, for example, about $0.5\sqrt{V_P}$ per generation for bristle number in *Drosophila*.

At present, the relative importance of preexisting variation due to alleles of small effect, as opposed to new mutations of larger effect, is unresolved. These two possibilities are not mutually exclusive. Mutations of large effect are likely to have deleterious **pleiotropic** side effects and so may make little long-term contribution to evolution. (We have already seen this at the level of a single gene: A substantial fraction of the variation in activity of alcohol dehydrogenase in *Drosophila* is caused by **transposable elements**, which are deleterious and hence rarely become fixed in a species [Figs. 13.11 and 14.27].) Thus, a large part of the standing variation that we see in a population may not in fact contribute to long-term change. There are, of course, many examples where adaptations are produced by major changes in single genes—most obviously resistance to insecticides and, in humans, resistance to malaria through change in hemoglobins and other genes. However, the drastic changes needed to make target molecules resistant usually interfere with their normal function (recall sickle-cell hemoglobin). Over time, alleles of smaller effect (known as **modifiers**) may evolve to ameliorate the unfortunate side effects of the genes responsible for the initial response to selection.

SELECTION ON MULTIPLE GENES

When Genes Are Associated with Each Other in Linkage Disequilibrium, Selection on One Gene Causes Changes at Others

So far, we have assumed that genes are randomly combined in the population. In that case, the chance of having any particular genotype is found by multiplying together the frequencies of all the individual alleles. The results are Hardy–Weinberg proportions of genotypes at each locus in a diploid (Box 1.1), and the alleles in each haploid genome are in linkage equilibrium (Box 15.5). Thus, the evolution of the population reduces to just the evolution of allele frequencies—a massive simplification, because we need follow far fewer variables. For example, with ten diploid loci, each with two alleles, we need only ten allele frequencies, instead of $2^{20} = 1,048,576$ different diploid genotypes.

This approach was derisively termed "beanbag genetics" by Ernst Mayr, who thought that it neglected the complex interactions between different genes in the whole organism. However, Haldane vigorously defended the approach. He pointed out that the effects of gene interactions on fitness (i.e., epistasis) *are* accounted for: As allele frequencies change, the additive effect of each allele also changes, because it interacts with a different set of alleles at other loci (see p. 389).

The key assumption of beanbag genetics is that genes are not statistically associated in the population. That is, deviations from Hardy–Weinberg proportions and associations among loci (i.e., linkage disequilibria) are negligible. This is a good approximation provided that the various forces that produce nonrandom associations among genes are weak relative to recombination and random mating. In this section, we will see how selection acts in cases where we *do* need to account for linkage disequilibria. We will return to these issues in more detail later, when we discuss the measurement of selection (Chapter 19), the origin of species (Chapter 22), and the evolution of genetic systems (Chapter 23).

If two alleles at different genetic loci are associated with each other in linkage disequilibrium, then selection for one of them will cause both to increase. This occurs for a simple reason: If one allele increases in frequency because it causes increased reproductive success, then another allele that tends to be found in the same genome will also increase, even if it has no direct effect on fitness. This spurious selection is known as **hitchhiking** (see p. 427). It reflects the difficulty of disentangling the causes of increased fitness from the many genetic differences that happen to be associated with it.

The phenomenon of hitchhiking is nicely illustrated by an experiment in which two strains of *Drosophila melanogaster* were crossed so as to create a complete association between a recessive lethal allele of the gene *Glued* and several closely linked markers (Fig. 17.35). The recessive allele quickly declined together with the markers that were initially associated with it. However, recombination breaks down linkage disequilibrium, so that linked markers were not completely eliminated with the lethal. As expected, the marker *Est-6* that was most distant from *Glued* was perturbed the least. Surprisingly, the markers tended to move back toward their original frequencies after the perturbation. This is presumably because selection on other genes, which had also been perturbed, tended to return them to their original equilibria and took the observed markers with them.

Several Processes Can Generate Linkage Disequilibrium

In the experiment shown in Figure 17.35, linkage disequilibria were generated by an artificial cross between two genetically distinct populations. Similar hitchhiking effects occur in nature, whenever there are linkage disequilibria with genes under selection.

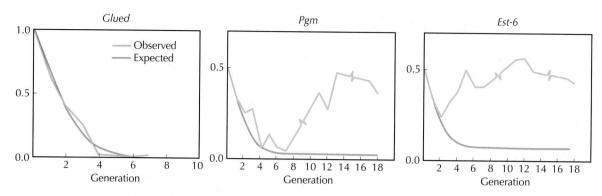

FIGURE 17.35. Hitchhiking causes selection at one locus to influence linked loci. The rate of recombination between *Pgm* and *Glued* is 3.5%, and between *Est*-6 and *Glued* is 8.6%. *Red lines* show expectations on the assumption that the only selection is against *Glued* homozygotes.

The most extreme case is a **selective sweep**, where a single favorable mutation occurs and rises to fixation in the whole population. Then, the surrounding region of chromosome will fix with it, greatly reducing genetic diversity (see p. 427).

Migration can also generate strong associations by mixing different populations (see p. 452). The consequences for a natural population can be seen in an experiment in which mice were transferred between two Scottish islands, from Eday to the Isle of May (Fig. 17.36A). The recipient population was highly inbred and had no variation at 69 enzyme loci. It also carried different chromosome arrangements and differed in skeletal morphology from the Eday mice. Although the introduced mice made up only a small fraction of the population, the genes they carried increased in frequency: Enzyme markers, chromosome arrangements, and morphology all shifted substantially toward those characteristic of Eday (Fig. 17.36B). The most likely explanation is that the original population was fixed for many deleterious recessive alleles. Selection to replace these by their Eday homologs increased the frequency of all the alleles from Eday in a hitchhiking event similar to that seen in the laboratory experiment (Fig. 17.35). This process, where selection acts on sets of genes introduced by migration, is important in understanding the barriers to gene exchange between populations that are diverging to become separate species (see p. 647).

Selection itself can build up linkage disequilibrium if it favors certain combinations of alleles—that is, if there is epistasis. One example is given by the **plasmids** that are carried by many bacteria. These often carry sets of genes that adapt the bacteria to particular conditions; such sets have been assembled and kept together as a result

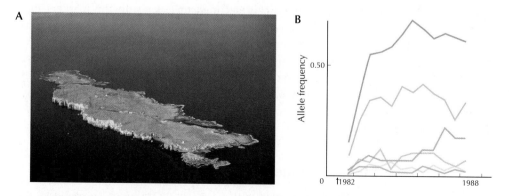

FIGURE 17.36. Hitchhiking in an island population. In 1982, 77 mice (*Mus domesticus*) from the Orkney island of Eday were released on the Isle of May, in the Firth of Forth (*A*). Introduced alleles at six allozyme loci increased in frequency (*B*), as did introduced chromosome arrangements and morphology.

of selection favoring particular gene combinations (see pp. 168–171). In the next chapter, we will see another striking example in which polymorphism for alternative butterfly wing patterns is controlled by a set of tightly linked loci, or a **supergene** (see p. 506). In all these cases, the strong linkage disequilibrium generated by epistatic selection has led to a reduction in recombination, which reduces the breakup of the favorable gene combinations. In Chapter 23, we will examine how the disequilibria generated by selection and drift influence rates of recombination.

SUMMARY

If some traits increase the reproduction of the organisms that carry them and if those traits are inherited, then they will increase in frequency. This process of natural selection acts on any kind of replicator, for example, RNA molecules replicating in vitro. The key idea that natural selection is due to inherited variation in fitness is made precise by Fisher's Fundamental Theorem: $\Delta\overline{W} = \text{var}_A(W)/\overline{W}$. Selection increases mean fitness $\overline{W}$ at a rate proportional to the additive genetic variance in fitness, $\text{var}_A(W)$. However, other processes tend to degrade it, so that mean fitness stays close to the value needed to keep population size within bounds.

Natural selection is the only process that leads to adaptation. Complex patterns may emerge from simple systems, but will only be directed to a function if they have been shaped by selection. Selection works step-by-step to to build highly improbable structures by bringing together alleles that are individually favored.

The effect of selection on allele frequencies or on quantitative traits is predicted by simple formulae. The rate of change of allele frequencies is proportional to the selection coefficient s, and the mean of a quantitative trait changes at a rate given by the product of the selection gradient and the additive genetic variance in the trait. In both cases, weak selection can produce changes that are very fast on an evolutionary timescale; in the case of allele frequencies, over times $\sim 1/s$.

Fitness depends on interactions between genes and with the environment. Dominance, epistasis, and frequency dependence all cause the selection on each allele to change as the composition of the population changes so that the outcome of selection can be complex. The effects of dominance and epistasis can be visualized by imagining that the population evolves across an adaptive landscape, climbing uphill toward alternative adaptive peaks.

Artificial selection on plants, animals, and microbes produces responses that can be sustained for a hundred or more generations. These sustained responses are based in part on variation in the base population, involving many genes of small effect, and in part on new mutations.

Different traits may be genetically correlated because some alleles affect both traits (pleiotropy) and because alleles are associated with each other in linkage disequilibrium. Selection on one trait then causes changes in genetically correlated traits. At the genetic level, selection on one allele changes the frequency of other alleles that are in linkage disequilibrium, a phenomenon known as hitchhiking.

FURTHER READING

Barton N.H. and Keightley P.D. 2002. Understanding quantitative genetic variation. *Nat. Rev. Genet.* **3:** 11–21.
  Reviews explanations for sustained responses to artificial selection.
Bell G. 1997. *Selection: The mechanism of evolution.* Chapman and Hall, New York.
  A textbook that focuses on the key process of selection and gives a more extended treatment of the material in this chapter.
Dawkins R. 1986. *The blind watchmaker.* Longman, London; Dawkins R. 1997. *Climbing Mount Improbable.* Penguin Science, London.
  Persuasive arguments for the power of Darwinian natural selection to generate complex adaptations.

Grant P.R. 1999. *The ecology and evolution of Darwin's finches.* Princeton University Press, Princeton, New Jersey.
  Summarizes many years' fieldwork on the Galápagos finches, including some of the best evidence for the action of natural selection in the wild.
Haldane J.B.S. 1932. *The causes of evolution.* Longman, New York.
  A classic work from the Evolutionary Synthesis and still one of the most readable accounts of the basics of selection on single loci.
Weiner J. 1995. *The beak of the finch.* Jonathan Cape, London.
  An excellent account of Grant's work, set in the wider context of evolutionary biology.

18

The Interaction between Selection and Other Forces

W E HAVE NOW EXPLAINED THE FUNDAMENTAL PROCESSES of mutation, recombination, random drift, gene flow, and selection. It is these processes that are responsible for changing the composition of populations—in other words, for evolution. In Chapters 15 and 16, we saw how the first four processes interact in the neutral theory, which provides a powerful **null model** for molecular evolution. In Chapter 17, we showed how natural selection makes organisms better able to survive and reproduce in their various environments. In this chapter, we will explore selection in more detail and will show how it interacts with the other evolutionary forces.

We emphasized in the previous chapter that selection is the only process that leads to adaptation. Correspondingly, other evolutionary processes tend to degrade adaptation (Fig. 17.6). Mutation randomizes genetic information, genetic drift randomizes genotype frequencies, and gene flow randomizes the positions of genes in space. The idea that selection is continually building up adaptation while other processes degrade it is a theme that links the various kinds of interactions that we examine in this chapter. We explore the idea further in the next chapter, where we move from explaining the basic principles using chosen examples to assessing the overall extent of selection.

Most New Mutations Are Lost by Chance, Even If They Increase Fitness

In Chapter 15, we emphasized the action of random drift in small populations. However, random drift has important effects even in extremely large populations whenever an allele is present in just a few copies. In the most extreme case, a new mutation arises as a single copy, but most likely will be lost by chance—regardless of its expected effects on fitness. For example, in a diploid sexual organism, a mutation first appears in a heterozygous individual. That individual may leave no offspring; even if it does, the muta-

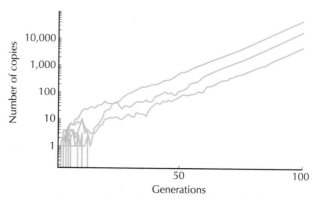

FIGURE 18.1. Most favorable mutations are lost by chance soon after they arise. The graph shows the number of copies of an allele with selective advantage $s = 10\%$ plotted against time. Even though the allele increases fitness, it survives in only 3 out of 30 replicates. In those cases, its numbers grow exponentially and so appear as straight lines on a log scale (*upper right*). The other 27 mutations are lost within a few generations (*lower left*). The probability of survival is $\sim 2s = 20\%$, so we would have expected 6 of 30 mutations to survive—rather more than the 3 that happened to survive in this example.

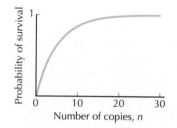

FIGURE 18.2. The probability of survival of an allele that increases fitness by $s = 10\%$ plotted against the number of copies, n. One copy has a chance of $\sim 2s = 20\%$ of surviving, and survival is almost certain once there are more than about 20 copies. The chance of survival depends on $n * s$, so that the graph has the same shape with weaker selection coefficients. For example, with $s = 1\%$, survival is almost assured with more than 200 copies.

tion has only a 50% chance of being passed on to each of them at meiosis. Then, even if the mutation does survive the hazards of the first generation, it may be lost later on. A simple mathematical argument shows that an allele with a selective advantage s has a chance of approximately $2s$ of being fixed in the population (Chapter 28 [online]). Moreover, if the **effective population size** is smaller than the census number ($N_e < N$), so that random drift is stronger, the **fixation probability** is reduced to $2s(N_e/N)$ (Fig. 18.1).

As the number of copies of a new allele increases, the chance of its ultimate fixation increases. However, this chance only becomes high when the number of copies is larger than $\sim 1/s$. For example, the chance that 100 copies of an allele with selective advantage of 1% will eventually be fixed is 86.5%; 1000 copies of an allele with advantage $s = 0.1\%$ or 10,000 copies with advantage 0.01% would have the same chance of survival (Fig. 18.2).

Populations Diverge Because They Pick Up Different Favorable Mutations

It is at first surprising that chance should play a role even in large populations and even when selection is strong. When populations are large enough, evolution does become deterministic. If there are so many genes that each mutation occurs in very many copies ($N_e\mu \gg 1$), then chance fluctuations average out, and any mutation that increases fitness is certain to be established. To see this, think back to the experiment on ribozyme evolution described in Figure 17.2. There, the initial population was extremely large ($\sim 10^{13}$ molecules), and mutations were generated in this starting pool at exceptionally high frequency (5% at each of 140 bases in the active region of the ribozyme). Thus, every possible single-base mutation was present in enormous numbers. Indeed, all possible *combinations* of several mutations were likely to have been present in many copies (e.g., ~ 1500 copies of every possible combination of four mutations). In the first few generations, a particularly fit combination of four mutations increased, and this increase was more or less deterministic (green curves in Fig. 17.2C). However, the final winner differed by seven mutations from the original and was probably not present in the original population. Instead, it was assembled by a series of random mutations that occurred later in the experiment—a process that did involve an element of chance.

This example of selection among RNA molecules is extreme. With typical mutation rates of 10^{-9} to 10^{-8} per base per generation in eukaryotes (see pp. 334 and 426), many species are not large enough to produce every single-base change in every generation (i.e., $N_e\mu \ll 1$). Moreover, favorable alleles may arise through multiple mutational changes or may involve complex rearrangements (e.g., duplications of large regions; Chapter 12). Such mutations are essentially unique and are subject to the hazards of random drift when they arise.

Often, many different mutations at the same or at different loci can increase fitness. For example, the anticoagulant poison warfarin acts by binding to vitamin K oxide reductase, an enzyme essential for blood clotting. Five years after warfarin was introduced into Britain in 1953, resistant rats were found in lowland Scotland, and 2 years after that, a different resistance allele started to spread in the Welsh Borders (Fig. 18.3). Many other independent outbreaks have since occurred involving at least three distinct alleles. Similarly, human populations have evolved partial resistance to malaria through a variety of mutations to several genes. The S allele of β-globin is common in Africa (recall Box 14.1), whereas the E allele is found in Southeast Asia (Fig. 18.4). In addition, thalassemias are caused by diverse rearrangements of the globin genes, and alleles at other loci, including glucose-6-phosphate dehydrogenase and the Duffy blood group, also cause resistance to malaria. The geographic distribution of these various alleles is likely to be largely a matter of chance.

Because the establishment of favorable mutations is a random process, populations diverge if they pick up different mutations in a different order. This is nicely illustrated by the long-term selection experiment that we saw in Figure 17.32. Over 10,000 generations, the fitness of the evolving populations increased relative to the ancestral population by about 70%. As the ability to grow on glucose increased, the growth rate on other carbon sources decreased (Fig. 18.5A). The loss of ability to

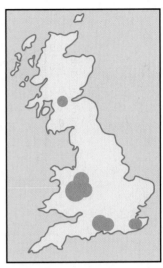

FIGURE 18.3. Resistance to the anticoagulant poison warfarin in rats (*Rattus norvegicus*) is due to multiple alleles. Areas where warfarin-resistant rats are found are in *red*. Here, resistant alleles are kept polymorphic by heterozygote advantage. The Scottish, Welsh, and English resistance outbreaks are due to different alleles.

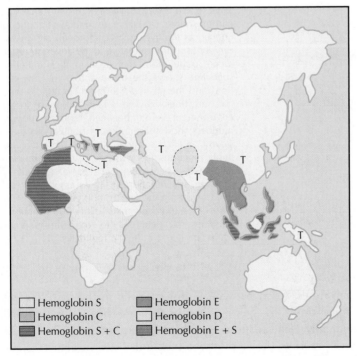

FIGURE 18.4. Human populations have evolved a variety of alleles that give partial resistance to malaria. The map shows the distribution of the common hemoglobin variants S, C, D, E; S is known as the sickle-cell allele. These alleles differ in amino acid sequence. There are also several hundred regulatory variants of α- and β-globin, which alter gene expression and cause thalassemia (T). These are almost all rare and have a restricted geographic distribution.

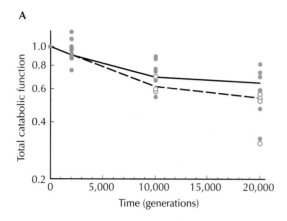

A

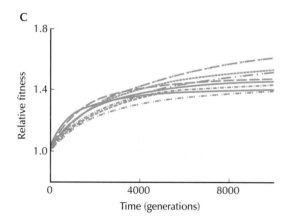

C

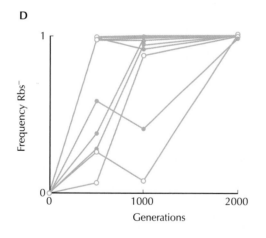

D

B

Carbon source	Time (generations)		
	2,000	10,000	20,000
Bromosuccinic acid	7	11	12
D-alanine	1	3	6
D-malic acid	5	12	12
D-ribose	12	12	12
D-saccharic acid	9	11	11
D-serine	12	11	10
D-sorbitol	12	11	11
Fructose-6-phosphate	11	10	9
Fumaric acid	9	12	12
Glucose-1-phosphate	12	11	10
Glucose-6-phosphate	11	12	8
Glucuronamide	0	4	8
L-asparagine	8	12	12
L-aspartic acid	9	12	12
L-glutamine	12	12	12
L-lactic acid	11	12	10
L-malic acid	7	12	12
Malic acid	9	12	12
Mono-methylsuccinate	2	12	12
Mucic acid	12	8	9
p-Hydroxyphenylacetic acid	5	12	11
Succinic acid	9	12	12
Uridine	12	12	10

FIGURE 18.5. Adaptation of 12 *Escherichia coli* populations to live on minimal medium with glucose as the sole carbon source. (*A*) Decrease in total catabolic function calculated as the average of the growth rate measured on 64 different substrates relative to the ancestor. *Open circles* indicate populations that evolved high mutation rate. The *solid line* shows the mean across the low mutation rate lines and the *dashed line* the mean across the high mutation rate lines. (*B*) The detailed response to life in a new environment varied between replicate populations. The numbers show the number of populations that grew more slowly on each substrate than the ancestor did. *Red* highlights functions that consistently decayed; *turquoise* indicates two cases where there was a significantly improved function. (*C*) Variation in fitness among the replicate populations during the first 10,000 generations. *Curves* are fitted to measurements made every 500 generations. (*D*) Frequencies of deletions to the ribose operon, which eventually fixed in all 12 replicates.

utilize other substrates varied between lines, especially in the first 2000 generations (Fig. 18.5B). For example, by then 7 of the 12 replicates had reduced growth on bromosuccinic acid, only 1 had reduced growth on D-alanine, and so on. By 20,000 generations, however, most of the 16 catabolic functions listed in Figure 18.5B had been lost.

Some functions were lost early on. By 2000 generations, every line had lost the ability to grow on D-ribose (Fig. 18.5C). All the populations carried deletions in the ribose operon, but the size and position of these deletions varied from population to population. Ribose catabolism was lost consistently and quickly because the ribose operon is flanked by repeated **insertion sequence** elements, which cause a high mutation rate (see pp. 342–343). Variants at other, less mutable, loci are established more erratically, leading to the wide variation in which functions are lost (Fig. 18.5B) and in overall fitness (Fig. 18.5C).

Given enough time, all 12 lines might accumulate the same set of favorable mutations and converge on a single optimal genotype. However, if there are interactions among sites (as we saw in the in vitro selection experiment of Fig. 17.2), then there may be multiple adaptive peaks, and divergence will be permanent (see pp. 466 and 472–475). The chance acquisition of different favorable mutations is one of several mechanisms of divergence, which we will examine when we consider the origin of new species on pages 642–644.

In Small Populations, Deleterious Alleles Can Be Fixed by Chance

Even deleterious alleles can be fixed by chance in a small population. Although an allele may, on average, slightly reduce fitness, there is some chance that the individuals who carry it will have slightly *more* offspring than expected, so that the deleterious allele drifts up to high frequency and, perhaps, displaces every copy of the supposedly superior allele.

The probability that a single copy of an allele will eventually be fixed, P, depends on the relative rates of drift ($\sim 1/N_e$) and selection ($\sim s$). The ratio between these rates is $N_e s$, and so this combined parameter determines whether selection or drift is the dominant process. We can think of this as the effective number of **selective deaths** in each generation. As long as $N_e s$ is not too negative, there is an appreciable chance that a mildly deleterious allele will fix. For example, if $s = -0.001$ and $N_e = N = 1000$, then $N_e s = -1$ and $P = 3.6 \times 10^{-5}$ (Fig. 18.6).

The chance fixation of weakly deleterious mutations could eventually cause the collapse of quite large populations. For the human population, a simple calculation suggests that fitness could collapse over approximately 300,000 generations or 6 million years (Box 18.1). Obviously, humans have many more urgent dangers to worry about. Even from a

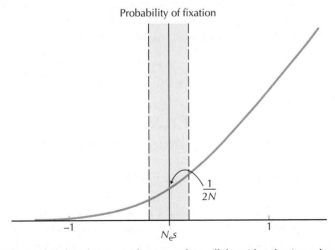

FIGURE 18.6. The probability that a single copy of an allele with selective advantage s will be fixed in a population of effective size N_e is $2s(N_e/N)/(1 - \exp(-4N_e s))$, where N is the actual number of individuals. The graph shows this probability plotted against $N_e s$, for $N_e = N$. If the allele is strongly favored ($N_e s \gg 1$), then $P \sim 2s(N_e/N)$. If $N_e s$ is small, then drift is much stronger than selection ($1/2N_e \gg s$), and so the allele is effectively neutral (*shaded strip*). Because each of the $2N$ genes in the population has the same chance of ultimately fixing, $P \sim 1/2N$ (see p. 425). Finally, if the allele is deleterious ($N_e s \ll -1$), then the probability of fixation becomes very small: $P \sim 2|s|(N_e/N)\exp(-4N_e|s|)$, where $|s|$ is the positive magnitude of selection (i.e., $-s$ if $s < 0$).

| Box 18.1 | Accumulation of Weakly Deleterious Mutations |

The total genome-wide rate of mutations that change amino acid sequence across the human genome is about 0.9 per haploid genome per generation, and there are about 90 mutations across the genome as a whole. (These estimates come both from direct estimates of mutation rates [see p. 334] and from the rate of divergence between neutral sequences [see p. 426].) If even a small proportion of these mutations were only mildly deleterious, then a significant fraction could accumulate. The probability of fixation of alleles with $N_e s = -0.5$ is approximately 0.3 times the neutral rate (Fig. 18.6), and so they will accumulate through random drift at a rate of approximately 0.3 times the mutation rate. Although the effect of each mutation would be small, the long-term rate of fitness decline could be substantial. The effective size of the human population required to account for observed levels of nucleotide diversity ($\pi \sim 0.0008$; see p. 368) is $N_e \sim 8000$, and so only very weakly selected variants have much chance of being fixed despite their deleterious effects (e.g., $|s| \sim 0.5/N_e \sim 6 \times 10^{-5}$). However, if 10% of the mutations affecting amino acid sequence had effects around this value and if noncoding regions contributed about the same number, then the net rate of fitness decline would be $2 \times 0.10 \times 0.9 \times 0.3 \times 6 \times 10^{-5} = 3.3 \times 10^{-6}$ or about 1/300,000 per generation.

genetic point of view, the increase in inherited diseases that are no longer countered by natural selection will be much more rapid than the very gradual erosion of function by random drift (see p. 773). However, this calculation does suggest that over evolutionary timescales, adaptations cannot be maintained by weak selection in species that have fallen to an effective size of, for example, a few thousand individuals. Whether such a pessimistic scenario is valid depends on whether adaptations are maintained by very weak selection ($s \sim 1/N_e$). It is possible that changes in amino acid sequence usually have large effects on proteins, so that selection against any change is strong (>0.1% say). However, the regulatory functions of noncoding DNA sequences (e.g., binding transcription factors) may be under weaker constraints and therefore more at risk from erosion by mutation and drift. If deleterious alleles do fix, then natural selection may be able to counteract the damage in two ways: The lost allele could be regenerated by mutation and then reestablished by selection, or changes in other genes could compensate for a loss of function.

There are few clear examples of adaptation that are maintained by very weak selection; it is hard to measure even strong selection coefficients. In the next chapter, we will examine evidence for this kind of weak selection and its interaction with mutation and drift. In particular, there is some evidence that slightly deleterious mutations have been fixed in the human population (see Web Notes).

Populations Cluster around Peaks on Wright's Adaptive Landscape

So far, we have discussed the chance that mutations will eventually either fix in the population or be lost. This gives a good description of evolution if mutation rates are low, so that populations are usually fixed for one or the other allele. Often, however, there are several alleles at a locus, and so we need to know how allele frequencies fluctuate.

In Chapter 17 (p. 472), we saw that selection tends to increase mean fitness $\overline{W}$, so that populations climb toward peaks in the **adaptive landscape**. Sewall Wright extended this idea to describe the interaction between several evolutionary processes. He showed that in a balance between random drift and selection, a collection of populations will tend to cluster around adaptive peaks, with a probability distribution proportional to $\overline{W}^{2N_e}$. Thus, the distribution will cluster more sharply around peaks in mean fitness as the effective population size increases (Fig. 18.7A,B). This gives a useful picture of the way selection pushes populations uphill, increasing mean fitness, whereas random drift disperses populations by randomizing allele frequencies.

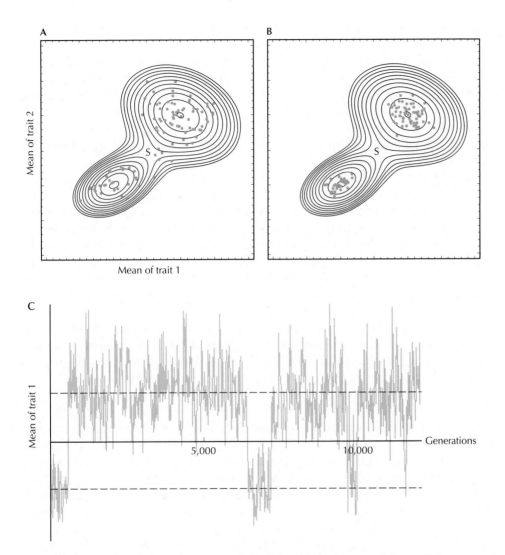

FIGURE 18.7. Under a balance between random drift and selection, populations will be scattered across the adaptive landscape in a distribution proportional to $\overline{W}\,^{2N_e}$. The contours show the mean fitness of a population plotted against the means of two quantitative traits. The traits are under **disruptive selection**, so that there are two adaptive peaks; genetic variance is assumed to be constant. *S* marks the saddle between the two peaks. Contours are plotted at $\overline{W} = 0.91, 0.92, ..., 0.99, 1$. (*A*) Populations of $N_e = 25$ are subject to strong random drift and so are scattered widely across the adaptive landscape. (*B*) In larger populations, with $N_e = 100$, selection is stronger than drift, and populations are clustered around the adaptive peaks. (*C*) A simulation of the time course of change in the first trait (x-axis in *A,B*) for $N_e = 100$. *Dashed lines* show the positions of the two adaptive peaks.

Wright derived a formula that includes the effects of mutation as well as random drift and selection. Suppose that there are two alternative alleles (P, Q) and that mutation rates between them are μ from Q to P and ν from P to Q. For a diploid population, Wright showed that the distribution of allele frequencies is proportional to

$$p^{4N_e\mu-1}\, q^{4N_e\nu-1}\, \overline{W}\,^{2N_e}. \tag{18.1}$$

The relative importance of mutation and drift depends on $4N_e\mu$ and $4N_e\nu$. These can be thought of as twice the numbers of mutations that are generated in every generation (twice, because in a diploid population there are effectively $2N_e$ genes that can mutate). Figure 18.8A shows how the distribution of allele frequencies changes as the number of mutations increases. For $4N_e\mu = 4N_e\nu = 1/2$, populations are mostly near fixation for one or the other allele, and so the distribution of allele frequencies is sharply peaked

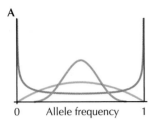

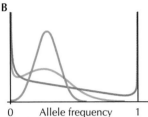

FIGURE 18.8. (*A*) The distribution of allele frequencies in a balance between mutation and random drift, from Wright's formula, $4N_e\mu = 4N_e\nu = 1/2, 2, 8$ (*purple, blue,* and *red,* respectively). (*B*) The distribution with selection favoring heterozygotes. Fitnesses of QQ:PQ:PP are $1 - sp_e$:1:$1 - sq_e$, with selection tending to maintain a polymorphism with $p_e = 0.3$ (see Box 17.1). $4N_e\mu = 4N_e\nu = 1/2$ and $N_es = 2, 8, 32$ (*purple, blue,* and *red,* respectively).

near 0 and 1 (purple curve in Fig. 18.8A). With a larger number of mutations per generation ($4N_e\mu = 4N_e\nu = 8$) (red curve in Fig. 18.8A), random drift is weaker relative to mutation, and so allele frequencies cluster around 0.5. In an infinite population, with equal mutation rates in each direction, allele frequencies will be at precisely $p = 0.5$.

Figure 18.8B shows the effect of selection favoring heterozygotes (Box 17.1). In this example, selection tends to maintain a polymorphism at allele frequency $p_e = 0.3$. When selection is weak relative to drift ($N_es = 2$), allele frequencies tend to be near 0 or 1; as N_es increases, selection dominates and the population clusters around the adaptive peak, at $p_e = 0.3$.

Wright's formula can describe migration in the **island model,** in which immigration of genes from a common gene pool is equivalent to mutation. However, it does not extend to other kinds of migration, for example, between neighboring demes in a one- or two-dimensional array. Similarly, with more than two alleles per locus, the formula applies only with particular patterns of mutation, and it cannot describe the effects of recombination on linkage disequilibrium. Nevertheless, even in these more general cases, it remains true that the behavior of populations depends on parameter combinations such as $N_e\mu$, N_es, and N_em, which describe the strength of different evolutionary forces relative to random drift.

Wright's formula shows how allele frequencies cluster around adaptive peaks when populations have reached a steady state between the different evolutionary processes. We can also ask about the rates of shifts between adaptive peaks. Typically, populations will fluctuate around an adaptive peak for a long time, but may occasionally jump to a new peak (e.g., Fig. 18.7C). Shifts from low to high peaks are more frequent than shifts in the opposite direction and so, on average, populations are more likely to be near the higher peaks. Populations are most likely to jump across the saddles that separate the adaptive peaks (S in Fig. 18.7A,B), because this path involves the least loss of fitness. The rate of shifts between peaks depends primarily on the chance of drifting to the saddle point relative to the chance of being at the peak; from Equation 18.1, this ratio is proportional to $\overline{W}_{\text{saddle}}^{2N_e} / \overline{W}_{\text{peak}}^{2N_e}$. Here, the crucial selection coefficient is defined by the loss of fitness needed to drop down from the original peak to the saddle: $1 - s^\star = \overline{W}_{\text{saddle}}/\overline{W}_{\text{peak}}$, so that the rate of shifts is $\sim(1 - s^\star)^{2N_e} \sim \exp(-2N_es^\star)$. Thus, random drift is only likely to knock populations from one peak to another if $N_es^\star$ is not too large. In the following chapters, we see how Wright used this process as the foundation of his **shifting balance** theory (pp. 607–609) and how shifts between alternative adaptive peaks caused by random drift may account for the evolution of new chromosome arrangements (Fig. 22.22).

SELECTION AND GENE FLOW

Favorable Alleles Spread Rapidly, Even If the Population Is Subdivided

Although favorable alleles are likely to be lost when present in just a few copies, they can spread rapidly once established. It takes approximately $\log(1/p_0)/s$ generations for an allele with advantage s to increase from frequency p_0 to 50% (see p. 469). This time can be quite short, even in populations so large that the initial frequency is very low. For example, in the experiment in *E. coli* illustrated in Figure 18.5, the ability to grow on ribose was lost as a result of fixation of deletions to the ribose operon (Fig. 18.5D). The selection favoring deletions was measured as $s = 1.4\%$, and the frequency of deletions in the ancestral population was 0.0005. Hence, deletions are expected to reach 50% in $(1/0.014)\log(1/0.0005) = 540$ generations. The actual time taken fluctuated widely, but averages around this value (Fig. 18.5D). An allele with a stronger selective advantage would spread more rapidly, in proportion to $1/s$. For example, with a 14% advantage, it would take approximately 54 generations to reach 50% from the same starting frequency.

In reality, populations are subdivided into local populations, or demes, that are connected by gene flow. This slows down the spread of adaptations; nevertheless, alleles that are favored everywhere can spread through the entire species quite rapidly, at least on an evolutionary timescale.

Remarkably, subdivision of the population into demes does not change the probability of fixation from the simple value $2s$ provided that gene flow is symmetrical. To understand why subdivision of this sort does not alter the fixation probability, consider an individual who produces offspring that then move to different places. As long as the distribution of offspring number (i.e., of **fitness**) is not affected by where the offspring live, the success of the alleles they carry will also be unaffected. In addition, subdivision into demes does not itself change the overall allele frequency. For example, exchange of genes between two populations of the same size leaves the overall allele frequency unchanged, even though it does reduce differences in allele frequency between the populations.

Subdivision does, however, slow the spread of an allele once it is established. If (as in most cases) the species is spread over a broad two-dimensional range, the favorable allele will increase to high frequency in one region and then spread out behind a **wave of advance**. Fisher showed that this wave settles down to move at a speed of about $\sqrt{2s\sigma^2}$. Here, σ^2 is the rate at which genes diffuse through the population (see p. 442) and s is the allele's selective advantage. For example, an allele with advantage $s = 0.5\%$ would spread at a rate $\sqrt{2 \times 0.005\sigma^2} = 0.1\sigma$, and so would take about ten generations to spread by one dispersal range σ (Fig. 18.9).

When genes or species spread by local diffusion, they expand behind a sharp wave front moving at a steady speed (e.g., Fig. 18.10A,B). In many cases, though, the spread is much faster than can be explained by local diffusion. For example, after the last ice age, many tree species spread as the glaciers retreated. Fossil pollen shows that the ranges often expanded faster than 100 m/year. Yet, the great majority of seeds fall within 30 m and take decades to grow to a mature tree. In such cases, expansion is through occasional long-range dispersal events. For example, a few seeds may be carried long distances by animals or, more recently, by humans. The pattern of spread is then quite different from the steady waves seen with local diffusion (contrast Fig. 18.10B with 18.10D). Occasional migrants found new colonies well away from the existing range. These colonies expand and eventually merge, leading to an erratic and patchy advance that tends to accelerate over time (Fig. 18.10C,D). These examples show the spread of species across the landscape, but exactly the same principles apply to the spread of genes through populations.

The Spread of Replicating RNA Molecules Can Be Studied in the Laboratory

Short RNA molecules will replicate in the presence of an RNA polymerase from the bacteriophage Qβ. If a small quantity of a particular RNA sequence is inoculated into one end of a capillary tube containing Qβ polymerase and the necessary substrates, then that sequence will increase exponentially at first and then spread out behind a wave of advance that settles to a steady speed (Fig. 18.11). These wave fronts spread at an average of 2.3 μms^{-1}, a speed that can be compared with the prediction from Fisher's formula, $\sqrt{2s\sigma^2}$. The rate of diffusion of the polymerase-RNA complex is $\sigma^2 \sim 7.6 \times 10^{-11}$ m^2s^{-1}. (This is defined as the variance of distance moved along the habitat per unit time, in just the same way as for genes in real organisms [see p. 442].) The rate of exponential increase of these RNA sequences was $s = 0.038$ s^{-1}, corresponding

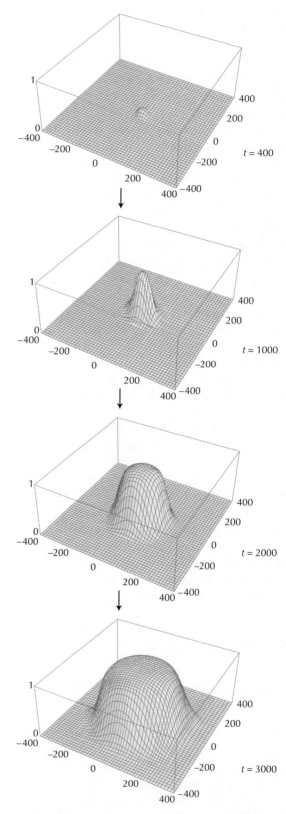

FIGURE 18.9. The spread of an allele with advantage $s = 0.5\%$. At first, numbers of the allele grow exponentially as $\exp(st)$ and spread out in a normal distribution with variance $\sigma^2 t$ in each direction. As the allele approaches fixation at the center, it spreads out behind a wave of advance, which has speed $\sqrt{2s\sigma^2}$—in this example, 0.1σ per generation. Graphs are for $t = 400$, 1000, 2000, and 3000 generations.

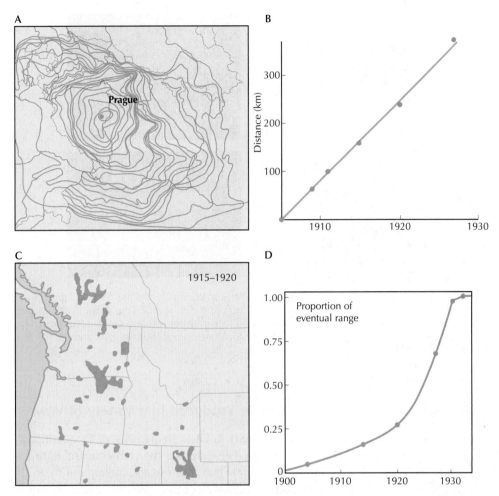

FIGURE 18.10. The muskrat (*Ondatra zibethica*) is native to North America, but spread rapidly through Europe after five muskrats escaped from a fur farm near Prague in 1905 (*A*). The wave of advance moved at a rate of about 14 km per year (*B*). In contrast, cheatgrass (*Bromus tectorum*) spread through rangelands in the western United States in a series of erratic long-distance movements, aided by the growth of the railroad system during the 19th century (*C,D*).

to a doubling roughly every 20 seconds. The predicted speed is thus

$$\sqrt{2(0.0380 \text{ s}^{-1})\, 7.6 \times 10^{-11} \text{ m}^2\text{s}^{-1}} = 2.4 \text{ μms}^{-1},$$

which agrees with the actual speed (Fig. 18.11).

We are following here the spread of a single type of RNA sequence, rather than competition between two different types (i.e., alleles). However, the mathematical description is the same in the two cases: Fisher's formula describes the spread of species through an empty habitat as well as the spread of new alleles that are replacing the previous (inferior) copy. In fact, Figure 18.11 shows evidence of the replacement of one allele by another. The arrow indicates a sharp increase in wave speed, which corresponds to the replacement of one sequence by another that replicates faster. Sequencing of the molecules before and after such changes showed that one sequence had indeed been replaced by another slightly fitter sequence. Even if no RNA is added, monomers will aggregate spontaneously to produce some random sequence. This sequence is then selected to replicate rapidly, so that the sequence that has the highest fitness under the experimental conditions emerges.

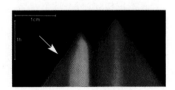

FIGURE 18.11. The rate of spread of replicating RNA molecules along a capillary tube. The experiment was seeded with a very low concentration of a particular 133-bp sequence. Wave fronts can be seen propagating at constant velocity to either side of the positions where the progeny of single RNA molecules established themselves. The *arrow* shows a sharp increase in growth rate caused by the evolution of a faster-replicating variant.

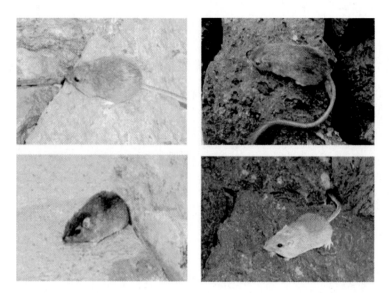

FIGURE 18.12. The rock pocket mouse, *Chaetodipus intermedius*, usually lives in light-colored rocks and has a correspondingly light coloration (*top left*). On several lava flows in the western United States, however, dark coat color has evolved, which better matches the dark volcanic rocks (*top right*). Mice on an inappropriate background are conspicuous (*bottom row*) and are likely to suffer greater predation by owls.

Geographic Patterns (Clines) Can Be Produced in a Variety of Ways

Often, we see systematic geographic gradients in allele frequencies or in quantitative traits. These may be over very short scales. For example, the coat color of mice can change from dark to light over less than a few hundred meters, corresponding to the changing color of the local rocks (Fig. 18.12). At the other extreme, gradients can be spread over entire continents; for example, the increase in body size at higher latitudes seen in many terrestrial animals.

Such gradients are referred to as **clines**. They are important for evolutionary studies for several reasons. The geographic patterns that Darwin observed during his voyage on the *Beagle* gave him a key piece of evidence for evolution (see pp. 16–18 and 70): Taxonomically similar organisms tended to be found in the same area, suggesting common descent, and the continuous transitions in form seen in space showed the possibility of an analogous continuity through time. Geographic patterns can also tell us about the mechanisms of evolution, because they represent the accumulation of differences that have evolved over long periods of time. Patterns in space can substitute for changes through time that cannot be observed directly.

Clines may simply be relics of the contact between divergent populations, which had been separated by climatic or geological changes. (Recall our discussion of the mixing of populations on pp. 452–455.) Such patterns are common in northern latitudes, where many species expanded from multiple refugia after the last glaciation, which ended less than 10,000 years ago. For example, the eelpout *Zoarces viviparus* is a fish found in both the Baltic and the North Seas. These two populations differ at several enzyme loci, which show smooth, parallel clines through the narrow straits that connect the two seas (Fig. 18.13). Concordance of clines at different genes is strong evidence that differences are a historic relic, rather than being influenced by selection.

Where selection is involved, clines may simply arise as a direct reflection of a changing environment. At each location, selection may maintain alternative alleles in a **balanced polymorphism** (e.g., by heterozygote advantage) with equilibrium frequencies that reflect local conditions. We have already seen one such example in our own species:

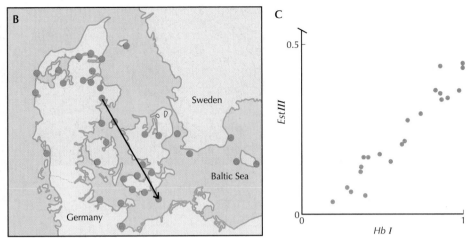

FIGURE 18.13. Concordant clines in the eelpout *Zoarces viviparus* (A). (B) The collecting sites, running from the North Sea through to the Baltic Sea. (C) Allele frequencies at the loci *EstIII* and *HbI* run parallel along this transect.

The sickle-cell hemoglobin allele is maintained because it gives partial resistance to malaria, and its frequency parallels the incidence of that disease (Fig. 18.4).

In these examples, the scale of the geographic pattern is too broad to be influenced by the movement of genes from place to place. However, sharper changes are opposed by the random movement of individuals and of the genes they carry. In the next section, we will see that by studying the interaction between gene flow and selection, we can better understand and quantify both processes.

Narrow Clines Can Be Maintained by a Balance between Selection and Gene Flow

Clines may reflect adaptation to abruptly changing environments. One of the best examples is the adaptation of the grass *Agrostis tenuis* to live on spoil heaps from mines that are contaminated with heavy metals such as lead, copper, and zinc. Few species can grow on such heavily polluted sites, and those that do can live there only because they have evolved tolerance to these toxic metals (Fig. 18.14A). Tolerance changes abruptly over just a few meters, and so the conflict between selection and gene flow can be seen clearly. Plants grown from seeds collected on the toxic soil are less tolerant than adults living on the toxic soil (right of Fig. 18.14B), presumably because they are fathered by pollen from tolerant plants off the mine. The opposite pattern is seen off the mine (left of Fig. 18.14B). Thus, the cline in resistance is shallower in seedlings than in adults (Fig. 18.14B) and is sharpened by selection in every generation.

A similar example of adaptation to changing environments shows how patterns of genotype frequencies in time and space can be used to estimate the strength of selection and gene flow. Every summer, areas within 20 km of the southern French coast are sprayed with organophosphate insecticide to control mosquitoes (*Culex pipiens*). In these areas, the insects have evolved resistance through changes in two genes. Resistance alleles at the *Ace 1* locus modify acetylcholinesterase, which is the target of the insecticide, and those at *Est 1* cause overproduction of detoxifying esterases. However, these resistance alleles remain confined to the coastal regions, showing that they are at a selective disadvantage in the absence of insecticide. Clines have formed at each locus, which shift back and forth in a regular way through the seasons. The strength of selection for and against resistance in different areas, and also the rate of dispersal, have been estimated by fitting a model to observed genotype frequencies (Fig. 18.15).

Clines can also be maintained by selection even in a homogeneous environment. The simplest example is where two alternative chromosome arrangements meet in a narrow boundary (e.g., Fig. 17.22). On either side, selection acts against whichever arrangement

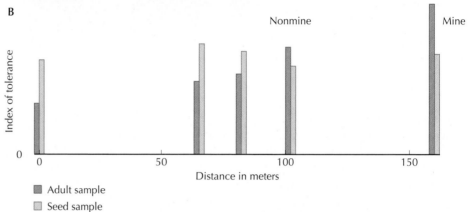

FIGURE 18.14. Sharp clines in the copper tolerance of the grass, *Agrostis tenuis*, at the edge of a mine in North Wales. (*A*) The sparse vegetation on spoil heaps polluted by heavy metals (*right*) contrasts with unpolluted areas (*left*). (*B*) Adult *Agrostis* taken from the mine are more tolerant than are plants grown up from seeds taken from the same location (compare *red* with *blue* at *right*). In contrast, adults are less tolerant than seedlings away from the mine (*lower left*).

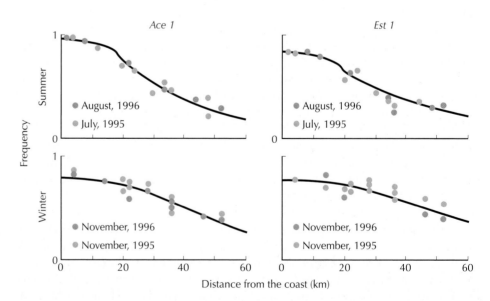

FIGURE 18.15. Clines in insecticide resistance alleles in the mosquito, *Culex pipiens*. Resistance alleles at both loci increase in the summer when insecticide is sprayed; over the winter the clines broaden out. The rate of gene flow during the summer was estimated as $\sigma^2 = 6.6$ km^2 per generation; during the autumn, it was higher, at 14.6 km^2 per generation. The selection coefficients favoring resistance alleles while insecticide was being sprayed were estimated as 0.33 and 0.19 for *Ace 1* and *Est 1*, whereas in unsprayed areas during summer, selection coefficients against them were 0.11 and 0.05, respectively.

is rarer, simply because the rarer arrangement is found mostly in heterozygotes, which are partially sterile (Box 12.2). Thus, a cline is maintained by a balance between selection against heterozygotes, which tends to fix one or the other chromosome arrangement, and gene flow, which broadens the cline. Figure 18.16 shows the boundary between two chromosomal rearrangements in the alpine grasshopper *Podisma pedestris*. Where these meet, they are separated by a cline approximately 800 m wide. This cline runs for more than 100 km through a wide variety of habitats and across the whole range of altitudes occupied by this species (~1500–3000 m). This pattern is strong evidence that the cline is maintained by steady selection against chromosomal heterozygotes rather than by selection that favors different arrangements in different environments.

The same kind of cline can be maintained whenever selection can maintain two alternative stable states (roughly speaking, when there are alternative adaptive peaks; see pp. 466 and 472–475). For example, butterflies of the species *Heliconius erato* and *Heliconius melpomene* are **Müllerian mimics**—they are distasteful to potential predators and advertise their distastefulness by bright warning patterns (see p. 475). In any one area, all butterflies from both species converge on one pattern. Any deviant from this pattern is more likely to be eaten, because it is not recognized as being distasteful. The range of these two *Heliconius* species is divided into a mosaic of different geographic races separated by narrow clines (Figs. 17.23 and 18.17).

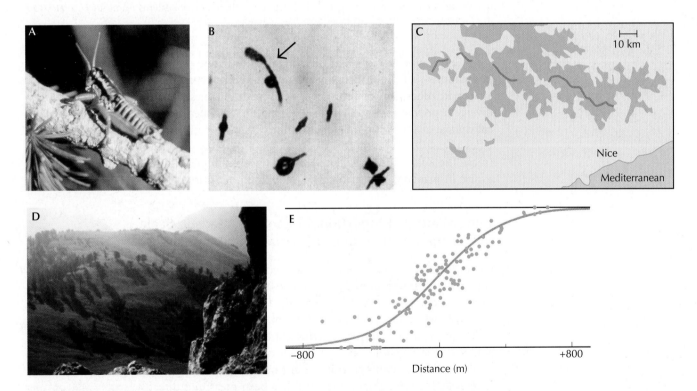

FIGURE 18.16. The chromosomal cline in the grasshopper *Podisma pedestris*. (*A*) An adult male. (*B*) Chromosome pairs at male meiosis; the cross-like structures are **chiasmata** that correspond to recombination events (see Fig. 12.24). The *arrow* shows a fusion between the X chromosome and one of the autosomes. This is paired up with the homologous autosome, which is now acting as a Y chromosome. (Males of this species carry one X; females carry two.) (*C*) This fusion between the X chromosome and the autosome is found only in the southern Maritime Alps. Mixed populations and heterozygous females are found only within a narrow cline about 800 m wide (*red line*). This species is common above 1500 m (*shaded area*). (*D*) Detail of the cline at its southeastern end. The frequency of the fusion runs from about 90% at *lower left* to about 10% at *upper right*. (*E*) The frequency of the fusion plotted against distance along this region. Each point represents a sample of about 20 males.

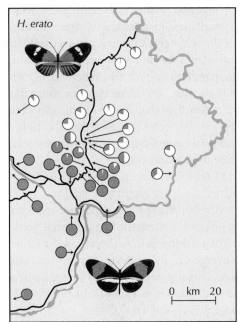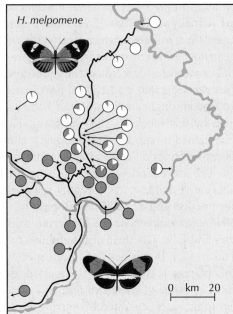

FIGURE 18.17. Parallel clines in the warningly colored butterflies *Heliconius erato* (*left*) and *Heliconius melpomene* (*right*) near Tarapoto in Peru. The pie-shaped symbols show the average frequency of alleles at the major genes that determine the different patterns (three in *H. erato*, four in *H. melpomene*). *Blue:* Rio Huallaga; *black:* roads.

The examples given so far have involved only one or a few genes or traits. Often, populations that differ in many ways meet and interbreed, forming what are known as **hybrid zones**, sets of more or less coincident clines. These are often found when divergent populations meet after having been separated by some climatic or geological change or following introduction into a new region. Such **secondary contact** provides a ready explanation for the coincidence of many genetic differences in one place (recall the *Zoarces* example, Fig. 18.13). However, in principle multiple differences can evolve in **primary contact** with no geographic separation.

Cline Width Is Proportional to a Characteristic Scale Set by the Relative Rates of Dispersal and Selection

Clines and hybrid zones can be described by a simple model of diffusion (see p. 442) combined with selection. The outcome is determined primarily by the rate of diffusion of genes, relative to selection. These determine a characteristic distance, $l = \sqrt{\sigma^2/2s}$, which determines the width of the cline. Strong selection (large s) and weak diffusion (small σ^2) lead to narrow clines (small l). The actual width of a cline depends on just how selection works, but typically the cline will have a width of $(2–4)l$.

For example, with selection against heterozygotes, we expect a cline with width $4l$. In *P. pedestris* (Fig. 18.16), dispersal has been measured directly as $\sigma^2 = 400$ m^2 per generation (i.e., the grasshoppers move with a standard deviation of 20 m in a year). The cline that separates the two alternative chromosome arrangements is approximately 800 m wide, and so selection of $s = 0.005$ against heterozygotes is needed to maintain such a cline ($800 \text{ m} = 4 \times \sqrt{400 \text{ m}^2 \text{ gen}^{-1}/(2 \times 0.005 \text{ gen}^{-1})}$).

The length scale $l = \sqrt{\sigma^2/2s}$ determines the minimum distance over which populations can adapt to localized selection. If an allele is favored within a region with diameter smaller than roughly l, then gene flow swamps selection. One interesting practical result of Thomas Lenormand and colleagues' study of insecticide resistance (Fig.

18.15) is that if about half the area were sprayed, resistance alleles could not be established—selection within sprayed areas would be overwhelmed by gene flow from elsewhere. This illustrates the general point that gene flow interferes with selection and can prevent the establishment of favored alleles if the area is relatively small.

We have given several examples of how populations can adapt and diverge over very short scales: rock pocket mice (Fig. 18.12), the grass *Agrostis tenuis* (Fig. 18.14), and insecticide resistance in mosquitoes (Fig. 18.15). It is hard to know how common these cases are, and still harder to know whether, by preventing local adaptation, gene flow limits the expansion of species into new habitats.

BALANCING SELECTION

Heterozygote Advantage Is Unlikely to Be the Main Form of Balancing Selection

What maintains genetic variation within populations? We have seen that there is extensive variation for almost all kinds of genes and traits and in almost all populations (see pp. 367–370 and 397–399). A major, and as yet unresolved, puzzle is to understand what causes this variation. In the following sections, we set out contrasting explanations. Then, in the next chapter, we will review the various ways of finding how variation is maintained and how it affects fitness.

As we saw in Chapter 1 (p. 33), evolutionary biology has been dominated by two kinds of explanation of variation. In their extreme forms, these reflect quite different views of the world. The **classical view** supposes that there is typically a single fittest wild-type genotype and that variation is primarily due to mutation. The **neutral theory** (see pp. 59 and 425–427) can be seen as a descendant of the original classical view to allow for the extensive genetic variation in protein and DNA sequence that was discovered in the 1960s. In contrast, the **balance view** supposes that most variation within species is maintained by selection and that differences between species are mostly adaptive. In the next three sections, we explore how **balancing selection** can operate. Then we set out the contrasting view that variation within populations is mainly due to deleterious mutations.

Variation can only be maintained by selection if rare alleles are favored. Alleles that are in danger of loss will tend to increase, thereby maintaining variation within the population. Thus, understanding how selection maintains variation amounts to understanding how an allele can become relatively fitter as it gets rarer.

If heterozygotes are fitter than homozygotes, then rare alleles gain an advantage, simply because they are more likely to find themselves in heterozygotes. We have already seen two examples of polymorphism maintained by heterozygote advantage: the Hb^S allele of β-globin (Box 14.1) and warfarin resistance in rats (Fig. 18.3). However, it is hard to judge the overall importance of this mechanism for maintaining balanced polymorphism from a few well-studied examples. A strong argument against it is that there is abundant genetic variation both at discrete genes and for quantitative traits in organisms such as fungi, which are predominantly haploid, or in organisms that usually self-fertilize and so are largely homozygous. Heterozygote advantage cannot act if heterozygotes are absent or rare.

Genetic Variation Can Be Maintained If Selection Favors Rare Alleles

Rare alleles can gain a fitness advantage in other ways. For example, many plants have evolved incompatibility systems, such that their ovules cannot be fertilized by their

FIGURE 18.18. The plant *Oenothera organensis* is restricted to a few canyons in the Organ Mountains of New Mexico and numbers fewer than 1000 plants in total. Despite that, it maintains many polymorphic self-incompatibility alleles.

own pollen. These systems are based on single genes and can act in two ways. In **gametophytic** systems, pollen carrying an allele that is shared with the ovule cannot fertilize, and in **sporophytic** systems, pollen derived from a parent that shares an allele with the ovule cannot fertilize. In either case, a rare allele has a strong advantage, because it can fertilize any plant in the population. Once an allele becomes common, plants carrying it will waste pollen in failed fertilizations of plants that share the same allele. Consequently, incompatibility loci show extremely high polymorphism. For example, 34 distinct alleles were found in a sample of only 135 *Oenothera organensis* (Fig. 18.18). Figure 18.19 shows the extent of diversity at the molecular level in another self-incompatible species. At the incompatibility locus of the flowering plant *Physalis longifolia*, 114 of 122 amino acids are polymorphic, and some sites have five or six different amino acid variants. In this example, it is not known which amino acids are actually responsible for determining mating compatibility. We consider the relation between the alleles that are actually under selection and the observed molecular variation in Chapter 19 (p. 506).

This kind of selection, in which the fitness of an allele decreases with its frequency, is known as negative **frequency-dependent selection**. The most familiar example is the ratio of males to females in a population. If the **sex ratio** in the population were biased toward males, then females would necessarily produce more offspring. Every child must have a male and a female parent, and so in a male-biased population, each female must on average produce more children than each male. Therefore, selection favors an increase in the proportion of females, which tends to restore equal proportions of the two sexes. This argument is associated with R.A. Fisher, but, in fact, was first made by Darwin and was analyzed mathematically in the late 19th century. The sex ratio is now one of the best-understood parts of evolutionary biology; we examine it in more detail in Chapter 21 (p. 603).

Another good example is **Batesian mimicry**, in which a palatable species gains protection from predators by mimicking distasteful models (Fig. 18.20). Such species often mimic a variety of model species. For example, in East Africa females of the swallowtail butterfly *Papilio dardanus* mimic several different model species that look quite different from one another. Any pattern that becomes common will tend to lose protection, because predators will no longer associate that pattern with prey; most of the butterflies that carry it will be from the mimic, not the model. Thus, rare alleles, coding for rare patterns, gain an advantage that maintains the polymorphism. It is remarkable that such a polymorphism can be maintained, because the patterns differ in many aspects of color pattern and wing shape. In *Papilio*, the different morphs are de-

```
DNTNT--RLMDCSPPPNYTNFQ-DKMLDDLDKHWTQLKIFKNKSKIDQSTWSYQYKKHGSCCQNLYNNQNMYFSLALHLKDKV
EKDKV--LQINCPPTPNYTNFQ-DKMLDDLDTHWTQLLLTKKTGLEEQRIWNYQFRKHGSCCREL-YNQSMYFSLALGLKAKV
DKNNS--LLMDCTPRPNYTYFPRNKMFADLDKHWTQLKITEDDAETDQSTWSRQYIKHGSCCRNL-YNQNMYFSLALHLKNRV
EKKGVD-KLTFCSAQPNYTIFKDKKMLDDLDKHWIQLMYSKENGLQKQEFWKSQYEKHGSCCLNR-YNQTAYFSLASHLKKKI
EKRGIK-MMVSCKPEVNYTLFQDRKMLDDLDKHWTQLKVSKDEGLEKQEAWKYQYEKHGACSQES-YNQNMYFSLALHLYERF
EKRGKN-IMVSCKPEVKYALFQDRKMLDDLDKHWIQLKVSKDEGLEKQEAWKHQYEKHGACSQES-YNQNMYFSLALHLYERF
DNFSA--KLNFCGPNTYDKTILKDYKKNKLYIHWPDLVVDEAKCKKDQKFWSDEYGKHGTCCEKT-YSQEQYFDLAMVLKDKF
DNIST--TLNFCKGVTYKNVTG--EKKNNLYIHWPDLLVEEANCKTYQTFWKKEYDKHGSCCEGT--NQEQYFDLAVALKDKF
DNIST--TLNKCKSIPYDKNMT-DDKKNMLYIRWPDLLVGEASCKKDQKFWKNEYEKHGTCCEES-YNQEQYFDLAMGLKDKF
DRNNS--VLVECEPFRGYTNFK-DNMLDELDKHWTQFKYDKTSGLKDQKTWRYQYRRHGTCCQEL-YNQDMYFSLALRLKRKF
DKNNS--VLVECQPLRGYTNFK-DNMLDELDKHWTQFKYDKTSGLKDQKTWRYQYRRHGTCCQEL-YNQDMYFSLALHLKRKV
              HVa                                    ,HVb
```

FIGURE 18.19. Frequency-dependent selection maintains high diversity at the self-incompatibility locus in the flowering plant, *Physalis longifolia*. The figure shows the amino acid sequences of 11 of 33 genes that were sampled from the eastern United States. (The 20 possible amino acids are denoted by their single-letter codes [Fig. 2.23].) All 33 amino acid sequences were different, and diversity is especially high in two hypervariable regions, HVa and HVb. *Red* indicates conserved amino acids.

FIGURE 18.20. Batesian mimicry in females of the swallowtail butterfly *Papilio dardanus*. The *left* column shows three different unpalatable model species in the family Danaidae, and the *right* column shows palatable *Papilio dardanus* mimics. All three patterns are found as polymorphisms within populations of *P. dardanus*.

termined by a cluster of tightly linked genes (a **supergene**) that carries several alternative combinations of alleles. This allows a clean switch between quite different patterns.

A well-understood example of frequency-dependent selection comes from the plant-colonizing bacterium *Pseudomonas fluorescens*. If this is grown in well-shaken liquid medium, then a single type is found. It is called SM, because it forms smooth colonies when grown on agar (Fig. 18.21A, left). However, if a genetically uniform SM clone is grown for a few days in small vials that are left undisturbed, the population diversifies in a reproducible way. The colonies now take varied forms when individual bacteria are seeded onto an agar plate. These can mainly be classed as the ancestral smooth (SM), as "wrinkly spreader" (WS), or as "fuzzy spreader" (FS) (Fig. 18.21A). SM grows within the medium, whereas WS forms dense mats at the surface (Fig. 18.21B), and so they occupy different **niches** within the vial. Competition experiments

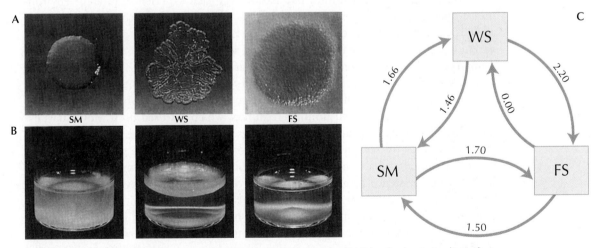

FIGURE 18.21. When *Pseudomonas fluorescens* is grown in small vials, the bacteria diversify in a reproducible way. (*A*) The three main types of colony: SM, smooth spreader (the ancestral type); WS, wrinkly spreader; FS, fuzzy spreader. (*B*) The distinctive growth patterns of these three types. (*C*) Summary of invasion experiments in which one type was introduced at low frequency (1%) into a population of another type. In all except one case (FS introduced into WS), the rare type invaded. The numbers on each arrow show the fitness of the rare type relative to the common type, measured over 7 days: numbers greater than 1 indicate successful invasion.

showed that in most cases each type has a fitness advantage when rare, which allows all three types to coexist at stable frequencies (Fig. 18.21C). (The one exception is that FS cannot invade WS. However, FS presumably can invade a population consisting of WS and SM, so that all three types can coexist.)

The mechanism of the frequency-dependent selection between WS and SM is understood, at least in outline. WS cells stick to each other and so can form dense mats at the air–broth interface. Although WS is rare, these mats allow it to exploit increased oxygen levels at the surface, but as the mat gets heavier, it sinks. A genetic analysis has shown that an **operon** responsible for synthesis of a cellulose-like polymer is essential for the WS phenotype, but other changes are also needed. Because reproduction is asexual, the three types can coexist without poorly adapted intermediates being produced by sex and recombination.

Often, negative frequency dependence arises because there is a variety of limiting resources available to a species and different genotypes tend to exploit different resources. Thus, a rare genotype will find its preferred resource relatively unexploited and will gain an advantage. One can view the previous examples in this way. For an incompatibility locus, the different resources are the various potential mates, whereas for Batesian mimicry, the resources are the various distasteful model species. In *P. fluorescens*, there is a spatial heterogeneity in the unstirred vials. On this view, the maintenance of genetic diversity within species and the coexistence of different species have the same fundamental cause: the advantage to a gene or a species that can more efficiently exploit an underused resource. We return to this issue when we discuss the origin of species (Chapter 22 [pp. 650–654]).

Only Certain Kinds of Fluctuating Selection Can Maintain Variation

Intuitively, one might expect that variation in selection from place to place or from time to time could maintain genetic variation. Diverse demands from the environment might maintain diverse populations. However, there is no direct or simple correspondence between variation in environment and variation in genotype. In this section, we show that fluctuating selection can maintain genetic variation, but only under particular circumstances.

Different individuals within a population will each experience different environments during their lifetimes. However, all that matters for the success of a particular allele is the *average* fitness of individuals who carry that allele. In a large population, variation between individuals will average out and, in a small one, it will lead to **random genetic drift**, which reduces genetic variation (Chapter 15).

We saw in the previous section that if different genotypes can exploit different resources, then rare types will be at an advantage and variation will be maintained. It is crucial that there be different limiting resources, which get used up if individuals exploit them. "Resources" could take many forms, but, to be concrete, think of an insect that can lay eggs on several different species of food plant. These grow intermingled within its habitat and are each capable of supporting a limited number of larvae (Fig. 18.22A). The existence of a variety of resources is not enough by itself. If different genotypes survive better or worse on different kinds of food plant, but exploit them at random, then they can in principle coexist, but only if selection is very strong (Fig. 18.22B). Polymorphism can be maintained with weak selection, and for a wide range of parameters, if genotypes use different resources—for example, if different genotypes choose to lay their eggs on different host plants (Fig. 18.22C).

What if the environment varies through space on a larger scale than the individual's dispersal range? We have already seen that as long as this scale is much larger than the characteristic scale *l* set by selection and dispersal, locally adapted alleles can be

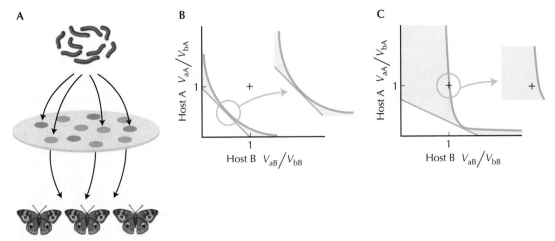

FIGURE 18.22. Polymorphism can be maintained in a heterogeneous environment only under special conditions. (*A*) Juveniles settle into separate patches. There is competition within each patch, so that each patch produces a fixed number of adults. There are two types of organism (a, b) and two kinds of patch (A, B); survival of type x on host Y is V_{xY}. (*B*) If juveniles settle at random on the two kinds of patch (A, B), then polymorphism is possible if one type survives much better on one patch and the other much better in the other patch (i.e., $V_{aA}/V_{bA} \gg 1$, $V_{aB}/V_{bB} \ll 1$ [*blue area at top left*] or $V_{aA}/V_{bA} \ll 1$, $V_{aB}/V_{bB} \gg 1$ [*blue area at bottom right*]). However, if selection is weak (*inset area near $V_{aA}/V_{aB} \sim 1$, $V_{aB}/V_{bB} \sim 1$*), polymorphism is possible only for a very limited range of parameters. (*C*) If the different types *choose* to settle in different habitats, polymorphism can easily be maintained even when selection is weak. Indeed, polymorphism is possible even when there is no selection within patches (around the *cross*, indicating equal survival).

maintained. (For example, melanic mice on dark rocks [Fig. 18.12] or grasses tolerant of heavy metal pollution [Fig. 18.14].) Such local selection can readily maintain variation within the species as a whole, but is less effective at maintaining variation *within* populations. Polymorphism is confined to the narrow boundaries between habitats.

Coevolution May Be an Important Source of Variation

Finally, what if selection varies over time? We have already seen that temporal fluctuations cannot by themselves maintain variation, because all that matters in the long run is the average fitness of each genotype (Fig. 17.17). However, a low rate of mutation can combine with fluctuating selection to generate some variability. If conditions change, so that previously deleterious alleles become advantageous, then new mutations can sweep through the population and will contribute to variation as they do so. Just as with spatial variation, the scale of fluctuations must be intermediate if substantial variation is to be generated. Rapid fluctuations will just average out, whereas if fluctuations are slow, populations will usually be fixed for one of the alleles.

Fitnesses are likely to vary with changing physical conditions (e.g., climate) and with changes in the biological environment as other species change in abundance and themselves evolve. The continual evolution of organisms in response to evolution in other species is termed the **Red Queen**, after the character in Lewis Carroll's *Through the Looking-Glass* who always had to run in order to stay in the same place. W.D. Hamilton argued that the struggle between parasites and their hosts is especially important. The evolution of resistance to the current parasite genotype may be countered by the evolution of new virulent genotypes in the parasite, leading in turn to new resistant genotypes in the host. In effect, rarer genotypes gain an advantage, because hosts will not have evolved resistance to rare parasite genotypes and because parasites will not have evolved the ability to infect rare host genotypes.

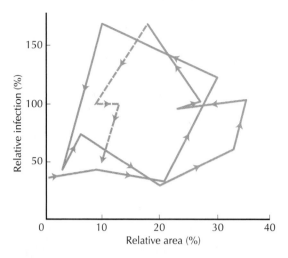

FIGURE 18.23. Evolution of virulence in powdery mildew (*Erysiphe graminis*), an asexual fungus that is a serious disease of barley. In the late 1960s, new varieties of barley were planted, which carried the resistance allele Mla12. By the end of the decade, mildew carrying the corresponding virulence genes Va12 began to spread. The varieties were withdrawn from cultivation, and by the mid-1970s infection rates had fallen. New varieties of barley carrying Mla12 and other resistance factors began to be planted extensively, but again lost their resistance. The rate of infection by mildew is plotted against the area planted by varieties carrying Mla12, for the years 1967–1983.

In Chapter 19 (p. 532), we will see indirect evidence of selection on genes involved in coevolutionary interactions. However, it is difficult to observe this kind of host–parasite **coevolution** in action. The genotypes responsible for virulence and resistance must be identified and their frequencies followed over many generations. In agriculture, pathogens often evolve to infect the commonly planted varieties of crops; farmers then plant new, resistant varieties that in turn become susceptible (Fig. 18.23). Lively and Dybdahl have studied a natural system, the coevolution between freshwater snails (*Potamopyrgus antipodarum*) and their trematode parasites (*Microphallus* sp.). In the populations illustrated in Figure 18.24, the snails reproduce asexually. The populations are divided into distinct clones, which can be identified by the combination of enzyme alleles they carry. The parasites are locally adapted, so that they best infect the snails found in the same lake (Fig. 18.24B). Moreover, parasites are most successful at infecting common clones, which demonstrates that rare genotypes have a frequency-dependent advantage (Fig. 18.24C). Finally, clones that become more common also become more highly infected, after a 1-year time lag (Fig. 18.24D). Together, these patterns suggest that there is a cycling of host and parasite genotypes, which maintains clonal variation in these populations.

Continual coevolution of this sort causes fluctuations in selection over just the right timescale to maintain substantial fitness variation; it may therefore be important in driving the evolution of mate preferences (see p. 583) and sexual reproduction (see p. 672). Although we have discussed hosts and parasites, the same kind of process arises in other interactions. For example, on page 591 we will see how conflicts between genes that are inherited in different ways or expressed in different contexts can also lead to rapid evolution and high levels of genetic variation.

MUTATION AND SELECTION

Deleterious Mutations Are Responsible for Much Variation

In contrast to balancing selection, the alternative explanation for maintenance of genetic variation is simple: Mutation generates variation, and selection eliminates it. This means that variation reduces fitness, rather than increasing it (e.g., by speeding up adaptation or by allowing diverse organisms to better exploit diverse resources). The very different implications of these two alternative explanations have led to strong controversy (see p. 33).

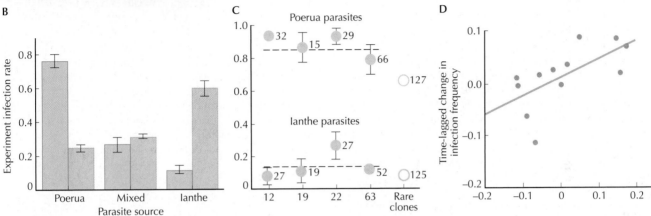

FIGURE 18.24. Coevolution between the freshwater snail *Potamopyrgus antipodarum* (*A*) and its trematode parasite (*Microphallus* sp.) studied over a 5-year period. (*B*) The rate of infection of snails by parasites from the same or from different lakes. Parasites infect snails from the same lake much more successfully. Snails and parasites came from two lakes in New Zealand, Poerua (*turquoise*) and Ianthe (*orange*). ("Mixed" parasites are offspring from a cross between parasites from different lakes.) (*C*) The Poerua snails are all asexual and can be classified into clones according to their allozyme genotype (12, 19, 22, or 63). There were four common clones, which were infected at a high rate by the parasites from the same lake (*top row*). However, the rare clones, taken together, were infected significantly less (*top right*). In contrast, parasites from a different lake infected the Poerua snails much less, regardless of whether they were members of common or rare clones (*bottom row*). (*D*) The rate of increase of the snail clone is correlated with the rate of increase of infection by parasites, but with a 1-year time lag. The *x*-axis shows the increase in host clone frequency from year *y* to *y* + 1, and the vertical axis shows the subsequent change in rate of infection, from year *y* + 1 to *y* + 2.

The variation generated by mutation is seen most clearly in inherited diseases, such as achondroplasia (short stature) or retinoblastoma (malignant eye tumors), which have a simple genetic basis. In the extreme case of a dominant lethal or sterile allele, all cases are directly due to a new mutation and their frequency in the population is twice the mutation rate. (Each diploid individual has two copies of each gene, either of which might mutate.) With weaker selection, where heterozygotes have fitness $1 - s$, the allele frequency reaches an equilibrium set by the balance between the rate of mutation and selection ($p = \mu/s$; Box 18.2). The frequency of affected heterozygotes is approximately twice this ($\sim 2\mu/s$), because at **Hardy–Weinberg** proportions, heterozygote frequency is $2pq$, and $q \sim 1$. We do not need to consider homozygote fitness here, because typically mutation is so much weaker than selection that the deleterious allele is rare and homozygotes hardly ever appear. The strength of selection can be estimated very simply for such diseases, because the fraction of cases due to new mutations (i.e., where neither parent is affected) is just equal to the selection coefficient (Table 18.1).

Many inherited diseases that are due to a single gene are caused by an autosomal recessive allele, which is manifested only in homozygotes. (These make up about one-

Box 18.2   Mutation/Selection Balance at a Single Locus

In Box 17.2, we saw how selection changes allele frequencies at a single locus. Here, we add the effects of mutation, so as to find the balance between selection eliminating deleterious alleles and mutation introducing them.

Suppose that there are two alternative alleles, P, Q, at frequencies p, q, respectively, and that allele P is deleterious. It is simplest to start by thinking of a haploid population, so that there are just two types. We set the fitness of P as $1 - s$ relative to a fitness of Q equal to 1. For weak selection ($s \ll 1$), we saw in Box 17.2 that this kind of selection will reduce the frequency of P at a rate $-spq$ per generation.

The effect of mutation is easily found. We will assume that mutation is in only one direction: Q alleles mutate to P at a rate μ, and mutation in the opposite direction is negligible. This is reasonable, because although there are many ways by which a functional allele can be disrupted, it is much harder for an allele that has been degraded to regain its function (recall Fig. 14.21; the "allele" P may be a set of many sequences with reduced fitness, rather than one particular sequence).

The frequency of Q after mutation is

$$q^{**} = q(1 - \mu),$$

because a fraction $1 - \mu$ of the Q alleles survive intact. The allele frequency therefore changes by $q^* - q = -\mu q$ or $\Delta p = p^* - p = +\mu q$. When selection is weak, we can just add this change to that due to selection (Chapter 28 [online]) so that

$$\Delta p = \mu q - spq.$$

At an equilibrium between mutation and selection, $p = \mu/s$.

This formula also applies to a diploid population provided that the deleterious allele is rare enough to be eliminated mostly in heterozygotes (rather than homozygotes). Then, Q alleles will almost all be in QQ homozygotes, and P alleles will almost all be in PQ heterozygotes. What matters is the difference in relative fitness between these two genotypes. Sometimes, fitnesses of QQ:PQ:PP are written $1:1 - hs:1 - s$, so that the equilibrium allele frequency is $p = \mu/hs$. We will keep the notation simple by denoting the heterozygote fitness $1 - s$, so that the equilibrium is still at $p = \mu/s$.

For a completely recessive allele (fitnesses of QQ:PQ:PP are $1:1:1 - s$), the rate of elimination of P by selection is $-sp^2q$. Thus,

$$\Delta p = \mu q - sp^2q,$$

and the equilibrium is at $p = \sqrt{\mu/s}$. This allele frequency can be much higher than when selection acts against heterozygotes. For example, if $\mu = 10^{-5}$ and $s = 10\%$, the equilibrium allele frequency is $\sqrt{10^{-5}/10^{-1}} = 1\%$. However, the PP homozygotes that express the deleterious effects are still very rare ($p^2 = 10^{-4}$).

TABLE 18.1. The proportion of cases of autosomal dominant disorders that are due to new mutations is equal to the selection coefficient against the deleterious alleles

Disorder	Proportion
Apert's syndrome	>0.95
Achondroplasia	0.80
Tuberous sclerosis	0.80
Neurofibromatosis	0.40
Marfan's syndrome	0.30
Myotonic dystrophy	0.25
Huntington's disease	0.01
Adult polycystic kidney	0.01
Familial hypercholesterolemia	<0.01

Reproduced from Vogel F. and Motulsky A.G. 1997. *Human genetics: Problems and approaches.* Springer-Verlag, Berlin, Table 9.8, p. 397 (modified from Goldstein J.L. and Brown M.S. 1977. *Annu. Rev. Biochem.* **46:** 897–930 [© Annual Reviews]).

third of the inherited diseases in humans that have a simple mode of inheritance, although this is probably an underestimate. Recessive inheritance is harder to confirm than dominant or X-linked and, in other species, spontaneous mutations with visible effects are predominantly recessive.) However, even when an allele is classed as "recessive" in its phenotypic effects, its frequency may still be determined primarily by selection against heterozygotes. Although this selection may be weak (so that there is no noticeable consequence to individuals carrying one copy of the allele), it can still eliminate more copies of the mutation, simply because there are so many more heterozygotes than homozygotes in a random-mating population. In some populations, however, recessive mutations are mainly eliminated in homozygotes that arise from matings between relatives. We consider the effects of inbreeding in more detail on pages 515–518.

We have already seen an example of mutation/selection balance at the molecular level. Recall that populations of *E. coli* grown on glucose lost the ability to use ribose as their carbon source (Fig. 18.5D). This loss was due to the fixation of deletions in the ribose operon, which were at a frequency of $p = 0.0005$ in the base population. The mutation rate to such deletions was measured directly, as $\mu = 5.4 \times 10^{-5}$ per generation. Because $p = \mu/s$ at equilibrium, a selective disadvantage of $s = 5.4 \times 10^{-5}/0.0005$ ~ 11% is needed to keep deletions at low frequency in the original population. This fairly strong selection contrasts with the weaker advantage to deleting the ribose operon when it is not required: $s = 1.4\%$ (see p. 492).

We have seen that in almost all organisms there is extensive variation in protein and DNA sequence. Typically, there is much less variation at functional sites than at putatively neutral sites. For example, **synonymous** variation, which does not alter amino acid sequence, is much more common than variations that do alter the protein (e.g., Figs. 13.21 and 18.25). The abundant variability that we see may be neutral, maintained in a balance between mutation and selection, or maintained by balancing selection. In Chapter 19, we will see how these three alternatives can be distinguished.

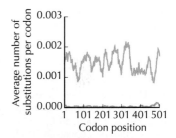

FIGURE 18.25. Synonymous (*blue*) and **nonsynonymous** (*red*) variation in the glycoprotein gene of vesicular stomatitis virus, an RNA virus that infects cattle. Variation is measured using a 20-bp sliding window. Overall, synonymous variation is 20 times higher.

Mutation Maintains Variation in Quantitative Traits

One of the most striking facts of biology is the widespread inherited variation that is seen in almost all traits in almost all organisms (see pp. 397–399). This variation is especially surprising because in most cases we expect it to affect fitness. It is hard to believe that growth rates, fruit weight, or flying ability are neutral. Although the number of bristles on a *Drosophila* abdomen may seem trivial to us, bristles are important sensory organs for the fly. We will see in the next chapter that it is surprisingly difficult to measure the effects of particular traits on fitness, but nevertheless it is widely accepted that these are usually under **stabilizing selection**, such that deviations from the optimum are detrimental. The key arguments are that, first, morphology typically remains constant for extremely long times (see p. 280 and Fig. 17.30) and, second, that an organism's function depends on the more or less precise coordination of many different traits (see p. 407). Yet, continual stabilizing selection should eliminate variation.

The simplest explanation for heritable variation in quantitative traits is that mutation balances stabilizing selection in much the same way that mutation and selection can balance at a single locus. Provided that deleterious alleles are rare, the genetic variation that is maintained depends in a simple way on the relative rates of mutation and selection: $V_G = 2UV_S$. In this formula, fitness declines away from the optimum following a **normal** curve, with variance V_S. Thus, small V_S implies strong stabilizing selection, so that only a narrow range of trait values gives high fitness. U is the total rate of mutations that affects the trait. The effects of individual alleles do not appear in the formula because mutations with large effects remain rare, whereas those of small effects can become more common. The amount of genetic variance contributed by alleles with different effects is therefore the same.

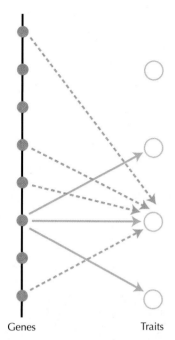

Genes   Traits

FIGURE 18.26. There is extensive pleiotropy. Each quantitative trait is influenced by many genes (*purple*); conversely, each gene typically influences many traits (*blue*).

For example, suppose that the strength of stabilizing selection is $V_S = 20V_E$. (Thus, an individual that deviates by one environmental standard deviation, $\sqrt{V_E}$, will have fitness reduced by $\sim 1/(2 \times 20) = 2.5\%$.) To maintain a heritability $h^2 = 50\%$ (i.e., $V_G = V_E$), we therefore need $U = 0.025$. The rate at which mutations affecting quantitative traits arise is not known with any accuracy, but this value is not unreasonable for traits that are affected by large numbers of genes (see p. 411).

This rough calculation shows that variation could be maintained by mutation if we look at just a single trait. However, a mutation/selection balance becomes much less plausible when we allow for the fact that very many traits influence fitness. If mutation rates are to be high enough to maintain variation (e.g., $U = 0.025$), then each trait must be affected by many genes. But in that case, each gene must affect many traits. In other words, there must be extensive **pleiotropy**, as we expect in any case from the complex relation between genes and traits (Fig. 18.26). Now, variation will be reduced by stabilizing selection on all the traits that are influenced by the gene, and the genetic variance V_G will be reduced by a factor equal to the number of traits affected. Once this effect of pleiotropy is appreciated, it becomes hard to see how mutation can account for high heritabilities.

Once we accept that alleles affecting quantitative traits are likely to show extensive pleiotropy, it becomes simpler to look at the problem in another way. We can think of each allele as affecting the trait of interest (e.g., bristle number) and also reducing fitness. Crucially, however, we do not need to assume that the allele reduces fitness *because* it changes the trait—the two effects could be quite separate. Now, genetic variance in the trait is maintained as a pleiotropic side effect of mutation/selection balance at the underlying loci. It is easy to show that this maintains a genetic variance $V_G = V_M/\bar{s}$, where V_M is the trait variance generated by mutation every generation and $\bar{s}$ is the average selection against the alleles involved. The **mutational variance** V_M has been measured in many organisms and is typically $0.001V_e$ to $0.01V_e$ (Table 14.3). Thus, high heritability ($h^2 = 50\%$, with $V_E = V_G$) can be maintained if the alleles involved reduce fitness by $\bar{s} = 0.001–0.01$. It is possible that the mutational variance is largely due to mildly deleterious alleles of this sort, which would then account for the high heritabilities we see. However, for this to be so, these alleles must have very small effects on traits under stabilizing selection, otherwise they would be eliminated too rapidly to contribute much genetic variance. At present, we know too little about the distribution of effects of mutations on traits and on fitness to know whether most variation could be due to a mutation/selection balance.

Variation in Quantitative Traits May Be a Side Effect of Polymorphisms Maintained for Other Reasons

What are the alternatives? Balancing selection acting directly on quantitative traits could maintain diversity by favoring rare phenotypes. Individuals with different trait values could exploit different resources. For example, birds with different beak sizes might handle differently sized seeds with varying efficiency. (Recall the example of selection in the Galápagos finches; Fig. 17.25.) Fluctuations in selection over time, combined with mutation, could also maintain variation, as discussed above for single genes. However, we do not know of any examples of these kinds of direct balancing selection.

Just as for mutation/selection balance, it may be better to think of quantitative variation as a pleiotropic side effect of variation that is maintained for reasons unrelated to the particular trait we happen to observe. Indeed, balanced polymorphisms are expected to have effects on all kinds of traits. For example, sickle-cell anemia is likely to reduce children's growth. In the polymorphic *Pseudomonas*, the growth pattern of colonies on agar plates is altered as a side effect of adaptation to different niches (Fig. 18.21). If balancing selection is widespread, it might account for substantial heritability.

How might these alternatives be distinguished? A key distinction is that mutation/selection maintains alleles at low frequency, whereas most forms of balancing selection maintain polymorphisms with alleles at high frequency. At present, evidence from studies of **quantitative trait loci** (QTLs) is conflicting. On the one hand, in several *Drosophila* studies, variation has been shown to be due to **transposable elements** that are at low frequency in the population and that are almost certainly maintained despite slightly deleterious effects. On the other hand, associations between common alleles at candidate loci and trait variation have been found, which suggest balancing selection (see p. 404). The question of whether variation is due to common or to rare alleles is crucial to determine the genetic basis of disease in human populations using **association studies** (see pp. 404–407 and 758–765*).*

Inbred Individuals Are Less Fit: Inbreeding Depression

When different populations of a crop plant or a domesticated animal are crossed, there is often a large increase in yield. This phenomenon is known as **heterosis**. It is exploited commercially in the production of F_1 hybrid seed from a cross between two inbred lines that individually would perform poorly. (Most maize is produced in this way.) The converse is that offspring from matings between relatives within a population often have reduced vigor—a phenomenon known as **inbreeding depression**. Both heterosis and inbreeding depression were well known long before other aspects of inheritance were understood. For example, Darwin performed extensive experiments on inbreeding depression in plants.

Although these phenomena were first observed in domesticated and laboratory populations, it is now clear that they are ubiquitous in nature. For example, a population of song sparrows (*Melospiza melodia*) on Mandarte Island, British Columbia, has been studied for many years and relationships between individual birds observed. A pedigree spanning 20 years or 16 generations allows the **inbreeding coefficient** F to be calculated for each individual (Box 15.3). The inbreeding coefficient for an individual correlates with its survival to adulthood. Inbreeding depression was measured by fitting a model in which survival is proportional to $\exp(-BF)$. Over the whole study, B averaged 2.7, corresponding to a reduction in survival of about 2.7% for every 1% increase in F. Although the degree of inbreeding depression varies considerably, values of this order have been found in many other studies. In humans, for example, data on matings between relatives have given estimates of inbreeding depression for survival (B) ranging from 0.57 to 2.55.

Heterosis is also seen in the Mandarte Island sparrow population. The offspring of immigrants from the mainland who mated with resident sparrows had higher fitness. Female F_1's laid earlier than the residents, whereas males were more likely to breed. Similar increases in fitness have been seen in several other cases in both plants and animals.

In most studies, the relationships between individuals within populations are not known, and it is not feasible to make experimental crosses to measure inbreeding depression. However, inbreeding depression can be evident as a correlation between fitness and heterozygosity at marker loci. In early studies using allozyme loci, such correlations were attributed to direct effects of **overdominance** at the enzymes themselves. These enzymes often have important metabolic roles, and we will see in the next chapter that they may often be under selection. However, correlations between heterozygosity and fitness components are also seen when heterozygosity is measured using markers such as **microsatellites**, which are most unlikely to have any measurable effect on fitness. Thus, heterozygosity is likely to reflect the overall level of inbreeding of the individual, so that correlations with fitness components are an indirect measure of inbreeding depression.

An example where something is known of the mechanisms involved comes from the

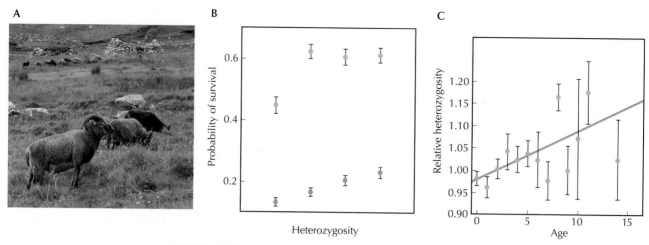

FIGURE 18.27. Overwinter survival of Soay sheep (*Ovis aries*) (*A*) is correlated with their heterozygosity at up to 14 microsatellite loci. (*B*) Survival has been corrected for environmental variables (e.g., weather) and is compared between four equal-sized quartiles of increasing heterozygosity. *Red circles*, overwinter survival of lambs; *blue circles*, adults. (*C*) As a result of the correlation between heterozygosity and survival, relative heterozygosity increases with age.

sheep that live on the isolated island of Hirta, in the St. Kilda archipelago northwest of Scotland. The Soay breed has lived on the archipelago for as far back as records exist, and the present Hirta population originates from 107 sheep, which were transferred from the island of Soay in 1932. The current Hirta population fluctuates between 500 and 2000 sheep. Overwinter survival of both lambs and adults increases substantially with heterozygosity at microsatellite loci (Fig. 18.27B) and, as a result, older sheep are more heterozygous (Fig. 18.27C). Inbred sheep are more susceptible to nematode gut parasites, and no relation was seen between heterozygosity and survival in sheep that had been cleared of parasites with antihelminthic drugs. This suggests that inbreeding depression in these sheep is largely mediated by variation in parasite resistance.

Inbreeding depression is often strong enough to affect population size and can even lead to extinction (Fig. 18.28). Thus, estimating the degree of inbreeding depression is important for conservation biology, both for designing captive breeding programs and for managing endangered populations in nature. For example, heterosis can be exploited by moving individuals between local populations, thereby enhancing their fitness.

Inbreeding Depression Can Be Caused by Deleterious Recessive Mutations or by Overdominance

There are two possible causes of inbreeding depression. Overdominance might maintain polymorphisms at multiple loci; thus, inbred individuals would be more homozygous at these loci and so would have reduced fitness. Alternatively, if deleterious recessive alleles are present in the population, then matings between relatives would tend to bring these together as homozygotes, again reducing fitness (Fig. 18.29). The same two explanations apply to heterosis. On the one hand, if two populations have fixed different alleles at overdominant loci, F_1 hybrids between them will be fitter. On the other hand, if deleterious recessives are at different frequencies in the populations, they are less likely to become homozygous in hybrids (recall the **complementation test**; Box 13.2).

These alternative explanations are hard to distinguish because if deleterious recessives are at tightly linked loci, they behave like a single overdominant locus and are unlikely to be separated by recombination in experimental crosses (Fig. 18.29, right).

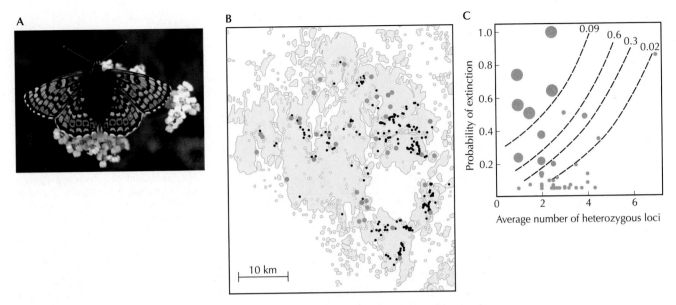

FIGURE 18.28. Small populations of the Glanville fritillary (*Melitaea cinxia*) (*A*) in Aland, south-western Finland, are less likely to go extinct if they are more heterozygous. (*B*) Meadows occupied by this butterfly are shown by *small black circles*; suitable but unoccupied meadows by *open blue circles*. Sampled populations shown in *green* survived, but those marked by *red circles* went extinct between 1985 and 1996. (*C*) The *y*-axis shows the probability of survival predicted from ecological factors; the *x*-axis shows the average number of heterozygous loci based on six allozyme loci and one microsatellite locus. Contours are the estimated survival probabilities: There is a significant effect of both ecological and genetic factors. The *red circles* are the populations that went extinct and the *green circles* are those that survived; the sizes of the circles are proportional to predicted extinction probability.

However, the alternatives do have different consequences. If inbreeding depression and heterosis are due to deleterious recessives, then an optimal homozygous genotype free of any deleterious alleles could be produced. This would be a practical benefit to farmers, because they would be able to plant true-breeding seed year after year, rather than depending on seed companies for F_1 hybrid seed.

If deleterious recessives are primarily responsible, then we might expect the fitter wild-type allele at each locus to be fixed in populations that are habitually inbred. Then, deleterious recessive alleles would be exposed to selection and thus purged from the population. There is evidence for purging of this sort. Populations that are inbred for

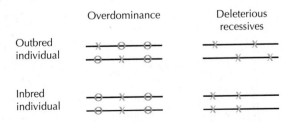

FIGURE 18.29. Alternative explanations for inbreeding depression. With overdominance, outbred individuals are fitter because they are more likely to be heterozygous for different alleles at over-dominant loci (different alleles indicated by *circles, crosses* on *left*). If deleterious recessives are responsible, then outbred individuals tend to be fitter because their two genomes carry different combinations of recessive deleterious mutations (*upper right*). In contrast, inbred individuals are more homozygous, so that the harmful effects of the recessive alleles are expressed.

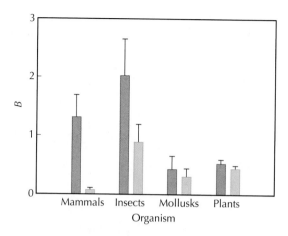

FIGURE 18.30. Evidence for purging of the deleterious recessive alleles responsible for inbreeding depression, *B*. *Red bars*, the extent of inbreeding depression, *B*, before experimental inbreeding; *blue bars*, the level after inbreeding for at least two generations.

several generations typically show a reduction in the degree of inbreeding depression (Fig. 18.30). However, such purging is inefficient. Plant populations that reproduce mainly by **self-fertilization** nevertheless still show substantial inbreeding depression. This is to be expected, because as we saw at the beginning of this chapter, deleterious alleles can increase by chance in small populations even if they reduce fitness (Box 18.1). Thus, if inbreeding depression is due to alleles with only mildly deleterious effects ($s \sim 1/N_e$), these alleles may persist to generate inbreeding depression within populations and may be fixed to contribute to heterosis in crosses between populations.

The existence of widespread inbreeding depression raises important issues, which we will examine in later chapters. Is the genetic variation that underlies inbreeding depression primarily due to balanced polymorphism or to deleterious recessives? Do these have large or small effects (Chapter 19)? What causes the divergence between populations that leads to heterosis (Chapter 22)? How have organisms evolved so as to avoid the harmful consequences of inbreeding (Chapter 23)?

SUMMARY

Random genetic drift interferes with selection even in large populations. A mutation that increases relative fitness by s has a chance of only approximately $2s$ of surviving the first few generations when it is present in just a few copies. Populations diverge because they pick up by chance different combinations of favorable mutations. Conversely, weakly deleterious mutations can be fixed if the product of effective population size and their selective disadvantage is not too large ($N_e s \sim 1$). In the long term, this process can substantially degrade fitness.

Wright quantified his metaphor of an adaptive landscape by showing that the distribution of allele frequencies under drift and selection is proportional to $\overline{W}^{2N_e}$. Thus, populations will cluster tightly around local adaptive peaks if $N_e s$ is large and only have an appreciable chance of jumping between peaks if $N_e s \sim 1$.

Favorable alleles spread rapidly, either behind a wave that advances at a steady rate of $\sqrt{2s\sigma^2}$ or via sporadic long-range jumps. If selection favors different alleles in differ-

ent places or just favors common alleles, then stable clines can form. These have widths proportional to $l = \sqrt{\sigma^2/2s}$.

Polymorphism is maintained if rare alleles tend to increase. Overdominance maintains variation because rare alleles are mostly found in the fitter heterozygotes. However, this cannot be a general explanation, because haploids and selfers are genetically variable. Frequency-dependent selection that favors rare alleles maintains an even sex ratio and maintains polymorphism in plant incompatibility systems and Batesian mimicry. Negative frequency-dependent selection arises when different genotypes exploit different limiting resources. Variation can also be maintained by coevolutionary interactions, where each species responds to evolution by others.

At single loci, a balance between mutation μ and selection s against heterozygotes maintains an allele frequency μ/s. Quantitative variation can also be maintained by a balance between stabilizing selection and mutation: $V_G = 2UV_S$, where U is the total rate of mutations affecting the trait and $\sqrt{V_S}$ measures the range of trait values that have high fitness. However, because pleiotropy is widespread, it may be better to think of quantitative variation as a side effect of polymorphisms that are maintained independently of the trait itself. Then, $V_G = V_M/\bar{s}$.

Inbred individuals are usually less fit, and outcrossed individuals tend to be fitter. Inbreeding depression and heterosis may be due to overdominance or to deleterious recessive alleles—possibilities that are surprisingly hard to distinguish.

FURTHER READING

Selection and Random Drift

Kimura M. 1983. *The neutral theory of molecular evolution.* Cambridge University Press, Cambridge.
   Chapter 6 gives a more detailed account of how selection, mutation, and drift interact.

Selection and Gene Flow

Endler J.A. 1977. *Geographic variation, speciation, and clines.* Princeton University Press, Princeton, New Jersey.
   A broad review of spatial variation.
Haldane J.B.S. 1932. *The causes of evolution.* Longman, New York.
   Still an excellent introduction to population genetics. The Appendix summarizes the basic theory for mutation and migration.
Roughgarden J. 1979. *Theory of population genetics and evolutionary ecology: An introduction.* Macmillan, New York.
   Gives a clear summary of the theory of selection in heterogeneous environments (Chapter 18.12 and 18.13) and of clines (Chapter 18.8–18.10).

Mutation and Selection

Falconer D.S. and Mackay T.F.C. 1995. *Introduction to quantitative genetics.* Longman, London.
   Chapter 20 summarizes arguments over how quantitative variation is maintained in nature.
Lynch M., Blanchard J., Houle D., Kibota T., Schultz S., et al. 1999. Spontaneous deleterious mutation. *Evolution* **53:** 645–663.
   A review of the evolutionary consequences of deleterious mutation.
Vogel F. and Motulsky A.G. 1997. *Human genetics: Problems and approaches.* Springer-Verlag, Berlin.
   A comprehensive review of the human examples used in this chapter. See Chapters 9 (mutation-selection balance), 12 (malaria resistance), and 13 (inbreeding).

C H A P T E R

19

Measuring Selection

I N THIS CHAPTER, WE TURN FROM GIVING examples of how selection can act to assessing its extent and magnitude. What causes differences between species and variation within them? How prevalent are different kinds of selection? How strong are typical selection coefficients? We will see that such issues are hard to answer, and so they continue to be perhaps the most fundamental open questions in evolutionary biology. At one extreme, variation in most coding sequences, and their regulatory elements, may be shaped by selection. Alternatively, the great majority of variants might be effectively neutral, their fate determined by random drift and mutation. These are very different views of the significance of variation for the functioning of the organism. Finding the fraction of sequence variation that significantly affects fitness and the consequence of that effect for the organism remain major challenges. Moreover, the answers to these questions are central to how we find the genetic basis of human disease and how we use artificial selection in agriculture.

Selection can be measured in a variety of ways. Different genotypes can be set to compete with each other in the laboratory; reproductive success can be observed in nature; selection can be measured indirectly, based on the interaction between different evolutionary processes; and inferences can be made from DNA sequence variation within and between species. At the end of the chapter, we bring all the evidence together to ask just how much selection typically acts on natural populations.

DIRECT MEASUREMENT OF SELECTION

Measuring Selection Directly Is Difficult

It is remarkably difficult to make accurate measurements of selection, whether on individual genes or on quantitative traits. There are two basic problems: first, measuring differences in **fitness** (i.e., in number of offspring or rate of reproduction) and, second, finding what causes these fitness differences—which trait, which gene, or, ultimately, which change in DNA sequence. The difficulties are really the same as those involved in finding the genes responsible for variation in quantitative traits (Chapter

14 [p. 402]). Here, we are focusing on a particular trait, fitness, that is itself hard to measure and that may be affected by many genetic differences.

To take the simplest case, suppose that two strains of some organism reproduce asexually. If the difference in growth rate (i.e., the **selection coefficient**) is s, then the ratio between the frequencies of the two strains changes exponentially, as e^{st}. For example, if growth rates differ by $s = 1\%$ per generation, after ten generations the ratio would change by a factor $e^{(0.01 \times 10)} \sim 1.1$ and after 100 generations by a factor $e^{(0.01 \times 100)} \sim 2.72$ (see pp. 467–469). In principle, all we need to do is count the numbers of the two types at the beginning and end of the experiment: The accuracy is limited by the total number we count, by the population size (which determines the rate of random drift), and by how long we can keep conditions constant. For example, if we measure rates of growth at low density, the experiment must stop when the population becomes so large that crowding alters growth rates.

In practice, microbial competition experiments of this sort can resolve fitness differences down to about 0.5% per generation. For example, Figure 19.1A shows the decline in frequency of a mutant strain of yeast relative to a competing wild-type strain.

It is harder to make direct measurements of selection in larger organisms, which reproduce more slowly, in smaller numbers, and under less easily controlled conditions. Figure 19.1B shows how fitnesses can be estimated from competition between three chromosome types in two replicate cage populations of *Drosophila melanogaster*. Note that the frequencies in the replicates tend to fluctuate in parallel as a result of slight changes in their common environment that alter selection. These fluctuations limit the accuracy of fitness estimates to about 5% in this experiment. In general, selection coefficients can be measured to an accuracy of only a few percent at best, which is a serious limitation. As we saw in Chapter 17 (p. 467), far smaller selection coefficients can have significant effects over relatively short times.

Genetic Manipulation Is Needed to Find the Effect of a Specific Genetic Difference

Even if the statistical difficulties of detecting small fitness differences are overcome, one only has a measurement of the net difference between two competing strains. The genomes of these strains may differ in many ways, including multiple point mutations, insertions and deletions, and other rearrangements. As we saw in Chapter 14, it is extremely difficult to find the effect of one particular genetic difference on fitness. Ideally, one would compare organisms that are genetically identical apart from the one difference of interest. In practice, it has been difficult to use genetic manipulation to make such specific changes. For example, markers used in the manipulation will have their own effect. Thus, direct genetic manipulation has so far been used to identify the precise causes of fitness differences in only a handful of cases (e.g., the alcohol dehydrogenase gene [*Adh*] in *Drosophila*; Fig. 14.27).

In the simplest kind of manipulation, a chosen gene is knocked out entirely. For a wide range of eukaryotes (yeast, nematodes, *Arabidopsis*, *Drosophila*, and the mouse), in about two-thirds of cases, this has no obvious effect on the phenotype. In yeast, competition experiments have been used to systematically measure the effects of gene knockouts on fitness in different environments (Fig. 19.2). Although this reveals genes that are essential only in certain environments, it still seems that only a minority of genes have a detectable effect on fitness under any of the experimental conditions.

Presumably, however, some selection does maintain such apparently nonessential genes—or, at least, did so under the natural conditions in which these model organisms lived until they were brought into the laboratory. The quantitative survey of Figure 19.2 is still fairly crude, because growth rates were measured for only a few generations under noncompetitive conditions. More sensitive measurements on 34 yeast deletion strains (as in Fig. 19.1A) have shown that all but seven of these did have sig-

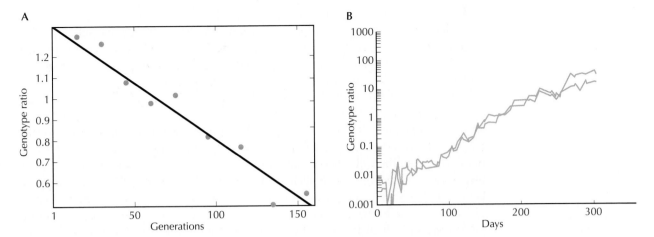

FIGURE 19.1. Selection coefficients can be measured as the rate of change of the ratio between two competing genotypes. (*A*) The frequency of a mutant strain of yeast relative to a competing wild-type strain. The ratio between the frequencies of the two strains is plotted on a logarithmic scale and so is expected to follow a straight line if fitnesses are constant; the slope of the line gives the selection coefficient. In this example, the ratio declines from 1.27:1 to 0.69:1 over 150 generations, and so the selection coefficient is estimated as $s = \ln(0.69/1.27)/150 = -0.004$ (or −0.4%) per generation. (*B*) The ratios between the frequencies of competing chromosomal genotypes in two replicate cages of *Drosophila melanogaster*. The slopes of the graphs give an estimate of relative fitness of 1.75, assuming a generation time of 15 days.

nificantly different fitness in rich growth medium. This suggests that although most genes may not be essential under benign conditions, their function may still be maintained by selection of a few percent or more.

Selection Can Be Measured by Correlating Fitness Differences with Genetic Variation

In a natural population that is reproducing sexually, selection can be measured by observing the fitness of individuals carrying different alleles at the gene of interest. We are still faced with the same difficulties as were seen in laboratory studies: Large samples are needed to detect small fitness differences and, more seriously, we do not know whether any differences are due to the alleles we observe rather than to linked variants. Again, we face the same difficulty as when finding genes responsible for varia-

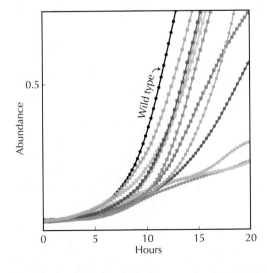

FIGURE 19.2. Genome-wide measurements of fitness in yeast. A series of strains was constructed; in each a single gene was knocked out. The set of 5916 strains (96% of all yeast genes) were then pooled into a single population and grown for several generations on rich glucose medium. This figure shows a subset of these strains compared with wild type. Overall, only 19% of genes were essential for growth on rich glucose medium and, of those deletions that were viable, only 15% grew more slowly than the wild type. Growth rates in several other environments were also measured. For example, 62 genes caused sensitivity to high-salt concentrations when deleted, and 128 caused sensitivity to high pH.

tion in quantitative traits or for human diseases (see Chapters 14 and 26). Indeed, **natural selection** itself faces this problem in picking out those variants that actually caused increased fitness, as opposed to those variants that happen to be associated with them. We will see in Chapter 23 that recombination and sexual reproduction may have evolved to allow natural selection to overcome this difficulty.

Following the discovery of extensive allozyme variation, many studies attempted to find whether this had detectable effects on fitness. In many cases, there were statistically significant differences between genotypes. However, it was usually not possible to show that the fitness differences were due to the alleles themselves. Convincing evidence requires that the fitness effects correlate with the known function of the gene. For example, *D. melanogaster* with the F allele of *Adh* have higher activity of this enzyme (partly because of the single amino acid difference associated with allele F, and partly because of increased gene expression; see pp. 405–406). In experiments with populations of flies held in cages, the F allele usually becomes more common when ethanol is present (Fig. 19.3A). Similarly, flies caught in wineries are more tolerant of ethanol than flies from the surrounding areas and, in some cases at least, the F allele is also more frequent in wineries. This association between alleles of the enzyme and the presence of its substrate, ethanol, could be due to variation at a closely linked site, but the most plausible interpretation is that the effects on fitness and on physiology are both due to *Adh* itself.

Further evidence that selection acts directly on the F/S allele comes from its geographic distribution. The frequency of the F allele systematically increases away from the tropics across five continents. It is implausible that consistent patterns seen over such wide areas could be due to a spurious association. Moreover, within the gene itself, the base responsible for the amino acid change characteristic of the F allele increases steadily in frequency from south to north along the U.S. coast. In contrast, closely linked DNA sequence variants show no such pattern (Fig. 19.3B). Thus, knowledge of the physiological effects of the allele and its spatial distribution supports a causal relation between fitness and the allele.

Although this example shows that selection acts on the *Adh* gene, it remains unclear exactly which sites within the gene are involved (see Fig. 14.27) or exactly how the gene causes differences in fitness. There are other examples of enzyme polymorphisms in which the selective mechanism is better understood. For example, different alleles of phosphoglucose isomerase (PGI) in *Colias* butterflies and of lactate dehydrogenase (LDH) in killifish (*Fundulus heteroclitus*) code for enzyme molecules that differ in their response to temperature; these differences are consistent with their effects on the physiology and behavior of the organism and with their geographic distribution.

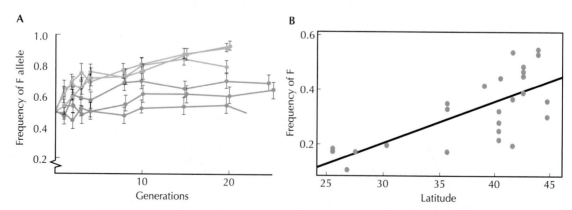

FIGURE 19.3. (*A*) The F allele of *alcohol dehydrogenase* increases in frequency in three replicate laboratory populations of *Drosophila melanogaster* that are exposed to ethanol. The *red lines* show control populations and the *blue lines* show populations with 15% ethanol added to the food. (*B*) The F allele, defined by a substitution of threonine for lysine, increases in frequency northward along the east coast of the United States.

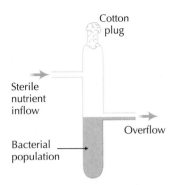

Enzyme Variation Is Often Selected

One of the best understood polymorphisms evolved spontaneously in laboratory populations of the bacterium *Escherichia coli*. A single cell founded a genetically uniform population, which was grown in a **chemostat**, with glucose provided as a carbon source (Fig. 19.4). After 773 generations, three distinct types were found together. This polymorphism was shown to be stable by experiments in which populations that started at different frequencies returned toward the same equilibrium (e.g., Fig. 19.5). One of the types was most efficient at taking up glucose and would be expected to displace the other two if glucose were the only limiting resource. The other two types survived because they specialized in using chemicals excreted by the first, namely, acetate and glycerol. This divergence was found to be due mainly to changes in the enzyme acetyl-coenzyme A synthetase (ACS), which scavenges acetate from the medium. The clone specializing on glucose carried a null mutation, which prevented it from using acetate, whereas the acetate specialist carried a constitutive mutation and so expressed ACS at a high level.

Polymorphisms of this kind often evolve in bacterial populations (e.g., Fig. 18.21). They are maintained by **frequency-dependent selection**, in which a rare genotype gains an advantage by exploiting resources that have not been exhausted by its competitors. These bacterial examples are rather different from the enzyme polymorphisms just discussed, because they involve asexually reproducing clones rather than individual genes in a sexual population. Because the *E. coli* polymorphism had recently evolved, it had a simple genetic basis. However, polymorphic clones might accumulate more genetic differences over time, allowing them to specialize further. In Chapter 22 (pp. 650–654), we will discuss how this kind of specialization to exploit different resources can lead to the evolution of new species, and how sex and recombination interfere with this process.

What fraction of the protein variation revealed by electrophoresis is under selection? It is hard to answer this question on the basis of a few examples such as *Adh* in *Drosophila*, *PGI* in *Colias*, or sickle-cell hemoglobin in humans, because successful detection of selection receives more attention than negative results. However, if we look at the ten polymorphisms in the first published survey of variation in *Drosophila melanogaster*, every one of the six that have been studied biochemically shows substantial kinetic differences between enzymes, in most cases with physiological consequences for the whole organism. In bacteria, chemostat experiments found no selective differences between natural variants at five enzyme loci under glucose-limited conditions. However, significant differences were found in about a third of cases when growth was limited by a different sugar. This strong dependence on the environment makes it difficult to know how selection is likely to act on *E. coli* in its natural habitat, the mammalian gut.

Overall, such surveys suggest that much of the variation in metabolic enzymes is maintained by selection, although this selection may act only in particular environments. However, we still do not know in detail how the mass of coding and noncoding DNA influences fitness. At each gene, are single amino acid differences maintained or are multiple amino acid and/or regulatory differences involved? Are the abundant metabolic enzymes that have been most intensively studied representative of other kinds of gene?

FIGURE 19.4. A chemostat is a simple device for maintaining a microbial population under constant conditions by introducing sterile nutrients at a constant rate.

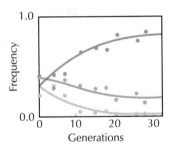

FIGURE 19.5. Coexistence of three genotypes in an *Escherichia coli* population. Over 30 generations, they evolved to a stable equilibrium with two types common and the third occurring at low frequency, but remaining present.

Selection Can Be Measured by Correlating Quantitative Traits with Fitness

Selection on quantitative traits can be studied in the same way as discrete Mendelian genes. One simply takes a sample of individuals and measures the trait, together with fitness or some component of fitness. The first such measurement was made by Herman Bumpus in 1898. He collected sparrows that had survived a severe New England

A

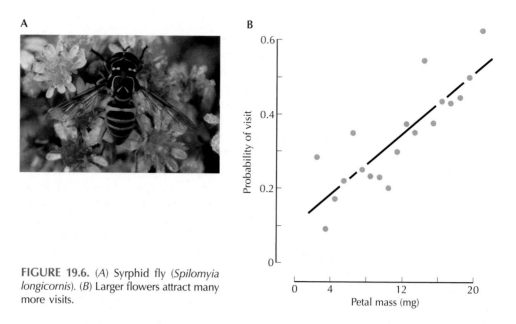

B

FIGURE 19.6. (A) Syrphid fly (*Spilomyia longicornis*). (B) Larger flowers attract many more visits.

storm, together with those that had died. The surviving male sparrows were significantly smaller than the rest of the population: This one episode of **directional selection** had reduced average body length by 1.8% or one standard deviation. In contrast, female sparrows were under **stabilizing selection**, which acts against extreme individuals (Box 17.3). Although the average length of the female sparrows did not change, the variance of the female survivors was 15% lower than before.

As we saw in Chapter 1 (p. 18), Bumpus' work was not followed up for many years. However, from the beginning of the 1980s, many such studies of "natural selection in the wild" were undertaken across a wide range of organisms. They used the same protocol, looking for an association between a set of quantitative traits and some measure of fitness. This simple approach often yielded estimates of strong selection (e.g., Fig. 19.6). In general, there might be a complicated relation between the trait and fitness, which can be visualized as a **fitness landscape** (Fig. 17.21). With enough data, it is possible to measure this landscape by estimating the average fitness of individuals as a function of their phenotype (Figs. 17.27 and 19.7).

Selection on a Correlated Trait Can Cause Apparent Selection on the Trait Being Measured

Just as for discrete genetic variation, the key problem with this approach is that we cannot easily tell whether fitness differences are caused by the measured trait or by something else that happens to be correlated with it. The difficulty can be avoided to some extent by measuring many traits and allowing for the correlation between them. In principle, this statistical approach allows the actual causal effects to be disentangled, but a very large sample size is needed. More fundamentally, one can never rule out, on purely statistical grounds, effects of traits that were *not* measured.

This problem is seen especially clearly in an experiment that attempted to measure selection on the number of abdominal bristles in *D. melanogaster*. Two experimental lines, which had been artificially selected for large and small bristle number, were crossed to produce an F_2 population. Larvae carrying chromosomes that caused intermediate bristle number survived better than those carrying chromosomes for small or large bristle number. That is, there is apparently strong stabilizing selection

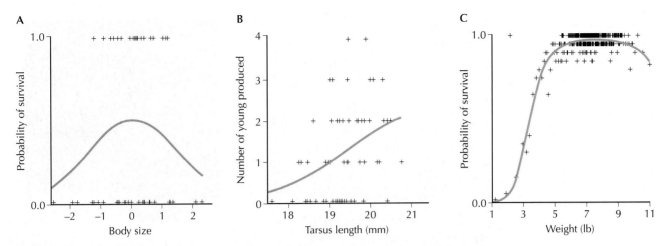

FIGURE 19.7. Examples of the relation between fitness and a quantitative trait (fitness landscapes). (*A*) Bumpus' data on survival of female house sparrows after a New England storm in 1898. The crosses at the top show the relative sizes of the survivors and those at the bottom show the relative sizes of the nonsurvivors. Stabilizing selection acts against extreme sizes. (*B*) Reproductive success in female song sparrows in British Columbia as a function of tarsus length. Here, directional selection acts to increase tarsus length. (*C*) Survival of male human infants as a function of birth weight. Small babies survive poorly, and unusually large babies survive slightly less well.

on bristle number acting via larval survival. However, bristles are only expressed in adults, and so this selection cannot be on bristle number itself. It may be due to **pleiotropic** effects of genes that affect both larval survival and adult bristle number, or it may be due to **linkage disequilibrium** between genes with separate effects on these two traits. (Recall from Fig. 14.26 that the gene *Achaete-scute* is associated with bristle number variation in natural populations. This gene is known to influence larval as well as adult development.)

Often, traits are found to be strongly correlated with fitness, suggesting that strong directional selection is acting to change them. Quantitative traits are almost always heritable (see p. 397), and so we expect to see rapid changes over time at a rate equal to the product of the **selection differential** and the **heritability** (see p. 478). Yet, such changes are often not observed in nature. There are many possible reasons for such discrepancies. For example, the change due to selection may be counterbalanced by some other process, such as immigration or a deteriorating environment. An important explanation, and one that illustrates the difficulty of measuring selection on quantitative traits, is illustrated by an example from the long-term study of red deer on the Scottish island of Rum. Antlers are used in contests between males, and stags with larger antlers produce significantly more offspring (Fig. 19.8A). Much of the variance in antler size is due to additive genetic effects ($h^2 = 0.36$). Thus, an increase in antler size of 0.165 standard deviations per generation is expected. Yet, over 29 years of the study (approximately four generations), antler size has in fact decreased. This apparent paradox has been resolved by separating out the two components of phenotype. The antler size of each deer is the sum of its **breeding value** (a measure of the average antler size of the sons of a stag) and the **environmental deviation**, which is due to random, noninherited variation (see pp. 387 and 393). The relation between fitness and breeding value is weak and is not statistically significant. The apparent selection is seen because selection acts on the environmental deviation, which by definition is not inherited and so cannot contribute to any response to selection.

Why should stags that have large antlers as a result of factors that are not inherited sire more offspring? The obvious explanation is that stags in good condition tend to leave more offspring and also tend to have larger antlers. Thus, a common factor—con-

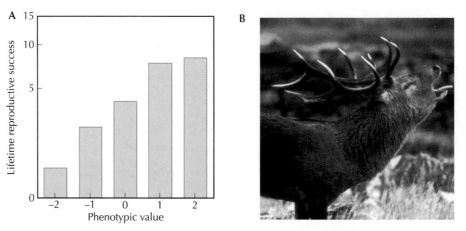

FIGURE 19.8. Antler size in red deer appears to be under strong selection, yet has not changed over a 29-year study period. (A) Lifetime reproductive success (i.e., fitness) is plotted against antler size measured in standard deviations from the mean. (B) One of the stags measured in this study.

dition—is correlated with both antler size and fitness, but large antlers *do not cause* an increase in fitness. The effects of selection on correlated traits such as condition can be compensated for statistically. However, it is usually not possible to find out which traits are, in fact, relevant and then to measure them. Selection on noninherited variation, arising from unknown traits, may be a common explanation for discrepancies between quantitative genetic predictions and the actual changes seen in populations.

As we saw for enzyme variation, convincing evidence that trait differences directly cause fitness differences comes from direct experimental manipulation. It is usually extremely difficult to manipulate a chosen quantitative trait in a sensible way. However, there are some exceptions. For example, manipulations of the tail streamers carried by barn swallows show that these are longer than their optimum with respect to flight ability because females prefer to mate with males that carry longer streamers (Fig. 19.9). As we will see in the next chapter, it is common for organisms to be shifted from their apparent optimum by mate preferences of this sort.

FIGURE 19.9. Experimental manipulation demonstrates selection on tail streamers in swallows. (A) A barn swallow, *Hirundo rustica*, feeding its young. Note the two tail streamers. (B) The flight time of male barn swallows, plotted against an experimental reduction in tail streamer length. (Results for females are similar.) The swallows fly fastest when streamers are about 12 mm shorter than their average in nature. Point size is proportional to number of birds.

Quantitative Traits Are Often Subject to Natural Selection

What is the typical strength of selection on quantitative traits? The strength of directional selection can be measured by the **selection gradient** β, standardized relative to the phenotypic standard deviation (see p. 478). This gives the increase in relative fitness associated with an increase of one standard deviation in the trait of interest. Thus, $\beta = 0.1$ indicates that an increase in the trait of one standard deviation is associated with a 10% increase in fitness. (As discussed above, although selection gradients are corrected for the effects of correlated traits that are measured in the same study, the fitness differences could still be caused by correlations with unmeasured traits.) An extensive survey of selection gradients shows that although some estimates are very high (up to $\beta \sim 1$), most are more modest, averaging $\beta = 0.22$ (Fig. 19.10A). Moreover, most individual estimates are not statistically significant, and so a large part of each estimate is due to sampling error (blue area in Fig. 19.10A). Indeed, estimates from large studies based on measurement of 1000 or more individuals tend to be much lower than those from smaller studies. This could be because negative results from small studies are less likely to be published, biasing the estimates upward in Figure 19.10A.

The strength of stabilizing or disruptive selection is measured by the standardized **quadratic selection gradient** γ (Fig. 19.10B). This is analogous to the linear selection gradient β and measures the change in variance in fitness caused by selection. For example, $\gamma = 0.1$ indicates disruptive selection, which increases the variance by 10%. Negative values indicate stabilizing selection, which decreases the variance. Overall, selection on the variance tends to be weaker than directional selection on the mean: Although some estimates of γ are large, the absolute value is around $\gamma = 0.1$; again, most individual estimates are not statistically significant. Remarkably, estimates are scattered symmetrically about $\gamma = 0$ (i.e., disruptive selection seems to be as common as stabilizing selection). This seems paradoxical because we expect most traits to evolve to an optimal value (Chapter 20 [pp. 556–560]) and because morphology is seen to change very slowly in the fossil record (Fig. 17.30). However, there are several reasons why this survey may be misleading. Most important, selection is likely to fluctuate, so that its long-term effect may be much weaker than our experiments suggest. This argument is supported by the slower rates of change seen over long timescales (Fig. 17.30) and by the changes in selection seen over just a few years in the Galápagos finches (Fig. 17.25).

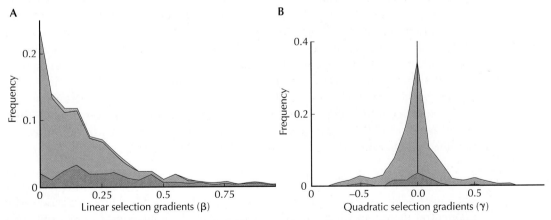

FIGURE 19.10. (*A*) The distribution of linear selection gradients (β) measured in a survey of 993 published estimates. The survey included roughly equal numbers of plants, invertebrates, and vertebrates. (*B*) The distribution of quadratic selection gradients (γ) based on 465 published estimates. Negative values indicate stabilizing selection, and positive values indicate disruptive selection. *Red areas* indicate individual estimates that are statistically significant (*p* < 5%). The *yellow area* in *A* indicates a few cases where significance was not stated.

In principle, it is straightforward to measure selection on a quantitative trait simply by finding the fitness of individuals with different values of the trait. Moderately strong selection is often seen (Fig. 19.10) and can easily account for the rates of change seen in the fossil record (Fig. 17.30). It is clear a priori that the very many characteristics of organisms that are required for their proper functioning must be under stabilizing selection. However, we are very far from knowing the actual distribution of selection or the number of traits that have significant effects on fitness.

INDIRECT MEASUREMENT

Selection Can Be Measured Indirectly through Its Interaction with Other Forces

How can we measure moderate or weak selection—selection that could cause change over hundreds rather than tens of generations, say? Our only hope is to use some indirect method based on understanding the interactions between different evolutionary processes.

Mutation and Selection

In the previous chapter, we saw how selection against deleterious mutations can be estimated (p. 513). The same approach has been used to show that selection against **null alleles** at enzyme loci is very weak. In a sample of *D. melanogaster*, nulls were found at an average frequency of $p = 0.0025$ at 20 autosomal loci. We expect an equilibrium between mutation and selection at $p = \mu/s$ (Box 18.2) and the rate at which mutation generates nulls was measured directly as $\mu = 3.9 \times 10^{-6}$. Thus, the average selection against heterozygotes is estimated as $s = \mu/p = (3.9 \times 10^{-6})/0.0025 = 0.0015$. Using similar methods, the selection against heterozygotes for recessive lethals has been estimated as approximately 2% (see Web Notes). This suggests that loss of function of one of the two copies of these enzymes reduces fitness by much less than for essential genes in general. Deleterious enzyme variants that change amino acid sequence rather than abolishing function altogether are likely to be under even weaker selection.

Dispersal and Selection

Selection can also be measured from the width of clines maintained by a balance between dispersal and selection. In Chapter 18 (p. 504), we saw that the width of such clines is proportional to a characteristic distance given by the ratio between the rate of dispersal and the square root of twice the selection coefficient: $l = \sigma/\sqrt{2s}$. Thus, if the dispersal rate σ and cline width are known, and the cline has reached an equilibrium, the strength of the selection maintaining it, s, can be estimated (e.g., Fig. 18.16). In practice, this method is limited by the difficulty of measuring dispersal rates directly. However, other information, such as information about changes through time, or linkage disequilibrium, enables estimates of both dispersal and selection to be determined purely from genetic data (as was done in Fig. 18.15).

Comparison with Neutral Theory

The most widely used methods for measuring selection are based on comparison with the **neutral theory**, in which variation is shaped by the interaction between mutation and random genetic drift (Chapter 15). The neutral theory serves as a well-understood **null hypothesis,** and deviations from it may be caused by various kinds of selection. In the following sections, we examine ways of detecting and measuring selection by comparison with the neutral theory.

Selection Can Be Detected by Comparison with Neutral Variation

The action of selection can be detected by comparing variants that are thought to affect fitness with variation that has negligible effects, that is, with neutral variation. This can be done by using differences *between* species or by using polymorphism *within* species. The most powerful methods compare variation both *within and between* species. Alternatively, the pattern of purely neutral variation may be distorted by selection at linked sites, which are not directly observed. These methods can be used to detect several kinds of selection: **purifying selection** against deleterious mutations, **balancing selection** that maintains variation, and **directional selection** that fixes favorable mutations (Box 17.3).

We begin with the first approach, comparison between different kinds of sequence variation between species. This comparison is usually based on the redundancy of the **genetic code** (Fig. 2.26). A change in the third position of a nucleotide triplet in the code usually does not alter the amino acid. Such a corresponding change is called **synonymous**. For example, CCU, CCA, CCC, and CCG all code for proline. In contrast, changes at either of the first two nucleotide positions of a triplet usually change what amino acid is coded for—a **nonsynonymous** change. Thus, synonymous changes can be used as a neutral benchmark. As we shall see later, this is only an approximation, because changes that do not alter protein sequence can nevertheless slightly alter fitness. Nevertheless, it is reasonable to assume that selection acts much more strongly on differences in amino acid sequence. Other kinds of noncoding sequence—for example, **introns** or **pseudogenes**—can be used as a neutral benchmark in the same way.

Deviations from the Molecular Clock Indicate Selection

A basic prediction of the neutral theory of molecular evolution is that the rate at which differences between species accumulate is equal to the rate of neutral mutation μ (p. 425). For the parts of the genome that have negligible effects on fitness, this equals the actual mutation rate. However, only a small fraction of mutations in functionally important sequences will be **effectively neutral** (Fig. 18.6), and so the rate of effectively neutral mutation (μ_N) will then be much lower than the total rate of mutation. Over a time t, a lineage will on average accumulate $\mu_N t$ changes as a result of mutation and random drift. The actual number is expected to follow a **Poisson distribution**, which has a variance equal to its mean (Fig. 13.30). Thus, we can measure variability in the rate of molecular evolution by the ratio between the variance and the mean of the number of changes, denoted by R. The neutral theory predicts that $R = 1$.

Does the **molecular clock** in fact fluctuate in rate as predicted by the neutral theory? Motoo Kimura first examined this question using protein sequences from mammals (e.g., Fig. 19.11). Kimura found that rates varied substantially more than this, and more extensive surveys of mammals have since confirmed this finding. However, these estimates are complicated by several factors. Especially for synonymous changes, divergence may be so great that several changes have occurred at a single site and yet would appear as only a single change (see Web Notes). Mammals did not all diverge at exactly the same time and, moreover, different lineages may vary in the expected number of **substitutions** (e.g., because of differences in generation time). These complications can all be corrected for statistically, and it remains clear that rates of amino acid substitution vary at least fivefold more among mammalian lineages than expected under the neutral theory.

This variation is surprising, because it implies that the underlying rate of molecular evolution is extremely variable. Substitutions must be clumped, so that several occur together in the same lineage. The most straightforward explanation for this pat-

	Human	Mouse	Rabbit	Dog	Horse	Cow
Human		27	14	15	25	25
Mouse			28	30	36	39
Rabbit				21	25	30
Dog					30	28
Horse						30

FIGURE 19.11. Variation in the number of amino acid differences between β-globins from mammalian species. Assuming that these species diverged at about the same time, under the neutral theory the number of differences between any two of them should follow a Poisson distribution, with a variance equal to the mean. From these data, the ratio between the variance and the mean is estimated as $R = 3.1$. (A correction for multiple substitutions is made; see Web Notes.)

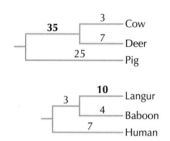

FIGURE 19.12. Bursts of substitutions in lysozyme (shown in boldface).

tern is that proteins occasionally undergo rapid bursts of adaptation, driven by positive selection. Many examples are known. For example, lysozyme is an enzyme that attacks bacterial cell walls and that functions as an antibiotic (e.g., in tears, saliva, and egg whites). However, lysozyme has independently acquired a new function in cloven-hoofed mammals (e.g., cows, deer) and in leaf-eating monkeys such as the langur, in which it digests bacterial remains in the stomach. Correspondingly, lysozyme in these animals has evolved to function in an acidic environment and to resist degradation by other digestive enzymes. Indeed, the enzyme evolved via bursts of amino acid substitution and, remarkably, many of these substitutions are the same in these two separate groups of animals (Fig. 19.12).

A High Rate of Amino Acid Evolution Relative to Synonymous Change Indicates Positive Selection: K_a/K_s

A simple and widely used way of detecting such episodes of positive selection is to calculate the ratio between the rate of substitutions that change the amino acid and the rate of synonymous substitutions. (This ratio is known as K_a/K_s, but is sometimes denoted d_N/d_S, where d_N refers to the rate of nonsynonymous substitutions.) Each rate is calculated per nucleotide site and so, in the absence of any selective constraint, it is expected to equal 1. The test is unaffected by variations in mutation rate from gene to gene, because it compares synonymous and nonsynonymous changes within the same gene. Most genes are at least somewhat constrained by selection and so usually K_a/K_s is smaller than 1. However, if selection has caused several amino acid changes between closely related species, then K_a/K_s can be larger than 1. This is strong evidence for the action of positive natural selection and has been observed many times. Most examples involve interactions between parasite and host or between molecules involved in sexual reproduction. In both cases, we expect a continual "arms race," in which parasites struggle to evade host defenses (see p. 509) and pollen or sperm compete to fertilize an egg (see p. 577). Rates of amino acid evolution may also exceed synonymous rates in genes that have recently become duplicated, so that one or both acquire new functions (see p. 710).

The extent of positive selection is greatly underestimated by simply counting cases where $K_a > K_s$ over the entire gene. The bulk of the gene may be under strong selective constraint with positive selection concentrated in a particular region. If sequences from

A

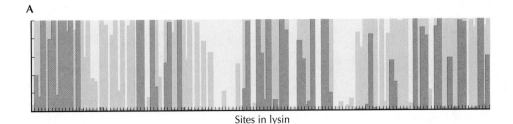

Sites in lysin

B

FIGURE 19.13. Rates of amino acid substitution vary greatly across the sperm lysin gene of abalone. (A) The *red bars* indicate that a codon evolves rapidly ($K_a/K_s \sim$ 3.1), the *blue bars* that a codon evolves at intermediate rates ($K_a/K_s \sim 0.91$), and the *gray bars* a low rate ($K_a/K_s \sim 0.09$). (B) Lysin crystal structure from the red abalone, with sites colored according to their rate of evolution (A). Sites that are likely to evolve rapidly (*red*) tend to cluster around the top and bottom, which are exposed.

several related species are available, it is possible to fit a model in which rates of amino acid evolution, K_a/K_s, vary from codon to codon. A good example comes from comparisons between the protein, sperm lysin, of 25 abalone species. These large marine mollusks release sperm into the sea. The species remain distinct in part because sperm lysin regulates species-specific interactions between sperm and egg. This membrane-bound protein binds to the envelope of the egg and unravels it, allowing fertilization. Figure 19.13 shows the probability that the amino acid at each codon evolves at a "fast," "medium," or "slow" rate (red, green, and gray, respectively). The rapidly evolving regions are concentrated where amino acids are exposed; regions buried in the membrane evolve more slowly and may be constrained to maintain the overall structure.

The Pattern of Variation within Populations Can Reveal Deviations from the Neutral Theory

The neutral theory predicts the pattern of variation within populations, and so deviations from this prediction indicate selection. This approach is more difficult, however, than one based on differences between species. Instead of sampling just one individual from each of several species, one must take large samples from within a population. Moreover, deviations from simple versions of the neutral model can be caused by population history (e.g., subdivision into **demes** or **population bottlenecks**), as well as by selection (see p. 427).

A simple example of this approach is shown in Figure 19.14, which gives the distribution of allele frequencies across 46 genes from a sample of plaice assayed by electrophoresis. The distribution fits the neutral theory quite well. As predicted, most alleles are rare, with a minority reaching intermediate frequency. However, there is a significant excess of rare alleles (leftmost frequency classes in Fig. 19.14).

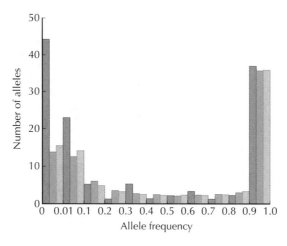

FIGURE 19.14. The distribution of allele frequencies at 46 genes in a sample of plaice, *Pleuronectes platessa*. The *red columns* show the frequency of alleles that were found in each frequency class (<0.01, 0.01–0.1, etc.). This is compared with the prediction from the neutral theory, either assuming the same neutral mutation rate at all genes (*blue*) or allowing for variation in rate (*orange*).

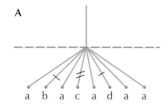

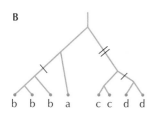

FIGURE 19.15. A population bottleneck (*A*) tends to produce an excess of rare alleles, compared with the pattern seen in a population of constant size (*B*). Assuming that each mutation generates a novel allele (the infinite allele model), the alleles observed in the sample are shown by the letters below each figure. In *A*, every new allele is present in only a single copy (labeled b–d). In contrast, in a sample from a population of constant size (*B*), mutations tend to be present in multiple copies (b–d) The *dotted line* in *A* indicates a sharp reduction in population size; all the lineages coalesce in a common ancestor who survived this bottleneck. The *black dashes* indicate mutations.

One of the first statistical tests of the neutral theory relies on this kind of comparison of observed with expected allele frequency distribution. It is based on a remarkable property of the neutral theory, which was discovered by Warren Ewens. As we saw in Chapter 15 (p. 427), a single parameter determines the evolution of a population of constant size under drift and mutation: $\theta = 4N_e\mu$, which is the product of the effective population size and the neutral mutation rate. The expected number of alleles in a sample increases with θ, but for a given number of alleles the frequency distribution of the alleles is independent of θ. Thus, it is possible to use this distribution to test the neutral theory without making any assumptions about the mutation rate or effective population size. For example, one study sampled 21 genes of the enzyme xanthine dehydrogenase in *Drosophila persimilis*. Ten different alleles were found, of which one was present in 12 copies and nine were found in only a single copy. This is the most extreme configuration possible and is extremely unlikely under the neutral **infinite-alleles** model; the chance of finding it is only 0.00012. As in the example of Figure 19.14, there are significantly too many rare alleles, present in just one copy in the sample.

This pattern of an excess of rare alleles has been found in several other surveys of electrophoretic variation. One interpretation is that these rare alleles are only slightly deleterious, so that they are kept at low frequency in the population. However, the same pattern can also be produced by expansion of a population after a sudden population bottleneck. In the most extreme case, the genealogy connecting the genes in a present-day sample will look like a **star**, in which all the lineages trace back to a common ancestor at the bottleneck (Fig. 19.15A). Then, any mutations that have accumulated since the bottleneck will be confined to one lineage, and so we will see one common ancestral allele and a set of more recent alleles, each in a single copy. In contrast, coalescence in a population of steady size occurs over a wide range of times (Fig. 19.15B). Mutations most often occur in the longer branches that lie deep in the genealogy and so produce descendant alleles that are at higher frequency.

Now that DNA sequences are available, these alternative explanations for an excess of rare alleles can be distinguished by comparing nonsynonymous with synonymous variation. In human populations, variants that alter amino acid sequence are rarer than those that do not, and amino acid changes that have major chemical effects are rarer than minor changes (Chapter 13 [pp. 370–371]). These different classes of polymorphism are all influenced by population history in the same way, and so the differences between them must reflect the stronger action of selection against mutations that change the protein's function.

These deviations from the simplest predictions are not fatal to the neutral theory. Species hardly ever consist of single, randomly mating populations of constant size, and selection does not cleanly distinguish between neutral alleles with no effect on fitness and strongly deleterious alleles, which are so rare that they are never seen in samples. It is difficult to use statistical tests alone to reject an extended version of the neutral theory that allows for the possibility of population structure and for weak selection against deleterious alleles. Yet, this extension retains the key feature that both differences between species and variation within them are caused by random drift and mutation, acting in opposition to selection.

Under the Neutral Theory, Polymorphism within Species Should Be Proportional to Divergence between Species

So far, we have looked separately at differences between species and at variation within them. However, under the neutral theory, these should be strictly proportional to each other, because both are due to mutation and drift. Both should vary with the rate of effectively neutral mutation, that is, with the fraction of mutations that escape elimination by purifying selection. We have seen examples of this pattern in Chapter 13 (pp. 370–371), where we showed that sequences under strong selective constraint tend to vary less both within and between populations. A broad survey of allozyme variation confirms this pattern: Proteins that tend to be highly heterozygous within species also tend to diverge more extensively between species (Fig. 19.16).

The neutral theory predicts that the ratio of divergence to polymorphism should be the same for all kinds of change, regardless of variation in mutation rate across the genome and regardless of the degree of purifying selection. Thus, if this ratio is higher for nonsynonymous changes than for synonymous changes, we infer that directional selection has established adaptive amino acid differences between the species.

The McDonald–Kreitman test quantifies this comparison of the ratio of divergence to polymorphism between different kinds of change (Box 19.1). This test can be applied systematically across many genes to give an overall estimate of the fraction of divergence caused by selection. In both *Drosophila* and in primates, there is an excess of nonsynonymous divergence between species relative to common polymorphism within species, which implies that a substantial fraction of amino acid differences between species have been established by selection. For example, 45% of the amino acid differences between *D. simulans* and *D. yakuba* are estimated to have been fixed by selection. Because there are about 13,600 genes in the *Drosophila* genome, with an av-

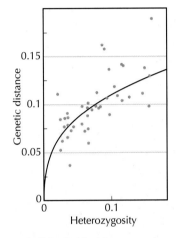

FIGURE 19.16. Divergence between species is greater for proteins that are more variable within populations. Genetic distance is plotted against heterozygosity for 42 proteins scored by electrophoresis in at least 50 animal species. The *curve* shows the expectation from the neutral theory.

Box 19.1	The McDonald–Kreitman Test

Under the neutral theory, both the divergence between species and the variation within them should be proportional to the neutral mutation rate. McDonald and Kreitman introduced a simple test based on this idea and applied it to data from the *alcohol dehydrogenase* polymorphism in *D. melanogaster*, *D. simulans*, and *Drosophila yakuba*.

	Fixed	Polymorphic
Synonymous	$D_s = 17$	$P_s = 42$
Nonsynonymous	$D_n = 7$	$P_n = 2$

Sites are classed as "polymorphic" if they show any variation within species and as "fixed" if they differ between species and are never seen to vary within them. Numbers of fixed differences that are synonymous and nonsynonymous are written as D_s and D_n, respectively. (These correspond to the K_s and K_a used elsewhere, but we keep to the slightly different notation used by McDonald and Kreitman.) The ratio of fixed differences to polymorphisms is much higher for nonsynonymous changes (i.e., $D_n/P_n = 7/2 \gg D_s/P_s = 17/42$). This suggests that selection has caused many of the amino acid differences between these species.

This approach can be extended to give an estimate of the overall proportion of nonsynonymous changes that are due to positive selection. If both synonymous and nonsynonymous changes were neutral, the ratio of divergence to polymorphism should be the same for both: $D_n/P_n = D_s/P_s$. Thus, we expect to see $D_s P_n/P_s$ nonsynonymous changes purely as a result of random drift. The difference $D_n - D_s P_n/P_s$ gives the excess number of nonsynonymous changes, which can be attributed to selection rather than drift. By averaging this quantity over many genes, the fraction of divergence due to selection can be estimated.

erage of 44 differences per gene, selection has established about 270,000 changes during the 6 million years for which these two species have been diverging. For primates, 35% of amino acid differences are estimated to be selected, which implies about 150,000 adaptive substitutions during the 30 million years since humans shared a common ancestor with old-world monkeys.

SELECTION ON LINKED LOCI

Selection Can Be Detected through Its Effects on Linked Neutral Variation

So far, we have examined the direct effects of selection on variation within and between species by looking at the selected variants themselves. However, even purely neutral variation is influenced by selection on linked loci, which might not be observed directly. This is because by chance a neutral variant can become associated with a site that directly influences fitness. In other words, random drift generates linkage disequilibria between the neutral variation that we observe and the variation that is actually under selection (see p. 432). (This is known as the **Hill–Robertson effect** and will be discussed further in Chapter 23.) Selection therefore influences linked neutral variation and can be detected by examining these patterns.

The simplest case is one in which a single favorable mutation arises and increases, eventually becoming fixed in the whole population. As the mutation increases in frequency, so does the entire stretch of chromosome in which it arose, a process known as **hitchhiking** or a **selective sweep** (see p. 486). Recombination will whittle away the size of the block that remains associated with the mutation, so that eventually the whole population will carry both the mutation and a small segment of the chromosome in which it happened to arise (Fig. 19.17A). The net result is that genetic variation is lost not just from the selected locus, but also in the surrounding region (Fig. 19.17C).

There are only a few cases where we know that a particular allele has recently been established by selection and so can demonstrate the effects of a selective sweep on variation at nearby sites. Some of the best examples are variants that give us resistance to malaria, which is thought to be a relatively recent disease. In sub-Saharan Africa, the Duffy allele FY*O is fixed, whereas in Europe and Asia it is rare. This allele differs from the ancestral FY*B allele by a single noncoding base change that prevents transcription. Homozygotes are resistant to one form of malaria, caused by *Plasmodium vivax*, because the gene product, a chemokine receptor, is required for the parasite to invade red blood cells. As expected for an allele that was established by a single mutation, variation in West Africans is substantially reduced over approximately 20 kb of sequence.

Selective Sweeps Cause a Sudden Burst of Coalescence

The effect of a selective sweep is to cause a sudden burst of **coalescence**. As one moves back in time, lineages will join to share a common ancestor at the time when the favorable mutation swept through the population. If we follow the ancestry of the selected locus itself, then *all* present-day copies trace back to a single ancestor on the chromosome that carried the new mutation. Loci that are less tightly linked to the selected site will show a less drastic pattern of coalescence. Some may trace back to the ancestral chromosome that carried the mutation, whereas other lineages may recombine away and coalesce much further back in time (Fig. 19.17B).

The concentration of coalescence caused by a selective sweep reduces variation, because the genes we sample will tend to share the same ancestral sequence. The pattern of variation will show an excess of rare variants. This is because many lineages

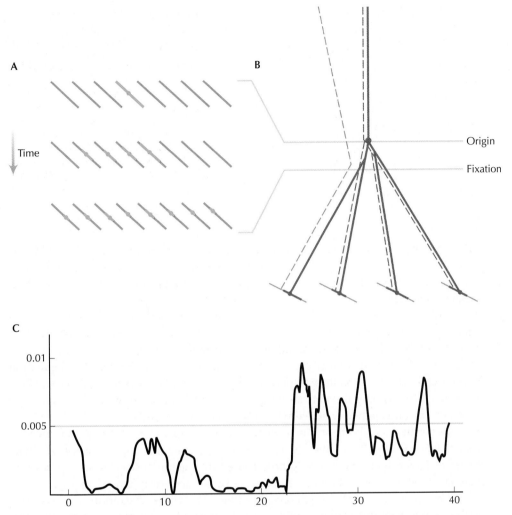

FIGURE 19.17. (*A*) Fixation of a favorable mutation at one locus will sweep variability out of a region of genome. Initially, a favorable mutation (*blue dot*) arises in one particular genome (*blue line, top row*). Eventually, the mutation fixes in the whole population, along with a fragment of the original genome. Thus, variation is swept out of a short region around the mutation. (*B*) The *solid lines* show the genealogy of four genes at the selected locus. These must coalesce during the sweep, because they all must carry the new mutation. The *dashed lines* show the ancestry of a sample of closely linked genes. Three of these stay with the selected mutation, but the leftmost lineage recombines away and has a different ancestry. (*C*) A simulation of genetic variability along 40 kb of genome soon after fixation of a favorable mutation at the center (20 kb), showing the nucleotide diversity π. Under the neutral theory, we expect $\pi = \theta = 4N_e\mu = 0.005$.

trace back to coalesce at the selective sweep, so that the genealogy looks like a star (Fig. 19.17B). Variants that arise after the sweep will therefore appear once, rather than in a cluster (recall Fig. 19.15A). As we saw in Figure 15.8B, genetic variation can also be reduced by a drastic reduction in population size, called a bottleneck. Such a population bottleneck has a similar effect on the genealogy, giving a star shape with coalescence concentrated in the short time when the population was small. Hence, a bottleneck also leads to an excess of rare variants in a similar way to a selective sweep. This makes it difficult to distinguish population bottlenecks from selective sweeps using data from a single locus. For example, human mitochondrial DNA shows an excess of rare variants, which has been attributed to a population expansion after humans migrated out of Africa, 50,000 to 100,000 years ago. However, this pattern could equally well be explained if an advantageous mitochondrial variant spread at that

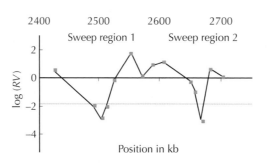

FIGURE 19.18. Selective sweeps can be identified from surveys of genetic variation. In this example, two regions of the *Drosophila melanogaster* X chromosome showed significantly lower variability at microsatellite loci in non-African than in African populations. These local reductions are probably due to the fixation of adaptive mutations, following the recent spread of *D. melanogaster* out of Africa. log(*RV*) is the ratio of variability in non-African populations relative to African populations on a log scale. The *gray line* shows the lower 95% confidence interval.

time. In that case, we could not infer the history of our entire species from the single mitochondrial genealogy. Nuclear genes do not show a consistent excess of rare variants; selection of some kind must be invoked to explain why different genes show different patterns.

To distinguish explanations based on selection on particular loci from those involving the whole population, we must compare patterns across many genes. The fixation of a favorable mutation should leave its trace in a localized reduction in genetic variation, confined to a particular section of the genome. In contrast, a population bottleneck is expected to affect all parts of the genome in the same way. Attempts have been made to estimate the net rate of adaptive evolution by counting the number of regions with unusually low genetic variation (e.g., Fig. 19.18).

One problem with these kinds of surveys is that local reductions may arise by chance (see Fig. 19.17C). However, such spurious reductions tend to occur over narrower regions of genome than do those caused by a genuine selective sweep, because they are generated over a longer timescale—by the drift of a neutral variant to fixation over approximately $2N_e$ generations, rather than through fixation of a mutation with selective advantage s, which takes approximately $1/s$ generations. Estimates of this kind are as yet tentative, but do hold the promise that the overall extent of adaptive evolution could be measured from genome-wide sequence data.

Deleterious Mutations Reduce Variation at Linked Sites: Background Selection

We have seen that the spread of a favorable mutation will reduce genetic variation in the surrounding region and distort the pattern of variation. Other kinds of selection also distort neutral variation at linked sites. In particular, deleterious mutations reduce variation, whereas balancing selection increases it. One of the most important problems in evolutionary biology at present is to find out not only how to detect selection, but also how to distinguish the different kinds of selection that may be acting.

When mutations have any appreciable effect on fitness, they are almost always harmful. This is simply because if they have any effect at all, it will be to disrupt functions that have been built up by natural selection. As we saw in Chapter 12, the total rate of deleterious mutations can be high, at least in complex multicellular organisms. If the population is to maintain high fitness, then every deleterious mutation that enters must eventually be removed by selection (Box 18.2). Crucially, every such **selective death** eliminates not only the deleterious mutation, but also the variation that was associated with it.

To understand this process in more detail, consider the simplest case—one in which there is no recombination (e.g., a very short region of the genome or an asex-

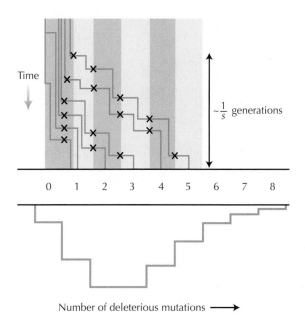

FIGURE 19.19. The ancestry of a population quickly traces back to the small fraction that is free of deleterious mutation. The histogram (*below*) shows the proportion of genomes carrying 0, 1, 2, ... deleterious mutations, and the diagram at the *top* shows how deleterious mutations (*crosses*) move individual genomes into less fit classes (i.e., to the *right*). The *blue lines* indicate ancestral lineages. Although most genes carry several mutations, they trace back to recent ancestors with fitter genomes. Ultimately, the ancestry of the whole population traces back into the fittest class, at left.

ually reproducing organism). If the total mutation rate in this region, U, is high, and selection against deleterious mutations, s, is weak, then most chromosomes will carry several deleterious mutations. (The average number is U/s, because U mutations are produced in each generation and each of them lasts on average $1/s$ generations before being eliminated by selection; recall Table 18.1.) In the long run, only members of the fittest class—those that carry no deleterious mutations—can leave offspring. All others will eventually be eliminated. Tracing back, the whole population must descend from the fittest class (Fig. 19.19). The pattern of variation is approximately the same as if the population consisted *only* of the small fraction that carry no deleterious mutations, and so neutral variation is considerably reduced. This reduction in the effective size of the population as a result of selection against deleterious mutations is termed **background selection**.

With recombination, the effect of background selection is greatly reduced, because associations between deleterious mutations and linked variation are broken up by meiosis. Nevertheless, the constant influx of deleterious mutations does reduce variation. To a good approximation, the effective population size is reduced by $e^{-U/R}$, where U/R is the rate of mutation divided by the rate of recombination. In most organisms, the effect is small. Microbes, which have only sporadic sex and recombination, tend to have low genomic mutation rates; but organisms that maintain larger genomes, and hence suffer more deleterious mutations per generation, usually have high rates of recombination. However, the effect may be important where recombination is restricted. For example, in self-fertilizing plants, individuals are predominantly homozygous, so that although recombination occurs, it has no effect on the composition of the population. Thus, the effective rate of recombination is greatly reduced. This may help explain the low levels of genetic diversity seen in many such species (e.g., Fig. 19.20)

The best evidence that selection does distort neutral variation comes from the observation that nucleotide diversity is lower in regions of the *D. melanogaster* genome in which recombination is reduced, namely, near the **centromeres** and **telomeres**, and in the tiny fourth chromosome. There is a good correlation between the nucleotide diversity at synonymous sites and the local rate of recombination (Fig. 19.21A). Moreover, the pattern of variation along the genome fits quite well with predictions from a model of background selection, given estimates of mutation and recombination rates

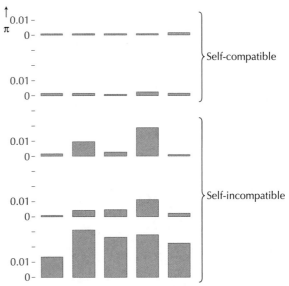

FIGURE 19.20. Diversity in self-fertilizing species tends to be lower than in their **outcrossing** relatives. This example shows levels of nucleotide diversity at silent sites, π, around five genes, sampled from five species of wild tomato. The three self-incompatible species are much more diverse than the two self-compatible species.

(Fig. 19.21B). This reduction is not due to a lower rate of mutation in regions with lower recombination, because divergence between *D. melanogaster* and *D. simulans* is not affected. (One might have supposed, for example, that recombination itself is mutagenic; but if it were, divergence between species should be increased by recombination to the same extent as variation within species.) The pattern, then, manifests itself as a reduction in the ratio of polymorphism to divergence below that expected under the neutral theory. However, unlike the examples we saw in the previous section, this is likely to be due to selection on linked loci, rather than on the synonymous sites themselves. At present, it is not clear whether this selection is due to deleterious mutations (i.e., background selection), to selective sweeps, or to a combination of these and other kinds of selection. All these may reduce diversity, and their effects are difficult to distinguish.

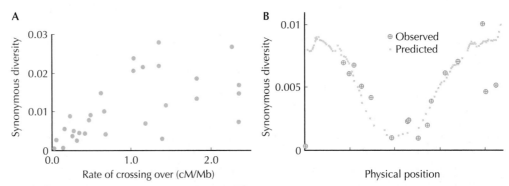

FIGURE 19.21. (*A*) Synonymous diversity is greatly reduced in regions of low crossing over in African populations of *Drosophila melanogaster*. (*B*) Observed and predicted diversity along the third chromosome of *D. melanogaster*, plotted against physical position. Predictions are based on the known pattern of recombination and assume selection against deleterious mutations with selection coefficient $s = 0.02$; mutation rate is taken as 0.4 over the whole chromosome. The fit is good, except near the tips of the chromosome.

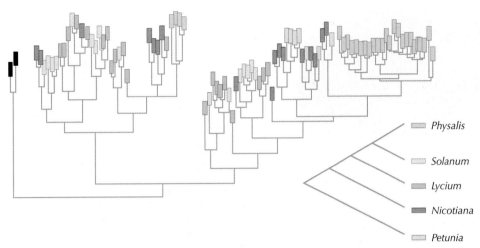

FIGURE 19.22. Genealogy showing the relationships between incompatibility genes in the Solanaceae. Closely related genes are often found in different species—and in this example even in different genera. *Black (left)* indicates an outgroup used to root the genealogy.

Balancing Selection Can Increase Neutral Variation

Balancing selection, which maintains two or more alternative alleles in a population, also increases the neutral variability in the surrounding regions of genome. The reason is simple. When different alleles are maintained for very long times, two separate gene pools persist, consisting of the sequences that are associated with one or the other allele. Over time, these sequences accumulate different neutral mutations and so diverge from each other. This divergence is reflected in increased neutral variability within the population. The clearest example is the incompatibility loci that, as we saw in Chapter 18 (p. 506), have evolved independently in several groups of flowering plants. Large numbers of alleles are maintained by frequency-dependent selection and these can persist for very long periods. Indeed, balancing selection can maintain the same allele over times longer than the species' life span, so that the relationship between genes shows little relationship with the phylogeny of the species in which those genes are found (Fig. 19.22). In other words, a gene can be more similar to a gene in another species than to other genes within the same species.

The effects of long-continued balancing selection can be seen in more detail in studies of *Brassica rapa* and *Brassica oleracea* (cabbages and turnip). Their incompatibility system involves two tightly linked genes: a pollen surface protein known as SP11 that interacts with a receptor kinase SRK on the surface of the stigma. If the stigma expresses an allele that is also expressed by the pollen grain, then germination and growth of the pollen tube is blocked. Different alleles within the same species show extreme divergence, with less than 50% amino acid identity; yet four pairs of alleles from different species are remarkably similar, despite the 2–3 million year separation of these species (red columns in Fig. 19.23). Apart from incompatibility alleles such as these, the only other well-understood example of a long-standing balanced polymorphism is the vertebrate **major histocompatibility complex** (**MHC**) genes, which present antigens to the immune system. Here again, polymorphisms are maintained across species, and so there is high amino acid divergence between alleles within species.

The effect of balancing selection on surrounding regions of the genome builds up over very long periods of time, because it depends on the gradual accumulation of different neutral mutations. Thus, the effect is confined to narrow regions, because even very low rates of recombination, of the same order as the mutation rate, will impede divergence between the alternative selected alleles. Fluctuations in the frequency of the

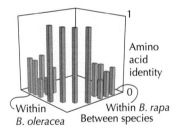

FIGURE 19.23. Protein sequences are more similar between homologous incompatibility alleles in different *Brassica* species (*red columns*) than they are between different alleles of the same species (*pink* and *blue columns*).

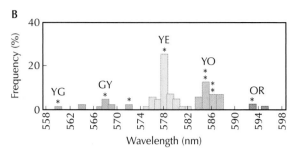

FIGURE 19.24. Sequence diversity gives evidence for selection on a color polymorphism in the Jamaican click beetle, *Pyrophorus plagiophthalamus*. (A) Two different luciferase genes are expressed in the dorsal and ventral light-producing organs (only thge dorsal organs are shown), and variation in each of these genes is responsible for separate polymorphisms. The ventral luciferase has three alleles, yellow-green (vYG), yellow (vYE), and orange (vOR). (B) The three alleles generate five phenotypes; vYG homozygotes (labeled YG), vYG/vYE heterozygotes (phenotype GY), and so on. The McDonald–Kreitman test shows a significant excess of amino acid change in the region that affects color. The vOR alleles show significantly lower sequence diversity than the vYE alleles (π = 0.00046 vs. π = 0.00129), suggesting that the vOR allele has recently increased.

balanced polymorphism over shorter timescales will tend to reduce variation in much the same way as does the selective sweep of a favorable mutation. For example, polymorphisms for chromosomal **inversions** in *Drosophila* are maintained by some kind of balancing selection (see Fig. 1.39). However, nucleotide diversity tends to be *lower* near the boundaries of these inversions, rather than being increased. This is thought to be because inversion polymorphisms are transient, so that there is no time for different chromosome arrangements to diverge much. Instead, variation is eliminated as inversions increase in a selective sweep. Figure 19.24 shows another example where a polymorphism is associated with reduced sequence diversity.

More generally, the complex histories of populations and of selection make interpretation of sequence variation much more complicated than was at first hoped. Although deviations from the simple null model of neutral evolution in a stable population can readily be detected, it is not at all easy to determine what causes these deviations. The strength and nature of selection cannot simply be read from sequence data.

SELECTION ON NONCODING DNA

Selection Acts on Some Noncoding Regions

The analysis of DNA sequence data is greatly aided by the genetic code. The code is essentially universal, so that the same methods can be used for all organisms and for all genes. Patterns of synonymous and nonsynonymous evolution can be compared by contrasting different kinds of change in each triplet codon—roughly speaking, by comparing rates at the first, second, and third positions of each codon. This leads to robust tests that are not affected by (for example) variation in base composition along the genome. Most important, coding regions from different species can be **aligned** with each other, making it possible to compare their sequences. Analysis of noncoding regions is much harder, because we have little idea of how such sequences function and therefore of how they might be constrained. Moreover, such sequences accumulate insertions and deletions at a much higher rate than those in coding regions, which makes alignment of divergent sequences difficult and complicates models of the mutational process.

Nevertheless, analysis of noncoding sequences is extremely important. Present-day organisms evolved from a primordial world in which biochemical functions were carried out primarily by RNA (Chapter 4); RNA molecules are still responsible for many key functions, such as translation of the genetic code. Noncoding regions regulate gene expression utilizing various mechanisms, including binding transcription factors and splicing introns out of messenger RNA molecules.

Organisms such as humans, flies, flowering plants, and nematodes have similar numbers of genes and yet very different morphologies (Fig. 13.17). Much of this complex and diverse organization must be due to different patterns of gene regulation as well as to functions based on RNA molecules rather than on proteins (Chapter 9). Some of these differences may be due to the amino acid sequences of regulatory proteins, but the majority are likely to be due to differences in noncoding regions of the genome. Yet, we have little idea at present of how much noncoding sequence is functional or just how these sequences function.

Codon Usage Bias Is Caused by Weak Selection for Translational Efficiency and Accuracy

The best-understood example of selection on synonymous variation (i.e., variation that does not alter amino acid sequence) is bias in the use of different triplet codons. Each amino acid can be coded by several alternative codons (Fig. 2.26), which one might expect to be equivalent and so used with equal frequency. However, there is often a strong bias toward the use of particular codons. For example, in both *Drosophila* and *Caenorhabditis elegans*, codons ending in C or G tend to be used preferentially. Such biases could be caused by either mutation or selection. Mutations might tend to produce, for example, C or G, causing codons involving these bases to accumulate. Selection for the particular amino acid sequences that are required for the protein to function would then tend to keep the first two positions more A-T rich but leave the third position to accumulate C and G as a result of biased mutation. Alternatively, selection might act to favor some codons over others; that is, organisms that use different codons might have different fitnesses even though they carry the same protein sequences.

Several lines of evidence suggest that in many organisms, **codon usage** is biased by weak selection, rather than simply being determined by mutation pressure. In *E. coli*, yeast, nematodes, and *Drosophila*, the degree of bias correlates with the level of gene expression (e.g., Fig. 19.25). This suggests that there is selection for translational efficiency. For example, a gene can be translated more quickly if the transfer RNAs (tRNAs) it requires are more abundant. This is supported by the observation that in *E. coli*, yeast, and other microbes the preferred codon corresponds to the most abundant tRNA. Most convincing is a comparison between genes in *D. melanogaster* whose transcript is spliced into mature mRNA in a variety of ways. Those sequences that are always translated show a higher codon bias than sequences that are translated less often.

As well as selection for efficient translation, other kinds of selection may also act. In *E. coli* and *Drosophila*, sites *within* genes that have greater functional importance (reflected in their slower divergence between species) show stronger bias. Selection for the overall speed of translation cannot explain this variation between sites within the *same* gene, because it should act equally on all codons. Selection may instead act on the *accuracy* of translation. A codon that calls for a rare tRNA has a greater chance of incorporating the wrong amino acid, because a more common, but incorrect, tRNA may bind before the correct one is found. Of course, selection does not act solely on translation rate and accuracy; even within coding regions, many sites have important regulatory functions at the RNA level. In the following section, we will examine such selection in more detail.

Fixation of the preferred codon is rarely seen even in highly expressed genes in *E. coli* and yeast, even though we expect it to be favored by selection. Now, the selection on any one codon must be quite weak. Indeed, codon usage bias is the best example we have of

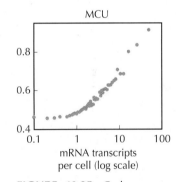

FIGURE 19.25. Codon usage bias increases with level of gene expression in yeast, *Saccharomyces cerevisiae*. The proportion of sites occupied by the preferred ("major") codon is plotted against the number of messenger RNA (mRNA) transcripts per cell.

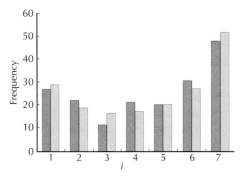

FIGURE 19.26. The observed frequency of alternative codons in *Escherichia coli* (*red bars*) fits with the prediction for an equilibrium between mutation, selection, and drift, with $N_e s$ = 0.8 (*blue bars*). Data are combined over 11 genes, each sampled for 8 sequences.

an adaptation maintained by very weak selection. So, this apparently imperfect adaptation may be due to random genetic drift, which interferes with selection and so reduces the strength of codon bias. We saw in Chapter 18 (pp. 494–496) that random drift interferes significantly with selection if $N_e s$ is small. (This key parameter is the ratio between the rate of selection and the rate of drift, $s/(1/N_e)$.) The extent of bias in *E. coli* implies $N_e s \sim 0.8$ (Fig. 19.26), and similar estimates have been obtained from a variety of other bacteria and from *Drosophila*. Interestingly, there is little evidence of selection for codon bias in mammals, which are likely to have smaller effective population sizes, $N_e s \ll 1$.

For *E. coli*, a rough calculation based on the effect of codon usage on translation rate and on the biochemical consequences of translation rate for reproduction suggests a selection coefficient proportional to protein production, with a maximum of approximately 10^{-4} per generation for the most highly expressed genes, such as those for ribosomal proteins. Although this selection is very weak, implying slow change over more than 10,000 generations, it is still much stronger than the mutation rate per nucleotide of approximately 10^{-9} per base per generation and much stronger than the rate of drift expected from the extremely large population size of *E. coli*. It is surprising that intermediate levels of bias (implying $N_e s \sim 1$) are seen across many different organisms and genes. We expect to see large $N_e s$—and hence complete codon bias—in highly expressed genes of abundant organisms and negligible bias for the opposite case.

A plausible explanation is suggested by the observation that in *Drosophila*, bias is lower in regions of low recombination. We saw in the previous section that selection for favorable mutations or against deleterious mutations reduces nucleotide variation at linked sites. The same process also reduces the response to selection by generating a kind of random drift. We investigate such effects in more detail in Chapter 23. For the moment, note that it may explain why both levels of genetic variability (see pp. 368–370) and values of $N_e s$ estimated from codon usage bias are not much larger in *E. coli* than in *Drosophila*, despite the very different census numbers of these two species. It may be that most random drift is caused by the random hitchhiking effects of linked loci, rather than by simple sampling (see p. 426).

Selection for Pairing in RNA Molecules Can Be Detected

Not all expressed gene sequences code for protein. Some produce RNA molecules having specific structures and functions. For example, ribosomal RNA has a highly conserved structure, which is constrained by interactions with ribosomal proteins and by its catalytic activity. Messenger RNA (mRNA) forms structures that are recognized as sites for splicing out introns, for transport into the cytoplasm, and for recognition by the translation machinery. These RNA structures, such as paired helical stems, are maintained primarily by base pairing, which requires proper alignment of complementary bases (Fig. 19.27). Paired helical stems can be identified by looking for correlated changes at different sites. A change at one site will tend to occur along with a

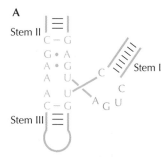

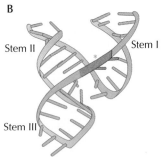

FIGURE 19.27. RNA structure. (*A*) Diagram showing how complementary base pairing (*lines*) maintains helical stems. (*B*) The corresponding three-dimensional structure.

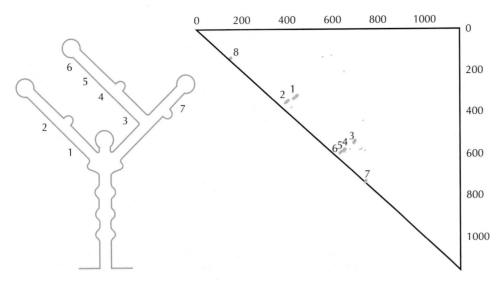

FIGURE 19.28. Correlations between changes at different positions in the *bicoid* gene sequence of *Drosophila* can be used to infer RNA secondary structure. The *triangle* shows pairs of positions at which changes occur together in the same part of the phylogenetic tree. Statistically significant regions are numbered 1 to 8. (For example, in region 3, changes at positions 534–542 tend to be accompanied by changes at the corresponding positions 687–695.) The structure at *left* shows the seven regions that have been independently confirmed.

corresponding change that maintains the pairing. In the Aim and Scope of this book, we saw how this method has been applied to find the structure of ribosomal RNA. Here, we give another example, which shows how this approach can be used systematically to identify pairing constraints in noncoding sequences and to estimate the strength of the selection that acts to maintain that pairing.

The *bicoid* gene is essential for the early development of *Drosophila* embryos. Bicoid mRNA produced by the mother must be localized at the anterior of the egg for the head and thorax to form. A conserved secondary structure in the 3′ untranslated region of the gene is responsible for this localization. By comparing bicoid sequences from nine *Drosophila* species, eight paired regions have been identified through correlations in changes at different positions (Fig. 19.28). This structure matches thermodynamic predictions, and seven of the regions have been confirmed by genetic manipulations. The advantage of the evolutionary method here is that it can be applied quite generally to identify previously unknown structures.

Selection to maintain RNA pairing can be seen within populations as well as between them. In *D. pseudoobscura*, two common **haplotypes** are seen in the introns of alcohol dehydrogenase (*Adh*). These have similar secondary structures, but quite different sequences (Fig. 19.29). These haplotypes are ancient and may be maintained by some kind of balancing selection. Linkage disequilibrium between sites involved in pairing is high and is probably maintained by selection, because recombinants between the two haplotypes would not pair properly. (This kind of selection, in this case favoring combinations that pair well, is referred to as **epistasis**; see Chapter 14 [p. 388].)

Functional Sequences Can Be Detected by Looking for Conserved Sequences

In these examples of codon usage bias and of pairing in mRNA, we believe we know the functional significance of the variation. In general, however, we have little idea of how noncoding sequences function or even of how much is maintained by selection. The idea that functional sequences can be identified simply by comparing DNA se-

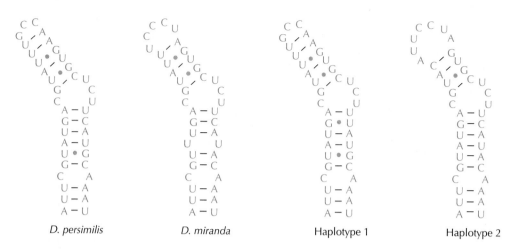

FIGURE 19.29. Two distinct haplotypes (labeled 1, 2) segregate within the first intron of *Adh* in *Drosophila pseudoobscura*. Although they form similar structures in the pre-mRNA, the sequences differ considerably. This polymorphism is ancient: Haplotype 1 is similar to the sequence in *Drosophila persimilis*, and haplotype 2 is similar to the sequence in *Drosophila miranda*.

quences is therefore attractive. We expect that functional sequences should diverge more slowly between species and so can be identified as regions of conserved sequence. One could compare sequence divergence with divergence at synonymous sites within coding regions, assuming that those are only weakly selected. Alternatively, one could look for variations in the degree of divergence along the genome and infer that islands of low divergence are functionally constrained.

Although this method is simple in principle, there are considerable difficulties in practice. The main problem is that without the standard reading frame provided by the genetic code, noncoding regions are free to accumulate insertions or deletions (termed **indels**). This makes it hard to align such sequences, especially when the species are highly divergent (Fig. 19.30). Slight errors in alignment can generate spuriously high rates of base substitution and spurious patterns of divergence. Moreover, it is very hard to derive reliable alignment algorithms that can work on large amounts of sequence data.

Most studies of noncoding divergence have used wide comparisons, where nonfunctional sequences have accumulated so many mutations that they are almost completely randomized. For example, comparisons between the genomes of the nematodes *C. elegans* and *Caenorhabditis briggsae* reveal short conserved regions, which are interspersed with regions with no significant similarity. Overall, about 18% of nucleotides are conserved in introns and intergenic regions, compared with 72% within exons. Because only 28% of the *Caenorhabditis* genome consists of exons, noncoding regions ac-

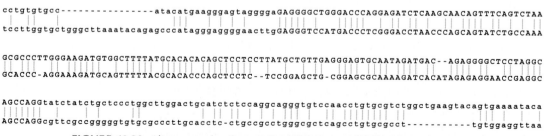

FIGURE 19.30. Alignment of a short region of mouse and human genomes. Here, insertions or deletions (*black dashed lines*) can be identified because enough sites have been conserved (*vertical solid blue lines*) to "anchor" the alignment. (*Lowercase letters*, regions with no significant similarity; *uppercase letters*, regions with >50% similarity.)

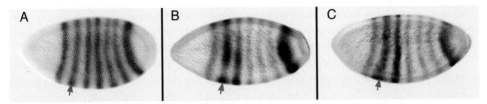

FIGURE 19.31. The functions of some genetic control elements are conserved between species of *Drosophila*. (*A*) Expression of endogenous *even-skipped* mRNA in *Drosophila melanogaster*. (*B*) *D. melanogaster* enhancers control the expression of LacZ mRNA (*purple stripes*) in *D. melanogaster*. (*C*) *D. pseudoobscura* enhancers control the expression of LacZ mRNA in *D. melanogaster* in the same way as do *D. melanogaster* sequences. In *B* and *C*, the *brown stripes* are endogenous *even-skipped* mRNA.

count for nearly half of all functionally constrained sites. A similar comparison of mouse with human sequences identified regions of around 100 bp in length that were more than 50% conserved, embedded in highly divergent sequences (e.g., Fig. 19.30). The degree of constraint was greater near the ends of genes, presumably because regulatory sequences are more common there. For each gene, approximately 1000 base pairs code for protein, and an additional 2000 bases per gene are also constrained by selection. Such studies suggest that in multicellular eukaryotes, at least as much noncoding as coding sequence is maintained by selection. One consequence is that the total rate of deleterious mutation is at least twice as great as expected based on coding regions alone.

Function can be conserved even if DNA sequence diverges considerably. Thus, the approaches just described may greatly underestimate the extent of function in noncoding sequences. We have seen that entirely different sequences can maintain the same RNA secondary structure, as long as pairing is maintained. Similarly, transcription factors may bind a wide range of sequences. Especially in eukaryotes, where regulatory sequences may be bound weakly by several transcription factors that themselves interact, the weakening or even loss of one binding site can be compensated for by the strengthening or gain of another. For example, the *even-skipped* gene is expressed in a pattern of seven stripes in all *Drosophila* species that have been examined. In *D. melanogaster*, detailed analysis has identified a specific enhancer sequence that regulates the expression of the second stripe, and this enhancer contains 12 binding sites, each of which binds one of the proteins encoded by the genes *giant*, *Krüppel*, *hunchback*, and *bicoid*. Remarkably, however, only three of these sites are completely conserved across *Drosophila* species; most sites show differences at individual bases, with some insertions and deletions, and new binding sites have arisen in some lineages. Despite this, the stripe 2 regulatory element from various *Drosophila* species will still produce an accurate stripe 2 expression pattern when introduced into *D. melanogaster* (Fig. 19.31). Examples such as this suggest that it may be hard to identify the selection acting on noncoding regions without knowing much more about how it actually functions. Changes in gene regulation cannot be inferred from sequence differences without detailed knowledge of the molecular mechanisms.

THE EXTENT OF SELECTION

It Is Difficult to Measure the Extent to Which Fitness Itself Is Inherited

We have seen that it is difficult to measure the effects of individual traits or genes on fitness. Only strong selection can be measured directly, and measurements are so laborious that they can be made for only a few examples. Indirect methods, which compare

DNA sequences with neutral expectations, can detect very weak selection and can be applied to genome-wide sequence data. However, such methods cannot easily measure the *strength* of selection. How then can we quantify the overall extent of selection?

One approach is suggested by Fisher's **Fundamental Theorem**: The rate of increase in mean fitness caused by selection on allele frequencies equals the **additive genetic variance** in fitness (Chapter 17 [p. 462]). Of course, the overall mean fitness stays roughly constant at the value needed to keep the population size steady. The decrease in fitness as the environment (physical and biotic) changes is countered by the increase due to selection. The additive genetic variance in fitness is therefore a good measure of the extent of natural selection.

In principle, the additive genetic variance in fitness can be measured in the same way as for any other kind of quantitative trait, by comparing the fitnesses of related individuals (Chapter 14 [p. 394]). This is relatively simple in laboratory experiments. For example, we saw in Figure 18.5C how the variance in fitness among different bacterial strains increased over time as mutations accumulated by selection and drift. However, such laboratory studies are not particularly helpful, because they require asexual reproduction and because fitness is determined by the experimenter or, at least, by the artificial conditions of the experiment. Measurement of the inheritance of fitness under natural conditions requires that we follow individual relationships in a population over several generations. This is straightforward for humans. Within European populations, for example, family size is strongly correlated across generations. This correlation is due to cultural as well as genetic factors, which are hard to separate. However, a recent study of Australian female twins found that the number of offspring produced is much more strongly correlated between identical than between nonidentical twins (41% vs. 18%). After allowing for level of education and religious background, these data suggest that about 40% of the variance in fitness is due to additive genetic factors.

Apart from humans, the heritability of fitness has been measured only in a handful of studies of birds and large mammals. The unmanaged population of red deer on the Scottish island of Rum has been studied intensively for more than 30 years. We have already seen how this dataset has given estimates of the relation between traits such as antler size and fitness (Fig. 19.8). It is possible to measure the fitness of individual deer (i.e., the number of offspring that they produce over their lifetime), and so the additive genetic variance in many traits, including fitness itself, has been estimated from the full set of relationships, making due allowance for environmental factors such as the weather and the degree of crowding. As is typical, a large fraction of the variance of morphological traits such as leg length and jaw size can be ascribed to additive genetic effects. That is, such traits show high **narrow-sense heritability**. However, traits more closely related to fitness, such as birth weight and longevity, show lower heritability (Fig. 19.32A), a pattern that is seen in wider surveys (Fig. 14.19). Most importantly, the heritability of fitness itself is not significantly different from 0. This pattern has traditionally been taken to show that selection has exhausted the additive genetic variation by fixing any allele that has a positive effect on fitness. However, it has recently been realized that this interpretation is wrong. The additive genetic variance of fitness-related traits, scaled relative to the trait mean, tends to be *higher* than that of other traits (e.g., Fig. 19.32B; p. 399). However, the environmental (nongenetic) component of divergence is higher still, leading to a lower heritability.

The variability of both the genetic and nongenetic components of fitness-related traits may be high because these kinds of traits are influenced by many features of the organism and its environment. Thus, they may be more sensitive to both genetic and nongenetic variation. Whatever the explanation, the consequence is that the additive genetic variance of fitness itself is difficult to estimate accurately. In the Rum study, the additive genetic variance in fitness is not significantly greater than zero because the

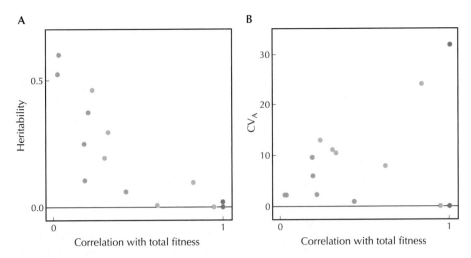

FIGURE 19.32. Quantitative genetics of fitness in the population of red deer on Rum. (*A*) The heritability of morphological and life history traits (*red* and *blue circles,* respectively) decreases with the correlation between the trait and fitness. The *magenta dots* show the heritability of fitness itself, which is estimated as near 0 in both sexes. (*B*) A similar plot for the additive genetic variance standardized relative to the square of the mean (CV_A). Total fitness has near-zero CV_A in females (*lower magenta dot*) but high CV_A in males (*upper magenta dot*).

nongenetic components of variance are so high. However, the actual value of the additive genetic variance might be substantial, corresponding to an appreciable increase in fitness as a result of selection in each generation.

If there is a high additive genetic variance in fitness such that natural selection is increasing mean fitness substantially in every generation, then some other process—essentially, either mutation, migration, or a deteriorating environment—must be counterbalancing the increase. Otherwise, fitness, and hence population size, would increase without limit. By artificially turning off selection, the rate of degradation of fitness due to such processes can be measured. This gives us an indirect way to measure the heritable variance in fitness required to counter such degradation. Natural selection can be prevented simply by ensuring that every individual produces the same number of offspring. In *Drosophila,* viability and fecundity declined by 0.2% per generation when measured under benign conditions, but fell an order of magnitude faster (about 2% per generation) when measured under competitive conditions. Because there is no migration and a (presumably) constant artificial environment, this decline is due to mutation. Thus, the rate is equal to the additive genetic variance in fitness associated with deleterious mutations, which would normally counterbalance the harmful consequences of mutation.

Similar rates of decline in fitness components have been seen in several experiments in which selection is prevented, either by using a highly inbred population or by use of **balancer** chromosomes in *Drosophila* (Box 13.2). In both the flowering plant *Arabidopsis thaliana* and the water flea *Daphnia pulex,* fitness components declined by around 1% per generation. In contrast, much lower rates of decline were seen in several protozoa, in the bacterium *E. coli,* and in the nematode *C. elegans.* As discussed in Chapter 12 (p. 349), these lower rates of decline may reflect a much lower genomic rate of mutation to deleterious alleles.

The Overall Extent of Selection Is Limited by Genetic Load

We have seen that it is difficult to measure the extent of natural selection either directly or indirectly. Another approach is to ask whether there is in principle an upper

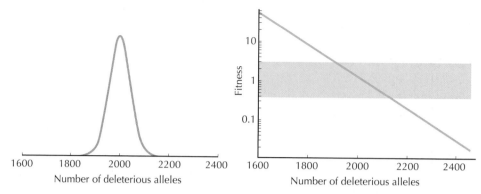

FIGURE 19.33. Selection on large numbers of genes generates only modest fitness variation. For example, selection $s = 1\%$ on alleles at frequency $p = 10\%$ at each of $n = 10,000$ diploid loci would give a total variance in fitness of $2ns^2pq = 0.18$. More than 98% of individuals have between 1900 and 2100 deleterious alleles and hence have fitness within a range 0.37–2.7 relative to the population mean (*shaded* region). (*Left*) The distribution of numbers of deleterious alleles per individual. (*Right*) How fitness declines with the number of deleterious alleles, assuming multiplicative effects; the average individual is assigned a fitness of 1.

limit to the power of selection. How many genes or traits can be selected and how fast? Natural selection molds the highly improbable combination of genes required to code for complex organisms by the repeated selection of random variations. Intuitively, one feels that there must be some limit to the rate at which this process can operate. From the perspective of a plant or animal breeder, this is a practical question: What is the fastest possible rate at which a domesticated species can be improved? Although such questions seem straightforward, they have proved remarkably difficult to quantify, and the limits are looser than one might guess.

One limit to the rate of increase in mean fitness is set by the variance in number of offspring—that is, by the total variance in fitness. The additive genetic variance in fitness cannot be larger than this and, as we just saw, is probably much smaller (i.e., the heritability of fitness is low). However, most organisms have a high variability in total fitness and so this hardly sets much of a constraint. Moreover, even moderately strong selection on very many loci gives reasonably low variance in fitness (Fig. 19.33).

The mean fitness relative to the maximum possible could set a much tighter limit. Such a reduction in mean fitness relative to the maximum possible is often termed a **genetic load**. More precisely, we can define the load as $L = 1 - \overline{W}/W_{\max}$, where $W_{\max}$ is the maximum possible fitness and $\overline{W}$ is the mean fitness. One can think of the genetic load as the proportion by which the average fitness of a population, $\overline{W}$, is reduced because it evolves by natural selection, rather than by some ideal process that always maintains the optimal genotype, with fitness $W_{\max}$.

There Are Several Kinds of Load, Corresponding to Different Kinds of Selection

The load associated with the selection of a favorable allele was defined by J.B.S. Haldane, who termed it the **cost of natural selection**; it is also known as the **substitution load**. Imagine that there is a sudden change in environment such that the population suffers an immediate drop in mean fitness. To take an example originally used by Haldane, think of the peppered moths (*Biston betularia*), which lost their camouflage when industrial pollution darkened their roosting sites (Figs. 1.38 amd 17.18). The ideal would be for the entire population to be replaced immediately by the fittest genotype—in this case, melanic moths. In fact, it took several generations for melanics to reach high frequency, and during these generations, many nonmelanic moths were eaten by birds. The substitution load in this example is the reduction in mean fitness that arose because the population did not immediately become melanic, summed over time (shaded area in Fig. 19.34). This represents a real load on the population, which (presumably) reduced its survival rate and hence its numbers.

Haldane showed that, assuming additive gene action, the total number of selective deaths required to increase an allele from a frequency p to fixation is approximately $2 \log_e(1/p)$. Remarkably, the load is independent of the strength and pattern of selection. This is because strongly selected alleles cause high mortality in each generation, but are fixed quickly, whereas weakly selected alleles cause little load in each generation, but take a long time to be fixed. The conclusion from a range of models is that the net load is likely to be in the range $L = 10–40$, assuming that the allele is initially rare. This number is the total number of **selective deaths** *per head of population*. For example, if the population contains 1 million individuals, then a substitution load of $L = 20$ implies that 20 million individuals must fail to reproduce because of the time taken for natural selection to complete an adaptation.

Haldane's argument suggests that there cannot be many amino acid substitutions per generation and was a key motivation for the neutral theory. How great is the substitution load required to fix the adaptive amino acid differences that we observe between species? Comparisons between synonymous and nonsynonymous divergence and variation yield an estimate of about 270,000 adaptive substitutions during 6 million years of divergence between *D. simulans* and *D. yakuba* (see pp. 535–536). Assuming ten generations per year, and allowing for the fact that divergence has occurred down both lineages, this implies roughly one substitution every $2 \times (6 \times 10^6)/270,000 \sim 450$ generations. Thus, the load per generation need only be in the range $(10–40)/450 = 2–10\%$. Similar estimates have been made for the divergence between humans and old-world monkeys. This level of selection could easily be borne by a rapidly reproducing organism such as *Drosophila* and does not seem excessive even for slower creatures such as ourselves. However, if adaptation is concentrated in short bursts (e.g., during species formation or in times of environmental change) or if a substantial fraction of noncoding differences are also adaptive, then the substitution load could become substantial.

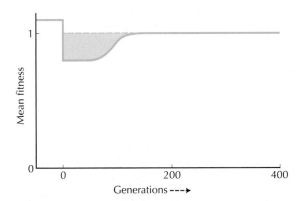

FIGURE 19.34. The replacement of one allele by another involves a substitution load. Initially, the population has a mean fitness of 1.1. The environment then changes suddenly, and mean fitness falls to 0.8. An allele P, which was previously unfavorable and was held at low frequency ($p_0 = 0.0001$) by selection, now becomes favorable: The fitnesses of genotypes QQ, PQ, and PP are 0.8, 0.9, and 1, respectively. After about 100 generations, this allele becomes common and sweeps to fixation. However, the population suffers a 20% drop in fitness while it waits for adaptation by natural selection. The net loss, indicated by the shaded area, is $2 \log_e(1/p_0) = 18.4$.

Deleterious Mutations and Balancing Selection Also Cause a Genetic Load

The constant influx of deleterious mutations also reduces the average fitness of a population. We saw in Box 18.2 that in a balance between mutation μ and selection s, the equilibrium frequency of deleterious alleles is μ/s. (Here, s is the reduction in fitness caused by a single deleterious allele; in diploids, it is the reduction in fitness of heterozygotes relative to wild-type homozygotes.) Now, the average loss of fitness due to each copy of the gene is $sp = \mu$. In a haploid population, therefore, the loss of fitness required to eliminate deleterious mutations, per head of population and per generation, is equal to the total mutation rate μ. In a diploid population, each individual carries two copies of every gene, and so the **mutation load** is twice as great: 2μ.

Just as for the substitution load, the result is independent of the strength of selection. Weakly deleterious alleles will have less effect on each individual that carries them, but will accumulate to higher frequency in the population. This result can be understood without recourse to algebra: Every deleterious mutation that enters the population by mutation must eventually be eliminated by a selective death.

If we assume that fitness effects multiply across genes, then the overall mean fitness is $(1 - \mu)^n$, which is, to a very good approximation, e^{-U}. Here, n is the number of gene copies in each individual that are subject to deleterious mutation and $U = n\mu$ is the total rate of deleterious mutation across the whole genome. Although mutation rates per gene are small, the *total* mutation rate U can be large (see Fig. 12.23) and consequently could have a substantial effect on the mean fitness of the whole population. (In diploids, both copies at a locus are subject to mutation. Thus, n is twice the number of loci and by convention $U = n\mu$ is the mutation rate per diploid genome.)

Other processes are associated with genetic loads in a similar way. Balancing selection also entails a **segregation load**. If heterozygotes are fitter than either homozygote, then a polymorphism will be maintained (Box 17.1). However, homozygotes will continually be generated, and so the population cannot contain more than 50% of the fittest genotype, the heterozygote. Migration into a population introduces alleles that are not well adapted to their new home (Chapter 16); random drift knocks allele frequencies away from their optimal values; and recombination breaks up gene combinations that have been built up by selection (see Chapter 23). All of these processes degrade fitness and so must be countered by natural selection. Together, they seem to set tight constraints on the extent of selection—on the number of adaptive changes that can be fixed, on the total rate of mutation in functionally important regions of the genome, and on the number of balanced polymorphisms.

Interactions between Genes Can Greatly Alleviate the Load

Simple load arguments of the kind outlined here have been strongly criticized, primarily on the grounds that they assume that the load on a single locus can be extrapolated up to the whole genome. In fact, genes could interact in such a way that the overall load may be very much less than expected from gene-by-gene calculations. Suppose that the environment changes so that many alleles, at present very rare, become favorable. If each of them has independent effects on fitness (one increasing early survival by 10%, another increasing male mating success by 1%, and so on), then the number of selective deaths needed to fix them all will be given by Haldane's calculation and will be proportional to the number of genes involved. However, selection acts much more efficiently if in every generation the individuals carrying the most favorable set of alleles are chosen to reproduce. This choice of the "best" in the population is known as **truncation selection** and gives the fastest response to selection for a given fraction of selected individuals. For this reason, it is the usual practice in artificial breeding programs. If the favorable alleles are rare, the situation in which the substitution load is most severe, then *every* favorable allele can be selected in each generation and the load is independent of the number of genes involved.

Similar arguments apply to the mutation and segregation loads. We have seen that truncation selection is much more efficient than selection that acts independently on each gene. If truncation selection is typical, then load arguments do not set any strong limit on the power of natural selection to fix good genes, to eliminate bad genes, or to maintain variation in the population. However, the question arises as to whether selection in fact acts in this way, so that populations are optimally **evolvable**. We return to this issue in Chapter 23 (pp. 689–692).

Gene interactions make it hard to estimate theoretically just how much the fitness of a population is reduced because organisms evolve by natural selection, rather than by some ideal alternative. Nevertheless, we know that genetic loads can be substantial. About 1.5% of newborn humans suffer from some inherited disease and a much larger fraction suffer from diseases with a strong heritable component. There are examples from the human population not only of loss of survival and reproduction caused by new mutations, but also of segregation load (e.g., sickle-cell anemia; Box 14.1) and substitution load (e.g., ailments associated with Western diets such as lactose intolerance; Chapter 26 [pp. 771–773]).

There may well be no "perfect" genotype. Some fraction of inherited disease and of natural senescence may be due to inevitable trade-offs (see p. 560). For example, deletions of components of the immune system can give resistance to some diseases, but might increase susceptibility to others; recall the examples of the Duffy allele that gives resistance to malaria (see p. 536) or the CCR5-Δ32 allele that gives resistance to HIV (see p. 434). Nevertheless, many of life's ills may be due to our dependence on natural selection.

SUMMARY

We have discussed many kinds of natural selection. There are well-understood examples of selection on visible Mendelian polymorphisms, on individual genes, and on quantitative traits. The action of selection can be seen in laboratory or natural populations or can be inferred from the traces it leaves in DNA sequence variation. Selection is clearly able to account for the fastest changes seen in the fossil record, including the evolution of our own species. Just as for Darwin, the strongest evidence for this comes from the impressive results of artificial selection on the farm and in the laboratory.

Nevertheless, there are many open questions. The argument between the **balance view** and **classical view** of genetic variation, which transformed into the **selectionist** versus **neutralist** debate of the 1970s, remains unresolved. It is widely accepted that much variation in DNA between species and within them has no significant effect on fitness and that functional sequences are primarily maintained against mutational degradation by purifying selection. However, there is disagreement about what fraction of differences in functional sequences between species are established by selection and about what fraction of polymorphism within species is maintained by balancing selection rather than by mutation.

Many more detailed questions about the nature and magnitude of selection remain open. Are adaptations typically due to selection of 10% or 0.1% or what? How much variation in fitness is inherited? Where balancing selection does act, how do rare alleles gain their advantage? To what extent are noncoding sequences shaped by selection? Why are there comparable levels of genetic variation across diverse organisms?

Indirect methods for detecting selection from samples of DNA sequences give some hope that we might be able to get answers to such questions that apply over the entire genome and over a range of organisms rather than in only a few examples. It is already feasible to identify functional sequences maintained by selection by comparing rates of divergence. Similarly, genome-wide estimates of the rate of adaptive evolution are beginning to be made. However, it is much harder to detect selection within present-day populations, to measure its strength, and to find out just how it acts. For this, a laborious combination of genetic manipulation and careful statistical analysis is necessary.

FURTHER READING

Barton N.H. and Partridge L. 2000. Limited to natural selection. *BioEssays* **22:** 1075–1084.

Discusses the various kinds of genetic load, and other constraints on the effectiveness of selection.

Bell G. 1997. *Selection: The mechanism of evolution*. Chapman and Hall, New York.

A detailed review of direct measurements of selection on genes and on quantitative traits.

Burt A. 1995. The evolution of fitness. *Evolution* **49:** 1–8.

Reviews evidence on the overall extent of selection, focusing on the additive genetic variance in fitness.

Eyre-Walker A. 2006. The genomic rate of adaptive evolution. *Trends Ecol. Evol.* **21:** 569–575.

Kimura M. 1983. *The neutral theory of molecular evolution*. Cambridge University Press, Cambridge; Gillespie J.H. 1991. *The causes of molecular evolution*. Oxford University Press, Oxford; Hey J. 1999. The neutralist, the fly, and the selectionist. *Trends Ecol. Evol.* **14:** 35–37.

Kimura and Gillespie present opposing views on the importance of selection in shaping molecular evolution. Hey (1999) gives a more recent view of the neutralist versus selectionist debate.

Kingsolver J.G., Hoekstra H.E., Hoekstra J.M., Berrigan D., Vignieri S.N., et al. 2001. The strength of phenotypic selection in natural populations. *Am. Nat.* **157:** 245–261.

Survey of "natural selection in the wild."

Kreitman M. 2001 Methods to detect selection in populations with applications to the human. *Annu. Rev. Genom. Hum. Genet.* **1:** 539–559; Fay J.C. and Wu C.I. 2001. The neutral theory in the genomic era. *Curr. Opin. Genet. Dev.* **11:** 642–646.

Reviews of methods that use sequence variation to detect selection.

Lynch M.J., Blanchard J., Houle D., Kibota T., Schultz S., et al. 1999. Perspective: Spontaneous deleterious mutation. *Evolution* **53:** 645–663.

Reviews the overall extent of selection due to deleterious mutations.

Nielsen R. 2005. Molecular signatures of natural selection. *Annu. Rev. Genet.* **39:** 197–218; Fay J.C. and Wu C.I. 2003. Sequence divergence, functional constraint, and selection in protein evolution. *Annu. Rev. Genom. Hum. Genet.* **4:** 213–235.

Reviews of the many recent studies that use sequence variation to infer the extent of adaptive evolution.

20

Phenotypic Evolution

W^E DO NOT KNOW THE DETAILED GENETIC mechanisms responsible for variation in morphology and behavior. How, then, can we understand the evolution of these kinds of traits? In this chapter, we discuss approaches that do not depend on genetics. We will not need to know how genes determine the phenotype or how much genetic variance there is. Much of evolutionary biology is successful exactly because it does not depend on knowing the precise mechanisms that determine phenotype. After all, Darwin developed the idea of evolution by natural selection in almost complete ignorance of heredity, and the basic framework of the **Evolutionary Synthesis** was built before the molecular basis of heredity was understood (Chapter 1).

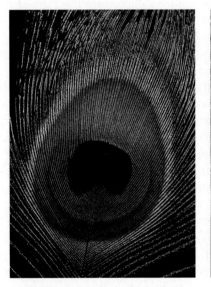 

We begin by showing how evolution can be seen as a problem of optimization. The classic example is the structure of the eye, which is optimized for visual function. Our main example will be the evolution of aging, which evolves as part of an optimal life history. We will also see how interactions between organisms can be seen as an **evolutionary game**. Here, our main example is the interaction between males and females that is **sexual selection**. We see that this can be understood by using ideas from game theory, but that the underlying genetics are also important.

What Kinds of Organisms Should Evolve?

We have seen how selection acts, given certain fitness differences between genotypes and given the way these genotypes are inherited (Chapter 17). We have also seen how selection interacts with other evolutionary forces (Chapter 18) and how selection can be detected and measured (Chapter 19). The action of selection is straightforward: Alleles that increase **fitness** will become more common. If fitness changes as the composition of the population changes, the outcome can be more complicated. For example, a polymorphism can be maintained by **balancing selection**. The process may also be highly stochastic, because of the randomness of mutation and reproduction in small populations. But if we know the fitnesses of different genotypes and the way they

FIGURE 20.1. Insect appendages can be put to diverse uses.

are inherited, then we can predict how the population is likely to evolve.

However, population genetic methods do not by themselves tell us *what* will evolve. To do that, we need to know what is fit and what is not. Obviously, this depends on circumstances—on the nature of the organism and its environment—and so we cannot expect to make detailed predictions. For example, the standard genetic code that maps the 64 triplet codons onto 20 amino acids may be largely arbitrary and not predictable in detail. The great diversity of insect species may be due to the many uses to which their appendages can be put (Fig. 20.1); but again, we could not hope to predict this diversity simply by knowing how insects develop. Nevertheless, there are ways of understanding in general terms what phenotypes will be fit and thus what kinds of organism we expect to evolve by natural selection. These various approaches form the theme of this chapter.

Although the methods we discuss are quite general, they have been developed in two contexts: the evolution of **life history** and the understanding of behavior as an **evolutionary game**. The life history of an organism is its pattern of survival and reproduction. We can ask such questions as: When will an organism reproduce? Will it reproduce once or over a prolonged period? How long will it live? For a **hermaphrodite**, how much resource will be put into male as opposed to female function? These kinds of questions can be understood as problems of optimization. What life history will produce the greatest overall fitness, given constraints on what can be achieved?

When there are interactions between individuals in a population, it is usually not possible to think of evolution as a simple optimization. There is no "best" solution, and even if there were, there is no guarantee that evolution would produce it. A fruitful approach has been to think of evolution as a game played between organisms. There may then be an **evolutionarily stable strategy**, which cannot be beaten by any alternative. We have already seen, in Chapters 17 and 18, how interactions between individuals cause selection to depend on the composition of a population. There, our focus was on how such **frequency-dependent selection** can maintain genetic diversity. In this chapter, we show how **game theory** can help us understand frequency dependence without needing to know the detailed genetics.

In the last part of the chapter, we turn to the interaction between males and females that leads to **sexual selection**. This is a key evolutionary process, which was first identified by Darwin, and which has been the focus of much research over the past 20 years. Here, we will see that both game theory and genetic arguments have been important. Indeed, many controversies have stemmed from a tension between ideas based on these two different approaches.

EVOLUTIONARY OPTIMIZATION

Much of Evolution Can Be Understood as Optimization

Before Darwin, the delicate complexities of nature were seen as being created to fulfill some function. For example, at the beginning of the 19th century, Richard Paley gave a careful account of how the vertebrate eye is well adapted to its purpose, in much the same way as the optical instruments of the time. The eye must do more than simply focus a clear image on the retina. The iris limits the amount of light that enters, the lens is built so that different colors focus in the same place, and muscles distort the eye so that it will focus over a wide range of distances. Such studies of **natural theology** were carried out to reveal the perfection of a divine designer (see p. 11). We now see natural selection, rather than any kind of "design," as being responsible for producing adaptive structures. Nevertheless, many features of organisms can still be understood

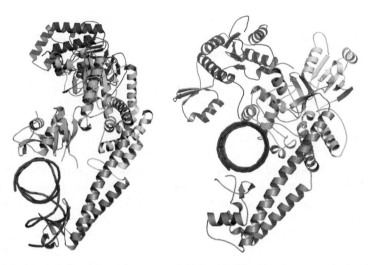

FIGURE 20.2. MutS (*left*) and topoisomerase II (*right*) bind DNA using very similar folds. However, the sequences of the proteins as a whole show that they are only distantly related. *Colored regions* show different domains.

as being nearly optimal, regardless of how they are produced. That is, we see them as maximizing either reproductive success itself or some component function that is necessary for reproductive success.

Later, in this chapter and the next, we will often emphasize *exceptions* to this simple view that natural selection leads to organisms that are in some sense "optimal." We will see cases where interactions between individuals, and conflicts between genes inherited in different ways, lead to apparently maladaptive outcomes. Such exceptions can tell us how the evolutionary process works. If organisms were simply the "best" that could be produced, then there would be no trace of their evolutionary history or of the processes that produced them (see Chapter 27 [online] for examples). Characteristics may be shared by distantly related groups because they are shaped by some common function. For example, bats, insects, and birds all have wings, but these are independent adaptations for flight rather than evidence of shared ancestry. Figure 20.2 shows how distantly related proteins have converged on the same structure at their active site. Such convergent features do not carry any information about ancestry; thus, when inferring evolutionary history from comparative analysis, we must be careful to exclude convergent features and focus on homologous ones (Chapter 28 [online]).

Nevertheless, it is important to remember that in most cases we can sensibly ask what some protein or organ or behavior is *for*. We can then find out how such features fulfill their function on the working assumption that the feature is close to optimal for that function. Our understanding of most of biology comes from the assumption that organs have a function and are optimized for it. When we do this, we are not studying the evolutionary process as such. Rather, we are assuming that evolution is dominated by straightforward natural selection, which maximizes fitness and hence optimizes the function we are studying.

The Optimum Is Defined Relative to Constraints on What Is Possible

The ideal organism would begin reproducing at birth and would sustain a high rate of reproduction throughout an indefinitely long life. Of course, such a **Darwinian Demon** is impossible; there are all sorts of constraints on reproduction and survival, and these constraints depend on the peculiar circumstances of each organism. We usually concentrate on some particular feature and seek the best solution within a limited range of possibilities. Moreover, within that limited range, it is usually sensible to choose a simple surrogate measure of fitness, based on the function in question, rather than trying to work out the exact consequences of the function for fitness. We illustrate the logic of optimization in Box 20.1, using two quite different examples.

This approach to understanding what kinds of organism will evolve is sometimes called the **adaptationist program**. It was strongly criticized in 1979 by Stephen Jay Gould and Richard Lewontin, on two main grounds. First, we cannot assume that natural selection will produce an optimal solution or that any particular feature is actually adaptive. Other evolutionary processes interfere with selection and, in any case, an organism's evolution is strongly constrained by its history. Gould and Lewontin began their paper by discussing the triangular spaces formed between two arches of a cathedral, which are known as **spandrels**. These are often elaborately decorated,

Box 20.1   The Logic of Optimization Arguments

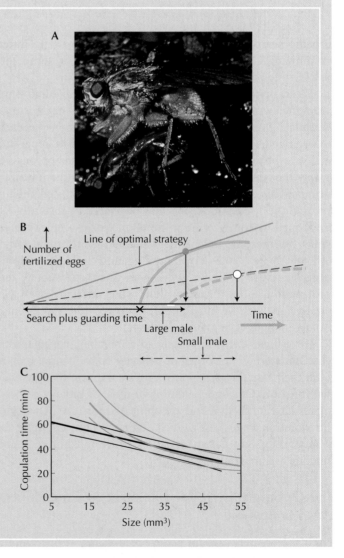

Optimization arguments share a simple logic. We ask what trait or combination of traits will give the highest fitness, subject to constraints on what is possible. In the first example, we focus on a single trait, the time dung flies spend copulating. In the second, we look at the entire metabolism of a bacterium.

Mating behavior of dung flies: Male dung flies (*Scatophaga stercoraria*) find fresh cattle droppings, and as females arrive, mate with them (Fig. 20.3A). They then guard the females as they lay eggs in the dung. A male must invest a fixed amount of time finding a mate and guarding her afterward (left of Fig. 20.3B). In this example, the average was 156.5 minutes. The females usually carry sperm from previous matings, which is displaced as the new mate copulates. The solid curve in Figure 20.3B shows how the proportion of eggs fertilized increases with copulation time, but with diminishing returns.

A male would maximize his fitness from a single mating by copulating indefinitely. However, his lifetime success depends on maximizing the number of eggs fertilized *per unit*

FIGURE 20.3. (A) Copulating pair of dung flies, *Scatophaga stercoraria*. (B) The *thick blue curves* show how the number of eggs that a male dung fly fertilizes increases with the time spent copulating. A male's fitness is proportional to the number of eggs he fertilizes per unit of time (i.e., proportional to the slope of the *straight lines*). This is maximized by the *straight line* that just touches the curve; the point where they touch shows the optimum copulation time. *Dashed lines* are for small males. (C) The observed copulation time as a function of male size is shown by the *thick black lines*. (The *upper* and *lower black lines* show the standard error of the estimate.) This matches the predicted optimum well (*red lines*, ±S.E.M.), except for very small males (*left*). Size is measured as the cube of hind tibia length.

Box 20.1. Continued.

time. This measure of fitness is given by the slope of a line from the origin (bottom left) to a point somewhere on the curve, because this slope is the ratio between number of eggs fertilized and time spent. Thus, the optimal strategy is given by the line of steepest slope. This just touches the curve, as shown in Figure 20.3.

Small males transfer sperm more slowly. They also take more time to successfully find a mate, because large males sometimes steal their females, forcing them to start searching again. Thus, the optimal copulation time for a small male is longer (dashed curve and line in Fig. 20.3B). The predicted decrease in optimal copulation time with male size, based on measured sperm transfer rates and search plus guard times, matches the actual relationship well (Fig. 20.3C). There is a discrepancy for very small males (left of figure), but such males are rare.

We are assuming here that the male controls copulation time, a plausible assumption in this case. In the next chapter, we will see the consequences of conflicts between male and female strategies.

***The metabolism of* Escherichia coli:** It would be extremely difficult to construct a full model of cellular metabolism. The rates of each reaction would need to be measured accurately and, moreover, measured as functions of the concentrations of all the substrates, products, and regulatory molecules. However, the structure of the metabolic network (what is converted to what) can be determined, and constraints can be put on it—reactions must conserve atoms and energy. This approach has been used to find the maximum growth rate of *E. coli* as a function of the rate of uptake of oxygen and of the carbon source.

This method does not determine the absolute growth rate, because that depends on the reaction rates, which are not specified. However, it does specify an optimal *relationship* between the rates of different reactions. For example, the red line in Figure 20.4A shows the relationship between the optimal rate of uptake of oxygen and of malate; the observed rates fit this prediction well (dots). Populations grown at higher temperatures or higher malate concentrations grow faster, but the relationship between oxygen uptake, malate uptake, and growth rate remains close to the optimum. If populations are grown on malate for many generations, their growth rate increases by about 20% (blue points in Fig. 20.4A), but the relative rates stay close to the predicted optimum.

Similar results were found for growth on succinate, acetate, and glucose, but bacteria grown on glycerol showed uptake rates that were scattered well away from the predicted optimum and grew more slowly than expected (Fig. 20.4B). However, when populations were grown for 700 generations on glycerol, they evolved the uptake rates predicted by the model (Fig. 20.4C).

In the dung fly example, a single trait (copulation time) was optimized, and everything else about the mating behavior was taken as given. The *E. coli* example is more ambitious, in that predictions were made about the entire metabolic network, involving 436 metabolites and 720 reactions. In both cases, however, experimental tests of the predictions were made more rigorous by testing predictions over a range of conditions: for the dung flies, over a range of male sizes, and for the metabolic network, over a range of temperatures and substrate concentrations.

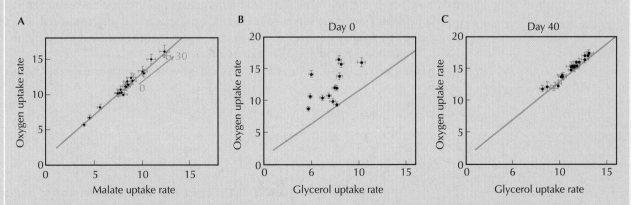

FIGURE 20.4. (*A*) The *red line* shows the predicted combination of oxygen uptake rate and malate uptake rate that will maximize growth rate. *Dots* show rates measured over a range of temperatures and malate concentrations. The two *blue points* are the start and end points of an evolution experiment that spanned 30 days (~500 generations). Units are mmole per gram dry weight per hour. (*B*) When grown on glycerol, the base population does not fit the predicted line of optima (*red*). (*C*) However, after 40 days (~700 generations) of selection on glycerol, the bacteria now follow the predicted optimum. Evolution toward the predicted optimum was seen in three replicate selection experiments. Units as in *A*.

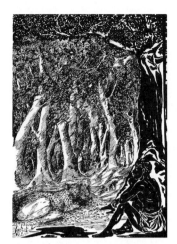

FIGURE 20.5. In one of his *Just So Stories*, Kipling explained how the leopard got its spots.

but are not present *because* of a need for decoration. Rather, they are the inevitable side effect of a roof supported by intersecting arches and, once present, have been exploited as decorative spaces. Extending the analogy, many features of organisms may have first evolved as side effects of other changes. Even if features have now acquired a function, their origin depended on quirks of history.

Gould and Lewontin's second line of argument was that the adaptationist view often led to unfalsifiable **just-so stories**—more or less plausible explanations, which could not be tested either experimentally or by observation of existing species. (The term comes from Rudyard Kipling's children's tales, which explained how the leopard got its spots and other such natural phenomena [Fig. 20.5].) This criticism has been influential and has led to considerably more care in making evolutionary arguments. More important, our knowledge of molecular evolution has revealed many nonadaptive features that are relics of ancient history and has brought a new emphasis on processes other than selection (see pp. 59–62).

Despite these criticisms, arguments about adaptation can be made rigorously. Any optimization argument *requires* that the constraints be spelled out, and these constraints do depend on the individual organism and its history. We will see examples of this approach throughout this chapter. Second, predictions can be falsified. If organisms are not as we predict them to be, then either we have chosen the wrong fitness measure, we have not specified the constraints properly, or other evolutionary processes may be interfering with natural selection. The advantage of the optimization approach is that it allows us to test whether we really do understand the problem at hand.

There Are Trade-offs between Components of Fitness

One way to think about optimization under constraint is to see that there are **trade-offs** between different components of fitness. Plants may divert limited resources of nutrients and energy into either vegetative growth or producing flowers; and it may be impossible to produce large showy flowers that attract pollinators *and* to produce many well-provisioned seeds. To take another example, when we think of the evolution of the **sex ratio** (see p. 506), we assume that the number of offspring is limited, so that more sons implies fewer daughters. Most fundamentally, it is not possible to attain maximum survival at the same time as attaining maximum reproductive output. Later, we will examine how these kinds of trade-offs can be detected and measured. For the moment, it is important to realize that trade-offs are another way of thinking about the constraints on what is possible (Fig. 20.6).

An intriguing example of how trade-offs can be used in an optimization argument

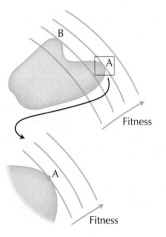

FIGURE 20.6. Optimization with constraints leads to trade-offs between different components of fitness. The figure shows the set of possible phenotypes (*green area*), together with contours of increasing fitness (*blue*). The two axes represent two traits that are each components of fitness. A shows the trait combination that gives the highest possible fitness. The *inset* shows the region around this global optimum. The boundary of the *shaded area* slopes downward, so that an increase in one component of fitness cannot be made without decreasing the other. This is what is meant by a trade-off. B is another local optimum, corresponding to a second **adaptive peak**, with lower fitness than A, but showing a similar trade-off between the two traits.

comes from the evolution of the genetic material itself. Why do we have four possible bases (A, T, G, and C) instead of 2 or 22? Clearly, organisms *could* evolve with DNA or RNA that uses only GC pairs or only AT pairs. Conversely, several alternative base pairs have been synthesized that fit into the Watson–Crick double helix and that have been shown to replicate in vitro (Fig. 20.7). Thus, it would be chemically possible for organisms to use three or four kinds of base pairs and so have six or eight letters in their "genetic alphabet."

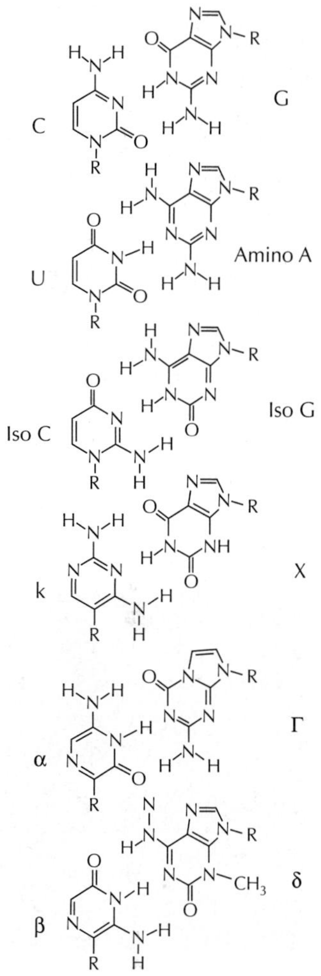

It has been argued that the present genetic system can be understood as a compromise between the need for accurate replication and the need for biochemical flexibility. The chance that the wrong base will pair during replication depends on the difference in energy between correct and incorrect pairs and on the number of alternatives available. Replication with only one base pair would be most accurate, because only G or C (say) would need to be distinguished. In our present world, a genetic code consisting solely of GC might therefore be preferable to one with GC or AT, because it would allow more accurate replication. (Information can still be coded with one base pair: Each strand would have a sequence such as GCCGGGG..., and each amino acid could be coded by four or five bases instead of three.)

However, the genetic system almost certainly evolved in an **RNA world** in which the molecules of heredity also carried out catalytic functions (Chapter 4 [p. 101]). Then, having more than one kind of base pair allows catalysis of many more reactions, because a greater variety of chemical groups are available. (In the same way, proteins can be adapted to catalyze a greater variety of reactions, because they are made up of 20 different amino acids.) By using some plausible chemical arguments about the rate of metabolism and accuracy of replication, it has been shown that a two-base-pair alphabet is most efficient and, moreover, that our actual two-base-pair alphabet is almost the most efficient possible, given the alternative bases that have been synthesized. As with all optimality arguments, this one can be criticized on the grounds that there may be other possibilities (other bases, say) or that the fitness measure is wrong. (In particular, we know very little about what kind of metabolism might need to be catalyzed.) However, this argument provides a way of understanding why we have the genetic alphabet that we do.

FIGURE 20.7. A synthetic set of six base pairs, which allows 12 "letters" in the genetic alphabet. The first two base pairs are those used by actual organisms. (Adenine has been modified so that all six pairs are maintained by three hydrogen bonds.)

AGING

Aging Is Not Inevitable

Why should rates of survival and reproduction decline in old age in so many organisms, humans included? Would it be optimal to maintain a functioning body that could reproduce indefinitely? Although **senescence** seems at first paradoxical, from an evolutionary point of view it is in fact one of the best-understood aspects of life history.

It is important to realize that senescence (or aging, a term we use in the same way here) is not the same as mortality. An organism that maintained the same physiological state throughout its life would still eventually die as a result of accidents of one sort or another—meeting a predator or catching a fatal disease, for example. Senescence is defined as a *decrease* in rates of survival or reproduction with age. It is not inevitable. Some fish, for example, continue growing throughout their lives, and as a result survival and reproduction increase for as long as a significant number escape accidental death. However, these are exceptions. Birds and mammals in nature show increasing mortality with age (e.g., Fig. 20.8), and even microbes age (Fig. 20.9).

Aging is not simply a consequence of physical constraints on what is possible. Presumably there is some limit to how long an organism can be maintained, because various kinds of damage accumulate. For example, cancers usually arise when two or

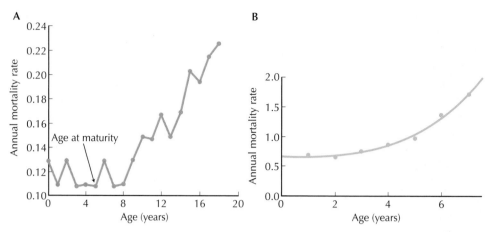

A

B

FIGURE 20.8. Mortality increases with age in natural populations of (A) the ring seal, *Phoca hispida* and (B) the pied flycatcher, *Ficedula hypoleuca*.

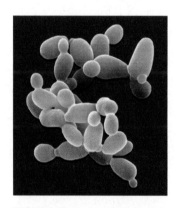

FIGURE 20.9. Asymmetric division in the yeast *Candida albicans*. Yeast cells reproduce by budding off a smaller daughter cell, but the mother cell cannot keep budding indefinitely.

more mutations occur in genes that control cell growth. Somatic mutations occur throughout life, and this is one reason why the rate of mortality through cancer increases with age. However, one can imagine that any particular kind of damage would be repaired or avoided, and, indeed, much cellular machinery is devoted to this kind of somatic maintenance (see Chapter 12). We know that organisms do not live as long as is physically possible, simply because some organisms do live much longer than others under benign conditions. Turtles, bats, and birds have longer life spans than comparable animals of the same body size and, conversely, salmon migrate into fresh water from the sea to breed just once and die soon after. Thus the degree of aging can evolve and might have been different from what it actually is.

Aging Evolves Because the Old Make Little Contribution to Fitness

The basic reason why survival and reproduction decline with age is that selection acts more weakly on later ages. Imagine an organism that does *not* senesce, that is, that maintains itself with the same rate of survival and reproduction indefinitely. It will still suffer accidental mortality and so, on average, reproducing early will produce more offspring than deferring reproduction until later; death might strike first. Therefore, natural selection acts more strongly on variations that act early in life and acts more and more weakly on late-acting variations (see Box 20.2). If, as a result, senescence starts to evolve, then there will be a feedback, so that selection on later ages becomes even weaker. In the extreme case, where an organism such as a salmon reproduces only once, there is absolutely no selection on subsequent survival.

Based on this evolutionary explanation of aging, we predict that in organisms with a low rate of accidental mortality, senescence should be less severe. This is consistent with the relatively long life spans of birds, bats, and turtles, who should suffer less predation because they can fly or have thick shells. Across species of birds and mammals, the rate of senescence is indeed correlated with baseline mortality (Fig. 20.12).

Perhaps the most striking example comes from social insects such as ants, termites, and honeybees. In such species, the queens live for an average of 10 years, whereas workers in the same species live for only a few weeks. The life spans of solitary insects average one month—100-fold less than a colonial queen. The long life span of queens is to be expected, both because they live within well-defended nests and because the rate of reproduction of the colony increases over time as it grows and

Selection Acts More Weakly on Later Life

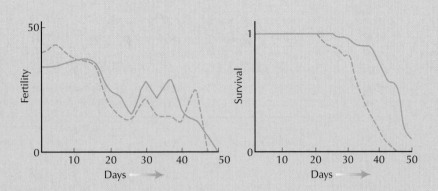

FIGURE 20.10. The fertility (*left*) and the proportion surviving (*right*) in males as a function of time since eclosion from the pupa. Male fertility is measured as the percentage of matings relative to a standard competitor. *Dashed lines* correspond to "early" lines, *solid lines* to "late" lines.

Here, we illustrate how selection on fertility and survival for the two sexes decreases with age. Lines of *Drosophila melanogaster* were selected for early or late reproduction. Fertility and survival shifted just as expected, in both sexes (Fig. 20.10).

The pair of graphs in Figure 20.11 shows how sensitive male fitness is to changes in fertility and survival (similar patterns were seen in females). This is a measure of the strength of selection on the life history. The key point is that

selection on later life is very weak. It is this diminished selection that is the ultimate cause of senescence.

In both graphs, selection is stronger in early life on the lines selected for early reproduction (dashed lines) and stronger in late life for the "late" lines (solid curves). The difference is greater for selection on fertility (left) than on survival (right). Thus, as selection shifts the life history toward earlier reproduction, selection itself becomes still more biased toward the young.

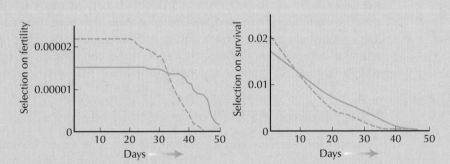

FIGURE 20.11. The sensitivity of fitness to survival and fertility in males. For fertility (*left*), units are as in Fig. 20.10. For survival (*right*), units are rate of increase per day per unit decrease in death rate. These figures are for populations with life history given by Fig. 20.10.

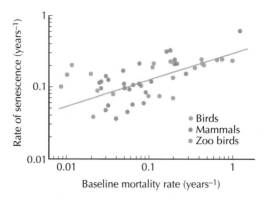

FIGURE 20.12. The rate of senescence increases with the baseline rate of accidental mortality in birds and mammals. The baseline mortality is the left-hand intercept of curves such as that in Fig. 20.8B, and the rate of senescence is a measure of how fast mortality increases with age.

establishes itself. Within the social insects, queens from species that have transient nests, and so must suffer the risks of moving the colony, live much shorter lives. It is remarkable that queens can live so long, despite the enormous metabolic effort required to produce eggs for the colony. (The mass of eggs they expel every day is greater than their own body weight.)

This evolutionary explanation of aging applies more generally. Traits that are selected more weakly will tend to degenerate and, in some cases, this can cause a feedback leading to complete collapse. Loss of completely unselected functions is seen in eyeless cave fish (Fig. 17.8), in the degeneration of **pseudogenes** that are no longer expressed (see pp. 218 and 371), in the loss of genes in bacteria (see p. 173), and in the loss of unnecessary metabolic functions (e.g., Fig. 18.5). Evidence for partial degeneration of more weakly selected traits is harder to find. We have seen one example in **codon usage bias**, which in yeast and *E. coli* is weaker in genes that are expressed less strongly (Fig. 19.26).

There are two main reasons why unselected traits might degenerate or, specifically in the case of aging, why survival and reproduction might decline with age. First, the optimal life history will be weighted toward early life, because that is when the greatest contribution to fitness is expected. Second, mutations that have deleterious effects on late life will not reduce fitness much and so will accumulate. Both the optimality and the mutation-accumulation theories must make some contribution to aging. However, we will see that present evidence supports the first more strongly than the second.

Aging May Evolve as Part of an Optimal Life History

Optimal life history involves a trade-off between early versus late survival and reproduction (Box 20.3). To the extent that late reproduction gives less return in fitness, the best compromise will emphasize performance in youth at the expense of old age. In genetic terms, alleles that are beneficial early in life may be favored even if they have harmful **pleiotropic** effects later on. The time when a gene is expressed (in the usual molecular biological sense) is irrelevant; what matters is when it affects fitness. (For example, a gene might be expressed in the developing heart of an embryo and could increase cardiac performance in youth. However, it might also increase the risk of heart failure in old age.) The argument is often phrased in terms of the pleiotropic effects of particular genes, and is referred to as **antagonistic pleiotropy**. However, the optimality theory is really independent of genetics. It simply depends on maximizing fitness, subject to constraints on life history.

Quite generally, reproduction reduces the rate of survival and reproduction later in life. Thus, the costs of reproduction give a general mechanism for the trade-off between early and late life. This has been demonstrated in a variety of ways. Female

Box 20.3   The Evolution of Senescence

Trade-offs and the evolution of senescence can be illustrated by a simple model of an organism with two age classes (Fig. 20.13). Each female produces one daughter at each of two ages. So, the life history is described by just two numbers: the chance of survival from birth to age 1 (J), and from age 1 to age 2 (A). The green curve shows the maximum possible survival probabilities, and represents the trade-off between adult and juvenile survival. The blue curves show contours of fitness, increasing to the upper right. The optimal life history is where the fitness contour from the origin just touches the trade-off curve (open circle, $J = 0.935$, $A = 0.505$). Note that fitness depends much more strongly on juvenile survival J than on adult survival A, simply because many individuals will have died before reaching adulthood. The arrows coming from the optimal point show the reduction of survival below the optimal value caused by deleterious mutations at a rate $U = 0.1$ per genome. The horizontal arrow shows the effect of mutations that reduce only adult survival. When the net rate of mutation exceeds a threshold ($U > 0.067$), these accumulate indefinitely, and reproduction at age 2 collapses, as shown here. The diagonal arrow shows the effects of mutations that reduce juvenile and adult survival equally. There is now no threshold; as the net mutation rate U increases, there is a steady decrease in survival.

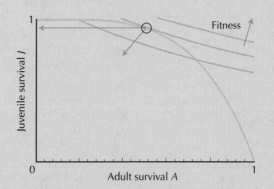

FIGURE 20.13. Optimal life history can be determined from fitness contours and a survival trade-off curve.

Drosophila that are sterilized, either by genetic mutations or by X rays, live longer. Conversely, if extra eggs are added to a blue tit's nest (*Parus caeruleus*), the mother is more likely to die over the following winter; if she does survive, she will produce fewer fledglings the following year. Artificial selection can also reveal the costs of reproduction. Populations of *D. melanogaster* can be propagated from either young parents or old parents. There is a direct response to such selection, so that "young" lines show higher fecundity in early life, whereas "old" lines show the opposite (Box 20.2). There is also an indirect response to selection in that "young" lines show lower survival in later life. This shows that the genetic changes that increased early survival had the pleiotropic side effect of reducing survival later on. Crucially, females from the "young" lines that were sterilized lived as long as sterilized females from the base population, showing that their shorter life span was a direct "cost of reproduction." Males also suffer costs to reproduction. Among the wild Soay sheep that live on the North Atlantic island of St. Kilda (Fig. 18.27), males rarely live beyond their first year, whereas females may live many years. Castrated males, which do not compete in the annual rut, live as long as females.

The Mutation Load May Be Concentrated on the Old

Senescence may also evolve because of the accumulation of deleterious mutations that affect late life. This explanation for aging was first proposed by J.B.S. Haldane in 1941. He pointed out that Huntington's disease, caused by a dominant allele that causes premature senility and death, is not eliminated by selection, because it acts in middle age, after most children have been raised. Late-acting mutations of this kind are expected to reach high frequency because they are only weakly constrained by selection. To the extent that there are deleterious mutations that act late in life, mutation accumulation will contribute to aging (see arrows in Fig. 20.13).

Several lines of evidence from laboratory populations of *D. melanogaster* suggest that mutation accumulation is of relatively minor importance. First, experiments in which mutations are allowed to accumulate (e.g., Fig. 14.30; p. 549) show little evidence of mutations that act *only* in late life. Mutated chromosomes suffer reduced survival and fecundity both early and late in life, suggesting that individual mutations will have deleterious effects both early and late and so will remain rare. Second, when laboratory populations are selected to reproduce at a younger age (Box 20.2), there is a decline in performance later in life that occurs at the *same* time as the increased fecundity earlier in life. If it were due to the accumulation of late-acting mutations that had been released from selection in the "young" lines, this decline would take many generations to appear. On the other hand, it has been predicted that the **additive genetic variance** for fitness components should increase with age if late-acting mutations accumulate; experimental observations on this prediction have given mixed results. However, all this evidence comes from *Drosophila*, and it may be that in organisms such as humans, with a higher total mutation rate, mutation accumulation makes a greater contribution.

Aging Is Influenced by Conserved Mechanisms for Optimizing the Life History

The evolutionary theories say little about *how* aging might occur. We might expect that many different factors would be involved, because selection on the functioning of the whole organism is relaxed in late life. The key point is that aging is expected to be a side effect of mutation and selection on other traits, rather than a tightly programmed function analogous to development of the embryo.

With this point in mind, the identification of specific mutations that greatly extend life span is surprising. Many such mutations have been found in yeast, nematodes, *Drosophila*, and mice (e.g., Fig. 20.14). These mutations are not completely paradoxical; they presumably have costs that would prevent them from being established in nature. For example, most long-lived mutants of mice are dwarfed and have small litters. What *is* surprising is that several of these genes are involved in a conserved signaling pathway, which in the multicellular organisms involves homologs of human insulin. It seems that this pathway senses nutritional conditions and adjusts metabolism accordingly. These recent observations fit with another previously puzzling observation: In organisms as diverse as yeast, nematodes, flies, and mammals, a reduction of food supply to about two-thirds of voluntary energy intake extends life span substantially. A plausible explanation is that when food is limited, life history is switched from current reproduction to survival, in the hope that reproduction can resume when conditions improve. Thus, there is a conserved mechanism that adjusts the life history toward the optimum for current nutritional conditions (see Web Notes for further details).

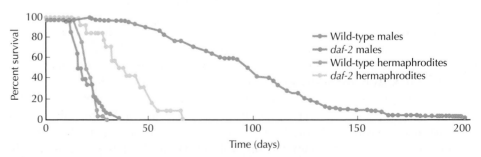

FIGURE 20.14. Mutations in the *daf-2* gene of the nematode, *Caenorhabditis elegans*, greatly extend life span, especially of males.

EVOLUTIONARY GAMES

Interactions between Individuals Can Be Understood as Evolutionary Games

So far in this chapter, we have shown how organisms can be optimized for a given function. However, in general, fitness depends on the biological environment, that is, on the abundance of different species and on the frequencies of different genotypes within species. Then, there is often no single "best" solution. Rather, we have to understand how all the organisms that interact will evolve in response to each other. To take the dung fly example, we must understand how female behavior responds to male tactics, and vice versa (Box 20.1).

When fitness depends on interactions between individuals, we cannot use simple optimization arguments to understand what will happen. In particular, natural selection will not necessarily lead to an increase in fitness. It may instead cause changes that are harmful to all the individual organisms involved, even though it is acting to favor fitter variants within each population. At first sight this seems to contradict Fisher's **Fundamental Theorem** (see p. 462), which states that natural selection causes an immediate increase in average fitness proportional to the additive genetic variance in fitness. However, the changes in gene frequency caused by natural selection may themselves change fitnesses in a way that is not taken into account by Fisher's algebra. Simple optimization arguments clearly fail in examples where evolving populations cycle endlessly (see Figs. 20.16–20.18).

We have already discussed frequency-dependent selection (pp. 475 and 505–508), concentrating on how it can maintain genetic polymorphism if rare variants do better (e.g., Figs. 18.19–18.21 and 19.5). Here, we introduce an important way of understanding the outcome of the frequency dependence that arises from the interactions between organisms: game theory. This can be seen as an extension of the optimization arguments that we discussed above; like them, it depends solely on phenotypes and not on the underlying genetics.

Imagine each organism as playing a game against the rest of its species or against other species. To understand this game, we have to know what each player can do and what the result of that action will be. That is, we have to define a set of possible **strategies** and a set of **payoffs** that depend on them. The strategies could be morphologies—body size or growth rate, say; behaviors, such as the time a dung fly spends copulating; or a set of behavioral or developmental rules—what to do in different circumstances, how to grow in different environments, and so on. The payoff is usually fitness, on the grounds that the "winner" in the game is the strategy that leads to the largest number of offspring and hence that propagates most rapidly through the population. The approach is close to the optimization framework we have seen already, in that it depends on a range of possible phenotypes (using the term in a broad sense) and on an optimization criterion. Crucially, however, the payoff depends on the *interaction* between individuals. If we imagine the simplest case—two players—then we must define the payoff to an individual playing one strategy as a function of what the opponent plays. This is often represented as a **payoff matrix** (Box 20.4).

Game theory was developed by John von Neumann in the 1950s as a tool in economics. The idea that evolution can be seen as a game played between individuals was first used by William Hamilton in 1967 and has since been developed primarily by John Maynard Smith. In many ways, it is more appropriate in evolution than in economics, because fitness provides a clearer payoff function than "utility" does for human behavior, and there is no need to assume that the players are behaving rationally. The key assumption is simply that a strategy that gives its player higher fitness will tend to spread by natural selection.

An Evolutionarily Stable Strategy Cannot Be Displaced by Any Alternative

The key concept in evolutionary game theory was introduced by Maynard Smith and George Price in 1973. To analyze a game, we seek an **evolutionarily stable strategy** or ESS, which cannot be displaced by any alternative. That is, the ESS must be at least as fit as any other feasible strategy, when almost all the population is playing the ESS (Box 20.4). There may be a single ESS, in which case we expect the population to evolve to it. If the current strategy is not an ESS, then fitter alternatives can invade (i.e., increase in frequency), and we expect that the population will eventually approach the ESS. There may be several ESSs, in which case the outcome will depend on where the population starts out. As we will see, there may even be no ESS, in which case the population may become polymorphic, or cycle endlessly. Understanding the ESS open to a population is a reliable, but not infallible, guide to what will actually evolve. In particular, the genetic system must be able to produce the ESS if it is to fix a single genotype that plays that strategy. For example, if the ESS can be produced only by a heterozy-

---

Box 20.4 Evolutionarily Stable Strategies in the Hawk–Dove Game

In the hawk–dove game, two individuals contest a reward with value V. The game is defined by the **payoff matrix** (a). The first row shows the payoff to a hawk that meets either another hawk (left) or a dove (right); the second row shows the payoff to a dove who meets a hawk or a dove. When a hawk meets a hawk, it has a 1/2 chance of winning and gaining the reward V, but also a 1/2 chance of losing and paying the cost of fighting C. The expected payoff is thus $(V - C)/2$. When a hawk meets a dove, it wins the reward V (top right). When a dove meets a hawk, it gains nothing, but because it does not fight, it also loses nothing (bottom left). Finally, when two doves meet, they share the reward (or have an equal chance of gaining it, which comes to the same thing).

(a)

payoff to	meeting	
	hawk	dove
hawk	$(V - C)/2$	V
dove	0	$V/2$

There are two distinct cases, depending on the cost of fighting relative to the reward. When the reward is worth more than the cost of fighting (payoff matrix (b), $V > C$; here, $V = 2$ and $C = 1$), then the "hawk" strategy is the ESS: A hawk can invade a population of doves (right column), but a dove cannot invade a population of hawks (left column). (But note that the population as a whole would be better off if it were all doves.)

(b)

payoff to	meeting	
	hawk	dove
hawk	1/2	2
dove	0	1

On the other hand, when the reward is worth less than the cost of fighting (payoff matrix (c), $V < C$; here, $V = 1$ and

$C = 2$), there is no "pure" ESS: Hawks can invade a population of doves, but doves can also invade a population of hawks. (Note that payoffs can be negative: We can think of them as changes in fitness relative to some arbitrary baseline.)

(c)

payoff to	meeting	
	hawk	dove
hawk	–1/2	1
dove	0	1/2

In this case (c), the ESS is a **mixed strategy** in which individuals all play hawk with probability V/C (here, 1/2), and otherwise play dove. To see this, think about the average success of an individual who plays hawk with some probability P, when the rest of the population plays the ESS (hawk half the time, dove the other half). If the individual plays hawk all the time ($P = 1$), then its payoff is –1/2 when it meets a hawk and 1 when it meets a dove, averaging 1/4. If it plays dove all the time ($P = 0$), then it wins nothing against a hawk and wins 1/2 against a dove, again averaging 1/4. So, when the population is at this mixed ESS, it makes no difference which strategy is played. Any strategy P will give the same payoff 1/4 when played against the current population.

In cases such as this, in which a range of possible invading strategies have the same fitness when played against the resident, an additional criterion is used to determine the ESS. Once the invader has reached an appreciable frequency, both invaders and residents will meet invaders. Given that invaders and residents have exactly the same payoff when they meet residents, the outcome is now determined by the outcome of contests with the invaders. We must ask what happens when the invader meets itself, compared with what happens when a resident meets the invader (see Web Notes for further details).

gote, then a sexual population will become polymorphic, and homozygotes will not play the expected ESS. In such cases, the detailed outcome depends on the genetics.

We will see examples of evolutionary games later in this chapter, when we examine interactions between males and females, and in the next chapter, where we discuss conflict and cooperation. First, however, we will see some examples of the diverse systems which can be understood using game theory.

Competition between Viruses Is an Evolutionary Game

When several viruses can infect the same host cell, they can interact by sharing the proteins that they need to replicate their genomes and to ensure that they are released as new infectious particles. With multiple infections, a virus can cheat, by contributing less to the collective pool than its competitors and by exploiting common resources from the pool for its own transmission. Thus, it is not surprising that almost all viruses that have been investigated are accompanied by defective viruses lacking essential genes that code for the viral coat or virus-specific polymerase. These **defective interfering viruses (DIVs)** gain an advantage in several ways. They may replicate faster simply because they are shorter; they may avoid transcription steps that would be needed only if they actually coded for protein; and they may encapsulate or replicate more often than their intact competitors.

The "game" played between DIV and wild-type virus can be represented by a payoff matrix that shows the fitness of "cooperating" (C) or "defecting" (D) virus (i.e., wild type or DIV) when the other viruses infecting a cell are either C or D (Fig. 20.15). We can see that neither C nor D is an ESS. If most viruses are wild type, the DIV has a selfish advantage, whereas if almost all are defectors, only the wild-type virus can replicate at all. Thus, when multiple infection is common, viral populations are typically polymorphic with respect to D versus C. They may contain a large fraction of DIVs, which reduce overall activity substantially. If viruses could play a mixed strategy and only *sometimes* cooperate, then there would be an ESS. A virus that sometimes cooperated and sometimes did not could not be displaced by either the pure C or the pure D strategy, and the ESS *probability* of cooperation would depend on levels of multiple infection. The idea of a mixed strategy is explained in more detail in Box 20.4. This game between viruses in fact follows the same logic as the hawk–dove game discussed there.

A slight variation on the game played between DIVs and their wild-type progenitor leads to one of the most intriguing games: the **Prisoner's Dilemma**. In the original version, two prisoners know that if neither confesses, neither can be convicted. However, if one confesses, he or she will get a reward, and the other will be given a heavy sentence. If both confess, they will both get the usual sentence. From a selfish point of view, it is always best to confess. If the other does not confess, there is a large advantage (a reward vs. prison). But if the other does confess, the prisoner will at least avoid a heavier sentence by confessing. Yet, both prisoners could go free if they stayed silent.

In terms of a payoff matrix, the Prisoner's Dilemma arises when "defection" is the fittest strategy even when defection is common. Then, defection is an ESS, because cooperation can never invade. Yet, a population at this ESS has lower fitness than one consisting solely of cooperators. We have already seen this situation in the hawk–dove game (Box 20.4, case (b)), where the structure of the payoff matrices is the same. A major advantage of a game-theory approach is that a wide range of examples can be described by a limited number of logical possibilities.

In the context of economics, a situation such as this is referred to as the **tragedy of the commons**: Self-interested exploitation of common resources leads to a worse outcome for all. Understanding this issue is especially important now that so many common resources are overexploited. (For example, it is in every fisherman's immediate interest to increase his catch of fish if others are competing for the same stocks; yet this has caused the collapse of many fisheries.) In economics and sociology, as well

	Defect (DIV)	Cooperate (wild type)
Defect (DIV)	0	$1 + s_2$
Cooperate (wild type)	$1 - s_1$	1

FIGURE 20.15. The payoff matrix for the game played between a defective interfering virus (DIV) and its wild-type form. In more general terms, this involves a cooperator (a virus that produces the proteins needed for replication and transmission), and defectors (viruses that "cheat" by using proteins coded by intact virus). If two DIVs infect a cell, they cannot replicate (*top left*), but if a DIV infects a cell together with a wild-type virus, it will outcompete the latter (*top right*, $0 < s_2$). Conversely, a wild-type virus will be at a disadvantage in a mixed infection ($0 < s_1 < 1$, *bottom left*).

as in biology, game theory has helped show how more sophisticated strategies (e.g., involving repeated interactions, or punishment of defectors by a social group) can avoid overexploitation and yet still be stable against invasion by more selfish alternatives. In the next chapter, we discuss these ideas (p. 613). We see that major transitions in evolution involve the evolution of cooperation between different replicating elements (pp. 614–606).

If There Is No ESS, Populations May Cycle: The Rock–Paper–Scissors Game

The children's game of **rock–paper–scissors** illustrates a slightly more complex game. Here, rock crushes scissors, scissors cut paper, and paper wraps rock (Fig. 20.16A). There is no single unbeatable strategy or, in other words, no "pure" ESS. Rock can be invaded by paper, paper by scissors, and scissors by rock. If there is a slight cost to a draw, then there is a mixed ESS—a strategy that is equally likely to play rock, scissors, or paper cannot be beaten. However, if there is a slight payoff from a draw, then there is not even a mixed ESS. In this situation, the population will tend to cycle endlessly.

Game theory alone cannot now predict what will happen. To understand how the population will change through time, we need to know how the strategies are inherited. That is, we need a genetic model. For example, suppose that the three strategies are coded by three alleles (R, S, P) at a single locus in a haploid; an R individual plays rock, an S individual scissors, and a P individual paper. Then, if there is a cost to a draw ($\varepsilon < 0$ in Fig. 20.16A), there is a stable polymorphism with the three alleles equally frequent. If, on the other hand, there is a payoff ($\varepsilon > 0$ in Fig. 20.16A), the

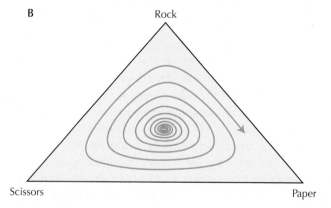

FIGURE 20.16. (*A*) A payoff matrix representing the children's game of rock–scissors–paper. The first row shows the payoff to an individual who plays rock when he or she meets each of the other strategies: Rock crushes scissors (payoff +1), but paper wraps rock (payoff –1). When there is a draw (e.g., rock meets rock), there may be a slight cost ($\varepsilon < 0$) or a slight payoff to a draw ($\varepsilon > 0$). (*B*) A genetic model of a haploid population in which three alleles of a single gene code for each of the three strategies. The rate of reproduction of each individual is proportional to the payoffs in *A*, with slight payoff to a draw ($\varepsilon = 0.1$). The polymorphic equilibrium where the three strategies are equally frequent is unstable, and the population cycles outward until one of the alleles is lost. The *triangle* represents the frequencies of the alleles that code for the three strategies.

population will tend to cycle between the three strategies. The cycles get larger and larger in amplitude, so that one of the alleles will eventually be lost by accident (Fig. 20.16B). R displaces S, and P displaces R, but by then S may already have been lost before it regains its selective advantage.

The outcome in this kind of situation depends on both the genetics and the initial conditions. In this example, the genetic model parallels the ESS. With a cost to a draw, the genetic model has a stable polymorphism with three types equally frequent, whereas the ESS is a mixed strategy in which every individual plays rock, scissors, or paper at random. With a payoff to a draw, there is no ESS, and in the genetic model, polymorphism is unstable. In general, however, genetic models may give different outcomes from ESS arguments.

Complex behavior, such as that seen in the rock–paper–scissors game, can occur quite readily in nature. The key requirement is that there should be no simple ranking of competitive ability, such that one type is always superior. One such situation is seen in an experiment with *E. coli*. Some strains of this bacterium can produce a **bacteriocin** that kills sensitive *E. coli*. Bacteriocin production is the product of a "col" plasmid, which also encodes a protein that gives the cell immunity to the toxin, and a protein that causes a fraction of bacteriocin-producing cells to lyse, releasing the toxin and killing its competitors. Sensitive strains occasionally mutate to become resistant as a result of changes in the genes encoding membrane proteins that bind and transport the bacteriocin. Thus, there are three viable kinds of bacterium: C (bacteriocin-producing and resistant), R (resistant), and S (sensitive). In the absence of bacteriocin, S tends to grow faster than R and R faster than C. This is because toxin production and resistance both involve costs: Producing bacteriocin diverts metabolic resources, and changing the membrane to give resistance interferes with its normal function. However, C will displace S by poisoning sensitive cells.

In a single well-mixed population (e.g., a stirred flask) that begins with all three at high frequency, S is eliminated rapidly, because it is poisoned by the bacteriocin. C cells are then outcompeted by R cells, which are then established permanently (Fig. 20.17A). (In principle, S could reinvade, but by the time C cells have been reduced enough to give S a net advantage, S have been lost from the population.) In contrast, if interactions are localized, the cycling characteristic of the rock–paper–scissors game emerges. Bacteria can be grown on agar plates and transferred onto new medium with a velvet pad that retains their spatial pattern. Starting with a uniform mixture, patches fixed for the three types emerge and move around in a cyclic way. The boundary between R and S moves in favor of S, that between S and C in favor of C, and that between C and R in favor of R (Fig. 20.17C). This example shows how diversity can be maintained by cycling between different types, provided that the spatial distribution of the habitat prevents loss of the rarer type from the population. (Recall that maintenance of diverse types in *Pseudomonas* also required spatial structure; Fig. 18.21.)

Behavioral Polymorphism in Lizards Gives a Natural Example of the Rock–Paper–Scissors Game

A quite different example of the rock–paper–scissors game can be seen in the competition between males of the side-blotched lizard, *Uta stansburiana*. These lizards are abundant in the deserts of the western United States; they mate in their first year and can be individually marked and recaptured. Three different male color morphs are seen, each with a distinct behavior (Fig. 20.18A). Orange-throated males are very aggressive, holding large territories in which they guard many females. Yellow-throated males do not hold territories and instead mimic females. They lurk in crevices and sneak matings with females within territories guarded by the orange-throated morph. Blue-throated males are less aggressive, and guard their mates in small territories; they

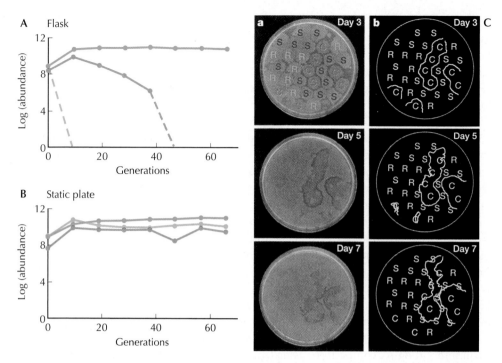

FIGURE 20.17. The rock–scissors–paper game in a bacterial population. (A) Abundance of three types of *Escherichia coli*, grown in a stirred flask: *green*, resistant; *red*, bacteriocin-producing; *blue*, sensitive. *Dashed lines* indicate that the type is undetectable. (B) When the bacteria are propagated on the surface of an agar plate, patches of the three types can coexist. The graphs show the average frequency over the whole plate. (C) An example showing how patches move over time. The letters show the initial distribution of the three types. The pictures on the *left* show the plates at days 3, 5, and 7: The patches inhabited by C cells are less dense and so can be distinguished by eye. The sketches on the *right* show how the boundaries between the three types move: C advances at the expense of S (*yellow lines*), but R advances at the expense of C (*pink lines*).

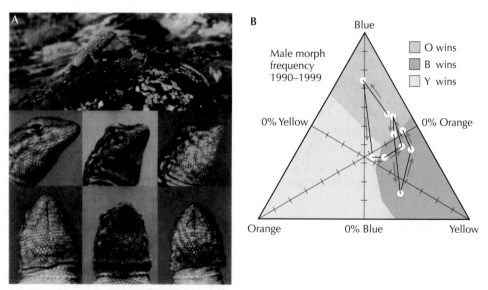

FIGURE 20.18. (A) Three color morphs of male side-blotched lizards (*Uta stansburiana*). Males with orange throats (*lower left* panels, and male at *left, top*) are "ultradominant" to males with blue (*lower center*, and male at *right, top*). Yellow-throated males (*lower right*) resemble females and sneak matings from orange-throated males. (B) Cycling of morph frequencies from 1990 to 1999 (direction indicated by *red arrows*). The *shaded areas* show frequency combinations in which orange, blue, or yellow are predicted to increase, based on measurements of how male fitness depends on the frequencies of other morphs in their neighborhood. (Fitnesses were measured by scoring offspring and adults for highly polymorphic **microsatellite** loci, allowing the father of each juvenile to be identified.)

can successfully defend against the yellow-throated morph. The inheritance of the three morphs is consistent with there being three alleles at a single locus. Male mating success depends strongly on the frequency of the morphs in the surrounding area, and the measured fitnesses imply a rock–paper–scissors relation. Correspondingly, the frequencies of the three morphs cycle over a 10-year period (Fig. 20.18B).

These various examples illustrate how game theory can help us understand evolution when the outcome depends on interactions between individuals. The fundamental principle is to seek one or more ESSs that cannot be displaced by any alternative. The strength of the method is that it does not involve any genetic assumptions and so can give a general understanding of phenotypic evolution. However, the outcome does depend on genetics if there is no ESS (as may happen in the rock–paper–scissors game) or if the ESS cannot actually be achieved by any feasible genotype. In the rest of this chapter, we will see that the interaction between males and females can be understood from both a genetic and a game theory perspective.

SEXUAL SELECTION

Sexual Selection Arises from Variation in Mating Success

In sexually reproducing populations, sexual selection arises from differences in the ability to find a mate. It is distinguished as a special kind of selection because it involves interactions between individuals of the same species, rather than with the external environment. Because there is no adaptation to definite externally imposed requirements, sexual selection can lead to arbitrary and apparently maladaptive evolution. The same is true of other kinds of interaction, such as the defective interfering viruses (Fig. 20.15) presented earlier in this chapter. However, sexual selection has pervasive effects in sexually reproducing species and, as we will see, is responsible for a large fraction of fitness variation. It therefore deserves special attention.

In organisms with separate sexes—male and female—there is often a strong asymmetry in mating behavior and in the associated morphology and physiology. Males often have elaborate structures that they use to compete with other males or to attract females. We have already seen several examples—in the luminescent click beetles (Fig. 19.24) and in the behavior of different color morphs of side-blotched lizards (Fig. 20.18). There are many other familiar examples, such as the extravagant plumage of the peacock, the large antlers used by stags to help them maintain harems of females, and the bright coloration of many male insects, to name only a few (Fig. 20.19). The consequences of sexual selection need not be so obvious. We will see that male genitalia and seminal fluid have evolved in remarkable ways so as to father more offspring, by ensuring fertilization in competition with other males (Fig. 20.20). As we will see in the next two chapters, this kind of conflict can contribute to speciation.

Sexual selection does not depend on each adult being a separate sex (Fig. 20.21). Nevertheless, male and female function can be distinguished. For example, individual flowering plants usually produce both pollen and ovules, which will grow to produce seed. The striking floral displays of insect-pollinated plants have evolved primarily as male structures that function to disseminate pollen, rather than to attract fertilizations. Larger flowers attract more insects and contribute more to the pollen pool (e.g., Fig. 19.6). However, smaller and much less costly flowers would usually suffice to ensure that ovules are fertilized.

Sexual Selection Can Be Maladaptive

Many of the male traits that evolve as a result of sexual selection seem maladaptive in that they reduce the survival of males. A peacock's tail or a stag's antlers must hinder

FIGURE 20.19. Examples of extravagant male traits. (*A*) Tail of a peacock (*Pavo cristatus*). (*B*) A male satin bowerbird (*Ptilonorhynchus violaceus*) with his bower decorated with brightly colored objects. (*C*) Stag beetles (*Lucanus cervus*) with their antler-like jaws. (*D*) Female swallowtail butterflies (*Papilio dardanus*) gain protection by mimicking a variety of other distasteful species. However, males have retained the ancestral pattern shown here, presumably to remain attractive to females. (*E*) A *Dasa nivea* flower, which has evolved to attract bees that will carry away its pollen.

escape from predators and divert resources from necessary functions. In some cases, we have direct evidence. For example, male sticklebacks develop bright red coloration in the breeding season, which has been shown to increase the rate of predation by trout (Fig. 20.22A). Moreover, in streams with higher predation pressure, male fish are less brightly colored (Fig. 20.22B). More generally, the absence of sexually selected traits in females (as in all the animal examples of Fig. 20.19), and their frequent absence in nonbreeding males (as in the stickleback and lizard examples, Figs. 20.18 and 20.22), imply that the traits are costly to fitness.

It is not so clear whether sexual selection impairs survival of the species as a whole. If males do not provide any parental care, then as long as there are enough males to

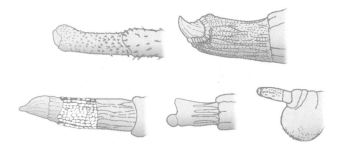

FIGURE 20.20. Male genitalia often have complex morphology, which has evolved both to transfer sperm more effectively and to stimulate the female to accept the sperm. These drawings show examples of primate penes.

fertilize the females, reduced male survival does not affect the reproduction of the species. Of course, this begs the question of why males are present at all or, indeed, why organisms go to the trouble of reproducing sexually. We take up these issues in Chapter 23. For the moment, we accept that there are distinct sexes, which are defined by the production of small male gametes and large female gametes. Here, we explore the consequences for sexual selection.

Sexual Selection Acts More Intensely among Males Than among Females

When Darwin first defined sexual selection, in *On the Origin of Species*, he described it as "a struggle between the males for the possession of the females." Why should there be such an asymmetry between the sexes? The basic reason is that females invest heavily in offspring, by producing large eggs and, in many cases, then nurturing the embryos and caring for the young. Thus, a female's fitness is limited by the number of eggs that she can produce and rear. In contrast, a male produces small gametes (sperm or pollen) and so can achieve high fitness by mating many times and fertilizing many females. Darwin described the difference between the sexes vividly in his book, *The Descent of Man and Selection in Relation to Sex*:

> The female has to expend much organic matter in the formation of her ova, whereas the male expends much force in fierce contests with his rivals, in wandering about in search of the female, in exerting his voice, pouring out odoriferous secretions, &c....On the whole the expenditure of matter and force by the two sexes is probably nearly equal, though effected in very different ways and at different rates.

Ultimately, the asymmetry in sexual selection arises from the asymmetry in gamete size that defines males and females (see pp. 685–687). Sometimes, the roles of the sexes are reversed (e.g., Fig. 20.23). Such cases confirm Darwin's explanation, because females then typically have the more conspicuous mating behavior.

The importance of sexual selection for the two sexes can be quantified. Typically, males show a higher variance in reproductive success than females, as a result of sexual selection. In the population of red deer from the Hebridean Island of Rum introduced in Chapter 19, in which individuals and their relationships have been monitored for many years, the variance in lifetime reproductive success among males is 22.0, compared with 6.4 for females. This difference between the sexes is largely due to the much greater variance in breeding success between males, compared with females (variances 51.3 vs. 8.8, respectively). (Here, lifetime reproductive success is the total number of offspring produced by a newborn individual, whereas breeding success is the number produced, given that an individual has survived to adulthood.)

Fitness differences are not just due to variation in finding a mate. Often, females mate with several males, and competition between pollen or sperm from different fathers can lead to large fitness differences. For example, in *Drosophila* there is remarkable variation in the ability of different male genotypes to displace sperm from

FIGURE 20.21. Examples of **hermaphrodites**. (*Top*) Most flowering plants produce both pollen and seed, usually from within the same flowers. (*Bottom*) When snails mate, they each inseminate their partner.

FIGURE 20.23. The red-necked phalarope (*Phalaropus lobatus*) is a rare example where the roles of the sexes are reversed: Males (*top*) tend eggs and chicks, and females (*bottom*) have brighter plumage and compete for males.

FIGURE 20.22. (*Left*) The bright red coloration that develops in male sticklebacks (*Gasterosteus aculeatus*) during the breeding season. Experimental exaggeration of the red coloration increases the rate of predation. (*Right*) The more cryptic silver color that is found at high frequency in streams with high predation pressure.

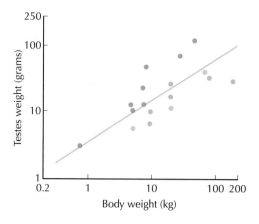

FIGURE 20.24. Testes are larger, relative to body size, in primates where many males compete to fertilize females. *Red dots*: multimale breeding system; *green dots*, monogamous male; *blue dots*, single male; the rightmost *green dot* is man.

earlier matings or to resist displacement by later matings.

Sexual selection is more intense in some species than others. If adults are strictly monogamous, then sexual selection is expected to be weak; and with equal proportions of the two sexes, all individuals will find a mate. In primate species in which females mate with several males when in breeding condition, the males have larger testes relative to their body size (Fig. 20.24), presumably because there is strong selection to produce more sperm when competing with other males. In birds, monogamous species generally show less difference in plumage between the sexes. Nevertheless, sexual selection can still act even with strict monogamy if more attractive males pair up earlier and so can raise more young (Fig. 20.25).

FIGURE 20.25. Sexual selection in a monogamous bird, the arctic skua (*Stercorarius parasiticus*). This is found in three forms—dark, intermediate, and pale—determined by two alleles at a single locus. The dark form (*A, top*) breeds earlier than the intermediate, which in turn breeds earlier than the pale (*A, bottom*). (*B*) Because earlier breeders have higher net fitness than late breeders, this gives the dark form a selective advantage—even though all birds pair up and mate. As a result, the dark form is spreading.

Sexual Selection Involves Competition between Males and Choice of Males by Females

Two kinds of sexual selection can be distinguished: competition between males and choice of males as mates by females. There are many obvious examples of the first category. Males of many animal species fight each other in order to keep control over females (e.g., Fig. 20.19C). It is also common for males to guard females once they have mated with them (recall the dung fly example in Box 20.1). Flowers compete to attract pollinators and thereby transfer more pollen to the surrounding **stigmas**. Competition between males may take less obvious forms—for example, when sperm or pollen compete. Male damselflies have a special structure that is used to sweep out sperm from previous matings (Fig. 20.26) and, conversely, many insects produce "mating plugs" that prevent females from taking in more sperm. In *Drosophila*, males produce seminal fluid, which causes females to re-mate less often and increases their immediate egg production at the expense of later survival.

In many cases, however, females appear to choose between potential mates. Such female choice is most obvious in species where breeding males aggregate in **leks** and females choose mates from among them. For example, male sage grouse of the western United States congregate in stony areas and hold territories there through the breeding season (Fig. 20.27). Females come to the lek to mate and mostly mate with the few males at the center. Male bowerbirds of New Guinea and Australia build elaborate bowers and decorate them with colorful objects. Females visit several bowers and seem to choose those males that have built the most attractive structures (Fig. 20.19B). This kind of mating system raises the question of why females should behave in this way: Why do they mate with the central male rather than at random?

It is not easy to distinguish female choice from competition between males, even in principle. (For example, how do we find out whether stag beetles with larger jaws gain mates because they are better at fighting with other males or are more attractive to females [Fig. 20.19C]?) However, if different females show *different* preferences, then female characteristics must be influencing mate choice. An unusually detailed example comes from sperm competition in *Drosophila*. There is wide variation between male genotypes in their ability to defend against displacement by sperm from a second mating, and the degree of sperm displacement varies greatly between female genotypes. Figure 20.28 shows that the relative success of different *Drosophila* males in fertilizing females depends strongly on which female genotype they mate with. There is no "best" male overall, despite strong differences between males in their ability to fertilize any one kind of female.

Female preferences such as those demonstrated in Figure 20.28 must impose strong sexual selection on males. We must ask, first, what kinds of males will be preferred and, second, why females should evolve behaviors that allow them to exert these preferences. That is, we must explain both female choice of one kind of male rather than another and the female choosiness that allows these preferences to be exerted. For example, why should female sage grouse prefer to mate with the males at the center of the lek, and why should females go to leks in order to choose a mate? In the following section, we explore the theory and evidence on the evolution of female choice and female choosiness.

Female Preferences May Evolve by Direct Selection on Females or as a Side Effect of Selection on Other Traits

Darwin's theory of sexual selection by female choice was neglected until recently. In part, this was because it was thought that female choice required some kind of conscious aesthetic sense. Both Darwin and Fisher argued that animals have an aesthetic sense that influences mate choice and that it is the evolutionary homolog of human aesthet-

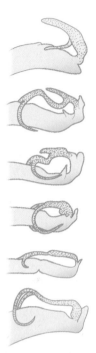

FIGURE 20.26. Male genitalia have evolved complex structures that aid competition with other males. Shown here are the penes of six species from the damselfly genus *Argia*. The *stippled areas* are used to scoop out sperm that have been stored by the female from previous matings.

FIGURE 20.27. Sage grouse.

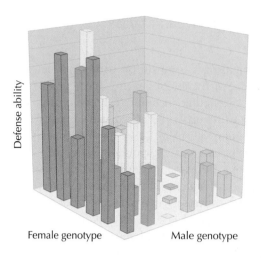

FIGURE 20.28. Genetic variation in sperm competition in *Drosophila*. The bars show the ability of a male's sperm to defend against displacement by a subsequent mating. This depends on an interaction between the genotypes of male and female. In this experiment, six male genotypes were tested against six female genotypes, giving 36 experiments in all.

ics. However, when we use the term "female preference," we simply refer to any features of females that make them more likely to mate with some males rather than with others. Female choice can occur whenever female traits lead to differences in mating success among males; it can occur in any sexually reproducing species, not just in animals.

Competition between males, and between male gametes, is easily understood. It may readily lead to behavior and morphology that reduce male survival. This is simply an example of a trade-off that shapes a life history optimal for overall fitness. Male traits that seem to be displays and are not directly involved in contests with other males are harder to understand. Why should peahens be attracted by the peacock's tail, or bowerbird females by elaborate bowers? Some displays function as signals to other males that a fight would be unduly costly. Often, however, males seem to have evolved extravagant displays that attract females, and females behave as if *choosing* between these displays.

There are three distinct kinds of explanation for the evolution of female preferences: **direct selection**, **sensory bias**, and **indirect selection**. Direct selection may act when females gain an immediate increase in fitness by choosing one male over another. Direct selection is clearly important when males provide parental care. In great tits, for example, females prefer to pair with males that have larger territories, and females nesting on such territories do raise more young. Many insects transfer nutrients as well as sperm and, again, females that receive larger "nuptial gifts" raise more young. There may be subtler benefits; for example, females may choose males who guard them against harassment by other males, they may mate with healthy males in order to avoid catching sexually transmitted diseases, or they may mate with males who will give them enough sperm to fertilize all their eggs.

The difficulty is to explain the existence of female preferences when there seems to be no immediate consequence for female fitness—for example, when a female sage grouse makes a brief visit to a lek and comes away with sperm from only her chosen male. The most straightforward explanation for this kind of female choice is simply that females have an innate sensory bias among males, for reasons that have nothing to do with the sexual selection caused by this bias. Females are under strong selection to find a mate and to ensure that this mate is of the right species. Thus, it is not surprising that females should prefer males that are more conspicuous. Bias may also be caused by other kinds of selection on the mating system (e.g., Fig. 20.29).

Evidence for sensory bias of this kind comes from phylogenetic comparisons across species. In the tungara frog (*Physalaemus pustulosus*, Fig. 20.30A), the male call consists of the basic "whine" that is found throughout this group of species, but also has additional "chucks" (Fig. 20.30B). This more complex call is costly, because it requires more energy to produce, and also is more likely to attract predators. However, female *P. pustulosus* strongly prefer calls with extra chucks. This preference arises because the chuck frequencies are tuned to stimulate the basilar papilla, one of two sound-sensing organs in the

FIGURE 20.29. Female preferences may be a side effect of sensory bias. Male nursery web spiders (*Pisaura mirabilis*) attract females with a prey gift, which they cover with silk so that it resembles a female's egg sac. Females of this species care for their eggs and so recognize egg-like objects more readily. Experimentally manipulated prey gifts were more attractive the more they resembled egg sacs.

frog's inner ear. (The basic whine only stimulates the amphibian papilla, the other sound-sensing organ.) Crucial evidence that the "chuck" evolved to exploit a preexisting sensory bias comes from the observation that in related species males do not have the distinctive "chuck," yet females have a strong preference for recorded calls containing this chuck. Other things being equal, this male trait would increase if it arose in those species. Examination of the distribution of trait and preference across the phylogeny shows that the preference evolved before the trait (Fig. 20.30B). Selection on the preference therefore cannot have anything to do with the trait that happens to be preferred, because the preference was present before the trait evolved.

Indirect Selection on Preferences Can Lead to Fisher's Runaway Process

Finally, female preference may evolve by indirect selection, caused by natural and sexual selection on the preferred traits. This is the most contentious of the three classes of explanation and has stimulated a great deal of theoretical and empirical work over the past 20 years. We will use arguments based on both population genetics and game theory.

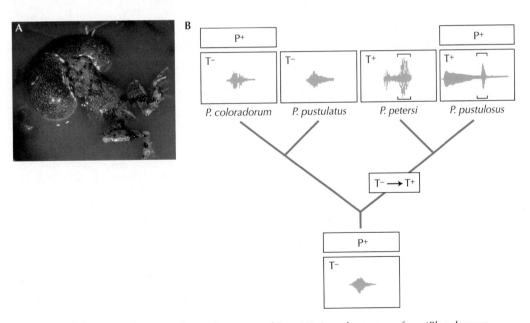

FIGURE 20.30. Phylogenetic evidence for sensory bias. (*A*) A male tungara frog (*Physalaemus pustulosus*) calling. (*B*) The phylogeny shows the relationships between four species in the genus *Physalaemus*. The four boxes (*top of the tree*) show the calls in the four species and the single box (*base of the tree*) is the inferred call of their ancestor. The sister species *P. petersi* and *P. pustulosus* have the chuck (*black bars*), but both *P. pustulosus* and *P. coloradorum* females prefer the chuck (P⁺), even though males of *P. coloradorum* do not express it. This shows that the female preference for the chuck evolved before the males evolved this trait. (Female preferences have not been tested in the other two species.)

These have both been important and have sometimes led to different conclusions.

As we saw in Chapter 17 (p. 479), a trait that does not itself affect fitness can still change if it is associated with traits that *are* selected. In quantitative genetic terms, this requires a **genetic correlation**; with discrete Mendelian genes, it requires **linkage disequilibrium** between the genes. Such linkage disequilibrium arises naturally as a consequence of mating preferences. Females who prefer to mate with certain kinds of male will produce offspring that carry a maternal genome containing genes for the female preference, and a paternal genome carrying genes for the preferred trait. Recombination then brings these genes together into the same genome. The net result is the generation of linkage disequilibrium in proportion to the strength of mating preferences, which allows selection on the male genes to cause a change in female preferences (Fig. 20.31).

Fisher first described this mechanism of indirect selection in 1915. Later, in his book *The Genetical Theory of Natural Selection* (1930), he argued that indirect selection could lead to a **runaway process** in which both a male trait and the female preference for it would be elaborated. Initially, a male trait might be increased by natural selection. Any female preference for it would become associated with the trait and would therefore increase with it, in a form of **hitchhiking** (see p. 485). As the preference increases, sexual selection gives the trait an additional advantage, and so the trait increases more rapidly. This in turn leads to a further strengthening of the preference. Fisher argued that this positive feedback could exaggerate the trait beyond the point where it is favored by natural selection, and that it would continue to evolve until halted by opposition from natural selection. This runaway process thus explains the elaboration of arbitrary and maladaptive traits, along with the female preferences for them.

Fisher's argument was not analyzed mathematically until the early 1980s, when Mark Kirkpatrick and Russell Lande independently set out models for the joint evolution of trait and preference. These models showed that a **runaway** could occur, but that it required a very strong association between trait and preference. A more likely outcome is a balance between natural and sexual selection on the trait, such that both trait and preference reach an equilibrium. In the absence of direct selection on the preference, there could be an infinite variety of outcomes. There might be no preference, in which case the trait would reach its optimum under natural selection, or there might be a strong preference, opposed by strong natural selection. The outcome depends on arbitrary sensory bias of the female. If direct selection acted on females, the preference would simply evolve to maximize the female's immediate fitness. These models thus support the two explanations discussed above (either an arbitrary sensory bias or direct selection) rather than indirect selection acting via a runaway process.

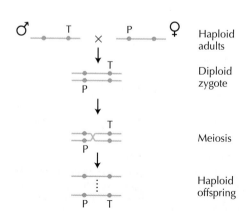

FIGURE 20.31. Mating preferences generate linkage disequilibrium between genes for male trait and female preference. This diagram shows a simple life cycle in which haploid adults mate to produce diploid zygotes, which then go through meiosis to give the next generation of haploid offspring. Some males carry an allele T, which causes an attractive trait (bright plumage, say). Some females carry an allele P, which causes a preference for this trait. The diagram shows the alleles P and T at different genetic loci along the chromosome. Because of the mating preferences, P and T tend to find themselves within the same diploid zygote. If there is crossing over, then P and T will be brought together on the same chromosome. In the next generation, therefore, the combination PT will be in excess, or in other words, in linkage disequilibrium. This linkage disequilibrium allows selection on the trait T to cause indirect selection on the preference P.

Sexual Characteristics May Evolve to Signal Genetic Quality

The conclusions from Kirkpatrick's and Lande's models, which emphasized the arbitrary outcome of sexual selection, jarred with the intuition of many naturalists, who suggested that females can choose **good genes**. Can a male signal his genetic quality to females, and can a female preference for the signal evolve because of its association with good genes? (By "genetic quality" or good genes, we mean genes that will increase fitness, for reasons that are not due to sexual selection on the signal. It is convenient to refer to such components of fitness as viability, although male fertility and mating success could also be included.) This is a special case of indirect selection, in which the female preference evolves because it is in linkage disequilibrium with genes that are selected.

There are two main difficulties for good genes models of this kind. First, the signal must remain honest. What prevents males of lower genetic quality from gaining mates by producing the attractive signal? Second, there must be enough variation in genetic quality among males to allow significant indirect selection on female preferences. As we will see, both problems can be avoided in principle. However, it remains unclear to what extent female preferences for good genes have in fact shaped sexual selection.

How can a reliable signal of male genetic quality evolve? This problem is best seen from the perspective of game theory. Is there an ESS in which males with genes that will give offspring of higher viability also tend to produce a particular signal? The solution was first proposed in 1975 by Amotz Zahavi, although it took some time for the theory to become clear. The crucial requirement is that the signal must be *costly* and, moreover, it must be *more* costly to males of lower genetic quality (i.e., to males with genes that reduce fitness for reasons unrelated to the signal trait). Zahavi termed this kind of trait a **handicap**. It is the *differential* cost that ensures that the handicap remains an honest signal of male merit. Inferior males could get more mates by displaying the signal, but would lose more in other components of fitness than they would gain by doing so. Hence, the best strategy, from the point of view of an individual male, is to display to an extent that increases with genetic quality. Once the association between signal and genetic quality is established, females will evolve to prefer the signal simply because they thereby gain fitter sons.

It Is Difficult to Find Evidence on Whether Females Choose Males with Good Genes

The most obvious test of the good genes hypothesis is to ask whether females who mate with attractive males bear offspring of higher "genetic quality"—for example, they survive better. Such correlations between male attractiveness and offspring quality have been observed in several studies. For example, in great tits (*Parus major*), males carrying larger black stripes (Fig. 20.32) are more successful in attracting mates, and the offspring of males with larger stripes survive better over winter. Thus, females who choose to mate with the more attractive males do gain an indirect advantage, because their sons will have higher viability. Another example comes from cockroaches (*Nauphoeta cinerea*), in which males that are more attractive produce offspring that develop faster (Fig. 20.33). These insects are live bearing, and so females who mate with attractive males gain a direct benefit by being able to produce broods more often, as well as an indirect advantage through faster growth of their offspring.

Another approach is to ask whether females who are allowed to choose their mates have offspring that survive better than those of females who are allowed no choice. Advantages to mate choice have been found in several cases. Recently, much attention has been focused on the surprising discovery that in many species of bird that appear monogamous, females frequently seek copulations from other males. Genetic methods have detected "extra-pair" matings in 90% of species studied, and even in socially monogamous species, 11% of offspring have come from extra-pair

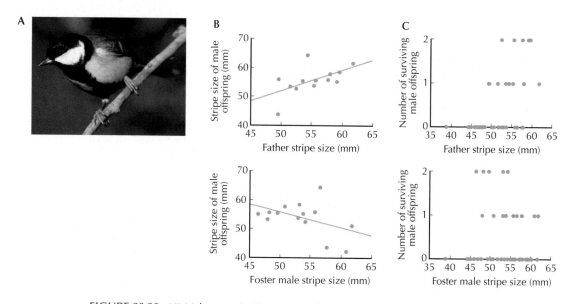

FIGURE 20.32. (*A*) Male great tits (*Parus major*) have a dark breast stripe that is attractive to females: Males with larger stripes are more likely to gain mates. In a cross-fostering experiment, clutches were swapped between nests. Thus, correlations with the biological parent could be compared with correlations with the randomly chosen foster parent as a control. There is substantial genetic variation in stripe size: The stripes of sons are strongly correlated with their biological father (*B, top*), but there is no significant correlation with the foster parent (*B, bottom*). Offspring whose biological fathers have larger stripes survive significantly better over winter (*C, top*), whereas there is no correlation with the foster father's stripe size (*C, bottom*).

fertilizations. In some cases, such offspring have been shown to have higher survival compared with their nestmates who were fathered by the male providing parental care. This suggests that females may be choosing males who can provide good parental care as their social mates, but also seek matings from males who can provide genes that will increase the fitness of their progeny (e.g., Fig. 20.34). Another possibility, however, is that some males are infertile, and that females seek multiple matings as an insurance against the failure of their first mate to fertilize their eggs.

Correlations of this kind can be hard to interpret. Although correlations between

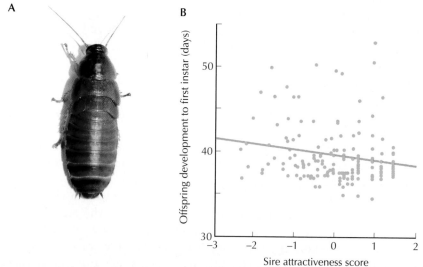

FIGURE 20.33. Female cockroaches (*Nauphoeta cinerea; A*) are attracted to males of high social status; this preference is mediated by a pheromonal signal. The offspring of males that score as being more attractive in mate-choice tests develop faster (*B*).

FIGURE 20.34. In a study of great reed warblers (*Acrocephalus arundinaceus*; *A*), females who obtain extra-pair fertilizations (EPFs) choose neighboring males with larger song repertoires: In all ten EPFs observed, the biological father had a larger repertoire than the cuckolded male. The male's song repertoire is correlated with the number of his offspring successfully recruited into the population (*B*). Thus, females who obtain EPFs get an indirect benefit through greater offspring survival.

attractive male traits and offspring quality are necessary if females are to evolve preferences for good genes, they do not imply that preferences evolved for that reason. We expect to see a correlation between male mating success and offspring survival for trivial reasons—vigorous males are more likely to win mates and are also more likely to have vigorous offspring if vigor is heritable. However, it is extremely difficult to demonstrate that female preferences for traits associated with vigor evolved *because* of the genetic benefits females gain by choosing to mate with such males.

Choice of Good Genes Requires Heritable Variation in Fitness

Another difficulty for the evolution of female preferences for good genes in males is that there must be a source of heritable variation in net fitness. In the original Kirkpatrick and Lande models, female preferences for good genes could not evolve, because there was no additive genetic variance in net fitness, which is inevitable if the only process acting is selection (see p. 462). However, as we saw in Chapter 18, there are many processes that can maintain additive fitness variance: mutation, migration, and some kinds of fluctuating selection, among others. Empirically, it is not clear how much heritable fitness variation is actually maintained by these processes (Chapter 19 [pp. 547–548]). Good genes models would become substantially more plausible if it were shown that there are typically high levels of additive variance for fitness.

In 1982, Hamilton and Marlene Zuk suggested that the coevolution between hosts and their parasites could be responsible for maintaining high levels of fitness variation (see p. 509). They went on to suggest that sexually selected traits could evolve to signal the ability of males to resist parasites. For example, the condition of a bird's plumage might indicate whether it is infected, and females might use this information to choose mates with genes that allowed them to better resist infection. Hamilton and Zuk supported their argument with a survey that showed that bird species with high levels of parasitism also tended to have brighter plumage and greater differences in plumage between the sexes. However, this cross-species correlation has not been found consistently in later surveys. Within species, correlations between heritable parasite resistance and sexual attractiveness have been found, but as noted in the previous section, it is not clear that female preferences have evolved as result of the indirect benefits caused by these correlations.

Selection can readily maintain nonadditive fitness variation—for example, by heterozygote advantage. In such cases, mating preferences may evolve that depend on female genotype. We have already seen an example of such preferences in sperm precedence in *Drosophila* (Fig. 20.28). These kinds of preference for genetic compatibility may be widespread. Indeed, we examine two especially important cases in later chapters: female choice for males of their own species (p. 648) and female preference for mating with nonrelatives, so as to avoid inbreeding (p. 687).

SUMMARY

Evolutionary biologists use a variety of approaches to understand what kinds of organisms are likely to evolve. If the fitness of an individual depends only on its own phenotype, then we expect the optimal phenotype (i.e., the phenotype that maximizes fitness, subject to constraints on what is possible) to evolve.

Senescence—the fall in rates of survival and reproduction with age—can evolve as part of an optimal life history. This is because older individuals contribute less to fitness, so that the selection that maintains late life is weaker. Survival and reproduction may be further reduced by deleterious mutations, which are opposed more weakly when they act later in life. These evolutionary arguments suggest that aging will involve many different components of fitness. It is surprising, therefore, that mutations have been found that substantially increase survival rates in old age. These influence a conserved pathway that reallocates effort from reproduction to survival when resources are scarce.

When fitness is determined by interactions between individuals, we expect an evolutionary stable strategy (ESS) (i.e., a strategy that cannot be displaced by any alternative) to evolve. However, there may be several such ESSs, in which case the outcome depends on the initial state of the population. If there is no ESS, the population may cycle endlessly. The ESS may be mixed, in which case a mixture of types is maintained. Even though each individual strives to maintain its fitness at the expense of others, the outcome may not be optimal for the population as a whole.

Much of evolution is driven by sexual selection, which arises from variation in mating success. Sexual selection acts most intensely between males. Male fitness increases by mating with more females, whereas a female's fitness is limited by her ability to raise offspring. Sexual selection involves both competition between males and choice of males by females. The latter poses special difficulties. Female preferences may be directly selected (e.g., if males provide parental care), in which case females should choose those males that will help them rear the most offspring. However, if males make no direct contribution to fitness, female preferences may evolve as a side effect of sensory bias, or through a runaway process, which amplifies preferences for male traits that are already preferred. Females may also prefer males that carry good genes that will make their offspring more fit. In this case, preferences must be for male traits that are costly, so that they can be borne only by males that genuinely do carry good genes.

FURTHER READING

Evolutionary Optimization

Parker G.A. and Maynard Smith J. 1990. Optimality theory in evolutionary biology. *Nature* **348**: 27–33.
    A clear statement of how optimization principles should be used.
Stearns S.C. 1992. *The evolution of life histories.* Oxford University Press, Oxford.
    A comprehensive textbook that covers much of the material in this chapter.

Aging

Guarente L. 2002. *Ageless quest: One scientist's search for genes that prolong youth.* Cold Spring Harbor Laboratory Press, Cold Spring Harbor, New York.
Partridge L. and Barton N.H. 1993. Optimality, mutation and the evolution of ageing. *Nature* **362**: 305–311.
    A review of alternative theories of aging, and of experimental evidence, primarily from *Drosophila.*

Evolutionary Games

Maynard Smith J. 1982. *Evolution and the theory of games.* Cambridge University Press, Cambridge.
    The classic account of evolutionary game theory.

Sexual Selection

Andersson M. 1994. *Sexual selection.* Princeton University Press, Princeton, New Jersey.
    A balanced and comprehensive text.
Cronin H. 1991. *The ant and the peacock.* Cambridge University Press, Cambridge.
    A readable history of ideas on sexual selection.
Maynard Smith J. and Harper D. 2003. *Animal signals.* Oxford University Press, Oxford.
    A recent survey of animal signaling, including signaling between males and females in sexual selection.

CHAPTER

21

Conflict and Cooperation

S OME OF THE MOST STRIKING FEATURES OF THE LIVING world involve cooperation between organisms. Most obvious to us is the intricacy of human society, in which no individual is self-sufficient. Among the social insects (ants, bees, wasps, and termites), a single queen produces all the offspring of the colony. In slime molds, cells aggregate to produce a fruiting body, but only a fraction of the cells contribute to dispersing spores (see Figs. 8.6H and 9.4). There are many less extreme examples of cooperative breeding, where some individuals help others to reproduce. Members of different species can also cooperate, in **mutualistic** or **symbiotic** relationships. Many flowering plants have evolved intimate relationships with their pollinators. Legumes produce root nodules that nourish nitrogen-fixing bacteria; lichens consist of a tight symbiosis between a fungus and either a photosynthetic alga or cyanobacterium; and many animals harbor specialized gut flora that supply essential biochemical functions (see pp. 163–166).

These examples are, for the most part, exceptional. Most insects are solitary; most flowering plants are pollinated by a variety of species; and most microbes do not disperse via fruiting bodies. Nevertheless, cooperation is fundamental to the evolution of complex organisms, as we will see when we discuss the **major transitions** in evolution at the end of this chapter. Until then, we will present more familiar examples, such as social insects and cooperatively breeding birds. Note, however, that the basic principles are quite general and illuminate an extraordinary range of apparently disparate phenomena.

It might at first seem surprising that cooperation could evolve through natural selection, which is fundamentally a competitive process: Genes compete with each other to replicate. Indeed, cooperation is always vulnerable to exploitation by "selfish" phenotypes. For example, worker bees may lay their own eggs instead of rearing the queen's eggs, and mitochondria may replicate faster rather than supplying energy to the cell. Thus, a key aim of this chapter is to show how cooperation can evolve despite conflict between the organisms involved. We begin by spending some time discussing examples of conflict, partly because these are fascinating phenomena, but also because they help us to understand how cooperation is maintained.

SOCIAL EVOLUTION

Selection Involves Conflict between Genes and Interaction between Relatives

In Chapter 17, we began by looking at selection in the simplest possible way. We focused on the **fitness** of individuals (i.e., the number of offspring that they produce) and initially made two key simplifications. First, we assumed that the spread of the genes carried by an individual depends only on that individual's fitness. Second, we assumed that fitness depends only on the individual's phenotype and not on the rest of the population. In this chapter, we relax both these assumptions.

First, we look at what happens when genes compete for transmission. The evolution of a population is determined solely by individual fitness only when every gene has the same chance of being transmitted. When Mendel's rules are broken, genes can increase in frequency without increasing the fitness of their bearer. We should emphasize here that this kind of advantage is possible only with sexual recombination. If offspring are genetically identical to their parents, then the fitness of individual and gene must be the same. Thus, we will return to some of the issues raised here when we look at the evolution of sex in Chapter 23.

Second, we will look at what happens when fitness depends on interaction between related individuals. We have already seen that the number of offspring an individual produces will almost always depend on the composition of the entire population, as well as on the phenotype of the individual. If the population is to remain within bounds, rather than going extinct or increasing without limit, then fitness must decrease as numbers increase (see p. 470). Often, this effect of other individuals will depend on their genotype, in which case we say that selection is **frequency dependent**. We have seen how negative frequency dependence can maintain genetic variation (see pp. 505–508), and then how game theory provides a way of understanding the (sometimes complicated) consequences of interactions that depend on genotype (see pp. 567–573). In this chapter, we will see how the outcome changes when interactions are between relatives, as is often the case. This will help us understand how both conflict and cooperation are modulated when the individuals involved share genes.

The study of the evolutionary consequences of interactions between individuals is sometimes termed **social evolution**. It overlaps with **evolutionary ecology**, which focuses on interactions between species and with their environment. Although most empirical work is on the more obvious and tractable examples such as social insects, it applies to a wide range of organisms—from viruses to humans.

In population genetic terms, conflict and cooperation can be understood simply by keeping track of how the numbers of different kinds of genes change as a result of frequency-dependent selection and **non-Mendelian inheritance**. However, much of the theory is verbal and involves a colorful terminology. For example, a gene that increases its own transmission but reduces the fitness of its host is termed **selfish**; and the opposite pattern, in which a gene increases the fitness of other individuals at the expense of its carrier, is termed **altruistic**.

A vivid terminology is helpful for understanding and remembering the theory and for guiding our intuition. However, it can mislead us in two ways. First, genes are obviously not conscious agents with "selfish" or "altruistic" motives. We do often think from the gene's point of view, and ask what the gene should do to maximize its transmission to future generations. But what we are really working out is whether alleles that cause various phenotypes would increase in the population. Usually (but not always) these two ways of looking at the problem give the same answer. Ideally, what is needed is a complete population-genetic model that tracks the numbers of all the different kinds of genes. Often, however, such a model is not very illuminating by itself. It is best complemented by a more intuitive interpretation.

Second, the use of everyday words suggests an immediate application of evolutionary principles to human behavior. As with any animal, the innate component of human behavior has evolved, in part through natural selection, and to that extent the principles of social evolution apply. However, language and culture have been at least as important as biological evolution in shaping our species. We discuss the issues arising from the interaction between biology and culture in Chapter 26.

CONFLICT BETWEEN GENES

Elements That Replicate Independently of the Host's Chromosomes May Be Genetic Parasites

Genetic elements can gain an advantage in several ways. They may be transmitted in excess through meiosis, leading to **segregation distortion** (Fig. 21.1); **transposable elements** may spread to new locations in the genome (Table 12.2); and elements transmitted through one sex may gain an advantage by biasing the **sex ratio** in favor of that

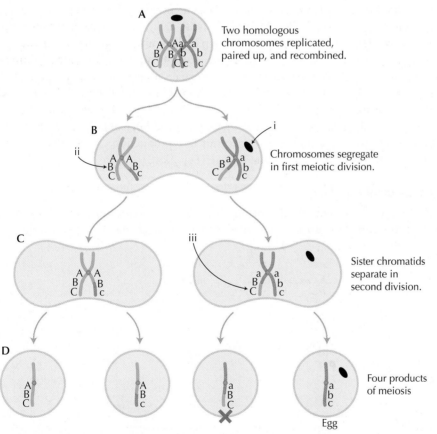

FIGURE 21.1. Segregation distortion can arise during meiosis in several ways (i, ii, iii). (*A*) Two homologous chromosomes have replicated into sister chromatids, paired up, and recombined. (The two genomes are shown in *red* and *blue*.) The homologous chromosomes then segregate in the first division of meiosis (*B*), and the sister chromatids separate in the second division (*C*) to produce the four products of meiosis (*D*). (i) A B chromosome (*black ellipse*) is shown segregating into the meiotic product that will form the egg (*bottom right*). Such preferential segregation is known as meiotic drive. (ii) A short stretch of sequence around the site of recombination forms a heteroduplex, in which base pairs are mismatched. Repair of these mismatches leads to gene conversion. Here, a B/b heteroduplex has been repaired to BB, giving an excess of B. (iii) Segregation distortion can also be caused by competition between meiotic products. One allele (C) is marked in heterozygotes, and haploid products carrying it are then destroyed (*red cross*).

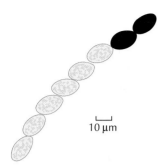

FIGURE 21.2. Gene conversion in fungi. In some species, the products of meiosis stay together, and so the segregation of alleles that affect spore color can be seen directly. This shows the products of a cross between wild-type *Sordaria brevicollis*, which has black spores, and a yellow-spored strain; the color difference is due to a single mutation. Occasionally, gene conversion changes one allele to the other, leading to a 6:2 segregation, as shown here, instead of the usual 4:4 segegration.

FIGURE 21.3. One of the maize cobs studied by Barbara McClintock, discoverer of transposable elements. The variegation is caused by transposition of the DNA-based transposon *Ds*, which causes chromosome breakage.

sex. We will look at some examples in detail in the following section, but here we begin with an overview of the variety of mechanisms that can be exploited.

Segregation Distortion

In plants and animals, only one of the four products of female meiosis becomes an egg cell. Thus, an allele can spread if it preferentially segregates into that product (Fig. 21.1, i). Such a bias is known as **meiotic drive**. A similar bias is produced by **gene conversion** (Fig. 21.1, ii), which can be seen directly in some fungi (Fig. 21.2). Biased gene conversion has only a weak effect (up to ~10^{-3} per site per generation) compared with the more extreme examples of segregation distortion that we will see below. However, it acts across all of the genome that can recombine and can have a significant long-term effect.

Segregation distortion can also be seen when gametes carrying one allele cause destruction of the other allele after meiosis (Fig. 21.1, iii). This is distinct from meiotic drive, but the outcome is the same: an excess of one allele over the other among the gametes produced by a heterozygote. (Recall that on pp. 575–578, we saw examples of competition between gametes from different individuals. Here, we are concerned with competition between gametes from the *same* diploid individual.)

Transposable Elements

Barbara McClintock discovered transposable elements in maize through the variegated kernels that they produced (Fig. 21.3; recall Table 12.2). Transposable elements do not gain any advantage merely by moving through the genome; they must also replicate and increase the number of copies present as well. Even then, there is no long-term advantage to elements that increase within strictly asexual lineages. Indeed, lineages that carry such elements will eventually go extinct if transposable elements have deleterious effects on the host genome (Fig. 21.4A). Transposable elements can survive only if there is at least some sexual reproduction, so that they can spread **horizontally** through the population (having spread within the genome prior to reproduction), rather than being confined by **vertical transmission** to one lineage (Fig. 21.4B).

Sex-biased Inheritance

Genes that are inherited strictly through one of the sexes gain an advantage if they promote reproduction of that sex. For example, the great majority of flowering plants

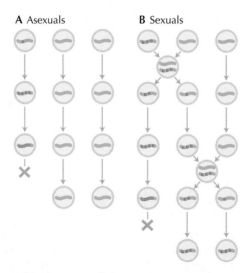

FIGURE 21.4. (*A*) Transposable elements (*red dots*) can increase within an asexual lineage, but if they have deleterious effects on their host genome, lineages that carry them will eventually go extinct. (*B*) In a sexual population, transposable elements can spread horizontally through the whole population. The diagram shows an organism such as yeast, which is usually haploid, but occasionally has sex.

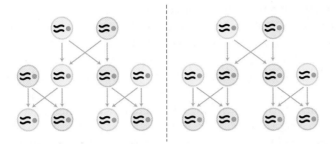

FIGURE 21.5. Genes inherited through one sex gain if they promote production of that sex. (*Left*) Mitochondrial DNA (mtDNA) in animals and flowering plants is inherited maternally. If each mother has one son and one daughter, the number of copies of her mtDNA in females (which is all that matters for the future) stays the same (*left*). However, if she carries mtDNA that causes only daughters to be produced (*right*), then the number of copies in females doubles every generation. The pairs of *wavy black lines* indicate the diploid nuclear genome, and the *green circles* indicate the mitochondrial genome. Males are *blue*; females are *pink*.

are **hermaphrodites**, that is, they produce both male and female gametes. However, in 5–10% of species, mixed populations are found, in which some plants are male sterile and so reproduce only as females. This male sterility is caused by mitochondrial variants, which are inherited down the female line. These variants gain an advantage if male-sterile plants allocate more resources to producing seed, rather than pollen, and so pass on more mitochondrial genomes (Fig. 21.5). (This phenomenon is termed **cytoplasmic male sterility**, or CMS. It is important for commercial plant breeding, because male-sterile plants cannot self-fertilize, and so crosses can be controlled more easily.) A similar, but more extreme, example is seen in the wasp *Nasonia vitripennis* (Fig. 21.6).

There Is No Clear-Cut Distinction between Genetic and Conventional Parasites

In ladybirds, a cytoplasmically inherited element occurs that kills males that carry it. Because siblings compete for resources (Fig. 21.7), this male killer gains an advantage in essentially the same way as with CMS in plants. However, this syndrome is caused by an intracellular rickettsia-like bacterium of the genus *Wolbachia* (Table 6.2) and can be eliminated by treatment with the antibiotic tetracycline. *Wolbachia* are widespread in arthropods, where they have evolved a number of strategies that aid their transmission down the maternal line. For example, *Wolbachia* recently infected populations of *Drosophila simulans* from Southern California, probably via a rare horizontal transmission from another species. Although infected males do not transmit

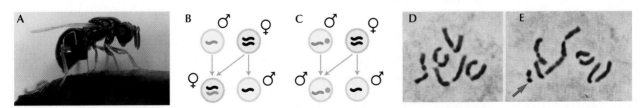

FIGURE 21.6. The wasp *Nasonia vitripennis* (*A*) provides an extreme example of **sex-biased inheritance**. Like other Hymenoptera (bees, wasps, and ants) this species is haplodiploid: Fertilized eggs develop as diploid females, but unfertilized eggs develop as haploid males (*B*). In many populations, a supernumerary B chromosome is found, called PSR (paternal sex ratio; *green dot* in *C*). Fertilized eggs that carry this chromosome eliminate the paternal genome (*red*), so that they develop as haploid males instead of diploid females (*C*). Thus, PSR gains an advantage by shifting reproduction toward males, the sex through which it is transmitted. (*D*) The normal haploid set of five chromosomes in a male; (*E*) the extra chromosome in a male carrying PSR (*arrow*).

FIGURE 21.7. Female ladybird larvae (*Adalia bipunctata*) hatching from a brood in which males have been killed by the cytoplasmically transmitted bacterium *Wolbachia*. The larvae are eating their dead brothers, which gives them and the *Wolbachia* they carry a growth advantage.

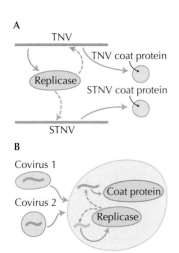

FIGURE 21.8. (*A*) Tobacco necrosis virus (TNV) is an RNA virus that produces both replicase and coat protein. Satellite tobacco necrosis virus (STNV) is an unrelated virus that produces its own coat protein but no replicase: Instead, it relies on the replicase produced by TNV. (*Solid arrows* indicate that the virus codes for the protein; *dashed arrows* indicate replication of the virus by the replicase.) (*B*) Tobacco rattle covirus consists of two kinds of particle, both of which are needed for successful infection. One codes for replicase, whereas the other codes for coat protein.

the bacterium, offspring from their matings with uninfected females die; in effect they sterilize their uninfected mates. Thus, the infection is spreading rapidly northward. Although sterility is caused by an intracellular bacterium, its pattern of inheritance is the same as for the mitochondrial genome. As a result, the mitochondrial DNA variant that happened to be associated with the original infection is **hitchhiking** northward with it.

The consequences are essentially the same, whether effects are caused by the mitochondrial genome as with CMS in plants, by supernumerary or **B chromosomes** (as with paternal sex ratio in *Nasonia*), or by an intracellular parasite such as *Wolbachia*. The continuum between parasites and selfish DNA is also apparent from the close relation between some viruses and transposable elements (Chapter 5 [pp. 133–136]). The key enzyme coded by transposable elements that replicate via an RNA intermediate is **reverse transcriptase**, which is homologous to the enzyme used by **retroviruses** to copy their RNA genomes into the DNA that they insert into the host's genome (see pp. 217–219). The close relationship between transposable elements and viruses is well illustrated by the *Ty* element of yeast, whose RNA intermediate is packaged in a protein coat within the cell. This is essentially a virus that cannot escape its host cell.

The difficulty in drawing a clear distinction between separate organisms on the one hand and competing genetic elements on the other is illustrated by the strange natural history of viruses. Usually (although not always), viruses are packaged in a protein coat when outside the cell. However, within the cell their genomes are typically naked and so share their replication machinery with other viruses. This makes parasitic relationships easy and also makes it hard to define the boundaries of the organism. For example, satellite tobacco necrosis virus (STNV) is a parasite of tobacco necrosis virus (TNV) that makes its own coat protein but relies on the RNA polymerase coded by TNV (Fig. 21.8). These two kinds of virus are distantly related and so are seen as separate organisms. Almost all viruses have **defective interfering viruses** (DIVs), which have simply lost some parts of the genome and so depend on intact virus for their transmission (see p. 569).

The terms "parasite" and "selfish DNA" clearly apply to the examples we have used so far. However, it is important to realize that replicating molecules can evolve mutualistic as well as parasitic relationships. For example, many plant RNA viruses are unable to complete their life cycles by themselves, because successful infection requires two different viruses with complementary functions (Fig. 21.8B). Such **coviruses** have been seen to evolve in the laboratory. In one primate cell line carrying the SV40 virus, two different kinds of DIV evolved, which together completely eliminated the parent SV40. Although each had lost essential functions, they could still replicate together. Later in this chapter, we will examine more examples of both symbiotic and parasitic relationships among molecules, cells, and organisms.

Selfish Elements Often Consist of Several Tightly Linked Components

Genetic elements can gain an advantage by breaking Mendel's rules. Some of the mechanisms are simple; for example, gene conversion gives a slight bias to one of the bases at a single site in a heterozygote. However, most cases of strong segregation distortion involve several tightly linked components. Typically, there are one or more distorter loci that can impair gametes that carry the sensitive allele at a responder locus.

The t-haplotype in wild house mouse populations (*Mus musculus*) is a well-understood example (Fig. 21.9A). Heterozygous (t/+) males transmit an excess of the t-haplotype. This distortion is due to impaired function of the +-bearing sperm produced by heterozygotes, rather than to an asymmetry at meiosis: Sperm recovered from the vagina of females mated to t/+ males show equal proportions of the two kinds of

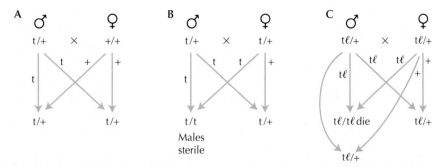

FIGURE 21.9. (A) Male mice heterozygous for the t-haplotype (t/+) pass on mostly t-bearing sperm, giving a strong transmission advantage. (B) In small, inbred local populations, t may be frequent enough that t/+ x t/+ matings become common. Then, nearly half the sons will be t/t, and therefore sterile. (C) Many t-haplotypes carry tightly linked recessive lethals (denoted ℓ here), which kill tℓ/tℓ homozygotes early in embryogenesis. These can be replaced by tℓ/+ heterozygotes, thus avoiding production of sterile sons and increasing net transmission of the tℓ combination.

DNA. The t-complex spans 30–40 Mb of genome on chromosome 17, which includes several distorter loci, and a single responder locus. Recombination is suppressed by multiple **inversions**. Interestingly, natural t-haplotypes carry a variety of different recessive lethals, which arose after the establishment of the driving haplotypes themselves. In families where both parents are t/+, nearly half of the sons will be t/t, and therefore sterile (Fig. 21.9B). If these die early on, because they carry a recessive lethal, then more offspring can be produced, and these will be mostly t/+. This gives lethal-bearing t-haplotypes a strong selective advantage (Fig. 21.9C). This is an example of **kin selection**, which we discuss later in this chapter.

The t-complex in mice is inherited on the autosomes. However, many more cases are known in which one of the sex chromosomes is transmitted in excess in the **heterogametic** sex (males in mammals and *Drosophila*, females in Lepidoptera). At least 23 examples of X-chromosome drive, and 6 of Y-chromosome drive, are known from insects, mammals, and flowering plants. The preponderance of examples involving the sex chromosomes may simply be because they are more easily detected through the strong bias in sex ratio that they produce. However, the lack of recombination between the sex chromosomes makes it easier for segregation distortion to evolve. Distorter loci on the X chromosome can target responder loci anywhere on the Y. In contrast, with autosomal systems the components must be tightly linked from the outset. The causes and consequences of sex chromosome evolution are explained in more detail in Chapter 23 (pp. 682–683).

Modifiers Evolve to Suppress Genes That Cheat

Genetic elements that evade Mendel's rules harm other genes in the organism. This harm can be the direct result of their preferential transmission (for example, t-bearing gametes spread by killing their wild-type brothers in heterozygous males) or it can be the indirect result of deleterious side effects (e.g., transposable elements may insert into the coding sequence). Thus, there is strong selection on the host genome for suppressors of non-Mendelian transmission.

Mitochondria that cause cytoplasmic male sterility in plants gain an advantage if they cause increased transmission through female gametes. This advantage may occur in any of several ways: Females avoid inbreeding depression because they cannot self-fertilize, or resources may be shifted from pollen production and from producing large showy flowers into production of seed (recall that flowers act primarily to export pollen [p. 573]). In the simplest case, the frequency of females would increase until pollen became so scarce that they failed to be fertilized. Hermaphrodites would then gain a counterbalancing advantage, because they could at least self-fertilize, rather than producing no offspring at all. In nature, restorer alleles that prevent male sterility arise in the nuclear genome. New mitochondrial sterility factors may then evolve, leading to a continual struggle between nucleus and cytoplasm. Typically, several different sterility cytotypes are found in a population, together with several more-or-less specific restorer alleles. In the moss campion, *Silene acaulis*, several highly divergent mitochondrial variants are found within each population, which suggests that a long-standing polymorphism has been maintained (see p. 541).

Suppressors of segregation distortion have been found in all well-studied systems. They also have been observed in laboratory experiments. An artificially constructed chromosome, in which the segregation distorter (*SD*) complex was fused to the Y chromosome, was introduced into populations of *D. melanogaster* (Fig. 21.10). This caused excess transmission of the Y, and hence an excess of males; indeed, several replicates went extinct for lack of females. However, the sex ratio returned toward equality over approximately ten generations, as suppressor alleles were established at multiple loci (Fig. 21.10).

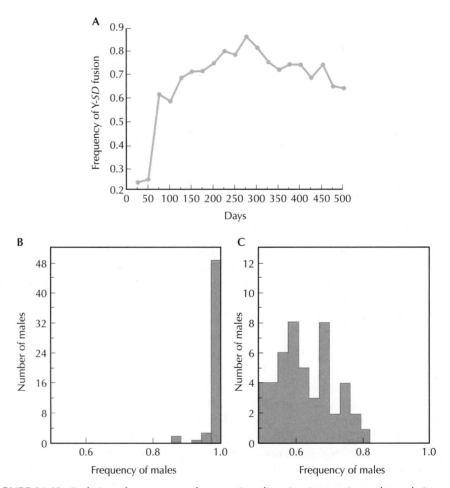

FIGURE 21.10. Evolution of suppressors of segregation distortion in experimental populations. (*A*) The frequency of a driving chromosome, constructed by fusing second chromosomes carrying *SD* to a Y chromosome (average over three replicates). The driving chromosomes increase rapidly at first, leading to a strongly male-biased population. However, they then decline, as suppressors evolve. Initially, offspring of males carrying Y-*SD* are almost all male (*B*). (*C*) However, by day 157, the male bias is much weaker and varies considerably between males. This histogram shows the distribution of offspring sex ratios for a cage that initially contained no genetic variation. Here, suppression of drive has evolved from variation generated de novo, by mutation.

Transmission distortion can be countered through behavioral changes. For example, female t/+ mice prefer to mate with +/+ rather than t/+ males, and so avoid producing inviable t/t offspring. In stalk-eyed flies, males have an exaggerated eye span. In two species, there is a driving X chromosome that causes a substantial excess of females, by destroying wild-type sperm in heterozygotes. There is also a Y-linked suppressor, which prevents destruction of sperm that carry it. In populations that were selected for increased eye span, the bias toward females decreased. The opposite was seen in replicate populations selected for decreased eye span. This shows that male eye span and the degree of drive are **genetically correlated** (see p. 479). Thus, females who choose to mate with males that have their eyes further apart gain both because their offspring will include more sons (valuable in a female-biased population) and because any son's sperm carrying suppressor alleles will be less sensitive to the destructive effects of the driving X. (This is a nice example of a preference for **good genes** [see p. 581].) Conversely, the females' mating preference tends to increase the frequency of suppressors of drive and so counters the spread of selfish X chromosomes.

Organisms have evolved sophisticated defenses against viruses and transposable elements. Perhaps the best understood is repeat-induced point mutation (RIP) in the fungus *Neurospora*. This process detects any duplicated sequence longer than approximately 1 kb at the stage after fertilization but before nuclear fusion and meiosis. Once recognized, duplicates are mutated at high frequency (predominantly, G:C to A:T transitions) and are also methylated. This is thought to be a defense against transposable elements. Although the *Neurospora* genome contains **transposon**-related sequences, these have been inactivated by RIP, and no active elements are known. Mechanisms that inactivate double-stranded RNA are widespread and may also be a defense against viruses and transposable elements that have double-stranded RNA as part of their life cycle.

When Sex Is Rare, All Genes Share the Same Interests

We have discussed mainly sexual eukaryotes, in which genes can spread horizontally, even if they harm the individual that carries them. In bacteria and archaea and in predominantly asexual eukaryotes, the situation is quite different. Genes stay together through many generations of asexual reproduction and so share the same evolutionary interests. Plasmids and various kinds of transposable elements are found in bacteria and archaea (see pp. 169–172) and do sometimes move between lineages by conjugation or other rare recombination events (see pp. 182–187 and 352). However, because their long-term propagation depends primarily on survival of the asexual lineage in which they live, they are expected to be benign (Fig. 21.4).

Plasmids often carry multiple genes for antibiotic resistance, which give a clear advantage to the host. Multiple drug resistance is of course a very recent phenomenon, caused by the excessive use of antibiotics. However, antibiotics such as penicillin are natural antibacterial defenses, and so resistance to them is ancient. Samples of bacteria taken from soil show a high diversity of plasmids carrying a variety of potentially favorable genes. We have already seen that in *Escherichia coli*, plasmids carry genes for the production of bacteriocins that are toxic to other bacteria. These also favor the host because they confer resistance to toxins that kill competitors (see p. 571).

Bacterial and archaeal transposons and insertion sequences also appear to be benign. In contrast to eukaryotes, no mechanisms are known that would confer resistance to them. (Restriction enzymes are thought to have evolved in order to destroy viral DNA, not transposable elements.) Moreover, these elements often have evolved ingenious mechanisms that limit their copy number, thereby reducing the harmful effects of random transposition (Fig. 21.11).

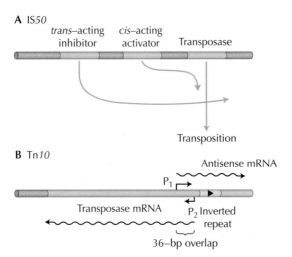

FIGURE 21.11. Because bacteria and archaea rarely indulge in sex, their transposons have evolved mechanisms that prevent excessive transposition. (*A*) The DNA-based transposon IS*50* produces a *cis*-acting activator of transposition and also a *trans*-acting inhibitor: As copy number increases, levels of the inhibitor rise and transposition ceases. (*B*) Another DNA-based transposon, Tn*10*, has two promoters near the start of the transposase gene. One initiates transcription of the transposase messenger RNA (mRNA), whereas the other initiates transcription in the opposite direction, producing an antisense RNA. At high copy number, this binds to the transposase mRNA and prevents its translation.

TABLE 21.1. Outbreeding is associated with the presence of B chromosomes in flowering plants

Breeding	B Chromosome	
System	Absent	Present
Inbred	52	3
Mixed	191	14
Outcrossed	66	27

Table shows the number of species in each class. The association is highly statistically significant.

In eukaryotes, we see a correlation between the rate of sexual reproduction and the prevalence of selfish elements. To take an extreme example, the bdelloid rotifers are one of the very few taxa that have apparently reproduced asexually for long periods (~100 million years [Myr]; Fig. 23.10). Correspondingly, they lack long interspersed nucleotide element (LINE) and gypsy-like retrotransposons that are found in all other eukaryotes that have been studied. Among flowering plants, outbreeding is associated with the presence of B chromosomes, as expected (Table 21.1).

Abundance of Selfish DNA Is Determined in Several Ways

The low frequency of individual insertions within populations provides evidence that selection limits the accumulation of transposons in *Drosophila*. We have already seen that such insertions are responsible for quantitative variation (p. 405); they are rare, and confined to the tips of genealogies, implying that they do not remain in the population for long. Counterselection may be due to the deleterious effects of the transpositions that they cause or to their disruption of gene function. A further possibility is that transposon numbers are limited by **ectopic recombination**, recombination between different elements, which leads to strongly deleterious chromosome rearrangements (see p. 339). This is consistent with the observation that transposons tend to accumulate in parts of the *Drosophila* genome that have low recombination (Fig. 21.12). However, this pattern could also be due to the lower efficiency of selection when linkage is tight (the **Hill–Robertson effect** [see p. 676]).

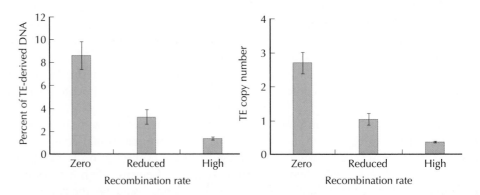

FIGURE 21.12. Transposable elements tend to accumulate in regions of low recombination. The *Drosophila melanogaster* genome was classified into regions with zero, reduced, or high recombination.

TABLE 21.2. Retrotransposons in *Drosophila* and humans show contrasting patterns

	Drosophila	Humans
Number of euchromatic copies (approximate)	1400	2,900,000
% of euchromatic DNA	2	42
Number of families (approximate)	60	100
non-LTR retrotransposons (LINEs)	20	3
SINEs	0	3
LTR retrotransposons	40	100
Insertions that are fixed	Few	Most
% of spontaneous mutations caused by retrotransposons	>50	<0.2

LTR, long terminal repeat; LINEs, long interspersed nucleotide elements; SINEs, short interspersed nucleotide elements.

The distribution of transposable elements in mammals and, in particular, in humans, is very different (Table 21.2). There are very large numbers of transposable elements, almost all fixed in the population and almost all inactive. Correspondingly, whereas most spontaneous mutations in *Drosophila* are caused by transposable elements, very few are in humans. The cause of the difference is not known. The lower effective population size of humans compared with *Drosophila* may mean that selection cannot counter the fixation of transposable elements. Ectopic recombination between transposons may be much rarer. (Indeed, it must be, because otherwise the very many repeated sequences would be disastrous.) Elements may tend to transpose into noncoding regions, where they have little harmful effect. All these factors contribute to explaining the difference, but their relative importance is unknown.

The spread of genetic elements can be limited by the increasingly deleterious effects of the elements themselves or by the accumulation of defective copies. Moreover, as we saw in the previous section, various kinds of suppressors can evolve. These will tend to increase as selfish elements increase, giving an indirect frequency dependence that can maintain a stable balance. Elements that spread by non-Mendelian transmission will not evolve to limit their own reproduction unless their fate is tied to that of their host by low recombination. However, they may well limit their own deleterious effects, for example, by preferentially inserting into heterochromatic or noncoding regions or by transposing only in the germ line (as with P elements).

Despite the variety of mechanisms that can limit the spread of selfish DNA, it remains unclear why the pressure to subvert Mendelian inheritance does not overwhelm organisms. Indeed, for all we know this may be a common cause of extinction. The simple fact that there are many more genes in the host genome than in the selfish element may make it much more likely that suppression will evolve. Indeed, we do typically see multiple suppressor alleles in nature (e.g., with cytoplasmic male sterility), and in the laboratory experiment on *SD* (Fig. 21.11), multiple suppressors arose de novo, by mutation, within a few tens of generations. To use a phrase coined by Egbert Leigh, a "parliament of genes" with a common interest in reproduction may successfully restrain an errant minority. Later in this chapter, we return to the broader issue of how cooperation can be stabilized.

Eukaryote Genomes Consist Largely of "Selfish" and "Junk" DNA

Typically, only a small fraction of eukaryotic genomes code for protein (Figs. 7.1 and 13.17 and pp. 216–219). Some of the non-protein-coding sequence functions in gene regulation, codes for ribosomal, transfer, and other functional RNAs, or maintains centromeres. However, estimates based on sequence conservation in *Drosophila* and mammals suggest that the amount of functional noncoding sequence is approximately the same as the amount of coding sequence (see pp. 542–547). This leaves a mass of sequence that is apparently not maintained by selection. Several lines of evidence suggest that this excess DNA indeed has no function in sustaining individual fitness. Rather, it consists of elements that replicate despite their harmful effects on the organism (**selfish DNA**) and of repeated sequences that accumulate as a by-product of the replication machinery (**junk DNA**).

The most powerful argument for this view of the genome is that similar species vary greatly in DNA content. **Genome size** (also known as **C value**) is expressed as the mass of a single haploid genome, measured in picograms (1 pg = 10^{-12} g), and corresponds to approximately 1000 Mb of DNA. Overall, eukaryote genomes vary from 0.0023 pg in the parasitic microsporidian *Encephalitozoon intestinalis* to 133 pg in lungfish; humans are intermediate, with 2.9 pg (Fig. 7.1). This range covers an enormous variety of organisms, from single-celled microbes to complex multicellular organisms. However, there is no relation between genome size and any subjective measure of organismal complexity. The smallest genome within a group of organisms is roughly correlated with their complexity, but within groups there is a huge range of genome size (Fig. 21.13A). Some of this variation is due to **polyploidy** (i.e., multiplication of the whole genome, both coding and noncoding; see pp. 326–328). However, most variation reflects a very wide range in the amount of noncoding DNA. This variation is seen within genera and even species. For example, DNA content varies 1.6-fold within the genus of kangaroo rats, *Dipodomys*; within maize (*Zea mays*), genomes vary by approximately 40% in size. Experimental deletion of two large segments of noncoding DNA from the mouse genome (1.5-Mb- and 0.85-Mb-long) had no detectable effect on survival and reproduction and hardly altered the expression of adjacent genes.

Most sequence evolves at approximately the mutation rate, which implies that it experiences little selective constraint (see pp. 425 and 546–547). As well as ruling out

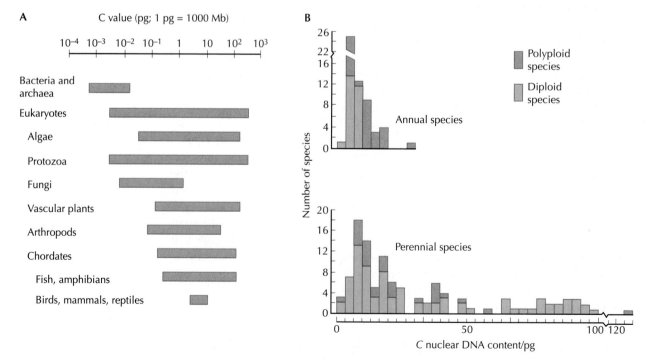

FIGURE 21.13. (*A*) Ranges of haploid genome size (C value) in major groups. (*B*) Annual species of flowering plants tend to have smaller genomes than perennial species. *Red bars* indicate species that are known to be polyploid. Overall, fast-growing "weeds" have genomes typically smaller than 10 pg, whereas almost all plants with genomes larger than 30 pg are perennial.

a precise function that benefits individual fitness, this also implies that most sequence is not selected for selfish replication. (Indeed, as we will see below, most genomes contain relatively few active transposons.)

Finally, it is very hard to see how natural selection could maintain very large amounts of DNA in a precise sequence. The **mutation load** would be excessive, and the fitness cost of establishing adaptive substitutions would be too high (see pp. 549–553). Note, though, that arguments from rate of sequence evolution or from genetic load do not rule out selection for the overall amount of DNA. (For example, it has been suggested that large genomes could be favored by selection for large nuclei.) However, as we have just seen, comparisons between species and experimental manipulations of *Drosophila* show that large genomes are not essential for building complex organisms.

We expect big genomes to be harmful because of the metabolic cost and the time taken to replicate large amounts of DNA, and because large cell size may constrain development and behavior. Comparisons between species support these intuitions. Plants that evolve annual, as opposed to perennial, life histories tend to have smaller genomes (Fig. 21.13B). In maize, artificial selection for early flowering causes a reduction in genome size. Similarly, salamanders that metamorphose from aquatic larvae into adults have genomes that are small relative to other salamanders (14–17 pg), whereas **neotenous** species, which avoid metamorphosis by reproducing as larvae, can have up to 76 pg of DNA per haploid genome. Insects show a similar pattern. **Holometabolous** insects, which metamorphose through a pupal stage (e.g., flies, butterflies, and beetles), have genomes smaller than 2 pg, whereas **hemimetabolous** species, which develop directly through a series of nymphal stages (such as grasshoppers and bugs), can evolve large genomes, up to 17 pg. Presumably, metamorphosis requires rapid cell division and is harder when massive amounts of DNA must be replicated.

Genome size is strongly correlated with cell size. Bolitoglossine salamanders have very large genomes (averaging 48 pg) housed within a small number of large neurons (Fig. 21.14). As a result, their brains have a simple structure. It has been proposed that this makes it impossible for them to actively capture prey, as do most adult salamanders; instead, they catch prey by ambush, using a fast projectile tongue. However, although such examples are suggestive, it is hard to establish a significant correlation between brain complexity and behavior, and still harder to disentangle cause and effect. Do large genomes lead to simpler brains and hence a change in behavior, or does a change in behavior alter selection on genome size?

Much of the Human Genome Consists of Defunct Transposable Elements

Given that selection is likely to favor smaller amounts of nonfunctional DNA, what pressures lead to large genomes? We focus on the selfish replication of transposable elements, but note that the accumulation of tandem repeats through unequal crossing over and other mutation-like processes is also important.

Overall, the number of copies of transposable elements, and the fraction of the genome derived from them, increases with genome size (Fig. 21.15). We discuss an especially well-understood example: the composition of our own genetic material. Nearly half of the human genome derives from transposable elements (Figs. 8.19 and 21.16), although almost all of this sequence is defunct. There is no evidence that DNA-based transposons have moved in the past 50 Myr, and long terminal repeat (LTR) retrotransposons are virtually all inactive. The third class of transposable elements, non-LTR LINE elements, has proliferated most remarkably. The only family still active is LINE-1, which makes up nearly 20% of our genome. However, only about 100 of these are intact and active 6-kb-long elements. This is because in most transposition events, reverse transcription ends prematurely, which leaves a fragment of approximately 1 kb.

FIGURE 21.14. The salamander *Bolitoglossa subpalmata* has a very large genome (69 pg), and hence it has large neural cells. Thus, it necessarily has a simple brain structure.

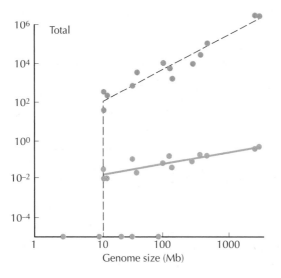

FIGURE 21.15. A substantial fraction of large genomes is derived from transposable elements. Both the number of copies (*red circles*) and the fraction of genome (*blue circles*) increase with genome size. Organisms with no mobile elements are plotted on the *x*-axis. (These data are from sequenced genomes of bacteria, archaea, and eukaryotes and so do not include large genomes.)

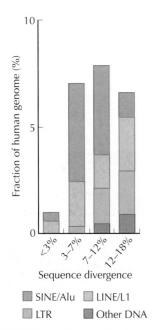

FIGURE 21.16. The distribution of ages of transposable elements in the human genome. This is measured by divergence from the human–mouse consensus sequence and is grouped into bins that correspond to about 25 Myr of divergence.

The transposase coded by intact LINE-1 elements has a reverse transcriptase activity that can cause the insertion of a variety of RNA molecules. This has led to the spread of several families of SINEs (short interspersed nucleotide elements), each a few hundred base pairs long. The most abundant is *Alu*, a 300-bp sequence derived from a structural RNA gene that makes up part of the ribosome. Now, *Alu* elements make up an extraordinary 10% of our genome. *Alu* began proliferating early in the primate radiation, about 40 Mya, and initially spread at a rate of around one insertion per generation. However, after insertion the poly(A) tail that is recognized by transposase accumulates mutations, so that the element becomes immobile. Therefore, the present rate of movement is very low, approximately 1 element per genome per 200 generations. Thus, only a small fraction of *Alu* actually transpose and these depend on the transposase coded by the few intact LINE-1 elements. So, only a small amount of the human genome is maintained by selection for selfish replication—the transposase and flanking sequences of the active LINE-1 elements and the poly(A) tails of those SINE elements that can still move. The bulk of the transposon-derived half of our genome consists of degenerate relics of insertions that occurred tens of millions of years ago (Fig. 21.16).

The extraordinary variation in genome size, and the vast excess of DNA in many species, is one of the most remarkable phenomena in biology. It has often been suggested that this DNA must have some function, such as maintaining large cell size or, less directly, facilitating chromosome rearrangements, forming novel genes, or generating other mutations that may occasionally be adaptive (see the next section). However, we will see in Chapter 23 (p. 659) that in sexual species, high mutation rates cannot be maintained because of their occasional positive effects, when the great majority of mutations are deleterious. More generally, the large variation between closely related species argues against any adaptive explanation (Fig. 21.13). It is clear that much of the eukaryote genome is a relic of the activities of self-replicating elements and of quirks of replication and recombination.

Transposable Elements Have Sometimes Been Co-opted to Aid Their Host

In many cases, transposable elements have acquired a function that aids host fitness. We can divide the known examples into two classes. First, transposable elements, or sequences derived from their activities, can acquire a new function that has nothing to do with selection for replication of the transposon. Given that, so much of many genomes is composed of transposable elements or their relics, it is hardly surprising that some should by chance acquire a function that aids the host. Second, the machinery for transposition, which originally evolved through selection for selfish replication, could be taken over to play a similar role for the host.

As we discussed in Chapter 8 (pp. 219–221), introns are thought to be derived from transposable elements. Although many genes can function without introns, proper gene regulation now often depends on how the full-length RNA transcript is spliced into messenger RNA (mRNA). In particular, many different proteins can be produced from the same set of exons via **alternative splicing**, and so protein diversity can be much greater than expected solely from the number of genes (Fig. 8.22). New genes may be produced by recombination between exons from different genes. Retrotransposon activity often causes reverse transcription of mRNA, leading to insertion of novel intronless genes. Usually these inserts degenerate into **pseudogenes**, but in some cases (eight in humans, for example) they acquire new functions. In these examples, the new functions are incidental by-products of the presence of transposable elements. We discuss the formation of novel genes in more depth in Chapter 24.

Sometimes, the host takes over a function that had originally been selected for propagation of the transposon. A few hundred human genes terminate transcription using signals derived from LTR retrotransposons, and other kinds of regulatory signal have been co-opted in a similar way. Forty-seven human genes are homologous to transposase (usually deriving from DNA-based transposons). In most of these cases it is not known what kind of selection maintains these functions. However, some are well understood, such as two key enzymes (RAG1 and RAG2), which are responsible for the somatic rearrangements that generate diverse antibodies in the vertebrate immune system (Fig. 9.10).

INTERACTIONS BETWEEN RELATIVES

Selection Acts on the Extended Phenotype

In the first half of this chapter, we considered the conflicts that arise when genes compete for transmission. In the following sections, we turn to examine interactions between relatives. Although these two issues may seem quite different, they are in fact closely related. In both, the spread of a gene cannot be predicted from the fitness of the individual that carries it. We have just seen how a gene with a transmission advantage can spread, even though it is harmful to the individual that bears it. Now, we will see that genes can spread if they help relations to reproduce, even if they are costly to their bearer.

Throughout, we have emphasized that typically the fitness of an individual depends not just on its own genotype, but also on the genotypes of others in the population. Put another way, a gene typically influences the reproduction not just of the individual that carries it, but also of its neighbors (Fig. 21.17). We expect such interactions when we think of social behavior in birds or mammals, but they exist much more widely. For example, soil microbes interact by consuming local nutrients and producing toxic waste products, as well as via specialized molecules that have evolved to influence others (e.g., bacteriocins and **quorum sensing**; p. 167 and Fig. 20.17). Richard Dawkins introduced the term **extended phenotype** to describe the net effects of a gene on all the individuals in the population.

There are many examples of apparent altruism, in which individuals behave so as to increase the reproduction of others, at some cost to themselves. The most striking example is the social insects (see Box 21.2), where, in the most extreme cases, all the individuals

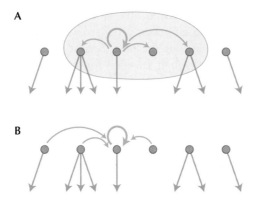

FIGURE 21.17. (*A*) An individual influences its own fitness (*heavy blue arrow*) but also influences its neighbors' fitness (*light blue arrows*). The total effect of an individual (or gene) on the rest of the population is called its extended phenotype. (*B*) Conversely, each individual is influenced by several neighbors: It falls within their extended phenotypes.

in a colony aid the reproduction of a single queen, who is fertilized by just one or a few males (Fig. 21.18B). There are many less extreme examples of cooperative breeding, in which helpers assist a reproductive pair. This is seen in approximately 220 species of birds (e.g., Fig. 21.18C). Alarm calls appear altruistic, because they draw attention to the caller (Fig. 21.18A). Similarly, it is hard to understand how the bright warning colors shown by many distasteful organisms could first evolve, because when they first arise, predators would not associate them with distastefulness. The altruism here lies in their training predators to avoid similarly patterned animals in the future (Fig. 21.18D). Finally, we have already seen that some bacteriocin-producing bacteria lyse, so as to release toxin that kills competing strains (Fig. 20.17). This benefits those bacteria of the same strain that do not lyse, because they are resistant to the bacteriocins (Fig. 21.18E).

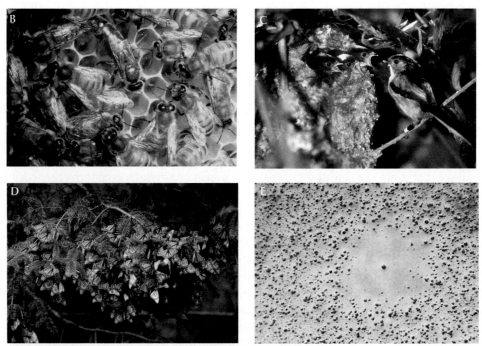

FIGURE 21.18. Examples of apparently altruistic traits. (*A*) Meerkat sentry (*Suricata suricatta*); (*B*) a colony of bees (*Apis mellifera*); (*C*) cooperatively breeding long-tailed tits (*Aegithalos caudatus*); (*D*) warningly colored monarch butterflies (*Danaus plexippus*) on their roost; and (*E*) bacteriocin production in *Escherichia coli*. The blank area of *E* in the center is the result of sensitive bacteria killed by a bacteriocin-producing colony.

Apparently Altruistic Traits Can Evolve by Kin Selection

Altruistic traits can evolve through **kin selection**. A gene can spread if it increases the reproduction of relatives, even if it reduces its own fitness. Charles Darwin suggested this explanation for the evolution of sterile workers in social insects, and R.A. Fisher invoked kin selection to explain the evolution of warning color (Fig. 21.18D). J.B.S. Haldane captured the essence of the idea in the characteristically pithy remark that he would be prepared to lay down his life for two brothers or eight cousins. However, the theory was not properly worked out until a series of papers by W.D. Hamilton, beginning in 1963.

Take the simplest case of a rare allele that reduces the fitness of its carrier by C; in other words, individuals that carry the allele, on average, have C fewer offspring (Fig. 21.19). Suppose that the allele also increases the fitness of neighboring individuals, so that their combined fitness increases by B (i.e., the neighbors leave, on average, a total of B more offspring). Then, the allele can increase if it tends to be carried by the neighbors. If it is sufficiently frequent in the neighboring individuals, then its cost C will be outweighed by its benefit B to others. More precisely, the allele will increase if $rB > C$, where r is twice the frequency of the allele among neighbors of an individual that carries the rare allele. This formula is known as **Hamilton's rule**. The same basic argument applies for a common allele or a quantitative trait.

Why should an allele be found in excess among those individuals that benefit from its effects? The most important cause of such clustering is that individuals that interact tend to be related—that is, they carry genes that are **identical by descent** (IBD) from a common ancestor. To a good approximation, the r that appears in Hamilton's rule is just equal to twice the coefficient of kinship that we defined in Box 15.3. It is twice the chance that a randomly chosen gene from one individual is IBD with a randomly chosen gene from the other. (The factor of 2 appears because a gene in one individual can be shared with either of the two homologous genes in another diploid individual.) For example, there is a chance of 1/4 that two genes at a diploid autosomal locus will be IBD between brothers and 1/16 between cousins—hence the relative weights that Haldane placed on his relatives' lives.

It is only an approximation to derive the coefficient r from the probability of identity by descent calculated from a pedigree relationship. There are two related difficulties. First, when a mutation arises, it is present in only one individual and gains no advantage by increasing the fitness of its blood relatives, because they do not yet carry the new allele (e.g., individuals a in Fig. 21.19). Second, the proportion of alleles carried in relatives can be distorted by strong selection. (For this reason, a

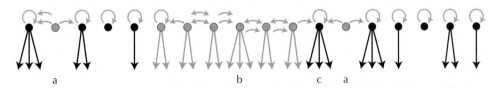

FIGURE 21.19. Alleles that are harmful to the individuals that carry them can nevertheless increase, provided that they benefit others who carry the same allele. In this example, the wild-type individuals (*black*) produce an average of one offspring each by their own efforts (*gray arrows*). Individuals that carry an "altruistic" allele (*red*) produce no offspring by their own efforts, but cause their immediate neighbors to produce an average of three extra offspring in total (*red arrows*). Thus, isolated "red" individuals (a) produce no offspring, but those in a cluster of the "altruist" alleles benefit (e.g., b): On average, the cluster of six individuals carrying the new allele produce 2.4 offspring each. Some wild-type individuals also benefit (e.g., c), but this does not directly affect the rate of increase of a rare "altruist" allele. In terms of Hamilton's rule, $C = 1$ (the reduction in number of offspring produced by the altruist's own efforts), $B = 3$ (the net increase in neighbors' fitness), and so the allele will increase if at least $R = 1/3$ of neighbors carry the "altruist" allele.

Box 21.1   Determining the Proportions of Shared Genes among Relatives

The sharing of genes between relatives can be calculated using the concept of identity by descent. Figure 21.20 shows a pedigree with two parents (generation 0, denoted G_0), who produce four siblings (G_1), who mate with unrelated individuals to produce four families that are related as cousins (G_2). Focus on an allele in one of the parents (marked red). This is passed on to two of the siblings, and to three of the next generation. In the first generation (G_1), an individual who carries the allele will find it in 1/3 of her siblings and therefore at an allele frequency of 1/6 in siblings. In the second generation, the leftmost individual will find that it shares the red allele with 2 of 10 cousins, whereas the two individuals on the right that carry the red allele will each find that they share it with 1 of 8 cousins. The average frequency of sharing with cousins is thus (1/3)(2/20 + [2 × 1/16]) = 0.075. This is just one random outcome of Mendelian segregation: On average, a rare allele found in one individual will be at frequency 1/4 in her siblings, and 1/16 = 0.0625 in her cousins. These frequencies can be derived from the probability of identity by descent and used to determine the strength of kin selection.

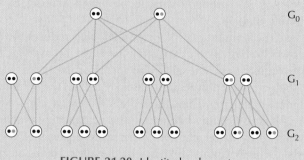

FIGURE 21.20. Identity by descent.

dominant lethal mutation can never be established, however much it increases the fitness of its neighbors.) Over several generations, a new allele with small effects will be found at the expected frequencies in relatives, and the coefficient r can then be identified with relatedness (Box 21.1). The simplest interpretation of Hamilton's rule, in which r represents relatedness, is valid only with weak selection and slow change, as is true for many results in population genetics.

Inclusive Fitness Describes the Effects of Kin Selection

We can think about the effects of kin selection in two ways. The more straightforward one is to count the number of copies of each kind of gene from one generation to the next; this is the typical population-genetic approach, and the one we have taken until now. Hamilton's rule is then exactly correct, provided that r measures the genetic similarity among individuals. As we have explained, this is only approximately equal to the relatedness expected from the pedigree.

Hamilton introduced a complementary approach, which can sometimes give us a useful intuition. He defined **inclusive fitness** by adding the effect of an allele on its carrier's fitness to the effect of that allele on other individuals' fitness, weighted by r. One can then say that selection on individuals tends to increase inclusive fitness. More precisely, individuals will act *as if* maximizing their inclusive fitness. They will help to increase their neighbors' fitness to the extent that those neighbors share genes, so that the focal individual's help is included in its inclusive fitness.

Although the basic concept is simple, a precise definition of inclusive fitness requires some care. The usual definition of fitness is simple: the number of offspring that an individual leaves in the next generation (see p. 418). However, it is not possible to define the inclusive fitness of individuals by taking a weighted sum of the number of offspring of the individual and its relatives. Each individual is expected to have a geometrically increasing number of relatives (one brother or sister, two cousins, . . .), and so the contribution of distant relatives would tend to infinity. Inclusive fitness focuses on the *causes* of individual fitness and attributes these causes to the actions of genes in other individuals. Thus, when we calculate an individual's inclusive fitness, we count only that part of its fitness that is due to the genes that the individual carries. Clearly, the inclusive fitness is not a quantity that could actually be measured. Rather, it serves to focus our attention on the effects of alleles on the extended phenotype—that is, on the reproduction of all the individuals that they influence. It is this that determines the fate of the allele.

Sex Ratio in Fig Wasps Is Determined by Both Competition and Relatedness

Sex ratio in fig wasps provides us with some of the best-understood examples of natural selection. There are about a thousand species of fig, each pollinated by a particular species of fig wasp. A number of female wasps enter the enclosed inflorescence of the fig almost simultaneously. These foundress females carry sperm and also pollen, which fertilizes the fig flowers. As in all wasps, fertilized eggs develop into diploid females, whereas unfertilized eggs develop into haploid males. Offspring mate within the fruit; males then die, and females disperse to start the next generation. This is an excellent experimental system, because the number of founding females and the sex ratio can be determined simply by examining the figs (Fig. 21.21).

The sex ratio is determined by the females' choice of whether or not to fertilize their eggs. If there is only a single foundress, the best choice is to produce mainly daughters, with just enough males to fertilize them. This maximizes the output of fertilized female offspring that emerge from the ripe fruit. However, if there are many foundresses, an individual female can increase her fitness by producing more males, which compete to fertilize females from other mothers. As explained on page 506, the sex ratio tends to 50% when there are so many foundresses that each fig contains a large randomly mating population. Thus, female-biased sex ratios arise when there is **local mate competition** (LMC), that is, when brothers compete with each other for matings.

The sex ratio that evolves also depends on the degree of relatedness within the population as a whole. To see this, think of the genetic contribution that a female makes to the next generation. Sons are produced from unfertilized eggs, and so each carries one genome from the mother. Daughters carry one genome from the mother, plus one from the father. Now, if the population is inbred there is a chance F that a gene from the father is IBD with a gene from the mother. On average, therefore, mothers make a genetic contribution of $1 + F$ via their daughters, relative to 1 via their sons: Inbreeding favors a female-biased sex ratio. To put this another way, an allele in the mother that increases the proportion of daughters by increasing the proportion of eggs that are fertilized is in effect aiding the transmission of genes from the father. Such an allele will tend to increase by kin selection if fathers tend to carry the same allele, because they are related to the mothers.

The effects of local mate competition and inbreeding combine to give a simple prediction for the equilibrium sex ratio. This prediction matches observations remarkably well (Fig. 21.22). Within a species, figs with fewer foundresses produce more daughters because of local mate competition. Thus, individual females adjust their sex ratios according to the theory, so as to optimize their genetic contribution to future generations. Moreover, some species have, on average, fewer foundresses than others and so are more inbred. As predicted, the more inbred species tend to produce more daughters, because kin selection favors females that fertilize their eggs and so help propagate the genes of the related fathers.

In general, it can be difficult to disentangle the effects of competition from those of relatedness. When relatives are clustered together, one might expect more cooperation as a result of kin selection. However, competition is also stronger in such circumstances and tends to reduce cooperation. We have just seen that in fig wasps, relatedness and competition can be separated, because relatedness depends on the average inbreeding in the population as a whole, whereas competition depends on the number of foundresses within individual figs.

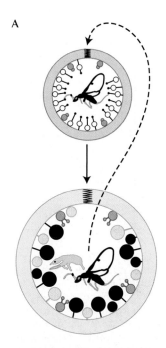

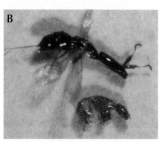

FIGURE 21.21. Figs and fig wasps are closely interdependent: Figs are pollinated by fig wasps, and wasps grow up and mate within the enclosed inflorescence of the fig. The wasp life cycle begins when a fertilized female wasp (*B, top*) enters through a narrow tunnel; this closes after one or a few foundress females have entered. The female wasps pollinate the female flowers of the fig, but also lay eggs in some fig ovaries (*blue*). The wasp larvae induce specialized **galls** (*black*). Wingless male wasps (*B, bottom*) emerge first and extend their genitalia to fertilize females while they are still inside their galls. The male wasps then bite holes in the fig that allow the winged females to disperse, taking with them both sperm from the wasp males and pollen from the male flowers of the fig (*blue*). The figs themselves are eaten by birds, which disperse the seeds (*yellow*).

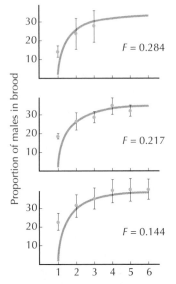

FIGURE 21.22. Local mate competition and **inbreeding** favor a female-biased sex ratio. The graphs show how the proportion of males within individual figs increases toward 50% as the number of foundress females increases, thus reducing competition between brothers for mates. The three graphs show results for three species, which have different overall levels of inbreeding: The more inbred species (*top*) have a stronger female bias. Levels of inbreeding for each species were calculated from the average number of foundresses. *Curves* show the theoretical expectation (see Web Notes), and error bars show standard deviations of sex ratio.

Parents and Offspring Have Different Evolutionary Interests

In many species, parents care for their offspring. The most familiar examples are birds and mammals, but many insects also show parental care, as do plants in which seeds are supplied with nutrients. The parent provisions the offspring, which compete with each other for resources.

In such cases, there is a conflict between offspring and parent over the level of provisioning and between siblings for access to parental resources. From the parents' point of view, there is a trade-off between producing more offspring in the current brood and conserving resources for future reproduction (see p. 560). This trade-off has been demonstrated in many experiments. For example, adding eggs to a bird's nest typically reduces survival and vigor of existing nestlings and also reduces the mother's future reproductive success. Offspring are selected to take more resources than is optimal for the brood as a whole or for the parents' lifetime fitness.

This conflict has led to extravagant begging displays in birds. These may initially have evolved as honest signals of the chicks' hunger, which elicit an adaptive response from the mother. However, once any behavior that can increase the parents' efforts is established, it will be exaggerated by selection on the offspring. There will be counterselection on the parents to resist such exploitative signals, and presumably the outcome will be a compromise between these opposing selective forces.

Evidence for the function of begging displays comes from studies of brood parasites such as cuckoos and cowbirds, which lay their eggs in other species' nests (Fig. 21.23A). Because the parasite's eggs are completely unrelated to their nestmates, selection for begging is not mitigated by any sharing of genes with parents or siblings. Often, the parasitic chick or its mother ejects all the hosts' offspring. However, in many cases they do not (Fig. 21.23B). At first sight, this is puzzling: Why should a cowbird tolerate unrelated nestmates that compete for food? One explanation is that extra begging by the hosts' chicks increases parental effort and leads to a net increase in provisioning of the parasite (Fig. 21.23C).

There is also a conflict between the evolutionary interests of offspring genes that are inherited from the father as opposed to the mother, which may explain the intriguing phenomenon of **genomic imprinting**. The great majority of genes are expressed in the same way, regardless of which parent they come from. However,

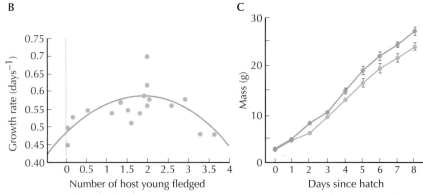

FIGURE 21.23. Begging by chicks stimulates the parents to provide more food. (*A*) A cowbird, *Molothrus ater*, begging in a nest containing chicks of its host, the yellow warbler (*Dendroica petechia*). (*B*) The cowbird does not eject all the host's offspring: A survey of 18 species shows that several host chicks usually fledge and that the growth rate of the cowbird is higher when an intermediate number of host chicks fledge. (*C*) Cowbirds grow faster when reared with two phoebe chicks (*red*) than when reared alone (*blue*).

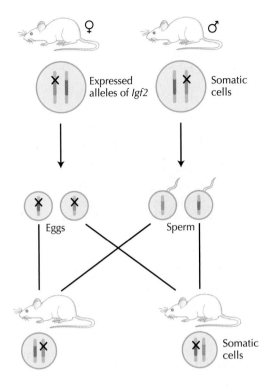

FIGURE 21.24. Some mammalian genes are imprinted, so that only the allele inherited from one parent is expressed. For example, the gene *Igf2* encodes an insulin-like growth factor. In mice and humans, only the copy inherited from the father is expressed in developing embryos. Conversely, the *Igf2r* gene (an insulin-like growth factor receptor) in mice is expressed only from the maternal allele. Imprinting is erased during meiosis; during formation of gametes, different imprints are established in eggs and in sperm. A *cross* indicates imprinting that suppresses gene expression. Note that different alleles (*purple* or *orange*) may be expressed as the genes pass down through the generations: All that matters is whether a gene was inherited from the mother or from the father.

approximately 50 genes in human and mouse are **imprinted**—that is, only one or the other copy is expressed (e.g., Fig. 21.24). A similar phenomenon is found in plants. A major obstacle to the cloning of individuals from somatic cells is that genes in such individuals may not be imprinted correctly.

One explanation for the origin of imprinting is based on an asymmetry in relatedness between alleles of different origin. Paternal alleles are not shared with the mother (assuming no inbreeding) and are only partly shared between siblings (Fig. 21.25). (The relatedness of siblings decreases with the number of fathers that contributes to the brood.) Thus, paternal alleles are selected more strongly to increase the growth of individual offspring. In contrast, kin selection acts on maternal alleles to mitigate the harmful effects of excessive competition on the mother and on other siblings. This explanation is supported by the observation that many imprinted genes have a growth-related function and, moreover, that paternally expressed genes tend to increase growth whereas maternally expressed genes tend to decrease growth (e.g., Fig. 21.24). However, imprinted genes do not evolve especially quickly (as measured by K_a/K_s; see p. 532), as might be expected if they were engaged in an evolutionary conflict.

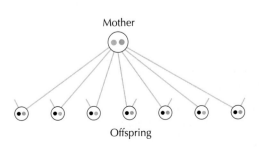

Mother

Offspring

FIGURE 21.25. Paternal alleles (*black*) are more likely to be selected to take more resources from the mother and from other siblings than are maternal alleles (*red, blue*). The diagram shows the most extreme case, where each offspring has a different father. Then, each paternal allele is unrelated to any other gene in the siblings or mother, whereas each maternal allele is identical by descent with one of the mother's alleles and with an allele in half of the siblings.

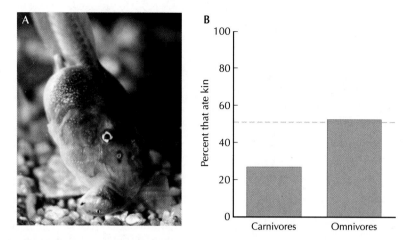

FIGURE 21.26. Tadpoles of the spadefoot toad, *Spea multiplicata*, can develop into a carnivorous morph when exposed to animal prey. These carnivores cannibalize conspecifics more often: (*A*) a carnivore about to eat an omnivore. However, they avoid eating kin (*left bar* in *B*). After a change in diet, these carnivores reverted to the omnivorous form and lost their kin discrimination (*right bar* in *B*). The *dashed line* shows the expectation in the absence of any kin discrimination.

Sometimes, Relatives Can Be Recognized

We have seen that alleles can spread if they increase the fitness of relatives. Clearly, there is strong selection to direct benefits specifically toward kin. Many examples of such **kin discrimination** are known. For example, long-tailed tits that fail to breed in a season may help another pair to rear their young; such help is almost always given to close relatives, both in nature and when a choice is offered experimentally. The carnivorous tadpoles of the spadefoot toad provide another example (Fig. 21.26). Kin discrimination is widespread in vertebrates. Across 18 studies, the average correlation between kinship and probability of helping is $r = 0.33$. Moreover, the degree of discrimination tends to be greater when the benefits of helping are greater, as shown in Figure 21.27.

A form of kin discrimination known as **quorum sensing** is widespread in bacteria (see p. 167). Many species secrete small molecules (homoserine lactones) and use lev-

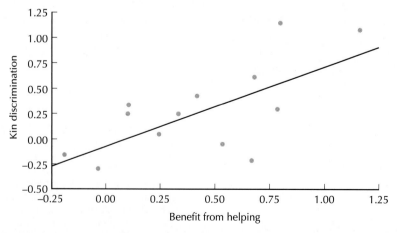

FIGURE 21.27. Across avian and mammalian species, kin discrimination increases with the benefits of helping. Kin discrimination is measured by the correlation between relatedness and amount of help given, and the benefit from helping is measured by the correlation between offspring survival/production and the amount of help.

els of these molecules to sense the density of relatives. For example, the potato pathogen *Erwinia carotovora* secretes antibiotics and extracellular enzymes that kill competing bacteria and digest the food plant. However, they are produced only when the signal molecule reaches high enough concentration. Such signaling systems are favored because the production of digestive enzymes benefits all the bacteria in the region. Clearly, strains that produce them can spread only if those same strains are common enough among the bacteria that benefit.

Selection among Groups Is Weaker Than Selection among Individuals

Until quite recently, arguments based on advantages to the group or to the species were widely accepted (see p. 9). However, during the 1960s, the many difficulties facing such arguments were made clear by evolutionists such as George Williams, W.D. Hamilton, and John Maynard Smith. In the following sections, we will see that although the most basic forms of **group selection** are unlikely to have contributed much to building complex adaptations, the coming together of previously independent replicators into larger units has been of fundamental importance.

In the simplest case, we can imagine a species that is subdivided into small local **demes**. These may go extinct, or they may send out colonists to found new demes. There is now a process of group selection that is analogous to natural selection between individuals: The unit of selection is the local deme rather than the individual organism. Because this selection acts through group fitness (the number of new colonies founded during a colony's lifetime), it can select for properties of the whole group, which may involve cooperation between individuals.

This model of group selection is clearly idealized. Real populations are not neatly subdivided into separate demes that suffer extinction and recolonization (**metapopulations**; see pp. 444–447). However, the same processes occur within viscous populations, which are distributed more or less continuously but with limited dispersal. What is important is that related individuals cluster together, so that selection can favor interactions that increase the fitness of the entire local population.

The fundamental difficulty with this simple form of group selection is that relationships between neighboring individuals are likely to be weak. In Chapter 16, we saw that F_{ST}, which measures the relatedness between nearby individuals relative to the population as a whole, is typically 10–20% (Fig. 16.10). For our purposes, $2F_{ST}$ corresponds to the coefficient r in Hamilton's rule; thus, selection on local groups is typically 2.5–5 times weaker than selection between individuals. Thus, it is both slow and likely to be overwhelmed by opposition from selection between individuals. A further difficulty is that the number of selective events is much smaller: The number of births and deaths of local demes must be far smaller than the number of individual births and deaths. Thus, random fluctuations (analogous to random genetic drift) strongly interfere with group selection.

Wright's Shifting Balance Involves Selection among Alternative Adaptive Peaks

In 1931, Sewall Wright proposed an ingenious theory that avoids some of the difficulties facing group selection. Wright realized that, typically, selection can push populations toward alternative stable equilibria, which he visualized as alternative peaks on an **adaptive landscape** (Fig. 21.28). (Recall that we saw several examples of alternative equilibria in Chapter 17, for example, different chromosome arrangements and butterfly warning patterns.) He saw this as posing a major problem for adaptive evolution. Populations would evolve toward the nearest **adaptive peak** and get stuck there, rather than finding the global optimum. Indeed, this is the main difficulty facing computer algorithms used

A Mass selection   **B** Wright's shifting balance

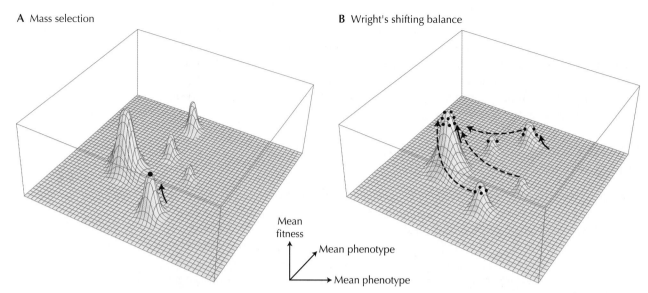

FIGURE 21.28. Sewall Wright argued that a network of local populations (demes) can evolve toward the best adaptive peak, through a shifting balance between random drift, selection within demes, and selection between demes. (*A*) Selection will push a single large population toward the nearest adaptive peak, which is unlikely to be the best possible. (*B*) If the population is subdivided into many local demes, then random drift scatters demes across the adaptive landscape. Selection within demes causes them to cluster around adaptive peaks (*solid arrows*), and selection between different adaptive peaks allows the whole population to cluster around the highest peak (*dashed arrows*). The diagrams show adaptive landscapes—graphs of mean fitness against the state of the population (e.g., allele frequencies, means of quantitative traits; see pp. 472 and 494). Bear in mind that populations can evolve in very many dimensions, not just two: These figures are caricatures of multidimensional reality.

for solving practical optimization problems (see p. 467), such as finding the optimal route for delivering goods or designing the most efficient layout of a microchip. The question becomes: How can one search efficiently among a large number of local optima?

Wright's solution was his **shifting balance theory** (SBT), so called because it involves a shifting balance between different evolutionary processes. Wright identified three phases. In the first, random drift scatters small local populations across the adaptive landscape (see pp. 494–496). (This requires that $N_e m$ and $N_e s$ be intermediate, so that drift, migration, and selection are all of similar strength.) In the second phase, selection pushes populations toward the nearest adaptive peak; and in the third phase, different adaptive peaks compete with each other to spread through the entire population. It is this final stage that involves a form of group selection. Populations at some adaptive peaks may do better than others, because they send out more migrants, go extinct more rarely, or reach greater population size. The shifting balance combines random drift with selection, so that populations can explore a variety of alternative adaptive peaks and can in principle establish a globally advantageous set of alleles.

Wright's scheme provides an elegant way for populations to explore the adaptive landscape. It avoids a key difficulty, in that selection within demes does not oppose selection between them; instead, selection keeps most populations near one or the other of the competing alternatives. Some computer algorithms use a similar method to search among alternative local optima (see p. 467). Wright's theory has had wide influence and has motivated many empirical studies of **population structure** and **epistasis**. Are random drift and population subdivision strong enough to allow demes to shift between adaptive peaks? Do genes interact in the right way to allow selection to sustain alternative equilibria? However, the SBT suffers from many of the same difficulties as simpler forms of group selection. Selection between demes at different adaptive peaks is inherently slow and subject to random events. Moreover, there is little evidence that local demes are in fact at different peaks. We do see examples (e.g., in *Heliconius* butterflies; Figs. 17.23 and 18.17), but these involve populations that cover wide areas and that often differ at very many traits. Thus, it is hard to see how Wright's shifting balance could build up complex adaptations in the way that selection between individuals in a sexual population could. Group selection cannot disentangle the many differences between local demes, whereas recombination within a sexual population can do so.

Selection Can Act among Species but Is Exceedingly Slow

Moving to a larger scale, there can be selection among species. Traits that cause a higher rate of speciation and a lower rate of extinction will tend to spread. Clearly, rates of species formation vary. The coelacanth, an ancient fish (one of the "living fossils"), has survived for the past 120 Myr, yet has never diverged into more than a few species. In contrast, the cichlid fishes of Lake Victoria have proliferated into approximately 300 species in only 500,000 years (see p. 650). The beetles (Coleoptera) have diversified on an extraordinary scale, with more than 350,000 species that vary greatly in morphology and ecological niche. (When Haldane was asked what he had learned about the Creator through his studies, he replied that the Creator had "an inordinate fondness for beetles.") In contrast, the Orthoptera (grasshoppers and crickets) comprise only approximately 20,000 species, having generally similar morphology and ways of life. Yet, the Orthoptera are older than the Coleoptera. The oldest fossil beetles are approximately 270 Myr old, whereas the oldest fossil Orthoptera are approximately 360 Myr old.

In these examples, it is hard to identify which traits are responsible for their differential proliferation. Indeed, it is quite difficult to show that two groups do in fact have intrinsically different rates of speciation and extinction. As we emphasized in Chapter 15, individual reproduction is a highly random process. However, rigorous comparative studies have demonstrated statistically significant associations between particular characteristics and net rates of speciation. For example, groups that show strong sexual dimorphism as a result of sexual selection show significantly higher rates of speciation (Fig. 21.29). Perhaps most important, asexual taxa have a high extinction rate and rarely diversify. Thus, **species selection** is an important factor in explaining the prevalence of sexual reproduction (see pp. 666–668). Similarly, long-term trends seen in the fossil record (e.g., the increasing size of horses through the Eocene) are partly due to a changing array of species, rather than to change within species.

However, although species selection plays a large part in shaping biological diversity, it is far too slow and sporadic to assemble complex adaptations. As for the shifting balance theory, selection between species must be based on their entire set of characteristics. Because there is no recombination, selection cannot pick out individual alleles or traits as can be done by individual selection within sexual populations. Logically, selection can operate at many different levels, but it is far more effective when acting at the lowest level on large numbers of individuals in sexual populations.

We are skeptical of the power of group or species selection. It is inherently slow, lacks the advantages of sexual reproduction, is susceptible to the random success of small numbers of groups or species, and is vulnerable to selection acting more rapidly at lower levels. Yet in a somewhat different sense, selection on groups of organisms has been fundamental to evolution. John Maynard Smith and Eörs Szathmáry have argued that the **major transitions** in evolution have involved the coming together of replicating individuals to form more complex entities (Table 21.3). The origin of life involved the aggregation of replicating molecules into cells and the assembly of separate molecules into chromosomes, thus forcing them to replicate together (see pp. 99–101). Eukaryotic cells arose from symbioses that established mitochondria, chloroplasts, and, perhaps, other organelles (Chapter 8). Once eukaryotes evolved obligate sexual

Polyandrous   Monandrous

FIGURE 21.29. Sexual selection is associated with high rates of speciation. For example, rates of speciation have been compared between sister groups of insects that differ in mating system. Over 25 independent comparisons, the rate of speciation was four times higher for **polyandrous** groups, in which females mate with many males, than for **monandrous** groups, where females mate with a single male. For example, the Drosophilidae (e.g., *A*) are polyandrous, whereas their sister group, the Culicidae (e.g., *B*) are monandrous; and within the butterfly genus *Heliconius*, the *numata* species (*C*) are polyandrous, in contrast with the monandrous sara/sapho clade (*D*).

reproduction, organisms reproduced as pairs rather than as single individuals (Chapters 8 and 23). Later still, multicellular organisms arose, with only a few cells contributing to reproduction (Chapter 9). In all these examples, previously independent replicators came together to produce a whole organism, so that selection thereafter acted at a higher level. Such organisms are vulnerable to exploitation by selection at lower levels (think of, e.g., transposable elements, mitochondria that cause male sterility, malignant tumors, competition for reproduction within slime molds or within beehives). Thus, the key to understanding these major transitions in biological organization is to see how cooperation can evolve and how competition can be suppressed.

TABLE 21.3. The major transitions in evolution all involve fundamental changes in the way hereditary information is passed on

Replicating molecules	→	Populations of molecules in compartments
Unlinked replicators	→	Chromosomes
RNA as gene and enzyme	→	DNA and protein (*genetic code*)
Prokaryotes	→	Eukaryotes
Asexual clones	→	Sexual populations
Single-celled organisms	→	Animals/plants/fungi (*cell differentiation*)
Solitary individuals	→	Social colonies (*nonbreeding castes*)
Primate societies	→	Human societies (*language*)

Apart from the evolution of the genetic code, all these transitions involve the coming together of previously independent replicators, to cooperate in a higher-level assembly that reproduces as a single unit.

EVOLUTION OF COOPERATION

Cooperation May Be to the Participants' Mutual Advantage

So far, we have emphasized kin selection, in which alleles help to propagate copies of themselves carried in other individuals. This is the only tenable explanation for extreme cases such as sterile workers in social insects, which make no direct contribution to future generations. But conversely, kin selection cannot explain mutualistic relations between different species, which do not share genes at all. We have already seen several examples of such mutualisms. Figs depend on fig wasps for pollination, whereas the wasps feed on the fruit. Squid provide a sheltered and nutrient-rich environment within which luminescent bacteria grow. Similarly, essential biochemical services are provided by endosymbiotic *Buchnera* within aphid guts or by nitrogen-fixing bacteria in the specialized root nodules of legumes (see p. 165). Eukaryotes depend on their ancient relationship with mitochondria and chloroplasts. In this section, we explore how cooperation between members of the same species can evolve through direct fitness benefits, in much the same way as in these examples of interspecific mutualisms.

We saw in the previous section that long-tailed tits prefer to help their kin; such kin discrimination is widespread in vertebrates (Fig. 21.27). However, in many cases there is no relation between the level of helping and kinship. For example, meerkats are cooperatively breeding mongooses that live in small groups in African deserts (Fig. 21.30). A dominant male and female produce most of the offspring and are aided by others in the group. Newborn meerkats are protected by "babysitters" for their first month of life, and these helpers suffer a substantial cost: Frequent babysitters may lose from 6% to 11% of their body weight. However, there is no tendency to preferentially help relatives.

Instead, helpers tend to be the larger and older individuals, who presumably suffer relatively less from their duties. It is thought that this behavior is maintained because helpers progress to eventually breed themselves. Thus, their actions eventually benefit their own individual fitness. Similarly, in *Polistes* wasps, about one-third of helpers are unrelated to the queen, but, by helping at the nest, increase their chances of eventually breeding (Fig. 21.31).

In these examples, help given now is rewarded by increased fitness later. Direct benefits can arise in other ways. For example, there may be a strong advantage to large group size, so that it pays every member of the group to accept unrelated incomers and to help rear each other's offspring. In many cases apparently altruistic behavior in fact brings an immediate benefit. For example, once they have finished feeding, Arabian babblers (Fig. 21.32) will find a high perch and look out for predators. Their watchfulness gives them an immediate advantage; although their warning cries may save the whole flock, this is a side effect of behavior that evolved by direct selection.

In many of these examples, cooperation helps all the individuals involved and increases overall fitness (think of nitrogen-fixing bacteria in legumes or the mutualism between squid and their luminescent bacteria). But in other cases, it is difficult to see any overall benefit. For example, figs must flower at just the time that wasps emerge from ripe fruit on another tree of the same species, so that there must be a succession of trees flowering throughout the year. This constrains them to flower in all seasons, and also makes survival at low density difficult. Even in social insects, the benefits of having large colonies are not obvious. Solitary ants may have essentially the same repertoire of behaviors as is found in a colony where these roles are divided across multiple castes (Box 21.2). For example, some colonial ants set up "gardens" in which fungi are reared for food, but similar agriculture is practiced by some solitary species. Overall, productivity per capita actually decreases with colony size. An advantage may lie in the sheer size of social colonies, which overwhelms competitors. Also, different activities (finding food, defense, etc.) can be performed simultaneously by different specialists; individuals do not need to switch between them.

FIGURE 21.30. Young meerkats are protected by unrelated "babysitters."

FIGURE 21.31. In *Polistes*, helpers are often unrelated to the queen.

FIGURE 21.32. An Arabian babbler (*Turdoides squamiceps*) watching for predators.

Box 21.2   Social Insects

Fully social (**eusocial**) insects include all 10,000 species of ant, all 2,200 species of termite, about 1,000 species of social wasps, and several thousand species of social bees. (Many species of bees and wasps are solitary.) Eusocial insects are characterized by cooperative care of the young, with one generation looking after the next, and by reproductive specialization, with eggs being laid by one or a few queens, and the young tended by sterile workers. Other than humans, the eusocial insects have evolved the most elaborate and diverse array of social organization. These range from simple groups of fewer than 100 individuals to colonies of army ants numbering 20 million or more. Research on insect social organization has greatly added to our understanding of social evolution in general.

Social insects have achieved extraordinary feats of construction and behavior. Termite mounds are built to allow a flow of air that regulates their temperature; grass-cutting ants build turrets that prevent flooding of their nests; and weaver ants build silk shelters for homopterans within their nests, which they farm for honeydew. Ingenious defenses include individuals who patrol the foliage around the nest, ant soldiers shaped to plug nest entrances when attacked, and guards who check the identity of those entering the nest. Defenses are often suicidal, as with barbed stings; some termites and ants literally explode, tearing their body wall open and spraying out large quantities of sticky fluid.

Typically, individuals pass through a series of specialized stages. For example, honeybee workers produce royal jelly, which is used to feed queen larvae between their 6th and 14th days. There may also be permanently differentiated morphological castes (e.g., large soldier ants), usually induced by differences in feeding early in life. Complex colonies may involve 20–40 distinct roles, carried out by different individuals, which can be coordinated to achieve complex tasks. For example, weaver ants build their nests by gluing together leaves using silk produced by larvae. One worker holds the larva above the seam while another glues the leaves. There may be complex foraging behaviors, in which individuals are recruited to food sources using foraging trails, or dances that encode the direction and distance to the source.

Why are 12 of the 13 eusocial insect groups Hymenoptera? The leading explanation is based on kin selection. Hymenoptera are **haplodiploid**, and so females are more closely related to their sisters than to their own offspring ($r = 3/4$ vs. $1/2$ if the queen is outbred and mates only once; see Fig. 21.43). Hence, female workers can gain more fitness by rearing their sisters than by rearing their own young. This also explains why workers are female rather than male; drones typically spend little time in the nest before leaving for their nuptial flight. The one exception to this pattern is the termite, which has normal diploid inheritance. Termites depend on intestinal microbes that break down their diet of wood and that must be passed between generations by anal feeding. This requires some degree of sociality.

We should not be surprised at the complex structures and behaviors that can emerge from the collective actions of simple individuals. There is a close analogy with multicellular organisms, in which relatively simple cells cause the development of complex individuals. What is surprising is that such complexity can evolve despite the genetic heterogeneity within a social colony, which makes it vulnerable to subversion from within.

Even where cooperation does benefit all parties, it may be unstable. This point is encapsulated in the **Prisoner's Dilemma** game (see p. 569), in which it is in each individual's interests to defect, even though once both have defected, both are worse off (Fig. 21.33). More generally, cooperative systems are always vulnerable to cheating. (For example, some fig wasps do not carry pollen between fig trees; or worker wasps sometimes lay their own eggs within colonies.) In the next section, we examine how cooperation can be protected from subversion by selection at a lower level.

	Opponent cooperates	Opponent defects
Cooperate	3	0
Defect	5	1

FIGURE 21.33. In the Prisoner's Dilemma game (see p. 569), the best strategy for each individual is to defect, even though both would do better by cooperating. The first column shows the payoff for each strategy when the opponent cooperates, and the second column shows the payoff when the opponent defects. Whatever the opponent does, it is best to defect; yet, when both defect, both do worse than when both cooperate (payoff 1 vs. 3).

Competition Can Be Suppressed by Reciprocation, Punishment, and Policing

When the same pair of individuals meet repeatedly, strategies can evolve that ensure that they cooperate. If pairs of individuals play the Prisoner's Dilemma game many times, a **tit-for-tat** strategy can be evolutionarily stable. Cooperate in the first round, but if a partner defects, then defect in the next round; if she cooperates, then cooperate in turn. However, although this kind of reciprocation is an attractive theoretical possibility, it is hard to find examples in nature. For one thing, interactions are usually between individuals that have different roles, and so are asymmetric.

Cooperation can be stabilized if other individuals prevent selfish behavior. For example, honeybee workers sometimes lay unfertilized eggs, but other workers almost always destroy the eggs. Intriguingly, such **policing** tends to be found in species where the queen mates many times. In such species, female workers are more closely related to the queen's sons than they are to other workers' sons.

The close relationship between yucca and its obligate pollinator, a moth, dates back approximately 40 Myr. The female moth pierces the yucca inflorescence and lays eggs there. She then climbs up to the stigma and deposits pollen. Moth larvae develop within the fruit; adult females mate and then disperse, taking pollen with them. It is in the moth's interests to pollinate, because that ensures that seeds will develop, thus providing food for her offspring. However, the moth is selected to lay more eggs than is optimal for the plant. Although emerging moths do distribute the plant's pollen, they also eat its seeds. In response, the plant produces an excess of flowers and aborts those that receive too little pollen or too many moth eggs. This **punishment** causes selection on the moths, in favor of a higher rate of pollination and lower egg loads (Fig. 21.34). A similar host response maintains the close association between legumes and the nitrogen-fixing bacteria that live in their root nodules (see p. 165). Again, this sets up a selection regime that enforces cooperation by the bacteria and shows a remarkable ability to distinguish cheaters from cooperators (Fig. 21.35).

Why do the different cells in a multicellular organism work together for the good of the whole? In animals, reproduction is confined to the germ line, so that selfish replication by somatic cells has no consequences beyond one generation—cancer cells leave no descendants. However, in plants there is no such separation of soma from germ line, and cells could compete to contribute to flowers. The most important factor

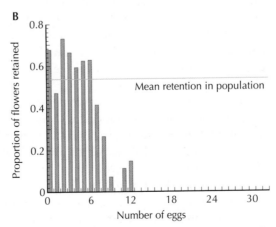

FIGURE 21.34. The yucca (*Yucca filamentosa*) is pollinated solely by the moth *Tegeticula yuccasella* (*A*). Moth larvae grow inside the inflorescence and eat some of the seeds. Host plants limit seed loss by aborting flowers in which too many eggs are laid (*B*): This imposes selection on the moths to limit the egg load. In addition, plants are less likely to abort flowers that receive more pollen from the moths.

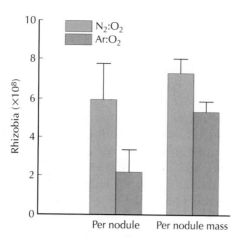

FIGURE 21.35. Soybeans (*Glycine max*) carry root nodules that host the symbiotic nitrogen-fixing bacterium *Bradyrhizobium japonicum*. When the bacterium is prevented from fixing nitrogen by growing the plants in an argon:oxygen atmosphere, the plant responds by reducing the oxygen supply. This greatly reduces bacterial numbers and mass.

maintaining cooperation is simply the close relationship between cells. The entire organism develops from a single fertilized egg, and so only somatic mutation, generated within an individual's lifetime, can have a disruptive effect. Because there is no reason why development should necessarily be from a single cell, it is thought that this unicellular stage evolved as a safeguard against competition within the organism. Indeed, we have seen that such competition does occur in slime molds and in social insects, where individuals are not genetically identical. In some colonial hydrozoans, relatedness is enforced in a different way, through an incompatibility system that allows only colonies that are close relatives to fuse.

Conflict and Cooperation Were Important in Early Evolution

We have emphasized that the major transitions in evolution involve the coming together of previously independent replicators, thus allowing more complex organisms to evolve (Table 21.3). How then is competition between components of the organism suppressed? This problem must have been solved at the origin of life itself. We saw in Chapter 4 (p. 99) that because the first replicating molecules must have copied inaccurately, they could not have been long (at most $\sim 1/\mu$ base pairs, where μ is the mutation rate per base per generation; see p. 552). However, that implies that they could not evolve more sophisticated replication machinery, which could reduce the mutation rate. Manfred Eigen suggested an ingenious solution. If different replicators depend on each other in a mutualistic relationship, then the system as a whole could encode much more information. He called such a system of interdependent replicators a **hypercycle** (Fig. 21.36). Although each component is limited in length by the mutation load, there is no upper limit to the number of components and hence to the complexity of the whole system.

The difficulty with Eigen's proposal is that it is vulnerable to selfish exploitation, especially because all the molecules share the biochemical products of each primordial enzyme molecule (Fig. 21.36B,C). Indeed, we have already seen that this is a serious problem for viruses, which are typically parasitized by defective interfering viruses. It seems that molecules must have aggregated together in cells (or possibly on the surfaces of minerals) so that products remain associated with the sets of molecules that produced them (see p. 100). It is then possible for a form of group selection to maintain cooperation, because protocells that contain efficient hypercycles will outcompete those that have been invaded by "selfish" variants. Indeed, once small groups of molecules have been partitioned into separate compartments that can compete with each other, hypercycles as envisaged by Eigen are no longer essential. For example, one molecule could have a replicase activity, which copies it

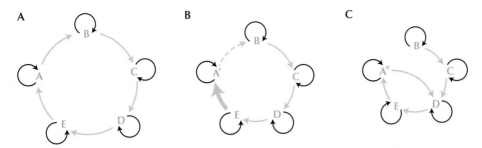

FIGURE 21.36. Hypercycles allow more genetic information to be replicated, but are unstable in several ways. (*A*) Several different replicating molecules (A, B, ... , E) can coexist in a population if they aid each others' replication: A helps B, B helps C, ... , and E helps A. This help could take many forms: provision of some metabolic product, directly aiding replication, etc. What is important is that B replicates faster as A increases in concentration, and similarly for the other links. Such diffuse mutualisms are a familiar feature of ecosystems (trees provide leaf litter for earthworms, which in turn improve silt structure and so aid trees) and are plausible for molecular ecosystems as well. (*B*) Hypercycles are vulnerable to invasion by variants. Compared with A, the variants (A') gain more from help by other species (E) and provide less help to others (B). (*C*) Hypercycles also tend to lose members (B, C), if variants (A") arise that sustain a shorter cycle.

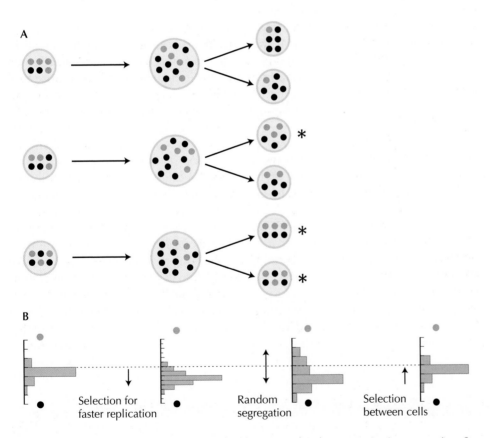

FIGURE 21.37. Selection on small groups of replicating molecules can maintain cooperation. Suppose that there are two kinds of replicating molecule (*red* and *black*), and that black tends to replicate faster than red. However, both types replicate faster when there are equal proportions of the two types. In a well-mixed population, black would take over, even though the population as a whole would then do worse. (This situation is essentially that described by the Prisoner's Dilemma [Fig. 21.33] and also is similar to the reproduction of coviruses [Fig. 21.8].) However, if the molecules are contained within small protocells, which divide when they become large, then the proportions of the two types after division will vary randomly. Selection for protocells with equal proportions (*) can then maintain near-optimal proportions. This kind of group selection can maintain hypercycles by preventing the invasion of selfish variants (Figs. 21.36B,C). However, it acts more generally: The different types do not each have to catalyze their own replication, as in Eigen's model. Instead, one could act as a replicase, whereas the others specialize in other metabolic and synthetic tasks.

and all the others, whereas other molecules specialize in catalyzing metabolic and synthetic reactions (Fig. 21.37). Once such a system is established, selection can favor other methods for ensuring cooperation—such as linking different genes onto a chromosome, which forces them all to replicate together. This ensures better control of the relative amounts of each gene after replication, but also makes it harder for one gene to replicate at the expense of the others.

We have already seen how conflict and cooperation are important for understanding the other major transitions (Table 21.3). The most recent transition identified by Maynard Smith and Szathmáry is the evolution of complex human societies. Apparently altruistic cooperation among unrelated individuals is a distinctive feature of humanity. In Chapter 26, we examine how language may have evolved to facilitate such cooperation.

SUMMARY

In this chapter, we have seen examples of the conflicts that can arise between genes that are transmitted in different ways and, conversely, how separate individuals can cooperate, either because it is in each others' interests to do so or because they share genes. Conflict and cooperation are intertwined, because cooperation is always liable to be subverted by genes that promote their own reproduction at the expense of the whole.

Genetic conflicts cause some extraordinary biological phenomena—elimination of the paternal genome or of competing sperm, mitochondria that cause male sterility, and transposable elements that can spread across the genome, to name a few. Much of biology is driven by conflict. A large fraction of eukaryotic genomes consists of defunct transposons; mutation is often due to active transposition and to recombination between repeated sequences; and natural selection is in large part due to the struggle between hosts and various kinds of parasites. In asexual organisms, the interests of all parts of the genome coincide, and so elements that are only occasionally transmitted horizontally are usually benign. As we see in Chapters 22 and 23, genetic conflict has been important for both speciation and the evolution of genetic systems.

Alleles that increase the fitness of neighbors can increase if those neighbors tend to carry that same allele. Hamilton's rule quantifies this kin selection. Kin selection tends to promote cooperation, but conflicts do arise even between close relatives—for example, between parents and offspring or between paternal and maternal genomes. Sharing of genes is most important between close relatives, but selection can also act between local demes or even between species. However, these forms of group selection are much weaker and less effective than selection between sexually reproducing individuals. Cooperation can evolve even when individuals are unrelated and do not share genes—the clearest examples being mutualisms between different species. Each individual may gain from cooperation, and cheating may be either prevented or punished by others. Ultimately, cooperation may become obligatory, so that eventually elements that previously replicated independently come together in a single, more complex organism. Over the long term, this process has been responsible for the major transitions in evolution.

FURTHER READING

Conflict between Genes

Hurst G.D. and Werren J.H. 2001. The role of selfish genetic elements in eukaryotic evolution. *Nat. Rev. Genet.* **2:** 597–606.
  Broad reviews of the various kinds of selfish element and their evolutionary consequences.
Kidwell M.G. and Lisch D.R. 2001. Transposable elements, parasitic DNA, and genome evolution. *Evolution* **55:** 1–24.
  A review of the natural history of transposable elements, which discusses how they are regulated and how they may be co-opted for host functions.
Nee S. and Maynard Smith J. 1990. The evolutionary biology of molecular parasites, pages 5–18 in *Parasitol. Today (Supplement)*, edited by A.E. Keymer and A.F. Read.
  A readable account of the fascinating biology of viruses and their parasites.
Partridge L. and Hurst L.D. 1998. Sex and conflict. *Science* **281:** 2003–2008.

Interactions between Relatives

Coyne J.A., Barton N.H., and Turelli M. 1997. A critique of Wright's shifting balance theory of evolution. *Evolution* **51**: 643–671.

Hamilton W.D. 1996. *Narrow roads of gene land. Vol. 1: Evolution of social behaviour.* W.H. Freeman, Oxford.

Includes Hamilton's collected papers on social evolution, accompanied by interesting (and idiosyncratic) autobiographical notes.

West S.A., Pen I., and Griffin A.S. 2002. Cooperation and competition between relatives. *Science* **296**: 72–75.

Discusses the opposing effects of competition and kin selection on the evolution of cooperation.

Williams G.C. 1992. *Natural selection: Domains, levels and challenges.* Oxford University Press, Oxford.

Includes a discussion of the inefficiency of group selection; see also Williams G.C. 1966. *Adaptation and natural selection.* Princeton University Press, Princeton, New Jersey.

Evolution of Cooperation

Anderson C. and McShea D.W. 2001. Individual versus social complexity, with particular reference to ant colonies. *Biol. Rev. Camb. Philos. Soc.* **76**: 211–238.

A fascinating review of the organization of ant colonies.

Clutton-Brock T. 2002. Breeding together: Kin selection and mutualism in cooperative vertebrates. *Science* **296**: 69–72; Cockburn A. 1998. Evolution of helping behaviour in cooperatively breeding birds. *Annu. Rev. Ecol. Syst.* **29**: 141–177.

These two reviews weigh the relative importance of the different ways cooperation can evolve.

Maynard Smith J. and Szathmáry E. 1995. *The major transitions in evolution.* W.H. Freeman, Oxford.

Brings together the major transitions in evolution in a common framework, which emphasizes cooperation among previously independent individuals. Szathmáry and Maynard Smith (1995, *Nature* **374**: 227–231) summarize the argument.

CHAPTER

22

Species and Speciation

S PECIES, ALONG WITH GENOMES, CELLS, AND ORGANISMS, are a fundamental unit
in biology. Most biological work is described as being on a particular species,
such as the fruit fly *Drosophila melanogaster,* brewer's yeast, *Saccharomyces cere-*
visiae, or the bacterium *Escherichia coli.* Ecologists are
primarily concerned with explaining the distribution and
abundance of species. For example, why are there more
species in the tropics than at high latitudes? In systematics,
the main task is to name each organism as a member of
some particular species. We can go beyond this basic classification and determine how species are related to each
other (see Chapter 27 [online]); but, even so, most phylogenetic work takes the species to be the basic unit.

It is generally accepted that species are real. There is
often disagreement about how to define higher taxa, such
as genera or families. Similarly, it is often unclear whether
different forms within a species should be described as geographic races or subspecies. In contrast, it is usually obvious how the organisms living in any one place should be
grouped into species. Think, for example, of the birds or
mammals with which you are familiar. The objective reality of species is demonstrated by the agreement between
professional taxonomists and the traditional classification

systems used by diverse cultures. For example, Ernst Mayr found that the local people
in New Guinea had named 136 of the 137 bird species that he had distinguished, and
several studies of "folk taxonomy" have found similarly close agreement. Organisms
do cluster into distinct types, which we call species.

Although we take for granted the clustering of organisms into species, the world
need not be this way. After all, the individuals from different species share a common
ancestor in the distant past and are connected by a continuous chain of inheritance.
Thus, in principle we could see a correspondingly continuous array of organisms, with
no sharp breaks in morphology or in DNA sequence. We will begin this chapter by
explaining how species are defined, and why they exist.

DEFINING SPECIES

The Continuity of Evolution Makes Species Hard to Define

FIGURE 22.1. Darwin's early thinking on evolution was stimulated by his observation that the giant tortoises on different islands in the Galápagos islands were distinct (see p. 1).

Although species are usually quite distinct at any one location, the picture becomes much less clear when we look at the whole geographic distribution. Often, populations on nearby islands are somewhat different, although still clearly related to each other (e.g., Fig. 22.1). Across a continuous habitat, there can be gradual changes that connect organisms that are quite distinct. Most of the examples of **clines** in Chapter 18 involved slight differences with a simple genetic basis, but often divergence can be more extreme. Sometimes, these extremes meet up and are recognized as distinct species or subspecies where they overlap. We have already seen an example of such a **ring species**: Two distinct forms of the greenish warbler meet in Siberia without interbreeding and yet are connected by a continuous intergradation south of the Tibetan Plateau (Fig. 16.18).

Overall, the relationships can be complex. For example, *Partula* is a genus of land snails found in Polynesia. On the island of Moorea, several species have been described, which differ substantially in morphology and preferred habitat, but show little divergence at **allozyme** loci. Each species coexisted with other species as a distinct form within local areas. Yet, most were connected with others through hybridization in other areas. For example, *Partula aurantia* coexisted with *Partula suturalis* in several places, but hybridized with it in others; and although *Partula tohiveana* coexisted with *P. suturalis* without interbreeding, it was connected with it through a continuous intergradation, via *Partula olympia* (Fig. 22.2). Sadly, *Partula* is now extinct in nature following the introduction of other snail species.

These difficulties should hardly surprise us. The apparently continuous connection between distinct forms seen in ring species can be thought of as mirroring the continuous evolutionary relationships with the common ancestor. In this chapter, we begin by asking what species are, and we see that no one definition can clearly divide organisms into separate species. We then go on to ask how species differ from each other, and finally we consider how they came to be different—that is, what evolutionary processes are involved in the origin of species?

There Are Several Ways of Defining Species

The most obvious approach, and the one used traditionally, is to define species by their appearance. Often, this raises no difficulties; recall the close agreement between Mayr's classification of the birds of New Guinea and that of the local inhabitants. However, appearances can be misleading, and our conclusions will depend on just what features we pay attention to.

In many species, there are striking differences between the two sexes (see p. 575). To take an extreme example, Darwin was the first to realize that the tiny creature found clinging to female barnacles is in fact the male of the species. Here, there is no difficulty provided that we treat the male and female together as different forms of the same species. The same argument applies to different stages in the life history (e.g., caterpillars and butterflies) or to the different forms within a polymorphic species. For example, the different mimetic forms of female swallowtail butterflies (Fig. 18.20) have very different wing patterns, and yet if we look at features such as genital structure, then we see that they are members of the same species. Here, we may be able to identify features that are selected to vary and then exclude them from our classification. The fundamental reason why these different forms are classified as members of the same species is that each line of descent crosses between the different forms as it passes

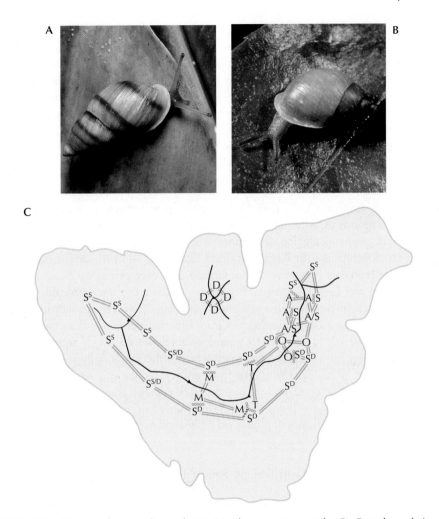

FIGURE 22.2. (*A*) *Partula suturalis* snail. (*B*) *Partula mooreana* snail. (*C*) Complex relationships among *Partula* snails. S, *P. suturalis*; A, *P. aurantia*; D, *P. dendroica*; M, *P. mooreana*; O, *P. olympia*; T, *P. tohiveana*. Superscript S indicates sinistral (left-coiling) *P. suturalis* and superscript D indicates dextral (right-coiling) *P. suturalis*. *Double lines* indicate intergradation through a continuous series of populations; A/S indicates populations where *P. aurantia* and *P. suturalis* hybridize; *hatching* indicates coexistence in nature without interbreeding. The *solid lines* indicate mountain ridges.

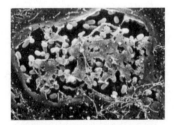

down the generations. Mothers bear sons, caterpillars grow into butterflies, one mimetic form has offspring with a variety of different forms, and so on.

The converse problem raises more difficulties in practice. Sibling species may appear almost identical and yet be distinct for at most a few cryptic characters. For example, *Drosophila pseudoobscura* and *Drosophila persimilis* were originally identified because F_1 hybrid males are sterile; later, they were found to differ in the number of teeth on the sex combs carried on the legs of male flies. This problem is especially serious in organisms with few clear-cut morphological features, such as parasites (e.g., Fig. 22.3). Even more extreme are microbes that cannot be cultured in the laboratory, which may be detected only through isolation of DNA fragments (see p. 148). Just how large a phenotypic difference, and of what kind, is needed to distinguish species?

Molecular divergence might seem to provide a more objective measure; we could use the DNA sequence to classify individuals into species. For want of a better alternative, this approach is used for **uncultured microbes**. In many studies, any ribosomal RNA sequence that is less than 97% identical with another sequence in the sample is

FIGURE 22.3. Microsporidians are intracellular parasites that have lost many of the basic features of eukaryotes: Their genome is smaller than in many bacteria, and they have lost much basic metabolism (see p. 198). They lack mitochondria, but their nuclear genome contains many mitochondrion-derived genes. Although originally thought to be distantly related to other eukaryotes, microsporidians are in fact highly modified fungi. The figure shows *Encephalitozoon cuniculi* within a human host cell.

deemed a different species. However, as we saw in Chapter 15 (p. 425), the degree of divergence in neutral sequence largely reflects the time since the sequences shared common ancestry. This divergence time varies greatly between what are generally accepted as species. We examine rates of speciation in more detail later in this chapter. But to appreciate the great diversity of speciation rates, contrast the divergence of the cichlid fishes within Lake Victoria within the past 100,000 years with "living fossils" such as the coelacanth, *Latimeria chalumnae,* which is separated from its closest living relative by at least 400 million years (Myr).

Using Genealogies to Define Species Is Problematic

Instead of using arbitrarily chosen characteristics to define species, we could try to use genealogical relationships. After all, any set of homologous genes has a definite relationship, which can (in principle) be determined from the pattern of DNA sequence variation. Such relationships are the basis of modern taxonomy and provide the evolutionary justification for Linnaeus' hierarchical classification. However, genealogy alone cannot define what species are. Genes in asexual organisms all share the same genealogy, but this can be divided up arbitrarily in many different ways (Fig. 22.4). For sexually reproducing organisms, different genes can be inherited from different parents and so have different ancestry. Thus, as we saw in Chapter 15, genealogical relationships vary from one gene to another (Fig. 15.11). Even bacterial genomes, which only occasionally recombine, show this mosaic ancestry (e.g., Fig. 15.16). For most eukaryotes, meiotic recombination is an obligatory part of the life cycle, and so ancestry varies greatly across the genome.

In the next section, we will see that the clearest definition of a species, at least for sexual organisms, is based on a special kind of character: structure, physiology, or behavior that prevents exchange of genetic material by sex and recombination.

Biological Species Are Defined as Being Reproductively Isolated

The most widely accepted definition of species is based on the **biological species concept (BSC)**. Although this idea has roots that trace back many centuries, it was first set out in its current form at the time of the **Evolutionary Synthesis** (see p. 30), primarily by Theodosius Dobzhansky, Mayr, and Sewall Wright. Mayr wrote in 1940 that:

> Species are groups of interbreeding natural populations that are reproductively isolated from other such groups.

A population is **reproductively isolated** if it possesses genetic differences that prevent successful interbreeding with other biological species under natural conditions. Two points must be made here. First, the definition does not include geographic isolation. Two populations that cannot interbreed because they are separated by some physical barrier such as a river or mountain range are not defined as separate species. This makes sense, because geographic isolation may be temporary (rivers change their course, mountains erode), whereas reproductive isolation caused by genetic differences is usually irreversible. Second, the genetic differences between species must prevent successful hybridization under natural conditions. For example, lions and tigers will hybridize in zoos (Fig. 22.5), but were never known to do so where their ranges previously overlapped in India.

The BSC applies only to organisms that reproduce sexually. Because the BSC cannot be applied to bacteria and archaea, which only exchange genes sporadically (see pp. 182–191), other methods are needed. Similarly, the BSC cannot be applied to fossils. As we trace each lineage back through time, ancestors appear more and more different from modern organisms. At some more or less arbitrary degree of morphological divergence, it is convenient to label these as distinct species.

Even among contemporary, sexually reproducing organisms, the BSC faces difficulties. As we would expect from the continuity of evolution, no single definition

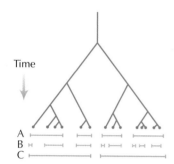

FIGURE 22.4. In asexual organisms, a species definition based on genealogy is arbitrary. The bars (A, B, C) show alternative groups of individuals that could be defined as species. Even if we demand that each group include all individuals that share a common ancestor, there are still several ways to divide up the genealogy. (The same argument applies if we think of the diagram as a phylogeny of species, rather than a genealogy of individuals: Classification into genera, families, and so on based on the phylogeny is to some extent arbitrary.)

FIGURE 22.5. A liger (a cross between a male lion and a female tiger).

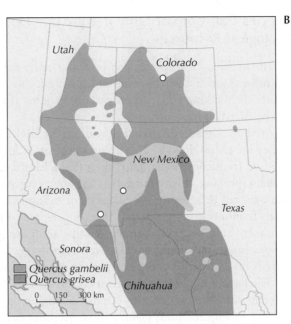

FIGURE 22.6. (*A*) *Quercus gambelii*. (*B*) The white oaks *Quercus grisea* and *Q. gambelii* have overlapping ranges in New Mexico and Arizona. In some places, genetically and morphologically intermediate trees are found, but despite this hybridization, most genetic markers remain distinct. In contrast, relationships between chloroplasts match geography more closely than they match phenotype, which indicates that these organelles have crossed taxonomic boundaries. The *lavender* region on the map indicates an overlap of the two species.

can divide living organisms into clearly separated species that correspond to our strong intuitions about what species are. Often, different kinds of organisms can remain clearly distinct in many features and yet can frequently exchange genes by hybridization. The oaks of North America, for example, are named as distinct species, which are found in different local habitats and which differ in many features that adapt them to those habitats (Fig. 22.6). Nevertheless, hybridization is frequent, and chloroplast genealogies do not reflect the clusters of adaptive traits or of nuclear genetic markers. Such groups of distinct yet interbreeding taxa are termed **syngameons** by botanists.

Many animal groups also contain populations that remain distinct despite hybridization (e.g., Figs. 18.15–18.17 and 22.2). Where the ranges of *D. pseudoobscura* and *D. persimilis* overlap, they occasionally hybridize; as we will see later, most regions of genome consequently show evidence of gene exchange (see Fig. 22.30). More generally, it is often observed that (as in the oak example) mitochondrial and chloroplast genomes flow between taxa that otherwise appear as good biological species, based on morphology and on nuclear genes.

If populations fail to produce fertile hybrids where they meet in nature, they can clearly be assigned to different biological species. Even if they do hybridize at a low level, in practice they may still be regarded as separate biological species. Just how much gene flow is allowed between biological species is debatable, and the status of populations separated by geographic barriers is hard to assess.

Despite these difficulties, the BSC still gives the clearest and most practicable species definition; thus, we will use it throughout the rest of this chapter. The fundamental idea is that a biological species forms a single gene pool, such that a favorable allele that arises in one individual can eventually spread through the entire species. In principle, any combination of the alleles found within a biological species can be produced by recombination. In contrast, different biological species evolve independently of each other in the sense that genetic variation in one biological species cannot contribute to evolution in another.

Species Can Coexist Only if They Use Different Ecological Resources

When two biological species compete, our naive expectation is that one will displace the other. Even if their fitnesses are identical, one will eventually become extinct by chance. In the long run, species can only coexist if they are maintained in a more or less balanced equilibrium.

We have already discussed how different **alleles** can coexist in a population only if they are maintained by **balancing selection** (Chapter 18 [pp. 505–510]). Selection must be **frequency dependent**, such that alleles tend to increase when they are rare. Broadly speaking, this requires that the alleles exploit different resources, so that their numbers can be separately regulated. For example, we saw how *Pseudomonas* bacteria diversify into distinct genotypes that are adapted to live in different parts of their environment (Fig. 18.21). We presented this as an example of different alleles at a single genetic locus, but could equally well regard the various types as different species; there is no clear distinction in a primarily asexual population.

The basic ecological principle of **competitive exclusion** holds that species must use different resources in order to coexist. A single population can be maintained within bounds only if the fitness of individuals declines as population numbers increase. This regulation is mediated by levels of some limiting resource, which declines as it is exploited more heavily. We might think of sheep grazing their pasture or predators consuming their prey. More generally, the resource might be space (sunny clearings required by a forest plant or territory required by a male bird) or indeed anything that facilitates reproduction and is consumed by it. Crucially, if several distinct populations use exactly the same resource, then their *relative* numbers cannot be regulated by this resource. To coexist stably, they must exploit different resources so that their numbers can be separately regulated; when any one population increases, its particular resource is depleted and its numbers stabilize.

The central aim of ecology is to understand what determines the distribution and abundance of species. Typically, species numbers are seen as being determined by the way birth and death rates depend on the abundance of all the species in the community; this in turn depends on how the species exploit, and respond to, the various resources that are available. Thus, species' numbers passively reflect the distribution of available resources; the processes of speciation and extinction are irrelevant. This view is supported by examples of ecological convergence, in which similar sets of species evolve independently in areas that apparently offer a similar range of resources. (For a classic example in nature, see Fig. 22.7, and recall the replicable divergence of *Pseudomonas* bacteria [Fig. 18.21].) In contrast, evolutionary biologists tend to see species' numbers as reflecting a statistical balance between the rates of origin of new species and their random extinction. At present, the relative importance of the resource distribution versus the dynamic processes of speciation and extinction is not well understood.

At the end of this chapter (p. 650), we see how these issues are linked with each other—that is, how the selection imposed by diverse resources can cause one species to split into two. First, however, we discuss how both reproductive isolation and diversifying selection play a role in generating the distinct clusters that we intuitively regard as "species."

Reproductive Isolation Can Arise in Many Ways

All sorts of genetic differences between populations can prevent successful hybridization and hence keep biological species separate. It is helpful to classify these according to when they act, from the first meeting between members of different species through to the production of hybrid offspring several generations later. (Sometimes, these different causes of reproductive isolation are referred to as "isolating mechanisms."

FIGURE 22.7. Marsupials in Australia and **placental mammals** on other continents have converged to occupy a similar range of ecological niches.

We avoid this term, because it implies that "mechanisms" have evolved *in order* to reduce gene exchange. As we shall see, this is rarely the case.)

The first form of isolation occurs when two species never meet each other, at least while they are able to mate. Obviously, we should not count organisms as different species simply because they happen to live in different places. However, genetic differences may cause individuals to *prefer* to live in different microhabitats. Although species occur together in the same general area, they may mate in different places (Fig. 22.8A). Similarly, adults may emerge at different seasons or be active at different times of day (Fig. 22.8B). One of the most remarkable examples of such temporal isolation

FIGURE 22.8. Temporal and habitat isolation. (*A*) The ladybird beetles *Epilachna nipponica* and *Epilachna yasutomii* feed and mate on thistles and blue cohosh, respectively. *E. nipponica* is shown here on its host plant. This kind of isolation is common in insects with specialized host preferences. (*B*) *Photinus* fireflies display at different times of day, thus contributing to reproductive isolation. Seasonal isolation, in which mating occurs at different times of year, is also common.

comes from North American cicadas. These spend many years in the soil as larvae and emerge in large numbers at regular intervals. The length of the life cycle varies, and so a 13-year and a 17-year race living in the same area will have the opportunity to mate with each other only every $13 \times 17 = 221$ years.

Even if males and females do meet at the same time and place, they may not choose to mate with each other. As we saw in Chapter 20, male traits and female preferences for those traits can evolve together in seemingly arbitrary ways. Thus, males of one species may not compete successfully with males of another or may not be chosen as mates by females of the other species. This reproductive isolation may be mediated by divergence in mating signals, such as different frog calls or flowers that attract different insects as pollinators (e.g., see Fig. 22.16). In many cases, different species do mate with each other, but fertilization fails, because sperm or pollen from different species may not be able to fertilize or may not compete successfully with conspecific gametes (e.g., Fig. 20.28).

Once fertilization has occurred, the F_1 zygote may be inviable, because the F_1 genotype fails to develop properly. Sometimes, there can be an incompatibility with the maternal genotype, so that crosses fail in one direction but not in the other. (Recall that early embryogenesis depends on gene products in the cytoplasm of the egg that were encoded by the mother's genome.) Often, F_1 hybrids are fully viable and yet completely sterile—for example, mules, which are first-generation hybrids between horses and donkeys. This may be because of a failure of the F_1 genotype to produce functional germ cells or may be because at meiosis chromosomal incompatibilities lead to **aneuploid** gametes that lack a full set of genes (Box 12.2 and p. 375).

Sometimes, F_2 or backcross individuals are born, but are inviable or sterile. If there is free recombination, the parental genotype could in principle be recovered, so that some recombinant hybrids would be viable. However, this may be highly unlikely, given the extremely large number of possible genotypes produced by crossing parents who differ at many genetic loci.

Reproductive Isolation May Be Prezygotic or Postzygotic

It is useful to classify isolating mechanisms into two classes, depending on whether they act before or after zygote formation. That is, we distinguish **prezygotic** from **postzygotic isolation** (Table 22.1). There are two reasons why we make this distinction. First,

TABLE 22.1. Classification of causes of reproductive isolation

Prezygotic barriers

   Potential mates live in the same place, but do not meet.

      habitat isolation

      temporal isolation (by time of day or time of year)

   Potential mates meet, but choose not to mate (behavioral isolation).

   Individuals copulate, but male gametes (sperm or pollen[a]) are not transferred.

   Male gametes are transferred, but the egg is not fertilized (gametic incompatibility).

Postzygotic barriers

   Zygote dies early in embryogenesis.

   F_1 hybrids are inviable.

   F_1 hybrids survive, but are sterile.

   Backcross or F_2 hybrids are inviable or sterile.

[a]In insect-pollinated plants, pollinators might visit a flower, but not pick up pollen successfully, or fail to deposit pollen on the recipient stigma.

natural selection can act to increase only prezygotic isolation, not postzygotic. Clearly, individuals who choose to mate with their own species, rather than wasting effort on mating with another, will be at a selective advantage. In contrast, natural selection cannot favor a further reduction in hybrid viability or fertility. Once a hybrid zygote is formed, a parent's interests are best served by maximizing its success. Later in this chapter, we examine how natural selection can increase prezygotic isolation.

Second, species that are not separated by prezygotic isolation cannot easily coexist. The rarer form will almost always mate with the other type and will therefore leave fewer offspring if there is some degree of postzygotic isolation. Hence the less common of the two will go extinct, as a result of this positive frequency-dependent selection. Whether two species will be able to coexist does not depend only on whether they have some prezygotic isolation. As discussed above, they must also occupy different ecological niches or, more precisely, must exploit different resources. If good biological species evolve without sufficient divergence in mating behavior or resource use, then they will remain in nonoverlapping regions, only meeting in a narrow contact zone (e.g., Fig. 17.22). (This is known as a **parapatric** distribution.) Further behavioral and/or ecological divergence must occur before they can coexist with each other over a wide area.

Molecular Divergence Can Measure the Rate at Which Reproductive Isolation Evolves

Even the most closely related pairs of species usually show substantial molecular divergence, which implies that they have been evolving separately for a long time: For example, *Drosophila simulans* has two sister species, *Drosophila sechellia* and *Drosophila mauritiana*, which are found on the islands of the Seychelles and on Mauritius, in the Indian Ocean. The **molecular clock** can be calibrated using the divergence of *Drosophila* species endemic to the Hawaiian island of Kauai, which formed 5.1 millions of years ago (Mya), leading to an estimated rate of divergence at **synonymous** sites of 1.14×10^{-8} per year. Using this method, *D. simulans* and *D. mauritiana* are estimated to have diverged ~0.93 Mya, and *D. simulans* and *D. melanogaster* ~5.4 Mya (Fig. 22.9). To take another example, many North American bird species were until recently thought to have diverged during the **Pleistocene**,

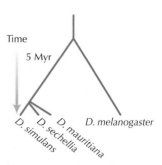

FIGURE 22.9. This phylogeny shows the relationships between flies in the *Drosophila melanogaster* species group, which will be discussed throughout this chapter.

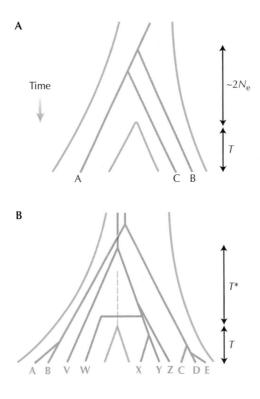

FIGURE 22.10. Using genetic divergence to estimate when populations separated. At time T, two species separate from a single ancestral population, which had effective size N_e. (A) The time to common ancestry of two lineages sampled from different populations averages $T + 2N_e$. This is because, tracing backward in time, even when the ancestral lineages (A, B) enter the same population, it will still take ~$2N_e$ generations for them to coalesce. When species are closely related (T less than ~$2N_e$), it is likely that an ancestral lineage (C) will not coalesce within its own species during time T, and that the genealogy will not correspond to the species' phylogeny. In this example, C is more closely related to A than it is to B. (B) If speciation is not instantaneous, there will be a period T^* during which some genes can flow between species. If there happens to be no gene flow in the ancestry of a gene, genealogies are likely to coincide with the phylogeny, with relatively ancient divergence (i.e., before $T + T^*$; A–E, which are a genealogy at one gene [*red*]). However, some loci will show gene flow and will have discordant genealogies (e.g., W–Y). V–Z are a genealogy at another gene (*green*).

within the last 250,000 years, when their ranges were split by successive glaciations. However, comparison of mitochondrial DNA (mtDNA) sequences suggests that divergence is often much more ancient, typically more than 1 Mya.

Estimates of divergence time need to be treated carefully for several reasons. We saw in Chapter 19 (p. 531) that the molecular clock varies more than the steady rate expected under the **neutral theory**; this may reflect the action of selection. Inaccuracies due to such variation in rate can be reduced by using noncoding sequences or synonymous sites that do not affect amino acid sequence, and by averaging across many loci (as was done for Fig. 22.9).

Another difficulty arises from polymorphisms within species. Even if two species separated very recently, sequences sampled from each would still differ; on average, they share common ancestry ~$2N_e$ generations back (Fig. 22.10; p. 422). In this situation, where species separated recently relative to the timescale of coalescence, the genealogies of individual genes will not match the species phylogeny [Fig. 22.10A], a phenomenon termed **lineage sorting**. Finally, even a low rate of gene exchange after separation can greatly reduce divergence. Even if two species have been distinct for an extremely long time, sequence divergence will plateau at an equilibrium that reflects the rate of gene flow rather than reflecting the time since their (partial) separation (Fig. 22.10B; p. 449).

Despite these complications, it is clear that even the most closely related species typically began to diverge long ago. To take the most familiar example, humans diverged from chimpanzees 5–8 Mya (see p. 729). We should bear in mind, however, that these comparisons are between *surviving* species. Far more species have been born than now survive, and so the long divergence times between extant species partly reflect the small fraction of species that are still living. In other words, the divergence times between the sister species that we have discussed depend on both speciation and extinction rates. This point is well illustrated by the recent discovery of fossils from many **hominin** species that branched off our own evolutionary tree, but later went extinct (Chapter 25).

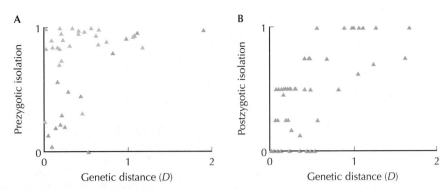

FIGURE 22.11. (*A*) Strength of prezygotic isolation plotted against genetic distance *D* for pairs of *Drosophila* species. Prezygotic isolation was measured by laboratory mate choice experiments. *Blue triangles*, sympatric pairs; *red triangles*, allopatric pairs. (*B*) Postzygotic isolation against Nei's *D*. There is no significant difference between sympatric and allopatric pairs, and so these are not distinguished. Crude measures of postzygotic isolation were given by the fraction of the four kinds of F_1 hybrids that survive or are fertile (male or female, from each of the two possible **reciprocal crosses**).

Reproductive Isolation Accumulates Slowly

We can measure the rate of formation of biological species by finding how rapidly reproductive isolation evolves. The best data come from comparisons between pairs of *Drosophila* species. Figure 22.11 shows a comparison between laboratory measures of prezygotic and postzygotic isolation and a measure of **genetic distance** *D*, which serves as a rough surrogate for divergence time. Prezygotic (behavioral) and postzygotic (survival and fertility) isolation between **allopatric** species (i.e., species that live in different places) accumulate at similar rates and reach 50% at a divergence time of rather more than 1 Myr (Fig. 22.11). Less extensive surveys in fish, birds, and Lepidoptera show similarly slow rates of accumulation of reproductive isolation. (Note that these crude measures tell us about isolation only in the laboratory and so may miss factors important in nature.)

This survey reveals a remarkable difference between allopatric and **sympatric** pairs of species—that is, between species that live in different places compared with those living together in the same area. Closely related species that live together show much stronger prezygotic isolation than those that live apart (0.83 vs. 0.29, for species with *D* < 0.5; Fig. 22.11A). No such difference is seen in the strength of postzygotic isolation. Species can live in sympatry only if they avoid interbreeding—that is, if they show strong prezygotic isolation. We consider the implication of this striking pattern toward the end of this chapter.

Reproductive isolation between bacteria that take up foreign DNA by **transformation** depends in a simple way on their genetic divergence. To be incorporated into the recipient's genome by homologous recombination, a fragment of DNA must be taken up from the medium; it must avoid being degraded by defense mechanisms such as the restriction endonuclease system; and it must form a heteroduplex with the host genome that allows homologous recombination to take place successfully. This last step depends primarily on the stability of the heteroduplex between divergent DNA sequences. There is a simple exponential relationship between the rate of transformation and the degree of sequence divergence (Fig. 22.12). Bacteria and archaea can take up genes from donors that are an order of magnitude more divergent than in crosses between eukaryotes.

The key point from this section is that reproductive isolation accumulates slowly. In other words, organisms that shared ancestry millions of generations back can nev-

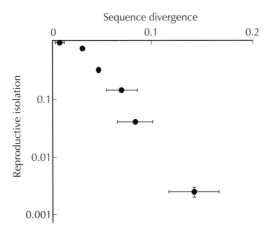

FIGURE 22.12. The rate at which *Bacillus subtilis* incorporates DNA by transformation decreases exponentially with the degree of sequence divergence. The rate at which DNA is incorporated into the recipient's genome, relative to control DNA from the same strain, is plotted against sequence divergence. *Bars* indicate standard errors.

ertheless successfully interbreed to produce viable and fertile hybrids. This partly reflects the slow rate at which organisms evolve. Perhaps the greatest surprise from molecular biology has been the extraordinary conservation of gene function and development system (Chapters 2 and 11). However, it also reflects the robustness of organisms to genetic change—a topic we return to in Chapter 23.

THE GENETICS OF SPECIATION

Occasionally, Species May Be Separated by a Single Genetic Change

In what ways do species differ from each other? The bulk of genetic differences between them are likely to be neutral and thus simply reflect the time since the species diverged (see p. 425). We are particularly interested in knowing which genetic differences actually define species by preventing them from interbreeding. Identifying the genetic differences that are responsible for reproductive isolation is perhaps the most straightforward task in the study of speciation, because it involves concrete genetic questions about present-day organisms, rather than about the past processes that produced them. However, as we will see in this section, it is still by no means easy. As yet, only a few such "speciation genes" have been identified at the molecular level.

The first geneticists emphasized the importance of single mutations as the cause of new species. They thought that new species were produced directly, by single mutations of large effect (see p. 25). This is, in fact, rarely the case, for two reasons. First, changes with large phenotypic effects almost always disrupt the organism and are deleterious (see p. 407). Second, a mutation that causes reproductive isolation would be selected against when rare. For example, a major chromosome rearrangement that made heterozygotes sterile would itself be sterile when it arose as a heterozygote; even if it could persist long enough to become homozygous, progeny formed by crossing with the rest of the population would be heterozygous and hence sterile. Similar arguments apply to prezygotic isolation. Females would not prefer a new male trait, whereas a new female preference might incur extra costs in searching for those rare males carrying the trait. Later, we see that such problems can be avoided when isolation is due to the accumulation of multiple genetic changes (pp. 642–644). However, it is hard to see how a major change that confers reproductive isolation in a *single* step could ever be established.

Despite these arguments, reproductive isolation can sometimes evolve in a single step. In this and the following section, we discuss the two most important examples: the origin of an asexually reproducing clone from sexual species, and the generation of **polyploid** species, where the number of copies of the whole genome increases abruptly.

If we keep strictly to the biological species concept, then every member of an asexual population is a separate "species," because it does not interbreed with any other individual. However, we can still regard the production of an asexually reproducing population from sexual progenitors as an instance of sudden speciation. In organisms that can reproduce vegetatively as well as by sexual reproduction (e.g., potatoes and bananas), asexuality can be achieved simply by loss of sexual function. The most interesting cases, however, involve development from unfertilized eggs—that is, by **parthenogenesis**. If females are produced, then continued asexual reproduction is possible. Often, successful parthenogens are produced when two different species hybridize. Then, the offspring are heterozygous at all autosomal loci and may gain a strong advantage because recessive deleterious alleles are masked (**heterosis**; p. 515). In addition, asexual species avoid the costs involved in finding a mate and gain a twofold advantage by producing exclusively daughters of identical genotype, rather than devoting half their effort to raising genes from another individual (see p. 669). As a result, parthenogens may greatly extend their range. In the long run, however, almost all asexual species go extinct and do not give rise to daughter species. We examine possible explanations for the rarity of asexual species in the next chapter.

The Best-Understood Mechanism for the Origin of Species Is through Polyploidy

The number of copies of a genome within each cell may increase in several ways and can occasionally produce a polyploid population, which is reproductively isolated from its progenitor. These kinds of whole-genome mutations are not uncommon; for example, 2–4% of human conceptions are **triploid** as a result of multiple fertilization (**polyspermy**), although these almost always abort spontaneously. In flowering plants, the rate of production of **tetraploid** individuals within a species is roughly 10^{-5}.

Once a tetraploid individual has formed, it must found a sustainable population if it is to found a new species. The key difficulty it faces is that it produces diploid gametes, whereas the ancestral population produces haploid gametes. Thus, crossing back to the original population will produce triploid, and hence sterile, individuals. As discussed above, the polyploid individuals may abandon sex altogether and found an asexual clone. A new sexual species can be produced if the tetraploid self-fertilizes and thus produces multiple offspring that can mate with each other to found a new population. This will be more likely if the parent has a long life span, and indeed polyploid speciation in plants is associated with self-fertilization and with a perennial life history. However, polyploid speciation is found even in organisms with separate sexes, which obviously cannot self-fertilize (e.g., in many fish and amphibians). In such cases, two tetraploid individuals may form by chance and mate with each other.

Polyploids may arise from the doubling of a genome from one species (**autopolyploidy**), as discussed above. However, the majority of polyploid species in flowering plants and animals arise through hybridization between two species (**allopolyploidy**). Allopolyploidy provides the best-documented examples of speciation, because experimental hybridization can recreate species that arose in nature (e.g., Fig. 22.13). The prevalence of allopolyploidy may be because cell division is disrupted in hybrids, so that polyploid mutations occur more often, and also because a hybrid polyploid gains high fitness through heterosis. Polyploidy makes this advantage permanent, because

FIGURE 22.13. *Galeopsis tetrahit* is a tetraploid species of mint, which was formed by hybridization between the diploid species *Galeopsis pubescens* and *Galeopsis speciosa*. In 1930, Arne Müntzing showed that artificial hybrid tetraploids appear similar to natural *G. tetrahit* and interbreed with them.

FIGURE 22.14. Chromosomes of highly polyploid plants. The record for the number of chromosomes among flowering plants is held by stonecrop, *Sedum suaveoleus* (shown here); which has a diploid number $2n = 640$ chromosomes and ~80 copies of the ancestral genome. Among plants as a whole, the highest chromosome number known is in the fern *Ophioglossum pycnostichum*, with $2n = 1260$ chromosomes and ~84 copies.

every member of the new species carries copies of both ancestral genomes. Therefore, wild-type copies permanently mask deleterious recessive alleles, and heterosis is not dissipated by Mendelian segregation.

How widespread is polyploidy? Many crop plants are polyploid, including coffee, cotton, tobacco, wheat, maize, and sugarcane. Some became polyploid in the distant past (~11 Mya for maize), whereas others, such as wheat, became polyploid during domestication. Some plants have evolved exceptionally high chromosome numbers through multiple rounds of polyploidization (Fig. 22.14). In ferns, it is estimated that 42% of recent changes in chromosome number have involved polyploidy (Fig. 22.15). Because about 16% of speciation events in ferns involve a change in chromosome number, approximately 7% (16% × 42%) of speciation events involve polyploidy. Similar arguments suggest that between 2% and 4% of speciation events in angiosperms (flowering plants) involved a doubling of the genome. Polyploid speciation is much rarer in animals than in plants, perhaps because animals commonly have separate sexes, so that at least two polyploid individuals are needed to found a new species. In contrast, plants usually combine male and female function in each individual, so that a single polyploid mutant can propagate by self-fertilization.

Doubling of the genome in the distant past has been detected by the presence of large numbers of duplicate genes. Stronger evidence of ancient polyploidy comes when

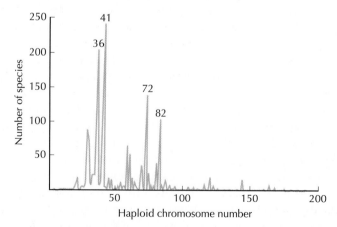

FIGURE 22.15. The proportions of odd and even chromosome numbers can be used to estimate the incidence of polyploidy. A doubling of the genome must give an even chromosome number, whereas other kinds of chromosomal change will cause random shifts between even and odd numbers. This figure shows the distribution of haploid chromosome numbers in ferns. (Fourteen species with more than 200 chromosomes in the haploid genome are omitted [e.g., Fig. 22.14].) Overall, there are 1092 even numbers and 637 odd numbers, which leads to the estimate given in the text.

duplicate blocks of genes retain the same linear order along the chromosomes. The sequencing of whole genomes has revealed several such cases, including brewer's yeast and *Arabidopsis thaliana*. It is likely that vertebrates trace back to two successive polyploidization events, although the evidence here is less clear. Following whole-genome duplications such as these, one or another copy of each gene may be lost, so that the species returns toward its original diploid gene number. However, duplicates are often retained and may diverge in function. We will examine the role of gene duplication in the evolution of novel function in Chapter 24.

Usually, Species Are Separated by Many Genetic Differences

There is a fundamental difficulty in studying the genetic basis of species differences: Reproductive isolation prevents the crosses that are needed for genetic studies. However, many taxa that are regarded as good species do occasionally hybridize, and, in any case, good biological species will often mate successfully in the laboratory even if they never do so in nature. We will now see how the methods described in Chapter 14 can be applied to investigate differences between species in the same way as for variation in quantitative traits within species.

We are most interested in certain kinds of difference: those responsible for the adaptations peculiar to each species, and those responsible for the reproductive isolation that defines them. The sudden origin of such differences through asexuality or polyploidy is exceptional. We will see that both adaptation and reproductive isolation are usually due to multiple genetic changes. We should bear in mind, however, that the differences we now see between species have accumulated over time. It is difficult to know *which* differences first caused the species to become reproductively isolated or how these crucial differences arose. We consider these more difficult questions concerning the mechanism of speciation later in this chapter.

The availability of molecular markers has allowed the **quantitative trait loci** (**QTLs**) responsible for species differences to be mapped in a variety of organisms (Chapter 14). A good example comes from two species of monkey flower, *Mimulus lewisii* and *Mimulus cardinalis* (Fig. 22.16). Even where their ranges overlap, these species rarely form hybrids. However, fertile F_1s can readily be produced by artificial pollination. Thus, there is strong prezygotic isolation, but little postzygotic isolation. There are differences in many floral traits that reduce cross-pollination, and for most of these, there is a major QTL that explains much of the genetic variance in the F_2 population and much of the difference in means between the species (Fig. 22.17). Most striking is carotenoid pigment production, which is almost entirely determined by a single Mendelian locus, *yup* (Fig. 22.17).

In this example, the finding that some alleles have large effects suggests that postzygotic isolation could have first evolved through simple genetic changes, which shifted flowers from pollination primarily by bees to pollination by hummingbirds. Once most flowers are hummingbird pollinated, selection would favor alleles of smaller effect, which would perfect the new adaptation. Of course, we have no evidence for this scenario. Indeed, it is possible that the alleles of large effect that we see now were built up by a series of substitutions at the same gene, each of smaller effect. In general, it is difficult to know which of many differences first led to reproductive isolation.

Excluding domesticated plants and animals, there have been rather few studies of phenotypic differences between species, and most of these have been in *Drosophila* and *Mimulus*. What is most striking is the wide range of genetic complexity underlying species differences. Some show simple Mendelian inheritance—recall the example in which *Ubx* caused differences in bristle pattern (Fig. 11.19)—whereas others involve large numbers of genes. This variation is not just due to differences in divergence time. Contrasting examples can be found even between the same pairs of species.

FIGURE 22.16. The flowers of *Mimulus lewisii* (*top*) and *Mimulus cardinalis* (*bottom*) with their typical pollinators.

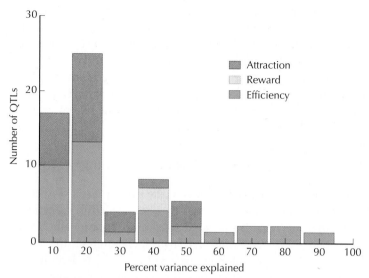

FIGURE 22.17. The distribution of quantitative trait locus (QTL) effects on pollination characters in *Mimulus*. These are measured as the proportion of the variance in the backcross population that is explained by the QTLs. The distribution pools estimates from backcrosses to *Mimulus lewisii* and to *Mimulus cardinalis*. Traits are classed as "attraction" (e.g., flower color), "reward" (nectar volume), and "efficiency" (e.g., stamen and pistil length).

Classical Genetics Has Allowed Detailed Analysis of Reproductive Incompatibilities in *Drosophila*

Most genetic analyses of species differences have focused on hybrid sterility and inviability in *Drosophila*. This is because these traits are key components of the reproductive isolation that defines biological species and because *Drosophila* genetics is so well developed. Early work by Dobzhansky and Muller showed that, typically, every chromosome has some effect on hybrid fitness. The development of molecular markers has made possible a much more detailed analysis. Here, we describe some examples that show the varied kinds of genetic interaction that can be found.

Incompatibilities between D. mauritiana and D. simulans

One of the most detailed studies involved 87 artificial lines, which each carried mainly *D. simulans* genes, but with a small segment derived from *D. mauritiana* at a random location in the genome (Fig. 22.18). When heterozygous, these introgressed segments had little effect on viability or fertility. Indeed, if they had severe effects, the experiment would not have succeeded. However, flies made homozygous for introgressed segments were inviable in about 5% of cases. A further 9% of lines showed maternal effects on viability, such that flies died when homozygous for the introgressed segment if their mothers were also homozygous. (Presumably, some essential gene product could be obtained either from the maternal cytoplasm or from the embryo.) About 5% of segments caused female sterility when homozygous, whereas 50% caused male sterility. These figures are for segments on the autosomes; segments carried on the single X chromosome of a male caused sterility in 75% of cases. We return to this example in the next section, where we discuss why male sterility is so common. For now, the main point to note is the very large number of regions of genome that lead to incompatibilities between these species.

The Bogotá Subspecies of D. pseudoobscura

The large number of genes responsible for reproductive isolation in this and other examples might just reflect the long time for which the species have been diverging.

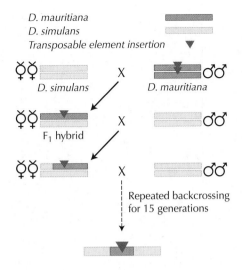

D. mauritiana
D. simulans
Transposable element insertion

D. simulans   X   D. mauritiana

F₁ hybrid   X

X

Repeated backcrossing
for 15 generations

FIGURE 22.18. Genetic analysis of incompatibilities between *Drosophila mauritiana* and *Drosophila simulans*. A white-eyed stock of *D. mauritiana*, homozygous for the recessive w^- allele, was transformed by injecting transposable elements into embryos. These elements inserted at random locations in the genome. The transformed flies were red-eyed because they had been engineered to carry the wild-type w^+ allele. Replicate lines were backcrossed to a white-eyed *D. simulans* stock, which also carried the w^- allele, for 15 generations, with red-eyed flies being selected in each generation. After 15 generations, each backcrossed line contained a segment from *D. mauritiana,* which represented, on average, 7% of the genome. (See p. 403 for more on this method.)

Although they are sister species, the molecular clock suggests that *D. simulans* and *D. mauritiana* separated about 0.9 Mya (Fig. 22.9). A study of two more closely related populations reveals less extensive divergence at an earlier stage in speciation. *D. pseudoobscura* is mainly distributed across the United States, but an isolated population (classified as a subspecies) is found around Bogotá, in Colombia. This is similar in **allozymes** and in DNA sequence to the main (USA) population and so is thought to have diverged relatively recently (150,000–230,000 years ago). When Bogotá subspecies females are mated with USA males, daughters are fertile but sons are completely sterile. In the opposite cross, all offspring are fertile. Three regions of the X chromosome interact in a complex way with genes on the autosomes to determine male fertility. Bogotá genes on the left arm of the X have no effect unless Bogotá genes from the right arm of the X are also present; conversely, third chromosomes from the USA genes have no effect on male fertility unless genes from the USA second chromosome are also present. The net consequence of these interactions is that out of 33 tested genotypes, only three showed significant sterility.

Crucially, several regions of the X chromosome and autosomes had no detectable effect at all. This suggests that only about 15 loci are responsible for the sterility of F₁ males. This contrast between the *D. pseudoobscura* subspecies and the wider comparisons between *Drosophila* species that we saw earlier is partly due to the different times over which the populations have been diverging. However, there is considerable variation in the numbers of genes involved in different traits, even when the populations have been diverging for similar times. For example, a single gene is responsible for differences in larval hairs between *D. sechellia* and *D. mauritiana*, yet many genes are involved in genitalic differences between *D. simulans* and *D. mauritiana* (Fig. 22.19). Reproductive isolation accumulates much less steadily than does DNA sequence divergence.

A striking difference between hybrid incompatibility and the phenotypic differences discussed earlier is that the former involve interactions between genes (i.e., **epistasis**), whereas the latter often involve effects that are more or less additive across genes. For example, complex interactions influence hybrid sterility in *D. pseudoobscura*, whereas the QTLs that influence the genitalia of *D. simulans* and *D. mauritiana* (Fig. 22.19) have roughly additive effects. Of course, hybrid sterility and inviability almost inevitably involve epistasis. An allele that causes problems on a foreign genetic background must work well on its own genetic background. However, in several cases (including crosses of *D. pseudoobscura* USA with the Bogotá subspecies), more complex interactions are seen, such that specific combinations of alleles must be present before incompatibility is expressed.

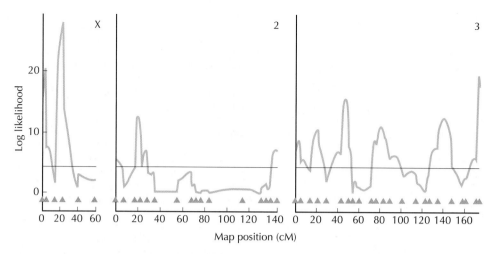

FIGURE 22.19. Polygenic basis of differences between the male genitalia of *Drosophila simulans* and *Drosophila mauritiana*. These species differ in the shape of the posterior genital lobe by ~35 environmental standard deviations ($\sqrt{V_e}$). At least 19 QTLs are responsible for this difference. The *blue lines* show the log likelihood score plotted against position along the three chromosomes (X, 2, 3). This gives statistical evidence for the presence of QTLs affecting genital shape; the *horizontal line* shows the threshold for statistical significance. The *red triangles* show the positions of the 63 genetic markers used in the backcross experiment.

Hybrid Rescue Alleles Show One of the Most Striking Kinds of Interaction

D. melanogaster and *D. simulans* have been diverging for ~5.4 Myr (Fig. 22.9). F_1 males of crosses between these species are completely inviable, and F_1 females are inviable or sterile. Thus, until recently the elaborate genetic tools available in *D. melanogaster* could not be used. However, five mutations have now been identified that rescue such F_1 males, and the recent discovery of an allele that rescues F_1 female fertility allows genetic analysis of this cross for the first time. The existence of such **hybrid rescue** alleles suggests that male viability has a simple genetic basis, despite the evidence for polygenic incompatibilities discussed above. We return to this apparent paradox below.

Some of the Genes Responsible for Reproductive Isolation Have Been Identified

What kinds of genes are responsible for hybrid sterility and inviability? So far, only a few "speciation genes" have been identified in molecular detail, but many more are likely to be discovered soon. Here, we describe two of the first examples.

The platyfish *Xiphophorus maculatus* is polymorphic for pigmented cells, which are not found in the swordtail *Xiphophorus helleri* (Fig. 22.20A). F_1 hybrids carry exaggerated spots and, when backcrossed to *X. helleri*, half the offspring lack spots, whereas the other half show spots ranging from F_1-like to malignant melanomas (Fig. 22.20B). This represents strong postzygotic isolation between these species. The basic explanation was worked out in the early years of classical genetics. *X. maculatus* carries a sex-linked *Tu* allele, which determines pigmented cells, and is also homozygous for a suppressor allele *R* at an autosomal locus (Fig. 22.20C). Both loci have now been isolated and characterized.

Our second example was found by using a powerful method for precisely mapping the genes responsible for the incompatibility between *D. simulans* and *D. melanogaster*. *D. simulans* males carrying the hybrid rescue allele *Lhr* were crossed to *D. melanogaster* females. Usually, the F_1 males would die early in embryogenesis, but in this experiment they were rescued by *Lhr*. However, the *D. melanogaster* females carried a small deficiency somewhere on their autosomes. Thus, the F_1 males carried only *D. simulans*

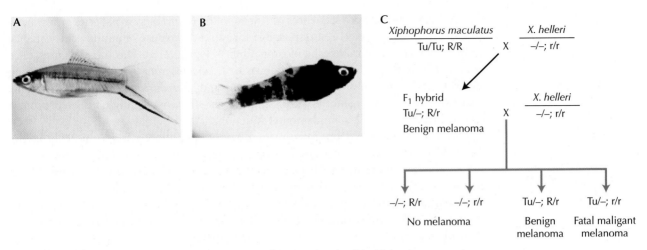

FIGURE 22.20. Genetics of hybrid inviability in crosses between the platyfish, *Xiphophorus maculatus*, and the swordtail, *Xiphophorus helleri* (shown in *A*). The extreme melanomas seen in F₂ fish and in the backcross to *X. helleri* (shown in *B*) are due to an interaction between two genes (*C*).

alleles in the region of genome uncovered by this deficiency and only *D. melanogaster* alleles on their single X chromosomes. A recessive incompatibility between the *D. melanogaster* X and this small region of *D. simulans* autosome would be revealed by the death of the F₁ males. A survey of approximately 7% of the genome revealed 20 small regions that contain incompatible alleles. The precise gene involved in one of these was determined using a **complementation test** (Box 13.1). Instead of using deficiencies, point mutations in candidate genes were used to uncover the exact location of the incompatibility. Any mutation that caused a loss of function of the *D. melanogaster* allele would unmask an incompatibility caused by the homologous *D. simulans* allele (Fig. 22.21A). The incompatibility was found to be in the amino terminus of *Nup96*, a structural component of the nuclear pore complex, which is responsible for transporting messenger RNA (mRNA) from nucleus to cytoplasm.

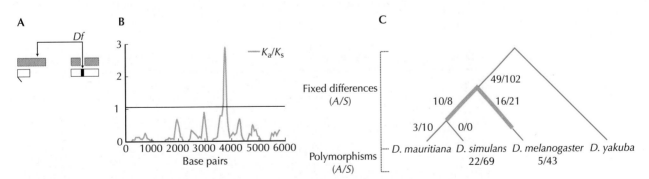

FIGURE 22.21. Genetics of incompatibility between *Drosophila simulans* and *Drosophila melanogaster*. (*A*) Hybrids are produced that carry an X chromosome from *D. melanogaster* (*red, left*) and a deficiency in a small region of an autosome (*red, right*). Any recessive allele in the corresponding region of the *D. simulans* genome (*black bar*) that interacts with recessive alleles on the *D. melanogaster* X will be unmasked by this deficiency, and these hybrid male genotypes will die. (*B*) The *sharp peak* shows an excess of amino acid substitution in the region of *Nup96* containing the incompatibility. The *red curve* shows the ratio of amino acid replacements relative to synonymous substitutions (K_a/K_s). (*C*) Comparisons between species show that the amino acid changes occurred in the lineage connecting *D. melanogaster* with *D. simulans* and *Drosophila mauritiana* (*heavy lines*), but before the latter two diverged. The figures give the numbers of replacement/synonymous substitutions (*A/S*) in *Nup96* for each branch of the phylogeny.

Comparisons between sequences of *Nup96* in the *D. melanogaster* species group show that natural selection has been responsible for rapid divergence of amino acid sequence in just the region that carries the incompatibility (Fig. 22.21B). This burst of divergence occurred in both the lineage leading down to *D. melanogaster* and the lineage leading down to the common ancestor of *D. simulans* and *D. mauritiana* (Fig. 22.21C). This selectively driven divergence is surprising for an otherwise highly conserved protein that is involved in such a fundamental function.

We can expect rapid progress in identifying "speciation genes," especially in the *D. melanogaster* group of species. Moreover, the methods described in Chapter 19 will tell us whether divergence has been driven by selection, as in the *Nup96* example. What is already clear, however, is that these "speciation genes" are not genes *for* reproductive isolation. Their effects on hybrids are unpredictable side effects of divergence, and their normal function (control of pigmentation or mRNA transport) has nothing to do with speciation.

Haldane's Rule Tells Us about the Genetic Basis of Species Differences

In 1922, J.B.S. Haldane wrote that:

> When in the offspring of two different animal taxa one sex is absent, rare or sterile, that sex is the heterozygous sex.

By "heterozygous," Haldane meant the sex that has heterozygous sex chromosomes. In mammals, and in many insect groups such as *Drosophila*, males are heterozygous (XY) for the sex chromosomes, whereas females carry two X chromosomes. Lepidoptera (butterflies and moths) and birds show the opposite pattern: The female genotype is denoted ZW and the male genotype ZZ. Thus, **Haldane's rule** states that in mammals and in *Drosophila*, F_1 hybrid males tend to be inviable or sterile, whereas in groups such as birds and butterflies, F_1 hybrid females tend to suffer more than males.

This is a remarkably strong and consistent pattern. It is obeyed in the great majority of those cases where one sex of hybrids suffers more than the other (Table 22.2). Moreover, such asymmetries are common in the early steps of speciation; the heterozygous sex loses fitness well before the homozygous sex. For example, in the survey of *Drosophila* species crosses (Fig. 22.11), pairs for which sterility or inviability appear only in male hybrids have Nei's genetic distance $D \sim 0.26$, whereas pairs in which

TABLE 22.2. Haldane's rule

Group	Phenotype	Asymmetric Hybridizations	Number Obeying Haldane's Rule
Heterogametic males			
Drosophila	Sterility	114	112
	Inviability	17	13
Mammals	Sterility	25	25
	Inviability	1	1
Heterogametic females			
Lepidoptera	Sterility	11	11
	Inviability	34	29
Birds	Sterility	23	21
	Inviability	30	30

From Table 1 of Orr H.A. 1997. *Annu. Rev. Ecol. Syst.* **28**: 195–218 (© Annual Reviews).

female hybrids suffer have an average $D \sim 0.88$. The large difference between these figures implies that as *Drosophila* species diverge, F_1 males become sterile and/or inviable relatively quickly; it takes much longer for F_1 females to begin to lose fitness. The strength of this effect can be appreciated by seeing that almost all the young species pairs in the lower left corner of Figure 22.11B represent cases where F_1 males die.

The likely explanation for Haldane's rule was first put forward by Muller, but has only recently been confirmed and clarified. It is now known as the **dominance theory**. The basic idea is that if incompatibilities are due to **recessive** alleles on the X chromosome, then these will be uncovered in the heterozygous sex. If one species carried an X-linked recessive allele *a* that causes problems when in an F_1 genetic background, this will be expressed in F_1 males, which carry only one copy. However, F_1 females have genotype *Aa*, and so will not suffer because the *a* allele will be masked. More precisely, if incompatibility alleles are on average recessive, then we will see Haldane's rule. (We assume that males have heterozygous sex chromosomes; obviously, a similar argument works when females are heterozygous.)

We have already seen direct evidence for this explanation from genetic analysis of species crosses in *Drosophila*. For example, small segments of the *D. mauritiana* genome introgressed into *D. simulans* have little effect when heterozygous, but frequently cause sterility or inviability when homozygous (see p. 634). Similarly, the screen for "speciation genes" shown in Figure 22.20A detected autosomal alleles that caused inviability or sterility when uncovered by a deficiency—just as the whole X chromosome is uncovered in male *Drosophila*.

Haldane's Rule Predicts the Large X Effect

The dominance theory neatly explains another striking pattern in the genetics of speciation. The X chromosome is involved in the incompatibilities responsible for Haldane's rule far more often than expected from its size alone. This **large X effect** has been known since the early work by Dobzhansky and Muller. Although most evidence comes from *Drosophila*, the sex chromosomes are disproportionately involved in postzygotic isolation in butterflies and mammals. Crucially, there is no evidence for any unusual role for the sex chromosomes in "ordinary" species differences or in quantitative genetic variation in general. The pattern is thought to reflect the uncovering of sex-linked recessive incompatibilities in the heterozygous sex, rather than any difference in the rate of evolution of the X itself.

In species where males have heterozygous sex chromosomes, another factor contributes to Haldane's rule: Male sterility evolves faster than female sterility. We have seen that genes involved in male reproduction evolve rapidly, as a result of sexual selection (e.g., Fig. 19.13). The importance of both faster male evolution and the dominance theory can be seen by contrasting two groups of mosquitoes. *Anopheles* mosquitoes have the same sex-determination system as in *Drosophila*, with a large X chromosome and a degenerate Y. In contrast, sex in *Aedes* mosquitoes is determined by a single locus (heterozygous in males, homozygous in females). The chromosomes carrying this locus are still called X and Y, but they are homologous, and both carry a full set of functioning genes. Thus, the dominance theory does not apply in *Aedes*, because a recessive incompatibility on the X chromosome would be complemented by the Y in males. However, sexual selection should act in the same way in both groups of mosquitoes. Therefore, a comparison between them can distinguish the contributions of sexual selection and of the dominance theory.

Table 22.3 shows that Haldane's rule is observed for sterility in *Aedes*. In 11 species crosses only males are sterile, whereas there are no cases where only females are sterile. In contrast, hybrid inviability is hardly ever seen in just one sex in *Aedes*, as expected. Neither sexual selection nor the dominance theory apply for *inviability* in species with

TABLE 22.3. The contributions of the dominance theory, and of faster male evolution due to sexual selection, can be distinguished by comparing two groups of mosquitoes

	Females Affected	Males Affected	Both Sexes Affected	Dominance Theory?	Faster Male Evolution?
Aedes (intact Y)					
Sterility	0	11	10	No	Yes
Inviability	1	1	11	No	No
Anopheles (degenerate Y)					
Sterility	0	56	20	Yes	Yes
Inviability	3	21	40	Yes	No

single-locus sex determination. In *Anopheles*, both factors contribute to hybrid sterility, and so male sterility is seen more often relative to both-sex sterility than in *Aedes* (56:20 vs. 11:10). Finally, *Anopheles* shows Haldane's rule for hybrid inviability, which can result only from the dominance theory (last row). This comparison shows that both the faster evolution of male sterility caused by sexual selection and the dominance theory contribute to Haldane's rule in mosquitoes. However, it is important to realize that sexual selection will act in the *opposite* direction to the dominance theory in groups where females have heterozygous sex chromosomes, such as birds and butterflies.

Haldane's rule and the large X effect provide strong evidence that the alleles responsible for postzygotic isolation are predominantly recessive. We do not at present understand *why* this should be. Deleterious mutations are usually recessive, because they cause a loss of function; however, it is not clear that the same explanation applies to incompatibility alleles, which are expressed only on a hybrid genetic background. We discuss the evolution of dominance in the next chapter (p. 690).

MECHANISMS OF SPECIATION

Natural Selection May Oppose the Evolution of Reproductive Isolation

It is relatively straightforward to find what kinds of genetic differences distinguish species and, in particular, what differences prevent successful interbreeding. The combination of new molecular tools with classical genetics is fueling rapid progress in this area. We now turn to a harder question: *How* does reproductive isolation evolve?

This question is hard to answer, simply because it concerns past events and processes rather than present-day differences. However, there is a particular difficulty in understanding the evolution of reproductive isolation as opposed to other past evolutionary events. As explained above, selection is expected to eliminate any allele that reduces hybrid fitness or makes interbreeding more difficult.

We have already seen several examples where selection opposes variants that cause some degree of reproductive isolation. For example, chromosome rearrangements cause meiotic problems when heterozygous, and, in **Müllerian mimicry**, patterns that warn of a butterfly's distastefulness are ineffective when rare. Such traits tend to be distributed in a mosaic, with selection against heterozygotes, recombinants, or rare alleles maintaining narrow **hybrid zones** that separate large, homogenous areas (Figs. 17.22, 18.16, and 18.17). In such examples, how do new chromosome arrangements or warning patterns ever get established?

The apparent difficulty in evolving reproductive isolation can be understood using Wright's metaphor of an **adaptive landscape** (Fig. 17.12). Populations are thought of as being at **adaptive peaks** of high mean fitness. Selection opposes any movement to new peaks, because this would involve crossing an adaptive valley of lower mean fitness. Species are seen as sets of coadapted genes, which have been selected to work well together, and which resist change—an idea emphasized by Dobzhansky and Mayr. The difficulty in seeing how reproductive isolation can evolve is essentially the same as the apparent difficulty in understanding the evolution of complex features, which require the coordinated action of many genes or traits (Chapter 24). We will first explain how **random drift** can act in opposition to selection to generate reproductive isolation. We then set out a different and, in many ways, simpler view, in which reproductive isolation is seen as the inevitable side effect of evolutionary change.

Random Drift Can Knock Populations from One Adaptive Peak to Another

In any population, allele frequencies fluctuate simply because the reproductive output of individuals is largely random (Chapter 15). We have seen that random drift can overcome selection and knock a population across an adaptive valley (pp. 494–496). For example, if heterozygotes for a chromosome rearrangement have their fertility reduced by a factor $1 - s$, then an intermediate population, which consists of 50% heterozygotes, will have mean fitness $\overline{W}_V / \overline{W}_P = 1 - s/2$. The chance that a chromosomal mutation will be fixed, taking the population to a new adaptive peak, is proportional to $(\overline{W}_V / \overline{W}_P)^{2N_e} = (1 - s/2)^{2N_e} \approx e^{-N_e s}$ (from Equation 18.1). A chromosomal mutation has an appreciable chance of being fixed, despite its selective disadvantage when rare, if $N_e s$ is less than about 10 (Fig. 22.22). For example, if $s = 1\%$, chromosomal evolution can occur through random drift in populations of 1000 or less.

A species cannot persist for long if it has such small numbers. Thus, random drift can effectively oppose moderate selection (e.g., $s \geq 1\%$) only within small local populations or when the entire species passes through a severe **bottleneck**. First, consider a species that is subdivided into local **demes**. A new chromosome arrangement

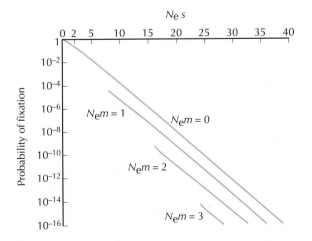

FIGURE 22.22. The probability of fixation of a chromosome rearrangement that reduces fitness by a factor $1 - s$, plotted against $N_e s$ on a logarithmic scale. Fixation becomes extremely unlikely when $N_e s$ is large. The lines for $N_e m = 1, 2, 3$ show how the chances of fixation of a new rearrangement are reduced if genes flow into the deme at a rate m per generation.

can arise by mutation and be established in one of these demes if $N_e s$ is small enough, and if not too many immigrants flow into the deme (Fig. 22.22). It can then spread if demes occasionally go extinct and are recolonized by neighboring individuals. By analogy with the **neutral theory** (see p. 425), the rate at which the entire species evolves is equal to the rate of change in any one deme. (If the new chromosome arrangement gives some advantage, either to the individual or to the deme that carries it, then the rate of spread can be considerably greater than this.) Observed rates of chromosomal evolution are consistent with this model, in which random drift within demes is followed by random extinction and recolonization. Here, reproductive isolation evolves as a side effect of Wright's **shifting balance** model of evolution (see pp. 607–609).

A basic problem with this scenario is that random drift cannot take populations across deep adaptive valleys; but on the other hand, shallow valleys cannot contribute much reproductive isolation. These opposing factors imply that isolation is most likely to be generated by weak incompatibilities, which reduce fitness by $s \approx 1/N_e$. Thus, isolation accumulates very slowly in drift-based theories.

What is the evidence that chromosomal evolution, and reproductive isolation more generally, are caused by random genetic drift? Across many groups of plants and animals, the rate of speciation correlates with the rate of chromosomal evolution (Fig. 22.23). For example, primates and horses have high rates of chromosomal change and have split into many species, whereas whales are chromosomally similar and are also much less speciose. Chromosomal change, however, does not directly cause speciation. Chromosomal differences make only a minor contribution to reproductive isolation (although speciation by polyploidy is an obvious exception). Rather, the correlations shown in Figure 22.23 have been thought to arise because a common factor—random drift—has caused both chromosomal evolution and speciation. However, this interpretation is not compelling. It is not clear that chromosomal evolution is in fact caused by drift, as in the model of Figure 22.22. Selection may well be responsible. Rearrangements may gain selective advantages (e.g., through **meiotic drive** or by changing genetic linkage [Chapter 23]) that outweigh their slight effects on fertility. Later in this chapter, we will see another way by which chromosomal rearrangements can facilitate speciation.

Reproductive Isolation Can Evolve without any Transient Reduction in Fitness

It is misleading to think of populations as being trapped on separate adaptive peaks and able to shift to new peaks only by random drift. Fitnesses fluctuate as the environment

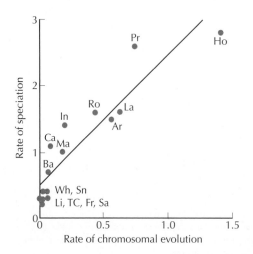

FIGURE 22.23. Correlation between rates of chromosomal evolution and speciation in vertebrates. Scales are per Myr. Ar, artiodactyls; Ba, bats; Ca, carnivores; Fr, frogs; Ho, horses; In, insectivores; La, lagomorphs; Li, lizards; Ma, marsupials; Pr, primates; Ro, rodents; Sa, salamanders; Sn, snakes; TC, turtles and crocodilians; Wh, whales.

changes and as other species evolve. If we picture the adaptive landscape itself as fluctuating, then it is easy to see how populations can move across it. More fundamentally, populations can evolve in many ways, which makes the usual one- or two-dimensional landscape (e.g., Fig. 17.12) inadequate. Even if we confine ourselves to point changes in coding sequence, there are three possible substitutions at each base, giving a total of approximately 10^8 possible single-step changes in the human coding sequence. The adaptive landscape has an enormous number of dimensions corresponding to the number of possible ways by which a population can alter its gene pool. If we imagine populations wandering through this complicated multidimensional space, it is clear that they will inevitably take different paths and eventually become incompatible.

Another way to understand this intuitive argument is to realize that only a small number of possible genotypes are ever tested by selection during evolution. Imagine two species that became separated approximately 1 Mya. Assuming 10,000 genes of 1000 base pairs each, and assuming a substitution rate of approximately 10^{-9} per base per year, we expect 20,000 differences between their coding sequences. (The factor of 2 arises because two lineages are diverging.) Now, if substitutions occur one after the other, then only about 20,000 different genotypes will have been produced at high frequency during divergence (Fig. 22.24). Yet, the F_1 genotype has never been produced, and in F_2 or backcross hybrids, an enormous number of new genotypes can be generated (~$3^{20,000}$ diploid genotypes). Because these have not been tested by natural selection, we expect them to be, on average, less fit than either parent species—in other words, we expect to see reproductive isolation.

Seen this way, it is remarkable that diverging genomes remain compatible for as long as they do. Individual genes can retain functional compatibility over much of evolutionary history. For example, human homologs can rescue yeast that lack essential cell cycle control genes. In another example, the gene *sonic hedgehog* plays the same role in organizing eye development in *Drosophila* and mouse (see Chapter 11 for other examples). At the level of the whole organism, taxa that have been diverging for millions of years can produce viable and fertile hybrids (see Fig. 22.11) even though their genomes have been extensively shuffled in F_1 meiosis.

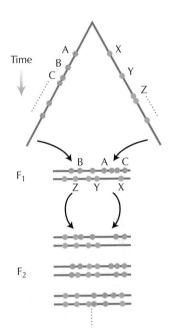

FIGURE 22.24. Two lineages diverge independently and accumulate substitutions at different loci (A, B, C, . . . in one, X, Y, Z, . . . in the other). The F_1 genotype is heterozygous for all these substitutions and has never been seen before. In the F_2, an enormous number of novel recombinant genotypes is generated; just three examples are shown here.

In the Dobzhansky–Muller Model, Speciation Is Not Opposed by Selection

The simplest genetic model to capture this argument is known as the **Dobzhansky–Muller model**. Initially, one locus carries allele a and a second locus carries allele b. In one lineage, allele A replaces a at the first locus, and in the other lineage, B replaces b. Individually, these new alleles cause no loss of fitness; they may give a slight advantage, so that they could be fixed by selection. When A and B meet for the first time in F_1 hybrids, they may prove incompatible. If the F_1 is completely inviable or sterile, then these two substitutions have caused the separation of two good biological species (Fig. 22.25).

The Dobzhansky–Muller model allows populations to diverge along a flat ridge in the adaptive landscape. This ridge surrounds a hole of low fitness into which hybrids fall (Fig. 22.25). The idea extends to large numbers of genes. Then, we imagine that diverging populations follow paths of high fitness, which are separated by regions of low fitness.

The analysis of hybrid sterility and viability in *Drosophila* fits with the Dobzhansky–Muller model, in that a few *specific* genes interact to cause severe loss of fitness. Moreover, these interactions are asymmetric. For example, in the platyfish *Xiphophorus*, the *Tu* and *r* alleles are incompatible because they cause malignant melanomas when brought together (Fig. 22.20). In contrast, the complementary recombinant *tu/R* is wild type, as predicted if this were the ancestral genotype under the Dobzhansky–Muller model.

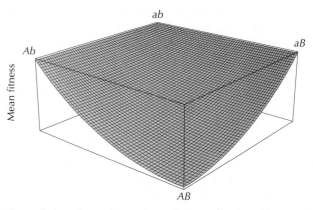

FIGURE 22.25. In the Dobzhansky–Muller model, an ancestral aabb population can substitute either allele A at one locus or B at another locus without any loss in fitness. However, if any genotype carrying both A and B dies or is sterile (i.e., has zero fitness), then there will be complete reproductive isolation. The adaptive landscape shows the mean fitness of a random-mating population as a function of allele frequencies at the two loci. The ancestral population fixed for a and b can evolve along ridges of high fitness, but when the derived populations (*Ab, aB*) cross, fitness is lost.

Chromosomal fusions provide a concrete case where strong reproductive isolation can evolve without ever being expressed during evolution. For example, the shrew *Sorex araneus* is divided into several chromosome races across Europe, each carrying different combinations of fusions between chromosome arms. F_1 hybrids between these chromosomal races are largely sterile, because multiple fusions pair up at meiosis (Fig. 22.26A). However, fusions could readily evolve from an ancestor with no fusions, because the heterozygotes would pair readily at meiosis, and so do not suffer much fitness loss. Indeed, this presumed ancestral genotype is found within hybrid populations, where it is selectively favored because it reduces the level of incompatibility (Fig. 22.26B).

THE GEOGRAPHY OF SPECIATION

Speciation Does Not Require Geographic Barriers

Historically, most arguments about speciation have focused on its geography. We can contrast three models (Fig. 22.27). So far, we have discussed the simplest: **allopatric speciation**, in which two populations diverge independently in complete geographic isolation. **Parapatric speciation** occurs when divergence occurs across a broadly continuous distribution, with diverging populations becoming separated by a series of clines. Finally, **sympatric speciation** occurs when a single randomly mating population splits into reproductively isolated groups, with no geographic isolation whatever.

Of course, speciation will in reality involve some mixture of these three idealized possibilities. The central question is the extent to which gene flow hinders divergence. Is strict geographic isolation required to allow divergence, or can species form even while the diverging populations continue to exchange genes?

In his early notebooks, Darwin emphasized geographic isolation as a cause of divergence. The importance of allopatric speciation was argued by Moritz Wagner in the mid-19th century, by David Starr Jordan at the turn of the 20th century, and by Mayr from the time of the Evolutionary Synthesis up to the present day. The key observation that supported this widely held view was that sister species often live in dif-

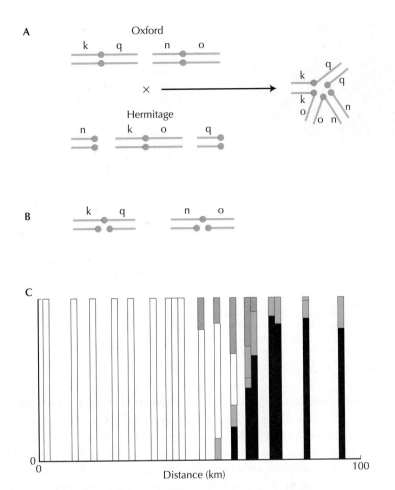

FIGURE 22.26. (*A*) The "Oxford" and "Hermitage" races of the shrew *Sorex araneus* differ by several chromosomal fusions: In the Oxford race, chromosome arm k is fused with q, and n with o, whereas in the Hermitage race, k is fused with o. (The dots represent the **centromere**.) In the meiosis of F_1 hybrids, the chromosome arms pair, which produces a tangle of five chromosomes. These fail to segregate properly, and so the F_1 is sterile. (*B*) In contrast, heterozygotes between **acrocentric** chromosomes and fused chromosomes pair to form simpler configurations that segregate correctly at meiosis. (*C*) Where the two chromosome races meet, few sterile individuals are found, because acrocentric chromosomes k and o have risen to high frequency (*red bars*). Thus, individuals tend to be heterozygous for simple combinations, which have high fertility (e.g., *B*). *White*, homozygous "Oxford" genotypes; *black*, homozygous "Hermitage" genotypes; *red*, homozygotes acrocentrics; *blue*, other hybrid genotypes.

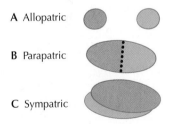

FIGURE 22.27. (*A*) In allopatric speciation, two populations are separated by some geographic barrier; they therefore diverge in complete independence. (*B*) In parapatric speciation, divergence occurs within a broadly continuous habitat; the diverging populations are separated by a set of clines (*dotted line*). (*C*) In sympatric speciation, there is no spatial separation whatever: nonrandom mating arises solely from genetic differences and not from spatial separation. These three cases represent extremes: in reality, speciation involves elements of each. The geographic pattern may change through time and may differ between genes and traits.

ferent places, separated by a geographical barrier (such as a river, mountain range, or ocean). This pattern suggests that divergence requires, or at least is encouraged by, some extrinsic barrier to gene flow.

We can quantify the association between geographic barriers and divergence. For example, fence lizards living on opposite sides of rivers differ more morphologically than those living on the same side, for a given distance between them (Fig. 22.28). In many groups, even the most closely related species tend to have an allopatric distribution, but overall the pattern is not as clear as was believed by earlier naturalists.

Yet, these observations apparently conflict with the examples we saw in Chapter 18, where divergence takes place despite gene flow. The processes responsible for divergence—selection and random drift—should be effective in the presence of some gene flow, especially if the population is spread over a broad area. We have seen that selection can maintain different alleles in different places, provided that it is sustained over dis-

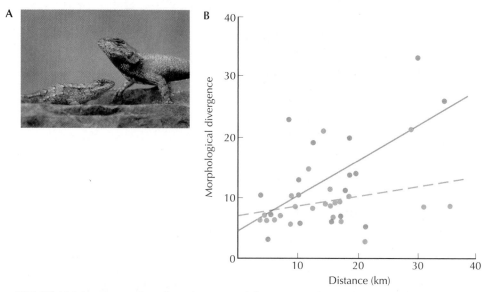

FIGURE 22.28. (*A*) Fence lizard (*Sceloporus undulatus*). (*B*) In the southeastern United States, populations that are separated by rivers (*red*) differ in morphology more than do populations separated by the same geographic distance, but with no barrier to movement (*blue)*.

tances larger than a characteristic scale, $l = \sigma/\sqrt{s}$ (where σ is the dispersal distance, and s is the strength of selection [see p. 504]). More directly, there are many examples where selection maintains clines that are far narrower than the species range (e.g., Figs. 18.14–18.17). Random drift can also cause fluctuations in allele frequency despite gene flow (see pp. 444–447). We saw earlier in this chapter how new chromosome rearrangements can be established by drift in a local deme and then spread out.

The Dobzhansky–Muller model of divergence is perhaps most sensitive to gene flow. Even here, however, incompatibilities can build up if the species range is large enough. Different favorable alleles will then be established in different places. (For example, think of the various ways in which human populations have evolved resistance to malaria [Fig. 18.4].) If these different alleles turn out to be incompatible with each other, they will remain separated by a narrow cline; over time, further differences will build up. (The chromosome races of shrews may be an example of this process.) In this scenario, barriers to gene flow will facilitate divergence, but are not essential for it.

It is difficult to distinguish whether populations diverged in allopatry, with no exchange of genes between them, or instead diverged in parapatry, while connected by gene flow. There has been a good deal of argument about whether present-day clines evolved in situ, with no geographic separation, or whether they are caused by **secondary contact** between divergent populations. There are clear examples of evolution in situ—for example, the local adaptation of the grass *Agrostis tenuis* to heavy metal pollution (Fig. 18.14). However, where many different clines coincide in a hybrid zone, it is likely that they were all brought together by secondary contact (as with red deer and sika deer [Fig. 16.19]). Indeed, in northern latitudes, hybrid zones must have formed after the last glaciation, less than 10,000 years ago. However, differences must have been evolving for far longer than this, and so the present distribution tells us little about how those differences originated. Most evolution must occur within abundant species, which are spread over wide areas, and so we expect that most divergence originally occurred in parapatry. The correlation that we now see between genetic divergence and the presence of barriers reflects the shuffling of species ranges over long periods of time and may not tell us much about the actual origin of the divergences (Fig. 22.29).

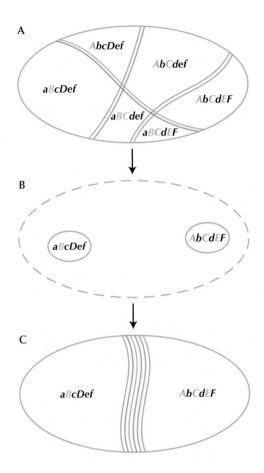

FIGURE 22.29. Changes in a species' range can bring together differences that evolved independently, in parapatry. (*A*) If a species is distributed across a broad area, divergence will build up. In this example, allele A is incompatible with allele B, C is incompatible with D, and E is incompatible with F; however, there is no interaction between these separate incompatibility systems, and so they may form clines in different places. If the species' range is reduced to two small refugia (*B*), which happen to contain gene combinations aBcDef and AbCdEF, then when the range expands again (*C*), all these independently evolved differences will be brought together in a single hybrid zone.

Genealogies and Chromosomal Rearrangements Give Evidence for Gene Flow during Speciation

Recent studies of sunflowers and *Drosophila* give strong, albeit indirect, evidence that barriers to gene flow do facilitate divergence, but that reproductive isolation has nevertheless evolved in the presence of gene flow.

D. pseudoobscura and *D. persimilis* are found together along the west coast of the United States. Although they are distinct species, they do occasionally hybridize. They differ by chromosomal inversions that cause little or no sterility by themselves, but that effectively prevent recombination in hybrids. Factors involved in many aspects of reproductive isolation have been mapped in crosses between *D. pseudoobscura* and *D. persimilis*. These include cuticular hydrocarbons and male wing vibrations involved in courtship, female mating preferences, hybrid inviability, and hybrid sterility. Remarkably, all these differences map to genes in or near the regions that differ in chromosome arrangement (Fig. 22.30A). The ancestry of genes in or near these regions follows the taxonomy, but the history of genes in the rest of the genome does not correspond to the species classification (Fig. 22.30B). (As we saw in Chapter 16 [p. 444], only a low rate of gene exchange is needed to shuffle genealogies in this way—roughly, one or more successful hybrids per generation.)

This study shows that genes for reproductive isolation are concentrated in regions with chromosomal differences, and that these effectively prevent gene exchange. In this example, the chromosomal inversions hold together multiple genetic differences. This allows distinct species to be maintained in sympatry, with different behaviors and ecological niches, despite gene flow across much of the genome. (At the end of this chapter, we will see an example of the same pattern in another genus of fly, *Rhagoletis*.) In-

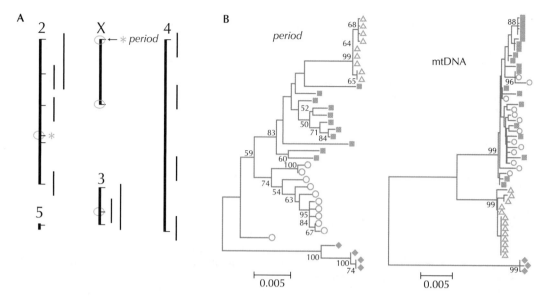

FIGURE 22.30. (*A*) All the QTLs responsible for reproductive isolation between *Drosophila pseudoobscura* and *Drosophila persimilis* map to two blocks of genome, marked by *red asterisks*. *Circles* show inversion differences fixed between the species. The position of the gene *period* is indicated. *Bars* to the *right* of the chromosomes indicate segments of genome that are known to have introgressed between the species. (*B*) Genealogies on the X (e.g., *period*) follow the species classification, with *D. persimilis* and *D. pseudoobscura* sequences forming separate groups. However, the nonrecombining locus (mitochondrial DNA) shows evidence of recent gene flow: The two species' sequences are intermingled on the genealogy and closely related. Other autosomal loci, outside the inverted regions, show evidence of recombination as well as gene flow. The scale represents nucleotide divergence per base. *D. pseudoobscura*, filled squares; *D. persimilis*, open circles; *Drosophila pseudoobscura bogotana*, open triangles; *Drosophila miranda*, filled diamonds.

deed, *Drosophila* species pairs that do not differ chromosomally are almost never found in the same place, whereas those with chromosomal differences often are.

The direct effects of chromosomal differences on meiosis contribute relatively little to reproductive isolation (see p. 642). Yet, chromosomal divergence is broadly correlated with the rate of speciation (Fig. 22.23), and in the *Drosophila* example, genes for reproductive isolation are found in rearranged regions of the genome. This may be because chromosomal differences greatly reduce recombination and so encourage divergence. In the previous section, we discussed one way in which this might happen. Favorable alleles located within the rearranged region would be trapped by the rearrangement for long enough for another incompatible allele to arise elsewhere. As multiple incompatibilities become associated with the rearrangement, its effect as a barrier would be strengthened, eventually leading to strong reproductive isolation.

In general, the evolution of biological species involves a reduction in recombination between diverging populations, which in turn accelerates divergence. In the next chapter, we will look at the converse factors, which maintain high levels of recombination within sexual species.

Selection against Crossbreeding Can Reinforce Reproductive Isolation

So far, we have seen reproductive isolation as a side effect of divergence that has occurred for other reasons, including local adaptation, sexual selection, random drift, or accumulation of favorable alleles that turn out to be incompatible. In principle, however, selection can directly favor stronger reproductive isolation. Individuals that choose to mate with others from their own population will pass on more of their genes because they will waste less reproductive effort on unfit hybrids. This process is known as **reinforcement**. It is important to realize that selection can favor prezygotic, but not postzygotic, isolation. A gene that further reduces the fitness of hybrids must be deleterious. (This argument assumes that offspring do not compete for resources. If they do, early death of a hybrid individual that has poor future prospects may be favored by **kin selection** [see pp. 601–603], because its siblings may then do better.)

Pied flycatchers (*Ficedula hypoleuca*) and black and white collared flycatchers (*Ficedula albicollis*) occur together across central and eastern Europe. They hybridize, although at a lower rate than would be expected if mating were random (2.6% of heterospecific matings were observed, compared to an expectation of 13.8%). There is strong postzygotic isolation—nearly three-quarters of eggs fail to hatch when one or both parents is hybrid, compared to approximately 5% otherwise. As one might expect for a bright plumage trait (Chapter 20), female collared flycatchers prefer males with black and white plumage, as do female pied flycatchers from outside the region of overlap (Fig. 22.31A). Correspondingly, males of both species are black and white outside the overlap region. However, male pied flycatchers in the overlap region have brown female-like plumage, and female pied flycatchers from that area prefer these duller males. This pattern strongly suggests that the duller male plumage and the female preference for it have evolved through selection to reduce crossbreeding. There

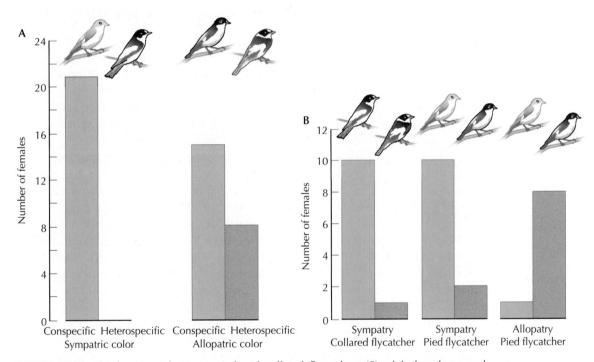

FIGURE 22.31. Reinforcement between pied and collared flycatchers (*Ficedula hypoleuca* and *Ficedula albicollis*). (A) Male pied and collared flycatchers from allopatric populations have similar black and white plumage (*top right*). However, where the ranges of the two species overlap, male pied flycatchers have a dull brown plumage, and male collared flycatchers have a more exaggerated white patch (*top left*). The greater difference in plumage between sympatric males reduces hybridization: Females of both species, taken from sympatric populations, always choose the conspecific male when given a choice between sympatric males (*left*), but often fail to distinguish when given a choice between allopatric males (*right*). (B) Female preferences have also diverged between allopatric and sympatric populations. Female collared flycatchers from the region of sympatry prefer males with larger patches of white (*left*), and female pied flycatchers from the region of sympatry prefer the dull plumage typical of males from that area (*middle*). However, female pied flycatchers from allopatric populations have the opposite preference, choosing black and white males over the duller males found in sympatry (*right*).

are several other such cases. For example, *D. pseudoobscura* and *D. persimilis* from sympatric populations show much stronger assortative mating in laboratory tests than do flies from allopatric populations.

Broader evidence for reinforcement comes from the observation that genetically similar *Drosophila* species pairs show much stronger prezygotic isolation when they are found together, in sympatry, than when they do not meet in nature. Because the strength of prezygotic isolation hardly overlaps between genetically similar sympatric and allopatric pairs (left of Fig. 22.11A), it is hard to argue that this is a "sieving" effect, with only those pairs that have strong prezygotic isolation able to live together.

These examples of reinforcement all involve species that overlap over large areas and are already strongly reproductively isolated. Here, once broad sympatry was achieved, selection for reinforcement would act over a large area and could be effective. However, it is much less clear that reinforcement could be effective when prezygotic isolation and ecological divergence are weak, so that the incipient species meet in a narrow set of clines. Then, selection for assortative mating is confined to a small region. The best example where reinforcement seems to have occurred in a narrow cline comes from *A. tenuis*, which has adapted to heavy metal pollution (Fig. 18.14). There, flowering time has diverged between plants at the edge of polluted areas, which reduces maladaptive gene flow; the rate of self-fertilization has also increased. However, there are few other examples where reinforcement has acted within narrow clines.

A Single Population Can Split into Two: Sympatric Speciation

Can species form even when there is no barrier whatever to gene flow? In other words, can a single randomly mating population split into two reproductively isolated gene pools? The possibility of such sympatric speciation has been suggested by examples where large numbers of species have radiated apparently from a single ancestral population and within a restricted geographic area. The best known example is the dramatic radiation of the cichlid fishes in many African lakes.

In each of the three great lakes (Malawi, Tanganyika, and Victoria; Fig. 22.32), species have evolved that vary widely in mating system, morphology, and feeding behavior. Some species pluck scales from other fish; some feed on large invertebrates; some comb algae for food; and some prey on other fish by ambush or chase. Lake Tanganyika is 9–12 Myr old and contains approximately 200 endemic cichlid species. Lake Malawi is 4–9 Myr old but is thought to have been colonized by cichlids only about 0.7 Mya; it now has more than 400 species. Lake Victoria is younger and shallower than the other two lakes (about 0.5 Myr old and less than 80 m deep). Nevertheless, it contains more than 300 endemic species. Each of the three species flocks originated separately from river-dwelling ancestors. Thus, these dramatic radiations into diverse ecological niches have occurred in parallel, within relatively short periods of time.

Did this dramatic diversification occur in allopatry, parapatry, or sympatry? In the larger lakes, fluctuations in lake level have separated different populations and thus provided the opportunity for allopatric speciation. In addition, the shore-dwelling species have limited dispersal and show considerable genetic divergence over short distances.

The best case for sympatric speciation comes from small lakes in West Africa. Lakes Barombi Mbo and Bermin in Cameroon are small (4.2 and 0.6 km², respectively) and have formed in homogeneous, conical craters of old volcanoes, which are now completely isolated from the surrounding river system. Thus, fluctuations in lake level would not have led to any splitting of populations. Yet, these lakes contain many cichlid species (11 and 9, respectively). These species are monophyletic, and the molecular clock suggests that they originated approximately 10,000 years ago. Interestingly, the deepest split in each case is between pelagic species that feed

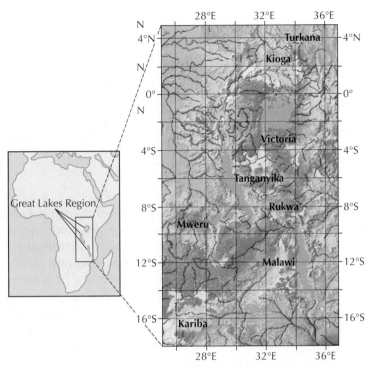

FIGURE 22.32. Locations of the African great lakes. *Light shading* on the main map indicates higher altitude.

on plankton and those that feed on the substrate. Most striking is Lake Ejagham, which is also small (~0.5 km²) and homogeneous and in which five morphs of ci-chlid can be distinguished. Two have been studied in detail and are associated with different microhabitats (Fig. 22.33). Although they do interbreed, there is assorta-tive mating by size (Fig. 22.34), which suggests incipient speciation in sympatry.

We have seen that two types can coexist provided that they exploit different resources. The rarer type then gains an advantage, because its own resource is used less; this generates negative frequency-dependent selection, which maintains a bal-anced equilibrium (pp. 505–508; Box 22.1). If the two types interbreed to produce in-termediate and maladapted offspring, then selection will act to reduce cross-mating. Thus, sympatric speciation combines selection for ecological diversification with rein-forcement of prezygotic isolation.

FIGURE 22.33. (*A–C*) The cichlid *Tilapia cf. deckerti* from Lake Ejagham, Cameroon, contains two sympatric forms: "little black," associated with shallow water, and "large black," associated with deep water. (*A*) Breeding pair of "little black" in shallow water. (*B*) Breeding female of the "large black" form, in front of a log hole (its typical breeding site). (*C*) Juvenile swarm, showing the long black mark at the base of the dorsal fin that distinguishes *T. cf. deckerti* from other cichlid species in this lake.

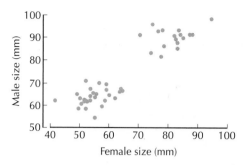

FIGURE 22.34. Assortative mating by size separates the two forms of *Tilapia cf. deckerti* in Lake Ejagham.

This scenario presents no difficulties once interbreeding is infrequent—for example, when two populations that have somewhat diverged meet each other. However, it is hard to see how a random mating population could maintain two distinct types when there is strong selection against intermediates between them (Box 22.1). This is because the *negative* frequency dependence due to use of different resources is opposed by *positive* frequency dependence arising from selection against intermediates. The rarer type will suffer more by being broken down through hybridization. A further difficulty is that if different sets of genes are needed to adapt to different resources and to choose mates, then these must become associated with each other (i.e., they must be in **linkage disequilibrium**) in order for the production of mal-

Box 22.1   Maintenance of Variation by Disruptive Selection

Disruptive selection on a quantitative trait can maintain variation in a sexual population, provided that there is competition between similar phenotypes. Suppose that resources are distributed along a continuous axis—for example, this might represent seed size. Resources are clustered around two peaks (small and large seeds, say). Individuals vary for a quantitative trait, so that individuals with a certain trait value are fittest when the corresponding resource is abundant (the trait might represent beak size, say, and birds with medium-sized beaks do best when medium-sized seeds are abundant). An asexual population will reach an equilibrium in which all phenotypes are equally fit, because the abundance of different individuals perfectly matches the abundance of the different resources (Fig. 22.35A). A sexual population may reach an equilibrium with a high genetic variance, which allows it to exploit both resources. However, quantitative traits in sexual populations have to follow a more or less normal distribution (see p. 385), so that many individuals will have poorly adapted intermediate phenotypes. Selection will favor assortative mating and, ultimately, sympatric speciation (Fig. 22.35B). If the population has low genetic variance, it will shift to exploit one or the other resource peak. Individuals that match the other peak will have high fitness, because that resource is underexploited. However, these individuals will mate with the common type, produce unfit intermediate offspring, and be eliminated. Thus, it is difficult for a sexual population to sustain genetic variation under disruptive selection (Fig. 22.35B).

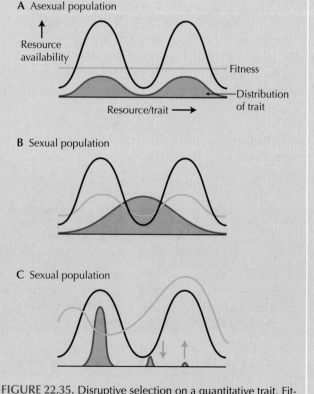

FIGURE 22.35. Disruptive selection on a quantitative trait. Fitness, *blue*; resource availability, *black*; trait distribution, *red*.

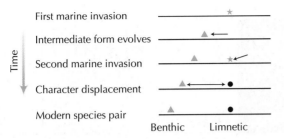

FIGURE 22.36. The benthic and limnetic forms of the stickleback *Gasterosteus aculeatus* each grow best in their own environment—benthic near the shore and limnetic in open water. Hybrids grow more slowly than the mean of the two forms and so do worse on average. The postglacial lakes of British Columbia were first invaded by sticklebacks from the sea, which have a limnetic-like diet. These invaders evolved an intermediate morphology—as is seen today in those lakes with a single morph. Subsequent invaders from the sea were to some extent reproductively isolated and evolved as limnetic specialists; the fish already present specialized as benthics.

adapted hybrids to be reduced. Unless the genes involved happen to be tightly linked (e.g., within chromosomal inversions), such associations will be broken down by recombination.

A good example where selection maintains different phenotypes despite interbreeding to produce unfit hybrids comes from the sticklebacks (*Gasterosteus aculeatus*) found in the lakes of British Columbia (see pp. 307–313). These lakes formed after the glaciers retreated less than 10,000 years ago. In many lakes, **benthic** and **limnetic** morphs are found, which tend to feed in shallow versus deep water, respectively. They differ in morphology, feeding behavior, and mate preference, but frequently hybridize in nature; hybrids have been shown to be less fit. However, although this example is consistent with the later stages of sympatric speciation, genealogies determined from nuclear genes show that the divergence did not arise in sympatry. Instead, there were multiple invasions of these lakes by the ancestral marine population. The first fish to establish in the lakes occupied a broad niche; fish that invaded later specialized in limnetic feeding, and the residents shifted from an intermediate form to specialize as the benthic form (Fig. 22.36).

Host Races Provide a Classic Example of Incipient Speciation

Many insects have diversified into **host races**, specialized to feed on different plant species. These have been taken as classic examples of sympatric speciation (indeed, the idea was first proposed in 1864). Multiple adaptations are needed to exploit different host plants, which leads to selection against intermediates. Moreover, if the insects mate on their preferred host, then assortative mating follows directly as a consequence of host choice. This makes it much easier to see how divergence and assortment could occur in sympatry.

The best-studied example is the apple maggot *Rhagoletis pomonella*. Flies of this species are found in Mexico and across much of the central and eastern United States, where they feed on hawthorn (Fig. 22.37A,B). Adults mate at the host plant and lay eggs on the young fruit. Larvae develop in the fruit, pupate in the soil, and then enter a winter **diapause**, emerging as adults in the following spring. Apples were introduced to North America by European settlers; about 150 years ago, *R. pomonella* began to exploit this new host species. Now, two host races (apple and hawthorn) coexist in the northeastern United States. These differ in host preference and in the time when they emerge as adults (apples flower and fruit several weeks earlier than hawthorn) (Fig. 22.37C). This causes some temporal isolation, because members of the different races tend not to be ready to mate at the same time. In addition, a behavioral preference

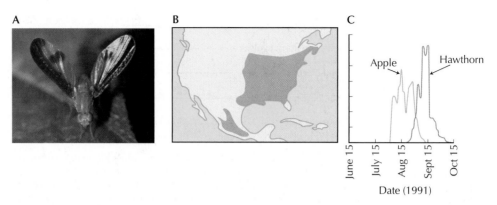

FIGURE 22.37. The apple maggot *Rhagoletis pomonella* (A) is distributed across the United States and Mexico (B). In the northeastern United States, two host races are found, which specialize on apple and hawthorn. Larvae emerge 3–4 weeks earlier from apple than from hawthorn (C), and so experience a warmer prewinter period and must diapause over winter as pupae for longer.

for different hosts reduces interbreeding between the races to approximately 5%. Finally, flies that develop on the wrong host plant may fail to enter diapause appropriately, thus causing severe mortality.

The two races differ in allele frequency at six allozyme loci, which map to three regions of the genome. This is surprising, because gene flow between the host races should rapidly eliminate differences, unless the allozymes are themselves under direct selection. This puzzle has recently been resolved by the discovery that the races differ by three inversions, which prevent recombination across roughly half the genome. The allozyme markers are thus linked to blocks of selected genes, and strong selection can act on each of the three linked blocks. The inversions found in the apple race derive from Mexico and may have been preadapted to early-fruiting apples by their southern origin. Perhaps, one of the inversions found in Mexico invaded the northeastern population by exploiting the new niche created by cultivation of apples. This initial polymorphism could then be strengthened by evolution of genes within the inversion to adapt to the new host and by recruitment of two further inversions to give the present polymorphism. This scenario includes elements of both parapatric and sympatric divergence.

The association between genes for reproductive isolation and inversions is remarkably similar to the *D. pseudoobscura/D. persimilis* example discussed above. In both cases, the suppression of recombination by the inversion facilitates the evolution of distinct combinations of alleles adapted to different niches.

In general, it is extremely difficult to show that divergence was truly sympatric, rather than involving some transient isolation of a local population or occurring within a broadly continuous distribution. The striking radiation of the cichlid fishes in the large East African lakes is likely to have involved population subdivision, at least of the shore-dwelling species. The most plausible case of sympatric speciation comes from the cichlid fishes of isolated crater lakes in Cameroon (Fig. 22.33). Here, the small size and uniformity of the lakes makes geographic isolation implausible.

SUMMARY

The continuity of evolution makes it hard to draw clear boundaries around species. Nevertheless, organisms do cluster into distinct forms, because of both their genetic relationship and the discrete ecological niches to which they adapt. Asexual organisms must be grouped into "species" according to this phenotypic clustering. However, with sexual reproduction, the clearest definition is the biological species, which corresponds to a separate gene pool.

Species can form in a single event that produces a parthenogenetic clone or a polyploid species. However, in the great majority of cases reproductive isolation accumulates slowly. It is caused by alleles that cause problems when introduced into a foreign genetic background, as a side effect of divergence. These incompatibilities are usually recessive, so they tend to be revealed when present in a single copy on a sex chromosome.

At first sight, it would seem that natural selection should oppose the evolution of reproductive isolation, so that random drift is necessary for speciation. However, speciation can occur without any transient reduction in fitness, as explained by the Dobzhansky–Muller model. If divergence is driven by positive selection, then it can occur despite gene flow, in parapatry.

In this chapter, we have been examining the relation between two kinds of diversity: genetic variation within species and the diversity of species that coexist in any one place. Both kinds of variation depend on the existence of a range of different ecological resources. Frequency-dependent selection can maintain polymorphism if genotypes exploit different resources, and it can allow the coexistence of different species by reducing competition between them. We have seen that, in principle, selection can reduce interbreeding between genotypes that exploit different resources, which leads to sympatric speciation. This possibility poses a major open question for evolutionary biology. Is speciation for the most part an arbitrary side effect of divergence that occurs for other reasons, or is it instead driven by selection, as an adaptation to reduce the disruption of well-adapted sets of genes by recombination? In the next chapter, we will look at the converse problem: Why is it that so many species maintain high levels of sex and recombination, even though this prevents them from adapting efficiently to diverse ecological niches?

FURTHER READING

Coyne J.A. and Orr H.A. 2004. *Speciation*. Sinauer Press, Sunderland, Massachussetts.

An excellent and up-to-date overview of the whole subject.

Howard D.J. and Berlocher S.H. 1997. *Endless forms: Species and speciation*. Oxford University Press, Oxford.

A collection of articles that show the diversity of views on the origin of species.

Orr H.A. 1997. Haldane's rule. *Annu. Rev. Ecol. Syst.* **28:** 195–218.

An excellent review, written soon after the basic explanation of Haldane's rule had become clear.

Orr H.A. and Presgraves D.C. 2000. Speciation by postzygotic isolation: Forces, genes and molecules. *BioEssays* **22:** 1085–1094.

A review of the genetic basis of reproductive isolation; complements the review of other kinds of species difference cited below (*Trends Ecol. Evol.* 16(7)). More recent references are given in the online notes.

Schluter D. 1996. Ecological causes of adaptive radiation. *Am. Nat.* **148:** S40–S64; Schluter D. 2000. *The ecology of adaptive radiation*. Oxford University Press, Oxford.

Schluter discusses cases of adaptive radiation, where many species have formed to fill diverse ecological niches.

Trends Ecol. Evol. 2001. **16(7):** 325–413.

A special issue devoted to speciation. Includes a discussion of species concepts by Hey, a review of the genetics of species differences by Orr, a review of the theory by Turelli et al., and an argument for the importance of ecological factors by Schluter.

The Cichlid System

Danley P.D. and Kocher T.D. 2001. Speciation in rapidly diverging systems: Lessons from Lake Malawi. *Mol. Ecol.* **10:** 1075–1086; Kornfield I. and Smith P.F. 2000. African cichlid fishes: Model systems for evolutionary biology. *Annu. Rev. Ecol. Syst.* **31:** 163–196.

Good recent reviews of the cichlid system.

Goldschmidt T. 1998. *Darwin's dreampond: Drama in Lake Victoria*. MIT Press, Cambridge, Massachusetts.

A nice popular account of the remarkable diversification of these fishes.

CHAPTER

23

Evolution of Genetic Systems

I N PART II OF THIS BOOK, WE TRACED THE HISTORY OF LIFE. We have seen how the first replicating molecules grouped together in chromosomes and cells; how a division of labor evolved between DNA molecules carrying genetic information and proteins, which are responsible for interpreting that information; how a symbiotic relationship led to eukaryotic cells, characterized by cooperation between organelles and by regular meiosis; and how multicellular organisms developed complex morphology and behavior, which allows diverse ways of life. Now, organisms such as ourselves have a sophisticated genetic system with many remarkable features: extraordinarily accurate replication, sexual union to produce new genotypes by recombination, fair segregation of alleles at meiosis, a largely diploid life cycle with separate sexes and diverse mechanisms that encourage **outcrossing**, and a robust developmental system that reliably generates complex organisms.

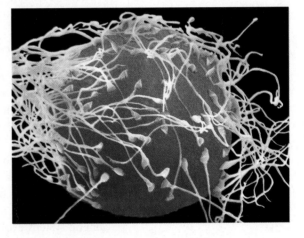

So far, we have mostly taken these features as given and worked out their consequences. Population genetics takes the rules of Mendelian inheritance and the fitness of different genotypes and uses them to work out how the composition of a population changes. For example, in Chapter 18, we saw how selection interacts with processes such as mutation and recombination to shape variation; in Chapter 20, we saw how sexual selection can lead to elaborate and apparently maladaptive behaviors; and in Chapter 22, we saw how populations diverge to become separate species. However, evolutionary biology goes beyond this and tries to understand the underlying genetic system itself. Why are the mechanisms of inheritance and development as they are?

STUDYING THE EVOLUTION OF GENETIC SYSTEMS

All Aspects of Genetic Systems Evolve

Every aspect of the **genetic system** raises an evolutionary question. What determines mutation rates? Why is sexual reproduction so widespread? Why are there separate sexes: males producing small sperm or pollen, females producing large eggs or ovules? What decides the structure of the networks that regulate gene expression? Much of evolutionary biology is now devoted to answering these and many similar questions.

657

However, investigating such questions is difficult. Many key features of the genetic system, such as meiosis and the genetic code, are unique, having evolved only once and in the distant past. This makes it almost impossible to use experimental or comparative methods to study them. Many aspects of the genetic system may be arbitrary. For example, Francis Crick argued that the genetic code is a "frozen accident," and exactly how a particular gene is regulated is largely fortuitous. Thus, it is hard to know how far we can go in seeking general explanations and how far the genetic system can be seen as well adapted rather than arbitrary.

Throughout the 20th century, it was widely accepted that genetic systems evolve "for the good of the species." Such arguments do have a part to play; we can understand the competition between asexual clones with different mutation rates, or between a sexual population and an asexual clone that arises from it, in terms of the average fitnesses of the competing populations. Also, **species selection** (see p. 609) does play some part in explaining the prevalence of sexual reproduction. However, such arguments cannot fully explain how features originated as variations between individuals within asexual populations or how they are now maintained. We know that most aspects of genetic systems (e.g., rates of mutation, recombination, and selfing) vary, and so we must explain how they can remain more or less stable despite that variation. It was the realization in the 1960s and 1970s (see p. 34), by John Maynard Smith, George Williams, and others, that arguments from **group selection** or species selection are inadequate that led to the present vigorous research in this area.

The Evolution of Genetic Systems Can Be Studied in Several Ways

Verbal arguments need to be laid out carefully when considering why some feature might be favored at the level of a group or individual. Often a mathematical model or computer simulation needs to be analyzed, in order to define the argument precisely and to work out its consequences. Theoretical arguments about the evolution of genetic systems go back to the early years of population genetics. Recently, much of the progress in the field has come from an understanding of how apparently diverse models relate to each other in a general way. As we show in this chapter, we now have a clear theoretical framework to guide us.

An important role for theory is to identify key quantities that influence how the genetic system evolves. For example, we will see that sexual reproduction can evolve because it aids the elimination of deleterious mutations, but only if the total rate of these mutations is sufficiently high. This has motivated many efforts to measure quantities such as the total rate of deleterious mutations, the additive genetic variance in fitness, and the rate of adaptive substitution (Chapter 19).

Although all organisms share the same basic genetic machinery, there is also extraordinary diversity. We will see that important evidence comes from comparisons between species and from research on organisms with unusual life cycles. Such comparisons have been made statistically rigorous with the aid of good phylogenies estimated from molecular data (Chapter 27 [online]). Moreover, with the recent availability of complete genome sequences, we are now able to make comparisons across different parts of the genome (e.g., between regions of high and low recombination; Fig. 19.21).

We can investigate the evolution of genetic systems experimentally, using rapidly reproducing organisms such as *Drosophila*, yeast, or *Escherichia coli* or even RNA molecules replicating in vitro (Fig. 17.2). We can either select directly on the genetic system (e.g., selecting for higher recombination rates), or we can examine the consequences of different rates of mutation, recombination, or sex. Although such experiments are confined to a few amenable organisms reared in the laboratory, they have at least confirmed the feasibility of several proposed theories.

In this chapter, we will illustrate all four approaches: theoretical, observational, comparative, and experimental. Indeed, we have already seen many examples of these approaches and have already discussed the evolution of the genetic code (Chapter 4), the evolution of mating preferences (Chapter 20), and the maintenance of a fair meiosis despite its exploitation by selfish genetic elements (Chapter 21). In the remainder of this chapter, we concentrate on three issues: the evolution of mutation rates, the evolution of sex and recombination, and the evolution of gene regulation and development. All three of these features of genetic systems influence the amount of genetic variation on which natural selection acts. Thus, throughout this chapter we will be asking how far the generation of genetic variation can be seen as an adaptation for facilitating the efficient action of selection.

EVOLUTION OF MUTATION RATES

The Cost of Preventing Harmful Mutations Drives the Evolution of Mutation Rates

Mutation is inevitable, because the genetic material cannot be replicated with perfect accuracy. However, mutation rates are under genetic control. We saw in Chapter 12 that there are elaborate proofreading and repair mechanisms that correct errors made during replication and that repair problems caused by DNA-damaging agents. Mutations in these error-correcting systems lead to **mutator** alleles, which have elevated mutation rates. However, there is also evidence that mutation rates are not as low as they could be. An experiment in *Drosophila melanogaster* successfully selected for lower mutation rates, and **antimutator** alleles have been found in *E. coli* and in bacteriophage. Broader evidence for the genetic control of mutation rates comes from comparisons across species. DNA-based microbes have rather similar genome-wide mutation rates, even though their genome sizes vary over four orders of magnitude (Fig. 12.23). Thus, the mutation rate *per base pair* varies greatly, suggesting that it could be much lower in microbes with small genomes than it actually is (Fig. 23.1).

So, we need to explain why mutation rates are as they are. Two main factors are involved. First, reducing mutation rate is costly, both in the metabolic energy needed for the proofreading and repair machinery and in the extra replication time involved. Second, mutations are almost always deleterious, because they disrupt a well-adapted organism (see p. 407). Aggregated over the entire genome, mutations cause a substantial reduction in fitness, known as the **mutation load**, and every deleterious mutation must be eliminated by a **selective death** (see p. 552). We have seen that the reduction in average fitness of a population due to deleterious mutations is independent of their selective effect. For asexual populations, average fit-

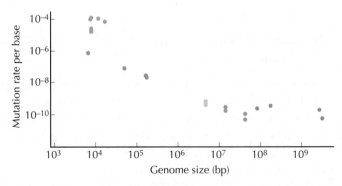

FIGURE 23.1. Mutation rates per base vary between organisms across about six orders of magnitude. (*Purple*) RNA viruses; (*red*) DNA-based viruses; (*orange*) *Escherichia coli*; (*green*) fungi; (*blue*) nematode; *Drosophila*; mouse; human (*left* to *right*).

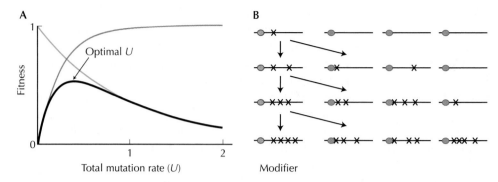

FIGURE 23.2. The mutation rate evolves as a compromise between the mutation load and the cost of lowering mutation rates. (*A*) In an asexual population, overall fitness (*black curve*) is the product of the effect of mutation on fitness (exp(–*U*); *blue curve*) and the physiological cost of reducing the rate of mutation (*purple curve*). (*B*) A sexual population will evolve a higher mutation rate than is optimal. The diagram shows the effect of a modifier allele that increases mutation rate (*blue circles*). Over time, deleterious alleles caused by this allele accumulate on the same genome (*left column*). However, recombination carries the mutations onto other genomes, which carry the low-mutation-rate allele at the modifier locus (*green circles*). Eventually, an equilibrium is reached in which the high-mutation-rate modifier may be associated with only a modest increase in the number of deleterious mutations (*bottom row*). It is this difference that causes indirect selection against high mutation rates.

ness is reduced by a factor e^{-U}, where U is the total genome-wide mutation rate. (In sexual populations, the load can be ameliorated somewhat.) Organisms with large functional genomes, such as humans, or with highly mutable genomes (e.g., RNA viruses such as HIV) may live close to the maximum sustainable mutation load.

In the simplest view, we expect a balance between selection to reduce the mutation load and the physiological cost of doing so. Strictly asexual organisms should evolve to a mutation rate that simply maximizes fitness, in a compromise between these two factors (Fig. 23.2A). However, because we do not know the cost to fitness of reducing mutation rates, it is difficult to test this theoretical prediction.

With sexual reproduction, mutation rates can evolve to be much higher than is optimal. This is because an allele that raises mutation rates gains an immediate and direct fitness advantage by avoiding the costs of proofreading and repair. It will suffer some additional mutation load from closely linked mutations that it generates, but it will recombine away from most of the harmful mutations that it causes. Thus, the extra load generated by mutator alleles is dispersed over the entire sexual population, rather than being borne entirely by the mutator allele itself (Fig. 23.2B). It is hard to know, though, whether sexual populations do in fact have higher mutation rates for this reason, because multicellular eukaryotes differ in so many ways from microbes.

Mutator Alleles Can Hitchhike to High Frequency with the Beneficial Alleles That They Cause

Mutation is the ultimate source of the variation required for adaptation by natural selection. Thus, one might expect that selection would favor increased mutation rates, because this would lead to faster adaptation. We will see that this can happen in asexual populations, but that it is unlikely in sexually reproducing populations.

Mutator alleles, which are often defects in DNA repair systems, are surprisingly common in natural populations of bacteria. Mutator alleles have often been found at high frequency in long-term selection experiments on bacteria. For example, when 12 bacterial populations were propagated for thousands of generations on glucose-limited medium, three of them acquired defects in mismatch repair, which allowed them

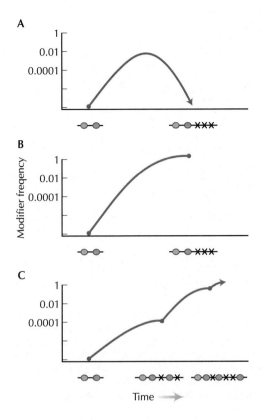

FIGURE 23.3. In asexual populations, mutator alleles can increase with the beneficial alleles that they cause. (*A*) If a modifier allele (*blue circles*) causes a large increase in mutation rates, then most beneficial mutations (*red circles*) may be caused by the modifier, even when the modifier allele is initially rare. The modifier will increase along with the beneficial mutation, but as it does so, will accumulate a load of deleterious mutations (*crosses*) and may be eliminated (*A, right*). However, other outcomes are possible. The modifier may fix in the population before being eliminated by the mutation load (*B*), or it may be boosted by successive beneficial mutations (*C*), so that it continues to increase despite the accumulating mutation load.

to accumulate favorable mutations more rapidly (Figs. 17.32 and 18.5). How is it that mutator alleles can rise to high frequency, even though almost all of the mutations that they produce must be deleterious?

Suppose that an asexual population finds itself in a new environment and must acquire new mutations in order to adapt. If mutator alleles greatly increase mutation rates—for example, by knocking out a repair pathway—then even if the mutator is rare, the first beneficial mutation is likely to be due to the mutator. The genome carrying the mutator and the beneficial allele will then increase, displacing the original, low-mutation-rate genotype. As it increases, it will also start to acquire deleterious mutations, and eventually the populations of mutators will suffer a load of e^{-U} (Fig. 23.3A). However, the allele may fix before that happens, or it may acquire a further boost from additional favorable mutations (Fig. 23.3B,C).

Figure 23.4 illustrates some of these points. A mutator strain of *E. coli*, produced by a defect at the *mutS* mismatch repair gene (see pp. 336–338 for a discussion of mismatch repair), showed a consistent advantage over wild-type (*mutS*$^+$) in the early stages of infection of initially germ-free mice, presumably because it accumulated the mutations needed to adapt to mouse guts more quickly (Fig. 23.4B). This interpretation was supported by the observation that the mutator lost its advantage when introduced at low frequency. In that case, the wild-type population would be more likely to acquire favorable mutations first, simply because it is more numerous. Moreover, it was shown directly that a mutator strain had acquired adaptive mutations during its time in the mouse gut. However, the mutator strain also acquired deleterious mutations, which prevented growth on minimal medium.

Although mutator alleles are found in nature and have been seen to increase as a consequence of the favorable mutations they produce (e.g., Fig. 23.4), they are nevertheless the exception. Most populations, most of the time, have extremely low mutation rates. Populations carrying mutator alleles must therefore either be eliminated as a result of the mutation load that they will eventually accumulate, or they must revert

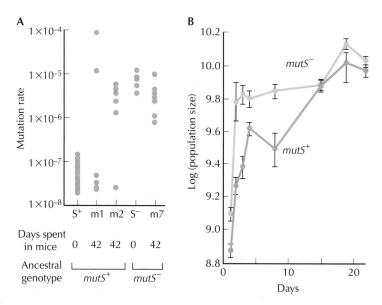

FIGURE 23.4. Mutator strains of *Escherichia coli* are at an advantage when reared in mouse guts. (*A*) A mutator strain (*mutS*⁻) was compared with wild type (*mutS*⁺); these differed in mutation rate by two orders of magnitude (S⁻, S⁺ in *A*). After 6 weeks in mouse guts, most strains retained their original mutation rates. However, clones with high mutation rates were isolated from two mice (m1, m2); these mutator strains arose from spontaneous *mutS*⁻ mutations. Conversely, one mouse inoculated with *mutS*⁻ bacteria contained clones that had evolved lower mutation rates (m7 in *A*). (*B*) The mutator strain (*mutS*⁻) has a significant advantage in the early stages of infection.

to low mutation rates. We have been discussing bacterial populations, which rarely exchange genes. In predominantly sexual populations, mutator alleles are unlikely to gain any significant advantage from the beneficial mutations they generate, because they will separate from them so rapidly.

Would higher mutation rates actually lead to faster adaptation? In plant and animal breeding, using mutagenesis to increase the rate of response to artificial selection is not considered worthwhile. Even in in vitro experiments, where the aim is to improve some particular molecule, mutagenesis is directed to particular sites (e.g., Fig. 17.2). As we shall see, mutation is much less effective than recombination as a means for generating useful variation.

There are several reasons why the rate of adaptation may not be limited by mutation rates. First, in a large population all possible single-base changes may occur many times in each generation. (This will be true for a population much larger than the reciprocal of the per-base mutation rate, >10⁹ individuals, say.) Second, the limiting factor may be the rate at which the biotic and physical environments change, rather than the number of mutations. Finally, in asexual populations we will see that mutations must be fixed in series (Fig. 23.18). If many different favorable mutations occur at about the same time, they will compete with each other so that only one can fix (this is called **clonal interference**). In sexual populations, recombination can bring different mutations together, which allows much faster adaptation.

Some Genes Have Evolved Mechanisms to Raise Their Mutation Rate

If organisms had reached an optimal phenotype, then under steady conditions the great majority of mutations would be deleterious, and selection would minimize mutation rates. However, the environment is not constant, and so at some genes selection favors random changes that make novel and adaptive variants. We will see that this can lead to the evolution of high mutation rates at these particular loci. We explore this topic by looking at examples in which diversifying selection arises from coevolution between bacterial pathogens and their hosts. There is a continual arms race between a host and a pathogen that leads to selection for novel variants in the pathogen that allow it to evade the host's defenses, and for counteradaptations by the host (see pp. 509–510). Such **Red Queen** coevolution leads both to rapid change in the genes involved in this interaction and to **indirect selection** for high mutation rates at specific locations in the genome.

Pathogenic bacteria have evolved a variety of mechanisms for increasing mutation rates at specific **contingency loci**. Most involve **microsatellites**, which are multiple repeats of short sequences that vary in copy number as a result of strand slippage during replication (see pp. 330–347). Other mechanisms include intragenomic recombination between nonhomologous sequences and **gene conversion**. The effect of these various mechanisms can be either to randomly switch gene expression or to generate random variation in RNA or protein sequence.

The most remarkable example of a specific mechanism for generating diversity is the vertebrate immune system, which was described in Chapter 9. This has two key components: B cells, which both carry **antibodies** on their surface and secrete them, and T cells, which carry special receptor molecules on their surface. Both antibodies and T-cell receptors are members of the **immunoglobulin** superfamily of proteins, and diversity in both is generated in a similar way.

This system relies on the production of an enormous diversity of antibodies, such that almost all foreign **antigens** can be recognized by a specific B cell. This diversity is generated by several mechanisms, which in effect cause a high rate of somatic mutation in just those parts of the antibody that recognize the antigen (Fig. 9.10). Together, these generate at least 10^{10} alternative antibody sequences among B cells. Further diversity is generated by somatic mutation when a clone of B cells proliferates during an immune response. The selection of cells that show greatest affinity for antigens causes an increased specificity. This maturation of the immune response is due to selection on variation generated by somatic mutation.

In the bacterial examples mentioned above, variation is advantageous either because it causes switching between alternative developmental states (e.g., fimbriae and capsule formation in *H. influenzae*), or because it causes variation in surface antigens, which helps evade the host's immune response (e.g., lipopolysaccharide in *H. influenzae*). (See Web Notes for further detail.) The diversity generated by the vertebrate immune system is advantageous because it provides material for selection between different B or T cells, which is responsible for the extraordinary specificity of the immune response. In the rest of this chapter, we will see how sex and recombination generate variation that also functions to facilitate adaptive selection.

EVOLUTION OF SEX AND RECOMBINATION

Almost All Organisms Have Some Form of Recombination

Much of this chapter is devoted to understanding why it is that almost all organisms have some kind of **sex** and **recombination**. Sex is the coming together of different genomes in the same individual. Usually, we have in mind the union of gametes to form a fertilized zygote, which is typical of eukaryotes. (This is sometimes termed **syngamy**; Fig. 23.5A.) However, we use the term broadly, to include the various ways by which bacterial and archaeal genomes can incorporate fragments of other genomes (Fig. 23.5B; pp. 182–191).

Recombination is the production of new combinations of genes. In **meiosis** this occurs in two ways: by **segregation** of whole chromosomes and by **crossing over** between homologous chromosomes. The consequences are similar—production of a new

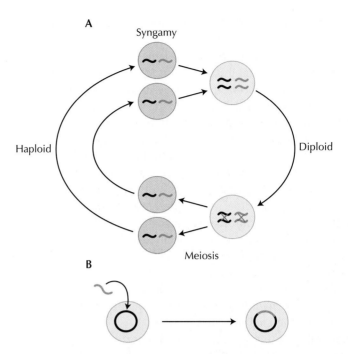

FIGURE 23.5. Sex in eukaryotes vs. bacteria and archaea. (*A*) The basic eukaryotic life cycle involves the coming together of haploid cells to form diploids (syngamy) and the production of haploid cells from diploid by meiosis. The diagram shows a genome arranged on two chromosomes. Recombination occurs at meiosis by both segregation of whole chromosomes and crossing over (*red* xs). (*B*) In bacteria and archaea, sex is asymmetric and does not involve reproduction (i.e., cell division). A fragment of DNA is incorporated into the chromosome by transformation, transduction, or conjugation.

genome containing some genes from one parent genome and some from another; and so we refer to both as forms of recombination. Again, although we will mostly be concerned with meiotic recombination, bacteria and archaea can also generate new recombinant genotypes (Fig. 23.5B; pp. 182–191).

A key feature of eukaryotes is a life cycle that includes sex and recombination (Chapter 8 [p. 222]). Necessarily, this involves an alternation of haploid and diploid phases: the sexual union of haploid gametes produces a diploid **zygote**, and meiosis of a diploid cell produces haploid offspring (Fig. 23.5A). Sex may be a regular part of the life cycle; in animals such as ourselves, sex and meiosis are required for reproduction (Fig. 23.6A). Other organisms can reproduce without sex (e.g., vegetative reproduction of plants, budding of offspring from the **cnidarian** *Hydra*, or mitotic division of single-celled microbes). Nevertheless, in the great majority of eukaryotes, sex does occur regularly, either at certain seasons or in certain environments (e.g., Fig. 23.6B–D).

Even where sex and meiosis are a regular part of the life cycle, the effective rate of recombination can vary considerably. The number of chromosomes determines the amount of recombination that occurs via segregation, and the location of crossing over determines the chance that two genes will be separated by recombination (Fig. 23.7). In addition, there is much heterogeneity in the rate of recombination per base pair (see pp. 435–436). Often, gametes from the same diploid individual unite, which leads to self-fertilization. (For example, self-fertilization occurs if a human host is infected by *Plasmodium* parasites from a single zygote, which then mate with each other when ingested by a mosquito [Fig. 23.6C].) If the diploid parent was heterozygous at more than one locus, these gametes will reproduce to produce new genotypes. However, repeated self-fertilization leads rapidly to fixation of a single homozygous genotype (see p. 420), after which recombination has no effect. Thus, continued selfing leads to a low effective recombination rate.

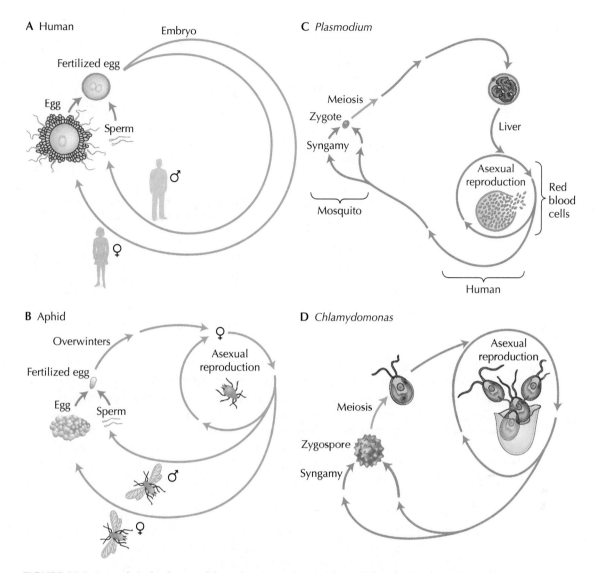

FIGURE 23.6. Examples of eukaryote life cycles. (*A*) Human males and females produce eggs and sperm by meiosis, which unite to give a diploid zygote that develops into a new male or female adult. Sex is required for reproduction, and the haploid stage (*red*) does not divide. (*B*) In aphids, the life cycle is similar, except that sex is facultative. Females reproduce parthenogenetically during the spring and summer, bearing genetically identical daughters. At the end of the season, diploid sexual males and females are produced. These produce haploid sperm and eggs by meiosis, which fuse to give a fertilized egg that overwinters. The egg hatches to give parthenogenetic females in the next season. (*C*) The sporozoan parasite responsible for malaria, *Plasmodium*, spends most of its life cycle as a haploid. It goes through several asexual stages in the host, where it reproduces asexually in the liver and then in the red blood cells. Some haploid asexuals differentiate into gametocytes, which are taken up by mosquitoes. These fuse in the mosquito gut to give a zygote, which immediately goes through meiosis to begin the life cycle anew. (*D*) The green alga *Chlamydomonas* has a similar life cycle to *Plasmodium*, again spending most of it life as a haploid. However, sexual reproduction is now triggered by nutritional conditions, which produce a highly resistant zygospore. When conditions improve, this zygospore goes through meiosis and hatches to produce new haploid cells. Haploid portions of each cycle are shown in *red*; diploid portions are shown in *blue*.

Meiosis I

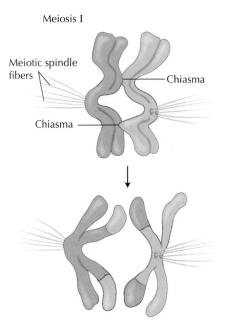

Meiotic spindle
fibers

Chiasma

Chiasma

FIGURE 23.7. At least one cross-over must occur
if the chromosomes are to segregate properly at
meiosis. During the first division of meiosis, homologous chromosomes attach to the meiotic
spindle and are pulled in opposite directions.
They remain attached to each other by *chiasmata* for some time; this attachment is necessary
for proper segregation. However, if this cross-
over occurs at the tips of the chromosome, almost no genes undergo effective recombination.

Although meiosis does not occur in bacteria and archaea, they nevertheless can occasionally transfer genes by **conjugation**, **transduction**, and **transformation** (see pp.
182–191). Indeed, we will see that occasional sex may be essential for their long-term
survival and brings clear benefits through the spread of advantageous traits such as
antibiotic resistance. However, sex in bacteria and archaea may be a side effect of other
processes, rather than itself being an adaptation that has evolved by selection. Conjugation and transduction are side effects of the propagation of plasmids and viruses
and do not seem to require further explanation. Transformation involves uptake of
DNA fragments, which may be primarily for nutrition as most ingested DNA is broken up into nucleotides, and in *Haemophilus*, for example, the genes responsible for
uptake of DNA are expressed when nucleotides are scarce. In the rest of this chapter
we concentrate on the evolution of eukaryotic sex.

Asexual Taxa Do Not Persist

The great majority of eukaryotes have sex as a regular part of their life cycle. Nevertheless, many taxa are entirely asexual or effectively so (Fig. 23.8). We saw in the previous chapter that asexual "species" are often generated when hybrids between sexual
species are able to produce genetically identical offspring and so can preserve the advantage of **heterosis**. (Examples include *Cnemidophorus* lizards and the minnows *Poeciliopsis monacha/lucida* [Fig. 23.8A,B].) Many flowering plants have become obligately
self-fertilizing; more rarely, they may be able to produce genetically identical seed (e.g.,
Fig. 23.8C), which avoids the loss of heterozygosity and **inbreeding depression** produced
by continued selfing. Many plants reproduce vegetatively, as do animals such as *Hydra*.

Asexuals may be successful in the short term and often spread over much wider
areas than their sexual relatives. However, on an evolutionary timescale, asexuals rarely
persist for long, which is shown most clearly by their phylogenetic distribution. Asexuals are found at the tips of phylogenetic trees and hardly ever give rise to daughter
species (Fig. 23.9). This pattern is seen clearly when we compare two kinds of
parthenogenesis. When unfertilized eggs produce haploid males and fertilized eggs
produce females, then we have a **haplodiploid** life cycle, which must involve sex. This

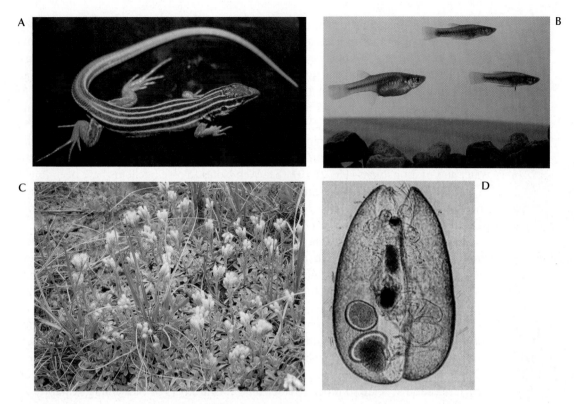

FIGURE 23.8. Examples of asexual reproduction in eukaryotes. (*A*) The all-female lizard *Cnemidophorus uniparens* and (*B*) the fish *Poeciliopsis monacha-lucida* are hybrids between two sexual species, which produce diploid eggs that go on to produce genetically identical offspring. In *B*, development still requires fertilization by sperm from *P. monacha*, but the sperm genome is discarded. (The larger fish is an asexual *P. monacha-lucida* and the two smaller fish are male *P. monacha*.) (*C*) The plant species *Antennaria parvifolia* contains both sexual and asexual females. (*D*) The parthenogenetic ostracod *Darwinula stevensoni*. Note the eggs in the brood pouch at *lower right*. All fossils of this species have such eggs, which implies that this species has been asexual for ~100 Myr. (Recently, however, living males have been found, suggesting that there may be occasional sexual reproduction.)

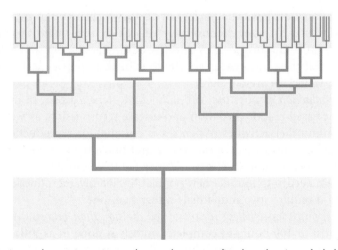

FIGURE 23.9. Asexual taxa (*green*) are almost always confined to the tips of phylogenetic trees, indicating that they are short lived on an evolutionary timescale.

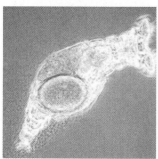

FIGURE 23.10. Adult bdelloid rotifers of two species. (*Top*) *Philodina roseola* (eating algae); (*bottom*) *Macrotrachela quadricornifera* (the large oval is a mature egg). Scale bar, 100 μm.

has evolved at least eight times, in most cases producing large and diverse taxa, such as the Hymenoptera (ants, bees, and wasps), Thysanoptera (thrips), and Monogononta (a class of rotifers). In contrast, the production of females from unfertilized eggs, which allows asexual reproduction, has evolved hundreds of times, but in almost all cases has led to a species that is very close to a bisexual progenitor. This form of parthenogenesis has hardly ever led to the diversification of a major taxonomic group.

There are a few exceptions, however, that, as we shall see, pose a major challenge to theories of sex that predict the long-term degeneration of asexuals. Examples include the ostracods (Fig. 23.8D) and the bdelloid rotifers, which are found worldwide in freshwater and moist habitats (Fig. 23.10). (See Web Notes.) Fossil bdelloids have been found in amber 35–40 Myr old, and estimates based on the molecular clock suggest a common ancestor approximately 100 Mya. Meiosis is absent, and neither males nor other evidence of sexual reproduction has ever been found.

This phylogenetic pattern, in which asexuals are confined to the tips of the tree, shows that species selection (Chapter 21 [p. 609]) plays some part in maintaining sexual reproduction. Asexual populations are likely to go extinct, because they accumulate deleterious mutations and they fail to adapt to changing conditions. Later in this chapter, we will elaborate on why this is so.

In general, however, we cannot appeal to selection between species to account for the prevalence of sex and recombination. Selection between species is an extremely weak evolutionary force, simply because rates of speciation and extinction are so slow relative to the reproduction of individuals, and because the numbers of species on which selection acts are so much smaller than the numbers of individuals (Chapter 21 [p. 609]). In examples such as those in Figure 23.8A–C, we must explain why sexual species persist in the short term in competition with asexual relatives. Within many species (e.g., *Plasmodium*, aphids, *Chlamydomonas*, and *Daphnia*), sex occurs at particular stages of the life cycle, often associated with a resting spore or transmission to a new host. In such cases, it is hard to see why a fully asexual life cycle does not evolve. Similarly, rates of recombination could be higher or lower; for example, artificial selection on recombination rates in *Drosophila* and other species produces a rapid response. Thus, we must explain what determines the current frequency of sex and recombination in much the same way as we did for mutation rates earlier in this chapter.

Sex and Recombination Have Physiological, Genetic, and Evolutionary Costs

Sex and recombination involve a variety of costs, which makes it especially hard to understand why they are so widespread. As we seek to understand their evolutionary persistence, we must not only identify selective advantages for sex and recombination, but also show that these advantages are strong enough to outweigh the more obvious disadvantages, which we present here.

Sexual reproduction requires that two individuals come together. In many cases, this involves an increased risk of predation and disease transmission, as well as diversion of resources from other activities. When sex is essential for reproduction, individuals may fail to produce offspring because they cannot find a mate. Sexual populations are vulnerable to extinction at low density and may fail to colonize new habitats if a male and a female are needed to found a new population. Plainly, reproduction is much less efficient when it requires two individuals rather than one.

Meiosis is a much less efficient means of cell division than mitosis. Meiosis typically requires from 10 to 100 hours to complete; mitosis is done in as little as 15 minutes to 4 hours. In meiosis, homologous chromosomes must pair up and the paired chromosomes must be broken and rejoined with extreme accuracy.

The cost of genome dilution. The cost that has received the most attention has been termed the "cost of meiosis," but it is more accurate to call it the **cost of genome dilution.** The idea is simple: The genes of a female who devotes all her resources to raising genetically identical offspring will have twice the fitness of those of a female whose offspring inherit half their genes from the father. Thus, if a parthenogenetic female arose within a sexual population, her genome would be expected to double in frequency in each generation and would soon displace her sexual progenitors. Crucially, this cost arises only when there are distinct sexes that contribute unequally (i.e., **anisogamy**). If two cells pooled their resources equally, then an asexual individual would gain no advantage. (We consider *why* sexes evolve later in this chapter.)

The cost of sexual selection. In the long term, sexual reproduction has deleterious consequences for the population as a whole. Once separate sexes have evolved, sexual selection causes males to produce elaborate traits and behaviors to attract females and compete with each other to find mates (Chapter 20). These can greatly reduce male survival. For example, male Soay sheep living wild on St. Kilda (Figs. 13.19 and 18.27) mostly die in their first winter, whereas ewes live for many years. Yet, if rams are castrated, they live for as long as the ewes. If males contribute no resources to offspring, reduced male survival may not harm the population. However, sexual conflict, which arises from the different reproductive interests of genes expressed in males and females, does reduce female fitness (recall the *Drosophila* example of p. 577).

The proliferation of selfish genes. Another long-term consequence of sexual reproduction is that it allows selfish genes to spread (e.g., organelles, B chromosomes, and transposable elements; see Fig. 21.4). These can substantially reduce the fitness of sexual populations; yet, with asexual reproduction all genes are inherited in the same way and so share the same interests. (Consistent with this, bdelloid rotifers lack potentially harmful transposable elements, which are found in related sexual taxa.) However, although sexual selection and the spread of selfish genes have important consequences, it is hard to see that they select against sex in the short term.

Recombination breaks up combinations of genes that have been built up by selection. We have concentrated so far on the costs of sex and meiosis. Recombination is also costly, because it breaks up combinations of genes that have been built up by selection. In addition, there may be a cost due to **ectopic recombination** between transposable elements, which generates lethal deletions and duplications (Chapter 21 [pp. 339–342 and 594]). Later in this chapter, we will see that in general, this recombination load leads to selection for reduced recombination. Experiments with *D. melanogaster* suggest that recombination does slightly reduce fitness; however, there is little other evidence on the magnitude of the **recombination load** (see p. 552).

Sex Affects a Population Only if Alleles Associate Nonrandomly

Sex and recombination shuffle genes into new combinations. In the simplest case, a heterozygote at a single locus, Aa, produces two kinds of genes, A and a. If these unite at random, a population with three distinct genotypes is produced. Here, a diverse population of haploid gametes is produced by meiosis, and a diverse population of diploid offspring is produced by the random sexual union of these gametes. With more than one locus, recombination can generate much more diversity. A heterozygote at two loci, AB/ab, produces four kinds of gametes (ab, aB, Ab, AB), from which ten distinct diploid offspring can be formed (counting AB/ab and Ab/aB as distinct). With $n = 10$ loci, there are $2^n = 1024$ different gametes and $2^{n-1}(2^n + 1) = 524,800$ distinct diploid genotypes. If we think of the thousands of sites that are typically heterozygous within an individual in an outcrossing sexual population ($n \approx 2.3 \times 10^6$ for humans; p. 355), it is clear that a vast number of genotypes can be produced.

Sex and recombination are powerful generators of diversity. Moreover, in contrast with mutation, this diversity is due to the reshuffling of alleles that function well in existing genotypes. From the perspective of a single gene, sex and recombination randomize the genetic background. With a single locus and two alleles, each gene can find itself paired with either A or a, and with many genetic loci, each gene can be combined with all possible sets of alleles at other loci. This randomization allows selection to act on the **average effect** of an individual gene, rather than on the combined effect of the whole genome (Chapter 14 [p. 389]). There is a close analogy here with experimental design. To evaluate the yield of a new crop variety under diverse conditions, it is best to grow it under all possible combinations of treatments (fertilizer, soil type, etc.).

These intuitive arguments, which appeal to the value of variation as material for natural selection, were first made by August Weismann in the late 19th century and were accepted without question until the 1970s. However, it has taken a great deal of effort to understand exactly how recombination can be selected in order to maintain variation, and the theoretical framework has become clear only recently. The key concept is that sex and recombination can alter the composition of a population only if there are deviations from **Hardy–Weinberg** proportions and from **linkage equilibrium**. In other words, there must be associations either between the two different alleles at a single locus or between alleles at different loci (Chapter 15 [pp. 432–435]). If we think of a single locus, then a population that is in Hardy–Weinberg proportions will stay in the same proportions even if reproduction is completely asexual (aa producing aa, Aa producing Aa, and so on). A shift to sexual reproduction will have no effect because although each individual produces more diverse offspring, the overall composition of the population stays the same (Fig. 23.11A). The same argument applies with multiple loci (Fig. 23.11B). Recombination acts to reduce linkage disequilibria (see p. 434), but once there is linkage equilibrium, it causes no further change.

We focus on how associations between loci (i.e., linkage disequilibria) are generated, and how recombination gains an advantage by breaking them up. Table 23.1 shows a simple classification, which will guide the rest of our discussion. The columns show the process responsible for generating the linkage disequilibrium on which recombination acts. Selection may favor certain combinations of alleles (termed **epistasis**; p. 389); migration may bring some gene combinations typical of one population into another (see p. 542); and random drift will cause random fluctuations in genotype frequencies and hence will produce random association between alleles (see pp. 434–435). The rows show the kind of selection that acts on the alleles involved. (Clearly, neutral alleles, with no effect on fitness, are irrelevant.) We may be concerned with the elimination of deleterious mutations, with response to selection that fluctuates in space or time, or with directional selection on favorable alleles or on quantitative traits. In the following sections, we discuss several models, which are classified within Table 23.1.

Competition between Relatives Can Favor Genotypic Diversity

Several theories focus on the advantage of diversity within families or larger groups of relatives. There are two ways in which diversity could be advantageous. First, a genetically diverse family could use diverse resources within its local habitat more efficiently and thus raise more offspring. In Chapters 18 and 19 we saw that substantial genetic variation can be maintained by **frequency-dependent selection**, when different genotypes exploit different resources. This is supported by the observation that mixtures of crop varieties often yield more than the average of their pure components and by comparing diversity between sexual and asexual populations. A second, slightly different possibility is that the survivors from a diverse family are more vigorous and so do better later in life. This has been likened to a lottery, in which the best strategy from the

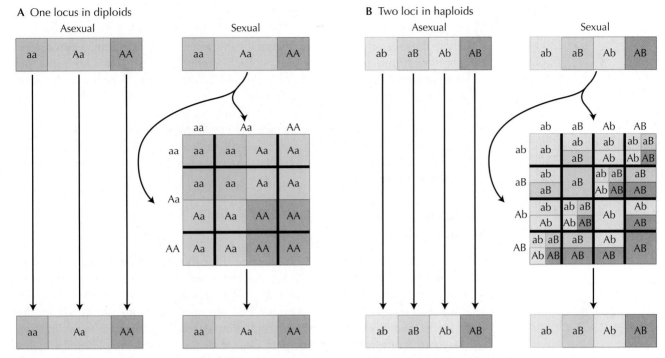

FIGURE 23.11. Sex and recombination alter the composition of a population only if there are non-random associations between genes. (*A*) With one locus, sex has no effect on a population that is in Hardy–Weinberg proportions. The *left-hand* diagram shows asexual reproduction, with each genotype producing identical offspring. The *right-hand* diagram shows sexual reproduction. The 4 x 4 diagram shows the 16 kinds of mating and their offspring: aa x aa produces all aa offspring (*top left*), Aa x Aa produces 1:2:1 proportions of aa:Aa:AA, and so on. The proportions in the next generation will be in Hardy–Weinberg proportions if mating is random. (*B*) Similarly, sex and re-combination do not alter the proportion of haploid genotypes if there is linkage equilibrium. The diagram is as in *A*, but now with two loci, giving four haploid genotypes (ab, aB, Ab, AB).

family's point of view is to choose different numbers, each of which might win, rather than all choosing the same number.

Although these theories seem plausible, they do not give a general explanation for the prevalence of sex. The difficulty is that siblings must compete with each other for the same resources. Genetic diversity may well increase net fitness in large populations, but in the absence of linkage disequilibria, sex and recombination would not increase diversity, and so modifiers that increase recombination would not increase through in-direct selection. Explanations that rely on competition between siblings can apply only

TABLE 23.1 Classification of models for the evolution of recombination

Form of Selection	Process Generating Linkage Disequilibrium	
	Epistasis	Random Drift
Deleterious mutations	Alleviation of mutation load	Muller's ratchet Background selection
Fluctuating selection	Fluctuating epistasis—e.g., host–parasite coevolution	Sibling competition
Favorable alleles	Sstabilizing election on a quantitative trait	Fisher–Muller argument Hitchhiking

to a fraction of species; moreover, the distribution of sexual reproduction across species does not suggest any correlation with the degree of sibling competition.

These theories rely on linkage disequilibrium generated by random subdivision of the population into small family groups. Each family has two parents with just four random genomes (assuming diploidy). Sex and recombination generate a very large number of genotypes among the offspring, in effect by breaking up linkage disequilibria produced by the extreme random drift that occurs within a "population" of two parents. Thus, the same theories operate whenever populations are subdivided into small groups. However, although models such as this seem plausible, a strong advantage to sex and recombination requires both strong population subdivision and strong frequency-dependent selection.

Fluctuating Environments Do Not Generally Favor Sex and Recombination

One might think that fluctuations in selection through space and time would favor recombination; perhaps adaptation to heterogeneous and unpredictable environments favors heterogeneous gene combinations. However, we saw in Chapter 18 that fluctuating selection does not itself maintain polymorphism. In much the same way, we will see that quite special conditions are needed for fluctuating selection to maintain recombination.

Maynard Smith made the argument clear with the aid of a simple model of two genes, each responsible for adaptation to two different environmental factors (e.g., cold vs. hot and wet vs. dry). Newborn individuals land on patches with different environments, are selected accordingly, and then come together as adults to mate randomly with each other (Fig. 23.12). If environments, and therefore fitnesses, fluctuate through time, then in the simplest case the allele at each locus with the highest geometric mean fitness will fix (see p. 508). If variation is to be maintained, selection must favor rare alleles. This could arise if the numbers emerging from each local patch are separately regulated. Even then, if environmental features vary independently across patches so that the selective advantage of alleles at one locus does not depend on which allele is present at the other locus, there will be no associations between alleles at the different loci (Fig. 23.13A). In other words, if there is no epistasis, there will be no linkage disequilibrium to drive the evolution of recombination. However, if there is a consistent correlation between environments (so that cold patches tend to be wet, and hot patches dry, say), then there will be consistent epistasis, and recombination will be disadvantageous (Fig. 23.13B). This is because recombination breaks up the favorable gene combination built up by selection ("cold" alleles with "wet," "hot" with "dry"). This recombination load causes indirect selection for *reduced* recombination.

For recombination to increase in this model, there must be frequency-dependent selection that maintains variation at multiple loci, and crucially there must be epistasis that *changes direction* every few generations (Fig. 23.13C). In one generation, selection must favor "cold" with "wet" and "hot" with "dry," but soon after the opposite association must be favored. Recombination then gains an advantage, because it breaks up gene combinations that *had* been favorable, but which are now *unfavorable* (Fig. 23.13D).

This model for the evolution of sex and recombination is odd, because recombination gains an advantage by *interfering* with selection. With fluctuating epistasis, a response to selection is *disadvantageous* because different gene combinations will soon be favored. It has been suggested that coevolution between parasites and their hosts could produce these perverse kinds of selection: Parasites adapt to host genotypes that are common now, so that those genotypes become more heavily infected in the future. This continual struggle between species is termed Red Queen evolution (see p. 509).

We saw an example of this kind of coevolution between host and parasite in Figure 18.24, which shows that in asexual populations of the snail *Potamopyrgus antipodarum*, clones that had been common are more susceptible to infection by the para-

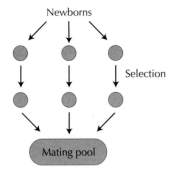

FIGURE 23.12. A simple model of a population that experiences a heterogeneous environment.

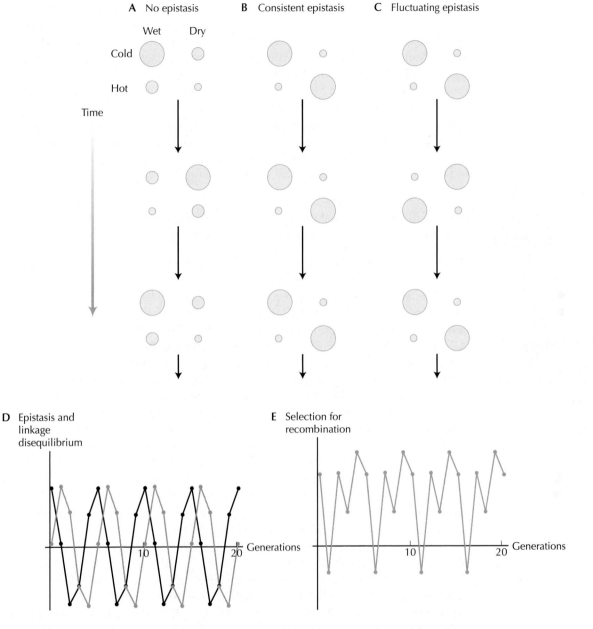

FIGURE 23.13. Fluctuating selection favors recombination only under restrictive conditions. Some kind of frequency-dependent selection must maintain polymorphism by favoring rare alleles. This example shows the **Levene model**, in which juveniles compete within patches with different environments, and the numbers emerging from each patch are fixed (indicated by the sizes of the *circles* in *A–C*). (*A*) Suppose that selection acts on two loci, which determine adaptation to two environmental factors. If there is no correlation between these factors, then there is no epistasis, no linkage disequilibria will be generated, and recombination makes no difference. (*B*) If there is a consistent correlation between the factors, then there is constant epistasis, and recombination is selected against because it breaks up favorable gene combinations. (*C*) Recombination can be favored if the correlation between environmental factors *changes* over time, so that epistasis fluctuates. (*D*) Then positive epistasis (*black*) builds up positive linkage disequilibria (*red*), but by the time these have become strong, epistasis has changed sign, and negative disequilibria are favored. (*E*) The strength of selection for recombination. Only relative values are shown. With no linkage, all allele frequencies at 1/2, and epistasis fluctuating between 0.1 and –0.1, the average selection against a modifier that reduces recombination by 0.1 is only 0.0000084.

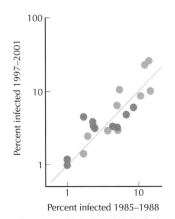

FIGURE 23.14. (*Top*) Populations of the freshwater snail *Potamopyrgus antipodarum* (*bottom*), which are heavily infected by trematode parasites, are more likely to include sexual individuals. (*Red circles*) Strictly asexual populations; (*blue circles*) mixed sexual and asexual populations. Infection levels remained similar across more than 10 years.

sites that live with them. These snails are interesting because in some populations, diploid outcrossing sexuals coexist with triploid asexual clones. The frequency of sexuals correlates strongly with levels of infection (Fig. 23.14), which suggests that host–parasite coevolution does favor sex in these snails. However, there is no evidence that this advantage acts through fluctuations in the gene combinations that are favored, as outlined above. More generally, computer simulations of host–parasite coevolution do often show an advantage for sex over asex, but it has not been shown that this is due to fluctuating epistasis. In the following sections, we will see how recombination and sex can gain an advantage by facilitating selection, as first proposed by Weismann. It may be that Red Queen coevolution is a significant contributor to the variance in fitness, which is needed by these more plausible models.

Recombination Increases the Additive Variance in Fitness When There Is Negative Linkage Disequilibrium

Fisher's **Fundamental Theorem** states that the increase in mean fitness due to selection is equal to the **additive genetic variance** in fitness divided by the mean (see p. 642). The additive genetic variance is thus the key measure of useful variation in a population (see pp. 547–549). Recombination will therefore facilitate natural selection only if it increases the additive genetic variance in fitness. We have explained that if there are no linkage disequilibria, then recombination has no effect on the population and, in particular, does not alter the variance in fitness. In this section, we explain how a special kind of linkage disequilibrium is required if recombination is to gain an advantage by increasing the additive variance in fitness that is available to selection.

We label alleles whose main (i.e., **additive**) effect is to increase fitness as "+" and those that reduce fitness as "–." If there are positive associations, so that ++ and – – are in excess, then the variance in fitness will be higher than at linkage equilibrium, because these extreme genotypes are overrepresented (Fig. 23.15A). Recombination will break up these associations and so will *reduce* the variance in fitness back toward its linkage equilibrium value. Conversely, if there are negative associations, so that +– and –+ are commoner than expected, the variance in fitness will be reduced by them, and recombination will tend to *increase* it (Fig. 23.15B).

Note that we define linkage disequilibria as positive or negative according to the additive effects of alleles on fitness. This labeling of alleles as "+" or "–" makes sense only if there is some kind of directional selection and, hence, some additive variance in fitness. In the previous section, we focused on epistasis and showed that recombination can be favored with epistasis alone only if epistasis fluctuates on the right time scale. The mechanism we are outlining in this section is quite different. It relies on an

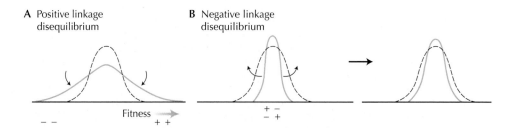

FIGURE 23.15. The effect of recombination on the variance in fitness depends on whether linkage disequilibria are positive or negative. (*A*) If ++ and – – combinations are in excess (i.e., positive linkage disequilibria), then the variance will be higher than when alleles are combined at random (*dashed curve*). Recombination therefore *reduces* the variance (*arrows*). (*B*) Conversely, if +– and –+ combinations are in excess (i.e., negative linkage disequilibria), then the variance will be lower than at linkage equilibrium (*dashed curve*), and recombination will *increase* the variance (*right*).

interaction between directional selection and negative linkage disequilibria.

We can now see a general way in which recombination can increase the fitness of a population and can itself increase by indirect selection (Fig. 23.16). If there are negative linkage disequilibria between favorable alleles, recombination will increase the additive variance in fitness, which will cause a faster response to selection. This gives an advantage to the whole population, which can explain why sexual populations are not outcompeted by asexual clones (as, for example, in the *Potomapyrgus* example of Fig. 23.14). Moreover, an allele that increases recombination will tend to find itself with the favorable combinations it generates and so will tend to increase with them. This gives an individual-level advantage to recombination, via indirect selection, which is greater if the modifier is tightly linked to the selected alleles. (Recall that the same kind of indirect selection acts on modifiers of mutation rate [Fig. 23.2B].)

What is crucial, then, is understanding why linkage disequilibrium tends to be negative, so that it interferes with selection. Why should favorable alleles be found on different genomes more often than expected by chance? Essentially there are two kinds of explanation: one based on the deterministic action of selection and one based on the random effects of genetic drift (Table 23.1).

Negative Linkage Disequilibria Can Be Generated by Epistasis or by Drift

The most straightforward possibility is that there is negative epistasis; in other words, that selection favors (+–) and (–+) combinations (as in Fig. 23.16). However, such epistasis must not be too strong, otherwise the immediate recombination load will outweigh the long-term advantage brought by an increased additive variance in fitness. There are some general reasons why we might expect negative epistasis. For example, if quantitative traits are typically under **stabilizing selection** (Box 17.3), then selection will act against extreme ++, – – combinations, and negative associations will develop. However, as yet, there is very little direct evidence on the extent and sign of epistasis (see p. 409; e.g., Fig. 23.17).

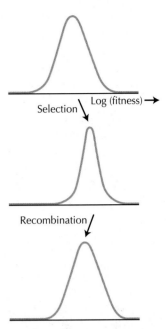

FIGURE 23.16. The distribution of fitness changes as a result of selection and recombination. Selection increases the mean log fitness by an amount equal to the additive genetic variance in fitness (p. 462). If selection favors negative associations, it generates negative linkage disequilibria, which reduce the variance in log fitness (Fig. 23.15B) and, hence, the future response to directional selection. Recombination causes an immediate reduction in mean log fitness by breaking up favored gene combinations, but facilitates future adaptation by increasing the variance in log fitness. Modifiers that increase recombination can be favored because they are associated with adaptive variation in fitness, even though they also are associated with an immediate recombination load.

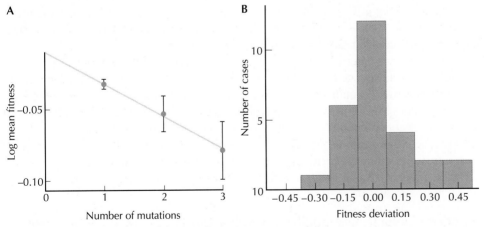

FIGURE 23.17. An experiment with *Escherichia coli* showed no systematic positive or negative epistasis. (*A*) The base population had adapted for 10,000 generations to glucose-limited minimal medium (Fig. 17.32). Transposable elements were inserted at random into one, two, or three sites, giving 225 mutated strains. The log mean fitness fell linearly with the number of mutations, which shows that the effects of each additional mutation are independent of the number of mutations already present. (*B*) In a separate experiment, combinations of pairs of individual mutations were constructed and the fitness of the pairs was compared with the fitness expected by multiplying their individual effects. There were significant deviations (indicating significant epistasis), but these showed no tendency to be systematically positive or negative.

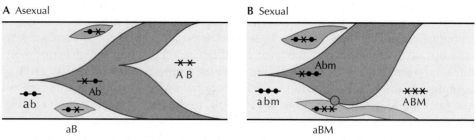

FIGURE 23.18. The Fisher–Muller argument. (*A*) Favorable mutations must be established sequentially in an asexual population. For example, if allele A is destined to replace a, then any favorable alleles that occur at other loci (B, for instance) can only be fixed if they occur within a genome that carries A. (*B*) With sexual reproduction, favorable mutations at different loci can be combined; this leads to an advantage to modifiers that causes sex and recombination. A favorable allele B that occurs with the unfavorable allele a can be fixed if it can recombine into association with A (*red circle*); if this requires that a modifier allele M be present, then allele M will also tend to increase by hitchhiking.

Random drift does tend to separate favorable alleles from each other, which generates negative linkage disequilibria. This is easiest to see if we think of the extreme case of new advantageous mutations entering an asexual population. R.A. Fisher and Hermann Muller both pointed out that such mutations must be accumulated one by one. If several favorable mutations arise at about the same time, they will almost certainly arise on different genomes, and so only one of them can be fixed (Fig. 23.18A). Sexual populations can evolve much faster, because mutations that arise on different genomes can be brought together by recombination (Fig. 23.18B). Moreover, modifiers that promote recombination will gain an advantage if they stay associated with the favorable combinations that they generate (Fig. 23.18B).

This advantage to sex and recombination is due to random drift. If the population were extremely large, then in every generation, double mutations would arise and recombination would not be needed to bring them together. Random drift alone generates random linkage disequilibria, with no tendency for associations to be positive or negative on average. However, positive associations (++, − −) are rapidly swept out of the population by selection, whereas negative associations (+−, −+) persist for much longer, because there is little difference in fitness between them. Thus, over time negative associations accumulate and slow down selection.

To summarize, negative linkage disequilibria are required if sex and recombination are to gain an advantage by increasing the efficacy of selection. Epistatic selection can generate negative disequilibria, but it is not clear why it should do so in general. Random drift does lead to negative associations that interfere with selection, but it is not clear that local populations are small enough, or selective sweeps frequent enough, for this to give a significant advantage across a wide range of organisms. In the following sections, we examine different ways in which negative associations can be generated; all these models fall into the framework described in Table 23.1.

Random Linkage Disequilibria Amplify the Effects of Genetic Drift

We can also understand why random linkage disequilibria interfere with selection by thinking of them as generating an additional source of random genetic drift. This idea was first put forward by W.G. Hill and Alan Robertson in 1966 and is known as the **Hill–Robertson** effect.

In Chapter 19 (p. 536), we saw that selection at one locus causes random perturbations at other loci and on average reduces genetic diversity. In a completely asexual population, all variation is lost when a favorable mutation sweeps to fixation, carrying with it the unique genome in which the mutation arose. In a sexual population, **selective sweeps** eliminate variation within a limited region of genome (Fig. 19.17). Conversely, the elimination of deleterious mutations reduces variation at linked loci in a process termed **background selection** (see p. 538). From the point of view of an individual gene, the random effects on fitness of the genetic background in which it happens to be embedded causes it to increase or decrease, regardless of its own direct effects on fitness.

The idea that selection at one locus in effect generates random drift at linked loci can be understood by recalling that fundamentally drift is caused by random variation in fitness (Chapter 15 [p. 418]). Usually, we think of this variation as being due to randomness in whether individual organisms reproduce, and whether individual genes are passed on at meiosis. In other words, thinking of fitness as a quantitative trait, we consider just the noninherited environmental component of its variance. However, the inherited component of fitness variance also contributes. Moreover, because the effects of being in a good or bad genetic background persist over many generations, this inherited variation in fitness produces a disproportionate amount of drift (Fig. 23.19A). Thus, even if the fraction of the variance in fitness that is inherited (i.e., the **heritability** of fitness) is small, it can still make a substantial contribution to random drift. Indeed, in Chapter 15 (p. 426), we discussed the possibility that at least in abundant species, *most* random drift is due to some kind of selection at linked loci.

So far, we have discussed **neutral** loci. However, the increased random drift caused by the Hill–Robertson effect also influences selected loci and tends to interfere with selection. If one thinks of selection as trying to pick out the genes that will on average give the highest fitness, any source of random noise must interfere with this measurement. Increasing the amount of random drift will reduce the probability that favorable alleles are fixed and will increase the chance that deleterious alleles are fixed (Fig. 18.6). Selection is indeed less effective in regions of the *Drosophila* genome that have low recombination (see p. 544).

Thinking of selection at one locus as inducing random drift at other loci is precisely equivalent to thinking in terms of negative linkage disequilibria between favorable alleles. The two approaches outlined in this and the previous section are just different ways of looking at the same process. We now turn to specific examples and evidence as to the importance of recombination in facilitating selection.

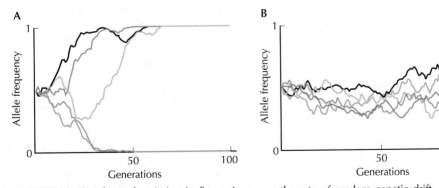

FIGURE 23.19. Inherited variation in fitness increases the rate of random genetic drift—a process known as the Hill–Robertson effect. (*A*) Neutral allele frequencies increase or decrease at random, as they become associated with fit or unfit genetic backgrounds. For example, one replicate (*green*) increases steadily between generations 30 and 50, because it happens to be in a particularly fit background. Across five replicates, neutral variation is lost by 70 generations. The simulation shows a population of 400 genomes, with a variance in relative fitness of 0.1 and a recombination rate of 0.05. (*B*) With no selection, allele frequency fluctuations are not correlated from one generation to the next, and so neutral variation persists for much longer.

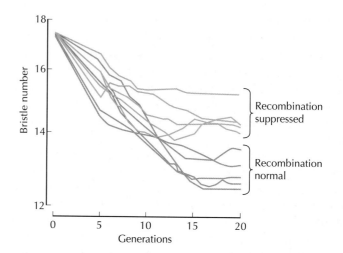

FIGURE 23.20. Selection for reduced bristle number in *Drosophila melanogaster* was less effective when recombination between the autosomes was suppressed by using **balancer chromosomes** (Box 13.2).

Favorable Alleles Are Brought Together by Recombination

If selection or drift produce negative linkage disequilibria, then the response to selection should be faster when recombination is also operating. This prediction has been tested experimentally both in *Drosophila* (Fig. 23.20) and in vitro (Fig. 23.21). Indeed, **genetic algorithms**, which evolve software by selecting among competing computer programs (e.g., Fig. 17.13), rely on recombination between programs to generate appropriate variation.

These examples do not show exactly how recombination increases additive genetic variance in fitness and hence increases the response to selection. This question has

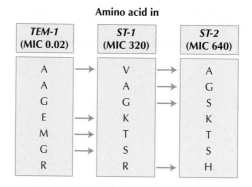

FIGURE 23.21. Recombination selects for much more efficient antibiotic resistance genes. In vitro the β-lactamase gene *TEM-1* in *E. coli* has poor activity against the antibiotic cefotaxime: The minimum inhibitory concentration (MIC) of antibiotic against *E. coli* carrying this gene is only 0.02 μg/ml. The gene was broken into fragments and amplified by **polymerase chain reaction**, which caused recombination between fragments and also introduced mutations. The recombined genes were inserted back into *E. coli*, and were then selected on increasing concentrations of antibiotic. This process was repeated three times and produced a gene *ST-1*, which conferred a 16,000-fold increase in resistance. This had four amino acid replacements, as well as four silent changes and a mutation that increased promoter strength. Two generations of recombination in the presence of excess wild-type DNA (a "backcross") eliminated mutations unnecessary for resistance, but also gave further amino acid substitutions that further doubled resistance. The final product was 64 times more resistant than any previously published *TEM-1*-derived gene; a similar experiment using mutation alone gave only a 16-fold increase in resistance. (*Letters* indicate amino acids at positions that evolved.)

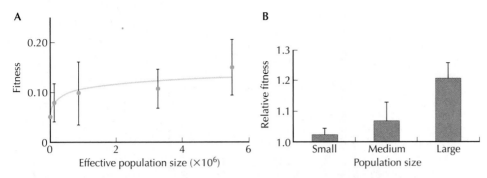

FIGURE 23.22. (*A*) The increase in fitness of asexual populations after adaptation to a new medium plotted against their effective population size. (*B*) The mean fitness of sexual lines relative to asexual populations with the same effective size.

been addressed in an experiment in which populations of the green alga *Chlamydomonas reinhardtii* were propagated through approximately 200 generations of growth. Initially these populations were genetically identical, so that adaptation depended on new mutations. One set of populations was allowed to reproduce sexually, whereas the other set remained strictly asexual. Both sets were passed through bottlenecks of varying sizes. The asexual populations adapted faster when passed through less severe bottlenecks, corresponding to larger effective population sizes, but with decreasing slope as population sizes increase (Fig. 23.22A). This is presumably because in small populations, favorable mutations rarely arise and so do not interfere with each other. However, as the rate of incorporation of new mutations increases, they begin to interfere with each other, so that there are diminishing returns to increased population size, as predicted by Fisher and Muller (Fig. 23.18A). Correspondingly, sexual reproduction has little advantage in small populations, but allows much faster adaptation in large populations (Fig. 23.22B), presumably by bringing together different favorable mutations, as in Figure 23.18B.

If recombination does improve the response to selection, then we expect higher recombination rates to evolve, as alleles that increase recombination **hitchhike** along with the favorable gene combinations that they produce. This is supported by the observations that domesticated mammals tend to have more cross-overs at meiosis than their wild relatives (Fig. 23.23) and that recombination rates tend to increase in laboratory populations subject to strong artificial selection (Fig. 23.24).

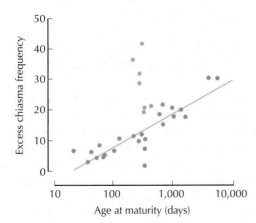

FIGURE 23.23. Domesticated species of mammal (*blue*) have a larger number of chiasmata than wild mammals (*red circles* and *gray regression line*) for a given age at maturity. The plot shows the number of chiasmata minus the number of haploid chromosomes, because for proper segregation one chiasma is needed for each chromosome.

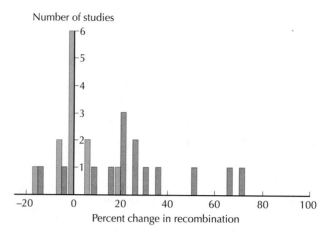

Number of studies

Percent change in recombination

FIGURE 23.24. Recombination rates tend to increase when populations are artificially selected for arbitrary traits (e.g., the ability of flies to move upward through a maze). Averaging across the several studies summarized here, selection was for 50 generations, on a population size of ~100 individuals. (*Blue*) Nonsignificant results; (*red*) significant results.

These various lines of evidence support the theoretical prediction that recombination and sex gain an advantage because they facilitate selection. However, whether this advantage is strong enough to outweigh the various costs of sex and recombination described on pages 668–669 depends on the net strength of directional selection, as measured by the additive variance in fitness—an issue on which at present we have little good evidence (see pp. 547–549).

In the Absence of Sex, Deleterious Alleles Accumulate

All populations must continually eliminate deleterious mutations if they are to survive. This elimination is much more efficient if there is recombination. In this section, we will see that the primary function of sex and recombination may be to prevent the fatal accumulation of mutations. Because all organisms suffer a mutation load, this is an attractive general explanation for the prevalence of sex and one that has received much attention in recent years.

Sexual reproduction can reduce the mutation load (see Web Notes). If mutation is always from good alleles to bad (a reasonable approximation) then the fitness of an asexual population is reduced by a factor e^{-U}, where U is the genomic mutation rate. Sexual reproduction makes no difference if the effects of different mutations on fitness multiply together, so that there are no linkage disequilibria. However, if there is negative epistasis, so that the effect of each additional mutation becomes more severe as the number of mutations increases, then negative linkage disequilibria will be generated. Recombination breaks these up and, by doing so, can substantially reduce the mutation load (Fig. 23.16). This gives a population-level advantage for sexuals over asexuals and also causes indirect selection for modifiers that increase the mutation rate.

Alexey S. Kondrashov has argued that this deterministic effect of the mutation load must be important in maintaining sex and recombination if the total genomic mutation rate U is high. In that case, populations could not survive unless there is both negative epistasis and sexual reproduction. The total mutation rate may be high enough for this argument to hold in organisms with large functional genomes, such as mammals. However, in the great majority of organisms, the mutation load is probably too small to cause substantial indirect selection for sex and recombination (Chapter 19 [p. 552]). Moreover, experiments with a variety of organisms have shown little evidence for negative epistasis between mutations (e.g., Fig. 23.17).

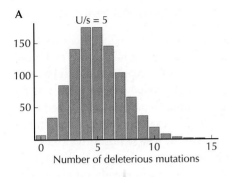

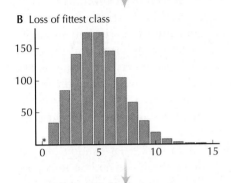

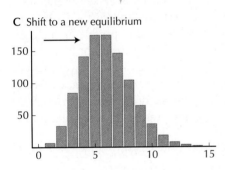

FIG. 23.25. Deleterious mutations accumulate in asexual populations via Muller's ratchet. (*A*) In a balance between mutation, at a total rate $U = 0.1$ and selection $s = 0.02$, an equilibrium is reached with, on average, $U/s = 5$ deleterious mutations per genome. However, in a population of 1000 genomes, there are only, on average, $1000\, e^{-U/s} = 6.7$ individuals who are free of mutations (*blue bar*). (*B*) This fittest class will eventually be lost by chance. If there is no recombination and no back mutation, then mutation-free individuals cannot be recovered. (*C*) The whole distribution shifts to the right in one click of Muller's ratchet and the process continues.

Random linkage disequilibria also interfere with selection, via the Hill–Robertson effect. We have seen that any kind of selection causes random fluctuations at linked loci, which may by chance fix deleterious mutations. This effect is especially severe in strictly asexual populations. To see this, consider a simple model of one-way mutations and multiplicative fitnesses. The population will approach an equilibrium distribution around an average of U/s mutations per genome (Box 18.2). For example, with selection of $s = 2\%$ against rare deleterious alleles and a genomic mutation rate of $U = 0.1$, the average number of mutant alleles carried by each individual is $U/s = 0.1/0.02 = 5$ (Fig. 23.25).

Muller pointed out that even in a very large asexual population, deleterious mutations must accumulate, in a process now known as **Muller's ratchet**. The fraction of the population that carries *no* deleterious mutations is $e^{-U/s}$, which can be very small if U/s is large. (For example, $e^{-U/s} = 0.0067$ if $U/s = 5$, and 0.000045 if $U/s = 10$.) Thus, the fittest genotype can be lost by chance, and once lost, it can never be recovered. As we saw in Figure 19.19, the whole population must trace its ancestry back to the fittest class, even if that class is rare. Once the fittest class is lost, the process begins again, but with all individuals carrying one extra mutation (Fig. 23.25). Even if the population size is in the millions, weakly selected mutants will still accumulate. Roughly speaking, asexual populations cannot prevent accumulation of alleles with fitness effects smaller than approximately $s = U/10$.

Y Chromosomes Degenerate Because They Do Not Recombine

The best evidence for the harmful long-term consequences of asexuality comes from sex chromosomes that do not recombine. To understand this evidence, we first need to see how sex chromosomes evolve. The ancestral state for plants and animals was **hermaphroditism** or **cosexuality**, where each individual produces both male and female gametes. (For example, most flowering plants produce both pollen and seed from the same individual.) The evolution of separate sexes (i.e., male individuals and female individuals) is a simple evolutionary transition, because it just requires the loss of either male or female function, rather than elaboration of a new, complex pathway. Separate sexes have evolved independently many times, by means of a wide variety of sex-determining mechanisms. In many groups, sex is determined by environmental cues. For example, alligators develop as females if the eggs are below a critical temperature threshold and as males otherwise. Genetic sex determination has arisen many times. Most commonly, one sex is homozygous at the sex-determining locus, and the other is heterozygous. Where the male is heterozygous (as in most insects and in mammals) the male and female genotypes are denoted XY and XX, respectively. When the female is heterozygous (as in birds and Lepidoptera), they are denoted ZW and ZZ. We will assume male heterozygosity and so refer to the chromosomes as X and Y, but the same arguments apply in the opposite case (see p. 638).

The first step toward separating the sexes is likely to have been fixation of a recessive mutation that causes male sterility. This would establish a **gynodioecious** population, consisting of both females and hermaphrodites. Such a male-sterile mutation could gain an advantage because females cannot self-fertilize, and so their offspring cannot suffer inbreeding depression. Also, if resources allocated to producing male gametes are diverted to female function, the number of offspring need not be reduced. Once a gynodioecious population is established, the sex ratio is skewed toward female function, and so males may gain an advantage because they are the rarer sex (see p. 506). Thus, a dominant *female* sterile allele can be established if it is tightly linked to the first male-sterility locus (Fig. 23.26). Although many present-day sex-determining mechanisms involve just a single switch gene (e.g., *SRY* in humans), or multiple genes, these are thought to be elaborations of an earlier two-gene system, as just described.

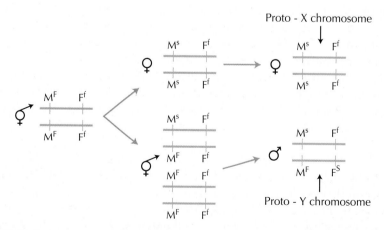

FIGURE 23.26. Separate sexes can evolve from an initially hermaphroditic population through the fixation of two mutations. First, a recessive allele M^s arises that causes male sterility when homozygous (M^sM^s), hence producing females. The other genotypes (M^sM^F, M^FM^F) remain as hermaphrodites, and so a polymorphic gynodioecious population is formed. Next, a dominant female sterility mutation, F^S, arises at a linked locus. Now, individuals homozygous for M^sF^f (*top right*) are female, and those heterozygous for M^sF^f/M^sF^S are male. There is strong selection for tighter linkage between the two loci, because recombinant genotypes such as M^sF^S/M^sF^f are completely sterile.

FIGURE 23.27. In guppies, bright coloration is favored by sexual selection in males, whereas dull color is favored in females. As a consequence, alleles for bright coloration are tightly linked to the sex-determining locus.

It is now easy to see how sex chromosomes can evolve. There is strong selection against recombination between the male- and female-sterility alleles, because this would produce sterile offspring (Fig. 23.26). This selection may suppress recombination over a wide region—as, for example, if a chromosomal **inversion** became associated with the male- and female-determining loci. Also, many genes may be polymorphic for alleles that work best in different sexes, and there will be strong selection for these to become linked to the sex-determining locus (e.g., Fig. 23.27). This is an example of the **reduction principle** by which consistent epistasis selects for reduced recombination.

Once recombination between X and Y chromosomes is suppressed, the Y chromosome is in a peculiar situation. It is always heterozygous, is confined to one sex, and, in effect, reproduces asexually. In contrast, genes on the X can recombine freely in females. The primary cause of degeneration of the Y is that in the absence of recombination, selection becomes ineffective and deleterious mutations accumulate. Genes may cease to function, or be lost altogether, and transposable elements may be fixed. Eventually, the Y chromosome accumulates repeated sequences, and may be lost altogether. Mechanisms of **dosage compensation** usually evolve to ensure that the level of gene expression is the same in males and females (at least for those genes that have not become sex specific). In **eutherian mammals**, one X is inactivated at random, whereas in *Drosophila*, genes on the X are expressed at twice the rate in males compared to females. New Y chromosomes can be generated if an X chromosome fuses with an autosome (recall the *Podisma* example [Fig. 18.16]), because the homologous autosome is then confined to males. The process of degeneration then begins afresh.

CONSEQUENCES OF SEX

We now turn from the evolution of sexual recombination to look at some of the consequences of sex. Sexual reproduction has led to many striking features, which we often take for granted. We consider some of the most basic: life cycles involving both haploid and diploid phases, evolution of the two sexes, and avoidance of inbreeding.

Diploid Life Cycles May Have Evolved to Mask Deleterious Alleles

Sex necessarily leads to an alternation between haploid and diploid phases of a life cycle. The fusion of gametes doubles the number of genomes, and meiosis halves the number. However, the relative lengths of the two phases vary greatly (Figs. 23.26 and 23.28). The haploid and diploid stages may look similar, as in the sea lettuce *Ulva* (Fig. 23.28C), or they may be specialized for different roles. For example, in ferns the large

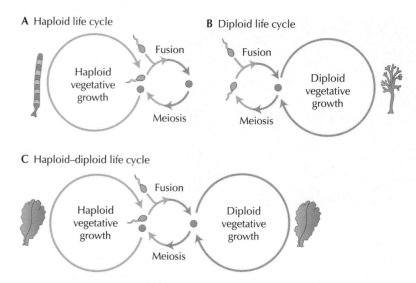

FIGURE 23.28. Eukaryotic life cycles vary in the time spent in haploid vs. diploid phases. (*A*) In many protists, mitotic divisions and development occur only in haploids: Fusion of gametes is followed immediately by meiosis. (*B*) Most animals have an almost entirely diploid life cycle, with no mitosis or development of haploids. (*C*) Many organisms spend substantial times as both haploids and diploids. These stages may be morphologically similar, as shown here, or may be distinct. The three possibilities are illustrated by species of alga: (*A*) *Ulothrix*; (*B*) *Fucus*; (*C*) *Ulva*. Larger cycles represent vegetative growth and smaller cycles represent sexual reproduction.

diploid sporophyte produces haploid spores by meiosis, which disperse and then differentiate into multicellular haploid gametophytes. These produce sperm and eggs, which fuse to begin the cycle anew.

How did eukaryotes move from asexual reproduction by mitosis to a life cycle with both syngamy and meiosis? It is plausible that a simple version of meiosis came first, and that genome doubling then took place by **endomitosis** (i.e., mitosis with no cell division; Fig. 23.29A). Indeed, some organisms have just such a life cycle. For example, the oxymonad *Pyrsonympha* has an asexual life cycle in which increases in genome number by endomitosis are followed by several reductional divisions. The question is then, why should organisms go to the trouble of alternating between haploid and diploid phases?

We have already seen that having two copies of a genome allows repair of double-stranded damage to one of them, by homologous recombination (Chapter 12 [p. 336]). Diploidy also allows development of large multicellular organisms, because somatic mutation in one genome is masked by functional copies in the other. (Cancer involves multiple mutations in regulatory genes; because these mutations are usually recessive, cancer would be fatally common if we were haploid rather than diploid.) However, single-celled diploids grow more slowly than haploids when uptake of nutrients is limiting, because they have a smaller surface area relative to their volume. Also, the long-term mutation load of diploids is twice that of haploids, essentially because they carry twice as many genes that can mutate (see p. 552). In the early evolution of haploid/diploid life cycles, the key factors were the ability of diploids to repair double-stranded damage, set against the faster growth rates of haploids. An alternation could be maintained if diploidy occurs at times when damage to DNA is likely and haploidy when growth is possible. Indeed, among unicellular eukaryotes, diploidy is associated with formation of resting cysts, which must survive harsh conditions.

Why should syngamy have largely replaced endomitosis as a means of genome doubling? We have already discussed the long-term advantages of sexual reproduction.

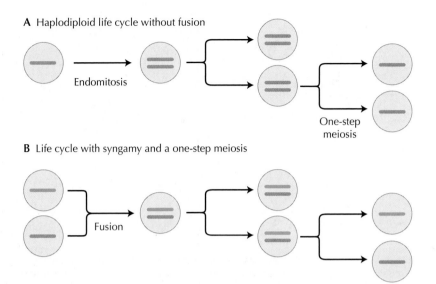

A Haplodiploid life cycle without fusion

Endomitosis

One-step
meiosis

B Life cycle with syngamy and a one-step meiosis

Fusion

FIGURE 23.29. (*A*) The first stage in the evolution of the modern eukaryotic life cycle may have involved a doubling of genome number by endomitosis. Haploidy could be restored by a simple one-step version of meiosis. (*B*) Fusion of genomes from different individuals (syngamy) then replaced endomitosis.

However, syngamy would give an immediate advantage by masking recessive deleterious mutations; two genomes brought together from different individuals would tend to carry different mutations (heterosis; p. 515). This masking has the consequence that deleterious alleles increase in frequency, because they are shielded from selection at the diploid stage. Eventually, the mutation load increases to twice that in haploids. This difference in mutation load would give an advantage to populations with a longer haploid phase, in competition with populations selected mainly as diploids. However, within a single sexual population, the long-term effects of mutation load do not favor modifiers that lengthen the haploid phase. This is because a longer diploid phase gives an immediate advantage from heterosis, whereas the harmful effects of an increased frequency of deleterious alleles is not confined to diploids, but is shared by the whole population. (Recall similar arguments about the evolution of mutation rates [Fig. 23.2].)

At present, it is not understood exactly why so many organisms have life cycles that include extended haploid and diploid stages (e.g., Fig. 23.28C). In general terms, such life cycles can be maintained if selection favors haploidy in one environment but diploidy in another (as suggested above for the earlier evolution of haploid/diploid alternation, before syngamy). We can, however, see some overall patterns. For example, multicellular eukaryotes spend most of their life cycle as diploids, which may be because diploidy masks somatic mutation.

The Two Sexes Evolved through a Division of Labor between Gametes of Different Size

Even when gametes are identical, there are usually distinct **mating types**. For example, in the yeast *Saccharomyces cerevisiae*, there are mating types α and **a**, and in the green alga *Chlamydamonas reinhardtii* the mating types are referred to as + and −. Only different types can mate successfully. The evolution of two mating types can be seen as the first stage in the evolution of the two sexes, male and female.

Mating types can evolve as a result of selection against mating with close relatives. The fusion of two single-celled haploid individuals presumably involves (at least) two

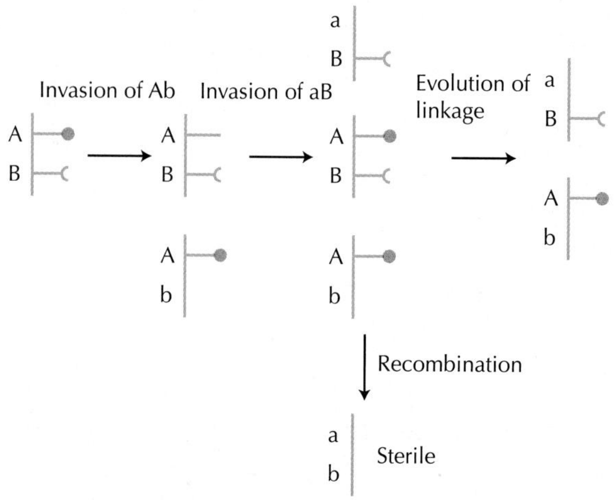

FIGURE 23.30. Mating-types may evolve via loss of function at each of two loci. Initially, two molecules are required for fusion (A, B). Individuals that lose one of these (Ab, say) may gain an advantage, because they cannot mate with close relatives. For the same reason, the complementary genotype can also invade. Recombination between Ab and aB produces ab cells, which cannot mate at all. Tight linkage between the genes therefore evolves, which leads eventually to a population containing just two alternative mating types, Ab and aB.

molecules on the cell surface (Fig. 23.30). We call the alleles coding for these A and B. Now, an individual that lost the function of one of these genes (aB, say) would still be able to mate with AB individuals, but could gain a considerable advantage, because it would be impossible for it to mate with close relatives that had the same genotype. Similarly, Ab genotypes that had lost the activity of the other molecules could also invade the population. Once both Ab and aB had become common, there would be selection for tight linkage between the genes because recombinant ab individuals could not mate at all. Once linkage became complete, the ancestral AB genotype would be lost, and two mating types (aB, Ab) would be established. (Note that this scenario is similar to the model for the separation of male and female function into two sexes [Fig. 23.26]. There, however, we were considering multicellular diploid individuals, which produce distinct male and female gametes. Here, we consider the much earlier evolution of two types of gamete in a haploid single-celled organism.)

Another feature of the differentiation of the two sexes involves the transmission of organelles. It is remarkable that in all known cases, mitochondria and chloroplasts are inherited from only one parent. This is not simply a consequence of the amount of cytoplasm transferred. In humans, approximately 50 mitochondrial genomes are transferred into the egg by the sperm, but are then specifically degraded. In mussels (*Mytilus*), there are two kinds of mitochondria, one of which is inherited from the father and one from the mother. Inheritance is still strictly uniparental, even though both sexes can transmit mitochondria. Even where the gametes are apparently identical, organelles are still inherited from just one mating type. For example, in *C. reinhardtii*, mitochondria are inherited from the − mating type, and chloroplasts from the + mating type.

Uniparental inheritance can be explained as an adaptation to prevent the spread of selfish organelles (recall Fig. 21.4). A nuclear gene, which causes elimination of organelles coming from the opposite mating type, will not suffer the effects of any selfish elements coming from the mate. Conversely, once organelles are inherited from one parent, they will be selected to be benign, because their success depends on that of their host. Interestingly, multiple mating types are found only in organisms such as fungi, in which mating involves transfer of nuclei, with no cytoplasmic fusion and hence no transfer of organelles. This is to be expected, because once transmission of organelles is coupled to mating type, it is impossible for more than two mating types to evolve. (For example, if type 1 transmits organelles, a mating between types 2 and 3 would give offspring with no organelles.)

Gender is defined by the production of two kinds of gametes: small male pollen or sperm and large female eggs. Why should so many species have evolved two sexes, rather than remaining **isogamous** or evolving several kinds of gamete? The most plausible explanation is that there is a trade-off between producing large numbers of small gametes and increasing the chances of fertilization by weight of numbers, as opposed to producing a small number of large gametes that will give large zygotes that have a better chance of survival. The key requirement is that there should be accelerating returns to zygote size, so that the fitness gained by increasing zygote size increases as the zygote gets larger. This explanation is supported by comparison between species in the Volvocales, a family of colonial algae (Fig. 23.31).

We have seen how an isogamous system can evolve two mating types, uniparental inheritance of organelles, and anisogamy. Once these features are established, there may be evolution of individuals that specialize in producing either male or female gametes; the sexes of these multicellular individuals may be determined environmentally or genetically (Fig. 23.36). Regardless of whether production of male and female gametes is separated into individuals of different sexes, sexual selection may elaborate costly traits that function to increase the success of male gametes (Chapter 20). Many of the most interesting phenomena in biology trace back to the evolution of two sexes.

Many Mechanisms Have Evolved for Avoidance of Inbreeding

We have seen that because most deleterious mutations are recessive, crosses between relations produce offspring that have low fitness (Chapter 18 [p. 515]). This phenomenon of **inbreeding depression** has profound consequences. We have already seen that it gives an immediate advantage to syngamy and to lengthening the diploid phase relative to the haploid. It also drives the evolution of mating types and the separation of male and female function into different individuals. In this section, we look at some other ways by which organisms avoid outcrossing.

Angiosperms (flowering plants) have an especially diverse range of mating systems, which have to a large extent been shaped as adaptations for avoidance of inbreeding depression. We have already seen that some groups have evolved incompatibility systems, such that pollen that shares alleles at the incompatibility locus with the female, or pollen from a plant that shares alleles, cannot fertilize (pp. 506 and 541). Incompatibility systems have evolved independently several times in angiosperms, and also in fungi and ciliates.

In the majority of angiosperms, male and female functions are combined within the same flower, so that there is a risk of self-fertilization. In animal-pollinated plants, anthers and stigmas may be separated, which makes selfing less likely. However, this also makes transfer of pollen less effective. This conflict between avoiding selfing and exporting pollen efficiently is resolved by **heterostyly**, where populations are polymorphic for distinct arrangements of anther and stigma. Heterostylous systems have evolved independently 28 times, and in several groups, **tristylous** systems have evolved with three distinct arrangements (e.g., see Fig. 23.33). As with **sex ratios** and **self-incompatibility**, frequency-dependent selection favors rare morphs, so that the polymorphism tends to equal frequencies. Other kinds of polymorphic systems are also found. Polymorphisms for left- versus right-handed flowers have evolved three times (Fig. 23.32A), and recently a remarkable example has been found in which two morphs change structure through the day, in opposite directions (Fig. 23.32B).

In many angiosperm species, flowers are specialized to function as male or female,

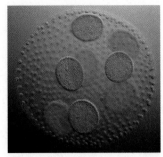

FIGURE 23.31. The Volvocales are an order of algae, some of which form large colonies with some specialization between cells (Fig. 9.1). Species with small colonies tend to be isogamous, whereas those with large, highly differentiated, colonies produce small motile gametes and large eggs. This is consistent with the theory that anisogamy is favored when there are accelerating returns from producing a large zygote, as one would expect if the zygote must develop into a complex colony. The figure shows a colony of *Volvox*, which includes up to 50,000 cells; daughter colonies can be seen developing within the parent colony.

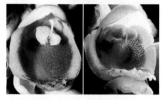

FIGURE 23.32. In *Cyanella alba* (*top*) and *Wachendorfia paniculata* (*middle*), populations are polymorphic for left- and right-handed flower structures; insects transfer pollen most efficiently between flowers with different handedness. In the tropical ginger *Alpinia* (*bottom*), one form acts as a male (i.e., exporting pollen) in the morning and as a female in the afternoon, whereas the other shows the opposite pattern. In the first morph (*left*), the style curves upward in the morning, so that the stigma cannot touch pollinating bees, but then grows downward in the afternoon, when anthers are depleted. The second morph (*right*) shows the opposite pattern; the figure shows the morphs in the morning.

but are still found on the same plant. Selfing may then be reduced by encouraging insect pollinators to move pollen between plants rather than within the same plant. For example, male and female flowers may open at different times. Where entire plants are male or female, selfing, of course, becomes impossible.

Many angiosperms have become obligate selfers. Such a transition is straightforward, because the initial step may be just the loss of one of the various mechanisms for inbreeding avoidance discussed above. For example, small populations of *Eichornia paniculata* have lost **tristyly** as a result of random drift (Fig. 23.33). Moreover, selfers gain an immediate advantage, because they transmit only their own genes to their offspring and can also still fertilize others in the population. (This advantage, termed **automatic selection**, is related to the cost of genome dilution.) Also, selfers do not suffer the risk of pollination failure (an advantage termed **reproductive assurance**). Thus, the evolution of selfing raises many of the same issues as the evolution of sex itself discussed above.

As with asexuals (Fig. 23.9), selfers tend to be distributed at the tips of phylogenies (Fig. 23.34A). This suggests that they are usually short-lived and go extinct relatively quickly. Consistent with this, self-fertilizing plants often retain floral structures, even though these are no longer needed to attract pollination (Fig. 23.34B). Selfing may evolve rarely because inbreeding depression outweighs the immediate advantages of selfing through automatic selection and reproductive assurance.

In this section, we have concentrated on flowering plants, because they have such a diversity of mating systems, and because they are so amenable to experimental manipulation. The same principles apply to other groups; a variety of mechanisms for avoidance of inbreeding have evolved in birds and mammals (e.g., Fig. 23.35). Often, only one sex disperses, an asymmetry that is believed to have evolved to ensure outcrossing. However, as is often the case in the study of adaptation (Chapter 20), it is hard to show unambiguously that a particular trait has evolved through selection to avoid inbreeding depression. For example, in humans there are strong taboos against marriage between close relatives. However, economic and social consequences of inbreeding may be at least as important as the genetic consequences as causes of incest avoidance (Chapter 26).

An intriguing possible mechanism for inbreeding avoidance is the preference of mice for mates that have different genotypes at the **major histocompatibility complex (MHC)**. These preferences are observed between mice that differ only in the MHC region and are based on odor: Female mice in estrus prefer the scent of males that are

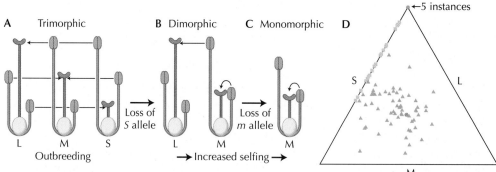

FIGURE 23.33. The aquatic plant *Eichornia paniculata* is tristylous. (*A*) It contains three flower morphs, and transfer of pollen is most efficient between different morphs (L, M, or S). In many populations in northeastern Brazil, polymorphism has been lost from one or both loci involved (*D*); this tends to occur in small populations subject to random drift. Thus, many populations are dimorphic for the M and L morphs (*blue triangles*) or monomorphic for the M morph (*red circle*). Dimorphic and monomorphic populations contain a selfing variant of the M form that has anther and stigma adjacent (as shown in *B* and *C*). In *D*, the frequency of each morph is indicated by the distance from the corresponding edge (labeled L, M, or S). *Orange squares* represent trimorphic populations.

A

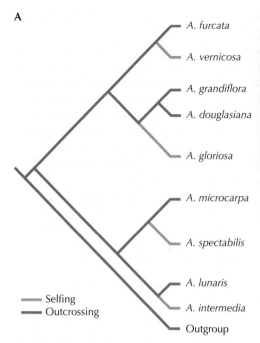

B

- A. furcata
- A. vernicosa
- A. grandiflora
- A. douglasiana
- A. gloriosa
- A. microcarpa
- A. spectabilis
- A. lunaris
- A. intermedia
- Outgroup

Selfing
Outcrossing

FIGURE 23.34. (*A*) Self-fertilization has evolved several times within the genus *Amsinckia*. However, selfing species are confined to the tips of the phylogeny, indicating recent origin. (*B*) Selfing species retain flowers, but these tend to be smaller. (*Left*) *Amsinckia furcata*, an outcrosser; (*right*) its sister taxon *Amsinckia vernicosa*, which is predominantly self-fertilizing.

dissimilar at the MHC. Similar MHC-based mate choice has been reported in other species, including humans. However, preferences for mating with different MHC genotypes may have evolved because they give offspring fitter genotypes at the MHC locus, rather than as a means for avoiding inbreeding depression per se. (Recall that MHC polymorphism is maintained by **balancing selection**, possibly because of greater disease resistance conferred by MHC heterozygosity and by rare MHC alleles [p. 541].) It is difficult to separate out inbreeding avoidance from other causes of selection on mate choice (see p. 581).

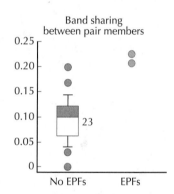

Band sharing
between pair members

No EPFs   EPFs

FIGURE 23.35. Extra-pair fertilizations (EPFs; Fig. 20.34) are more common when partners are genetically similar, as measured by sharing of bands in multilocus DNA fingerprints (Box 13.3). This allows females who are paired with a related mate to avoid producing offspring that suffer inbreeding depression. Full siblings share 50% of bands. For broods without extra-pair young, data are presented as median, bars showing 10th and 90th percentiles, and data points outside these percentiles (*red dots*). The number of broods is given beside the box. For broods with EPFs, the individual data points are shown (*blue dots*). Data are for western sandpiper (*Calidris mauri*); similar patterns are seen in Kentish plover (*Charadrius alexandrinus*) and common sandpiper (*Actitis hypoleucos*).

EVOLUTION OF EVOLVABILITY

Evolution Depends on the Relation between Genotype and Phenotype

Recently, there has been much interest in the evolution of **evolvability**. There are various definitions of this term, but in this section we take it to mean the ability of a population to generate phenotypic variation that is useful material for natural selection. The term is sometimes used in a narrow sense, to refer to the additive genetic variance measured relative to the trait mean (see p. 399). Here, however, we use it more broadly and in a qualitative rather than a quantitative sense. Thus, evolvability includes evolution of mutation rates, and of sex and recombination, as discussed above. However, most recent discussions of evolvability have come from developmental and molecular biologists, whose focus is on the way genetic variation is translated into phenotypic variation.

So far in this chapter, we have concentrated on the way the genetic system shapes variation in genotype frequencies. However, as biologists, we are most interested in characteristics of the whole organism (body shape, behavior, and so on), which are the outcome of a complex developmental process and which depend on the combined action of many genes. In Chapter 14, we saw how the evolution of quantitative traits can be described quite simply, despite the complex processes that lead from genotype to phenotype. There, we took that relationship as given; here, we ask why it is as it is. To what extent has the relation between genotype and phenotype been shaped by selection? Has it evolved in order to generate adaptive variation?

If variation is to be "useful material for natural selection," then it must cause changes that improve some function, without disrupting the rest of the organism. Thus, there are two complementary issues in the evolution of evolvability: (1) How can an organism be buffered against the potentially deleterious effects of genetic changes? and (2) How can phenotypic changes that improve function be produced? These two aspects to the problem mirror our discussion of the evolution of mutation rates and of recombination. There, we were concerned both with the load due to deleterious mutations and with the incorporation of favorable mutations (Table 23.1). In the rest of this chapter, we concentrate on the robustness of organisms to deleterious mutations; in the next chapter, we consider the complementary question of how organisms can change despite this robustness.

Dominance of Wild-type Alleles Is Only Weakly Selected

Consider the simplest case of a single locus, with two alleles: wild-type and mutant. What determines the phenotypes of the three diploid genotypes? Specifically, why are the great majority of mutant alleles recessive and the wild-type allele dominant (Chapter 13 [p. 358])? Fisher proposed that because rare alleles appear mainly in heterozygotes, selection acts to ameliorate their effects on heterozygotes. Although mutations may have an additive effect when they first arise, they eventually evolve to become recessive. (This does not much reduce the mutation load, because as alleles evolve smaller effects on heterozygotes, they become more common. As long as heterozygotes are slightly less fit, one selective death of a heterozygote is needed to eliminate each mutation [see p. 552].)

Sewall Wright opposed Fisher's theory primarily because selection on modifiers of dominance at a single locus is extremely weak, being of the same order as the mutation rate. Wright argued that in reality, modifiers would have **pleiotropic** effects on fitness that would be far stronger than the weak selection arising from their effects on dominance. Also, Fisher's theory requires an enormous number of modifier alleles, with effects on specific loci or even on specific mutant alleles. Therefore, Wright proposed the physiological explanation that is now widely accepted. If the rate at which a reaction proceeds shows diminishing returns with enzyme activity, then alleles causing loss of function will tend to be recessive (Fig. 23.36).

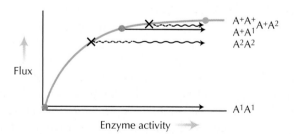

FIGURE 23.36. Dominance of wild-type alleles arises naturally if the flux through a metabolic pathway shows diminishing returns with the level of enzyme activity. The *blue circle* shows the wild-type homozygote (A^+A^+), which produces enough enzyme activity that it is not the limiting factor under normal conditions. *Red circles* show the effect of an allele A^1, which completely abolishes activity in the homozygote. It causes only a small reduction in flux in the heterozygote and so appears approximately recessive. In contrast, an allele of small effect (A^2) has an approximately additive effect, with the heterozygote A^+A^2 being close to halfway between the two homozygotes A^+A^+ and A^2A^2.

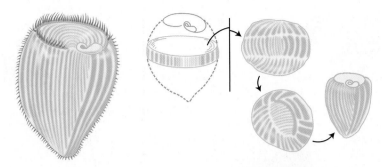

FIGURE 23.37. The single-celled ciliate *Stentor coerulus* can recover from gross disruption. The normal cell (0.5- to 1-mm long) has a series of pigmented stripes (*left*). These change smoothly in width around the cell, except along a line where thin stripes meet thick (*front right of cell*). If a ring of tissue is excised surgically, the normal pattern is restored within a few days, even though the manipulated cell is much smaller.

Why would enzymes be produced in such excess that halving their activity has little phenotypic effect? J.B.S. Haldane argued that this is a consequence of selection for a "safety factor" against fluctuating conditions. In general, the flux through a metabolic pathway tends to be insensitive to small changes in the activity of any one enzyme within it; this provides a quantitative justification for Wright's key assumption. These arguments apply to enzymes involved in metabolism, but it seems plausible that they carry over to regulatory networks. Evidence in support of the physiological explanation comes from the observation that in *Drosophila*, alleles with small effects on viability show little dominance—just as expected under the physiological theory (Fig. 23.36). Most convincing, spontaneous mutations in *C. reinhardtii* are mostly recessive. Because this alga is haploid through most of its life cycle, selection cannot have acted to modify dominance.

Recent debate has concerned the broader properties of the networks of interacting genes responsible for development. How can a precisely coordinated form develop, despite environmental and genetic perturbation (e.g., Fig. 23.37; Chapters 9 and 11)? Conversely, how can the novel variants required for evolutionary change be produced without fatal disruption of the current organism? We saw in Chapter 13 (p. 358) that tightly regulated phenotypes conceal extensive genetic variation, which can be revealed in extreme environments or on new genetic backgrounds. For example, C.H. Waddington's classic experiments showed that expression of the cross-veinless mutant, *cv*, in *D. melanogaster* is highly variable. Selection for a larger gap in wing variation on a homozygous *cv/cv* background eventually produced flies with gaps on a wild-type background even though wing variation is normally highly constrained (Fig. 14.20A). We will examine how novel phenotypes can be selected in this way in the next chapter.

A striking example of such cryptic variation in *Drosophila* involves the heatshock protein Hsp90, a **chaperone** that assists the correct folding of denatured proteins (Fig. 23.38A). It is one of the most abundant proteins in the cell and is produced in large quantities under stressful conditions (e.g., at high temperatures). Flies with reduced expression of Hsp90 show a remarkable range of phenotypic variation, which is due to polymorphism in large numbers of genes; normally the effects of this polymorphism are hidden by the buffering effect of *Hsp90* (Fig. 23.38). Reduced Hsp90 in *Arabidopsis* reveals similar variation. In *E. coli*, overexpression of another chaperone, GroEL, was found to greatly increase the fitness of lines that had accumulated deleterious mutations over 3000 generations. Thus, chaperones are an example of a mechanism that buffers organisms against both environmental and genetic perturbations.

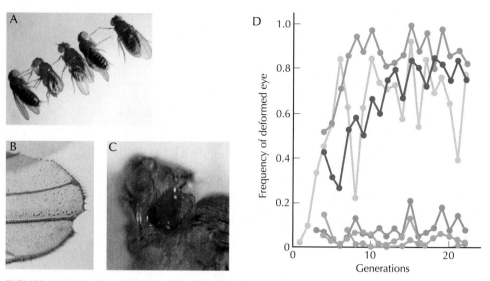

FIGURE 23.38. Reduced levels of the heat-shock protein Hsp90 cause a wide variety of abnormalities (e.g., (*A*) small wings; (*B*) notched wings; (*C*) deformed eye). Particular abnormalities are found in crosses with particular wild-type stocks, showing that Hsp90 interacts with genetic variation that is not normally expressed. (*D*) Selection for (*red, deep purple,* and *orange lines*) deformed eye in flies with reduced Hsp90 expression causes a rapid response, which confirms that cryptic genetic variation is involved. As a control, selection was applied to flies that did not exhibit the eye mutation (*blue, green,* and *light purple lines*).

Robustness Can Evolve in Several Ways

In general terms, we can identify several causes of the remarkable robustness of development. Specific systems have evolved to reduce disruption—for example, chaperones such as Hsp90, temperature compensation in biological clocks, and DNA repair systems. There is extensive redundancy, which may involve duplicate genes or which may involve nonhomologous genes with overlapping function. We have seen one such example in the set of overlapping transcription factors that regulate expression of *even-skipped* (Fig. 19.31). More generally, it is remarkable that in all eukaryotes that have been examined, most genes can be eliminated with no obvious phenotypic effect (see p. 522). Finally, many functions are not specified precisely, but rather develop by a trial-and-error process analogous to selection. **Exploratory systems** such as this are essential if the limited amount of information coded in the genome is to specify complex organisms; they also make development robust to environmental and genetic perturbations. We consider them in more detail in the next chapter.

How has robustness evolved? To some degree, robustness may be a property of any large functional network. However, selection will strongly favor systems that are robust to random noise and that function across the range of environments experienced by the organism. (In the context of the evolution of dominance, this is Haldane's argument for the evolution of a "safety factor" in the production of metabolic enzymes.) Similarly, there may also be selection for robustness toward *genetic* perturbations. This is more plausible for changes that affect the action of many genes (e.g., Hsp90), rather than for the modifiers of single loci postulated by Fisher, simply because the selection pressure is proportional to the net mutation rate at all the genes whose action is being modified. On this argument, the relative importance of robustness to environmental versus genetic variation may depend on the relative magnitudes of these perturbations.

SUMMARY

The evolution of genetic systems involves those traits that affect the way that genetic information is passed from one generation to the next: rates of mutation and recombination, the proportion of the life cycle spent as diploid rather than haploid, the level of inbreeding, the degree of sexual reproduction, and so on. Even if alleles that modify the genetic system have no direct effect on fitness, they may facilitate the generation of fitter variants in the future and so can themselves hitchhike to higher frequency. This is known as **indirect selection**. Theoretical predictions made by studying hypothetical modifier alleles can be tested by making broad comparisons across species, by observing natural populations, or by experiments on rapidly reproducing organisms in the laboratory.

Most mutations are deleterious, and so selection acts primarily to reduce mutation rates. In asexual populations, mutator alleles can hitchhike to high frequency along with the beneficial alleles that they produce. However, such indirect selection for higher mutation rates is ineffective in sexual populations. Genes that experience fluctuating selection (especially those involved in interactions between host and pathogen) may evolve special mechanisms for generating higher levels of random variation.

Sexual reproduction has a variety of costs—most important, the cost to a female of providing resources to raise the male's genes. (This is known as the **cost of genome dilution**.) Recombination is also costly, because it breaks up fit combinations of alleles that have been built up by selection. Nevertheless, sex and recombination are widespread. Asexual taxa are almost always short-lived, but the differential survival of sexual and asexual species (**species selection**) cannot explain the prevalence of sex. Many species have facultative sex, and recombination rates vary; thus, we must explain how individual alleles that increase sex and recombination can be favored, despite the obvious costs.

Sex and recombination only affect populations if there are nonrandom associations between alleles which they can break up. Thus, any population genetic explanation depends on the existence of deviations from Hardy–Weinberg proportions and/or linkage disequilibria. In the most plausible theories, recombination breaks up negative associations between favourable alleles, generating additive genetic variance in fitness that fuels natural selection. Negative associations can be generated by certain kinds of epistatic selection or by random drift. They may involve deleterious mutations, fluctuating selection, or multiple selective sweeps. As we saw in Chapter 19, the relative importance of these different kinds of selection remains an open question.

Sex has many consequences—most directly, an alternation between haploid and diploid parts of the life cycle. Diploidy may be favored because it facilitates repair of double-stranded breaks in the DNA, and because it masks deleterious recessive alleles. Many features of the genetic system arise from selection to avoid inbreeding, which unmasks deleterious recessives, eliminating the advantage of diploidy. Selection against inbreeding drives the evolution of separate mating types and separate sexes and is responsible for the extraordinary devices that reduce selfing in flowering plants.

Evolution depends on **evolvability**: the ability of populations to generate variation that can be used by natural selection. The evolution of genetic systems is thus essentially about evolvability, but the term is currently used mainly to refer to the relation between genotype and phenotype: Can the genetic variation produced by mutation and recombination produce fitter phenotypes without disrupting existing function? We take up these issues in the following chapter, on the evolution of novelty.

FURTHER READING

Evolution of Mutation Rates

Sniegowski, P.D., Gerrish P.J., Johnson T., and Shaver A. 2000. The evolution of mutation rates: Separating causes from consequences. *BioEssays* **22:** 1057–1066.
    Explains the evolutionary forces that shape mutation rates and the evidence for their importance.

Evolution of Sex and Recombination

Barton N.H. and Charlesworth B. 1998. Why sex and recombination? *Science* **281:** 1986–1990; Burt A. 2000. Sex, recombination and the efficacy of selection—Was Weissman right? *Evolution* **54:** 337–351;

Otto S. and Lenormand T. 2002. Resolving the paradox of sex and recombination. *Nat. Rev. Genet.* **3:** 252–261; Rice W.R. 2002. Experimental tests of the adaptive significance of sexual recombination. *Nat. Rev. Genet.* **3:** 241–246.
    Some recent reviews of the central problem: Why is sexual recombination so widespread?

Consequences of Sex

Barrett S.C. 2003. Mating strategies in flowering plants: The outcrossing-selfing paradigm and beyond. *Philos. Trans. R. Soc. (Lond.) B Biol. Sci.* **358:** 991–1004.

A broad review of the ingenious devices that flowering plants use to avoid inbreeding.

Charlesworth B. 1991. The evolution of sex chromosomes. *Science* **251:** 1030–1033. Charlesworth D., Charlesworth B., and Marais G. 2005. Steps in the evolution of heteromorphic sex chromosomes. *Heredity* **95:** 118–128.

These two reviews cover the theory and evidence relating to sex chromosome evolution.

Felsenstein J. 1974. The evolutionary advantage of recombination. *Genetics* **78:** 737–756. Maynard Smith J. 1978 *The evolution of sex.* Cambridge University Press, Cambridge.

Classic works that helped establish our modern understanding of the evolution of sex and recombination.

Evolvability

Kirschner M. and Gerhart J. 1998. Evolvability. *Proc. Natl. Acad. Sci.* **95:** 8420–8427. Kirschner M. and Gerhart J. 2005. *The plausibility of life: Resolving Darwin's dilemma.* Yale University Press, New Haven, Connecticut.

The authors argue that gene regulation and development have evolved so as to facilitate further evolution—a controversial argument.

24

Evolution of Novelty

ARLY IN THE HISTORY OF LIFE, STRIKING NEW FEATURES appeared that are now shared by all living organisms—the basic machinery for replicating and translating genetic information and the fundamental metabolic mechanisms that provide the energy and materials to run these processes. Building on this common foundation, an astounding variety of novelties arose that were critical to the diversification of life. Among the thousands of examples are specialized metabolic processes, such as the ability to degrade and metabolize cellulose; the development of organelles, such as the endoplasmic reticulum and the mitochondrion; the formation of organs, like the eye and the heart; and the establishment of new behaviors, such as storage of seeds by birds and dam building by beavers. More fundamental than these individual examples are the novelties that permit the diversification of major taxa. For example, the presence of introns in eukaryotic genomes enables alternative RNA splicing to produce a diversity of proteins (Fig. 8.22) and has facilitated re-

combination between different protein domains (Fig. 8.23). The specialization of individual segments and associated appendages in various animal groups has allowed them to exploit a wide variety of ecological niches (Chapter 11 [pp. 289–305]).

In this chapter, we discuss how novelty evolves. We have already seen how imperfect inheritance generates new genetic variants (Chapter 12), how different genetic variants interact to determine phenotype (Chapter 14), and how selection shapes this genetic and phenotypic variation to build up complex adaptations (Chapter 17). This chapter will bring together ideas and examples that have been encountered before, but will focus on understanding those novelties that at first seem the hardest to explain. Evolution has produced an astonishing variety of novel adaptations, but there are some general principles: small changes accumulate step-by-step; existing functions are co-opted for new purposes; new gene combinations are generated by **lateral transfer** and **symbiosis**; adaptations are robust, functioning despite environmental and genetic change; and adaptive systems evolve without requiring explicit blueprints to specify complex functions. These principles make it possible to see how even the most complex features evolved.

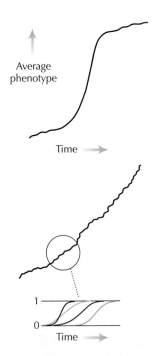

FIGURE 24.1. A change in phenotype may be due mainly to the fixation of a major mutation of large effect (*top*). In contrast, it might instead be due to the accumulation of very many substitutions of small effect. The vertical axis in both figures shows the average phenotype. This changes continuously as the frequencies of the alleles that affect the phenotype change (*bottom, inset*).

THE BASIC FEATURES OF NOVELTY

Quantitative Changes Can Lead to Qualitatively New Features

It is tempting to suppose that the evolution of novelties involves major mutations that establish the novelties' key features in a single step. Indeed, the first geneticists emphasized major mutations as the basis for speciation (see p. 25). **Quantitative trait loci** (QTLs) of large effect do sometimes contribute to phenotypic differences within and between species (see pp. 298–318, 408–409, and 633–638), and, in some instances, major mutations may indeed be involved in establishing a novel feature. Nevertheless, what are now distinct forms are, in fact, the end points of a long evolutionary process; even sister species typically diverged millions of years ago, leaving no trace of the intervening generations. Thus, what exists now is typically due to the accumulation of many small changes. What now seems to be a discontinuous transition may actually be the result of a more gradual and continuous process (Fig. 24.1). Similarly, a QTL of major effect could have arisen by a series of mutations of small effect over a long period of time (e.g., Fig. 14.27).

A qualitative change may arise in two ways. First, a sufficiently large continuous change may be seen as a qualitative shift. (For example, a change in plumage color from red to green consists of a shift in the wavelength of reflected light.) Different forms may be connected by a continuous transformation involving changes in just a few key growth rates. Differential growth (known as **allometry**) and changes in the timing of developmental events (**heterochrony**) provide simple routes that can lead to radical changes in body shape (Fig. 24.2). Similar possibilities exist for other aspects of phenotype; for example, major changes in the way ants forage for food might be caused by slight shifts in the rules that they use to lay down and follow pheromone trails.

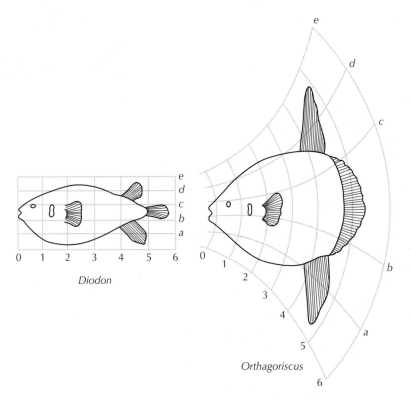

FIGURE 24.2. In his book *On Growth and Form*, D'Arcy Thompson (1917) suggested that quite different morphologies could be produced by simple geometrical transformations caused by differences in rates of growth. This example shows how a porcupinefish (*Diodon*) transforms to a sunfish (*Orthagoriscus*).

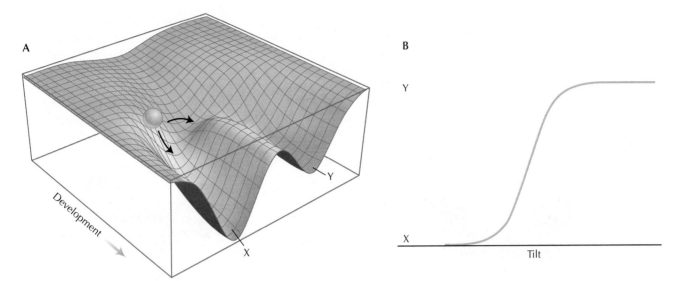

FIGURE 24.3. Slight changes in the development of an organism can cause sudden shifts to a new outcome. (*A*) Development can be seen as analogous to a ball rolling down a gradient. Slight tilts to the surface cause a shift to a new channel. (*B*) The chance of developing to phenotype X rather than Y suddenly changes as the tilt passes a critical point.

Second, and more fundamentally, small and continuous changes can flip a system from one state to another, thus producing a substantial qualitative shift. A continuous change in some underlying variable, such as a gene's level of expression, may cause a sudden switch to a qualitatively new outcome. C.H. Waddington illustrated this basic point by depicting the development of an organism as analogous to a ball rolling down a channel on a slope with multiple channels. A slight perturbation will switch its path to a new channel, thereby producing a distinct outcome (Fig. 24.3). Chapter 11 provided examples involving *Hox* genes, in which changes in level and timing of *Hox* gene expression can lead to major morphological changes (pp. 293–307). Waddington's analogy assumes that development is **canalized** into a limited number of possible paths—each of which is robust to small perturbations. Why this might be was discussed at the end of Chapter 23, and we will return to this key issue later in this chapter.

The implication of Figure 24.3 is that a continuous change in the development of a single organism can lead to a sudden shift to a qualitatively different phenotype. However, even if the *individual* phenotype changes continuously, a change in the direction of selection can cause the *population* to suddenly shift to a new state. This is best illustrated in terms of the **adaptive landscape**, which shows how the average fitness of the population depends on the averages of several quantitative traits (see p. 472). If this landscape contains two **adaptive peaks**, then selection will take the population uphill to the nearest peak. A population on the lower peak can evolve to a higher one if a slight change in environment temporarily removes the valley. Figure 24.4 shows an example involving mimicry in burnet moths, in which both the genetics and the selective forces are well understood.

We saw in Chapter 22 that populations may move to a new adaptive peak in a variety of ways. As well as a shift in environment, random genetic drift may knock a population across a valley of reduced fitness, despite opposition from natural selection. There will be only an appreciable chance that this will happen if the product of effective population size and the depth of the valley in mean fitness is not too large (see p. 641). We also saw that populations can evolve along "ridges" in the adaptive landscape, so that they can come to be separated by a deep valley, even though this divergence was never opposed by strong selection (see pp. 642–644). In the example of Figure 24.4, a shift to a new warning pattern was made easier because a shift to a new allele at one gene gives a reasonable resemblance to the new pattern, which later can be improved by a substitution at a second gene. Let us examine this example in more detail.

The burnet moth *Zygaena ephialtes* exists in two forms, each mimicking a different toxic model species: *Zygaena filipendulae* (Fig. 24.4A, left) and *Amata phegea* (Fig. 24.4A, right). This is an example of Müllerian mimicry, in which several distasteful

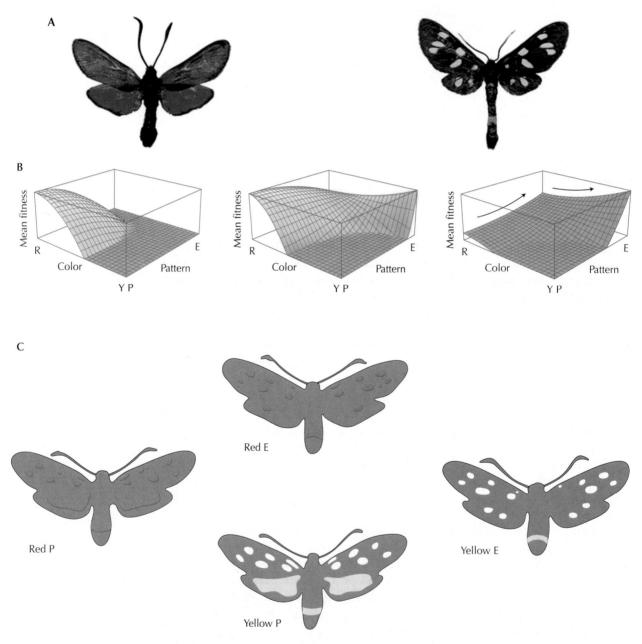

FIGURE 24.4. Müllerian mimicry in the burnet moth *Zygaena ephialtes* illustrates how populations can diverge onto different adaptive peaks as a result of varying selection pressures. (*A*) Two distasteful model species, *Zygaena filipendulae* (*left*) and *Amata phegea* (*right*). (*B*) Plots of mean fitness against the frequency of the alleles determining color and pattern in *Z. ephialtes* ("adaptive landscapes"). The *left* panel is for a region where *Z. filipendulae* is common, and the *right* panel is for a region where *A. phegea* is common; the *middle* panel shows the two alternate adaptive peaks that exist where both models are present. (*C*) The two mimetic forms of *Z. ephialtes* (Red P, *left*, and Yellow E, *right*), and the recombinant forms (*center*).

species gain an advantage by sharing the same warning pattern, so as to better advertise their distastefulness to potential predators (see p. 475). The two forms of *Z. ephialtes* differ in color and pattern, with each aspect of their phenotype controlled by alternative alleles at a single locus. A dominant allele at one locus results in red instead of yellow spots, and a dominant allele at another unlinked locus gives the "P" pattern, with a large patch on the hindwing, instead of the "E" pattern (Fig. 24.4C). Thus, the red P form mimics *Z. filipendulae* (in Fig. 24.4, compare panel A left and panel C left) and the yellow E form mimics *A. phegea* (in Fig. 24.4, compare panel A right and panel C right).

The two forms of *Z. ephialtes* meet in a narrow **hybrid zone** in Austria, where recombinant phenotypes, Red E and Yellow P, are produced (Fig. 24.4C, middle). These recombinants do not carry either warning pattern, and so selection against them keeps the two parental gene combinations separate. Thus, the two forms represent different adaptive peaks: A new allele of either gene could not invade. In Figure 24.4B, the middle diagram shows the adaptive landscape in a region where both models are present, so that there are two adaptive peaks.

How could a new mimetic form arise despite this selection? A plausible hypothesis is that in southern Italy, *A. phegea* is much more common than *Z. filipendulae*. Therefore, even a poor mimic would be at an advantage compared to the original pattern, which resembles a model that is very rare. The E-pattern allele could therefore increase from low frequency. Once it became common, the yellow color allele would be at a strong advantage and would fix, giving the new yellow-E form (note the arrows in Fig. 24.4B, far right).

In this example, the diverging populations can each evolve uphill on the adaptive landscape, and yet come to be separated by an adaptive valley. The question of how adaptive valleys can be bypassed becomes central when we consider how multiple interdependent changes can be established, a subject that is explored next.

Novelty Often Requires Multiple Interdependent Changes

The hardest novelties to explain are those requiring the coordination of many features. Complex adaptations such as the vertebrate eye, the seed-storing behavior of some birds, or the molecular machinery of the ribosome are remarkable because they all involve many components, which depend on each other in an intricate way. The interaction between multiple interdependent components makes it likely that any change will disrupt the whole system and, conversely, makes it hard to see how a new set of components can be brought together. Examples of such problems recur throughout this chapter. In this section, we summarize the general issues and the relevant arguments that were encountered in earlier chapters.

In Chapter 14 (p. 407), we set out Fisher's geometrical argument that complex functions are most likely to evolve via changes of small effect. Specifically, if selection is to coordinate n components, then the typical size of a favorable mutation will decrease with the square root of that number ($\sim 1/\sqrt{n}$). This argument is not decisive, however. There are several reasons why the effective number of dimensions, n, in which organisms evolve may be smaller than first appears, so that substantial changes can occur without undue disruption. First, changes in just a few factors (e.g., growth rates in different parts of the embryo or subtleties in the expression of key transcription factors during development) can have radical effects on form (e.g., Figs. 24.2 and 11.28). Second, even a rough approximation to the optimal form may be advantageous (e.g., Fig. 24.4); and, third, if the organism is organized into more or less independent modules, then one module can change without disrupting the others.

Empirically, we know that although different parts of the phenotype are tightly correlated, selection nevertheless can rapidly produce new phenotypes that are quite

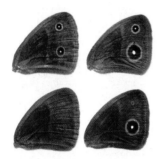

FIGURE 24.5. Selection can readily break down constraints to produce novel phenotypes. Within populations of the butterfly *Bicyclus anynana*, there is a strong **genetic correlation** between the sizes of the anterior and the posterior eyespots. Nevertheless, 25 generations of artificial selection produced all four possible combinations of small vs. large anterior and posterior eyespots.

different from anything seen before. We have already seen one such example in Chapter 14. Within populations of flies, different measures of wing shape vary together, and yet artificial selection can rapidly alter this relationship, producing wings quite unlike any shape seen before (Fig. 14.23). This is because although there is much more genetic variation for changes in wing size than there is for changes in wing shape, there is still *some* variation in shape, and so selection can still change shape. A similar example is seen in *Bicyclus* butterflies, in which there is normally a tight relationship between the relative sizes of different eyespots. Again, artificial selection can readily break down these constraints, thereby generating novel phenotypes (Fig. 24.5).

The real issue, then, is not whether variation that takes the right direction across a continuous space of possibilities can be produced; rather, it is whether such variation can be favored by selection, given that it may disrupt current function and may not give any advantage until several components of the new function have been assembled. Different species may have evolved different complex functions, which are separated by adaptive valleys. However, the existence of these valleys may have been irrelevant to the evolution of those species from their common ancestor. In the previous section, we saw that there are several ways by which such adaptive valleys can be bypassed. In the rest of this chapter, several concrete examples are presented as we focus on the evolutionary mechanisms and processes that produce novelties.

These may act at any biological level: from the gene to development to morphology to behavior. What is important is how novelties originate: what kinds of genetic change are involved and how they are caused by evolutionary processes (primarily, by selection). Although in many cases **inventions** require multiple complex genetic changes, this is not always the case. Conversely, although **innovations** are frequently the result of simple, small genetic changes, they can sometimes require many evolutionary steps.

The remainder of this chapter has been divided into six sections. We begin with a discussion of how novelty can arise directly from changes in the activity of individual gene products. Then we discuss how the origin of novelty is fostered by a variety of biological features: regulation, redundancy, modularity, and **symbioses**. We close the chapter with a discussion of the role natural selection plays in the origin of novelty and in particular how natural selection can allow simple steps to be pieced together to yield major changes to the organism.

CHANGES IN THE ACTIVITY OF GENE PRODUCTS

The simplest way that novelty arises is when one or a few mutations alter the functions of a single gene product, which directly changes some important phenotypic property of the organism. For example, an alteration in a biosynthetic enzyme that produces a pigment might directly change feather color, or a change in the activity of a digestive enzyme might shift the eating habits of an organism. In this section, we will see how simple changes in protein or RNA sequence may directly change the activity of a macromolecule, indirectly change its conformation, or cause complete loss of function. All of these kinds of change can lead to novel organisms.

It is very important to remember that mutation is random with respect to adaptation (see pp. 344–347). Thus, most mutations will either be inconsequential or deleterious. Nevertheless, some will produce effects that lead to novel features that increase fitness.

Small Sequence Changes in the Active Site Can Greatly Alter Catalytic Functions of Enzymes

A straightforward way for enzyme function to change is through changes in its **active site**. The active site is the center of catalytic activity and usually consists of only a few amino acids. When these amino acids are altered, the substrate specificity may change.

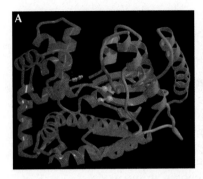
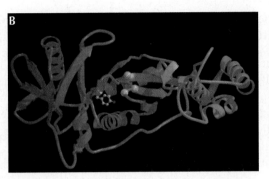
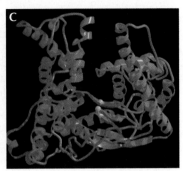

FIGURE 24.6. Changes in specificity of nucleic acid polymerases. The crystal structures of three different types of nucleic acid polymerase are shown. (*A*) Klenow fragment DNA-dependent DNA polymerase, (*B*) human immunodeficiency virus (HIV) RNA-dependent DNA polymerase (also known as **reverse transcriptase**), and (*C*) T7 phage DNA-dependent RNA polymerase. The three-dimensional structures of these proteins are very similar, indicating a common evolutionary origin. Each enzyme is considered to have three main domains much like a hand: a palm domain (shown in *blue*), a thumb domain (*green*), and a finger domain (*purple*). The sugar-specificity region of each enzyme is highlighted in *yellow* and *red*. Changes in these few amino acids are all that is required to change the specificity of these enzymes.

A good example involves the nucleic acid polymerases, which catalyze the production of either RNA or DNA using an existing nucleic acid as a template. Although the functions of these different enzymes are distinct, one form can be converted to another with relative ease—sometimes by changing only a single amino acid (Fig. 24.6).

Changes in the active site can also lead to changes in enzyme kinetics (i.e., reaction rates) and these too can have major biological effects. Some excellent examples come from experimental studies initiated by the discovery of structural polymorphisms in metabolic enzymes (see pp. 368 and 522–525). For example, in *Colias* butterflies, slight differences in the amino acid sequence of the enzyme phosphoglucose isomerase (PGI) lead to different rates of glycolysis. These rate variations in turn affect a variety of fitness-related traits in the butterflies, such as their ability to fly at various temperatures and their patterns of activity during the day (see p. 525). The differences in PGI represent small amino acid sequence variations relative to the overall structure of the protein, but nevertheless have major consequences for the butterflies: They alter patterns of behavior and influence the ability to reproduce successfully (Fig. 24.7).

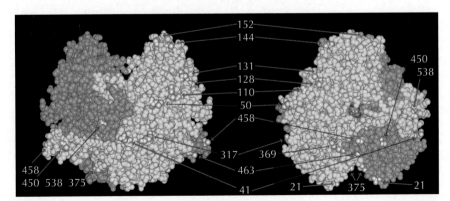

FIGURE 24.7. Amino acid differences in the phosphoglucose isomerase (PGI) enzyme of the butterfly *Colias eurytheme* cause temperature-sensitive differences in kinetics, which in turn cause substantial differences in behavior and physiology (see p. 525). The enzyme is a dimer, made up of two monomers shown here in *green* and *yellow*. Several alternative alleles are maintained in the population; these differ by, on average, four amino acids, which typically have different charge. *Numbers* indicate the positions of polymorphic amino acids in the protein sequence. *Left* and *right* panels show views from different angles.

Protein Conformation (and Hence Function) Can Be Greatly Altered by Small Changes in Amino Acid Sequence

The nucleic acid polymerase example (Fig. 24.6) illustrates phenotypic effects that result from amino acid changes within the active site of an enzyme. Slight changes can also alter the overall shape of a protein. A protein's shape (also called its conformation) is critical for all of its functions. Even single amino acid changes can dramatically alter the shape of a protein by disrupting existing structural features (such as α helices, β sheets, folds, and hydrophobic cores) or by creating new features (e.g., generating a new **hydrophobic core** by replacing a charged residue with a hydrophobic one in a region of the protein dominated by hydrophobic amino acids).

Changes in conformation can affect protein function in many ways. For example, such changes can affect the **allosteric interactions** that allow one region of a protein to regulate the activity of other regions (see pp. 54–56). A conformational change can also affect an enzyme's interaction with its cofactors. For example, changes to just three amino acids in the retinal binding region of red opsin can shift its light absorption properties to green (Fig. 24.8), which in turn affects color vision.

The potential for changes in conformation to affect function extends to all proteins, not just enzymes. Perhaps the best examples are from proteins involved in **molecular recognition** (Fig. 24.9). Here, the shape of a protein regulates its interaction with other molecules (e.g., by binding to them), which in turn triggers a response. An example of

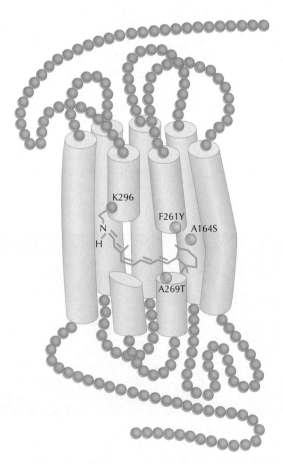

FIGURE 24.8. In humans, the color shift between red and green opsins is due to only three amino acid differences, at positions 164, 261, and 269. These are involved in binding of the retinal molecule (*orange*).

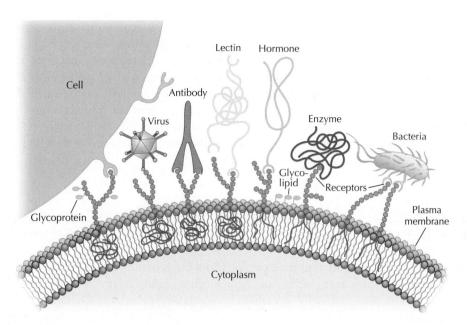

FIGURE 24.9. Molecular recognition. Examples include antibody–antigen interactions, receptors binding to hormones, and enzymes binding to inhibitors.

how changes lead to novelty comes from **bacteriophage** that gain entry to their host by recognizing and attaching to molecules on the surface of their bacterial target. Bacteria can evolve resistance to the phage through genetic changes that alter the topology and electrochemical nature of their targeted surface molecules. Bacteriophage can become infectious again if they, in turn, undergo genetic changes that permit binding to the modified bacterial surface molecules. The same type of "arms race" occurs between the human immune system and human pathogens. In each case, simple genetic changes, including single-nucleotide substitutions, can change the ability of a pathogen to infect a host or the ability of a host to mount an immune response to the pathogen.

Loss of Function Can Produce Novel Phenotypes

The novelties presented so far are generated by converting one activity or function into another. However, novelty can also arise when an existing function is lost. The evolution of pathogenicity provides one such example. In some pathogens, genes have evolved that suppress the harmful effects of the pathogen on the host organism—an advantage if transmission of the pathogen requires survival of the host. Inactivation of these genes thus can lead to the emergence of a novel pathogen (Fig. 24.10). Conversely, deletion of a host protein that is recognized by the pathogen may confer resistance; recall the example of the CCR5-Δ32 deletion, which gives resistance to HIV (see p. 434).

A conceptually similar phenomenon is seen in viruses that cause cancer. The pathway to the development of cancer usually requires the inactivation of tumor suppressor genes. Such genes delicately regulate cell growth so that overall function is maintained. When enough of them are inactivated, cell growth becomes unregulated, leading to cancer. Many viruses target these genes for inactivation, which in turn means that the viruses can directly or indirectly cause tumors to form—which may enhance their transmission. From the point of view of the virus, the creation of tumors is a novelty, but one that may involve the relatively simple inactivation of tumor suppressor genes in the host.

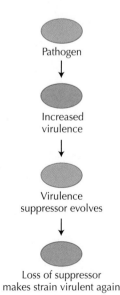

FIGURE 24.10. Pathogenicity suppression.

How does loss of function occur? Point mutations or frameshifts can inactivate a gene, for example, by adding a stop codon early in a protein-coding sequence. Deletion of a gene can also lead to loss of function. Larger deletions can inactivate many genes at once, resulting in a series of changes in an organism's biology, possibly creating major novelties. (Recall that insertion and deletion polymorphisms are surprisingly common in human populations [Fig. 13.32].) Finally, one of the more common ways to inactivate a gene is through the insertion of a transposable element (see p. 342). When transposable elements are highly active, they repeatedly generate novelty by altering gene expression or by gene inactivation.

CHANGES IN GENE REGULATION AND INTERACTIONS WITHIN A NETWORK: TARGETING, DIFFERENTIATION, AND DEVELOPMENT

In biological systems, function is determined by what activities molecules can perform and by the details of when, where, and how much of a particular molecule is present. The latter are determined by a variety of regulatory processes. Changes in these processes are a major source of novelty. In this section we first discuss how changes within a gene can affect its regulation. Then we discuss a more pervasive source of evolutionary novelty: changes in single gene products or single environmental conditions that can create a cascade of changes in a regulatory network.

Function Depends Not Just on the Potential for Biochemical Activity But Also on When, Where, and How Much of That Activity Is Present

In general, the "when, where, and how much" of a gene product is answered by a combination of three factors: production (e.g., transcription and translation), degradation, and targeting. Each of these processes is controlled in part by the sequence of the gene, which in turn allows for changes in these features to produce novelty. For example, the stability of RNAs and proteins depends on the presence of particular motifs that make them the target of degradation or protection processes. If a mutation adds a degradation motif, the final product of that gene (be it RNA or protein) will be diminished. Similarly, changes in a promoter can alter transcription through changes in transcription factor binding motifs (Fig. 24.11).

The ultimate destinations of RNA and protein molecules are also controlled by sequence motifs within these molecules. Mutations that alter these targeting signals can affect where a protein or RNA ends up, which can initiate the creation of a novel function or process. For example, a single mutation has been identified that causes a fatty acid desaturase (an enzyme that creates carbon–carbon double bonds in fatty acids by removing hydrogen atoms) in *Arabidopsis thaliana* to move from the cytoplasm to the chloroplast. This change in location leads the enzyme to catalyze a different reaction, because the substrates are different in the two locations.

Two key features of the systems described above allow novelty to be generated more readily. First, the regulatory signals usually consist of short simple sequence patterns. For example, protein-targeting signals are often less than ten amino acids long. In addition, the processes at work are rarely "on/off" switches but instead work on sliding scales determined by how well a particular sequence is bound by some other molecule. For example, a perfect match between a transcription factor and its binding site might result in strong transcriptional activation. If, however, there are a few mismatches between the protein and the binding motif, which results in weaker binding, transcription might still be activated, albeit at a lower level (recall the example in-

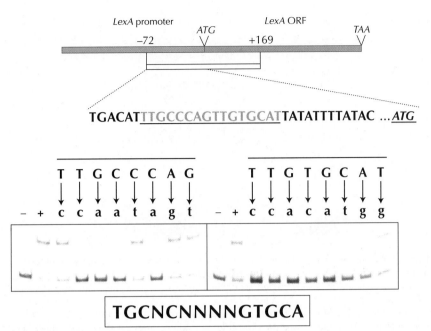

FIGURE 24.11. Binding of the LexA transcription repressor to DNA depends on a simple sequence motif. Results are shown for a gel mobility shift experiment using the LexA protein from the bacterium *Fibrobacter succinogenes*. The protein was mixed with different DNA fragments and subjected to electrophoresis. Stronger LexA/DNA binding correlates with more of the total DNA fraction in each lane appearing near the top of the gel. A control is shown on the *left* without (–) and with (+) LexA mixed with the wild-type LexA box found upstream of the *LexA* gene from this species (in addition to regulating many other genes, LexA protein regulates its own expression). The other lanes represent single-nucleotide substitutions in the LexA box and the mobility of these fragments when mixed with LexA protein. Note that many mutations disrupt binding completely (indicated by most of the DNA running at the bottom of the gel), whereas others have little or no effect on binding. ORF, open reading frame.

volving *even-skipped* [Fig. 19.31]). These two features (small motifs and sliding scales) make it relatively easy to evolve new motifs and change the specificity of others.

Regulatory Processes Allow for Responses to Changing Environmental Conditions

A common feature of almost all organisms, and even all cells within multicellular organisms, is their ability to respond rapidly to changes in their environment. These responses frequently involve numerous parts of a cell or organism and interconnected mechanisms such as activation and repression of metabolic pathways, changes in cell shape, changes in heart rate, and movement (cell or organismal), to name but a few. Thus, an organism can be characterized as having not just one phenotype, but a set of phenotypes appropriate for the various conditions in which it finds itself. Such a complex, environment-dependent phenotype is known as a **reaction norm** (e.g., Fig. 14.5B).

Adaptability to varying conditions can facilitate the origin of novelty in two ways. First, a robustness to environmental perturbations makes organisms robust to genetic perturbations as well, so that mutations of large phenotypic effect do not interfere with an organism's function as much as might be expected (recall Chapter 23 [pp. 689–692], and see below). Second, because a single genome can generate multiple phenotypes depending on the conditions, new phenotypes can be generated more readily by combining elements of existing phenotypes. In a sense, the organism contains cryptic phenotypes that can be elicited by changes in the regulatory networks.

Control of transcription is one of the best-understood regulatory processes. In the

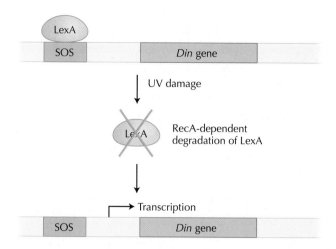

FIGURE 24.12. The SOS response in *Escherichia coli* works via the negative regulation of genes by the LexA protein. LexA binds to an upstream region (known as an SOS box) of a suite of genes known as *Din* genes (for DNA *d*amage *in*ducible, because one of the main ways to activate the SOS response is through DNA damage). When cells encounter an activating signal such as DNA damage (in this case, due to UV radiation), the LexA protein cleaves itself (with the assistance of the RecA protein). This causes derepression (i.e., activation) of all of the genes with SOS boxes.

SOS system in *Escherichia coli*, a set of approximately 20 genes contains a binding motif for the protein LexA in their promoter regions (Fig. 24.11). When LexA is bound to these promoters, transcription of these genes is repressed. When *E. coli* is exposed to certain environmental stresses, such as UV irradiation, a series of steps is triggered that ultimately cause the LexA repressor to cleave itself in half. Once cleaved, LexA falls off its repressor sites, which leads to the activation of transcription of all the "SOS" genes that were inhibited by LexA. Among the genes induced are DNA repair genes, which aid in the cells' response to UV irradiation (Fig. 24.12).

Comparative analysis of SOS-like systems in various bacteria shows how this system has evolved. One way is by changing the targets of LexA repression. Two features of regulatory systems, outlined above, facilitate the evolution of novelty. The LexA-binding motif is relatively small (~20 nucleotides), and its degree of repression changes continuously, rather than as an on/off switch. Thus, minor nucleotide changes can affect the strength of repression in any gene that is already repressed by LexA, and only a few mutations are needed to create a new LexA-binding motif for other genes. This in turn allows the suite of genes induced (and their level of induction) to change relatively rapidly.

Novelty also arises when the DNA-binding sites within LexA change. The LexA-binding motif is determined by the general structure of the LexA protein and the particular amino acids that interact with the DNA. Thus, in one or a few mutational steps, the pattern of gene repression by LexA can change for an entire organism (Fig. 24.13).

Although the SOS system is relatively simple, it serves as an excellent model for understanding how novel regulatory patterns can be generated: Changes in the regulatory sequences allow target genes to change how they are regulated, and changes in the regulators can change whole pathways at once. The LexA example is one of transcriptional repression, but, of course, transcription can be activated instead of repressed, and numerous examples of evolutionary changes in transcriptional activators and activator binding sites are known. Many other components of gene expression can also be altered—in fact, every step of hierarchical regulation can serve as a target for the evolution of novelty. This is true for translational regulation by initiation factors and small RNA molecules (**microRNAs**), phosphorylation cascades by kinases, control of protein folding by chaperonins, and regulation of the flow of metabolites through metabolic pathways by feedback inhibition. In Chapter 9 (pp. 244–250), we gave examples of how

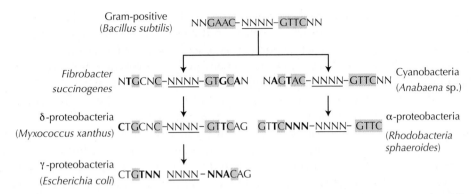

FIGURE 24.13. LexA recognition sites in different bacterial species. The diagram shows the different consensus sequences that are recognized by different species' LexA proteins. Bases that are similar to those in the gram-positive species are highlighted in *orange*. N is any of the four nucleotides.

changes in messenger RNA (mRNA) splicing, mRNA stability, transcriptional regulation, protein phosphorylation, and protein localization play important roles in the development of multicellular organisms. Chapter 11 focused particularly on the role of transcriptional regulation in morphological evolution (pp. 293–318). All of these levels of genetic change, and many additional ones as well, constitute the multitude of mechanisms that lead to evolutionary changes through alterations in gene regulation.

Differentiation and Development Allow Programmed Regulatory Changes and the Division of Labor

Regulatory changes are not always simple responses to a changing environment. In most species, there are preprogrammed patterns of gene regulation and regulation of gene products. Examples include the mechanisms that drive the cell cycle, the differentiation processes by which single-celled organisms sporulate or change mating type, the differentiation of cells within multicellular organisms (Chapter 9 [pp. 230–236]), and the generation of castes within social organisms, such as ants (Box 21.2).

In all of these examples the use of regulatory processes allows for a division of labor, a term associated with the economist Adam Smith (see Fig. 1.13). Smith emphasized that a process is more efficient when it is separated into components, each of which can be better performed by a specialized individual. A gene duplication can be advantageous if the ancestral gene performed two functions, which can each be carried out more effectively under the control of separate genes. Division of labor can also involve temporal separation of events (such as the sexual vs. asexual phases of the life cycle of many single-celled eukaryotes [Fig. 23.28C]) or allocation of tasks to different cells or individuals. Examples of the latter occur in developmental pathways in plants and animals, as we saw in Chapters 9 and 11. Examples are also evident in life history strategies, such as in cicadas, where the larval stage feeds, whereas the short-lived adults mate and disperse. The existence of these differentiation processes not only provides an immediate advantage through specialization, but also opens up qualitatively new possibilities. The evolution of color vision provides an example of how differentiation can play a role in the evolution of novelty.

In animals, rhodopsin molecules are able to absorb photons and then create an electrical signal in the neural photoreceptor cell in which they reside. During the course of evolution, rhodopsin genes have duplicated and subsequently evolved to encode rhodopsins that show maximal absorption for different wavelengths (i.e., colors) of light. To have useful color vision, however, differentiated photoreceptor neurons must also evolve. For example, within each eye facet of the fly *Drosophila*, there is a specialized photoreceptor cell that expresses a UV-sensitive rhodopsin and projects its axon connection to a different part of the brain than those photoreceptors that express rhodopsins that respond to visible wavelengths of light. In this way, flies can distinguish a UV radiation source from a visible light source. In general, these same types of recruitment and replacement processes can occur in all manner of developmental and physiological events and thus create novelties in processes ranging from cell-cycle progression to organ development.

Studies of developmental mechanisms and the way in which they evolve (some of which were described in Chapters 9 and 11) reveal some key features that have been found to contribute to the complexity and novelty that can be generated through regulatory processes. These include, but are not limited to, the following features: (1) the existence of distinct subpopulations of cells and tissues in which regulation can differ; (2) specific regulatory genes that are multifunctional and can be used in different ways in different contexts (e.g., at different time points and in different tissues); (3) the targets of regulation are often controlled by multiple interacting regulators; and (4) many regulators overlap in function, resulting in greater evolutionary flexibility. In fact, studies of both simple and complicated regulatory networks provide some key insights into how regulatory networks in general could have evolved. Crucially, a relatively simple set of rules (e.g., if protein A binds, transcription is increased; if protein B binds, transcription is reduced) can be combined to generate immense diversity of patterns. This is similar to the way that arbitrarily complex computations can be performed by combining a simple set of logical rules (and, or, not, ...).

REDUNDANCY

Most of the examples given above refer to changes in the function of some gene (e.g., changing the specificity of an enzyme or changing the location and timing of gene expression). Although it may be beneficial to an organism to have a new function, this cannot evolve if it interferes with an existing, and more important, function. Even when the old function is not lost entirely, conflicting selective pressures may prevent a single gene from evolving multiple functions. For example, the enzyme lactate dehydrogenase (LDH) has gained an additional function, acting as a crystallin that forms the lens of bird eyes. So, in this case, the protein has two functions: as an enzyme and as a structural molecule. Evolutionary studies have shown in birds this dual-function protein has undergone amino acid changes at positions that are otherwise conserved in all other vertebrate LDH proteins that are involved only in metabolism. This suggests that its metabolic activity has been diminished as a consequence of its new function in vision (Fig. 24.14). One way around this constraint is for new functions to evolve from scratch in regions of the genome with no function. Although this is clearly possible (and is discussed in more detail in Box 24.1), it is rare. An alternative pathway, and one that is pervasive in evolution, is through gene duplication.

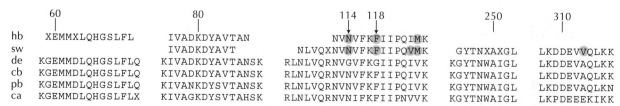

FIGURE 24.14. Conflicting pressures in one gene. An alignment of a region of lactate dehydrogenases from various vertebrates (cb, chicken LDH-B; pb, pig LDH-B; ca, chicken LDH-A) and lens crystallins from two bird species (hb, hummingbird ε-crystallin; sw, swift ε-crystallin). Also shown is a protein from ducks (labeled de) that serves as both an LDH-B and a crystallin. Note that some residues (labeled with *arrows*) are conserved in the other LDHs (including many not shown) and in some crystallins but are different in the dual-function enzyme. This may represent conflicting selection pressures that could diminish the activity of the LDH while improving the function of this protein as a crystallin. The *shaded* residues indicate differences between ε-crystallin in swift or hummingbird and duck.

Box 24.1   New Processes Can Evolve from Scratch

Here we consider a few examples of how new processes can evolve from nonfunctional entities.

One way this can occur is when noncoding DNA converts to coding DNA. For example, when transcription and translation signals are incorporated into noncoding regions of the genome, the transcription machinery is able to synthesize mRNA from that region of DNA, and the mRNA can be translated, which in some cases will generate novel proteins. Although the insertion of transcription and translation signals into noncoding regions can occur randomly, the generation of novel proteins is not an entirely random process. It is more likely to occur in parts of the genome that are high in guanine and cytosine, because these regions contain fewer stop codons in any random stretch of DNA (the stop codons—TAA, TAG, and TGA—are rich in adenine and thymine). Furthermore, this leads to differences among species, because organisms with high G + C content have a greater probability of inventing new genes than do organisms with lower G + C content.

Noncoding DNA is more easily converted to coding DNA in regions of the genome where genes are present. The elimination of a stop codon, either by a point mutation or a frameshift (Fig. 12.2), can lead to the creation of new **open reading frames** and possibly new or modified gene products. Elimination of a stop codon adds amino acids to the carboxyl end of the primary protein sequence. Although the majority of frameshift mutations lead to loss of protein function, in some cases they can generate new functions, for example, by generating new domains in the three-dimensional structure of the protein (see pp. 219–221 and 712–713 and Figs. 24.19 and 24.20 for additional discussion about domains).

Frameshifts and stop codon readthrough do not result only from changes in the DNA sequence within the gene. Certain frameshifts, known as programmed frameshifts, which are found in a large number of species, are controlled by translation factors that cause the translation apparatus to "jump" to a new reading frame. By using programmed frameshifts in a regulated fashion, an organism can produce two separate proteins from a single gene. When programmed frameshifts can occur in multiple regions of a genome, then changes in the translation factor(s) that controls the frameshifts can alter several protein sequences at once. In other cases, a stop codon can be read through by the presence of alternative transfer RNAs (tRNAs) that recognize the stop codon instead of one of the "normal" 61 codons. Alternative tRNAs capable of recognizing a stop codon will (like normal tRNAs) insert and amino acid into the growing polypeptide. The latter example is a situation where the genetic code itself has been changed (Table 5.3). This can occur in a variety of ways (e.g., changes in aminoacyl-tRNA synthetases or in tRNAs), which all can lead to significant novelty. It is also possible for new genes to be created through modification of noncoding regions in other genes and conversion of these into coding regions. An example of this involving antifreeze proteins in fish is shown in Figure 24.15.

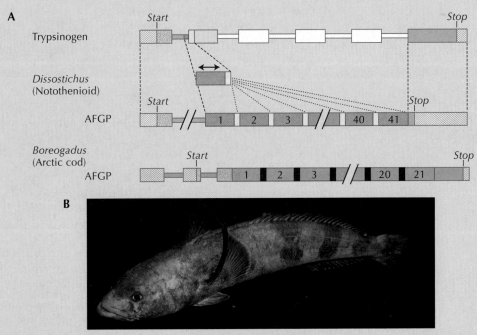

FIGURE 24.15. Origin of new genes. (A) Schematic diagram of the gene structures of three protein-coding genes including two antifreeze glycoproteins (AFGPs) and trypsinogen. *Large boxes* correspond to exons and *thin boxes* to introns. *Hatching* shows untranslated regions. Signal peptides are *stippled*. Regions of similar sequence between the genes are in similar colors. Regions of homology between the trypsinogen and AFGP genes from the Antarctic fish *Dissostichus* are connected by *dashed lines*. It has been proposed that the AFGP of *Dissostichus* evolved from the trypsinogen through an expansion of a DNA sequence element that can encode the protein repeat found in the AFGP protein. The nonhomologous AFGP from another fish is shown below. (B) *Dissostichus mawsoni.*

Box 24.1. Continued.

The yeast **prion**, *PSI⁺*, provides an intriguing example of the effects of incorrect termination of translation. This is an aberrant form of a protein that is involved in recognizing stop codons. The prion form converts normal protein into the aberrant form and sequesters it in inactive complexes (Fig. 24.16). (Prions are a remarkable form of inheritance that is not based on nucleic acids: Diseases such as kuru and scrapie that are caused by prions are transmitted by proteins that have taken on an abnormal conformation.) As a result, proteins are expressed with a variety of additional sequences, thereby causing wide phenotypic variation.

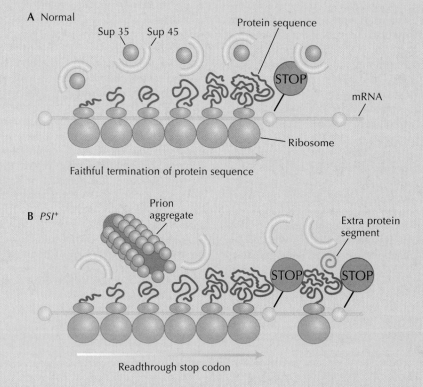

FIGURE 24.16. The yeast prion *PSI⁺* is an abnormal form of the protein Sup35, which normally binds the protein Sup45 to form a complex that assists termination of translation (*A*). The prion form aggregates, which interferes with normal termination, and causes extra protein segments to be added (*B*).

Gene Duplication and Divergence Allows for Functional Diversification without Loss of Previous Functions

Gene duplication generates redundancy. As long as a duplicate retains the regulatory information of the original, a gene copy is free to evolve new functions while the original gene maintains the original function. Duplication is a frequent event in evolution. It can occur at the gene level, where it is most common for a duplicate to be generated adjacent to the original gene. Larger segments of a genome or entire chromosomes can be duplicated, and, in the extreme case, duplication of the whole genome leads to **polyploids** in which every gene is present in multiple copies (see p. 328 and Chapter 22).

As discussed on pages 125–128, the duplicates within a genome are known as **paralogs**, because they evolve in parallel within a species. Examples include the different globin genes that make up hemoglobin and the two forms of elongation factor (Tu and G) used to root the Tree of Life (pp. 126–127). After a gene duplication event occurs, one of the two duplicates may evolve new functions while the original functions are maintained by the other (Fig. 24.17). It is for this reason that paralogous genes frequently have different functions, whereas orthologous genes frequently have the same function. (Recall that orthologous genes are the same form of a gene in different species, e.g., elongation factor Tu in humans and mice.)

Although gene duplication is common, the generation of new functions by divergence of a gene copy is rare. It is far more likely that a gene copy will accumulate deleterious mutations and decay into a **pseudogene**. Conversely, if an increased level of the original function is favored, both copies may be maintained in their original functional state. Moreover, duplicated genes evolve in different ways from nonduplicated ones—**gene conversion** between them impedes their divergence, and deletion by recombination may eliminate one of them (see Chapter 12 [p. 339]).

If the original gene had more than one function (as in the lens crystallin example), then selection will favor specialization of function by the product of each gene copy. This is a relatively rapid route to divergence. In contrast, de novo evolution of a new function by a duplicate copy requires a period of neutral evolution in which the most likely outcome is loss. A straightforward and common means to gene specialization is through changes in regulatory sequences that cause the copies to be expressed at slightly different times or locations. For example, human embryos express one set of hemoglobin genes. After birth, the embryonic genes are shut down and a separate set of hemoglobin genes is activated. This temporal pattern of gene expression reflects regulatory evolution. Similarly, duplications of the *Hox* genes have led to copies that have different expression domains and bind to different target sequences (see Chapter 11 [pp. 290–293]).

The existence of duplicate genes allows the functions of one or both of the duplicates to diversify. Through positive selection for specialization of each gene and through the accumulation of neutral or nearly neutral mutations, one or both of the duplicates is able to traverse the adaptive landscape of gene fitness and avoid adaptive valleys in ways that a single gene could not. This then accelerates the development of novel functions.

Although gene duplication followed by divergence is the best-studied way in which redundancy contributes to novelty, the same general schema is seen in any redundant system. For example, when an organism possesses multiple metabolic pathways that interconvert the same substrates, one of the pathways can diverge without compromising the function of the organism. For the most part, this redundancy does not rely on the presence of duplicate genes, but instead reflects the general robustness of organisms. (For example, recall how multiple transcription factors can allow gene expression in a variety of combinations [p. 547].)

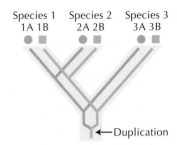

FIGURE 24.17. Duplication and divergence. The figure shows an evolutionary tree of three species (indicated by the *thicker, gray* branches). At the base of the tree, in the ancestor of all three species, a gene underwent a duplication event generating paralogous genes and the paralogs diverged into two distinct functional types, indicated in *red* and *blue*. As the branches for species 1, 2, and 3 separated, each retained both paralogs and their distinct functions. Duplication events such as those shown here lead to redundancy that can be the basis of functional diversification.

ROBUSTNESS, MODULARITY, AND COMPARTMENTALIZATION

Another feature of biological systems that contributes to the origin of novelty is the existence of systems with distinct subparts. Many terms have been applied to describe such subdivision of systems. For example, **compartmentalization**, which was discussed in Chapter 4 (pp. 99–102), is usually used for the division of a system into *physically* distinct subparts. (There, we were concerned with the origin of independently replicating organisms; here, we restrict attention to compartments that differentiate parts of a single organism.) When systems can be divided into *functionally* distinct subparts, this is frequently referred to as **modularity**. For simplicity, here we use the term modularity to refer to any type of subdivision into distinct parts that are each capable of operating with a great degree of independence. Modularity occurs at every level of biological systems, from proteins to cells to multicellular organisms. As we explained above, the key feature of modularity in terms of novelty is that it allows different parts of an organism (modules) to evolve somewhat independently. That is, changes in one part will have only limited direct effects on the functions of other parts. This is not to say that changes will not affect the whole organism, but rather that one part may not be significantly constrained by changes to another part. In the following sections, we discuss some of the different types of modularity and how these contribute to the origin of novelty.

Mixing and Matching of Protein Domains Allow Creation of Diverse Activities

When the sequences of homologous proteins are examined, it is found that some regions are highly conserved and others poorly so (recall p. 532). One of the main reasons for this is the existence of distinct domains in their three-dimensional structures. For example, in many enzymes, the amino acids at or near the active site are highly conserved across species, whereas those in other regions of the protein tend to be less conserved. In multifunctional enzymes, multiple domains are frequently conserved, interspersed among poorly conserved regions. (It should be pointed out that in some proteins, even though three-dimensional structure may be highly conserved across species, the primary sequence does not show conserved domains, because the amino acids that interact in three-dimensional space are not neighbors in the primary sequence. However, in the majority of cases there are conserved sections of the primary sequence that can be considered distinct domains [Fig. 24.18].)

The existence of distinct domains in the primary sequence facilitates the origin of novelty in two ways. First, changes at one domain can influence the activity of another via changes in protein conformation. As we saw in Chapter 2, such **allosteric** interactions allow regulatory links between domains with arbitrary functions (pp. 54–56). Second, domains can be relatively easily recombined. This mixing and matching can take many forms. In some cases a single core domain is shared among a group of proteins and additional domains come in from various sources (e.g., Fig. 24.19). In other cases, great variation is seen across species in the mixing of domains (Fig. 24.20).

How does this mixing and matching actually occur? One process is known as **exon shuffling** (see p. 222), where domains in proteins are encoded by exons in the DNA sequence and introns act as regions of recombination. The longer the intron, the more likely it would be for a random recombination event to mix around the domains in the exons. Because it is known that introns frequently share some sequence similarity even between different proteins, homologous recombination between the introns of different genes could generate new proteins. Although exon shuffling can occur, in many cases it appears that protein domains are not congruent with exons, so that recombination in the intron would actually disrupt domains. Thus, other explanations are required to explain the diversity of domain patterns.

In eukaryotes, alteration in gene splicing patterns can generate novelty. This can happen in two ways. First, the splicing pattern of a gene can vary under different conditions, such as at different temperatures or in different cell types. In theory, **alternative splicing** (see pp. 220–221) allows a single gene to code for thousands or even millions of different proteins by recombining the spliced sequences in a multitude of combinations. In practice, however, alternative splicing produces a relatively small number of different proteins in an organism. Nevertheless, it opens up the *possibility* of creating a vast number of novel proteins. Second, a gene's splicing patterns can be altered when mutations change the location of its intron–exon boundaries. Changes in splicing patterns provide a more straightforward mechanism of generating diversity than does exon shuffling.

The existence of proteins with multiple domains that have distinct functions also allows other types of evolution. For example, a simple domain deletion would lead to a loss of one of the functions without necessarily compromising another function. If

A
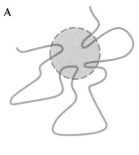

B
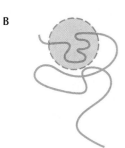

FIGURE 24.18. (A) The active site of a protein (*red*) may be made up of several different segments of the primary sequence. Usually, however, it consists of a single stretch of sequence (B), which will tend to be conserved.

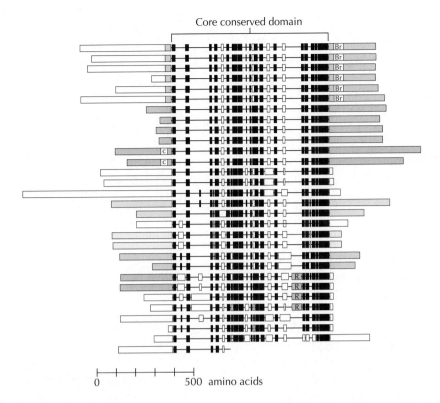

FIGURE 24.19. Schematic diagram of an alignment of proteins in the SNF2 family of DNA-dependent ATPases. The figure shows different proteins in *horizontal bars* with distinct protein domains in different colors. All proteins in the family share a core SNF2 domain (*black*) that carries out the conserved activity found in all these proteins. Many of the proteins have additional domains (*other colors*) that provide specificity to the reaction.

this were to occur after a gene duplication event, the organism would likely survive and the domain loss could allow for the specialization of the two duplicates.

Compartmentalization Is an Important Contributor to Novelty

One of the easiest ways to create modules within an organism is to create separate compartments: for example, nucleus versus cytoplasm, different organs in plants and animals, and outer versus inner membrane in gram-negative bacteria. Such compartments are distinguished by master signals that control the expression of many different genes, such as the *Hox* genes that determine which genes will be expressed in each segment. Compartmentalization makes it relatively easy for systems to evolve that influence an individual compartment. All that is needed is for particular genes or gene products to respond to the signals that define compartments. Two examples are the crystallins expressed in the lens of the eye (see above) and changes in segment identity (Chapter 11 [pp. 293–318]). We give some additional examples here that highlight how compartmentalization contributes to the origin of novelty.

Organelles partition a eukaryotic cell into membrane-bound substructures capable of maintaining unique internal environments. The ability of an organelle to maintain internal conditions that are very different from those in the rest of the cell has certainly facilitated the generation of novelty. For example, lysosomes are full of digestive enzymes that degrade macromolecules (Fig. 24.21). Without the existence of a separate compartment for digestion of macromolecules, the acid-pH-based degradative pathways present there probably could not have evolved.

Most organisms have some control over where they live. Bacteria can use their fla-

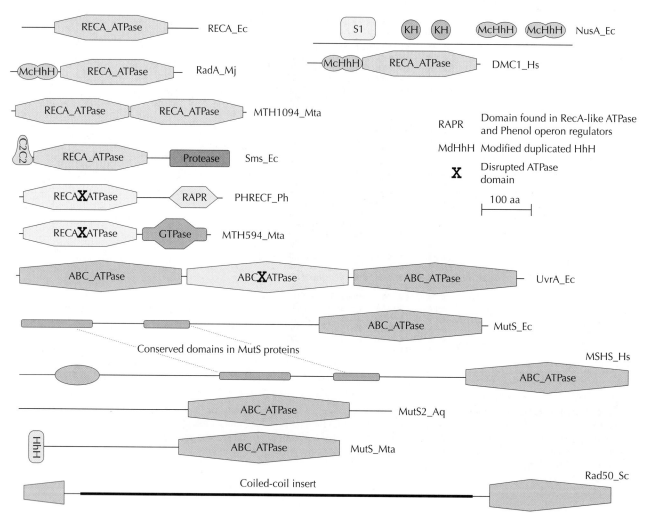

FIGURE 24.20. Domains found in known ATPases show an enormous diversity of patterns. This is just a small subset of the proteins found in only a small number of species. The total diversity of domain structures across all proteins is therefore immense. Proteins are shown approximately to scale. Possible nonfunctional motifs are indicated by *crossed* symbols. The protein IDs refer to gene names or names from the Swiss-Prot database. Species abbreviations: Af, *Archaeoglobus fulgidus*; Aq, *Aquifex aeolicus*; Ce, *Caenorhabditis elegans*; Dm, *Drosophila melanogaster*; Ec, *Escherichia coli*; Hs, *Homo sapiens*; Mj, *Methanococcus jannaschii*; Mta, *Methanobacterium thermoautotrophicum*; Mtu, *Mycobacterium tuberculosis*; Ph, *Pyrococcus horikoshii*; Sc, *Saccharomyces cerevisiae*.

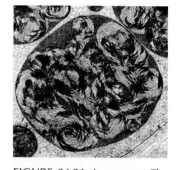

FIGURE 24.21. Lysosomes. The pH of 4.8 within lysosomes (more acidic than the cytosol at pH 7) aids greatly in the digestion process. The amino acid content and sequence of lysosomal proteins have been selected to work better at low pH.

gellae to move along chemical and physical gradients, plants can influence where their pollen and seeds disperse, and animals can choose to exploit different hosts. Differences in such habitat preferences may have a simple genetic basis but may cause the selection experienced by the whole genome to change. For example, in Chapter 22, we saw how a change in the host plant preferred by *Rhagoletis* flies has led to multiple adaptations to those plants and may eventually generate two separately specialized species (p. 653).

Modularity of Regulatory and Developmental Networks

Just as we can think of modularity and compartmentalization in gene structure, so we can think about it for developmental mechanisms. Networks of genetic interactions and pathways form modules that can be used repeatedly in development. For example, specific cell–cell signaling systems involve cascades of molecular interactions and feedback loops (see Chapter 9 [pp. 248–250]). Any one signaling system can, and usually is, used repeatedly during development, but because of the different contexts in which these signals occur, the same signaling pathway can orchestrate different outcomes each time it is used. This allows a relatively small amount of genetic information to generate many different outcomes. It also means that mutations that cause these signals to initiate in new locations can have profound effects on development, as they will be able to activate entire cascades of genetic interactions. Although in most cases the result will be deleterious, in some cases these changes can create advantageous alterations in development.

The ways in which developmentally important genes are regulated also leads to modularity. As described in Chapter 11 (pp. 307–313), different domains of *Pitx-1* expression in sticklebacks—that is, different tissues in which *Pitx-1* is expressed—are presumed to be controlled by separate *cis*-regulatory elements. Because of this, mutations can arise that alter the expression of *Pitx-1* in the pelvis, without affecting expression in the head. This avoids potentially widespread pleiotropic effects and presumably allowed for *Pitx-1*-mediated evolutionary changes in the stickleback pelvic spines, without disrupting development of the head of the fish. *Hox* genes provide another example of this type of evolution. Duplications of *Hox* genes during evolution have allowed *Hox* gene functions to be subdivided and diversified. At the same time, subtler changes in the regulation of individual *Hox* genes have also played a role in morphological evolution. For example, the evolutionary changes in leg hair patterns between *Drosophila* species appear to be controlled by relatively subtle changes in the *Ubx* gene that result in differences in the levels of Ubx protein expression during leg development in different species (see Fig. 11.19). These types of regulatory modularity, in which there is independent control of regulatory networks in different parts of an organism, appear to be a critical component of many developmental genes. At the same time, they make *cis*-regulatory mutations a prime mechanism for evolutionary change and facilitate the independent evolution of different body parts.

The examples given in Chapter 11 help to highlight the concepts of both modularity and compartmentalization in development. The module that initiates insect appendage development is composed of the intersection of specific signaling cascades, which in turn activates a suite of transcription factors; the *Dll* expression shown in Fig. 11.13 is one component of this module. Appendage development has the potential to be activated in every segment, but the action of *Hox* genes prevents this activation from occurring in the abdomen. In segments where appendage development is initiated, *Hox* genes influence the type of appendage that will develop (e.g., wing vs. **haltere**). As explained in more detail below, this system of organization allows development to be robust, but also allows mutations to produce potentially advantageous variations.

Exploratory Systems Offer Robust Routes to Novelty

It is remarkable that complex organisms can be specified by so little information. Even the human genome, which is larger than most, codes only for a hundred or so megabytes of information—comparable with the operating system of a personal computer. As we noted in Chapter 14, the organism is not determined entirely by its genome: The DNA sequence must be interpreted by the cellular machinery (p. 381). Nevertheless, the ability of a fertilized egg to develop into an organism with a complex morphology and behavior is extraordinary. Development must unfold from the limited information contained in the zygote and, moreover, must do so reliably, despite random variation in the environment.

The developmental process is made possible by a variety of regulatory systems that reliably produce complex outcomes from limited inputs. In particular, some scientists emphasize the importance of **exploratory systems** that constrain and shape initially random variation so as to produce a final well-coordinated outcome. There is a loose analogy with natural selection, which shapes random variation over many generations of replication to produce organisms with high fitness. Indeed, in some cases these systems actually depend on natural selection between replicating lineages within a developing organism. Furthermore, in this type of system, developmental steps are closely knit together by feedback loops that allow events to be coordinated without having to specify them all individually. This allows variation to be tolerated so that outcomes are not thrown off by random variations in the process, but, at the same time, the coordination makes it possible for substantial changes to occur if key steps are altered.

A simple example is the development of the vertebrate limb. This involves not just growth of a bone in the right place, but also the correct placement of muscles together with their blood supply and innervation. If each component tissue were specified independently, by some kind of precise blueprint, then a great deal of information would be needed. Indeed, in this view, we would require a kind of **homunculus** within the egg, which explicitly contained all the information needed to describe the mature organism, and that homunculus would itself need to be specified in the same way (see p. 11). If development proceeded in this way, then not only would an enormous amount of information be needed within the egg, but also, if different components got out of step, a great deal could go wrong.

In fact, a simple molecular gradient determines the initial differentiation of the primordial limb. Once the initial limb outgrowth is established, muscles are laid down in response, and then followed by growth of nerves and blood supply stimulated by the presence of muscle. A general process that applies to all parts of the body ensures that each part of the limb skeleton contains a complete set of the necessary structures to produce a functional appendage. These events can also be modulated by local signals that cause the relatively small differences between different limbs and limb parts.

In this example, coordination between different tissues ensures that one signal can specify a complex structure. Similar examples were provided in Chapter 11. For example, the single transcription factor, Pax6, when ectopically expressed in a larval imaginal disk of *Drosophila*, can set off the entire cascade of events required to produce an eye structure (Figs. 11.33 and 11.34). Thus, ectopic Pax6 expression can create organized eye tissue on the antennae, legs, or wings. The result is not just random photoreceptor cells, but instead correctly patterned ommatidia (eye facets) containing the correct photoreceptor types, all properly arranged, and accompanied by the appropriate lens and pigment cells. The neurons produced will even grow into the central nervous system and attempt to make appropriate connections with targets in the brain.

When examined in detail, the developmental process often achieves this coordination in a particular way: A generic structure is first initiated and is then refined into the final, tightly specified form. For example, a mature muscle cell receives signals from a single motor neuron. However, an immature muscle cell is innervated with multiple neurons, which compete with each other. Those neurons that transmit more signals to the developing muscle receive as feedback growth factors required for neuron survival and as a result are more likely to survive, so that eventually each muscle cell is innervated by a single motor neuron. Similar competitive processes occur throughout neural development. During development of the brain, many more neurons and synaptic connections between them are formed than eventually survive, and their survival is shaped by the signals that they transmit. Thus, although the overall structure of the brain is rigidly determined, its fine structure develops through interplay with the outside environment (Fig. 24.22).

The vertebrate immune system relies on natural selection between developing cell lineages to generate diverse antibodies that recognize foreign antigens, but not antigens from the organism itself. During early development, a combination of recombination and mutation generates an enormous variety of genetically distinct cells, each producing a unique antibody (Fig. 9.10). Any cell producing an antibody that reacts against the organism is eliminated, but this still leaves a great diversity of cells that can respond to

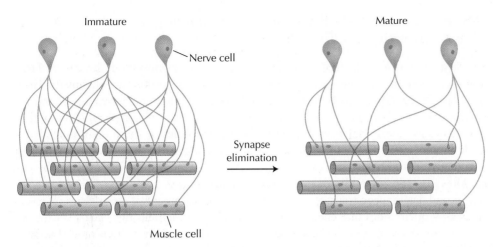

FIGURE 24.22. Competition between motor neurons ensures that eventually only a single neuron instructs each muscle cell.

foreign antigens (proteins or other macromolecules from a foreign source, such as an invading bacterium). Subsequently, when a cell recognizes a specific antigen, the cell is stimulated to divide, so that after a few days a large quantity of specific antibody is produced. That particular cell lineage remains, and so reaction to future challenges by the same antigen is much faster. This phenomenon of **immune memory** is the basis for the success of vaccination. The key point here is that if the system were not so flexible, then such an enormous number of antibodies would require an equally enormous amount of information to be specified in advance. Moreover, such an inflexible system would not be able to react to challenges within the lifetime of the individual. This kind of exploratory system, based on natural selection within the individual organism, allows a limited amount of genetic information to generate a complex and adaptable system.

In this section, we have emphasized processes that allow complex organisms to be coded by compact genomes and to develop reliably despite unpredictable perturbations. These features also make it much easier to generate novelties, because any change is likely to at least maintain existing functions and may specify an advantageous set of coordinated phenotypic changes without requiring an improbable coincidence of multiple changes to the genome. Thus, for example, changes in the expression pattern of a *Hox* gene can set off an entire cascade of events resulting in a segmental transformation that produces coordinated changes in all tissues, including ectoderm, mesoderm, and the nervous system. This greatly enhances the possibility that evolutionary modulations in *Hox* gene expression can produce a phenotypic outcome that is potentially beneficial. This same principle applies to the multitude of developmental genes that coordinate the great variety of structures, organs, and cell types during development.

ACQUIRING NEW FUNCTIONS FROM OTHER SPECIES: GENE TRANSFER AND SYMBIOSIS

Modules cannot be completely independent, because they all are encoded by genes that evolve within the same genome. This potential constraint is avoided when processes that evolved in separate lineages are brought together. This is exemplified by two processes: lateral gene transfer and symbiosis. Lateral gene transfer was discussed in Chapters 7 and 12, and symbiosis was discussed in Chapters 8 and 21. Here, we consider how these two processes contribute to the origin of novelty.

Lateral Gene Transfer Allows Species to Sample the Genetic Diversity of Other Lineages

Lateral gene transfer facilitates the creation of evolutionary novelty in three ways. First, it allows an organism to acquire entire new functions from another organism. Thus, in a way, the genetic diversity of the planet serves as a functional breeding ground for species that are able to acquire and use DNA from others. Examples include antibiotic-resistance genes that spread between bacterial species in the human gut, pathogenicity and virulence factors that are shared among pathogens, the acquisition of degradative pathways by soil-dwelling microbes, and the acquisition of genes by eukaryotes from the ancestors of mitochondria and chloroplasts.

Second, lateral gene transfer contributes to the replacement of some genes in an organism by other genes that encode the same process. Sometimes this occurs via the acquisition of a homolog of the gene in question, thus creating a situation much like gene duplication, in which the species possesses two forms of the same gene. This redundancy permits functional diversification, just as with duplication. Alternatively, the newly acquired gene may encode an endogenous biochemical function, even if the gene is not homologous to any native genes. Subsequent deletion or mutation within the host genome, such that the associated biochemical function is now encoded exclusively by the donated gene, is known as **nonhomologous gene displacement** (Fig. 24.23). This process generates diversity in biochemical pathways, because although the new gene may encode a protein that performs the same general reaction, it may have distinct properties from the native protein. For example, it may be regulated differently or it may carry out additional biochemical reactions.

Third, lateral transfer allows completely new processes to be created by combining newly donated functions with processes already active in the host organism. In this way lateral transfer works much like symbiosis, which we discuss next.

Symbioses between Organisms Are a Breeding Ground for Novelty

Lateral gene transfer is a complicated means of acquiring functions present in another organism. Many obstacles impede the transfer process, and successful transfers often

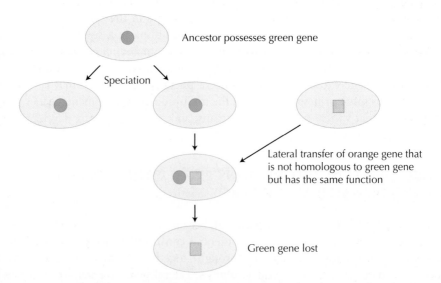

FIGURE 24.23. Nonhomologous gene displacement. A gene that has the same function as an endogenous gene but that is not homologous to the endogenous gene is acquired from another species by lateral transfer. The endogenous gene is subsequently lost, resulting in an organism that carries out the same functions as its ancestor and its relatives but does so using a nonhomologous process.

do not lead to functional products in the new species because of incompatibilities, such as differences in codon usage and promoter signals (see pp. 187–188). There is, however, a way of acquiring the functions of another organism that avoids the obstacles to incorporating genes into the host genome: **symbiosis**.

Symbiosis here refers to two distinct species that live in close association with each other. This does not have to be a mutually beneficial association: One species could be parasitizing the other or one could be indifferent. All that is required is that there be some close association. Symbiosis provides benefits of both compartmentalization and gene transfer. As with gene transfer, an organism can acquire functions without having to evolve them itself. Because symbioses can be established much more readily than gene transfers and because they can occur between any types of organism, symbiosis allows for more rapid and diverse types of functional acquisition. Symbiosis also provides the benefits of compartmentalization because once a symbiosis forms, the different partners can evolve processes independently of each other. Because their genomes are separate, they are not subject to the same constraints as when different compartments are encoded by the same genome within one organism.

Given that symbiosis provides the benefits of both gene acquisition and compartmentalization, it is not surprising that it is one of the most pervasive and effective means of producing novelty. Symbioses have contributed to the origin of eukaryotes, the diversification of algae and green plants, the ability of animals to survive in the deep sea, the origin of many eukaryotic kingdoms, and the existence of viruses.

Some very interesting symbioses are those in which the partners form consortia that are able to carry out biochemical processes together that would be impossible as separate entities. For example, it is common to find bacteria that grow on carbon monoxide and that produce hydrogen gas as waste living in association with methanogenic archaea, which take the hydrogen from the carbon monoxotrophs and use it to produce methane. The carbon monoxotrophs benefit greatly because their reactions are quite costly: Having another organism rapidly utilize their waste product helps drive the reactions forward. The methanogens benefit because they get free hydrogen for their production of methane. It is likely that the particular details of the processes in each species have been fine-tuned because of the symbiosis and that some of the reactions present might never have been able to evolve if the symbiosis did not exist.

Mitochondria and chloroplasts are extreme examples of how symbiosis can produce novelties. Eukaryotes were able to become photosynthetic only through the acquisition of chloroplasts, so that symbiosis was required for the diversification of all plants and algae we see today, as well as by the eukaryotic lineages that have secondary and tertiary symbioses with algae. Most of the **major transitions** in evolution (Table 21.3) involve the coming together of previously independent **replicators** to produce new, more complex organisms. In the origin of eukaryotes, the symbiotic relationship with the bacterial ancestors of the mitochondria and chloroplasts ultimately became so close that these organelles lost almost all of their genes by lateral transfer to the nucleus. A similar kind of symbiosis may have been involved in the early evolution of life, when replicating molecules associated within the original cells, subsequently becoming joined together into chromosomes.

NATURAL SELECTION OVER LONG PERIODS OF TIME LEADS TO THE ORIGIN OF NOVELTY

At the beginning of this chapter, we emphasized that the key obstacle to the evolution of complex novelties is that they involve multiple interdependent changes. Despite this difficulty, such changes can evolve if successive changes are favored by selection: A population can be seen as evolving across an adaptive landscape along ridges of increasing average fitness. A variety of examples were presented showing how genetic mechanisms can aid the development of novelty primarily by allowing variants to contribute to new functions without disrupting existing functions, allowing them to be picked up by selection.

We now consider how the evolutionary process—primarily selection—can aid novelty. As was explained at the start of this chapter, for these features to evolve they must either be created in large jumps involving rare co-occurrences of genetic changes or they must be created in a series of steps. Both types of event are greatly aided by two things: the passage of time and the cumulative action of natural selection. Unlikely events will eventually occur, but far more important is that selection assembles multiple changes step-by-step.

Multiple Favorable Mutations Are Unlikely to Occur Together

Over many generations, and many reproducing individuals, even very improbable mutations will occur and may be picked up by selection. Nevertheless, it is unlikely that more than a few favorable mutations will occur together by chance. Instead, multiple changes must be assembled by selection. We can quantify this argument by referring back to the in vitro selection experiment on RNA molecules that opened Chapter 17. This experiment began by generating mutations with very high frequency (5% per site) in a 140-bp region. The initial population was so large (around 10^{12} molecules), and the mutation rate so high, that it is likely that every four-base variant of the original sequence was present in multiple copies. However, the chance that a specific seven-base variant was present is very small. Thus, it is likely that the four variants that increased quickly (green curves in Fig. 17.2) were present in the original population, and that the other three mutations (blue and yellow curves) arose later, by mutation of the four-variant sequence. This example is an extreme one, because such an enormous population of molecules is involved, and because the mutation rate was artificially high. Even here, however, it is vanishingly unlikely that more than a few favorable variants would arise together by chance. Moreover, there is negligible recombination in this experiment, so that the fittest sequence can build up step-by-step with one mutation following another (Fig. 23.18). In a sexual population, a favorable combination of variants would be broken up, even if it did arise by accident.

Even Partial Function Can Give a Strong Advantage Relative to What Went Before

Complex novelties require many changes and so cannot be generated in a single step or by chance conjunction of several mutations. Moreover, in a sexual population, multiple changes are broken up. Therefore a route is required that assembles changes step-by-step. For example, even if a new organ made use of only preexisting genes and proteins, getting them to work together in just the right way would require a series of steps, which would virtually never occur by chance.

Although time and population numbers may allow for a vast number of rare mutational events to occur, this can hardly explain all complex novelties that have arisen during the course of evolution. All but the simplest structures are far too unlikely to occur by chance mutation. What is crucial is that individual variants occur that increase fitness, even if slightly. The population can then climb steadily upward along ridges in the adaptive landscape (e.g., Fig. 24.4).

Individual steps on the way to producing a complex and novel function may themselves be advantageous (see Chapter 17 [p. 464]). As John Maynard Smith put it, poor vision is better than no vision at all (Fig. 24.24). Thus, the precursor to eyes probably began with simple photoreceptors that detected light. The rhodopsin proteins used for detecting light are still found today in many species without eyes and are also used to detect and/or respond to light in those species. In fact, the biochemical function of these proteins is highly conserved across all organisms and their role in eye function is more of an issue of context than of novel biochemistry. After a time, the simple photoreceptor may have become surrounded by light-blocking pigment cells so that or-

FIGURE 24.24. John Maynard Smith.

ganisms could determine where the light was coming from. The rhodopsins also diversified to allow color discrimination, and the combination of the physical arrangement of greater numbers of photoreceptors, the elaboration of lenses and reflective surfaces, and the growing complexity of the nervous system eventually led to eyes that could produce accurate and detailed images of the world. Each step toward the complex structure led to slightly improved vision. Conversely, even though any change to the current (finely tuned) eye may be deleterious, there still can be a path of increasing fitness that led to the current state.

One way to show that such step-by-step evolution of novelty has happened is to use the fossil record or **character state reconstruction** (Chapter 27 [online]) based on modern organisms to infer the history of features. This is the approach that is being used in the study of feather-like structures on many dinosaur fossils, as scientists seek to understand the origin of bird flight. The discovery of feathered but flightless dinosaurs provides strong evidence that the feathers used to assist in the flight of birds did not arise by selection for flight, but for some other purpose. Once present, however, they could be molded into the essential component for flying that they now are (Figs. 3.17 and 10.33).

Likewise, the fossil record reveals in some detail how early vertebrates evolved from swimming creatures to ones that could walk on land: Much of the transition from fins to limbs actually occurred in the water. Fish living in shallow waters gradually evolved limb-like structures that increasingly acquired most of the skeletal elements we associate with the limbs of extant land creatures. These fish used their limbs to crawl about in the shallow water, slowly becoming better able to support their weight and then to eventually venture out onto land (Fig. 24.25).

The evolution of insect wings also appears to have occurred gradually (see p. 276). Some suggest that insect limbs may have evolved from the gills of an aquatic crustacean-like ancestor. As the insect ancestors ventured out of the water, they evolved a tracheal system for gas exchange, but the gills were retained and instead occasionally used to sail or glide. This behavior is still seen today in mayflies. Eventually, these appendages were elaborated into the wings of flying insects. An alternative hypothesis is that wings elaborated from lobes on the legs that were initially used for aerial maneuvering when ancestral insects glided down from vegetation. This aerial behavior is still retained by many wingless insects.

In Vitro and in Silico Selection Experiments Show How Quickly Novelty Can Be Generated

Experiments in which RNA and DNA molecules replicate in vitro have attracted much attention (both academic and commercial) in recent years. In vitro selection experiments aim to produce new molecules that will carry out particular functions, such as binding a drug target or catalyzing specific reactions. Although the chemical principles that govern nucleic acids are well understood, it is still beyond our ability to design such molecules from first principles. In vitro selection offers a general way to produce novel molecules that could not be designed a priori.

Several examples have already been presented in which novel biochemical functions have evolved over just a few generations. At the start of the book, we saw how selection has been used to produce novel fluorescent proteins (p. 5). In Chapter 4, we saw that in vitro selection has been used to produce **ribozymes** that will catalyze replication of short RNA sequences—a key step in the origin of life (p. 103). In Chapter 17, we saw that 12 generations of selection produced a ribozyme that would work efficiently in the presence of calcium ions rather than the magnesium ions it originally required. Of course, such experiments cannot produce much complexity, because they involve single short nucleic acid sequences. However, they do show that selection can rapidly produce new functions that involve coordinated changes at multiple sites.

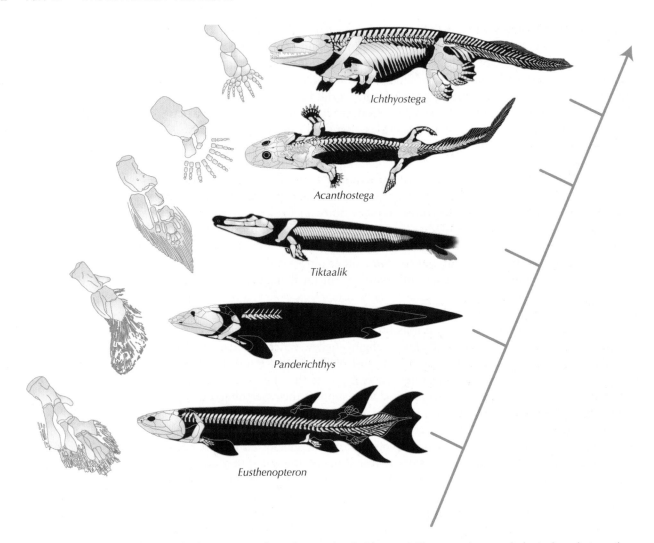

FIGURE 24.25. Limb evolution. The fossil record illustrates the morphological evolution of appendages in the lineage leading from ancient fish to modern tetrapods (four-limbed animals like ourselves). The illustrated fossils all come from the middle to late Devonian period (398–359 Mya; an isolated forelimb skeleton is shown next to each animal, except for *Ichthyostega*, where a hindlimb is shown). *Ichthyostega* and *Acanthostega* were early tetrapods with distinct bones of the appendages that are similar to those found in modern tetrapod wrists, ankles, and digits, whereas *Eusthenopteron* was a lobe-finned fish that possessed limbs with a fish-like fin organization. *Panderichthys* and *Tiktaalik* were aquatic predatory fish with crocodile-like skulls that "walked" on the river bottom and whose appendages reveal the intermediate steps in the transition from fins to limbs.

FIGURE 24.26. The placement of components on a computer chip is often designed by a **genetic algorithm**, which uses natural selection to search for the optimal solution. This microprocessor contains 15 million transistors.

Selection is also used to solve complex computational problems, such as finding the optimal placement of millions of components on a silicon chip (Fig. 24.26) or finding the best way to route packets of information through a network. These problems are beyond human capabilities, and computer algorithms designed by humans are often ineffective. In contrast, selection among alternative algorithms is straightforward to implement and frequently leads to solutions at least as effective as could be obtained by a laborious process of rational design. For example, in Chapter 17, we saw how artificial selection produced several ingenious designs for electronic circuits that had earlier been thought novel enough to patent (Fig. 17.13).

The aim of such **evolutionary computation** is to make selection for the desired function as efficient as possible. Typically, available computer power limits populations

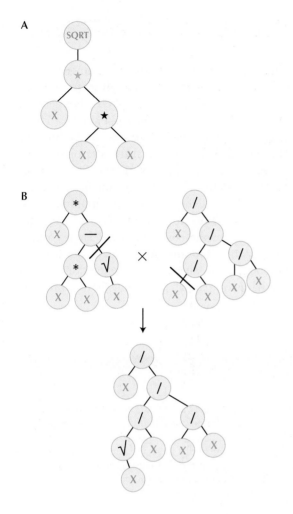

FIGURE 24.27. Kepler's third law states that the square of the orbital period of a planet P is proportional to the cube of its average distance from the sun, X, to within a constant: $P = X^{3/2}$. In a conventional programming language such as Fortran, this relation would appear as a string of characters, P = SQRT(X*X*X). Making random changes to each character (e.g., changing SQRT to SRT or deleting a parenthesis) would produce uninterpretable code. However, the relation also can be represented as a tree, with each node representing a computation on the two descendant branches. Now, random changes between legitimate values at the nodes (e.g., from * to +) or recombinations that splice together different trees will still make sense. (B) For example, how recombination between expressions representing $X*(X^2 - \sqrt{X})$ and $X/((X/X)/(X/X))$ gives an "offspring" representing $X/((\sqrt{X}/X)/(X/X))$. If fitness increases as the expression becomes closer to the correct function (A), then selection, mutation, and recombination will cause a population of initially random trees to approach the solution shown in A.

to a few hundred competing programs. The best are chosen to reproduce in each generation. As with any form of selection, there is a trade-off between choosing a small number (and hence increasing the immediate response to selection) and propagating a population large enough to maintain the variation necessary to sustain the response to selection. Variation is generated by mutation (by making random changes to the programs) and by sexual recombination (by combining two or more successful solutions). For the reasons explained in Chapter 23, sexual reproduction improves performance; most importantly, it allows advantageous variants that arose in different lineages to be brought together by recombination. The key to success, however, lies in choosing the relation between genotype and phenotype. That is, the problem must be coded such that mutation and recombination have an appreciable chance of generating some improvement and do not destroy existing functions. Conventional programming languages (e.g., Fortran or C) are extremely fragile; splicing together two working programs, or making random changes, is usually disastrous. Therefore, special languages are used, which are robust to mutation and recombination (Fig. 24.27). The issues that arise in designing effective evolutionary algorithms are close to those involved in understanding the evolvability of biological systems.

SUMMARY

In this chapter we have shown examples of how novelty can be created by natural selection acting on simple mutational events. Qualitatively new features can evolve through the accumulation of many small changes. The central issue is to understand how features with many interdependent components can evolve without disruption of existing function. Several features of biological systems greatly aid this process. At the level of single genes, small changes in sequence can alter catalytic and regulatory functions; sometimes, a simple loss of function can lead to novel phenotypes. Developmental and regulatory networks have evolved to be robust to varying environments and have a modular structure—both features that facilitate novelty. Gene duplication and redundancy allow diversification while existing function is retained; compart-

mentalization and differentiation allow functions expressed at different times and places to diversify; and "exploratory systems" make it more likely that genetic changes will still lead to functioning phenotypes. Functions that evolved in different organisms can be brought together in one organism by lateral transfer or by symbiosis. All these features of the genetic and developmental mechanism make it easier for organisms to evolve novel functions. Even if the number of mutational steps required to generate some novel feature is great, such features can occur given the combination of large population numbers, large amounts of time, and the force of natural selection. Thus, although the generation of novelty may be hard to imagine, it is not as hard a problem as once thought.

FURTHER READING

General Readings

Carroll S.B., Grenier J.K., and Wetherbee S.D. 2005. *From DNA to diversity: Molecular genetics and the evolution of animal design.* Blackwell Publishing, Oxford.

Coen E. 2000. *The art of genes: How organisms make themselves.* Oxford University Press, Oxford.

Argues that complex organisms develop through an interaction between the genes and the developing organism, rather than by simple execution of a "genetic blueprint." This makes it easier to see how novel features can evolve without disrupting existing function.

Kirschner M.W. and Gerhart J. 2005. *The plausibility of life.* Yale University Press, New Haven, Connecticut. Kitano H. 2004. Biological robustness. *Nat. Rev. Genet.* **5:** 826–837.

General discussions of the features that make biological systems robust, and thereby facilitate the evolution of novelties. Kitano gives examples from cellular regulatory networks, whereas Kirschner and Gerhart focus more on the developmental process.

Long, M., Betran E., Thornton K., and Wang W. 2003. The origin of new genes: Glimpses from the old and young. *Nat. Rev. Genet,* **4:** 865–875. Otto S.P. and Yong P. 2002. The evolution of gene duplicates. *Adv. Genet.* **46:** 451–483.

Reviews of the role of gene duplication in the origin of novelty.

Maynard Smith J. and Száthmáry E. 1995. *The major transitions in evolution.* W.H. Freeman, Oxford. Száthmáry E. and Maynard Smith J. 1995. The major evolutionary transitions. *Nature* **374:** 227–231.

A wide-ranging argument that the major innovations in evolution have arisen when previously independently replicating individuals come together in a more complex organism.

Symbioses

Moran N.A. 2006. Symbiosis. *Curr. Biol.* **16:** R866–R871.

Smith J.M. 1989. Evolution: Generating novelty by symbiosis. *Nature* **341:** 284–285.

Origin and Diversification of Genes

Long M., Betrán E., Thornton K., and Wang W. 2003. The origin of new genes: Glimpses from the young and old. *Nat. Rev. Genet.* **4:** 865–875.

Ohta T. 1989. Role of gene duplication in evolution. *Genome* **31:** 301–310.

Ohta T. 1991. Multigene families and the evolution of complexity. *J. Mol. Evol.* **33:** 34–41.

PART IV

HUMAN EVOLUTION

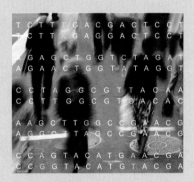

I n these final chapters, we focus on our own species. Chapter 25 begins by tracing our ancestry since the human lineage separated from the ancestors of the great apes. Here, remarkable progress has been made, largely from the study of a series of newly discovered fossils. It is now clear that, for much of our history, diverse hominin species have coexisted, although only our lineage has survived. Genetic evidence has helped reveal how our species spread after anatomically modern humans left Africa. Most recently, comparisons between human and great ape genome sequences have begun to reveal some of the genetic differences that may be responsible for uniquely human features. However, it remains very hard to discern the evolutionary history of features, such as language, that leave no direct physical trace.

In the final chapter, we illustrate how evolutionary biology helps us to understand present-day humanity. We concentrate on two issues: the application of genetics to medicine and the use of evolutionary ideas to understand human nature. The massive effort that has gone into determining genome sequences has been motivated primarily by possible medical applications. Most disease is influenced by complex interactions between multiple genes and the environment; thus, the methods described in Chapter 14 can be applied directly. Our human nature—especially our social behavior—is based on our biology and has no doubt evolved, at least to some degree, as a set of adaptations that promote individual survival and reproduction. However, understanding exactly what selective forces may have been involved, and disentangling biological from cultural factors, remains extremely difficult.

CHAPTER

25

Human Evolutionary History

HOW, WHERE, AND WHY DID HUMANS EVOLVE? What is our relationship with the great apes, and with extinct lineages that we see as fossils? What selective pressures led to upright walking and language? What has led to a species that has harnessed so many of the planet's resources for its own ends and that is able to contemplate questions such as these?

In this chapter we review the evolution of **hominins**, primate lineages that led to humans but not to the other great apes. Our story begins with the split between the hominin and chimpanzee lineages, approximately 5–8 million years ago (Mya). We then follow the divergence of hominins into multiple forms throughout Africa and the rest of the Old World, with the eventual appearance of physically modern humans beginning about 200,000 years ago. The majority of the hominin story is visible only through fossils or artifacts. For the vast majority of hominin history, we have no genetic information whatsoever and we probably never will. Genetic analyses have, however, made decisive contributions concerning the appearance of modern humans, largely settling what was long viewed as an irresolv-

able debate among anthropologists about human origins. We shall also comment, where possible, on when and where hominins acquired the characteristics now thought of as human. Relatively little is known about how and why hominin lineages evolved human characteristics, such as our remarkable cognitive abilities. At the end of the chapter, we look at several new genetic and linguistic studies that hint at novel directions for addressing perhaps the biggest question in human evolution: What makes us different from the other great apes?

PLACING HUMANS ON THE TREE OF LIFE

The capacity of humans for symbolic thought and the complexity of our societies so far outstrip those of our nearest relatives that there is a tendency to view "humanness" as a unique, integrated package. In terms of what makes us human, we see ourselves

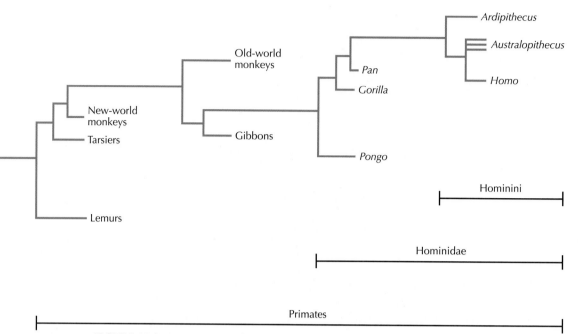

FIGURE 25.1. Organization of the order Primates. This tree represents only the branching order among the lineages. The lengths of the branches are not meaningful.

as equally far from all of our closest relatives, not recognizing, for example, any more indications of humanness in the chimpanzee than in the gorilla. Perhaps for these reasons the morphologies of living species have often been interpreted as providing evidence of a close relationship among the African apes and a much more distant relationship between them and humans. More recent evidence based on molecular data indicates that humans and African apes are closely related. Let us take a moment to explore the primate phylogenetic tree.

The order Primates contains all of the species commonly related to monkeys, lemurs, tarsiers, and apes (Fig. 25.1). On one branch of the Primate tree are three families, the old-world monkeys (Cercopithecidae), gibbons (Hylobatidae), and humans and other great apes (Hominidae). For most scientists, the terms Hominidae and great apes are interchangeable. The family Hominidae contains several branches: orangutans, chimpanzees, gorillas, and humans (Fig. 25.2). This family is divided into tribes.

FIGURE 25.2. (*Left to right*) Chimpanzee (*Pan*), gorilla *(Gorilla)*, orangutan *(Pongo)*, and human (*Homo*; Charles Darwin).

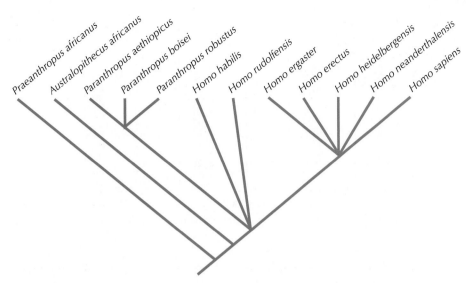

FIGURE 25.3. Phylogenetic relationships among the hominins. Branches with more than two descendants indicate relationships that cannot be resolved. Note that although all taxa are shown at the same level, they lived at different times.

The tribe containing humans and our extinct ancestors is called the Hominini (Fig. 25.3). Members of this tribe are termed hominins.

Chimpanzees Are Our Closest Relative

Until the 1960s, limited, mostly nonmolecular data (e.g., morphology, anatomy) were available for identifying our closest relatives and determining how long ago we shared a common ancestor with them. Using the data that were available, no agreement could be reached concerning the relationships among the great apes.

With the development of molecular techniques for evaluating phylogenies (see Box 5.1 and Chapter 27 [online]), the relationships among the great apes, including us, were established. One of the first applications of molecular data estimated the separation time between humans, apes, and old-world monkeys using a measurement known as **immunological distance**. The premise of this approach is that as proteins diverge, antibody binding will be reduced. If an antigen from species A is used to produce antibodies, then these antibodies will bind strongly to the protein from species A, but less strongly to a homologous protein from species B because of amino acid substitutions. The magnitude of change in binding affinity can then be used as a rough evolutionary clock. Using this technique, it was estimated that humans diverged from other apes approximately 5 Mya (Fig. 25.4). Prior to this immunological distance analysis, most scientists believed the divergence had begun at least 20 Mya, with *Ramapithecus*, which had been considered a direct ancestor to modern humans, estimated to have lived about 14 Mya. This dramatic result using molecular data forced a major reworking of ideas about early human evolution.

Extensive analyses of multiple gene genealogies have confirmed arguments based on the **molecular clock** that the separation between human and chimpanzee lineages occurred 5–8 Mya. Thus, at most loci, genes in humans are more closely related to their homologs in chimps than they are to any other living primate species. Among the great apes, chimpanzees are our closest relatives (Fig. 25.5). Note, however, that the split from the gorilla lineage occurred not long before the human–chimpanzee split, so a substantial fraction of the human genome contains genes that are more closely related to their gorilla homologs than to chimp homologs (**lineage sorting**; see p. 628).

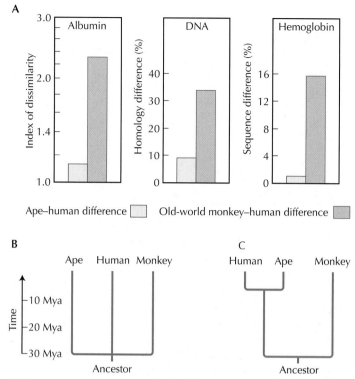

FIGURE 25.4. (A) Quantitative comparisons of the difference between human macromolecules and those of African apes and old-world monkeys. The results of these comparisons showed that the phylogenetic relationship of humans to African apes and old-world monkeys was as shown in C rather than as in B. "Index of dissimilarity" and "Homology difference" are measures of divergence based on antigenic cross-reaction and on DNA–DNA annealing temperature, respectively.

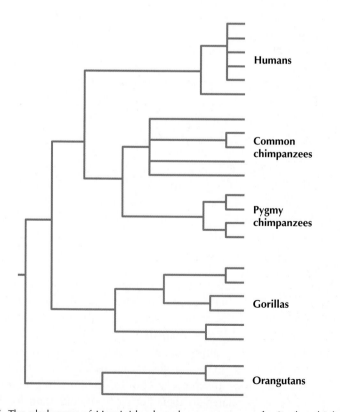

FIGURE 25.5. The phylogeny of Hominidae based on sequences of mitochondrial COII genes.

Taxonomy Based on Fossils Is Technically and Conceptually Difficult

Theories on the pattern of human evolution have ranged from a single linear progression beginning with the common ancestor of the great apes and leading to ourselves to a highly branched phylogeny containing many diverse species who, except for *H. sapiens*, are all extinct. (Recall that, in general, the phylogenies that connect present-day species are far simpler than the complete phylogeny that includes all extinct lineages [Fig. 5.11].)

As discussed in Chapter 22, species are hard to define and recognize, even when we deal with living organisms. When dealing with sexually reproducing groups, most biologists use the **biological species concept**, which defines species as sharing a common gene pool separated from that of other species by barriers to gene exchange (see p. 622). However, for human evolution, fossils give us little clue as to likely mating behavior, and even morphological data are often scarce. Thus, recognizing species boundaries at a given point in time in the past presents a considerable challenge. More fundamentally, lineages change over time, and the biological species concept does not tell us how to divide a continuous line of descent into distinct species. In practice, hominin taxonomy is based on differences that exist both among biological species at a single point in time and within the same lineage at different points in time. Division of a single evolving lineage at an arbitrary time point is made for taxonomic convenience and does not have biological significance. Despite the taxonomic ambiguity, it is of interest to trace through time the "human" traits that distinguish us from our closest relatives, which we pursue in the next section.

THE EVOLUTION OF HOMININS

Two recent finds have rekindled the debate about the overall bushiness of the hominid evolutionary tree. The most primitive *Homo* fossil known outside Africa was recently discovered at the same site as other, more modern forms. This suggests that for part of its history the **hominin** lineage may have comprised only one very diverse species, perhaps with extensive geographic structuring. On the other hand, the very early hominin *Sahelanthropus*, which was found in Chad in Central Africa (dated 6–7 Mya), has a surprising mixture of primitive and more modern hominin features. This suggests that there has been a high level of **homoplasy** at different times in the hominin lineages: That is, similar forms may have appeared in different species because of convergent and parallel evolution rather than phylogenetic relationship.

This section presents an overview of how the "human" traits that distinguish us from our closest relatives have changed through time. Because of the inherent ambiguity of taxonomy based on fossil evidence, the focus will be less on the names and ranges of species and more on the appearance of those traits that distinguish us from other primates. Up until approximately 20,000 years ago, the hominins very likely included more than one species, and perhaps many more for much of the last 2 million years. In addition, many of these species had traits that, among the living primates, are now unique to us.

The oldest known hominin along the human lineage was discovered in West Africa and lived about 6 Mya. *Sahelanthropus* is known only from cranial fragments (Fig. 25.6). Nonetheless, changes in dentition and the structure of the basic cranium make a compelling case that *Sahelanthropus* is closer to us than to the chimpanzee, although perhaps with little else to distinguish it from the common ancestor of humans and chimps, who was also living approximately 6 Mya. Although paleoanthropologists believe that *Sahelanthropus* is on our lineage, with so few traits to work with, the group may have been on the lineage of the ancestor of humans and chimps, and the features that now appear hominin-like were lost along the lineage leading to the chimp. The discovery of

FIGURE 25.6. *Sahelanthropus* skull. (*Left*) Facial view. (*Right*) Lateral view. This specimen, found in Chad in Central Africa, is between 6 and 7 million years old. Its distance from the East African Rift Valley suggests that the earliest hominids may have been widely distributed. Scale bar, 5 cm.

Sahelanthropus makes clear that the groups near the base of the human lineage did not acquire a set of hominin characteristics as a package, which is just as we would expect from evolution. We have been evolving separately from lineages leading to all other living primates for more than 5 million years. It is to be expected that features unique to humans have been acquired throughout this period and are distributed throughout the various hominin species that have appeared and disappeared (Fig. 25.7).

Australopithecus Was One of the Earliest Bipedal Hominins

Some spectacular discoveries of fossil hominins have been made, including the famous "Lucy," a partial skeleton of an *Australopithecus afarensis* dated at 3.18 Mya (Fig. 25.8). However, most finds are nothing more than the fragment of a skull or a limb, leaving anthropologists to infer the details of the remaining skeleton. They do so by analogy to other species, or on the basis of engineering principles, although the conclusions are often ambiguous and contentious.

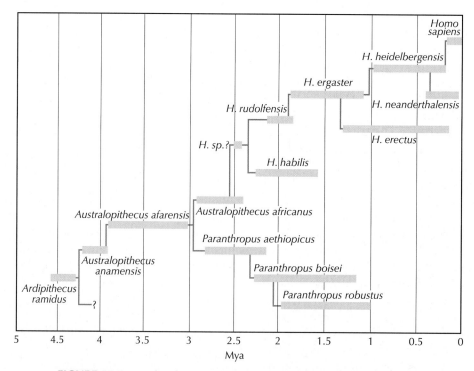

FIGURE 25.7. Postulated time spans and relationships of hominin species.

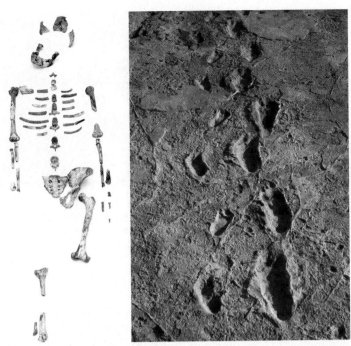

FIGURE 25.8. *Australopithecus afarensis* skeleton (*left*) of the famous "Lucy" found in the Afar region of Ethiopia and dating from about 3.3 Mya. (*Right*) Trail of hominin footprints fossilized in volcanic ash. This 70-m trail was found by Mary Leakey's expedition at Laetoli, Tanzania in 1978. It dates to 3.6 Mya and shows that hominins had acquired the upright, bipedal, free-striding gait of modern man by this date. The footprints show a well-developed arch to the foot and no divergence of the big toe. They were made by two adults, probably belonging to *A. afarensis*, with possibly a third set belonging to a child who walked in the footsteps of one of the adults.

The earliest member of the genus *Australopithecus* is *A. anamensis*, first documented in Kenya and believed to have lived beginning approximately 4.2 Mya (Fig. 25.9; see also Fig. 25.7). Although *A. anamensis* is one of the many forms known only from small bone fragments, even at this early stage there was clearly diversity among hominids, with the more ape-like genus *Ardipithecus* living slightly before and perhaps overlapping with *A. anamensis*. However, *Ardipithecus* has **derived features** that do not appear in the hominin line or its inferred ancestors. Thus, it is likely to be the earliest of the many side branches apparent in the hominin family.

Because walking upright on two limbs—bipedalism—is a key human adaptation, there are many theories concerning its origin. Darwin thought that it evolved in order to free the hands for fine manipulations, and this view retains some currency today. Other suggestions include greater stamina for long distance travel, more effective hunting, and improved ability to carry food. It is possible, however, that there is no single overriding reason for bipedalism, but rather a mosaic of different advantages and uses. Many theories place the advantages to bipedalism in a savanna context—for example, the ability to see long distances. The sites where *A. anamensis* fossils have been found are inferred to have been relatively dry woodland when *A. anamensis* was alive. However, the picture is complicated somewhat by *Ardipithecus*. Although thought on less secure evidence to have been bipedal as well, we know that its habitat was heavily wooded. Although the fossil evidence is very poor for *A. anamensis*, what exists shows few differences from *A. afarensis*, for which the fossil evidence is much richer. With the arrival of *A. afarensis*, we have the first well-known hominid species in whom bipedalism is now a certainty, evocatively attested to by the footprints of three individuals immortalized in volcanic ash (see Fig. 25.8).

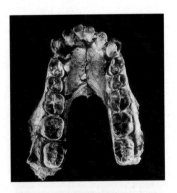

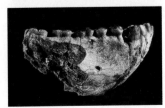

FIGURE 25.9. *Australopithecus anamensis* mandible dating from 4.1 Mya. The mandible has both primitive and derived features, suggesting that this species may be an ancestor of *Australopithecus afarensis*.

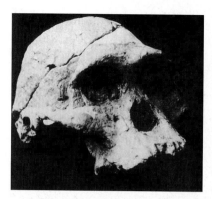

FIGURE 25.10. Gracile (*left*) and robust (*right*) skull forms of Australopithecines. These groups make clear that at least at some points in the hominin past there was considerable morphological variation among groups. Different hominin groups at comparable times, or the same time, appear to have had very different lifestyles, socially and/or ecologically.

As far as can be discerned from fossil evidence, the shift to bipedalism was not accompanied by the development of other modern human features. Relative to body size, the brain of *A. anamensis* was within the range of modern apes. There are no indications of development of a modern vocal apparatus, no indications of tool use, and sexual dimorphism appears to have been extreme.

It is easy to view the modest developments within *A. afarensis* as a mere transition between something ape-like and something more human. In fact, *A. afarensis* was a stable and successful form, persisting for much more than a million years with only modest modifications. The *A. afarensis* period does, however, mark an increase in hominin diversity. Approximately 3 Mya, the *A. afarensis* lineages split into forms with marked differences in bone structure, referred to as **gracile** and **robust** forms (Fig. 25.10).

Homo habilis Was the First Hominid to Use a Complex Set of Tools

The earliest evidence of human-made tools dates to approximately 2.5 Mya in Gona, Ethiopia. Paleoanthropologists distinguish between tools and a complex set of tools fashioned for different purposes, termed a tool industry. The earliest known tool industry was discovered at Olduvai Gorge in 1971, amid a mixture of robust and gracile hominin fossils; it dates to approximately 1.8 Mya. Although the tools were found among australopithecine fossils, the australopiths were probably not the toolmakers, because they had been evolving for more than a million years without any apparent development of tools, and because their relative brain size was not very different from that of modern apes. The first toolmakers were by definition humans and would therefore be more modern in form than the australopiths. Instead, the tools were ascribed to the more gracile form found at Olduvai, eventually introduced as *H. habilis*. The brain size of *H. habilis* was only modestly larger than the australopithecine brain, and was much less than the arbitrary thresholds that had been previously suggested for admittance to the *Homo* genus (Fig. 25.11).

Whereas information about australopithecine species is very limited, a great deal more is known about *Homo* species. There is evidence of hunting, the use of relatively complex tools, the use of fire, and prolonged childhood (Fig. 25.12). However, few details of the social organization and foraging habits of early *Homo* species are known.

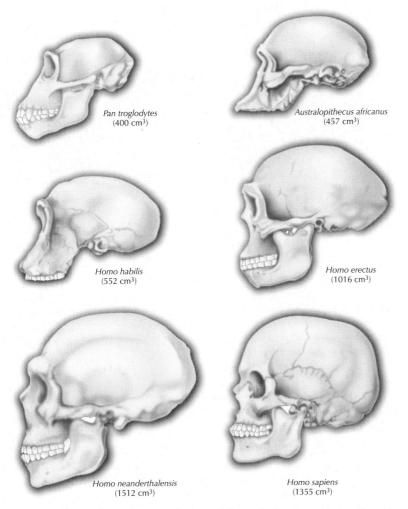

Pan troglodytes
(400 cm³)

Australopithecus africanus
(457 cm³)

Homo habilis
(552 cm³)

Homo erectus
(1016 cm³)

Homo neanderthalensis
(1512 cm³)

Homo sapiens
(1355 cm³)

FIGURE 25.11. Series of hominid skulls showing different brain sizes. Notice that *Homo neanderthalensis* had a greater volume that that of modern humans.

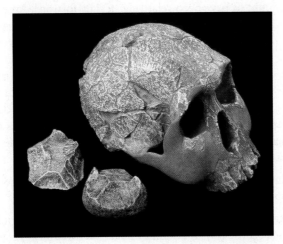

FIGURE 25.12. *Homo habilis* skull, dating from about 1.8 Mya and found in Kenya. The complexity of *Homo* tools indicates a substantially more complex culture and social organization than that of earlier hominins. Two shaped rocks found with the fossil are seen at *lower left*.

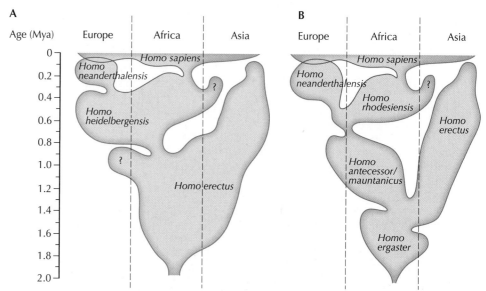

FIGURE 25.13. Two different views of the geographic and temporal ranges of key hominin groups based on fossil discoveries and different taxonomic schemes. (A) This distribution posits that Neanderthals and modern humans both derived from *Homo heidelbergensis*. (B) This distribution shows an alternative view, with a *Homo rhodesiensis* ancestor for *Homo sapiens*, and postulates more species deeper in our ancestry.

It seems likely that *Homo* marked a dramatic increase in cognitive abilities over earlier hominids. Enhanced cognitive abilities, and the concomitant development of complex social organization, probably allowed for an increase in population density, which may explain the greater amount of fossil evidence with *Homo* as compared with earlier hominins. It is also likely that these advances permitted the dramatic increase in geographic range achieved by *Homo* beginning as early as 1.8 Mya (Fig. 25.13).

The Details of Hominin Dispersal from Africa Remain Unresolved

Exactly *who* came out of Africa 1.8 Mya is an open question. Until recently, the evolution of the *Homo* line after the appearance of *H. habilis*, and its spread through the Old World, was thought to have been relatively straightforward. An early *Homo* species differentiated in Africa into *Homo erectus*, who subsequently spread throughout the Old World (Fig. 25.13A). This picture has been complicated, however, by the data that indicate that there may be two distinct early lineages (*Homo rudolfensis* and *H. habilis*), and it is not clear which of these may have led to later forms that are sometimes now recognized as two groups: *Homo ergaster* and *H. erectus*. Moreover, although it appears fairly clear that *H. ergaster* evolved in Africa, *H. erectus* finds have been dated between 1.6 and 1.8 Mya in central and Southeast Asia. These dates are indistinguishable from the oldest African ones, which are from East Africa. The fossil evidence is therefore consistent with two alternative histories.

1. The descendant of *H. habilis/H. rudolfensis* evolved into *H. erectus* in Africa and soon after dispersed throughout the Old World.

2. The descendant of *H. habilis/H. rudolfensis* dispersed from Africa and somewhere outside of Africa—probably in Southeast Asia—evolved into *H. erectus*, which subsequently dispersed back through the Old World, including Africa.

Either way, after dispersing throughout the Old World, whether originally from Africa or eastern Asia, *Homo* populations subsequently differentiated into different forms in different geographic areas.

Box 25.1   Distinguishing *Homo* Species Remains Controversial

Most paleoanthropologists recognize at least four "*H. erectus*-like" forms of hominins: *H. erectus, H. heidelbergensis, H. neanderthalensis,* and *H. sapiens.* Others, however, have recognized earlier differences, such as an earlier divide between *H. ergaster* and *H. erectus.* The oldest *H. heidelbergensis* fossils are European, and similar fossils have been found in various African sites (Fig. 25.14). *H. heidelbergensis* has both ancestral *H. erectus* features and some derived *H. sapiens* features, suggesting that it, or something like it, gave rise to modern hominids. Agreement breaks down completely, however, for groups that appeared after *H. heidelbergensis,* as paleoanthropologists continue to debate the origin of anatomically modern humans. At its heart, this dispute

concerns which, if any, groups subsequent to *H. heidelbergensis* were reproductively isolated from one another, and so form separate **biological species.** In discussing this debate, we shall generally refer simply to anatomically modern humans and lump the earlier forms together as either *H. erectus* or archaics. As will be seen, the principal features of this debate can be addressed without worrying about exactly what to call the various *H. erectus* groups or even whether all those extant hominids except modern humans were reproductively isolated from one another. The very recently discovered *H. erectus* from a 1.8-Mya site, on the eastern shores of the Black Sea, has suggested to some that many different forms are, in fact, members of a single species.

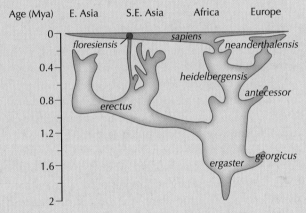

FIGURE 25.14. Distribution of *Homo erectus* and *H. erectus*-like forms outside of Africa.

The Earliest Appearance of Anatomically Modern Humans Was in Eastern Africa

Throughout its range, the descendants of *H. rudolfensis* evolved into different forms, including *H. ergaster, H. erectus, H. antecessor, H. heidelbergensis,* and *H. neanderthalensis* (Box 25.1) For simplicity, these forms are often referred to collectively as archaics (i.e., archaic *Homo* species). Within some of these groups there are strong regional differences, and it seems likely that more types will be recognized. There are no convincing arguments, however, that can be marshaled based on morphological data to determine which forms were distinct biological species. Despite the diversity in hominin forms, none of them was truly modern, although different groups had modern features to a greater or lesser extent. For example, some *H. neanderthalensis* had cranial volumes even larger than we do.

Forms that are recognized by anthropologists as fully modern are collectively called **anatomically modern humans** and are known from sites around what is now Ethiopia. Radiometric analyses of two skulls, Omo 1 and 2, found near Kibish in Ethiopia in the 1960s (Fig. 25.15), revealed that these two people lived nearly 200,000 years ago. The antiquity of this collection of old sites near one another in Ethiopia and Eritrea suggests that this may have been our earliest home. In all features that

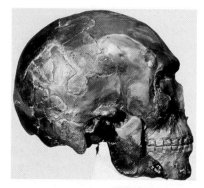
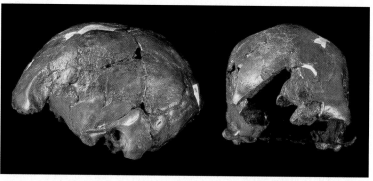

FIGURE 25.15. Omo I (*left*) and Omo II (*right*). Anatomically modern, these skulls were found in the Omo Basin in Ethiopia and were originally dated to 130,000 years ago. More recent radiometric dating, however, has pushed the date back to 200,000 years ago, making them the oldest known examples of modern humans. Omo I is a reconstruction; Omo II is a cranium only.

can be discerned from the fossil record these people were largely indistinguishable from modern humans. There is often a tacit assumption, therefore, that early anatomically modern humans also possessed fully modern cognitive functions (including a highly developed language), but this is probably something that will never be known with certainty.

In some cases archaeology has done much more than just show us *where* our ancestors lived. On lucky occasions evidence persists that gives us fascinating glimpses into *how* they lived. One of the most interesting of these early finds is from modern-day Eritrea. We know very little about these people, but we do know that they lived by and off of the sea. This seems to have been a new habitat for modern hominins, and the finding has bolstered claims of a coastal route out of Africa. Some anthropologists have postulated that the earliest large dispersals of modern humans from Africa may have been from the Horn of Africa and along the southern coasts of Asia (Fig. 25.16). This is certainly

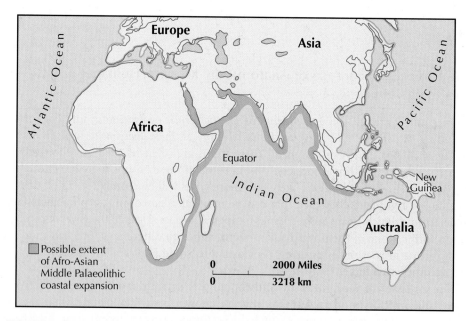

FIGURE 25.16. Representation of shorelines as they were around 65,000 years ago and possible coastal routes for early movements of modern humans out of Africa. *Red lines* are the current shorelines.

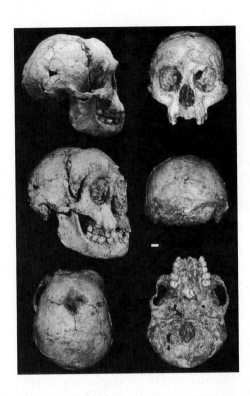

FIGURE 25.17. *Homo floresiensis* cranium and mandible. This species, which existed from about 38,000 to 18,000 years ago and was found in Flores, Indonesia, has a combination of primitive and derived features. It overlapped with *Homo sapiens* in this region and was strikingly small bodied, standing only 1 m tall.

possible, as neither genetic nor archaeological data make clear the route out of Africa, although no conclusive evidence in support of the coastal route has yet been found either. In any case, it raises the broader question of the origin and spread of anatomically modern humans, one of the most fiercely debated topics in human evolution.

Have *H. sapiens* Always Been Alone?

The earlier hominin forms did not just step aside for the modern ones. There was a long period of coexistence that may have included living in close proximity. Until very recently it was thought that we were the only hominins in existence since the Neanderthal disappearance approximately 20,000–30,000 years ago. But in 2003 an extraordinary find from Flores, an Indonesian island near Bali, introduced what may be a new close relative. As recently as 18,000 years ago a diminutive **archaic human** form inhabited this island, apparently undergoing the rapid reduction in body size that often occurs on islands. Named *Homo floresiensis*, these hominins may have weighed only about 50 pounds (Fig. 25.17). It has been argued, however, that the main cranium used to describe the species could be an example of a microcephaly, a genetic disorder characterized by severe reductions in cranial size and other characteristics. At the time of writing, this dispute is unresolved.

The Genetic Origin of Modern Humans: Out-of-Africa or Multiregional Evolution?

How and when did anatomically modern humans spread across the world? The oldest reliable date associated with modern humans is approximately 200,000 years ago in East Africa. Outside Africa, the oldest site containing fossils of modern humans is in modern-day Israel and dates to approximately 100,000 years ago. The site in Israel is of particular interest because moderns may have lived there concurrently with *H. neanderthalensis*, who had moved there much earlier (Fig. 25.18). For unknown reasons there

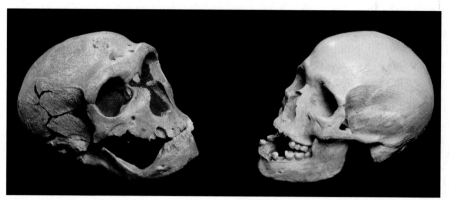

FIGURE 25.18. A Neanderthal skull (*left*) and a modern skull (*right*). Neanderthals were extant in Europe around 150,000–30,000 years ago.

was a considerable gap before modern humans reached other areas beyond Africa. Intriguingly, many of the earliest sites are in insular Southeast Asia and Australasia, consistent with the possibility of a coastal route of expansion from East Africa along the Arabian peninsula and Asian subcontinents and finally to Southeast Asia (see Fig. 25.16). The question this raises is, "What happened between the appearance of modern humans and their eventual colonization of both the old and new worlds?" This question has been the subject of highly polarized debates, out of which have arisen two models, now known as the **out-of-Africa model** and the **multiregional evolution model**. Note that this debate refers only to the appearance of modern humans. The much earlier dispersal from Africa (see p. 736) involving archaic *Homo* species is not controversial.

The multiregional model in its most extreme version posits that the hominin species and populations evolving throughout the Old World, such as *H. neanderthalensis* in Europe or *H. erectus* in East Asia, evolved in situ into the modern populations now living in these regions. Although rarely explicitly discussed, this model implies that the species ancestral to modern humans were not true biological species (Chapter 22 [p. 622]). The strongest morphological evidence for multiregionalism comes from a perceived continuity of form ranging from the archaic to the modern periods within specific geographic regions, including Europe, China, and Java.

The out-of-Africa hypothesis, on the other hand, proposes that modern humans emerged recently in Africa and spread to the rest of the world, where they replaced the archaic hominins. Objective and quantitative analysis of patterns in the fossil record, however, is a complicated business. For example, samples from geographic areas are rarely considered exhaustively, which inevitably biases their interpretation. Claims of regional continuity are strongly contested by a number of anthropologists, many of whom see evidence in the fossil record of a general population replacement associated with the advent of modern forms in different geographic regions.

Although both models are sometimes represented in the literature in highly specific forms, they are in fact each more appropriately thought of as classes of models. At its heart, the out-of-Africa/multiregional debate is about whether non-African archaic *Homo* populations made genetic contributions to the modern gene pool. Although debate about the origin of modern humans has traditionally been highly polarized, the out-of-Africa model and multiregional evolution model are best seen as the extreme ends of a continuum focusing on the degree of archaic genetic input into the modern gene pool. This perspective is of particular importance for interpretation of the genetic evidence considered in the following sections. We will see that it is quite difficult to reconcile the existing genetic evidence with large-scale genetic contributions from differentiated archaic groups into the modern gene pool, as is postulated by the multiregional model.

GENETICS AND HUMAN EVOLUTION

The earliest genetic studies in human evolution long preceded the era of recombinant DNA and instead focused on protein level variation at loci determining blood groups, immunoglobulins, and the HLA (human leukocyte antigen) complex, which determines **antigenic diversity** (see pp. 234 and 541). However, it was the development of protein electrophoresis in 1966 that made it possible to systematically screen protein variants at a broad range of loci (see pp. 59 and 360). But the availability of detailed data is only part of what is needed to study evolution. It is also necessary to know what kind of questions should be addressed with the data, and, in the early stages, the interpretation of genetic data was not well grounded in evolutionary theory.

In the development of evolutionary models, there is always a trade-off between realism and tractability. If a model is too simple, inferences based on it will have little relevance to biological reality. But if the model is too complex it may be impossible to use, either because of mathematical intractability, because it requires the use of parameters that are not known and cannot be inferred, or because it weaves together everything that is known into a compelling but hopelessly complicated story. (Recall our discussion of phylogeography in Chapter 16 [pp. 451–455].) When developing an evolutionary model, the goal is to capture enough of the reality to represent the processes of interest while being specific enough to be testable (see Box 25.2; Fig. 25.19). As a compromise between simplicity and realism, the Italian geneticist Luca Cavalli-Sforza and others assumed that, on a global scale, as human populations colonized different geographic regions, the new populations became effectively isolated from the parent populations. Within regions, gene flow might combine with selection and random drift to generate **clines** and **isolation by distance**, as we discussed in Chapters 16 and 18. On a large scale, however, human population history is approximated by a branching tree of effectively isolated populations. This might seem a reasonable assumption, but it must be remembered that movement of just a few individuals per generation greatly reduces neutral divergence between populations (see Chapter 16 [p. 445]). As we will see below, the assumption of limited migration between populations is critical.

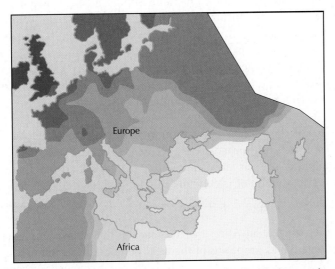

FIGURE 25.19. As early farmers spread into Europe, they carried with them a characteristic set of alleles. As they interbred with the resident populations, the frequencies of these alleles declined, leaving a cline running from east to west. The map shows the average frequency of alleles characteristic of early farming populations. The *lightest shade* correlates with areas in which farming was present 8500 years ago, and the *darkest shade* correlates with areas in which farming was present no more than 5500 years ago.

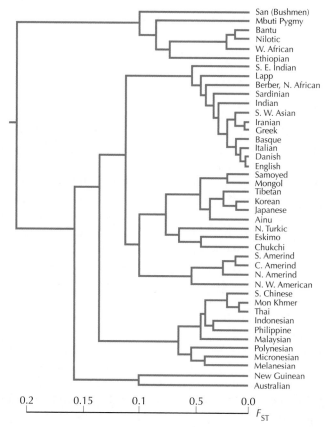

FIGURE 25.20. A **phenogram** showing the similarity of different populations in terms of the frequencies of 120 alleles.

Human Demography Can Be Modeled as a Hierarchical Structure

Assuming that colonization of new geographic regions leads to complete genetic isolation, relationships among human populations should form phylogenetic trees, similar to those that describe relationships among different species. Under this model, relationships among populations from the different regions could be appropriately represented as a tree, with nodes in the tree representing points at which an ancestral population split to form two genetically isolated populations. Efforts to understand the evolutionary history of human populations on a global scale have therefore focused on estimation of the times when ancestral populations split in two.

Methods for estimating phylogenetic trees are discussed in detail in Chapter 27 (online). Studies of the human population have often been based on the use of **genetic distances**, which are measures of divergence that are expected to increase linearly with time under the relevant demography (e.g., constant size following complete isolation). However, the use of genetic distances to estimate phylogenetic trees rests on the key assumption that human populations were, in fact, genetically isolated from one another.

It must be recognized, however, that trees such as that shown in Figure 25.20 do not themselves validate the assumptions of the basic demographic model. If human populations never actually became reproductively isolated, but exchanged migrants at some rate, then the pattern of genetic distances, and the estimated phylogeny, would reflect the pattern of gene flow, rather than their actual history. The basic assumption that some human groups split into isolated populations can be tested by comparing the inferred separation times with those estimated from archaeological data. For example, modern humans are

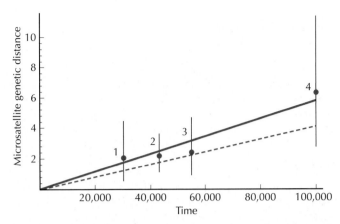

FIGURE 25.21. The *x*-axis shows the estimated separation times of major population groupings (based on archaeological data) and the *y*-axis shows the estimated genetic distance based on microsatellite variation. The population comparisons are *1,* Amerind vs. East Asian; *2,* European vs. East Asian and Amerind; *3,* Melanesian vs. Sahulland; *4,* African vs. non-African, based on archaeological estimates of the arrival time of anatomically modern humans into various continental regions. (*Solid line*) The best fitting straight line through the points and the origin. (*Dotted line*) The slope of the genetic distance expected based on mutation rate estimated from the pedigrees.

first recorded outside of Africa in the Levant (i.e., countries bordering the eastern Mediterranean) about 100,000 years ago, in Australasia about 70,000 years ago, and in the New World about 15,000–20,000 years ago. If the genetic distances among human populations on different continents were actually determined mainly by migration rates, as opposed to separation times, there would be no reason to expect a correlation between genetically inferred separation times and the estimated times when humans arrived on new continents (Fig. 25.21). Although data are sparse, these analyses suggest that at least for groups of people living in different continents, the history of population separation has not been greatly obscured by subsequent migration of populations from different continents.

Most geneticists today are unwilling to assume a priori that different human populations have been isolated for much of their histories, even for intercontinental comparisons. In many ways, however, the validity of the model is less important than its existence. By specifying a precise demographic model, and using formal population genetics to estimate the parameters of that model, Cavalli-Sforza and his colleagues ushered in the modern era of the study of human genetic variation. Fortunately, the development of new inferential frameworks is allowing the evaluation of more realistic, and thus more complicated, models (Box 25.2).

Genetic Evidence Helps Us Understand the Origin of Anatomically Modern Humans

When trees were first constructed with large data sets, the position of the root separated sub-Saharan African populations from all non-African populations—a pattern that has been confirmed by all recent studies. Assuming that populations split into separate lineages, this suggests that the first population was in Africa, and that all non-African populations are derived from a group that split off from the African group and differentiated into different groups throughout the non-African world. If the assumption of population bifurcations is incorrect, however, these results tell us nothing about who was where first; instead, they would reflect mainly the pattern of migration.

Even ignoring the question of whether migration might have affected the analysis, this result is insufficient on its own to reject multiregionalism in favor of the out-of-Africa model. All anthropologists agree that *Homo* has an African origin, and so even

Box 25.2   Statistical Inference in Genetic Anthropology

The availability of abundant genetic data and of fast computers has made possible new approaches to inference from genetic data. Instead of summarizing the data by statistics such as F_{ST} (see p. 446), we determine the probability of obtaining the data given some evolutionary model; this is known as the **likelihood** of the model. In most cases, this probability cannot be determined analytically, but rather is estimated by computer simulation. Using simulation to estimate these probabilities has become easier because of the efficiency and relative simplicity of **coalescent** approaches for simulating the evolutionary process (see Chapter 15).

For example, Lounès Chikhi and Mark A. Beaumont developed a method for estimating the proportion of a population (P_1, P_2) that came from each of two parental populations (see Fig. 25.22). The model takes as input the genotypes of the descendants of each parental population and of the admixed one. It then estimates P_1 and P_2, and the amount of drift that has subsequently occurred in each of the three populations is represented by the combined parameter t/N_e, where t is the time since the admixture event and N_e is the effective population size. Just as important as what an explicit model such as this one can tell us about human history is what it can tell us about what cannot be inferred reliably. For example, in the area of work often referred to as **phylogeography** (see pp. 451–455) some researchers have attempted to infer details about the historical movements of populations directly from the currently observed frequencies of lineages in different geographic areas. For example, some have argued that lineages that are observed in Europe but not in the Near East must have been resident in Europe before the arrival of Near Eastern farmers from Europe (see Fig. 25.19), and that the current proportions of such lineages can be used to infer the original proportions of migrants and nonmigrants. The Chikhi–Beaumont model, however, makes clear that because of genetic drift that has occurred since the colonization of Europe, current proportions of lineages, even if these were correctly identified as migrant and nonmigrant, provide a very poor estimate of the original proportions (Fig. 25.22). When these more detailed analyses are carried out, they often show that a great many of the claims that have emerged from phylogeographic studies using single-locus systems, such as mtDNA (**mitochondrial DNA**) and Y chromosomes, are suspect (see also pp. 451–455).

In human evolution, this approach for developing a specific and tailored model is highly effective when external historical information makes clear exactly what sort of model is appropriate. In this case, the model was developed to estimate the proportions of different prehistoric Eurasian populations in the modern European gene pool. The reason that highly specific models such as this one are used is that we do not currently have available general models that are appropriate for all settings. For example, we do not currently have models that could be used appropriately in any kind of admixture scenario, which is itself a fairly specific situation. This is one of the reasons why population genetic inference should not be done at the push of a button on a keyboard. It is also the reason why there is a continuing role for more ad hoc methods of analysis in human genetic history (see pp. 769–770).

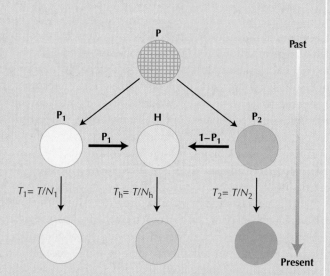

FIGURE 25.22. Model to estimate admixture in populations. P is the founder population. H is a hybrid population formed by members of the P_1 and P_2 parental populations. The impact of genetic drift on allele frequencies in all populations depends on the time elapsed and the effective population sizes (N_1, N_h, N_2). Models such as these can be used to estimate the time since a hybrid population was formed (scaled by population size) and/or the proportion of input from each parental population.

the multiregional model would predict an African/non-African divide as the deepest split in the human population tree. The difference, however, is in *when* the split occurred. *H. erectus* emerged from Africa more than 1.7 Mya, whereas modern humans by any estimate are much more recent. Estimates made using a wide range of marker systems and methods date the deepest split in the tree of human populations at well under 200,000 years ago. Such a recent date is difficult to reconcile with many, but not all, versions of the multiregional model. Thus, if populations were established throughout the world more than 1.7 Mya, and if gene flow among them was limited thereafter, the deepest split should be much deeper than 200,000 years—in fact, closer to 1.7 million years. Although estimated dates are uncertain, there has been clear consistency among genetic marker systems, from classical blood groups and allozymes to microsatellite and single-nucleotide polymorphisms. Clearly, if the model of population bifurcations is correct, then we are a demographically young species that only recently emerged from Africa.

The multiregional model can be ruled out with analyses such as these, however, only if there are external reasons for believing that the model of population bifurcations is accurate on a global scale. If there is considerable gene flow among regions, then the "separation times" estimated in these analyses have little meaning. For example, regular genetic exchange between African and non-African populations would lead to estimated separation times much more recent than the time of first emergence from Africa.

The Genealogies of Y Chromosomes and mtDNA Can Be Estimated

The Y chromosome is inherited **patrilineally** (from fathers to sons) and therefore carries information about the evolutionary past of males, complementing information carried by the matrilineal **mtDNA** molecule. The nonrecombining part of the Y chromosome is about 22 Mb long, or about 43% of its total length.

Because of the size of the nonrecombining segment, it harbors a tremendous number of polymorphic sites that can be used to infer the genealogical relationships among human Y chromosomes. Critically, within this long stretch of DNA there are a variety of markers present with different mutation properties. For example, single-nucleotide polymorphisms and insertions and deletions on the Y chromosome occur so rarely that polymorphisms of this sort usually trace back to a single mutation in human evolutionary history. Multiple mutations at the same site are unlikely. For this reason, these markers are known as **unique event polymorphisms**, or UEPs. Because the chromosomes sharing a mutation in a sample will share a common origin, they form a related group. Consequently, UEPs break the Y chromosome into unambiguous genealogical groups (Fig. 25.23; recall Fig. 13.12 and see Chapter 27 [online]).

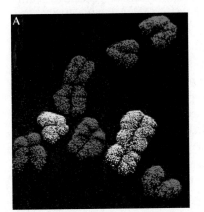
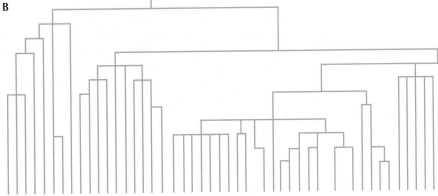

FIGURE 25.23. A portion of the Y chromosome does not recombine. (*A*) Human chromosomes showing an example of the X chromosome (*blue*) and the Y chromosome (*pink*). A long stretch of the Y-chromosome DNA sequence shares a single genealogical history, so the many mutations that have occurred throughout the sequence help to reveal the single genealogy. (*B*) A portion of the genealogy of the Y chromsome, based on a large number of mutations that occur sufficiently infrequently in human evolution that they can be considered effectively unique.

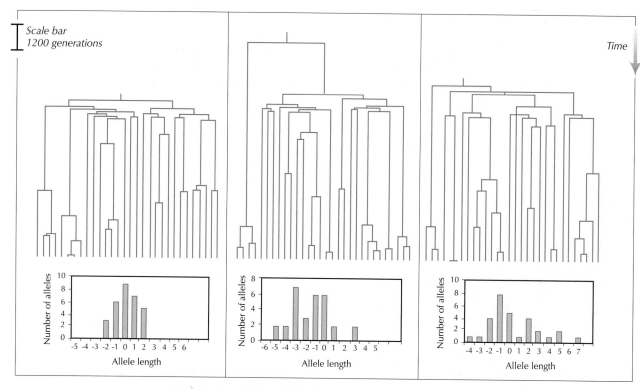

FIGURE 25.24. Growth affects the shape of gene genealogies and thereby affects the distribution of variants on alleles sampled from the population. The impact of demography on the shape of the gene genealogy can be seen using different kinds of genetic markers. Here three examples of genealogies are shown from a simulation of a rapidly growing population. The genealogy tends to show multiple deep branches, and in consequence the distribution of microsatellite allele sizes is relatively even.

In addition to UEPs there are also many rapidly evolving microsatellites. In approximately 2 out of every 100 parent–offspring transmissions, on average, a Y chromosome microsatellite will increase or decrease by one or more repeat units. This is a sufficiently rapid rate of evolution that differences emerge among Y chromosomes within the major genealogical groupings described by UEPs, allowing the reconstruction of fine-scale genealogical relationships and the estimation of genealogical time depths. This is also the reason that microsatellites are the markers of choice for Y-chromosome forensic work (p. 768).

Genetic markers on the Y chromosome can tell us about the past history of the human population. One approach is based on the **mismatch distribution**. In a growing population, most coalescence occurred in the distant past when the population was small. Thus, most pairs of chromosomes have been separated from one another for similar amounts of time, and genealogies tend to have deep branches (recall Fig. 15.8). For this reason, the distribution of sequence differences between pairs has a single peak, centered on the amount of divergence, that is in proportion to the total depth of the gene tree (Fig. 25.24). In contrast, in a population of constant size, coalescence times are exponentially distributed, and so the mismatch distribution is correspondingly broad.

It is important to realize, however, that inferences based on single genetic loci are highly variable. As we saw in Chapter 15, variation at a single locus results from a single realization of the evolutionary process. That is, the genealogy is a random draw from the distribution of genealogies associated with a given evolutionary situation. For this reason a single genealogy will inherently lack statistical power in any effort to infer population parameters. An even more fundamental problem, however, is that selection may act on the genealogy under study, which will make it atypical of other genealogies and would lead to systematic biases in the estimation of population parameters. Thus, although the statistics such as the mismatch distribution are used to infer the rate of growth of the whole population, they really tell us about the single genealogy in question. As we discussed before (pp. 451–455 and 536–538), the only way to distinguish the effects of overall demography from the selection and drift on individual loci is to analyze multiple genetic loci.

Closely Related Y-Chromosome and mtDNA Lineages Are Inconsistent with the Multiregional Model

Most versions of the multiregional model predict deep genealogies, with genes sampled from different continents tracing back through the archaic populations that lived in the same region. For example, archaic humans have lived in Europe and East Asia for approximately the last million years, and it is hard to imagine extensive gene flow between these two regions. Thus, lineages that trace back into these two populations are expected to have had a common ancestor no less than 1 Mya. Yet, the estimated time to the most recent common ancestors of the human Y chromosome and mtDNA are nowhere near this old, ranging from 100,000 to 150,000 years ago for the mtDNA and somewhat less for the Y chromosome. Although current estimates of genealogical depth are uncertain, these very young dates are statistically inconsistent with a coalescence time anywhere near 1 million years.

These relatively recent dates have not changed despite the fact that extremely large and diverse sets of mtDNA and Y chromosomes have now been evaluated. We cannot be sure that there are *no* archaic lineages among any human populations for these two genetic systems, but they certainly cannot be very common. Thus, the strongest current evidence against the multiregional model is also the simplest: No ancient mtDNA or Y-chromosome lineages have been found.

Unfortunately, data from these two genetic systems do not rule out the multiregional model entirely. If there were a modest rate of gene flow between the archaic populations, then even if most lineages traced back for a million or more years within their local populations, some loci would show much more recent coalescence through occasional migration events (Fig. 22.10); it is possible that both Y chromosomes and mtDNA show that pattern. This problem would be exacerbated if both uniparental systems happened to have experienced worldwide **selective sweeps**, which could spread recently derived favorable alleles even with very low migration rates.

Although the absence of old lineages at the Y chromosome and mtDNA does argue against many versions of the multiregional model, the presence of old lineages elsewhere in the genome would not necessarily argue in favor of multiregionalism. In a panmictic diploid population that is at equilibrium, the expected time to the most recent common ancestor for all of the alleles in the population is $4N_e$ (see Chapter 15). Most estimates place the human effective population size at 10,000, implying that the expected time to the most recent common ancestor for an autosomal locus is 40,000 generations, or about 800,000 years, assuming a generation time of 20 years. However, this is only the average value over many realizations of the evolutionary process (or, equivalently, the average over many unlinked loci in a single population). A given locus might have a much deeper, or much shallower, time depth (see p. 423). For this reason, an autosomal locus at which some pairs of genes diverged from one another hundreds of thousands of years ago (i.e., an "old" set of lineages) would not by itself be strong evidence in favor of the multiregional model. Moreover, **balancing selection** can maintain separate alleles for very long times, as shown by the polymorphic major histocompatibility complex alleles that we share with chimpanzees (see p. 541). A decisive test between the alternatives will require

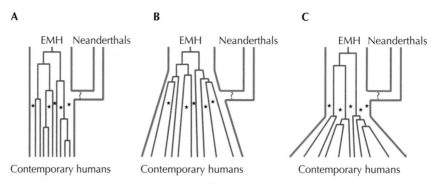

FIGURE 25.25. Failure to find Neanderthal sequences in a sample of ancient modern humans does not preclude Neanderthal–human interbreeding, but rather only places a boundary on the extent of the interbreeding, which is dependent on the assumed demography. Under the constant-size model (*left*), the data are consistent with a 25% Neanderthal input, whereas under population growth (*middle*), the data constrain the Neanderthal input to a lower value. If the population was constant until interbreeding with Neanderthal and expanded thereafter, the Neanderthal contribution could have been larger. EMH, early modern human.

good genealogies at multiple autosomal loci. If the majority of lineages sampled from different continents coalesce relatively recently (as is seen for the Y chromosome and mtDNA), then any multiregional model becomes implausible.

Analysis of Ancient DNA Refutes the Multiregional Model

Analyses of ancient mtDNA have also been used to argue against the multiregional model. Under that model we expect a closer genetic relationship between current inhabitants of Europe and the ancestral populations that were there before the arrival of anatomically modern humans than between current inhabitants of Europe and moderns from elsewhere. Therefore, a simple approach would be to ask whether any mtDNA from modern Europeans shows closer similarity to Neanderthal than to other non-European modern sequences. Mitochondrial DNA sequences from Neanderthal specimens not only show no affiliation with modern European samples (Fig. 25.25), but are also highly divergent from all modern sequences, with an estimated average divergence time from modern human sequences of about 600,000 years. Recent comparisons of Neanderthal nuclear sequences give a similar date.

It does not immediately follow from this result, however, that Europeans and Neanderthals did not interbreed at all when they overlapped in Europe 30,000–40,000 years ago. There could have been significant exchange between archaics and moderns, but the archaic lineages may simply have been lost by drift during the long period following the overlap between Neanderthals and moderns. The absence of archaic mtDNA lineages does not necessarily rule out the presence of archaic lineages elsewhere in the genome.

However, this is an inappropriate argument for assessing the multiregional model. In a single population it is quite possible that there were Neanderthal lineages in the modern human gene pool 30,000 years ago as a result of interbreeding that have since been lost by genetic drift. However, under the multiregional model, different geographic areas would have been at least partially isolated. Therefore, if we do not see archaic mtDNA lineages in any present-day population, we would have to argue that they have been lost independently in all those populations. Thus, ancient mtDNA evidence would weigh much more heavily against the multiregional model than when only a single panmictic population is considered. It becomes more difficult to argue that the archaic lineages were lost by chance if they must be lost repeatedly in semi-isolated populations; the more likely explanation is that the "old" lineages were not there in the first place. That is, humans are a demographically young species.

GENOMICS AND HUMANNESS

As was discussed at the beginning of this chapter, we remain largely ignorant about many of the big questions related to human evolution. We know that burial customs and art have been around for tens of thousands of years. But we do not know when language evolved, when our current cognitive abilities developed, or which other hominins may have been self-aware, although many and varied theories have been proposed on all of these issues. One view is that rapid biological changes occurred about 50,000 years ago and that these led to a sudden increase in both the cultural complexity and the range of modern humans as reflected in the fossil record.

Although we do know some details of the behavior of our ancestors and primate cousins, what can be seen or inferred in the fossil record is clearly only a glimpse of reality. In addition, we know next to nothing about what selection pressures were most important in the development of hominin characteristics. In the remainder of this chapter, we describe two promising directions for shedding light on the evolution of modern human characteristics. First, we discuss how modern genetic and genomic advances might help us address some of these questions. Then, we close by discussing how language may eventually be dissected into component parts whose evolution can be modeled and tested.

Gene Expression Patterns Appear to Have Evolved Faster in the Hominid Lineages Than in Other Primate Lineages

Despite obvious phenotypic differences, humans and chimps have fairly similar DNA sequences, differing only in 1 out of every 100 bases. As we saw in Chapter 22, such similarity is typical of sister species, which are generally separated by a few million years and show correspondingly low levels of divergence. This similarity led to the hypothesis that evolution is driven more by changes in gene expression than by structural differences in proteins. This is possible, as the relative importance of structural and regulatory changes within proteins and DNA is poorly understood (see pp. 542–547). Moreover, selection conserves the sequence of at least as much noncoding DNA as coding sequence, most likely because it functions to control gene expression (see p. 547). Nevertheless, although such arguments are plausible, we should bear in mind that any homologous pair of human and chimp proteins is distinguished by an average of two amino acid differences (see p. 536), giving ample scope for divergence in protein sequence to play a role.

The role of regulatory variation in hominid evolution can be quantified by using **microarrays** to compare patterns of gene expression in humans, chimpanzees, and other species. In one experiment, the levels of expression of transcripts for 12,000 different genes from liver, blood leukocytes, and brain samples from various primates were measured (Fig. 25.26). Faster rates of change in gene expression were found in the human lineage for all three tissue types. However, there is a methodological bias here, because the oligonucleotides used to construct the microarray were human, which would result in low hybridization rates and hence low expression levels in the nonhuman representatives. The hominid brain tissues, however, appeared to evolve fivefold faster than the chimpanzee lineage—a striking difference.

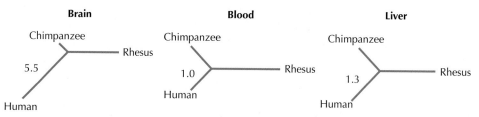

FIGURE 25.26. The relative extent of changes in gene expression among three primate species and three tissues. Numbers give the extent of change in the lineages leading to humans relative to those leading to chimpanzees.

Subsequent studies have shown that across primates, patterns of gene expression in the brain change more slowly than in most tissues; in contrast, gene expression in the testes diverges faster than in other tissues. Amino acid sequence divergence shows a similar pattern: Genes expressed in the brain diverge more slowly, whereas those in the testes diverge more rapidly. This is consistent with greater constraint on gene expression in the brain and with **positive selection** being prevalent on genes expressed in the testes. (Recall that rapid divergence of male sexual function is seen in other species [e.g., Fig. 19.13].) Crucially, however, the rate of change in gene expression and sequence divergence in the human lineage relative to that along the lineage leading to chimpanzees has been found to be faster for genes expressed in the brain than for other tissues, as in the example shown in Figure 25.26. These results are a long way from identifying the changes that are responsible for human cognitive abilities, but do show that the hominid line stands out from other primates in terms of the rate at which transcript levels have changed in the brain.

It Is Difficult to Identify Specific Genes Responsible for Language

Language is one of the most impressive and distinctive of all human traits. Humans have an innate capacity to develop a sophisticated, symbolic system of communication, as evidenced by the example of isolated groups of deaf people, who spontaneously develop sign languages that compare to spoken ones in complexity and flexibility. Unfortunately, language is also one of those human traits that leaves behind almost no trace in the fossil or archaeological record. We have virtually no evidence, for example, about when humans first developed a complex language. Some argue that it developed early in our history—millions of years ago—whereas others believe it is more recent, reaching its full development only about 100,000 years ago with the appearance of modern humans. Currently, we cannot even answer the more limited question of whether all extant human languages derive from a single ancestral one.

Another perspective on the evolution of human language has emerged from studies of a rare genetic disease, called speech-language disorder 1, that involves severe impairments of articulation and grammar. In studying a family with multiple affected individuals, scientists used linkage analyses (see Chapters 14 and 26) to track down the responsible mutation to the *FOXP2* gene. This encodes a **transcription factor** that represses gene expression, although the set of genes it influences is poorly defined. This gene is important in the development of lung epithelium and also may be involved in neuronal development. Patterns of *FOXP2* gene diversity have been studied in humans, several primates, and mice. These studies indicate that there is an excess of replacement substitutions in the lineage leading to humans, suggesting that this gene has been under selection at some point after the human and chimpanzee lineages split (Fig. 25.27). The pattern of variation among humans also suggests a history of positive selection. When selected mutations sweep to fixation, linked neutral alleles are either fixed or are dragged up to high frequency (see pp. 427 and 537). As a result, reduced

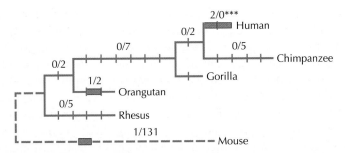

FIGURE 25.27. The *FOXP2* gene is the first gene identified that carries a mutation that causes a specific language deficit in humans. The silent and replacement nucleotide substitutions in this gene as mapped on a primate phylogeny are shown. (*Red bars*) Amino acid changes; (*blue tick marks*) nucleotide changes. Data suggest that the *FOXP2* gene has been the target of selection during recent human evolution after the separation of the human lineage from the common ancestor with the chimpanzee. Numbers show how many nonsynonymous/synonymous changes have occurred along each branch.

variation and an excess of high frequency–derived mutations flanking a fixed difference provide evidence for positive selection. Such methods led to estimates that a selective sweep occurred at *FOXP2* some time in approximately the past 120,000 years. Because anatomically modern humans appeared at about the same time as the upper limit on the date of the sweep, it has been suggested that amino acid changes in *FOXP2* may have played a role in the development of language.

Given that mutations in *FOXP2* result in speech impairment, there is a possibility that any such selection on *FOXP2* may have been associated with the emergence of language ability. However, the case is at best circumstantial. Although patterns of sequence variation within and between species do indicate selection, bear in mind that large numbers of genes show such patterns (see p. 536). Just because a mutation has effects on speech, it does not follow that this is the primary role of the gene itself, or that its evolution has been driven by selection for linguistic ability.

So how do we move beyond conjecture? Probably the most important thing is to assess the biological function of FOXP2. For example, which genes have their expression patterns modified by FOXP2, and in what way? Learning more about the biology of FOXP2 might enable us to assess the idea that this selection was associated with language in particular. Functional analysis of FOXP2 could ultimately strengthen or weaken the case, although it is hard to devise a method for finding the effect of subtle changes in the *FOXP2* gene on language ability; the usual approach of studying transgenic mice would hardly work.

Cis-regulatory Variation in the Prodynorphin Gene Suggests Another Approach to Identifying "Humanness"

Recent studies of the gene *PDYN*, which encodes the neurotransmitter prodynorphin, provide a useful model for studies of brain-related phenotypes in humans. Prodynorphin is an endorphin, which binds to opiate receptors. As such, it mediates our experience of pain, social attachments, learning, memory, and addiction. A 68-base-pair (bp) **tandem repeat** polymorphism in the human *PDYN* promoter, located 1250 bp upstream from the start of transcription, has been shown to influence the inducibility of the gene. Association studies (Chapters 14 and 26) have implicated this tandem repeat in a range of disorders, including schizophrenia, cocaine addiction, and epilepsy.

Given these attributes, *PDYN* is an excellent candidate for understanding several components of human nature. It harbors variation in its *cis*-regulatory region capable, in principle, of exerting subtle quantitative changes in gene expression. Thus, we can ask whether patterns of sequence variation suggest that selection has acted, as we did for the example of *FOXP2*.

Three different kinds of evidence were used to test for selection on *PDYN*: population genetic, phylogenetic, and functional. First, the **frequency spectrum** of linked neutral mutations was assessed. In a sample of 74 haplotypes from an Austrian population, an excess of high frequency variants was seen; in contrast, a chimpanzee sample exhibited no such departure. In addition, excess geographic variation was seen in variants upstream of the gene, as measured by an elevated F_{ST} statistic across six old-world populations (see p. 446). This is a signature of geographically heterogeneous positive selection, which causes allele frequencies to differ among populations more rapidly than if genetic drift and migration were acting alone. Linked microsatellite loci also showed reduced variation in three of the six populations, as would be expected following a selective sweep.

Comparisons between species also give evidence that selection has acted at the *PDYN* locus, although over a longer period of time. There is a substantial excess of five substitutions in just one part of the phylogeny—specifically, along the lineage leading to humans. Remarkably, positive selection appears to be confined to the regulatory region. These patterns all suggest recent selection, but do not implicate any particular sites or function.

The third piece of evidence is functional. Changes in the upstream variant are already known to modify inducibility of the gene, but we can ask specifically about the effects of the putatively selected nucleotides that were fixed in the human lineage. When a human neural cell line was transfected with constructs bearing 3 kb of human or chimpanzee *PDYN* *cis*-regulatory DNA, the human 68-bp element was found to drive significantly higher *PDYN* expression than the chimpanzee element, regardless of whether the flanking sequence came from chimp or human. In contrast, no differences between chimp and human were observed when the same experiment was performed in a nonneural cell line, suggesting that the functional effects of the selected sequence are brain specific.

Cases for selection will always be circumstantial. But here, the marshaling of three lines of evidence and the fact that the gene appears to be strongly associated with a set of traits that are regarded as particularly "human" make *PDYN* worthy of further investigation.

The Evolution of Language

As we argued in Chapter 24, complex adaptations typically evolve through a number of distinct steps, each relatively simple. This appears to also be the case for human language. Our linguistic abilities can be parsed into more or less discrete abilities, each of which may have been selected in its own right. This subtle view of language evolution enables us to imagine how natural selection might have led to increased language abilities, by selecting for specific improvements in communication that are each well short of modern language abilities. This viewpoint parallels ideas about the evolution of other complex structures, such as the origin of eyes, and associated perceptive abilities, such as vision (see Chapter 24).

Some of these distinct aspects in the evolution of language ability can still be seen today in specific features of modern languages. Let us consider several possible steps in the evolution of language (Fig. 25.28). The first is the establishment of a limited set of symbols that can be used in a nonspecific way, which distinguishes them from the largely situation-specific communication of other animals—for example, distress signals. Getting from this point to modern language requires two major innovations, which appear to be largely independent. One is the capacity to combine sounds to produce an essentially unlimited number of different symbols or words. Some researchers see, in modern one-word expressions, fossils of an earlier one-word stage in human language. Thus, words such as "wow" or "ouch" are examples of words that can stand entirely alone to express something largely independent of context.

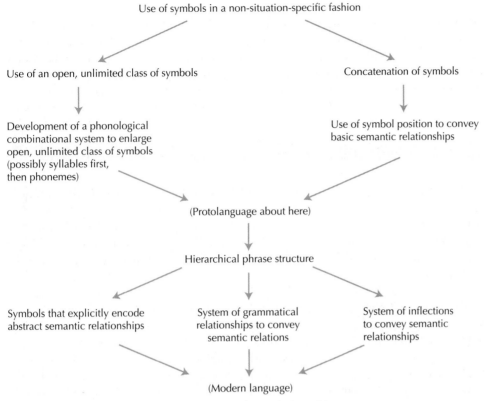

Use of symbols in a non-situation-specific fashion

Use of an open, unlimited class of symbols

Concatenation of symbols

Development of a phonological combinational system to enlarge open, unlimited class of symbols (possibly syllables first, then phonemes)

Use of symbol position to convey basic semantic relationships

(Protolanguage about here)

Hierarchical phrase structure

Symbols that explicitly encode abstract semantic relationships

System of grammatical relationships to convey semantic relations

System of inflections to convey semantic relationships

(Modern language)

FIGURE 25.28. Steps in the evolution of language.

The second step is putting these symbols together, following some set of rules or **syntax**, to express more complicated ideas. Even before syntax, the symbol stage of human language would seem to have been useful and more flexible than the communications of untrained primates. With training, other apes manage no more than several hundred different symbols. In contrast, without syntax humans have language abilities that far outstrip those of other apes, and it is not hard to imagine that this expanded vocabulary would itself be advantageous and subject to elaboration by natural selection. In support of the idea that this ability may have evolved at least partially independently of syntax, speakers of second languages late in life can easily be counted on to master the vocabulary, whereas the grammar frequently remains beyond their capabilities.

Finally comes the evolution of syntax itself. Here progress in identifying clearly distinguishable elements is less clear, but some steps seem obvious. For example, an early stage could consist of just combining words, with context controlling the interpretation. One simple rule for clarifying ambiguities would be word order, which is, of course, still used in various ways throughout modern languages. An innovation on word order would be the structural replacement of a word with a phrase, as is easily accommodated by all modern and historically known languages. Final aspects of syntax that make up modern language include syntax rules governing words and phrases that are separated in sentences, and the inflection of words.

From the perspective of evolutionary biology, efforts to dissect language are both important and useful. Their importance lies in clarifying how something as complex as human language could evolve from much simpler ingredients. These efforts are useful because they suggest ways of understanding the evolution of language ability by analyzing the structure of modern language.

Predicting the future is always hazardous, but it seems likely that the next generation of human evolutionists will turn their attention to understanding phenotypic evolution (as, for example, is being done with the role of prodynorphin in human brains). What genetic changes make us different from other primates and what genetic differences make us different from one another? These are questions that we have barely begun to answer, but the tools needed to address these questions are slowly becoming available. In the next chapter, we turn from the history of our species to focus on variation within it, and the implications for medicine and for wider social issues.

SUMMARY

Together, evolutionary genetics and paleoanthropology have outlined the broad patterns of human evolution. Humans separated from a branch leading to the chimpanzee about 5 Mya. The subsequent bulk of the evolutionary changes that created modern humans occurred in Africa. Modern humans derive the vast majority, if not all, of their genetic heritage from a relatively small population that lived in Africa between 50,000 and 100,000 years ago. The broad outlines of our demographic history are well established. So what is left for the field of human evolution?

The traits now thought of as human first appeared at different times, and many also appeared in species that are not in our direct line of ancestry. Although we think of human traits as an integrated set, because no other hominin species survived to the present, depending on which trait is considered (e.g., toolmaking, brain size, or bipedalism), different extinct species would be afforded the human moniker. Even among linguists, there is a tendency to dissect the human capacity for language into separable steps, the existence of which is hinted at in various features of modern languages, in the nature of communication in other animals, and by the characteristics of language-impaired humans. Such mosaic patterns are exactly what we would expect of the evolutionary process. Different traits appear at different times in response to different environments for different members of a phylogeny. Some of these traits will be specific to single groups in the phylogeny, and some, which occur at early branching points, will appear in multiple members. Today, *Homo sapiens* is the only surviving hominin, and so it is our features that we see as a distinctive package of traits that separates us from other primates.

FURTHER READING

The Fossil Record

Lewin R. and Foley R. 1995. *Principles of human evolution.* Blackwell Science, London

Tattersall I. 1995. *The fossil trail: How we know what we think we know about human evolution.* Oxford University Press, Oxford.

Genetics of Humanness

Cavalli-Sforza L.L., Menozzi P., and Piazza A. 1994. *The history and geography of human genes.* Princeton University Press, Princeton, New Jersey.

A comprehensive review of how variation in allele frequencies among human populations has been used to infer evolutionary relationship. It also provides succinct and accessible overviews of the histories of human populations on each of the continents.

Enard W. and Pääbo S. 2004. Comparative primate genomics. *Annu. Rev. Genomics Hum. Genet.* **5:** 351–378; Khaitovich P., Enard W., Lachmann M., and Pääbo S. 2006. Evolution of primate gene expression. *Nat. Rev. Genet.* **7:** 693–702.

These papers review our increasing knowledge of primate genomes and genomics, emphasizing approaches that may help to identify the changes responsible for unique human attributes.

Tishkoff S.A. and Verrelli B.C. 2003. Patterns of human genetic diversity: Implications for human evolutionary history and disease. *Annu. Rev. Genomics Hum. Genet.* **4:** 293–340.

An up-to-date review of how genetic analyses have informed our understanding of human demographic history. It describes the implications of patterns of genetic variation for medical genetics, the subject of Chapter 26.

Evolution of Language

Pinker S. 1995. *The language instinct: The new science of language and mind.* Penguin, New York; Jackendoff R. 2003. *Foundations of language: Brain, meaning, grammar, evolution.* Oxford University Press, Oxford.

Pinker gives a highly readable accout of the argument that humans have an innate "language instinct." Jackendoff makes the case in greater depth; both discuss how our ability to use language might have evolved.

CHAPTER

26

Current Issues in Human Evolution

EVOLUTIONARY BIOLOGY HAS MANY AND DIVERSE IMPLICATIONS for humanity—both in the way that we deal with practical problems and for our understanding of ourselves. In this chapter, we focus first on the genetic variation that influences people's susceptibility to disease and the way that they respond to treatment. Here, an evolutionary perspective is essential to make sense of the mass of genetic data available from human populations and to find how best to use it. We then move on to wider questions. We ask about the nature of human variation and the extent to which this has been, and is being, shaped by selection. We then discuss the extent to which human disease and human psychology can be understood from an evolutionary point of view in the same way as for other biological phenomena.

THE GENETIC BASIS OF DISEASE

Why Study Disease Genetics?

The search for the genetic causes of disease has been central to human genetics for decades. It concentrated first on Mendelian diseases, which are those that depend largely or entirely on mutation at a single gene. Cystic fibrosis is one well-known example in which two mutant copies of the *CFTR* gene will cause disease. This is the most common autosomal recessive disease in northwest European populations, but is nevertheless quite rare, with an incidence of about 1 in 2000. More recently, attention has begun to shift toward more common diseases that have more complicated modes of inheritance. The motivations for uncovering the genetic causes of these two types of diseases are somewhat different. In the case of Mendelian diseases it is often appropriate to think of the mutation as directly causing the disease. Therefore, knowledge of the mutation may directly aid treatment or may even open the door to **gene therapy**, in which functional copies of defective genes are provided.

The genetic basis of common diseases is less clear-cut. Most such diseases are strongly inherited; that is, relatives of individuals with the disease are more likely to get the disease themselves. However, this genetic variation in susceptibility is very rarely due to a single gene (Chapter 14); instead, multiple alleles are involved, each somewhat increasing risk. The two primary reasons for identifying risk-conferring alleles are that they may point the way toward therapies and that they may allow estimation of individual risk of specific common diseases. A good example illustrating the first motivation is PPARg, a transcription factor that has a coding polymorphism strongly associated with the risk of developing type 2 diabetes. In fact, it has been calculated that if the high-risk allele were not present, there would be 20% fewer cases of type 2 diabetes in the United States. The protein encoded by PPARg is, in fact, the target of an important class of medicines used in the treatment of diabetes. Thus, if these medicines did not already exist, then the discovery of the PPARg polymorphism would have suggested this gene product as a good target for new drugs.

The case for the clinical relevance of risk prediction is much harder to make, both because risks cannot be predicted accurately and because it is unclear how knowledge of risk might be used. We discuss these issues in more detail after first explaining how genes for Mendelian disease and for complex disease can be identified.

The Genes Responsible for Mendelian Disease Can Be Mapped

Soon after the rediscovery of Mendel's work at the start of the 20th century, it became clear that several inherited diseases are caused by defects in single genes. These produce characteristic patterns of segregation, depending on whether the disease allele is dominant or recessive, autosomal or X-linked (Fig. 26.1). It was appreciated early on that human disease could be caused by "inborn errors of metabolism" (see p. 42), that is, by failure of crucial biochemical steps, as a result of defects in the genes that code for the necessary enzyme. In some cases, the biochemical phenotype allows the defective gene to be identified. For example, in 1934 Asbjørn Følling, a Norwegian doctor, noticed that the urine of some mentally retarded patients had a musty or "mousey" odor. This was shown to be due to a buildup of phenylpyruvic acid, which led to the discovery of the biochemical anomaly now known as phenylketonuria (PKU).

Classical genetics provides another route to finding the genes responsible for disease by establishing linkage between the disease gene and a polymorphic molecular marker. With multiple markers, a linear genetic map can be constructed that corresponds to the physical arrangement of genes along the linear chromosome. The first evidence for linkage in humans was found in the 1930s, through associations between inherited disease and markers such as blood groups and color blindness; statistical methods for estimating the rate of recombination between markers and the disease-causing allele were developed during the following years (Fig. 26.2). We would expect that affected members of the pedigree share the same allele at markers that are linked to the disease gene, but there would be no increased sharing of alleles among affected individuals at unlinked loci. Such associations allow the disease-causing gene to be localized to a specific chromosomal region. However, until recently such methods were of little practical use, because it was unlikely that a disease allele would be tightly linked to one of the very few genetic markers that were available. Moreover, even if the disease allele could be accurately located, there would have been no way to identify and isolate the actual gene responsible.

By the 1980s, the development of techniques for genetic manipulation (see p. 56) made it possible to apply classical genetic methods to locate and identify disease genes. First, an essentially unlimited number of polymorphic genetic markers became available, which could be placed on a dense genetic map (Box 13.3). Second, it became possible to directly manipulate a gene once its position was identified; the relevant region of genome could be put into microbial clones and sequenced (see p. 56). The new method became known for a time as **reverse genetics**, and then more commonly as **positional cloning**.

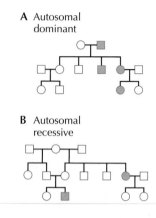

A Autosomal dominant

B Autosomal recessive

C X–linked recessive

FIGURE 26.1. Typical patterns of inheritance for rare diseases caused by single alleles. *Circles*: females; *squares*: males; *red symbols*: affected. (*A*) Autosomal dominant: Affected individuals must have an affected parent; when one parent is affected, half of its offspring will be affected. (*B*) Autosomal recessive: Affected individuals typically have two heterozygous parents, each with normal phenotype; one-quarter of offspring from such matings are affected. Offspring from matings between relatives (e.g., at *left*) are much more likely to be affected, because they may inherit the disease allele from both their mother and their father (see Box 15.3). (*C*) X-linked recessive: Heterozygous mothers (e.g., *top left*) are normal, but one-half of their sons inherit the disease allele on their single X chromosome and are affected. Sons of affected males are not affected, but their daughters are all heterozygous, and so one-half of their grandsons are affected (e.g., *bottom row*).

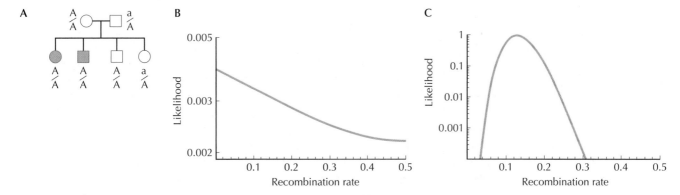

FIGURE 26.2. The rate of recombination between a marker and a disease locus can be estimated from pedigrees. (A) Two siblings are affected by a disease caused by a rare autosomal recessive allele (*shaded symbols*). The parents must both be heterozygous for the disease allele. The family is genotyped for a marker (A, a); the father and one daughter are heterozygous (*circles*: females; *squares*: males). (B) The probability that this pattern would be observed as a function of the recombination rate between the marker and the disease locus. This is known as the **likelihood** of the recombination rate. For this very small set of data, most likely there is complete linkage ($r = 0$, at *left*), with the marker A linked to the disease allele; but it is not much less likely that there is no linkage ($r = 0.5$, *right*). (C) With more data (here 200 families of the same size as in *A*), a better estimate can be made: The most likely estimate is that $r = 0.13$ (*peak of graph*). The data were simulated with $r = 0.1$, so this is reasonably accurate. Likelihood is scaled relative to the maximum, at 1. See Chapter 14 (pp. 399–402) for further details of this method in the context of QTL mapping.

The key resource required for positional cloning is a map of genetic markers that covers all of the human chromosomes; initially, this was built using **restriction fragment length polymorphisms** (Box 13.3). This idea was to become a major driving force behind the human genome project, which had as a central aim the generation of both **physical** and **genetic maps**. Construction of the physical map required knowledge of unique sequence stretches throughout the genome so that researchers could use sequence to determine unambiguously what part of the genome they were examining. Genetic maps required the development of a dense set of genetic markers and estimation of the recombination rates among them.

Before a dense set of genetic markers was available, finding genes for Mendelian disease was exceptionally hard work, just as was the case for any **quantitative trait locus (QTL)** (Chapter 14). Linkage analyses of data from pedigrees allow localization of disease genes to within, at best, 1 centiMorgan (cM). (This is because it is unlikely that 1-cM regions of genome will be broken up by meiosis within a pedigree that spans a few generations. Therefore, a higher density of markers would give little extra information.) In the human genome, 1 cM translates to approximately 1 megabase of sequence. Even by today's standards, this is still a huge amount of DNA to sequence exhaustively in the search for mutations that cause disease. Typically, there would be roughly ten genes in a region of this size. In the early days, researchers would work laboriously along the implicated region, looking for some obvious cause of disease—for example, a mutation that led to the presence of a stop codon in the middle of a region coding for an amino acid sequence. However, if nothing obvious turned up, then it was necessary to continue trying to localize the gene, either by using more families or through association methods, which are discussed below.

One of the primary motivations for the complete sequencing carried out by the Human Genome Project (HGP) was to facilitate the final identification of disease genes once a gene had been localized to a genomic region. The HGP has identified and assigned some known or predicted functions to most human genes. Therefore,

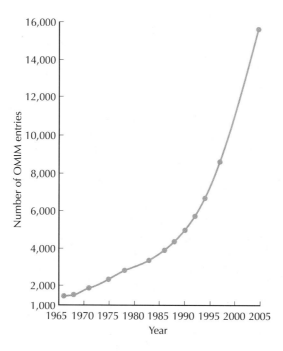

FIGURE 26.3. Total number of Mendelian genes listed in the *Online Mendelian Inheritance in Man* (OMIM) database, from 1965 onward. By January 2007, 17,238 genes had been identified. Of these, approximately 1200 genes for Mendelian disease have been identified.

once a disease gene has been localized to a megabase of DNA, it is possible to look up what genes are there, sequence just the exons of those genes in affected persons, and, often, find the causal mutation. Whereas graduate students once spent years trying to find their gene, and careers would be made or lost on whether they happened to find it, it is now common for laboratories to acquire a new family, show that there must be a single gene causing the disease, and then find the responsible gene within a month. Thanks to the tools provided by the HGP, what was once cutting-edge science is now, in most cases, routine technical work. In consequence, the genes for more than 1200 Mendelian diseases have now been identified (Fig. 26.3).

Associations between Genetic Markers and Phenotype Can Tell Us the Genetic Basis of Complex Traits

The success of human geneticists in identifying disease-causing genes has been remarkable and is the clearest immediate benefit from the HGP. Unfortunately, these advances have had little impact on the overall health of the human population, simply because diseases with clear-cut Mendelian inheritance are rare, affecting only a minority of people. In contrast, most people are affected by common, but complex, diseases such as asthma, diabetes, obesity, and cardiovascular and psychiatric disorders. Susceptibility to these diseases is strongly heritable, but they show no simple Mendelian pattern. Instead, they are influenced by multiple genes in a complex and poorly understood interaction with the environment. The challenge for human geneticists is to find ways to determine the genetic basis of such complex traits.

We saw in the last section that it is now straightforward to map the genes responsible for Mendelian diseases essentially through associations between genetic markers and the disease within pedigrees. In Chapter 14, we explained how this approach can be used to map the QTLs that are responsible for variation in complex traits: by looking for associations between genetic markers and phenotype within experimental crosses or within families. However, several factors make this approach much harder in humans or, indeed, in any natural population: Observations are on small families, rather than on controlled crosses with large numbers of offspring; different alleles segregate in different families; and environments vary uncontrollably. This makes it practically impossible to map the QTLs responsible for diseases with a complex genetic basis solely from observations on pedigrees in the manner described above. In addition, there are only a limited number of recombination events within families, and so it is impossible in principle to use such studies to map genes to closer than about 1 cM.

Association studies are now widely viewed as the most promising alternative for studying the genetic basis of common disorders, because, in principle, they have much more statistical power to detect and localize alleles of modest effect (Fig. 26.4). In this approach, one looks for associations within an entire population rather than within families of close relatives. Typically, a **case–control** design is used, in which the genotypes of a sample of cases are compared with the genotypes of a control sample that is chosen as far as possible from the same population; that is, the control sample matches by social class, ethnicity, and so on.

There are two alternative methods—direct and indirect. The direct approach assumes that the genetic variation we observe actually causes the disease. Thus, we might focus attention on variants in exons that cause amino acid substitutions, on changes at intron–exon boundaries that might alter splicing of mRNA, or at regulatory sequences known to influence gene expression. However, there are serious problems with this approach. First, there are an enormous number of polymorphisms of these kinds, and studies of QTLs have shown that it is very hard to determine which of several variants in a gene is actually responsible (see pp. 402–403). Indeed, as we saw in the example of *Adh* polymorphism in *Drosophila*, several closely linked variants may interact to influence the trait (Fig. 14.27). Second, a substantial fraction of the noncoding genome—at least comparable in quantity to the coding sequence—is evolutionarily constrained (see p. 547). This implies that many of the variants that influence disease susceptibility may be in noncoding regions and cannot be identified given our limited understanding of gene regulation. Consistent with this, in the few cases where the actual sequence variants that affect quantitative traits have been identified, they frequently do not alter protein sequence and would not be identified as likely candidates in advance (e.g., see pp. 312, 317, and 403–406).

The direct approach is motivated by experience of Mendelian disease, in which the vast majority of causal mutations have been found to alter protein sequence (Table 26.1). However, the Mendelian experience may not apply to the common, complex diseases. The phenotypic effects of Mendelian mutations can be severe and extremely debilitating, often resulting in early death. On the other hand, alleles that influence common diseases may only slightly alter the chances of disease, perhaps only in specific

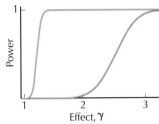

FIGURE 26.4. Studies based on associations between disease and markers in the population as a whole have more power to detect linkage than studies that use only associations within families. The graph shows the chance of detecting an allele that increases disease risk by a factor γ in a survey of 1000 families of affected sibling pairs (this chance is termed the statistical power). The *red curve* shows results based on whether the affected siblings share an allele. This test only involves associations within families, and so is relatively weak: The allele is likely to be detected only if it increases disease risk by ~2.5 times. The *blue curve* shows results from the **transmission disequilibrium test** (**TDT**), which looks for alleles that are transmitted in excess to affected offspring. This requires a consistent association (i.e., linkage disequilibrium) in the population as a whole and can in principle detect alleles that increase disease risk by ~25%.

TABLE 26.1. Relative frequency of types of mutations underlying disease phenotypes

Change	Number	Percent of Total
Deletion	6,085	21.8
Insertion/duplication	1,911	6.8
Complex rearrangement	512	1.8
Repeat variations	38	0.1
Missense/nonsense	16,441	58.9
Splicing	2,727	9.8
Regulatory	213	0.8
Total	27,027	100.0

*Data are from the Human Gene Mutation Database (June 2002); http://www.hgmd.cf.ac.uk.

environments. Variants acting in this way may be systematically different from those responsible for Mendelian disease. We discuss this issue in more detail below (pp. 763–765).

Association Studies Are Based on Linkage Disequilibrium

The indirect approach to association studies does not depend on assumptions about which polymorphisms are most likely to cause disease. Rather, it relies on the fact that closely linked polymorphisms are often statistically associated with one another; in other words, they are in **linkage disequilibrium**. We saw in Chapter 15 that because closely linked sites tend to share the same ancestry, they also share the same pattern of polymorphism (Fig. 26.5). Typically, just a few **haplotypes** will segregate, rather than the very large number of possible combinations of markers that would be seen in the absence of linkage disequilibrium.

The presence of strong linkage disequilibria between neighboring polymorphisms has two consequences for association studies. First, it makes the direct approach discussed above impracticable. Typically, we will see several markers that are all strongly associated with disease (Fig. 26.5). As we saw in Chapter 14 (pp. 402–405), it is impossible in principle to tell which of these actually *causes* disease from purely statistical arguments—some kind of experimental evidence is needed. Second, because just a few haplotypes segregate in each region of a genome, we do not need to type all of the markers contained within them. By choosing an appropriate subset of markers, known as **tagging single-nucleotide polymorphisms (tSNPs)**, we can distinguish the haplotypes much more efficiently (Fig. 26.5; recall Fig. 15.15). This idea was the main motivation for the HapMap Project, which seeks to describe patterns of linkage disequilibrium throughout the human genome so as to facilitate the selection of tSNPs.

This approach has recently been assessed by using the HapMap data to select a set of tagging polymorphisms that represent genetic variation throughout the entire human genome. Samples drawn from northern Europe show that just 300,000 SNPs represent the vast majority of polymorphisms very well. For example, more than 80% of polymorphisms are represented by at least one of the tagging SNPs, with a correlation coefficient of at least 0.8. This means that the tagging SNPs represent 8 million of the total number of 10 million SNPs in the human genome, with only modest loss of statistical power. (Here, SNPs are defined as having minor allele frequency >1%.) However, results for the sample from West Africa are much worse. This is because African populations are known to have less linkage disequilibrium than any non-African population. Thus, many more polymorphisms would be required to carry out mapping studies in these African populations with a similar statistical power.

The efficiency of association studies that are based on the indirect approach depends on how far linkage disequilibrium extends across the genetic map. This question has received intense study over recent years. As we saw in Chapter 15, it has become clear that linkage disequilibrium extends much further than expected from the **standard neutral model** of a single well-mixed population with constant population size and with constant recombination rates (Fig. 15.20). This extensive linkage disequilibrium reflects shared ancestry across long stretches of genome such that blocks of several kilobases share the same genealogy. This discrepancy may be partly due to population subdivision, to past population bottlenecks, and to recurrent selective sweeps. However, it is clear that a key factor in shaping the pattern of linkage disequilibrium is the presence of **recombination hot spots**: Cross-overs are concentrated at particular points and so do not break up linkage disequilibrium within the long regions between these hot spots (p. 436).

The presence of extensive linkage disequilibrium both helps and hinders association studies (see p. 436). On the one hand, it makes it possible to describe variation between individuals using far fewer markers than otherwise, as explained above. On the

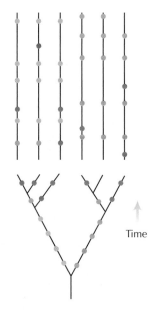

FIGURE 26.5. Any short region of genome will share the same ancestral genealogy (*bottom*). Six sampled sequences are shown on the *top*. These carry any mutations that occurred on the lineages leading down to them (*colored dots*). Every mutation is assumed to occur at a unique place in the genome (i.e., the **infinite-sites model**). The sequences fall into two distinct haplotypes corresponding to the two main branches of the genealogy. The haplotype could be determined by scoring any one of the *red* or *light blue* mutations, which could therefore act as tagging single-nucleotide polymorphisms (tSNPs).

Time

other hand, it makes it impossible to locate the responsible alleles any closer than the average size of a nonrecombined haplotype block. Within such blocks, there typically will be several candidate sites that cannot be distinguished. Later in this chapter, we will see how such limitations influence the potential usefulness of association studies.

Searching the Genome for Associations with Quantitative Traits Is Difficult

Association studies can easily test the hypothesis that a given polymorphism influences a given disease. This approach would be straightforward if there were only a few such hypotheses to test. But there are estimated to be 10 million polymorphisms in the human genome. Indeed, given that there are 6 billion people on the planet, and the rate of point mutations is approximately 5×10^{-9} per site per generation, most sites of the genome will differ among at least some individuals unless such differences are lethal. How do we find which of these many variable sites actually influence disease?

In principle, the loci that influence a quantitative trait can be found simply by searching for associations with markers scattered randomly across the genome. The idea of using associations between markers and disease in the whole population is attractive, because family studies of the kind we have just discussed cannot accurately locate the genes responsible for complex traits. However, there are formidable, and perhaps insuperable, obstacles.

The first difficulty is that when statistical tests are made for associations with very large numbers of markers, many apparently "significant" results will arise by chance. By definition, we expect that 1 in every 20 tests will be significant at a level of $p < 5\%$, even under the **null hypothesis** of no association. This means that when very many tests are performed, the threshold for significance of any one of them must be much more stringent; thus, we are less likely to detect a real association amid the statistical "noise." This point is nicely illustrated by a study that attempted to replicate associations between **candidate genes** and Alzheimer's disease. A sample of 121 cases of Alzheimer's disease was matched with 152 control individuals from the Scottish population and was scored for markers at 54 genes that were thought likely to influence the disease. Thirteen of these genes had previously been reported as showing significant associations with Alzheimer's disease in separate studies. Overall, 2.8% of tests were significant at $p < 5\%$ (slightly fewer than one would expect purely by chance) and, in a follow-up study based on a new sample from the same population, all but one of these initially significant tests became insignificant (Fig. 26.6). What is most disturbing about this study is that it included the *ApoE*4* allele, which has been shown to be strongly associated with an increased risk of Alzheimer's disease in many studies of both families and whole populations. This allele did indeed show a strong association with the disease in this study, but when allowance was made for multiple testing, this association became barely significant. The key point is that this statistical problem will become still more severe if studies are extended from 54 candidate genes to many thousands of markers scattered across the genome.

The limiting factor in association studies is not the number of genetic markers but rather the number of individuals in the sample. Thus, knowing the complete DNA sequence of each individual would hardly solve our difficulties. In a homogeneous population, weak correlations could be detected by taking a sufficiently large sample. However, the heterogeneity of real populations makes this impractical. On the one hand, the actual genetic variants responsible for complex diseases vary from population to population and may also depend on the environment. This is exploited in family studies, where families that show clear inheritance of some trait can be used to identify the gene responsible. However, the gene and allele involved will vary from family to

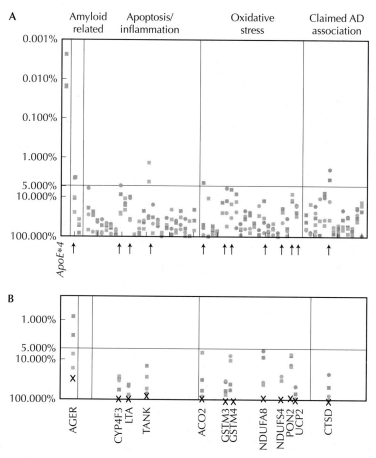

FIGURE 26.6. A search for associations between genetic markers and Alzheimer's disease (AD) illustrates the difficulties of combining multiple statistical tests. (*A*) The statistical significance of associations considered one by one. The 54 genes studied are divided into different functional classes; the *right-hand* section shows 13 genes that showed significant associations in previous studies. *A* shows the initial survey and *B* shows a follow-up using a separate sample and a subset of markers (indicated by *arrows above*). Of the genes that initially showed significance, only one (*AGER*) remained so in the follow-up. The crosses in *B* show the *p* values after proper correction for multiple testing: None are significant.

family, and each particular allele may make a negligible contribution to variation in the population as a whole. On the other hand, spurious associations can be generated if the population is actually a mixture from genetically different sources (Chapter 16 [pp. 452–455]). For example, in the United States, the Hb^S sickle-cell allele is associated with any marker that is common in the African–American population.

The ability to hunt systematically for gene variants is relatively recent, and most people consider genetic studies of common diseases to be in the very early stages. The field also suffers considerably from methodological weaknesses and from severe publication biases that favor positive over negative results. Promising associations usually cannot be replicated, as we saw in the Alzheimer's study just discussed. This is to be expected, both because of the statistical errors that are inevitable in even large studies and because there may be genuine heterogeneity between study populations. Nevertheless, an analysis of multiple studies shows that associations are replicated much more often than expected by chance alone and suggests that about half of these well-studied associations do genuinely reveal markers that are correlated with elevated risks of specific common diseases (Table 26.2). Similarly, there are about 70 variants for which a strong case can be made that they influence response to one or more pre-

TABLE 26.2. How replicable are association studies?

Associated Gene, Phenotype	Number of Studies (Total)	Number of Studies ($p < 0.05$) Same Direction as Original Report	Number of Studies ($p < 0.05$) Opposite Direction as Original Report
ABCC8, type 2 diabetes[a]	9	1	1
ABCC8, type 2 diabetes[a]	4	2	0
ADD1, hypertension	18	5	2
APOE, schizophrenia	12	0	1
BLMH, Alzheimer's disease	5	0	0
COL1A1, osteoporotic fracture	12	5	0
COMT, bipolar disorder	12	0	0
COMT, schizophrenia	9	0	1
CTLA4, type 1 diabetes	20	8	0
DRD2, schizophrenia	8	2	1
DRD3, schizophrenia	48	5	2
GSTM1, breast cancer	15	0	0
GSTM1, head/neck cancer	25	3	1
GYS1, type 2 diabetes	3	0	1
HTR2A, schizophrenia	28	3	1
INSR, type 2 diabetes	4	1	0
INSR, type 2 diabetes	4	0	0
KCNJ11, type 2 diabetes	6	0	0
NTF3, schizophrenia	7	0	0
PON1, coronary artery disease	14	5	0
PPARG, type 2 diabetes	14	4	1
SERPINE1, myocardial infarction	13	1	0
SLCA1, type 2 diabetes	3	2	0
SLCA2, type 2 diabetes	3	0	0
TPH, bipolar disorder	5	0	0
Total	301	47	12

From Table 1 of Lohmueller K.E. et al. 2003. *Nat. Genet.* **33:** 177–182 (© Macmillan Publishers).

[a]The two entries for *ABCC8* in the first and second rows are for variants in the intron and exon, respectively.

A survey of 25 associations between markers and complex diseases found a total of 301 follow-up studies. Of these, 59 were significant at the 5% level, far higher than the number expected if the initial reports had been "false positives" (301 x 0.05 ~ 15), but still a minority. Moreover, many of the subsequent significant findings were in the opposite direction. The significant replications were concentrated among 11 of the associations (in bold type), which suggests that about one-half of these associations actually exist.

scription medicines. Given how little work has been done relating genome-wide variation to disease risk, it seems fair to conclude that as large-scale studies are implemented, many more variants will soon be identified. Later in this chapter, we discuss how useful such variants might actually be.

Are There Common Variants for Common Diseases?

It is hard to devise optimal strategies for finding disease-causing variants without knowing about the nature of those variants. One of the key questions is whether such variants tend to be rare or common in the human population. We can imagine two alternative scenarios. At one extreme, the mutations that contribute most to the burden of common disease may be similar to those responsible for rare Mendelian disease. They may be harmful and thus rapidly eliminated by selection. Therefore, each mutation will remain rare and will be specific to the local population in which it arose. Alternatively, the gene variants that contribute to common diseases may be much more common in the population than those responsible for the rarer Mendelian diseases. They might be effectively neutral; they might be in the process of sweeping in or out of the population; or they might be maintained by some kind of **balancing selection** (see pp. 476 and 541).

Of course, the questions of what kinds of variants are responsible for genetically complex traits and how selection acts on them are general ones in evolutionary biology. As we saw in Chapter 19, we know very little in general about what maintains the widespread heritable variation that we see for most traits in most populations. In particular, the contrast between the **classical** and the **balance views** of variation goes back to the early years of population genetics (see p. 34) and is still unresolved. However, several of the clearest examples we have of balancing selection evolved as a response to disease (e.g., sickle-cell anemia [Box 14.1] and the **major histocompatibility complex** [**MHC**; pp. 541 and 688]), and much natural selection arises from the continual coevolution between pathogens and their hosts (see p. 509).

One might argue that variation that increased susceptibility to disease is different from quantitative variation in general, because it reduces fitness. That is plainly true when a single dominant allele is responsible for a severe disease with early onset. However, it is not at all clear that the majority of alleles that predispose to disease do reduce fitness or did so in past generations. Even if an allele substantially increases the risk of a disease that affects, say, 5% of the population, it may cause only a modest reduction in fitness, which may be outweighed by other unknown effects. Effects with onset after reproductive age are not directly selected at all (see pp. 562–564). For example, Huntington's disease is due to dominant mutations that cause premature senility in middle age. However, single mutations may persist for many generations and cause large numbers of cases, thus showing that mutations do not greatly reduce reproductive fitness. Where disease is caused by recessive alleles, detrimental effects in the homozygote may be outweighed by advantages to the heterozygote, which will typically be much more numerous (from the **Hardy–Weinberg formula**, $2p(1 - p) \gg p^2$, when the allele frequency p is small). The obvious example here is the Hb^S hemoglobin allele, for which the deaths of homozygotes from sickle-cell anemia are balanced by the resistance to malaria of heterozygotes (Fig. 14.29).

Modern humans live in a markedly different environment from their recent ancestors; thus, common disease might result from a mismatch between alleles and their environments. Alleles that were once well adapted may be in the process of being replaced by a fitter alternative, but may still be common in the population. An example often cited to support the "common disease–common variant" hypothesis is the *ApoE*4* allele, which appears to be an example of such a mismatch. *ApoE*4* is associated with increased risk of coronary artery and Alzheimer's disease and ranges in frequency from approximately 10% to approximately 40% in different human populations. Strikingly, this allele is rarest in those populations that have long practiced agriculture. It is thought that *ApoE*4* may have been advantageous in our ancestral, nutrient-poor environment but deleterious with the current, high-nutrient diet of the developed world. However, we have few examples of alleles that underlie the common diseases; thus, it is currently hard to know which class of variants is most important.

One might also argue that an allele that destroys the function of a gene is likely to be deleterious; after all, in the long term any functional gene must have been maintained by positive selection. However, although most of the alleles responsible for rare Mendelian disease do destroy gene function (Table 26.1), the alleles that underlie common disease are likely to have subtler effects, as is the case for alleles that affect quantitative traits (Fig. 14.29). Also, loss of function may actually be advantageous under changed circumstances: Recall the *CCRΔ5* deletion, which protects against **HIV** (**human immunodeficiency virus**) infection by eliminating the receptor molecule that the virus uses to gain entry to cells of the immune system. This allele may be common in European populations because it protected against past disease in a similar way (see p. 434).

Beyond their inherent interest, the characteristics of variants that cause human disease have practical implications. Population-based gene mapping will be far easier if a small number of common variants are involved (Fig. 26.7A), rather than a myriad of heterogeneous mutations that are individually rare (Fig. 26.7B). Although we know very little about genetic control of the common diseases, we know a great deal about the genetic control of simple diseases that are caused by a single locus. Most loci underlying Mendelian disease have a diverse allelic spectrum; many different mutations at the same locus cause disease. As a typical example, many different mutations in the *SCN1A* gene have been documented as causing Mendelian forms of epilepsy: One study of 93 infants with severe epilepsy found that 33 had mutations that altered the protein sequence, almost all of them different. This heterogeneity is to be expected. All random mutations that have any effect at all are almost certain to degrade existing functions that have been built up by selection (see pp. 510–514).

The diversity of mutations underlying diseases with simple Mendelian inheritance does not rule out the possibility that common diseases are influenced by common variants. If variation in traits of medical importance is primarily a side effect of balancing selection rather than of deleterious mutations, then only a few alleles are likely to be found at each locus. These alleles must have special properties that allow them to coexist in the population; thus, although each allelic class may have accumulated substantial neutral variability over time, it may have homogeneous biological properties—in particular, all members of an allelic class may have a similar effect on disease risk (Fig. 26.7A). (Note that this is not inevitable. Some kinds of balancing selection can maintain very many distinct alleles at a locus, e.g., at the MHC locus or at plant **self-incompatibility** loci [see p. 541].)

It is also possible that even if common diseases are primarily influenced by deleterious alleles, these alleles might be much commoner than those responsible for rare Mendelian diseases. A mutation that causes a serious disease will quickly be eliminated by strong selection and will only contribute a small number of copies before it is lost (~$1/s$, where s is the selection coefficient against it). In contrast, a mutation that has a smaller effect on mortality may be less strongly selected and so will persist for longer and reach higher frequency in the population. Moreover, the recent expansion of the human population may have accentuated this difference. Weakly deleterious alleles (with $s < 1\%$, say) typically trace back some hundreds of generations (their average age being ~$1/s$ generations) to a time when the population size was much smaller. Therefore, the numbers of these deleterious alleles will have increased along with the population as a whole. The net result is that weakly deleterious alleles will be genetically more homogeneous, reflecting the smaller population in which they originated, than will strongly deleterious alleles, which will be younger and more heterogeneous (Fig. 26.8).

What Can We Do with Genetic Information?

At the start of this chapter, we identified two justifications for mapping disease genes: Knowing the causes of diseases helps us to treat them, and associations with genetic markers may help us to estimate the risk of disease. In this section, we return to assess these justifications and ask more broadly how useful genetic information of this kind really is.

The first justification is straightforward. The genome projects have provided the framework for identifying the complete set of genes that we and other species carry. This knowledge is enormously valuable for understanding biology in general and is likely to lead to many medical advances. More specifically, knowing the causes of rare diseases that show Mendelian inheritance tells us much about the causes of common

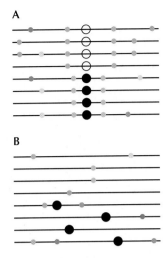

FIGURE 26.7. (*A*) Common diseases might be influenced by common variants that are maintained by balancing selection. In that case, it is likely that a small number of causal alleles would be involved. The diagram shows two alternative alleles that are maintained by balancing selection (*open* and *closed circles*) and that influence disease risk. Other neutral variation will also be present (*colored dots*) and, at tightly linked sites, will be in strong linkage disequilibrium with the causal allele (*red* and *blue dots*). However, for long-standing balanced polymorphisms only a very small region of genome will be influenced in this way ($r \sim \mu$ [p. 541]). (*B*) In contrast, common diseases might be influenced by many different deleterious mutations at each gene (*filled circles*): There are many ways to disrupt gene function, and so we might expect this to be typical.

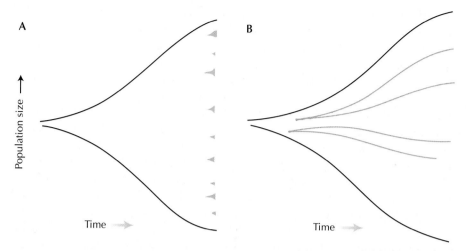

FIGURE 26.8. Population expansion influences allelic heterogeneity. (*A*) Strongly deleterious alleles (*blue*) must have originated recently, within the large current population. Therefore, many different mutations are likely to contribute. (*B*) More weakly deleterious alleles (*red*) tend to be older and so trace back to the smaller ancestral population. Most mutations will be lost by random drift, but those that survive will have expanded in numbers with the population as a whole. Here, two mutations are shown as contributing.

diseases, in the same way that the study of gene knockouts has told us so much about the normal functioning of organisms. Recent advances also make possible not only specifically genetic approaches—most obviously gene therapy, in which defective genes are replaced—but also sensitive diagnostic tools that distinguish different pathogens or different types of cancer, enabling more appropriate treatment. We might worry about the costs of these medical advances, about whether they will be available to all, and whether limited resources are best spent on such sophisticated technologies. However, these are general concerns and involve no specific evolutionary or population genetic considerations.

When we come to the second justification, where we consider how genetic markers might be used in medicine and beyond, there are more complex and contentious issues. An understanding of the evolutionary causes of population genetic variation does not settle these issues, but is essential to an informed debate. Here, we briefly outline the difficulties raised by the coming availability of detailed genetic information in different areas.

Implications for Medicine

One of the reasons often given for funding extensive genetic surveys of the human population and for the development of techniques for rapidly genotyping individuals is the prospect of **personalized medicine**—that is, tailoring treatment to fit the individual. For example, individuals vary considerably in how they metabolize drugs, which influences both the drugs' effectiveness and toxicity. Determining genotype at the genes responsible for drug metabolism could be a cost-effective way to make treatments safer and more successful. (This approach is similar to the characterization of tumors, with the aim of finding the most appropriate treatment for individual cancers. The difference is that here it is the individual rather than the tumor that is characterized.)

Perhaps the most direct benefit from genotyping the general population might be that screening for diseases such as cancer could be focused on those most at risk. (For example, at present, women with a family history of breast cancer may be typed for the *BRCA1* allele. The presence of the allele correlates with a substantial increase in the risk of developing breast cancer.) However, it is far from clear what other benefits there would be of assessing disease risk for healthy people. For example, suppose that we could accurately predict the risk that an individual would suffer heart disease or diabetes. Most of the changes in lifestyle that would reduce risk from such diseases (e.g., more exercise and improved diet) would be beneficial for everyone. Those predicted to be at higher risk might be more inclined to lead healthier lives, but would those predicted to be at lower risk do the opposite? This problem—whether genetic information is worth having when it leads to no specific cure—is seen most starkly for those with a family history of late-onset disease. Children and siblings of those with Huntington's disease themselves have a 50% chance of carrying the dominant allele responsible for the disease; if they do, they will inevitably suffer premature senility in middle age. Because there is no treatment, most of those at risk choose not to be tested.

Prenatal Screening

At present, genetic information is most widely available through prenatal screening. Some defects can be detected from samples of the mother's blood. For example, an extra copy of chromosome 21 causes Down syndrome, or trisomy 21. This leads to lower levels of α-fetoprotein and unconjugated estriol and to higher levels of human chorionic gonadotropin in the mother's bloodstream. However, tests based on this are uncertain. The embryo's genotype can be determined only by **amniocentesis**, in which a sample of placental tissue is taken. For those prepared to consider abortion, genetic information could be valuable if there is a high risk of some serious problem. (For example, older mothers have a higher risk of trisomy 21 and so may be offered amniocentesis.) Where there is an exceptionally high risk (e.g., if a sibling has a disease caused by homozygosity for a recessive allele or if both parents are known to be heterozygous for such an allele), then in vitro fertilization could be considered. Surplus embryos are produced in the normal course of this technique, and so healthy genotypes can be chosen from them before implantation. For the general population, however, the risks of amniocentesis are high enough that routine genetic screening of embryos is not justified. That might change if it becomes feasible to genotype the small numbers of fetal cells that circulate in the mother's bloodstream.

As the genetic causes of human variation become better understood, far more detailed information could be derived from an embryo's genotype. Rather than just knowing that there is an increased risk of one of a few serious inherited defects as at present, parents might in principle be given estimates of the risks of a large number of diseases occurring both early and late in life. Inevitably, every individual will be at risk from several more or less serious problems. It is hard to see how parents could deal with such a mass of statistical information or how much information should be given.

The issues become still more obscure when we consider the possibility of knowing about variation in general. It is already feasible to determine discrete traits such as eye and hair color. It is unlikely, but not inconceivable, that useful information could be given about quantitative traits such as height or intelligence, because very many genes influence these traits, in interaction with the environment. But still, there is an issue of principle about whether parents who use in vitro fertilization should be allowed to choose on the basis of such characteristics. Most countries where in vitro fertilization is regulated do not allow such choice. However, there is already in practice extensive choice of offspring on the basis of sex by selectively aborting daughters (Fig. 26.9).

Insurance

A person may take out life insurance so that the person's family will be supported if he or she dies or is incapacitated. In those countries where health care is privately funded, insurance is also needed to cover the unpredictable costs of medical treatment.

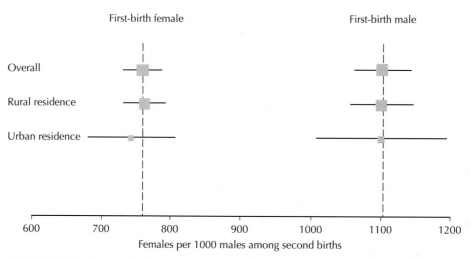

FIGURE 26.9. In India and China, the sex ratio is skewed toward males when the first birth is female. A survey of more than 1.1 million households in India showed that the proportion of daughters in second births is much lower when the firstborn was a daughter.

It is a basic principle that any information available to the insured is also available to the insurer—people who know that they will become seriously ill should not be able to take out insurance at the same rates as others. However, although insurers may require a medical examination, they do not at present demand any genetic tests. Moreover, insurers do not on the whole vary premiums according to factors that are known to have a strong association with mortality, for example, weight or social class. This contrasts with other kinds of insurance: Premiums for car insurance are carefully adjusted according to make of car, age of driver, and so on.

There are serious concerns that if genetic information were to give accurate estimates of future health and life expectancy, then insurers might demand to know the genotype of their clients. Some might benefit from lower premiums, but others might be unable to obtain any insurance at all. However, it is not clear how far insurers would actually vary their premiums according to estimated risk, given that they have been reluctant to do so until now.

Identifying Individuals

We now have an extraordinary ability to identify individuals from minute traces of their DNA. In the United Kingdom, the National DNA Database, which has been compiled from samples taken from suspects and witnesses by police, now covers around 3.4 million people, including more than 500,000 children. Individuals can be identified even if they are not in the database, because close relatives of known genotypes can be identified as sharing alleles at multiple loci. It is now becoming possible to use genetic markers to identify ethnic origin, eye color, and other such features. These advances have transformed police work and have prevented many miscarriages of justice. Other benefits include the ability to identify victims of natural and man-made disasters. However, these unprecedented abilities raise concerns about individual privacy and about whether they give states too much power.

In all these applications, there is a general worry about how accurate estimates based on genetic information would actually be. In the popular media, we often read that genes have been found "for" this or that trait. What this usually means is that in a survey of a few hundred individuals, a statistically significant association has been found between some genetic variant and the trait. As we discussed above, we expect results to vary between populations and between environments, and, indeed, such associations often cannot be replicated (e.g., Fig. 26.6; Table 26.2). It is instructive to compare the recent enthusiasm for genetic markers with our long experience in epidemiology. Despite great efforts, involving many large surveys, there are very few robust and widely accepted associations between dietary factors and disease. It is extremely difficult to reliably detect an association between disease and any factor—genetic or not—in a heterogeneous population and a heterogeneous environment. Moreover, any such association is likely to be modulated by interactions with the environment and with other genetic factors.

A somewhat different concern is that even if there is a real association between an allele and some particular disease, that allele may be associated with other diseases as well. We have already seen one such example in the *ApoE*4* allele, which predisposes to both Alzheimer's disease and to coronary heart disease. Such **pleiotropic** effects do not matter if we are concerned about only one disease—for example, in deciding whether to screen for cancer. However, it is not at all clear that the overall risks to health and life expectancy can be estimated by considering the effects of only particular alleles on susceptibility to particular diseases. This is a special concern if balancing selection is widespread, so that common variants are maintained by a balance of selective forces, each acting via different biological routes. For example, the Hb^S allele causes severe anemia when homozygous, but confers partial resistance to malaria in the heterozygote; and although alleles of the major histocompatibility loci are known to be associated with particular rare diseases, they may also affect susceptibility to many other conditions that we know nothing about. Both the statistical difficulties that are already apparent from conventional epidemiology and the potentially complex effects of alleles should make us extremely cautious in using genetic information.

UNDERSTANDING HUMAN NATURE

In the previous chapter, we set out what is known about how our species evolved after its separation from our closest surviving relatives. So far in this chapter, we have discussed human population genetics, focusing on the definite issues of how the genes that influence disease can be found. Now, we turn to broader questions. We begin by summarizing the background pattern of genetic variation within the human species and then move on to consider characteristics that have been shaped by selection and the implications for medicine. We end this chapter by discussing the limitations of attempts to use evolutionary biology to understand our own species.

Human Variation Cannot Be Subdivided into Discrete Races

There is a long history of attempts to classify mankind into separate races with distinct characteristics. What does our modern knowledge tell us about the nature of genetic differences between human populations?

There are significant differences in allele frequency among human populations. These are best summarized by the statistic F_{ST}, which measures the variance in allele frequency among populations relative to the total variation in the entire population ($F_{ST} = \text{var}(p)/p(1 - p)$; p. 446). Worldwide, $F_{ST} \sim 0.15$, which is typical of animal species (see Fig. 16.10). Roughly speaking, 80% of variation is found within populations and 20% between populations; typically, there will be modest allele-frequency differences between populations. Morphological variation (e.g., skull shape) follows a similar pattern, with the statistic Q_{ST} being about the same as F_{ST} (p. 446; see Web Notes).

We also can think in terms of the ancestry of genes. A pair of lineages have some chance of coalescing in a common ancestor within the local population, but are more likely to trace back many thousands of generations to the ancestral population that lived in Africa before anatomically modern humans left that continent (see pp.

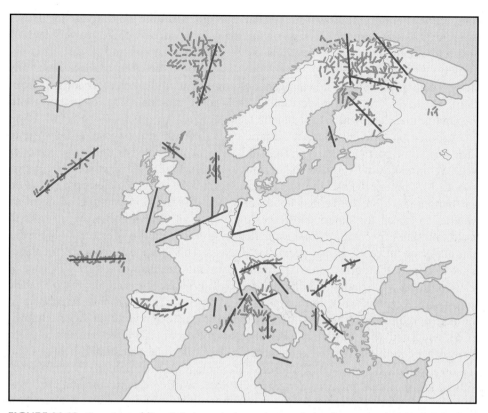

FIGURE 26.10. Genetic and linguistic boundaries correspond closely across Europe. The *red* lines indicate average gradients in allele frequency from a survey of 60 alleles at 3119 locations. These were used to estimate 33 zones of sharp genetic change, indicated by the *purple lines*. Of these, all but two correspond to linguistic boundaries. Boundaries shown in the sea are between adjacent land masses (e.g., Ireland vs. Iceland).

743–748). (Recall that two randomly chosen sequences are likely to differ with probability $\pi \sim 0.001$. Assuming a mutation rate of $\mu \sim 5 \times 10^{-9}$ per site per year, this implies a time to common ancestry of around $2N_e = (\pi/2\mu) = 10^5$ years [p. 426].) If we think instead of our pedigree ancestors (i.e., our great-great…grandparents), and if we go back just 40 generations (~1000 years), then we all share essentially the same set of ancestors (see p. 428).

This moderate level of variation between populations does not reflect sharp boundaries between races, but neither does it consist entirely of smooth **clines**. We have seen that tree-like models can be fitted to human genetic data (see p. 72) and that individual genotypes can be assigned reliably to their continent of origin (see p. 454). Within Europe, regions can be identified where genetic differences are especially sharp, and these correspond to linguistic boundaries (Fig. 26.10). However, none of these patterns involves large allele-frequency differences; they are a statistical aggregate of many modest differences. If we look at the frequencies of individual alleles, we see geographic patterns that are a mixture of variation at all spatial scales. Recall also that genetic evidence makes the **multiregional hypothesis** unlikely. Present-day populations on different continents all trace back to an anatomically modern population that evolved in Africa rather than to more ancient roots within each continent (see pp. 748–749). All this is consistent with what we know about human population structure. Our species is somewhat divided by geographic and cultural barriers, but there is nevertheless sufficient **gene flow** to make our species well mixed over evolutionary timescales.

Natural Selection Has Shaped, and Is Shaping, Human Variation

We have described patterns of variation that are likely to be largely neutral. By definition, such variation has no significant effect on fitness and so is also unlikely to affect characteristics that actually matter to us. Neutral variation reflects our ancestry, which is largely shared, but does not tell us about differences that affect survival and reproduction.

We saw in Chapter 16 that very low levels of gene flow, acting over the very long timescales characteristic of random drift (for humans, roughly $N_e \sim 10^4$ generations), can greatly reduce genetic divergence. The worldwide F_{ST} value of approximately 0.15 is consistent with exchange of roughly one individual between continents in every generation (i.e., $Nm \sim 1$ [Box 16.2]). Such low rates of gene flow pose no significant obstacle to changes caused by even weak selection ($s \sim 10^{-3}$, say).

Some genes do indeed show unusually high geographic divergence. We have already noted several examples. For example, the *ApoE*4* allele, which increases the risk of heart disease and Alzheimer's disease, is much rarer in populations that have long practiced agriculture (see p. 764); and the *CCRΔ5* allele, which is much more common in European than African populations, confers partial resistance to HIV (see p. 434). Considerable variation in F_{ST} is expected between loci even in the absence of selection, but, nevertheless, some loci can be identified as having significantly higher F_{ST} than expected under a simple **island model** (Fig. 26.11). We have already seen that, in *Drosophila*, regions that show lower genetic diversity in the European population relative to the African indicate recent fixation of alleles that were favored in the new northern environment (Fig. 19.18). In humans, systematic surveys of genome-wide variation within and between populations are likely to identify many further candidates for diversifying selection (Fig. 26.11).

One of the best-understood examples of natural selection in humans is lactose tolerance in European populations. Lactose is the only nutritionally significant carbohydrate in milk, but must be broken down into glucose and galactose by an intestinal enzyme, lactase, before it can be used. Lactase is active up until weaning, but in almost all mammals it then declines to a low level. Thus, most humans cannot digest lactose and are intolerant of large amounts of unfermented milk. However, persistence of lactase into adulthood is common in northern European populations, allowing them to digest milk. Two polymorphisms upstream of the *LCT* gene that encodes lac-

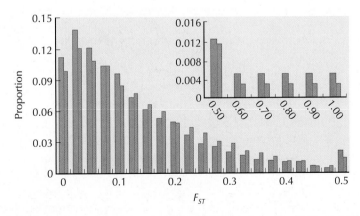

FIGURE 26.11. Distribution of F_{ST} across loci. The *blue bars* show the distribution of F_{ST} observed between three populations: East Asian, African–American, and European–American (42 individuals each), for 26,530 SNP (single-nucleotide polymorphism) markers. *Red bars* show the pattern expected assuming an island model with three populations of the same constant size. The inset shows the tail of the distribution, for $F_{ST} \geq 0.5$. Loci with unexpectedly high F_{ST} may have been subject to diversifying selection, although a complex demography also may have been responsible.

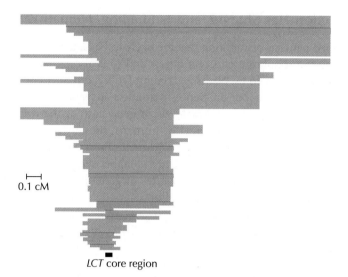

0.1 cM

LCT core region

FIGURE 26.12. Evidence of selection for lactose tolerance. Long stretches of homozygosity give evidence for recent selection on the allele of the *LCT* gene that causes lactose tolerance in European populations. Each row shows the length of a haplotype that has identical sequence. The haplotypes that carry the tolerance allele (*red*) are, on average, much longer than those that do not (*blue*), and extend up to ~1 cM.

tase are strongly associated with each other and with lactose tolerance and have a frequency that matches the distribution of dairy production, reaching approximately 77% in Scandinavia. This allele is found in haplotypes that are homogeneous over long stretches of the genome (up to 1 cM, or about 1 Mb [Fig. 26.12]). This indicates that the allele arose recently as a single mutation, so that recombination has not had time to break up the ancestral sequence (p. 434). The length of the haplotype is consistent with an origin around 10,000 years ago, roughly when dairy farming became established; selection of a few percent would be needed to raise it to its present frequency.

Another plausible candidate for a selected trait is skin color. Although this has traditionally been used to define human races, it varies more or less continuously worldwide (Fig. 26.13). It is thought that dark skin is ancestral, both because our primate rel-

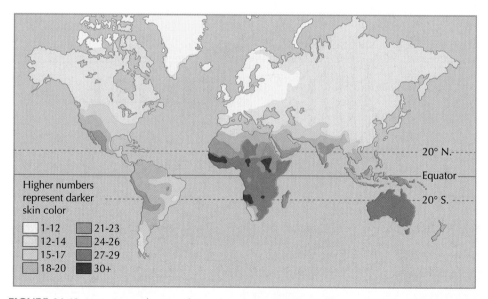

Higher numbers represent darker skin color

- 1-12
- 12-14
- 15-17
- 18-20
- 21-23
- 24-26
- 27-29
- 30+

20° N.

Equator

20° S.

FIGURE 26.13. Variation in human skin color is associated with levels of UV irradiation, which are higher near the equator.

atives have dark skin and because of the recent African ancestry of all humans (see pp. 743–749). Dark skin protects against the damaging effects of UV radiation. However, UV light is required for synthesis of vitamin D, and in higher latitudes, where UV levels are lower, dark-skinned people suffer from vitamin D deficiency. Thus, it is plausible that pale skin has evolved in higher latitudes to avoid this problem; indeed, skin color is closely correlated with levels of UV irradiation (Fig. 26.13). Human skin color has a complex genetic basis involving several genes. Recently, the human homolog of a gene that influences pigmentation in zebra fish was found to explain about one-quarter of the difference in pigmentation between Europeans and Africans. An allele in which a threonine has replaced an alanine is closely associated with lighter pigmentation in mixed African–American populations; this allele is absent in dark-skinned populations, but is at very high frequency among light-skinned Europeans. As in the lactase example above, the allele that is associated with paler skin is flanked by a region with low sequence variation, which implies that selection has recently raised it to high frequency.

Relaxed Selection Will Eventually Lead to an Increased Incidence of Inherited Disease

Some have argued that selection no longer acts in affluent populations, because they live in a benign environment. Selection has clearly changed its nature. In many countries, infectious disease and starvation are now minor causes of mortality (although new diseases such as AIDS and influenza may change that picture). However, it is plainly not the case that there is *no* selection. Some people have more offspring than others, and much of this variation in fitness is likely to be inherited; quantitative traits associated with fitness have high genetic variance (see p. 547). Indeed, we can see examples of particular alleles that are currently under strong selection (e.g., Fig. 26.14).

How quickly will new selective conditions take effect? In particular, how quickly will the relaxation of selection against many inherited diseases that is allowed by modern medicine cause them to increase in frequency? This depends very much on their mode of inheritance. Recessive deleterious alleles will increase very slowly when selec-

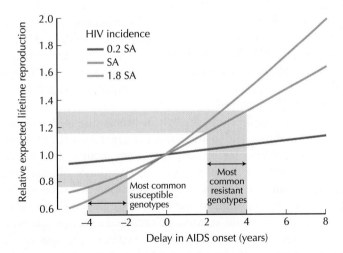

FIGURE 26.14. The *CCRΔ5* gene strongly influences susceptibility to AIDS and is therefore under strong selection (see p. 434). The graph shows the relative fitness of genotypes that alter the onset of AIDS plotted against the delay in onset; the *middle line* is based on data from South Africa, where the incidence of HIV infection was ~20% in 1999. The *upper* and *lower curves* show fitness in populations with higher (~1.8 times) and lower (~0.2 times) incidence of HIV infection, respectively. The most common resistance genotypes have about 50% greater fitness than the most common susceptible genotypes, based on South African data. Comparing the three curves, we see that the strength of selection increases with increasing incidence of AIDS.

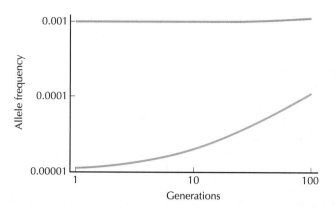

FIGURE 26.15. Changes in allele frequency following relaxation of selection. The frequency of a recessive lethal allele (*red*) barely changes after 100 generations in which homozygotes survive and reproduce. This is because most copies are hidden from selection in heterozygotes. In contrast, a dominant allele (*blue*), which initially has disadvantage $s = 10\%$ in the heterozygote, increases relatively faster if selection is relaxed: It doubles in frequency after ten generations and increases 11-fold after 100 generations.

tion against homozygotes is relaxed, simply because these alleles are mostly carried in the much more numerous heterozygotes. If the heterozygote has the same fitness as the wild-type homozygote, then the allele will increase exceedingly slowly, at a rate equal to the mutation rate (Fig. 26.15, red line). It is likely that the heterozygote is under some selection, in which case the short-term outcome depends almost entirely on that selection and hardly at all on what happens to the rare homozygotes.

However, recessive alleles are an extreme case. With other forms of inheritance, change can be much more rapid (e.g., Fig. 26.15, blue line). A dominant lethal allele is never passed on, and so its frequency equals the mutation rate in that generation. If medical treatment allows individuals carrying the allele to reproduce, then the allele again increases at a rate equal to the mutation rate. However, unlike the case of a recessive allele, this increase is immediately reflected in the incidence of the disease, which will double in the first generation and increase by a factor t after t generations. As a concrete example, deletions occur on the Y chromosome at a substantial rate and cause infertility. However, men with such deletions can often father children by in vitro fertilization because although their sperm have low motility, they can still fertilize the egg. The deletion will then be passed on to their sons, and so the frequency of this form of infertility will increase rapidly, at a rate comparable with the rate of in vitro fertilization to treat male infertility.

We have seen that at equilibrium, the loss of fitness due to deleterious mutations depends on the total mutation rate per genome per generation, U; in the simplest case, average fitness is reduced by a factor $\exp(-U)$. (This assumes that the effects of different mutations multiply. In a sexual population, some kinds of gene interaction may reduce the **mutation load** [see p. 552].) Because the total mutation rate in humans is high (possibly $U > 1$), the standing mutation load is also likely to be high, and yet more deleterious mutations would accumulate rapidly if selection against them ceased. If selection were entirely relaxed (admittedly, an extreme assumption), then the load would increase by approximately Us per generation, where s is the average selective effect of deleterious mutations. It is likely that most deleterious mutations have small effects, and so this decline would occur over many generations ($t \sim 1/s$). Note that this calculation gives the fitness loss in the absence of medical intervention; the assumption is that under modern conditions, there would be no actual loss of fitness. Nevertheless, there is cause for concern as to the consequences over timescales of some hundreds of generations—long from a human perspective, but very short in evolutionary terms.

It is unclear how effective natural selection has been in our recent history. On the one hand, our relatively low effective population size ($N_e \sim 10^4$) makes it impossible for us to avoid accumulating mildly deleterious mutations (see p. 493), and there is evidence that more such mutations have accumulated along our lineage than along that of our sister species, the chimpanzee (see Web Notes). However, it could be that our low effective population size actually reflects the effects of a large number of **selective sweeps**, and therefore the *success* of natural selection. (Recall that N_e is really a measure of the inverse rate of genetic drift. It can be reduced by both low population size and selective sweeps [p. 427].) When a favorable mutation is swept to fixation, it may carry with it more weakly deleterious alleles that happen to be tightly linked to the original mutation. Thus, we can make two radically different interpretations of the observed low genetic diversity within our species: on the one hand, that it reflects a low population size in the past, implying inefficient selection, or on the other, that it is a side effect of intense adaptive selection.

Application of Genomic Medicine Depends on the Extent of Racial Variation

Advances in genomic medicine are raising a number of questions about race, ranging from whether all ethnic and racial groups will benefit equally from genomic advances to whether identifying differences among racial or ethnic groups could provide a new foundation for racism. The interface of race and genetics has been one of the most controversial areas in modern science. Some of the worst excesses of the 20th century were fueled in part by claims of racial superiority, with academic research in some cases playing directly supporting roles.

However, in the decades following World War II, a clear consensus emerged among virtually the entire community of human and population geneticists that genetic differences among racial groups are slight. This was strongly confirmed by patterns of molecular variation, as discussed above. Genetic variation is largely shared; although some alleles do differ more between populations as a result of selection, these geographic patterns vary from gene to gene and do not correspond to traditionally defined racial groups.

However, the question of how geographic ancestry correlates with genetics has recently reemerged as a contentious topic, with many now arguing that it is irresponsible to deny the differences among groups because of the potential medical implications. For example, the U.S. Food and Drug Administration (FDA) has recently approved BiDil, a combination therapy for congestive heart failure, for use specifically in the African–American community. BiDil is a combination of two older generic medicines that together reduce blood pressure and possibly protect the heart and blood vessels from damage associated with certain cardiovascular diseases. The combination was first used in drug trials in an ethnically mixed population of patients with heart failure, but the drug did not work well enough to win approval for treating this condition. Subsequent analyses, however, suggested that the drug worked better among African–Americans, leading to a trial exclusively in that community in 2001. In 2005 the FDA approved the drug for use in African–Americans, the first drug for which ethnicity is part of the indication.

There are two reasons why a medicine may work differently, on average, among Americans of European and African descent. First, it may be that the environments for the two groups are different. African–Americans are exposed to more lead and are more likely to eat high-fat foods and less likely to eat fresh fruit and vegetables, on average. They also have poorer access to health care. The other reason is that the two groups may have genetic differences that influence drug response. The classic position that genetic differences among racial or ethnic groups are slight would seem to rule out a genetic explanation for any differences in efficacy. As noted, however, this ignores the fact that many polymorphisms are in fact expected to show geographic structuring. Indeed, an analysis of gene variants that have been shown to be associated with drug response makes clear that average differences between African–Americans and European–Americans are common. This certainly does not mean that the reason that BiDil works differently is genetic, but it does mean that genetic explanations cannot be ruled out on first principles. However, even if genetics is a contributing factor, it does not follow that it makes sense to treat on the basis of racial or ethnic labels.

Clinically, the aim is to ensure that patients get the right medicine. If we knew the precise underlying genetic or environmental factor(s) responsible, we could do a much better job of identifying who is likely to respond well and who is not. A good illustration of this point involves codeine, a widely used analgesic. Codeine itself does nothing to relieve pain. It is only after an enzyme called CYP2D6 changes codeine into the related chemical morphine that pain is relieved. Approximately one in ten North Europeans lacks this enzyme, and so for them codeine offers no pain relief. In the Arabian Peninsula, however, at least 97% of people produce the enzyme. On average, therefore, codeine is more effective in people from the Arabian Peninsula than in people from Europe. Our knowledge that CYP2D6 determines response to codeine, however, is no reason to consider whether people are from Sweden or Saudi Arabia—it would be far more informative to look directly at CYP2D6. For BiDil, this is currently impossible because we do not know what to look for. In this case, knowledge of the underlying genetic causes of the variable response would help to optimize the use of the medicines.

Many researchers worry that the growing acceptance of genetic differences among racial and ethnic groups could contribute to a resurgent racism underpinned by modern genomics. Particular concerns have been raised about hypothetical scenarios in which gene variants that are associated with behavioral traits, such as cognitive measures, show allele-frequency differences among racial or ethnic groups. It is too early to know what gene variants might be found that are associated with cognition and other sensitive traits, but the evidence so far gives little cause for concern that these variants will show geographic patterns that significantly favor one group over another. Because of both selection and genetic drift, there do appear to be differences among people from different parts of the world, on average. (Recall the examples of lactose tolerance and skin color described above.) Hopefully it is not overoptimistic to predict that in the coming years geneticists will increasingly view genetic differences among groups as a matter of both historical interest and potential medical significance and not as something to fear.

Darwinian Medicine Seeks Ultimate Explanations

The role of evolutionary biology in explaining the ultimate causes of human disease was articulated most forcefully by George C. Williams and is often referred to as **Darwinian medicine**. The basic idea is that disease is shaped by evolution, and that the treatment of disease can be improved by an understanding of the evolutionary forces that have operated on it. More or less plausible evolutionary explanations have been offered for a range of different conditions. For example, nausea and peculiar food preferences among pregnant women may be an adaptive mechanism that protects fetuses from the toxins found in normal diets (Fig. 26.16). Fever may be an evolutionary adaptation that creates an inhospitable environment for pathogens. Similarly, the global epidemic of obesity may result from our long-term adaptation to a hunter–gatherer's diet, and the vulnerability of all mammals, including humans, to choking is a direct result of ancient constraints that appear to make it developmentally impossible to separate the openings leading to our stomachs and our lungs.

Darwinian medicine has paid particular attention to the study of aging. As we saw in Chapter 20, the evolutionary explanation of aging is that survival and reproduction in late life are under weaker selection than in early life. This is simply because most individuals are likely to die before they become old, and so alleles that influence only late life are rarely expressed and hence exposed to selection. The reduced selection on older individuals means that where there is a trade-off between early reproduction and maintenance of the organism for future reproduction, the former will receive greater weight (see p. 564). In addition, deleterious mutations will place a greater burden on late life, because they will be eliminated less effectively by selection (see p. 565).

What does this evolutionary explanation tell us about human senescence? First, aging is not inevitable. This is shown most directly by comparisons between species. For example, birds, bats, and tortoises, which are relatively safe from predators, live much longer than other species of similar size, because they suffer less extrinsic mortality (see p. 562). However, this does not mean that it will be easy to extend the human life span. Evolutionary theory implies that there are many different pathways that could cause aging, because in principle any physiological process could be subject to selection that favors early reproductive output or survival over later reproductive output or survival. Thus, if some pharmacological intervention fixed one problem of old age (e.g., a neuroprotective agent that prevented dementia), then, shortly thereafter, aging individuals would suffer deterioration of some other function, such as cardiovascular function. In short, "magic bullets" for senescence are unlikely to be found.

The argument may be correct, but it is unclear whether knowing that aging is "tough" will be helpful in finding new strategies to thwart it. More fundamentally, it is also unclear whether aging really is "tougher," in that it has a more complicated pathophysiology than other common conditions afflicting humans. As described above, it may well be that both rare and common diseases are caused primarily by a heterogeneous mass of deleterious mutations, each individually rare and each degrading function in a different way. On that view, all common diseases would be equally "tough," and the underlying causes of a given disease would differ from person to person. A different "magic bullet" would be required for each patient.

The view that the mechanism of senescence is likely to be complex, involving the failure of many functions, is apparently contradicted by the discovery in a wide range of organisms of mutations that greatly extend life span. These act via a conserved signaling pathway, which acts to shift resources from reproduction to survival when food is scarce (see p. 566). This pathway may have evolved to allow organisms to maintain themselves until conditions improve and reproduction is more likely to be successful. It is responsible for the observation (in organisms from yeast to humans) that reduced intake of food energy extends life span. Indeed, this is the only environmental intervention that has been reliably shown to extend life span.

Although evolutionary explanations provide a satisfying framework for understanding human response to pathogens and the nature of senescence, by and large they do not offer an obvious route to improved treatment of diseases. In contrast, an evolutionary approach has been of great practical use in understanding the evolution of disease organisms themselves. Perhaps the clearest example is the evolution of antibiotic resistance. Without an understanding that adaptive evolution among bacteria has caused the increase of resistance over the past 50 years it is hard to see how anyone would have come to the conclusion that prophylactic use of antibiotics should be curtailed. Similarly, understanding the evolution of resistance to antiviral drugs is crucial in treating HIV infections, and phylogenetic methods have been essential in tracing the history and spread of infectious diseases (e.g., Fig. 15.19). In general, medicine is most concerned with proximate causes as opposed to ultimate explanations, and here evolutionary approaches have emerged as a key component in efforts to determine the genetic bases of common diseases.

Evolutionary Psychology Attempts to Understand Human Nature

The greatest impact of Darwin's theory of evolution by natural selection was on our understanding of our own species—the knowledge that we share ancestry with our primate relatives and beyond, and that our distinct adaptations evolved by natural selection. Although Darwin did not discuss humans in the *Origin*, he keenly appreciated the implications of his theory for our understanding of our place in the world and went on to write two books that dealt specifically with human psychology and its continuity with animal behavior. However, as we saw in Chapter 1, others did not take up or develop the concept of natural selection. The idea of evolution was widely applied to humanity, but with only the most tenuous connection with Darwin's work. The strongest influence in the century following the *Origin* was the idea of progress, of evolution directed toward humans. Even among biologists, evolution was often seen as acting "for the good of the species" (p. 34).

The importance of Darwinian natural selection, acting on individuals or on genes, only reemerged around 1970, through the work of Williams, Maynard Smith, and William Hamilton. Its application to animal behavior—including human behavior—was emphasized in Edward O. Wilson's book *Sociobiology: The New Synthesis* (1975) and by Richard Dawkins' *The Selfish Gene* (1976). The study of animal behavior was transformed by application of evolutionary principles and grew into a thriving field (e.g., see Chapter 20). The extension to humans was highly controversial and led to bitter criticisms. These came in part from social scientists, who argued against any application of biological principles to humans. However, there was also strong criticism from the evolutionary biologists Stephen Jay Gould and Richard Lewontin, who objected to the assumption that human behavioral traits are adaptations built up by natural selection rather than arbitrary side effects of the evolutionary process (see p. 557).

More recently, several research programs have emerged that use evolutionary principles to study human behavior. The most prominent is **evolutionary psychology**, which focuses on universal human traits, especially those involved in social cooperation, mate choice, and parental investment. Evolutionary psychologists argue that we are not necessarily adapted to current conditions; that is, we do not necessarily behave so as to maximize the reproductive success of our relatives and ourselves in our present environment. Rather, humans are thought to be adapted to conditions in the **Pleistocene**, 0.1–2.5 Mya, when we lived as hunter–gatherers. This is termed the **environment of evolutionary adaptation** (EEA).

Evolutionary psychologists also suppose that we have evolved specific modules for tasks such as detecting cheating or choosing a mate. Such modules are thought of as analogous to subroutines in computer software, which are dedicated to specific functions that enhance fitness. Basic cognitive processes are known to be organized into modules (e.g., in vision, there are groups of neurons that can detect edges, or even faces), but there is considerable debate over evolutionary psychology's hypothesis of "massive modularity." However, this argument over psychological mechanism is not central to the evolutionary question of whether specific behaviors are adaptations that have evolved by natural selection. As we saw in Chapter 14, selection will shape phenotypes provided that there is inherited variation for the selected traits; the genetic and physiological mechanisms are not directly relevant.

Glossary

absolute fitness: *See* fitness.

acritarch: An organic-walled microfossil, found in ancient rocks, that is interpreted as the reproductive cyst of a eukaryote.

acrocentric: A chromosome with the centromere close to one end.

adaptation: A trait that functions to increase fitness and that evolved for that function.

adaptationist program: An approach to understanding evolution that assumes that traits are adaptations.

adaptive landscape: *See* fitness landscape.

adaptive peak: A local maximum in the fitness landscape.

adaptive radiation: Diversification of a single lineage into species that exploit diverse ecological niches.

additive genetic variance: The total variance, V_A, contributed by the additive effects of each gene (in other words, the variance of breeding values). The response to selection on a quantitative trait is proportional to V_A.

additive genetic variance in fitness: The additive genetic variance, V_W, of fitness. Fisher's Fundamental Theorem states that the increase in mean fitness is proportional to this quantity.

additive model: A model in which a quantitative trait is the sum of effects of all the genes involved and of a random environment.

additive tree: A phylogenetic tree in which the branch lengths are proportional to the evolutionary distance between nodes; also known as a phylogram.

ADH: The enzyme alcohol dehydrogenase.

aging: A decrease in survival or reproduction with age. It is equivalent to senescence.

align, aligned, alignment: To line up different DNA, RNA, or protein sequences. In most cases alignments include gaps where one molecule has an insertion or deletion relative to others. For phylogenetic analysis, each position (i.e., column) in the alignment should include homologous residues (bases or amino acids) from the different molecules.

allele: A particular form of a gene.

allele frequency: The frequency of a particular allele in a population.

allelic: Two variants are allelic if they are alleles at the same genetic locus.

allometry: Change in proportions with body size. For example, the size of a stag's antlers relative to its body size increases with body size.

allopatric speciation: The formation of reproductively isolated species due to the divergence of populations that are geographically isolated from each other.

allopatry: The complete separation of populations by geographic barriers.

allopolyploid: A polyploid in which the multiple genomes are derived from different populations or species.

allosteric interaction: *See* allostery.

allostery: A change in the shape of a protein or ribozyme due to binding of a molecule at one site, which then changes activity at a distant site.

allozyme: One of several variant forms of an enzyme coded by alternative alleles at a single genetic locus.

α-helix: Common structural motif of proteins in which a linear sequence of amino acids folds into a right-hand helix stabilized by internal hydrogen bonding between backbone atoms.

α-proteobacteria: A major class of bacteria that includes many photosynthetic species, many pathogens (e.g., Rickettsias), and many mutualistic symbionts, including the ancestors of mitochondria.

alternative splicing: The process by which the initial RNA made from a single gene can be spliced into different mature messenger RNAs, which in turn produce different proteins.

altruistic: A gene, trait, or behavior that reduces the fitness of its bearer but increases the fitness of other individuals.

alveolate: Member of Alveolata, one of the major kingdoms of eukaryotes. It includes apicomplexans, dinoflagellates, and ciliates.

amelioration: The process in which DNA that has been acquired by lateral gene transfer changes in composition (e.g., G + C content and codon usage) to resemble the genome in which it resides.

AMH: *See* anatomically modern human.

amniocentesis: Removal of amniotic fluid that surrounds the embryo. The fluid contains fetal cells that can be used for prenatal diagnosis.

amoeba: A single-celled eukaryote that has no fixed shape. This phenotype is found in many eukaryotic lineages.

Amoebozoa: One of the major kingdoms of eukaryotes. Most species are heterotrophic. It includes some species that do not have mitochondria.

amphipathic: A molecule that has both hydrophobic and hydrophilic components (e.g., the phospholipids that make up membranes).

analogy: Similarity that is not due to homology (i.e., common ancestry).

analysis of variance: The separation of the variance into a sum of components, a widely used statistical technique that is the basis for quantitative genetics.

anatomically modern human (AMH): A member of the human lineage that is recognized by anthropologists as having essentially the same anatomy as present-day *Homo sapiens*.

ancestral characteristic: A trait found in both an organism being studied and the common ancestor of a group to which the study organism belongs.

aneuploid: A cell or organism having an abnormal set of chromosomes.

angiosperm: Flowering plant.

anisogamy: Differentiation of gametes into two (or more) sizes.

annelids: Member of Annelida, a phylum within the Lophotrochozoa. This group of segmented worms includes earthworms, leeches, and bristle worms.

antagonistic pleiotropy: Describes alleles that increase one fitness component but decrease another. It is also a term for theories of aging that involve such alleles.

anther: The male reproductive organ where pollen is produced in a flowering plant.

antibody: A protein that binds to a specific antigen.

antigen: A chemical that triggers an immune response by binding to a specific antibody.

antimutator: An allele or genotype that reduces mutation rates.

antisense oligonucleotide: A short synthetic nucleic acid sequence that is complementary to an mRNA sequence. Through a variety of mechanisms, antisense oligonucleotides can be used to inactivate gene function by interfering with the ability of an mRNA to produce its corresponding protein product.

apicomplexan: Member of Apicomplexa, a phylum of eukaryotes, in the alveolate kingdom. Many species in this phylum are parasites (e.g., *Plasmodium falciparum,* the causative agent of malaria).

apicoplast: An organelle found in apicomplexan species that is involved in metabolic processes. It is derived from photosynthetic chloroplasts even though it is not now involved in photosynthesis.

archaea: One of the three domains of life. These are species that do not have nuclei and thus were originally grouped with bacteria into the "prokaryotes." They were identified as a separate domain by analysis of rRNA sequences.

archaic human: A hominin that does not have the modern anatomy of *Homo sapiens*. Contrast with anatomically modern human.

argument from design: The argument that the order seen in the living world implies that it was created by a divine power.

arthropod: Member of Arthropoda, a major phylum within the ecdysozoa. This group, which is characterized by jointed legs, includes the insects, centipedes, millipedes, spiders, and crustaceans.

association study: A survey of associations between genetic markers and quantitative traits (including human disease) that aims to locate the QTLs that cause trait variation.

ATP: Adenosine triphosphate.

autocatalytic network: A chemical system that outputs a chemical that is a catalyst for the original reaction or that leads to other reactions that eventually output a catalyst for the original reaction.

autoinducer: A chemical used in quorum sensing that is secreted by cells and then used to quantify cell density.

autoinduction: The induction of a regulatory cascade in quorum sensing. It is triggered in response to the crossing of a threshold concentration of autoinducer in the environment.

automatic selection: The increase in frequency of an allele that increases the rate of selfing. It is closely related to the twofold cost of genome dilution.

autopolyploid: A polyploid that carries multiple genomes derived from within the same population.

autosome: A chromosome that is inherited in the usual Mendelian way, in contrast to sex chromosomes and mtDNA.

average effect: The effect of an allele on a quantitative trait as estimated by regression. For a random-mating population that is in linkage equilibrium, this is equivalent to the average excess.

average excess: The difference between the average trait value of individuals who carry a particular allele and the average of the population. For a random-mating population that is in linkage equilibrium, this is equivalent to the average effect.

B chromosome: An extra chromosome that is not required for normal function in either sex and is present in only some individuals.

backcross: A cross between a hybrid individual and one of the parental genotypes.

background selection: The reduction in genetic diversity caused by selection against deleterious alleles at linked loci.

bacteria: One of the three domains of life. These are species that do not have nuclei and thus were originally grouped with archaea into the "prokaryotes."

bacteriocin: A toxin produced by a bacterium that kills its competitors.

bacteriophage: A virus that infects bacteria. The term is commonly used to refer to viruses that infect archaea as well.

balanced polymorphism: A stable polymorphism maintained by balancing selection.

balancer chromosome: A *Drosophila* chromosome that carries multiple inversions, recessive lethals, and dominant markers. Such chromosomes cannot recombine and cannot become homozygous and so can be used to keep individual wild-type chromosomes intact.

balance view: The view that genetic variation is mostly maintained by balancing selection. This contrasts with the classical view that most variation is due to deleterious mutations.

balancing selection: Selection that maintains polymorphism.

Batesian mimicry: A palatable mimic evolves to resemble a distasteful model species and thereby suffers less predation.

benthic: Living at the bottom of a body of water.

β: *See* selection gradient.

β-sheet: Common structural motif of proteins in which linear amino acid sequences ("strands") located in different regions of the polypeptide chain align adjacent to each other and are stabilized by hydrogen bonding between backbone atoms located in different strands.

bilaterian: Member of the group that includes the majority of animal phyla and includes all animals with bilateral (left/right) symmetry.

binomial distribution: If n genes are sampled from a population in which the P allele has frequency p, then the chance of finding j P alleles is

$$\frac{n!}{j!(n-j)!} \; p^j (1-p)^{n-j}.$$

binomial nomenclature: Naming scheme for species in which there is a genus name and a species name.

biological species concept (BSC): Definition of species as groups of individuals that can successfully interbreed with each other in nature but that are reproductively isolated from other such groups.

biometry: The application of statistical methods to biology.

body plan: An organism's overall morphology that is created by the reproducible spatial positioning of differentiated cell types.

bootstrap: A statistical method for measuring consistency in datasets in which new simulated datasets are generated by sampling with replacement. In phylogenetic inference, bootstrap values represent the percentage of times that a particular clade is seen in trees generated with different simulated datasets.

bottleneck: *See* population bottleneck.

brachiopod: Member of the phylum Brachiopoda within the lophotrochozoa, composed of a group of marine animals that superficially resemble clams but are only distantly related to the mollusks.

branch: Portion of an evolutionary tree diagram connecting two nodes.

branch length: The length of a particular branch in an evolutionary tree. In certain types of evolutionary trees, the branch length is used to represent the amount of evolutionary change or time. A scale bar is used to represent the units of measurement.

breeding value: The sum of the average effect of each gene. If an individual is mated at random to others in the population, its offspring will deviate from the population average by one-half the breeding value.

broad-sense heritability (H^2): *See* heritability.

bryozoan: Member of a phylum (Ectoprocta or Bryozoa) of sessile colonial animals, commonly referred to as sea mats or moss animals, that are superficially similar to corals but are instead members of the Lophotrochozoa.

BSC: *See* biological species concept.

C-value: *See* genome size.

canalization: The buffering of development such that the same form is produced despite genetic and environmental perturbations.

candidate gene: A gene that is thought likely to influence the trait of interest, usually because major mutations at the gene affect the trait.

canonical code: The genetic code that is used almost universally.

capsid: A protein casing that makes up the outside of a virus particle.

case-control study: A form of association study in which "case" individuals with, for example, a disease are compared with "control" individuals without it.

catalysis: The facilitation of a chemical reaction by a molecule that is not itself altered by the reaction.

cDNA: *See* complementary DNA.

centiMorgan (cM): A distance on the genetic map that corresponds to a 1% recombination rate. *See* Morgan.

Central Dogma: Information can pass from nucleic acid to protein but not in the opposite direction.

centromere: The region of chromosome that attaches to the spindle at mitosis and meiosis.

chaetognath: Any worm of the phylum Chaetognatha, commonly called an arrowworm. This is a transparent marine worm with horizontal lateral and caudal fins and a row of movable curved spines at each side of the mouth.

chaperone: A protein that assists other proteins in achieving a properly folded state.

character state reconstruction: A method used to infer ancestral and derived character states and traits.

chemical evolution: Chemical reactions that could have generated complex compounds from simple ones prior to the origin of life.

chemostat: A device that allows populations of microorganisms to be maintained in a steady state.

chert: Very fine grained silica (SiO_2) that forms layers or nodules in sequences of sedimentary rocks.

chiasma: The cross-like structure formed by crossing over during meiosis.

chlorophyte: Member of a phylum of eukaryotes that are all single-celled green algae and closely related to green plants.

chloroplast: A photosynthetic organelle found in many plants, algae, and other microbial eukaryotes that is evolutionarily derived from cyanobacteria.

choanoflagellate: Member of a phylum of eukaryotes including single-celled flagellated species. This is the sister phylum to animals.

chordate: Member of a major phylum (Chordata) within the deuterostomes, which includes the vertebrates and closely allied invertebrates such as tunicates and amphioxus. All chordates contain a solid rod, called a notochord, along the length of the body during embryogenesis and a dorsal nerve cord and pharyngeal pouches.

chromatid: One of the two copies of a chromosome after it has been replicated.

chromatin: A compact structure of DNA and protein found in eukaryotic nuclei.

chromosomal fusion: *See* fusion, chromosomal.

chromosomal inversion: *See* inversion, chromosomal.

ciliate: Member of a phylum of eukaryotes including single-celled species. Most are coated on the outside with cilia, which are used for movement and cellular functions. They are in the kingdom Alveolata.

***cis*-regulation:** Refers to the regulation of when and where a gene is transcribed by DNA sequences that lie to either side (5′ or 3′) of the gene or within the introns of the gene.

clade: A group of species or genes that includes all descendants of an ancestral species or gene. *See* monophyletic.

cladistics: A method of classification that is based on the order of branching in a phylogenetic tree rather than on phenotypic similarity.

cladogram: A phylogenetic tree in which the only information given is about the relationships among taxa (i.e., the length of the branches is not meaningful).

classical view: The view that genetic variation is mostly due to deleterious mutations. This contrasts with the balance view that most variation is maintained by balancing selection.

cline: A smooth change from place to place across a spatially continuous habitat. The term refers to a spatial gradient in any measurable characteristic—for example, allele frequency or the mean of a quantitative trait.

clonal interference: In an asexual population, different clones, each favored by selection, compete with each other so that only one can succeed.

clone: A set of genetically identical individuals. In genetic engineering, a line of microbes that carry a particular sequence from another species.

cM: *See* centiMorgan.

CMS: *See* cytoplasmic male sterility.

cnidarian: Member of a major animal phylum (Cnidaria) that includes corals, sea anemones, hydra, and jellyfish. The group is characterized by the presence of stinging cells called nematocysts.

coalescence: The merging of two genetic lineages into a single common ancestor.

coalescence time: The time back to when two genes share a common ancestor.

coalescent process: A model in which as one moves back in time, each pair of lineages coalesces at a rate $1/2N_e$.

coancestry: A measure of the relatedness of two individuals. It is the chance that two randomly chosen genes, one from each individual, are identical by descent.

codon: Three bases that code for a single amino acid.

codon usage: The frequency with which each of the alternative codons that code for an amino acid is used.

codon usage bias: A bias toward use of one of several alternative codons that code for the same amino acid.

coefficient of kinship: *See* coancestry.

coevolution: The joint evolution of two species, with each responding to selection imposed by the other.

colloid: A substance that contains components in different phases (e.g., minute solid particles within a liquid).

compartmentalization: Subdivision of molecules, cells, or genetic functions into discrete spatial or temporal units. For example, this can refer to the grouping together of a protein into distinct parts of the cell or of networks of genetic interactions into distinct functional units.

competence: The uptake of DNA directly from the environment.

competitive exclusion: Species that use exactly the same resources cannot coexist in a stable equilibrium.

complementary DNA (cDNA): DNA that is complementary to messenger RNA. By producing cDNA using reverse transcription from mRNA, actively expressed genes can be identified.

complementation test: A test for determining whether two mutations are in the same gene (they do not complement) or different genes (they complement).

component of fitness: *See* fitness component.

component of variance: *See* variance component.

concerted evolution: The evolution of repeated sequences, which tend to remain homogeneous because of processes such as unequal crossing over and gene conversion.

confidence interval: The range of parameter values that do not deviate significantly from a null hypothesis.

conjugation: The transfer through a pilus of DNA from one bacterium or archaeon to another.

conservative DNA transposons: A DNA-based transposable element that moves itself to a new place in the genome but does not leave a copy in the original location.

constitutive: Referes to a gene that is always expressed.

contingency loci: Loci made up of microsatellite repeats (e.g., ATATATAT) in which, when the number of copies of the repeat changes, the phenotype of the cell changes drastically. These are common in the genomes of some bacterial pathogens.

convergence (also convergent evolution): The process by which features with no common ancestry become similar as a result of selection.

correlation coefficient: The most commonly used measure of correlation between two variables (x, y). It was devised by Karl Pearson and is defined as the ratio between the covariance and the square roots of the variances:

$$\mathrm{cov}(x, y)/\sqrt{\mathrm{var}(x)\mathrm{var}(y)}.$$

correlation, genetic: *See* genetic correlation.

cosexual: Producing both male and female gametes. It is equivalent to hermaphrodite.

cost of genome dilution: The disadvantage to a female who reproduces sexually and therefore devotes resources to propagating her mate's genes.

cost of meiosis: *See* cost of genome dilution.

cost of natural selection: *See* substitution load.

covariance: A measure of association between two variables (x, y). It is defined as the expected product of their deviations from the mean:

$$\mathrm{cov}(x, y) = \mathrm{E}[(x - \bar{x})(y - \bar{y})].$$

Note that $\mathrm{cov}(x, x) = \mathrm{var}(x)$.

covariance matrix: An $n \times n$ matrix giving the covariances between a set of n variables. The diagonal gives the covariances of each variable with itself (i.e., the variances).

covirus: One of a pair of viruses that have complementary functions and that must coinfect a cell for successful viral transmission.

cristae: Infolded internal membranes, such as that seen in mitochondria and plastids.

crossover: A recombination event within a chromosome at meiosis.

crown group: That part of a clade of living and fossil organisms that includes the last common ancestor of all the living forms and all of its descendants. *See* stem group.

ctenophore: Member of a major animal phylum of solitary gelatinous marine animals commonly called comb jellies or sea gooseberries (phylum Ctenophora). Members of this group have rows of small combs composed of cilia that are used for locomotion.

cultural evolution: Change in culture (i.e., information passed on by learning and imitation rather than by biological inheritance).

culturing: The growth of a particular microorganism in the laboratory in isolation from other organisms.

cyanobacteria: One of the major phyla of bacteria. Most species are photosynthetic, and chloroplasts are derived from this group.

cytology: Study of cells.

cytoplasmic male sterility (CMS): Loss of male function due to a cytoplasmically inherited factor in flowering plants.

cytoskeleton: The system of protein filaments in the cytoplasm of eukaryotic cells that gives the cell its shape and its capacity for directed movement and that participates in the directed transport of molecules within a cell.

D: This symbol is used in several ways. *See* genetic distance and linkage disequilibrium. D_n and D_s are also used as alternative notations for K_a and K_s.

Darwin (d): A unit for the rate of change of morphology. It was introduced by J.B.S. Haldane and is equal to a change by a factor e = 2.718... per million years or 1 part per million per year.

Darwinian Demon: A hypothetical organism with indefinitely high survival and fertility. This is an idea used to emphasize that all real organisms must face constraints on their fitness.

Darwinian medicine: The application of evolutionary principles to medicine.

defective interfering virus (DIV): A virus that has lost some function and that depends on coinfection with intact virus for transmission.

deism: The view that God works through fixed laws of nature; that is, that events occur without supernatural intervention.

deme: A discrete local population.

dendrogram: A phylogenetic tree in which the branch lengths are constrained to all be equidistant from the root; also known as an ultrameric tree.

density-dependent selection: Selection that occurs when relative fitness depends on population density.

derived characteristic: A trait found in an organism that was not present in the common ancestor of a group of organisms being studied.

descent with modification: Darwin's term for evolution.

deuterostome: Member of one of the two large groups of bilaterian animals including the echinoderms, hemichordates, and chordates. In this group, the initial embryonic opening becomes the anus. *See* protostome.

developmental constraints: Limits on what kinds of organism can develop.

diapause: A resting stage that allows organisms to survive harsh conditions.

differentiation: The process by which a cell becomes more and more specialized in its function and morphology through the regulation of gene expression and biochemical activities.

diffusion: Spreading due to the cumulative effect of small random movements.

diffusion approximation: A mathematical approximation that describes diffusion using a differential equation. It describes the spread of chemical concentration or allele frequency through physical space and also the spread of a probability distribution through the space of allele or genotype frequencies.

dimer: Two molecules that are bound together. Similarly, trimers and tetramers are clusters of three and four molecules, respectively.

dinoflagellate: Member of a phylum of eukaryotes in the Alveolate kingdom. All are single-celled organisms and many are photosynthetic.

diploid: Carrying two copies of each chromosome.

direct selection: *See* selection, direct.

directional selection: Selection favoring one allele over another or favoring increased values of a quantitative trait. It is equivalent to positive selection.

discicristate: A kingdom of eukaryotes. It includes the kinetoplastids (e.g., trypanosomes) and euglenoids.

disruptive selection: Selection favoring extreme values of a trait.

distyly: A polymorphism with two different arrangements of anther and stigma that promotes outcrossing.

DIV: *See* defective interfering virus.

divergence: The acquisition of differences after evolutionary separation (e.g., of species).

dizygotic twins: Twins formed from separate zygotes and therefore related in the same way as siblings.

DNA ligation: Chemical connection of two DNA strands. It is used in DNA repair, replication, and other molecular processes.

Dobzhansky–Muller model: A simple model for the evolution of reproductive isolation in which two populations accumulate different alleles. These alleles cause no loss of fitness within the genetic background in which they arose, but derived alleles from different populations are incompatible with each other.

dominance: If the heterozygote is precisely intermediate between the two homozygotes, there is no dominance. Any deviation from this additive model is described as dominance.

dominance deviation: The difference between the trait value of a genotype and the value expected with no dominance.

dominance theory: An explanation for Haldane's rule, which assumes that F_1 sterility or inviability is caused by recessive alleles.

dominance variance (V_D): The variance in a quantitative trait that is caused by dominance. It is defined as the variance of dominance deviations.

dominant: An allele is completely dominant with respect to a certain phenotype if it produces that phenotype when present in either one or two copies.

dominant lethal: *See* lethal.

dosage compensation: A mechanism that ensures that sex-linked genes are expressed at the appropriate level in both males and females.

drift: *See* random genetic drift.

ecdysozoan: Member of a major subdivision within the protostomes that includes the arthropods, nematodes, and several smaller phyla. Members of this group possess an external covering called a cuticle that molts as the animal grows.

echinoderm: Member of a major phylum within the deuterostomes that includes sea urchins, starfish, crinoids, and sea cucumbers. Although they possess bilateral symmetry initially, adults usually show a pentaradial symmetry.

ecotype: A genotype adapted to a particular environment.

ectoderm: One of three cell layers found in bilaterian embryos (the other two being the endoderm and mesoderm). The ectoderm goes on to form structures such as the external skin, mouth, and the nervous system.

ectopic recombination: Recombination between repetitive DNA elements found in different regions of the genome (e.g., between transposable elements at different sites). This can lead to chromosome rearrangements and deletions.

EEA: *See* environment of evolutionary adaptation.

effectively neutral: An allele that alters fitness by less than approximately $1/N_e$. Its probability of fixation is approximately the same as if it were absolutely neutral.

effective population size (N_e): The size of the ideal Wright–Fisher population that would give the same rate of random drift as the actual population in question.

electrophoresis: A technique in which molecules are pulled through a porous medium by an electric field and so are separated according to their charge and mobility.

endocytosis: The engulfment of material found outside a cell by surrounding it with the cell membrane.

endoderm: One of three cell layers found in bilaterian embryos (the other two being the ectoderm and mesoderm). The endoderm goes on to form structures such as the lining of the digestive system and portions of organs such as the liver, lung, and pancreas.

endomembrane system: Series of intracellular membrane compartments found in eukaryotic cells.

endomitosis: Mitosis with no cell division, which leads to a doubling of ploidy.

endoplasmic reticulum: Eukaryotic membrane compartment involved in translation, folding, and transport of proteins.

endosymbiosis: A symbiosis in which one organism lives within cells of another.

entropy: A quantitative measure of disorder. The Second Law of Thermodynamics states that the entropy of a closed system can never decrease.

environmental deviation: The difference E between the expected trait value of a given genotype and its actual value.

environmental genomics: Large-scale sequencing of DNA isolated directly from environmental samples (e.g., soil, air, and water).

environmental variance (V_E): The variance of the environmental deviation, $\text{var}(E) = V_E$.

environmental variation: Variation between genetically identical individuals.

environment of evolutionary adaptation (EEA): In evolutionary psychology, the Pleistocene environment in which adaptive human traits evolved.

epigenesis: The development of an organism from a zygote through cell differentiation and formation of morphology. The term also refers to the theory that organisms develop in this way rather than by preformation.

epistasis: Interaction between alleles in their effect on a trait. If a quantitative trait is given by adding up contributions from different loci, then we say that there is no epistasis.

ESS: *See* evolutionarily stable strategy.

euchromatin: The part of the eukaryote genome that is not condensed and that contains most active genes. It contrasts with heterochromatin.

eugenics: The belief that the human gene pool can be improved by selective breeding.

euglenoid: Member of a class of eukaryotic microorganisms bearing flagella. Most members live in freshwater and many possess chloroplasts.

eukaryote: One of the three domains of life. Its species are characterized by the presence of a nucleus.

eusocial: Fully social organisms in which only one or a few individuals in a colony reproduce.

eutherian mammal: A mammal having a placenta. This includes all mammals except monotremes and marsupials.

evolutionarily stable strategy (ESS): A strategy that cannot be displaced by any alternative. Individuals that play the ESS must be at least as fit as any feasible alternative when the rest of the population is playing the ESS.

evolutionary character state reconstruction: *See* character state reconstruction.

evolutionary computation: Using evolutionary processes—especially natural selection—to solve computational problems. *See* genetic programming, genetic algorithm.

evolutionary ecology: Study of the evolutionary consequences of interactions between species and between a species and its environment.

evolutionary game: An interaction between individuals in which the payoff depends on the strategy played by each of them.

evolutionary psychology: A field that applies evolutionary principles to understand universal human traits. It is usually assumed that humans are adapted to their past "environment of evolutionary adaptation."

Evolutionary Synthesis: The synthesis during the 1930s and 1940s of population genetics with other fields of biology (e.g., paleontology, systematics, and botany).

evolvability: Ability to generate heritable variation that can be exploited by selection.

excavate: Member of a kingdom of eukaryotes. All are single-celled species and none are known to have mitochondria.

exon: A protein-coding region of a protein-coding gene.

exon shuffling: Recombination events that mix exons from two different genes.

expected heterozygosity: *See* gene diversity.

exploratory system: Systems that shape initially random variation so as to produce a well-coordinated functional outcome.

exponential: Growing at a constant rate r, so that numbers increase with e^{rt}.

extended phenotype: The phenotype of all the individuals affected by a gene.

extremophile: An organism that thrives in environments that are at the extremes of conditions where life is normally found.

fitness: The number of offspring left by an individual after one generation. The fitness of an allele is the average fitness of individuals that carry that allele.

fitness component: Traits, such as survival, mating success, and reproduction, that combine to determine fitness.

fitness landscape: Either a graph of fitness as a function of individual genotype or phenotype or of population mean fitness as a function of allele frequencies or trait means.

fitness, mean: The average fitness of a population.

fitness, relative: The fitness divided by the mean fitness or by the fitness of a reference genotype.

fixation probability: *See* probability of fixation.

fixed (also fix): When all copies of a gene carry the same allele, that allele is said to be fixed.

fluctuation test: An experimental method for measuring mutation rates and for determining whether mutations arise prior to or in response to selection.

founder effect: The loss of genetic variation and the consequent change in genotype frequencies that occurs when a small number of individuals found a new population. It is equivalent to a burst of random genetic drift.

frameshift mutation: An insertion or a deletion mutation that leads to a change in the reading frame in a protein-coding gene.

frequency-dependent selection: Selection that occurs when relative fitness depends on genotype frequencies.

frequency spectrum: The distribution of allele frequencies.

F_{ST}: A measure of the genetic diversity between subpopulations relative to the total:

$$F_{ST} = \mathrm{var}(p)/\overline{p}\,\overline{q}.$$

It was devised by Sewall Wright.

Fundamental Theorem (of Natural Selection): The increase in mean fitness due to selection on allele frequencies is equal to the additive genetic variance in fitness divided by the mean fitness:

$$\Delta \overline{W} = \mathrm{var}_A(W)/\overline{W}.$$

fusion, chromosomal: A mutation in which two chromosomes are connected at their ends. The product is a metacentric chromosome.

galls: Structures induced in a plant by a parasite (e.g., a bacterium or an insect) that nurture that parasite.

game: *See* evolutionary game.

game theory: *See* evolutionary game.

gametophyte: In plants, the haploid structure that produces gametes. It contrasts with sporophyte.

γ: *See* selection gradient, quadratic.

gas hydrate: Frozen deposits rich in hydrocarbons that occur in the deep ocean basins.

Gaussian distribution: *See* normal distribution.

gene: A stretch of DNA (or, in some viruses, RNA) sequence that codes for a protein or RNA molecule, together with associated regulatory elements.

genealogy: The tree-like ancestral relationship that connects a set of genes at a single genetic locus.

gene conversion: A meiotic process in which nonreciprocal exchange of genetic information occurs as a result of heteroduplex formation between non-sister chromatids. Thus, heterozygous sites are converted to sites homozygous for one or the other allele.

gene diversity (*H*): The probability that two randomly chosen genes will carry different alleles. It is equal to the heterozygosity for a randomly mating diploid population with no selection.

gene dosage: The number of copies of a gene within an individual.

gene flow: The movement of genes from place to place. The term usually refers to movement in space but can also refer to movement between microhabitats or to introgression between distinct populations or species.

gene therapy: A treatment in which functional genes are introduced into the patient.

genetic algorithm: An algorithm that applies selection, mutation, and recombination to a population of computer programs in order to solve computational problems.

genetic assimilation: The process in which a phenotype, normally expressed only in a specific environment, through selection shows increased expression in that environment, which may cause the phenotype to be expressed under normal conditions as well.

genetic background: The set of genes with which a gene of interest is associated. Recombination moves genes from one genetic background to another.

genetic code: The code that translates 64 possible triplet codons into amino acids and translation stop signals.

genetic correlation: A correlation between the breeding values for different traits. This may be due to linkage disequilibrium between genes affecting the different traits or to pleiotropy, in which alleles affect both traits.

genetic distance: A measure of the difference in allele frequencies between populations. The most widely used measure of genetic distance is Nei's *D*.

genetic drift: *See* random genetic drift.

genetic engineering: The manipulation of organisms by the artificial introduction of DNA sequence in order to change their characteristics.

genetic load: The loss of mean fitness relative to some ideal fitness:

$$L = (1 - \overline{W}/W_{max}).$$

It includes mutation, substitution, recombination, and segregation loads.

genetic locus: *See* locus.

genetic map: A map of the linear order of genes constructed by measuring the rates of recombination between them.

genetic marker: A polymorphic locus that is used to observe genetic variation but that is not itself of primary interest.

genetic programming: Generating computer programs by selection on a population of variant programs. It is similar to the use of genetic algorithms.

genetic recombination: *See* recombination.

genetic system: The system immediately responsible for transmission of genetic information.

genome size (C-value): The size of a single haploid genome. It is sometimes expressed as the mass of a single haploid genome measured in picograms ($1 \text{ pg} = 10^{-12}$ g) and corresponding to approximately 1000 Mb of DNA.

genomic imprinting: *See* imprinting, genomic.

genotype: The set of alleles carried by an individual.

genotypic value: The average trait value G of individuals with a particular genotype.

genotypic variance (V_G): The variance of the genotypic value: $\text{var}(G) = V_G$.

geometric mean: An average defined by the nth root of the product of n values:

$$\bar{x}_G = \left(\prod_{i=1}^{n} x_i \right)^{1/n}.$$

germ line: In a multicellular organism, the lineage of cells that will generate gametes via meiosis. It contrasts with soma.

glaucocystophyte: Member of a phylum of eukaryotes that contains single-celled photosynthetic species.

good genes: Genes that increase fitness for reasons that are not due to sexual selection on a signal trait. Such components of fitness are most conveniently thought of as viability, but fertility and mating success could also be included.

gracile: Graceful, slender, and delicate. The term is used to describe the large rear teeth of certain Australopithecenes. It contrasts with robust.

Gram stain: A stain that specifically detects the type of cell wall and membrane structure found in some bacterial species. These are known as "gram-positive" species.

great plate count anomaly: A phenomenon in which the number of cells from natural environments that can be grown in culture is much less than what can be seen through a microscope. This was one of the first indications of the difficulties in culturing many of the microbes found in the environment.

group selection: Selection of traits that increase survival and proliferation of groups of individuals.

gynodioecious: A population that contains both females and hermaphrodites.

H: *See* gene diversity.

H^2: *See* heritability.

h^2: *See* heritability.

Haldane's rule: When in the offspring of two different animal taxa, one sex is absent, rare, or sterile, that sex is the sex with heterozygous sex chromosomes. Defined by J.B.S. Haldane in 1922.

halophilic: Describes an organism with a preference for growth in high-salt environments.

haltere: Sense organs found in Diptera on the second thoracic segment, evolutionarily derived by modification of the wings. They are used to help balance during flight.

Hamilton's rule: An allele that increases the fitness of its neighbors by B but reduces the fitness of its carrier by C will increase if $rB > C$, where r is a measure of the genetic similarity of neighbors to the focal individual.

handicap: A trait that signals a male's genetic quality. Its association with good genes is maintained because it is less costly to males of higher quality.

haplodiploid: A system of sex determination in which fertilized eggs develop as diploid females and unfertilized eggs develop as haploid males.

haploid: Carrying one copy of each chromosome.

haplotype: A particular combination of alleles in a haploid—that is, a haploid genotype.

Hardy–Weinberg proportions: The frequencies of diploid genotypes produced after random mating. In the simplest case of two alleles at frequencies $q + p = 1$, the three genotypes QQ, PQ, PP are at frequencies $q^2, 2pq, p^2$.

harmonic mean: An average defined by

$$\bar{x}_H = 1/\left(\frac{1}{n} \sum_{i=1}^{n} \frac{1}{x_i} \right).$$

It gives greatest weight to small values. The effective population size is the harmonic mean of the population sizes in each generation.

hemichordate: Member of a diverse phylum of marine animals including the acorn worms and pterobranchs (phylum Hemichordata).

hemimetabolous: Developing directly through a series of nymphal stages with a similar morphology to the adult (e.g., as in grasshoppers and bugs). It contrasts with holometabolous.

heritability: The fraction of phenotypic variance that is inherited. The broad-sense heritability refers to the total genotypic variance ($H^2 = V_G/V_P$), whereas the narrow-sense heritability refers to the additive genetic variance ($h^2 = V_A/V_P$).

hermaphrodite: An individual that produces both male and female gametes.

heterochrony: The change in the relative timing or duration of events during development achieved by altering the relative onset or ending of particular developmental processes. (24)

heterogametic: The sex that carries distinct sex chromosomes. For example, in mammals, males are the heterogametic sex.

heterokont: One of the kingdoms of eukaryotes. It includes a diverse collection of single-celled species including diatoms.

heterosis: The increase in fitness seen in a cross between different populations.

heterostyly: A polymorphism for distinct arrangements of anther and stigma.

heterozygosity: The proportion of heterozygotes in a population.

heterozygote: A diploid individual that carries two different alleles at a locus.

heterozygote advantage: *See* overdominance.

hierarchical classification: A system of classifying organisms in a nested series of levels—domain, kingdom, phylum, class, order, family, genus, species.

Hill–Robertson effect: The interference between selection at linked loci. It was first analyzed by Hill and Robertson in 1966.

hitchhiking: The increase in a neutral allele that happens to be associated with a selectively favorable allele at another locus. It is sometimes used more broadly to refer to any kind of indirect selection.

HIV: *See* human immunodeficiency virus.

HLA: Human leukocyte antigen. *See* major histocompatibility complex.

Holliday junction: The cross-like structure formed by crossing over between two DNA double helices.

holometabolous: Metamorphosis through a pupal stage (as in flies, butterflies, and beetles). It contrasts with hemimetabolous.

homeobox: A sequence, approximately 180 nucleotides long, that is translated into a DNA-binding domain called the homeodomain. This sequence is found in many transcription factors that play a role in pattern formation and cell differentiation.

homeodomain: A sequence, approximately 60 amino acids long, that is encoded by a homeobox DNA sequence. This protein motif forms a DNA-binding domain that is found in many transcription factors that play a role in pattern formation and cell differentiation.

homeotic: Describes a class of mutations that transforms one part of an organism into another part. For example, a particular homeotic mutation in *Drosophila* transforms the antennae into legs.

hominid: Member of the great apes (family Hominidae), which now include human, gorilla, orangutan, and chimpanzee.

hominin: All taxa closer to humans than to chimpanzee. Apart from ourselves, all these taxa are now extinct.

homologous recombination: The process by which two pieces of DNA, identical or nearly identical in sequence (e.g., two copies of a chromosome), align and exchange a portion of DNA.

homology: A similarity attributable to its derivation from the same ancestral feature. In genetic studies, it refers to genes that are present in the same locus in the genome. In phenotypic studies, it refers to character traits or states that are present in a set of species and their common ancestor.

homoplasy: A similarity of traits that is not due to homology but instead to convergence or parallel evolution.

homozygote: A diploid individual that carries the same allele at a genetic locus.

homunculus: A "little man" that was supposedly introduced into a fertilized egg by the sperm and that guided its development. *See* preformation.

horizontal transmission: Transmission of genetic information between different individuals other than from parent to offspring.

host races: Genetically distinct populations that specialize on different hosts.

hot spots, recombination: *See* recombination hot spots.

human immundeficiency virus (HIV): The virus responsible for AIDS.

hybrid rescue allele: An allele that alleviates hybrid sterility or inviability.

hybrid zone: A narrow region in which genetically distinct populations meet, mate, and hybridize.

hydrogenosome: An organelle of some eukaryotes that produces hydrogen gas and ATP. It is possibly derived from mitochondria.

hydrophilic: A molecule or portion of a molecule that readily dissolves in water via the formation of hydrogen bonds.

hydrophobic: A molecule or portion of a molecule that does not readily dissolve in water.

hydrophobic core: A portion of a protein that avoids dissolution in water and is composed of a set of hydrophobic amino acids.

hypercycle: Cooperation between a set of replicating molecules (A, B, ..., Z), in which A aids replication of B, B aids C, ..., and Z aids A. It was proposed as a way for early life to replicate genetic information despite high mutation rates.

hyperthermophile: An organism that thrives at temperatures above 80°C.

IBD: *See* identity by descent.

identical twins: Twins formed from a single zygote and therefore genetically identical. *See* monozygotic twins.

identity by descent (IBD): Genes that are inherited from the same gene in an ancestral population are identical by descent.

immune memory: The increased response of the immune system to an antigen that had been encountered before. This is the basis of vaccination.

immunoglobulin: A family of proteins involved in the immune system. It includes antibodies.

immunological distance: A measurement of phylogenetic relationship based on differences in antibody–antigen binding affinity between species. The magnitude of the difference is used as a rough evolutionary clock.

imprinting, genomic: A phenomenon in which the phenotype of a heterozygote depends on which allele came from the father and which from the mother.

inbred: Produced by mating between relatives.

inbred line: A population produced by continued self-fertilization or mating between close relatives. After several generations of inbreeding, an inbred line becomes genetically homogeneous.

inbreeding: Mating between relatives.

inbreeding coefficient: The chance that two homologous genes in a diploid individual are identical by descent.

inbreeding depression: Reduced vigor of inbred individuals.

inclusive fitness: A measure of individual fitness that includes the effects of that individual on its neighbors' fitness and discounts the effects of neighbors on the individual's own fitness.

incomplete dominance: A condition in which heterozygotes express a trait that is distinguishable from both homozygotes.

indel: An insertion or a deletion mutation involving a small number of bases.

indirect selection: Selection on a trait that arises from its association with other traits rather than because of its direct causal effect on fitness. The term may also refer to selection that arises from the effects of a gene in one individual on the fitness of other individuals.

induction: Increase in gene expression in response to a regulatory signal.

infinite-alleles model: A model that assumes that every mutation generates a new allele.

infinitesimal model: A model that assumes that quantitative trait variation is caused by a very large number of loci, with infinitesimally small additive effects. Selection alters the mean, but the genetic variance remains constant. It is equivalent to assuming that offspring are normally distributed around the mid-parental value, with constant variance.

infinite-sites model: A model that assumes a very large number of sites, each with a very low mutation rate, so that every mutation is unique.

informational genes: Genes involved in core "informational" processes including DNA replication and repair, transcription, and translation. They are thought to be less prone to lateral gene transfer than operational genes.

inheritance of acquired characteristics: Transmission of characteristics acquired during an organism's lifetime to its offspring. This view is associated with Lamarck, but it was widely held until the establishment of genetics.

innovation: A change to a preexisting feature.

insertion sequence (IS): A class of transposable elements found in bacteria and archaea.

in situ hybridization: A labeled DNA or RNA probe is hybridized to a tissue section or whole embryo and viewed under the microscope to determine when and where a specific mRNA is expressed. Alternatively, the probe is hybridized to a chromosome spread and viewed under the microscope to determine the location of a specific genomic sequence.

intelligent design: The argument that organisms are irreducibly complex and so must have been created by an intelligent designer. It is a version of the argument from design.

interaction variance: The variance in a quantitative trait, V_I, caused by epistatic interactions between loci. It is the difference between the genotypic and the additive plus dominance variance and is $V_I = V_G - V_A - V_D$.

intercalating agent: A chemical that resembles DNA bases and can insert into DNA backbones during replication, leading to insertion or deletion mistakes in replication.

introgression: Movement of genes from one genetic background to another, as a result of hybridization between individuals from distinct populations.

intron: A noncoding sequence that interrupts the coding sequence.

invention: A fundamentally new feature of an organism.

inversion, chromosomal: A mutation in which a section of DNA sequence in the offspring is inverted relative to its orientation in the parent. The process involves excision and reinsertion of the DNA.

irreducibly complex: A system that cannot function if any one of its components is missing.

IS: *See* insertion sequence.

island model: The simplest model of population structure. A fraction *m* of the genes in a deme comes from outside. Immigration may be from the mainland population or from other demes.

isogamous: Producing a single kind of gamete.

isolation by distance: Divergence between allele frequencies in different places within a spatially continuous population. The term usually refers to divergence caused by random genetic drift.

isotope: Forms of an element that differ in atomic mass.

isozymes: Enzymes with different amino acid sequences that catalyze the same reaction. This includes variants coded by different genetic loci as well as allozymes, which are coded by homologous genes at the same locus.

junk DNA: Sequences that accumulate by mutation and that are neutral or deleterious.

just-so stories: Untestable explanations for adaptations.

K_a: The rate of nonsynonymous substitutions, which alter amino acid sequence. This is sometimes denoted D_n.

kerogen: A class of organic compounds found in some sedimentary rocks and thought to be derived from organic molecules from living organisms.

kinase: An enzyme that adds phosphate groups onto other molecules.

kin discrimination: The ability to distinguish between related and unrelated individuals.

kinetoplast: An independently replicating organelle lying near the base of the flagellum in kinetoplastids.

kinetoplastid: Member of a phylum of eukaryotes in the Excavata kingdom characterized by the presence of a kinetoplast organelle.

kinorhynch: Member of a phylum of tiny spiny animals (phylum Kinorhyncha).

kin selection: A change in the frequency of an allele, caused by the effect of that allele on the fitness of other individuals who carry the allele.

K_s: The rate of synonymous substitutions that do not alter amino acid sequence. It is sometimes denoted D_s.

labyrinthulid: Member of a phylum of eukaryotes in the heterokont kingdom.

lagging strand: During DNA replication, the strand that is synthesized in the 3′ to 5′ direction by ligation of short DNA strands synthesized discontinuously in the 5′ to 3′ direction.

Lamarckism: *See* inheritance of acquired characteristics.

large X effect: The observation that the X chromosome is involved in hybrid incompatibilities more often than would be expected from its size.

last universal common ancestor (LUCA): The common ancestor of all modern life-forms.

lateral gene transfer (LGT): The transmission of DNA from one evolutionary lineage to another. Also known as horizontal gene transfer.

leading strand: During DNA replication, the strand that is synthesized in the 5′ to 3′ direction by continuous polymerization at the growing 3′ tip.

lek: An arena where males gather and are chosen as mates by females.

lethal: A recessive lethal allele kills its bearer when homozygous, whereas a dominant lethal allele kills when present in a single copy.

Levene model: A model of a structured population in which individuals from a single gene pool spend part of their lives competing within small patches.

LGT: *See* lateral gene transfer.

life history: An organism's pattern of survival and reproduction.

life-history trait: Traits, such as mortality rate, fertility, or age of reproduction, that are closely associated with fitness.

likelihood: Given a hypothesis, the probability of observing certain data.

limnetic: Occurring in the deeper open water of lakes or ponds.

LINE: *See* long interspersed nucleotide element.

lineage sorting: The process by which, following separation of two species, the ancestry of every gene converges to the overall phylogeny of the species. This process takes about $2N_e$ generations, and incomplete lineage sorting implies discordance between genealogies.

linkage: Genes that are carried on the same chromosome are said to be linked.

linkage disequilibrium: Nonrandom associations between alleles at two or more genetic loci.

linkage equilibrium: Absence of linkage disequilibrium, so that haplotype frequencies are equal to the product of allele frequencies.

liposome: A spherical lipid bilayer.

LMC: *See* local mate competition.

load: *See* genetic load.

load, mutation: *See* mutation load.

load, recombination: *See* recombination load.

load, segregation: *See* segregation load.

load, substitution: *See* substitution load.

lobopod: The fossil members of a phylum of caterpillar-like animals. The living forms are terrestrial and commonly referred to as velvet worms. The fossils are marine and were diverse in the Cambrian (phylum Onychophora).

local mate competition (LMC): Competition for mates within a local group (e.g., between fig wasps within a single fig).

locus: A location on the genome. It may refer to a single nucleotide site or to a substantial stretch of DNA sequence.

long-branch attraction: A phenomenon in phylogenetic analyses when rapidly evolving lineages are inferred to be closely related, regardless of their true evolutionary relationships.

long interspersed nucleotide element (LINE): A class of transposable element.

lophotrochozoan: Member of a major subdivision within the protostomes that includes the annelids, mollusks, bryozoans, brachiopods, and several additional small phyla.

LUCA: *See* last universal common ancestor.

lycophyte: Member of a diverse group of early land plants including the lycopods and zosterophylls.

lycopod: Member of a group of plants that includes giant trees in the Carboniferous coal swamp forests and the living club mosses.

lycopsid: *See* lycopod.

lysogenic: Describes bacteria and archaea that have bacteriophage integrated into their genomes. The bacteriophage may be activated, leading to lysis of the cell.

macroevolution: Evolutionary change at or above the species level.

macronucleus: The larger of the two nuclei found in ciliate cells. It functions in ways similar to the somatic cells of animals.

major histocompatibility complex (MHC): A set of closely linked genes in vertebrates that play a key role in the immune response. In humans, it is known as the HLA (human leukocyte antigen) complex.

major transitions: Identified by Maynard Smith and Szathmáry as major changes in the way hereditary information is transmitted. Almost all involve the coming together of previously independent replicators to form more complex entities.

mating types: A polymorphism in which individuals can mate only with a different type. *See* self-incompatibility.

mean: Usually refers to the arithmetic mean: for n values, $z_1, ..., z_n, \bar{z} = (\Sigma_i z_i)/n$. *See also* the harmonic mean and the geometric mean.

mean fitness: *See* fitness, mean.

meiosis: A cellular division process that is involved in sexual reproduction in eukaryotes in which gametes are produced having half the number of copies of each chromosome as the parents.

meiotic drive: Preferential movement of a chromosome during meiosis toward the pole of the cell that will go on to produce gametes.

mesoderm: One of three cell layers found in bilaterian embryos (the other two being the ectoderm and endoderm). The mesoderm goes on to form structures such as the muscles and internal bones and portions of organs such as the kidney and reproductive system.

mesophile: An organism that prefers to live at moderate temperatures.

Mesoproterozoic: Division of time from 1600 to 1000 Mya.

messenger RNA (mRNA): The RNA molecule that is transcribed from the DNA and takes sequence information to the ribosome, where it is translated into protein.

messenger RNA processing: The collective term for the modifications to eukaryotic RNA that are necessary before the RNA can be transported to the cytoplasm for translation.

metagenomics: Large-scale sequencing of DNA isolated directly from environmental samples (e.g., soil, air, and water).

metapopulation: A collection of discrete demes. The first metapopulation models involved extinction and recolonization, but the term is now used more broadly.

metazoa: The group that includes all animal phyla, including sponges, ctenophores, cnidarians, and the bilaterians.

MHC: *See* major histocompatibility complex.

microarray: An array of short oligonucleotides, bound to a substrate, that can be used to simultaneously measure the concentration of large numbers of different DNA or RNA sequences.

microfilament: Minute fibers composed primarily of actin involved in the structural integrity and movement of eukaryotic cells. Together with microtubules these make up the cytoskeleton.

micronucleus: The smaller of the two nuclei found in ciliate cells. It functions in ways similar to the germ cells of animals.

microRNA: A family of RNA molecules, approximately 22 nucleotides long, that regulates the expression of some eukaryotic genes.

microsatellite: A short array of repeated sequences, each a few base pairs long. Microsatellites tend to be highly polymorphic and so are widely used as genetic markers.

microsporidia: A group of single-celled eukaryotes that were once considered to be their own phylum but are now considered part of the fungal phylum.

microtubule: A major component of the cytoskeleton, composed of the protein tubulin. It is used by eukaryotic cells to regulate their shape and control their movements.

migration: Movement from place to place. Here, it is used synonymously with gene flow.

mimicry: An adaptive resemblance between one organism and another. *See* Batesian mimicry and Müllerian mimicry.

minisatellite: Multiple copies of short sequences, from 9 base pairs up to several hundred base pairs. They are scattered over many sites, with 10–100 repeats, and are highly variable and hence useful as genetic markers.

mismatch distribution: The distribution of numbers of differences between random pairs of sequences sampled from a population.

mismatch repair: The process of repairing damage to DNA that has already been replicated.

missense mutation: A nucleotide substitution within a protein-coding region of a gene that leads to the replacement of one amino acid by a different amino acid.

mitochondria: The eukaryote organelles responsible for aerobic respiration. Mitochondria are derived from α-proteobacteria.

mitochondrial DNA (mtDNA): The genome contained within mitochondria. Animal mtDNA is widely used as a genetic marker, because it does not recombine.

mitosis: Cellular division process that is involved in asexual reproduction in eukaryotes in which each daughter cell gets a copy of the chromosomes of the parent.

mixed strategy: Where individuals play two or more stategies at random.

modifier: In the strict sense, an allele that has no direct effect on fitness but does have some other effect on the genetic system that may be indirectly selected. More broadly, it refers to an allele that alters the expression of alleles at other loci.

modularity: Subdivision into distinct parts or modules that can function independently. In evolutionary psychology, the term refers to independent mental functions.

molecular clock: The constant rate of accumulation of amino acid or DNA sequence differences.

molecular recognition: The binding of two molecules though noncovalent bonds in which the shape of the molecules plays a key role in the strength of binding.

molecular recombination: The physical cutting and joining of DNA molecules.

mollusks: A major phylum within the lophotrochozoa. This group of animals includes clams, mussels, chitons, octopus, squid, and nudibranchs.

monandrous: Where females mate with a single male.

monophyletic: Describes a set of taxa that all descend from a common ancestral taxon—that is, a group of organisms or genes that share a common ancestor to the exclusion of all other entities.

monosomy: The presence of only one chromosome instead of a pair in a cell's nucleus.

monozygotic twins: *See* identical twins.

Morgan: The unit of distance on the genetic map. Over short map distances, the distance between loci in Morgans equals the recombination rate between them. Over long distances, the recombination rate is lower than the map distance because of multiple cross-overs.

mRNA: *See* messenger RNA.

mtDNA: *See* mitochondrial DNA.

μ: *See* mutation rate.

Müllerian mimicry: A distasteful species evolves to resemble another distasteful species. Both gain increased protection, because predators learn to avoid the common pattern more quickly.

Muller's ratchet: The degeneration of an asexual population that arises from the random and irreversible loss of the fittest genotype.

multiple testing: If many significance tests are carried out, then some will reject the null hypothesis just by chance. Thus, significance levels must be reduced when multiple tests are performed.

multiregional model: The hypothesis that different hominins found across the Old World (such as *Homo erectus* and *Homo neanderthalensis*) evolved in situ into modern human populations. It contrasts with the out-of-Africa model.

mutation: A heritable change in the genetic material of an organism that does not involve reciprocal combination.

mutational heritability: The ratio, V_M/V_E, between the mutational variance and the environmental variance.

mutational variance (V_M): The variance in a quantitative trait caused by new mutations in each generation.

mutation load: The loss of mean fitness caused by deleterious mutations:

$$L = 1 - (\bar{W}/W_{max}),$$

where W_{max} is the fitness of the wild-type genotype that carries no mutations.

mutation rate (μ): The rate at which mutations are generated.

mutator: An allele that causes an increased mutation rate.

mutualism: An interaction between species from which all involved gain.

NADPH: Nicotine adenine dinucleotide phosphate (in its reduced form). It is used as an energy and redox carrier in all organisms.

narrow-sense heritability (h^2): *See* heritability.

naturalistic fallacy: The mistaken argument that what *is* justifies what *should be*—for example, that evolution, or the mechanism of evolution, justifies particular moralities.

natural selection: The process by which genotypes with higher fitness increase in frequency in a population.

natural theology: A theology based on reason and ordinary experience instead of on special or supposedly supernatural revelation.

N_e: *See* effective population size.

nearly isogenic line (NIL): A line produced by continued crossing of one line back to another, combined with selection. An NIL contains small segments of genome derived from one line, which are enriched for QTLs that influence the selected trait.

neighborhood size: Proportional to the product of the rate of gene flow, σ^2, and the effective population size per unit area (i.e., the effective density), ρ, which determines the relative rate of gene flow and random drift in a spatially continuous population. In two dimensions, it is defined as $Nb = 4\pi\rho\sigma^2$.

nematode: Member of a major phylum within the Ecdysozoa of very diverse and abundant worm-like animals, including the round worms and thread worms (phylum Nemata).

Neoproterozoic: Division of time from 1000 Mya to the base of the Cambrian at 542 Mya.

neoteny: Reproduction by juveniles. For example, axolotls reproduce without metamorphosis into adult salamanders.

neutral: Having no effect on fitness.

neutral evolution: Evolving without the influence of natural selection.

neutralist: One who holds that most molecular divergence and variation has negligible effect on fitness, at least at the time when it evolved.

neutral mutation: A mutation that does not affect fitness.

neutral theory: The theory that genetic variation is neutral and is shaped primarily by mutation and random genetic drift.

niche: The set of ecological environments in which a species can survive and reproduce.

NIL: *See* nearly isogenic line.

node: The point in a phylogenetic tree where one branch splits into two.

nonhomologous gene displacement: A lateral gene transfer event in which a gene that carries out a particular function is replaced by a nonhomologous gene that carries out a similar function.

non-Mendelian inheritance: Inheritance that does not follow Mendelian patterns. (Linkage and sex linkage, which were not known to Mendel, are not included.)

nonsense mutation: A point mutation in a protein-coding region that produces a stop codon, prematurely truncating the protein sequence.

nonsynonymous mutation: A point mutation in a protein-coding region that changes a codon such that it alters the resulting amino acid sequence of the protein.

normal distribution: The bell-shaped curve that describes the distribution of the sum of a large number of independent variables.

nucleolus: Subcompartment within the nucleus that is involved primarily in making ribosome components

nucleomorph: A highly reduced relic of a nucleus. It is found in many eukaryotic species that have undergone secondary symbioses in which one eukaryote became the endosymbiont of another eukaryote.

nucleotide: A nitrogenous base attached to a ribose or deoxyribose sugar and a phosphate molecule. It is the basic unit of a nucleic acid.

nucleotide diversity (π): The chance that two randomly chosen copies of a nucleotide site will carry different bases. It is similar to gene diversity and applies to single nucleotides.

nucleotide site: A particular nucleotide in the DNA or RNA sequence.

null allele: An allele of a gene in which function has been completely abolished. It is generally considered equivalent to the complete deletion of the gene.

null hypothesis: A hypothesis that is presumed true and against which alternative hypotheses are tested statistically.

null model: *See* null allele.

Occam's razor: The general principal that if all else is equal, the simplest explanation is best. It is applied when using parsimony methods during phylogenetic reconstruction.

Okazaki fragments: Small fragments of DNA made during DNA replication of the lagging strand.

oligonucleotide: A short piece of DNA, no more than about 20 base pairs long.

onychophoran: Member of a phylum of caterpillar-like animals. The living forms are terrestrial and are commonly referred to as velvet worms; the fossils are marine and were diverse in the Cambrian (phylum Onychophora).

open reading frame (ORF): A section of a genome that contains the codons used to make a protein. In bacteria and archaea open reading frames are usually found in contiguous stretches in the genome; in eukaryotes they are frequently interrupted by introns.

operational gene: Genes involved in metabolic and other peripheral processes. They are meant to contrast with informational genes and are thought to be more prone to lateral gene transfer than informational genes.

operational taxonomic unit (OTU): Any entity (e.g., an organism, gene, species, or population) used for phylogenetic study.

operon: A set of adjacent genes whose transcription is regulated as a single unit.

opisthokont: Member of a kingdom of eukaryotes including the metazoa, fungi, and choanoflagellates.

ORF: *See* open reading frame.

orthogenesis: An inherent tendency for lineages to change in a particular direction.

orthologous: *See* orthologous genes.

orthologous genes: Genes that are homologous (share a common ancestry) and have diverged from each other due to the separation of the species in which the genes are found (e.g., α-globin from humans and mice). Contrast this with paralogous genes.

outcrossing: Mating with unrelated individuals.

outgroup: An organism or gene from an evolutionary lineage that separated from those lineages being studied prior to the existence of their common ancestor.

out-of-Africa model: The hypothesis that modern humans evolved recently in Africa and spread from there, replacing archaic hominins. It contrasts with the multiregional model.

overdominance: Describes heterozygotes that have higher trait values (usually higher fitness) than either homozygote.

ovule: The female gamete in a flowering plant. It is analogous to the egg in animals.

Paleoproterozoic: Division of time from 2500 to 1600 Mya.

panmictic: Describes a population in which every individual has the same chance of mating with every other: in other words, where there is no population structure.

paracentric inversion: A mutation that involves a chromosomal inversion that does not span the centromere.

parallel evolution: The process by which features that once were different become similar by experiencing the same changes in different evolutionary lineages.

paralogous genes: Genes that are homologous (share a common ancestry) and have diverged from each other after gene duplication events (e.g., α- and β-globins). Contrast this with orthologous genes.

parapatric: A geographic distribution in which different types are found in different places and meet only in a narrow zone.

parapatric speciation: The evolution of new species within a spatially extended population in the presence of gene flow.

paraphyletic: Describes a group of organisms or genes that share a common ancestor to the exclusion of all other entities but in which some members of the group are excluded.

parsimony: General approach to evolutionary reconstructions in which the goal is to identify theories (e.g., evolutionary branching patterns) that require the fewest number of evolutionary events (and thus might be considered the simplest).

parthenogenesis: The production of offspring from unfertilized eggs.

pathogenicity islands: Contiguous sections of a pathogen's genome that contain a disproportionate (relative to number of base pairs) number of the factors that cause pathogenicity.

patrilineal: Inherited from the father (e.g., the Y chromosome in mammals).

payoff: In an evolutionary context, the increase in fitness due to a contest.

payoff matrix: A matrix showing the payoff that is won by each possible strategy when played against each other strategy.

PCR: *See* polymerase chain reaction.

pedigree: The family relationships between individuals in a sexual population.

pericentric inversion: A mutation involving a chromosomal inversion that spans the centromere.

peroxisome: A membrane-bound organelle in eukaryotes involved in detoxification.

personalized medicine: Use of information on an individual's genotype to improve its health.

phage: *See* bacteriophage.

phenocopy: A phenotype induced by an environmental agent (e.g., a temperature shock), which is similar to that produced by a genetic mutation.

phenogram: A branching diagram that links entities by estimates of overall similarity.

phenotype: The observed characteristics of an individual.

phenotypic value: The actual value P of a quantitative trait, which is made up of contributions from both genotype and environment ($P = G + E$).

phenotypic variance (V_p): The variance of the phenotypic value, $var(P) = V_P$.

phosphorite: A sedimentary rock rich in phosphate.

phylogenetic anchor: The use of the phylogeny of a gene to infer the organismal source of a small piece of DNA. It is used in metagenomics.

phylogenetic tree: A diagram showing the evolutionary history of organisms or genes.

phylogeny: The evolutionary history of organisms or genes.

phylogeography: Inference of population history from the genealogy that connects genes sampled from different geographic locations. It is often based on genealogies inferred from mitochondrial DNA.

phylogram: A phylogenetic tree in which the branch lengths are proportional to the evolutionary distance between nodes. It is also known as an additive tree.

phylotype: The phylogenetic type of an uncultured organism as inferred from analysis of its ribosomal RNA sequence.

physical map: A map that gives the physical location of a genetic variant on the DNA sequence. It contrasts with a genetic map, which is determined using classical genetics.

π: *See* nucleotide diversity.

placental mammals: *See* eutherian mammals.

plasmid: A genetic element that can replicate autonomously but is usually smaller than the chromosome. It is frequently required only for specialized conditions and is commonly found in bacteria and archaea and sometimes in eukaryotes.

plastid: A specialized organelle found in plants, algae, and a variety of single-celled eukaryotes. It comes in a variety of forms including chloroplasts and apicoplasts.

plate tectonics: The mechanism by which the plates that make up the surface of the earth interact with one another, including the formation and subduction of oceanic crust.

pleiotropy: When one allele affects two or more traits.

Pleistocene: The geological period between 1.8 Mya and ~11,000 years ago. It includes major glaciations.

ploidy: The number of copies of each chromosome in the organism.

pluripotency: The ability to differentiate into multiple cell types.

point mutation: A mutation that involves a change from one base to another.

Poisson distribution: The probability that j independent events occur is $(\lambda^j/j!)e^{-\lambda}$, where λ is the expected number of events.

policing: Where selfish behavior is prevented by other individuals.

polyandrous: Describes females that mate with many males.

polygenic: Influenced by multiple genes.

polymerase chain reaction (PCR): A method for amplifying as little as a single copy of a specific nucleic acid molecule, which is recognized because it binds to a pair of primer sequences.

polymorphism: The presence of multiple alleles or inherited phenotypes at appreciable frequency.

polyphyletic: Describes a group of organisms or genes that do not share a common ancestor.

polyploid: A cell or chromosome carrying more than two genomes (e.g., triploid, tetraploid).

polyspermy: Fertilization of an egg by more than one sperm.

polytene chromosome: A chromosome that consists of large numbers of parallel DNA strands, making their structure clearly visible.

polytomy: A portion of a phylogenetic tree in which more than two branches emerge from a single node. It can be used to represent radiation events as well as ambiguities in knowledge.

population bottleneck: A brief reduction in population size, which causes a burst of random genetic drift. It may be caused by a founder effect.

population genetics: Study of the processes that change the genetic composition of populations.

population structure: Any deviation from the ideal state of a single panmictic population. Individuals may be more likely to mate with those in the same place or in the same microhabitat, for example.

positional cloning: Process by which data from genetic crosses are used to identify a DNA fragment that contains a desired gene sequence.

positive selection: *See* directional selection.

postzygotic isolation: Reproductive isolation that acts after production of an F_1 zygote through hybrid inviability or sterility.

prebiotic synthesis: The naturally occurring synthesis of organic compounds before there was life on Earth.

preformation: The view that an embryo develops through the unfolding of preexisting form. This view contrasts with epigenesis.

prenylation: The addition of isoprenoid groups to proteins. It is common in many eukaryotes.

prezygotic isolation: Reproductive isolation that stops production of an F_1 zygote by preventing cross-mating.

priapulid: Member of a phylum of worm-like animals (phylum Priapulida).

primary contact: Where populations have been in contact throughout their divergence. It contrasts with secondary contact.

primase: Enzyme used to initiate replication of DNA.

prion: A protein that can take on alternative stable conformations. It is the infectious agent in diseases such as scrapie and bovine spongiform encephalopathy (BSE).

Prisoner's Dilemma: A game in which both players have a lower fitness when they play the ESS than when they both play the alternative.

probability of fixation: The chance that a single copy of an allele will ultimately fix throughout the entire population.

prokaryotes: Organisms that do not have nuclei. This includes both bacteria and archaea, two distinct phylogenetic domains of life, and thus the term may represent a polyphyletic grouping of organisms.

proofreading: Correction of DNA replication mistakes by the DNA polymerase enzyme.

protein electrophoresis: A method of analyzing a mixture of proteins by separating the molecules based on physical characteristics such as size, shape, or isoelectric point.

protostome: Member of one of two large groups of bilaterian animals including ecdysozoans and lophotrochozoans (*see* deuterostome). In this group, the initial embryonic opening becomes the mouth.

pseudogene: A gene that has lost its function and is degenerating under mutation and drift.

psychrophiles: Organisms that prefer to grow at low temperatures.

punctuated equilibrium: A theory devised by Eldredge and Gould (in 1972) based on the observation that in the fossil record species often show long periods of stasis, punctuated by rapid morphological change associated with speciation.

punishment: Where individuals who behave in ways that reduce others' fitness are made to suffer reduced fitness as a result.

purifying selection: Selection against deleterious alleles.

purines: A class of nucleic acid bases including adenine (A) and guanine (G).

pyrimidines: A class of nucleic acid bases including thymine (T), cytosine (C), and uracil (U).

QTL: *See* quantitative trait locus.

quadratic selection gradient: *See* selection gradient, quadratic.

quantitative genetics: The study of the inheritance of genetically complex traits.

quantitative trait: Some characteristic that may be influenced by multiple genes and that is studied by the methods of quantitative genetics.

quantitative trait locus (QTL): A region of genome that influences a quantitative trait.

quorum sensing: A mechanism that allows individual bacteria to sense the density of their population.

radioisotope dating: *See* radiometric dating.

radiolarian: Member of a phylum of radially symmetric eukaryotes in the Rhizaria kingdom. They have a strong fossil record.

radiometric dating: A method of dating samples based on analysis of radioactive isotopes and the products of their decay.

random (genetic) drift: The random change in genotype frequency caused by random variation in individual reproduction. Looking backward in time, this process causes coalescence of lineages.

random segregation: During meiosis, the two chromosomes of a pair are distributed randomly to the gametes, each gamete having an equal chance of receiving either chromosome.

rDNA: *See* ribosomal DNA.

reaction norm: The set of phenotypes expressed by a single genotype across a range of environments.

recessive: An allele is recessive with respect to a certain phenotype if it produces that phenotype only when present in two copies, that is, as a homozygote.

recessive lethal: *See* lethal.

reciprocal cross: If a cross is made between A males and B females, then the reciprocal cross is between B males and A females.

reciprocal translocation: A translocation mutation in which parts of two different chromosomes are exchanged.

recombination: The generation of new combinations of genes.

recombination hot spot: A localized region with exceptionally high recombination rate.

recombination load: The loss of mean fitness caused by recombination breaking up combinations of alleles that are favored by epistasis:

$$L = (1 - \bar{W}/W_{max}),$$

where W_{max} is the fitness of the fittest combination of alleles.

recombination, molecular: *See* molecular recombination.

recombination rate: The proportion of recombinant gametes.

Red Queen: Continual coevolution between two species (e.g., between host and parasite).

reduction principle: If selection is the only process acting, then the recombination rate will tend to decrease.

regression: The way in which a variable y depends on another variable x can be represented by a simple regression model,

$$y = \alpha + \beta x + \varepsilon,$$

where ε is a random deviation. β is known as the regression coefficient.

reinforcement: The strengthening of prezygotic isolation through selection against cross-matings that produce unfit hybrid offspring.

relative fitness: *See* fitness, relative.

replicase: An enzyme that copies any form of genome (i.e., in the origin of life, the genome may not have been DNA or RNA based).

replicative DNA transposon: A DNA-based transposable element that moves itself to a new place in the genome and also leaves a copy in the original location.

replicator: Any entity that can replicate. It usually refers to a DNA-based genome but can include prions and words passed on by cultural transmission.

reproductive assurance: The assurance that an individual can fertilize its eggs or ovules by selfing.

reproductive isolation: The separation of distinct gene pools, as a result of genetic differences that prevent successful interbreeding.

rescue, hybrid: *See* hybrid rescue allele.

restriction enzyme: An enzyme that cuts DNA at specific sites, typically four to six bases long. The natural function of restriction enzymes is to destroy foreign DNA that enters the cell.

restriction fragment length polymorphism (RFLP): A method for detecting DNA sequence variation that detects variation in the length of fragments cut by restriction enzymes.

retroposon: A DNA transposable element that replicates through an RNA intermediate and does not have long terminal repeats.

retrotransposon: A DNA transposable element that replicates through an RNA intermediate and has long terminal repeats on both sides.

retrovirus: A virus that has an RNA genome and replicates it through a DNA intermediate. DNA can be inserted into the genome of host species.

reverse genetics: Term used to describe any of a variety of molecular methods that allow a wild-type allele of a gene to be targeted and replaced by an engineered mutant allele. *See* positional cloning.

reverse transcription: Some viruses produce enzymes that reverse the transcription process by copying RNA back into a complementary DNA sequence. This process produces a complementary DNA copy of an RNA molecule and is used by retroviruses and retrotransposons.

RFLP: *See* restriction fragment length polymorphism.

rhizaria: Another name for the kingdom of eukaryotes known as Cercozoa.

rhyniophyte: Member of an early group of vascular plants.

ribosomal DNA (rDNA): The DNA sequence that codes for the ribosomal RNAs, which form the core of the ribosome.

ribosomal RNA (rRNA): The highly conserved RNA molecules that are found within ribosomes. They are widely used for estimating phylogenies.

ribosome: The protein–RNA complex responsible for translating the genetic code.

ribozyme: An RNA with catalytic activity.

ring species: A chain of interbreeding populations whose ends overlap without interbreeding.

RNA-mediated interference (RNAi): Mechanism of RNA-based regulation of gene function that results from the inhibition of gene expression through the formation of double-stranded RNA.

RNA world: The stage before the evolution of the genetic code when RNA was responsible for both heredity and catalysis.

robust: Used to describe early human fossils—specifically, the large rear teeth of certain Australopithecines. It contrasts with gracile.

rock–paper–scissors game: A game in which strategy A beats B, B beats C, and C beats A.

root: The most ancient branch in a phylogenetic tree.

runaway process: Arises from the positive feedback between the evolution of a male trait and the female preference for that trait.

S: *See* selection differential.

s: *See* selection coefficient.

saltation: A variation of large effect; also, a major mutation.

satellite DNA: Highly repeated DNA sequence, which was originally detected as a "satellite" component with a density distinct from the rest of the genome. *See* microsatellite and minisatellite.

SBT: *See* shifting balance theory.

secondary contact: Contact between populations that had previously been geographically separate (i.e., allopatric). It contrasts with primary contact.

segregating sites: Sites that are polymorphic in a sample of sequences.

segregation: The movement of two homologous chromosomes during meiosis, one to each pole of the cell. Also, the production of different genotypes in the offspring of heterozygotes, as a result of this random meiotic process.

segregation distortion: Deviation from the expected 1:1 segregation of alleles from a heterozygote at meiosis. It includes both meiotic drive and differential survival of the haploid products of meiosis.

segregation load: The loss of mean fitness caused by the segregation of homozygotes, when polymorphism is maintained by heterozygote advantage:

$$L = (1 - \bar{W}/W_{max}),$$

where W_{max} is the fitness of the fittest heterozygous genotype.

selection, background: *See* background selection.

selection coefficient (*s*): Difference in relative fitness.

selection differential (*S*): The difference in mean trait value between those that reproduce and the original population.

selection, direct: A change in genotype frequency that is caused by the effects on fitness of the alleles themselves.

selection gradient (β): The gradient of a regression of fitness on trait value.

selection gradient, quadratic (γ): The coefficient of a regression of squared deviation of a trait against fitness. Negative values indicate stabilizing selection, which reduces the variance of the trait.

selectionist: One who holds that most molecular divergence and variation is shaped by selection. It contrasts with neutralist.

selection, purifying: *See* purifying selection.

selection response: The change in mean trait value over one generation.

selection, truncation: *See* truncation selection.

selective death: A failure to survive or reproduce; also, a loss of fitness attributable to differences in genotype.

selective sweep: Increase of neutral alleles by hitchhiking with a favorable mutation. This sweeps variation out of a region of the genome surrounding the favorable mutation.

self-fertilization: When a hermaphroditic organism mates with itself.

self-incompatibility: When individuals cannot self-fertilize.

selfing: *See* self-fertilization.

selfish: A gene that increases its own transmission but reduces the fitness of the individual that carries it.

selfish DNA: Sequences that replicate faster than the rest of the genome and that reduce the fitness of the individual carrying them.

self-replication: Used to describe a molecule or other structure that can cause its own replication.

semiconservative: Describes the replication of double-stranded DNA, where the two new molecules each carry one strand from their parent and a complementary strand that has been newly synthesized.

senescence: *See* aging.

sensory bias: An innate preference for particular male traits, which did not evolve as a result of the sexual selection caused by that preference.

sex: Production of offspring that are a mixture between two different parental genotypes.

sex-biased inheritance: Transmission of genes through one or the other sex (e.g., maternal inheritance or sex linkage).

sex chromosome: A chromosome that is inherited differently by the two sexes. In mammals, males carry an X and a Y chromosome, and females carry two X chromosomes. In birds and butterflies, males carry two Z chromosomes, and females carry a Z and a W chromosome.

sex ratio: The ratio of males to females in the population.

sexual selection: Selection arising from variation in the ability to find a mate.

shifting balance theory (SBT): A theory developed by Sewall Wright, in which species evolve toward the best among many alternative adaptive peaks.

short interspersed nucleotide element (SINE): A class of transposable element.

shotgun sequencing: A method of sequencing genomes and environmental samples in which random fragments of DNA are sequenced and then computational methods are used to "reassemble" genomes from the sample.

simple sequence repeats (SSRs): Tandem repeats of a short sequence.

simulated annealing: An optimization algorithm that chooses random changes that improve the desired trait.

SINE: *See* short interspersed nucleotide element.

single-nucleotide polymorphism (SNP): A SNP (pronounced "snip") occurs when members of a population vary in which base they carry at a single nucleotide site.

sister chromatid: The two copies of a chromosome after it has been replicated.

site: *See* nucleotide site.

slime mold: Eukaryotes, from multiple phyla, that normally exist as single-celled amoeba-like organisms but that some-times gather together into "slugs," which move together as a unit.

slip-strand mispairing (SSM): A process in which a DNA polymerase adds one too many or one too few copies of a repetitive sequence during replication.

SNP: *See* single-nucleotide polymorphism.

Social Darwinism: The idea that, by analogy with natural selection, societies evolve through competition between individuals or groups.

social evolution: The study of the evolutionary consequences of interactions between individuals.

soma: Those parts of a multicellular organism that will not directly produce gametes. It contrasts with germ line.

spandrels: Triangular spaces formed where two arches intersect. Used by Gould and Lewontin to refer to structures that are necessary by-products of traits that arose for reasons quite unrelated to any present function.

speciation: The process by which new species are formed.

species selection: Selection between species arising from differences in the rate of speciation and/or extinction of lineages.

species tree: A phylogenetic tree showing the relationships among species. It is usually shown to contrast with gene trees, which may include events such as gene duplication and lateral gene transfer.

specificity: Where individual molecules take up a stable conformation with specific biological functions.

sphenopsid: A member of a group of plants that includes trees in the Carboniferous coal swamp forests as well as the living horsetail (*Equisetum*).

sponges: Common name for members of the phylum *Porifera*, which are thought to be the earliest branching lineage of animals. Sponges feed by moving water through their bodies using specialized cells with cilia called choanocytes.

sporangium (*pl.* sporangia): A structure containing spores.

sporophyte: The diploid phase of the life cycle of plants that gives rise to the production of spores by means of meiosis. It contrasts with gametophyte.

SSM: *See* slip-strand mispairing.

SSR: *See* simple sequence repeats.

stabilizing selection: Selection that favors intermediate trait values.

standard (genetic) code: *See* canonical code.

standard neutral model: The simplest version of the neutral theory, in which mutation and random genetic drift act in a single panmictic population of constant size.

star genealogy: A genealogy in which all lineages coalesce in a common ancestor at the same time. It is produced by a population bottleneck or by a selective sweep.

statistical power: The chance that the null hypothesis will be rejected when the data are generated by a different model.

stem group: The series of extinct organisms within a clade of living and fossil organisms that lie below the crown group. *See* crown group.

stem-loop structure: A hairpin structure in an RNA molecule that is maintained by complementary base pairing.

steranes: Chemical derivatives of sterols that have been used as chemical fossils.

stereoisomers: Molecules whose atoms are connected with each other in the same way but are arranged differently in space. This includes enantiomers, which are mirror-image reflections of each other.

sterol: Amphipathic molecules found in the membranes of many organisms, especially eukaryotes.

stigma: The female reproductive organ in a flowering plant, which receives pollen.

strategy: A general term used in game theory. It is equivalent to a phenotype or reaction norm. It refers to the morphology or behavior expressed by an individual under a range of circumstances.

structure: *See* population structure and spatial structure.

structured coalescent: An extension to the coalescent process in which lineages move from place to place as they trace backward in time while they coalesce.

structured population: *See* population structure.

style: The long structure between the stigma and ovule in a flower. The pollen must grow through the style in order to fertilize the plant.

substitution: The replacement in a population of one nucleotide or amino acid by another.

substitution load: The total loss of mean fitness caused because favorable alleles substitute gradually by selection rather than instantaneously. It is the integral of

$$1 - \overline{W}/W_{\max}$$

over time, where the mean fitness $\overline{W}$ increases toward the maximum fitness $W_{\max}$.

supercoiling: Higher-order twisting of DNA strands.

supergene: A cluster of tightly linked genes, which allow distinct alternative morphs to be maintained as a polymorphism within one population.

suppressor mutation: A secondary mutation that can cancel the effect of a primary mutation, resulting in a wild-type phenotype.

symbiosis: A close association between two organisms.

sympatric speciation: The separation of a single population into two or more reproductively isolated species in the absence of any geographic barriers.

sympatry: Coexistence in the same place.

synapsis: The lining up of homologous chromosomes in meiosis.

syngameon: A botanical term, referring to a cluster of taxa that are morphologically distinct and yet exchange genes.

syngamy: The union of two genomes, which leads to a doubling of ploidy level.

synonymous mutation: A point mutation in a protein-coding region that changes a codon such that it does not alter the resulting amino acid sequence of the protein.

syntax: The rules that determine how words combine to make phrases and sentences.

systematics: *See* taxonomy.

tagging single-nucleotide polymorphisms (tSNPs): A set of single-nucleotide polymorphisms that distinguish different haplotypes segregating in a population.

tandem (DNA) repeats: A series of repeated sequences, arrayed next to each other.

tandem duplication: A duplication mutation in which the duplicated DNA is found next to the original DNA.

tautomerization: The spontaneous isomerization of a nitrogen base from its normal *keto* (or *amino*) form to an alternative hydrogen-bonding *enol* (or *imino*) form.

taxon (*pl.* taxa): A unit of classification (e.g., species, genus).

taxonomy: Classification of living organisms (also known as systematics).

TDT: *See* transmission disequilibrium test.

telocentric: A chromosome with the centromere in the middle.

telomerase: An enzyme complex found in most eukaryotes that maintains the length of telomeres through successive divisions.

telomere: The end of a chromosome.

tetraploid: Carrying four copies of each chromosome gene.

theistic evolution: A view in which religious teachings are seen as compatible with biological evolution. Species share common ancestry and change through natural selection and other evolutionary processes but with divine guidance and intervention.

theory: A set of interconnected hypotheses that leads to testable predictions.

thermophile: An organism with a growth temperature optimum between 50°C and 80°C.

θ: The product $4N_e\mu$, which measures the relative rates of random drift ($1/2N_e$) and mutation (μ) and hence determines the variation maintained in a balance between these processes.

threshold model: A model that states that a discrete trait is present only when the quantity of an underlying continuous trait is greater than some threshold.

tip: The terminal node on a phylogenetic tree.

tit-for-tat: The winning strategy in the repeated Prisoner's Dilemma game.

tracheid: A cell with strengthened walls that functions to transport fluid within plants.

trade-off: A situation where one trait cannot be increased without decreasing another. The term is used to describe constraints in optimization arguments.

tragedy of the commons: Where self-interested exploitation of common resources leads to a worse outcome for all. *See* Prisoner's Dilemma.

trait: *See* quantitative trait.

transcription: Replication of an RNA strand complementary to a DNA sequence.

transcription factor: A molecule that binds to the promoter and regulates transcription.

transduction: The movement of genes from a donor cell to a recipient cell with a virus as the vector.

transfer RNA (tRNA): An RNA molecule that couples a specific amino acid to a specific sequence of three bases. It is responsible for translating the genetic code.

transformation: The introduction of a fragment of DNA into a genome. Transformation occurs naturally in some bacteria and archaea; in the laboratory, it is the basis of genetic engineering.

transgenic: Genetically manipulated to carry genes from another individual or species.

transition: A mutation in which a purine replaces another purine or a pyrimidine replaces another pyrimidine.

translation: Synthesis of protein with amino acid sequence encoded by an RNA sequence.

translocation: A rearrangement mutation in which part of one chromosome breaks away and joins another.

transmission disequilibrium test (TDT): A statistical test that detects associations between genetic markers and disease alleles by looking for marker alleles that are transmitted in excess to affected offspring.

transposable element: A genetic element that can move from one location in the genome to another.

transposon: *See* transposable element.

transversion: A mutation in which a pyrimidine replaces a purine, or vice versa.

Tree of Life: A phylogenetic tree showing the relationships among all cellular organisms.

trigonotarbid: Member of an order of extinct terrestrial spider-like animals (order Trigonotarbida).

trimerophyte: Member of an early group of vascular plants.

triploid: Carrying three genomes.

trisomy: Possession of three copies of one chromosome (in humans this is abnormal because there are normally only two copies).

tristyly: A polymorphism with three different arrangements of anther and stigma. It promotes outcrossing.

tRNA: *See* transfer RNA.

trochophore: Larval type characteristic of many protostomes including annelids and many mollusks.

true breeding: A population or individual that produces genetically identical offspring.

truncation selection: Selection that eliminates those with the largest (or smallest) trait values.

UEP: *See* unique event polymorphism.

ultracentrifuge: A very high-speed centrifuge used to separate macromolecules.

ultrameric tree: A phylogenetic tree in which the branch lengths are constrained to all be equidistant from the root. It is also known as a dendrogram.

uncultured microbes: Microbes that have never successfully been grown in isolation in the laboratory.

underdominance: Describes heterozygotes that have lower trait values (usually lower fitness) than either homozygote.

unequal crossover: The outcome when two tandemly repeated sequences do not pair correctly. A cross-over between them produces one genome with fewer repeats and one with more.

uniformitarianism: The assumption that the same natural processes acted in the past as are observed to operate now.

unique event polymorphism (UEP): An allele for which all copies derive from a single mutational event.

universal genetic code: *See* canonical code.

universal homology: A homologous trait found in all cellular organisms.

unrooted tree: A phylogenetic tree in which the root is not shown (frequently because it is not known).

V_A: *See* additive genetic variance.

variable number tandem repeat (VNTR): A general term for tandemly repeated sequences. It includes satellite, microsatellite, and minisatellite sequences.

variance: The mean squared deviation from the average:

$$\text{var}(x) = \text{E}[(x - \bar{x})^2].$$

variance component: A contribution to the phenotypic variance due to a specific kind of variation. It includes environmental, additive, dominance, and interaction variances.

var(x): *See* variance.

vascular plants: Division of plants with vascular tissues, which function in transporting fluids.

V_D: *See* dominance variance.

V_E: *See* environmental variance.

vertical descent: The evolution of species by a branching pattern.

vertical evolution: *See* vertical descent.

vertical inheritance: The transmission of traits from parent to offspring.

vertical transmission: *See* vertical inheritance.

V_G: *See* genotypic variance.

virulence: The degree of pathogenicity of a parasite.

V_M: *See* mutational variance.

VNTR: *See* variable number tandem repeat.

V_W: *See* additive genetic variance in fitness.

wave of advance: A favorable allele advances behind a moving cline, known as a wave of advance.

wild type: The commonest allele at a locus or the most common genotype.

wobble pairing: The ability of tRNAs to hybridize with codons even when only the first two nucleotides follow the standard G-C and A-U base-pairing rules.

Wright–Castle estimator: A method for estimating the number of genes that influence a quantitative trait. It is based on a comparison between the variance in an F_2 population and the difference in means between the parent populations.

Wright–Fisher model: A standard model of random genetic drift, in which each gene is drawn at random from $2N$ genes in the previous generation.

xerophiles: Organisms that prefer to grow in very dry conditions.

zosterophyll: A type of early vascular plant that carried spores laterally along the stem.

zygote: The diploid cell formed by union of two haploid gametes.

Figure Credits

Note: Permissions for use of the images shown in these sample pages are in process and provisional. Every effort has been made to contact the copyright holders of figures and tables in this text. Any copyright holders we have been unable to reach or for whom inaccurate information has been provided are invited to contact Cold Spring Harbor Laboratory Press. Figure credits, pending permission, are listed below.

Abbreviations: AAAS, American Association for the Advancement of Science; APS, American Philosophical Society; ASBMB, American Society for Biochemistry and Molecular Biology; ASM, American Society for Microbiology; CSHL, Cold Spring Harbor Laboratory; CSHLP, Cold Spring Harbor Laboratory Press; GSA, Genetics Society of America; NAS USA, National Academy of Sciences, U.S.A.; NLM, National Library of Medicine; PNAS, Proceedings of the National Academy of Sciences; PR, Photo Researchers, Inc.; VU, Visuals Unlimited. BL, bottom left; BR, bottom right; ML, middle left; MR, middle right; TL, top left; TR, top right.

Cover and Title Pages

1st Image: A map of the genome of the bacterium *Deinococcus radiodurans* R1. A portion of Fig. 1 from White O. et al., Genome sequence of the radioresistant bacterium *Deinococcus radiodurans* R1. *Science* **286:** 1571–1577, © 1999 AAAS. **2nd Image:** Mineral skeleton of a radiolarian (marine). The hard skeleton is composed of silica. SEM x145. © Dr. Dennis Kunkel/Visuals Unlimited. **3rd Image:** Grasshopper embryo (image duplicated and rotated fivefold). Courtesy of Sabbi Lall and Nipam H. Patel, University of California, Berkeley. **4th Image:** Photo of enhanced brain MRI image, © Scott Camazine.

Preface

Earthrise—Apollo 8. Courtesy of NASA.

Part Openers

Part I left, *HMS Beagle*, Mt. Sarmiento, reprinted from Darwin, Charles, *The Voyage of the Beagle*; **Part I middle,** fossil lotus leaf, © Ken Lucas/VU; **Part I right,** microscope view DNA SEM, © Science VU/LL/VU; **Part II left,** trilobite fossil, © Ken Lucas/VU; **Part II middle,** enterobacter SEM, © Dr. David Phillips/VU; **Part II right,** humpback whale, © Masa Ushioda/VU; **Part III left,** cichlid fish, © Fredrik Hagblom; **Part III middle,** small ground finch, © Gerald and Buff Corsi/VU; **Part III right,** rotifer, © Science VU/VU; **Part IV left,** stone tool, © Javier Trueba/Madrid Scientific Films/PR; **Part IV middle,** school assembly, © David R. Frazier/ PR; **Part IV right,** people, U.S. Department of Energy Human Genome Program.

Aim & Scope

A&S.1 TL, Antarctic dry valley, Laura Connor and Effie Jarret, © 1999; **A&S.1 TR,** tubeworms, © Science VU/VU; **A&S.1 ML,** moth antennae Rippel Electron Microscope Facility, Dartmouth College; **A&S.1 MR,** screech owl, © Bowers Photo; **A&S.1 BL,** crow, Gavin Hunt, Department of Psychology, University of Auckland; **A&S.1 BR,** honeybees, © Simon Fraser/SPL/PR; **A&S.2,** apicoplasts, courtesy of Geoffrey McFadden, University of Melbourne; **A&S.3 B,C,** RNA folding, adapted from Wimberly B.T. et al., *Nature* **407:** 327–339, © 2000 Macmillan; **A&S.4,** Fluorescent proteins, reprinted from Crameri A. et al., *Nat. Biotechnol.* **14:** 315–319, © 1996 Macmillan.

Chapter 1

1.0 left, Darwin, reprinted from Darwin C., 1859, *On the Origin of Species by Means of Natural Selection,* title page, John Murray, London from facsimile edition published by Harvard University Press, 1966, courtesy of CSHL Library and Archives; **1.0 middle,** Mendel, reprinted from Mendel G.J., *Versuche über Pflanzen-Hybriden. Vorgelegt in den Sitzungen vom 8. Februar und 8. März 1865.* Verhandlungen des naturforschenden Vereines in Brünn, Band IV, Heft 1 (1865): 3–47, title page, Brünn: Verlag des Vereines, 1866, courtesy of Special Collections, Falvey Memorial Library, Villanova University; **1.0 right,** Watson–Crick double helix, courtesy of CSHL Library and Archives; **1.1,** adapted from Crick F., *Nature* **227:** 461–563, © 1970 Macmillan; **1.3,** redrawn from http://web.clas.ufl.edu/users/rhatch/images/greatChain.gif, credit: Robert A. Hatch; **1.5,** engravings of fossils reproduced in Robert Hooke's *Posthumous Works,* Museum of the History of Science, Oxford; **1.6 top,** Regional Nature Park of the Auvergne Volcanoes, located in France, © Mark Boulton/PR; **1.6 bottom,** National Portrait Gallery, London; **1.7,** NLM; **1.8,** reprinted from Buckland W., *Transactions of the Geological Society of London,* series 2, V.1 (1824), pp. 390–396, courtesy of Linda Hall Library of Science, Engineering and Technology; **1.9,** courtesy of the Syndics of Cambridge University Library; **1.10,** reprinted from *The Endeavour Journal of Joseph Banks 1768–1771,* Sydney, Public Library of New South Wales, 1962, edited by J.C. Beaglehole, available as part of the "Sir Joseph Banks Electronic Archive" Web site; **1.11,** National Portrait Gallery, London; **1.12A,** NLM; **1.13,** courtesy of Adam Smith Institute, http://www.adamsmith.org; **1.16,** courtesy of the Syndics of Cambridge University Library; **1.18,** Wellcome Institute Library, London; **1.20,** reprinted from Drawings by Ape from Vanity Fair (**left**) July 24, 1869, (**center**) January 28, 1871, (**right**) Sept. 30, 1871; **1.21,** engraving, "Adventure with curl-crested toucans," frontispiece to Vol. I of Bates H.W., 1863, *The Naturalist on the River Amazons,* image courtesy of James Mallet, University College London, http://www.ucl.ac.uk/taxome/; **1.22,** courtesy of Nipam H. Patel; **1.23,** British Marine Life Study Society, © Chris Rowe; **1.24,** reprinted from Desmond A. et al., 1992 (©1991), *Darwin,* Fig. 53, "Fancy pigeons" (facing p. 301), Warner Books, New York (first published in Great Britain), collection of CSHL Library and Archives; **1.25,** modified from Galton F., 1885, Regression towards mediocrity in hereditary stature. *J. Anthropol. Inst.* **15:** 246–267; **1.26 left,** Courtesy of CSHL Library and Archives; **1.26 middle,** NLM; **1.26 right,** Galton F., 1908, *Memories of My Life,* Methuen & Co., London, collection of CSHL Library and Archives; **1.27,** courtesy of APS; **1.28,** data from Mendel's 1866 paper; **1.29,** reprinted from Boveri T., 1910, *Die Potenzen der Ascaris-Blastomeri bei abgaederter Furchung, Zugleich ein Beitrag zur Frage qualitativungleicher Chromosomen-Teilung. Festschrift zum sechzig-*

sten Geburtstag Richard Hertwigs, Band III, Gustav Fisher, Jena; **1.30,** courtesy of APS; **1.32,** redrawn from Nilsson-Ehle H., 1914, *Vilka erfarenheter hava hittills vunnits rörande möjligheten av växters acklimatisering? Kunglig Landtbruksakademiens Handlingar och Tidskrift,* after Gould J. et al., *Biological Science, 6e,* Fig. 1.32, © 1996 W.W. Norton and Co.; **1.33,** Reproduced from Castle W.E. et al., 1914, *Piebald rats and selection/An experimental test of the effectiveness of selection and of the theory of gametic purity in Mendelian crosses,* Plate 1 (facing p. 56), The Carnegie Institute of Washington, Washington, D.C., courtesy CSHL Library and Archives; **1.35 top,** courtesy of APS, Bronson Price papers; **1.35 middle,** courtesy of APS, Wright papers; **1.35 bottom,** reprinted from Clark R.W., *J.B.S.: The Life and Work of J.B.S. Haldane,* © 1969 Coward-McCann; **1.36,** The Rockefeller University Archives, photo by Don C. Young; **1.38 top,** © Rob & Ann Simpson/VU; **1.38 bottom,** © Leroy Simon/VU; **1.39A,** reprinted from Lewontin R.C. et al., *Dobzhansky's "Genetics of Natural Populations,"* Fig. 34, © 1981 Columbia University Press; **1.39B–D,** redrawn from Dobzhansky T., *Genetics* **28:** 162–186, © 1943 GSA; **1.40 top,** courtesy of APS; **1.40 bottom,** courtesy of Ernst Mayr Library of the Museum of Comparative Zoology, Harvard University.

Chapter 2

2.0, Hemoglobin, reprinted from Barrick D. et al., *Methods Enzymol.* **379:** 28–54, © 2004 Elsevier, image courtesy of Chien Ho, Carnegie Mellon; **2.1,** Annenberg Rare Book and Manuscript Library, University of Pennsylvania; **2.2 top,** NLM; **2.3,** Edgar Fahs Smith Collection, University of Pennsylvania Library; **2.4,** Science Museum Pictorial; **2.5,** adapted from Hunter G.K., *Vital Forces: The Discovery of the Molecular Basis of Life,* © 2000 Elsevier; **2.6,** redrawn from Svedberg T. et al., *Proc. R. Soc. Lond. A, Math. Phys. Sci.* **170:** 40–79, © 1939 Royal Society; **2.7,** reprinted from Sumner J., 1964, *Nobel Lectures, Chemistry 1942–1962,* p.117, Elsevier, Amsterdam, © The Nobel Foundation 1946; **2.8,** © 1939 Cornell University Press; **2.9,** © 1953 Los Angeles Times, reprinted with permission; **2.10,** © Dr. James W. Richardson/VU; **2.11,** © Dr. Dennis Kunkel/VU; **2.12,** © Department of Chemistry, University of Cambridge; **2.13,** Courtesy of APS; **2.14,** Courtesy of The Rockefeller University Archive Center; **2.16,** © A.C. Barrington Brown/PR; **2.17 left,** courtesy of the James D. Watson Collection, CSHL Library and Archives; **2.17 right,** © King's College Archives, King's College London; **2.18,** courtesy of the James D. Watson Collection, CSHL Library and Archives; **2.19,** adapted from Watson J.D. et al., *Nature* **171:** 964–967, © 1953 Macmillan; **2.20,** reprinted from Meselson M. et al., *PNAS* **44:** 671–682; **2.21,** reprinted from Ingram V., *Nature* **178:** 792–794, © 1956 Macmillan; **2.22 top,** courtesy of CSHL Library and Archives; **2.22 bottom,** redrawn from Gamow G., *Nature* **173:** 318, © 1954 Macmillan; **2.23,** redrawn from http://www. biochem.ucl.ac.uk/bsm/dbbrowser/jj/aastruct.html with permission of Nancy Watson; **2.24,** adapted from Stryer L., 1981, *Biochemistry,* 2e, Fig. 12-12c,© 2004 John Wiley, based on a drawing by Dr. Sung-Hou Kim; **2.25 top,** Collection of NLM, courtesy of Marshall W. Nirenberg; **2.25 bottom,** reprinted from Jones O.W. et al., *PNAS* **48:** 2115–2123; **2.26,** adapted from Crick F.H.C., *Cold Spring Harbor Symp. Quant. Biol.* **31:** 3–9, © 1966 CSHLP; **2.28,** © Institut Pasteur; **2.29, 2.30A,** adapted from Judson H.F., *The Eighth Day of Creation; Makers of the Revolution in Biology,* expanded edition pp. 372, 392, CSHLP, © 1996 by Horace Freeland Judson; **2.30B,** adapted from Pardee A.B. et al., *J. Mol. Biol.* **1:** 165–178, © 1959 Academic Press; **2.31,** adapted from Crick F., *Nature* **227:** 461–563, © 1970 Macmillan; **2.32,** © BBC; **2.33 top,** reprinted from Perutz M.E., 1962 *Nobel Lectures, Chemistry 1942–1962,* p. 665, Elsevier, Amsterdam, © The Nobel Foundation 1946; **2.33 bottom,** adapted from Judson H.F., *The Eighth Day of Creation: Makers of the Revolution in Biology,* expanded edition, p. 497, CSHLP, © 1996 by Horace Freeland Judson; **2.34,** redrawn from Voet D.D. et al., *Biochemistry, 3e,* Fig 10.16, © 2004 John Wiley; **2.35,** reprinted from Wimberly B.T. et al., *Nature* **407:** 327–339, © 2000 Macmillan; **2.36,** modified from

Michel F. et al., *J. Mol. Biol.* **216:** 585–610, © 1990 Academic Press Limited; **2.38,** adapted from Kimura M., *The Neutral Theory of Molecular Evolution,* Fig. 4.4, © 1983 Cambridge University Press; **2.39,** reprinted from Kimura M., *Genetics* **140:** 1–5, © 1995 GSA; **2.40,** © Dr. Gopal Murti/VU; **2.41,** © Joe McDonald/VU; **2.42,** Kipling R., *Just So Stories,* "The Elephant's Child," p. 53, 1926 Macmillan, London; **2.43,** Rousseau H., *Fight between a Tiger and a Buffalo,* 1908, gift of the Hanna Fund, © The Cleveland Museum of Art.

Chapter 3

3.0, Rheinhold's monkey, courtesy, University of Edinburgh; **3.1,** courtesy, CSHL Archives; **3.3A left,** © Arthur Morris/VU; **3.3A right,** © Bill Beatty/VU; **3.4 A–C,** photos courtesy of Nipam H. Patel; **3.6,** adapted from *Spice Island Voyage,* University of Limerick, Ireland Project; **3.7A,** © Dr. John Cunningham/VU; **3.7B,** © David Sieren/VU; **3.7C,D,** © Inga Spence/VU; **3.7E,** © Wally Eberhart/VU; **3.8A,** modified from Carroll S.P. et al., *Evolution* **46:** 1052–1069, © 1992 Society for the Study of Evolution; **3.9,** redrawn from Molla A. et al., *Nat. Med.* **2:** 760–766, © 1996 Macmillan; **3.10 left, middle,** Web source, no longer available; **3.10 right,** Nikola Repke, Institute of Systematic Botany and Botanic Garden, University of Zurich; **3.11 top,** © Adam Jones/VU; **3.11 bottom,** from http://staff.science.nus.edu.sg/~scilooe/srp_2003/sci_paper/botanic/research_paper/ lim_yifan.pdf; **3.13,** modified from Sheldon P.R., *Nature* **330:** 561–563 and references therein, © 1987 Macmillan; **3.14 top,** © Natural History Museum, London; **3.14 bottom,** © William J. Weber/VU; **3.15,** Woodburne M.O. et al., *Science* **218:** 284–286, © Robert A. Hicks, rahicksphoto.com; **3.16,** © Dr. John Cunningham/VU; **3.17,** reprinted from Xu X. et al., *Nature* **431:** 680–684, © 2004 Macmillan; **3.18,** redrawn from de Muizon C., *Nature* **413:** 259–260, © 2001 Macmillan; **3.19,** courtesy, Fisher Papers, Barr Smith Library, University of Adelaide; **3.21,** adapted from Eldredge N., *Darwin: Discovering the Tree of Life,* © 2005 W.W. Norton, New York, London.

Chapter 4

4.0, Yellowstone hot spring, © Inga Spence/VU; **4.2,** © Kevin and Betty Collins/VU; **4.3,** reprinted from Buick R., *Paleobiology 2e,* Briggs D. et al., eds., p.14, © 2001 Blackwell Publishing; **4.4,** modified from Joyce G.F., *Nature* **418:** 214–221, © 2002 Macmillan; **4.6A,** photo courtesy of NASA; **4.8,** redrawn from Orgel L.E., *Trends Biochem. Sci.* **23:** 491–495, © 1998 Elsevier; **4.9,** photo courtesy of the Canadian Scientific Submersible Facility; **4.10,** redrawn from Hazen R.M., *Sci. Am.* **284:** 76–85, © 2001 Scientific American; **4.11,** redrawn from Lee D.H. et al., *Curr. Opin. Chem. Biol.* **1:** 491–496 © 1997 Elsevier; **4.12,** reprinted from Lewis R.J. et al., *Nature* **298:** 393–396, © 1982 Macmillan; **4.13A,B,** redrawn from Maynard Smith J. et al., *The Major Transitions in Evolution,* Figs. 4.1, 4.2, © 1998 Oxford University Press; **4.14A,** redrawn from Web source, no longer available; **4.14B,C,** adapted from Berg J.M. et al., *Biochemistry 5e,* Figs. 12.10, 12.12, © 2002 W.H. Freeman; **4.16A,B,** redrawn from Lodish H. et al., *Molecular Cell Biology,* Fig. 4.12, © W.H. Freeman; **4.17,** adapted from Alberts B. et al., *Molecular Biology of the Cell, 4e,* Figs. 6.92, 6.99, © 2002 Garland Science; **4.18,** redrawn from http://www.lpi.usra.edu/publications/MSR/Bada/ Fig3.GIF; **4.19A,B,** redrawn from Brown T.A., *Genomes 2e,* Fig. 15.3, © 2002 Wiley-Liss.

Chapter 5

5.0, 70S ribosome, http://rna.ucsc.edu/rnacenter/ribosome_images. html; **5.1,** reprinted from Darwin C., *On the Origin of Species;* **5.10,** redrawn from Morrison C.L. et al., *Proc. R Soc. Lond. B* **269:** 345–350, © 2002 The Royal Society; **5.15,** redrawn from Whitaker R.H., *Science* **163:** 150–160, © 1969 AAAS; **5.17,** redrawn from Pace N.R., *Science* **276:** 734–740, © 1977 AAAS; **5.20A,B,** redrawn from Eisen J.A., *Genome Res.* **8:** 163–167, © 1998 CSHLP; **5.21,** redrawn from Brown J.R. et al., *Microbiol. Mol. Biol. Rev.* **61:** 456–502,

© 1997 American Society for Microbiology; **5.22C,D,** redrawn from Forterre P. et al., *Bioessays* **21**: 871–879, © 1999 Wiley-Liss, Inc.; **5.23A,** adapted from Penny D. et al., *Curr. Opin. Genet. Dev.* **9**: 672–677, © 1999 Elsevier; **5.23B,** redrawn from Doolittle W.F., *Sci. Am.* (Feb) 2000: 90–95, © Scientific American, Inc.; **5.24A,** redrawn from Eisen J.A., *Curr. Opin. Microbiol.* **3**: 475–480, © 2000 Elsevier; **5.24B,** redrawn from Huynen et al., *Science* **286**: 1443, © 1999 AAAS; **5.25,** modified from Rohwer F. et al., *J. Bacteriol.* **184**: 4529–4535, © 2002 ASM.

Chapter 6

6.0 left, *Escherichia coli* 0157:h7, © Gary Gaugler/VU; **6.0 right,** *Methanococcus jannaschii,* © B. Boonyaratanakornkit, D.S. Clark, G. Vrodoljak/EM Lab, University of California, Berkeley/VU; **6.1A,** © Dr. David Phillips/ VU; **6.1B,C,** © Dr. Dennis Kunkel/VU; **6.1D,** © Gary Gaugler/VU; **6.1E,** © Tina Carvalho/VU; **6.1F,** © Gary Gaugler/VU; **6.1G,** © Science VU/VU; **6.1H,** © Dr. David Phillips/VU; **6.2 top,** © Arthur Siegelman/VU; **6.3A,** © Dr. Fred Hossler/ VU; **6.4 top, middle,** © Dr. Gopal Murti/VU; **6.6A,** http://www. uoguelph.ca/~gbarron/MIS-CELLANEOUS/ nov00.htm; **6.6B,** reprinted from Wang J. et al., *Appl. Environ. Microbiol.* **68**: 417–422, © 2002 ASM; **6.6C,** reprinted from Schulz H.N. et al., *Science* **284**: 493–495, © 1999 AAAS; **6.7 top,** http://uninews.unimelb.edu.au/mediaComms/Collage_C23_Jun14_fig3. jpg; **6.7 middle,** © B. Boonyaratanakornkit, D.S. Clark, G. Vrodoljak/EM Lab, University of California, Berkeley/VU; **6.7 bottom,** http://www. wissenschaft-online.de/sixcms/media.php/591/nanoarch2.102460.jpg, © Karl Stetter; **6.8 top,** reprinted from Pereiera S.L., *PNAS* **94**: 12633–12637, © 1997 NAS USA; **6.8 bottom,** source unknown; **6.9,** unattributable; **6.10,** redrawn from Embley T.M. et al., *Syst. Appl. Microbiol.* **16**: 25-29, © 1993 Elsevier; **6.11,** redrawn from Hugenholtz P., *Genome Biol.* **3**: reviews0003.1–0003.8, © 2002 BioMed Central; **6.13, 6.14,** redrawn from Amann R.I. et al., *Microbiol. Rev.* **59**: 143–169, © 1995 ASM; **6.16,** redrawn from Madigan M.T. et al., *Brock Biology of Microorganisms, 9e,* Figs. 5.12, 5.13, © 2000 Prentice Hall; **6.17,** adapted from Web source, no longer accessible; **6.18,** redrawn from Suhre K. et al., *J. Biol. Chem.* **278**: 17198–17202, © 2003 ASBMB; **6.19,** redrawn from Madigan M.T. et al., *Brock Biology of Microorganisms, 9e,* Fig. 5.19, © 2000 Prentice Hall; **6.20,** redrawn from Battista J.R. et al., *Trends Microbiol.* **7**: 362–365, © 1999 Elsevier; **6.21D,** Web source no longer available; **6.22,** redrawn from Whitmarsh J. et al., *Concepts in Photobiology: Photosynthesis and Photomorphogenesis,* Singhal G.S. et al., eds., Fig. 1, © 1999 Narosa Publishers and Kluwer Academic; **6.23,** adapted from Whitmarsh J. and Govindjee, *The Photosynthetic Process,* Fig. 6, http://www.life.uiuc.edu/govindjee/paper/fig6.gif and from Madigan M.T. et al., *Brock Biology of Microorganisms, 9e,* Fig. 13.6, © 2000 Prentice Hall; **6.24,** adapted from Madigan M.T. et al., *Brock Biology of Microorganisms, 9e,* Figs. 9.10, 9.11, © 2000 Prentice Hall; **6.25 top,** reprinted from http://www.ascidians.com/families/didemnidae/Lissoclinum_ patella/lissoclinumpatella.htm; **6.25 bottom,** photo courtesy of Jacques Ravel; **6.27,** redrawn from Taga M.E. et al., *PNAS* **100**: 14549–14554, © 2003 NAS USA.

Chapter 7

7.0, *Deinococcus radiodurans* genome, White O. et al., *Science* **286**: 1571–1577, © 1999 AAAS; **7.1,** adapted from Bentley S.D. et al., *Annu. Rev. Genet.* **38**: 771–791, © 2004 Annual Reviews, based on data from DOGS http://www.cbs.dtu.dk/databases/DOGS/; **7.2,** adapted from Brown T.A., *Genomes, 2e,* Fig 2.2, © 1999 Bios Scientific Publishers Ltd, by permission of Taylor & Francis Books U.K.; **7.4,** adapted from Ptashne M. et al., *Genes and Signals,* Fig. 1.2, © 2002 CSHLP, and from Brown T.A., *Genomes, 2e,* Fig. 2.20, © 1999 Bios Scientific Publishers Ltd, by permission of Taylor & Francis Books U.K.; **7.5,** redrawn from Moran N.A. et al., *Genome Biol.* **2**: research0054.1–0054.12, © 2001 Nancy A. Moran; **7.7,** redrawn from Welch R.A. et al., *PNAS* **99**:

17020–17024, © 2002 NAS USA; **7.8A–C, 7.9B–E,** modified from Helfman J. 1996. *Dotplot Patterns: A Literal Look at Pattern Languages, Theory and Practice of Object Systems (TAPOS),* special issue on Patterns, V2(1), pp. 31–41 and Helfman J., *Similarity Patterns in Language, Proceedings of the 1994 IEEE Symposium on Visual Languages,* St. Louis, Missouri, pp. 173–175, October 1994; **7.11,** redrawn from Casjens S., *Annu. Rev. Genet.* **32**: 339–377, © 1998 Annual Reviews; **7.12,** redrawn from Barloy-Hubler F. et al., *Nucleic Acids Res.* **29**: 2747–2756, © 2001 Oxford University Press; **7.13,** redrawn from Wu M. et al., *PLoS Biol.* **2**: 0327, © 2004 Public Library of Science; **7.14A,B,C,** redrawn from Eisen J.A. et al., 2000. *Genome Biol.* **1**: research0011.1–0011.9, © 2000 BioMed Central Ltd; **7.15A,** Web source no longer available; **7.15B,** © Science VU/Visuals Unlimited; **7.16, 7.17,** adapted from Bushman F., *Lateral DNA Transfer,* Figs. 1.4, 1.3, © 2002 CSHLP; **7.18A,** © Science VU/VU; **7.19, 7.20,** adapted from Collignon P.J., *Med. J. Australia* **177**: 325–329, © 2002 Australasian Medical Publishing Co; **7.21,** adapted from Schmidt H. et al., *Clin. Microbiol. Rev.* **17**: 14–56, © 2004 ASM; **7.22,** redrawn from Lawrence J.G. et al., *J. Mol. Evol.* **44**: 383–397, © 1997 Springer; **7.23,** redrawn from Ochman H. et al. *Nature* **405**: 299–304, © 2000 Macmillan; **7.24, 7.25,** redrawn from Lerat E. et al., *PLoS Biol.* **3**: E130, © 2005 Public Library of Science; **7.26,** reprinted from Wu M. et al., *PLoS Genetics* **1**: e65, © 2005 Public Library of Science.

Chapter 8

8.0, Liver cell, © Dr. Gopal Murti/VU; **8.1A,** © Dr. T.J. Beveridge/VU; **8.1B,** © Dr. Gopal Murti/VU; **8.1C,D,** adapted from Madigan M.T. et al., *Brock Biology of Microorganisms,* 9e, © 2000 Prentice Hall; **8.2A,** adapted from http://biology.kenyon.edu/courses/biol114/Chap01/chrom_struct. html; **8.4A,** adapted from Keeling P.J. et al., *Annu. Rev. Microbiol.* **56**: 93–116, © 2002 Annual Reviews; **8.4B,** redrawn from Phillippe H. et al., *Mol. Biol. Evol.* **17**: 830–834, © 2000 Oxford University Press; **8.5,** redrawn from Baldauf S.L. et al., *Science* **300**: 1703–1706, © 2003 AAAS; **8.6A,** © Jerome Paulin/VU; **8.6B,C,** © Wim van Egmond/VU; **8.6D,** © CSIRO Marine Research/VU; **8.6E,** © Joe Scott/VU; **8.6F,** © Dr. David M. Phillips/VU; **8.6G,** © Linda Sims/VU; **8.6H,** © Dr. Richard Kessel and Dr. Gene Shih/VU; **8.6I,** © Ken Lucas/VU; **8.7,** reprinted from Javaux E.J. et al., *Nature* **412**: 66–69, © 2001 Macmillan; **8.8,** redrawn from Marechal E. et al., *Trends Plant Sci.* **6**: 200–205, © 2001 Elsevier; **8.9,** adapted from Dyall S.D. et al., *Science* **304**: 253, © 2004 AAAS; **8.10,** redrawn from Gray M.W., *Curr. Opin. Genet. Dev.* **9**: 678–687, © 1999 Elsevier; **8.11A(a),** © Michael Abbey/VU; **8.11A(b),** © Dr. James W. Richardson/VU; **8.11A(c),** © Dr. Philip Size/VU; **8.11A(d),** © Michael Abbey/VU; **8.11B,** redrawn from Archibald J.M. et al., *Trends Genet.* **18**: 577–584, © 2002 Elsevier; **8.11C,** redrawn from Gilson P.R., *Genome Biol.* **2**: 1022.1–1022.5, © 2001 BioMed Central Ltd.; **8.12,** redrawn from Doolittle W.F., *Nature* **392**: 15–16, © 1998 Macmillan; **8.13,** redrawn from Akhmanova A. et al., *Nature* **396**: 527–528, © 1998 Macmillan; **8.14,** redrawn from Roger A.J. et al., *PNAS* **93**: 14618–14622, © 1996 NAS USA; **8.15A,** reprinted from Wang J. et al., *Appl. Environ. Microbiol.* **68**: 417–422, © 2002 ASM; **8.15B,** reprinted from Fuerst J.A. et al., *PNAS* **88**: 8184–8188, © 1991 NAS USA; **8.16,** modified from Lake J.A. et al., *PNAS* **91**: 2880–2881, © 1994 NAS USA; **8.17A,** modified from Brown T.A., *Genomes, 2e,* Fig 13.24, © 2002 Wiley-Liss Publishers; **8.17B,** based on Alberts B. et al., *Molecular Biology of the Cell, 4e,* Fig 5.34, © 2002 Garland Science; **8.18A,** adapted from Alberts B. et al., *Molecular Biology of the Cell, 4e,* Fig. 4.50, © 2002 Garland Science; **8.18B,** redrawn from Lin X. et al., *Nature* **402**: 761–768, © 1991 Macmillan; **8.19,** redrawn from International Human Genome Sequencing Consortium, *Nature* **409**: 860–921, © 2001 Macmillan; **8.20,** redrawn from Brown T.A., *Genomes, 2e,* Fig 2.26, © 2002 Wiley-Liss Publishers; **8.22,** modified from Lees-Miller J.P. et al., *Mol. Cell. Biol.* **10**: 1729–1742, © 1990 ASM; **8.24,** posttranslational modification section, modified from Seet B.T. et al., *Nat. Rev. Mol. Cell. Biol.* **7**: 473–483, © 2006 Macmillan.

Chapter 9

9.0, Zebrafish embryo, photo courtesy of Nipam H. Patel; **9.1A,C,D,** photos courtesy of Aurora M. Nedelcu, University of New Brunswick; **9.1B,** M.J. Wynne, University of Michigan; **9.2, 9.3,** sources unknown; **9.5A,** redrawn from King N., *Dev. Cell.* **7:** 313–325, © 2004 Elsevier; **9.5B,D,** reprinted from Kent W.S., *A Manual of the Infusoria,* Plates 1V-13, X-2, © 1880, David Brogue, London; **9.5C,** photo courtesy of Nicole King, University of California, Berkeley; **9.6,** human figure, redrawn from da Vinci L., *Vitruvian Man;* alveoli, © Dr. Fred Hossler/VU; red blood cells, © Dr. David M. Phillips/VU; stomach, © Dr. Michael Webb/VU; small intestine, © Dr. Dennis Kunkel/ VU; skeletal muscle, © Science VU/VU; **9.8,** adapted from Reyer R.W., *Q. Rev. Biol.* **29:** 1–46, © 1954 University of Chicago Press; **9.9 left,** © Naylah Feanny/CORBIS; **9.9 right,** reprinted from Chan A.W.S. et al., *Science* **287:** 317–319, © 2000 AAAS; **9.10,** redrawn from deVillartay J.-P. et al., *Nat. Rev. Immunol.* **3:** 962–972, © 2003 Macmillan; **9.11,** modified from Alberts B. et al., *Molecular Biology of the Cell, 4e,* Fig. 7.59, © 2002 Garland Science; **9.12,** redrawn from http://homepages.strath.ac.uk/~dfs99109/Brain/Calcitoningene.gif; **9.13,** reprinted from Hall A., *Science* **279:** 509–514, © 1998 AAAS; **9.14A,B,C,** adapted from Web source, no longer available; **9.14D,** modified from www.northland.cc.mn.us/biology/AP2Online/Fall2001/Nervous/images/motor_sensory_neuron.gif; **9.15,** adapted from Stewart T.A. et al., *PNAS* **78:** 6314–6318, © 1981 NAS USA; **9.19 left,** photo courtesy of Nipam H. Patel; **9.19 middle,** photo courtesy of Nipan H. Patel; **9.19 right,** © James R. McCullagh/VU; **9.20,** photos courtesy of Nipam H. Patel; **9.21 A,B,D,E,F,** photos courtesy of Nipam H. Patel; **9.21C,** © Alex Kerstitch/VU; **9.21G,I,** © Tom Adams/VU; **9.21H,** © James McCullagh/VU; **9.22A–D,** photos courtesy of Nipam H. Patel; **9.22E,** © Steven Haddock, haddock@lifesci.ucsb.edu; **9.22F,** © Erling Svenson; **9.22G,** © Dr. James Castner/VU; **9.23A,** redrawn from www.devbio.com/images/ch11/1101fig1.jpg; **9.23B,** redrawn from www.ls.berkeley.edu/images/divisions/bio/gallery_mcb/fish_embryo>lg.jpg, © Sharon Amacher; **9.23C top,** © Mike Noren; **9.23C bottom,** redrawn from Parichy D.M., *Heredity* **97:** 200–210 © Macmillan; **9.24,** photo courtesy of Nipam H. Patel; **9.25,** modified from McManus M.T. et al., *Nat. Rev. Genet.* **3:** 737– 747, © 2002 Macmillan; **9.26A–C,** modified from Lai E.C. et al., *Dev. Biol.* **269:** 1–17, © 2004 Elsevier; **9.26D–F,** reprinted from Roegiers F. et al., *Nat. Cell. Biol.* **3:** 58–67, © 2001 Macmillan; **9.27 top,** redrawn from Hall D.H. et al., C. elegans *Atlas,* Fig. 8.27B, © CSHLP; **9.27 bottom,** redrawn from www.wormatlas.org/handbook/reproductivesystem/reproductivesystem1.htm.

Chapter 10

10.0, de la Beche, *Duria Antiquor—A More Ancient Dorsetshire,* National Museum of Wales; **10.1,** based on data in Gradstein F. et al., *A Geologic Time Scale,* © 2004 Cambridge University Press; **10.2,** adapted from Smith A.B., *Systematics and the Fossil Record,* p. 62, © 1994 Blackwell Publishing; **10.3,** reprinted from Ostrom J.H., *Bulletin of the Peabody Museum of Natural History,* frontispiece, © 1969 Linda Hall Library; **10.4,** adapted from Wegener A., *Die Entstehung der Kontinente und Ozeane,* © 1915; **10.5A,B,** adapted from Hill R.S. et al., *Palaeobiology 2,* p. 457, © 2001 Blackwell Publishing; **10.5C,** photo courtesy of Joy and Bob Coghlan, Australian Plants Society Tasmania Inc; **10.7A,** photo courtesy of Dr. Shuhai Xiao, Virginia Polytechnic Institute; **10.7B,C,** reprinted from Yuan X. et al., *Doushantuo Fossils: Life on the Eve of Animal Radiation,* pp. 95, 88, © 2002 University of Science and Technology of China Press; **10.8,** reprinted from Bengtson S. et al., *Science* **277:** 1645–1648, © 1997 AAAS; **10.9,** adapted from Selden P.A. et al., *Evolution of Fossil Ecosystems,* p. 10, © 2004 Manson, London; **10.10 top,** redrawn from Fedonkin M.A. et al., *Nature* **388:** 868–871, © 1997 Macmillan; **10.10 bottom,** photo courtesy of Derek Briggs; **10.11,** photo courtesy of Yale Peabody Museum; **10.12,** reprinted from Bengston S. et al., *Science* **257:** 367–369, © 1992 Macmillan; **10.13,** adapted from Fortey R.A., *Fossils: The Key to the Past, 2e,* p. 150, © 1991 Natural History Mu-

seum, London; **10.14 left,** modified from Selden P.A. et al., *Evolution of Fossil Ecosystems,* p. 20, © 2004 Manson, London; **10.14 right,** reprinted from Whittington H.B., *The Burgess Shale,* p.12, © 1985 Yale University Press; **10.15,** modified from Briggs D., *Am. Sci.* **19:** 130–141, © 1991 Sigma Xi, The Scientific Research Co.; **10.16,** adapted from original drawing by Susan Butts, based on design by Matthew Wills; **10.17,** adapted from Blaxter M., *Nature* **413:** 121–122, © 2001 Macmillan; **10.18,** reprinted from Shu D.G. et al., *Nature* **421:** 526–529, © 2003 Macmillan; **10.19,** adapted from Benton M.J., *Palaeobiology 2,* Briggs D. et al., eds., p. 215, © 2001 Blackwell Publishing; **10.20A,C,** adapted from Gould S.J., *Wonderful Life,* p. 46, © 1989 Norton; **10.20B,D,** adapted from Wills M.A. et al., *BioEssays* **22:** 1142–1152, © 2000 John Wiley & Sons; **10.21,** modified from Cracraft J. et al., eds., *Assembling the Tree of Life,* p. 505, © 2004 Oxford University Press; **10.22,** MacNaughton R.B. et al., *Geology* **30:** 391–394, © 2001 Geological Society of America, photo courtesy of Robert MacNaughton, Geological Survey of Canada; **10.23A,B,** reprinted from Wilson H.M. et al., *J. Paleontol.* **78:** 169–184, © 2004 Paleontological Society; **10.23C,** adapted from Gray J. et al., *Am. Sci.* **80:** 444–456, © 1992 Sigma Xi, The Scientific Research Co.; **10.24,** adapted from McKerrow W.S., *An Illustrated Guide,* p. 132, © 1978 MIT Press; **10.25,** adapted from Selden P.A. , *Palaeobiology 2,* Briggs D. et al., eds., p. 73, © 2001 Blackwell Publishing; **10.26,** adapted from Willis K.J. et al., *The Evolution of Plants,* p. 58, © 2002 Oxford University Press; **10.27,** adapted from Selden P.A., et al., *Evolution of Fossil Ecosystems.* p. 50, © 2004 Manson, London; **10.28,** adapted from Dimichele W.A., *Palaeobiology 2,* Briggs D. et al., eds., p. 79, © 2001 Blackwell Publishing; **10.29,** adapted from Coates M., *Palaeobiology 2,* Briggs D. et al., eds., p. 75, © 2001 Blackwell Publishing; **10.30 top,** modified from Engel M.S. et al., *Nature* **427:** 627–630, © 2004 Macmillan; **10.30 bottom,** reprinted from Engel M.S. et al., *Nature* **427:** 627–630, © 2004 Macmillan; **10.31,** reprinted from Benton M.J., *The Book of Life,* p. 23, © 1993 Hutchinson; **10.32,** reprinted from Mayr G. et al., *Science* **310:** 1483–1486, © 2005 AAAS; **10.33A,** reprinted from Xu X. et al., *Nature* **421:** 335–340, © 2003 Macmillan; **10.33B,** reprinted from Chang M.M., ed., *The Jehol Biota,* p. 124, © 2003 Shanghai Scientific; **10.34,** modified from Ridley M., *Evolution,* p. 562, © 1996 Blackwell Publishing; **10.35,** adapted from Sepkoski J.J., *Paleobiology* **10:** 246–267, © 1984 Paleontological Society; **10.36,** adapted from Taylor P.D., ed., *Extinctions in the History of Life,* p. 14, © 2004 Cambridge University Press; **10.37,** © John Sibbick; **10.38,** adapted from Benton M.J., *When Life Nearly Died,* pp. 182, 118, © 2003 Thames and Hudson; **10.39,** modified from the Intergovernmental Panel on Climate Change (IPCC) report, Climate Change 2001: The Scientific Basis, Web source, no longer available.

Chapter 11

11.0, *Hox* gene mRNA expression in crustacean embryo, Danielle Liubicich and Nipam H. Patel; **11.1,** Nipam H. Patel; **11.2,** redrawn from Gilbert S.E., *Developmental Biology, 6e,* p. 366, © 2000 Sinauer Associates; **11.3A,** modified from Lawrence P.A., *The Making of a Fly. The Genetics of Animal Design,* p. 112, © 1992 Blackwell Science; **11.3B,** Nipam H. Patel; **11.4,** image courtesy of Flybase http://flybase.org, credit Rudi Turner, Indiana University; **11.5B,** redrawn from http://www.biosci.ki.se/groups/tbu/homeo.html; **11.6,** modified from Carroll S.B. et al., *From DNA to Diversity: Molecular Genetics and the Evolution of Animal Design, 2e,* p. 115, © 2005 Blackwell Publishing; **11.7,** redrawn from Gilbert S.E., *Developmental Biology, 6e,* p. 366, © 2000 Sinauer Associates; **11.8A,** modified from Ramirez-Solis R. et al., *Cell* **73:** 279–294, © 1993 Elsevier; **11.8B,** modified from Wellik D.M. et al., *Science* **301:** 363–367, © 2003 AAAS; **11.9A–C,** Nipam H. Patel; **11.9D–F,** modified from Carroll S.B. et al., *From DNA to Diversity: Molecular Genetics and the Evolution of Animal Design, 2e,* p. 148, © 2005 Blackwell Publishing; **11.10A–C,** Nipam H. Patel; **11.11A-H,** Nipam H. Patel; **11.12,** adapted from Carroll S.B. et al., *From DNA to Diversity: Molecular Genetics and the Evolution of Animal Design, 2e,* p. 155, © 2005 Blackwell

Publishing; **11.13**, Nipam H. Patel; **11.14**, Nipam H. Patel; **11.15**, redrawn from Lewis D.L. et al., *PNAS* **97**: 4504–4509, © 2000 NAS USA; **11.16**, **11.17**, **11.18**, modified from Averof M. et al., *Nature* **388**: 682–686, © 1997 Macmillan; **11.19**, based on Stern D.L., *Nature* **396**: 463–466, © 1998 Macmillan; **11.20A**, redrawn from http://www.beesies.nl/stekelbaarsje. htm; **11.20B**, redrawn from http://mednews.stanford.edu/mcr/archive/ 2004/04_21_04.html; **11.20C**, redrawn from http://www.speciesatrisk.gc. ca/search/speciesDetails_e.cfm?SpeciesID=554; **11.20D**, redrawn from Bell M. et al., *The Evolutionary Biology of the Three Spine Stickleback*, pp. 1–27, © 1994 Oxford University Press; **11.22**, redrawn from Peichel C.L. et al., *Nature* **414**: 901–905, © 2001 Macmillan and also Shapiro M.D. et al., *Nature* **428**: 717–723, © 2004 Macmillan; **11.23A**, reprinted from Logan M. et al., *Development* **125**: 2825–2835, © 1998 Company of Biologists Ltd; **11.23B,C**, Lanctot C. et al., *Development* **126**: 1805–1810, © 1998 Company of Biologists Ltd; **11.23D**, Marcil A. et al., *Development* **130**: 4555, © 1998 Company of Biologists Ltd; **11.24**, Shapiro M. et al., *Nature* **428**: 717–723, © 2004 Macmillan; **11.25**, photo by John Doebley; **11.26**, Nipam H. Patel, photo by John Doebley; **11.27**, modified from Doebley J. et al., *PNAS* **87**: 9888–9892, © 1990 John Doebley; **11.28**, reprinted from Doebley J., *Annu. Rev. Genet.* **38**: 37–59, © 2004 Annual Reviews; **11.29**, reprinted from Hubbard L. et al., *Genetics* **162**: 1927–1935, © 2002 GSA; **11.30**, modified from http://teosinte.wisc.edu/; **11.31**, reprinted from Jaenicke-Despre V. et al., *Science* **302**: 1206–1208, © 2003 AAAS; **11.32A**, Julia Serrano; **11.32B**, courtesy of Maria del Pilar Gomez and Enrico Nasi; **11.32C**, Nipam H. Patel; **11.32D**, BIODIDAC, biodidac.bio.uottawa.ca; **11.32E,F**, modified from Gehring W.J., *J. Hered.* **96**: 171–184, © 2005 Oxford University Press; **11.33A**, Halder G. et al., *Science* **267**: 1788–1792, ©1995 AAAS; **11.33B**, Gehring W.J., *J. Hered.* **96**: 171–184, © 2005 Oxford University Press.

Chapter 12

12.0, *Escherichia coli* chromosome, reprinted from Cairns J.P., *Cold Spring Harbor Symp. Quant. Biol.* **28**: 44, © 1963 CSHLP; **12.2**, based on Lodish H., *Molecular Cell Biology*, Fig. 8.4, © W.H. Freeman; **12.3B**, modified from Griffiths A.J.F. et al., *Modern Genetic Analysis, 2e*, Fig. 12.4, © 1999 W.H. Freeman; **12.4**, source generated by author, using the MUMMER program, Comprehensive Microbial Resource; **12.6**, modified from http://www.uic.edu/classes/bms655/lesson9.html, Fig. 8; **12.7**, adapted from Madigan M.T. et al., *Brock Biology of Microorganisms, 8e*, Fig. 6.17, © 2006 Pearson Prentice Hall; **12.8A**, modified from Griffiths A.J.F. et al., *Modern Genetic Analysis, 2e*, Fig. 10.6, © 1999 W.H. Freeman; **12.8B**, redrawn from Gardner E.J. et al., *Principles of Genetics, 5e*, © 1984 John Wiley & Sons; **12.9**, modified from Goldstein D.B. et al., *Microsatellites: Evolution and Applications*, © 1999 Oxford University Press; **12.10**, adapted from Friedberg E.C., *DNA Repair*, © 1985 W.H. Freeman; **12.11**, modified from Nelson D.L. et al., *Lehninger Principles of Biochemistry, 3e*, Fig. 10.34, © 2000 Worth Publishers; **12.12**, redrawn from Griffiths A.J.F. et al., *Modern Genetic Analysis, 2e*, Fig. 8.16 © 1999 W.H. Freeman; **12.13A**, modified from Brown T.A., *Genomes, 2e*, Fig. 14.19, © 1999 Wiley-Liss; **12.13B**, redrawn from Watson J.D. et al., *Molecular Biology of the Gene, 3e*, © 1976 W.A. Benjamin; **12.13C**, adapted from Alberts B. et al., *Essential Cell Biology*, Fig. 6.30, © 1998 Garland Publishing; **12.14**, adapted from Alberts B. et al., *Essential Cell Biology*, Fig. 6.18, © 1998 Garland Publishing; **12.15**, Lodish H., *Molecular Cell Biology*, Fig. 12.24, © W.H. Freeman; **12.16A–D**, adapted from http://www.medgen.ubc.ca/ wrobinson/mosaic/tri_how.htm; **12.17A**, redrawn from Griffiths A.J.F. et al., *An Introduction to Genetic Analysis, 8e*, Fig. 15.24, © 2005 W.H. Freeman; **12.17B**, redrawn from Griffiths A.J.F. et al., *Introduction to Genetic Analysis, 8e*, Fig. 15.23, © 2005 W.H. Freeman; **12.18A**, http://www. ncbi.nlm.nih.gov/books/bv.fcgi?rid=cooper.figgrp.823; **12.18B**, Strachan T. et al., *Human Molecular Genetics*, Figs. 9.7, 9.8, © 1999 Garland Science; **12.19**, modified from Griffiths A.J.F. et al., *Modern Genetic Analysis, 2e*, Fig.

8.16, © 1999 W.H. Freeman; **12.21**, Web source, no longer available; **12.22**, reprinted from Baron S., *Medical Microbiology, 4e*, Fig. 5.4, © 1996 University of Texas Medical Branch; **12.23**, modified from Sniegowski P. et al., *BioEssays* **22**: 1057–1066, © 2000 John Wiley & Sons; **12.24B**, modified from Griffiths A.J.F. et al., *An Introduction to Genetic Analysis, 8e*, Fig. 4.6, © 2005 W.H. Freeman; **12.24C**, modified from Griffiths A.J.F. et al., *An Introduction to Genetic Analysis, 8e*, Fig. 4.9, © 2005 W.H. Freeman; **12.25**, based on http://www3.niaid.nih.gov/news/focuson/flu/illustrations/ antigenic/antigenicshift.htm.

Chapter 13

13.0, *Colias* phosphoglucose isomerase, redrawn from Wheat C.W. et al., *Mol. Biol. Evol.* **23**: 499–512, © 2005 Oxford University Press; **13.4**, adapted from Carlson E., *Mendel's Legacy: The Origin of Classical Genetics*, p. 188, © 2004 CSHLP; **13.5A**, adapted from material supplied by Peter Mordan, Natural History Museum, London; **13.5B**, redrawn from Jong P.W. et al., *J. Exp. Biol.* **199**: 2655–2666, © 1996 Company of Biologists Ltd., and other sources; **13.7**, photo courtesy of Nipam H. Patel; **13.9**, modified from Vogel F. et al., *Human Genetics: Problems and Approaches, 3e*, Fig. 12.4, courtesy of U. Barthe-Witte, © 1997 Springer-Verlag; **13.10**, redrawn from Valdes A.M. et al., *Genetics* **133**: 737–749, © 1993 GSA; **13.11**, **13.14**, adapted from Aquadro C.F. et al., *Genetics* **114**: 1165–1190, © 1986 GSA; **13.15A**, reprinted from Hubby J.L. et al., *Genetics* **54**: 577–594, © 1966 GSA; **13.15B**, reprinted from Patil N. et al., *Science* **294**: 1719–1723, © 2001 AAAS; **13.15C**, courtesy of Mark Blaxter, IEB, University of Edinburgh; **13.15D**, data supplied by Silvia Perez-Espona; **13.16**, adapted from Nevo E. et al., in *Lecture Notes in Biomathematics*, Levin S., ed., *V 53: Evolutionary Dynamics of Genetic Diversity*, Mani G.S., ed., pp. 23, 25, © 1984 Springer-Verlag; **13.17A,B**, modified from Lynch M. et al., *Science* **302**: 1401–1404, © 2003 AAAS; **13.18**, © Richard Herrmann/VU; **13.19**, photo courtesy of Ian Stevenson; **13.20**, redrawn from Gillespie J.H., *The Causes of Molecular Evolution*, Fig. 1.17, © 1991 Oxford University Press; data from Nei M. et al., *Evol. Biol.* **17**: 73–118, © 1984 Kluwer Academic Publishers; **13.21**, modified from Cargill M. et al., *Nat. Genet.* **22**: 231–238, © 1999 Macmillan; **13.22**, modified from Livingston R.J. et al., *Genome Res.* **14**: 1821–1831, © 2004 CSHLP; **13.23**, redrawn from The International SNP Map Working Group, *Nature* **409**: 928–933, © 2001 Macmillan; **13.24**, drawn from data from Table 2 in The International SNP Map Working Group. *Nature* **409**: 928–933, © 2001 Macmillan; **13.26A,B**, modified from Kimura M., *The Neutral Theory of Molecular Evolution*, Figs. 4.2, 4.4, © 1983 Cambridge University Press; **13.28A,B**, modified from Sibley C.G. et al., *Sci. Am.* **254**: 82–93, © 1986 Scientific American; **13.29**, redrawn from Kimura M., *Philos. Trans. R. Soc. Lond. B* **312**: 343–354, © 1986 The Royal Society, **13.31**, redrawn from Dobzhansky T. et al., *Genetics* **23**: 28–64, © 1938 GSA; **13.32**, modified from Sebat J. et al., *Science* **305**: 525–528, © 2004 AAAS; **13.33**, reprinted from Eichler E.E. et al., *Science* **301**: 793–797, © 2003 AAAS.

Chapter 14

14.0, Dog breeds, © Carolyn A. McKeone/PR; **14.1**, redrawn from Rutherford S., *Nat. Rev. Genet.* **4**: 263–274, © 2003 Macmillan; **14.2A**, graph, modified from Quackenbush J., *Nat. Rev. Genet.* **2**: 418–427, © 2001 Macmillan; **14.2A**, photo, courtesy of U.S. Dept. of Energy Genomics: GTL Program; **14.2B**, graph, redrawn from Falconer D.S. et al., *Introduction to Quantitative Genetics*, Fig. 6.2C, © 1995 Longman, London; **14.2B**, illustration, redrawn from Patterson J.T., *Studies in the Genetics of Drosophilae of the Southwest*, Plate 1V, © 1943 University of Texas; **14.2C**, graph, redrawn from Berthold P., *Bird Migration: A General Study*, Fig. 7.3, © 2001 Oxford University Press; **14.2C**, photo, © Johann Oli Hilmarsson; **14.3**, **14.5A**, data from Powers L., *Biometrics* **6**: 145–163, © 1950 International Biometric Society; **14.5B**, data courtesy of Sara Via, University of Maryland; **14.5C**, data from Pearson K. et al., *Biometrika* **2**: 357–462, © 1903 Biometrika Trust, University College London; **14.9A,B**, redrawn from Clark A.G. et al., *Genetics* **147**: 157–164, © 1997 GSA; **14.11, 14.12**, based on data from Allison A.C., *Ann. Hum. Genet.* **21**: 67, © 1956 Cambridge University Press, and Allison A.C., *Council for International Organisations for Medical Science Symposium on Abnormal Haemoglobins*, © 1964 Blackwell Publishing; **14.15**, redrawn from McClearn G.E. et al., *Science* **276**: 1560–1563, © 1997 AAAS; **14.19**, redrawn from Mousseau T.A. et al., *Heredity* **59**: 181–198, © 1987 Macmillan; **14.20A,B**, reprinted from Gibson G. et al., *BioEssays* **22**: 372–380, © 2000 John Wiley & Sons; **14.23A–D**, redrawn from Weber K. et al., *Genetics* **153**: 773–786, © 1999 GSA; **14.25, 14.26**, reprinted from Frary A. et al., *Science* **289**: 85–88, © 2000 AAAS; **14.27A,B**, redrawn from Stam L.F. et al., *Genetics* **144**: 1559–1564, © 1996 GSA; **14.28**, redrawn from Barton N.H. et al., *Nat. Rev. Genet.* **3**: 11–21, © 2002 Macmillan; **14.29A**, redrawn from Hayes B. et al., *Genet. Sel. Evol.* **33**: 209–230, © 2001 EDP Sciences; **14.29B**, redrawn from Shrimpton A.E. et al., *Genetics* **118**: 445–459, © 1988 GSA; **14.30**, redrawn from Azevedo R.B.R. et al., *Genetics* **162**: 755–765, © 2002 GSA.

Chapter 15

15.0, Genetic drift; **15.3A–C**, redrawn from Buri P., *Evolution* **10**: 367–402, © 1956 Society for the Study of Evolution; **15.4**, redrawn from Clayton G.A. et al., *J. Genet.* **55**: 131–151, © 1957 Indian Academy of Sciences; **15.9**, data courtesy of Leigh-Brown A.J., Centre for HIV Research, University of Edinburgh; **15.12B,C**, redrawn from Derrida B. et al., *Phys. Rev. Lett.* **82**: 1987–1990, © 1999 American Physical Society; **15.15**, reprinted from Patel N. et al., *Science* **294**: 1719–1723, © 2001 AAAS; **15.16**, modified from Maynard Smith J., *Annu. Rev. Ecol. Syst.* **21**: 1–12, © 1990 Annual Reviews; **15.18**, from data in Stephens J.C. et al., *Am. J. Hum. Genet.* **62**: 1507–1515, © 1998 University of Chicago Press; **15.19**, modified from Phillips M.S. et al., *Nat. Genet.* **33**: 382–387, © 2003 Macmillan; **15.20A,B**, redrawn from Reich D.E. et al., *Nat. Genet.* **32**: 135–142, © 2002 Macmillan.

Chapter 16

16.0, Glanville fritillary, courtesy of Tapio Gustafsson; **16.1A**, adapted from material supplied by Peter B. Mordan, Natural History Museum London; **16.1B,C**, adapted from Ochman H. et al., *PNAS* **80**: 4189–4193, © 1983 Ochman et al.; **16.4A**, reprinted from Dobzhansky T.G., *Dobzhansky's Genetics of Natural Populations*, Fig. 37, © 1981 Columbia University Press; **16.4B**, drawn from data in Dobzhansky T.G. et al., *Genetics* **28**: 304–340, © 1943 GSA; **16.5**, photo courtesy of Nicholas Barton; **16.10**, redrawn from Morjan C.L. et al., *Mol. Ecol.* **13**: 1341–1356, © 2003 Blackwell; **16.16**, redrawn from Hewitt G., *Nature* **405**: 907–913, © 2000 Macmillan; **16.17A,B**, adapted from Avise J.C., *Molecular Markers, Natural History and Evolution*, pp. 243, 244, © 1994 Chapman & Hall; **16.18A**, photo courtesy of Darren E. Irwin, University of British Columbia; **16.18B,C**, redrawn from Irwin D.E., *Evolution* **56**: 2383–2394, © 2002 Society for the Study of Evolution; **16.19A**, photo courtesy Josephine Pemberton, University of Edinburgh; **16.19B,C**, drawn from data from Goodman S.J. et al., *Genetics* **152**: 355–371, © 1999 GSA.

Chapter 17

17.0, Deepsea tubeworms, © Science VU/VU; **17.2A–C**, redrawn from Lehman N. et al., *Curr. Biol.* **3**: 723–734, © 1993 Elsevier; **17.3A**, courtesy Louise Matthews, CTVM Edinburgh; **17.3B**, modified from Woolhouse M. et al., *Nature* **411**: 258–259, © 2001 Macmillan; **17.4**, courtesy CSHL Archives; **17.7A**, NASA/JPL/Malin Space Science Systems; **17.7B**, image from Journal of Chemical Education software; **17.7C**, © Dr. Harold Fisher/VU; **17.7D**, reprinted from Hofer T. et al., *Proc. R. Soc. Biol. Sci.*

259: 249–257, © 1995 Royal Society; **17.8**, © Ken Lucas/VU; **17.10**, Museum Boijmans-van Beuningen, Rotterdam; **17.11**, adapted from Bell G., *Selection: The Mechanism of Evolution*, © 1997 Chapman and Hall, with permission of Springer Science and Business Media; **17.13**, Redrawn from Koza J.R. et al., *Evolution as Computation*, Figs. 6, 9, 10, 13, © 2002 with permission of Springer Science and Business Media; **17.18**, © Rob and Ann Simpson/VU, © Leroy Simon/VU; **17.22**, redrawn from White M.J.D., *Modes of Speciation*. Fig. 16, © 1978 W.H. Freeman; **17.23A,B**, modified from Schappert P., *A World For Butterflies: Their Lives, Behavior and Future*, p. 98, © 2000 Key Porter Books, photo by John Lightfoot, Light Art & Design Inc.; **17.24A**, reprinted from Grant P.R. et al., *Science* **313**: 224, photo courtesy of B. Rosemary Grant; **17.24B**, photo, © D. Parer and E. Parer-Cook; **17.24C,D**, reprinted from Grant P.R. et al., *Science* **313**: 224, photo courtesy of B. Rosemary Grant; **17.25, 17.26**, redrawn from Grant P., *The Ecology and Evolution of Darwin's Finches*, Figs. 55, 56, 60, © 1986 Princeton University Press; **17.27B**, data from Moore A.J., *Evolution* **44**: 315–331, © Society for the Study of Evolution; **17.29A**, photos by John Doebley; **17.29B right**, © Beth Davidow/VU; **17.29B left**, © Inga Spence/VU; **17.30**, modified from Gingerich P.D., *Science* **222**: 159–161, © 1983 AAAS; **17.31A**, redrawn from Dudley J. et al., *Plant Breeding Reviews*, V. 24, Part 1, © 2004 John Wiley & Sons; **17.31B**, redrawn from Barton N.H. et al., *Nat. Rev. Genet.* **3**: 11–21, © 2002 Macmillan; **17.32**, redrawn from Lenski R.E. et al., *PNAS* **91**: 6608–6618, © 1994 NAS USA; **17.33**, redrawn from Barton N.H. et al., *Nat. Rev. Genet.* **3**: 11–21, © 2002 Macmillan; **17.34**, redrawn from Yoo B.H., *Genet. Res.* **35**: 1–17, © Cambridge Unversity Press; **17.35**, redrawn from Clegg M.T. et al., *Genetics* **83**: 793–810, © 1976 GSA; **17.36A**, P&A Macdonald/SNH; **17.36B**, redrawn from Berry R. et al., *J. Zool.* **225**: 615–632, © 1991 Zoological Society of London.

Chapter 18

18.0, Trelogan mine clines, photo courtesy of Janis Antonovics, University of Virginia; **18.3**, Web source, no longer available; **18.4**, modified from Vogel F. et al., *Human Genetics: Problems and Approaches*, Fig. 12.19, © 1997 Springer; **18.5A,B**, redrawn from Cooper V.S. et al., *Nature* **407**: 736–739. © 2000 Macmillan; **18.5C**, redrawn from Lenski R.E. et al., *PNAS* **91**: 6608–6618, © 1994 NAS USA; **18.5D**, redrawn from Cooper V.S. et al., *J. Bacteriol.* **183**: 2834–2841, © 2001 ASM; **18.10A–D**, redrawn from Shigesada N. et al., *Biological Invasions*, Figs. 2.1, 2.11, © 1997 Oxford University Press; **18.11**, modified from McCaskill J.S. et al., *PNAS* **90**: 4191–4195, © 1993 NAS USA; **18.12**, reprinted from Nachman M.W. et al., *PNAS* **100**: 5268–5273, © 2003 NAS USA; **18.13A**, © Rudolf Svensen; **18.13B,C**, Christiansen F.B., *Lecture Notes in Biomathematics: Measuring Selection in Natural Populations*, Christiansen and Fenchel, eds., **19**: 21–49, © 1977 Springer Verlag; **18.14A**, photo courtesy of Janis Antonovics, University of Virginia; **18.14B**, redrawn from McNeilly T., *Heredity* **23**: 99–108, © 1968 Macmillan; **18.15**, redrawn from Lenormand T. et al., *Nature* **400**: 861–864, © 1999 Macmillan; **18.16A,B**, photos courtesy of Nick Barton; **18.16C**, redrawn from Barton N.H. et al., *Nature* **341**: 497–503, © 1989 Macmillan; **18.16D**, photo courtesy of Nick Barton; **18.17**, modified from Mallet J.L.B. et al., *Evolution* **43**: 421–431, © 1989 Society for the Study of Evolution; **18.18**, photo courtesy of Walt and Mimi Miller; **18.19**, redrawn from Lu Y., *J. Mol. Evol.* **57**: 784–793, © 2002 Springer-Verlag; **18.20**, photo courtesy of Nipam H. Patel; **18.21A–C**, reprinted from Rainey R.B. et al., *Nature* **394**: 69-72, © 1998 Macmillan; **18.23**, redrawn from Bell G., *Selection: The Mechanism of Evolution*, p. 550, © 1997 Chapman and Hall; **18.24A**, http://www.esg.montana.edu/aim/taxa/mollusca/pag1044b.jpg; **18.24B,C**, redrawn from Lively C.M. et al., *Nature* **405**: 679–691, © 2000 Macmillan; **18.24D**, redrawn from Dybdahl M.F. et al., *Evolution* **52**: 1057–1066, © 1998 Society for the Study of Evolution; **18.25**, redrawn from Jenkins G.M. et al., *Mol. Biol. Evol.* **18**: 987–994, © 2001 Oxford University Press; **18.27A,** photo courtesy of Kathie Miller, www.soayfarms. com; **18.27B,C,** redrawn from Coltman D.W. et al., *Evolution* **53:** 1259-1267, © 1999 Society for the Study of Evolution; **18.28A,** photo courtesy of Tari Haahtela; **18.28B,C,** redrawn from Saccheri I. et al., *Nature* **392:** 491–494, © 1998 Macmillan; **18.30,** redrawn from Crnokrak P. et al., *Evolution* **56:** 2347– 2358, © 1999 Society for the Study of Evolution.

Chapter 19

19.0, Barn swallows, photo courtesy of Matthew Evans, University of Exeter in Cornwall; **19.1A,** modified from Thatcher J.W. et al., *PNAS* **95:** 253–257, © 1998 NAS USA; **19.1B,** modified from Fowler K. et al., *Proc. R. Soc. Lond. B* **264:** 191–199, © 1997 Royal Society; **19.2,** modified from Giaever G. et al., *Nature* **418:** 387–391, © 2002 Macmillan; **19.3A,** modified from van Delden W. et al., *Genetics* **90:** 161–191, © 1978 GSA; **19.3B,** modified from Berry A. et al., *Genetics* **134:** 869–893, © 1993 GSA; **19.4,** modified from Ridley M., *Evolution, 1e,* p. 77, © 1993 Blackwell; **19.5,** modified from Rozenzweig R.F., *Genetics* **137:** 903–917, © 1994 GSA; **19.6A,** © Gary Meszaros/VU; **19.6B,** modified from Bell G., *Proc. R. Soc. Lond. B* **224:** 223–265, © 1985 Royal Society; **19.7,** modified from Schluter D., *Evolution* **42:** 849–861, © 1988 Society for the Study of Evolution; **19.8A,** modified from Merila J. et al., *Genetica* **112–113:** 199–222, © 2001 Springer; **19.8B,** photo courtesy of Dr. Loeske Kruuk; **19.9A,** photo courtesy of Matthew Evans, University of Exeter in Cornwall; **19.9B,** redrawn from Rowe L.V. et al., *Behav. Ecol.* **12:** 157–163, © 2001 Oxford University Press; **19.10A,** redrawn from Kingsolver J.G. et al., *Am. Naturalist* **157:** 245–261, © 2001 University of Chicago Press; **19.10B,** redrawn from Butlin R.K. et al., *Philos. Trans. R Soc. Lond. B* **334:** 297–308, © 1991 Royal Society; **19.11,** data from Kimura M., *The Neutral Theory of Molecular Evolution,* © 1983 Cambridge University Press; **19.12,** modified from Gillespie J.H., *The Causes of Molecular Evolution,* © 1991 Oxford University Press; **19.13,** redrawn from Yang Z. et al., *Mol. Biol. Evol.* **17:** 1446–1455, © 2000 Oxford University Press; **19.14,** redrawn from Kimura M., *The Neutral Theory of Molecular Evolution,* Fig. 8.3, © 1983 Cambridge University Press; **19.16,** modified from Skibinski D.O.F. et al., *Genetics* **135:** 233–248, © 1993 GSA; **19.17C,** redrawn from Kim Y. et al., *Genetics* **160:** 765–777, © 2002 GSA; **19.18,** modified from Harr B. et al., *PNAS* **99:** 12949–12954, © 2002 NAS USA; **19.20,** data from Baudry E. et al., *Genetics* **158:** 1725–1735, © 2001 GSA; **19.21A,** modified from Andolfatto P. et al., *Genetics* **158:** 657–665, © 2001 GSA; **19.21B,** modified from Hudson R.R. et al., *Philos. Trans. R. Soc. Lond. B* **349:** 19–23, © 1995 Royal Society; **19.22,** redrawn from Richman A., *Mol. Ecol.* **9:** 1953–1964, © 2000 Blackwell, and references therein; **19.23,** redrawn from Charlesworth D., *Curr. Biol.* **12:** R424–R426, © 2002 Elsevier; **19.24A,** photo courtesy of Jeffrey L. Feder, University of Notre Dame; **19.24B,** redrawn from Stolz U. et al., *PNAS* **100:**14955–14959, © 2003 NAS USA; **19.25,** redrawn from Akashi H., *Curr. Opin. Genet. Dev.* **11:** 660–666, © 2001 Elsevier; **19.26,** redrawn from Smith N.G.C. et al., *J. Mol. Evol.* **53:** 225–236, © 2001 Springer; **19.27A,B,** redrawn from Ke A. et al., *The RNA World, 3e,* p. 124, Gesteland R.F. et al., eds., © 2006 CSHLP; **19.28,** redrawn from Parsch J. et al., *Genetics* **154:** 909–921, © 2000 GSA; **19.29,** redrawn from Kirby D.A. et al., *PNAS* **92:** 9047–9051, © 1995 NAS USA; **19.30,** redrawn from Shabalina S.A. et al., *Trends Genet.* **17:** 373–376, © 2001 Elsevier; **19.31,** courtesy of Nipam H. Patel; **19.32,** redrawn from Kruuk L.E.B. et al., *PNAS* **97:** 698–703, © 2000 NAS USA.

Chapter 20

20.0 left, Peacock feather, © John Gerlach/VU; **20.0 right,** northern saw whet owl, © Joe McDonald/VU; **20.1 top, middle,** © Gary Meszaros/VU; **20.1 bottom,** © Richard Walters/VU; **20.2,** reprinted from Timsit Y. et al., *J. Mol. Biol.* **284:** 1289–1299, © 1998 Elsevier, photo courtesy of

Youri Timsit; **20.3A,** © Stephen Cresswell; **20.3C,** reprinted from Parker G.A. et al., *Nature* **370:** 53–56, © 1994 Macmillan; **20.4A–C,** modified from Ibarra R.U. et al., *Nature* **420:** 186–189, © 2002 Macmillan; **20.5,** Kipling R., *Just So Stories,* © 1926 Macmillan; **20.7,** redrawn from Szathmary E., *Proc. R Soc. Lond. B* **245:** 91–99, © 1991 The Royal Society; **20.8A,** redrawn from Promislow D.E.L., *Evolution* **45:** 1869–1887, © 1991 Society for the Study of Evolution; **20.8B,** redrawn from Ricklefs R.E., *Am. Nat.* **152:** 24–44, © 1998 University of Chicago Press; **20.9,** © Dr. David Phillips/VU; **20.10, 20.11,** redrawn from Partridge L. et al., *Nature* **362:** 305–311, © 1993 Macmillan; **20.12,** redrawn from Ricklefs R.E., *Am. Nat.* **152:** 24–44, © 1998 University of Chicago Press; **20.13,** redrawn from Partridge L. et al., *Nature* **362:** 305–311, © 1993 Macmillan; **20.14,** redrawn from Partridge L. et al., *Nat. Rev. Genet.* **3:** 165–175, © 2002 Macmillan; **20.17A–C,** modified from Kerr B. et al., *Nature* **418:** 171–174, © 2002 Macmillan; **20.18A,** reprinted from Sinervo B., *Nature* **380:** 240, © 1996 Macmillan; **20.18B,** redrawn from Sinervo B., *Genetica* **112:** 417–434, © 2001 Kluwer Academic Publishers; **20.19A,** © Tom J. Ulrich/VU; **20.19B,** © Gerald and Buff Corsi/VU; **20.19C,** © Rob and Ann Simpson/VU; **20.19D,** © Rick and Nora Bowers/VU; **20.19E,** reprinted from Anderson B., *Am. J. Botany* **92:** 1342–1349, © 2005 Botanical Society of America; **20.20,** redrawn from Eberhard W.G., *Sexual Selection and Animal Genitalia,* Fig 1.4, © 1985 Harvard University Press; **20.21 top,** © Wally Eberhart/VU; **20.21 bottom,** © Dr. Dennis Kunkel/VU; **20.22,** reprinted from Hagen D.W. et al., *Evolution* **34:** 1050–1059, © 1980 Society for the Study of Evolution; **20.23 top, bottom,** © Arthur Morris/VU; **20.24,** redrawn from Harcourt A.H. et al., *Nature* **293:** 55–57, © 1981 Macmillan; **20.25A top,** © Gerard Fuehre/VU; **20.25A bottom,** © Arthur Morris/VU; **20.25B,** redrawn from O'Donald P. et al., *Heredity* **33:** 1–16, © 1974 Macmillan; **20.26,** redrawn from Eberhard W.G., *Sexual Selection and Animal Genitalia,* Fig. 2.3, © 1985 Harvard University Press; **20.27,** Dave Menke, U.S. Fish and Wildlife Service; **20.28,** redrawn from Clark A.G., *Heredity* **88:** 148–153, © 2002 Macmillan; **20.29,** © Ray Coleman/VU; **20.30A,** © Gregory G. Dimijian/PR; **20.30B,** redrawn from Kirkpatrick M. et al., *Nature* **350:** 33–38, © 1991 Macmillan; **20.32A,** © Arthur Morris/VU; **20.32B,** redrawn from Moore A.J., *Behav. Ecol. Sociobiol.* **35:** 235–241, © 1994 Springer-Verlag GmbH & Co; **20.32C,** redrawn from Norris K., *Nature* **362:** 537–539, © 1993 Macmillan; **20.33A,** http://gallery.insect.cz; photo courtesy of Josef Dvorak; **20.33B,** redrawn from Moore A.J., *Behav. Evol. Sociobiol.* **35:** 235–241, © 1994 Springer-Verlag GmbH & Co; **20.34A,** http://www.camacdonald.com/birding/GreatReed-Warbler(PD).jpg, photo courtesy of Pascal Dubois; **20.34B,** redrawn from Hasselquist D. et al., *Nature* **381:** 229–232, © 1996 Macmillan.

Chapter 21

21.0, Rufous hummingbird, © Charles Melton/VU; **21.1,** modified from Watson J.D. et al., *Molecular Biology of the Gene, 5e,* Fig. 10.14, © 2004 Pearson Education Inc; **21.2,** redrawn from Whitehouse H.L.K, *Towards an Understanding of the Mechanics of Heredity,* Plate 16.1B, © 1973 E. Arnold, London; **21.3,** APS Library, photo courtesy of NLM; **21.6A,** photo courtesy of Michael E. Clark; **21.6D,E,** reprinted from Nur U. et al., *Science* **240:** 512–514, © 1988 AAAS; **21.7,** reprinted from Partridge L. et al., *Science* **281:** 2003–2008, © 1998 AAAS; **21.10A–C,** modified from Lyttle T.W., *Genetics* **91:** 339–357, © 1979 GSA; **21.12,** modified from Bartolome C. et al., *Mol. Biol. Evol.* **19:** 926–937, © 2002 Oxford University Press; **21.13A,** redrawn from Cavalier-Smith T., *Annu. Rev. Biophys. Bioeng.* **11:** 273–302, © 1982 Annual Reviews; **21.13B,** modified from Bennett M.D., *Proc. R Soc. Lond. B* **181:** 109–135, © 1972 Royal Society of London; **21.14,** photo courtesy of Alessandro Catenazzi; **21.15,** reprinted from Lynch M. et al., *Science* **302:** 1401–1404, © 2003 AAAS; **21.16,** redrawn from International Human Genome Sequencing Consortium, *Nature* **409:** 860, © 2001 AAAS; **21.18A,** © Wendy Dennis/VU; **21.18B,** © E. S. Ross/VU;

21.18C, Courtesy of Ben Hatchwell; **21.18D,** © Fritz Polking/ VU; **21.18E,** reprinted from Chao L. et al., *PNAS* **78:** 6324–6328, © 1981 NAS USA; **21.21A,B,** redrawn from Cook J.M. et al., *Trends Ecol. Evol.* **18:** 243, © 2003 Elsevier; **21.22,** modified from Herre E.A., *Science* **228:** 896–898, © 1985 AAAS; **21.23A,** © Arthur Morris/VU; **21.23B,** redrawn from Kilner R.M. et al., *Science* **305:** 877–879, © 2004 AAAS; **21.24,** based on Watson J.D. et al., *Molecular Biology of the Gene, 5e,* Fig. 21.27, © 2004 Pearson Education Inc; **21.26A,** photo courtesy of David Pfennig, University of North Carolina; **21.26B,** modified from Pfennig D.W., *Proc. R. Soc. Lond. B* **266:** 57–61, © 1999 Royal Society of London; **21.27,** modified from Griffin A.S. et al., *Science* **302:** 634–636, © 2003 AAAS; **21.28,** based on data from Wright S., *Evolution: Selected Papers,* Fig. 4, © 1986 University of Chicago Press; **21.29A,** © Dr. David M. Phillips/VU; **21.29B,** © Dr Dennis Kunkel/ VU; **21.29C,** courtesy of Mathieu Joron, © 2006; **21.29D,** © Greg Vandeleest/VU; **21.30,** reprinted from Griffin A.S. et al., *Trends Ecol. Evol.* **17:** 15–21, © 2002 Elsevier, photo courtesy of Ashleigh Griffin; **21.31,** © Charles Melton/VU; **21.32,** reprinted from Clutton-Brock T., *Science* **296:** 69–72, © 2002 AAAS; **21.34A,** © Ken Wagner/VU; **21.34B,** redrawn from Pellmyr O. et al., *Nature* **372:** 257–260, © 1994 Macmillan; **21.35,** redrawn from Kiers E.T. et al., *Nature* **425:** 78–81, © 2003 Macmillan; **21.37,** adapted from Maynard Smith J. et al., *The Major Transitions in Evolution,* Fig 4.11, © 1996 W.H. Freeman.

Chapter 22

22.0, American Southwest desert species; **22.1 top,** © Ken Lucas/VU; **22.1 bottom,** © Gerald & Buff Corsi/VU; **22.2A,** © Rod Williams/ naturepl.com; **22.2B,** © David Fox/www.osfimages.com; **22.2C,** redrawn from Murray J. et al., *Proc. R. Soc. Lond. B* **211:** 83–117, © 1980 Royal Society of London; **22.3,** reprinted from Keeling P.J., *Nature* **414:** 401–402, © 2001 Macmillan; **22.5,** © Tom Brakefield/CORBIS; **22.6A,** © Doug Sokell/VU; **22.6B,** modified from Howard D.J. et al., *Evolution* **51:** 747–755, © 1997 Society for the Study of Evolution; **22.8A,** photo courtesy of Takayuki Ohgushi, University of Kyoto; **22.8B,** © Adam Jones/ VU; **22.11,** modified from Coyne J.A. et al., *Evolution* **51:** 295–303, © 1997 Society for the Study of Evolution; **22.12,** modified from Roberts M.S. et al., *Genetics* **134:** 401–408, © 1993 GSA; **22.13,** Connecticut Botanical Society, photo courtesy of Eleanor Saulys; **22.14,** www.sedumphotos.net, photo courtesy of Wayne Fagerlund; **22.15,** redrawn from Otto S.P. et al., *Annu. Rev. Genet.* **34:** 401–437, © 2000 Annual Reviews; **22.16,** photo courtesy of Doug Schemske and Toby Bradshaw; **22.17,** modified from Bradshaw H.D. et al., *Genetics* **149:** 367–382, © 1998 GSA; **22.18,** redrawn from True J.R. et al., *Genetics* **142:** 819–837, © 1996 GSA; **22.19A,B,** redrawn from Zeng Z.B. et al., *Genetics* **154:** 299–310, © 2000 GSA; **22.20A–C,** reprinted from Orr H.A. et al., *Bioessays* **22:** 1085–1094, © 2000 John Wiley & Sons; **22.21A–C,** redrawn from Presgraves D.C. et al., *Nature* **423:** 715–719, © 2003 Macmillan; **22.22,** modified from Lande R., *Evolution* **33:** 234-251, © 1979 Society for the Study of Evolution; **22.23,** redrawn from Bush G.L. et al., *PNAS* **74:** 3942–3946, © 1977 NAS USA; **22.26B,** redrawn from Searle J.B., *Proc. R. Soc. Lond. B* **229:** 277–298, © 1986 Royal Society of London; **22.28A,** © Rob and Ann Simpson/VU; **22.28B,** modified from Pounds J.A. et al., *Evolution* **35:** 516–528, © 1981 Society for the Study of Evolution; **22.30A,** redrawn from Noor M.A.F. et al., *PNAS* **98:** 12084–12088, © 2001 NAS USA; **22.30B,** redrawn from Machado C.A. et al., *Proc. R. Soc. Lond. B* **270:** 1193–1202, © 1986 Royal Society of London; **22.31A,B,** redrawn from Stre G.-P. et al., *Nature* **387:** 589–592, © 1997 Macmillan; **22.32,** modified from http://www.jason.oceanobs.com/ html/actualites/image_du_mois/200303_uk.html; **22.33, 22.34,** reprinted from Schliewen U.K. et al., *Mol. Ecol.* **10:** 1471–1488, © 2001 Blackwell Publishing; **22.36,** modified from Pritchard J.R. et al., *Evol. Ecol. Res.* **3:** 209–220, with permission of authors; **22.37A,** © Bill Beatty/VU; **22.37B,** redrawn from Feder J.L. et al., *PNAS* **100:**

10314–10319, © 2003 NAS USA; **22.37C,** redrawn from Filchak K.E. et al., *Nature* **407:** 739–742, © 2000 Macmillan.

Chapter 23

23.0, Human egg and sperm, © Dr. John Cunningham/VU; **23.1,** redrawn from Sniegowski P. et al., *BioEssays* **22:** 1057–1066, © 2000 John Wiley & Sons; **23.4A,B,** modified from Giraud A. et al., *Science* **291:** 2606–2608, © 2001 AAAS; **23.8A,** photo courtesy of Carl Lieb, Laboratory of Environmental Biology, University of Texas, El Paso; **23.8B,** photo courtesy of Robert C. Vrijenhoek, Monterey Bay Aquarium Research Institute; **23.8C,** © Robert Bielesch; **23.8D,** reprinted from Butlin R., *Nat. Rev. Genet.* **3:** 311–317, © 2002 Macmillan; **23.9,** redrawn from Rice W.R., *Nat. Rev. Genet.* **3:** 241–246, © 2002 Macmillan; **23.10A,B,** reprinted from Butlin R., *Nat. Rev. Genet.* **3:** 311–317, © 2002 Macmillan; **23.14A,** redrawn from Lively C.M. et al., *Evol. Ecol. Res.* **4:** 219–226, © 2002 Evolutionary Ecology Ltd; **23.14B,** photo courtesy of Dr. Daniel L. Gustafson, Montana State University; **23.16,** redrawn from Barton N. et al., *Science* **281:** 1986–1990, © 1998 AAAS; **23.17A,B,** modified from Elena S.F. et al., *Nature* **390:** 395–398, © 1997 Macmillan; **23.18,** redrawn from Barton N. et al., *Science* **281:** 1986–1990, © 1998 AAAS; **23.20,** modified from Burt A., *Evolution* **54:** 337–351, © 2000 Society for the Study of Evolution; **23.21,** modified from Smith G.P., *Nature* **370:** 324–325, © 1995 Macmillan; **23.22A,B,** redrawn from Colegrave N., *Nature* **420:** 664–666, © 2002 Macmillan; **23.23,** redrawn from Blirt A. et al., *Nature* **326:** 803–805, © 1987 Macmillan; **23.24,** redrawn from Otto S.P. et al., *Nat. Rev. Genet.* **3:** 252–261, © 2002 Macmillan; **23.26,** modified from Chadwick D. et al., *The Genetics and Biology of Sex Determination,* p. 211, © 2002 J. Wiley; **23.27,** reprinted from Lahn B.T. et al., *Nat. Rev. Genet.* **2:** 207–216, © 2001 Macmillan; **23.28,** redrawn from Mable B.K. et al., *BioEssays* **20:** 453–462, © 1998 John Wiley & Sons; **23.29,** redrawn from Maynard-Smith J. et al., *The Major Transitions in Evolution,* Fig. 9.1a,b, © 1995 W.H. Freeman, **23.31,** © Wim van Egmond/VU; **23.32A,B,** reprinted from Barrett S.C.H., *Nat. Rev. Genet.* **3:** 274–284, © 2002 Macmillan, photo credit L.D. Harder, University of Calgary; **23.32C,** reprinted from Barrett S.C.H., *Nat. Rev. Genet.* **3:** 274–284, © 2002 Macmillan, photo credit Q.-L. Li, Xishuangbanna Tropical Botanical Garden, Mengla, China; **23.33A–C,** redrawn from Otte D. et al., *Speciation and Its Consequences,* Barrett, Figs. 3 and 4, © 1989 Sinauer Associates; **23.34A,** redrawn from Barrett S.C., *Nat. Rev. Genet.* **3:** 274–284, © 2002 Macmillan; **23.34B,** reprinted from Barrett S.C., *Nat. Rev. Genet.* **3:** 274–284, © 2002 Macmillan, photo courtesy of Daniel J. Schoen, McGill University, Canada; **23.35,** redrawn from Blomqvist D. et al., *Nature* **419:** 613–615, © 2002 Macmillan; **23.37,** redrawn from Kirschner M. et al., *Cell* **100:** 79–88, © 2000 Elsevier; **23.38A–C,** reprinted from Rutherford S.L. et al., *Nature* **396:** 336–346, © 1998 Macmillan; **23.38D,** redrawn from Rutherford S.L. et al., *Nature* **396:** 336–346, © 1998 Macmillan.

Chapter 24

24.0, Western U.S. beaver, beaver dam, © Thomas & Pat Leeson/PR; **24.2,** redrawn from Thompson D., *On Growth and Form,* Bonner J.T., ed., p. 301, © 1961 Cambridge University Press; **24.4A,** photo courtesy of Nipam H. Patel; **24.4C,** modified from Turner J.R.G., *Ecological Genetics and Evolution,* E.R. Creed, ed., Fig.11.1, © 1971 Oxford, Blackwell Scientific; **24.5,** reprinted from Brakefield P.M., *Zoology* **106:** 283–290, © 2003 Elsevier; **24.6,** reprinted from Joyce C.M., *PNAS* **94:** 1619–1622, © 1997 NAS USA; **24.7,** reprinted from Wheat C.W. et al., *Mol. Biol. Evol.* **23:** 499–512, © 2006 Oxford University Press; photo courtesy of Christopher Wheat; **24.9,** adapted from http://www.slic2.wsu.edu:82/hurlbert/micro101/ images/lock_key.gif; **24.11,** redrawn from Mazon G. et al., *Microbiology* **150:** 3783–3795, © 2004 MAIK Nauka/Interperiodic Publishing; **24.12,** adapted from Fry R.C. et al., *Annu. Rev. Microbiol.* **59:** 357–377,

© 2005 Annual Reviews; **24.13,** redrawn from Mazon G. et al., *Microbiology* **150:** 3783–3795, © 2004 MAIK Nauka/Interperiodic Publishing; **24.14,** redrawn from Wistow G. et al., *PNAS* **87:** 6277–6280, © 1990 NAS USA; **24.15,** redrawn from Logsdon J.M. et al., *PNAS* **94:** 3485–3487, © 1997 NAS USA; photo courtesy of Paul A. Cziko; **24.16,** modified from Partridge L. et al., *Nature* **407:** 457–458, © 2000 Macmillan; **24.17,** redrawn from Eisen J.A., *Genome Res.* **8:** 163–167, © 1998 CSHLP; **24.19,** modified from Eisen J.A., *Nucleic Acids Res.* **23:** 2715–2723, © 1995 Oxford University Press; **24.20,** redrawn from Aravind L. et al., *Nucleic Acids Res.* **27:** 1223–1242, © 1999 Oxford University Press; **24.21,** © Dr. Don W. Fawcett/VU; **24.22,** modified from Kirschner M.W. et al., *The Plausibility of Life: Resolving Darwin's Dilemma,* Fig. 24, © 2005 Yale University Press; **24.23,** adapted from Koonin E.V. et al., *Curr. Opin. Genet. Dev.* **6:** 757–762, © 1996 Elsevier; **24.24,** photo courtesy of University of Sussex; **24.25,** Ichthyostega limb skeleton, http://www.geocities.com/gilson_medufpr/icthiostega.html; **24.25,** skeletal limbs, redrawn from Shubin N.H. et al., *Nature* **440:** 764–771, © 2006 Macmillan; **24.25 top two images,** adapted from Coates M., *Palaeobiology II,* p. 75, Briggs and Crowther, eds., © 2001 Blackwell Science; **24.25 bottom three images,** adapted from Ahlberg P.E. et al., *Nature* **440:** 747–749, © 2006 Macmillan; **24.26,** photo Tom Way, courtesy of International Business Machines Corp; **24.27,** modified from Mitchell M., *An Introduction to Genetic Algorithms,* pp. 36–39, © 1998 MIT Press.

Chapter 25

25.0, *Australopithecus* skulls, © Daniel Herard/PR; **25.1,** Based on data at http://www.tol.org; **25.2,** chimp, © Fritz Polking/VU; gorilla, © Joe McDonald/VU; orangutan, © Theo Allofs/VU; Darwin, © Library of Congress/PR; **25.3,** adapted from Wood B. et al., *Science* **284:** 65–71, © 1999 AAAS; **25.4,** adapted from Wilson A.C. et al., *PNAS* **63:** 1088–1093, © 1969 NAS USA; **25.5,** adapted from Ruvolo M. et al., *PNAS* **91:** 8900–8904, © 1994 NAS USA; **25.6,** reprinted from Brunet M. et al., *Nature* **418:** 145–151, © 2002 Macmillan; **25.7,** adapted from Johanson D. et al., *From Lucy to Language,* p. 38, with permission of Nèvraumont Publishing Co.; **25.8,** © John Reader/PR; **25.9,** Johanson D. et al., *From Lucy to Language,* p. 123, with permission of Nèvraumont Publishing Co., photo by Robert I.M. Campbell, courtesy of National Museums of Kenya; **25.10,** Lewin R., *Human Evolution, An Illustrated Introduction, 3e,* p. 107, © 1993 Blackwell Scientific; **25.11,** adapted from http://www.scientific-art.com/portfolio%20palaeontology%20pages/skulls.htm, © 1994 Deborah Maizels; data for brain volumes from Carroll S., *Nature* **422:** 849–857, © 2003 Macmillan; **25.12,** © Pascal Goetgheluck/PR;

25.13, adapted from Hublin J.-J., pp. 99–121 and Rightmire G.P., pp. 123–133, *Human Roots: Africa and Asia in the Middle Pleistocene,* L. Barham et al., eds., © 2001 Western Academic & Specialist Press; **25.14,** adapted from Lahr M.M. et al., *Nature* **431:** 1043–1044, © 2004 Macmillan; **25.15 left,** The Natural History Museum, London, © Michael Day; **25.15 right,** © The Natural History Museum, London; **25.16,** adapted from Ferris M. et al., *Nature* **396:** 226–228, © 1998 Macmillan; **25.17,** reprinted from Brown P. et al., *Nature* **431:** 1055–1061, © 2004 Macmillan; **25.18,** © Pascal Goetgheluck/PR; **25.19,** adapted from Menozzi P. et al., *Science* **201:** 786–792, © 1978 AAAS; **25.20,** adapted from Cavalli-Sforza L. et al., *The History and Geography of Human Genes,* © 1994 Princeton University Press; **25.21,** adapted from Goldstein D.B. et al., *PNAS* **92:** 6723–6727, © 1995 NAS USA; **25.22,** adapted from Barbujani G. et al., *Annu. Rev. Genomics Hum. Genet.* **5:** 119–150, © 2004 Annual Reviews; **25.23,** photo, © Dr. K.G. Murti/VU; **25.23,** illustration, adapted from Underhill P.A. et al., *Nat. Genet.* **26:** 358–361, © 2000 Macmillan; **25.24,** adapted from Reich D.E. et al., *PNAS* **95:** 8119–8123, © 1998 NAS USA; **25.25,** adapted from Serre D. et al., *PLoS Biol.* **2:** E57, © 2004 Public Library of Science; **25.26,** adapted from Enard W. et al., *Science* **296:** 340–343, © 2002 AAAS; **25.27,** adapted from Enard W. et al., *Nature* **418:** 869–872, © 2002 Macmillan; **25.28,** reprinted from Jackendoff R., *Trends Cogn. Sci.* **3:** 272–279, © 1991 Elsevier.

Chapter 26

26.0, Human crowd, © Mark Burnett/PR; **26.3,** modified from Hamosh A. et al., *Nucleic Acids Res.* **33:** D514–D517, © 2005 Oxford University Press; **26.6,** redrawn from Emahazion T. et al., *Trends Genet.* **17:** 407–413, © 2001 Elsevier; **26.9,** adapted from Jha P. et al., *Lancet* **367:** 211–218, © 2006 Little, Brown & Co.; **26.10,** modified from Barbujani G. et al., *PNAS* **87:** 1816–1819, © 1990 NAS USA; **26.11,** redrawn from Akey J.M. et al., *Genome Res.* **12:** 1805–1814, © 2002 CSHLP; **26.12,** redrawn from Bersaglieri T. et al., *Am. J. Hum. Genet.* **74:** 1111–1120, © 2004 University of Chicago Press; **26.13,** modified from Barsh G.S., *PLoS Biol.* **1:** 019, © 2003 Public Library of Science; **26.14,** redrawn from Schliekelman P. et al., *Nature* **411:** 545, © 2001 Macmillan; **26.16,** adapted from Flaxman S.M. et al., *Q. Rev. Biol.* **75:** 113–148, © 2000 University of Chicago Press.

Back Endpaper

Courtesy of Sandie Baldauf, University of York.

Index

A

Abalone, 533, 533f
Abdominal-A (abd-A) gene, 288–289, 289f, 290f, 299–300, 300f
Abdominal-B (Abd-B) gene, 288–289, 289f, 290f
Absolute fitness, 419, 461, 462, 468, 470
Acanthostega, 276, 277f
Acetyl-coenzyme A synthetase (ACS), 525
Achaete-scute region, *Drosophila,* 404, 405f
Achondroplasia, 511
Acritarchs, 202, 202f, 260, 260f
Acrocephalus arundinaceus, 583f
Actin, 235, 236, 236f
Active site, enzyme, 700–701, 701f
Adaptation
    alleles of small effect, influence of, 407–408, 407f
    diversity of, 2f
    induction compared, 52
    natural selection as cause of, 10, 23, 457, 463–464
Adaptation and Natural Selection (Williams), 34
Adaptationist program, 558, 560
Adaptive landscape
    description, 79f, 472–473, 473f
    limitations of metaphor, 473–475
    novelty evolution, 697, 697f
    random drift and selection, balance between, 494–496, 495f
    shifting balance theory (SBT), 607–608, 608f
    speciation and, 641–642
Adaptive peaks, 473, 479, 480f, 493, 494–496, 495f, 560f, 607–609, 641–642, 697, 698f
Adaptive radiations, 70
Adaptor hypothesis, 47, 49
Additive genetic variance
    definition, 394
    estimation of variance components, 394–397, 395f, 397f
    extent of natural selection, as measure of, 548–550
    late acting mutations, 566
    narrow-sense heritability, 391, 399
    in negative linkage disequilibrium, 674–675, 675f
    quantitative traits, 446, 478–479
    random drift, effect of, 418, 418f

recombination and, 678
selection effect on, 482–484
sickle-cell hemoglobin, 391
Additive genetic variance in fitness, 462–463, 462f
Additive model, 385, 392, 402, 409
Additive selection, 471, 471f, 471t
Additive tree, 112
Adenosine triphosphate (ATP), use in Calvin cycle, 161
Adh gene. *See* Alcohol dehydrogenase *(Adh)* gene
Aedes mosquito, 639–640, 640t
Aequorea victoria, 5
AFGPs (antifreeze glycoproteins), 709f
Age of Enlightenment, 13
Aggregation, of unicellular individuals, 226
Aging
    Darwinian medicine, 777
    evolution of senescence, 565, 565f
    life span, mutations extending, 566, 566f
    mortality
        baseline, 562, 564f
        increase with age, 561, 562f
    mutation load, 565–566
    noninevitability of, 561–562, 562f
    as part of optimal life history, 564–565, 565f, 566
    selection, weakness of, 562–564
Agrobacterium tumefaciens
    conjugation in, 183, 185f
    linear chromosomes, 171–172
Agrostis tenuis, 501, 502f, 646, 650
Alanine, synthesis of, 92–93
Alarm calls, 600
Alcohol dehydrogenase *(Adh)* gene, 361, 363–366, 363f, 366f, 405–406, 406f, 524, 524f, 535, 545, 546f
Algae
    charophyte algae, 201
    chlorophyte algae, 226
    land forms, emergence of, 272
Allele frequencies
    adaptive landscapes and, 472–475, 473f
    definition, 356
    fitness and, 467–469, 467f–469f
    in island model, 441–442, 441f, 444–445, 444f, 445f
    linkage disequilibrium and, 433
    random drift of, 415–420

random drift/selection balance, 494–496, 496f
    selection and, 470–471, 471f, 471t
    variation among demes, 444–446
    variation described by, 363–364
Alleles
    age of new, 434, 434f
    allozymes, 360, 361
    average effect of, 389, 391–393, 392f, 462, 670
    average excess of, 389–393, 392f
    definition, 356
    deleterious, fixation of, 493–494
    dominance of wild-type, 690–691, 690f
    favorable, spread of, 496–497, 498f
    fitness of, 419, 467–469, 467f–469f, 470
    hybrid rescue, 636
    linkage disequilibrium, 432–437
    modifiers, 484
    neutral, 419
    purging of deleterious, 517–518, 518f
    rare, excess of, 533–535, 534f, 537
    recessive, 357, 358
    survival probability of, 490, 490f
    wild-type, 357
Allelism, tests for, 359, 359f
Allometry, 401, 696
Allopatric speciation, 629, 644–646, 645f
Allopolyploidy, 631
Allostery, 55–56, 55f, 102, 702
Allozymes
    definition, 360
    divergence between species, 535
    Drosophila species, 361, 524, 635
    host races, 654
Alphabet, genetic, 561, 561f
α-globin
    molecular clock and, 371–372, 372f, 373f
    rate of amino acid change, 425
α-proteobacteria, mitochondrial origin and, 207, 209, 210
Alternative splicing
    cell differentiation and, 235–236, 236f
    novelty creation by, 712
    protein diversity from, 220–221, 220f, 599
Altman, Sidney, 57
Altruism
    evolution of apparent by kin selection, 601–602, 601f
    examples of apparent, 599–600, 600f

Page references followed by f denote figures; those followed by t denote tables.

Altruistic gene, 586
Alu elements, 217, 218f, 598, 598f
Alvarez, Walter, 284
Alveolates, 200, 200f, 201f
Alzheimer's disease, 761, 762f, 764, 769, 771
Amelioration, of genes, 188, 189f
Amino acid changes
    enzyme active site, 700–701, 701f
    protein conformation, 702–703, 702f
Amino acids
    composition and optimal growth
        temperature, 154–155, 155f,
        156, 156t
    differences, 371–375, 372f–375f, 374t
    evolution, 532–533, 533f
    genetic code and, 104–105
    polymer creation by chemical evolution,
        97
    prebiotic synthesis experiments, 93, 94t
    RNA interaction with, evolution of,
        105–106, 105f
    stereoisomers, 104f
    structure, 48f
    substitutions, 531–533, 532f, 533f, 551
    universality of handedness, 66–67
Amniocentesis, 767
Amoebas, 195
Amoebozoa, 200f, 201, 201f
Amphibian eye, regeneration in the, 232, 233f
Amphipathic structure, 100
Analogous structures, 67, 68f
Analogy
    convergence, 114
    homology compared, 67, 113t
    parallel evolution, 114, 115f
Analysis of variance, 30
Ancestral character state, 119, 119f, 128
Ancestry
    genome blocks, 427–428
    pedigree, 428–429, 429f
    random drift and, 420–425
Aneuploid gametes, 340, 626
Angiosperms, 632, 687–689
Animal-cloning experiments, 232–233, 233f
Animals. *See also specific animals; specific*
        groups
    body plans, 239, 240f–243f, 241–242
    phylogenetic relationships, 240f
Anisogamy, 669
Annelids, 241–242, 242f
Anomalocaris, 264, 264f
Anopheles mosquito, 639–640, 640t
Antagonistic pleiotropy, 564
Antarctica, marsupial fossils, 74, 74f
Antennapedia (Antp-C), 288–289, 289f, 290f,
    291f
Antibiotic resistance
    conjugation, role of, 184, 185f
    in *Neisseria meningitidis*, 432, 432f
Antibodies, 663
Antifreeze glycoproteins (AFGPs), 709f
Antigenic diversity, 741
Antigenic shift in influenza, 351f
Antigens, 663
Antimutator alleles, 659
Antisense oligonucleotides, 247
Antler size, 527–528, 528f

Apes, 728–729, 728f, 730f, 734
Aphids, 165, 665f
Apicomplexans, 200, 204f, 207
Apicoplasts, 3, 3f
ApoE*4 allele, 761, 762f, 764, 769, 771
Apple maggot, 653, 654f
Aquifex pyrophilus, 155
Aquificales, 153, 154, 159
Aquinas, Saint Thomas, 11
Arabian babblers, 611, 611f
Arabidopsis thaliana
    alternative splicing, 221
    development in, 245, 246
    nuclear transfer of plastid genes, 205
    transposons, 216, 216f
Archaea
    Bacteria and Eukarya compared, 138t,
        141, 142f
    biological diversification, 151–168
        biochemical diversity, 158–163, 159t,
            160f, 162f–163f
        extreme environments, 151–158, 152t,
            153f–155f, 153t, 156t,
            157f–158f
        interactions with other species,
            163–168, 164t, 167f
    conjugation, 352
    differentiating features of, 126t
    environmental studies, 147–151, 149f, 151f
    features characteristic of, 141, 142f
    genome, 169–180
        gene content, 175–176, 178
        gene order, 178–180
        size and density, 172–173, 172f
        streamline, 173–175
        variations in, 169–172
    habitat, 140–141
    lateral DNA transfer, 182–191
    naming, 121
    phylogenetic diversification, 142–151
    phylogenetic tree, 144f
    similarity to eukaryotes, 128
    subgroups, 145, 147f
    in Tree of Life, 123f
Archaebacteria, naming of, 121
Archaeopteryx, 278, 279f
Arctic skua, 576f
Ardipithecus, 733
Argument from design, 11, 78
Aristotle, 11
Armadillos, 74f
Artemia salina, 300–301, 302, 303, 303f
Arthropods
    classification of, 242
    embryo similarity in, 69f
    flight, evolution of, 276–277, 278f
    in fossil record, 265, 266f
    interrelationships of groups, 267f
    parallel evolution, 114, 115f
    terrestrial forms, appearance of, 271, 271f
Artificial selection
    costs of reproduction, 565
    Darwin and, 71
    novelty generation, 700
    of pigeons, 19, 20f
    power of, 480–482, 481f, 482f
    recombination rates, 679, 680f

Asexual populations
    bottlenecks and, 679
    deleterious allele accumulation, 680–681,
        681f
    lack of persistence of taxa, 666, 667f, 668
    mutation rate evolution, 660, 660f
    species definition in, 622, 622f
Asexual reproduction
    examples of, 667f
    mutator alleles, 661, 661f
    parthenogenesis, 631, 666, 667f, 668, 669
    random genetic drift and, 415
    Volvox, 227, 227f
Association studies
    in human disease genetics, 515, 751,
        759–763, 763t
    linkage disequilibrium and, 760–761
    quantitative traits, 761–763
    statistical power of, 759, 759f
Asteroid impact, 284, 284f
ATPases, 713f, 714f
ATP (adenosine triphosphate), use in Calvin
    cycle, 161
Audureau, Alice, 41
Australopithecus, 732–734, 733f–735f
Autobiography (Darwin), 16f
Autocatalytic network, 97, 97f
Autoinducer, 167, 167f
Autoinduction, 167
Automatic selection, 688
Autopolyploidy, 631
Autosomal dominant disorders, 512t
Autotrophic, 159, 159t
Average effect of an allele, 389, 391–393,
    392f, 462, 670
Average excess of an allele, 389–393, 392f
Avery, Oswald, 41, 43, 44f

B

Bacillus subtilis, 630f
Backcross, 385
Background selection, 538–540, 677
Bacteria
    Archaea and Eukarya compared, 138t,
        142f
    biological diversification, 151–168
        biochemical diversity, 158–163, 159t,
            160f, 162f–163f
        extreme environments, 151–158, 152t,
            153f–155f, 153t, 156t,
            157f–158f
        interactions with other species,
            163–168, 164f, 164t, 165f,
            167f
    cell compared to eukaryotic cell, 196f
    conjugation, 352
    differentiating features of, 126t
    diversity of, 137–138, 139–140, 139f, 141f
    environmental studies, 147–151, 149f,
        151f
    features characteristic of, 138–140, 139f,
        140f
    genome, 169–182
        gene content, 175–176, 178
        gene order, 178–180, 179f–182f
        size and density, 172–173, 172f–173f

streamline, 173–175, 175f
   variations in, 169–172, 170t
lateral DNA transfer, 182–191
   amelioration of genes, 188, 189f
   barriers to, 187–188
   conjugation, 183–184, 184f, 185f
   core genes and, 188–191, 190f
   transduction, 184, 186–187, 186f, 187f
   transformation, 183, 183f
phyla, table of, 145t–146t
phylogenetic diversification, 142–151
phylogenetic tree, 144f
reproductive isolation, 629, 630f
in Tree of Life, 123f
Bacteriocins, 571, 572f, 600, 600f
Bacteriophage
   in fluctuation test, 345, 346f
   genetic research on, 42, 42f
   novelty evolution, 703
Bacteriorhodopsin, 162–163, 163f
Balanced polymorphism, 500, 514–515, 541, 541f
Balancer chromosomes, 359, 359f, 409, 549
Balance view, 34, 505
Balancing selection
   description, 476, 476f
   detecting, 531
   genetic load from, 552
   heterozygote advantage and, 505
   human disease genetics, 764–765
   in human evolution, 747
   MHC polymorphism, 689
   neutral variation, effect on, 541–542, 541f, 542f
   quantitative traits, 514–515
   variation maintenance by, 34, 505–510, 531
Base pair, definition, 357
Base substitution, 326, 326t, 327f, 328
Bates, Henry Walter, 18, 19f
Batesian mimicry, 506–507, 507f
Bateson, William, 21, 22, 25, 288
Bats, flight evolution in, 278
B chromosomes, 589f, 590, 594, 594t
Bdelloid rotifers, 668, 668f
Beadle, George, 42, 42f, 313–315, 314f
Beagle, H.M.S., 16
Beanbag genetics, 485
Beaumont, Mark A., 744
Beetles, *Hox* genes in, 300, 302f
Begging displays in birds, 604, 604f
Belozerskii, Andrei, 54
Benzer, Seymour, 42
Berg, Paul, 56
Bernal, John Desmond, 54
Berzelius, Jöns Jakob, 38, 38f
β-galactosidase, regulation of, 50, 51f, 52–53
β-globins, molecular clock and, 372, 373f
Bias, sensory, 578–579, 579f, 580
Bible, 82
bicoid gene, 545, 545f
Bicyclus butterflies, 700, 700f
BiDil, 775–776
Bilateral symmetry, 241
Bilaterians, in fossil record, 265, 266f
Binary fission
   in archaea, 141

in bacteria, 138, 140f
Binomial distribution, in Wright-Fisher model, 416, 417
Binomial nomenclature, 119
Biochemistry
   of bacteria and archaea, 158–163, 159t, 160f, 162f–163f
   role in emergence of molecular biology, 42
   shared, 66–67
Biodiversity
   evolution of, 62–63, 63f
   extinction, 281, 283–285, 283f
   in fossil record, 254, 280–285, 280f, 282f, 283f
Biofilms, 168
Biological nomenclature, 12
Biological species concept (BSC), 622–624, 731
Biomarkers, archaea membrane lipids as, 141
Biometrics, 21, 24, 30
Biomineralization, appearance in fossil record, 262–263
Bipedalism, origin of, 733
Birds
   begging displays, 604, 604f
   divergence of North American, 627–628
   flight, evolution of, 278, 279f, 721
   kin discrimination in, 606, 606f
   sexual selection, 574f–577f, 576
Biston betularia, 31, 32f, 470, 470f, 550–551
Bithorax (BX-C) complex, 288–289, 289f, 290f, 295, 303
Blending inheritance, 20, 32
Body plan, definition, 225
Bootstrap value, 113
Borges, Jorge Luis, 465
Bottlenecks, population
   asexual populations, 679
   coalescence bursts, 424
   distinguishing from selective sweep, 537–538
   drastic, effect of, 419
   rare allele excess, 534, 534f, 537
   sika deer, 454
Boveri, Theodor, 23, 24f
Bowerbirds, 577
Brachet, Jean, 48
Brachiopods, in fossil record, 265, 266f
Bradyrhizobium, 165
Bragg, William Lawrence, 43
Branches, phylogenetic tree, 111–112, 111f, 112f
Branch length, 112, 112f, 114f
Brassica oleracea, diverse varieties of, 71f
BRCA1 allele, 766–767
Breast cancer, 766–767
Breeding value, 390–391, 391f, 393, 393f, 401, 527
Brenner, Sydney, 47
Bristle patterns, fly, 305–307, 306f
Broad-sense heritability, 391, 394, 397
Brood parasites, 604, 604f
Bryozoans, in fossil record, 265, 266f
BSC (biological species concept), 622–624, 731
Buchnera aphidicola, 169–171, 174–175, 175f

Buchner, Eduard, 38, 39f
Buchner, Hans, 38
Buckland, William, 13, 13f
Buddhism, 66
Buffon, Georges-Louis Leclerc, Comte de, 14, 14f
Bumpus, Herman, 18, 525–526
Burgess Shale, 263, 264f, 269
Burnet moth, 698–699, 698f
Butterflies
   Batesian mimicry, 506, 507f
   eyespots, 700, 700f
   inbreeding depression, 517f
   Müllerian mimicry, 503, 504f, 640
   warning coloration, 600f

C

Cadherins, 230
Caenorhabditis elegans
   body size, mutational variance for, 409f
   codon usage bias, 543
   daf-2 gene, 566f
   development in, 244, 249, 250f
Calcitonin, 236, 236f
Calcitonin gene-related peptide (CGRP), 235, 236f
Callus, of cells, 232, 232f
Calvin cycle, 159, 160f, 161
Cambrian explosion, 244, 263–272
   Burgess Shale, 263–264, 264f, 266f, 269
   disparity of organisms, 269, 269f
   diversity of organisms, 267–270, 268f, 269f, 282f
   metazoan phyla, 264–265, 266f
   terrestrial forms, 271–272, 271f, 272f
   timing of, 270
Cambrian period, in geological time scale, 255, 255f
cAMP (cyclic adenosine monophosphate), aggregation in *Dictyostelium* and, 228
Cancer, 237–238, 238f, 703
Candida albicans, 562f
Candidate genes, 403–404, 406–407
Canonical (standard) genetic code, 105
Capsids, 184
Carbon-14 dating, 89
Carbonaceous shales, 259
Carbon dioxide, global warming and, 281, 285
Carbon fixation, by bacteria and archaea, 159, 160f, 161
Carboniferous period, 275, 276, 276f
Carsonella, 377
Case-control design, 759
Caspersson, Torbjörn, 48
Castle, William, 28, 28f, 399
Catalysis, definition of process, 38, 38f
Catalytic RNA, 57–58, 57f
Catastrophists, 12
Catholic Church, 66, 82
The Causes of Evolution (Haldane), 30
Cavalli-Sforza, Luca, 741, 743
CCR5 gene, 434, 434f, 765, 771, 773f
cDNA (complementary DNA), insertion into genome, 218

Cech, Thomas, 57
Cell–cell interactions, 249, 250f
Cell communication, multicellularity and, 229–230
Cell lineage, in embryogenesis, 248–249
Cell membrane, 141, 142f
Cellulose, digestion of, 165–166
Cell wall, bacterial, 138, 139f
Cenozoic era, 255, 255f
Central Dogma, 10f, 53, 53f, 344, 381
Centromeres
    function of, 215
    inversions, 328, 329f
    recombination reduction near, 216, 539
    transposon accumulation in, 216, 216f
Cepaea nemoralis, 440, 440f
Cercozoa, 200, 200f
CFTR gene, 755
CFUs (colony-forming units), 148
CGRP (calcitonin gene-related peptide), 235, 236f
Chaetognaths, in fossil record, 265, 266f
Chambers, Robert, 15–16, 15f, 17
Changeux, Jean-Pierre, 55
Chaperone, 691
Character state reconstruction, 117–119, 118f, 156, 721
Chargaff, Erwin, 42, 43, 43f
Charniodiscus, 263f
Charophyte algae, 201
Chatton, Edouard, 120
Cheatgrass, spread of, 499f
Chemical evolution, 96–97, 96f
Chemical fossils, 141
Chemoautotrophic symbioses, 164
Chemostat, 525, 525f
Chemosynthesis at hydrothermal vent, 95
Chemotaxis, 138
Chemotrophic, 158, 159t
Chengjiang fossils, 263, 265, 266f, 267
Chert, 259
Chetverikov, Sergei, 31
Chicxulub Crater, 284, 284f
Chikhi–Beaumont model, 744
Chikhi, Lounès, 744
Chimeric organisms, 131–132
Chimpanzee, 80–81, 728f, 729, 730f, 749–750, 750f, 775
Chlamydia, 143–144
Chlamydomonas, 226
    C. reinhardtii, 679, 685–686, 691
    life cycle, 665f
Chlorarachniophytes, 201
Chlorophyll, 161
Chlorophytes, 201, 226
Chloroplasts
    apicoplast relationship to, 3
    definition, 196
    DNA sequence divergence, 372
    endosymbiotic origin of, 3
    gene transfer to nuclear genomes, 132
    genome flow between taxa, 623, 623f
    as mutualistic symbiosis, 163–164
Choanocytes, 229, 229f, 241
Choanoflagellates, 201, 229–230, 229f
Chondromyces crocatus, 139, 141f
Chordates, 241, 242f, 265, 266f

Chorthippus parallelus, 452f
Christian churches, 66
Chromatids, 222
Chromatin, description, 195
Chromosomal fusions, 644
Chromosomal inversions, 359
Chromosomal rearrangements, 375–377, 376f–378f, 376t
Chromosomes
    in archaea, 141, 142f
    B, 589f, 590, 594, 594f
    bacterial, 138
    balancer, 359, 359f, 409, 549
    Boveri's experiments with, 23, 24f
    cytological examination, 375
    inversions, 542
    linear, 171–172
    linkage, 24
    number of, 632, 632f
    plasmids compared, 170–171
    ploidy, 222–223
    polytene, 375, 376f
    rates of evolution, 375t, 376
    sex-linked genes, 25
    structure of eukaryotic, 214–216
Cichlid fishes, 650–651, 651f, 652f
Ciliates, 200
Cis-regulation, of *Pitx1,* 311–312
Clade, 111, 364
Cladistic analyses, 268
Cladograms, 111–112, 257f
Classical view, 34
Classification of organisms
    eukaryotes, 198, 199f, 200–202, 200f, 201f
    hierarchical structure as evidence for evolution, 67–70, 69f
    mistakes/revisions in, 143–144
    molecular support of, 68
    schemes
        Linnaean, 119
        Tree of Life, 119–121, 120f–123f
        viral, 133, 134t
Clays
    role in chemical evolution, 96f
    self-organization of monomers into polymers, 99
Clines
    alpine grasshopper, 503, 503f
    butterfly mimicry, 503, 504f
    concordant, 500, 501f
    copper tolerance in grass, 501, 502f
    eelpout, 500, 501f
    gene flow and, 443–444, 443f
    human populations, 741, 741f, 770
    hybrid zone, 646
    insecticide resistance in mosquito, 501, 502f
    maintenance of narrow, 501–504
    production of, 500–501
    scale of, 500, 501
    width, 440, 504–505, 530
Clonal interference, 662
Clone, 56, 415
Cloning
    animal, 232–233, 233f
    positional, 756
Closed cycles, 96–97
Closed system, 79

Cloudina, 262, 262f
Cnemidophorus lizards, 666, 667f
Cnidarians, 241, 241f, 265, 266f
Coalescence, 420–425, 423f, 424f
    due to selective sweep, 536–538, 537f, 538f
    in human evolution, 748, 769
    in island model, 449–450, 449f, 450f
    lineage, 365f
    linkage disequilibrium and, 436
    mismatch distribution and, 746
    recombination and, 429, 430, 430f
Coalescence time, 449–450, 450f
Coancestry, 421
Cockroaches, 581, 582f
Codeine, 776
Codons, variation in, 104
Codon usage bias, 188, 543–544, 543f, 544f
Coefficient of kinship, 421
Coelacanth, 609, 622
Coevolution, 509–510, 510f, 511f, 672, 674, 674f
Cofactors, 702
Cohn, Melvin, 50
Colias butterflies, 701, 701f
Colloid, 39
Colony-forming units (CFUs), 148
Color vision, evolution of, 707–708
Comets, organic molecules in, 94–95
Common ancestor, 115–117, 116t. *See also* LUCA (last universal common ancestor)
Compartmentalization, 99–101, 711, 713–714
Competence, 168, 183, 183f
Competition
    between biological species, 624
    local mate, 603, 604f
    between motor neurons, 716, 717f
    between relatives, 670–672
Competitive exclusion principle, 624
Complementary DNA (cDNA), insertion into genome, 218
Complementation test, 357, 357f, 358, 359, 516, 637
Complex traits, human disease genetics, 758–760
Computation, evolutionary, 467, 467f, 722–723
Concatenation-based trees, 190
Concerted evolution, 342
Confidence interval, 423
Conflict between genes, 587–599
    segregation distortion, 587, 587f, 588, 588f, 591, 592, 592f
    selfish DNA, 590–599
        abundance in eukaryotes, 594–598, 594f, 594t, 595t, 598f
        asexual reproduction and, 593, 594f
        co-opted, 598–599
        kin selection, 591
        modifiers to suppress, 591–593, 592f, 595
        t-haplotype in mice, 590–591, 591f
    sex-based inheritance, 588–589, 589f
    transposable elements, 588, 588f
Conjugation, 183–184, 184f, 185f, 352
Conservation, molecular, 58, 58f, 59

Conservative DNA transposons, 217, 218f
Constitutive mutations, 52
Continental drift, 258f. *See also* Plate tectonics
Contingency loci, 663
Continuous variation, 25–29, 27f
Convergence
    ecological, 624, 625f
    phenotypic features prone to, 142
    of proteins, 557, 557f
Convergent evolution, 114, 172, 278f
Cook, James, 14, 14f
Cooksonia, 273–274, 274f
Cooperation, evolution of, 611–616
    competition, suppression of, 613–614,
        613f–614f
    importance in early evolution, 614, 615f,
        616
    mutual advantages of, 611–612
    social insects, 611, 612
Cooperative breeding, 600, 600f
Copernicus, Nicolaus, 11
Copper tolerance, clines in, 501, 502f
Coral reefs, Darwin's view on, 16, 17f
Core genes, 132, 175–176, 175f, 190, 190f
Corey, Robert, 43
Correlation analysis, 155–157
Correlation coefficient, 386
Correlations, between traits, 480, 480f
Correns, Carl, 24
Cosexuality, 682
Cost of genome dilution, 669
Cost of natural selection, 550
Covalent bonds, Pauling's work with, 41
Covariance, 386–387, 394–396
Coviruses, 590
Cowbirds, 604, 604f
CPD (cyclobutane pyrimidine dimer), 333f
Crab-like forms, parallel evolution of, 114,
        115f
Crabs, natural selection in, 18, 20f
Cretaceous, 277–278, 278f, 279f
Crick, Francis, 43, 44f, 45–47, 49, 53, 54, 101,
        658
Cristae, 200
Crossbreeding, selection against, 648–650
Crossing over
    in meiosis, 222, 328, 663, 666f
    between repetitive DNA, 341–342, 342f
    unequal, 341, 341f, 375
Crustaceans, macroevolution of morphology,
        301–305, 303f–305f
Cryptic genetic variation, 398–399, 398f
Ctenophora, 241, 241f, 265, 266f
CTP2D6 enzyme, 776
Cultural evolution, 780
Culture, of organisms, 148
Cuvier, Georges, 13, 13f, 15
Cyanobacteria
    ancient, 91, 91f
    endosymbiosis and, 3
    plastid origin and, 203, 207, 208f
    stromatolites, 90–91, 90f
    symbiosis, 164
Cyclic adenosine monophosphate (cAMP),
        aggregation in
        Dictyostelium and, 228
Cyclobutane pyrimidine dimer (CPD), 333f

CyIIIa gene, 235
Cystic fibrosis, 755
Cytology
    of chromosomes, 375
    development of, 23
Cytoplasmic male sterility, 589, 591
Cytosine
    deamination, 331–332, 333f, 347
    methylation of, 347
Cytoskeleton, 196, 198, 221–222

D

daf-2 gene, 566f
Darwin, Charles
    artificial selection and, 71, 397, 480–481
    bipedalism, evolution of, 733
    on common ancestor, 116
    on co-option of structures, 75
    on diversity, 82
    early years, 16
    embryology and development, 244, 287
    embryo similarities and, 67
    geographic patterns observed by, 70, 500
    inherited variation, documentation of,
        358
    kin selection, 601
    Law of Succession, 74
    on natural selection, 458
    on origin of life, 92
    *On the Origin of Species by Means of
        Natural Selection*, 9, 9f,
        16–18, 20–21, 65, 70, 75, 78,
        93, 110, 110f, 116, 244, 287,
        356, 458, 575
    photograph of, 66f
    sex ratio, 507
    on sexual selection, 575
    Sketch, 82
    on varieties, 70
    vertebrate eye and, 78
    vertical inheritance and, 109–110
Darwin, Erasmus, 14, 16
Darwinian Demon, 557
Darwinian medicine, 776–778
Dating, radiometric, 88–89
Daughter species, 109
Dawkins, Richard, 599, 778
Deamination of DNA bases, 331–332, 333f,
        347
Defective interfering viruses (DIVs), 464,
        569, 569f, 590
Degradation motif, 704
Deinococcus radiodurans, 143, 143f, 155,
        157–158, 158f, 183
Deinonychus, 256, 257f
Deism, 11, 20
Delbrück, Max, 41, 42, 345, 346f
Deletions, 326, 340, 342f
Delta gene, 405
de Maupertuis, Pierre, 14, 14f
Demes
    coalescence in, 449–450, 449f
    definition, 441
    extinction and recolonization, 447, 447f
    genetic variation, 444–451, 444f, 445f
    group selection, 607–609

island model, 441–442, 441f, 444–445,
        444f, 445f
    subdivision into, 607, 641
Demography, human, 742–743, 742f
Dendrogram, 112
Density dependence, 461
Density-dependent selection, 470
Deoxyribonucleotides
    formation from ribonucleotides, 134
    synthesis of, 102, 102f
Derived character states, 119, 119f, 128
Derived features, of *Ardipithecus*, 733
Descartes, René, 11, 12, 38
*Descent of Man and Selection in Relation to
        Sex, The* (Darwin), 575
Descent with modification
    body plans as example of, 238
    definition, 110
    hierarchical classification explained by,
        67, 70
    homologous structures, 67, 68f
Desiccation, resistance to, 157–158
Design, natural selection as cause for
        appearance of, 74–75,
        465–467
Deuterostomes, 240f, 241, 242f, 265, 266f
Development
    anterior–posterior patterning, 288–293
    body plans, 238–250
    Cambrian radiation and, 270
    cell–cell interactions, 249, 250f
    cell lineage, role of, 248–249
    control of patterning, 249–250
    differentiation, 230–238
    genetic screens, 244–246, 245f, 246f
    reverse genetic studies of, 246–248, 247f
    transcriptional control, 235
Developmental constraints, 464
Developmental programs, evolution of,
        287–321
    eye evolution, 319–321, 319f–320f
    Hox (homeotic) genes
        conservation of function and
            structure, 292–293
        crustacean morphology, involvement
            in, 301–305, 303f–305f
        discovery of, 288, 288f
        evolutionary history in metazoan
            phyla, 293f
        expression of, 288–290, 289f, 290f
        fly bristle patterns, role in, 305–307,
            306f
        gene organization, 289f, 292–293, 294f
        homeobox sequence, 290–292, 291f
        larval appendage evolution,
            involvement in, 298–301,
            300f–302f
        ubiquity in animals, 292, 293f
        variation within populations, 307
        wing evolution, involvement in,
            293–298, 296f, 297f, 299f
    maize evolution from teosinte, 313–318,
        314f–318f
    skeletal evolution in sticklebacks,
        307–313, 308f, 310f–312f
    universality of developmental systems,
        318–321

de Vries, Hugo, 24
Diabetes, PPARg allele and, 756
Diapause, 653–654
Dicer enzyme, 247f
Dickinsonia, 262, 263f
Dictyostelium discoideum, 228–229, 228f
Differentiation, 230–238
    cancer as loss of, 237–238, 238f
    division of labor, 230–231, 231f
    evolution of, 226
    genetic information, constancy of,
        231–233, 232f–233f
    genomic rearrangements, 233–234, 234f
    maintenance of, 236
    mechanisms of, 234–236
    progressive nature of, 236, 237f
Diffusion approximation, 442
Digestive symbioses, 165–166
Dimer, 361
Dinoflagellates, 200, 201f, 207
Dinosaurs
    extinction, 284
    feathered, 76, 76f
    Megalosaurus, discovery of, 13, 13f
Diploids
    alternation with haploid state, 222, 224
    diploid life cycles, evolution of, 683–685,
        685f
    selection on, 470–472, 471f
Directional selection, 476–477, 476f, 526,
        527f, 531, 535
Direct selection, 578
Discicristates, 200, 200f, 201f
Disease
    association studies, 759–763, 763t
    Darwinian medicine, 776–778
    genetic basis of, 755–769
    genomic medicine, 775–776
    incidence of inherited, increase in,
        773–775, 774f
    mapping disease genes, 756–758, 757f
    Mendelian genetics, 756–758, 756f, 758f
    patterns of inheritance, 756f
    QTLs, 758–759
    reasons for study of disease genetics,
        755–756
    use of genetic information, 765–769
        forensics, 768–769
        insurance, 767–768
        medical implications, 766–767
        prenatal screening, 767, 768f
    variants for common diseases, 763–765,
        765f, 766f
Disparity, 269
Dispersal rates, 530
Disruptive selection, 476, 476f, 652, 652f
Dissostichus mawsoni, 711f
Distal-less gene, 299–301, 300f
Divergence
    definition, 110
    Dobzhansky–Muller model, 643, 644f, 646
    McDonald–Kreitman test and, 535–536
Diversifying selection, 771
Diversity, of adaptation, 2f
Division of labor, 707
DIVs (defective interfering viruses), 464, 569,
    569f, 590

DNA
    chemical reactivity, 102, 106
    RNA replaced by, 106–107, 107f
    structure, research on
        Watson and Crick, 43, 44f, 45–46
        Wilkins and Franklin, 43, 44f, 45
DNA ligation, 335, 335f
DNA polymerase, 46, 328, 330, 334, 336, 337f
DNA repair
    desiccation resistance, 158
    mutator alleles, 660
    processes, 335–336, 335f
    radiation resistance, 158
DNA replication
    autocatalytic network, 97, 97f
    error-correcting mechanisms, 336–337,
        337f, 338f
    errors in, 1, 328, 330
    origin of replication, 179
    process, 330, 330f
    semiconservative, 46, 46f
    symmetric inversions and, 179–180, 182f
    telomeres, 214–215, 215f
    Watson and Crick's speculation on, 46
DNA sequencing
    Maxam–Gilbert method, 56
    Sanger method, 56
Dobzhansky–Muller model, 643, 644f, 646
Dobzhansky, Theodosius, 1, 31, 31f, 32f, 34,
        59, 442, 443f, 622, 634, 639,
        641
Dog, breeds of, 71
Domains, 121, 128
Dominance
    description, 388–389, 388f, 389f
    fitness and, 472, 472f
    of wild-type alleles, 690–691, 690f
Dominance deviation, 391
Dominance theory, 639, 640, 640t
Dominance variance, 394, 395–397, 397f
Dominant character, 22
Dosage compensation, 682
Dotplots, genome, 176–177, 176f–177f
Double mutations, 676
Double-strand breaks, DNA, 334, 334f, 335
Double-stranded RNA, 246, 247f
Doushantuo Formation, 256, 259–260, 259f,
    260f
Down syndrome, 328, 767
Drosophila. See also specific species
    artificial selection experiments, 482, 482f
    balancer chromosomes, 359, 359f, 549
    bicoid gene, 545, 545f
    bristle number, 484, 484f, 526–527, 678f
    codon usage bias, 543–544
    crossing over in, 222
    development in, 247–249, 248f
    genetic control element conservation,
        547, 547f
    genetic distance between species, 629, 629f
    genetic variation, 369–370, 370f
    heat shock proteins, 691, 692f
    hybridization, 623
    inversion polymorphism, 340
    McDonald–Kreitman test, 535–536
    polymorphism in, 368, 535
    polytene chromosomes, 375, 376f

    reproductive incompatibilities in,
        634–639, 635f–637f
    reproductive isolation, 647, 648f, 650
    sexual selection in, 577, 578f
    species differences, 621
    species divergence, 627, 627f
    substitution load, 551
    transposons in, 343, 594–595, 594f, 595t
    Ubx gene, 715
    variance in bristle number, 404–405, 405f,
        408f
Drosophila mauritiana, 634–635, 635f–636f,
        638–639, 639
Drosophila melanogaster
    Adh gene, 361, 363–366, 363f, 366f,
        405–406, 406f, 524
    aging, 563, 563f, 565, 566
    bristle number, 526–527, 678f
    development in, 244–245, 246, 246f
    effective population size, 432
    even-skipped gene, 547, 547f
    eye phenotypes, 357, 357f
    genetic diversity in, 426–427
    hitchhiking phenomenon, 485, 486f
    Hox (homeotic) genes, 288–296, 288f–291f,
        294f, 297f, 299–301, 300f,
        303–307, 306f
    Morgan's work with, 24–25, 25f
    nucleotide diversity, 432, 539–540, 540f
    polymorphisms in, 525
    QTL mapping and wing shape, 401–402,
        402f
    random drift experiments, 417f, 418, 418f
    recombination rates, 432
    reproductive incompatibility, 636–638
    segregation distorter *(SD)* complex, 592,
        592f
    selection coefficient measurement, 522,
        523f
    selective sweep, 538f
    venation pattern in, 398–399, 398f
Drosophila persimilis
    assortative mating, 650
    hybridization, 623, 647, 648f
    xanthine dehydrogenase alleles in, 534
Drosophila pseudoobscura
    Adh gene haplotypes, 545, 546f
    assortative mating, 650
    Bogotá subspecies, 634–635
    dispersal rate, 442, 443f
    hybridization, 623, 647, 648f
Drosophila sechellia, 635
Drosophila simulans, 589–590, 634–639,
        635f–637f, 639
Drosophila willistoni, recessive variants,
        frequency of, 359
Drug development, anti-malaria drugs, 3
Dung flies, mating behavior of, 558–559, 558f
Duplication
    mutations, 326, 328f, 342f
    of transposable elements, 343
Dybdahl, Mark F., 510

E

Earth, age of, 12, 20, 73, 88, 90
Ecdysozoa, 240f, 241, 242, 243f, 265, 266f

Echinoderms, 241, 242f, 265, 266f
Ectoderm
    actin expression in, 235
    in ctenophores and cnidarians, 265
Ectodysplasin (Eda) gene, 313
Ectopic recombination, 339, 343, 594, 595, 669
Edentates, 74f
Ediacara organisms, 260–262, 261f, 262f
EEA (environment of evolutionary
        adaptation), 778
Eelpout, 500, 501f
Effective population size, 417, 419, 420,
        426–427, 432, 444, 747
Eichornia paniculata, 688, 688f
Eigen, Manfred, 99, 614
Eigen's paradox, 99, 100
Eldredge, Niles, 80, 280, 280f
Electron transfer chains, 161
Electrophoresis, invention of, 40
Elephant seal, 368, 369f
Elongation factor G (EF-G), 127, 127f
Elongation factor Tu (EF-Tu), 127, 127f
Embryogenesis, patterning in, 248–250
Embryos. *See also* Development
    fossil, 260, 260f
    similarity between species, 67, 69f
Emerson, Ralph, 313
Encephalitozoon cuniculi, 198, 216, 377, 621f
Encephalitozoon intestinalis, 596
End–Cretaceous (K–T) extinction, 284, 284f
Endocytosis, 203
Endoderm, 235, 265
Endokaryotic models, of nucleus origin,
        212–213, 213f
Endomembrane system, 196
Endomitosis, 684, 685f
Endoplasmic reticulum, 196
Endosymbiosis, 202–214
    chloroplast origin, 3
    evidence for, 203
    gene content of endosymbiont, 204
    mitochondria, origin of, 203, 207,
        209–210, 209f–210f
    nucleus, origin of, 210, 211t, 212–213,
        213f
    organelle genomes, reduction in,
        204–207, 206f
    phylogenetic studies, 204–207, 206f, 207,
        209
    plastids, origin of, 203, 207, 208f
        secondary, 207, 208f
Energy of metabolism, diversity of, 158–163,
        159t
Entamoeba, 209–210, 210f
Entropy, 79
Environmental deviation, 387–388, 393, 527
Environmental genomics, 150
Environmental variance, 387–388, 394, 399,
        483, 483f, 677
Environmental variation, 387–388
Environment of evolutionary adaptation
        (EEA), 778
Environment, study of microbes in the,
        147–151, 149f, 151f
Enzymatic adaptation, 50
Enzymes
    active site, changes in, 700–701, 701f

allozymes, 360, 361, 524, 535
colloid theory, 39
design, de novo, 465
discovery of, 38
one gene–one enzyme principle, 42, 42f
polymorphisms, 524–525, 525f
variation, selection for, 525
Eocene, 278
EPF (extra-pair fertilization), 581–582, 583f,
        689f
Epigenesis, 11
Epilachna nipponica, 626f
Epilepsy, 765
Epistasis
    adaptive landscapes and, 475
    description of, 388–389, 388f, 389f
    fitness and, 472, 472f, 475
    linkage disequilibrium, 671t
    negative linkage disequilibrium, 675, 675f,
        676
    recombination and, 672, 673f, 674, 676
    reduction principle, 682
Epulopiscium fischelsoni, 140
Erwinia carotovora, 607
Escaped transcript model, 135–136
Escherichia coli
    adaptation to growth on glucose minimal
        media, 491–493, 492f
    artificial selection experiments, 482, 482f
    bacteriocin production, 571, 572f, 600f
    chaperone, 691
    codon usage bias, 543–544, 544f
    core genes, 175–176, 176f
    enzyme polymorphisms, 525, 525f
    epistasis, 675f
    evolutionary optimization, 559, 559f
    genetic diversity in, 427
    genomic islands, 178, 178f
    mutator strains, 661, 662f
    operons, 172, 174f
    plasmids in, 169–170
    polymorphism in, 368
    SOS response in, 706, 706f
ESS (evolutionarily stable strategy), 556,
        568–573
Essay on Population (Malthus), 17
Ether, as mutagen, 358, 360f
Eugenics, 82
Euglenoids, 200
Eukarya, in Tree of Life, 123f
Eukaryotes
    definition, 195
    differentiating features of, 126t
    diversification, 221–224
        regulatory networks, 222, 223f
        sexual reproduction, 222, 224, 224f
        single-celled eukaryotes, 221–222
    endosymbiosis, 202–214
        evidence for, 203
        mitochondria, origin of, 203, 207,
            209–210, 209f–210f
        nucleus, origin of, 210, 211t, 212–213,
            213f
        organelle genomes, reduction in,
            204–207, 206f
        phylogenetic studies, 204–207, 206f,
            207, 209

plastids, origin of, 203, 207, 208f
    secondary, 207, 208f
features of, 195–198, 196f, 197f
fossils, early, 202, 202f
genome size and density, 172–173,
        172f–173f
intracellular symbioses, 203
life cycles, 664, 664f, 665f, 684f, 685f
lineages, 198, 199f, 200–202, 200f, 201f
nuclear genome
    chromosome structure, 214–216, 215f,
        216f
    genome size, 216
    introns, 219–221, 219t, 220f, 221f
    pseudogenes, 218–219, 219f
    tandem DNA, 217
    transposable elements, 216, 216f,
        217–219, 218f
in tree of Life, 120
Eusocial insects, 612
even-skipped gene, 547
Evidence for evolution
    direct observation, 71–73
    fossil record, 73–74
    natural selection as cause of appearance
        of design, 74–75
    patterns of relationships, 66–71
        geographic distribution, 70–71, 70f
        hierarchical classification, 67–70, 69f
        universally shared features, 66–67
Evolutionarily stable strategy (ESS), 556,
        568–573
Evolutionary biology
    history, 9–36
    molecular biology and, 5, 35, 60–61
    neutral theory, effect of, 59–60
    practical uses, 3, 5
Evolutionary character state reconstruction,
        117–119, 118f
Evolutionary computation, 467, 467f,
        722–723
Evolutionary developmental biology
        (EvoDevo), 287
Evolutionary ecology, 586
Evolutionary psychology, 778–780
Evolutionary Synthesis, 30–35, 37, 60, 62, 65
Evolutionary trees. *See* Phylogenetic trees;
        Tree of Life
Evolution in Mendelian Populations (Wright),
        30
Evolution, summary of, 10, 10f
Evolvability, 464, 689–692, 690f–692f
Ewens, Warren, 534
Excavates, 200, 200f, 201f
Excision repair mechanisms, 335f, 336
Exon shuffling, 221, 221f, 712
Exon theory of genes, 221
Expected heterozygosity, 363
Exploratory systems, 692, 715–717
Extended phenotype, 599–600, 600f
Extinction
    inbreeding depression, effect of, 516, 517f
    K–T (end-Cretaceous), 284, 284f
    Modern, 285
    Permian, 281, 283, 283f
Extra-pair fertilization (EPF), 581–582, 583f,
        689f

Extremophiles, 151–158, 152t, 153f–155f,
153t, 156t, 157f–158f
Eyeless cave fishes, 464, 464f
eyeless mutation, 320
Eyes, *Pax6* and evolution of, 319–321, 320f

F

Fact, 81
Fatty acid desaturase, 704
FDA (Food and Drug Administration), 775
Feathered dinosaurs, 76, 76f
Feldspar, 96f
Female preferences, sexual selection and,
577–584, 579f, 580f, 582f,
583f
Fence lizards, 645, 646f
Fermentation, 38
Ferments, 38
Ferns, 632
Fertilization
extra-pair fertilizations (EPF), 581–582,
583f, 689f
polyspermy, 631
self-fertilization, 664
Fibrinopeptides, 374, 374f, 425
Ficedula, 649, 649f
Fig wasps, sex ratio in, 603, 603f
Filopodia, 236
Finger ridges, heritability of, 397
Fischer, Emil, 39
Fisher, R.A., 30, 30f, 407–408, 407f, 416, 462f,
497, 548, 580, 601, 676, 679
Fish, limb evolution, 721, 722f
Fitness
absolute, 419, 461, 462, 468, 470
adaptive landscape, 472–475, 473f
additive genetic variance, 462–463, 462f,
674–675, 675f, 678
allele, 467–469, 467f–469f, 470
components of, 460–461
definition, 418–419, 460
Fisher's geometric argument, 407–408, 407f
in Fundamental Theorem of Natural
Selection, 462–463, 462f
geometric mean, 469, 469f
inbreeding depression, 515–518
inclusive, 602–603
inherited variation in, 356
interactions influencing, 470–472, 470f,
472f
mean, 461, 462–463, 462f, 463f, 472–474,
473f, 494–496, 495f
random drift, effect of, 419
relative, 419, 461, 468–469, 475
reproductive isolation and, 642–643
selection measurement and, 521–530
Fitness component, 461
Fitness landscape, 466f, 526. *See also* Adaptive
landscape
Fixation
of deleterious alleles, 493–494
of favorable mutation, 536, 537f, 538
probability of chromosome
rearrangement, 641f
Fixation probability, 490, 493, 493f, 497
Flagella

in archaea, 141
bacterial, 138, 140f
choanoflagellates, 229, 229f
eukaryotic, 198
Flight, evolution of, 114, 276–278, 278f, 279f
Fluctuation test, 345, 346f
Fly bristle patterns, 305–307, 306f
Flycatchers, 649, 649f
Focal adhesion kinase, 236
Folk taxonomy, 619
Følling, Asbjørn, 756
Food and Drug Administration (FDA), 775
Foot-and-mouth disease virus, fitness change
in, 460f
Football team model of evolution, 63
Foraminifera, 201, 280, 284
Ford, E.B., 31, 31f, 59
Forensics, 768–769
Formose reaction, 93, 94f
Fossil record
body plans, history of, 244
Cambrian explosion, 263–272
Burgess Shale, 263–264, 264f, 266f, 269
disparity of organisms, 269, 269f
diversity of organisms, 267–270, 268f,
269f, 282f
metazoan phyla, 264–265, 266f
terrestrial forms, 271–272, 271f, 272f
timing of, 270
data from, 256
diversity, patterns of, 268f
Ediacara organisms, 260–262, 261f, 262f
embryos, 260, 260f
eukaryotes, appearance of, 202, 202f
as evidence for evolution, 73–74, 73f–74f
evolutionary phenomena revealed by, 254
extinctions, 281, 283–285, 283f
human evolution, 731–740, 732f–740f,
749
Law of Succession, 74
Neoproterozoic Doushantuo Formation,
259–260, 259f, 260f
origin of life, 90–91
pattern of diversity, 280–285, 280f, 282f,
283f
shells, appearance of, 262–263, 263f
species identified in, 254
terrestrial forms
Cambrian, 271–272, 271f, 272f
Devonian, 272–276
diversity, development of, 273f
flight, emergence of, 276–279
Neoproterozoic, 272
Ordovician, 271
plants, 272–275, 274f, 276f
Silurian, 271, 273
vertebrates, 275–276, 277f
Fossils
age of, 73
chemical, 91, 141
microfossils, 256
molecular, 102, 202, 219
stromatolites, 90–91, 90f
taxa, evolutionary relationships of
Archaeopteryx, 256, 257,
257f
trace, 267

trackways, 278
unlikelihood of fossilization, 253–254
FOXP2 gene, 750–751, 751f
Frameshifts, 326, 327f, 709
Franklin, Rosalind, 43, 45, 45f
Frequency-dependent selection, 470, 472,
472f, 475, 506–508, 506f,
507f, 525, 670, 672
Frequency spectrum, 752
Frozen accident hypothesis, 106
F_{ST}, 445–451, 447t, 448t, 450f, 769, 771, 771f
Fundamental Theorem of Natural Selection,
462–463, 462f, 548
Fungi
gene conversion, 588, 588f
microsporidia, 198, 199f

G

Galápagos finches
beak shape, 477, 478, 478f
directional selection in, 477
Galápagos Islands, adaptive radiations in, 70
Galeopsis tetrahit, 631f
Galilei, Galileo, 11
Galton, Francis, 20, 21, 21f, 22f, 387
Games, evolutionary, 567–573
defective interfering viruses (DIVs), 569,
569f
evolutionarily stable strategy, 568–573
hawk–dove game, 568
interactions between individuals, 567
Prisoner's Dilemma, 569, 612, 612f
rock–scissors–paper, 570–573, 570f, 572f
tragedy of the commons, 569–570
Gametes, 222, 685–687
Game theory, 567
Gametophytes, 275, 506
Gamow, George, 46–47, 47f
Garrod, Archibald, 42
Gas hydrates, 283
Gasterosteus aculeatus, 308, 653, 653f
GC content
pathogenicity islands, 166, 187
of thermophiles, 155
Gemmata obscuriglobus, 140, 141f, 212, 213f
Genealogy, 364–366
genetic divergence and, 628f
human Y chromosome, 745–748, 745f
inferring from sequences, 424, 424f
lineage sorting, 628
random drift, influence of, 420–424, 423f,
424f
species definition and, 622, 622f
in structured populations, 449–455
Gene conversion, 588, 588f, 663, 711
Gene displacement, nonhomologous, 718, 718f
Gene diversity, definition, 426
Gene dosage, 180
Gene duplication
description, 126
paralogs, 126–127, 127f
plasmid genes, 171
redundancy generation, 710–711, 711f
use in rooting Tree of Life, 125–128, 127f
Gene expression patterns, in primate
lineages, 749–750, 750f

Gene flow
    clines and, 443–444, 443f
    definition, 440
    diffusion rate, 442–444, 443f
    geographic barriers, 644–646
    homogenizing effect of, 441–442, 444
    human populations, 770
    island model, 441–442, 441f, 444–445, 444f, 445f
    linkage disequilibrium, 452, 454–455
    random genetic drift, effect of, 444–449
    rate estimation from F_{ST}, 448–449
    selection and, 496–505
    speciation and, 644–648
    structured populations, 439–440
    wave of advance, 497, 499f
Gene knockout, effect on fitness, 522–523, 523f
Gene loss
    in bacteria and archaea, 174–175, 175f, 191
    organelle, 206
Gene order, 178–180, 179f–182f, 207f, 378f
Gene regulation
    β-galactosidase, 50, 51f, 52–53
    cell differentiation and, 234–236, 235f, 236f
    changes in, 704–708, 705f–707f
    in eukaryotes, 222, 223f
    lac system, 50, 51f, 52
    λ system, 52
Genes
    definition, 356, 357f
    redundant, 57
    split, 62
Genesis, 65, 82
Gene targeting, 248
Gene therapy, 755
Genetic algorithms, 467, 678
Genetic alphabet, 561, 561f
The Genetical Theory of Natural Selection
    (Fisher), 30, 407, 580
Genetic assimilation, 398
Genetic backgrounds, 393, 432, 670
Genetic code
    adaptor hypothesis, 47, 49
    changes in, 67, 67f
    deciphering of, 47–49
    degenerate nature of, 53
    Gamow's model, 46–47, 47f
    mitochondria, 205t
    origin of, 104–106
    redundancy, 531
    standard (canonical) code, 105, 118–119
    table, 49f
    universality of, 66, 105, 381
    variations in, 118–119, 118f
Genetic correlations, 480, 480f
Genetic distances
    among human populations, 742–743
    reproductive isolation and, 629, 629f
Genetic divergence between populations
    (F_{ST}), 445–451, 447t, 448t, 450f, 769, 771, 771f
Genetic diversity, 363, 368–369, 426–427
Genetic drift, 60, 413–437
    allele frequencies, 415–420
    coalescence, 420–425, 423f, 424f

gene flow and, 444–449
geographic variation generated by, 444–445
in human evolution, 748
linkage disequilibrium, 432–437, 671t
natural selection compared, 460
negative linkage disequilibrium, 676
neutral theory of molecular evolution, 425–427
random linkage disequilibrium, 676–677, 677f
rate as determined by variance in fitness, 418–419
recombination, 427–437
selection and, 489–496
in shifting balance theory (SBT), 608, 608f
speciation and, 641–642
variance, effect on, 416–420
Wright–Fisher model, 416–418, 419, 420, 422
Wright's work on, 30
Genetic load, 75, 549–553
    from balancing selection, 552
    definition, 550
    gene interactions, 552–553
    mutation, 552, 565–566, 597, 659–660, 660f, 680, 684–685, 774
    segregation, 552
    substitution load, 550–551, 551f
Genetic markers, 35, 361–363
    for disease genes, 756–757, 757f
    QTL mapping and, 404–405, 405f
Genetic programming, 467
Genetic recombination, variation generated by, 325
Genetics and the Origin of Species
    (Dobzhansky), 31
Genetics, role in emergence of molecular biology, 41–42
Genetic systems, evolution of, 657–692
    evolvability, evolution of, 689–692, 690f–692f
    mutation rates, evolution of, 659–663, 659f–662f
    sex and recombination, evolution of, 663–683, 664f–668f, 671f–683f, 671t
    sex, consequences of, 683–689, 684f–689f
    studying, 658–659
Genetic terminology, 356–357
Genetic variance, random drift effect on, 418, 418f
Genetic variation, 355–379. *See also* Variation
    abundance of, 367–371
    in abundant species, 426–427
    allele frequencies, 363–364
    cline, 440, 443–444, 443f
    cryptic, 358, 360
    evolution, necessity for, 355–356, 356f, 358
    functional significance and, 370–371, 371f
    genealogies, 364–366
    gene arrangement, 375–377, 376f–378f, 376t
    genetic markers, 361–362
    maintenance of, 33–34
    molecular clock, 371–375, 373f, 375f

nature of, 33
nucleotide diversity, 364, 368, 369f, 370, 370f, 371f, 372f
polymorphisms, 358, 360–361, 363, 368, 368f
population size, effect of, 368–370, 426–427
spatial patterns, 440
terminology, 356–357
types, 367–377, 367f
Gene transfer, *Hfr*, 52
Genome density, 172–173, 173f
Genome dilution, cost of, 669
Genome sequencing, uncultured microbes, 150–151, 151f
Genome size, 172–173, 172f, 216, 348, 348f, 596–597, 596f
Genome streamlining, 136, 173–175, 175f
Genomic imprinting, 604–605, 605f
Genomic island, 178, 178f
Genomic medicine, 775–776
Genotype
    definition, 356, 357, 358
    distinction from phenotype, 25
    fitness of, 419
    haplotype, 357
    self-replication of, 98
Genotypic diversity, from competition between relatives, 670–672
Genotypic value, 387, 388, 390, 391f, 393
Genotypic variance, 390, 394
Geographic distribution, as evidence for evolution, 70–71, 70f
Geological time, 255, 255f
Geology, 12–13, 12f. *See also* Fossil record; Fossils
Geometric mean fitness, 469, 469f
Germ line, 21
Germ Plasm (Weismann), 24
GFP (green fluorescent protein), molecular breeding experiments with, 5, 5f
Giardia, 209–210
Gilbert, Walter, 56
Glanville fritillary, 517f
Glaucocystophytes, 203, 207
Gliding mammals, parallel evolution of, 114
Global warming, 281–283, 284, 285, 285f
Glucose, stereroisomers of, 39
Glued gene, 485, 486f
Glycosylation, 222
Glyptodon, 74f
God of the Gaps, 81
Golgi apparatus, 196
Gondwana, 71, 259
Gonium, 226
Good genes, 581–584, 593
Gorge, Olduvai, 734
Gorilla, 728, 728f, 729, 730f
Gould, Stephen Jay, 80, 82, 269, 280, 280f, 558, 560, 778
Gradualists, 22
Gram stain, 138, 139f
Grant, Robert, 16
Graphical alignment, for comparing circular genomes, 176–177, 176f–177f

Grasshopper, clines in alpine, 503, 503f
Great Chain of Being, 11–12, 11f, 15
Great reed warblers, 583f
Great tits, 581, 582f
Green fluorescent protein (GFP), molecular
        breeding experiments with,
        5, 5f
Griffith, Frederick, 42–43
Griffith, John, 43
GroEL chaperone, 691
Group I and Group II introns, 220
Group selection, 607–609
Growth temperature, 152–157, 153f–155f,
        153t, 156t
Guide RNAs, 106
Guppies, 683f
Gynodioecious population, 682, 682f
Gypsy-like retrotransposons, 594

H

Habitat preferences, 714
Haeckel, Ernst, 18, 21, 120, 120f
Haemophilus influenzae, 169, 175, 175f, 179f,
        183, 666
Haikouichthys, 265, 268f
Haldane, J.B.S., 30, 30f, 31, 32f, 34, 42, 74, 93,
        485, 550–551, 565, 601, 609,
        638–640, 638t, 692
Haldane's rule, 638–640, 638t
Half-lives, 88–89, 89t
Haloferax volcanii, 170
Halophiles, 157–158, 157f, 162
Halteres, 358, 360f
Hamilton's rule, 601–602
Hamilton, W.D., 34, 509, 567, 583, 601–602,
        607
Handicap, 581
Haplodiploids, 223, 612, 666, 668
Haploid, 222–223
Haplotype
    Adh gene introns, 545, 546f
    in association studies, 760–761
    definition, 357, 364
    structure of human chromosome 21, 429,
        431, 431f
HapMap Project, 760
Hardy, G.H., 28
Hardy–Weinberg law, 28–29, 29f, 363, 470,
        670, 764
Hawk–dove game, 568
Heat shock proteins, 58f, 691, 692f
Heavy metal pollution, 646, 650
Hedgehog, 452f
Helicobacter pylori, 179f
Heliconius butterflies, 475–476, 475f
Hemichordates, in fossil record, 265, 266f
Hemimetabolous insects, 240f, 597
Hemoglobin
    allele evolution and partial resistance to
        malaria, 491, 491f
    allele fitness, 419
    allosteric interactions in, 55, 55f
    gene regulation, 234–235, 235f
    heterozygote advantage, 470, 472
    orthologs and paralogs, 126–127
    Perutz's experiments with, 54–55, 54f

sickle cell, 46, 47f, 390–392, 390t, 391f,
        470, 472
    structure, 37f, 39
Heritability
    of bill shape, 477, 479f
    broad-sense, 391, 394, 397
    of fitness, 548
    mutational, 409–411, 410t
    narrow-sense, 391, 394, 397, 398f, 399,
        446, 477, 548
    quantitative traits, 527
    as slope of regression line, 477–478, 479f
Hermaphrodite, 556, 575f, 589, 592, 682, 682f
Herschel, John, 78
Heterochrony, 696
Heterogametic sex, 591
Heterokonts, 200, 200f, 201f
Heterosis, 515
Heterostyly, 687
Heterotrophic, 159, 159t
Heterozygosity
    divergence relationship to, 535, 535f
    expected, 363
    at microsatellite loci, 515–516, 516f
    population size and, 369, 370f
Heterozygote advantage, 359, 470, 471, 471f,
        472, 505
Heterozygotes, 24
Hfr, 52, 53f
HGP (Human Genome Project), 757–758
Hierarchical classification system, 119
Hill–Robertson effect, 536, 595, 676–677,
        677f
Hill, W.G., 676
Hinduism, 66
Histone proteins, 141, 142f, 197f, 373–374,
        374f, 425
Hitchhiking
    description, 427, 485–486, 486f
    mitochondrial DNA variant, 590
    recombination, effect of, 536, 679
    sexual selection, 580
HIV. See Human immunodeficiency virus
        (HIV)
HLA (human leukocyte antigen) complex,
        741
Hoagland, Mahlon, 48–49
Hofmeister, Franz, 39
Holliday junction, 341
Holometabolous insects, 239, 240f, 597
Homeobox, 290–292, 291f
Homeosis, 288
Homeotic genes, 288–290. See also Hox
        (homeotic) genes
Hominidae, 728, 730f
Hominin evolution, 731–740
    anatomically modern humans, 737–740,
        738f, 743–745
    Ardipithecus, 733
    Australopithecus, 732–734, 733f, 734f, 735f
    dispersal from Africa, 736–740, 738f
    geographic and temporal ranges, 736f
    Homo erectus, 735f, 736, 737, 737f, 740, 744
    Homo ergaster, 736
    Homo floresiensis, 739, 739f
    Homo habilis, 734, 735f
    Homo heidelbergensis, 736f, 737

Homo neanderthalensis, 735f, 737, 739,
        740, 740f, 748, 748f
    Homo rhodesiensis, 736f
    Homo rudolfensis, 736, 737
    multiregional evolution model, 740,
        743–745, 747–749, 770
    out-of-Africa model, 740, 743
    phylogenetic relationship, 729f
    Sahelanthropus, 731–732, 732f
    skull size and brain volume, 735f
    time spans and relationships of species,
        732f
Homo erectus, 735f, 736, 737, 737f, 740, 744
Homo floresiensis, 739, 739f
Homo habilis, 734, 735f
Homo heidelbergensis, 736f, 737
Homologous chromosomes, 24
Homologous genes, ancestral gene, 427
Homologous recombination
    crossing over, 349
    DNA repair, 336
    mechanism of, 339, 341, 341f
    in microbes, 187–188
    for organelle to nuclear gene transfer, 205
    targeted mutation, 246
Homologous structures, 67, 68f, 110, 113t
Homology
    analogy compared, 67, 113t
    manifestations of, 110
    orthologs, 126
    paralogs, 126
    universal homologies, 115–116, 116t, 120
Homo neanderthalensis, 735f, 736, 739, 740,
        740f
Homoplasy, in human lineages, 731
Homo rhodesiensis, 736f
Homo rudolfensis, 736, 737
Homunculus, 716
Hooker, Joseph, 92
Hooke, Robert, 12
Horizontal gene transfer. See Lateral gene
        transfer
Host races, 653–654
Housekeeping genes, 132, 171, 370
Hox (homeotic) genes, 68, 75, 697
    conservation of function and structure,
        292–293
    crustacean morphology, involvement in,
        301–305, 303f–305f
    discovery of, 288, 288f
    duplication and divergence, 711, 715
    evolutionary history in metazoan phyla,
        293f
    expression of, 288–290, 289f, 290f
    fly bristle patterns, role in, 305–307, 306f
    gene organization, 289f, 292–293, 294f
    homeobox sequence, 290–292, 291f
    larval appendage evolution, involvement
        in, 298–301, 300f–302f
    ubiquity in animals, 292, 293f
    variation within populations, 307
    wing evolution, involvement in, 293–298,
        296f, 297f, 299f
Hsp90, 691, 692f
Human evolution, 725–781
    disease, genetic basis of, 755–769
        association studies, 759–763, 763t

Darwinian medicine, 776–778
genomic medicine, 775–776
mapping disease genes, 756–758, 757f
Mendelian genetics, 756–758, 756f,
758f
patterns of inheritance, 756f
QTLs, 758–759
reasons for study of disease genetics,
755–756
use of genetic information, 765–769
variants for common diseases,
763–765, 765f, 766f
genetics, 741–749
Chikhi–Beaumont model, 744
clines, 741, 741f
hierarchical structure of human demography, 742–743, 742f, 743f
mismatch distribution, 746–747, 746f
mitochondrial DNA genealogy, 745,
747, 748–749
statistical inference in population
genetics, 744, 744f
Y-chromosome genealogy, 745–748,
745f
hominin evolution, 731–740
anatomically modern humans,
737–740, 738f, 743–745
Ardipithecus, 733
Australopithecus, 732–734, 733f, 734f,
735f
dispersal from Africa, 736–740, 738f
geographic and temporal ranges, 736f
Homo erectus, 735f, 736, 737, 737f,
740, 744
Homo ergaster, 736
Homo floresiensis, 739, 739f
Homo habilis, 734, 735f
Homo heidelbergensis, 736f, 737
Homo neanderthalensis, 735f, 737, 739,
740, 740f, 748, 748f
Homo rhodesiensis, 736f
Homo rudolfensis, 736, 737
multiregional evolution model, 740,
743–745, 747–749, 770
out-of-Africa model, 740, 743
phylogenetic relationship, 729f
Sahelanthropus, 731–732, 732f
skull size and brain volume, 735f
time spans and relationships of
species, 732f
human nature, 778–781
language, 750–751, 751f, 752–754, 753f
natural selection, 771–773, 771f, 772f
lactose tolerance, 771–772, 772f
relaxation in, 773–775, 774f
skin color, 772–773, 772f
psychology, evolutionary, 778–780
races, 769–770, 775–776
tree of life, human position on, 727–731
Human genome
chimpanzee compared, 80–81
number of genes in, 80
transposable elements in, 597–598, 598f
Human Genome Project (HGP), 757–758
Human immunodeficiency virus (HIV)
CCRΔ5 gene and, 765, 771, 773f
evolution of, 71, 72f, 74

extinction and recolonization in HIV
populations, 447
genealogy, 424
mutation rate, 348
Human nature, understanding, 778–781
Hummingbirds, pollination by, 633, 633f
Huntington's disease, 565, 767
Hutton, James, 12–13, 12f
Huxley, Thomas Henry, 18, 19f, 21
Hybridization
biological species concept (BSC) and,
622–623, 622f–623f
introgression, 452, 454, 454f
species origin from, 73, 73f
Hybrid rescue alleles, 636
Hybrid sterility, 634–636, 638–640
Hybrid zones, 452f, 474f, 504, 640, 646, 699
Hydrogen bonds
base pairing in DNA, 45, 45f
Pauling's work with, 41
Hydrogenosomes, 196, 209, 209f
Hydrophilic, 100
Hydrophobic, 100
Hydrophobic core, protein, 702
Hydrothermal vents
chemoautotrophic symbiosis, 164
metal sulfides at, 96
oasis, 95, 95f
as origin-of-life site, 95, 96
as reducing environment, 95
Hydroxypropionate pathway, 159, 160f
Hypercycle, 614, 615f
Hyperthermophiles, 152–154, 155f

I

Ichthyostega, 275, 276, 277f
Identical twins, variance analysis, 394, 395f
Identity by descent (IBD), 420–422, 421f,
601, 602, 602f
Immune memory, 717
Immune system
diversity in, 663
genomic rearrangements, 233–234, 234f
natural selection, 716–717
Immunoglobulin, 663
Immunoglobulin gene, rearrangement,
233–234, 234f
Immunological distance, 729
Imprinting, genomic, 604–605, 605f
Inborn errors of metabolism, 42, 756
Inbred lines, variance analysis, 394
Inbreeding
female-biased sex ratio and, 604f
identity by descent, 421–422, 421f
in pedigree, 429f
Wright's work on, 30
Inbreeding coefficient, 421, 515
Inbreeding depression, 358, 422, 515–518,
687–689
Incest avoidance, 688
Inclusive fitness, 602–603
Incompatibility loci, balanced selection and,
541, 541f
Incompatibility systems, in plants, 505–506,
506f
Incomplete dominance, 24

Indels, 326, 327f, 337, 340, 546
Indirect selection, 578–580, 663, 675
Induction, introduction of term, 52
Inference, evolutionary, 114
Infinite-alleles model, 534
Infinitesimal model, 483
Infinite-sites model, 424, 760f
Influenza, 61, 61f, 351, 351f
Information, in human genome, 1
Informational genes, 191
Ingram, Vernon, 46, 47f
Inheritance of acquired characteristics, 15,
21, 344
Innovations, 700
*An Inquiry into the Nature and Causes of the
Wealth of Nations* (Smith), 15
Insecticide resistance, clines in, 501, 502f
Insects. *See also specific insects*
aging, 562–564
appendages, 556f
flight, evolution of, 276–277, 278f
larval appendage evolution, 298–301,
300f–302f
phylogenetic relationships, 239, 240f
social, 599–600, 600f, 611, 612
wing evolution, 293–298, 296f, 297f, 299f,
721
Insertions, 326, 340
Insertion sequences, prokaryotic, 593
In situ hybridization, 149, 149f, 311
Insurance, 767–768
Integrins, 236
Intelligent designer, 66, 81
Interaction variances, 394
Interbreeding, Neanderthal–human, 748,
748f
Intercalating agents, DNA, 332–333
Intracellular Pangensesis (de Vries), 24
Introgression, 452, 454, 454f
Introns
alternative splicing, 599
in bacteria and archaea, 172, 174
discovery of, 56–57
evolution of, 62
origin of, 599
self-splicing from *Tetrahymena*, 458, 459f
types of, 219–221, 219t
variations in, 361
Inventions, 700
Inversion polymorphisms, 542
Inversions
in balancer chromosomes, 359
chromosomal, 31, 32f
creation of, 342f
description, 328
effect on meiotic, 340
recombination suppression by, 591
In vitro evolution experiments
catalytic RNA, 102
RNA ligases, 103
In vitro selection experiments, on ribozymes,
458–459, 459f
Iridium, 284
Irreducibly complex system, 78
Island model, 441–442, 441f, 444–445, 444f,
445f, 449–451, 449f, 450f,
496, 771

Isogamous system, 687, 687f
Isolation by distance, 741
Isomorphous replacement, 54–55
Isotopes
    definition, 88
    heavy, 40
    radioactive, 40
    radioisotope dating, 88–90, 89t
Itano, Harvey, 46

J

Jacob, François, 50, 51f, 52–53, 53f, 54, 55
Jamaican click beetle, 542f
Jenkin, Fleeming, 20
Johannsen, Wilhelm, 25
Jordan, David Starr, 644
Junk DNA, 173–174, 216, 220, 595–598
Just-so stories, 62, 62f, 560, 560f

K

Karyogenic models, of nucleus origin, 210,
    212, 213f
Kelvin, Lord, 20
Kendrew, John, 54f, 55
Kepler's third law, 723f
Kerogen, 91
Kimberella, 262, 263f
Kimura, Motoo, 59, 59f, 408, 425, 531
Kinase, in pre-tRNA processing, 219
Kin discrimination, 606–607, 606f
Kinetoplast, 200
Kinetoplastids, 200, 207
Kinorhynchs, in fossil record, 265
Kin selection
    definition, 601
    haplodiploid eusocial, 612
    identity by descent, 601, 602, 602f
    inclusive fitness, 602–603
    t-haplotype in mice, 591
Kipling, Rudyard, 62, 62f, 560, 560f
Kirkpatrick, Mark, 580, 583
Kondrashov, Alexey S., 680
Kornberg, Arthur, 46
K–T extinction, 284, 284f

L

Labyrinthulids, 200, 201f
Lac operon from *Escherichia coli,* 174f
Lac system, 50, 51f, 52, 345–346
Lactose permease, 52
Lactate dehydrogenase (LDH), 75, 524, 708f
Lactose tolerance, 771–772, 772f
Ladybird beetles, 626f
Ladybirds, *Wolbachia* infection, 589, 590f
Lagging strand, 330
Lamarck, Jean-Baptiste de, 14–15, 14f
Lamarckism, 21, 41
λ system, 52
Lamellipodia, 236
Lande, Russell, 580, 583
Land organisms, emergence of, 271–272,
    271f, 272f
Language, 750–751, 751f, 752–754, 753f, 770,
    770f

Large X effect, 639–640
Last universal common ancestor. *See* LUCA
Late acting mutations, 565–567
Lateral DNA transfer, 182–191
    amelioration of genes, 188, 189f
    barriers to, 187–188
    conjugation, 183–184, 184f, 185f
    core genes and, 188–191, 190f
    transduction, 184, 186–187, 186f, 187f
    transformation, 183, 183f
Lateral gene transfer (LGT)
    in eukaryotes, 214
    importance of, 191
    LUCA concept and, 116
    novelty creation, 718, 718f
    plasmids, 171
    sexual recombination compared, 183
    Tree of Life, effect on, 131–132, 131f, 133f
    as variation source, 351–352
    in viruses, 134
Lavoisier, Antoine-Laurent, 38, 38f
Law of Succession, 74
LCT gene, 771–772, 772f
LDH (lactate dehydrogenase), 75, 524, 708f
Leading strand, 330
Le Bel, Joseph-Achille, 38
Lederberg, Esther, 52
Lederberg, Joshua, 41
Leibniz, Gottfried Wilhelm, 38
Leks, 577
Lenormand, Thomas, 505
Lens cystallin, 708f
Lenski, Richard E., 482
Let-3 gene, 249, 250f
Levene model, 673f
Lewis, Edward, 288, 295
Lewontin, Richard, 558, 560, 778
LexA, 705f, 706, 707f
LGT. *See* Lateral gene transfer (LGT)
The Library of Babel (Borges), 465, 465f
Lichens, 164
Life cycle
    eukaryotic, 664, 664f, 665f
    prokaryotes, 664f
Life, definition of, 87
Life history, definition, 556
Life-history traits, 399
Life span, mutations extending, 566, 566f
Ligase, in pre-tRNA processing, 219
Ligation, in self-replicating system, 99
Liger, 622
Likelihood, plot of, 401
Lilium longiflorum, 216
Limb evolution, 721, 722f
Lineage sorting, 628
LINEs (long interspersed nuclear elements),
    217, 218f, 594, 597–598,
    598f
Linkage, 24
Linkage disequilibrium
    adaptive landscapes and, 474
    association studies and, 760–761
    description, 432–435
    epistasis, 671t
    generation by mixing populations, 452,
    454–455
    hitchhiking phenomenon, 485–486, 486f

    negative, 674–676, 675f
    QTLs, 436–437
    random drift, 671t
    rate of decay, 434
    recombination, effect of, 670–672, 671t
    selection on multiple genes, 485–487
    sexual selection, 580, 580f, 581
    variation, 435–436, 435f
Linked loci, selection on, 536–542
Linnaean classification scheme, 119
Linnaeus, Carolus, 12, 119
Lipids
    archaea, bacteria, and eukaryotic
        compared, 141, 142f
    compartmentalization function, 100
    structure, 100, 100f
Liposomes, 100, 100f
Lithotrophic, 159, 159t
Lively, Curt, 510
Lizards, behavioral polymorphism in, 571,
    572f, 573
Lobopods, in fossil record, 265, 266f
Local mate competition (LMC), 603, 604f
Locus
    definition, 356
    mutation/selection balance at a single
        locus, 512
Long-branch attraction, 130–131, 130f
Long interspersed nucleotide element. *See*
    LINEs (long interspersed
    nuclear elements)
Long-tailed tits, 600f, 606
Long terminal repeats (LTRs), 217, 597, 598f,
    599
Lophotrochozoa, 240f, 241, 243f, 265, 266f
Loss of function, 703–704
Low-pass filters, selection of, 467, 467f
LTRs (long terminal repeats), 217, 597, 598f,
    599
LUCA (last universal common ancestor)
    description, 116–117
    features of, 117–119
    gene families, inferred, 129t
    hyperthermophilia of, 154
    inferring properties of, 124, 125f, 128,
        129, 129t
    lateral gene transfer and, 131–132, 131f
    protein translation in, 129–130
    in Tree of Life, 116, 117f
Lucy, 732, 733f
Luria, Salvador, 41, 345, 346f
Lwoff, André, 41, 42, 50, 51f, 52
Lycophytes, 272–273
Lycopods, 274, 275
Lyell, Charles, 13, 13f, 16
Lysogenic phage, 52
Lysogeny, 50
Lysosomes, 196, 713, 714f
Lysozyme, 532, 532f

M

Macroevolution
    fossil record, 254
    punctuated equilibrium, 80
Macromolecules, origin of term, 39
Macronucleus, 200

Maize
    artificial selection, 481, 481f, 482f
    evolution from teosinte, 313–318,
        314f–318f
    heterosis in, 515
    transposable elements in, 343, 588, 588f
Major histocompatibility complex (MHC),
        541, 688–689, 747, 764
Major transitions in evolution, 609–610,
        610t, 719
Malaria, 3, 3f, 470, 472, 491, 491f, 536
Male genitalia, 573, 574f, 577, 577f, 635, 636f
Male sterility, 589, 591–592
Malthus, Thomas Robert, 17
Mammals
    adaptation to life in the sea, 76, 77f
    convergence of marsupial and placental,
        625f
    X inactivation, 682
Map
    genetic, 757
    physical, 757
Mapping
    disease genes, 756–758, 757f
    quantitative trait loci (QTLs), 399–409, 401f
Marginal effect of a gene, 389
Markuelia, 260, 260f
Marsupials
    convergence with placental mammals, 625f
    fossils in Antarctica, 74, 74f
Massive modularity hypothesis, 778
Materials for the Study of Variation (Bateson),
        21
Maternal lineage, 214
*Mathematical Theory of Natural and Artificial
        Selection* (Haldane), 30
Mating behavior of dung flies, 558–559, 558f
Mating plug, 75
Matings, extra-pair, 581–582, 583f
Mating types, 685–687, 686f
Matrilineal inheritance, 745
Matthaei, Johann, 49
Maxam, Allan, 56
Maynard Smith, John, 63, 78, 567, 568, 607,
        609, 616, 672, 720, 720f
Mayr, Ernst, 32, 33f, 485, 619, 620, 622, 641
McCarty, Maclyn, 43
McClintock, Barbara, 588, 588f
McDonald–Kreitman test, 535, 542f
McLeod, Colin, 43
Mean, 385
Mean fitness, 461
    in adaptive landscape, 494–496, 495f
    adaptive landscape and, 472–474, 473f
    genetic load and, 550–551, 551f
Medicine
    Darwinian, 776–778
    genomic, 775–776
Meerkat, 600f, 611, 611f
Megalosaurus, 13, 13f
Megatherium, 17f
Meiosis
    crossing over, 222
    efficiency, 668
    polymorphisms, effect on recombination,
        340, 340f
    process, 222, 224f, 349, 350f

randomness in, 428
recombination in, 349, 350f, 663–664
segregation errors, 338–339, 339f
segregation of chromosomes, 349
Meiotic drive, 588, 642
Melting temperature of double-stranded
        DNA, 373, 373f
Mendel, Gregor, 22f
    continuous variation and, 26
    heredity in pea, 22–23, 23f
    photograph of, 22f
    rediscovery of work, 9, 22
Mendelian genetics, human diseases,
        756–758, 756f, 758f
Meselson, Matthew, 46, 46f
Mesoderm, 235
Mesophiles, 152, 153f, 155f, 156t
Mesozoic era, 255, 255f
Messenger RNA (mRNA)
    antisense, 247
    processing, 235
    role of, 54
    transcription of, 9
    translation of, 9
Metagenomics, 150
Metal sulfide compounds, 96
Metapopulation, 444, 607
Metazoa
    appearance in fossil record, 270
    Ediacara, 261–262, 262f
    opisthokonts, 201
    phylogenetic relationships, 240f
Meteorites
    in Earth age estimation, 90
    organic molecules in, 94–95
Methanogenesis, 166
Methanogens, 161
Methanosarcina acetivorans, 172
Methylation, of DNA, 57
Methylcytosine, deamination of, 332, 333f
MHC (major histocompatibility complex),
        541, 688–689, 747, 764
Microarrays, 362, 749
Microbiology, role in emergence of molecular
        biology, 41
Micrococcus radiodurans, 143
Microfilaments, 198
Micronucleus, 200
Microorganisms
    environmental studies, 147–151, 149f,
        151f
    in Tree of Life, 120–121, 124
    unculturable, 3, 148–151, 148f
Microraptor, 278, 279f
MicroRNAs, 236, 706
Microsatellite DNA
    description, 362
    as heterozygosity marker, 515–516, 516f
    markers, 309, 361
    mutation rates, 663
    slip-strand mispairing, 330, 347
    tandem repeat size, 217
    Y chromosome, 746
Microsporidia, 198, 199f, 621f
Microtubules, 198
Migration, of human populations, 742–743
Miller, Stanley L., 34, 93

Miller–Urey experiments, 93–95, 93f, 94t
Mimicry
    Batesian, 506–507, 507f
    Müllerian, 18, 19f, 475–476, 475f, 503,
        504f, 640, 698–699, 698f
Mimivirus, 133, 136
Mimulus, 633, 633f, 634f
Minerals
    role in chemical evolution, 96f
    self-organization of monomers into
        polymers, 99
Minisatellite DNA
    description, 362
    slip-strand mispairing, 347
    tandem repeat size, 217
    use in identifying relationships, 362, 362f
Mint, tetraploid, 631f
Mismatch distribution, 746–747, 746f
Mismatch repair, 188, 336–337, 338f
Missense mutation, 326, 327f
Mitochondria
    amitochondrial lineages, 209–210,
        209f–210f
    cristae, 200
    as mutualistic symbiosis, 163
    origin of, 203, 207, 209–210, 209f–210f
    paternal inheritance in mussels, 686
    similarity to bacteria, 203, 207f
Mitochondrial DNA (mtDNA)
    genetic code, 118, 205t
    gene transfer to nuclear genomes, 132,
        204–207, 206f
    genome flow between taxa, 623
    genome reduction, 204–207, 206f
    human genealogy, 745, 747, 748–749
    male sterility, 589, 591–592
    North American bird species divergence,
        628
    phylogeographic studies of, 451–452,
        452f, 453f
    rate of evolution in genes, 214
Mitochondrial Eve, 427
Modifiers, 484
Modularity, 711–712, 714–715
Molecular biology
    Central Dogma of, 53, 53f
    emergence, technical advances crucial for,
        40
    evolutionary biology and, 5, 35, 60–61
    origin of, 37–64
    universality of mechanisms, 66
Molecular breeding, green fluorescent protein
        (GFP) example, 3, 5, 5f
Molecular clock
    calibration with fossil record, 254
    description, 59, 59f, 270, 270f, 371–375,
        373f, 375f
    deviations as measurement of selection,
        531–532, 532f
    Drosophila species divergence, 627
    human–chimpanzee lineage separation,
        729
    neutral theory and, 425
Molecular divergence, species definition by,
        621–622, 627–628, 627f–628f
Molecular fossils, 202, 219
Molecular recognition, 702, 703f

Molecular systematics
    description, 120
    history of use, 120–121
    in phylogenetic studies of archaea and
        bacteria, 142–145, 147–150
    universal genes, 124
Mollusks, 241, 242f
Molothrus ater, 604f
Monandrous groups, 610f
Monera, 120
Monkey flower, 633, 633f, 634f
Monkeys, old-world, 728, 729, 730f
Monod, Jacques, 40, 41, 50, 51f, 52, 53f, 54,
    55–56, 75
Monophyletic groups, 111, 111f, 239, 254
Monosomy, 326
Morgan, Thomas Hunt, 24–25, 25f, 34
Morning sickness, 779–780, 779f
Mortality
    baseline, 562, 564f
    increase with age, 561, 562f
Mosquito, insecticide resistance clines in,
    501, 502f
Moss campion, 592
Mosses (bryophytes), 272–273
Mouse
    Hox genes, 294f, 295f
    imprinting, genomic, 605, 605f
    MHC-based mate choice, 688–689
    Pitx1 gene, 309, 311f
    t-haplotype, 590–591, 591f
mRNA. *See* Messenger RNA (mRNA)
Mules, 626
Müller, Fritz, 18
Muller, Hermann, 34, 82, 634, 639, 676, 679
Müllerian mimicry, 475–476, 475f, 503, 504f,
    640, 698–699, 698f
Muller's ratchet, 681, 681f
Mullis, Kary, 56
Multicellularity
    body plans, 238–250
        of animal groups, 239, 240f–243f,
            241–242
        as defining group characteristic,
            238–239, 239f, 240f
        in fossil record, 244
        genetics of, 244–250
    differentiation, 230–238
        cancer as loss of, 237–238, 238f
        division of labor, 230–231, 231f
        genetic information, constancy of,
            231–233, 232f–233f
        genomic rearrangements, 233–234,
            234f
        maintenance of, 236
        mechanisms of, 234–236
        progressive nature of, 236, 237f
    emergence of, 225–230
        aggregation, 226, 228–229, 228f
        cell recognition and communication,
            229–230
        choanoflagellates, 229–230, 229f
        environmental inputs, involvement of,
            230
        fossil record, 259–262
        in Volvocales, 226–227, 226f–228f
Multiple genes, selection on, 485–487

Multiple sequence alignment, 120, 122f
Multiregional model, 740, 743–745, 747–749,
    770
Müntzing, Arne, 631f
Murchison meteorite, 95
Murchison, Roderick, 13
Muskrat, spread of, 499f
Mussels, mitochondria of, 686
Mutation
    accumulation of weakly deleterious, 494
    age of new allele, 434, 434f
    background selection against deleterious,
        538–540, 539f
    base substitutions, 326, 326t, 327f, 328
    categories of, 325–326
    definition, 325
    in DNA replication, 328, 330, 334,
        336–337, 337f, 338f
    duplications, 326, 328f, 342f
    favorable, establishment of, 490–493,
        491f, 492f
    fixation of, 536, 537f, 538, 641, 641f
    frameshift, 326, 327f
    indels, 326, 327f, 337, 340
    inversions, 328, 329f, 340f, 342f
    late acting, 565–567
    loss by chance, 489–490, 490f
    missense, 326, 327f
    multiple favorable, 720
    mutagenesis screens for development
        genes, 244–246, 245f, 246f
    neutral, 326, 415, 425
    nonsense, 67f
    physical damage to DNA, 330–332, 334
        deamination of bases, 331–332, 333f
        double-strand breaks, 334, 334f
        intercalating agents, 332–333
        UV irradiation, 332, 333f
    pleiotropic side effects, 484
    ploidy, 328, 339
    point, 326
    polymorphisms, effects of, 340, 340f
    protection, prevention, and correction
        mechanisms, 334–337
        DNA repair processes, 335–336, 335f
        error correction, 336–337, 337f, 338f
    quantitative trait variation, maintenance
        of, 513
    randomness of, 413, 414, 415
    rates and patterns, 343–349
        in different genomic locations, 347
        species differences, 348–349
        by type of mutation, 347
    recombination-generated, 339–342,
        340f
    segregation errors, 338–339, 339f
    selection, interaction with, 510–518
    slip-strand mispairing (SSM), 330, 332f,
        336, 337, 347
    of small effect, 407–408, 407f
    somy, 326, 328, 329
    speciation and, 630
    suppressor, 67f
    tautomerization, 328, 331f, 347
    translocations, 328, 329f, 340, 340f, 342f
    from transposition, 218, 342–343
    types underlying disease phenotypes, 759t

Mutational heritability, 409–411, 410t
Mutational variance, 409–411, 409f, 410t,
    483f, 484, 514
Mutation load, 552, 565–566, 597, 659–660,
    660f, 680, 684–685, 774
Mutation rate
    evolution of, 659–663, 659f–662f
    molecular clock, 425
    rate of neutral divergence, 425, 425f
Mutator alleles, 659–662, 661f, 662f
Mutualism, 163–166, 165f, 611
Mycoplasma genitalium, 172, 175, 175f
Myoglobin, 54f, 55
Mysid shrimp, 302–303, 304, 304f
Myxobacteria, 172

N

NADPH, 161, 203
Nanoarchaeum equitans, 172
Narrow-sense heritability, 391, 394, 397, 399,
    446, 477, 548
Nasonia vitripennis, 589, 589f
National DNA Database, United Kingdom,
    768
Naturalistic fallacy, 82
Natural selection, 457–487. *See also* Selection
    acceptance of, 18–20
    action on replicating molecules, 458–460,
        459f
    adaptive landscape, 472–475, 473f
    alternative evolutionary mechanisms to,
        20–21
    as cause for the appearance of design,
        74–75
    as cause of adaptation, 65, 457, 463–464
    cost of, 550
    Darwin's discovery of, 457–458
    definition, 2
    density-dependent, 470
    on diploids, 470–472, 470f, 472f
    directional selection, 470–471, 471f
    frequency-dependent, 470, 472, 472f,
        475–476
    Fundamental Theorem of Natural
        Selection, 462–463, 462f
    genetic load, 75
    human populations, 771–775, 771f, 772f,
        774f
    as imperfect mechanism, 75
    mean fitness, 462–463, 462f, 472–474,
        473f
    novelty creation, 719–723, 722f
    objections to as the cause of adaptation,
        78–81
    *On the Origin of Species by Means of
        Natural Selection,* 9, 16–18,
        20–21
    on quantitative traits, 529–530
    random genetic drift compared, 414, 460
    reproductive isolation, opposition to
        evolution of, 640–641
    step-by-step nature of, 464–465, 465f
    timescale of, 472
    variation in fitness, dependence on,
        457–458, 458f, 460
Natural theology, 11, 78, 556

Natural Theology (Paley), 16
Naturphilosophie, 15
Nauphoeta cinerea, 581, 582f
Nausea and vomiting during pregnancy (NVP), 779–780, 779f
Nearly isogenic lines (NILs), 403, 404f
Neighborhood size, 451
Nei's genetic distance, 638
Neisseria meningitidis, 432, 432f
Nematodes, 242, 265, 266f
Neo-Lamarckian inheritance, 21
Neoproterozoic era, 202, 259–260, 262, 272
Neotenous salamanders, 597
Neptunism, 12, 12f
Networks, modularity of regulatory and developmental, 714–715
Neural tube formation, 236, 237f
Neurons, differentiation of, 236, 237f
Neurospora crassa, 42, 42f, 593
Neutral evolution, 414
Neutral mutations, 326, 415, 425
Neutral theory, 59–60, 59f, 366, 425–427, 505, 530, 533–536, 534f, 642
Neutral variation
    balancing selection, effect of, 541–542, 541f, 542f
    linked loci and, 536–542
    selection compared to, 531
Newton, Isaac, 11
NILs (nearly isogenic lines), 403, 404f
Nilsson-Ehle, Herman, 26–27
Nirenberg, Marshall, 49, 49f
Nitrogen fixation, 164–165, 165f, 613, 614f
Nodes, phylogenetic tree, 111, 111f, 112f
Noncoding DNA
    abundance in eukaryotic genomes, 595–598, 598f
    conserved sequences, 545–547, 546f
    functional sequences, detection of, 545–547
    insertions and deletions in, 542, 546, 546f
    selection on, 542–547
    species differences in, 543
Nonhomologous end joining, 335
Nonhomologous gene displacement, 718, 718f
Non-Mendelian inheritance, 586
Nonsense mutations, 67f
Nonsynonymous change, 370, 374, 374f, 513f, 531, 532, 535
Normal distribution
    and gene flow, 442
    of variation, 384f, 385–387, 385f–388f
Northrop, John, 39
Nothofagus, 259, 259f
Novelty, evolution of, 695–723
    compartmentalization, 711, 713–714
    exploratory systems, 715–717
    features of novelty, 696–700
    gene product activity, changes in, 700–704
    gene regulation, changes in, 704–708, 705f–707f
    lateral gene transfer, 718, 718f
    modularity, 711–712, 714–715
    natural selection and, 719–723, 722f
    redundancy, 708–711, 708f, 711f
    robustness, 715–717
    from scratch, 709–710, 709f–710f

symbioses and, 718–719
Nuclear membrane, evolutionary history of, 128
Nucleation, 98–99, 98f
Nucleic acid polymerases, changes in, 700–701, 701f
Nucleoid, 138
Nucleomorph, 201, 207
Nucleosomes, 197f
Nucleotide diversity
    description, 364, 368
    difference among organisms, 369f
    effective population size inferred from, 432
    HIV, 447, 447t
    in humans, 370, 370f, 371f, 372f
    within a population, 424
    recombination reduction and, 539–540, 540f
    variation in a genome, 435–436
Nucleotides
    definition, 357
    polymerization by UV cross-linking, 97, 97f
    prebiotic synthesis, lack of, 95, 96
    stereoisomers, 104f
Nucleotide site, 357
Nucleus
    origin of, 210–213
        endokaryotic models, 212–213, 213f
        karyogenic models, 210, 212, 213f
        summary of models, 211t
        prokaryote–eukaryote distinction, 120
Null alleles, 530
Null hypothesis, 530, 761
Number of genes, estimating, 399, 399f
Numb mutant, 248–249, 248f
Nup96 gene, 637–638, 637f
Nursery web spiders, 579f
Nutritional symbioses, 165
NVP (nausea and vomiting during pregnancy), 779–780, 779f

O

Oaks, 623, 623f
Objections to evolution
    to the fact of evolution, 76–78
    to natural selection as cause of adaptation, 78–81
Occam's razor, 117
Oenothera lamarckiana, 24
Oenothera organensis, 506, 506f
Offspring, evolutionary interests of parents compared, 604–605, 604f–605f
Okazaki fragment, 214, 330
Oligonucleotide, array, 361
Oltmannsiella, 226
Omo 1 and 2, 737, 738f
Online Mendelian Inheritance in Man (OMIM) database, 758f
On the Origin of Species by Means of Natural Selection (Darwin), 16–18, 20–21, 78, 93
    animal embryology and development, 244
    common ancestry, 117
    co-option of structures, 75
    embryology and development, 287

geographic distribution, 70
natural selection explained in, 458
publication of, 9, 9f, 65
sexual selection, 575
variation documentation in, 356
vertical inheritance illustration, 110, 110f
Onychophorans, 265, 300
Oparin, Alexander, 34, 93
Open reading frames, creation of new, 709
Open system, 79
Operational genes, 191
Operational taxonomic units (OTUs), 111, 111f
Operons
    in archaea, 141
    description, 139
    genome compaction and, 172, 174
    Jacob and Monod, 54
    lac, 174f
    lateral transfer, 174
    in *Pseudomonas fluorescens*, 508
Opisthokonts, 200f, 201, 201f
Optimal growth temperature, 152–153, 153f
Optimization, evolutionary, 556–566
    aging, 564–565, 565f, 566
    constraints on what is possible, 557–558, 560
    logic of argument, 558–559, 558f–559f
    trade-off, 560–561, 560f
Orangutan, 728f, 730f
Orchids, hybridization and polyploidy in, 73, 73f
Organelle genomes
    gene transfer to nuclear genomes, 204–207, 206f
    reduced content of, 204
Organelles. *See also specific organelles*
    apicoplast, 3, 3f
    movement, 196
Organic molecules, production from inorganic compounds, 92–95
Organotrophic, 159, 159t
Origin of life
    how, 91–107
        chemical evolution, 95–97
        compartmentalization, 99–101
        DNA replacement of RNA, 106–107
        genetic code, origin of, 104–106
        organic molecules, generation of, 92–95
        RNA world, 101–104
        self-replication, 98–99
        steps involved, 91f, 92
        strategies for study of, 92
    when, 87–91
Origin of Life (Oparin), 34
Origin of new species
    after hybridization, 73, 73f
    geography, 644–654
    mechanisms, 640–644
    speed of, 71
Origin of replication, symmetric inversions around, 179–180, 181f
Orr, H. Allen, 408
Orthogenesis, 21
Orthologous genes, 176

Orthologs, 126–127, 127f, 291f, 292
Osmotic pressure, 157
Ostrom, John, 256
OTUs (operational taxonomic units), 111, 111f
Outgroup, 125, 126f, 365
Out-of-Africa model, 740, 743
Overdominance, 389f, 471, 471f, 471t, 515, 516, 517f
Oxygen radicals, 334, 335

P

PaJaMo experiment, 52–53, 53f, 54
Paleocene, 284
Paleoproterozoic era, 202
Paleozoic era, 255, 255f
Paley, William, 16
Pandorina, 226
Pangea, 256, 258, 258f
Panmictic population, 439
Panmixis, 439
Papilio dardanus, 506–507, 507f
Paracentric inversion, 328, 329f
Parallel evolution, 114, 115f
Paralogs, 126–127, 127f, 710
Parapatric distribution, 473, 474f, 476, 627
Parapatric speciation, 644, 646, 647f
Paraphyletic group, 111, 111f, 112f
Parasites
    brood, 604, 604f
    coevolution with host, 509–510, 510f, 511f, 672, 674
    coviruses, 590
    distinction between genetic and conventional, 589–590
    resistance signaled by sexually selected traits, 583
    of self-replicating systems, 99
Parasitic RNA, 103
Paratartaric acid, 38, 40f
Pardee, Arthur, 52, 53f
Parents, evolutionary interests of, 604–605, 604f–605f
Parsimony, 117, 124
Parthenogenesis, 631, 666, 667f, 668, 669
Particulate inheritance, 23, 24
Partula, 620, 621f
Parus major, 581, 582f
Pasteur, Louis, 38, 39f
Paternal alleles, selection on, 605, 605f
Paternal lineage, 214
Pathogenicity, 166
Pathogenicity islands, 166, 187, 187f
Pathogenicity suppression, 703, 703f
Pathogens, 166–167
Patrilineal inheritance, 745
Pauling, Linus, 39, 41, 41f, 42f, 43, 46, 54
Pax6 gene, 319–321, 320f
Payoff matrix, 567–570, 569f, 570f
PCR. *See* Polymerase chain reaction (PCR)
PDYN (prodynorphin), 751–752
Pearson, Karl, 21, 22f
Pedigree, 428–429, 429f
P elements, 595
Peppered moth, 470, 470f, 550–551
Pericentric inversion, 328, 329f
Permian period

extinctions, 281, 283, 283f
    wing evolution, 276
Peroxisomes, 196
Personalized medicine, 766
Perutz, Max, 54–55, 54f
PGI (phosphoglucose isomerase), 524, 701, 701f
Phage proteomic tree, 134, 135f, 184, 186–187, 186f–187f
Phalaropus lobatus, 575f
Phanerozoic era, 255, 255f, 262
Phenogram, human populations, 742f
Phenotype
    components of, 384, 387–388, 387f, 393
    description, 10, 358
    distinction from genotype, 25
Phenotypic variance, 394, 399, 446, 478
Phenylketonuria (PKU), 756
Philosophie Zoologique (Lamarck), 14, 15
Phosphoglucose isomerase (PGI), 524, 701, 701f
Phospholipids
    compartmentalization function, 100
    structure, 100, 100f
Phosphorites, 259–260, 260f
Photinus fireflies, 626f
Photoautotrophic symbioses, 164
Photoautotrophy, 161–162
Photoheterotrophy, 162
Photoreactivation, 335
Photosynthesis, by bacteria and archaea, 161–163, 162f, 163f
Photosystems, 161
Phototaxis, 138
Phototrophy, 158, 159t, 161–163, 162f, 163f
Phylloscopus trochiloides, 453f
Phylogenetic anchor, 150, 151f
Phylogenetic diversification of bacteria and archaea, 142–151
Phylogenetic studies, of eukaryotes, 198, 200f
Phylogenetic trees. *See also* Tree of Life
    additive, 112
    archaea, 144f
    asexual taxa in, 667f, 668
    bacteria, 144f
    bootstrap values, 113
    branch length, 112, 112f, 114f
    cladograms, 111–112
    components, 111, 111f
    concatenation-based, 190
    dendrogram, 112
    eukaryotic groups, 200f
    extinct lineages, 117f
    genetic distances, use of, 742
    inference from, 113, 114
    long-branch attraction artifact, 130–131, 130f
    of major evolutionary groups of eukaryotes, 200f
    microsporidia, 199f
    monophyletic group, 111, 111f, 112f
    outgroup, 125, 126f
    paraphyletic group, 111, 111f, 112f
    parsimony principle, 124
    phylogram, 112, 112f
    polyphyletic group, 111, 111f, 128
    polytomy, 112–113, 112f

rooted, 113, 113f, 125, 126f
    Tree of Life, 123f, 128f
    ultrameric, 112
    vertical and horizontal, 111f
Phylogeny, genealogy compared, 364
Phylogeography, 451–452, 452f, 453f, 744
Phylogram, 112, 112f
Phylotypes, 149–150
Physalaemus pustulosus, 578–579, 579f
Physalis longifolia, 506, 506f
Physics, randomness in, 414
Pitx1 gene, 309, 310f, 311–312, 311f, 715
Pitx2 gene, 309, 311
PKU (phenylketonuria), 756
Placental mammals, convergence with marsupials, 625f
Plaice, 533, 534f
Plantae, subgroups of, 200f, 201
Plants
    B chromosomes, 594, 594t
    Carboniferous coal swamp forest, 275, 276f
    chromosome number, 632, 632f
    extinctions, 284
    imprinting in, 605
    innovations, evolutionary, 274f
    male sterility, 589, 591–592
    mosses, 272–273
    phylogenetic relationships, 239f
    polyploidy, 631f, 632, 632f
    regeneration of, 232, 232f
    seed, 275
    self-fertilizing, 688, 689f
    vascular, 272–275
Plasmids
    in archaea, 141, 170–171
    in bacteria, 138, 140f, 169–171
    copy number, 171
    functions of, 170–172, 171t
    lateral gene transfer, 171
    multiple drug resistance, 593
    selection on, 486–487
    size, 170, 170t
Plasmodium, 3, 133, 204f, 665f
Plastids. *See also* Chloroplasts
    gene transfer to nuclear genomes, 204–207, 206f
    genome reduction, 204
    primary and secondary, 207, 208f
    similarity to cyanobacteria, 203
Plate count anomaly, 148, 148f
Plate tectonics, 256, 258–259
Plato, 11
Platyfish, 636, 637f, 643
Pleiotropy
    antagonistic, 564
    bristle number in *Drosophila*, effect on, 527
    linkage disequilibrium, 527
    modifiers of dominance, 690
    of mutation side effects, 484
    in quantitative traits, 514, 514f
Pleistocene era, 627, 778
Pleodorina, 226–227
Pleuropod, 299–300, 301f
Pluripotency of differentiated cells, 232
Pneumodesmus newmani, 271, 271f
Podisma pedestris, 442–443, 443f, 503, 503f

Poeciliopsis minnows, 666, 667f
Point mutations, 326
Poisson distribution, 375, 375f, 419, 420, 531
Policing, 613
Polistes wasps, 611, 611f
Pollination
    figs by fig wasps, 603, 603f
    sexual selection, 573, 577
    species differences in, 633, 633f, 634f
    yucca, 613, 613f
Polyandrous groups, 610f
Polychaetes, 241–242, 243f
Polygenic variation, 383–384, 399. *See also*
        Quantitative genetics;
        Quantitative variation
Polymerase chain reaction (PCR)
    invention of, 56
    use in environmental genomics, 150–151,
        151f
Polymers
    chemical evolution of, 97, 97f
    formation by nucleation, 98, 98f
    guided polymerization, 99
Polymorphism
    Adh gene intron haplotypes, 546f
    balanced, 500, 514–515, 541, 541f
    behavioral in lizards, 571, 572f, 573
    for chromosomal inversions in
        Drosophila, 542
    color in Jamaican click beetle, 542f
    description, 358
    enzyme, 524–525, 525f
    examples, 358f
    flowers, left- and right-handed, 687, 688f
    frequency of, 360, 368, 368f
    linkage disequilibrium, 760–761
    maintenance
        by frequency-dependent selection,
            506–508, 506f, 507f
        in heterogeneous environment, 508,
            509f
        by heterozygote advantage, 505
    neutral theory and, 535–536
    PPARg allele and, 756
    quantitative trait variation and, 514–515
    single-nucleotide polymorphisms (SNPs),
        349, 361, 362, 362f, 423, 760
    species divergence and, 628
    unique event (UEPs), 745–746
Polyphyletic group, 111, 111f, 128
Polyploidy
    allopolyploidy, 631
    autopolyploidy, 631
    description, 73, 377
    genome size and, 596, 596f
    incidence of, 632, 632f
    orchids, 73f
    polyspermy and, 631
    speciation from, 631–633, 631f–632f
Polyspermy, 631
Polytene chromosomes, 375, 376f
Polytomy, 112–113, 112f
Population bottlenecks. *See* Bottlenecks,
        population
Population structure
    definition, 439
    genealogies, 449–455

gene flow, 439–449
Porifera, 241, 241f
Positional cloning, 756
Positive eugenics, 82
Positive selection, 750
Posttranscriptional regulation, 235–236
Postzygotic isolation, 626–627, 627t, 629
Potamopyrgus antipodarum, 672, 674f
Powdery mildew, virulence in, 510f
PPARg, 756
Prasinophytes, 201
PRD domain, 319–320
Prebiotic synthesis, 93–96, 93f–95f, 94t, 100
Precambrian period, 255, 255f, 270
Preformation, 11
Prenatal screening, 767
Prenylation, 222
Prezygotic isolation, 626–627, 627t, 629, 629f
Priapulids, in fossil record, 265, 266f,
        269–270
Price, George, 568
Primary contact, 504
Primase, 330
Primates
    divergence estimate, 536
    gene expression patterns, 749–750, 750f
    organization of order, 728f
    penes, 574f
    testes, 576, 576f
Primula, hybridization of species, 73f
Principia (Newton), 11
Principles of Geology (Lyell), 13, 13f, 16
Prions, 57, 98, 710, 710f
Prisoner's Dilemma, 569, 612, 612f
Prochloron, 164, 164f
Prodynorphin *(PDYN)*, 751–752
Progenote hypothesis, 130, 132
Programmed frameshifts, 709
Programming languages, 723, 723f
Prokaryotes. *See also* Archaea; Bacteria
    life cycle, 664f
    as polyphyletic group, 128
    in Tree of Life, 120
Promoters
    in archaea, 141
    in transposable elements, 343
Proofreading, 326, 336, 337f
Protein
    colloid theory, 39
    diversity, 368
    mixing and matching of protein domains,
        712–713, 712f, 713f
    size determination by ultracentrifuge, 39,
        41f
    structure, 39
    synthesis, general mechanism of, 50f
    universality of handedness, 66–67
Protein electrophoresis, 360
Protein microspheres, 100
Protista, photosynthesis in, 200
Protostomes, 240f, 241, 265, 266f
Pseudogenes, 75, 218–219, 219f, 371, 374,
        425, 599, 711
Pseudomonas fluorescens, frequency-
        dependent selection in,
        507–508, 507f
PSR (paternal sex ratio) chromosome, 589f

Psychology, evolutionary, 778–780
Psychrophiles, 152, 153f
Pterosaurs, 277–278, 278f
Punctuated equilibrium, 12, 80, 80f, 280, 280f
Punishment, 613
Pure culture, 148
Pure line theory, of Johannsen, 25
Purifying selection, 476, 531, 535
Purines, 326
Pyrimidines, 326
Pyrite, 281
Pyrsonympha, 684
Python, vestigial structures in, 75f

Q

Quadratic selection gradient, 529, 529f
Quantitative genetics
    additive model, 385, 388, 388f, 392, 402,
        409
    alleles of small effect, 407–408, 407f
    applications in human genetics, 384
    average effect of an allele, 389, 391–393,
        392f
    average excess of an allele, 389–393, 392f
    breeding value, 390–391, 391f, 393, 393f,
        401
    candidate genes, indicators for, 403–404,
        406–407
    cryptic variation, 398–399, 398f
    dominance, 388–389, 388f, 389f
    environmental variance, 387–388, 399
    epistasis, 388–389, 388f, 389f, 401
    genetic variance, calculating components
        of, 390–391, 390t, 391f
    heritability, 391, 394, 397, 398f, 399,
        409–411, 410t
    mutational variance, 409–411, 409f, 410t
    normal distribution of traits, 384f–388f,
        385–387
    number of genes, estimating, 399, 399f
    phenotype, components of, 384, 387–388,
        387f, 393
    polygenic traits, 383–384
    quantitative trait locus (QTL) mapping,
        315, 315f, 399–409, 401f
    threshold model, 386, 386f, 399
    variance components, estimating,
        394–397, 395f
    variation between individuals, 382–383
Quantitative trait loci (QTLs), 35
    definition, 356
    description, 399
    disease-associated, 758–759
    linkage disequilibrium and, 436–437
    mapping, 315, 315f, 399–409, 401f
    novelty evolution and, 696
    reproductive isolation and, 648f
    speciation and, 633, 634f
    stickleback skeletal evolution, 309, 310f,
        311, 312–313
Quantitative traits
    in association studies, 761–763
    coefficient of kinship, 421
    heritability, 527
    heritable variation in, 397–399
    mutation, 513–514

Quantitative traits (*continued*)
  pleiotropy, 514, 514f
  Q$_{ST}$, 446, 769
  random drift, effect of, 418, 418f
  selection on, 476–484
  selection on, measuring, 525–530,
      526f–529f
  variation maintenance in, 513–515
Quantitative variation
  analysis of, 385–399
  generation of, 409–411
  genetic basis of, 399–409
Quercus, 623f
Quetzalcoatlus, 277, 278f
Quorum sensing, 167–168, 167f, 606–607

R

Races, human, 769–770, 775–776
Radial symmetry, 241
Radiation, resistance in *Deinococcus
      radiodurans*, 143, 158, 158f
Radioisotope dating, 88–90, 89t
Radiolarians, 200
Radiometric dating, 88–89
Ramapithecus, 729
Random genetic drift, 60, 413–437
  allele frequencies, 415–420
  coalescence, 420–425, 423f, 424f
  gene flow and, 444–449
  geographic variation generated by,
      444–445
  linkage disequilibrium, 432–437, 671t
  natural selection compared, 460
  negative linkage disequilibrium, 676
  neutral theory of molecular evolution,
      425–427
  random linkage disequilibrium, 676–677,
      677f
  rate as determined by variance in fitness,
      418–419
  recombination, 427–437
  selection and, 489–496
  in shifting balance theory (SBT), 608,
      608f
  speciation and, 641–642
  variance, effect on, 416–420
  Wright–Fisher model, 416–418, 419, 420,
      422
Random linkage disequilibria, 676–677, 677f
Random segregation, 349
Rates of evolution, 114, 114f
Rat, warfarin resistance in, 491, 491f
Ray, John, 11, 11f
Reaction center, 161, 162f
Reaction norm, 705
Reactive acetyl CoA pathway, 160f
Rearrangements, genomic, 233–234, 234f
Receptor tyrosine kinases (RTKs), 230
Recessive alleles, 357, 358
  inbreeding depression and, 516
  purging of deleterious, 517–518, 518f
  selection against, 513
Recessive character, 22
Recessive lethals, 358, 483, 484f
Reciprocal translocation, 340, 342f
Recombination

additive genetic variance, 674–675, 675f,
    678
  ancestry and, 427–432
  artificial selection and, 679, 680f
  background selection, effect on, 539–540
  coalescence and, 429, 430, 430f
  costs, physiological, genetic and evolu-
      tionary, 668–669
  definition, 663
  discovery in bacteria, 41
  ectopic, 339, 343, 594, 595, 669
  epistasis, 672, 673f, 674, 676
  evolution of, 663–683, 664f–668f,
      671f–683f, 671t
  favorable alleles, 678–680
  in fluctuating environments, 672, 672f,
      673f, 674
  gradient within eukaryotic chromosomes,
      216
  homologous, 187–188, 205, 246, 336, 339,
      341, 341f
  hot spots, 760
  inversion suppression of, 591
  lateral DNA transfer, 182–191
  linkage disequilibrium, 432–437,
      670–672, 671t
  between marker and a disease locus, 756,
      757f
  meiosis, 663–664, 666f
  in molecular breeding, 5
  negative linkage disequilibrium, 674–676,
      675f
  nonrandom associations of alleles,
      669–670, 671f
  polymorphism effects on meiotic, 340, 340f
  random drift and, 427–437
  sexual, 182–183, 349–351, 350f
  ubiquity of, 663–664
Red deer, 61, 61f, 454, 454f, 527–528, 528f,
      548, 549f, 575
Red-necked phalarope, 575f
Red Queen evolution, 509, 663, 672, 673f, 674
Reduction principle, 682
Reductive tricarboxylic acid (rTCA) cycle,
      159, 160f
Redundancy, 708–711, 708f, 711f
Regeneration
  amphibian eye, 232, 233f
  plant, 232, 232f
Regression, 20, 21f, 389, 392, 392f
Regression line, statistical, 477–478, 478f
Regulation
  gene
    β-galactosidase, 50, 51f, 52–53
    cell differentiation and, 234–236, 235f,
        236f
    changes in, 704–708, 705f–707f
    in eukaryotes, 222, 223f
    lac system, 50, 51f, 52
    λ system, 52
    posttranscriptional, 235–236
    transcriptional, 249
Reinforcement, 649–650, 649f
Relative fitness, 419, 461, 468–469, 475
Relatives, interactions between, 599–610
  altruism, apparent, 599–601, 600f
  competition between, 670–672

covariances between, 396, 396f
  genomic imprinting, 604–605, 605f
  group selection, 607–609
  identity by descent, 601, 602, 602f
  kin discrimination, 606–607, 606f
  kin selection, 601–603
  quorum sensing, 606–607
  selection on extended phenotype,
      599–600, 600f
  sex ratio in fig wasps, 603, 603f
  shifting balance theory (SBT), 607–609,
      608f
Relative survival, 391, 391f
Religious beliefs, 65–66, 82
Repeat-induced point mutation (RIP), 593
Repetitive DNA
  in bacteria and archaea, 173, 173f
  crossing over between, 341–342, 342f
  in eukaryotes, 172–173, 173f
  LINEs (long interspersed nuclear ele-
      ments), 217, 218f, 594,
      597–598, 598f
  microsatellite DNA
    description, 362
    as heterozygosity marker, 515–516, 516f
    markers, 309, 361
    mutation rates, 663
    slip-strand mispairing, 330, 347
    tandem repeat size, 217
    Y chromosome, 746
  minisatellite DNA
    description, 362
    slip-strand mispairing, 347
    tandem repeat size, 217
    use in identifying relationships, 362,
        362f
  satellite DNA, 217
  SINEs (short interspersed nuclear
      elements), 217, 218f, 598,
      598f
  tandem DNA arrays, 217
  in telomeres, 214
  transposable elements, 216, 216f,
      217–219, 218f
  variable number tandem repeats
      (VNTRs), 347, 362
Replica plating, 344–345, 344f
Replicase, 99, 103f
Replication
  accuracy, 99
  compartmentalization of, 99–101
  hypercycles, 614, 615f
  parasitism of self-replicating systems, 99
  RNA, 103–104, 103f
  self-replication, evolution of, 98–99, 98f
  spread of replicating RNA molecules, 497,
      499, 499f
  viral, 135
Replicative DNA transposons, 217
Replicators
  cooperation among, 614, 615f, 616
  fitness of, 460
  guided polymerization in, 99
  simplicity of first, 78
Reproduction. *See also* Asexual reproduction;
      Sexual reproduction
  costs, 564–565

random, 413–415
Reproductive assurance, 688
Reproductive isolation, 622–630. *See also*
    Speciation
    accumulation of, 629–630, 630f
    definition, 622
    fitness reduction and, 642–643
    forms of, 624–627, 626f, 627t
    gene flow, in presence of, 647
    genes responsible for, 636–638, 637f
    hybridization, 622–623, 622f–623f
    natural selection in opposition to,
        640–641
    QTLs, 648f
    rate of evolution of, 627–628
    reinforcement, 649–650, 649f
    selection against cross-breeding, 648–650
    from single genetic change, 630–631
    species definition by, 280
Restorer allele, 592
Restriction enzymes
    as barriers to lateral gene transfer, 187
    RFLP creation by, 361
Restriction fragment length polymorphisms
    (RFLPs), 142, 361–362,
        362f, 757
Retinoblastoma, 511
Retrotransposons, 217, 599
Retroviruses, 217
Reverse genetics, 246–248, 756
Reverse transcriptase
    crystal structure, 701f
    discovery of, 57
    pseudogene creation by, 218, 219f
    from transposable elements, 590
Rhagoletis pomonella, 653, 654f
Rhizaria, 200–201, 201f
Rhizobium, 165, 165f, 166
Rhodophytes, 201
Rhodopsin, 707–708, 720–721
Rhynia, 274, 274f
Rhynie Chert, 272f, 274, 275f, 276
Rhyniognatha, 276, 278f
Rhyniophytes, 274
Ribonucleotide reductase, 102f, 134
Ribonucleotides
    conversion to deoxyribonucleotides, 134
    deoxyribonucleotide synthesis from, 102,
        102f
Ribose
    half-life of, 95
    prebiotic synthesis experiments, 93, 94f, 95f
Ribosomal RNA (rRNA), 53–54
    as phylogenetic anchor, 150, 151f
    phylogenetic trees of bacteria and
        archaea, 144f
    structure, 3, 4f
    transcription of, 9
    Tree of Life, 128, 128f, 133f, 154f
    use in environmental sampling, 149–151,
        149f, 151f
    use in molecular systematics, 120–121
    in vitro selection experiments, 458–459,
        459f
Ribosomes
    RNA association, discovery of, 48
    role in protein synthesis, 48–49, 54

structure, 56, 56f
    translation by, 9
Ribozymes, 57f, 102t, 458–459, 459f
Ribulose bisphosphate carboxylase (rubisco),
    161
Rickettsia, 143–144
Ring species, 620
RIP (repeat-induced point mutation), 593
Ritonavir, HIV resistance to, 72f
RNA
    amino acid interactions with, evolution
        of, 105–106, 105f
    ancient roles of, 102
    base pairing in, 544–545, 544f
    catalytic, 57–58, 57f, 101, 102, 102t, 106,
        220
    as genotype and phenotype, 101, 103
    guide, 106
    parasitic, 103
    in protein synthesis, 48–49
    replication, 103–104, 103f
    roles of, modern, 101–102, 101t
    selection for pairing in RNA molecules,
        543–544
    spread of replicating RNA molecules, 497,
        499, 499f
    structure, 102, 103f, 544–545, 544f, 545f
RNA interference (RNAi), 246, 247f
RNA ligase, 103
RNA polymerase, 103, 141, 348
RNA world, 63
    catalytic RNAs, 67
    DNA replacement of RNA, 106–107, 107f
    replication system, evolution of, 103–104,
        103f
    RNA as genotype and phenotype,
        101–103
Robertson, Alan, 676
Robustness, evolution of, 692
Rockefeller Foundation, 42
Rock–paper–scissors game, 570–573, 570f,
    572f
Rock pocket mouse, 500f
Rocks
    primitive metabolism on, 96, 96f
    radioisotope dating of, 88–90
Rooted trees, 113, 113f, 124–129, 125f, 127f,
    128f
rRNA. *See* Ribosomal RNA (rRNA)
RTKs (receptor tyrosine kinases), 230
Ruminant digestion, 166
Runaway process, 580

S

Saccharomyces cerevisiae, 219, 220, 543f, 685
Sage grouse, 577, 577f
Sahelanthropus, 731–732, 732f
Salamanders, genome size, 597, 597f
Saltation, 21
Saltationists, 22
Sanger, Fredrick, 42, 43f, 56
Satellite DNA, 217
Satellite tobacco necrosis virus (STNV), 590,
    590f
Scabrous gene, 405
Scala Naturae, 11, 11f

Sceloporus undulatus, 646f
Schizosaccharomyces pombe, 172
SCN1A gene, 765
Scott-Moncrieff, Rose, 42
Scr orthologs, 291f
Seasonal isolation, 626f
Secondary contact, 504
Secondary structures, RNA, 102, 103f
Sedgwick, Adam, 13
Sedum, 632f
Segregating sites, 364
Segregation distorter *(SD)* complex, 592, 592f
Segregation distortion, 587, 587f, 588, 588f,
    591, 592, 592f
Segregation errors, 338–339, 339f
Segregation, in meiosis, 663–664, 666f
Segregation load, 552
Seilacher, Adolf, 262
Selection, 521–553. *See also* Natural selection
    adaptive landscapes, 472–475, 473f, 479,
        480f
    additive, 471, 471f, 471t
    additive genetic variance, effect on,
        480–482, 481f, 482–484,
        482f
    among species, 609–610
    artificial, 480–482, 481f, 482f
    automatic, 688
    background, 677
    balancing, 476, 476f, 505–510, 514–515,
        531, 541–542, 541f, 552,
        689, 747, 764–765
    correlations between traits, 480, 480f
    density-dependent, 470
    direct, 578
    directional, 476–477, 476f, 526, 527, 527f,
        531, 535
    disruptive, 476, 476f, 652, 652f
    diversifying, 771
    extent of selection, 547–553
        additive genetic variance and, 548–549
        genetic load and, 549–553, 551f
    frequency-dependent, 470, 472, 472f,
        475–476, 506–508, 506f,
        507f, 525, 670, 672
    gene flow and, 496–505
    genetic load, 549–553
    group, 607–609
    indirect, 578–580, 663, 675
    on linked loci, 536–542
        background selection, 536–538, 537f,
        538–540, 538f
        balancing selection, 541–542, 541f,
        542f
        selective sweeps, coalescence from,
        536–538, 537f, 538f
    measurement, direct, 521–530
        difficulty of, 521–522
        fitness correlation with quantitative
        traits, 525–526, 527f
        fitness correlation with variation,
        523–524
        genetic manipulation, use of, 522–523,
        523f
        on quantitative traits, 525–530,
        526f–529f
    measurement, indirect, 530–536

Selection (*continued*)
  amino acid evolution rate, 531–532,
    532–533, 532f, 533f
  dispersal/selection interaction, 530
  molecular clock deviations, 531–532,
    532f
  mutation/selection interaction, 530
  neutral theory, comparison to, 530,
    533–536, 534f, 535f
  neutral variation, comparison to, 530
  modes of, 476, 476f
  on multiple genes, 485–487
  mutation, interaction with, 510–518
  on noncoding DNA, 542–547
  overdominance, 471, 471f, 471t
  positive, 750
  purifying, 476, 531, 535
  on quantitative traits, 476–484
  random drift and, 489–496
  rational design compared, 465–467, 466f
  on replicating molecules, 458–460, 459f
  sexual, 573–584
  species
    punctuated equilibrium, 80
    sexual reproduction prevalence and,
      609–610
    slowness of, 609
  stabilizing, 476, 476f, 513–514, 526, 527,
    527f, 675
  timescale of, 472
  truncation, 476, 476f, 552–553
  underdominance, 471, 471f, 471t, 476
  in vitro experiments, 458–459, 459f
Selection coefficients, 461, 469, 472, 511,
    512t, 522, 523f, 544
Selection differential, 477, 479, 527
Selection gradient, 478–479, 478f, 529, 529f
Selection response, 477
Selective deaths, 493, 538, 551, 552, 659
Selective sweep
  coalescence from, 536–538, 537f, 538f
  description, 427, 486
  distinguishing bottlenecks from, 537–538
  in human evolution, 747
  recombination, effect of, 536
Selenocysteine, 119
Self-fertilization
  diversity in species, 539, 540f
  evolution of, 688, 689f
  repeated, 664
Self-incompatibility, 505–506, 506f, 539, 540f
Selfish DNA, 173–174, 216, 669
Selfish genes
  definition, 586
  proliferation of, 669
The Selfish Gene (Dawkins), 778
Self-organization, 463–464
Self-replication, evolution of, 98–99, 98f
Semiconservative DNA replication, 46, 46f
Senescence, 561–566, 777. *See also* Aging
Sensory bias, 578–579, 579f, 580
Sepkoski, Jack, 268–269, 281, 282f
Sex. *See also* Sexual reproduction
  consequences of, 683–689
  definition, 663
  in fluctuating environments, 672, 672f,
    673f, 674

nonrandom association of alleles,
    669–670, 671f
  physiological, genetic, and evolutionary
    costs, 668–669
Sex-biased inheritance, 588–589, 589f
Sex chromosomes, evolution of, 682–683,
    682f
Sexes, evolution of separate, 682–683, 683f
Sex-linked genes, 25, 419, 420
Sex ratio
  in fig wasps, 603, 603f
  human offspring choice, 767, 768f
  local mate competition (LMC), 603, 604f
  population, 507
  trade-offs, 560
Sexual recombination. *See also*
    Recombination
  description, 182
  lateral DNA transfer compared, 183
  variation generation by, 349–351
Sexual reproduction
  Dictyostelium, 228f
  inbreeding, avoidance of, 687–689
  as key invention in eukaryotic evolution,
    222, 224, 224f
  mating types, 685–687, 686f
  mutation rate evolution, 660, 660f
  variation generated by, 27–28
  Volvox, 227, 228f
Sexual selection, 75, 573–584
  competition between males, 577–578,
    577f
  cost of, 669
  female preferences, 577–584, 579f, 580f,
    582f, 583f
  good genes, choice of, 581–584
  maladaptive, 573–575
  males, action on, 574f–576f, 575–576
  mating success, variation as cause of, 573
  rates of speciation and, 609, 610f
  species differences in intensity of, 576
Sheep
  inbreeding depression, 516, 516f
  Soay, 369, 370f, 516, 516f, 565, 669
Shells, appearance in fossil record, 262–263,
    263f
Shifting balance model of evolution, 30, 496,
    642
Shotgun sequencing, environmental,
    150–151, 151f
Shrew, 644, 645f
Sibling competition, 671–672
Sickle-cell anemia, 46, 47f, 326, 327f,
    390–392, 390t, 391f, 470,
    514
Side effects, evolution of, 79
Sika deer, 454, 454f
Silene acaulis, 592
Silicon chip, 722, 722f
Simpson, George Gaylord, 32, 33f
Simulated annealing, 467
SINEs (short interspersed nuclear elements),
    217, 218f, 598, 598f
Single-celled eukaryotes, diversity of,
    221–222
Single-nucleotide polymorphisms (SNPs),
    349, 361, 362, 362f, 423, 760

Sister chromatids, 349
Site, 357
Skeletal evolution, in sticklebacks, 307–313,
    308f, 310f–312f
Skin color, human, 772–773, 772f
Slime molds, 200, 201, 228–229, 228f
Slip-strand mispairing (SSM), 330, 332f, 336,
    337, 347
Small-eye mutation, 320
Small interfering RNAs (siRNAs), 247f
Smith, Adam, 15, 15f, 17, 707
Snails
  coevolution with parasites, 672, 674, 674f
  species relationships, 620, 621f
SNF2 family of DNA-dependent ATPases,
    713f
SNPs (single-nucleotide polymorphisms),
    349, 361, 362, 362f, 423,
    760
Soapberry bugs, 71, 72f
Soay sheep, 369, 370f, 516, 516f, 565, 669
Social Darwinism, 82
Social evolution, 586–587
Social insects, 599–600, 600f, 611, 612
Sociality, 167–168, 167f
Sociobiology: The New Synthesis (Wilson),
    778
Soma, 21
Sonic hedgehog gene, 643
Sorangium cellulosum, 172, 377
Sordaria, 588f
Sorex araneus, 644, 645f
SOS response in *Escherichia coli*, 706, 706f
Spadefoot toad, 606, 606f
Spandrels, 558, 560
Sparrows
  heterosis, 515
  inbreeding depression, 515
  selection on quantitative traits, 525–526,
    527f
Spatial structure, 439–440
Spea multiplicata, 606f
Special creation, 81
Speciation
  genetics of, 630–640
    Drosophila, reproductive
      incompatibilities in,
      634–639, 635f–637f
    Haldane's rule, 638–640, 638t
    polyploidy, 631–633, 631f–632f
    quantitative trait loci (QTLs), 633,
      634f
    speciation genes, 630, 636, 638, 639
  geography of, 644–654
    allopatric speciation, 644–646, 645f
    host races, 653–654
    parapatric speciation, 644, 646, 647f
    sympatric speciation, 644, 650–653
  mechanisms of, 640–644
    Dobzhansky–Muller model, 643, 644f,
      646
    natural selection, 640–641
    polyploidy, 631–633, 631f–632f
    random genetic drift, 641–642
  rate, sexual selection and, 609, 610f
Species
  competitive exclusion principle, 624

defining, 620–630
   asexual organisms, 622, 622f
   biological species concept (BSC), 622–624
   genealogy, 622, 622f
   by molecular divergence, 621–622, 627–628, 627f–628f
   by morphology, 620–621
   reproductive isolation, 622–630
  as a fundamental unit in biology, 619
  numbers, 624
  ring, 620
  selection among, 609
Species selection
  punctuated equilibrium, 80
  sexual reproduction prevalence and, 609–610
  slowness of, 609
Species tree, 190, 191f
Specificity, role of hydrogen bonds in, 41
Speech-language disorder 1, 750
Spencer, Herbert, 21
Sperm competition, 577, 578f
Sperm lysin, 533, 533f
Sphenopsids, 272
Spirin, Alexander, 54
Spliceosomal introns, 220
Splicing
  alternative, 220–221, 220f, 235–236, 236f, 599, 712
  exon shuffling, 712
  RNA catalysis and, 102
Sponges, 229, 229f, 241, 241f, 264–265, 264f, 266f
Sporangia, 274
Sporophytic systems, 506
Sports, 20, 21
Sporulation, 191, 192f
SSM (slip-strand mispairing), 330, 332f, 336, 337, 347
Stabilizing selection, 476, 476f, 513–514, 526, 527f, 675
Stag beetles, 577
Stahl, Franklin, 46, 46f
Standard (canonical) genetic code, 105, 118–119
Standard neutral model, 760
Star genealogy, 537
Statistical inference, in human population genetics, 744, 744f
Statistical power, 759, 759f
Staudinger, Hermann, 39
Stebbins, G. Ledyard, 32
Stentor coerulus, 691f
Steranes, 202
Stercorarius parasiticus, 576f
Stereochemistry theory, 106
Stereoisomers, 39, 40f, 104, 104f
Sterols, 202
Sticklebacks
  hybrids, 653, 653f
  Pitx1 expression, 715
  sexual selection, 574, 575f
  skeletal evolution in, 307–313, 308f, 310f–312f
STNV (satellite tobacco necrosis virus), 590, 590f

Stonecrop, 632f
Stop codon readthrough, 709
Streptococcus pneumoniae, 183
Stromatolites, 90–91, 90f
Structured coalescent, 449
Structured populations, 439–440, 461
Structure, spatial, 439–440
su1 gene, 317
Substitution load, 550–551, 551f
Substitutions, amino acid, 531–533, 532f, 533f
Sugars
  half-life of, 95
  prebiotic synthesis experiments, 93, 94f, 95f
Sumner, James, 39
Supercoiling, DNA, 195, 197f
Supergene, 507
Suppressor mutations, 67f
Survival, relative, 391, 391f
Sutton, Walter, 24
Svedberg, Theodor, 39, 40, 41f
Swallows, 528, 528f
Swordtail, 636, 637f
Symbiosis
  mutualism, 163–166, 165f
  novelty creation, 718–719
  parasitism, 166–167
  sociality, 167–168, 167f
  types of, 164t
Symmetric inversions, genome, 179–180, 181f, 182f
Sympatric speciation, 629, 644, 650–653
Synapsis, 328
Syngameons, 623
Syngamy, 663, 664f, 684–685, 685f
Synonymous changes, 59, 370–371, 374, 374f, 513, 513f, 531, 532, 535
Synonymous diversity, 539–540, 540f
Synonymous sites, 627
Syntax, 753
Syrphid fly, 526f
Systema Naturae (Linnaeus), 12
Systematics and the Origin of Species (Mayr), 32
Szathmáry, Eörs, 609, 616

T

Tagging single-nucleotide polymorphisms (tSNPs), 760
Tandem DNA arrays, 217
Tandem duplication, 326, 328f
TATA-binding protein, 141
Tatum, Edward, 42, 42f
Tautomerization, 328, 331f, 347
Taxonomy, folk, 619
Tbx4 gene, 309, 311, 311f
TDT (transmission disequilibrium test), 759f
Telomerase, 171, 214–215
Telomeres
  recombination reduction near, 539
  repetitive DNA in, 214
  replication of, 171
TEM-1 gene, 678f
Temperature, life at extreme, 152–157, 153f–155f, 153t, 156t

Tempo and Mode in Evolution (Simpson), 32
Temporal isolation, 625–626, 626f, 653
Teosinte branched 1 (tb1) gene, 315–318, 315f, 316f
Teosinte, evolution of maize from, 313–318, 314f–318f
Teosinte glume architecture 1 (tga1) gene, 317–318
Terminology, genetic, 356–357
Termites, 612
Terrestrial organisms, evolution of
  Cambrian, 271–272, 271f, 272f
  Devonian, 272–276
  diversity, development of, 273f
  flight, emergence of, 276–279
  Neoproterozoic, 272
  Ordovician, 271
  plants, 272–275, 274f, 276f
  Silurian, 271, 273
  vertebrates, 275–276, 277f
Tetrahymena ribozyme, 458, 459f
Tetranucleotide hypothesis, 39
Tetraploidy, 631, 631f
Tetrapods, evolution of, 275–276, 277f
t-haplotype, mouse, 590–591, 591f
Theistic evolution, 66
Theory, 81
Theory of Mutation (de Vries), 24
Thermodynamics, second law of, 79
Thermophiles, 152–156, 155f, 156t
Thermotogales, 153, 154
Thermotoga maritima, 155
Thermus, 138, 143, 143f, 155
Thiomargarita namibiensis, 140, 141f
Thompson, D'Arcy, 696f
Thomson, William, 20
Threshold, 386, 386f, 399
Thymine dimer, 333f
Tinbergen, Nikolaas, 308
tinman gene, 247
Tips, phylogenetic tree, 111, 111f, 112f
Tiselius, Arne, 40
Tit-for-tat strategy, 613
Tobacco necrosis virus (TNV), 590, 590f
Tomato, QTL responsible for weight in, 403, 404f
Toothless mammals, 74f
Tortoises, giant, 620f
Trace fossils, 267
Tracheids, 273, 274f
Trackways, 271, 278
Trade-offs, between components of fitness, 560–561, 560f
Tragedy of the commons, 569–570
Transcription
  control of, 234–235, 705–707
  description, 1, 9
  elucidation of process, 54
  multicellularity and, 226
  translation coupled to, 138–139, 141, 197f
Transcriptional regulation, 249
Transcription factors, 235
  binding motifs, changes in, 704, 705f
  diabetes-associated, 756
  homeobox-containing genes, 290, 292
  language-associated, 750
  maize *tb1* gene, 316

Transduction, 184, 186–187, 186f, 187f
Transfer RNAs (tRNAs)
  as adaptor molecule, 47, 48f
  evolution of, 105–106
  introns and, 219
  organelle, 204
Transformation
  Avery's experiments, 43, 44f
  description, 183, 183f
  Griffith's experiment, 42–43
  reproductive isolation and, 629, 630f
Transitions, 326, 326t, 328, 347
Translation
  description, 1
  efficiency and accuracy, selection for, 543–544
  in eukaryotes, 196, 197f
  evolution of system, 104–106, 105f
  in LUCA, 129–130
  transcription coupled to, 139, 141, 197f
Translocation
  description, 328, 329f
  effect on meiotic recombination, 340, 340f
  reciprocal, 340, 342f
Transmission disequilibrium test (TDT), 759f
Transposable elements
  co-opted, 598–599
  defense against, 593
  density at centromere, 216, 216f
  discovery in maize, 588
  DNA transposons, 217–218, 218f
  dominance and epistasis effects, 388, 389f
  in *Drosophila*, 594–595, 594f, 595t
  frequency in genome, 57
  horizontal versus vertical spread, 588, 588f
  in human genome, 597–598, 598f
  intron origin and, 599
  movement between species, 188
  mutational events from, 218
  number of copies of, 173
  relationship to viruses, 590
  retrotransposons, 217
  selfish, 35
  types, 342t
  virus resemblence to, 135–136
Transposition
  as mutation source, 342–343
  replicative and nonreplicative, 343f
Transposons
  benign nature of prokaryotic, 593, 594f
  promoters in, 343
  types of, 342t
  universality of, 343
Transversions, 326, 326t, 347
Tree of Life
  history of, 119–121, 120f–121f, 123f
  human position on, 727–731
  lateral gene transfer, effect of, 131–132, 131f, 133f
  LUCA in, 116, 117f
  molecular systematics, use of, 120–121, 122f, 123f
  photosynthesis in, 161
  ribosomal RNA (rRNA), 128f, 133f, 154f
  rooted, 124–129, 125f, 127f, 128f

viruses and, 133–136
Trematode parasites, 510, 511f, 674f
Tribolium castaneum, 300, 302f
Trichomonas, 209–210, 210f
Trigonotarbids, 271, 272f, 274
Trilobites, 74f, 80f, 265
Trimerophytes, 274
Triops, 303f
Triploidy, 631, 631f
Trisomy, 326, 328, 767
Tristyly, 687, 688, 688f
TRNA. *See* Transfer RNAs (tRNAs)
Trochophore, 265
True-breeding, 22
Truncation selection, 476, 476f, 552–553
Trypsinogen, 709f
Tschermak von Seysenegg, Erich, 24
Tu allele, 636, 637f, 643
Tulerpeton, 276, 277f
Tungara frog, 578–579, 579f
Twins, variance analysis, 394, 395f
Ty element, 590

U

UEPs (unique event polymorphisms), 745–746
Ultrabithorax (Ubx) gene
  crustacean morphology, role in, 301–305, 303f–305f
  cyptic polymorphism, 359–360, 360f
  discovery of, 288, 288f
  expression, 288–289, 289f
  fly bristle pattern, role in, 305–307, 306f
  larval appendage evolution, role in, 298–301, 300f, 301f, 302f
  variation within populations, 307
  wing evolution, role in, 294–296, 296f, 297f, 298, 299f
Ultracentrifuge, 39–41, 41f
Ultrameric tree, 112
Uncultured microbes, 148–151, 148f
Underdominance, 471, 471f, 471t, 476
Uniformitarianism, 12, 13, 16
Unique event polymorphisms (UEPs), 745–746
Universal ancestor, 1
Universal genes, 124
Universal genetic code, 105
Universal homologies, 115–116, 116t, 120
Universally shared features, of organisms, 66–67
Uranium, 89, 89t
Urea, synthesis of, 92, 92f
Urey, Harold C., 93
Ursus arctos, 452f
Use and disuse of organs, 21
Ussher, James, 12
Uta stansburiana, 571, 572f
UV cross-linking, for nucleotide polymerization, 97, 97f
UV irradiation, DNA damage from, 332, 333f

V

Vandiemenella grasshoppers, 474f
Van Leeuwenhoek, Antoni, 119

van't Hoff, Jacobus Hendricus, 38
Variable number tandem repeats (VNTRs), 347, 362
Variance
  additive genetic, 391, 394–397, 397f, 399, 418, 418f, 446, 462–463, 462f, 478–479, 482–484, 548–550, 566, 674–675, 678
  of allele frequency in the island model, 444–445, 445f
  components, estimating, 395–397, 395f
  covariance, 386–387, 394–396
  defined, 30
  dominance, 394, 395–397, 397f
  environmental, 387–388, 394, 399, 483, 483f, 677
  estimating components of, 394–397, 395f
  genotypic, 390, 394
  interaction, 394
  mutational, 409–411, 409f, 410t, 483f, 484, 514
  in normal distribution, 385
  phenotypic, 394, 399, 446, 478
Variation
  cryptic, 398–399, 398f
  environmental, 387–388
  genetic
    abundance of, 367–371
    in abundant species, 426–427
    allele frequencies, 363–364
    cline, 440, 443–444, 443f
    cryptic, 358, 360
    evolution, necessity for, 355–356, 356f, 358
    functional significance and, 370–371, 371f
    genealogies, 364–366
    gene arrangement, 375–377, 376f–378f, 376t
    genetic markers, 361–362
    maintenance of, 33–34
    molecular clock, 371–375, 373f, 375f
    nature of, 33
    nucleotide diversity, 364, 368, 369f, 370, 370f, 371f, 372f
    polymorphisms, 358, 360–361, 363, 368, 368f
    population size, effect of, 368–370, 426–427
    spatial patterns, 440
    terminology, 356–357
    types, 367–377, 367f
  in genetically complex traits, 381–412
  by lateral gene transfer, 351–352
  maintenance by disruptive selection, 652, 652f
  by mutation, 325–349
  mutation/selection balance and, 513–515
  neutral
    balancing selection, effect of, 541–542, 541f, 542f
    linked loci and, 536–542
    selection compared to, 531
  normal distribution, 384f, 385–387, 385f–388f
  polygenic, 383–384, 399

quantitative
    analysis of, 385–399
        generation of, 409–411
        genetic basis of, 399–409
    in quantitative traits, 513–515
    relatives, resemblance between, 396, 396f
    selection on, 457–487
    separation of causes, 393–394
    by sexual recombination, 349–351
Variation and Evolution in Plants (Stebbins), 32
Varieties, as incipient species, 70
Vascular plants, 273–275
Vertebrates
    classification of, 68, 69f
    correlation between rates of chromosomal
        evolution and speciation,
        642f
    flight, evolution of, 277–278, 278f, 279f
    immune system, 233–234, 234f, 663,
        716–717
    kin discrimination, 606
    limb development, 716
    monophyletic and paraphyletic groups, 112f
    pentadactyl limb, 68, 75
    polyploidy, 633
    terrestrial, emergence of, 275–276, 277f
Vertical evolution, in bacteria and archaea,
        188–194
Vertical inheritance
    definition, 110
    illustration in *On the Origin of Species*,
        110, 110f
Vesicular stomatitis virus, 513f
Vestiges of the Natural History of Creation
        (Chambers), 15–16, 15f, 17
Vestigial genes, 174
Vestigial structures, 75, 75f, 77f
Vibrio cholerae, 171
Vibrio harveyi, 168
Virulence, 166
Viruses
    cancer and, 703
    competition between, 569, 569f
    coviruses, 590
    defective interfering viruses (DIVs), 464,
        569, 569f, 590
    diversity of, 133, 134t
    genetic exchange, 351, 351f
    lateral gene transfer, 134
    mutation rates, 348
    origin of, 134–136
    parasite viruses, 590, 590f
    phylogeny of, 134, 135f
    rate of evolution, 134

restriction enzyme evolution and, 593
    transposable elements and, 590
Vital force, 38
Vitalism, theory of, 92
Vitamin D, 773
VNTRs (variable number tandem repeats),
        347, 362
Volcanic eruptions, 281
Volvocales, 687, 687f
Volvox, 226–227, 226f–228f
Vulcanism, 12, 12f

W

Waddington, Conrad, 307, 358, 691, 697
Wagner, Moritz, 644
Walcott, Charles, 264f
Wallace, Alfred Russel, 17, 17f, 18, 70–71, 70f,
        79
Warfarin, resistance to, 491, 491f
Warning colors, 600, 600f
Wasp
    cooperation in, 611, 611f
    sex-biased inheritance, 589, 589f
    sex ratio in fig wasps, 603, 603f
Watson, James, 43, 44f, 45–46
Wave of advance, 497, 499f
Weaver, Warren, 42
Weber, Ken, 482
Wedgwood, Josiah, 16
Wegener, Alfred, 256, 258f
Weinberg, Wilhelm, 28
Weismann, August, 21, 23, 24, 670, 674
Weldon, W.F.R., 18, 20f, 21, 22f
Whales, evolution of, 76, 77f
Whittaker, Robert, 120, 121f
Wilberforce, Bishop, 19f
Wild-type allele, 34, 357
Wilkins, Maurice, 43, 45, 45f
Williams, George C., 34, 607, 776
Wilson, Edward O., 778
Wings
    evolution of insect, 721
    Hox gene role in evolutionary change,
        293–298, 296f, 297f, 299f
    structure of, 68f
*The Wisdom of God Manifested in the Works
        of the Creation* (Ray), 11
Wobble pairing, 204
Woese, Carl, 106, 120–121, 130
Wolbachia, 589–590, 590f
Wollman, Elie, 50, 52
Wright–Fisher model, 416–418, 419, 420,
        422

Wright, Sewall, 30, 30f, 399, 416, 442, 443f,
        472–475, 494–496,
        607–609, 608f, 622, 642,
        690
Wright's F_{ST}, 445–451, 447t, 448t, 450f
Wynne-Edwards, V.C., 34

X

Xanthine dehydrogenase, 534
X chromosome
    inactivation, 682
    incompatibilities and recessive alleles on,
        639
    large X effect, 639–640
X-chromosome drive, 591, 593
Xenopus, antisense oligonucleotides, 247
Xerophiles, 158
Xiphophorus helleri, 636, 637f
Xiphophorus maculatus, 636, 637f, 643
X-ray crystallography, development of, 39,
        40, 41, 42f
X-ray diffraction, of DNA, 43, 45, 45f

Y

Y chromosome Adam, 427
Y-chromosome drive, 591
Y chromosomes
    degeneration of, 682–683
    human genealogy, 745–748, 745f
Yeast
    aging, 562f
    codon usage bias, 543f
    fitness, genome-wide measurements of,
        522–523, 523f
    mating types, 685
    prion, 710, 710f
    Ty element, 590
Yucca, pollination of, 613, 613f
Yule, George Udny, 26

Z

Zahavi, Amotz, 581
Zamecnik, Paul, 48–49
Zebrafish mutagenesis screen, 244, 245f
Zircon crystals, 89, 90
Zoarces viviparus, 500, 501f
Zootype, 293
Zosterophylls, 274
Zuk, Marlene, 583
Zygaena, 698–699, 698f
Zygotes, 222